The Spanish Coastal Systems

Juan A. Morales
Editor

The Spanish Coastal Systems

Dynamic Processes, Sediments and Management

Springer

Editor
Juan A. Morales
Department of Earth Sciences
University of Huelva
Huelva, Spain

ISBN 978-3-030-06605-5 ISBN 978-3-319-93169-2 (eBook)
https://doi.org/10.1007/978-3-319-93169-2

This Springer imprint is published by the registered company Springer Nature Switzerland AG
The registered company address is: Gewerbestrasse 11, 6330 Cham, Switzerland

*I dedicate this book to my grandfather,
Juan González, who made me questioning
the dynamics of my surrounding world.*

Preface

During the 30 years of my profession, I was interested in visiting all the coasts of Spain. My training as Doctor in Coastal Geology allowed me to observe and interpret many of the features I saw in these coasts, because I read about its currents, waves, and tides before visiting them. In addition, it was in the year 2000 when the Spanish Littoral Geomorphology Conference (GeoLit) began to be held every two years. Since then there was a meeting point for researchers and professionals who investigate the sedimentological and geomorphological processes that shape the coast.

In the framework of these conferences, the coastal researchers had every two years scientific encounters to make a tuning about the investigations conducted during that period. The place of each conference was elected among the different tracks, trying to complete the entire Spanish Coast. In the conference field trips, not only we have gone through many coasts of Spain which have been explained by scientists who have studied them, but that we have also created a family among all those who love the coastal dynamic. The members of this family are the authors of the chapters that built this book today.

This book synthesizes the scientific experience of investigators of many universities, research centers, and enterprises in the last 40 years of studying the dynamic, geomorphology, and sedimentology of the Spanish coastal systems.

Huelva, Spain Juan A. Morales

Acknowledgements

I would like to thank the *Family of the Coast*, my colleagues and friends, who are the authors of the chapters of this book. You know that without your contributions and aims this book would not have been possible.

The final quality in the edition of this book was possible thanks to the invaluable work of Alexis Vizcaino and Femina Joshi Arul Thas.

The knowledge contained in the chapters of this book was acquired by each of us by means of the continuous financial support from the Spanish Ministry of Science (under different denominations along years), regional and local research plans, and always supported by the funds from EU. All these administrations created in the last decades' programs to finance research projects allowed a great scientific development during all these years.

Personally, I would like to acknowledge the guides in my first years of career from my coaches, Federico Vilas, Germán Flor, and Richard A. (Skip) Davis; all of them, are now retired, but always available for their scientific counsels and help in life.

The planning of this book was possible by the help and knowledge of my direct colleagues, José Borrego, Mercedes Cantano, Berta Carro, José Manuel Gutiérrez-Mas, and Antonio Rodriguez-Ramírez, during my decades with them. Also, my co-authors for years merit a special attention: thanks Irene Delgado, Claudio Lozano, Germán Flor-Blanco, Mouncef Sedrati, Erwan Garel, and Gabriel Pendón. With all of them, I got many sediment and water samples, many current measurements, too much sand in my shoes, and too much mud in my hands and face.

Finally, thanks to my wife, who has the suggesting name of *Mar* (Sea), for being always available to visit new coasts with me, for the encouragement in getting my studies, developing my science, and editing this book.

Contents

Chapter 1
Introducing the Spanish Coast

Juan A. Morales and Augusto Pérez-Alberti

1.1 Introduction

Spain is definitely a coastal country. The Spanish coast extends along more than 9000 km and 22 of the 50 Spanish provinces have coastal edges over the Atlantic Ocean or the Mediterranean Sea (Fig. 1.1). In general terms it is a low altitude coast. A topographic analysis of the 1000 m near the shoreline lets know that the most part of this fringe is below 50 m of altitude. Exactly, a 76.06% have less than 50 m, 13.70% between 50 and 100 m; 6.44% between 100 and 150 m and only 3.79% exceeds 150 m.

The Spanish coast is too diverse displaying different littoral environments. Beaches, dunes, rias, estuaries, deltas, fan-deltas, lagoons, rocky coasts and cliffs with different geologic and dynamic controls are distributed along this extension. But are the beaches the most valuable environments from an economic point of view. A big part of the Spanish Economy is based on the tourism and the tourist come to Spain looking for "sun and beaches". For that, is essential for the political authorities have beaches in a good "state of health", including the sand amount and the beach services. In Spain the main part of political tasks are not centered for the Government of the nation, but distributed between regional autonomic responsibles, but a proof of the importance of the beach maintenance is that this task depends directly on the National Government (National Directorate of Coasts).

In the last six decades the entire coast of Spain has experienced a deep human modification and is actually under an extreme urban pressure. During the Decades of the 80s and early 90s, the main way of action for maintenance and the man-

J. A. Morales (✉)
Department of Earth Sciences, University of Huelva, Huelva, Spain
e-mail: jmorales@uhu.es

A. Pérez-Alberti
Department of Geography, University of Santiago de Compostela, Galicia, Spain
e-mail: augusto.perez@usc.es

J. A. Morales (ed.), *The Spanish Coastal Systems*,
https://doi.org/10.1007/978-3-319-93169-2_1

agement of the Spanish coast was the building of fixed structures (groins, jetties and breakwaters). The main purpose of these actions was to modify the marine dynamics, thus reducing the rate of erosion. These structures, linked with punctual replenishments, allowed the restoration of some damaged coastal areas. However, in the late 90s the authorities noted that this type of work also interrupted the transport of sediment in the coast, generating erosion downstream of the modified area. Today, the trend towards so-called soft solutions is clear. Regeneration of the coast is, fundamentally, through the contribution of a volume of sand able to compensate the erosion.

According with the data of Greenpeace (July, 2012) the costs during the period 2009–2011 of the beach restoration was: 552 million euros in Galician Coast, 211 million euros in Cantabrian Coast (Asturias, Cantabria and Vasque Country), 168 million euros in eastern Mediterranean coast (Cataluña, Valencia and Murcia), 42 million euros in the Balearic Islands, 474 million euros in south Andalousian coast and 242 million euros in Canary Islands. Nevertheless, the studies oriented to the understanding of the coast commenced just three decades ago, when the coastal evolution started affecting in an unexpected way to many badly planned human structures.

This book reflects the scientific knowledge acquired by different research teams over the last 30 years, which have contributed to an applied understanding. So today, the coastal management can be boarded in an integrated way.

1.2 Factors Affecting the Spanish Coast

1.2.1 Winds and Waves

A coast can be defined as fringe of terrain where the continent meets the sea, although in this encounter also the atmosphere plays an important role. The wind regime controls the wave provenance and littoral drift acting on a coast, making it an important element to consider in order to understanding the dynamics of this coastal sector. Normally, the wind dynamic is the responsible to generate waves and these waves are one of the most important coastal actors in a coast. Winds can originate waves on the same coastal front (sea waves) or in oceanic areas located offshore (swell waves). In this case, the waves are propagated to reach the coast. On the other hand, the wind is also an actor itself, because is the responsible of the dune dynamics.

The dominant wave provenance in relation with the orientation of the coastline is a characteristic of fundamental importance to sediment transport. The existence of an angle between the sense of wave propagation and the direction of the coast is responsible of the existence of a littoral drift component in the sediment transport, also called longshore current. Normally, fair weather waves are related with a swell propagation and contribute to prograde the coastal profile to get a slight slope.

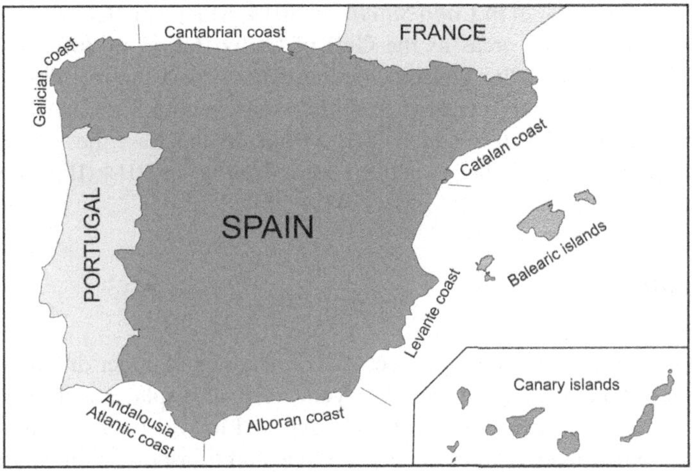

Fig. 1.1 Spanish coasts according with the division observed in this book

Waves higher than 1.5 m are associated with storm sea waves. These waves tend to be destructive for the coastal profile because they generate a movement of sediments towards the sea, which results in a general erosional effect. The influence of these storms is more important during the winter, causing erosion episodes on the beaches and cliffs so as marine surges in the coastal front and overwash phenomena in the back barrier areas. In most of Spanish coasts, these episodes destroy periodically the urban esplanades located in front of the coastal towns.

Wave conditions have a cyclical behavior resulting in varying onshore and offshore transports over the coastal profile. Usually, the conditions vary seasonally from a depositional situation in summer to become erosional in winter. The net balance of the coastal sedimentary budget depends on the relationships between wave dimensions, coastal slope, sedimentary supply and coastal environments. This balance can be also cyclical from a multi-year point of view and normally responds to climatic cycles.

Fair weather waves are of moderate dimensions (Menendez et al. 2009). The significant heights (Hs) are less than 0.5 m in the Mediterranean coast, dominating the provenance of East component. On the Atlantic coast (north and southwest), Hs rises to 1.5 m, reaching a maximum of 2.0 m on the Galician coast. During storm times Hs is also higher in the Galician coast, reaching values of 5.0 m. These values decrease to 4.0 in the northern coast, 3.0 m in the southwestern Atlantic coast and have its minimum of 1.8 m in the Mediterranean coast. Maximum waves are recorded in the northern (7.0 m) and northeastern (8.0 m) coasts. This record of wave dimensions are related with the provenance of waves, being the higher waves those coming from north Atlantic area, which arrive to the coast in form of swell from N and NW.

Maximum waves recorded in the Atlantic coast have been measured in Galicia, specifically in *Cabo Vilán*, with 13.5 m in 2009, reaching 12.9 m in *Estaca de*

Bares in 2008 and 12.0 m in *Cabo Silleiro* in 2014. Much smaller values have been in measured in southern areas as the Gulf of Cadiz (6.6 m in 2003) or 4.8 m in *Tenerife Sur* in 1999. In the Western Mediterranean coast the recorded values are similar to those of the southwestern Atlantic coasts, with 6.3 m in Alboran Sea in 1963 to 6.6 m in *Cabo de Gata* in 2015, while to the East the values increase recording 7.4 m in *Cap de Begur* and 8.1 m in *Mahón*, in 2014 (Balearic Islands).

1.2.2 Tides

The influence of tides on coastal processes is fundamental under different points of view. From a dynamic perspective, the cyclic vertical displacement in water level results in a variation of the foreshore tract exposed to wave action throughout the time. On the other hand, tidal currents can erode and transport sediment themselves in tide-dominated environments. That can be highly important in terms of sedimentation, especially in systems drained by tides like inlets, bays and fluvial mouths where the tide is the main responsible of the sedimentary infilling. Spanish coast is divided in two types from its tidal regime. The Atlantic coast has a mesotidal character with a mean range of about 2 m influencing the presence in this coast of important tidal systems like the rias of Galicia or the estuaries developed along the Cantabrian and the western Andalousian coasts. On the contrary, the Mediterranean coasts are microtidal, with astronomic tidal ranges minor to 25 cm and do not develop tidal systems at all.

1.2.3 Sediment Availability

The availability of sediment to be transported by waves and tides in the littoral exerts a first-order control on the dynamics of a coastal zone (Wright 1977). In Spain, the main part of the coastal sediment arrived to the coast through the 6 longer fluvial systems. These rivers (*Miño*, *Guadiana*, *Guadalquivir*, *Ebro*, *Júcar* and *Segura*) drained historically the central area of the Iberian Peninsula, transporting sedimentary material along hundreds of kilometers to reach the coastal area. This central peninsula is also drained by the rivers Duero and Tajo, but these finally reach the sea in Portuguese coasts.

Available sediment grain size also play a critical role in coastal evolution, by controlling the morphodynamic processes and the sediment depositional patterns at different temporal-spatial scales (Orton and Reading 1993). The rivers *Guadiana* and *Ebro* transported big amounts of sand able to build in their mouths important deltas, being the main suppliers of sand in each respective coast. On the contrary, the *Guadalquivir River* supply mainly suspended matter that infilled a wide estuary located in the southwest of Spain which developed the extensive marshes that presently constitute the *Doñana* National Park. Nevertheless, this system, do not supply sand material to feed the adjacent coast. The other three main rivers (*Miño* in the northwest, *Jucar* and *Segura* in the southeast) provided to their respective coasts sandy material to be distributed along the nearer coastal systems, but only the Segura was able to build a true depositional coast, whereas *Miño* and *Júcar* do not supply enough sand to feed the potential wave transport, remaining a sedimentary deficit in their adjacent open coasts that maintains a rocky character.

Additionally, it turns out that sediment supply to a coastal region is very closely linked to the tectonic setting of the coastal zone. In the main part of Spanish coastal systems uplifted mountains constitute denudation areas providing a good source of sediment to be transported by short rivers to near coasts. Normally, these short rivers transport gravels and sands, that infilled partially the estuaries developed in their mouths and occasionally some pocket beaches located in the nearest open areas. This is the case of all the coastal systems located in the northern, eastern and southeastern coasts of Spain. On the contrary, there are no substantial uplifted mountainous areas that can supply sediment near the southwestern coast of Spain. Nevertheless this is the best sand-supplied coast of this country and the only that maintains a prograding character. The long-term sediment balance between shoreface and shelf is also of crucial importance to coastal evolution (Carter et al. 1987) and this is the only littoral tract in Spain where the coastal agents are directly in contact with a sandy tertiary basin that continues underwater to a silty sand continental shelf. Both, continental and marine materials supply all the sand that tides and waves are able to rework. In addition, the Guadiana River is so excellent sand supplier that not only was able to build a delta but also to feed of sand the littoral drift acting in the western portion of this coast.

1.2.4 Climate Change and Sea Level Movements

The quantitative relationships between the rate of sea-level change, the rate of sediment supply and the capability of coastal agents to rework mark the nature of the long-term coastal evolution (Curray 1964). In this sense, Global warming causes sea level rise while stronger waves and currents eat away at the coastline. In a parallel way, a modification of the amount of mean rain water is expectable, impacting on the amount of available material transported by continental waters to reach the coast. According to the Spanish Environment Ministry, in the south-eastern coast the expected rise in the sea level is 4.0 mm/year, whereas the northern coast 2.5 mm a year. At same time, in Spain a decreasing of the rains will imply a shortage of sedimentary supply from rivers.

After a study done by the EU (Policy Research Corporation 2016), in the case of a general Sea Level Rise, the most vulnerable areas are the estuaries, deltas and enclosed beaches. Research studies have indicated that a Sea Level Rise of 0.4 m is considered to be a reasonable scenario for Spain by 2100. According with this studio, the climatic changes could result in the disappearance of 40% of the beaches in the Cantabrian coast. The inundation effect would also result in the disappearance of about 50% of the *Ebro Delta* in Cataluña or *Doñana Estuary* in Andalousia. In addition, significant increases in wave height as well as changes in the direction of the waves and an increasing of the number of storms per year have been observed along the Cantabrian and Galician coasts over the past 50 years. This effect has also been observed in other areas, like northern Cataluña, leading to an increased risk of flood events along the low slope coasts. The total coastline subject presently to erosion is 757 km^3 (11.5%), but the most vulnerable regions in this respect are Andalucía with erosion along 41% of its coastline, Cataluña with 33% and Valencia with 26%, despite these regions having protection structures and nourished beaches along 25% their coastline.

1.2.5 Tsunamis

Tsunamis are waves of large wavelength caused by eventual disturbances of the sea floor during seisms, volcanic eruptions or submarine landslides (Dawson and Smith 2000). From a human point of view, a tsunami is a low frequency high-magnitude event, but within a geological timescale is a frequent phenomenon. The impact of these events at the coastal areas is usually catastrophic. In open coasts the effect of a tsunami is destructive, not only for the human constructions, but also for the proper coastal sedimentary environments, representing a breakpoint in the coastal evolution. In protected low-energy coastal environments, such as lagoons, estuaries, or coastal lakes, the impact of a tsunami can create a recognizable sedimentary record difficult to distinguish from storm facies. This record may include surficial formations as cheniers and washover fans (Rodríguez-Ramírez and Camacho 2008) or subtidal layers which will be conserved interbedded into the ordinary coastal sediments (Ruiz et al. 2005).

The coast of the Gulf of Cadiz has been repeatedly struck by tsunamis during the late Holocene and the historic period. Recent papers identified from a geological point of view the record of, at least, fourteen catastrophic tsunamis affecting this coast (Morales et al. 2008; Gutierrez-Mas et al. 2009; Lario et al. 2011; Ruiz et al. 2013). Seven of these events were recorded in historical documents (Campos 1991). Pherhaps, the most famous is the tsunami that followed the November 11, 1755 Lisbon Earthquake, that destroyed the city of Lisbon and many coastal settlements along the southwestern Iberian coast. The geological effects of this event on the evolution of these coasts were also widely documented (Dabrio et al. 1998; Luque 2008; Lima et al. 2010).

Less known is the action of tsunami in the Mediterranean coast of Spain. Recent papers demonstrated the actions of tsunamis in the eastern coast of the Iberian Peninsula and on the Balearic Islands by studying large boulders present on coastal rocky terraces and cliffs (e.g. Roig-Munar et al. 2015). The hard nature of these coasts minimized the influence of these phenomena on the coastal evolution, but the within sedimentary environments also experienced the events as destroying breakpoints.

1.2.6 Human Actions

Dams along the rivers disrupt the natural sediment process to their mouths and coastal adjacent areas. The 6 long fluvial systems with mouths in Spanish coasts (*Miño*, *Guadiana*, *Guadalquivir*, *Ebro*, *Júcar* and *Segura*) are regulated by 211 dams and many of them are located just in the main courses. In addition, also numerous short rivers coming to the coasts just from coastal mountains are also dammed. These dams cut the bypass of sediments causing a sedimentary deficit on the adjacent coastal systems being one of the human actions with higher impact on the coast.

Other important human action is the caused by the alteration of the tidal network. In Atlantic coasts artificial salt pans and fishing farms occupy wide areas of ancient tidal flats. In parallel, extensions of salt marshes were reclaimed by agricultural activities. These facts generated a great decreasing of tidal prism in tidal systems, with a consequent diminution of the tidal currents developed in their mouths.

Finally, in open coasts, many hard structures were built to stabilize tidal channels (jetties) or avoid the coastal erosion (groins and seawalls). In this sense, more than 60% of estuarine mouths in the Atlantic coasts are presently controlled by jetties, whereas the main part of urban beaches along the Mediterranean coastline was rigidized with rocky structures. Many chains of coastal dunes were dismantled to transform the in urban promenades. These structures contributed to a leak of flexibility of the coast.

1.3 History of Human Occupation

As in the rest of coasts all over the world, the starting point of the Holocene coastal evolution in Spain was the beginning of the Flandrian Transgression that followed the melting of the Würm glaciations. Since then, the Sea Level was rise at a varying rate, with an estimated average of 7.8 m each thousands of years. The arrival of the sea at the present level is estimated occurred around 4,000 years BP (Friedman et al. 1992). In the northeast coast of Spain the Sea Level curve indicates a stabilization of the relative level 5 m under the present position occurred around 7,000 years ago (Alonso and Pagés 2010). This stand maintained during at least 3,000 years, then, it started to go up again to reach 2 m under the present level around 3,700 years ago. In southwestern Spain the relative movements are completely similar (Delgado et al. 2012), existing evidences of the oldest surficial sandy formations around 3,700 years B.P.

Evidences of early human occupations of the coast during pre-roman civilizations are abundant in Spain. In southwestern Spain, foundries of copper metallurgy were located in the coast of the 3rd millennium before Christ. Previous works of Chalcolithic ages evidenced in the continental area of this coast would be sunk by the last Sea Level Rise of the Flandrian Transgression. In the 3rd and 2nd millennium before Christ other civilization named *Argar* was established in the southeastern Spain. Some of their villages were fortified cities located in positions that allowed a control of the adjacent coast. At the end of this period the first chains of sandy barriers were already build by the coastal agents in the sedimentary littorals.

Later, in the southwestern area Tartessos civilization commenced in Spain the commercial relationships with Greeks (since 1,200 B.C.) and Phoenicians (since 1,000 B.C.). These relationships were based on the Spanish metals (bronze and silver) and implied a wide harbor development. Phoenicians also established commercial villages in all the Mediterranean coast of Spain. In northern Spain, Celtic villages (named *castros*) occupy privileged positions near estuaries and elevated points of the coast since VI century B.C. In the same ages, towers made in stone named *Talayots* were build in the Balearic Islands to domain the coast.

After a brief Carthaginian domination of Iberia that finished with the 2nd Punic war, Roman times signified the establishment of many port cities. 16 of these cities were located in the Mediterranean coast whereas only 6 were in the Atlantic coast. This fact suggests that the main part of these cities were found with the commercial objective of catering Rome with Spanish products. During Roman domination, there was a climatic optimum and some authors suggested that the Sea Level was located some centimeters above its present position. In facts some present estuaries are described in the roman texts as "*pallus*" (palustrine environments), giving an idea of the state of clogging of these systems.

The Visigoth domination implied a recession of the harbour activities that reborn with the Arab invasion. In general terms, the middle age was a warm period with an agricultural peak and the peoples from the Arabian Peninsula and the northern Africa colonized and used the incipient deltaic systems for agriculture, developing also systems to seize the force of the sea. In this sense, the marshes were modified to build the first tidal mills. Arabs also fortified many Mediterranean coastal towns to certificate their military domain of these coasts. Some of these coastal cities were progressively conquered by the Christians during all the middle age wars.

After the final conquest of Spain by Christians, the Renaissance arrived to this land. This fact coincided in time with the discovery of America in 1492. The colonization of this new world was done by using ships even bigger and stronger and these boats were constructed using oak wood. The use of a high amount of wood signified an intense deforestation of wide continental areas that remained unprotected and submitted to erosive processes. During centuries, the rivers transported amounts of sediments to coasts that evolved quickly, developing coastal plains and prograding deltas like *Guadiana* or *Ebro*. This is a moment where the coasts were not very populated, with the exception of important port cities like Bilbao, Barcelona, Valencia or Cadiz. The piracy was usual in this epoch. That was why a defensive system constituted by towers was built along the entire coast of Spain.

During the centuries XVIII–XIX the little ice age allowed the complete colonization of the coastal plains. A bad time for the agriculture induced the development of the intensive fishery and many small fisher villages were born in all the coasts.

The industrial revolution arrived to Spain with an enormous port development and the location on the coastal surroundings of wide industrial complexes. More than half of the Spanish population move to live on the coast. That was the real beginning of the coastal human transformation, but not only by the industrial port infrastructures. This transformation was accompanied by a human invasion of the coastal wetlands that were extensively exploited in agricultural or industrial uses like rice fields or salt pans.

In the XX century the extensive widening of the harbor areas signify a new dimension for the coastal occupation, but in parallel, a new tansformative element arrived to the Spanish coast: the sun-and-beaches tourism. Wide coastal plains were transformed in new urban areas. Airports, marinas and similar infrastructures were created to satisfy the tourism needs. But the most serious situation was that all this urbanization was done without a real plan or with a plan that ignored the natural functioning of the coast. That was how arrived the speculative operations, the excessive coastal modification or the uncontrolled wastes. In this context the Law for coasts was dictated in 1988 to try to avoid all of this uncontrolled use of the littoral areas. Since then, the coastal fringe is considered as public domain. In consequence, many houses built in previous epochs were destroyed and now all the projects to modify the coast need strong environmental studies and an integral plan of coastal management.

In the future, new challenges are raised, especially regarding Global Change and Sea Level Rise, but also a new vision must be done to include in the coastal management elements as the new energy sources.

1.4 Galician Coast

Extends between the mounts of the rivers *Miño* and *Eo* (Fig. 1.2). This is a rocky coast including cliffs (Fig. 1.3a) and low-slopped coasts, where a 86% are under 100 m of altitude and only a 0.76% exceeds the 300 m. It is composed by igneous and metamorphic paleozoic materials deformed by the Variscan Orogeny, but tectonically re-elevated during the Alpine orogeny when the Atlantic Ocean was opened (Gómez-Pujol et al. 2014). In the building of coastal forms, the tectonic played an important role, just by the superposition of these two tectonic cycles. The coast with the highest altitude corresponds with the plutonic rocks of *A Capelada* Complex (A Coruña) where are the highest cliffs of the Iberian Peninsula, reaching 600 m in *Cabo Ortegal*.

The most characteristic elements are the rias, that are estuaries with a state of infilling incipient or null (Fig. 1.3b), located in the mouth of rivers coming directly from the range and with a very short course. A part of these estuaries (*Rias Baixas*) are related with a wide strike-slip fault system that also originated grabens in the continent. Other part (*Rias Altas*) are also related with the presence of a lithologic

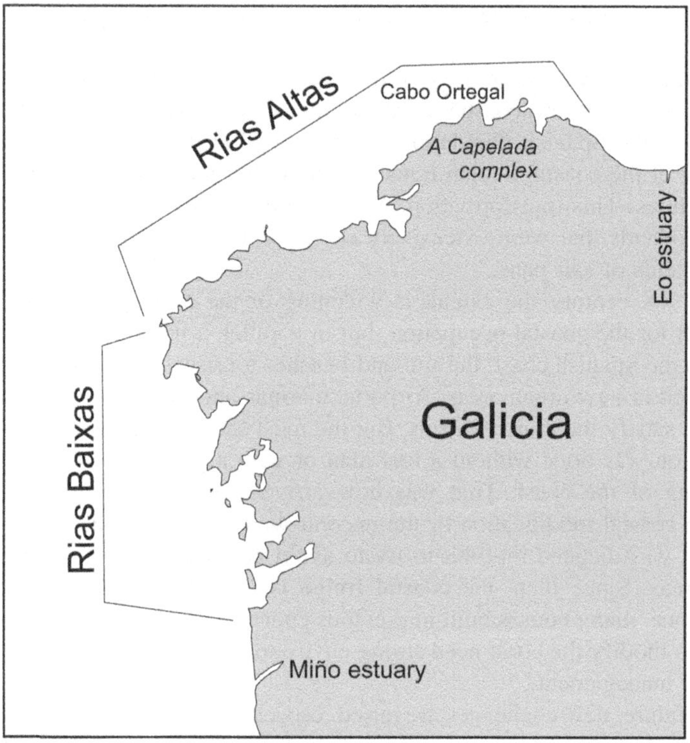

Fig. 1.2 Coastal configuration of the Galician coast, indicating the main coastal systems

banded oriented in N-S direction. This process favored the differential erosion and the fluvial incision of the less competent materials.

Normally the rivers transport too small amounts of sandy sediments, for that, this coast do not developed important beaches, but short beaches located in the middle part of estuaries or in the bottom of cliff bays, as the case of Corrubedo Beach-Lagoon system (Fig. 1.3c). Pebble beaches are also common (Fig. 1.3d). The climate of Galicia is too rainy, so the beaches are too far and hide and are not extensively used by tourists, for that are the most natural and less managed beaches of Spain.

In the 5 years comprised between 2001 and 2005 in Galicia, the most important performances were held at beaches in the provinces of Coruña and Pontevedra (Pérez-Alberti et al. 2013). Galicia received an investment of 450 million Euros, which were employed to improve 55 km of promenades and nourished only 20 km of beaches.

Nevertheless, on the Galician coast, the General Directorate of Coasts has carried out a series of geophysical surveys and marine drilling campaigns. The main aim of these studies was the evaluation of sand existing in the seabed and its possible use as zones of loan in the replenishment of beaches. In these campaigns, the 10 km next to the coast, along the Galician shelf, between the years 1987 and 1994 are prospected.

Fig. 1.3 Some examples of coasts in Galicia. **a** Cliff coast in *As Catedrais*. **b** *Ria de Pontevedra*. **c** Dune-Marsh in *Corrubedo*. **d** Pebble beach in *Coido*

1.5 Cantabrian Coast

The north coast of Spain extends between the Eo River estuary and the border with France. This coast presents a high degree of similitude with the previously described coast. A 82.7% of the Cantabrian coast is lower than 100 m of altitude and only the 1.1% exceeds the 300 m. In the western sector, metamorphic rocks domain the lithologies, whereas towards the East from the *Nalón Estuary*, the coastar rocks are mainly composed by limestones. The Cantabrian Coast, as the Galician one, is controlled by the rising of the *Cantrabrian Range* during the Variscan and Alpine orogenies.

The relief is characterized by a chain of coastal cliffs. Between Eo and Nalon Rivers, cliffs with a flat top and a very sloped front are dominant. These flat tops of the cliffs extend to the continent by means of a coastal plain named *Cantabrian rasa* (Fig. 1.4a). Eastward, from *Cabo de Peñas*, sectors of flat-topped cliffs alternate with slight-sloped cliffs, including sectors with a strong karstification (Fig. 1.4b).

Fig. 1.4 Some examples of coasts in the Cantabrian Coast. **a** Cliff coast in the surroundings of *Cabo de Palos*, where different *rasa* plains can be seen. **b** Small beach developed by karstification in a doline in *Gulpiyuri*. **c** Sandy barrier closing the *Tina Menor* Estuary (Cantabria). **d** Wide tidal flats and marshes in the Natural Park of *Urdaibai* (Basque Country). **e** Small beach developed in a cliff front in Mixota (Asturias)

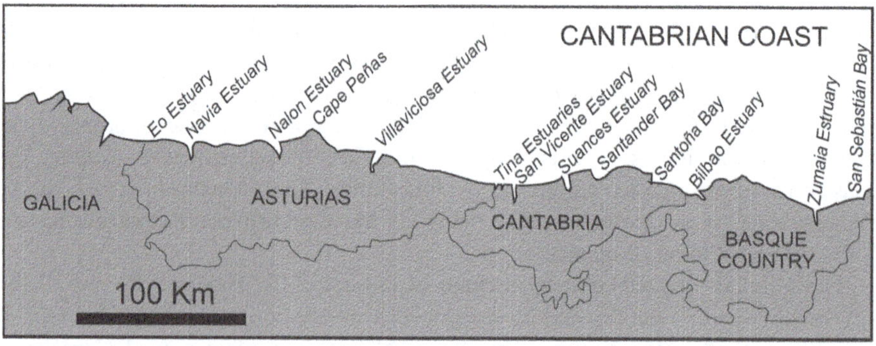

Fig. 1.5 Cantabrian Coast, indicating the administrative divisions and the main coastal features

A set of small estuaries cutted the coast in a perpendicular way. The cantabrian estuaries are also named rias by the local people; nevertheless, these are not rias from a geological point of view. The most significant of these estuaries are (Fig. 1.5): *Navia*, *Nalon*, *Villaviciosa*, *Tina Mayor*, *Tina Menor*, *San Vicente de la*

Barquera, *Suances*, *Bilbao*, *Mundaka*, *Deva*, *Zumaia*, *Orio* and *Pasajes*, existing also valleys occupied by bays as *Santoña*, *Santander* and *San Sebastián*. The rivers draining these estuaries are as short as the coast of Galicia and also come from a very near range, but in this case, the estuarine environments developed in their mouths are smaller and shallower and, in consequence, the degree of infilling of these systems is higher. So the main part of these estuaries developed sandy barriers in their mouths (Fig. 1.4c) that normally are constituted by wide beach-dune systems, developing ecologically rich tidal flats and salt marshes in their inner parts (Fig. 1.4d).

The main part of Cantabrian beaches are just developed in the front of these barriers, but also small beaches can also be developed in very small bays of the cliff front (Fig. 1.4e). The human use of these minor beaches is very restricted, because the access is normally difficult and crossing highly scarped roads. In consequence, these beaches are very natural and normally less managed.

The most popular beaches in the Cantabrian coast are those located just in the estuarine mouths, because these are most extensive and the access by car is easy. The problem of these beaches is that the main part of them was modified by the human action in relation with the control of the navigation to the harbors located in the inner estuaries. So, many of these inlets are bordered by jetties. Some of these jetties were built so early as the 40s decade, when the Spanish coastal engineers experimented firstly these kind of solutions. The main part of these beaches suffered a quick evolution after the building of the jetties and many of them experienced a rotation process, with progradation in the areas closer to the jetties and erosion in the farter areas. Now, the main part of these beaches has a new equilibrium and all of these systems have a cyclic behavior with strong erosive periods during stormy winters. In many cases, these storms eroded the dune fields located in the backshore area, which act as a sand reserve. Nevertheless during 80s and 90' many of these dune fields were transformed or destroyed to build. For that, the present efforts of the authorities were centered in the maintenance and development of the dune fields, that can avoid the necessity to nourish the beach after important storms.

Nevertheless, in many occasions during the last decade, some beaches have to be nourished. An example was the action developed in 1996 in the western beach (*Playa de Poniente*) in *Gijón* (Asturias) that was nearly disappeared after the building of a marina. The actions included the building of a limiting groin, the building of a promenade and the replenishment of the beach. For that 120,666.8 tons of rocks were used to build the groins and 451,493 m^3 of sand were employed to nourish a beach surface of 300,000 m^2. The sand was taken from a submarine place located 6 miles to the northwest of *El Musel* harbor. The total cost was 24 million Euros and an important study of environmental impact was aboarded.

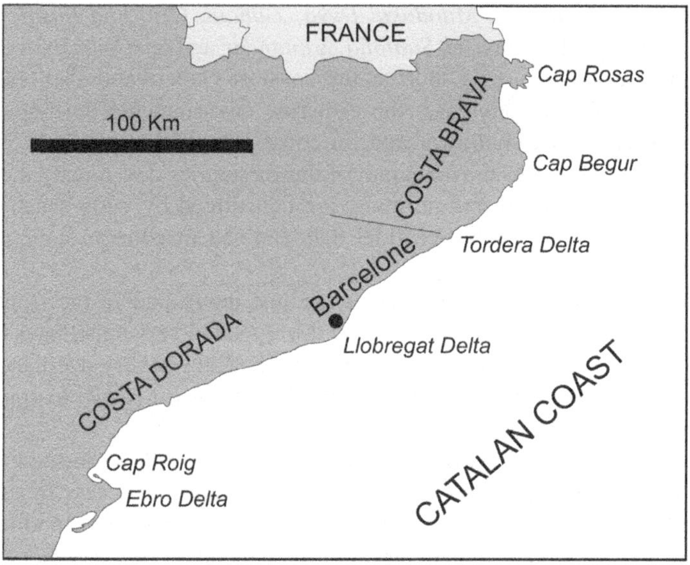

Fig. 1.6 Catalan Coast, indicating the main coastal features

1.6 Catalan Coast

The eastern coast of Spain, from the border with France to the *Ebro Delta* in the SE point of Iberia is a coast with a bigger diversity from a geomorphological point of view. Two well-defined sectors can be differentiated in the coastal front. The first one, named *Costa Dorada*, extends from the *Ebro* mouth to the *Tordera Delta* and the other from here to the border with France. The first sector is a low coast in which a dominance of deltas (*Ebro*, *Llobregat* and *Tordera*). The second, the so-called *Costa Brava*, is a generally low rocky coast with a predominance of low-slopped cliffs (Fig. 1.6). In general, a 90% of this coast do not exceeds the 100 m of altitude and only a 0.2% his higher than 300 m.

From a geological point of view, plio-quaternary sediments are characteristic of the southern flat areas (*Ebro Delta*, *Campo de Tarragona* or surroundings of *El Prat de Barcelona*). To the North, mesozoic rocks of the *Catalan Range* conditions the coastal morphology. In the surrounding of *Cap Roig* the jurassic and cretacean limestones domains, whereas from the surroundings of Barcelona to *Cap de Begur* domain the granitic rocks. Near the border with France, terrigenous sedimentary rocks as sandstones, shales and conglometates are the most significant lithologies. This variety of rocky formations exerts an important control on the coastal morphodynamics.

The most significant factor controlling the sedimentation is an active littoral drift acting in a north-to-south sense. From a geological point of view all this coast is dominated by the presence of three Alpine ranges (Pirineos, Iberian and Coastal

Fig. 1.7 Some examples of coasts in Catalonia. **a** Beach developed on a rocky platform in *Capellans* Beach (Photo by Jorge Franganillo). **b** Small beach in a rocky coastal cell in Tossa de Mar (Photo by Alberto Gonzalez-Rovira). **c** Aerial view of the Ebro Delta. **d** Coastal defenses in Cambrills Beach. **e** Barcelona Beach during the touristic season (Photo by Manuel Martin)

Catalan ranges), composed by Mesozoic materials deformed during the Alpine cycle. The main structural axis are orientated with a component W-E, then the main rivers flux along valleys longer than in the northern coast despite to have a rocky dominated coast. So, the southern part of this coast is flooded by one of the longer and most discharging rivers of Iberia: the Ebro River.

The geological configuration of this coast induced the existence of rocky tracks separated by more or less open embayments partially covered by long beaches developed on a rocky subtidal platform (Fig. 1.7a). The morphological configuration of the entire coast is very similar, but the presence of many caps and bays divide the coast in many sedimentary cells (Fig. 1.7b). The only exception is the Ebro river mouth, which developed the biggest delta in Spain and one of the most important of Europe (Fig. 1.7c), considered as the most important supplier of sediments of the eastern coast of Spain. Regrettably, this river was regulated by more than 40 dams and presents now an important sedimentary deficit, suffered not only by the beaches located in the delta front, but by all the coastal cells located downdrift.

On the other hand the eastern coast of Spain is the most exploited from a touristic point of view and the cost of beach maintenance is the higher of Spain. In this sense, many coastal defenses were built along the beaches (Fig. 1.7d). Only the

5000 km of urban beaches of the city of Barcelona (Fig. 1.7e) received a budget of 33 million Euros for restoration between 2006 and 2010. The works of beach management included the building of submerged breakwaters, the beach replenishment with 700.000 m^3 and a continuous monitoring by cameras, studied and analised by the Superior Scientific Bureau of Spain (CSIC).

1.7 Levante and Alboran Coasts

Between the Ebro Delta and the Gibraltar Strait, the Levantine-Alboran coast stretches. This coast enlarges the Mediterranean coast of three Spanish regions: Valencia, Murcia and Andalucía (Fig. 1.8). In this coastal fringe, a 91% is lower than 100 m of altitude and less 1% exceeds the 300 m. From this point of view, this littoral is a low and ravined coast, where alternate rocky tracks with wide coastal plains. Good examples of the rocky tracks can be observed in *Sierra Helada* (Alicante), the segment between *Cabo de Palos* and *Cabo de Gata* (Almeria). Examples of the coastal plains are the flats of Castellón, Valencia, Alicante or Cartagena, some of them associated to coastal lagoons like the *Albufera de Valencia* or the *Mar Menor* in Murcia.

The proximity of the *Betic Range*, generates two distinctly different broad sectors in the configuration of the relief (Fig. 1.9). The first, from the *Ebro Delta* to *Cabo de Palos*, is constituted by a sedimentary coast with wide sandy plains. The second, much more controlled by the geological nature of the rocks, is constituted

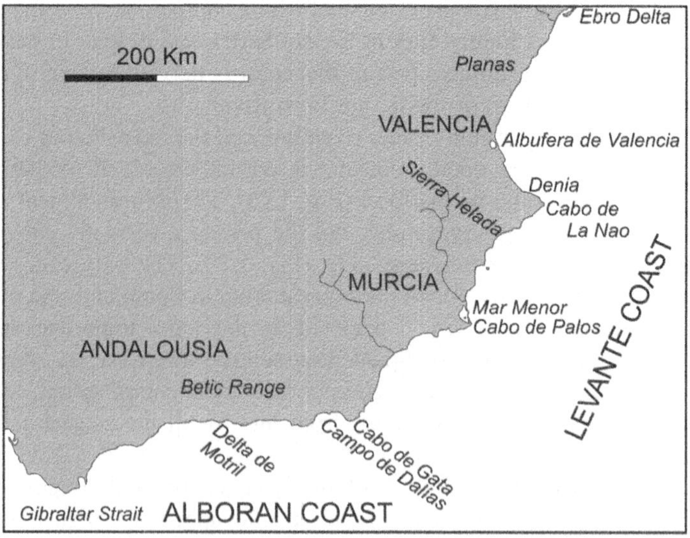

Fig. 1.8 Levante and Alboran Coasts, indicating the main coastal features

Fig. 1.9 Some examples of coasts in Levante and Alboran. **a** Double tombolo completely urbanized from *Calpe Rock* (Levante Coast). **b** Cliffs excavated in volcanic rocks in *Cabo de Gata*. **c** Betic Range see from the sea in Marbella (Alboran Coast)

by low-slopped cliffs cutted by narrow valleys where ephemeral streams named *ramblas* are developed. In this second track the flat sectors are less abundant and narrower, with the exception of the track *Campo de Dalias-El Ejido* (Almería) and *Motril* (Granada), where extensive fan deltas were developed.

Very tourist beaches stretch along this coast and all of them have a clear tendency to the erosive processes. As proof of this fact, they are constant nourishment that the Coastal Administration is obligated to do in these beaches. An example of nourishment in this coast was the replenishment of *Denia* (Alicante). This restoration try to resolve a problem of continuous erosion caused downdrift of a jetty. The replenishment was done in 2006, with a volume of sand of 953,307 m^3 taken by a suction dredge from an area of seabed located in the front of the cliff of *Sierra Helada*, 35 m under the sea level. The sand had a medium size of 0.22 mm and was distributed in three beaches with a total of 2,810 m long, covering an area of 106,000 m^2, with a medium slope of 3.5%. In order to avoid further losses of sand, three groins perpendicular to the coast were built, but the previous groin of *Punta del Molino* was 60 m shortened. Since then other restoration were necessary in the same beach. The last of them in July 2011, when 300 m of beach were replenished with 800 tons of sand, which disappeared in only a week, with a cost of near 12,000 Euros.

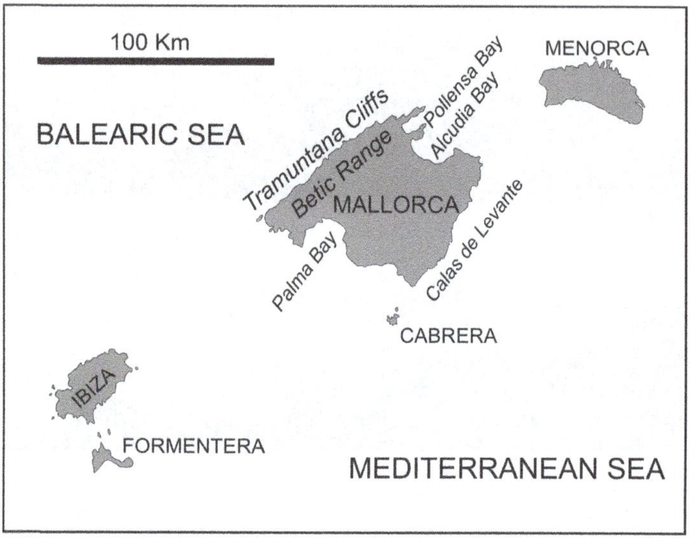

Fig. 1.10 Balearic Islands, indicating the main coastal features

Fig. 1.11 Some examples of coasts in Balearic Islands. **a** *Tramuntana Cliffs* along the northern coast of *Mallorca* (Photo by M. Rosa Ferre). **b** *Cala* in the eastern coast of Mallorca (Photo by Vicente Miramón). **c** Endokarstic cavity connected with the coast (Photo by Roberto Lumbreras). **d** Tsunamigenic blocks in the southern coast of Menorca (Photo by M. Mar Mateo)

1.8 Balearic Coasts

The Balearic Islands are described as an extension of the Betic Range (Fig. 1.10). 80% of their coast does not exceed 100 m of altitude and a little more than 5% is higher than 300 m. These are composed mainly by Mesozoic carbonatic rocks in the northern flank of the islands and Neogene sediments to the rest of littoral tracks. This geological distribution conditioned the presence of high cliff without beaches to the northwest (*Tramuntana cliffs*, Fig. 1.11a) and abundant bays and beaches in the remaining coasts of the islands.

The main part of these rocks (Mesozoic and Neogene) has an important carbonatic component. This fact conditioned the development of significant karstic processes. In the surface, short rivers were able to excavate narrow valleys being inundated by the sea water after the last rise, forming a characteristical geometry of the fluvial mouths named *calas* (Fig. 1.11b). There is also an important action of the karstic process in the interior of the rocks, constituting an important endokarst system (Fig. 1.11c). A characteristic feature of these rocky coasts are the presence of huge blocks on top of the cliffs (Fig. 1.11d), recent studies attributed a tsunamigenic origin to the same.

From a general point of view there is an important deficit of sand in all the beaches and all of them are configurated as beaches on rocky subtidal platforms. All of these beaches are very important from a touristic point of view, and are very visited specially by Germans. The deficit of sand is so important that the amount of sand lost in the feet of the tourist was evaluated by the scientists and the green algae eliminated from the beaches are previously dried extracting as maximum sand as possible before to be finally removed. Are also important the relationships between

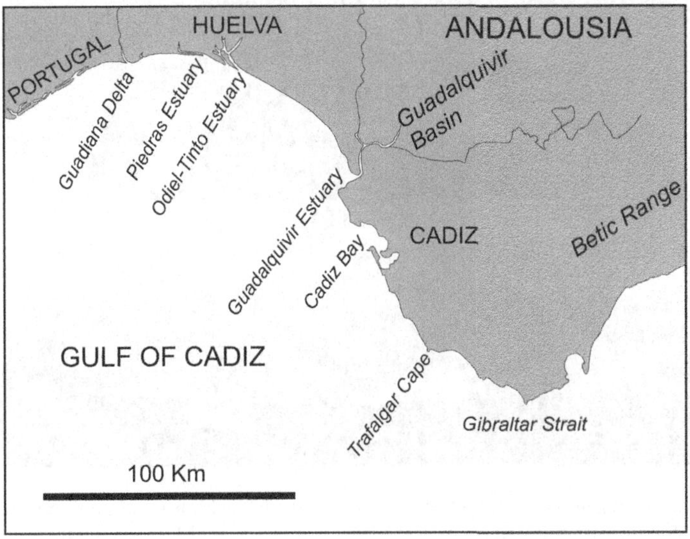

Fig. 1.12 Andalousia Atlantic Coast, indicating the main coastal features

dune systems and beaches and many programs of dune protection and wind traps for new dune creation were developed.

1.9 Andalousia Atlantic Coast

Extends to the west from the Gibraltar Strait. It is mainly a low sedimentary coast with rocky tracks (Fig. 1.12). 98% of this coast does not exceed 50 tall and only 2.57% up to 100 m.

This mesotidal coast is divided in two sectors. The first, developed along the coasts of Cadiz province is completely similar to those described for the Mediterranean coast. This sector is a rocky coast interrupted by little sedimentary systems, located in the margin of the *Betic Range*. It develops some short rivers with small estuaries and a clear deficit of sand. The beaches are small and located on rocky platforms (Fig. 1.13a). The second sector, located in the province of Huelva, clearly differs from all the other coasts of Spain, since is the only located in the front of a cenozoic basin (Guadalquivir Basin) and is also the only coast in the entire Iberian Peninsula which is well supplied of sand. This big budget of sand is based in the previous coastal erosion of the front of the neogene formations (Fig. 1.13b) and also because it receives all the sedimentary supply arrived from the central peninsula transported by the Guadiana river. This coast evolved from a previous barrier island-lagoonal system that was completely infilled with inlets that

Fig. 1.13 Some examples in the Atlantic Coast of Andalousia. **a** Cliffs of Conil, in the Cadiz Coast, developing a beach on the rocky platfom. **b** Cliff of Mazagon in the Huelva Coast, excavated in Cenozoic sediments developing a wide beach on the front. **c** Guadiana Delta in the border with Portugal. **d** Marshes of the Piedras Estuary (Photo by Wolfgang Michalsky)

were completely closed forming a continuous beach during near 100 km. This sandy body also extends seawards and do not presents rocky outcrops as a continuous base, but as punctual and disperse locations under 14 m depth.

The continuity of this long beach is based in a strong littoral drift acting west-to-east, distributing the sand from the Guadiana River (Fig. 1.13c) along the remaining coast and it is only interrupted by three fluvio-tidal systems that presently constitute estuaries (Fig. 1.13d), which developed closing barrier with significant beach-dune systems in their frontal areas.

In spite of this enormous amount of sand, many of the beaches of this coast had to be replenished in the last decades. The cause is a strong regulation of the Guadiana River and the building of many port structures that interrupted the longshore transport since late 70s, united to the building of urban settlements too near to the coast destroying the dune systems. In the Atlantic coast of Andalousia are abundant the cases of beach nourishment.

1.10 Canarian Coasts

The Canary Islands have a total longitude of coasts of 1,583 km (Fig. 1.14). It is a predominantly rocky coastline, characterized by vertical cliffs, modeled on volcanic rocks, separated by a wide network of streams. Around 70% of the coasts of the Canary Islands reached the 100 m high and 12% exceed 300 m.

It's a very recent volcanic archipelago, whose outcropping rocks are Cenozoic, although the main volcanic buildings are supported by oceanic crust of Jurassic age. In some parts of the coast a clear unconformity separates the volcanic materials (Fig. 1.15a) from plio-quaternary sedimentary formations (Fig. 1.15b). These formations can be constituted by alluvial deposits, shallow-water marine deposits (normally beach rocks) or aeolianites. The location of the islands in a tropical zone

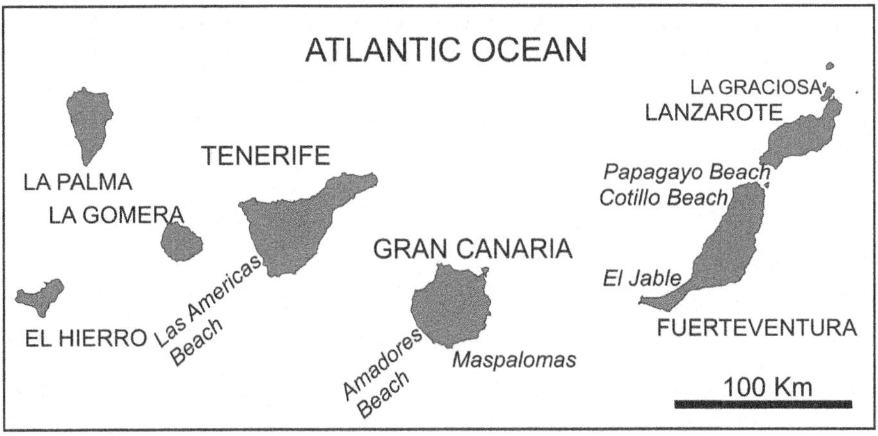

Fig. 1.14 Canary Islands, indicating the main coastal features

Fig. 1.15 Some examples in the Canary Islands Coast. **a** Cliffs excavated in volcanic rocks in the southwestern coast of *Gran Canaria*. **b** Cliffs excavated in stratified volcanic-sedimentary rocks in *La Graciosa*. **c** Beach-dune system in *Maspalomas* (Gran Canaria). **d** Artificial beach of *Las Teresitas*

influenced by trade winds and strong currents infer to these coasts an erosional character, being the low sedimentary coasts exceptional cases. Nevertheless, some of the Canarian Islands have some spectacular natural beach-dune systems like *El Jable* and *Cotillo-Tostón* in Fuerteventura, *Maspalomas* in Gran Canaria (Fig. 1.15c), *La Francesca* in La Graciosa or *Papagayo* in Lanzarote. These systems normally consist of a thin layer of sand located on a rocky flat, being very sensible to the processes of sand remobylisation.

The Canary Island coast is one of the most visited and touristically occupied in Spain. The inversion of the Spanish Government in coastal nourishment is very high. Some of the best examples of artificial beaches located in front of cliff systems are located in these islands (Fig. 1.15c). Perhaps the best example is *Playa de las Américas* (Tenerife), existing other well known examples like *Playa de Amadores* (Gran Canaria).

References

Campos, M. L. (1991). Tsunami hazard on the Spanish coasts of the Iberian peninsula. *Science of Tsunami Hazards, 9,* 83–90.

Carter, R. W. G., Johnston, T. W., McKenna, J., & Orford, J. D. (1987). Sea-level, sediment supply and coastal changes: Examples from the coast of Ireland. *Progress in Oceanography, 18,* 79–101.

Curray, I. R. (1964). Transgressions and regressions. In R. L. Miller (Ed.), *Papers in marine geology* (pp. 175–203). New York: Macmillan.

Dabrio, C. J., Goy, J. L., & Zazo, C. (1998). The record of the tsunami produced by the 1755 Lisbon earthquake in Valdelagrana spit (Gulf of Cádiz, southern Spain). *Geogaceta, 23,* 31–34.

Dawson, S., & Smith, D. E. (2000). The sedimentology of Middle Holocene tsunami facies in northern Sutherland, Scotland, UK. *Marine Geology, 170,* 69–79.

Delgado, J., Boski, T., Nieto, J. M., Pereira, L., Moura, D., Gomes, A., Sousa, C., & García-Tenorio, R. (2012). Sea-level rise and anthropogenic activities recorded in the late Pleistocene/ Holocene sedimentary infill of the Guadiana Estuary (SW Iberia). *Quaternary Science Reviews, 33,* 121–141.

Friedman, G. M., Sanders, J. E., & Kopaska-Merkel, D. C. (1992). *Principles of sedimentary deposits: Stratigraphy and sedimentology* (pp 717). New York: Macmillan.

Gómez-Pujol, L., Pérez-Alberti, A., Blanco-Chao, R., Costa, S., Neves, M., & Del Río, L. (2014). The rock coast of continental Europe in the Atlantic. *Geological Society, London, Memoirs, 40,* 77–88.

Gutiérrez-Más, J. M., Juan, C., & Morales, J. A. (2009). Evidence of high-energy events in shelly layers interbedded in coastal Holocene sands in Cádiz Bay (south-west Spain). *Earth Surface Processes and Landforms, 34,* 810–823.

Lario, J., Zazo, C., Goy, J. L., Silva, P. G., Bardají, T., Cabero, A., et al. (2011). Holocene palaeotsunami catalogue of SW Iberia. *Quaternary International, 242,* 196–209.

Lima, V. V., Miranda, J. M., Baptista, M. A., Catalao, J., González, M., Otero, L., et al. (2010). Impact of a 1755-like tsunami in Huelva, Spain. *Natural Hazards and Earth System Sciences, 10,* 139–148.

Luque, L. (2008). El impacto de eventos catastróficos costeros en el litoral del Golfo de Cádiz. *Revista* Atlántica-Mediterránea *de Prehistoria y* Arqueología *Social, 10,* 131–153.

Menéndez, M., Méndez, F. J., Izaguirre, C., Luceño, A., & Losada, I. J. (2009). The influence of seasonality on estimating return values of significant wave height. *Coastal Engineering, 56,* 211–219.

Morales, J. A., Borrego, J., San Miguel, E. G., López-González, N., & Carro, B. (2008). Sedimentary record of recent tsunamis in the Huelva estuary (southwestern Spain). *Quaternary Science Reviews, 27,* 734–746.

Orton, G. J., & Reading, H. G. (1993). Variability of deltaic processes in terms of sediment supply, with particular emphasis on grain size. *Sedimentology, 40,* 475–512.

Pérez-Alberti, A., Pires, A., Freitas, L., & Chaminé, H. (2013). Shoreline change mapping along the coast of Galicia, Spain. *Proceedings of the Institution of Civil Engineers, 166,* 125.

Policy Research Corporation. (2016). *The economics of climate change adaptation in EU coastal areas* (153 pp). European Commission. Report for the Directorate-General for Maritime Affairs and Fisheries.

Rodriguez-Ramirez, A., & Camacho, C. M. (2008). Formation of chenier plain of the Doñana marshland (SW Spain): Observations and geomorphic model. *Marine Geology, 254,* 187–196.

Roig-Munar, F. X., Rodríguez-Perea, A., Martín Prieto, J. A., Vilaplan, J. M., & Gelabert, B. (2015). Morfometría de bloques de tsunami en las costas rocosas del Este de Mallorca (Islas Baleares). *GeoTemas, 15,* 229–232.

Ruiz, F., Rodríguez-Ramírez, A., Cáceres, L. M., Rodríguez-Vidal, J., Carretero, M. I., Abad, M., et al. (2005). Evidence of high-energy events in the geological record: Mid-Holocene evolution of the southwestern Doñana National Park (SW Spain). *Palaeogeography, Palaeoclimatology, Palaeoecology, 229,* 212–229.

Ruiz, F., Rodríguez-Vidal, J., Abad, M., Cáceres, L. M., Carretero, M. I., Pozo, M., Rodríguez-Llanes, J. M., Gómez-Toscano, F., Izquierdo, T., Font, E., Toscano, A. (2013). Sedimentological and geomorphological imprints of Holocene tsunamis in southwestern Spain: An approach to establish the recurrence period. *Geomorphology, 203,* 97–104.

Wright, L. D. (1977). Sediment transport and deposition at river mouths: A synthesis. *Geological Society of America Bulletin, 88,* 857–868.

Part I
Rocky Coasts and Cliffs

Chapter 2
The Rocky Coasts of Northwest Spain

Augusto Pérez-Alberti and Alejandro Gómez-Pazo

2.1 Introduction

The coastline of Galicia is more than 2100 km long (POL Galicia 2010) (Fig. 2.1). Two broad types of coast can be differentiated in the region: zones with rías and zones without rías. Marine inlets dominate in the former, whereas rectilinear stretches dominate in the latter and only small coves or estuaries occur. The megaforms of coastal relief in northwest Spain are clearly determined by the tectonic structure, whereas lithological differentiation has played a predominant role in the genesis of meso and microforms (Pérez-Alberti and Blanco-Chao 2005). In general, different factors are involved in shaping the coastline: the overall structure is determined by tectonic processes; the lithology causes differential erosional processes that define the broad features of the coastal front; and, finally, the succession of geomorphological processes that have taken place over time have determined the specific forms and distribution of the different environments. In addition, human activity has affected many areas, particularly the low-lying coastline.

The Galician coast is classified as mesotidal, with a mean tidal range of 2.5 m and a maximal range of around 4 m. According to Davies (1972), the tidal regime occurs in the transition zone between areas mainly affected by swell waves and those affected by storm waves. The swell waves mainly arise from the northwest and less often from the west and southwest. High waves (>3 m) mainly occur in winter and are generated by depressions moving from northwest and western directions. The largest waves have been recorded at Cabo Vilán (13.5 m in 2009), Estaca de Bares (2.9 m in 2008) and Cabo Silleiro (12.01 m in 2014).

A. Pérez-Alberti (✉) · A. Gómez-Pazo
Department of Geography, Faculty of Geography and History, University of Santiago de Compostela, Praza da Universidade 1, 15782 Santiago de Compostela, Galicia, Spain
e-mail: augusto.perez@usc.es

A. Gómez-Pazo
e-mail: a.gomez@usc.es

J. A. Morales (ed.), *The Spanish Coastal Systems*,
https://doi.org/10.1007/978-3-319-93169-2_2

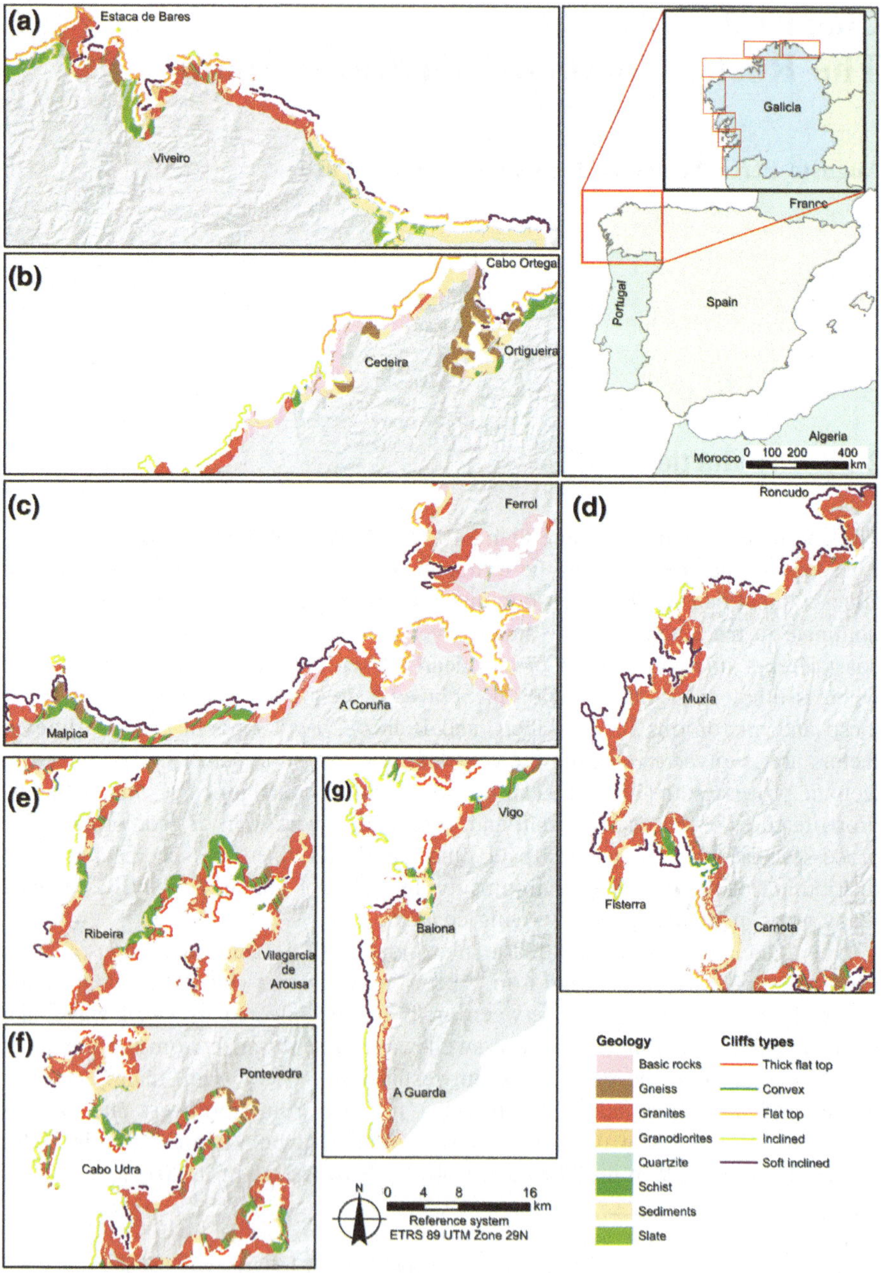

Fig. 2.1 Lithological map and types of cliffs on the Galician coast

2.2 Structural and Lithological Control of the Genesis of Rocky Coasts

The structure of coastal landforms is closely related to the geotectonic evolution of Galicia and can be observed at different scales, ranging from megaforms, which demarcate the overall design of the coast, to microforms, which introduce subtle, small-scale differences. The structural configuration of Galicia in general, and of the coast in particular, is the result of a complex evolutionary process characterized by the overlapping of two orogenic cycles: the Variscan and the Alpine orogenies. The first gave rise to a group of paleozoic rocks originating between the Precambrian and the Upper Carboniferous and whose structure is marked by the effects of the Variscan orogeny, although some structures were altered during the Alpine cycle, giving rise to the Cantabrian Range (Alonso et al. 1996).

Generation of an intense fracture network led to compartmentalization of the coast and the genesis of a sawtooth design with numerous inlets and outlets. Obviously not all of the structures are related to the fracture network as the lithological banding also introduces differentiating elements derived from the presence of different types of rock and as a result of this, differential weathering processes.

A large part of the Galician coastline is formed on granitic rock, although other types of rock appear in some areas (Fig. 2.1). This is the case of the basic and ultrabasic rocks that dominate in the Cabo Ortegal Complex within the Serra da Capelada, and of the slates and schists in the surroundings of the Ría de Ortigueira and the northern part of the Ría de Vigo. Thus, the type of rock, its mineralogical composition and degree of fracturing are, along with the degree of weathering, of vital importance in shaping the coastline. For any scale of analysis, the arrangement of the rocky outcrops establishes many of the lines traced by vegetation on the coast and, similarly, may establish a relationship between the shape of the vertical profile and the structural arrangement of the materials. However, the effectivity and action of the erosive processes not only depend on the type of rock, but also on the geometry and on patterns of discontinuities, which establish lines of weaknesses that favour erosion (Trenhaile 1987; Sunamura 1992).

2.3 The Coastal Cliffs

Sea cliffs are understood to comprise rock faces that occur at the edge of the sea and that are directly or indirectly affected by wave dynamics. However, the multiple factors involved both in the past and at present make it very difficult to generate a clear, unequivocal classification of cliffs. The shape of a cliff is the result of interactions between diverse variables: geological, climate, oceanic and biogeographical, together with the depth of the water and the amount and type of base material, the topography of the cliff surface and changes in height relative to sea level. As a result of the combined actions of these factors, the cliff profile is related to erosion by marine or subaerial agents and the time during which these operate (Emery and Kuhn 1982). Cliff profiles are therefore determined by the interaction between marine and terrestrial processes, the influence of which is more closely related to the relative than to the absolute effectivity, i.e. there is no dominant process but rather a balance between processes, which together determine the dynamics that shape the profile.

Marine erosion includes abrasion processes, weathering, mechanical attack and biological activity. In turn, the effects of marine erosion also initiate other processes, such as mass movements, which may also occur as a direct result of terrestrial factors. Littoral morphogenetic processes are therefore not linear in nature and are not accumulative but vary in accordance with the existing system (Cowell and Thom 1994). Although these processes may act at the same time, overlap or occur in spatial juxtaposition, they may also give rise to temporal succession. Thus, although the profiles of many cliffs are inherited, they are continually being affected by other phenomena. Most cliffs can therefore be characterized as polygenic forms.

Cliffs can therefore be differentiated by their shape, the dominant type of rock, the composition of the cliff base, the height, slope and orientation and the degree of stability. The number of types of cliff will therefore depend on the number of analytical elements considered.

2.4 Classification of Galician Cliffs by Their Shape

Analysis using a Geographic Information System (GIS) tool including a 2 and a 5 m Digital Elevation Model (DEM) and an orthophoto of spatial resolution 0.25 m, together with field validation of the data has enabled differentiation of 3 broad categories of cliffs. These also include numerous slight differences at local levels. The main categories are (a) cliffs with convex or concave slope; (b) flat-topped cliffs with no associated posterior plain; and (c) flat-topped cliffs with associated posterior plain.

The profiles shown in Figs. 2.2 and 2.3 were constructed using a DEM of spatial resolution 5 m, except for those in Figs. 2.2(9) and 2.3(3), which were constructed using a 2 m DEM to enable more detailed analysis of the shapes of these sectors. Different elevational and longitudinal scales were used to improve the accuracy of representation of the morphological characteristics of the cliff profiles.

2.4.1 Cliffs with Convex or Concave Slopes

These cliffs have a profile with two well-differentiated segments: a basal escarpment or sea cliff and an upper slope that tends to be either convex or concave, depending on the form, height and slope of each of the segments. Although all authors

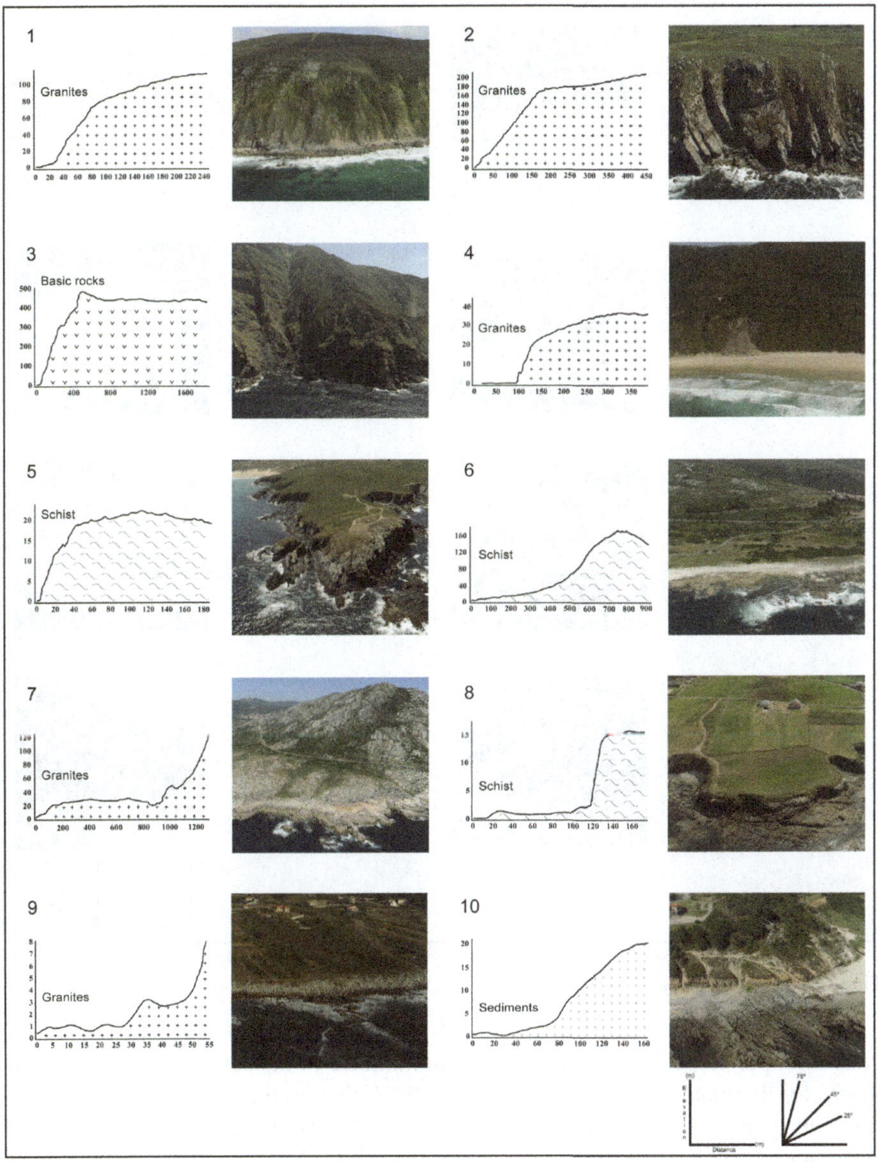

Fig. 2.2 Types of cliffs and associated profiles (I)

generally consider that these cliffs are recent forms, opinions about their origin vary. Cliffs are thus considered to be the result of the synchronous interaction of subaerial and marine processes, but are also viewed as undulating relief with gentle slopes, shaped under continental conditions and later attacked by the sea (Guilcher 1954). Emery and Khun (1982) consider that the compound-type profiles, with

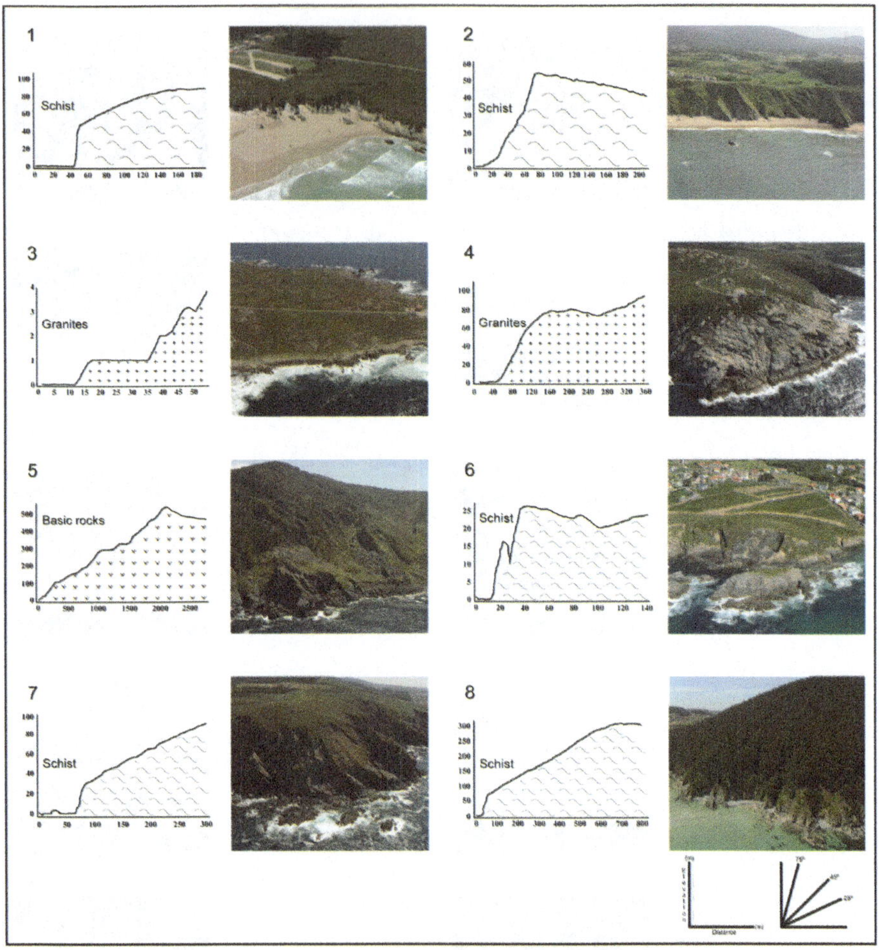

Fig. 2.3 Types of cliffs and associated profiles (II)

convex upper slope, merely reflect a predominance of subaerial processes that are more closely related to the resistance of the rock than to the temporal succession of different types of climate. The scheme proposed by Trenhaile (1987) suggests that the shape of a cliff depends on the balance between subaerial and marine processes, mainly due to the variations in sea level, as a consequence of climate variations during the Quaternary.

It is also possible to differentiate between cliff slopes with steep gradients (Fig. 2.2(2)) and those with gentle gradients (Fig. 2.3(8)). Concave (Fig. 2.2(4)) and convex (Fig. 2.2(6)) slopes can also be distinguished. These types of cliffs and profiles are mainly associated with granitic rock, in the case of the concave profiles, and with schists, as observed in the profile used as an example of a convex form.

These types of cliffs are widely distributed along the Galician coast, between the Rías Baixas in the south and Ferrol in the north (A Coruña).

2.4.2 *Flat-Topped Cliffs Without Associated Plain*

Sea-facing slopes vary greatly, ranging from almost horizontal flanks to slopes of less than 20°. The profile may also be concave or convex and is therefore somewhat similar to the previous category. The difference lies in the existence of a small plateau in the upper part of slope less than 8°, which does not favour continental dynamics and therefore a marine influence is essential.

This type of cliff is represented in Fig. 2.3(6) and is characteristic of zones where schists dominate, as in northern coastal areas such as Valdoviño (A Coruña).

2.4.3 *Flat-Topped Cliffs with Associated Plain*

These cliffs are associated with wide inland plateaus. These plateaus have traditionally been considered ancient marine features. The best examples of this is the Rasa Cantábrica. However, apart from the obvious difficulties in explaining the conditions necessary for shaping a coastal platform, of more than 1000 m, there is also no erosive (grottos/caves, paleocliffs, potholes etc.) or sedimentary (beach) evidence. These plateaus may therefore correspond to ancient, flattened continental surfaces that are currently beside the sea.

The profile of these cliffs varies depending on the location, ranging from almost vertical (Fig. 2.2(8)) to gentle slopes (Fig. 2.3(5)). This, along with the lithology or degree of fracture has a greater influence on the dynamics than the existence of large surfaces, as flat surfaces dissipate the energy of the continental waters, favouring the influence of wave action on cliff evolution. The best examples of this type of cliff is found in areas dominated by schists and where the predominant material is classified as basic or ultrabasic rock. The most representative examples of these cliffs appear in eastern Galicia, on the Cantabrian coast.

2.5 Classification of Cliffs by Height and Slope

Guilcher (1954) classified sea cliffs on the basis of height as follows: low cliffs (less than 2 m), medium cliffs (between 2 and 10 m), high cliffs (more than 10 m), and mega-cliffs (higher than 500 m). Bird (2008) used the same terms for cliffs higher than 500 m, but applied the term high cliffs to those of between 100 and 500 m. These classifications are theoretical and do not have a clear empirical basis.

Table 2.1 Relationship between elevation and cliff slope on the Galician coast

Slope \ Elevation	0	< 50	50-100	100-150	150-200	200-250	250-300	> 300
0	5.79	93.88	0.25	0.05	0.01	0.01	0.00	0.00
< 4	0.00	88.39	8.32	2.16	0.69	0.29	0.06	0.10
4 - 8	0.00	70.09	20.16	6.05	2.12	0.66	0.29	0.63
8 - 16	0.00	47.15	31.29	13.24	4.84	1.80	0.76	0.92
16 - 32	0.00	31.25	30.39	20.30	9.86	4.43	2.09	1.69
32 - 64	0.00	43.06	20.22	10.79	7.26	5.56	4.73	8.38
> 64	0.00	18.73	37.48	19.78	9.89	4.24	1.77	8.12
Total	0.87	67.42	18.04	7.78	3.19	1.32	0.62	0.76

The red colors are the highest percentages and the green colors indicate the lowest values

Examination of the Galician coastline, in this case using a DEM of spatial resolution 5 m, indicates a predominance of low-lying cliffs, also with low slopes (Table 2.1). For example, 94% of the cliffs lower than 50 m have a slope of less than 4°. However, 8% of the higher cliffs have slopes higher than 64°. It appears evident that the importance of higher slopes increases with increasing elevation.

2.6 Classification of Cliffs on the Basis on Lithology

2.6.1 Cliffs Formed by Granite Rock

The composition of granite is not uniform and granite rocks do not all respond in the same way to weathering processes. Two mica fine-grained granite may give rise to very different forms than those formed by granodiorites and coarse-grained granite. There is not, therefore a clear relationship between the type of granite and the cliff profile. The degree of fracturing is also important. Rectilinear cliffs tend to be abundant in sites where the granite rocks are associated with intense fracture networks. When flat surfaces occur towards the interior, flat-topped cliffs or with a gently sloping posterior plateau are formed. By contrast, when the relief is more abrupt, with steeper slopes and a higher degree of weathering, the cliffs tend to be sub-vertical, with numerous signs of detachments and landslides. Cliffs formed on granodiorites are usually curved and covered by boulders.

A third element to consider is the height of the profile. The change in height of a cliff and profile may increase the area and slope of the basin that receives water, leading to a higher concentration of moisture in the upper part of the seaward slopes. This explains the presence of detachments associated with hydrostatic overloading and the existence of accumulations of boulders at the cliff base. The change in energy as a consequence of changes in the sea level will influence the importance of marine processes.

Rock weathering should not be overlooked as in intensely weathered zones with abundant detachments, the cliffs are undulating, with circular inlets and significant changes in vegetation cover. The presence of weathered material and numerous fractures enables water to penetrate and saturate the soil, thus increasing the weight and favouring detachments at the cliff face. In cliffs formed on granodiorites, edaphogenic processes have given rise to differential weathering of the rock. The wave action causes gradual washing away of the weathered material, thus favouring the detachment of boulders or the in situ disinterment of forms carved in the rock interior. These weathering and/or disinterment processes cause the appearance of rounded, convex forms and of accumulations of boulders at the base of the cliff. When the erosion causes the disappearance of the weathering layer, the cliff landscape is characterized by large boulders.

Cliffs formed on granite rocks are widely distributed along the Galician coast. Those on the northern coast are particularly noteworthy (Fig. 2.1a, c), in contrast to those in areas of the southern coast between Cabo Silleiro and A Guarda (Pontevedra) (Fig. 2.1g) where the granite forms have less influence and are often found associated with accumulations of heterometric debris at the cliff base (Fig. 2.2(9)).

2.6.2 Cliffs Formed on Metamorphic Rocks

The mineralogical composition, the relative amount of quartz, the degree of stratification, schistosity and weathering generate very different types of cliffs, leading to coastline with numerous inlets and outlets marked by fracturing. Cliffs formed on slates and schists are highly conditioned both by the presence of dipping strata and by the orientation relative to wave attack. This explains the presence of vertical, inclined or sheer cliffs.

Intense weathering changes the shape of the coastline. Rotational detachments or landslides usually predominate, depending on the degree of weathering, and rocky platforms usually appear in front of cliffs, as seen in the interior of the Galician rías. In high energy zones, the dynamics of the cliffs formed on metamorphic rocks are marked by detachments, rotational landslides or by sedimentary cover.

Cliffs that form on metamorphic rock are widely distributed along the Cantabrian coast in Galicia. The most notable examples of this flat-topped type of cliff are found on schist material in the zone between Viveiro and Ribadeo (Lugo) (Fig. 2.1a). In other areas such as in northern Galicia, the metamorphic rocks are associated with variable slopes with numerous detachments (Fig. 2.3(1, 2)).

2.6.3 Cliffs Formed on Recent Sediments

Numerous ancient deposits formed by variable depositional sequences are found on the coast and give rise to heterogeneous granulometric composition and well-defined morphosedimentary evolution (Pérez-Alberti et al. 2009). These areas reflect the changes that the coast has undergone during at least 40 000 years (Pérez-Alberti et al. 2009), as a result of the interactions of marine regressions and transgressions with variations in climate and local depositional conditions. At an individual level, the characteristics defining these cliffs depend on the location, the source area, the association of sedimentary faces and the cliff shape.

Coastal deposits are currently being subjected to intense erosion, making it difficult to evaluate the area occupied at the moment of maximum accumulation. In general, depositional facies of continental origin are frequent and are acting as active cliffs with a high degree of mobility and which are affected by erosive processes, mainly of marine and to a lesser extent continental origin. The destruction of these deposits uncovers ancient littoral forms, such as platforms and caves.

The appearance of this dynamic cliff type is more localized that that of other cliff types by the more frequent presence of low sedimentary zones relative to well-defined cliffs. The profiles in these areas correspond to forms with gentle slopes and low to intermediate heights (Fig. 2.2(10)).

2.7 Classification of Cliffs According to the Composition of the Base

Information about the composition of the cliff base is fundamental for understanding cliff dynamics and evolutionary behaviour as the presence of material at the cliff base can slow down some erosion processes.

2.7.1 Cliffs with Sandy Areas at the Base

Sandy beaches dominate at the base of cliffs as a result of the action of sedimentary processes, which mobilize the material from nearby areas. The beach is converted into a type of defence at the cliff base so that the cliff is only affected by very high energy wave action. In this context, the coastal front mainly develops as a result of the action of continental run-off as large detachments or landslides, of potential risk to beach users.

Cliffs with beaches at their base are relatively common on the Galician coast. These types of cliffs may have different forms and composition, ranging from areas dominated by schists with vertical slopes and with a beach at the base (Fig. 2.3(1)) to granite cliffs with large accumulations of sediment at the base (Fig. 2.3(4)). Examples of the first type are found in the surroundings of Rinlo (Lugo) (Fig. 2.1a), whereas examples of the second type occur in the area of Ferrol (A Coruña) (Fig. 2.1b, c).

2.7.2 Cliffs with Accumulations of Boulders at the Base

This type of cliff is relatively abundant in Galicia. The boulders are of different sizes, with the longest axis reaching more than 2 m. They usually appear together with small abrasive platforms or stacks, which indicate that the cliff is retreating. The blocks effectively protect the cliff wall from the wave action by dissipating the energy of the waves.

Accumulation of boulders at the cliff base is almost exclusively associated with granitic areas of the Galician coast. The cliffs at Corrubedo (A Coruña) (Fig. 2.1e) and in the area between Cabo Silleiro and A Guarda (Pontevedra) (Fig. 2.1g) are clear examples of this type of cliff. In general, two types of profile are possible: one in which the height increases gradually with the boulders height that are present at the base (Fig. 2.3(3)) and the other in which the boulder-covered base is succeeded by an almost vertical cliff (Fig. 2.2(9)).

2.7.3 Cliffs with Platforms at the Base

These are found in highly localised areas of the coast, e.g. on the Cantabrian coast (Fig. 2.1a). They are formed on metamorphic rocks, in which stratification of the material or the degree of weathering favour retreat of the cliff.

2.8 Cliff Dynamics

Cliff stability depends on (a) the type of rock; (b) the structure; (c) the degree of weathering; (d) location in a high or low energy site; (e) orientation relative to the passage of storms and (f) the gradient, which favours the different types of movement of the existing material. These movements basically comprise detachments, collapses and rotational landslides, in addition to leaching of the weathered layer in granite zones (Pérez-Alberti et al. 2009).

2.8.1 Detachments

Detachments are basically fallen rocks. They can be differentiated into those that affect the cliff face and those produced in specific parts of the cliff. The latter type can be further divided into those produced in the upper part of the cliff and those at the bottom of the cliff. They are generally produced in deeply fractured, weathered rocks and in recent deposits. In the former case, detachments are most abundant in the highest sites. The best examples are found in coastal areas between Estaca de Bares and Ortigueira, although they also occur in other sites such as the sea wall on the coasts of Valdoviño, Narón and Ferrol (A Coruña) (Fig. 2.1b, c).

2.8.2 *Collapses*

These are produced as a result of a basal sapping. They can be differentiated into those produced in rocks and those affecting only the sedimentary cover. They may give rise to variable forms, the most spectacular of which are the blowholes (called *ollos* in Galicia) located on the coasts of Ribadeo (Lugo), Arteixo, Malpica and Laracha (A Coruña) (Fig. 2.1c).

2.8.3 *Rotational Landslides*

These are circular, spoon-shaped movements that cause overall retreat of the cliff. They have been observed in highly weathered rock, especially in the basic rocks in the Cabo Ortegal complex in the surroundings of Cariño (A Coruña) (Figs. 2.1a, b and 2.4).

2.9 Coastal Platforms

These are generally narrow platforms located in the intertidal zone. The term most commonly used at present is *shore platform*. However, various synonyms are used in the relevant literature, including *coastal platform, marine bench, high-water rock platform, abrasion platform, storm-wave platform* and *abrasion platform* (Sunamura 1992; Stephenson and Kirk 2005; Feal-Pérez et al. 2009).

Use of the term *shore platform* is generally adopted because it does not indicate the origin, as in the term abrasion, which in the context of littoral geomorphology, describes the erosion of a rocky surface by the movement of particles of sediment by wave action (Sunamura 1992; Trenhaile 1987, 1997). The movements produced by bearing, friction and saltation may potentially be produced from the fracture line to the maximum level reached by the *wave run-up*, whereas some sediment cannot be mobilized. However, although abrasion is not directly related to the tidal range, their efficacy tends to decrease rapidly at depth, outside of the *foreshore* zone (Robinson 1977a, b; Trenhaile 1987, 1997).

Platforms are very abundant in the northwest of the Iberian Peninsula and occur both in the interior of the rías and outside of these, in high and low energy zones,

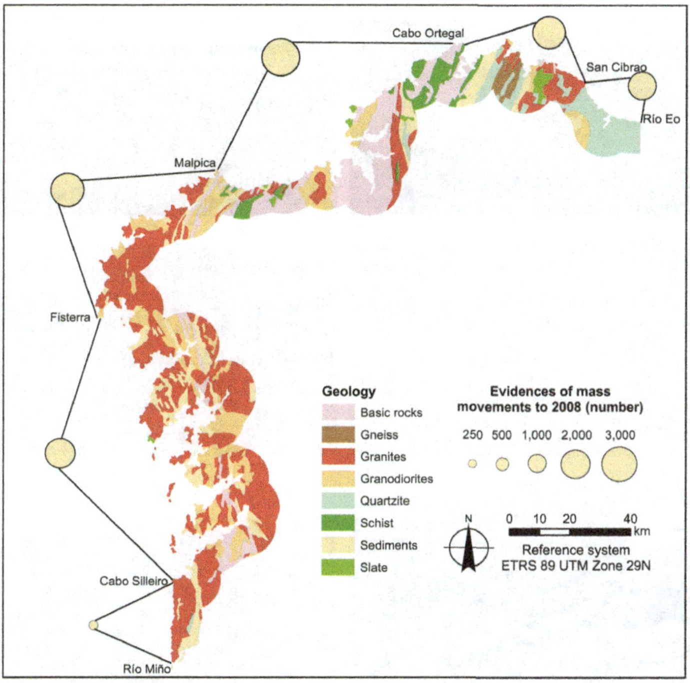

Fig. 2.4 Evidence of mass movements on the Galician coast since 2008 (modified from Pérez-Alberti et al. 2009)

and are formed on either metamorphic rock (Fig. 2.5(1, 2 and 4)) or granitic rock (Fig. 2.5(3)). In some sites, narrow platforms (5 m wide) exist: these are subhorizontal and of low uniformity, sometimes even with ledges of up to 1 m high. In other sites, the platforms are more highly developed, of width up to 50–100 m, notably uniform and with slope of between 0° and 2° covered by drift material that forms deposits of variable size and power, comprising material ranging in size from gravel to metre-wide boulders. Although the platforms are sometimes caused by abrasion, there is some evidence showing that they are formed as a result of weathering processes and therefore differential dissection.

2.9.1 *Platform Formation*

The question arises as to whether the platforms are formed by the same processes as cliff retreat or by other processes not necessarily consistent with the former (Trenhaile 1982). Various authors have suggested that the platforms are developed close to a level of weathering defined by saturation of the rock, either by leaching of

Fig. 2.5 Coastal platforms: (1) Caamaño (A Coruña); (2) Sanxenxo (Pontevedra); (3) Punta Corrubedo (A Coruña); (4) Ridadeo (Lugo)

weathered material or by concentration of wave energy (Dana 1849; Bartrum 1916; Bell and Clarke 1909 cited by Trenhaile 1987).

Trenhaile (1973, 1974, 1978, 1980, 1982, 1987) attempted to establish a relationship between the diverse parameters of the geometry of the platforms based on mathematical models and statistical techniques, observing a correlation between the tidal range, resistance of the rock and wave energy. According to this author, of all parameters considered, the tidal range is the main factor determining the slope of the platforms and is more important than others such as structure and lithology. The slope of the platform will also decrease with the tidal range, as the wave energy is spatially and temporally concentrated in a narrower area. Thus, almost horizontal platforms may be formed when the tidal range is less than 3 m (Trenhaile 1974). Other aspects, such as the mean elevation of the platform, or its uniformity or width will depend to a greater extent on local conditions, such as the type of rock, height of the cliff or amount and size of the rubble accumulated at the base.

In light of the evidence from the Galician coast, it is considered that, as for cliff retreat, the formation and evolution of coastal platforms is largely determined by the balance between previous weathering of the rock, structural factors and wave energy. The formation of platforms involves a balance between the rock resistance and the wave energy, where the rock resistance is understood to result from the different factors that determine the fragility of a material (mineralogical and petrological characteristics, degree of fracturing, etc.) for example boulder beaches, considered below.

The existence in some sites of ancient deposits on the platforms (Trenhaile et al. 1999) indicate that these are polygenic forms formed during the Eemiense interglacial period (Perez-Alberti et al. 2010) and are currently being reshaped.

2.10 Boulder Beaches (*coídos*)

Areas characterised by large accumulations of boulders are relatively abundant on the Atlantic coast. These areas are called *coidos* in Galician. They can be differentiated into those resulting from (i) weathering of the layer of granite rock, (ii) the dismantling of granite platforms, and (iii) the remobilization of deposits generated by movement of material on the cliff faces or of ancient deposits, mainly of periglacial or snow origin (Blanco-Chao et al. 2002; Pérez-Alberti and Bedoya 2004).

In the first case, an initial phase of subsurface weathering of the rock is followed by leaching by marine or continental waters. Granite weathering has been marked by the existence of an extensive network of fractures generated as the weathering advances and by the genetic predisposition of some types of granite, such as granodiorites, to undergo spheroidal decomposition. Moreover, the presence of nuclei of unaltered granite within the weathered material has favoured the accumulation of boulders of variable size. Good examples of these are found at the mouths of the rías of Ferrol and O Barqueiro (A Coruña) (Fig. 2.1c), at the Punta do Couso, in Corrubedo (A Coruña) and on the coast of O Grove (Pontevedra). The size of the boulders does not favour their movement as the wave energy is generally dissipated by these massive macro-supported accumulations with numerous cavities, leading to differential accumulation. The hollow spaces between the largest blocks act as sediment traps for the smaller blocks. By contrast, accumulations formed by smaller clasts are highly dynamic. Independently of their origin, boulder beaches can be differentiated into 5 main morphological types: longitudinal, double-peaked, bow-shaped, channel-type and simple-peaked (Fig. 2.6).

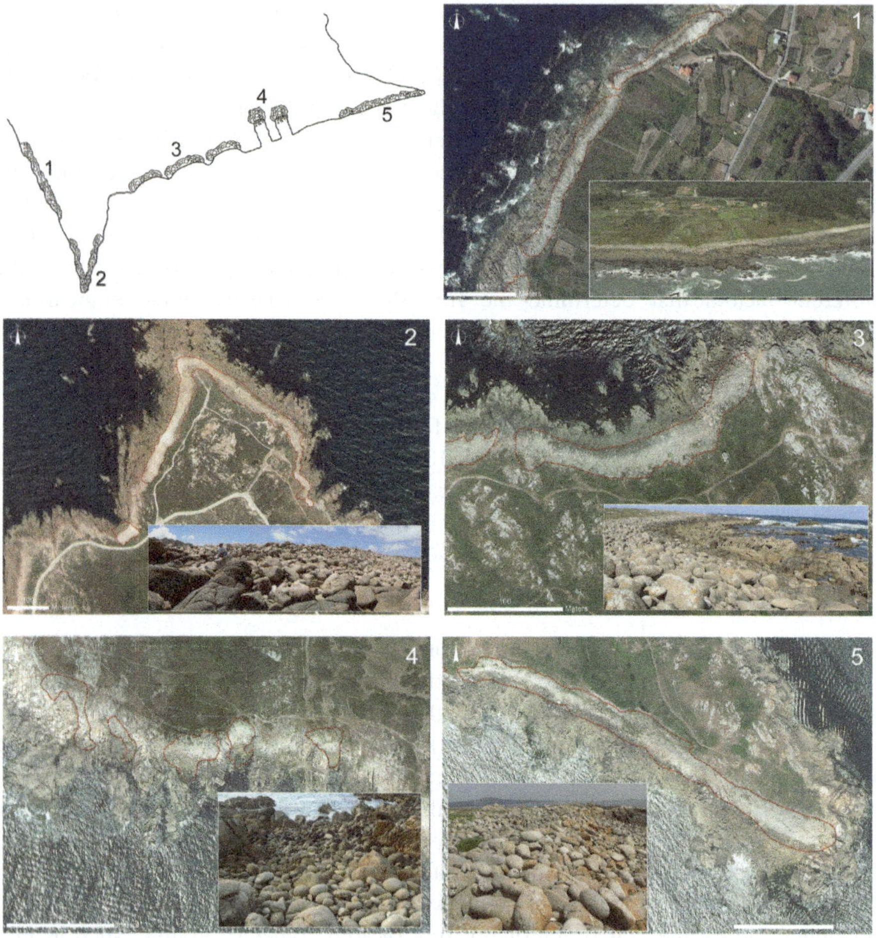

Fig. 2.6 Types of boulder beaches (*coidos*)

2.10.1 Types

2.10.1.1 Longitudinal

Boulder beaches are formed at the upper level of narrow and irregular coastal platforms of width 50–75 m (depending on the area). These beaches become progressively longer between Cabo Silleiro and A Guarda (Fig. 2.1g). They are comprised of heterometric granite boulders of diameter up to 4 m with many specimens larger than 1 m in diameter. The boulders are rounded or angular and form crests of width less than 80 m and height between 5 and 6 m (Pérez-Alberti et al. 2012).

Their origin is related to processes driven by the fracturing of the material on the platforms, favouring weathering at the discontinuities and thus contributing to the fragmentation. Channels or small cavities then gradually open up along lines of weakness in the structure or lithology, gradually forming an extensive network both horizontally and vertically. The high wave energy from NW or SW directions has driven the mobilization and accumulation of material.

2.10.1.2 Double Peaked

Good examples of this type of boulder beach are found in the area of the coast between Cabo Vilán and Camelle (A Coruña) (Fig. 2.1d). The extensive fracture network favours the formation of small boulders, generally of diameter less than 50 cm. The size favours the mobility of the boulders making them rounder than otherwise. On the other hand, the sawtooth arrangement of the coastline gives rise to numerous inlets and outlets. The area is thus affected by a combination of high energy waves of SW, W and NW components that transport the boulders to one face or other from arrowhead-shaped outlets.

Boulder beaches of width 20–40 m and height 4–5 m are located (as in the previously described type) at the upper level of narrow platforms not usually wider than 60–70 m.

2.10.1.3 Bow-Shaped

Shorter than the previous type, bow-shaped boulder beaches can reach lengths of 200–300 m (depending on the area) and are separated by rocky outlets. The accumulations can cover areas of width up to 80 m. Unlike the previously described types, these beaches are not always associated with platforms and usually extend from the lower tidal level to 8 m above the tidal levels in areas such as Laxe Brava (Ribeira, A Coruña) (Fig. 2.6(3)).

The granulometric properties of the material are more variable and the boulders are rounder than on other beaches. The largest diameter of some of the boulders reaches 2 m, although it is not usually greater than 1 m. They overlap towards the land, pushed by waves from a NW or SW direction (Pérez-Alberti et al. 2012; Pérez-Alberti and Trenhaile 2015a, b).

2.10.1.4 Channel-Type

Boulder beaches can also form narrow channels, of width 30–50 m and length 70–80 m, opening up from a wide network of fractures that follow a N-S direction (Fig. 2.6(4)). These beaches are very similar to the bow-type beaches regarding the clast size and distribution from low tide level. However, they are characterized by being mushroom shaped, with a lower level delimited by rocky walls and an upper level on the coastal plateau (Pérez-Alberti et al. 2012).

2.10.1.5 Simple Peaked

This type of beach is formed by accumulations of boulders that have been moved in the direction of the sedimentary materials. To some extent, they comprise the final sectors of long beaches situated at the upper level of the platform (Fig. 2.5(3)). The only example in Galicia is at Punta Corrubedo (Ribeira, A Coruña) (Fig. 2.6(5)). The beach is of maximum width 40 m and maximum height, 5 m. It is formed by rounded boulders usually of diameter less than 1 m.

2.10.2 Dynamics

Boulder beaches are surprisingly dynamic. Their movements have been monitored in two areas of the Galicia coast: Laxe Brava (Barbanza Peninsula) (Fig. 2.1f) and Oia (south of Pontevedra) (Fig. 2.1g). Pérez-Alberti and Trenhaile (2015a, b) have demonstrated movement of 80% of the boulders located at between 0.5 and 6 masl at Oia (Pontevedra) and between 58 and 69% of the boulders at Laxe Brava (Ribeira, A Coruña) located at of between 1 and 4.5 masl.

RFID (Radio-Frequency IDentification) sensors have recently been used to monitor the movements of boulders at Oia. Gómez-Pazo and Pérez-Alberti (2017) have reported that 81% of the boulders moved more than 50 cm in the winter of 2016–2017, with an average displacement of 5.06 m. These changes in position are associated with winter storms and the arrangement of the boulders on the beach.

Although boulder beaches in Galicia are usually associated with granite platforms, in other parts of the world they have been reported to be associated with other types of rock. However, they are always associated with intensely fractured, high energy areas (Blanco-Chao et al. 2007; Knight et al. 2009; Hall 2011; McKenna et al. 2011; Knight and Burningham 2011; Stephenson and Naylor 2011). Interest in their study has, however focused on discussing whether boulder beaches are formed by paleo-tsunami or storm waves (Nott 2004; Paris et al. 2011), and most studies have concentrated on almost bare rock surfaces without much associated sediment (Trenhaile 1987, 2011). Scarce interest has been shown in rocky coasts as sedimentary environments (Felton 2002; Noormets et al. 2004). Thus, despite the evident dynamic relationship between terrestrial and depositional forms and processes, boulder beaches are generally considered to be different entities.

As already mentioned, the boulders present on boulder beaches have either arisen as a result of platform or cliff erosion, or by washing of deposits of glacial, periglacial or nival origin or of other Quaternary deposits (Oak 1984; McKenna 2005; Chen et al. 2011; Blanco-Chao et al. 2007; Pérez-Alberti et al. 2009). However, independently of their origin, it has been shown that the boulder beaches in Galicia (Pérez-Alberti and Trenhaile 2015a, b) are moving on coastal platforms pushed by large storm waves, to sites located well above the high tide level. This gives rise to intense erosion in some areas and has a protective role in the evolution of the platform edges, especially in sites where the beaches are adjusting to the current sea level.

References

Alonso, J. L., Pulgar, J. A., García-Ramos, J. C., & Barba, P. (1996). Tertiary basins and Alpine tectonics in the Cantabrian Mountain (NW Spain). In P. Friend & C. Dabrio (Eds.), *Tertiary basins of Spain* (pp. 19–22). Cambridge: Cambridge University Press.

Bird, E. (2008). *Coastal geomorphology: An introduction* (p. 436). Chichester: Wiley.

Blanco-Chao, R., Costa Casais, M., Martínez Cortizas, A., Pérez-Alberti, A., & Vázquez Paz, M. (2002). Holocene evolution in Galician coast (NW Spain): An example of paraglacial dynamics. *Quaternary International, 93–94,* 149–159.

Blanco-Chao, R., Pérez-Alberti, A., Trenhaile, A. S., Costa Casais, A., & Valcarcel-Diaz, M. (2007). Shore platform abrasion in a para-periglacial environment, Galicia, northwestern Spain. *Geomorphology, 83,* 136–151.

Chen, B., Chen, Z., Stephenson, W., & Finlayson, B. (2011). Morphodynamics of a boulder beach, Putuo Island, SE China coast: The role of storm and typhoon. *Marine Geology, 283,* 106–115.

Cowell, P. J., & Thom, B. G. (1994). Morphodynamics of coastal evolution. In R. W. G. Carter & C. D. Woodroffe (Eds.), *Coastal evolution. Late Quaternary shoreline morphodynamics*. Cambridge University Press.

Davies, J. L. (1972). *Geographical variation in Coastal development* (p. 204). Edinburgh: Oliver and Boyd.

Emery, K. O., & Kuhn, G. G. (1982). Sea cliffs: Their processes, profiles and classification. *Geological Society of America Bulletin*, 93.

Feal-Pérez, A., Blanco-Chao, R., & Valcárcel-Díaz, M. (2009). Influencia de formas y procesos heredados en la evolución reciente y en los procesos morfodinámicos actuales en un sector de costa rocosa: Punta Gallín, costa cantábrica gallega. *Revista de la Sociedad Geológica de España, 22*(1–2), 67–78.

Felton, E. A. (2002). Sedimentology of rocky shorelines: 1. A review of the problem, with analytical methods, and insights gained from the Hulopoe Gravel and the modern rocky shoreline of Lanai, Hawaii. *Sedimentary Geology, 152,* 221–245.

Gómez-Pazo, A., & Pérez-Alberti, A. (2017). Initial evaluation of the use of RFID sensors for monitoring the geomorphological dynamics of a boulder beach. (Oia, Galicia). *Geo-Temas, 17,* 143–146.

Guilcher, A. (1954). *Morphologie littorale et sousmarine* (p. 126). Paris: University Press.

Hall, A. M. (2011). Storm wave currents, boulder movement and shore platform development: A case study from East Lothian Scotland. *Marine Geology, 283,* 98–105.

Knight, J., & Burningham, H. (2011). Boulder dynamics on an Atlantic-facing rock coastline, northwest Ireland. *Marine Geology, 283,* 56–65.

Knight, J., Burningham, H., & Barrett-Mold, C. (2009). The geomorphology and controls on development of a boulder-strewn rock-platform. *Journal of Coastal Research SI, 56,* 1646–1650.

McKenna, J. (2005). Boulder beaches. In M. L. Schwartz (Ed.), *Encyclopedia of Coastal Science* (pp. 206–208). Berlin: Springer.

McKenna, J., Jackson, D. W. T., Andrew, J., & Cooper, G. (2011). In situ exhumation from bedrock of large rounded boulders at the Giant's Causeway, Northern Ireland: An alternative genesis for large shore boulders (mega-clasts). *Marine Geology, 283,* 25–35.

Noormets, R., Crook, K. A. W., & Felton, E. A. (2004). Sedimentology of rocky shoreline: 3. Hydrodynamics of megaclast emplacement and transport on a shore platform Oahu, Hawaii. *Sedimentary Geology, 172,* 41–65.

Nott, J. (2004). The tsunami hypothesis: Comparisons of the field evidences against the effects, on the Western Australian coast, of some of the most powerful storms on Earth. *Marine Geology, 208,* 1–12.

Oak, H. L. (1984). The boulder beach: A fundamentally distinct sedimentary assemblage. *Annals of the Association of American Geographers, 74,* 71–82.

Paris, R., Naylor, L. A., & Stephenson, W. (2011). Boulders as a signature of storms on rock coasts. *Marine Geology, 283,* 1–11.

Pérez-Alberti, A., & Bedoya, L. J. (2004). Caracterización de las playas de Cantos y bloques (coídos) en el noroeste de la Península Ibérica. In R. Blanco-Chao, J. López Bedoya, A. Pérez-Alberti (Eds.), *Procesos geomorfológicos y evolución costera: actas de la II Reunión deGeomorfología Litoral* (pp. 371–400). Santiago de Compostela.

Pérez-Alberti, A., & Blanco-Chao, R. (2005). Controles y balances geomorfológicos en costas rocosas de macizos antiguos. El ejemplo de Galicia (Noroeste de la Península Ibérica). *Geomorfología litoral i Quaternari.* Publicaciones de la Universitat de Valencia.

Pérez-Alberti, A., Blanco-Chao, R., Otero, M., Macias, I., Bedoya, J. L., & Valcarcel-Díaz, M. (2009). Cambios ambientais detectados na costa de Galicia durante o Plistoceno e Holoceno e dinámica actual. In P. Muñuriz, F. Cañamero, & G. Gesteira (Eds.), *Evidencias e Impactos do cambio climático en Galicia.* Xunta de Galicia. Santiago de Compostela.

Pérez-Alberti, A., Cunha, P. P., Murray, A. S., & Buylaert, J. P. (2010). On the coastal evolution of NW Iberia (Galicia) during the late Pleistocene. In *18 International Sedimentological Congress.* Mendoza (Argentina).

Pérez-Alberti, A., & Trenhaile, A. S. (2015a). An initial evaluation of drone-based monitoring of boulder beaches in Galicia, north-western Spain. *Earth Surface Processes and Landforms, 40,* 105–111.

Pérez-Alberti, A., & Trenhaile, A. S. (2015b). Clast mobility within boulder beaches over two winters in Galicia, northwestern Spain. *Geomorphology, 248,* 411–426.

Pérez-Alberti, A., Trenhaile, A. S., Pires, A., López-Bedoya, J., Chaminé, H. I., & Gomes, A. (2012). The effect of boulders on shore platform development and morphology in Galicia, north west Spain. *Continental Shelf Research, 48,* 122–137.

POL Galicia. (2010). http://www.xunta.es/litoral/web/index.php/descargables.

Robinson, L. A. (1977a). The morphology and development of the northeast Yorkshire Shore Platform. *Marine Geology, 23,* 237–255.

Robinson, L. A. (1977b). Marine erosive processes at the cliff foot. *Marine Geology, 23,* 257–271.

Stephenson, W., & Kirk, R. (2005). Shore platforms. In *Encyclopedia of coastal science* (pp. 873–875). Netherlands: Springer.

Stephenson, W., & Naylor, L. A. (2011). Within site geological contingency and its effects on rock coast erosion. *Journal of Coastal Research, 64,* 831–835.

Sunamura, T. (1992). *The geomorphology of rocky coasts* (302 pp). Wiley.

Trenhaile, A. S. (1973). The geometry of shore platforms in England and Wales. *The Transactions of the Institute of British Geographers, 62,* 129–142.

Trenhaile, A. S. (1974). The morphology and classification of shore platforms in England and Wales. *Geografiska Annaler, 56A,* 103–110.

Trenhaile, A. S. (1978). The shore platformas of Gaspé, Quebec. *Annals of the Association of American Geographers, 68*(1), 95–114.

Trenhaile, A. S. (1980). Shore platform: A neglected coastal feature. *Progress in Physical Geography, 4,* 1–23.

Trenhaile, A. S. (1982). The width of shore platforms; a theoretical approach. *Geografiska Annaler. Series A. Physical Geography*, 147–158.

Trenhaile, A. S. (1987). *The geomorphology of rock coasts* (p. 384). Oxford: Oxford University Press.

Trenhaile, A. S. (1997). *Coastal dynamics and landforms* (p. 366). Oxford: Clarendon Press.

Trenhaile, A. S. (2011). Cliffs and rocky coasts. In Wolanski & McLusky (Eds.), *Treatise on estuarine and coastal science*. Academic Press, Waltham.

Trenhaile, A. S., Pérez-Alberti, A., Costa Casais, M., Martínez Cortizas, A., & Blanco-Chao, R. (1999). Rock coast inheritance: An example from Galicia, Northern Spain. *Earth Surface Processes and Landforms, 24,* 605–621.

Chapter 3
Cliff Coast of Asturias

María José Domínguez-Cuesta, Pablo Valenzuela, Laura Rodríguez-Rodríguez, Daniel Ballesteros, Montserrat Jiménez-Sánchez, Laura Piñuela and José Carlos García-Ramos

3.1 Introduction

The Principality of Asturias (or simply Asturias) is a region located in the North of Spain (SW of Europe) limited by the Cantabrian Sea in the North and the Castilla y León, Cantabria and Galicia regions in the South, East and West respectively. The Asturias Coast represents around 30% of the Cantabrian Coast, the northern limit of the Iberian Peninsula, and is surrounded by the Cantabrian Sea. This sea represents the transition of the Atlantic Ocean to the Biscay Gulf, between Spain and France. Towards the South, the Cantabrian Coast is limited by the Cantabrian Mountains, up to 2,648 m altitude, which axis is located only at 20–50 km from the sea.

M. J. Domínguez-Cuesta (✉) · P. Valenzuela · L. Rodríguez-Rodríguez · D. Ballesteros
M. Jiménez-Sánchez
Departamento de Geología, Universidad de Oviedo, C/Jesús Arias de Velasco s/n,
33005 Oviedo, Spain
e-mail: dominguezmaria@uniovi.es

P. Valenzuela
e-mail: pvalenzuela@geol.uniovi.es

L. Rodríguez-Rodríguez
e-mail: laurarr@geol.uniovi.es

D. Ballesteros
e-mail: ballesteros@geol.uniovi.es

M. Jiménez-Sánchez
e-mail: mjimenez@geol.uniovi.es

L. Piñuela · J. C. García-Ramos
Museo del Jurásico de Asturias (MUJA), Rasa de San Telmo s/n,
33328 Colunga, Spain
e-mail: lpinuela.muja@gmail.com

J. C. García-Ramos
e-mail: jcgramos.muja@gmail.com

J. A. Morales (ed.), *The Spanish Coastal Systems*,
https://doi.org/10.1007/978-3-319-93169-2_3

The Asturian Cantabrian coast presents a general E-W trend along ca. 200 km between 43° 20′–40′N latitude and 7° 2′/4° 30′W longitude (Fig. 3.1a, b). The official coastal limit determined by the IHM (Instituto Hidrográfico de la Marina-Hydrographic Institute of the Navy) exceeds 660 km in real length, considering the numerous islets scattered along the fragmented Asturian coast. The IHN differentiates a double line constituted by the high and the low tide limits. In one hand, because the Asturian coast is an eminently a steep cliff coast, it is very frequent that both lines overlap themselves. On the other hand, the IHN distinguishes between a natural and an artificial coastline. In this work, the high tide limit is considered as the Asturian coast boundary, which mostly corresponds to the natural coastline (570 km). The remaining stretch of coast consists of artificial constructions, breakwaters, piers or bridges (86 km) and estuaries or fluvial transition environments (11 km).

In this work, the coastal area is considered as the inland territory closer than 5 km from the coastline that is analysed to better understand the geomorphological processes in a wider context. Respect to the altitude (Fig. 3.1c), more than 77% of the coastal area is below 200 m above the sea level (asl), being the 50% under 100 m. Only 0.15% of the coastal area exceed 1000 m elevation, which together with the slope distribution (more 40% is above 20°; Fig. 3.1d) show the roughness of the relief. The coastal area includes two towns with regional relevance, Gijón and Avilés with >270,000 and >80,000 inhabitants respectively, and more than 250 villages (Fig. 3.1b).

In general, the Asturian coast is eminently rocky and abrupt (cliff-bound coast), with a predominance of north-facing cliffs dotted with small coves and beaches (Flor and Flor-Blanco 2014a, b). The cliffs have steep slopes often showing dipping angles close to 90°, sometimes overturned, and frequent 5–50 m heights, with maximum height values exceeding 120 m West of Villaviciosa (Fig. 3.1d). From a geomorphological point of view, the coast can be divided into three parts (numbered as 1, 2 and 3 in Fig. 3.1c). The part 1 corresponds to the westernmost area of the Asturian coast, extending from Eo Estuary (in the boundary with Galicia) to the Avilés Estuary. It is characterized by an approximately 90 km-long-rectilinear outline interrupted by capes such as Busto and Vidio capes. Estuaries related to the Navia and Nalón rivers and others watercourses are fitted in this cliff-line coast. The central part (number 2 in Fig. 3.1c) is about 25 km length and corresponds mainly with Peñas Cape and its surroundings. The cape shows triangular shape from the plan view, with two sections with SW-NE and NW-SE directions, on both sides of the cape. The most important coast infrastructures of Asturias, such as the seaports of Avilés and Gijón, are located in this area (Flor et al. 2006). The easternmost part (number 3 in Fig. 3.1c) is the longest one with 100 km length. It presents an initial section with a W-E trend, followed by another section trending NW-SE, and other one with a slight S-SE orientation. This third part comprises the Villaviciosa, Ribadesella and Tina Mayor estuaries, corresponding the latter one with the Deva River that is the limit with the Cantabria region.

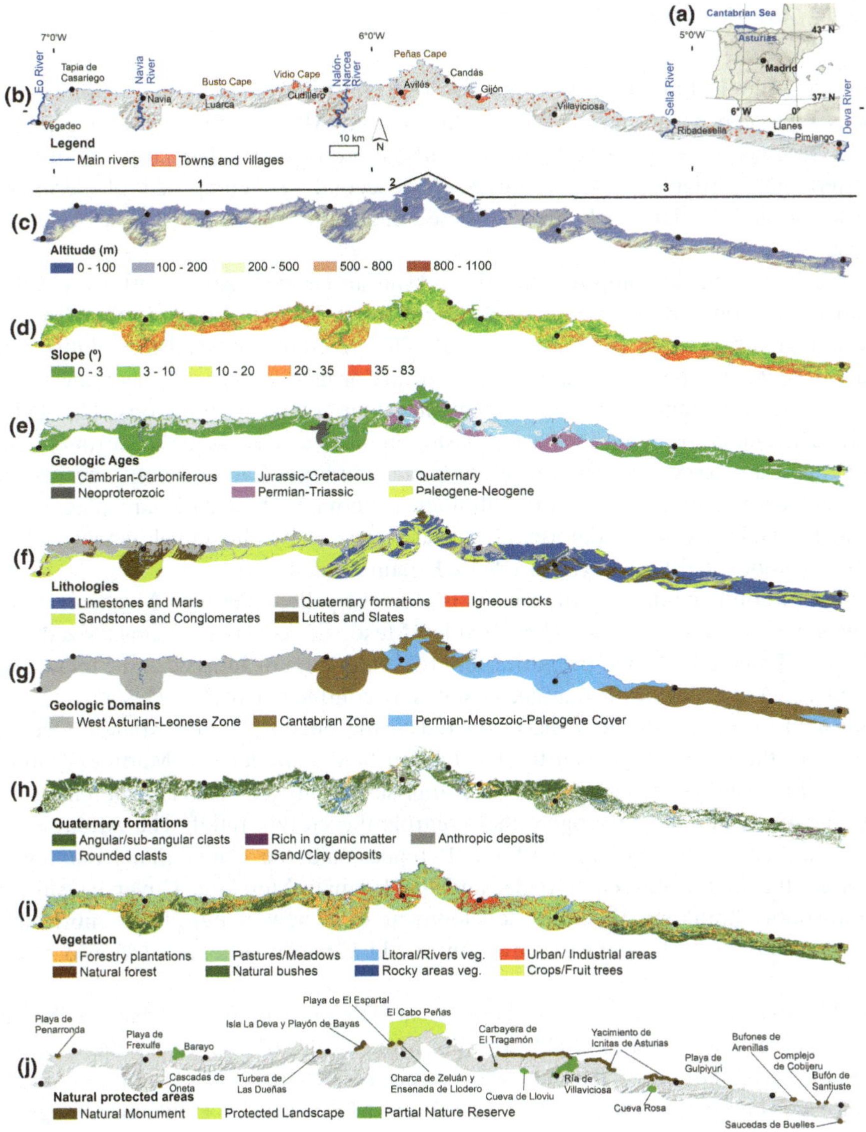

Fig. 3.1 Coastal area of Asturias (closer than 5 km from the high tidal coastline), **a** Location of the Asturian coast. **b** Main rivers, capes and towns. **c** Altitude. **d** Slope. **e** Distribution of rocks and surficial deposits according to their age. **f** Bedrock lithology. **g** Main geologic domains. **h** Quaternary formations. **i** Vegetation. **j** Natural protected areas. *Source of data* Digital terrain model (**b**, **c**, **d**, **j**) from Principado de Asturias, 2008; geologic information (**e**, **f**, **g**) from González Menéndez et al. 2008; Merino Tomé et al. 2013; Quaternary formations (**h**) and vegetation (**i**) from Principado de Asturias, 2002; and Natural protected areas (**j**) from Ministerio de Agricultura y Pesca, Alimentación y Medio Ambiente 2017

3.1.1 Geological Setting

From a geological point of view, the Asturian coast is formed by two main sets of rocks separated by an unconformity (Bastida and Aller 1995) (Fig. 3.1e–g). The first set consists of pre-Cambrian to Carboniferous rocks that crop out in the western and eastern part of Asturias. The second set, composed of Permian, Mesozoic and Tertiary rocks, is distributed along the central part of the Asturian coast.

The first main set comprises the West-Asturian Leonese Zone (ZAOL) and the Cantabrian Zone (ZC), included into the Iberian Massif (Lotze 1945). On the western Asturian coast, the rocks of the ZAOL are eminently sandstone, lutite and slate that present the typical deformation of the orogen hinterlands, showing large recumbent folds, regional metamorphism and intense tectonic foliations. The rocks of the ZC that crops out in the eastern Asturian coast as well as in the surroundings of Peñas Cape, constitute an almost complete Cambrian-Carboniferous succession in which carbonate and siliciclastic lithologies alternate. These rocks are affected by a thick-skinned tectonic deformation, characterized by the development of an E-direct imbricated thrust system (Pérez-Estaún et al. 1988).

The second main set of rocks is the so-called Permian-Mesozoic and Paleogene-Neogene covers. The Permian-Mesozoic cover is characterized by Permian-Triassic lutite and evaporite, and the alternation of Jurassic and Cretaceous carbonate and siliciclastic marine, costal and continental rocks. This cover was deposited in a extensional scenario related to the opening of the Atlantic Ocean, including the Biscay Gulf and the Cantabrian Sea (Lepvrier and Martínez-García 1990). The Paleogene-Neogene Cover comprises mainly siliciclastic materials from the erosion of the alpine orogen, that controls the current relief. The Alpine orogeny, linked to the formation of the Pyrenees and the Cantabrian Mountains, affected the first main set of rocks and the Permian-Mesozoic Cover causing a north-south shortening with the development of a new South-direct imbricated thrust system as well as the reactivation of old variscan structures (Alonso et al. 1996).

The spatial analyses of the Geological Digital Continues Map of Spain (González Menéndez et al. 2008; Merino Tomé et al. 2013) provides the chrono-lithologic distribution of the Asturian coastal rocks (Fig. 3.1f, g). 74.2% of the geological bedrock is Paleozoic in age, while Mesozoic and Paleogene-Neogene rocks occupy 24.5% and Neoproterozoic rocks represent 1.3% of the coastal area. Siliceous sandstone, conglomerate, lutite and slate occupy more than 61.75% of the 5 km buffer area, while limestone and marl constitute 39% and igneous rocks less than 1% of the geological bedrock.

3.1.2 Geomorphological Setting

In the Asturian coast, flat erosional surfaces locally known as “rasas”, located between 5 and 285 m asl, constitute the most outstanding elements of the relief in this area which run parallel to the coast (Flor 1983; Mary 1983). In general, the rasas dip less than 5° to the sea, descending steeply to the coast where the lower rasas are preserved. The emersion of the marine terraces occurred during, at least, the Pliocene-Quaternary by coastal uplifting estimated in around 1–2 mm a^{-1} (Jiménez-Sánchez et al. 2006; Álvarez-Marrón et al. 2008). A sole terrace with 100 km long is visible in the western part of the Asturian coast, modified probably by Quaternary faults (Álvarez-Marrón et al. 2008). Towards the East, it is possible to distinguish up to 12 levels of staggered rasas between the Nalón Estuary and the eastern boundary of Asturias (Flor and Flor-Blanco 2014a, b). The development of rasas and marine terraces depends strongly on the lithology of the bedrock, since rasas modelled on siliceous bedrocks are better preserved than terraces formed in calcareous substratum (Domínguez-Cuesta et al. 2015). Quartzite rasas are easily recognized in the western Asturias (Álvarez-Marrón et al. 2008), although rasas sculpted on Ordovician quartzite sandstone are also distinguished in eastern Asturias (Flor 2000; Jiménez-Sánchez et al. 2006; Adrados 2011). In Carboniferous and Jurassic limestone bedrocks, the delimitation of limestone rasas can not be carried out so easily because their original geometry was strongly modified by intensive karstification since at least Middle Pleistocene (Jiménez-Sánchez et al. 2006; Ballesteros et al. 2017). All rasas are crossed by the most important rivers of Asturias and also an incipient fluvial network can be identified on these flat surfaces; thus small streams that flow directly into the sea can be observed, some of which are presently hanging valley landforms.

Most of rasas, especially lower than 150 m altitude, are marine terraces originated by the emersion of marine abrasion platform (Álvarez-Marrón et al. 2008). The upper rasas could correspond ancient continental surfaces (Flor and Peón 2004). The marine origin is evidence by the identification of “rasa deposits” in the coastal surfaces (Flor 1983; Mary 1983; Adrados 2011), comprising around 60% of sand, 20% of silt and 20% of pebbles of sandstone or quartzite, which are frequently rod or disc shaped. Marine fauna was also recognized in these deposits (Flor 1983; Hoyos Gómez and Herrero 1989).

On their northern front, the rasas closest to the sea are delimited by steep cliffs. At the foot of these cliffs, abrasion platforms are preserved with different development degrees and slope angles depending on lithology. The current geomorphological evolution of cliffs takes place through slope instability processes, conditioned by the action of marine scouring and undercutting by waves, which helps to fall down and rework all materials detached from the cliff front (Adrados 2011; Flor-Blanco et al. 2015). The activity of these processes is an indicator of the importance that the progressive retreat of cliffs can achieve. According to the marine interpretation of the rasas, the limits between them would correspond to paleocoastlines (Álvarez-Marrón et al. 2008). These limits usually dip 20°–60° to the North.

With respect to fluvial and marine geomorphological process interactions, it is remarkable that to the West of Peñas Cape, the Nalón River meets the sea after draining 5,442 km^2, almost two thirds of the Asturian region. It supplies a great amount of sediments that are distributed towards the East in favor of the littoral drift (Flor and Flor-Blanco 2005). The presence of Peñas Cape causes a screen effect that drives to an unequal distribution of the sandy sediments on both sides. Thereby, the western coast of the Peñas Cape captures a large amount of these sediments. On the contrary, a shade area deficient of sandy sediments is created towards the East.

According to the thematic mapping of the Principado de Asturias (2002), the surficial formations of the coastal area (Fig. 3.1h) are mainly deposits of angular and sub-angular clasts rich in matrix (46%), followed by deposits of angular and sub-angular clasts with scarce matrix (36%). Most of these deposits come mainly from gravity and ancient fluvial processes. Other Quaternary formations are composed by rounded clasts (9%), sands and clays (5%) mainly related to fluvial, estuaries, dunes and beach deposits, anthropic deposits (3%) or deposits rich in organic matter (1%).

3.1.3 Climate, Coastal Dynamics and Vegetation Setting

The Asturian territory shows a characteristic Oceanic climate. More deeply, the coastal belt shows a Temperate climate Cfb, without dry season and with a temperate summer, according to the Köppen-Geiger classification (García Couto 2015; Olivares Navarro 2015). Precipitation and temperature are quite homogeneous, although with slight variations along the Asturian coast. Datasets from the AEMET (Agencia Estatal de Meteorología-Spanish Meteorological Agency) show average annual precipitation and temperature values that range from 997 mm $year^{-1}$ and 14.2 °C in the western coast (Lois de Castropol weather station), to 1122 mm $year^{-1}$ and 13.2 °C in the center of the region (Avilés-Airport weather station), and to (iii) 1152 mm $year^{-1}$ and 13.2 °C in the eastern zone (Llanes weather station). The average number of rainy days with accumulated values >1 mm in 24 h is of 123 (Botey et al. 2013), mainly concentrated between October and May, while the minimum precipitation values are recorded between June and September.

Two main precipitation patterns are the most frequent in Asturias during autumn and winter: (i) frontal rainfall associated with low-pressure systems, and (ii) orographic rain due to northern maritime air masses. In contrast, brief episodes of heavy rainfall, due to strong instability of air masses are typical during spring and early summer (Valenzuela et al. 2017a). Due to the orographic effect of the Cantabrian Mountains that produce the discharge of moisture-laden fronts when they crash against the first row of mountain foothills, precipitation is lower in the littoral band than in the nearby inland areas. Temperature also shows differences between the coast belt and the rest of the region, with softer values in areas closer to the coastline, which is attributed to the thermal regulation role of the sea.

Wind regime shows a strong seasonal behavior, with prevailing SW winds during autumn and winter related to the arrival of Atlantic depressions, and cold and dry N-NW winds during summer. Meanwhile maritime W-NW winds are common the rest of the year. In general, most intense winds are those with a NW-SW component, although only 15% of the existing records show speeds exceeding 18 km h^{-1}. Moreover, the periods of calm (wind slower than 6 km h^{-1}) are more frequent during autumn (Flor and Flor-Blanco 2009).

The action of the tides combined with the waves give shape to the Asturian coast in which the effect of coastal drift involves a transport of sediment to the East (Flor and Flor-Blanco 2005). The Asturian coast, like the entire Cantabrian coast, experiences semi-diurnal and mesomareal tidal ranges (Flor and Flor-Blanco 2014a, b), showing an increase and jet lag towards the East. The main component of the coastal drift is toward the East and it is altered by the presence of Peñas Cape, in the central area of Asturias producing upwelling phenomena (Flor and Flor-Blanco 2009). Although the coastline morphology is mostly linear, the presence of capes, coves and bathymetric irregularities led to great variations in the action of wind and waves. Thus, storm devastation often concentrates along cape and shallow areas. More usual waves are 2–3 m in height produced by all direction wind. However, maritime storms caused by winds from the W and the NW may lead to waves over 7 m in height. These storms show strong seasonality with peaks in winter and significant inter-annual variability (Izaguirre et al. 2011). During the period December 2013–March 2014, a set of maritime storms reached the Cantabrian Coast like many other Atlantic coasts of Europe, causing relevant structural damages, floods, important erosion in the littoral areas and the stoppage of fishing activities (Menéndez et al. 2014).

The mild present-day climate in the Asturian coast explains the survival of plants especially sensitive to low temperatures that do not appear in areas that are more interior. Vegetation in the littoral belt is closely related to its geological characteristics and climatic conditions lightly drier and warmer than inland. The distribution of vegetation closely reflects the four sections established attending to lithological criteria (Fig. 3.1g), each of which has a vegetation cover (plants and groups of communities) with different facies depending on whether it is exposed or protected from the sea activity (Álvarez Arbesú 2008). Thus, they vary from communities constituted by a few number of small-size plants with scattered distribution along the rock joints most exposed to the waves, to dense shrubs that thrive in wind-sheltered-areas. The effect of wind and salinity transported in aerosols determines the appearance of new vegetation waists: the most affected is formed by dense meadows (aerohalophilous grasslands) and the less affected by aerohalophilous bushes, usually dominated by *Ulex* sp. According to previous explanations, the plant communities recognized in the Asturian cliffs can be systematized in five vegetation belts: (i) Halocasmofitic communities; (ii) Aerohalophilous grasslands; (iii) Aerohalophilous scrubs; (iv) Shrub formations; and (v) other halophilic communities (Álvarez Arbesú 2008).

In the vegetation cover of the 5 km buffer coastal area mapped by Principado de Asturias (2002) (Fig. 3.1i), pastures and meadows are predominant (44%), while forestry plantations and natural bushes represent 26 and 13% of total surface, respectively. Remaining surface percentages includes residual vegetation from urban/industrial areas (7%), natural forest (5%), crops and fruit trees (<3%), vegetation from rocky areas (<2.5%), and vegetation from littoral/fluvial areas (<1.5%).

3.1.4 Geoheritage in Natural Protected Areas in the Asturian Coast

From an environmental point of view, the Asturian coast is one of the best preserved in Spain. The regional government has carried out an intense policy of protection since the 90s, with the creation of several protection figures. The Law of the Principality of Asturias 5/91, of April 5th, on the Protection of Natural Areas, establishes four categories for the protection of natural spaces. From higher to lower level of protection, these categories are (i) Natural Parks, (ii) Natural Reserves, (iii) Natural Monuments, and (iv) Protected Landscapes. The wide variety of environments along the Asturian coast as well as its high degree of conservation have led to the creation of nineteen figures of protected areas in the 5 km buffer area inland from the coastline (Fig. 3.1j). These spaces show interest from the geological, geomorphological, geoheritage and touristic points of view, among others. Near the coast, but not on the coastline itself, there are several protected areas, such as the natural monuments Cascadas de Oneta (Oneta Waterfalls), Carbayera de El Tragamón (The Tragamón Oak Grove) or Saucedas de Buelles (Buelles Willow), and the partial nature reserves as Cueva de Lloviu (Lloviu Cave) or Cueva Rosa (Rosa Cave). In adition, there are thirteen protected areas in the Asturian coastline. Six of them correspond to non-cliff areas such as the natural monuments of Playa de Penarronda (Penarronda Beach), Playa de Frexulfe (Frexulfe Beach), Playa de El Espartal (El Espartal Beach), Charca de Zeluán y Ensenada de Llodero (Zeluán Pond and Llodero Cove) and two partial nature reserves named Barayo and Ría de Villaviciosa (Villaviciosa Estuary). Up to eight protected figures are strictly located in cliff areas and will be described ahead. One of them is a protected landscape, a protection figure that arises to preserve specific areas of the natural environment that for their aesthetic and cultural values deserve special protection. The remaining seven natural monuments are spaces or nature elements constituted basically by formations of notorious singularity, rarity or beauty that deserve, for the same reason a special protection.

(i) Cabo Peñas (Peñas Cape) Protected Landscape (80/1995 Decree, of May 12th, BOPA 135 of June 13th, 1995) is the northernmost territory of Asturias (Figs. 3.1j, 3.2c). The 19.26 km^2 of protected area consist of 19 km-long by 3 km-wide at the most prominent point on the coastal strip and the corresponding inland waters of the maritime area. Together with the landscape

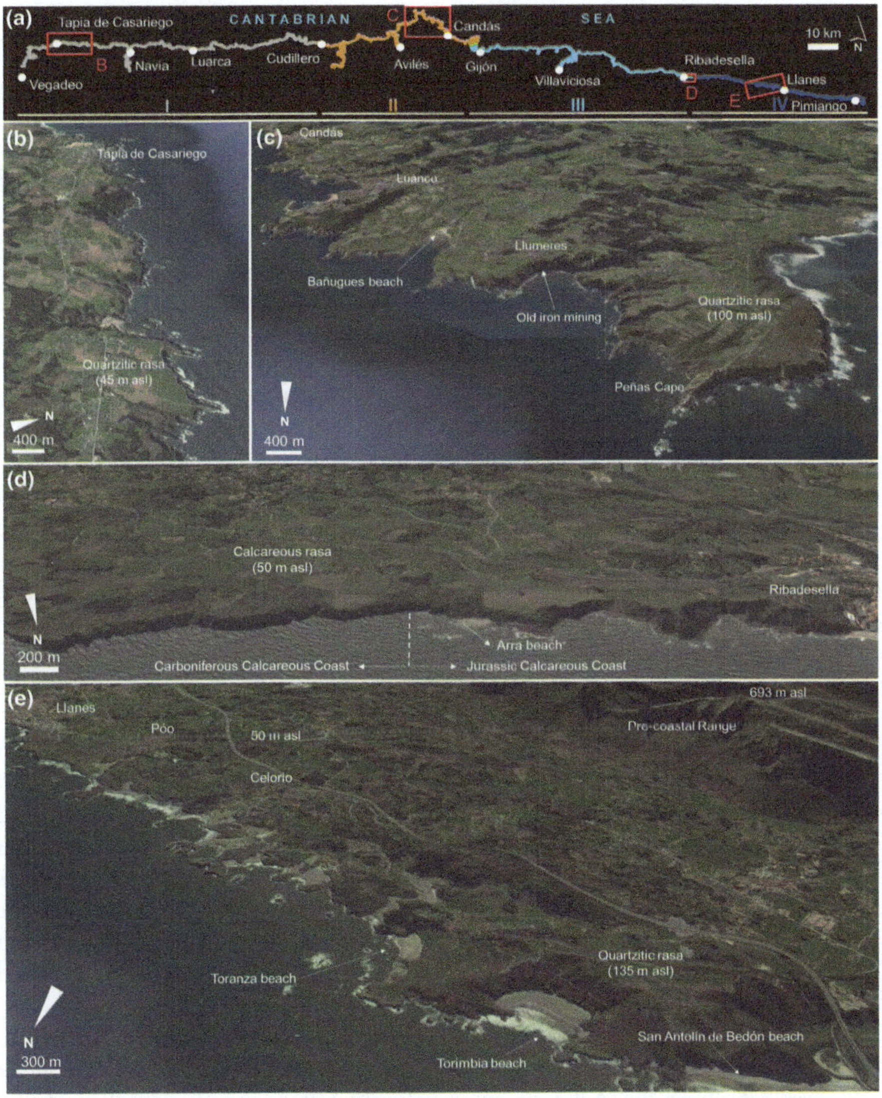

Fig. 3.2 **a** Location of the four lithological sections distinguished in this work along the Asturian coast. **b** Example of the coast section I. **c** Example of the coast section II. **d** Example of the coast section III. **e** Example of the coast section IV. *Source of data* Google Earth (**b–e**)

value, the scenic beauty, the location close to the largest population centers and the maintenance of ecosystems of great natural value, make Peñas Cape a privileged place in the predilections of the Asturians. Besides, the local landscape includes beaches such as Xagó or Bañugues, and cliffs and islets, among which La Erbosa islet stands out as, the second in size of the Asturian coast, after La Deva island. The Peñas Cape area is the point of the coast where the rasa is best preserved, sculpted on the Ordovician quartzite cliffs at an altitude of 100 m a.s.l. This landscape has partially been modified by fluvial channels, being possible to observe in this area only small streams, some of them are hanging valleys with respect to the current coastline.

(ii) Turbera de Las Dueñas (Las Dueñas Peatbog) Natural Monument (99/2002 Decree, of July 25th, BOPA 192 of August 19th, 2002) is a peatbog of 0.26 km^2 with natural and patrimonial interest located West of Cudillero. It is the largest one present along the Asturian coast, since peatbogs are most frequent in mountain areas where because low temperatures inhibit microbial organic matter degradation. Its formation is attributed to the presence of a smooth surface relief rasa supplied by three small streams installed on an impermeable Ordovician quartzite bedrock that hinders water permeability, giving rise to an anoxic environment.

(iii) Isla La Deva y Playón de Bayas (La Deva Island and Bayas Big Beach) Natural Monument (20/2002 Decree, of February 14th, BOPA 49 of February 28th, 2002) includes in its 1.1 km^2 area the biggest beach and island of Asturias plus a maritime surface of 0.32 km^2 between the island and the coast. It should be noted that the Asturias Airport is located just 400 m from Playón de Bayas on a 120 m a.s.l. rasa. La Deva Island is an Ordovician quartzitic promontory of approximately 800 m-long and 400 m-wide and ca. 90 m-high, located about 350 m from the coast at the eastern end of Playón de Bayas. Its vegetation cover includes three vegetation belts than can be differentiated in the Cantabrian coast cliffs: (i) a first band of very scarce cover of herbaceous species that root along the cliff joints; (ii) a second band of grassland of *Festuca rubra* ssp. *Pruinosa*; and (iii) a third band of bushes, mainly gorses and heathers. The main interest of La Deva Island is that it is the nesting area of some legally protected species classified as of special interest such as the shag (*Phalacrocorax aristotelis*), the European storm petrel (*Hydrobates pelagicus*), or the peregrine falcon (*Falco peregrinus*). It is also remarkable the presence of a rock lizard endemic subespecies (*Podarcis muralis rasquinetti*).

(iv) Yacimientos de Icnitas de Asturias (Asturian Tracksites) Natural Monument (45/2001 Decree, of April 19th, BOPA 106 of May 9th, 2001) includes three sectors along the Asturian coast between Gijón and Ribadesella, around 44.2 km in length. It is also popularly known as "La Costa de los Dinosaurios" (The Dinosaur Coast), an exceptional area of the Cantabrian coastline with a remarkable palaeontological heritage of scientific, cultural and touristic interest. This littoral, characterized by the abundance of dinosaur and other Jurassic reptile footprints and bones, shows different rocks (sandstone, siltstone, conglomerate, limestone, marl,…) that evidence changes in the Jurassic sedimentary environments. The invertebrate body fossils and trace fossils are very frequent and diverse, both in Early and Late Jurassic times. Plant fossils are also abundant and taxonomically diverse including ferns, horsetails, cycads, bennettitaleans, conifers, ginkgoals and others. There are also multiple logs of large coniferous trees, some of them with oil impregnations like the famous Asturian jet, a semi-precious stone exploited and worked during many centuries and known from old archaeological sites. The abundant inorganic sedimentary structures are preserved in the rocks by mud cracks, ripple marks, groove-, flute- and rill casts, Liessegang rings, honeycomb weathering, cross-stratification, nodules, septarian concretions, water-escape deformations, convolute-beds, raindrop marks, halite pseudomorphs, etc. The high number and the wide morphological variety of reptile footprints and its excellent preservation, make these Asturian Jurassic fossils between the best of the world. Furthermore, the Jurassic coast of Asturias is a great tourist attraction due to the environmental quality and the beauty of landscape. It is also remarkable the presence of the Museo del Jurásico de Asturias-MUJA (Jurassic Museum of Asturias) near Colunga locality, between Villaviciosa and Ribadesella towns, at 160 m asl, 700 m away from the coast, with almost 160.000 visitors in 2017.

The Carboniferous calcareous stretch of coast near Llanes constitutes an old rasa heavily karstified that gives it an irregular imprint on surface with karren development, a great amount of islets, tombolos and caves (Fig. 3.2e). Some cavities are presently connected to the sea originating geomorphological landforms of great beauty such as beaches in karst depressions or blowholes. The following four natural monuments are linked to this scenery.

(v) Playa de Gulpiyuri (Gulpiyuri Beach) Natural Monument (Decree 139/2001, of December 5th, BOPA 297 of December 26th, 2001) is a 0.038 km^2 area located between Ribadesella and Llanes. Despite its small size (49 m in length and 39 m in width), it is a geomorphological singularity that consists of a beach located into a karst doline sited one hundred meters inland and connected with the sea by karst conduits. The formation of this closed beach that seems a pool of marine waters during the high tide is due to the flood of a karst sinkhole by seawaters, very common in this area. The beach is made of quartzite cobbles and pebbles and sand underground transported by seawaters, especially during maritime storms. Its formation would be related to a collapse doline captured by seawaters during Holocene marine transgression (Flor-Blanco et al. 2015).

(vi) Complejo de Cobijeru (Cobijeru Complex) Natural Monument (Decree 140/2001, of December 5th, BOPA 297 of December 26th, 2001) is a 0.0873 km^2 area sited between Llanes and Pimiango. The Cobijeru complex includes the karst depression in which Cobijeru or Las Acacias Beach is located, the karst depression known as El Molín beach (or Marimuerto depression), the Cobijeru cave, the surroundings of both karst depressions and the closest section of the cliff. This complex is one of the most interesting geomorphological singularities of the Asturian coast and it will be described in detail in Sect. 3.3.3.

(vii) Blowholes are karst conduits connected to the sea and that have their upper entrance exposed at the top surface of cliffs. These conduits have up to 5 m diameter and usually dip more than 40°. During high tides and storms, blowholes work as water-air canyons that result in blasts of water and characteristic snorts, up to 30–50 m height. Oftentimes, and particularly during storms, they are visible and audible from hundreds or even thousands of meters distance. The blowholes produce characteristic erosion in the karren, as rillenkarren or pinnacles, due to the fall of the watersea. In the Asturian coast there are more than a hundred blowholes being two of them protected as Natural Monuments: Bufones de Arenillas (Arenillas Blowholes) Natural Monument (Decree 143/2001, of December 5, BOPA 297 of December 26, 2001) and Bufón de Santiuste (Santiuste Blowhole) Natural Monument (Decree 141/2001, of December 5, BOPA 297 of December 26, 2001). Santiuste Blowhole is the largest blowhole in the Asturian coast and its location in an environment of recognized quality, advice-taking measures to ensure its maintenance and conservation, given the touristic interest it arouses.

3.2 Current Geomorphological Processes in the Asturian Coast

To carry out the description of the cliff coast, we will consider four stretches of coastline divided according to lithological criteria (Figs. 3.1g, and 3.2a). From the West to the East, the sections considered are: (i) Western Siliciclastic Coast (34% of Asturias coastline), (ii) Central Palaeozoic Mixed Coast (19%), (iii) Eastern-Central Jurassic Coast (27%), and (iv) Eastern Carboniferous Calcareous Coast (20%). The western section corresponds to the siliciclastic rocks of the ZAOL while Palaeozoic limestone, sandstone, quartzite and lutite of the Cantabrian Zone in the surroundings of the Peñas Cape are considered as Cabo Peñas mixed section. The eastern-central Mesozoic section mainly includes limestones and marls. Finally, Carboniferous limestones outcropping in the eastern coast of Asturias constitute the eastern section.

Currently, the main geomorphological processes involved in modelling the Asturian cliff coast correspond to gravity, coastal and karst processes. The greater or lesser influence of each of them is strongly related to bedrock lithology and local structural geology. The text below describes in detail each lithological section, including some singular examples.

3.2.1 *Section I: The Western Siliceous Palaeozoic Coast*

Located in the western coast of Asturias, between the Eo Estuary and the town of Cudillero (Fig. 3.2a), the predominant bedrock lithology mainly consists of quartzite rocks of Paleozoic age belonging to the geological domain of the ZAOL. Rocky cliffs in this area reach heights over 100 m to the West of Cudillero. The highest slopes are reached in the eastern part of this section (>65°), from Luarca eastward to Cudillero. Bedrock in this zone shows a predominant orientation NE-SW both in stratification and tectonic structures, conditioning a series of rocky headlands with the same direction (Fig. 3.2b). Headlands shelter scarce and discontinuous beaches of small entity, mostly composed of pebbles, except for ten sand beaches of greater entity, placed at the following river mouths (from W to E): Penarronda, Navia, Frejulfe, Barayo, Otur, Luarca, Cave, San Pedro, Oleiros, and Concha de Artedo. Rasas are particularly well preserved along this section, being possible to recognize an almost continuous surface gradually rising from the Eo Estuary, at 25–30 m a.s.l., to Cudillero, at 120 m a.s.l., where Las Dueñas Peatbog is located. Las Dueñas Peatbog was formed during the early Holocene, ca. 11500 cal years BP (López-Merino et al. 2006). In addition, rasas in some places (i.e. near Tapia de Casariego, Busto Cape or Vidio Cape) show on top patches of sediments that are known as "rasa deposits", which consist of sands and rounded pebbles most commonly ascribed to marine origin. Absolute dating of marine terraces through a combination of cosmogenic nuclides (^{21}Ne, ^{10}Be, ^{26}Al) yielded minimum exposure ages of 1–2 Ma (Álvarez-Marrón et al. 2008).

The presence of penetrative tectonic foliations in the siliciclastic bedrock of this section induces a rock cracking that weakens rocks and predisposes them to destabilization. The current geomorphological activity in this section is mainly linked to littoral and gravity processes, mainly rockfalls and landslides (see Sect. 3.3.1). Damages to human constructions are usual, like it occurred since 2014 to 2017 in the access road to Arnao Beach in Castropol, west of Tapia de Casariego. Occasionally, stabilization works have been carried out to minimize the impact of cliff regression on human infrastructures, as it was the case of works carried out in Las Represas Beach in 2015, east of Tapia de Casariego.

3.2.2 Section II: The Central Palaeozoic Mixed Coast

The coastline of section II presents a NE-SW trend from the mouth of the Nalón River (East of Cudillero), to Peñas Cape (Fig. 3.2a, c). From this point on, this stretch of coastline turns to a NW-SE trend and extends to El Musel Port, the most important anthropic construction of the entire Asturian coast. The shape of the coastline, together with a lithological change with alternating sedimentary lithologies of Paleozoic age and varied nature and resistance, results in a coastal pattern with inlets and promontories. In several of these coves some of the longest beaches in Asturias have been developed, exceeding 1.5 km in length, like Playón de Bayas, Salinas and Xagó. Eastward, the screen effect of Peñas Cape on coastal drift favoured beach developed in small coves like Bañugues or Antromero, among others. Cliff slopes above 65° are usual in this stretch of coastline. The highest cliff is located in front of La Deva Island which reaches 117 m-high. The surroundings of Peñas Cape presents cliffs about 100 m-high. It is worth mentioning the presence of the Avilés Canyon as the most outstanding bathymetric feature of the Asturian coast. Despite its location 12 km far away from the coast, it is a remarkable feature because it is a deeply incised submarine canyon oblique to the E-W trending coast. The submarine canyon throws littoral sediments to the abyssal bottom of the Biscay Gulf (Gómez-Ballesteros et al. 2014). It drops more than 4600 m-deep in a 40 km-wide continental platform, being one among the deepest in the world in terms of net relief (Fernández-Viejo et al. 2014).

Geomorphological activity that currently occurs in this section of coast is, like in section I, mainly associated with littoral and slope processes, with minor participation of karst processes. The mass movements are very varied, being possible to find from rockfalls of small entity to a wide variety of landslides involving larger volumes of sediments and rocks. Thus, in the surroundings of Candás town, it can be observed translational sliding mass movements affecting some buildings of the village. Likewise, the action of sea undercutting at the base of Punta de San Antonio Cape, made of Devonian ferruginous sandstone bedrock (Naranco formation) has caused the backward of the cliff by translational slides. This process has endangered the San Antonio Chapel (17th century), which was moved several meters inland in order to avoid future damages due to coastal retreat (works extended from 2011 to 2015). At the same location, coastal undermining is affecting the toe of three big mass movements affecting little consolidated sandstones and siltstones of Triassic age. Maritime erosion processes along ca. 500 m escarpments of the toe of the movements are putting under hazard the position of the cemetery.

Near of Salinas town, an old cliff is currently disconnected from the sea due to the construction of a road, which had once been the passage of the tram. However, gravity processes are still shaping the ancient cliff in the entire area known as "Pinos Altos", which mainly evolves through debris flow processes. A little further westward, gravity processes also affect a cliff developed on Devonian limestone bedrock. It is an emblematic site due to its paleontological interest, as it is a fossil reef known as the Arnao Devonian reef (Arbizu et al. 2012). This point, included in the global Spanish Inventory of Geosites (http://info.igme.es/ielig/LIGInfo.aspx?codigo=CA003), shows elements of interest not only paleontological but also mining and architectural (Sect. 3.3.4).

To the west of Gijón city, there is a district called El Muselín that shows a particular steepped topography that reveals the existence of ancient terrain movements, particularly multiple hectometre size rotational landslides (Farias et al. 2012). El Muselín mass movements were developed affecting Ordovician quartzite sandstone that are unconformably overlain by Permian lutite. Assuming that sea under-cutting erosion at the base of the cliff was the most probable triggering factor, El Muselín mass movements seems to remain mostly stable since the construction of El Musel seaport.

3.2.3 Section III: The *Eastern-Central* Jurassic Coast

The third section extends from Gijón to Ribadesella and T is internationally known for the reptile fossil remains, mainly dinosaur tracks found in it (Sect. 3.3.2). The morphological pattern of this coastline changes with respect to two previous sections, being much more straight without pronounced promontories (Fig. 3.2a, d). In this section, the higher cliffs are more than 120 m height with slopes higher than 65°, mainly in the sector between Gijón and Villaviciosa. The activity of the cliffs is very high in this area evidenced continuous presence of gravels reworked by the sea at the bottom of the cliff. As in the previous sections, there are emblematic beaches in the third section. From West to East, the best examples of Sandy beaches are Gijón, Peñarrubia, Serín, Estaño, La Ñora, España, Merón, Tazones, Rodiles, Lastres, La Griega, La Isla, La Espasa, Arenal de Morís, Vega and Ribadesella. In addition, there are also numerous gravely beaches as Aranzón, La Conejera, El Sable, Huerres, La Atalaya, Arbidel, Xicu and Arra, among others. The presence of a diverse lithology together with a different disposition of the stratification respect to the coastline implies a different geomorphological activity (García-Ramos 2013). In the coastal areas with alternation of centimetric to decimetric layers of limestone and marl from the Lower Jurassic, the activity due to rockfall is large, although the coastal dynamics quickly incorporates them to pebble beaches. In other areas where the Late Jurassic siliceous conglomerates (La Ñora Formation) and sandstones (Vega and Lastres formations), falls of large blocks are very frequent being usual that they remain at the foot of the cliffs for long time (García-Ramos 2013). It is common that this lithology causes risk situations, which have forced to carry out sometimes large investments in stabilization and signalling works; it is the case of the high scarp located on Peñarrubia Beach (Gijón), where cleaning of the cliff, creation of berms, and mesh placement, among others measurements have been carried out. On the other hand, in this section of coast, it can also been recognized areas in which the coastal activity has undermined the lower part of the coastal edge, evolving later by flows or landslides, both translational and rotational. Deposits of ancient landslides with tens of meters in length are recognized, for example in the sea-cliffs of Oles and Tazones (Villaviciosa), Luces and Huerres (Colunga), Tereñes and Arra (Ribadesella). It is also common that these slope deposits are affected by surface flows.

3.2.4 Section IV: The Eastern Carboniferous Limestone Coast

The last section of the Asturian coast corresponds to the easternmost area of the Cantabrian Zone, where different types of Paleozoic rocks appear with predominate of Carboniferous limestones (Barcaliente, Valdeteja and Picos de Europa formations) and Ordovician quartzite sandstone (Barrios Formation). The imprint of the coast line changes again and, unlike the previous section (Mesozoic limestones), it presents numerous promontories and coves (Fig. 3.2a, e). The pattern change takes place in Arra Beach, which is the latest pebble beach eastward from Ribadesella in section III. Further East, pebble beaches almost disappear and the calcareous cliff is close to vertical and directly in contact with the sea. Although cliffs do not overcome 100 m in height in this section, slopes steeper than 65° are very frequent. This stretch of Asturian coast can be considered a singular unique place in the world for the great profusion of landforms of mixed, karst and marine origin giving rise to some emblematic landforms such as blowholes or closed beaches, some of which are subject to different protection figures (see Sect. 3.1.3). The whole area can be considered a very interesting place where it is possible to enjoy an exceptional geological heritage.

Cliffs developed over Carboniferous limestone bedrock are subvertical and bedrock remains fresh, only interrupted by dissolution hollows and caves, created also by marine abrasion, as well as the possibility to form sea corrosion notches, usually without abrasion platforms. Coastal stretches of Picos de Europa Formation limestone are straightest, while limestone of Valdeteja and Barcaliente formations have a cut profile that has allowed the formation of many coves, islands and hemitombolos.

Beaches are very scarce along coast section IV and can be divided between those ones developed over quartzite bedrock and those formed from calcareous bedrock. Beaches developed over quartzite bedrock are the typical small coves of the Cantabrian coast, like Torimbia and Andrín coves. Moreover, cliffs and hillsides on Ordovician quartzite materials display gentler slopes, developing many landslides as well as old perigacial and talus gravel deposits. A good example of an active quartzite cliff is the backshore of San Antolín de Bedón Beach, halfway between Ribadesella and Llanes, where evidence of recent coast retreat through debris slides are visible.

Different patterns beaches formed on calcareous bedrock can be distinguished: (i) small beaches and coves developed from former sinkholes captured by the sea (e.g. Borizu, Celorio or Toró beaches near Llanes); (ii) narrow beaches far from the coastline at the back of narrow, long and straight crevices that penetrate up to 500 m inland (e.g. Guadamía, Villanueva or Cuevas del Mar beaches); and (iii) closed beaches in karst depressions (like Gulpiyuri or Cobijeru) that formed from sea-captured sinkholes preserving their close-depression shape (Sects. 3.1.3 and 3.3.3).

3.3 Geomorphologic Singularities of the Asturian Coast

3.3.1 Landslide Instability Risks in the Asturian Rocky Coast

Landslides constitute one of the most important geomorphological hazards in Asturias causing mainly material damages and some human victims. They produce relevant annual costs linked to the repairing works of basic infrastructures and indirect economic losses derived from the closure of roads, the evacuation of people, the suspension of transport services and the stoppage of industrial production.

In particular, rocky coastal cliffs are one of the most landslide prone areas in the whole Asturian territory. This fact is due to the combination of preparatory factors for the occurrence of instabilities, such as the presence of slopes steeper than 65°, the alternation of lithologies with different mechanical behaviors and the scarce vegetation cover, with the recurrence of natural phenomena that are recognized as landslide triggers, such as strong waves or intense rainfall episodes.

Furthermore, the littoral belt of Asturias is nowadays one of the most populated areas in the region, which is linked to the presence of industrial activity (fishing industry, factories and harbours) and the touristic exploitation of the natural resources (landscape, beaches, estuaries and tracksites). The municipalities of the littoral belt have a total population of 486,298 inhabitants (INE 2015), showing a very uneven and disperse distribution from the low population density of the rural areas, in the eastern and western coasts (25–100 inhab. km^{-2}), to the densely populated cities of Gijón and Avilés, located in the central coast (>1,000 and 3,000 inhab. km^{-2}, respectively). The population dispersion and the production centers results in a very dense infrastructure network (roads, railway lines, water supply, electricity networks, etc.) running along the coast that is very vulnerable to slope instability dynamics (Fig. 3.3a, b). For all these reasons, the population and infrastructures settled in the coastal belt of Asturias are exposed to relevant levels of landslide risk.

During the period 1980–2016, a total of 224 individual landslides affecting rocky coast areas have been recorded in the BAPA database, a landslide inventory developed for the Asturian territory (www.geol.uniovi.es/BAPA/) (Valenzuela et al. 2017b) (Fig. 3.3). Events recorded include all landslide types, with a predominance of rockfalls (50%) over flows (14%) and slides (13%), while the instability type could not be determined in the remaining records (23%). In the majority of cases, natural slopes and cliffs have been the most affected areas (63%), while different kinds of human alteration might have exerted some influence in the configuration of the remaining affected slopes (47%). Moreover, a natural factor has been pointed out as the main trigger in 64% of the recorded landslides. However, and despite the proximity of the sea, the majority of these events have been related to precipitation (54%), occurring on the top of the cliffs or in areas where the cliff base is protected

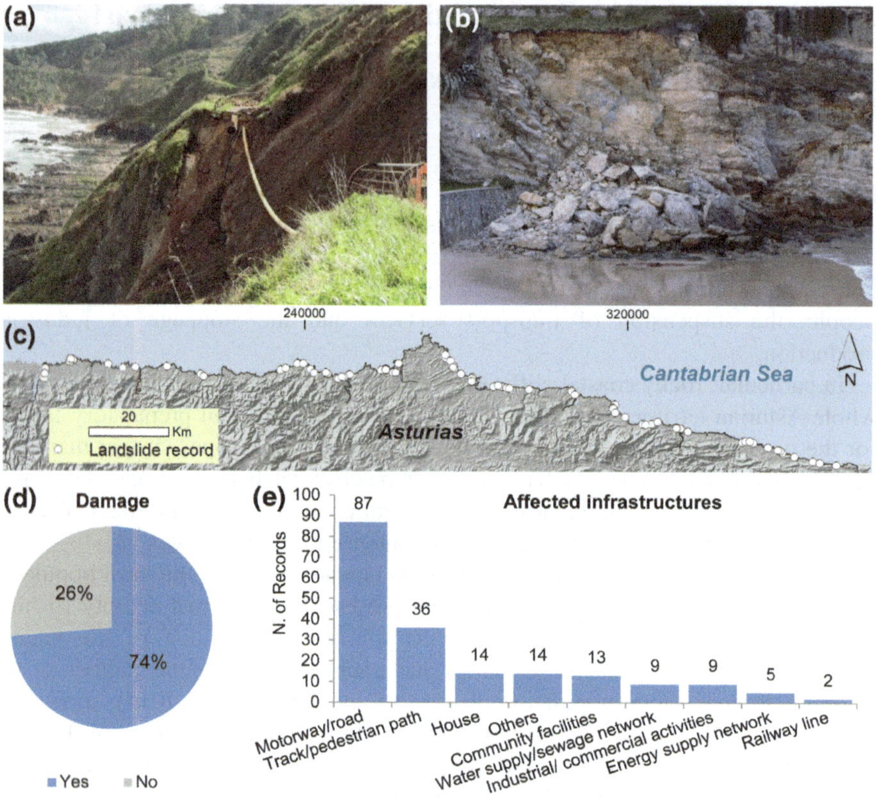

Fig. 3.3 **a** Sewage network affected by a landslide near Candás (section I). **b** Rockfall affecting the Sablón Beach in Llanes (section IV). **c** Landslide distribution in rocky coast of Asturias recorded between 1980 and 2016 in the BAPA landslide database (Valenzuela et al. 2017b). **d** Percentage of landslides causing some kind of damage. **e** Affected infrastructures

from wave action due to the presence of seaport facilities, roads or other infrastructures.

As a demonstration of the high landslide risk present in the area, at least 74% of the recorded landslides produced some kind of damage (Fig. 3.3d). In the vast majority of the cases (73%), landslides produced material losses, while only 1% of the recorded events caused human victims. Landslides mainly affected the communication network (46%), paths and pedestrian access (16%) and houses and community facilities (14%), followed by water supply or sewage networks (5%), industrial/commercial activities (5%), energy supply networks (3%) and railway lines (1%) (Fig. 3.3e). The aforementioned high exposure to landsliding often makes it necessary the implementation of protection measures, especially in populated areas, roads and port facilities, which implies major investment, even reaching figures higher than 1,000,000 €.

3.3.2 A Paleontological Unique Site

Along "The Dinosaur Coast" there are nine signposted tracksites where it is possible to see a large variety of dinosaur footprints. Maybe, the most complete and interesting tracksites along the coast are Tazones lighthouse sea-cliffs (Villaviciosa), La Griega Beach (Colunga) and Tereñes sea-cliffs (Ribadesella).

In Tazones lighthouse sea-cliffs, it is possible to observe tridactyl tracks attributed to theropods with different orientations (García-Ramos et al. 2004). There is also an elongated and sinuous impression due to the tail of a dinosaur, maybe produced by a theropod (García-Ramos et al. 2004). In another surface there are tridactyl tracks belonging to a small ornithopod (*Anomoepus)* (Piñuela Suárez 2015).

La Griega Beach is the most famous and accessible tracksite in the Asturian coast being the first one discovered in this region (Fig. 3.4a). One of the largest dinosaur footprints in the world, with 1.25 m in length, can be seen on this tracksite (Mensink and Mertmann 1984; García-Ramos et al. 2004, 2006; Lockley et al. 2007; Piñuela Suárez 2015). The same surface preserves another quadrupedal dinosaur trackway (García-Ramos et al. 2004, 2006; Lockley et al. 2007) attributed to a stegosaur (*Deltapodus*) by Piñuela Suárez (2015).

In Tereñes sea-cliffs, it is possible to see small ornithopod footprints (*Anomoepus*) and a trackway produced by a medium to large-size theropod (García-Ramos et al. 2004). In addition, it is remarkable a stegosaur trackway (Fig. 3.4b) assigned to *Deltapodus* ichnogenus, relatively scarce in the world. The dynamic of this coast aforementioned is very high, mainly in winter when the Cantabrian Sea can be very stormy. During a storm, a very interesting surface with four parallel trackways belonging to medium to large-size ornithopods (Piñuela et al. 2016), unique for the Jurassic, was broken and destroyed by the sea.

The collection of dinosaur tracks from the Jurassic Museum of Asturias (MUJA), recovered from this littoral, is the most complete in Europe and the third all over the world. Together with them appear numerous tracks and trackways belonging to other reptiles as pterosaurs (maybe the best on the fossil record, in which it is possible to observe skin impressions and evidences of the interdigital webs), crocodiles, turtles and lizards (the unique trackways for the Jurassic), now stored at MUJA. It also frequently appears reptile bones, in this sense it should be pointed out the most complete skeletons of ichthyosaurs and plesiosaurs in the Iberian Peninsula.

3.3.3 **A Geomorphological Singularity: Cobijeru Karst Complex *Natural Monument***

The Cobijeru Complex Natural Monument sited in eastern Asturias is a small karst area developed in a limestone rasa since, at least, the Middle Pleistocene. This area

Fig. 3.4 **a** La Griega Beach tracksite (Colunga). The yellow arrow indicates the sauropod trackway, including the largest footprints for the Jurassic in the world, with 1.25 m in length; the green arrow indicates the stegosaur trackway. **b** A view of Tereñes sea-cliffs and location of (**c**). **c** Stegosaur trackway preserved on the top of a sandstone bed near Ribadesella

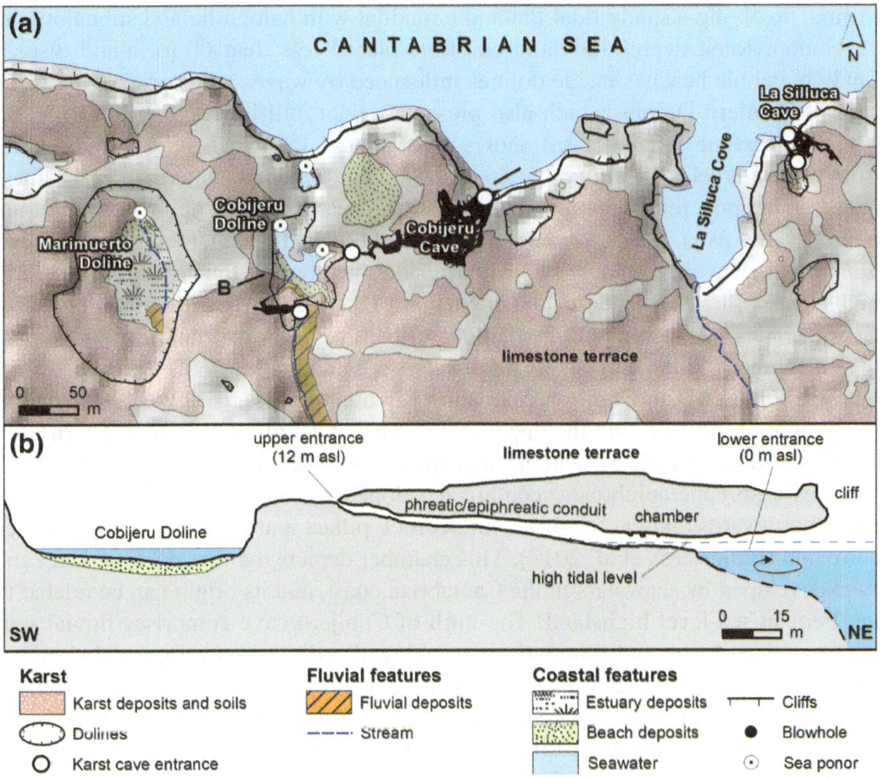

Fig. 3.5 **a** Geomorphological map of Cobijeru Natural Monument. **b** Sections of Cobijeru Doline and Cobijeru cave

comprises singular surface karst depressions and karst conduits affected by marine processes, which development was controlled by sea-level oscillations (Flor 1999; Ballesteros et al. 2017).

Singular littoral karst landforms involve depressions permanently occupied by the sea and dolines temporally invaded by seawaters. In relative age terms, the oldest ones correspond to depressions located in the coastline and captured by the sea, most likely during the Holocene, forming shallow coves with a rounded shape (Flor-Blanco et al. 2015). This is the case of La Silluca cove, East of the Cobijeru Complex (Fig. 3.5a). Dolines temporally invaded by seawaters are located 50–100 m inland from the coastline. These dolines show frequent subvertical walls, evidencing that collapse processes were key for local doline formation, maybe between the MIS 4 and 1. These depressions are connected directly by karst conduits with the sea, allowing the temporal invasion of the depressions by seawaters during high tides. The Marimuerto doline is placed ca. 80–100 m inland from the coastline and preserves an old tidal mill for grain grinding installed in the north end of its outlet. This restricted environment is sheltered and almost unaffected by storms, involving a sandy tidal flat and a mudflat with halophilic and subhalophilic vegetation. Karst depressions and coastline placed less than 60 m inland display sandy to cobble beaches inside dolines influenced by waves and storms. This is the case of Cobijeru Doline, which also presents a tidal mill in its eastern part.

The limestone terrace (rasa) shows more than 1 km of karst caves extending from the sea-level to an altitude of 40 m a.s.l, including the Cobijeru and La Silluca caves as the most relevant examples. La Silluca cave is located at the bottom of the cliff and involves bone remains of a straight-tusked elephant (*Palaeoloxodon antiquus*) that were probably deposited inside the cave by the sea during a warm highstand sea-level episode (Flor 1999; Pinto Llona and Aguirre 1999). The cave also includes archaeological remains ascribed to the Asturian technocomplex (epipaleolithic), characterized by mollusks shells and lithic tools used to collect marine mollusks. The Cobijeru cave shows two entrances, the lowest one sited at the bottom of the cliff, and the upper one located at 12 m a.s.l., in the eastern limit of the Cobijeru doline, evidencing that the doline intercepted the cave. The cave includes a phreatic/epiphreatic conduit developed by ancient freshwater flow oriented northwards, and a chamber with bedrock pillars and flat roofs formed by sea corrosion (Ballesteros et al. 2017). This chamber depicts the first documented karst cave developed by seawaters in the Cantabrian coast, and its origin can be related to the Eemian sea-level highstand. The infill of Cobijeru cave comprises fluvial sediments, speleothems and others deposits. Fluvial sediments are formed by subangular quartzite cobbles that came from the erosion of an upper quartzite terrace, placed at 150 m altitude toward the South. U/Th dates obtained from speleothems points out that the lower limestone marine terrace would be covered by alluvial fans at least at 140–150 and 60–70 ka, introducing quartzitic sediments that almost fulfilled the caves. This situation occurred during cold and sea level lowstand conditions, matching MIS (Marine Isotope Stage) 6 and 4. Besides, a molar of wild horse (*Equus ferus*) found into the Cobijeru Cave suggests an episode of steppe conditions with scarce vegetation covering the fluvial fans around 65 ka ago (Ballesteros et al. 2017).

3.3.4 Human Uses and Geology in the Asturian Rocky Coast

Coastal cliffs in Asturias are the setting of part of the most prominent cultural heritage of Asturias. Paleolithic sites, Iron Age hillforts, and the medieval Santiago Route are part of this extraordinary legacy, partly preserved in the cliff coast. However, marine erosion, with active gravity processes derived from wave impact and undercutting, is one of the main natural processes affecting the preservation of archaeological sites. Therefore, cultural heritage management should consider their influence and effects.

Asturias shows more than 600 Paleolithic sites, including 273 caves and shelters, and 350 open air sites (Turrero et al. 2013). More than 65% or these sites are located in marine terraces, which could have been preferential places for human occupation because of the smooth relief and the Cantabrian shoreline proximity. In addition, their relative continuity along the coast, from West to East, explains that they had acted as natural corridors for prehistoric human and animal populations from Europe. However, prehistoric occupation would have also been conditioned by the ancient shoreline position, which during the Last Glacial Maximum, was set more than 100 m below the current sea level and located some kilometres to the N. Subsequent coastal retreat and coeval sea level rise would have hampered the preservation of other Paleolithic sites located close to the shoreline.

Among the Paleolithic sites of Asturias, it is worth mentioning five prehistoric caves included in the List of World Heritage in 2008, as a part of the property known as "Cave of Altamira and Paleolithic cave art of Northern Spain" (http://whc.unesco.org/en/list/310/). One of them is El Pindal Cave, located to the East of Asturias (4° 30′W, 43° 23′N). El Pindal Cave shows some of the most prominent rock art representations of the Cantabrian coast. Known since 1908, art in this cave includes paintings and engravings of, at least Magdalenian age, being remarkable a mammoth, a fish and several symbolic paintings (González-Pumariega Solís 2008). However, El Pindal Cave is also relevant because of its geomorphological interest (Jiménez-Sánchez et al. 2006). The cave was formed in a karst massif modelled in Carboniferous limestone with a recharge area located in a marine terrace at 50–61 m asl (57 m average altitude). The cave trends from East-West direction showing 619 m length. The entrance, oriented to the East, is placed in a coastal cliff at 24 m asl. The development of the cave was controlled by East-West trending bedding that would have favoured the emplacement of phreatic conduits probably toward the Deva River, towards the East, which would have marked the base level of the karst system, coevally with marine terrace uplifting. The entrance of alluvial deposits from the marine terrace reliefs located to the South (Pimiango terrace, 160–175 m), slope processes (roof block collapse and subsidence evidence of cave floor), climatic and sea-level oscillations and precipitation of speleothems controlled the geomorphologic evolution of the cave at least during the Late Pleistocene (Jiménez-Sánchez et al. 2006; Stoll et al. 2013).

In Asturias, twenty-nine hillforts attributed to 9th BC to 1th AD centuries have been described, especially to the West of the region (Villa Valdés 2007). Hillforts located by the side of the coast are termed "Sea hillforts" or "maritime hillforts" (Camino Mayor 1995). Sea hillforts were strategic locations during Roman times (1st and 2nd centuries) because they allowed Romans to control both marine and terrestrial routes, and also to exploit the Cantabrian coast natural resources. One of the most relevant examples of sea hillforts is El Castiellu o Castro de Podes, located in Punta'l Castiellu Cape (San Martín de Podes, placed West of Peñas Cape). Castro de Podes site has more than 75,000 m^2 surface and exhibits some archaeological structures, as two ditches and one parapet (Camino Mayor 1995). The cape is modelled in Devonian rocks affected by faults, cut by an ancient marine terrace ranging from 25 to 35 m a.s.l. and limited by marine cliffs to the W, N and

E. Marine wave action undercuts the base of the cliffs, leading to its instability by rockfalls, topples, slides and flows that are also destroying part of the archaeological structures (Jiménez-Sánchez and Ballesteros 2017). Another hillfort, named La Campa Torres, involved probably a Roman lighthouse (Fernández-Ochoa and Morillo 2010) associated to the Roman settlement of Gijón, constructed in ancient tombolo connected to the land by a sandy bar. The settlement was defended by walls and towers in its south part, and by a natural cliff 30 m height in the North (Fernández-Ochoa 2006). This privileged position was also used by the Spanish Army to install a canyon battery in 18–20th centuries. At present, Gijón continues being the most density town of Asturias with the major industry and commercial activity due to its location at the coastline.

Moreover, Asturias includes 560 km of the Santiago de Compostela Route (Camino de Santiago), a relevant pilgrimage route since the Middle Age, inscribed in the List of World Heritate in 1993. Particularly, the Coastal Route was included in the List in 2015 as an extension of the original route (http://whc.unesco.org/en/list/669). The Coastal Route extends several kilometres along of the Asturian cliffs, being at present one of the main touristic resources in the area.

Fishing is also a traditional activity linked to the Asturian coast that was already developed by prehistoric human, as evidenced archaeological excavations in which shellfish and other marine remains have appeared (Turrero et al. 2013). At present, eighteen fishing seaports operate along the entire Asturian coast. All of them are strategically located taking advantage of the guard position that cliffs represent, being Viavélez, Ortiguera, Puerto de Vega, Luarca, and Cudillero seaports perfect examples for being especially linked to cliffs that originally constituted real natural seaports. In addition, minor structures as small cranes or stairs sculpted in the rock were built in the cliffs to access to the sea. These structures allow to collect seafood and shellfish in cliff-line coast and to descend boats to the sea, which are latterly raised to save far from the shore. This kind of structures is especially common in the Carboniferous limestone coast, where the cliffs are very continued.

On the other hand, mining activities linked to the extraction of coal, metallic elements or fluorite, among others, have been very important and widespread throughout the Asturian territory. This activity has also traditionally been present along the coast. Thus, mining activities developed by the side of the Asturian rocky coast already started with the Roman Empire, who extracted metals such as gold in Salave (next to Tapia de Casariego) or iron in Llumeres (southeast of Peñas Cape). Fluorite has been also extracted since the 1920s, being the Berbes Mine, placed west from Ribadesella, the closest to the sea fluorite exploitation.

Black jet, also known as "azabache", has been traditionally extracted to produce jewels along the Jurassic Calcareous Coast. Asturian black jet derives from Jurassic woods and is known worldwide because of its easiness to be carved and high quality together with Whitbey's jet black in United Kingdom. Mostly of azabache mines are located between Gijón and Ribadesella, in La Marina area and in Oles (Villaviciosa). Azabache mines often were conceived as small galleries excavated from the coast cliffs. Mining exploitations also include multiple open cast aggregate mines that are placed a few kilometres inland from the coastline, such as Perecil (Perlora, next to Candás) or El Estrellín (Avilés).

However, the most emblematic mine along the Asturian coastline is Arnao (northwest of Avilés), a coal mine initially exploited in 1591 and considered the oldest vertical undermine of the Iberian Peninsula and, at the time, the unique mine of Europe excavated under the sea. Its activity peaked from 1833 to 1915 with the funding of the Asturian Royal Company of Mines (Real Compañía Asturiana de Minas or RCAM) and ended when the seawaters started to fill the deepest extraction

Fig. 3.6 **a** Panoramic view of the Arnao Mine, now turned into a museum. **b** The outer lining made of bricks and zinc scales has preserved the original wooden tower of the Arnao well, unique in the entire region. **c** The Asturiana de Zinc factory at Arnao was built on Devonian limestone bedrock included as part of the Arnao reef geosite (Arbizu et al. 2012), visible at the image front. **d** Panoramic view of 'El Plano' rocky cliff, where a railway was once bringing quartzite boulders from the wave-cut bench to the top of the cliff to produce crucibles used in the extraction line of zinc in Arnao

galleries at 80 m depth. Nowadays, the engine building and the tower of the well extraction (both built in 1834) have been turned into a museum (Fig. 3.6a, b) and it is possible to visit the shallowest level of galleries placed 20 m below surface. Moreover, the RCAM had a strong importance in the industrial development of the whole region. Taking advantage of the strategic location of Arnao, close to Avilés seaport, and given the energetic cost of production, they built a metallurgic plant in Arnao to produce zinc oxide from the mineral that was being extracted more to the East in Guipúzcoa (Basque Country). Arnao factory was built on infill materials in front of Jurassic conglomerate cliffs of La Ñora Formation (Fig. 3.6c). Zinc production in Arnao started in 1855 and involved a singular type of exploitation in the rocky cliff placed in front of La Deva Island, located ca. 4 km West from the factory. This area (Fig. 3.6d) is still known as El Plano (The Plane) because a railway truck was installed in the cliff to lift Cambrian-Ordovician quartzite sandstone boulders accumulated at the base of the cliff up to its top and transported by road to the Arnao factory. The high purity of quartz in El Plano bedrock made it ideal to produce the crucibles that were used in the extraction line of zinc through the use of coal ovens. Nowadays, zinc production applies the electrolytic extraction line at the San Juan de Nieva factory, built in 1960 by RCAM and Asturiana de Zinc SA (AZSA) on top of El Espartal dunes and Avilés estuary deposits, by the sideof Avilés town.

Other industrial activities developed close to the Asturian rocky coast that have had a strong environmental footprint are the Aboño thermal power station built close to Gijón in 1974, which have completely distorted both the Aboño inlet and beach. Also, the steel factory Ensidesa, built in Avilés in 1950 and later named Aceralia (Arcelor-Mittal multinational) poured about two million tons of coke scoria residues in the Cabo Negro area, using a ca. 7 km-long cable railway as transport method.

References

Adrados, L. (2011). *Costa Oriental de Asturias. Un paisaje singular*. Adrados Ediciones Oviedo, 243 pp.

Alonso, J. L., Pulgar, J., García-Ramos, J. C., & Barba, P. (1996). Tertiary basins and Alpine tectonics in the Cantabrian Mountains (NW Spain). In P. F. Friend & C. J. Dabrio (Eds.), *Tertiary basins of Spain*. New York: Cambridge University Press.

Álvarez Arbesú, R. (2008). *La cubierta vegetal del litoral asturiano*. Documentos del Jardín Botánico Atlántico (Gijón) 5. Ediciones Trea, 128 pp.

Álvarez-Marrón, J., Hetzel, R., Niederman, S., Menéndez, R., & Marquínez, J. (2008). Origin, structure and exposure history of a wave-cut platform more than 1 Ma in age at the coast of northern Spain: A multiple cosmogenic nuclide approach. *Geomorphology, 93*, 316–334.

Arbizu, M., Méndez-Bedia, I., Busquets, P. A., Pérez-Estaún, A., Álvarez-Laó, D. J., & Turrero, P. (2012). El patrimonio natural y cultural de Arnao (Castrillón): geología, fósiles e historia minera. In J. R. Bahamonde Rionda et al (Eds.), *Geo-Guías Excursiones VIII Congreso Geológico de España* (Vol. 9, pp. 9–38).

Ballesteros, D., Rodríguez-Rodríguez, L., González-Lemos, S., Giralt, S., Álvarez-Lao, D. J., Adrados, L., et al. (2017). New evidence of sea-level lowstands and paleoenvironment during MIS 6 and 4 in the Cantabrian coastal karst: the Cobiheru cave (North Iberia). *Earth Surface Processes and Landforms, 42,* 1704–1716.

Bastida, F., & Aller, J. (1995). Rasgos geológicos generales. In C. Aramburu C & F. Bastida (Eds.), *Geología de Asturias*. Trea.

Botey, R., Guijarro, J. A., & Jiménez, A. (2013). *Valores Normales de Precipitación Mensual 1981-2010.* Ministerio de Agricultura, Alimentación y Medio Ambiente-Agencia Estatal de Meteorología, 55 pp.

Camino Mayor, J. (1995). *Los castros marítimos en Asturias* (p. 252). Oviedo: Real Instituto de Estudios Asturianos.

Domínguez-Cuesta, M. J., Jiménez-Sánchez, M., González-Fernández, J. A., Quintana, L., Flor, G., & Flor-Blanco, G. (2015). GIS as a tool to detect flat erosional surfaces in coastal areas: a case study in North Spain. *Geologica Acta, 13*(2), 97–106.

Farias, P., Domínguez-Cuesta, M. J., López, C., & Pando, L. (2012). Deslizamientos rotacionales en la vertiente oriental del cabo de Torres (Gijón, España). *Geotemas, 13,* 524.

Fernández Ochoa, C. (2006). The army and the urban walls in Late Roman Spain: Defence and strategy. In A. Morillo Cerdán & J. Aurrecoechea (Eds.), *Roman army in Hispania: An archaeological guide*. León: Universidad de León.

Fernández Ochoa, C., & Morillo, A. (2010). Roman ligthhouses on the Atlantic Coast. In C. Carreras, R. Morais (Eds.), *The Western Roman Atlantic Façade*. Publishers of British Archaeological Reports.

Fernández-Viejo, G., López-Fernández, C., Domínguez-Cuesta, M. J., & Cadenas, P. (2014). How much confidence can be conferred on tectonic maps of continental shelves? The Cantabrian-Fault case. *Nature Scientific Reports, 4* (Article number 3661).

Flor, G. (1983). Las rasas asturianas: ensayo de correlación y emplazamiento. *Trabajos de Geología, 13,* 65–81.

Flor, G. (1999). Estudio geológico de la Cueva de La Silluca (Concejo de Llanes). In *Excavaciones arqueológicas en Asturias 1995-98*. Servicio de Publicaciones del Principado de Asturias. Oviedo, Spain.

Flor, G. (2000). El entorno geológico del río y la playa de Bedón. *Bedoniana, 2,* 55–74.

Flor, G., & Flor-Blanco, G. (2005). An introduction to the erosion and sedimentation problems in the coastal regions of Asturias and Cantabria (NW Spain) and its implications on environmental management. *Journal of Coastal Research, 49,* 58–63.

Flor, G., Busto J.A., & Flor-Blanco, G. (2006). Morphological and sedimentary patterns of ports of the Asturian Region (NW Spain). *Journal of Coastal Research, 48,* 35–40.

Flor, G., & Flor-Blanco, G. (2009). *Aspectos morfológicos, dinámicos y sedimentarios del sector costero: desembocadura del Nalón-playa de Bañugues. Guía de Campo.* 6º Simposio sobre el Margen Ibérico Atlántico, Oviedo, 67 pp.

Flor, G., & Flor-Blanco, G. (2014a). Raised beaches in the Cantabrian Coast. In F. Gutiérrez, & M. Gutierrez Elorza (Eds.), *Landscapes and landforms in Spain*. Springer.

Flor, G., & Flor-Blanco, G. (2014b). Componentes de viento generadores de morfologías y campos de dunas costeras en Asturias (NO de España). *Geomorfología y Cuaternario, 28*(3–4), 47–68.

Flor, G., & Peón, A. (2004). Rasas y superficies de erosión continentales en el relieve alpídico del noroeste peninuslar y los depósitos terciarios. In M. Araújo & A. Gomez (Eds.), *Geomorfologia do NW da Península Ibérica*. Porto.

Flor-Blanco, G., Bruschi, V., Adrados, L., Flor, G., Trenhaile, A. S., & Domínguez-Cuesta, M. J. et al. (2015). Propuesta de un modelo de evolución de costas acantiladas calcáreas: Asturias y Cantabria (Norte de España). *Geo-Temas, 15,* 13–16.

García Couto, M. A. (Ed.). (2015). *Iberian climate atlas*. Agencia Estatal de Meteorología (España) and Instituto de Meteorología (Portugal). Madrid, Spain, 79 pp.

García-Ramos, J. C. (2013). El Jurásico de la costa centro-oriental de Asturias. Un Monumento Natural de alto interés patrimonial. *Geo-Temas, 14,* 19–29.

García-Ramos, J. C., Piñuela, L., & Lires, J. (2004). *Guía del Jurásico de Asturias*. Zinco Comunicación. Gijón, Asturias, 118 pp.

García-Ramos, J. C., Piñuela, L., & Lires, J. (2006). *Atlas del Jurásico de Asturias* (p. 225). Oviedo: Ediciones Nóbel.

Gómez-Ballesteros, M., Druet, M., Muñoz, A., Arrese, B., Rivera, J., Sánchez, F., et al. (2014). Geomorphology of the Avilés Canyon System, Cantabrian Sea (Bay of Biscay). *Deep Sea Research, 106,* 99–117.

González Menéndez, L., Heredia, N., & Marcos, A. (2008). Mapa Geológico Digital continuo E. 1:50.000, Zona Asturoccidental-Leonesa (Zona-1100). In *GEODE. Mapa Geológico Digital continuo de España.* http://info.igme.es/cartografiadigital/geologica/geoiezona.aspx?!d=Z1100 (online). Accessed September, 2017.

González-Pumariega Solís, M. (2008). *Guía del Arte Rupestre Paleolítico en Asturias* (p. 127). Pola de Siero: Ménsula Ediciones.

Hoyos Gómez, M., & Herrero, N. (1989). El karst en la Cornisa Cantábrica. In J. Durán Valsero & L. López Martínez (Eds.), *El karst en España*. Madrid: Sociedad Española de Geomorfología.

INE-Instituto Nacional de Estadística. (2015). *Cifras oficiales del Padrón municipal*. http://www.ine.es/dynt3/inebase/es/index.html?padre=517&dh=1 (online). Accessed September 15, 2017.

Izaguirre, C. F. J., Méndez, M., Menéndez, I. J., & Losada. (2011). Global extreme wave height variability based on satellite data. *Geophysical Research Letters, 38,* L10607.

Jiménez-Sánchez, M., & Ballesteros, D. (2017). Análisis preliminar del riesgo geoarqueológico en Castros Marítimos de Asturias: el ejemplo de El Castiellu de Podes. *Geogaceta* (in press).

Jiménez-Sánchez, M., Bischoff, J. L., Stoll, H., & Aranburu, A. (2006). A geochronological approach for cave evolution in the Cantabrian Coast (Pindal Cave, NW Spain). *Zeitschrift für Geomorphologie, Supplementary, 147,* 129–141.

Lepvrier, C., & Martínez-García, E. (1990). Fault development and stress evolution of the post-Hercynian Asturian Basin (Asturias and Cantabria, northwestern Spain). *Tectonophysics, 184,* 345–356.

Lockley, M., Lires, J., García-Ramos, J. C., Piñuela, L., & Avanzini, M. (2007). Shrinking the world's largest Dinosaur tracks: Observations on the Ichnotaxonomy of Gigantosauropus asturiensis and Hispanosauropus hauboldi from the Upper Jurassic of Asturias, Spain. *Ichnos, 14,* 247–255.

López-Merino, L., López Sáez, J. A., & López García, P. (2006). Estudio palinológico de la turbera litoral holocena de Las Dueñas (Cudillero, Asturias, España). *Revista Española de Micropaleontología, 38*(2–3), 299–308.

Lotze, F. (1945). Zur gliederung der varisziden der Iberischen Meseta. *Geotektonische Forschungen, 6,* 78–92.

Mary, G. (1983). Evolución del margen costero de la Cordillera Cantábrica en Asturias desde el Mioceno. *Trabajos de Geología, 13,* 3–35.

Menéndez, M., Espejo, A., Pérez, J., Méndez, F. J., & Losada, I. J. (2014). Análisis climatológico de los temporales marítimos en la Cornisa Atlántica durante el invierno 2013-14. *Publicaciones de la Asociación Española de Climatología serie A, 9,* 489–498.

Mensink, H., & D. Mertmann. (1984). Dinosaurier-Fährten (*Gigantosauropus asturiensis n.g. n. sp.; Hispanosauropus hauboldi n. g. n. sp.*) im Jura Asturiens bei La Griega und Ribadesella (Spanien). *Neus Jahrbuch für Geologie und Paläontologie Monatshefte, 1984*(7), 405–415.

Merino Tomé, O., Suárez Rodríguez, A., & Alonso, J. L. (2013). Mapa Geológico Digital continuo E. 1:50.000, Zona Cantábrica (Zona-1000). In *GEODE. Mapa Geológico Digital continuo de España.* http://info.igme.es/cartografiadigital/geologica/geoiezona.aspx?!d=Z1000 (online). Accessed September 15, 2017.

Ministerio de Agricultura y Pesca, Alimentación y Medio Ambiente, Banco de Datos de la Naturaleza. (2017). http://www.mapama.gob.es/es/biodiversidad/servicios/banco-datos-naturaleza/informacion-disponible/ENP_Descargas.aspx (online). Accessed September 15, 2017.

Olivares Navarro, C. (2015). *Climatología en Asturias en el periodo 1981-2010*. Master dissertation, Universidad de Oviedo, 55 pp.

Pérez-Estaún, A., Bastida, F., Alonso, J. L., Marquinez, J., Aller, J., Álvarez-Marrón, J., et al. (1988). A thin-skinned tectonics model for an arcuate fold and thrust belt: The Cantabrian Zone (Variscan ibero-Armorican Arc). *Tectonics, 7,* 517–537.

Pinto Llona, A., & Aguirre, E. (1999). Presencia del elefante antiguo Elephas (Plaleoxodon) antiquus en la Cueva de La Silluca (Buelna, Asturias). *Excavaciones arqueológicas en Asturias 1995-98* (pp. 225–232). Oviedo: Servicio de Publicaciones del Principado de Asturias.

Piñuela, L., García-Ramos, J. C., Romano, M., & Ruiz-Omeñaca, J. I. (2016). First record of gregarious behaviour in medium-large robust Jurassic ornithopods: evidence from the Kimmeridgian trackways of Asturias (N Spain) and some general considerations on large ornithopod tracks in the Mesozoic record. *Ichnos, 23*(3–4), 298–311.

Piñuela Suárez, L. (2015). *Huellas de dinosaurios y de otros reptiles del Jurásico Superior de Asturias*. Tesis Doctoral, Universidad de Oviedo, 326 pp.

Stoll, H. M., Moreno, A., Méndez-Vicente, A., González-Lemos, S., Jiménez-Sánchez, M., Domínguez-Cuesta, M. J., et al. (2013). Paleoclimate and growth rates of speleothems over the last two climatic cycles and relationship with climate in NW Iberian Peninsula. *Quaternary Research, 80*(2), 284–290.

Turrero, P., Domínguez-Cuesta, M. J., Jiménez-Sánchez, M., & García-Vázquez, E. (2013). The spatial distribution of Palaeolithic human settlements and its influence on palaeoecological studies: A case from Northern Iberia. *Journal of Archaeological Science, 40,* 4127–4138.

Valenzuela, P., Domínguez-Cuesta, M. J., Mora García, M. A., & Jiménez-Sánchez, M. (2017a). Rainfall thresholds for the triggering of landslides considering previous soil moisture conditions (Asturias, NW Spain). *Landslides on line*, 1–10.

Valenzuela, P., Domínguez-Cuesta, M. J., Mora García, M. A., & Jiménez-Sánchez, M. (2017b). A spatio-temporal landslide inventory for the NW of Spain: BAPA database. *Geomorphology, 293,* 11–23.

Villa Valdés, A. (2007). Mil años de poblados fortificados en Asturias (siglos IX a.C.-II d. C.). In J. Fernández-Tresguerres (coord): *Astures y Romanos: nuevas perspectivas*. Real Instituto de Estudios Asturianos. Oviedo.

Chapter 4
The Cantabrian Rocky Coast

Viola Bruschi and Juan Remondo

4.1 Introduction

The Cantabrian Coast is located in the Spanish north littoral, in the southeastern sector of the Bay of Biscay (Cantabrian Sea), conforming the northern border of the Regions of Cantabria and Basque Country, the latter matching Biscay and Guipúzcoa provinces (Fig. 4.1). This coastal strip has an obvious strategic and economic interest and concentrates a variety of activities (urban and industrial development, fishing, tourism, commerce, etc.). This has led to a very intense transformation of coastal areas, particularly during the last decades.

This sector of the Spanish Coast is a rectilinear and emergent coastline with abrupt cliffy trace, predominantly erosive (Rivas 2000). It is dominated by steep cliffs (typically more than 20 m high), interrupted by drowned river valley estuaries. The Cantabrian Rock Coast has been scarcely studied so far so there are only few scientific contributions on literature devoted to the characteristics and dynamics of the rock coast and cliffs in this area.

The Cantabrian littoral is a mesotidal coast with a semi-diurnal tide character (Puertos del Estado 2018). It is subject to intense surface processes, such as strong waves and high tides. Waves come mainly from NW while wind principal components are NW, SW and NE. This coastal area is exposed to large storms from the NW, produced by North Atlantic low pressure systems. During sea storms, strong NW swell waves dominate, reaching up to 26 m high, with a significant height of more

V. Bruschi (✉)
Departamento de Ciencia e Ingeniería del Terreno y de los Materiales,
Universidad de Cantabria, Avenida Los Castros s/n, 39005 Santander, Spain
e-mail: viola.bruschi@unican.es

J. Remondo
Departamento de Ciencias de la Tierra y Física de la Materia Condensada,
Universidad de Cantabria, Avenida Los Castros s/n, 39005 Santander, Spain
e-mail: juan.remondo@unican.es

J. A. Morales (ed.), *The Spanish Coastal Systems*,
https://doi.org/10.1007/978-3-319-93169-2_4

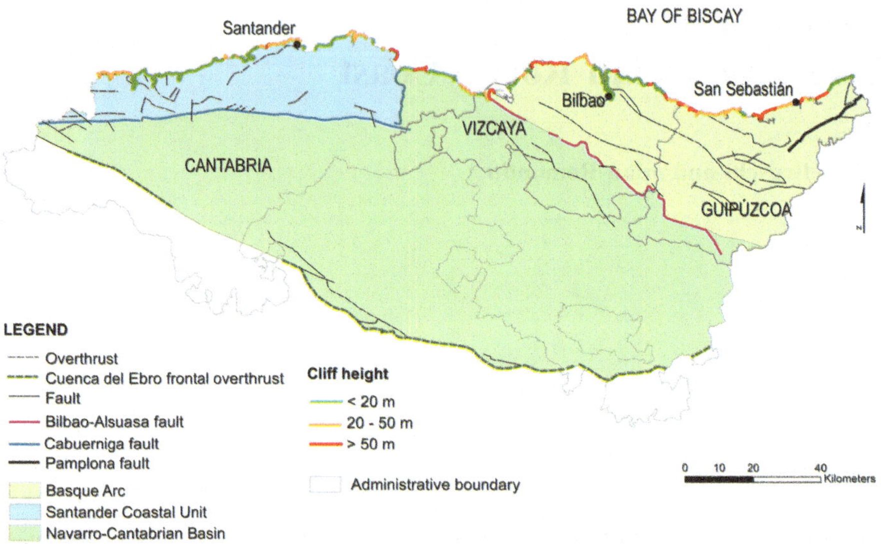

Fig. 4.1 Main geological domains and structures of the Basque-Cantabrian basin (based on Vera 2004). Coastal cliff heights, in classes, are also represented

than 14 m (Puertos del Estado 2018). Meteorological tides and associated strong waves can elevate sea-level about 60 cm during several days (Marcos et al. 2009).

The flora of the Cantabrian cliffs includes halophyte species exclusive of the littoral ecosystems, with very singular adaptations to the marine environment (thickened parenchyma, rosette leaves, etc.) (Diaz de Terán et al. 2016). *Crythmum marítimum*, *Limonium vulgare*, *Anthyllis vulneraria*, *Silene vulgaris* subsp. *marítima* and *Lylium pirenaicum*, are some representative examples. Cliff margins are covered by *Erica vagans* heath, dominant specie, with *Erica cinerea*, *Ulex europaeus*, *Daboecia cantabrica*, etc. In shallow soils grow saline grass of *Festuca rubra*, with *Plantago marítima*, *Plantago coronopus*, *Daucus carota*, etc. Associated to massive limestone formations, with very limited soils, holm oak woods develop. The scarcity and singularity of these cliff ecosystems justify their inclusion in the Nature 2000 Network, as High-Priority Habitats for Conservation (European Commission 1992).

4.2 Geology

From the geological point of view (Fig. 4.2 and Table 4.1), the Cantabrian Rock Coast is located in the eastern sector of the commonly named Cantabrian Range (also Cantabrian Mountains or Cantabrian Cordillera), corresponding to the geological Basque-Cantabrian basin or domain. It is the western extension of the

Fig. 4.2 Geological sketch map of the Cantabrian Rocky Coast (based on Rodríguez Fernández et al. 2014). A more detailed explanation on the lithostratigraphic units are presented in Table 4.1. Graphs in the right represent the cumulated length of both, coastline and geological structure directions

Table 4.1 Main lithostratigraphic units in the Cantabrian Rocky Coast (based on Rodríguez Fernández et al. 2014)

Lithology	Epoch-age
Limestone, dolostone, marl, sandstone, shale and occasionally conglomerate	Paleocene
Alternation of well-stratified limestone, clayey limestone, marls and white and red marlaceous lime. Calcareous sandstone and shale (Flysch facies)	Upper cretaceous–paleocene
Dolostone, limestone, calcarenite, massive dolomitic breccia, calcareous sandstones, marls, bioclastic limestones	Upper cretaceous
Well-stratified limestone, marl and massive dolostone, occasionally with sand at the base	Upper Cretaceous (Cenomanian–Turonian)
Sandy limestone, bioclastic limestone, sandstone, marl and shale	Upper cretaceous (Cenomanian)
Limestone, limestone with *Toucasias*, clastic limestone, marl and sandstone. Urgonian complex	Lower cretaceous (Albian–Aptian)
Red and green shales with gypsum, occasionally with sandstone intercalations (Keuper facies)	Upper triasic

Pyrenees, resulting from the convergence between the Iberian and the European plates during Eocene-Miocene, which produced a large regional E-W Alpine structure which is parallel to the coast of the Bay of Biscay (Fig. 4.1). This litho-structure controls the coastline trace, although there is a minor divergence between the structural strike and the direction of the coastal line.

4.3 Geomorphology

The main factors that control the geomorphology of this coast are the eustatic changes (sea level oscillations and continental uplift), the hard lithology and the geological structure, the relative orientation of prevailing waves and, with increasing importance, human activity.

The Cantabrian Rock Coast is characterized by the predominance of steep cliffs, developed on hard rocks mainly (Fig. 4.3). Raised wave cut benches (marine terraces) are also attached to this emerged coast. Embayed (pocket) beaches and, principally, fluvial valleys submerged by the sea (rías, estuaries and bays) interrupt the abrupt shoreline.

Marine abrasion surfaces are well developed at the bottom of the cliffs (Fig. 4.3a, b). The continental shelf of the southeastern part of the Bay of Biscay is

Fig. 4.3 Examples of coastal cliffs in the Cantabria Rock Coast area. **a** Panoramic view of the cliffs of the Basque Coast Geopark on Flysich facies. The picture has been taken from the Zumaia municipality toward the west. Coastal cliff evolves and retreat mainly through mass movements caused by sea wave erosion. There is a pebble beach at the bottom of the cliff and a well-developed abrasion platform. Note that the waves (from NW) impact obliquely to the coast (W-E). Several marine terraces can be observed in the background. **b** Landslides affecting the upper part of the cliff in La Arnía, where houses are exposed to hazard. There is an abrasion platform at the foot of the cliff. **c** San Julián cove in Liendo. Steep cliffs developed on massive limestone of the Urgonian Complex, which is affected by a halokinetic processes. The diapir outcrops in the lower part (red colours), pushing and destabilizing the massive limestone and producing rock falls. **d** San Juan de Gaztelugatxe area. It is a quite representative hard rock coast from the northern coast of Biscay, next to the Machichaco Cape. Islets formed in Cretaceous limestone, detached from the cliff (retreat) and corresponding to an ancient coastline

quite narrow and is also controlled by structural features (Galpasoro et al. 2010). According to these authors, different morphological features can be identified on the seafloor (0–100 m below mean sea level): rock erosional outcrops, 8 shore terraces up to −92 m, paleo river channels, sandy depositional features prograding wedges, sorted bedforms, ripple sandbars and anthropogenic structures.

Stepped sequences of ancient marine abrasion surfaces (marine terraces or rasas) are also relevant landforms in this rock coast sector. They can be interpreted as wave-cut platforms perched above the sea level and locally covered by a thin veneer of deposits (Gutiérrez et al. 2014). In a review provided by Flor and Flor-Blanco (2014), six continental erosional surfaces and twelve marine terraces (rasas) with ages ranging from Upper Eocene and Holocene and heights between 285 and 4 m are described. The marine origin of some of those terraces is to some extend questionable and only the surfaces bellow 65 m contain beach sediments (the lowest four terrace levels) while the rest present only alluvial and slope deposits and can be considered as continental. According to these authors, they can be traced from the French border to the central coast of the Lugo province (Galicia), although different sectors can be distinguished: in eastern Galicia there is only one terrace level which gradually reduces the height to the west; in the western part of Asturias, terraces appear at lower elevations and are less numerous (five levels) in comparison with other parts of the Cantabrian Coast; in eastern Asturias and in Cantabria, 12 terrace levels with quite good continuity appear. Finally, in the Basque Coast erosion levels are worse preserved and scarcely studied.

Hernández-Pacheco et al. (1954) described a 7, 20, 40, 60 m high rasas in the surroundings of Bermeo. In the right margin of Nervión River (northwestern Gran Bilbao) other discontinuous marine terraces can be seen (230 and 95 m). Hazera (1968) described an 80 m high surface in Algorta. In eastern Guipúzcoa, rasas with altitudes of 285 and 230 m have been reported in the Jaizquibel-Ulía Mountain by Edeso and Ugarte (1990) and Flor and Flor-Blanco (2014).

Marine terraces are higher in the Cantabrian Coast than in other sectors of the northern Spanish coast due to a more intense continental uplift in this sector (Fillon et al. 2016). For instance, the rasa with a mean elevation of 35 m in central Asturias corresponds to more than 40 m in Biscay. Nevertheless, lowest terraces (less than 10 m high) remain at a constant elevation along the northern coast of Spain. Lower terraces frequently present beach deposits on top (Flor and Flor-Blanco 2014). For instance, the lowest rasa are covered by gravel and sand beach deposits in Pechón, Suances, Guernica, La Navia, Zarauz, Bidasoa Pasajes, Jerra, Oyambre, etc., although they are scarce and not well preserved (Rivas 2000).

According to Mary (1983), the oldest marine terrace formed either during the Aquitanian-Langhian transgression at Early Miocene (although upper terraces match a continental origin) or in the Lower Pliocene, indicating long term uplift rates of 0.013 and 0.05 mm/year, respectively. Álvarez-Marrón et al. (2008) dated the 160 m high terrace in Asturias by cosmogenic nuclides in a minimum age of 1–2 Ma, obtaining an uplift rate of 0.07–0.15 mm/year. Extrapolating this estimate, the highest marine terrace (285 m) formed 4–2 Ma ago. Jiménez-Sánchez et al. (2006) have suggested a tentative maximum elevation rate of the Cantabrian coast of about 0.19 mm/year on the basis of U/Th geochronological dates obtained from speleothems within a cave located in a sea cliff located in the eastern sector of Asturias. They consider that the uplift is still an active process at present. Álvarez-Marrón et al. (2005) estimated a quite similar rise of 0.5–0.2 mm/year from the 100 m marine terrace in Western Asturias.

The existence of uplifted marine paleosurfaces is the result of the coupling of sea level rise and epeirogenic uplift. Sea level have considerably oscillated during the Cenozoic, being the late Oligocene the last time sea reached eustatic level levels noticeably higher than current ones (Kominz et al. 1998) and continental uplift seems to be also still active. Despite the principal Alpine overthrusting ceased before Late Miocene in the Cantabrian range, there are still clear evidences of contemporary tectonic activity. In this way, seismicity in the region is quite frequent, although normally of low magnitude (López-Fernández 2007; Gibson et al. 2007; Pando et al. 2007). Moreover, the Cantabrian Range uplift has continued for the last 1–2 Ma, as registered in the marine terraces covered by Pleistocene deposits with evidences of neotectonic activity (Gutiérrez-Cheverol et al. 2006).

Regarding sea level change, Leorri et al. (2012) present a review in which authors provide a new sea level change curve for the Basque coast, covering the last 12,000 years. The curve shows two main phases: a rapid relative sea level rise period (from 27 to 5 m bmsl) at ca 10,000–7,000 cal year BP; and a relative slow sea level rise from ca 7,000 cal year BP until present. Accordingly, they provide rates for sea level rise between 0.7 and 0.3 mm/year, which are close to the glacioeustatic contribution from land ice melting for the last 7,000 years. However, these figures are lower than the rates obtained northward at the French coast and higher than the ones from Portugal, showing a clear north-south gradient. This fact are explained by decreasing rates of vertical land motion associated with the subsidence of the peripheral bulge due to Eurasian deglaciation (glacio-isostasy). Superimposed to this effect there would be another signal induced by the ocean (hydro-isostasy) that produces a gradient perpendicular to the shoreline, due to the ocean loading, with a broad uplift of the continent which surpasses 2 m, according to the isostatic sea level model proposed in Leorri et al. (2012). In more recent times, relative sea level rise has been reported by Leorri et al. (2013) for the eastern Cantabrian Coast, with an estimated rate of 1.9 mm/year for the 20th century, the greatest sea level rise in SW Europe.

However, more precise paleo-environmental reconstructions are constrained by several deficiencies, above all: the infrequent and discontinuous sedimentary records, almost exclusively circumscribed to estuaries, the little known continental motion and deformation produced by current and recent neotectonics and the limited number of dates obtained by absolute dating techniques. In any case, it is remarkable the scarcity of research works on the rock coast of this sector of the Spanish littoral. This fact is particularly pertinent in the case of cliffs due to accessibility limitations.

The Cantabrian emergent coast has been affected by several geomorphic processes, mainly fluvial and alluvial, karstic, halokinesis and, especially, slope processes (Figs. 4.3 and 4.4). All these surface processes have produced different and

Fig. 4.4 Examples of erosional processes and landforms in the Cantabrian Rocky Coast, illustrating cliff retreat and coastal evolution resulting from sea wave undermining. **a** Mass movement caused by sea wave erosion on marl and limestone materials of the Upper Cretaceous in Costa Quebrada. **b** Differential erosion depending on lithology (from marls to marly limestone). Rock (block) fall deposit. It is one of the principal mechanisms of coastal cliff retreat. **c** Active climbing dunes in Sonabia pocket beach where dunes move uphill the cliff of Monte Candina, consisting of Aptian massive limestone. **d** Puente del Diablo (Santander), an ancient karstic pipeline exhumed by cliff erosion and retreat, which is conditioned by the litho-structure. Puente del Diablo felt down in November 2010 (right). The image in the left shows the former state

characteristic types of landforms and deposits, which are not specific of this sector of rock coast. The incidence of the erosive processes largely depends on the lithological strength to stress: hard rock coastline (rock mechanics) and soft rock coastline (soft rock and soil behavior).

Fluvial processes are particularly identifiable from fluvial incisions. Obviously, diffuse fluvial and alluvial deposits and landforms are common in the rocky coast. The best examples of that fluvial incision can be seen in the main river mouths and estuaries (rías and bays). Cantabrian rivers and valleys are short, abrupt and mighty. Several fluvial terrace sequence have been identified in the eastern Cantabrian Range with a similar geomorphological arrangement all over the area.

Fluvial incision rates are considerably high in the Pas and Besaya rivers, located in the central-eastern sector of the Cantabrian Mountains (González-Díez et al. 1996a, b). On the basis of radiocarbon dating and archaeological remains, these authors calculate incision rates ranging from 0.2 to 12.5 mm/year, with an average of 0.5 mm/year for the last 100,000 years. Extrapolating the average incision rate, they could obtain ages of 1.6, 1.2, 0.5 and 0.3 Ma for the erosive surfaces, previous to the fluvial terraces, located at 125, 250, 600 and 800 m, respectively. The highest surface could correspond to the paleosurfaces more to the west described by Álvarez-Marrón et al. (2003) as relict peneplain surfaces uplifted and dissected mainly by fluvio-glacial erosion. Arriolabengoa et al. (2015) found 8 fluvial terrace levels in the Deba Valley (Guipúzcoa), correlated with dated endokarstic sediments and slope erosion intensity. Fluvial incision rate estimates are smaller than those obtained for the Cantabria area (González-Díez 1996a).

Mass movements constitute the predominant geomorphic process affecting this sector of coastline, being the principal cause of cliff retreat and coastal evolution (Fig. 4.4a, b). Therefore, these slope processes are relevant cause of geomorphic hazard and risk. The Cantabrian Coast is dotted with landslides caused by the erosion and undermining of the cliffs (Corominas et al. 2018). The outcrops of Cretaceous and Eocene flysch in the Basque Country are particularly susceptible due to the unfavorable arrangement of the layers and the low strength of the lithology. Undermining produced by marine erosion are favored by sea level rise and recurrent strong sea storms. Falls and slides affecting the coastal cliffs, are especially frequent in those made up of soft rocks, such as the flysch in Zumaya. A recent synthetic review of landslide activity in Spain, with specific references to the Cantabrians have been presented by Bonachea et al. (2014). Despite most of the contributions in literature focus on recent landsliding, González-Díez et al. (1996a, 1999) studied the landslide activity in the Cantabrian Range for the last 100,000 years, in connection with climate, seismicity and human intervention and analyzing its contribution to landscape evolution.

Karstic processes and landforms are also important, as limestone (and dolostone) lithology that is one of the most frequent material in the Cantabrian Rock Coast (Fig. 4.4c). The evolution of the cave systems in the karst massifs must be related with base level lowering by fluvial incision and therefore with negative eustatism. An extensive review of the karst landscape evolution in the coastal area of the Bay of Biscay is presented by Aranburu et al. (2015). Associated with the same type of lithology, Elorza and Higuera-Ruiz (2015) described quite recent erosive forms (ventifacts), generated by sand abrasion on the Urgonian limestone, affected by minor weathering forms, produced by chemical solution, postdate the erosive forms.

Aeolian geomorphic landforms can be seen in Barrika (NW Biscay) (Cearreta et al. 1990), where there is a thick cliff-top sand deposit located 50 m a.s.l., formed by a climbing wind dune and probably related to the 40 m high rasa. A good example of active climbing dunes occurs in Sonabia (Fig. 4.4d) where dunes move from the beach uphill to the cliff of Monte Candina. On the other hand, a good example of halokinesis affecting coastal cliffs (Fig. 4.3c) can be seen in the beach of San Julian (Cantabria).

Anthropic activity has become a crucial and growing geological agent during the Holocene and particularly in recent times, producing new anthropo-geoforms. A noticeable example of these new forms are the breakwaters of the coastal port of

Bilbao, anchored in the rock cliffs and seafloor. Of course there are many other sedimentary and erosive anthropic landforms all over the coastal area. Moreover, human activity impacts the coast, increasing severely the erosive and sedimentation rates and modifying the behavior of the geomorphic processes (Remondo et al. (2005). Human intervention increase is coupled with climatic warming, sea level rise and other worldwide environmental changes, constituting the global change that, for sure, will affect rock coast and cliffs. Among all those worldwide changes, sea level rise will be probably the most decisive. Leorri et al. (2012) estimated an average of 1.9 ± 0.3 mm a^{-1} for the 20th century, particularly after 1950. A projection of 10–68 cm for the end of the present century, with 50 cm as the most likely scenario, was made in a report for the Ministry of the Environment (Moreno et al. 2005). The assessment of the effects of climate change on the Spanish Coasts to design adaptation strategies for 2050–2100 is presented in Medina et al. (2004) and Méndez et al. (2004). In any case, erosive effects of global change are already being taking place. Rivas and Cendrero (1994) obtained retreat rates of 2 m a^{-1} in the Oligocene-clay cliffs in Oyambre (Cantabria), although they only can be considered local estimates. Even more noticeable are the increasing erosive effects and damage caused by recurrent and more and more frequent and intense sea storms.

The evolution of landforms and deposits over time, resulting from some of those geomorphic processes acting simultaneously, has shaped the current coastal landscape. An example of the evolution and retreat of a sector of Cantabrian Rock Coast in Costa Quebrada Geologic Park (Cantabria) is shown in Fig. 4.5, which is based on the works carried out by (Díaz de Terán et al. 2016). The shape of this coastline is determined by differential erosion that is controlled by the alternation of soft and hard materials (Cretaceous marls and limestones) laid sub-parallel to the coastline (Fig. 4.4b). The model of evolution begins with a rectilinear coastline constituted by hard and fractured Aptian limestone, very resistant to erosion. The a in Fig. 4.5 represents a former coast line some undetermined-time ago. The action of the sea waves undermines the cliffs, running the fractured limestone through and eroding softer marls located immediately behind it (b in Fig. 4.5). Once the erosion is progressing at the bottom of the marl body, removing materials, the upper part of the cliff collapses, falling down and producing a progressive emptying of the sinkhole (b in Fig. 4.5). The next step in the model is the development of a small abrasion platform (c in Fig. 4.5). After the front of the cliff dismantled, sea waves can impact directly on the softer materials, which produces a progressive retreat of the cliff, facilitated by the erodibility of the materials which are prone to erosion (d in Fig. 4.5). The last stage of evolution observable in this sector of the coast is illustrated in e (Fig. 4.5), where it is possible to differentiate two coves, one containing beach sediments and the other characterized by the development of an abrasion platform. In the case of the eastern inlet, some hard-rock blocks of an early coastline remains, making it possible the protection from sea wave erosion and the beach formation. In the other case, there are smaller hard-rock blocks protecting the soft materials. Therefore, the cliff is more exposed to sea wave action and the abrasion platform can be formed, but sand deposits don't preserve.

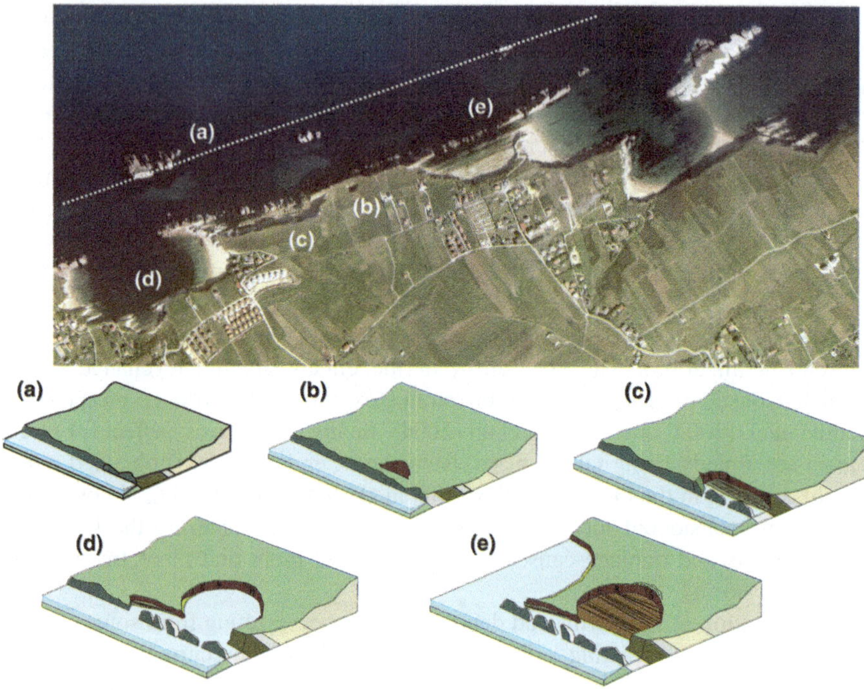

Fig. 4.5 Model explaining the evolution of the coastline in the Costa Quebrada Geologic Park (within the Protected Natural Space of Dunas de Liencres y Costa Quebrada). Block-diagrams represent different phases of evolution (letters (a), the former, to (e), the most advanced). Current landforms in the picture shows these stages of evolution ((a–e) on the image above). For a more detailed explanation of the figure, please see the main text. This model is based in the works carried out by Díaz de Terán et al. (2016)

All these examples of processes and landforms observable along the whole coastline have a clearly geologic and natural interest from the point of view of the conservation (Figs. 4.3a, d and 4.4c, d). From Tina Mayor (west end of the Cantabrian Coast) to the Bidasoa estuary (east) there are many geological elements declared of regional, national or international interest. On the one hand, there are numerous sites recorded in the National Inventory of Geosites (http://info.igme.es/ielig/), due to their local, regional or national merit, such as the Tina Menor Estuary (coastal landforms of high scientific interest), Santa Justa Anticline (anticline in the coastal cliff with a chapel in its core), Abrasion Platform at La Arnía, Liendo diapir or, the Black Flysch and the Slump of Armintza, the Gernika Anticline or clay and ophite of the Bakio diapir (Duque and Elizaga 1983). All these features, inventoried at the regional level, although not necessarily correspond to any conservative figure, are reflected within the corresponding Land Plans for the Littoral. On the other hand, there are other examples of high international interest, which are catalogued in the Global Geosites Programme, as the Cretaceous-Cenozoic boundary, the flysch facies (recognized by the IUGS and UNESCO as one of the greatest geological outcrops in the Earth) in the Geoparque de la Costa Vasca (Basque Coast Geopark), or the estuary and littoral in the Biosphere reserve of Urdaibai (Global Geisites Project; Carcavilla and Palacio 2010; García-Cortés 2009). Those areas are plenty of examples of spectacular coastal erosion processes with large cliffs, mass movements, abrasion platforms and rasas. Finally, there are other protected natural spaces, such as National Parks, sites of community importance (SIC) and the above mentioned biosphere Reserve, subject to high degree of protection with specific regulations.

References

Álvarez-Marrón, J., Hetzel, R., Niedermann, S., Menéndez, R., & Marquínez, J., (2008) Origin, structure and exposure history of a wave-cut platform more than 1 Ma in age at the coast of northern Spain: a multiple cosmogenic nuclide approach. *Geomorphology, 93*, 316–334.

Álvarez-Marrón, J., Menéndez, R., & Marquínez, J. (2003). Intramontane landscape evolution in the initial stages of coastal mountain growth (Cantabrian Mountains, Spain). *Geophysical Research Abstracts, 5*, 36–50.

Álvarez-Marrón, J., Menéndez, R., Niedermann, S., Hetzel, R., & Marquínez, J. (2005). Structure and exposure history of a large marine terrace along the Cantabrian shoreline of northern Spain. *Geophysical Research Abstracts, 7*, 73–76.

Aranburu, A., Arriolabengoa, M., Iriarte, E., Giralt, S., Yusta, I., & Martínez-Pillado, V., et al. (2015). Karst landscape evolution in the littoral area of the Bay of Biscay (North Iberian Peninsula). *Quaternary International, 364*, 231–243.

Arriolabengoa, M., Iriarte, E., del Val, M., & Aramburu, A. (2015). Terrazas fluviales y registros endokárstico del valle del río Deba (Guipúzcoa): una visión integrada de la evolución de los valles cantábricos. In J. P. Galve, J. M. Azañón, J. V. Pérez Peña, & P. Ruano (Eds.), *Una visión global del Cuaternario. El hombre como condicionante de procesos geológicos* (Vol. XIV). Granada: Reunion Nacional de Cuaternario.

Bonachea, J., Bruschi, V. M., Fernández-Maroto, G., Remondo, J., González-Díez, A., & Díaz de Terán, J. R., et al. (2014). Geomorphic hazards in Spain. In F. Gutiérrez & M. Gutiérrez (Eds.), *Landscapes and landforms of Spain. World geomorphological series*. Springer.

Carcavilla, L., & Palacio, J. (2010). *Proyecto Geosites: aportación española al patrimonio geológico mundial*. Madrid: Instituto Geológico y Minero de España.

Cearreta, A., Edeso, J. M., Merino, A., Ugalde, T., & Ugarte, F. M. (1990). *Las Dunas litorales de Barrika (costa occidental de Bizcaya)*. Kobie, 1977–1983.

Corominas, J., Mateos, R. M., & Remondo, J. (2018). Review of landslide occurrence in Spain and its relation to climate. In K. Ho, S. Lacasse, & L. Picarelli (Eds.), *Slope safety preparedness for impact of climate change*. London: Balkema.

Día de Terán, J. R., Bruschi, V., Cendrero, A., Francés, E., Flor, E., & Gonzáles Lastra, J. R. (2016). *El litoral entre Santander y Liencres*. Ed. Estudio. Santander.

Duque, L. C., & Elizaga, E. (1983). *Puntos de interés geológico en el sector oriental de la Cordillera Cantábria*. Madrid: IGME.

Edeso, J. M., & Ugarte, F. M. (1990). Algunos datos sobre la paleo-geografía litoral cuaternaria de la Costa: Jaizkibel-Bahía de Txingudi (Golfo de Vizcaya). *Cuad Sec Eusko Ikaskuntza, 16*, 27–76.

Elorza, J., & Higuera-Ruiz, R. (2015). Morfologías escalonadas de erosión eólica y disolución química sobre las calizas urgonianas en la playa de Valdearenas (Sonabia, Cantabria). *Revista de la Sociedad Geológica de España 28*, 25–39.

European Commission. (1992). Directiva 92/43/CEE del Consejo, de 21 de mayo de 1992.

Fillon, C., Pedreira, D., van der Beek, P. A., Huismans, R. S., Barbero, L., & Pulgar, J. A. (2016). Alpine exhumation of the central Cantabrian Mountains, Northwest Spain. *Tectonics, 35,* 1–18.

Flor, G., & Flor-Blanco, G. (2014). *Raised beaches in the Cantabrian Coast.* In F. Gutiérrez & M. Gutiérrez (Eds.), *Landscapes and Landforms of Spain. World geomorphological series.* Springer.

Galparsoro, I., Borja, A., Legorburu, A., Hernández, I., Chust, C., & Liria, G., et al. (2010). Morphological characteristics of the Basque continental shelf (Bay of Biscay, northern Spain): Their implications for integrated coastal zone management. *Geomorphology, 118,* 314–329.

García-Cortés, A. (Ed.). (2009). *Spanish geological frameworks and geosites: An approach to Spanish geological heritage of international relevance* (p. 234). CD-ROM: Geological Survey of Spain (IGME).

Gibson, M., Sinclair, H. D., Lynn, G. J., & Stuart, F. M. (2007). Late- to post-orogenic exhumation of the Central Pyrenees revealed through combined thermochronological data and modelling. *Basin Research, 19*(3), 323–334.

González-Díez, A., Díaz de Terán, J. R., Barba, J., Remondo, J., & Cendrero, A. (1996a). A proposal of a chronological model for the fluvial terraces of the rivers in the northern Cantabrian Range. *Geogaceta, 20*(5), 1096–1099.

González-Díez, A., Remondo, J., Díaz de Terán, J. R., & Cendrero, A. (1999). A methodological approach for the analysis of the temporal occurrence and triggering factors of landslides. *Geomorphology, 30,* 95–113.

González-Díez, A., Salas, L., Díaz de Terán, J. R., & Cendrero, A. (1996b). Late quaternary climate changes and mass movement frequency and magnitude in the Cantabrian region, Spain. *Geomorphology, 15,* 291–309.

Gutiérrez, F., Gutiérrez, M., & Martín-Serrano, A. (2014). The geology and geomorphology of Spain: A concise introduction. In F. Gutiérrez & M. Gutiérrez (Eds.), *Landscapes and landforms of Spain.* World geomorphological series. Springer.

Gutiérrez-Claverol, M., López-Fernández, C., & Alonso, J. L. (2006). Procesos neotectónicos en los depósitos de rasa en la zona de Canero (Occidente de Asturias). *Geogaceta, 40,* 75–78.

Hazera, J. (1968). La region de Bilbao et son arrière-pays. Étude géomorphologique. Thèse de Doctorat. Sociedad Ciencias Naturales de Aranzadi. San Sebastian. Munibe, XX (1/4).

Hernández-Pacheco, F., Llopis Lladó, N., Jordá-Cerdá, J., & Martínez, J. A. (1954). Guía de la excursion n° 2. El cuaternario de la region Cantábrica. V Congreso Internacional INQUA. Diputación Provincial de Asturias. Oviedo.

Jiménez-Sánchez, M., Bischoff, J. L., Stoll, H., & Aramburu, A. (2006). A geochronological approach for cave evolution in the Cantabrian Coast (Pindal Cave, NW Spain). *Zeitschrift für Geomorphologie NF, 147,* 129–141.

Kominz, M., Miller, K. G., & Browning, J. V. (1998). Long-term and short-term global Cenozoic sea level estimates. *Geology, 26,* 311–314.

Leorri, E., Cearreta, A., García-Artola, A., Irabien, M. J., & Blake, W. H. (2013). Relative sea-level rise in the Basque coast (N Spain): Different environmental consequences on the coastal area. *Ocean and Coastal Management, 77,* 3–13.

Leorri, E., Cearreta, A., & Milne, G. (2012). Field observations and modelling of Holocene sea-level changes in the southern Bay of Biscay: Implications for understanding current rates of relative sea-level change and vertical land motion along the Atlantic coast of SW Europe. *Quaternary Science Reviews, 42,* 59–73.

López-Fernández, C. (2007). Actividad sísmica, zonación sismotectónica y peligrosidad sísmica en el noroeste de la Península Ibérica. *Serie Nova Terra, 35,* 278.

Marcos, M., Tsimplis, M. N., & Shaw, G. P. (2009). Sea level extremes in southern Europe. *Journal of Geography Research: Oceans, 114*(C1), C01007.

Mary, G. (1983). Evolución del margen costero de la Coordillera Cantábria de Asturias desde el Mioceno. *Trabajos de Geología, 13,* 3–35.

Medina, R., Losada, I. J., Méndez, F. J., Olabarrieta, M., Liste, M., & Menéndez, M., et al. (2004). Impacto en la Costa Española por efecto del Cambio Climático. Ministerio de Medio Ambiente, Oficina Español de Cambio Climático-Dirección General de Costas, Madrid.

Méndez, F. J., Medina, R., Losada, I. J., Olabarrieta, M., Tomás, A., & Liste, M., et al. (2004). estudio de los impactos en la costa española por efecto del Cambio Climático. In J. C. García Codrón, C. Diego-Liaño, P. Fernández de Arróyabe-Hernáez, C. Garmendia Pedraja, D. Rasilla Álvarez (Eds.), el clima entre el mar y la montaña. Asociación Española de Climatología and Universidad de Cantabria, Serie A, No. 4, Santander.

Moreno, J. M., Aguiló, E., Alonso, S., Álvarez Cobelas, M., Anadón, R., Ballester, F., et al. (2005). *A preliminary assessment of the impacts in Spain due to the effects of climate change* (ECCE Project – Final Report). Ministry of the Environment and the University of Castilla La Mancha.

Pando, L. A., López Fernández, C., de la Rubia Mir, L. (2007). Actividad sísmica, zonación sismotectónica y peligrosidad sísmica en el noroeste de la Península Ibérica. Nova Terra 35.

Puertos del Estado. (2018). http://www.puertos.es/.

Remondo, J., González-Díez, A., Soto, J., Díaz de Terán, J. R., & Cendrero, A. (2005). Human impact on geomorphic processes and hazards in mountain areas. *Geomorphology, 73,* 185–206.

Rivas, V. (2000). Clima y nivel del mar: reconstrucción de las posiciones marinas cuaternarias a través de las evidencias en el Litoral Cantábrico. In: García Codrón JC (coord.) La reconstrucción del clima en época preinstrumental. Pub. Universidad de Cantabria. Santander.

Rivas, V., & Cendrero, A. (1994). Human in fluence in a low-hazard coastal area: An approach to risk assessment and proposal of mitigation strategies. *Journal of Coastal Research, 12,* 289–298.

Rodríguez Fernández, L. R., López Olmedo, F., Oliveira, J. T., Medialdea, T., Terrinha, P., & Matas, J., et al. (2014). Mapa Geológico de España y Portugal, 1: 1,000,000. IGME and LNEG (Spanish and Portuguese Geological Surveys).

Vera, J. A. (2004). Geología de España. Pral (ed) Madrid.

Chapter 5
Rocky Coast in Catalonia

Isabel Montoya-Montes, María José Sánchez-García, Ignacio Alonso, Mariona Casamayor and Inmaculada Rodríguez Santalla

5.1 Introduction

Rocky coasts represent the majority of the world shorelines (Stephenson et al. 2013). Relevant advances have been made in the knowledge and understanding of rocky coast morphodynamics using different techniques both in the field and in laboratory (Stephenson et al. 2010, 2013; Lim et al. 2010). Morphology and the numerous processes that operate according to geological and climatic conditions have been widely studied by several authors (Trenhaile 1987, 2011; Sunamura 1992, 2015 among others), but on the same way, these studies are limited if they are compared with the research literature on low-lying coasts.

Going through the Mediterranean and Black Sea, the rocky coast bordering these basins corresponds to more than 50% of the coastline (Furlani et al. 2014). The current physiography of these rocky regions is closely related to sea level variations along the history (Zazo et al. 2003; Pirazzoli 2005; Lambeck et al. 2010; Benjamin et al. 2017).

The Catalan littoral, in turn, counts on an extension of rocky coast that represents almost 40% of the Catalan coastline (GENCAT 2015). Geomorphological studies

I. Montoya-Montes (✉) · M. J. Sánchez-García · I. Alonso · M. Casamayor
Instituto de Oceanografía y Cambio Global, IOCAG, Universidad de Las Palmas de Gran Canaria, ULPGC, Campus de Tafira, 35017 Las Palmas de Gran Canaria, Spain
e-mail: isabel.montoya@ulpgc.es

M. J. Sánchez-García
e-mail: mariajose.sanchez@ulpgc.es

I. Alonso
e-mail: ignacio.alonso.bilbao@ulpgc.es

M. Casamayor
e-mail: mariona.casamayor101@alu.ulpgc.es

I. R. Santalla
Universidad Rey Juan Carlos, c/Tulipán sn, 28933 Móstoles, Madrid, Spain
e-mail: inmaculada.rodriguez@urjc.es

J. A. Morales (ed.), *The Spanish Coastal Systems*,
https://doi.org/10.1007/978-3-319-93169-2_5

on the whole coast at 1:250.000 scale (Calvet and Gallart 1973), and specifically on the northern sector at 1:200.000 (Butzer 1964; Barbaza 1970) had been tackled. More recently the geomorphological map of Spain (1:1.000.000) (IGME 2005), presents the subaerial and submerged morphologies allowing the understanding of sedimentary environments as continuous systems. Moreover, a detailed coastal typology of the Catalan coast has been also carried out by Ballesteros et al. (2007), Mariani et al. (2014) and GENCAT (2015). They revealed the importance of geomorphological factors influencing the development of biological communities at the Catalan littoral, through the partition of the rocky shoreline in several sectors, each one characterized by a different biological community.

One of the most outstanding aspects when considering rock coasts are those related to how these coasts evolve. Cliff retreat has been largely studied, and the driving mechanisms of this process are quite well understood. There is a clear agreement that sea cliff retreat is often linked to large waves, heavy precipitation and seismic events, although some authors have also considered other factors such as rock hardness, structural aspects of the rock, like the presence of faults, fractures and joints, and mechanical abrasion mechanisms (Kline et al. 2014).

Regarding precipitation, Young et al. (2009) found a high correlation between the timing of rainfall and coastal cliff erosion in southern California. Nevertheless, rainfall runoff directed onto specific cliff areas can create erosion hot spots unrelated to cliff hardness or regional rainfall quantities (Young 2018).

Even though in the last decades several papers have greatly contributed to give some light into the understanding of rock coasts, there are still many uncertainties. Some of them are related to rock resistance to erosion processes, including weathering and biological processes, as well as the erosional effectiveness of wave processes. Other aspects, which remain unclear, are the role of inheritance in long term evolution of cliffs and shore platforms, and the response of rock coasts to climate change and sea-level rise (Stephenson et al. 2013).

5.2 Study Area Background

5.2.1 Climate, Precipitation and Sediment-Water Discharge

Mediterranean climate is characterised by seasonal variability and conditioned by physiography and sea presence. Catalan river systems show relatively small size and relatively high slope due to the short distance between mountain headwaters and river mouths, so gullies and streams are prominent features in the Catalan area, which experiences the typical torrential, seasonal and sporadic rainfall. Most of the precipitation over the drainage basins occurs in autumn and, to a lesser extent, in spring (Martin-Vide and López-Bustins 2006). Due to its specific local and regional characteristics, the Catalan coast is particularly exposed to flash floods (Llasat et al. 2008). Observing the Catalan watersheds discharging in the Mediterranean Sea, the water

Table 5.1 Mean water discharge (MWD) of Catalan river systems, maximum monthly-mean discharge (mMWD), and standard deviation (σ) calculated from the normalized monthly-mean water discharge values

River system	MWD ($m^3 s^{-1}$)	Period	Max. mMWD recorded ($m^3 s^{-1}$)	σ
1. Muga	3.26	1972–1997	71.96	2.17
2. Fluvià	1.34	1912–2001	11.07	1.09
3. Ter	12.13	1912–2001	113.85	1.11
4. Tordera	0.73	1958–1999	6.05	1.29
5. Besós	6.85	1967–2001	91.41	1.29
6. Llobregat	16.64	1967–2001	125.39	1.26
7. Foix	0.39	1917–1995	22.31	3.43
8. Gaià	0.26	1928–2001	3.85	1.54
9. Francoli	0.60	1929–2001	6.25	1.34
10. Riudecanyes	0.32	1930–1999	5.70	1.69
11. Ebro	410.42	1912–2000	2171.17	0.69

Data from Amblás et al. (2006). Location of rivers is shown in Fig. 5.1

discharge of the Ebro river represents 89% of the total fluvial water contribution to the northeast Iberian margin (Table 5.1). However, during flash flood events the sediment load transported by small rivers is not negligible, as long as higher water discharge tends to imply higher erosive capacity (Tucker and Bras 2000; Amblás et al. 2006).

Annual rain precipitation along the coast shows average values that are slightly larger along the northern half of the Catalan coast (around 600–700 mm/year) than in the southern half (500–600 mm/year), aspect that will have influence on the shaping of the rocky coast morphology. This difference is related to the higher elevations of the Transversal Chain in comparison to the Littoral and Pre-Littoral Chains (Martínez et al. 2007). Maximum values of annual rain are close to 1,300 mm/year in the Pyrenees.

When looking at daily intensity in precipitations, the eastern side of the Transversal Range and the Ebro delta are the zones with the largest intensities of daily rain for all percentiles. For instance, the average intensity achieves values close to 80 mm/day for the 95th percentile in these areas. These daily amounts are very important, since they represent a high percentage of the annual amount (Martínez et al. 2007). The reason for the highest daily intensities in the Catalan coast lies in the fact that autumn precipitations are usually more intense because of high values of the sea-surface temperature, which acts as a moisture source in the Mediterranean cyclogenesis (Millán et al. 1995; Pastor et al. 2001).

Soil cover is one of the factors controlling the amount of water and sediment reaching the coast and the erosive processes modelling the littoral physiography. Most of the watersheds of the region are densely vegetated. Nevertheless, Gaià and Francoli basins show highly erodible land such as dry-farmed crops, while Llobregat and Besós rivers are conditioned by the important industrial and urban development around Barcelona city (Liquete et al. 2009) (Table 5.1).

Large seasonal variability in wind patterns as well as spatial distribution of wind can be explained because of the complex environmental characteristics of the NW Mediterranean Sea and due to its semi-enclosed character (Campins et al. 2011). Low to medium average wind intensity is characteristic of this region, with some extreme events occurring (Sánchez-Arcilla et al. 2008).

5.2.2 Wave Climate, Storms and Sea Level

Wave storms are important because of their potential damage to the coast. At the Catalan coast, they are normally associated to two different situations (Cateura et al. 2004): (i) The initial positioning of an intense high-pressure area on the British islands, leading to NE and E air fluxes on the Catalan littoral. (ii) Mediterranean cyclogenesis due to a high level of cold air pool deepening and the passage of the resulting low in front of the Catalan littoral. This generates E winds, except in the Ebro delta area where the wind comes from the NW due to orographic effects.

Seasonal wave climate is present, with low-energy conditions during summer and high-energy situations during autumn and winter. Mean significant wave height (Hs) is 0.7 m and mean wave periods are 4–6 s. However, during extreme storms the situation is significantly different, since significant wave height up to 7 m and periods of 13 s have been recorded (Bowman et al. 2009) (Fig. 5.1). Episodes of extreme situations are common in the Mediterranean Sea. The most severe wave storm events come from ENE, and the frequency and intensity of the storm events decrease southwards (Mendoza and Jiménez 2008). For the south coast of Catalonia, the mean wave climate shows that the yearly mean significant wave height is about 0.8 m. The maximum recorded Hs was close to 6 m, corresponding to maximum wave heights of 10 m. The maximum recorded wave peak period was 14.3 s, with a yearly mean of 5 s (Bolaños et al. 2009).

An analysis of 10 years (1990–2002) of wave data from the Ebro XIOM (Oceanographic and Meteorological Instruments Network from Generalitat de Catalunya) buoy, considering the mean wave climate and storm events, revealed two main groups of wave storms with NW and E directions (Bolaños and Sánchez-Arcilla 2006). Over 300 storms were recorded in the Ebro delta during 10 years, showing a mean duration of 20 h. The storm period with largest Hs was mainly from October to March. The most persistent storms were found to be from the East with a mean Hs of 2 m while the shortest were from the NW with a mean Hs of 1.7 m. It was found that storms with a maximum Hs of about of 3.5 m occurred once a year.

Regarding the future evolution of storms, Casas-Prat and Sierra (2012) performed an extrapolation of waves for 2050 based on the 44-year hindcast wave climate database (1958–2001) from the European HIPOCAS project (Guedes Soares et al. 2002). As a result of their analysis, comparison between present and future scenarios reveals that the magnitude of the wave climate energy either does not change significantly or that it slightly decreases.

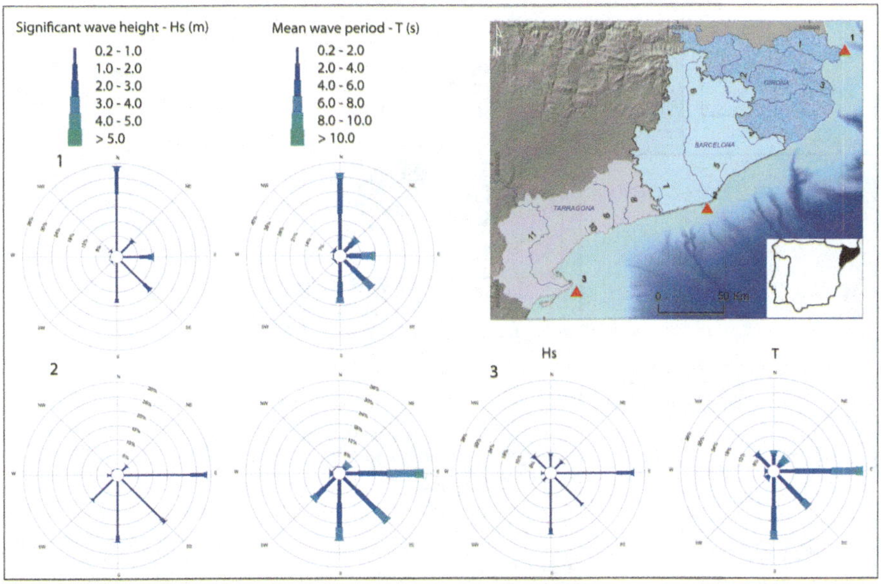

Fig. 5.1 Rose diagrams representing significant wave height (Hs) and mean wave period (T) for three SIMAR points along Catalan coast: (1) 2124148, Creus Cape; (2) 2109135, Barcelona; (3) 2095128, Ebro delta (Database: Puertos del Estado). Map presenting position of SIMAR points, main rivers cited in Table 5.1 and the three Catalan coastal provinces: Girona, Barcelona and Tarragona

Tidal range in Catalan coast is classified as microtidal. Nevertheless, the most important variations in sea level along the Catalan coast are due to meteorological conditions and the resonant effect in bays and harbours. Storm surges may be of the order of 1 m, a magnitude much larger than tidal range (Bolaños et al. 2009). This sea level variation has a very important effect on storm risk and coastal flooding.

5.2.3 Geology

The coastline of Catalonia has a complex geological history and a variety of lithologies can be found there (schist, gneisses, slates, limestone and granites). At the end of Palaeozoic, the Hercynian orogeny happened and a sedimentation period of limestone and conglomerates took place. After that, two main episodes occurred, one compressional Alpine (Early Oligocene to early Late Oligocene), with associated thrusts and folds E-W oriented; and a second extensional phase (Late Oligocene to Early Miocene), related to the Valencia Trough opening. A complex faults-system NE-SW oriented was developed forming horsts and grabens (Vegas 1992; Roca et al. 1999), distinguishing the Pre-Coastal Range, the Coastal Range and the Littoral Depression. From Early Miocene to present, a post-rift phase occurs and thick sedimentary sequences were deposited (Clavell and Berastegui 1991).

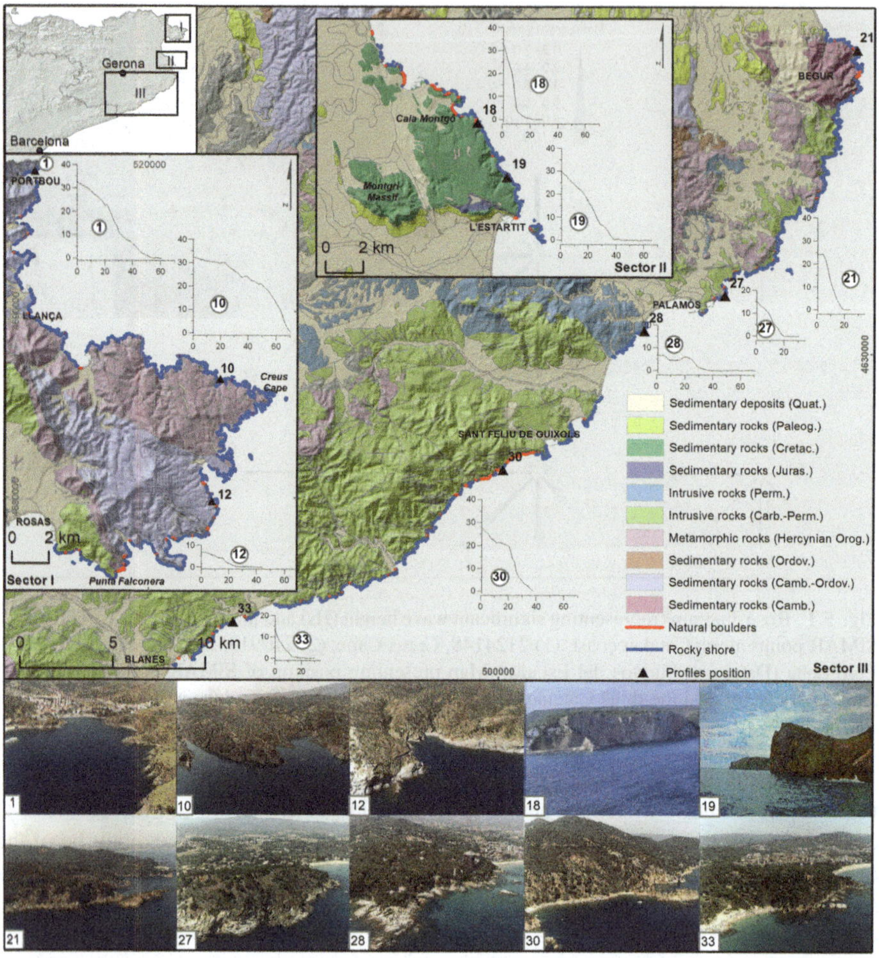

Fig. 5.2 Geological map of Girona province (ICC et al. 2006), including the cartography of the rocky sections of the coastline (GENCAT 2015), which has been distributed in three sectors (I, II and III). Topographic profiles are presented (obtained from LiDAR data from ICGC mentioned in the text. Horizontal and vertical distances measured are metres. At the lower part, oblique photographs of the different spots are presented (MAPAMA 2013). Numbers in photographs corresponds with numbers in profiles and in turn with numbers along the coast (Base topography: MDE SRTM30)

Thus, geological and structural backgrounds are controlling factors on the coastal configuration. According to geological criteria from N to S, it is possible to discern different geological areas (Figs. 5.2, 5.3 and 5.4) (ICC 2002; ICC et al. 2006; ICGC 2010): (i) Palaeozoic metamorphic and granitic materials in the eastern border of Pyrenees forming the Creus Cape Peninsula and the coast North of it. (ii) Empordà Depression, which is filled by sediments transported by Muga, Fluvià and Ter rivers. (iii) Mesozoic rocks are found on the Roses shelf and consist on

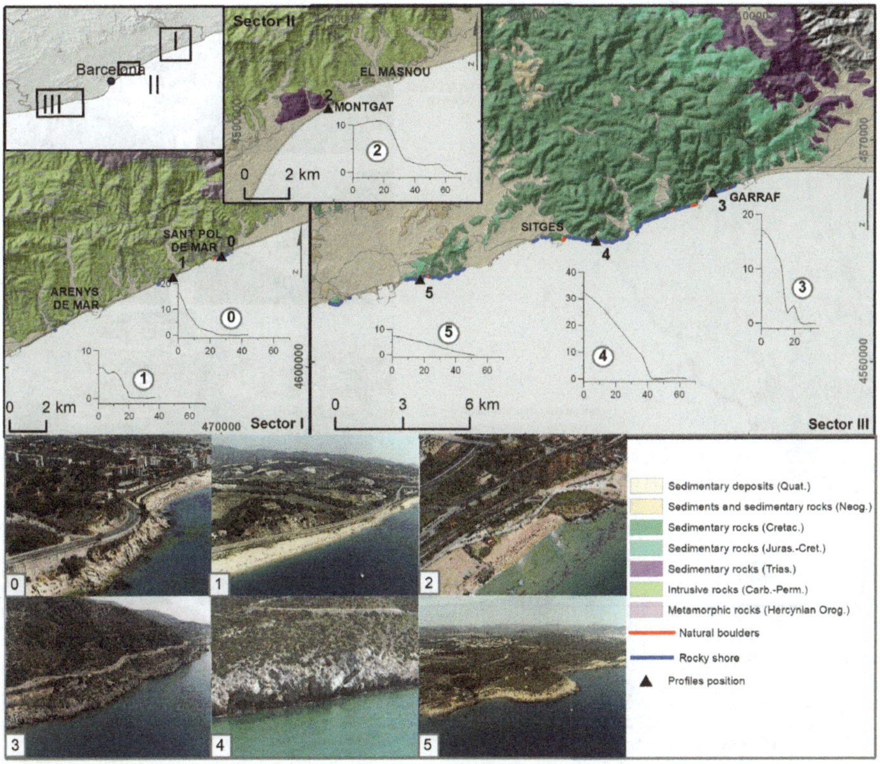

Fig. 5.3 Geological map of Barcelona province (ICC et al. 2006), including the cartography of the rocky sections of the coastline (GENCAT 2015), which has been distributed in three sectors (I, II and III). Topographic profiles are presented (obtained from LiDAR data from ICGC mentioned in the text. Horizontal and vertical distances measured are metres. At the low part, oblique photographs are presented (MAPAMA 2013). Numbers in photographs corresponds with numbers in profiles and in turn with numbers along the coast (Base topography: MDE SRTM30)

bioclastic mudstones that form the Montgrí Massif and the Medas Islands (Duran et al. 2014). (iv) Cape Begur which also is formed by Palaeozoic materials. (v) Coastal Range with NE-SW orientation, and Hercynian granitic rock at Maresme and Barcelona region. This area corresponds to a 40 km long beach region interrupted only by a series of harbours and groins, developed over granitic rocks, with sandy sediments supplied by Tordera River. Barcelona region, counts on artificial beaches and a harbour (15 km long), and the southern part is developed over the delta plain of the Llobregat River, which generates a sandy coastline over 18 km long. (vi) Limestone at Garraf Massif forming low calcareous cliffs. (vii) Tarragona coast that consists on beaches sometimes bounded by groins and harbours with small calcareous coastal relief that corresponds to the end of

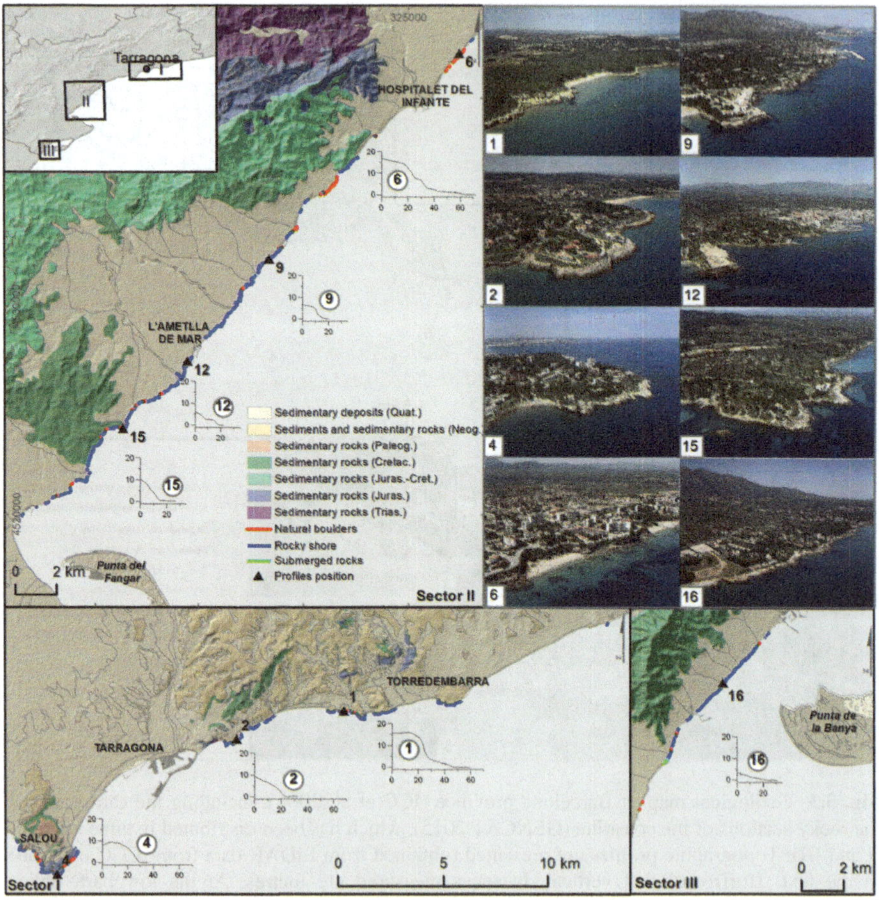

Fig. 5.4 Geological map of Tarragona province (ICC et al. 2006), including the cartography of the rocky sections of the coastline (GENCAT 2015), which has been distributed in three sectors (I, II and III). Topographic profiles are presented (obtained from LiDAR data from ICGC mentioned in the text. Horizontal and vertical distances measured are metres. At the low part, oblique photographs are presented (MAPAMA 2013). Numbers in photographs corresponds with numbers in profiles and in turn with numbers along the coast (Base topography: MDE SRTM30)

Pre-Coastal range (Salou Cape region). (viii) Ebro delta sedimentary region with a delta plain of about 320 km^2.

5.2.4 *Natural Hazards*

Analysis of natural hazards has been playing a key role through the last decades, since there is an increasing concern about vulnerability of the territory.

Consequently, governments, public and private institutions make efforts in prevent the fatal consequences of catastrophic events. Catalonia government has implemented a risk management plan in order to protect population and infrastructures from eventual natural disasters.

Catalonia appears among the regions with the highest losses related with floods and seismic hazards for the period 1987–2002 and it also remains on the first positions in the estimations for 2004–2033 period (Ferrer Gijón et al. 2004). Initiatives as the project RISKCAT offer a global view of the problems in Catalonia related to each one of the hazards considered, such as snow avalanches, gravitational processes, sinking and subsidence, floods, littoral events, earthquakes and volcanism (Vilaplana 2008).

The map for prevention of geological risks in Catalonia carried out at (1:25,000 scale) (ICGC 2017) includes the hazard cartography for the following processes: landslides, rock falls, sinking and torrential fluxes. It also includes an inventory of seismic activity indicators, areas potentially flooded according to geomorphological criteria, inventory of terrain movements and other documents. At the coast, the region southwards Sant Feliu de Guíxols (Fig. 5.2) has been mapped as hazardous because of rock falls.

RISKCAT project, in turn, identifies the coastal rocky sectors as areas with moderate landslides susceptibility. The whole coast presents high risk of flood hazard excepting some points at Begur region (Fig. 5.2), where the risk degree turns to moderate. The seismic zonation indicates that seismic risk is slightly higher on the northern half of Catalonia (Vilaplana 2008).

5.3 Physiography and Types of Rocky Coastline

The Catalan coastline extends approximately for 1,100 km (at a scale of 1:1000). Ballesteros et al. (2007), Mariani et al. (2014) and GENCAT (2015), through the study of habitats, obtained a detailed cartograph of types of substrate along the Catalan coast, in which they estimated the rocky substrate extension over 400 km, that represents around 36% of the littoral (Table 5.2), compared to 30.5% of beach systems and 30% corresponding to the anthropogenic coast. There are many pocket beaches along Catalan rocky coast. Natural pocket beaches are especially abundant at Girona coast and Garraf Massif regions, while artificial pocket beaches are frequent at Tarragona coast and at Barcelona region (Bowman et al. 2009).

Using data from the Digital Elevation Model (DEM) generated by the National Geographic Institute of Spain with data from 2008 to 2011, with 0.5 m of resolution, heights and slopes were calculated for the coastal fringe considering a 25 m wide buffer, from coast to inner land. Reclassification of the DEM and slope, allowed the organization of the topographic information (height and slope) in five classes. The area occupied for each class is shown in Table 5.3 for the whole Catalonia as well as for province.

Table 5.2 Types of coast and extension along Catalan coastline

Coast typology	Extension by province (km)			Total extension (km)	Total extension (%)
	Girona	Barcelona	Tarragona		
Beach	60.7	78	200.4	339	30.5
Harbour docks	41	102.1	66	209	18.8
Concrete walls	4.9	17.7	9	31.6	2.8
Breakwaters	13.7	43.8	36.4	94	8.4
Rocky shore	336.4	14.2	48.8	400.4	36
Natural rocky boulders	24.4	0.9	5.4	30.7	2.8
Rocky shores whose emerged position did not reach the supralittoral level	1.8	0.3	1.1	3.1	0.3
Infralittoral not emerging rocks	0.9	0	0.2	1.1	0.1
Caves	1.4	0.1	0.4	1.9	0.2
Rocky pools	2.1	0.0	0.4	2.5	0.2

Data obtained from GENCAT (2015)

The results of the DEM classification enabled to determine that around 43% (approx. 765 Ha) of the Catalan coast could be considered as high coast (with heights higher than 2 m). On one hand, Girona province presents the bigger extension of high coast and also the highest values of height and slope (123 m and 85° respectively), which are located in Creus Cape area (Fig. 5.2). On the other hand, the bigger surface with low coast is in Tarragona province. The spatial distribution of both variables stands out because of the increasing height of the coast from S to N, and it also shows that the highest values are related with the oldest geologic materials: granites and schists from Paleozoic period. Higher and steepest coastal sections concentrate on the rocky coast.

Table 5.3 Surface (Ha) classified by heights (in meters) and slopes (in degrees) of the coastline at Catalonia, by provinces, considering a buffer fringe of 25 m

Height (m)	<2		2–5		5–10		10–20		>20	
	(Ha)	%	(Ha)	%	(Ha)	%	(Ha)	%	(Ha)	%
Girona	316.6	30.9	114.4	29.1	99.5	60.1	98.7	76.8	71.4	91.2
Barcelona	197.8	19.3	159.7	40.6	26.3	15.9	10.2	8.0	5.5	7.0
Tarragona	508.7	49.7	119.1	30.3	39.8	24.0	19.5	15.2	1.4	1.8
Total	1023.2	100	393.3	100	165.5	100	128.5	100	78.2	100
Slope (°)	<10		10–20		20–40		40–60		>60	
	(Ha)	%	(Ha)	%	(Ha)	%	(Ha)	%	(Ha)	%
Girona	288.7	23.8	109.9	76.0	179.4	76.0	93.2	85.3	29.4	93.6
Barcelona	328.8	27.1	40.3	8.9	20.9	8.9	7.7	7.1	1.7	5.5
Tarragona	596.9	49.2	47.3	15.1	35.7	15.1	8.3	7.6	0.3	0.8
Total	1214.4	100	197.6	100	236.1	100	109.2	100	31.4	100

5.3.1 Girona Coast

Rocky coast at Girona represents 75% of the province coastline (Table 5.2). Both, waves and precipitations constitute two important erosive agents modelling the landscape of Girona coastline. The most frequent waves in that region are those coming from N, but the highest significant heights and mean period correspond to those waves coming from E (Fig. 5.1). Precipitation in this region is high if it is compared with the one registered on the south of the Catalan coastline (Martínez et al. 2007). From a geological-geomorphological point of view, three main rocky sectors separated by two long beaches can be differentiated. These sectors are described by Barbaza (1970), in a detailed work about the morphology of rocky coast.

At sector I (Portbou—Creus Cape Peninsula) the coast is very irregular and crenulated with lots of indentations with infrequent but well developed shore platforms (Furlani et al. 2014). The coastline is N-S oriented in the northern sector and turns to NW-SE at Llançà. Then, at Creus Cape it changes to NE-SW and finally at Punta Falconera it turns again to NW-SE until Roses city. Lithologically, Palaeozoic greywacke, schists and granites are the predominant materials. All of them were affected by regional metamorphism during Carboniferous-Permian period (ICC et al. 2006) (Fig. 5.2).

Height of cliffs was inferred from LiDAR data downloaded from the Catalonian Cartographic and Geologic Institute website, corresponding to flights carried out in 2008, 2010 and 2011, all of them with 0.5 m of resolution. Topographic cross-shore profiles show that this northern sector counts on two main typologies of profile morphology (Fig. 5.2). On one hand profiles with heights between 30 and 40 m, slopes around 45° and rectilinear or convex surface are found. On the other hand much lower profiles (almost 10 m) and smoother slopes also appear in the Creus Cape Peninsula. These profiles indicate that both subaerial and marine processes have influence modelling these surfaces, since the convexity of some of them is indicating the importance of aerial erosion. However the steep slopes found in this section make the marine processes not negligible, especially taking into account that action of continental episodic water fluxes, wind and waves are in particular relevant at Creus Cape Peninsula.

Indentations or inlets are erosive morphologies related with streams and gullies mouths. Many pocket beaches are found along this sector, but not so much as the numerous examples founded at the third sector (Costa Brava). Moreover, honeycomb and tafonis are also geomorphological features commonly found in this area (Barbaza 1970).

The following rocky sector, sector II, extends from L'Escala to L'Estartit, and it includes the Montgrí Massif region (Fig. 5.2). It is a relatively short section, with around 15 km length and is constituted by a rocky formation whose materials are mainly limestones from Cretaceous Period. Moreover, there are dolomites and limestones from Jurassic, and clays with conglomerate levels from Paleogene (ICC et al. 2006). Oriented NW-SE, the cliffs at this zone are quite subvertical.

The profiles and especially the photographs of the region, show the clear influence of marine processes eroding the front even though it is true that the materials are quite rigid and resistant. Geomorphologic features like notches can be observed in some points of the base of the cliff (Fig. 5.2).

Sector III, runs from Begur to Blanes and it counts on with around 51 km of cliffed coast. The orientation of the littoral changes 90° with respect the previous sector and turns to NE-SW. Thus, tectonics is controlling the coast design. It is formed by intrusive rocks (granites) from Carboniferous to Permian Periods, as well as metamorphic rocks affected by Hercynian Orogeny.

The alternating composition of materials gives the coast a characteristic configuration regarding the cliff structure. Differential erosion is observed since there are differences in resistance to marine action between shales and schists (Calvet and Gallart 1973). Furthermore, several Pleistocene abrasion surfaces indicating different positions of sea level have been described (Solé 1962; Butzer 1964).

This nearly longitudinal sector is completely rocky except at Palamós Bay and Platja d'Aro Beach at north of Sant Feliu de Guíxols, which are long sandy beaches. Moreover, countless pocket beaches truncating the coast characterise this sector (Calvet and Gallart 1973). Bowman et al. (2009) indicated that many rocky pockets beaches lack the substantial sediments contributing to reshaping, and asymmetric morphosedimentary patterns were noticed through the observation of alongshore sediment flux between opposite promontories.

Topographic profiles are in general quite vertical excepting those in the granites between Palamós and Sant Feliu the Guíxols, which are much more horizontal though irregular. Cliff heights range from 30 to 10 m and a mixed character between subaerial and marine processes is observed.

5.3.2 Barcelona Coast

Coastline at Barcelona province is completely different to the one in Girona. Rocky coast at Barcelona represents only 6% of the province coastline (Table 5.2).

Incident waves come mainly from E, SE and S, but the eastern ones are the most frequent and energetic ones, with mean periods between 8 and 10 s and significant height around 3 m (Fig. 5.1). Precipitation in this region ranges from 500 to 600 mm (data from Catalonian Meteorological Service 2018). Water discharge is also considerable because Llobregat river is the largest river after Ebro (Table 5.1), and if joined to other minor ones, it contributes with a mean discharge of almost 25 $m^3 s^{-1}$.

Coastal orientation, from N to S, is ENE-WSW until Llobregat delta, where it turns to NNE-SSW, and then at the south of Llobregat river mouth turns again to ENE-WSW.

Three main outcrops of rocky coast have been described (Fig. 5.3). Sector I, at north, is formed by granites and diorites from Carboniferous-Permian period. It outcrops in the coast next to Sant Pol de Mar and Arenys de Mar (ICC et al. 2006).

Oriented ENE-WSW, coastline in this section is moderately high, ranging between 5 and 20 m with a fractured aspect in contact with the sea at Sant Pol de Mar, or protected by beaches at Arenys de Mar.

Profiles appear concave, with steeped slopes on the top and smooth in the foot. On the other hand, convex profiles also appear with sheer slopes and with irregular surface as result of erosion and weathering processes modelling the superficial shape of the rock.

Montgat outcrop constitutes the second sector described. Orientation of coastline is ENE-WSW at this point, but from here to southwards it turns to NNE-SSW, as result of the reshaping of the coast because of sedimentary materials supplied by Llobregat river.

Miocene weathered conglomerates with sandy or clay matrix form the small outcrop (ICC et al. 2006), in contact with the sea and bordered by sandy beaches. A breakwater placed there as continuity of the rock mass interrupts littoral dynamics increasing beach width northwards. As consequence of this breakwater, erosive features appear southwards. Rocky boulders appear in the toe and it is vegetated in the top. The cliff height reaches 10 m and its transversal convex profile shows the top and a step in the middle of the cliff both dipping landward.

The third sector, at Garraf-Sitges region, is ENE-WSW oriented and it shows three rocky sections. From N to S, appear Garraf Mountains, a rock massif formed by limestones with dolomite intercalations from Cretaceous and Jurassic period (ICC et al. 2006). Moreover, at Sitges town Pleistocene piedmont alluvial fan deposits are found. Cliff height reaches 30 m in some points and quite vertical and convex slopes are observed. They appear in direct contact with the sea and notches in the base are found. Thus, marine processes have strong influence in this section.

Karstic morphological features are characteristic of this massif and elements as blowholes, basin pools, pinnacles, etc. have been described (Daura et al. 2014; Furlani et al. 2014).

Still in this sector III, southwards Sitges town, two more rocky outcrops are found. The same calcareous materials described before compose them, but there are differences regarding height and slope of the cliff front. Heights between 5 and 10 m, with quite smooth constant slopes (nor convex nor concave) are observed. They are exposed to the sea but with a clear influence of subaerial processes.

5.3.3 Tarragona Coast

Tarragona presents the largest coastline and the highest percentage of urbanized land, if compared with the rest of Spanish Mediterranean provinces (Ministerio de Fomento 2011). Tarragona province has a littoral mainly covered by beach systems, which extend 200 km long: open beaches of reduced width backed in most cases by infrastructures of transport, industries and especially housing. Several pocket beaches are found, and count with extensions ranging from few meters to more than 1 km. They appear surrounded by small cliffs with steep or vertical topographies, interspersed in rocky promontories. Rocky coast at Tarragona represents 15.5% of the province coastline (Table 5.2), reshaped by waves and superficial flows.

Waves approach from three different directions, NW, ENE, and S. During storm events the wave heights do not exceed 6 m in any case, but it is observed that those storms with more than 3 m (Hs) are more frequent in the waves coming from the ENE and E (XIOM 2007). However the periods are always greater than 4 s, with the waves of the first quadrant (ENE to E) also presenting the highest periods. The only effective waves will be those of the ENE and S, being the first, as already mentioned, more frequent (Jiménez et al. 2000).

Precipitation regime on the coast of Tarragona is characterised by annual average values of 500–600 mm (Font Tullot 2000), being September and October the months in which the maximum values are recorded. Maximum daily rainfall in the province of Tarragona is around 200 mm while the maximum rainfall recorded in one hour exceeds 80 mm in Ebro delta (Font Tullot 2000).

Coastal orientation from N to S slightly varies from ENE-WSW (Torredembarra–Salou), to NNE-SSW (Salou Cape–North Ebro delta). Southwards Ebro delta with its characteristic deltaic morphology, the coast keeps going with NNE-SSW orientation. The 50 km of rocky coast are distributed along these three sections (Fig. 5.4).

Northern Sector (Sector I in Fig. 5.4), counts on five rocky sections formed by Neogene sedimentary materials (conglomerates, calcarenites, clays), and Pleistocene piedmont deposits characterised as proximal facies of alluvial fans. At Salou Cape, Jurassic limestones, dolomites and marls with abundant fossil content are found (IGME 1972; Fernández-López 2001; Moreno-Bedmar et al, 2017). Heights observed are around 10–15 m, and convex profiles are dominant. Furlani et al. (2014) described the presence of Low Jurassic stacks at Salou and Tarragona.

Southwards, there are around 30 km of rocky materials (Sector II at Fig. 5.4). Sedimentary lithologies from Cretaceous and Jurassic outcrops, such as limestones, marls, dolomites and gravitational deposits typical from proximal parts of alluvial fans are also found. Heights of 10 m are rarely exceeded and again convex profiles are observed. Subaerial processes are dominant; however, Furlani et al. (2014), reported nips shaped in alluvial Quaternary deposits.

Finally, on the southern limit of the province, a short rocky section composed by Quaternary alluvial fan deposits is identified (Sector III). Heights are lower than 5 m, and slopes are very smooth.

5.4 Case Study of Landslide Susceptibility at Coastal Cliffs: Mont-Roig del Camp Cliffs (Tarragona)

Mont-Roig del Camp cliffs are located on the coastal central part of Tarragona province (Fig. 5.5a), and extend for 2.5 km and reach up to 19 m high. The high levels of weathering and degradation as well as the intensive recreational use of the region during summer months led the local authorities to establish some restricted zones because of the rock fall hazard that affects the beaches. Several narrow pocket beaches protect the base of the cliffs, which present inactive parts alternating with active ones.

The materials forming the cliffs correspond to alluvial fan deposits (Upper Pleistocene, 125–300 Ky BP) (CSN 2001). Stratigraphically, they consist of three lithological units, which are (Fig. 5.5b): (i, ii) conglomerates (units I, III), associated with proximal zones and showing evidences of high energy, and (iii) clay (unit II), related to distal zones of the alluvial fan. Evidences of periods without any deposition also appear. Moreover, undulated and erosive contacts have been identified. This area was affected by tectonic deformation during Pliocene and Quaternary (Masana 1995), so the maximum compression axis oriented from NNE-SSW to ENE-WSW during Pliocene, changed during Quaternary from N-S to

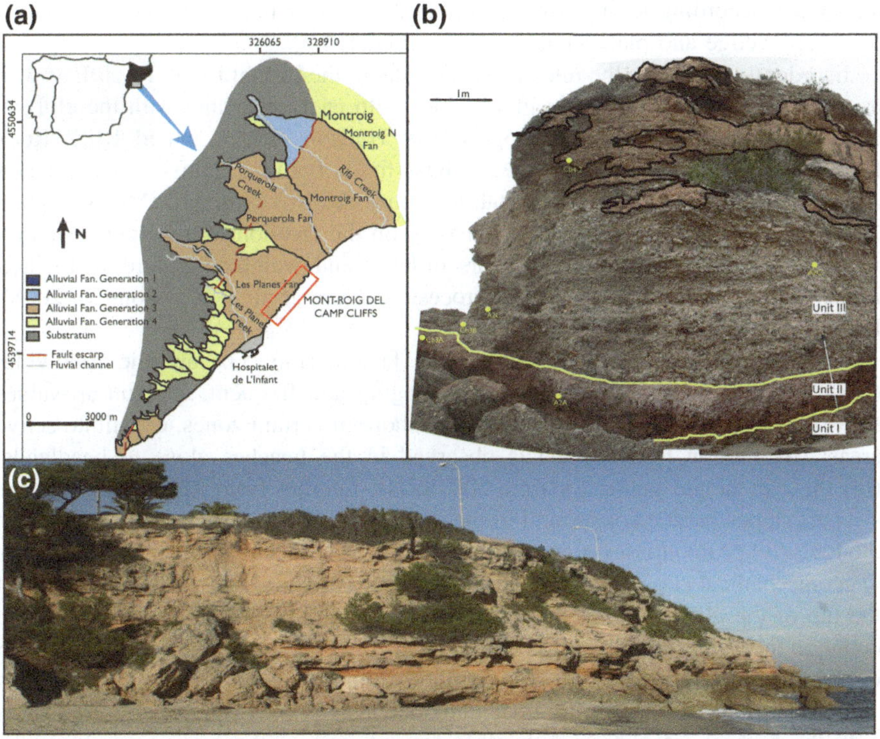

Fig. 5.5 **a** Location and geomorphological setting of Mont-Roig del Camp cliffs (adapted from CSN 2001). **b** Field scheme: Units I and III are conglomerates, with large boulders, size sorting, and occasional fine-grained bodies; Unit II, clay with some small lenticular bodies, and even some conglomeratic fragments. Sharp and straight contact between clays and conglomerates, especially with unit C; Yellow lines corresponds to the contact between units. Brown shaded area indicates the presence of calcareous crust. Yellow points indicate samples position. **c** General view of the fractured and weathered cliff

NE-SW coinciding with main fractures described by Montoya et al. (2012) for this cliff front. Sporadic and seasonal rainfall joined to short course streams cause eventually floods and high runoff rates, eroding and supplying sediments to the coast (López-Bermúdez and Gomáriz-Castillo 2006).

The stability of this coastal section is affected by subaerial and marine processes, in addition to the physical, chemical, and mechanical characteristics of the material forming these cliffs (Fig. 5.5c). Thus, susceptibility of the cliff to potential instabilities was evaluated through the analysis of the influence on the cliff of the following factors: (i) wave exposure: storm waves and run-up analyses were conducted; (ii) shoreline variations: analyses at the long/medium term through the comparison of aerial photographs; (iii, iv) cliff height and slope: data obtaining from Digital Elevation Model; (v) geotechnical quality of the rock mass: field and laboratory description of rock mass (Fig. 5.5b), calculation of resistance parameters and quality indexes; (vi) superficial runoff: application of method of MOPU (1990) modified by Témez (1991) and Ferrer (1993); and (vi) cliff orientations favouring landslides: according to the cases of preferred joints orientation, different scenarios for likely wedge and planar failures were simulated.

In order to determine the relevant parameters in the instability of the cliff, as well as the interactions of the different variables with each other and with the cliff, an approximation of the method proposed by Hudson (1992), called RES (Rock Engineering Systems), was applied. Thus, first of all, a rating of each factor according to its influence in the instability processes was proposed (Montoya et al. 2012) (Fig. 5.6a). Following stage consisted on the creation of a weighting matrix with the previously mentioned factors ordered and weighted according to their influence in the landslide formation process (Carrara et al. 1995; Montoya 2008). Finally, a susceptibility map was obtained.

Main results from the analysis of the factors (Fig. 5.6) are the following: (i) easterly storm events are the most damaging and frequent, and run-up values show that both headlands and cliffs at backshore of certain zones, constitute active areas; (ii) erosion (0.2 m/year) is observed in the beaches close to headlands; (iii) the cliff height decreases from SW to NE ranging from 19 to 2 m, and the steepest slopes (>50°) correspond to the top of the cliffs and the smoothest ones to the foot; (iv) promontories present poor quality of the rock mass (estimated with the Q geomechanical classification after Barton et al. 1974); (v) runoff flow estimated at the top of the cliff is higher southwestwards; (vi) cliff fronts NE and SE oriented count on the highest number of potential slides cases, and fronts NW, W, SW and S oriented, are scenarios with low probability of mass movement.

Slope, orientation favouring rock failures and rock mass quality are the factors with the highest interaction with other factors, and so the higher weight in the failure generation. Surface runoff and shoreline variations are highly dominant over the others, what means that the rest of the factors do not have much more influence over them. This can be explained because both easterly waves and runoff can be understood as external and triggering factors non-dependant of intrinsic characteristics of the rock mass.

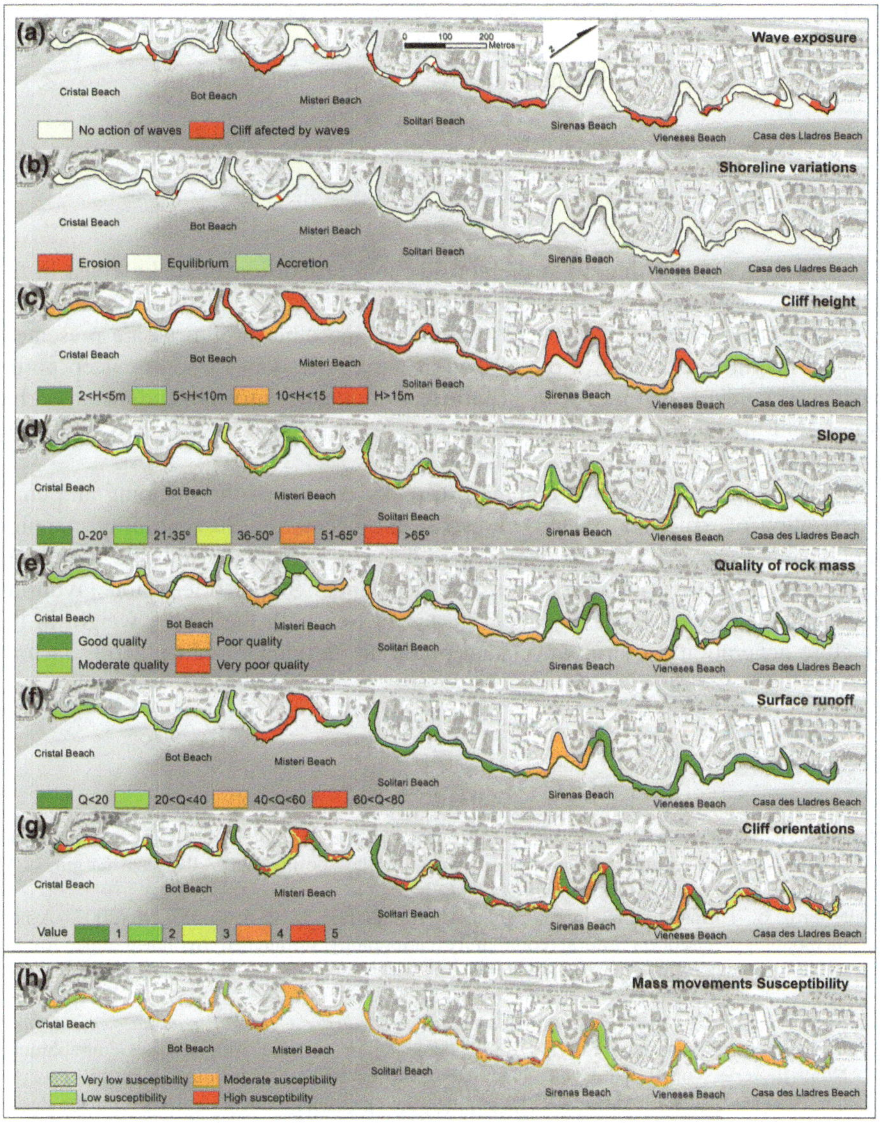

Fig. 5.6 Upper figures: Factors classification, where classes with lower values will contribute to landslide occurrence in a lower grade than classes with higher values. **a** Wave exposure; **b** Shoreline variations; **c** Cliff height; **d** Cliff slope; **e** Quality of rock mass; **f** Surface runoff; and **g** Cliff orientations favouring the landslides; Lower figure: Map representing susceptibility to mass movements [Modified from Montoya et al. (2012); Orthophotograph base: ICC et al. (2006)]

The cartography of susceptibility to mass movements shows that 61% of the cliff surface is moderately susceptible, 33% has low degree and just 6% are highly susceptible (Fig. 5.6h). Headlands along the central part of Mont-Roig del Camp cliffs present the highest degree of susceptibility, corresponding to steep slopes, significant degradation by subaerial weathering and wave action over the cliff controlled in some cases by the protective role of shoreline.

5.5 Final Remarks

To conclude, we will highlight the considerable proportion of rocky coast in the Catalan coast as well as the variety of lithologies, heights and slopes found.

Remarkable differences between Girona and the other provinces are presented: (i) cliff heights are higher at the N than at the S; (ii) slopes also are steep northwards; (iii) older lithologies with a character dominantly intrusive and metamorphic are found mainly northwards, while younger sedimentary materials dominate southwards.

Rock fall events are quite common in rocky coasts, so the mapping and characterization of materials and processes became necessary. Susceptibility mapping and analysis carried out on Mont-Roig del Camp beaches indicate that main instability problems are concentrated in headlands. Slope, orientation favouring rock failures and rock mass quality, are the factors with the higher weight in the fracture generation. In turn, waves and runoff are dominant and triggering factors.

References

Amblás, D., Canals, M., Urgeles, R., Lastras, G., Liquete, C., & Hughes-Clarke, J. E., et al. (2006). Morphogenetic mesoscale analysis of the northeastern Iberian margin, NW Mediterranean Basin. *Marine Geology, 234,* 3–20.

Ballesteros, E., Torras, X., Pinedo, S., García, M., Mangialajo, L., & de Torres, M. (2007). A new methodology based on littoral community cartography dominated by macroalgae for the implementation of the European Water Framework Directive. *Marine Pollution Bulletin, 55,* 172–180.

Barbaza, Y. (1970). *Morphologie des secteurs rocheaux du littoral catalan septentrional.* Memories et documents, Service de Documentation et de Cartographie Géographiques. Nouvelle série, vol. II. Ed. Centre National de la Recherche Scientifique. Paris, 150 pp.

Barton, N., Lien, R., & Lunde, J. (1974). Engineering classification of rock masses for the design of tunnel support. *Rock Mechanics, 6,* 189–236.

Benjamin, J., Rovere, A., Fontana, A., Furlani, S., Vacchi, M., & Inglis, R. H., et al. (2017). Late Quaternary sea-level changes and early human societies in the central and eastern Mediterranean Basin: An interdisciplinary review. *Quaternary International, 449,* 29–57.

Bolaños, R., Jorda, G., Cateura, J., Lopez, J., Puigdefabregas, J., & Gomez, J., et al. (2009). The XIOM: 20 years of a regional coastal observatory in the Spanish Catalan coast. *Journal of Marine Systems, 77,* 237–260.

Bolaños, R., & Sáchez-Arcilla, A. (2006). A note on nearshore wave features. Implications for wave generation. *Progress in Oceanography, 70,* 168–180.

Bowman, D., Guillén, J., López, J., & Pellegrino, V. (2009). Planview geometry and morphological characteristics of pocket beaches on the Catalan coast (Spain). *Geomorphology, 108,* 191–199.

Butzer, K. (1964). Pleistocene geomorphology and stratigraphy of the Costa Brava Region, Catalonia. Abhandlungen, Akademie der Wissenschaften und der Literatur (Mainz). *Mathematisch Naturwissenschaftliche Klasse, 1,* 1–51.

Calvet, J., & Gallart, F. (1973). Esquema morfologico de la Costa Catalana. *Acta Geologica Hispanica, 8,* 125–130.

Campins, J., Genovés, A., Picornell, M. A., & Jansà, A. (2011). Climatology of Mediterranean cyclones using the ERA-40 dataset. *International Journal of Climatology, 31,* 1596–1614.

Carrara, A., Cardinali, M., Guzzetti, F., & Reichenbach, P. (1995). GIS technology in mapping landslide hazard. In A. Carrara & F. Guzzetti (Eds.), *Geographical information systems in assessing natural hazards*. The Netherlands: Kluwer Academic Publisher.

Casas-Prat, M., & Sierra, J. P. (2012). Trend analysis of wave direction and associated impacts on the Catalan coast. *Climatic Change, 115,* 667–691.

Catalonian Meteorological Service. (2018). *Historical data.* http://www.meteo.cat/. Accessed February 28, 2018.

Cateura, J., Sanchez-Arcilla, A., Bolaños-Sanchez, R., & Sairouni, A. (2004). Atmospheric conditions during severe wave storms in the north-western Mediterranean. Generation, evolution, decay and implications. In *37th CIESM Congress*, Barcelona, Spain.

Clavell, E., & Berastegui, X. (1991). Petroleum geology of the Gulf of Valencia. In A. M. Spencer (Ed.), *Generation, accumulation and production of Europe's hydrocarbons* (Spec Publ Eur Assoc Pet Geosci Vol. 1, pp. 355–368).

CSN. (2001). *Proyecto Datación* (p. 159). Madrid: Consejo de Seguridad Nuclear.

Daura, J., Sanz, M., Fornós, J. J., Asensio, A., & Julià, R. (2014). Karst evolution of the Garraf Massif (Barcelona, Spain): Doline formation, chronology and archaeo-palaeontological archives. *Journal of Cave and Karst Studies, 76*(2), 69–87.

Duran, R., Canals, M., Sanz, J. L., Lastras, G., Amblas, D., & Micallef, A. (2014). Morphology and sediment dynamics of the northern Catalan continental shelf, northwestern Mediterranean Sea. *Geomorphology, 204,* 1–20.

Fernández-López, S. R. (2001). Upper Bathonian ammonites of the Catalan Basin (Tivissa and Cap Salou, Spain). *Hantkeniana, 3,* 25–39.

Ferrer, F. J. (1993). *Recomendaciones para el cálculo hidrometeorológico de avenidas* (p. 76). Madrid: CEDEX. Ministerio de Fomento.

Ferrer Gijón, M., García López-Davalillo, J. C., González de Vallejo, L. I., Rodríguez Franco, J. A., Estévez Martín, H., & Trimboli, M. (2004). *Análisis del impacto de los riesgos geológicos en España. Evaluación de pérdidas por terremotos e inundaciones en el periodo 1987-2001 y estimación para el periodo 2004-2033.* Technical Report. Instituto Geológico y Minero de España, Madrid, Spain, 103 pp.

Font Tullot, I. (2000). *Climatología de España y Portugal* (p. 422). Salamanca: Ediciones Universidad Salamanca.

Furlani, S., Pappalardo, M., Gomez-Pujol, L., & Chelli, A. (2014). The rock coast of the mediterranean and black seas. In D. M. Kennedy, et al. (Eds). *Rock coast geomorphology: A global synthesis.* London, UK: Geological Society, London Memoirs 40.

GENCAT—Generalitat de Catalunya. (2015). *Manual dels hàbitats litorals de Catalunya.* Dept. de Territori i Sostenibilitat. Generalitat de Catalunya. Barcelona, 251 pp.

Guedes Soares, C., Carretero-Albiach, J. C., Weisse, R., & Alvarez-Fanjul, E. (2002). A 40 years hindcast of wind, sea level and waves in European waters. In *Proceedings of the 21st International Conference on Offshore Mechanics and Arctic Engineering* (pp. 669–675).

Hudson, J. A. (1992). *Rock engineering systems: Theory and practice* (p. 185). Chichester: Ellis Horwood.

ICC—Intitut Cartogràfic de Catalunya. (2002). *Mapa geològic de Catalunya 1:250.000.* Institut Cartogràfic de Catalunya. Barcelona, Spain.

ICC-IGC-GENCAT-IGME. (2006). *Mapa geològic de Catalunya 1:50.000*. Institut Cartogràfic de Catalunya – Institut Geològic de Catalunya – Generalitat de Catalunya – Instituto Geológico y Minero de España, Barcelona, Spain.

ICGC. (2010). *Atles geològic de Catalunya* (p. 463). Barcelona, Spain: Institut Geològic de Catalunya.

ICGC. (2017). *Mapa per a la Prevenció de Riscos Geològics 1:25.000*. Dept. de Territori i Sostenibilitat. Generalitat de Catalunya, Barcelona.

IGME. (2005). *Mapa geomorfológico de España y del margen continental. Escala 1: 1.000.000*. Instituto Geológico y Minero de España, Madrid, Spain.

IGME—Instituto Geológico y Minero de España. (1972). *Mapa Geológico de España. 1: 200000. Síntesis de la Cartografía existente. Tarragona*. Instituto Geológico y Minero de España, Madrid, Spain.

Jiménez, J. A., Valdemoro, H. I., Gracia, V., & Sánchez-Arcilla, A. (2000). *Estudio sobre la situación del tramo costero Cambrils-Vandellós (T.M. Mont-Roig del Camp). Evolución costera y dinámica litoral*. Technical Report. Universitat Politécnica de Catalunya. Barcelona.

Kline, S. W., Adams, P. N., & Limber, P. W. (2014). The unsteady nature of sea cliff retreat due to mechanical abrasion, failure and comminution feedbacks. *Geomorphology, 219,* 53–67.

Lambeck, K., Woodroffe, C. D., Antonioli, F., Anzidei, M., Gehrels, W. R., & Laborel, J., et al. (2010). Paleoenvironmental records, geophysical modeling, and reconstruction of sea-level trends and variability on centennial and longer timescales. In J. A. Church, et al. (Eds.), *Understanding sea-level rise and variability*. Oxford, UK: Wiley.

Lim, M., Rosser, N. J., Allison, R. J., & Petley, D. N. (2010). Erosional processes in the hard rock coastal cliffs at Staithes, North Yorkshire. *Geomorphology, 114*(1–2), 12–21.

Liquete, C., Canals, M., Ludwig, W., & Arnau, P. (2009). Sediment discharge of the rivers of Catalunya, NE Spain, and the influence of human impacts. *Journal of Hydrology, 366,* 76–88.

Llasat, M. C., López, L., Barnolas, M., & Llasat-Botija, M. (2008). Flash-floods in Catalonia: The social perception in a context of changing vulnerability. *Advances in Geosciences, 17,* 63–70.

López-Bermúdez, F., & Gomáriz-Castillo, F. (2006). Las ramblas, agentes reguladores del litoral mediterráneo ibérico. El ejemplo de la rambla de Las Moreras. Murcia. In E. Sanjaume & J. Mateu (Eds.), *Geomorfología Litoral i Quaternari. Homenatge al Prof. V.M. Rosselló i Verger*. Universitat de València, Valencia.

MAPAMA—Ministerio de Agricultura y Pesca, Alimentación y Medio Ambiente. (2013). Guía de Playas de España.

Mariani, S., Cefali, M. D., Terradas, M., Chappuis, E., & Ballesteros, E. (2014). Using catenas for GIS-based mapping of NW Mediterranean littoral habitats. *Estuarine, Coastal and Shelf Science, 147,* 56–67.

Martínez, M. D., Lana, X., Burgueño, A., & Serra, C. (2007). Spatial and temporal daily rainfall regime in Catalonia (NE Spain) derived from four precipitation indices, years 1950–2000. *International Journal of Climatology, 27,* 123–138.

Martín-Vide, J., & López-Bustins, J. A. (2006). The Western mediterranean oscillation and rainfall in the Iberian Peninsula. *International Journal of Climatology, 26,* 1455–1475.

Masana, E. (1995). *L'activitat neotectònica a les Cadenes Costero Catalanes*. Ph.D. thesis, Universidad de Barcelona, 444 pp.

Mendoza, E., & Jiménez, J. A. (2008). Coastal storm classification on the Catalan littoral (NW Mediterranean). *Ingeniería hidráulica en México, 23*(2), 21–32.

Millán, M., Estrela, M. J., & Caselles, V. (1995). Torrential precipitations on the Spanish east coast: The role of the mediterranean sea surface temperature. *Atmospheric Research, 36,* 1–16.

Ministerio de Fomento. (2011). SIOSE—Sistema de Información sobre Ocupación del Suelo de España. http://www.siose.es/. Accessed February 27, 2018.

Montoya, I. (2008). *Análisis de la susceptibilidad a los movimientos de ladera en los acantilados de la franja costera de Mont-Roig del Camp (Tarragona)*. Ph.D. Thesis. Universidad Rey Juan Carlos, Madrid, 198 pp.

Montoya, I., Rodríguez, I., Sánchez, M. J., Alcántara-Carrió, J., Martín, S., & Gómez, D., et al. (2012). Mapping of landslide susceptibility at coastal cliffs. The Mont-Roig del Camp study case. *Geologica Acta, 10*(4), 439–455.

MOPU. (1990). *Instrucción 5.2-IC. Drenaje Superficial*. Centro de publicaciones del Ministerio de Obras Públicas y Urbanismo, Madrid, 84 pp.

Moreno-Bedmar, J. A., Robert, E., Matamales-Andreu, R., & Bover-Arnal, T. (2017). Review of the early Albian ammonites of the Montmell Formation near Marmellar (Salou-Garraf Basin, Tarragona, Catalonia, Spain). *Carnets de Géologie, 17*(1), 1–10.

Pastor, F., Estrela, M. J., Peñarrocha, D., & Millán, M. M. (2001). Torrential rains on the Spanish Mediterranean Coast: Modelling the effects of the sea surface temperature. *Journal of Applied Meteorology, 40,* 1180–1195.

Pirazzoli, P. A. (2005). A review of possible eustatic, isostatic and tectonic contributions in eight late-Holocene relative sea-level histories from the Mediterranean area. *Quaternary Science Reviews, 24,* 1989–2001.

Roca, E., Sans, M., Cabrera, L., & Marzo, M. (1999). Oligocene to Middle Miocene evolution of the Central Catalan margin (northwestern Mediterranean). *Tectonophysics, 315,* 209–233.

Sánchez-Arcilla, A., González-Marco, D., & Bolaños, R. (2008). A review of wave climate and prediction along the Spanish Mediterranean coast. *Natural hazards and earth system sciences, 8,* 1217–1228.

Solé, L. (1962). Le Quaternarie marin des Baléares et ses rapports avec les côtes méditerranéennes de la Péninsule Iberique. *Quaternaria, 1,* 309–342.

Stephenson, W. J., Dickson, M. E., & Trenhaile, A. S. (2013). Rock coasts. In J. Shroder, & D. J. Sherman (Eds.), *Treatise on geomorphology* (Vol. 10), Coastal Geomorphology. San Diego, CA: Academic Press.

Stephenson, W. J., Kirk, R. J., Hemmingsen, S. A., & Hemmingsen, M. A. (2010). Decadal scale micro erosion rates on shore platforms. *Geomorphology, 114*(1–2), 22–29.

Sunamura, T. (1992). *Geomorphology of rocky coasts* (p. 302). England: Willey.

Sunamura, T. (2015). Rocky coast processes: With special reference to the recession of soft rock cliffs. In *Proceedings of the Japan Academy. Series B, Physical and Biological Sciences* (Vol. 91, No. 9, pp. 481–500).

Témez, J. R. (1991). Extended and improved rational method. In *Proceedings of XXIV IAHR Congress* (Vol. A, pp. 33–40).

Trenhaile, A. S. (1987). *The geomorphology of rock coasts* (p. 344). Oxford: Oxford University Press.

Trenhaile, A. S. (2011). Cliffs and rock coasts. *Treatise on Estuarine and Coastal Science, 3,* 171–191.

Tucker, G. E., & Bras, R. L. (2000). A stochastic approach to modeling the role of rainfall variability in drainage basin evolution. *Water Resources Research, 36*(7), 1953–1964.

Vegas, R. (1992). The Valencia trough and the origin of the Western Mediterranean basins. *Tectonophysics, 203,* 249–261.

Vilaplana, J. M. (2008). *RISKCAT – Els riscos naturals a Catalunya. Informes del CADS 6.* Consell Assessor per al Desenvolupament Sostenible. Generalitat de Catalunya, Barcelona, 74 pp.

XIOM—Xarxa d'Instruments Oceanogràfics i Meteorològics de Catalunya. (2007). Boies d'onatge. Dades obtingudes l'any 2007. Technical Report. Generalitat de Catalunya, Barcelona, Spain.

Young, A. P. (2018). Decadal-scale coastal cliff retreat in southern and central California. *Geomorphology, 300,* 164–175.

Young, A. P., Guza, R. T., Flick, R. E., O'Reilly, W. C., & Gutierrez, R. (2009). Rain, waves, and short-term seacliff evolution. *Marine Geology, 267,* 1–7.

Zazo, C., Goy, J. L., Dabrio, C. J., Bardají, T., Hillaire-Marcel, C., Ghaleb, B., et al. (2003). Pleistocene raised marine terraces of the Spanish Mediterranean and Atlantic coasts: Records of coastal uplift, sea-level highstands and climate changes. *Marine Geology, 194,* 103–133.

Chapter 6
The Rocky Coasts of Balearic Islands

Pablo Balaguer, Guillem X. Pons and Miquel Mir-Gual

Abstract Balearic Islands are located in the centre of Western Mediterranean and is composed by five major islands (from biggest to smallest): Mallorca, Menorca, Ibiza, Formentera and Archipelago of Cabrera. Mallorca located at the centre of archipelago has an extension of 3,636 km^2, Menorca, located at the Northeast, has an extension of 695 km^2, Ibiza and Formentera, called Pitiuses Islands and located at the South, has an extension of 572 and 82 km^2 respectively. Mallorca is the island which have a higher coastal length with 842 km (including Archipelago of Cabrera), Menorca follows with 433 km, Ibiza has 334 km and Formentera 115 km. These coasts, and specially the rocky shores, are the result of the geological and geomorphological characteristics and modelling processes, as fluvial, karst, bioerosion and mass movements, between others conditioned by maritime climate which have configured the structural units of each island. According to general results for the entire archipelago, the percentage between rocky coasts or coasts formed by cohesive materials and beach coasts or coasts formed by non-cohesive materials is 80–20.

6.1 Introduction

Balearic Islands are located in the centre of Western Mediterranean and is composed by five major islands (from biggest to smallest): Mallorca, Menorca, Ibiza, Formentera and Archipelago of Cabrera (Fig. 6.1). Mallorca located at the centre of

P. Balaguer (✉)
SOCIB, Balearic Islands Coastal Observing and Forecasting System (MINECO-CAIB), ParcBit, Ed. Naorte, Cra. Valldemossa km 7.4, 07121 Palma, Balearic Islands, Spain
e-mail: pbalaguer@socib.es

G. X. Pons · M. Mir-Gual
Departament de Geografia, Universitat de les Illes Balears, Cra. Valldemossa km 7.5, 07122 Palma, Balearic Islands, Spain
e-mail: guillemx.pons@uib.es

M. Mir-Gual
e-mail: mirgual@gmail.com

J. A. Morales (ed.), *The Spanish Coastal Systems*,
https://doi.org/10.1007/978-3-319-93169-2_6

archipelago has an extension of 3,636 km^2, Menorca, located at the Northeast, has an extension of 695 km^2, Ibiza and Formentera, called Pitiuses Islands and located at the South, has an extension of 572 and 82 km^2 respectively. Mallorca is the island which have a higher coastal length with 842 km (including Archipelago of Cabrera), Menorca follows with 433 km, Ibiza has 334 km and Formentera 115 km.

Balearic Islands emerge from a submerged promontory (Balearic Promontory) as a prolongation towards ENE of the Beticas Mountain Rage which are located at Southeastern of the Peninsula Iberica (Fig. 6.1). Thus, Balearic Promontory and Balearic Islands are limited by the units of Valencia Channel at NW, Argelo-Balear Basin at S and SE and Liguro-Provenzal Basin at NNE. Balearic Promontory is where the continental shelf of the Balearic Islands is developed, at bathymetry of −200 m the platforms presents a strong slope (more accentuated on SE margin) that reach the abyssal plans of the Valencia Channel, Argelo-Balear Basin and Liguro-Provenzal Basin (Fig. 6.1).

Balearic Islands are constituted by materials corresponding to the last 400 million years of the geological history. The oldest materials are mainly located at northern area of Menorca (Tramuntana Region) and corresponds to the Silurian and Devonian periods of Paleozoic or Primary era, there is also a specific area of NW coast of Mallorca with an outcrop of this era, located close to middle of Tramuntana Mountain Range (Fig. 6.1). The Mesozoic or Secundary era are the most abundant materials and it could be found in the major part of the mountain ranges of Mallorca

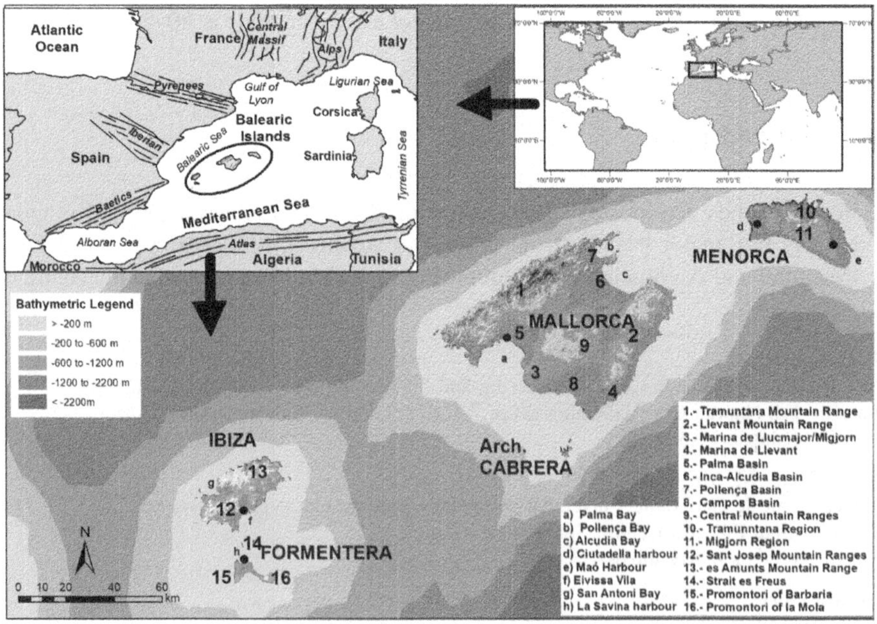

Fig. 6.1 Location of study area, Balearic Islands, located at the centre of the Western Mediterranean

(Tramuntana and Llevant) and Ibiza (Amunts and Serres de Sant Josep) and in some outcrops of Northern area of Menorca. Tertiary materials are very significant in the areas of Southern Menorca (Migjorn Region), Eastern and Southern Mallorca (Marina de Llevant and Marina de Migjorn respectively) and in Formentera Island. The Quaternary are represented mainly by eolian and littoral deposits, most of them, located close to the shoreline and are highly affected by eolian and wave-induced processes. The archipelago exists in a complex geological setting in the Western Mediterranean (Gelabert 1997). The islands are affected by two tectonic phases which configures the current geomorphological appearance. The first compressive phase, occurred between Paleogene and Middle Miocene (related to Alpine Orogeny), and the second one, extensive phase, occurred during the Upper Miocene which generated a structure of horsts and grabens. The major part of the lithology of the islands has a calcareous origin except the Northern region of Menorca (Tramuntana Region) that is mainly constituted by silicic materials.

Concerning the maritime climate the archipelago have a micro-tidal regime with oscillations that not used to be greater than 0.25 m, the sea level is conditioned to barometric changes, wind induced and storms, astronomic and gravitational influence (Basterretxea 2004). This fact conditions that the swash, splash and spray zones in rocky coasts are very stable and only varying its amplitude during marine storms events. According to this, in low rocky coasts (generally with heights less than 3 m) the variation of the amplitude of these 3 zones will be most appreciable than in high and cliffed coasts where especially the limits of swash and splash zones will be quite constant.

In general terms, the position of the Balearic Islands implies considerable fetch length towards most of the cardinal points (N, NE, E, SE, and SW). between North Balearics and Gulf of Lion there are more than 350 km, Between the North and Northeast of Balearics and Gulf of Ligurian exists more than 600 km, in the eastern side, between Balearics and Sardinia, there are around 350 km length and towards ESE the distance of the fetch is more than 800 km (until Sicily island). The Fetch towards SW, from Balearics to Alboran Sea, is longer than 600 km length (Fig. 6.1). The considerable length of the fetch surrounding the archipelago implies the possibility of creating big waves with long period which arrives to the coasts, and in this way. In this way the great energy provided by waves during the storms from N, NE, E, SE and SW contribute to model the coasts and their dynamics.

Regarding the distribution and frequency of significant wave height (Hs), is completely related with the high ciclogenetic activity of the Mediterranean sea region, highly conditioned by the mountains range of the surrounding area (Cañellas et al. 2007; Cañellas 2010). In our case the role of the Pyrenees and the Alps are decisive for the wind and pressure distribution over the basin of the Western Mediterranean (Ibit.). According to Cañellas (2010) the calculation of mean values of the Hs values, from the 44-year SIMAR database (former HIPOCAS data base) vary throughout the year, presenting the highest values during autum-winter seasons. Mean annual values of Hs in the area of Balearic islands varies between 0.8 m height in Ibiza, Formentera and Mallorca, and 0.9–1.1 m in southern and northern Menorca respectively. The values concerning the Hs for a

Table 6.1 Rocky coasts of the Balearic Islands in figures. Absolute values has been calculated according with the last edition of the Balearic Islands topographic map (2008–2012) of the Balearic Territorial information system (Sitibsa)

Coastal classification by height (%)							
		Mallorca	Menorca	Ibiza	Formentera	Arch. Cabrera	Balearic Islands
Rocky coasts	0–3 m	14.2	1.4	10.3	31.6	3.1	12.12
	3–5 m	10.5	32.5	16.9	16.6	19.2	19.14
	5–15 m	9.7	25.9	18.2	11.2	21.8	17.36
	15–30 m	20.8	13.3	16.5	11.9	17.8	16.06
	30–50 m	9.5	5.8	11.4	6.3	15.8	9.76
	50–100 m	5	0.5	5.5	4.5	15.8	6.26
	>100 m	5.9	0	2.8	1.1	2.1	2.38
Overall rocky %		75.6	79.4	81.6	83.2	95.6	83.1
Man-made		14.8	8.5	8.6	2.2	0.5	6.9
Beaches		9.6	12.1	9.8	14.6	3.9	10
Overall %		100	100	100	100	100	100
Absolute values (km)							
Overall coastal length (km)		783	433	334	115	59	1,724

return period of 50 years show maximum heights around 11 m in the North and Northwestern quadrant of the archipelago. The maritime climate show a relative constant behaviour throughout the year with a predominance of wave directions from NE and SW during winter months and during summer season this predominance is less accentuated and waves directions from E acquire greater importance.

The coasts of the Balearic Islands are the result of the geological and geomorphological characteristics and modelling processes, as well as the configuration of the structural units that constitute each one of the islands. From a macroscale perspective, the percentage between rocky coasts or coasts formed by cohesive materials and beach coasts or coasts formed by non-cohesive materials is 80–20 very (without taking into account man-made modified coasts) (Table 6.1). This percentage is very similar to the world coastal distribution of 80% for rocky coasts and 20% for beach coasts (Emery and Kuhn 1982; Granja 2009; Stephenson 2015). The configuration of rocky coasts of the Balearics at present are mainly conditioned by the characteristics of the maritime climate of the western Mediterranean combined with fluvial processes, karst, bioerosion and mass movements.

According to just commented, the aim of this chapter is to provide a general description of the rocky coasts of the Balearic Islands according to the geomorphological and structural units of each island, main characteristics (heights, shapes and materials), and modelling processes which configures the major part of the coastal zone of the Islands.

6.2 Methods

The description of the rocky coasts of the Balearic Islands has been based on the analysis of the main characteristics of these type of coasts through the field work, detailed observations of the several imagery of the coastline and in the analysis of the cartography 1/5000 scale. The analysis and management of all this information has been carried out with Geographic Information Systems (GIS) programs as ArcMap and gvSIG.

Maps used have been the official topographic maps of the Balearic Islands (Mapa Topogràfic Balear) scale 1/5000 corresponding with the last series edited in 1995 and 2008. The map series edited in 2008 is the newest and is used to calculate the official data regarding surface and coastal length of the islands offered by the Balearic Statistics Institute (Institut Balear d'Estadística. IBESTAT. www.ibestat.es). Coastal heights for Mallorca have been established according to direct observations during field work and previous works of Balaguer (2005, 2007) regarding coastal classification of Mallorca. Coastal heights for the rest of the islands (Menorca, Ibiza, Formentera and Archipelago of Cabrera) have been used the cartography and Digital Elevation Model (DEM) scale 1/5000 scale and the creation of contours of 3, 5, 15, 30, 50 and 100 m. At the same time two buffers of 25 and 40 m have been calculates from the coastline of the cartography of 2008. The coastal height has been determined from the join of contours of 3, 5, 15 m with the buffer of 25 m and joining the contours of 30, 50 and 100 m with the contour of 40 m. The values obtained have been expressed with percentages.

The thematic cartography regarding Environmental Sensitivity (ES) of the Coastline of the Balearic Islands of the Balearic Islands Coastal Observing and Forecasting System (SOCIB) has been used for check some aspects and to complete the classification. ES determines the behaviour of the materials of the coastline in case of an oil spill or marine pollution event. The classification of ES for the coasts of the Balearic Islands (Balaguer et al. 2017) is available through a web-based Map Viewer (http://gis.socib.es/sacosta/composer). At the same time, open source tools have been used in order to check some aspects and take more information about the coastline of the Balearic Islands, this is the case of Google Earth and Web Map Service of Sitibsa (Territorial Information System of the Balearic Islands. http://ideib.caib.es/visualitzador/visor.jsp).

The study and analysis of the Balearic coasts regarding the lithology, digital geological maps from Geological and Mining Institute from Spain (IGME) and Social and Environmental Observatori of Menorca (OBSAM) have been crossed with the digital maps of the coastline.

For the islands of Mallorca, Menorca and Formentera, the description of the shoreline has been discussed according with the structural units which conforms the islands. In these islands, the structural units condition clearly the landforms and the coasts are quite different depending on the units where are developed. In Mallorca there are 4 structural units (IGME) with shoreline, these are: (1) Tramuntana Mountain Range, (2) Llevant Mountain Range, (3) Post-orogenic reliefs of Upper

Miocene and (4) Post-orogenic basins. In Menorca there are two clear structural units or regions (Bourrouilh 1983), these are Trammuntana region (northern island) and Migjorn region (southern island). Formentera is structured according 2 main units and a third one could also be considered according to the objective of this work. These are: (1) Promontories of la Mola and Barbaria, (2) Central Barrier and (3) the third unit considered are the Outer Islands located in the strait of Ibiza called es Freus. The rest of the islands, Ibiza and the archipelago of Cabrera don't have structural units well differentiated and clearly manifested in the land forms. Cabrera is considered a part of the Llevant Mountain Range of Mallorca (Sàbat 1986) and in Ibiza 3 units are defined (Rangheard 1984) but are units thrust-folded to one another and each independent unit don't reflect a well differentiated landform and or landscapes as occurs in the islands of Mallorca and Menorca, in this way the geological landscape of Ibiza is quite uniform.

The shape, whether for low or high rocky coasts, is indicated in the general description of the rocky coasts. The shapes commented has been vertical, convex, concave and stepped according with proposed by Emery and Kuhn (1982), Trenhaile (1987), Carter (1988), Sunamura (1992) and Woodroffe (2003). This shapes are developed from the prevalence of marine processes (vertical shape), combination between marine and aerial processes (convex shape) and prevalence of aerial processes (concave shape). The case of stepped shaped coasts are relate with the presence of material with different resistance front weathering processes, especially marine processes, and in the particular case of Balearic Islands these type of shaping is closely relate with the variation of sea-level during the lasts glacial and interglacial Quaternary periods.

The determination of coastal roughness has been based with the simplification of the coastline of the digital map scale 1/5000 of 2008. For this calculation it had developed with ArcGis 10.3 using the tool of generalisation (simplify line). Coastal roughness show the degree of articulation of the shoreline and could offer information about the main processes modelling the coast as fluvial processes, relation with structure, marine erosion, etc. In the studies related with Environmental Sensitivity of the coasts, data about coastal roughness indicates the degree of difficulty for cleaning and recovery tasks of coasts affected by an oil spill or marine pollution event. When the shoreline have a bigger roughness implies that much more difficult will be the cleaning and recovery tasks.

6.3 Description of Rocky Coasts of the Balearic Islands

Rocky coasts has been defined by several authors, but the common feature is that rocky coasts are the ones which are composed by cohesive and consolidated materials (Trenhaile 1987, 2002; Sunamura 1992). But Carter (1988) differentiates two main types of coasts according with the main modelling processes, these are erosive and accumulation coasts. Erosive coasts are rocky coasts, these coasts always use to have a negative sedimentary balance and the erosion processes is an irreversible process, in the other hand, beach coasts, formed by unconsolidated materials tend to have a dynamic equilibrium between erosion and accretion (Sunamura 1992).

This chapter is divided in two parts, first referred to the main morphologies and processes which characterize the rock coasts of the Balearic Islands and a second part concerning the results obtained from the analysis of rocky coast classification and main landforms.

6.3.1 Morphologies and Processes

The major part of dynamics and weathering studies on rocky coasts in the Baleraric Islands have done in the island of Mallorca and something less in the island of Menorca. The homogeneity in terms of geology, geomorphology and physiography in all the islands (with the exception of the region of Tramuntana of Menorca) the intensity of the modelling and weathering processes are quite uniform. This fact, together with the homogeneity of the substrate and appearance of some coastal structural units in different islands (i.e. Reef Unif of the Upper Miocene and folded reliefs in Mallorca, Ibiza and Cabera), the studies about dynamics carried out in one island can be extrapolated to the rest of the islands of the archipelago.

In this chapter main landforms and morphologies observed in the rocky coasts off the Balearic Island will be exposed considering the process which have lead their evolution. The coasts of the Balearic Islands are mainly constituted by calcareous substrate and in this way karst processes will be important creating many kinds of landforms and morphologies in the coastal zones (Fig. 6.2a) (Gómez-Pujol and Fornós 2001). Main morphologies on calcareous rocky coasts are shown in Fig. 6.2a (Gómez-Pujol and Fornós 2001), this scheme was created according to karst processes and littoral karren morphologies and landforms. The influence of the biological action on the erosion of rocky coasts (biological weathering/bioerosion) could be considered as a part of karst processes because, the major part of this weathering is based on the solution of the calcareous rock. Figure 6.2a shows a general morphological scheme for the case of Majorcan coasts but extrapolated to the rest of the islands relating the slope, marine dynamics and morphological distribution and biological zonation (Gómez-Pujol and Fornós 2001). The thick black lines of Fig. 6.2a represents different rocky coast profiles: low-flat rocky platforms, low coasts with supra-tidal platforms, stepped coasts, low-medium and high cliffs and the swash, splash and spray zone controls the distribution of morphologies. According to Fig. 6.2a, there are presence of lichen and cyanobacteria, basic organisms in order to justify the major part of bioerosion processes and are all around the calcareous coasts. These contribute to weathering processes of the substrate by means of their own metabolism (Moses and Smith 1994; Chen et al. 2000; Fiol et al. 1996; Peyrot-Clausade et al. 1995) and attracting grazing organisms as gastropods, crustaceans and echinoderms. The distribution of grazing organisms is mainly located along swash and splash areas (Melaraphe, Cirripeda

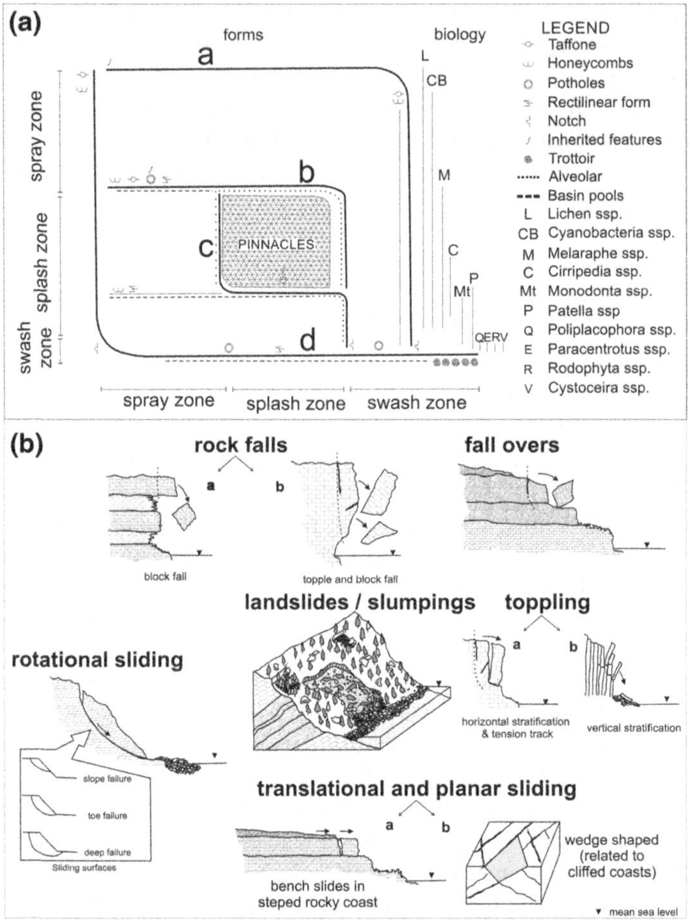

Fig. 6.2 **a** Scheme of the zonation of main forms and microforms and their distribution in rocky coasts. Morphological scheme according to rocky coasts of Mallorca of Gómez-Pujol and Fornós (2001) also applicable to the coasts of the entire archipelago. **b** Main types of mass movements and rock falls of the rocky coasts of Balearic Islands, scheme from Ayala et al. (1986) and Balaguer (2005)

ssp., Monodonta ssp., Patella ssp., Poliplacophora ssp. and Paracentrotus ssp.), and some species as Melaraphe ssp. also can be found in spray zone areas relatively far from the shoreline. The activity of these organisms leads the development of alveolar forms (Spencer 1988; Torunski 1979; Kelletat 1980) and contribute the development of other forms much more controlled by dissolution processes as basin pools. According to above explained all the forms and morphologies in Fig. 6.2a are controlled by several processes, this is the case of basin pools which are controlled by the presence of lineal fractures, dissolution processes and something by bioerosion. Other forms indicated in Fig. 6.2a as taffone, honeycombs, potholes combine wind action, dissolution processes and salt weathering, between others (Trenhaile 1987). The presence of pinnacles are areas where the rock substrate is very weathered, these kind of forms normally appear over the main sea level and just to be the product of dismantled former forms (i.e. basin pools) (Miller and Mason 1994; Sunamura 1992). Finally, Fig. 6.2a indicates the existence of trottoir which are biogenic constructions developed by calcareous algae and vermetid snails (Dalongeville 1995), trottoir are developed very closed to the main sea level and constitutes narrow organic protusions (Trenhaile 2014) as "micro shore-platforms" and conferring some kind of protection to the rock substrate, and also this features grow better under exposed conditions and vermetids as filter feeders are benefited (Spencer and Viles 2002).

Regarding erosion rates and according to the processes and morphologies above mentioned, mean erosion rates calculated for the role of bioerosion on the rocky coasts of Mallorca varies between −0.00709 and −0.587 mm/year (Villanueva et al. 2000; Vidal et al. 2001, 2013; Gómez-Pujol and Fornós 2002; Fornós et al. 2006; Gómez-Pujol 2006). Erosion rates estimated for calcarenite rock decay for a period of 10 years reach values between −1.76 and −2.35 mm/year (Gómez-Pujol et al. 2002; Balaguer et al. 2013), these values have the same order of magnitude as erosion rates obtained by experiments based on exposure of rock tablets along the rocky coasts of Mallorca (Gómez-Pujol and Fornós 2004).

Mass movements in rock coasts usually are related to catastrophic events happened in a short period of time which results in significant changes of the coasts and represents big erosion rates in a short period of time. Mass movements are commonly related to cliffed and high coasts and also depends on the characteristics of fractures, type of material, weather conditions and marine climate. Figure 6.2b shows a compilation of some of the major types of mass movements which can be observed in the rocky coasts of the Balearic Islands. Common rock falls, toppling processes, fall overs can be observed both in shorelines formed by folded materials and shorelines of the post-orogenic relief units but not common in post-orogenic basins. Rotational slides, landslides, slumps and planar slides are more related to high coasts formed by folded materials (structural units mainly affected by Alpine Orogeny). Translational slides according with the sketch showed in Fig. 6.2b, are common processes observed in post-orogenic coasts and stepped shaped where the stratification is horizontal with presence of sand, mud and relict soils between strata, and affected by an intense fracturation which allows the liberation of big rocky blocks.

In cliff coasts not only exists catastrophic events based on mass movements that represents a type of discontinuous weathering, continuous processes also have been described for cliffs of S, SE and E of Mallorca. These kind of continuous weathering processes are called granular disintegration, and is a continuous dismantling processes that affects cliffs and is directly related to the presence of marine salts, wetting and drying cycles, and influence of wind and rainfall which produces the liberation of particles continuously.

Weathering processes on cliffs coasts based on mass movements (mainly rock falls) and granular disintegration have been studied in rocky coasts of Mallorca

(Balaguer and Fornós 2003; Balaguer 2005). The study and calculation of both processes provides estimated erosion rates for the coasts of Mallorca/Balearic Islands of −0.78 mm/year for mass movements and rock falls of cliffed coasts, and −0.022 to −0,082 mm/year for processes of granular disintegration.

Rock quarrying processes depends on the presence of different types of fractures and also depends on the dominant process in each zone presented in Fig. 6.3b. Quarrying processes are described in all types of rocky coasts, but are more related to low rocky coasts (0–3 m) and low cliffed coasts (3–5 m) (Table 6.1). In this way lost blocks by quarrying processes tend to be greater close to the shoreline and tend to be smaller because the only weathering processes is exercised by salts without the power of waves. At the same time in the zones close to the shoreline appear other processes (dissolution and disintegration) related to higher humidity. In this zones the surface is rough and there are a massive presence of pinnacles and coalescence of basin pools together with quarrying processes induced by wave action (Balaguer 2005). Studies on rock quarrying process in Balearic Islands represents a loss of 0.055 and 2.246 m^3 but the period when this volume have been released have not been calculated (Balaguer 2005).

One of the noticeable general features of rocky coasts of Balearic Islands are the print of inherited forms, processes and sedimentary deposits from the Quaternary. The study of the Quaternary in Balearic Islands basically has been focused on the island of Mallorca but the rest of the islands have also had a notable role on the studies of processes, landforms and coastal deposits (Butzer and Cuerda 1962, Pomar and Cuerda 1979; Cuerda 1989; Cuerda and Sacarès 1992; Gràcia et al. 2001; Ginés et al. 2012; Del Valle 2016; Pomar-Bauzà 2016). The succession of episodes of marine regression and transgression related to glacial and interglacial periods, has left many footprints, traces in the coast that demonstrate the effects of the variations of sea level during this period. In this way, rocky coasts are the type of coasts which show better the geomorphological features inherited from the glacial (low sea levels) and interglacial (higher sea levels) periods, and specifically high and cliff rocky coasts are the ones which shows several traces corresponding to many periods. Figure 6.3a shows a graphic representation of several geomorphological evidences of changes in the sea level during the Quaternary for the coasts of the Balearic Islands (Gràcia et al. 2001). This work (ibid.) reviews the geomorphological evidences of sea level changes in the island of Mallorca and it could also be extrapolated to rest of the archipelago, Fig. 6.3a summarizes most of the geomorphological evidences in rocky coasts due to sea level changes. The tracks and features of Fig. 6.3a can have different origins as: (1) karstic, (2) marine, (3) combination between aerial and marine processes. (1) Karstic evidences are mainly justified by the calcareous nature of the island, the main features and morphologies are caves, speleothems, corrosion morphologies, formation of varves. Phreatic paleo-sea-level on speleothems (feature "E" in Fig. 6.3a) are very useful to stablish the oscillation curve of sea level and Dorale et al. (2010) have used the examples from Mallorca to determine with high precision the oscillation curve for the last 81,000 years. (2) Marine evidences can be erosion forms as notches, shore platforms or surf benches, water level notch marks, paleo-cliffs, marine abrasion

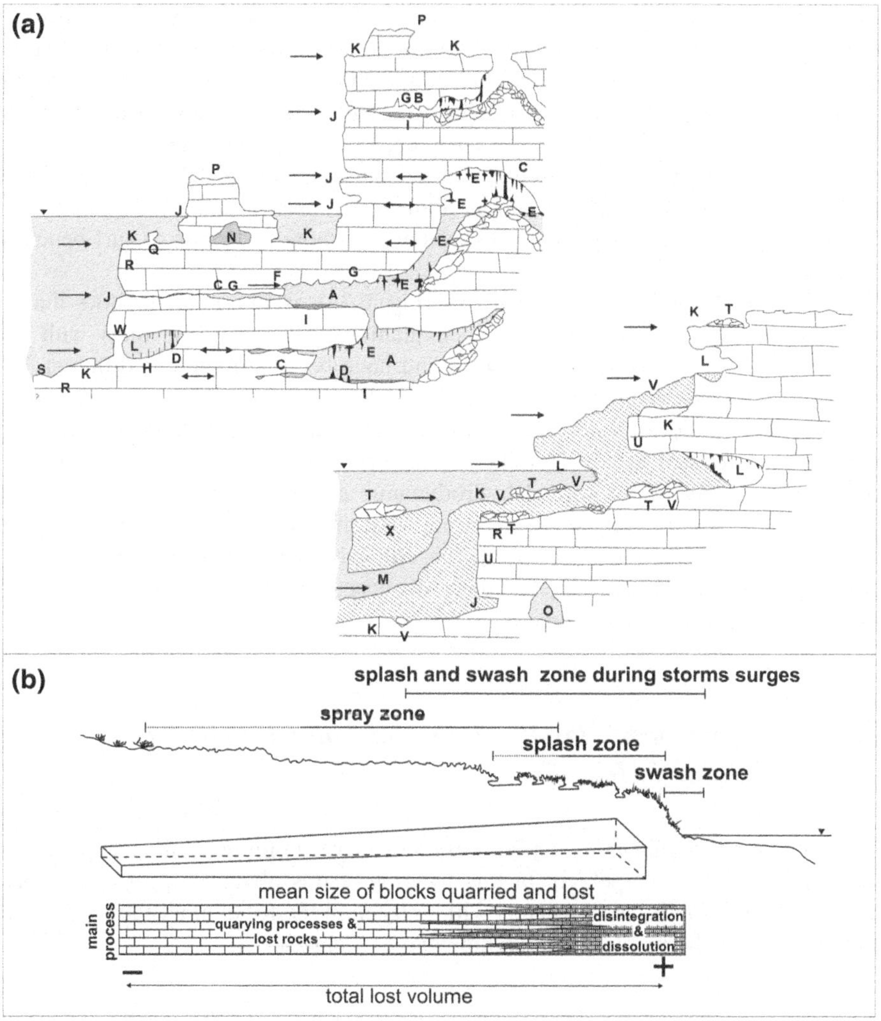

Fig. 6.3 **a** Graphic representation of the main geomorphologic evidence of sea-level changes by Gràcia et al. (2001). (a) Flooded phreatic caves, (b) fossil phreatic caves, (c) stable galleries, (d) submerged speleothems, (e) phreatic paleo-sea-level indicators on speleothems, (f) water-level notch marks, (g) corrosion morphologies, (h) quaternary fossils in non-inundated cavities, (i) formation of varves, (j) notches, (k) shore platforms or surf benches, (l) marine abrasion caves, (m) blowholes, (n) arches, (o) tunnels, (p) rock peak, (q) pedestal, (r) headland, (s) submarine canyons, (t) fields of loose blocks, (u) paleo-cliffs, (v) pothholes, (w) abrassion columns, (x) fossil dune ramps. **b** General scheme of the mechanical weathering, based on lost blocks and pebbles, in the rocky coasts, specially for low rocky coasts, of Mallorca and Northeastern coasts of Catalunya from Balaguer (2005). Processes observed happens in all Balearic coasts

caves, blowholes, arches, tunnels and abrasion columns ("columns" formed by abrasion inside the marine abrasion caves), between others. Quaternary marine evidences on rocky coasts also can be sedimentation forms and the most relevant are fields of loose blocks and fossil deposits. (3) Quaternary evidences according to a combination aerial and marine processes, according with our criteria, are the ones related with presence of fossil-cliffs, littoral paleontological data (Vicens 2015), fossil beaches and fossil dune ramps. Sometimes fossil beaches and dunes contribute to fossilize other morphologies as notches, shore platforms or surf benches, arches, caves, etc.

Regarding the fields of loose blocks (Fig. 6.3a), is important to note that in recent times some of this kind of accumulations, especially the ones with big weight, have been studied and analysed and have been attributed to different episodes of tsunamis (Roig-Munar et al. 2013; Roig-Munar 2016). These kind of accumulation of blocs usually are imbricated, are heavies (more than a tonne), and normally are located at considerable distance from the cliff edge or shoreline in order to be displaced by normal hydrodynamic conditions including big storms/ storm surges. This kind of studies consider that the size of these big blocks, their arrangement, and their distance from the cliff edge and the values of the water columns (run-up) necessaries for their displacement are enough evidences to associate these deposits to tsunamis.

6.3.2 Rocky Coast Classification and Main Landforms of Balearic Islands

In this chapter the main characteristics of rocky coasts in figures will be exposed for each island of the archipelago. In this way the results obtained from rocky coast classification of the Balearic Islands will be listed in the following order: Mallorca, Menorca, Ibiza, Formentera and archipelago of Cabrera.

6.3.2.1 Mallorca

The island of Mallorca have a shoreline length of 783 km, 75.6% are rocky coasts, 14.8% are shorelines modified by man-made structures and the rest 9.6% are beaches (sand and gravel) (Table 6.1). According to this, the cohesive coasts of Mallorca are around the 90% of the island (rocky and man-made structures). More than 60% of the coasts of Mallorca can be considered high coasts (higher than 3 m) and prevails the coasts with heights between 3 and 50 m. Mallorca is the island with major high and cliffed coasts higher than 100 m (Table 6.1), this reason is justified by the presence of Tramuntana Mountain Rage which have the higher reliefs of the archipelago and have a big littoral domain (Figs. 6.1 and 6.4). Mallorca has an index of coastal roughness of 1.6 (Table 6.2), this implies that the Majorcan coasts

Table 6.2 Index of coastal roughness of the Balearic Islands. Values for the whole Balearics, for each island and for some coastal relief units

Coastal roughness (coastal length/simplified coastal length)			
Menorca	1.9	Mallorca and Cabrera	1.6
Region of Tramuntana (Menorca)	2.1	Mallorca	1.6
Region of Migjorn (Menorca)	1.6	Marina of Migjorn (Mallorca)	1.4
Pitiüses Islands	1.5	Marina of Llevant (Mallorca)	1.6
Ibiza	1.6	Archipelago of Cabrera	1.4
Formentera	1.3	Balearic Islands	1.6

have a moderate roughness with some coastal stretches very sinuous with presence of important natural harbours (Andratx, Soller, Portocolom) and big presence of inlets (-calas- as Portopetro, Calador), enclosed bays and coves (Fig. 6.5).

In Mallorca there are 4 structural units (IGME) with littoral domain which confers somehow, these are: (1) Tramuntana Mountain Range, (2) Llevant Mountain Range, (3) Post-orogenic reliefs of Upper Miocene and (4) Post-orogenic basins (Fig. 6.4).

The major part of rocky coasts of Mallorca are formed by limestones and dolomites of the Lower Jurassic (Lias) with around 25% of the overall coastal length, rocky coasts formed by Quaternary materials are around the 21% and the

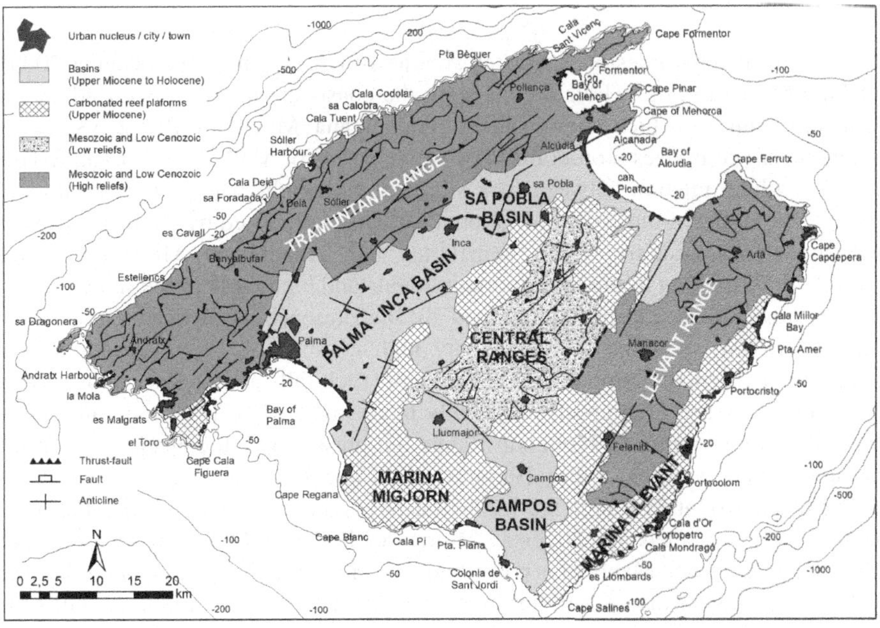

Fig. 6.4 Main characteristics of relief units, geological and geomorphological features of the island of Mallorca. Schemes and representation from Fornós et al. (2005)

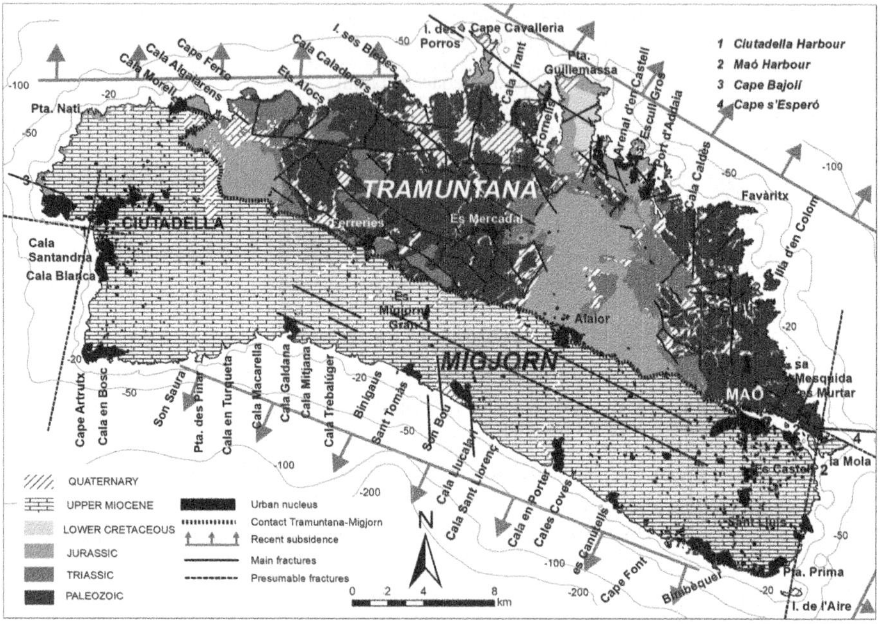

Fig. 6.5 Main characteristics of relief units, geological and geomorphological features of the island of Menorca. Schemes and representation from Bourrouilh (1983), Fornós and Obrador (2003) and Gelabert (2003)

19% are formed by limestones and calcarenites of Upper Miocene (Balaguer 2005). In Mallorca and in the whole archipelago, the coasts are highly influenced by the structural unit to which they belong. Tramuntana Mountain Range (Serra de Tramuntana) comprises around the 40% of the coasts of Mallorca and rocky coasts presents a high relation between rock strength and height. The general scheme of the NW shoreline of the coast of Mallorca is highly influenced by the disposition of the thrust-fold systems which constitutes the Tramuntana Mountain Range (Fig. 6.4). According to this, is important to outstand high cliffs of the sa Dragonera Island, higher than 100 m, conditioned by the presence of powerful thrust-fold. Along the NW shoreline, dominated by high coasts with concave and vertical shape, the existence of slope mass movements, rock falls and boulder talus base is frequent. Cliffs and high coasts, especially those higher than 100 m are all around the littoral of Tramuntana Mountain Range but present a high concentration in the coastal stretch between Soller Harbour and Cape Formentor. Lower coasts of Tramuntana Mountain Range are located at NE and SW extremes inside the bays, natural harbours, inlets and coves as Andratx or Pollença. Coastal stretch between Cape Pinar and Cape Menorca is constituted by Mesozoic limestones and dolomites with heights higher than 30 m, but in the surroundings of Cape Menorca, cliffs are higher than 100 m.

Llevant Mountain Range are close to the 9% of the total coastal length of Mallorca (Fig. 6.4), relliefs are softer and lower than in the case of Tramuntana Range and for this reason more than the 50% of the coasts are lower than 3 m and the general shape is stepped. Higher coastlines (more than 100 m) are located in the areas of Cape Ferrutx, Cape Vermell and Cape Pinar, two last capes are located between Cape Capdepera and Bay of Cala Millor (Fig. 6.4). Lower cliffs are step and concave shaped and are frequent at E of Cape Ferrutx and NE coasts in the surroundings of Cape Capdepera (Fig. 6.4).

Coasts of the post-orogenic reliefs of Upper Miocene are close to the 30% of the total coastal length of Mallorca. There are three main principal areas constituted by this unit with littoral domain (Fig. 6.4), first at the E is the Marina de Llevant, following at the S is the Marina de Migjorn and the third one is located at SW of the Bay of Palma. The general scheme of coasts of these units is characterized by the presence of vertical, concave and stepped cliffs and low coasts, very influenced by fractures and the high density of incised valleys at the shoreline giving place to inlets and caves. This inlets and caves are commonly called "calas" and it formation is conditioned by several factors as fluvial, system of fractures and karst processes (Gómez-Pujol et al. 2013). Marina de Llevant, Marina de Migjorn and the reliefs of the SW of Bay of Palma have around has roughly 100, 40 and 25 km coastal length. Major part of the coasts have vertical, concave and stepped shapes and close to the 90% of these shoreline are high coasts. In the area between el Toro and Cape Cala Figuera (SW Bay of Palma, Fig. 6.4), coasts have concave and vertical shape and some stretches are higher than 100 m. Other areas as Cape Blanc and Cape Regana the dominant type of coast are plunging cliffs with heights between 50 and 100 m. In these units, especially in Marines of Llevant and Migjorn there are an important development of karstic caves Most of them are very important and provides valuable information to determine the oscillation of sea level during the Quaternary (i.e. in Llevant: caves of Genovesa, Pirata-Pont, sa Piqueta, Cala Varques, Gleda, Camp des Pou, dets Ases, des Coll, D'en Bassol, Drac de Cala Santanyí. In Migjorn: Cave of Vallgornera). Some of these caves have been destined to tourism (i.e. Drac Caves) (Gràcia et al. 2011).

The shoreline regarding to post-orogenic basins, basically located in the centre of the big bays of the island (Fig. 6.4) are around the 20% of the total coastal length of the island. Major part of rocky coasts located at bays of Alcudia, Pollença and Palma and developed at Campos Basin (Fig. 6.4) are low coasts (less than 3 m height) and have stepped shape due to the influence of marine transgressions and regressions during the Quaternary. The materials are limestones, conglomerates and calcarenites (eolianites). These characteristics makes these shorelines more accessible and are the most populated areas of the island (especially the Bay of Palma) and are the most modified shorelines by man-made structures of the island.

6.3.2.2 Menorca

The island of Menorca have a shoreline length of 433 km, 79.4% are rocky coasts, 8.5% are shorelines modified by man-made structures and the rest 12.1% are beaches (sand and gravel) (Table 6.1). According to this, the cohesive coasts of Menorca are the 88% of the island (rocky and man-made structures). 78% of the coasts of Menorca can be considered high coasts (higher than 3 m) and prevails the coasts with heights between 3 and 15 m, maximum cliff heights are comprised between 50 and 100 m (Table 6.1). Menorca is composed by two different geological and lithological regions, Tramuntana at N and Migjorn at S (Fig. 6.5), this fact conditions the main characteristics of the coastline. In this way, Menorca has an index of coastal roughness of 1.9 (Table 6.2), this implies that the Menorcan coastline is very articulated (Fig. 6.5). The coastline of Tramunatana is highly articulated with presence of many inlets, coves, headlands and natural harbours (Fig. 6.5), its index of coastal roughness is 2.1, this implies that this coastal area is the most rough of the whole archipelago (Table 6.2). The region of Migjorn is formed by post-orogenic reliefs of Upper Miocene, as the cases of Marina of Migjorn and Llevant of Mallorca (Figs. 6.4 and 6.5), the index of coastal roughness is quite similar to the values of Marina the Llevant, those are highly influenced by presence of fluvial incised valleys forming calas (Table 6.2). In this way coastal characteristics on the same structural units in different islands are repeated.

The coastline of Menorca is conditioned by the existence of two structural and physiographic units of Tramunatna (N) and Migjorn (S). Coastline of Tramuntana has around 259 and 49 km corresponds to islets, in the other hand, coastline of Migjorn has 174 and 14 km corresponds to islets. The main difference is that the northern one (Tramuntana) is mainly formed by folded materials affected by Hercynian and Alpine Orogeny. The coasts are variated and has high coasts concave, convex or vertical shaped of coasts, plunging cliffs, low rocky coasts and pocket beaches shorter than 1 km. This region or unit is condition by the presence of faults with directions WNW-ESE and two thrust-fold systems with directions NE-SW and NW-SE (Gelabert 2003). According to just mentioned, shoreline is highly articulated and sinuous and presents many inlets, coves and important natural harbours as Fornells and Addaia. The Southern unit (Migjorn) is formed by post-orogenic materials (Fig. 6.5), this unit corresponds with the post-orogenic reliefs of Upper Miocene of Mallorca (Marina de Llevant, Marina de Migjorn and unit of SW of Bay of Palma) and Formentera (Promontories of Barbaria and la Mola) (Figs. 6.4 and 6.6). MIgjorn rocky coasts are stepped and vertical shaped and an important density of inlets (calas). Materials are carbonates corresponding with reef facies from Upper Miocene (Obrador 1998; Fornós and Obrador 2003; Obrador and Pomar 2004). According to Gelabert (2003), region of Migjorn can be defined as a "soft" anticline due by the presence of an inverse fault NNE-SSW in the central area. The crest of this anticline has the higher coasts. The fracture system of this unit has direction E-W at eastern and western sides, and N-S in the central area, in this way, these fracture systems controls the development of fluvial system, in this context fluvial system together with karst processes create canyons and capture dolines in coastal areas. This concurrence of factors in the coastline contribute to the development of "calas" (Gelabert 2003; Segura and Pardo 2003; Rosselló 2004; Gómez-Pujol et al. 2013).

According to Balaguer et al. (2017) the coastline higher than 3 m for Tramunatana and Migjorn are the 82% (aprox. 212 km) and 72% (aprox. 174 km) respectively. In Tramuntana, the higher coastlines (heights between 50 and 100 m) are located in the surroundings of Pta. Guillemassa and Cala en Caldés (Limestones and dolomites from Jurassic), in the coastal stretch between ets Alocs and Cape Ferro (quartz-sandstones from Permian-Triassic). Coasts with heights comprised between 30 and 50 m are the most repeated, these are formed by calcareous turbidites from Carboniferous located between sa Mesquida and Cape Esperó, formed by limestones and dolomites from Jurassic surroundings of Pta Guillemassa (Mola of Fornells), Cape Cavalleria, la Mola (E Maó Harbour) and formed by sandstones and limestones and dolomites from Triassic and Jurassic at NW of the island close to Cala Morell (Fig. 6.5). In Migjorn the prevailing lithologies which constitutes the rocky coasts are limestones and calcarenites from Upper Miocene and eolianites, conglomerates and sandstones from Quaternary. The higher coastlines are located in the central sector of the Migjorn coastline with vertical and plunging cliffs with heights between 30 and 50 m in the area of Cala Llucalcari and Cala Sant Llorenç, between Cala Turqueta and Cala Macarella, in the coastal stretch between Cala Morell and Punta Nati (NW of the island, with Jurassic limestones at the base), between Cales Coves and es Canutells at Eastern sector of this coastline (Fig. 6.5). In the Eastern and western extremes of the region of Migjorn, prevailing coasts are low cliffs with heights less than 15 m formed by limestones and clacarenites of the Upper Miocene and sandstones from Quaternary, at the same time, this kind of cliffs are the most common rocky coasts in the central sector.

Concerning the last type of cohesive "rocky" coast, coastlines modified by man-made structures, these are less than 9% of the coastal length and are mainly concentrated in the natural harbours of Maó (E), Ciutadella (W) and Addaia (N), Fornells (N) and the outer dike at Son Blanc, Ciutadella (W) (Fig. 6.5).

6.3.2.3 Ibiza

The island of Ibiza have a shoreline length of 334 km, 81.6% are rocky coasts, 8.6% are shorelines modified by man-made structures and the rest 9,8% are beaches (sand and gravel) (Table 6.1). The total coastal length of Ibiza is divided in 68 km of coastal length corresponding to the islets and 266 km corresponding with the "main" island. The cohesive coasts of Ibiza are around the 90.2% of the island (rocky and man-made structures). More than 70% of the coasts of Ibiza can be considered high coasts (higher than 3 m) and prevails the coasts with heights between 3 and 30 m (more than 50% of the total of coastline). Ibiza has an index of coastal roughness of 1.6 (Table 6.2), this implies that its coasts have a moderate roughness with some coastal stretches very sinuous with presence of important bays, natural harbours, inlets and coves as Bay of Sant Antoni, Port Sant Miquel,

Benirràs, Portinatx in the W and NW coastline and bays of Sta Eulalia and Ibiza Habour at E and SE of the island (Fig. 6.6). There are some coastal stretches with high presence of minor islands and islets, this condition increases the coastal roughness. This is the case for some coastal areas of the W (Conillera), SW (Vedrà and Vedranell) and E (Cala Pada and es Canar) of the island.

Ibiza has 3 geological units defined by Rangheard (1984) but are units thrust-folded to one another and each independent unit don't reflect a well differentiated landform and or landscapes as occurs in other islands of the archipelago, and also the Quaternary layers covers a considerable part of the island, in this way, the geological landscape of Ibiza is quite uniform (Fig. 6.6).

The major part of low rocky or cohesive coasts of Ibiza are concentrate at E and SE coastlines (areas of es Cavallet, surroundings of the cities of Ibiza and Sta Eulalia), these are also remarkable at N and NE coasts as es Figueral or Cape Moscarter and inside of some coves and natural harbours of the NW and W coasts as the Bay of Sant Antoni. Low rocky coasts usually have stepped shapes, a good example of this is the coastline of es Cavallet at the S of Ibiza. The case of Sta Eulalia, low rocky coasts are extended between Cala Blanca and es Canar and the major part is formed by Quaternary materials (Fig. 6.6). The cases of Ibiza, Sant Antoni and Sta Eulalia concentrates de major part of the shorelines modified by man-made structures.

In the S and SW of Ibiza higher coasts generally are vertical and concave shaped. Usually have boulder talus at the base of the cliffs and in the areas dominated by Quaternary materials landslides are observed. In this area cliffs with heights up to 30 m are located in es Codolar-ses Salines. The shoreline of es Codolar is formed by a pebble and gravel beach which is provided by the weathering of western cliffs of the area. Higher coastlines with heights between 50 and 100 m are in the coastal stretch between Cape Llentrisca and Cala Carbó and in the S coast of islet of es Vedrà (Fig. 6.6).

The area of SE of Ibiza, the higher coastlines are generally convex shaped alternating with vertical and concave forms with boulder at the base especially those located in the surroundings of the city of Ibiza and in the coastal stretch between Talamanca and Cala Blanca. Higher cliffs are located in the area of Cape Llibrell with heights between 50 and 100 m (Fig. 6.6).

In the NE the higher cliffs (heights between 50 and 100 m) are located at Cap Roig with vertical and concave shapes, concave cliffs usually have boulders at the base and in some hillshades some scree and areas plenty of blocks are observed.

In the N of the island higher coastlines are located at N of Cala Sant Vicenç and are formed by limestones and dolomites of the Jurassic, there are a variety of shapes but concave cliffs with boulder at the base are the most frequent, this is the case of the area of ses Caletes with cliffs formed by conglomerates, marlstones and calcarenites of the Lower and Medium Miocene. The coastline between Cala d'en Serra y ses Caletes has maximum cliff heights between 50 and 100 m (Fig. 6.6).

In the NW and W coast of Ibiza has the higher cliffs of the islands, this area is the littoral domain of es Amunts Mountain Range. The roughness of the NW coastline of Ibiza is due to the presence of complex inlets and coves (i.e. Portinatx,

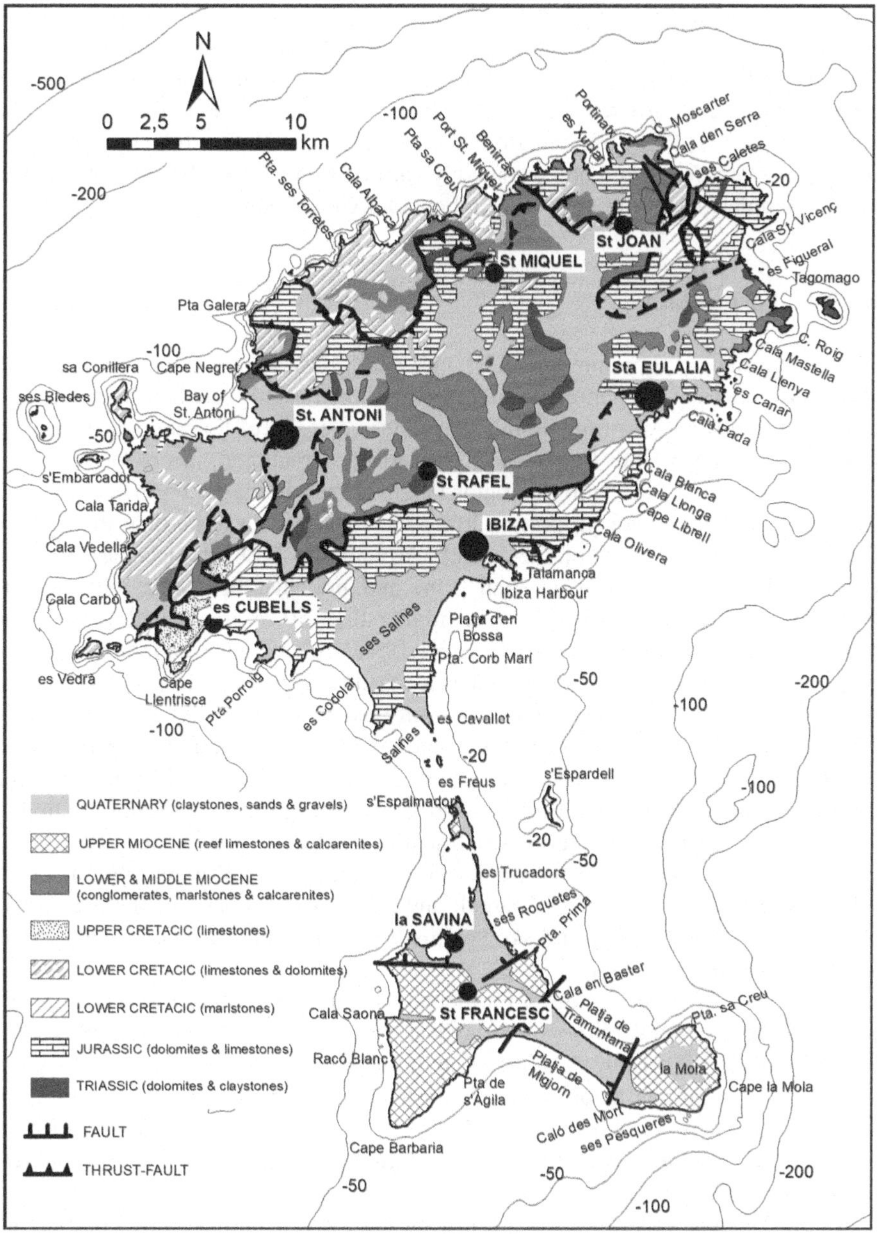

Fig. 6.6 Main characteristics of relief units, geological and geomorphological features of Pitiüses Islands (Ibiza and Formentera). Schemes and representation for Ibiza are drawn from Rangheard (1984) and for the case of Formentera from Rangheard (1984)

Fig. 6.6). In the coastal stretch between Port de St Miquel and Pta Galera is the area with the higher cliffs, these are plunging cliffs formed by limestones and dolomites from the Jurassic and Cretaceous, in this area the maximum heights (>100 m) are located in the coastal stretch between Albarca and Pta sa Creu (Fig. 6.6). In the coastal areas at S of Bay of Sant Antoni coasts and some cliffs with heights between 3 and 15 m are developed and are mainly formed by materials from Quaternary, a good examples are in the area of s'Embarcador/Cala Compte.

6.3.2.4 Formentera

Formentera have a shoreline length of 115 km, 83.2% are rocky coasts, 2.2% are shorelines modified by man-made structures and the rest 14.6 are beaches (sand and gravel) (Table 6.1). According to this, the cohesive coasts of Formentera are around the 85% of the island (rocky and man-made structures), according to this, Formentera is the island with the major length of beach coasts. Formentera is also the island with the major proportion low rocky coasts with more than the 31% of the overall coastal length of the island. Around the 50% of the coasts can be considered high coasts (higher than 3 m) and prevails the coasts with heights between 3 and 30 m (Table 6.1). Cliff higher than 50 m are around 11% of the coastline and those higher than 100 m are 1.1%. Formentera has an index of coastal roughness of 1.3 (Table 6.2), this implies a coastline very rectilinear with few inlets and coves, there aren't natural harbours, only exists s'Estany des Peix close to la Savina (Fig. 6.6) and acts as a shallow lagoon enclosed by a barrier formed by sand and Pleistocene eolianites. The limestones and calcarenites the of Upper Miocene constitutes the units of Cape Barbaria and la Mola (Fig. 6.6), these unit are highly controlled by system of fractures which justify the lineal shape of the coasts.

Low rocky coasts (0–3 m) are located at Pta des Trucadors, W and S of s'Espalmador, W of islet of s'Espardell and all the rocky coasts of the areas of la Savina and Platja des Migjorn have heights between 0 and 3 m. Low rocky coasts usually have stepped shape.

Major part of high coasts of Formentera have stepped, vertical and concave shape. Steeped shaped cliffs are located at the W of the island alternating with vertical cliffs in the area of Cala Saona, in this area boulders at the base are frequent due to the effect of marine processes. In the coastal stretch between Cala Saona and Pta de s'Àguila is formed by vertical cliffs with heights between 30 and 100 m (Fig. 6.6). In the surroundings of Pta de s'Àguila some accumulations of boulders at the base are observed.

Upper Miocene promontories of Formentera (Barbaria and la Mola, Fig. 6.6) are controlled by fracturation with orthogonal direction, and according to this, some coastal stretches are very articulated describing right angles, this is the case of the shoreline of the area of Pta de s'Àguila.

The highest coastlines of Formentera are located at the Promontory of la Mola, cliffs usually are plunging cliffs and vertical or concave cliffs with boulders at the base. In this promontory heights are comprised between 30 and 100 m. In Caló des

Morts and ses Pesqueres there are concave and vertical cliffs with accumulation of debris at the base. Highest coastlines are plunging cliffs at Cape la Mola.

Coastal stretch of Platja de Tramuntana, the place name makes reference to a beach, but most of the 90% of it is a stepped rocky coast no higher than 5 m composed by eolianites and calcarenites from Quaternary. Coastal stretch between Cala en Baster and Punta Prima is formed by Upper Miocene calcarenites with heights between 5 and 15 m. Rocky coasts of the area of ses Roquetes until es Trucadors is stepped and formed by Quaternary eolianites.

6.3.2.5 Archipelago of Cabrera

The archipelago of Cabrera have a shoreline length of 59 km, 46.6 km belongs to the main island and 12.4 km belongs to the northern islets (Table 6.1, Fig. 6.7). 95.6% are rocky coasts, only around 0.5% are shorelines modified by man-made structures and the rest 3.9% are beaches (sand and gravel) (Table 6.1). According to this, the cohesive coasts of the archipelago of Cabrera are the 96% of the total coastal length, according to this, this island is by far the one with major concentration of rocky/cohesive coasts. More than 92% of the coasts of the archipelago of Cabrera can be considered high coasts (higher than 3 m), high coasts with heights between 15 and 100 m represents around the 50% (Table 6.1). This archipelago has an index of coastal roughness of 1.4 (Table 6.2), this implies that this coasts have a low-moderate roughness with presence of some headlands (Cape Llebeig and Cape Enciola), natural harbour (Cabrera Harbour), inlets and coves (Cala Santa Maria, es Codolar and l'Olla) (Fig. 6.7).

Cabrera is a prolongation towards SW of the Llevant Mountain Range of Mallorca (Sàbat 1986). The archipelago of Cabrera is composed by a main island, Cabrera and several islets and minor islands, most of them, localized between Cabera and Mallorca. The lithologies, structures and the landscapes composed by soft reliefs are very similar with the Llevant Mountain Range.

Low rocky coasts usually are located inside the coves, inlets (Cala Santa Maria), Cabrera Harbour, in some sectors of Cape Falco, at W and E coastlines of islet Conillera and islets Pla and Foradada with high. Higher rocky coasts with elevations between 5 and 30 m are located at NNW of the island of Cabrera (surroundings of Cape Morobutí), central coasts of Cala Santa María and surroundings of Cape Xoriguer (Fig. 6.7).

Major part of Cape Llebeig are vertical plunging cliffs of heights between 30 and 50 m formed by lower Jurassic dolomites, same heights are observed in the entrance of the harbour under the castle, but in this case this have concave shape localized in sheltered marine conditions with boulders and cobble at the base and presence of some pebble and gravel beaches (Fig. 6.7). Plunging cliffs at N, NE and W of Cala Santa Maria and major part of headland of Cape Ensiola formed by plunging cliffs alternating with convex shapes until the area of es Codolar including the islet and strait of s'Imperial at SE of Cabrera (Fig. 6.7).

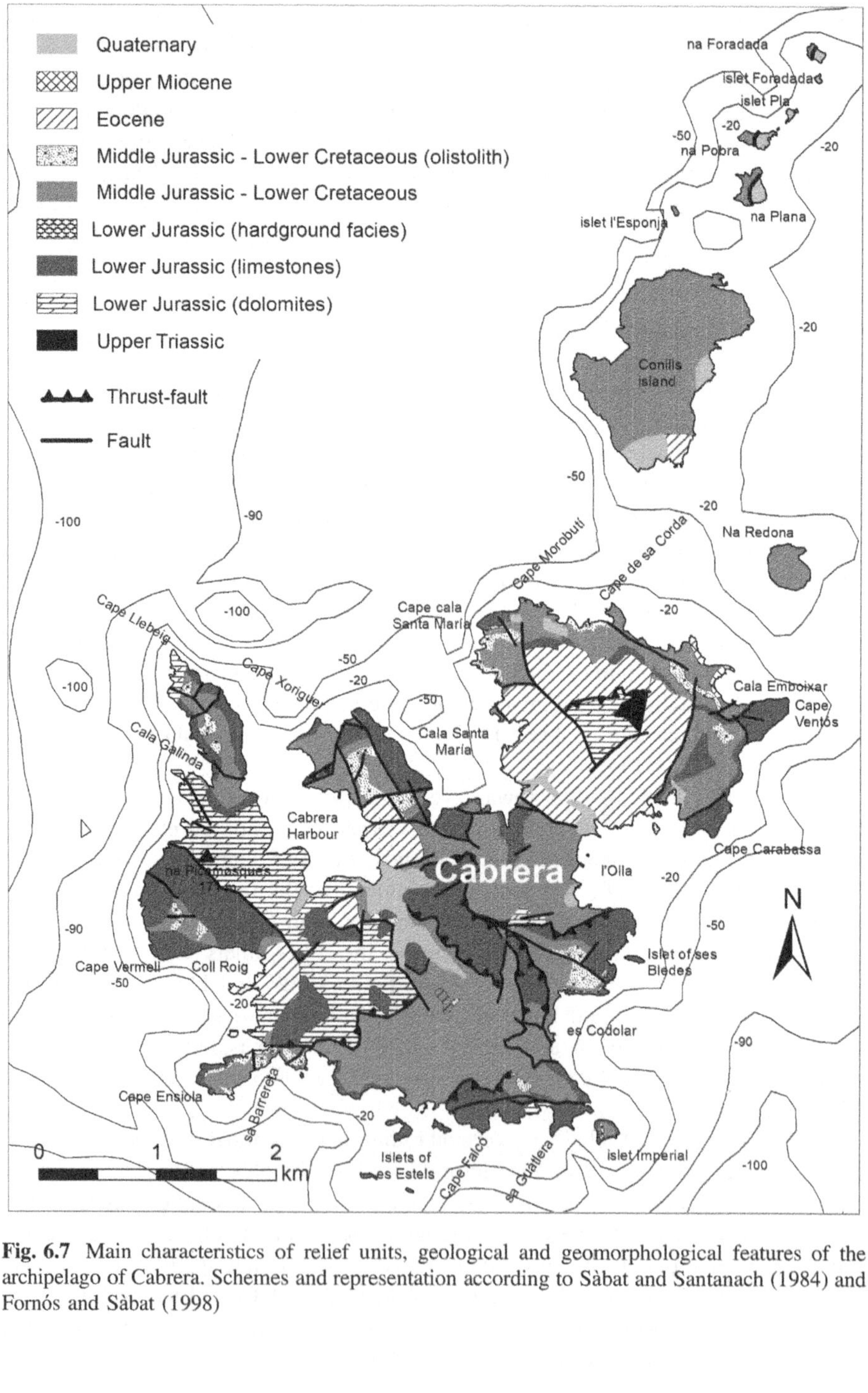

Fig. 6.7 Main characteristics of relief units, geological and geomorphological features of the archipelago of Cabrera. Schemes and representation according to Sàbat and Santanach (1984) and Fornós and Sàbat (1998)

In the area of es Codolar and central part of l'Olla higher cliffs have heights between 30 and 100 m, these are vertical cliffs with boulder and pebbles at the base with presence of pebble and gravel beaches (Fig. 6.7). These coasts are formed by dolomites and limestones of Lowe Jurassic.

6.4 Conclusions

Balearic Islands have a coastal length of 1,724 km, 83% are rocky coasts, around 7% of coastlines modified by man-made structures and 10% are beaches. In this way 22% of the coastline of the Balearic Islands can be considered as low cohesive/ rocky coasts (low rocky coasts and man-made structures) and the remaining 70% are high rocky coasts higher than 3 m. Higher coasts of the archipelago are located in Mallorca at littoral domains of Serra de Tramuntana Mountain Range and post-orogenic reliefs of Upper Miocene that exceed 100–120 m. In each island there are good examples of cliffs and high coasts around 100 m, in Menorca highest coasts are located at the northern coastline, in Ibiza at NW coasts in the littoral domains of es Amunts mountain range, in Formentera highest cliffs are locate at W at Cape la Mola and in Cabrera higher cliffs are located at SW and NE extremes of the main island.

Rocky coasts of Balearic Islands reflects a repetition of some types of coastal landscapes as result of the repetition of the same structural units in every island. Is the case of the post-orogenic reliefs of the Upper Miocene which are localized in the islands of Mallorca (Marinas of Migjorn and Llevant), Menorca (region of Migjorn) and Formentera (promontories of Barbaria and la Mola) (Figs. 4, 5 and 6). Coasts developed on this structural unit have same shapes and the vegetation (communities and associations) are the type or quite similar conforming very similar landscapes. In this way processes which affects these rocky coasts in one island, will also be assumed for other coastlines developed on the same structural unit. Similarities between the Balearic coasts developed on the reliefs of Upper Miocene limestones and calcarenites in each island can be observed. The case of the coasts developed on the relief of the Tramuntana mountain range in Mallorca have also many similarities with the coasts of the NW of Ibiza (Amunts mountain range), the organization, disposal and orientation of the thrust-faults are very similar and constitutes the same type coastal landscape. In the other hand the coasts developed at NE of Mallorca corresponding to the littoral domains of Llevant mountain range are very similar to the coasts of the Archipelago of Cabrera. These coasts belongs to the same geological unit and some of the characteristic features are soft reliefs, more frequency of convex coastal shapes and lower heights in comparison with Tramuntana mountain range. Regarding the coasts developed on post-orogenic basins in Mallorca are coasts mainly formed by Quaternary materials (sands, conglomerates, calcarenites, eolianites) usually have stepped shapes and repeated in many shorelines of the islands. Some of best examples of this type of shoreline are located inside the big bays of Mallorca, S Ibiza and Quaternary deposits of Formentera.

The current look of the rocky coasts of the Balearic Islands is a result of many processes imbricated, some of them act simultaneously (current processes) (Fig. 6.2a) and others have acted in the past, fundamentally during the Quaternary (Fig. 6.3). Weathering processes are the most common process acting on the Balearic rocky coasts, these kind of processes can be continuous and persistence over time (Fig. 6.2a) and discontinuous processes, only acting during specific times (Fig. 6.2b). Dissolution processes (karst), eolian erosion, salt weathering and the paper developed by organism eroding the rocky coasts can be considered as continuous weathering processes, the changes occurred in rock surface is not perceptible. The granular disintegration is another continuous process and its effects are not perceptible on the face of the cliffs either. Erosion rates obtained from several studies oscillates between −0.00709 and −0.587 mm/year for bioerosion processes, −1.76 and −2.35 mm/year for wind and salt weathering processes, −0.022 and 0.082 mm/year for granular disintegration processes. Despite the reduced magnitude of these erosion rates, its weathering effects during long periods (i.e. geological periods) could justify the evolution of a coastal cliff at SE of Mallorca (Fornós et al. 2005). Discontinuous erosion processes are usually catastrophic events which usually produce big erosion rates in a short period of time. Recurrence of these kind of processes can be of decade or century order. This processes are mass movements in cliffs or high coasts activated by aerial and/or marine processes (Figs. 2b and 3b). Erosion rates of cliff retreat in the post-orogenic reliefs of Upper Miocene (Figs. 4, 5 and 6) has been estimated in −0.78 mm/year. Trottoir or biogenic constructions, is the only continuous "construction" process described in the coasts of Balearic Islands (Gómez-Pujol and Fornós 2001). Accumulation of boulders or debris at the cliff base or the accumulation or redistribution of boulders due to the action of tsunamis (Roig-Munar 2016) are visible processes but accumulation rates not yet been established.

References

Ayala, F. J., Andreu, F. J., Marques, M., Ferrer, M., De Simón, A., Fernández, I., Olalla, C., Gómez-Puyuelo, J., Sampedro, J., & Cienfuegos, F. J. (1986). Manual de Taludes. Instituto Geológico y Minero de España (IGME) & Estudios y proyectos Técnicos Industriales, S.A. (EPTISA) (Eds.). Madrid. 450 pp.

Balaguer, P. (2005). *Tipus i evolució de les costes rocoses de Mallorca*. Ph.D. unpublished. Department of Geography of the Univeristy of the Balearic Islands, 373 p.

Balaguer, P. (2007). Inventari quantitatiu de les costes rocoses de Mallorca. In G. X. Pons & D. Vicens (Eds.), *Homenatge a D. Joan Cuerda Barceló. Monografia Societat d'Història Natural de les Balears* (Vol. 14, pp. 201–230).

Balaguer, P., & Fornós, J. J. (2003). Processos erosius als penya-segats costaners del SE de Mallorca (Illes Balears, Mediterrània Occidental): dades preliminars de l'erosió contínua per disgregació granular. *Boll. Soc. Hist. Nat. Balears, 46*, 37–50.

Balaguer, P., Ll, Gómez-Pujol, Fuster, M., & Fornós, J. J. (2013). Alteración Alveolar en areniscas carbonatadas y tasa de erosión asociada, 10 años de obsevaciones en las canteras litorales de es Carnatge (Mallorca, Illes Balears). *Geotemas, 14,* 79–82.

Balaguer, P., March, D., Frontera, B., Vizoso, G., Ruiz, M. & Tintoré, J. (2017). Sensibilidad Ambiental de la Líneade Costa (SALC), una tarea en continua revisión. IX Jornadas de Geomorfología Litoral. *Geotemas, 17,* 195–198.

Basterretxea, G., Orfila, A., Jordi, A., Casas, B., Lynett, P., Liu, P. L. F., et al. (2004). Seasonal dynamics of a microtidal pocket beach with *Posidonia oceanica* seabeds (Mallorca, Spain). *Journal of Coastal Research, 20,* 1155–1164.

Bourrouilh, R. (1983). Estratigrafía, sedimentología y tectónica de la isla de Menorca y del Noreste de Mallorca (Baleares). La terminación Nororiental de las Cordilleras Béticas en el Mediterráneo Occidental. Madrid, Memoria del Instituto Geológico y Minero de España (Vol. 2, 672 pp).

Butzer, K. W., & Cuerda, J. (1962). Coastal stratigraphy of Southern Mallorca and its implications for the Pleistocene chronology of the Mediterranean Sea. *Journal of Geology, 70,* 398–416.

Cañellas, B. (2010). *Long-term extreme wave height events in the Balearic Sea: characterization, variability and prediction.* Tesis Doctoral. Palma, Universitat de les Illes Balears. 85 pp.

Cañellas, B., Orfila, A., Méndez, F. J., Menéndez, M., Gómez-Pujol, L., & Tintoré, J. (2007). Application of a pot model to estimate the extreme significant wave height levels around the Balearic-sea (western mediterranean). *Journal of Coastal Research, 50,* 329–333.

Carter, R. W. G. (1988). *Coastal environments* (p. 617). London, UK: Academic Press.

Chen, J., Blume, H. P., & Beyer, L. (2000). Weathering of rocks induced by lichen colonization, a review. *CATENA, 39,* 121–146.

Cuerda, J. (1989). Los tiempos cuaternarios en Baleares. Conselleria de Cultura, Educació i Esports. Direcció General de cultura. Govern Balear. Palma de Mallorca. 310 pp.

Cuerda, J., & Sacarès, J. (1992). El Quaternari al Migjorn de Mallorca. Direcció General de Cultura. Conselleria de Cultura, Educació i Esports. Govern Balear. 130 pp. Palma de Mallorca.

Dalongeville, R. (1995). Le role des irganismes constructeurs dans la morphologhie des littoraux de la mer méditerranée: algues clcaires et vermetidés. *Norois, 42,* 73–88.

Del Valle, L. (2016). *El registre sedimentari eòlic del Plistocè Litoral d'Eivissa.* Ph.D. Unpublished. Department of Geography of University of the Balearic Islands.

Dorale, J. A., Onac, B. P., Fornós, J. J., Ginés, J., Ginés, A., Tuccimei, P., et al. (2010). Sea-level highstand 81,000 years ago in Mallorca. *Science, 327*(5967), 860–863.

Emery, K. O., & Kuhn, C. G. (1982). Sea cliffs: Their processes, profiles and classifications. *Geological Society of America Bulletin, 93,* 644–654.

Fiol, Ll., Fornós, J. J., & Ginés, A. (1996). Effects of biokarstic processes on the development of solutional rillenkarren in limestone rocks. *Earth Surface Processes and Landforms, 21,* 447–452.

Fornós, J.J. & Sàbat, F. (1998). Aspectes geològics de l'arxipèlag de Cabrera. In: Fornós, J.J. (ed.) *Aspectes geològics de les Balears.* Universitat de les IllesBalears, 67–93. ISBN: 84-7632-368-9.

Fornós, J. J., & Obrador, A. (2003). Geología de Menorca. In V. M. Rosselló, J. J. Fornós, & L. Gómez-Pujol (Eds.), *Introducción a la Geografía Física de Menorca. Mon. Soc. Hist. Nat. Balears* (Vol. 10, pp. 31–38). Palma, AGE, Universitat de València, Universitat de les Illes Balears, Societat d'Història Natural de les Balears.

Fornós, J. J., Balaguer, P., Gelabert, B., & Gómez-Pujol, L. (2005). Pleistocene formation, evolution, retreat, rates and processes in a carbonate coastal cliff (SE of Mallorca Is., Western Mediterranean). *Journal of Coastal Research SI, 49,* 15–21.

Fornós, J. J., Pons, G. X., Gómez-Pujol, L., & Balaguer, P. (2006). The role of biological processes of downwearing due to grazing organisms of Mallorca carbonate coasts (Western Mediterranean). *Zeitschrift für Geomorphologie, 144,* 161–189.

Gelabert, B. (1997). *La estructura geológica de la mitad occidental de la isla de Mallorca.* Instituto Tecnológico Geominero de España. Colección memorias, 129 pp.

Gelabert, B. (2003). La estructura geológica de Menorca: Las zonas de Tramuntana y de Migjorn. In V. M. Rosselló, J. J. Fornós, & L. Gómez-Pujol (Eds.), *Introducción a la Geografía Física de Menorca. Mon. Soc. Hist. Nat. Balears* (Vol. 10, pp. 39–48). Palma, AGE, Universitat de València, Universitat de les Illes Balears, Societat d'Història Natural de les Balears.

Ginés, A., Ginés, J., Gómez-Pujol, L., Onac, B. P., & Fornós, J. J. (Eds.). (2012). Mallorca: A Mediterranean benchmark for quaternary studies. *Mon. Soc. Hist. Nat. Balears, 18*, 13–53.

Gómez-Pujol, L. (2006). *Patrons, taxes i formes d'erosió a les costes rocoses carbonatades de Mallorca*. Ph.D., Unpublished. Department of Geography of the Univeristy of the Balearic Islands.

Gómez-Pujol, L., & Fornós, J. J. (2001). Les microformes de meteorització del litoral calcari de Mallorca: Aproximació de la seva sistematització. *Endins, 24*, 169–185.

Gómez-Pujol, Ll., & Fornós, J. J. (2002). Estudio integrado del lapiaz costeroo de Mallorca dentro del Proyecto ESPED: Metodología y resultados preliminares. Bol. *Soc. Esp. Espeleol. Y Ciencias del Karst, 3*, 106–115.

Gómez-Pujol, L., & Fornós, J. J. (2004) Tasas de meteorización química en la costa carbonatada de Mallorca. Evaluación preliminar mediante tests de exposición (weight-los rock tablets). In R. Blanco, J. Bedoya, & A. Pérez-Alberti (Eds.), *Procesos geomorfológicos y evolución costera. SPUSC, Santiago de Compostela. Comunicaciones* (pp. 401–410).

Gómez-Pujol, L., Balaguer, P., & Fornós, J. J. (2002). Meteorización del patrimonio histórico en ambientes costeros marinos: el caso de la Torre d'en Beu (Santanyí, Mallorca). In E. Serrano, et al. (Eds.), *Estudios recientes (2000–2002) en Geomorfología, PUV, Valladolid. Comunicaciones* (pp. 403–413).

Gómez-Pujol, L., Gelabert, B., Fornós, J. J., Pardo-Pascual, J. E., Rosselló, V. M., Segura, E. S., et al. (2013). Structural control on the presence and character of calas: Observations from Balearic Islands limestone rocks coasts macroforms. *Geomorphology, 194*, 1–15.

Gràcia, F., Clamor, B., Landreth, R., Vicens, D., & Watkinson, P. (2001). Evidències geomorfològiques dels canvis del nivell marí. In G. X. Pons & J. A. Guijarro (Eds.), *El Canvi Climàtic: passat present i futur* (pp. 91–120). ISBN: 84-87818-34-X.

Gràcia, F., Clamor, B., Gamundí, P., Fornés, J. J., & Watkinson, P. (2011). Cavitats subaquàtiques de la franja litoral de Mallorca. Endins 35/Mon. *Societat d'Història Natural de les Balears, 17*, 103–132.

Granja, H. M. (2009). Rocky coasts. In F. I. Islaay (Ed.), *Coastal zone and estuaries* (pp. 135–163). ISBN: 978-1-84826-016-0 (eBook) UNESCO-EOLSS.

Kelletat, D. (1980). Formenschatz und Prozessgefüge des Biokarstes an der Küste von Nordost-Mallorca (Cala Guya). Beiträge zu regionalen Küstenmorfhologie des Mittelmeerraumes VII. Berliner Geographische Studien (Vol. 7, pp. 99–113). Berlin.

Miller, W. R., & Mason, T. R. (1994). Erosional features of coastal beachrock and aeolianite outcrops in Natal and Zululand, South Africa. *Journal of Coastal Research, 10*, 374–394.

Moses, C. A., & Smith, B. J. (1994). Limestone weathering in the supra-tidal zone: An example from Mallorca. In D. Robinson, & R. G. B. Williams (Eds.), *Rock Weathering and Landform Evolution* (pp. 433–451). Wiley Ltd.

Obrador, A. (1998). Un cop d'ull a la geologia menorquina. In Fornós, J.J. (ed.) *Aspectes geològics de les Balears*. Universitat de les IllesBalears, 39–66. ISBN: 84-7632-368-9.

Obrador, A. & Pomar, L. (2004). El Miocè del Migjorn de Menorca. In J.J Fornós, A. i Obrador, & V.M. Rosselló (Eds.), *Història Natural del Migjorn de Menorca*: el medi físic i l'influx humà. *Mon. Soc. Hist. Nat. Balears, 11*, 73–92. ISBN: 84-609-3585-X.

Peyrot-Clausade, M., Le Campion, T., Harmelin, M., Romano, J. C., Chazottes, V., Pari, N., et al. (1995). La bioérosion dans le cycle des carbonates: essais de quantification des processus en Plynésie française. *Bulletin de la Societe Geologique de France, 1*, 85–94.

Pomar, L., & Cuerda, J. (1979). Los depósitos marinos pleistocenos en Mallorca. Acta Geológica Hispánica. *Homenaje a Luís Solé i Sabarís, 14*, 505–513.

Pomar-Bauzà, F. (2016). *Arquitectura i fàcies deposicionals de la interferència entre sedimentació al·luvial, col·luvial i eòlica a les Illes Balears durant el plistocè superior: implicacions paleoclimàtiques*. Ph.D. Unpublished. Department of Geography of University of the Balearic Islands.

Rangheard, Y. (1984). The geological history of Eivissa and Formentera. In H. Kuhbier, J. A. Alcover, & C. Guerau (Eds.), *Biogeography and ecology of the Pityusic Islands* (pp. 25–104). DR. W. Junk Publishers.

Roig-Munar, F. X. (2016). *Blocs de tempesta i tsunami a les costes rocoses de les Illes Balears. Anàlisi geomorfològica i morfomètrica.* Ph.D. unpublished. Departament de Geodinàmica i Geofísica. Universitat de Barcelona. 410 p.

Roig-Munar, F. X., Rodríguez-Perea, A., Martín-Prieto, J. A., Vilaplana, J. M., Gelabert, B., Pons, G. X., & Mir-Gual, M. (2013). Presència de blocs acumulats a terrasses i penya-segats marins a les illes de Menorca i Mallorca: tsunamis o tempestes? VI Jornades de Medi Ambient de les Illes Balears. Ponències i Resums. Societat d'Història Natural de les Illes Balears (pp. 62–64).

Rosselló, V. M. (2004). El litoral y les cales. In J. J. Fornós, A. Obrador, & V. M. Rosselló (Eds.), *Història Natural del Migjorn de Menorca: el medi físic i l'influx humà. Mon. Soc. Hist. Nat. Balears* (Vol. 11, pp. 177–122). Palma, Societat d'Història Natural de Balears, Institut Menorquí d'Estudis.

Sàbat, F. (1986). *Estructura geològica de les Serres de Llevant de Mallorca (Balears).* Ph.D. Universitat de Barcelona. 2 Vol. Unpublished.

Sàbat, F., & Santanach, P. (1984). Tectònica extensiva d'edat juràssica a l'illa de Cabrera (Balears). *Acta Geologica Hispànica, 19*(4), 227–234.

Segura, F., & Pardo, J. E. (2003). Los Barrancos del Migjorn de Menorca. In V. M. Rosselló, J. J. Fornós, & L. Gómez-Pujol (Eds.), *Introducción a la Geografía Física de Menorca. Mon. Soc. Hist. Nat. Balears* (Vol. 10, pp. 71–86). Palma, AGE, Universitat de València, Universitat de les Illes Balears, Societat d'Història Natural de les Balears.

Spencer, T. (1988). Limestone coastal morphology: the biological contribution. *Porgress in Physical Geography, 12,* 66–101.

Spencer, T., & Viles, H. (2002). Bioconstruction, bioerosion and disturbance on tropical coasts: Coral reefs and rocky limestone shores. *Geomorphology, 48,* 23–50.

Stephenson, W. (2015). Rock coasts, in coastal environments and global change. In G. Masselink & R. Gehrels (Eds.). Chichester, UK: Wiley Ltd.

Sunamura, T. (1992). *The geomorphology of rocky coasts* (p. 302). Chichester, UK: Wiley.

Torunski, H. (1979). Biological erosion and its significance for morphogenesis of limestone coasts and for nearshore sedimentation (Northern Adriatic). *Senckenbergiana Maritima, 11,* 193–265.

Trenhaile, A. S. (1987). *The geomorphology of rock coasts.* Oxford: Clarendon Press (Ed.).

Trenhaile, A.S. (2002) Rock coasts, with particular emphasis on shore platforms. *Geomorphology, 48,* 7–22.

Trenhaile, A. S. (2014). Climate change and its impact on rock coasts. In D. M. Kennedy, W. J. Stephenson, & L. A. Naylor (Eds.), *Rock coast geomorphology*. Geological Society.

Vicens, D. (2015). *El registre paleontològic dels dipòsits litorals quaternaris a l'Illa de Mallorca (Illes Balears, Mediterrània occcidental).* Ph.D. Universitat de les Illes Balears, 985 pp.

Vidal, M., Fornós, J. J., Palmer, M., & Pons, G. X. (2001) Quantificació de la capacitat erosiva de *Melaraphe neritoides* (Linnaeus, 1758). In G. X. Pons (Ed.), *III Jornades del Medi Ambient de les Illes Balears*: 78. Palma de Mallorca.

Vidal, M., Fornós, J. J., Gómez-Pujol, L., Palmer, M., Pons, G. X., & Balaguer, P. (2013). Exploring rock coast bioerosion: Rock fragment intestine transit time and erosion rates computation of the gastropod *Monodonta articulata* (Lamarck, 1822). *Journal of Coastal Research, 65*(Special Issue), 1704–1709. ISSN: 1551-5036.

Villanueva, G., Balaguer, P., Baldo, M., Fornós, J. J., Gómez-Pujol, L., & Pons, G. X. (2000). Importancia de la bioerosión en la evolución del litoral rocoso: efecto de Melaraphe neritoides y Monodonta turbinata en el litoral de Mallorca y norte de Cataluña. *Geotemas, 1*(4), 229–302.

Woodroffe, C. D. (2003). *Coasts. Form, Processes and Evolution.* Cambridge: Cambridge University Press.

Chapter 7
Littoral Endokarst from Mallorca Island (Western Mediterranean)

Joan J. Fornós, Àngel Ginés, Joaquín Ginés, Francesc Gràcia and Antoni Merino

7.1 Introduction

Karst, in a wide sense, constitutes a series of complex intermingled processes evidenced by a series of forms and features leading to a distinctive landscape. Among these set of processes and the resulting landforms, karst can be divided into exokarst and endokarst (Bögli 1980). The former takes place at the surface where carbonate rocks are directly exposed to the meteoric processes, or under a soil cover, being characterised by a series of forms as depressions, karren features, etc. The endokarst is related to the underground circulation of water (vadose and phreatic) inside the geologic carbonate formations, where the development of caves and conduits as well as an important series of precipitates (speleothems) take place. However, it must be mentioned that some authors reserve this term to explain the endogenous void generation mechanisms occurring at much greater depths (see a

J. J. Fornós (✉) · À. Ginés · J. Ginés · F. Gràcia · A. Merino
Grup de Ciències de la Terra, Universitat de les Illes Balears (UIB),
Crta. Valldemossa, km 7,5, 07122 Palma (Mallorca), Spain
e-mail: joan.fornos@uib.es

À. Ginés
e-mail: agines.gracia@yahoo.es

J. Ginés
e-mail: jginesgracia@yahoo.es

F. Gràcia
e-mail: xescgracia@yahoo.es

A. Merino
e-mail: tonymerinoj@gmail.com

J. J. Fornós · À. Ginés · J. Ginés · F. Gràcia
Societat Espeleològica Balear (SEB), C. Margarida Xirgú, 16,
07011 Palma (Mallorca), Spain

J. A. Morales (ed.), *The Spanish Coastal Systems*,
https://doi.org/10.1007/978-3-319-93169-2_7

detailed discussion in Klimchouk 2017), which are not the case of the phenomena described in these pages. So, the term endokarst will refer now exclusively to the subterranean forms of limestone landscapes (Bögli 1980), mainly generated within the meteoric domain in the upper stories of the Earth crust.

The content of this paper will focus on the endokarst carved in the coastal fringe. Among the involved processes, the inputs of marine waters in the littoral aquifers generate a broad set of geochemical interactions, that are mainly linked to different hydrological horizons like the water table and the haloclines. In this way, the mixing of freshwater, brackish or even marine waters, is key for the development of dissolution/precipitation karst processes in these littoral areas.

In the case of Mallorca Island, the fact that carbonate rocks are omnipresent together with an appropriate climatology implies that karstic processes are the main responsible factor in the landscape development as well as the littoral shaping. The special geomorphological and hydrologic conditions of this area, along with sea level fluctuations, ecology, or paleoclimate, produce a littoral complex interaction that give rise to a characteristic coastline.

The objective of the present paper deals with the brief description of all related aspects of the endokarstic phenomena that affect the littoral fringe, trying to summarize their effects in the morphologic development of the coastline.

7.2 Geographical and Geological Setting

7.2.1 Physiography of Mallorca Island

Located in the western Mediterranean and forming part of the Balearic archipelago (Fig. 7.1a), Mallorca is the seventh greatest island in extension of the Mediterranean Sea, having a perimeter around 560 km, a surface area of about 3650 km^2 and a maximum height of 1445 m (Ginés et al. 2012a). The maximum height, the Puig Major, is located in the Serra de Tramuntana, a structured mountain range nearly 90 km in length and with a mean width of 15 km, which forms an alignment that goes from the SW to the NE. This alignment mostly composed of Jurassic limestones and dolostones has a slight asymmetry, showing steeper slopes in the north-western side characterising this part of the coast by high and spectacular cliffs. The south-eastern side shows slopes with less inclination giving way to the central area of the island, characterized by a smooth topography and composed of softer Cenozoic marly materials.

The above mentioned central region ends toward the east in the Serres de Llevant range, an alignment of structured Mesozoic deposits with a less pronounced relief, which is surrounded by a post-orogenic reefal carbonate platform built during the Upper Miocene times. This platform characterizes the relief of the eastern and southern coasts causing a continuous vertical cliff 20 m in mean height, only cut by the incision of ephemeral creeks that develop the classical *"cala"* (cove) morphology.

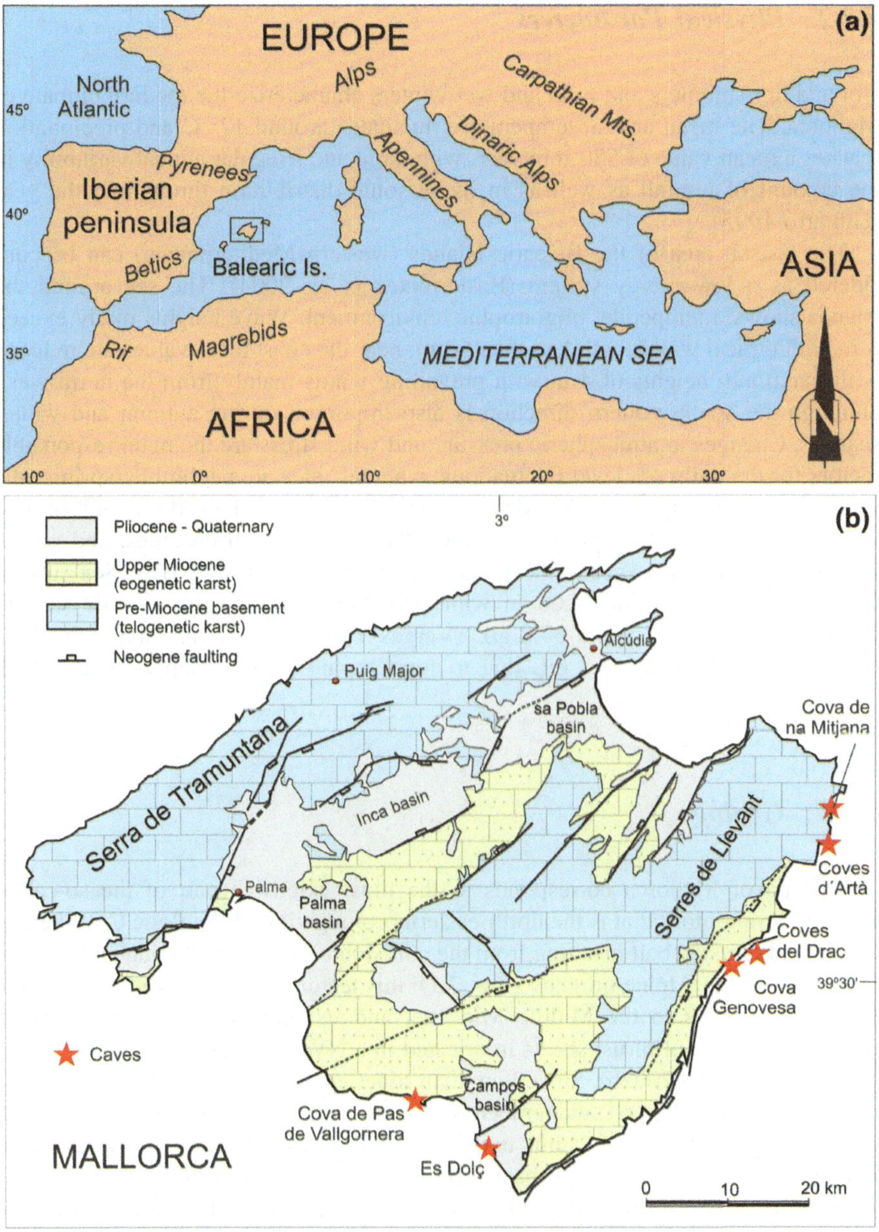

Fig. 7.1 **a** Location of Mallorca Island in the western Mediterranean. **b** Simplified geological map and main karst regions in Mallorca showing the location of the caves cited in the text

7.2.2 Physical Parameters

Hot and dry summers and mild and wet winters characterize the modern climate of Mallorca. The mean annual temperature fluctuates around 17 °C and precipitation reaches a mean value of 580 mm/year, with large and irregular annual variability in the amount of rainfall as well as in its seasonal distribution throughout the year (Guijarro 1995).

The coastal area of the Balearic Islands (western Mediterranean) can be considered as a low-energy system (Basterretxea et al. 2004). The sea around the islands shows a temperate, oligotrophic, environment. Wave heights rarely exceed 8 m with typical wavelength less than 50 m; near the coast these values are reduced with maximum heights of 4 m with prevailing winds mainly from the north-west, although the south-western direction is also important during autumn and winter seasons. Changes in atmospheric pressure and wind stress are the main responsible factors for the daily sea level fluctuations, whereas tides are negligible (spring tidal range of less than 0.25 m). The Mediterranean Sea is a landlocked sea with limited exchange, characterized by high salinities, temperatures and densities, and with a net evaporation that exceeds the precipitation (Tanhua et al. 2013). Sea surface temperature ranges from 14.08 in winter to 24.9 °C in summer, while salinity fluctuates between 37.5 and 37.8 g/L (Vargas-Yáñez et al. 2017) although it can reach values of up to 38.3 g/L due to topographic effects (López-Jurado et al. 2015).

7.2.3 Geology

The island of Mallorca corresponds to the most extensive part of the so-called Balearic Promontory that is the north-eastern prolongation of the Betic Cordillera, a folded and thrusted belt resulting from the collision between the Iberian and African plates during the Alpine orogeny (Fig. 7.1); this tectonic event took place from the Upper Cretaceous to the Middle Miocene, and its main deformation structures consist of a series of thrust sheets imbricated in a NW transport direction (Gelabert 1998; Fornós et al. 2002b). Since Langhian times extensional deformation prevails, causing a series of horst and graben structures clearly evidenced by alternating geomorphological and structural units (Fig. 7.1b), corresponding respectively to mountain ranges and subsident basins (Fornós and Gelabert 1995).

With an approximate stratigraphic sequence of 3000 m in thickness, the geological history of Mallorca evolves continuously from the Carboniferous to the Quaternary, with a gap at the base of the Tertiary (Palaeocene—early Eocene). Although it shows a great variation regarding the sedimentary environments, ranging from lacustrine to deep pelagic facies, the carbonate lithology is predominant. Leaving aside the small Carboniferous outcrop of pelites with weak metamorphism at the foot of the Serra de Tramuntana, the Mesozoic deposits (more than

1500 m thick) show a basal sequence characterized by the classical German Triassic. This, in turn, exhibits continental, shallow marine and evaporitic facies, with some volcanism, that evolves to a shallow marine deposition during the Lower Jurassic, constituting a massive micritic limestone slab up to 400 m in thickness giving rise to the main reliefs of the mountain ranges. Later, the sequence changes to deep pelagic sedimentation characterized by marls and marly limestones that embraces the rest of the Mesozoic. After the emersion of the area, sedimentation resumes in the upper Eocene with littoral and lacustrine deposits until the deposition of turbidite facies of the middle Miocene, that are coetaneous with the main alpine orogenic phase in the area. Post-orogenic sedimentation starts with the development of a carbonate platform in a tropical environment resulting in a reefal deposition characterized by limestone and calcarenite deposits. The Pliocene and Quaternary facies, also with a calcarenitic and calcisiltitic lithology, fill the main basins while the depressed areas, placed at the foot of the ranges, accumulate their denudation materials. On top of the sequence, a Middle Pleistocene aeolian system unconformably overlies both the Upper Miocene and the Pliocene sequences (Nielsen et al. 2004). In littoral areas, recent Pleistocene sediments consist of aeolian calcarenite and beach-related deposits, whereas alluvial fan deposition takes place along the foothills of the mountain ridges. The whole Cenozoic sequence generally exceeds 1500 m in thickness (Gibbons and Moreno 2002).

Special attention should be drawn to the evolution of the Upper Miocene to Quaternary sequences, that constitute the post-orogenic platform building up the eastern and southern coasts of Mallorca. In this lithological context, the littoral endokarst is conspicuous and its morphogenetic evolution is controlled by the different carbonate facies present (Ginés et al. 2014). Sea-cliffs of the eastern and southern coasts display nice sets of the following lithostratigraphic sequences. The base of the Upper Miocene consists of marly limestones corresponding to an open platform deposition. Above these deposits, limestones belonging to a Reef complex sedimentation, crop out all along the decameter scale height of littoral cliffs. Its architecture shows complex accretional geometric relationships related to oscillations in relative sea level (Pomar 1991). These lithofacies present an elevated primary porosity which brings about a high permeability; these levels grade to marly limestones slightly dolomitized showing a conspicuous horizontal lamination (Pomar and Ward 1994). Over an important erosion surface, the Terminal Complex (Esteban 1979), a restricted coastal environment deposition with oolitic sand shoals and stromatolites (Pomar et al. 1996), give way to the major erosion surface corresponding to the Uper Messinian low-stand and the subsequent Pliocene transgression.

7.2.4 *The Mallorcan Karst*

Karstic features are present over more than two thirds of the island favoured by the omnipresence of carbonate rocks, both limestone and dolomites as well as calcarenite and marly limestones deposits. These rocks have been subjected to long-term karstification processes generating a great variety of landforms as well as caves and conduits that may contain a wealth of speleothems and sediments, related to their speleogenetic evolution. Traditionally, three different areas have been identified linked to the distribution of geologic and climatic factors (Ginés and Ginés 1989), two rather parallel mountain ranges (Serra de Tramuntana and Serres de Llevant) and the Migjorn coastal plain (Fig. 7.1b).

In Serra de Tramuntana, the main karstic region sculptured principally over Jurassic and early Miocene limestones (telogenetic karst), karren fields show the most impressive features, although karst depressions and karstic canyons are also present. Subterranean cavities, mostly vertical shafts, represent more than 70% of the caves present in this karst region (Ginés and Ginés 1987). In Serres de Llevant the predominance of marls and marly limestones restricts the evolution of exo-karstic landforms; exhumed subsoil karren features or incipient superficial minor sculptures are mainly present. Vadose shafts with vertical pits are the more common endokarst features, including some interesting littoral caves as well, as the touristic Coves d'Artà that outstands for its great dimensions and impressive speleothem decoration. Caves are relatively scarce in this karst area and only represent around 5% of total caves in the island.

In the case of the Migjorn area, the great permeability of the calcarenites appertaining to the Upper Miocene, as well as their relationship with the coastal processes, allow an important development of horizontal caves being specially characterized by their richness in speleothem accumulations (i.e. Coves del Drac). Instead, exokarst is limited to some subsoil karren features or scant karstic depressions of medium to large size. Most of these caves are carved in diagenetically immature calcarenites (eogenetic karst), and can be considered as caves developed in the littoral fringe (Ginés and Ginés 2011) constituting a relevant example of coastal endokarst. The general morphology of these caves consists of a succession of chambers that have been shaped by collapse processes favoured by the fluctuation of the Pleistocene sea level. In fact, the sea level history can be inferred from the deposition of phreatic overgrowths on speleothems, occurring in the Mallorcan coastal caves which are partially drowned by brackish waters (Tuccimei et al. 2006, 2010; Ginés et al. 2012b). The interferences between the endokarst and the coastline evolution are also controlled by the Quaternary sea level history, producing a wide range of morpho-sedimentary effects, as for example, the presence of marine and/or aeolian sedimentary infillings into some littoral caves (Ginés et al. 2011). The involvement of hypogene speleogenetic processes recently documented adds new insights to the development of this subterranean world (Fornós et al. 2011; Ginés et al. 2017).

7.3 Coastal Karst Speleogenesis

7.3.1 Development and Geochemical Characteristics

One of the most outstanding processes in the coastal karst speleogenesis, apart from the common karst dissolution potential from carbonic acid derived from atmospheric and soil CO_2, corresponds to the hydrology and geochemistry related to the complex interaction between freshwater and marine saltwater (Fratesi 2013). This occurrence, as well as the one that mixes phreatic with vadose waters, produces some chemical effects influencing the carbonate solubility, usually inducing undersaturation of the mixed waters depending on the chemistry of the solutions and their proportions (Plummer 1975; Palmer 1991). In that sense, many variables as ion pairs, ionic strength, presence of different acids or organic matter, can act provoking different chemical effects. This solutional potential is remarkable at the contact between the freshwater lens and the marine water, especially if this contact is sharp (halocline). Needless to say, that variability of partial pressure of CO_2 or temperature, among others, can have a direct effect on the dissolution processes of the different carbonate minerals (Back et al. 1986; Fratesi 2013).

In that geochemical context, a general model has been proposed for the development of coastal caves: the *flank margin caves* of Mylroie and Carew (1990). The Carbonate Island Karst Model (CIKM) of Mylroie and Mylroie (2007) was an attempt to characterize the complexity, both geochemically and geomorphological, that accounts for carbonate coasts in insular settings. This model tries to integrate not only the geochemical processes, but also sea level changes and their relationship with the carbonate substrate (eogenetic vs. telogenetic) or the presence of non-carbonate rocks and their relationship with sea level (Mylroie 2013). Some recent contributions (Ginés and Ginés 2007; Ginés et al. 2014) have proposed the need of broadening and generalising this model as a Carbonate Coast Karst Model.

7.3.2 The Mallorcan Eogenetic Karst

The coastal karst developed in the eastern and southern areas of Mallorca can be considered as an eogenetic karst in the sense of Vacher and Mylroie (2002). It corresponds to a pervasive highly porous medium formed by relatively young shallow-water carbonates, that constitutes the lithological substrate present in these zones of the island, having undergone a meteoric diagenesis concurrently with the process of karstification. Caves located in this karst region are carved in calcarenites from Upper Miocene (reef and shallow carbonate platform deposits) and Pleistocene (eolianites). They correspond to a post-orogenic belt which fringes the reliefs built up, during the alpine orogeny, by the folded and thrusted Mezosoic carbonate deposits.

From a hydrological point of view the young calcarenite deposits, that show a great degree of porosity (eogenetic rocks), can be considered as an unconfined coastal aquifer, although the great variability that show the carbonate sedimentary facies present, give way to important lateral and vertical variations in permeability. This hydrological parameter is enhanced by the preservation of large-scale reefal porosity during the early diagenesis stages. This fact is clearly visible in the pattern, morphology and development of voids and conduits (Ginés et al. 2014) and favours the generation of caves showing extensive collapses.

The first speleogenetic explanations about the littoral endokarst in Mallorca were formulated by Martel (1896) in connection with the exploration of Coves del Drac, a world-wide famous touristic cave. In his opinion that littoral cave could be considered as developed under marine erosion processes. Later authors, based their explanations on conventional karst mechanisms, although influenced by the proximity of the coastline and invoking an important role played by underground water flows (among others, Faura y Sans 1926). Modern knowledge and cave exploration stimulate a critical discussion on speleogenesis (Ginés and Ginés 1989, 1992) according to the models developed to explain caves of the Caribbean area, with some special insights as the relevant role played by the breakdown collapses (Ginés and Ginés 2007; Gràcia et al. 2006; Ginés et al. 2013). A better understanding of endokarst evolution rises after the progress of subaqueous exploration of these littoral caves, that has permitted to outline a more accurate and global vision of the cave development, morphological features, as well as the chronological framework that embraces the endokarst evolution (Gràcia 2015).

According to the present state of knowledge (Ginés et al. 2013, and references included therein), we can consider that the speleogenetical processes of the Mallorca littoral endokarst start with the creation of the initial voids and passages in phreatic conditions (Fig. 7.2a), being related to the freshwater-seawater mixing zone proposed by Mylroie and Carew (1990) for the flank margin caves model. Fractures and joints, although present everywhere but without any evident visibility, are not important in its development. Only locally, can they play an important role especially regarding the geomorphological features of the coastline, or the plan development of some caves (Ginés et al. 2014).

The removal of rock in those high porosity aquifers, seems not only be related to the general drainage of the island but with the periodic tidal flushing that provokes significant movement of large masses of groundwater near de coast (Ginés and Ginés 2007).

Collapse chambers that are conspicuous in the Mallorcan littoral endokarst (Ginés 2000; Ginés et al. 2013), are closely related, in all probability, apart from the double type of porosity both depositional and karstic, to recurrent rising and falling of the sea level and consequently of the water table during the Quaternary times. In this regard, the continuous development and evolution of the karstic voids could explain the upward growth of the breakdown domes favoured by the mechanical instability and local stress caused by joint presence (Fig. 7.2b, c). Coalescence of

Fig. 7.2 Subterranean landscapes in the coastal karst of Mallorca Island, developed in young carbonates (Upper Miocene to Pleistocene). **a** Phreatic solutional conduit in Cova des Pas de Vallgornera (*Photo* A. Cirer). **b** Breakdown processes affecting the underwater passages of Es Dolç (*Photo* A. Cirer). **c** The vadose evolution of underground passages includes extensive collapse of the ceilings as well as spectacular speleothem deposits; the presence of large brackish pools corresponds to the present-day sea level (*Photo* A. Merino)

the collapsed zones would promote connectivity between chambers, thus enhancing the water flow and finally reaching the topographical surface giving way to external cave entrances. This fact could favour the presence inside the cave chambers and conduits of a series of entrance facies, as aeolian sands or fine-grained material coming from the surficial runoff (Fornós et al. 2009), in addition to the accumulation of rock blocs and boulders due to the roof collapse, mostly favoured during the Quaternary sea level low-stands.

From a general point of view the Mallorcan endokarst can be viewed as a series of large collapse chambers that evidence the significant role that breakdown processes play in the development of the littoral eogenetic endokarst (Fig. 7.3). The vadose flow is not necessary to explain the speleogenetic evolution of these caves (Ginés and Ginés 1992; Ginés et al. 2013). Mixing solution and sea level fluctuation at different scales (tidal and/or eustatic movements) can explain the cave patterns that show the coastal karst caves of the island.

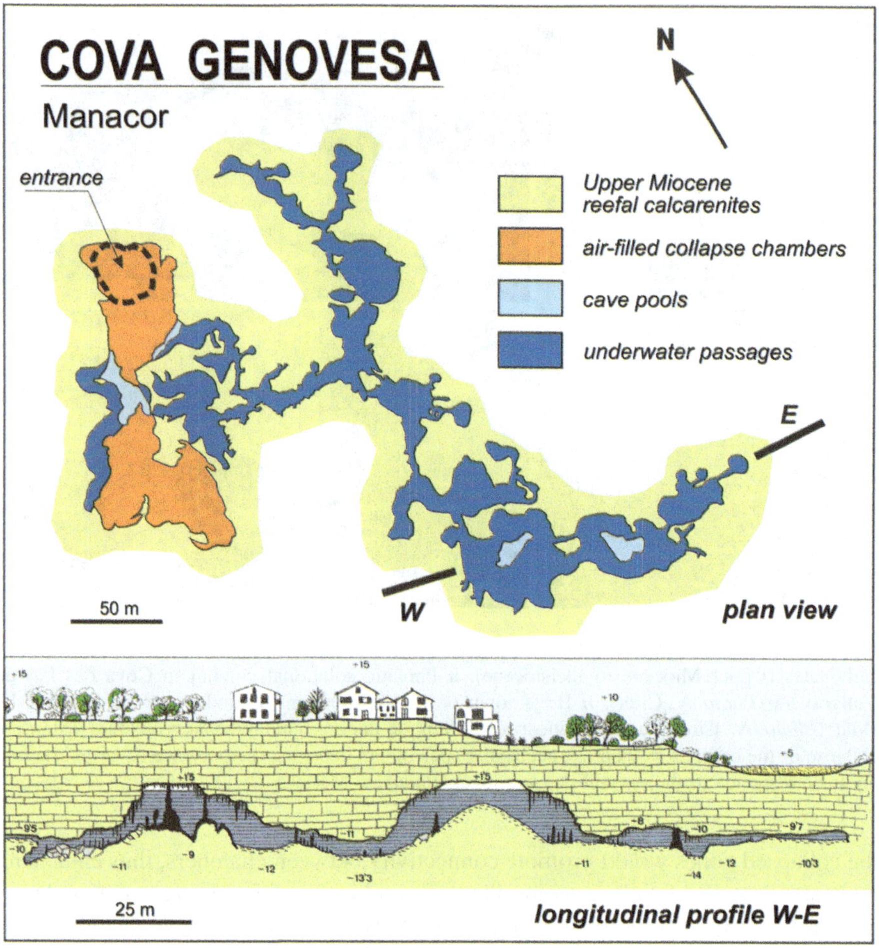

Fig. 7.3 Simplified topographical survey of Cova Genovesa. A longitudinal profile of some submerged passages is included; note the collapse chambers represented in the profile and the presence of speleothems deposited during low sea-stands

7.3.3 *Hypogenic Influences*

Recent research has evidenced the presence of solutional features and deposits related to a hypogene basal recharge occurring in littoral karst caves of Mallorca (Merino and Fornós 2010; Fornós et al. 2011; Ginés et al. 2017), that can be related to the hydrological studies that reported a series of low-grade geothermal anomalies (López 2007). These hypogene phenomena are associated to an extensional fault system showing a SW–NE trend that affected the post-orogenic deposits (López and Mateos 2006) probably since the Middle Pliocene.

In this context, there are many evidences pointing toward the involvement of hypogene speleogenesis in the formation of the Mallorcan coastal caves. Several authors have described the presence of hypogene features as well as some specific cave deposits (Fornós et al. 2011; Ginés et al. 2017); they mainly correspond to solutional rising forms and blackish crusts and fine sediments. Among the more habitual features it is possible to distinguish small ascending channels with a great variability in morphology and dimension (Ginés et al. 2009; Merino and Fornós 2010). Besides these rising wall channels, feeding points in the floor of the passages as well as channels and outlets in their ceilings characterise this hypogene morphological suite, being also possible to observe other morphologies such as dead ends, partitions, and solutional ascending hollows or alveoli.

The presence of ascending vents, located at the floor of the galleries, is also very common, showing some related speleothems, such as crusts and cave rims of uncommon mineral composition in the Mallorcan karst (huntite, strontianite, nordstrandite, celestine, or barite, among others). In addition, the accumulation of fine grained black sediments, enriched in Mn and Fe oxides and/or hydroxides and alternating with silty material and reddish clays, is especially abundant. Although gypsum is scarcely present, there are no proofs about sulfuric acid speleogenesis (SAS).

Nowadays it lacks any clear evidence supporting that hypogene processes are currently active. However, there is a big amount of data suggesting that hypogene processes have been involved in the morphogenesis of some coastal caves, at least producing some relevant hypogene imprints (Ginés et al. 2017).

While explaining the speleogenetic mechanisms that have formed the coastal endokarst of Mallorca, we must take into consideration a series of additional mechanisms, that have left their imprints in modelling the coastal karst acting in quite different degrees; these mechanisms show a great complexity due to their interaction, both in time and in space. Apart from the classical meteoric recharge with very variable CO_2 inputs, together with the more relevant and generalised coastal mixing processes, we must also consider a deep recharge rising and mixing with the littoral groundwater.

7.4 Littoral Features (Rocky Coast Expressions of Endokarst Features)

7.4.1 Karstic-Marine Caves and Calas

The morphological expression of endokarstic processes in the Mallorcan rocky coast is conspicuous, specially where the calcarenite rocks appertaining to the Upper Miocene crop out along the littoral cliffs of the island (Fig. 7.4).

Generally speaking, it corresponds to a very articulated coast characterised by a series of entrances and projections, with a great range of dimensions, that are related

(a)
(b)
(c)

◂**Fig. 7.4** Marine erosion features along the eastern rocky coasts of Mallorca Island. **a** Sea cave and coastal arch in the Mondragó natural park area, formed exploiting paleokarst features existing in the Upper Miocene carbonates. **b** The wide entrance in the center of the picture gives access to Cova des Coloms, a karst cave captured by means of sea erosion and coastal cliff retreat. **c** Cova Marina des Pont is a karstic collapse chamber almost totally dismantled by marine erosion processes; the height of the arch is 19 m (*Photos* J. Ginés)

to the complex interaction between marine dynamics, endokarst and fluvial processes. Frequently, coastline indentations are an outcome linked to the presence of littoral caves, explaining the notable coastal complexity if considered at a mesoscale level. In this regard, we can encounter several meso- to macroscale features such as the karstic-marine caves (Montoriol-Pous 1971) that result, on the one hand, from the interaction between marine dynamics and the evolution of the endokarst. And on the other, it is worth to mention the presence of *calas* (Gómez-Pujol et al. 2013) due to the interaction of fluvial processes, marine dynamics and sea-level oscillation during the Quaternary times, with some imprints related to the karst processes acting in the littoral fringe.

The structural control caused by the presence of joints and fractures must be taken into account, not only related to the endokarst development along master passages, but also representing weakness points where marine dynamics can act in preference. In that sense, the density and spaced out of fractures, as well as their preferential direction, play a fundamental role in the coastline configuration, together with other factors such as lithology and the characteristics of the surficial drainage network.

Needless to say, that the progressive evolution of the endokarst features along with the collapse of the resulting cavities can boost marine erosive action (cave captures), modifying the topographical treats by disrupting the rectilinear character caused by the structural control of the sea cliffs. This process can be extended to the macroscale when coalescence of collapses occur as is the case of some coastal bights. The so-called *calas* (Fig. 7.5), are phenomena present in the carbonate plateaus of the Mediterranean area, corresponding to embayed (flooded) rocky coastline landforms related to steep-sided drowned valleys that were deeply incised during low sea level stages (Gómez-Pujol et al. 2013). Their formation is related to the absence of structural trends parallel to the coast, and their development depends also on the catchment size of the fluvial drainage basin, lithology (presence of impervious rocks), degree of stream incision at its mouth, and the possible existence of valley infilling associated with Quaternary sea-level oscillations.

Taking into account all of these variables, seems clear the decisive dependence of the morphological trend of the coastline on fracture densities and spacing, as well as on the development of karstic phenomena (caves, surficial collapses, etc.) and their capture by the marine erosion dynamics.

Fig. 7.5 Indentations of the coastline, locally known with the term of *calas*, result from the flooding of a non-functional surficial drainage network together with the frequent capture of caves, and karst landforms in general, by marine erosion processes

7.4.2 *Evolutionary Model*

The presence of dissolution cavities showing a parallel trend with the coastline, together with the existence of karstic conduits under the present-day fluvial network associated to the *calas* landforms (Gràcia et al. 2000), have permitted to suggest an evolutionary model for the Mallorcan carbonate littoral.

Preferential dissolution related to the underwater mixing zone (Smart and Whitaker 1991) would create a series of voids, mainly developed at the haloclines of the coastal aquifers owing to marine water intrusion (Whitaker and Smart 1990; Mylroie and Carew 1990). Apart from this fact, the occurrence of fractures and joints transversal to the coastline would favour the dissolution by preferential flow through the enhanced permeability paths that represent those structural discontinuities (Back et al. 1984). This greater discharge of phreatic water, and therefore the intersection of structural discontinuities with the intrusion wedge of sea water, would boost the generation of karstic porosity that eventually could be the subject of marine erosion processes leading to the capture of exo- and endokarst morphologies created along the littoral fringe. Obviously, other discontinuities as the penetrative stratification (alternating calcarenites with calcisiltites) or the depositional slopes due to the progradation of the Miocene platform, can also act as preferential dissolution horizons during sea level stabilization periods.

All these coastal features are finally conditioned by glacioeustatic oscillations of sea level during Quaternary times. The glacioeustic variability of this period, would result in the superposition of several levels of dissolution that would favour the formation of extensive littoral caves, as well as exokarstic landforms that eventually can collapse, mainly during the moments of low sea-stands due to the loss of hydraulic support.

The final morphology of the coast would then be the result of a complex scenario including, apart from the geochemical processes that result from the underwater interference between meteoric waters and marine intrusion, the involvement of fluvial processes, the structural tectonic conditionings, the variability of lithological factors and, evidently, the sea dynamics.

7.5 Littoral Endokarst and Sea Level History

7.5.1 Sea Level Controlled Phreatic Speleothems

An outstanding aspect of the littoral caves of Mallorca is the presence of phreatic crystallizations (Ginés and Ginés 1974; Pomar et al. 1979) commonly found in the subterranean brackish pools corresponding to the current sea level (Tuccimei et al. 2010; Ginés et al. 2012b). The relationship with sea level oscillations becomes evident as the water surface of these pools (the coastal water table) fluctuates daily related to minor tidal and/or barometric variations, as it does the western Mediterranean current sea level (Boop et al. 2017).

Crystal morphology of these encrustations is very variable (acicular, bladed, and prismatic) forming diverse types of aggregates, that range from millimetre to centimetre in size, that are organized in different fabrics resulting in an enormous morphological variety for this kind of speleothems (Ginés et al. 2005, 2012b). Their mineralogy corresponds to calcite and/or aragonite, being the calcite the dominant phase mostly represented by high-Mg calcite. Although aragonite precipitation was previously attributed to warmer periods (Pomar et al. 1976), isotopic analysis seems to indicate the role of the meteoric recharge in the precipitation of the different calcium carbonate phases (Vesica et al. 2000). Calcite would precipitate associated to higher meteoric recharge, while during conditions of lower rainfall aragonite would be the phase present (Csoma et al. 2006). Other factors as Mg and Sr content in water, pCO_2, salinity, Mg/Ca ratios and evaporation, among others, can also be involved in the precipitation of different phases (Hill and Forti 1997).

Encrustations, that commonly develop over a vadose speleothem support or along the cave walls (POS: Phreatic Overgrowths on Speleothems), form spectacular bulky coatings that show their maximum thickness corresponding to the mean position of the underground water table, thus indicating the mean sea level at

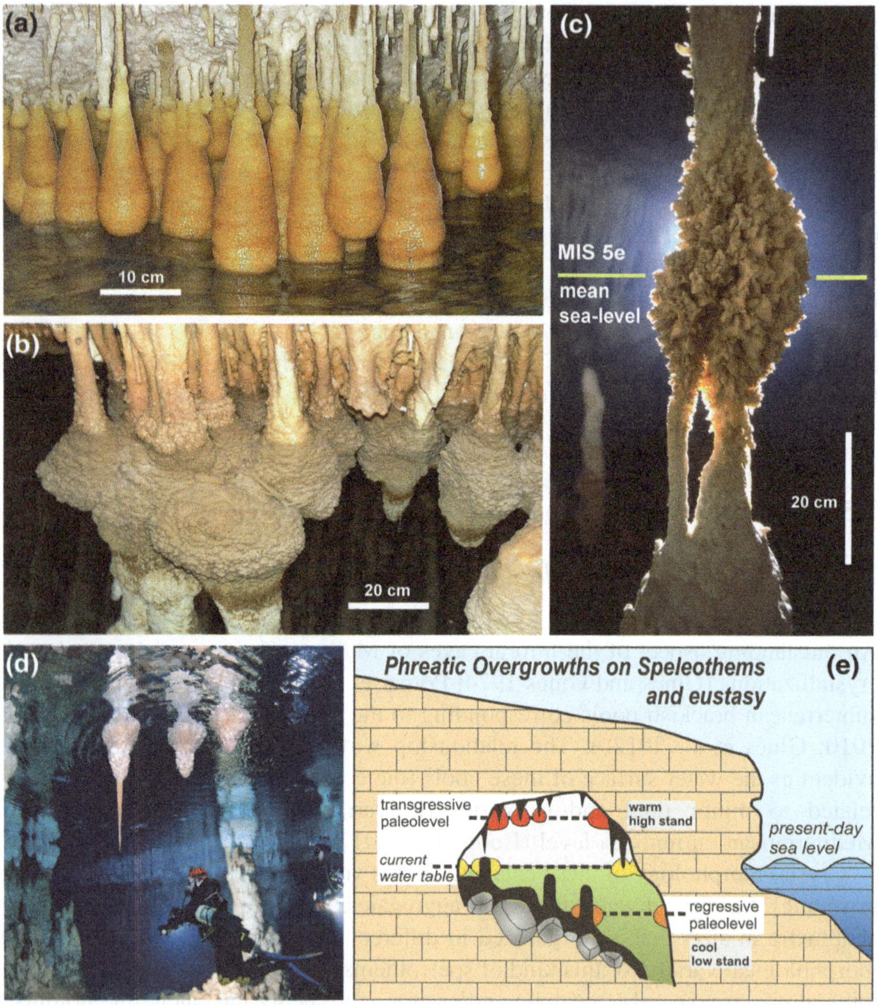

Fig. 7.6 Phreatic overgrowths on speleothems (POS) that act as a register of sea level history. **a** Aragonite encrustations in Cova des Pas de Vallgornera corresponding to present-day sea level (*Photo* A. Merino). **b** Calcite overgrowths in Cova A de Cala Varques, developed in the today's fluctuation range of the brackish pools (*Photo* B. P. Onac). **c** Phreatic overgrowths paleolevel deposited 2.6 m above the current sea level during the Last Interglacial (*Photo* A. Merino). **d** Late Holocene POS observable in the underwater sections of Coves del Drac (*Photo* A. Cirer). **e** Conceptual sketch of a karst littoral cave of Mallorca hosting present-day and ancient POS deposits; broken lines represent the mean elevation of the ground water table during each recorded sea-stand

the moment of their precipitation (Fig. 7.6). In that sense, these phreatically encrusted speleothems can be used as ideal sea level indicators in terms of both age and elevation (Ginés et al. 2012b; Onac et al. 2012).

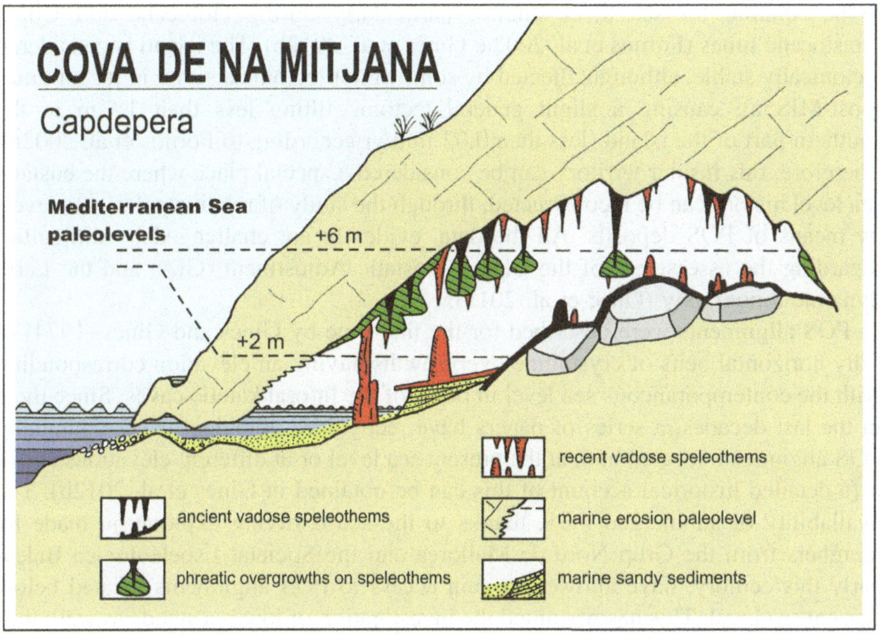

Fig. 7.7 Schematic representation of a coastal cave in Jurassic limestones, located at the north-eastern area of Mallorca, showing conspicuous morphologies and infillings related to Quaternary sea level oscillations. The paleolevel of phreatic overgrowths on speleothems (+6 m asl) may correspond to some Middle Pleistocene sea high-stand, whereas the marine capture of the cave presumably occurred during the Last Interglacial (marine erosion paleolevel at +2 m asl)

Repeated flooding of coastal caves, controlled by glacioeustatic sea level oscillations during the Plio-quaternary, have produced the encrustation of these strictly horizontal bands at different elevations both above and below the present-day sea level. The observed horizontal alignments at those different levels in the caves are a record and a consequence of ancient lower or higher sea-stands, that once dated will bring about important palaeoclimatic information and very precise sea level elevations. These data can be correlated with other features along the cliffy coasts where the Pleistocene sea level fluctuations left marks as marine abrasion notches, marine boring organisms found in the entrances of some coastal caves, or patches of marine and aeolian fossiliferous sands (Fig. 7.7).

7.5.2 Dating Pliocene to Holocene Sea Level Oscillations

POS encrustations represent accurate proxies of ancient sea level stands in the western Mediterranean basin. In this respect Mallorca, and specially its eastern littoral part where the endokarst is conspicuous all along the coastline, has proven to be suitable for sea level studies, particularly during Holocene and Upper Pleistocene times (Fornós et al. 2002a; Ginés et al. 2012b). The island is considered tectonically stable, although affected by some minor tectonic activity in recent times (post-MIS5a), causing a slight general tectonic tilting less than 1.5 m in the southern part of the island (less than 0.02 mm/yr according to Fornós et al. 2002a). Therefore, this insular territory can be considered a special place where the eustatic sea level history can be reconstructed, through the study of relative paleo- sea levels by means of POS deposits. All the data, evidently are challenged by difficulties regarding the assessment of the Glacial Isostatic Adjustment (GIA) and the Earth dynamic topography (Onac et al. 2017).

POS alignments were described for the first time by Ginés and Ginés (1974) as fully horizontal belts of crystalline overgrowths having an elevation corresponding with the contemporaneous sea level in pools of the littoral karstic caves. Since then, in the last decades, a series of papers have recognized abundant caves containing POS alignments located both at the present sea level or at different elevations above it (a detailed historical account of this can be obtained in Ginés et al. 2012b). The availability of submerged POS, thanks to the scuba diving exploration made by members from the Grup Nord de Mallorca and the Societat Espeleològica Balear early this century, have allowed to gain access to POS alignments located below present sea level. This has permitted the investigation of low sea stands as well, thus completing a precise sea level curve for the Balearics area.

Until now, up to 30 POS alignments have been reported at different heights ranging from +40 m apsl to −23 m bpsl. U/Th, U/Pb and ^{14}C datings give ages ranging from Pliocene times (4.07 Ma, Onac et al. 2017) to the present (Tuccimei et al. 2010). The most well represented period is the MIS5 stage that corresponds to the Last Interglacial (Tuccimei et al. 2006; Dorale et al. 2010).

The chronological information obtained have permitted, not only to determine the patterns characteristics of the eustatic curve, but the rates of the sea level change based on U-series ages performed on the POS alignments. In that sense, it has been specially interesting the data obtained during the sea level fluctuations that occurred during the Last Interglacial period, where intervals of stable sea stand (up to 10 ka long) alternate with rapid positive and negative sea level changes greater than 18 m in amplitude (Tuccimei et al. 2010). This has conducted to the deduction of sea level change rates that can fluctuate between a minimum value of 2.9 mm/yr and a maximum of 20.0 mm/yr. The average value corresponds to 5.9 mm/yr (Tuccimei et al. 2006; Dorale, et al. 2010). Similar values (3.5–6.0 mm/yr) have been obtained by Harmon (1985) in Bermuda Islands.

The understanding of the sea level behaviour during past high sea stands, when climate was warmer than today, can help to refine the predictions of the magnitude of sea level rise in the coming times due to anthropogenic causes. This is a crucial focus point in the framework of the observed climatic change, and it is about to become a great societal problem for the next centuries.

7.6 Threats, Management and Protection

The coastal endokarst of Mallorca is threatened in several ways mainly related to the high urbanization and touristic development indexes, occurring in our limited insular territory from the fifties of the past century. Besides the relatively important extension of the island (3,650 km^2) it hosts a population exceeding today 850,000 inhabitants, together with more than 13 millions/year of visitors attracted currently by the existing touristic offer. This very high human pressure is clearly concentrated toward the littoral areas, due to the summer-time touristic model of Mallorca, focused basically to sun and beach holidays.

The main hazards are directly linked to urban development, within the two-fold scenario of intensive building development along the coastal fringe of the island, as well as the menaces affecting its coastal aquifers. From the hydrogeological point of view, the overexploitation of the underground resources is a generalized phenomenon, which produce serious problems of salt water intrusion in the littoral karstic aquifers. Another aspect of the hydrogeological issues endangering the endokarst is the sewage contamination linked to extensive disseminated housing in the coastal areas. The sewage collection infrastructures, when existing, are insufficient during the summer months; furthermore, the presence of cesspits, out of regulation, produces abundant cases of fecal contamination in the coastal aquifers. These circumstances represent an important threat to the habitats of the rich crustacean stygofauna populating the brackish pools of the Mallorcan coastal caves (Gràcia and Jaume 2011).

A quite different menace to the littoral caves consists in the excessive frequentation of some cavities, as a consequence linked to the practice of the so-called "adventure sports", specially the cave-diving activities which have dramatically grown the last decades. The visits to coastal caves are frequently developed by local people, as well as by increasing amounts of tourists that are recruited and guided by small companies devoted to cave-diving, caving, canyoning, and other activities performed in natural environments.

The management and protection of the littoral endokarst is a matter not well-solved in Mallorca, nor in the Balearic Islands in general. The municipal and regional authorities have abundant regulations in order to minimize the impact of urban and housing development in the underground water resources, as well as to avoid possible problems produced by the collapse of subjacent cavities; in both cases, the regulations are not sufficiently strict being usually applied with excessive tolerance.

The caves as a general subject are not protected by the public administration, with the exception of those sites containing archaeological remains. The only protection status that applies to some Mallorcan caves is the existence of a list of 30 localities that were proposed, and registered in 2006, as *Sites of Community Importance*, within the European *Natura 2000 Network* (Jaume et al. 2001). Among these protected sites, up to 14 significant coastal caves were included in the list, that are currently under the protection of the "Conselleria de Medi Ambient, Agricultura i Pesca" (the Regional Environmental Authority) as *Special Areas of Conservation* according to the Habitats Directive. As an example, the most outstanding littoral cave in the island "Cova des Pas de Vallgornera, with a development exceeding 78,000 m" is today strictly managed by the Regional Environmental Authority, being only accessible for exploration and investigation purposes. Nevertheless, the management and protection of the coastal karst areas of Mallorca, generally speaking, are far from reaching adequate and effective conservation measures.

Acknowledgements The research of this work was supported by grants from the Spanish Government, corresponding to different MINECO projects: CGL2013-48441-P and CGL2016-79246-P (AEI-FEDER, EU). Thanks are due to the team of cave divers from Grup Nord de Mallorca (Pollença, Mallorca) and Societat Espeleològica Balear for their tasks related to underwater sampling, and particularly to Bernat Clamor and Antoni Cirer.

References

Back, W., Hanshaw, B. B., & Van Driel, J. N. (1984). Role of groundwater in shaping the Eastern Coastline of the Yucatan Peninsula, Mexico. In R. G. LaFleur (Ed.), *Groundwater as a Geomorphic Agent* (Vol. 13, pp. 281–293). Allen and Unwin, Inc. London. The Binghamton Symposia in Geomorphology: International Series.

Back, W., Hanshaw, B. B., Herman, J. S., & Van Driel, J. N. (1986). Differential dissolution of a Pleistocene reef in the ground-water mixing zone of coastal Yucatán, México. *Geology, 14,* 137–140.

Basterretxea, G., Orfila, A., Jordi, A., Casas, B., Lynett, P., Liu, P. L.-F., et al. (2004). Seasonal dynamics of a microtidal pocket beach with *Posidonia oceanica* seabeds (Mallorca, Spain). *Journal of Coastal Research, 20,* 1155–1164.

Bögli, A. (1980). *Karst hydrology and physical speleology* (284 pp). Berlin: Springer-Verlag.

Boop, L., Wynn, J. G., Thompson, G., Fornós, J. J., & Onac, B. P. (2017). Interactions between surface conditions, the Mediterranean Sea, and cave climate within two littoral caves in Mallorca: Implications for the formation of phreatic overgrowths on spleothems. *Journal of Cave and Karst Studies, 79*(1), 59–72.

Csoma, A. E., Goldstein, R. H., & Pomar, L. (2006). Pleistocene speleothems of Mallorca: Implications for paleoclimate and carbonate diagenesis in mixing zones. *Sedimentology, 53,* 213–236.

Dorale, J. A., Onac, B. P., Fornós, J. J., Ginés, J., Ginés, A., Tuccimei, P., et al. (2010). Sea- level 81,000 years ago in Mallorca. *Science, 327,* 860–863.

Esteban, M. (1979). Significance of the upper Miocene reefs of the Western Mediterranean. *Palaeog, Palaeoeco, Palaeoclim, 29,* 169–188.

Faura y Sans, M. (1926). *Cuevas de Mallorca* (78 pp). S.A, Madrid: XIV Cong. Geol. Intern., Inst. Geol. España, Gráficas Reunidas.

Fornós, J. J., & Gelabert, B. (1995). Litologia i tectònica del carst de Mallorca/ Lithology and tectonics of the Majorcan karst. In A. Ginés & J. Ginés (Eds.), *El carst i les coves de Mallorca/ Karst and caves in Mallorca* (Vol. 3, pp. 27–43). Palma de Mallorca: Endins, 20/Mon. Soc. Hist. Nat. Balears.

Fornós, J. J., Gelabert, B., Ginés, A., Ginés, J., Tuccimei, P., & Vesica, P. L. (2002a). Phreatic overgrowths on speleothems: A useful tool in structural geology in littoral karstic landscapes. The example of eastern Mallorca (Balearic Islands). *Geodinamica Acta, 15,* 113–125.

Fornós, J. J., Ginés, J., & Gràcia, F. (2009). Present-day sedimentary facies in the coastal karst caves of Mallorca Island (Western Mediterranean). *Journal of Cave and Karst Studies, 71*(1), 86–99.

Fornós, J. J., Merino, A., Ginés, J., Ginés, A., & Gràcia, F. (2011). Solutional features and cave deposits related to hypogene speleogenetic processes in a littoral cave of Mallorca Island (western Mediterranean). *Carbonates and Evaporites, 26,* 69–81.

Fornós, J. J., Pomar, L., & Ramos-Guerrero, E. (2002b). Balearic Islands, Tertiary. In W. Gibbons & T. Moreno (Eds.), *The geology of Spain* (pp. 327–334, 405–406). London: The Geological Society.

Fratesi, B. (2013). Hydrology and geochemistry of the freshwaters lens in coastal karst. In M. J. Lace & J. E. Mylroie (Eds.), *Coastal karst landforms* (Vol. 5, pp. 59–75). Springer, Coastal Research Library.

Gelabert, B. (1998). *La estructura geológica de la mitad occidental de la isla de Mallorca* (129 pp). Madrid: Memorias del Instituto Tecnológico Geominero de España.

Ginés, A., & Ginés, J. (1974). Consideraciones sobre los mecanismos de fosilización de la Cova de sa Bassa Blanca y su paralelismo con formaciones marinas del Cuaternario. *Bolletí de la Societat d'Història Natural de les Balears, 19,* 11–28.

Ginés, A., & Ginés, J. (1987). Características espeleológicas del karst de Mallorca. *Endins, 13,* 3–19.

Ginés, A., & Ginés, J. (1992). Las Coves del Drac (Manacor, Mallorca) Apuntes históricos y espeleogenéticos. *Endins, 17–18,* 5–20.

Ginés, A., & Ginés, J. (2007). Eogenetic karst, glacioeustatic cave pools and anchialine environments on Mallorca Island: A discussion of coastal speleogenesis. *International Journal of Speleology, 36*(2), 57–67.

Ginés, A., Ginés, J., Gómez-Pujol, L., Onac, B. P., & Fornós, J. J. (2012a). *Mallorca: A Mediterranean Benchmark for Quaternary Studies.* Mon. Soc. Hist. Nat. Balears 18. Palma, 219 pp.

Ginés, A., Ginés, J., & Gràcia, F. (2013). Cave development and patterns of caves and cave systems in the eogenetic coastal karst of southern Mallorca (Balearic Islands, Spain). In M. J. Lace & J. E. Mylroie (Eds.), *Coastal Karst Landforms* (Vol. 5, pp. 245–260). Springer, Coastal Research Library.

Ginés J (2000) *El karst litoral en el levante de Mallorca; una aproximación al conocimiento de su morfogénesis y cronología.* Ph. D. Departament de Ciències de la Terra. Universitat de les Illes Balears.

Ginés, J., & Ginés, A. (1989). El karst en las Islas Baleares. In J. J. Durán et al. (Eds.), *El karst en España* (Vol. 4, pp. 163–174). Madrid: Monografía, Sociedad Española de Geomorfología.

Ginés, J., & Ginés, A. (2011). Classificació morfogenètica de les cavitats càrstiques de les Illes Balears, *Endins/Mon Soc Hist Nat Balears, 17*(35), 85–102.

Ginés, J., Fornós, J. J., & Ginés, A. (2005). Els espeleotemes freàtics del Quaternari de Mallorca: aspectes morfològics, mineralògics i cristal·logràfics. In E. Sanjaume & J. F. Mateu (Eds.), *Geomorfologia litoral i Quaternari. Homenatge al professor Vicenç M. Rosselló i Verger* (pp. 151–165). Valencia: Universitat de València.

Ginés, J., Fornós, J. J., Ginés, A., Merino, A., & Gràcia, F. (2014). Geologic constraints and speleogenesis of Cova des Pas de Vallgornera, a complex coastal cave from Mallorca Island (Western Mediterranean). *International Journal of Speleology, 43*(2), 105–124.

Ginés, J., Fornós, J. J., Gràcia, F., Merino, A., Onac, B. P., & Ginés, A. (2017). Hypogene imprints in coastal Karst Caves from Mallorca Island (Western Mediterranean): Morphological features and speleogenetic approach. In A. Klimchouk, et al. (Eds.), *Hypogene Karst Regions and Caves of the World* (pp. 99–112). New York: Cave and Karst Systems of the World, Springer International Publishing AG.

Ginés, J., Ginés, A., Fornós, J. J., Merino, A., & Gràcia, F. (2009). On the role of hypogene speleogenesis in shaping the coastal endokarst of southern Mallorca (western Mediterranean). In A. B. Klimchouk & D. C. Ford (Eds.), *Hypogene speleogenesis and karst hydrogeology of artesian basins* (Vol. 1, pp. 91–99). Simferopol, Ukraine: Ukrainian Institute of Speleology and Karstology, Special paper.

Ginés, J., Ginés, A., & Fornós, J. J. (2011). Dades sobre paleocarst i espeleocronologia de les Illes Balears. *Endins/Mon Soc. Hist. Nat. Balears, 35*(17), 213–226.

Ginés, J., Ginés, A., Fornós, J. J., Tuccimei, P., Onac, B. P., & Gràcia, F. (2012b). Phreatic overgrowths on Speleothems (POS) from Mallorca, Spain: updating forty years of research. In A. Ginés et al. (Eds.), *Mallorca: A Mediterranean benchmark for Quaternary studies* (Vol. 18, pp. 111–146). Mon. Soc. Hist. Nat. Balears.

Gómez-Pujol, L., Gelabert, B., Fornós, J. J., Pardo-Pascual, J. E., Rosselló, V. M., Segura, F. S., et al. (2013). Structural control on the presence and character of calas: Observations from Balearic Islands limestone rock coast macroforms. *Geomorphology, 194,* 1–15.

Gràcia, F. (2015). *Les cavitats subaquàtiques de les zones coataneres del llevant i Migjorn de Mallorca* (PhD Thesis. Universitat de les Illes Balears). 984 pp.

Gràcia, F., Clamor, B., & Lavergne, J. J. (2000). Les coves de Cala Varques (Manacor, Mallorca). *Endins, 23,* 41–57.

Gràcia, F., Clamor, B., Fornós, J. J., Jaume, D., & Febrer, M. (2006). El sistema Pirata—Pont—Piqueta (Manacor, Mallorca): geomorfologia, espeleogènesi, hidrologia, sedimentologia i fauna. *Endins, 29,* 25–64.

Gràcia, D., & Jaume, D. (2011). La fauna aquàtica dels hábitats anguihalins i dolçaquícoles de les cavitats balears. *Endins, 35,* 257–268.

Gibbons, W., & Moreno, T. (2002). *The geology of Spain* (649 pp). London: The Geological Society.

Guijarro, J. A. (1995). Aspectes bioclimatics del karst de Mallorca/Bioclimatic aspects of karst in Mallorca. In A. Ginés & J. Ginés (Eds.), *El carst i les coves de Mallorca/ Karst and caves in Mallorca. Endins/Mon.Soc.Hist.Nat. Balears, 3,* 17–26.

Harmon, R. S. (1985). Late Pleistocene sea level history of Bermuda: A review. In *Geomorfología Litoral y Cuaternario. Homenaje a Juan Cuerda* (pp. 53–60). València: Univ. de València.

Hill, C. A., & Forti, P. (1997). *Cave minerals of the world* (2nd ed., 463 pp). Huntsville, Alabama: National Speleological Society.

Jaume, D., Pons, G. X., Gràcia, F., & Vicens, D. (2001). *Atles de cavitats càrstiques de les Balears elevades a Lloc d'Interès Comunitari (LICs)* (157 pp). Palma: Conselleria de Medi Ambient, Govern de les Illes Balears.

Klimchouk, A. (2017). Types and settings of hypogene karst. In A. Klimchouk, et al. (Eds.), *Hypogene Karst Regions and Caves of the World* (pp. 1–39). New York: Cave and Karst Systems of the World, Springer International Publishing AG.

López, J. M. (2007). *Las manifestaciones hidrotermales del sur de Llucmajor, Mallorca* (Unpublished investigation report). Palma de Mallorca: Departament de Ciències de la Terra, Universitat de les Illes Balears, p. 132.

López, J. M., & Mateos, R. M. (2006). Control estructural de las anomalías geotérmicas y la intrusión marina en la plataforma de Llucmajor y la cubeta de Campos (Mallorca). In *Las aguas subterráneas en los países Mediterráneos* (Vol. 17, pp. 607–613). Instituto Geológico y Minero de España, Madrid. Serie Hidrogeología y Aguas Subterráneas.

López-Jurado, J. L., Balbín, R., Amengual, B., Aparicio-González, A., Fernández de Puelles, M. L., García-Martínez, M. C., et al. (2015). The RADMED monitoring program: Towards an ecosystem approach. *Ocean Sci Discuss, 12,* 645–671.

Martel, E. A. (1896). Sous Terre. Cueva del Drach, a Majorque. *Ann Club Alpin Franc, 23,* 1–32.

Merino, A., & Fornós, J. J. (2010). Los conjuntos morfológicos de flujo ascendente (morphologic suite of rising flow) en la Cova des Pas de Vallgornera (Llucmajor, Mallorca). *Endins, 34,* 85–100.

Montoriol-Pous, J. (1971). Estudio de una captura kárstico-marina en la isla de Cabrera (Baleares). *Acta Geologica Hispánica, 6*(4), 89–91.

Mylroie, J. E. (2013). Coastal karst development in carbonate rocks. In M. J. Lace & J. E. Mylroie (Eds.), *Coastal Karst landforms* (Vol. 5, pp. 77–109). Springer, Coastal Research Library.

Mylroie, J. E., & Carew, J. L. (1990). The flank margin model for dissolution cave development in carbonate platforms. *Earth Surface Processes and Landforms, 15,* 413–424.

Mylroie, J. E., & Mylroie, J. R. (2007). Developments of the Carbonate Island Karst Model. *Journal of Cave and Karst Studies, 69,* 59–75.

Nielsen, K. A., Clemmensen, L. B., & Fornós, J. J. (2004). Middle Pleistocene magneto-stratigraphy and susceptibility stratigraphy: Data from carbonate aeolian systems, Mallorca Western Mediterranean. *Quaternary Science Reviews, 23,* 1733–1756.

Onac, B. P., Ginés, A., Ginés, J., Fornós, J. J., & Dorale, J. A. (2012), Late Quaternary Sea-level History: a Speleothem Perspective. In A. Ginés et al. (Eds.), *Mallorca: A mediterranean benchmark for quaternary studies* (Vol. 18, pp. 147–162). Mon. Soc. Hist. Nat. Balears.

Onac, B. P., Polyak, V. J., Fornós, J. J., Dumitru, O. A., Asmeron, Y., Ginés, J., & Ginés, A. (2017). Timing of Pliocene sea-level high-stands in western Mediterranean using U-Pb ages of cave deposits from Mallorca Island. In *17th International Congress of Speleology*. Sidney 22–28 July.

Palmer, A. N. (1991). The origin and morphology of limestone caves. *Geological Society of America Bulletin, 103,* 1–21.

Plummer, L. N. (1975). Mixing of sea water with calcium carbonate ground wáter. In E. H. T. Whitten (Ed.), *Quantitative studies in geological sciences* (Vol. 142, pp. 219–236). Colorado: Memoir Geol. Soc. Am.

Pomar, L. (1991). Reef geometries, erosion surfaces and high-frequency sea-level changes, Upper Miocene Reef Complex, Mallorca, Spain. *Sedimentology, 38,* 243–269.

Pomar, L., Ginés, A., & Fontarnau, R. (1976). Las cristalizaciones freáticas. *Endins, 3,* 3–25.

Pomar, L., Ginés, A., & Ginés, J. (1979). Morfología, estructura y origen de los espeleotemas epiacuáticos. *Endins, 5–6,* 3–17.

Pomar, L., & Ward, W. C. (1994). Response of late Miocene Mediterranean reef platform to high-frequency eustasy. *Geology, 22,* 131–134.

Pomar, L., Ward, W. C., & Green, D. G. (1996). Upper Miocene reef complex of the Llucmajor area, Mallorca, Spain. In E. Franseen et al. (Eds.), *Models for carbonate stratigraphy from Miocene reef complexes of the Mediterranean regions* (Vol. 5, pp. 191–225). SEPM Concepts in Sedimentology and Paleontology.

Smart, P. L., & Whitaker, F. F. (1991). Karst processes, hydrology and porosity evolution. In M. Wright et al. (Eds.), *Palaeokarsts and Palaeokarstic reservoirs* (Vol. 2, pp. 1–54). P.R.I.S. Occ. Publ. Series, University of Reading.

Tanhua, T., Hainbucher, D., Schroeder, K., Cardin, V., Álvarez, M., & Civitarese, G. (2013). The Mediterranean Sea system: A review and an introduction to the special issue. *Ocean Science, 9,* 789–803.

Tuccimei, P., Ginés, J., Delitala, M. C., Ginés, A., Gràcia, F., & Fornós, J. J. (2006). Last interglacial sea level changes in Mallorca island (Western Mediterranean). High precision U-series data from phreatic overgrowths on speleothems. *Zeits für Geomorph, 50,* 1–21.

Tuccimei, P., Soligo, M., Ginés, J., Ginés, A., Fornós, J. J., Kramers, J., et al. (2010). Constraining Holocene sea levels using U-Th ages of phreatic overgrowths on speleothems from coastal caves in Mallorca (Western Mediterranean). *Earth Surface Processes and Landforms, 35*(7), 782–790.

Vacher, H. L., & Mylroie, J. E. (2002). Eogenetic Karst from the perspective of an equivalent porous medium. *Carbonates and Evaporites, 17*(2), 182–196.

Vargas-Yáñez, M., García-Martínez, M. C., Moya, F., Balbín, R., López-Jurado, J. L., Serra, M., et al. (2017). Updating temperature and salinity mean values and trends in the Western Mediterranean: The RADMED project. *Progress in Oceanography, 157,* 27–46.

Vesica, P., Tuccimei, P., Turi, B., Fornós, J. J., Ginés, A., & Ginés, J. (2000). Late Pleistocene paleoclimates and sea-level change in the Mediterranean as inferred from stable isotope and U-series studies of overgrowth on speleothems. *Quaternary Science Reviews, 19*(9), 865–879.
Whitaker, F. F., & Smart, P. L. (1990). Active circulation of saline ground waters in carbonate platforms: Evidence from the Great Bahama Bank. *Geology, 18*, 200–203.

Chapter 8
The Rocky Coastlines of the Canary Islands

Amalia Yanes-Luque

8.1 Introduction

The Canary geographic area, an archipelago of 7,447 km^2, consists of seven islands, four islets and several seamounts on the eastern edge of the Mid Atlantic (27°–30°N and 13°–16°W), and has a coastal morphology with its own features. This is result of the interference between eruptive activity and marine dynamics. In the Canary Islands, as in Hawaii, the Azores, Madeira and many other oceanic areas of active volcanism, the persistence of eruptive manifestations increases the coastal front. The arrival of lava flows to the sea and the location of emission centers in the sea or its immediate vicinity counteract the retreat of the coastline caused by erosion. Therefore, the areas where the waves have blurred the original volcanic features coexist with sectors where they are still fresh (Guilcher 1981; Mitchell 1998; Etienne and Paris 2007; Ramalho et al. 2013). This interference coincides, moreover, with sea level oscillations. Quaternary environmental changes are an aspect to be assessed, as is the case in many coastlines of a marked polygenic character (Trenhaile 1987).

In this context, the relief of the island front is abrupt, since almost 78% of the 1,554 km of its perimeter is rocky. Cliffs and abrasion platforms in volcanic constructions, such as the basal complex, ancient massifs, ridges, cones and lava flows, are examples of this abruptness; besides which escarpments in terrestrial and marine sedimentary accumulations have reached the shoreline as part of their development process. Both forms are geomorphological units of a specific physiognomy, since the volcanic and sedimentary structures that the sea transforms differ in chronology, size, internal constitution, lithological nature and degree of weatherization of the

A. Yanes-Luque (✉)
Department of Geography and History, University of La Laguna,
Plaza del Rector D. José Carlos Alberto Bethencourt, Campus de Guajara,
38200 La Laguna, Santa Cruz de Tenerife, Spain
e-mail: ayanes@ull.edu.es

J. A. Morales (ed.), *The Spanish Coastal Systems*,
https://doi.org/10.1007/978-3-319-93169-2_8

rock. The spatial continuity of many cliffs and the reduced presence of beaches of considerable size reinforce the vigour of coastal modeling. The recording of great depths very close to the shore and the scarcity of materials susceptible to transport and accumulation limit the beaches to 17% of the insular coastline contour. Manmade works make up 5% of the coastline.

Rocky shorelines are also particular biotopes. A pronounced tendency to verticality, lack of soil and the incidence of marine spray condition this on the cliffs. Tidal changes and the splashes of the waves act on the abrasion platforms. Such aspects favour the occupation of these biotopes by halophytic communities, which give way to a sub-desert scrub of *Euphorbiaceas* at the top of the cliffs (Yanes and Beltrán 2009). The undeniable natural values that converge in large areas of these coasts present, however, symptoms of alteration which are due to changes in land use, thereby favouring an increasingly intense socioeconomic occupation of the coastal environment.

8.2 The Geographical Framework

The analysis of the rocky coastlines of the Canary Islands requires, in the first place, a consideration of the magmatic processes that create the archipelago, since the lavas and the sea are the essential factors in their geological history. This refers to the alkaline volcanism of a *hotspot* in the interior of the African lithospheric plate, which moves slowly from west to east. The eastern islands (Lanzarote, Fuerteventura and Gran Canaria) are therefore older than the western islands (Tenerife, La Gomera, La Palma and El Hierro) (Carracedo 2011).

The abovementioned *hotspot* throws materials to the bottom of the sea through a system of fissures of the oceanic crust, by way of structural axes in rectilinear layout and with dominant NE-SW, NW-SE and N-S courses. This process has been occurring for between 30 and 35 million years, and the last volcanic manifestation recorded in the Canary Islands until now was in 2011 (La Restinga, S of El Hierro). The expulsion of these materials is, however, intermittent, considering the existence of several eruptive cycles separated by rest stages, which are sometimes prolonged. It should also be borne in mind that the beginning and end of these cycles vary from one island to another and that the emission mechanisms, the volume and lithology of the expelled materials and the type of edifice constructed change, depending on the period in question. The Canary insular geology is characterized, then, by the development of an initial cycle of oligo-miocene underwater volcanism, followed by a first subaerial miocene cycle and a second plio-quaternary cycle. The progression of volcanism to the west explains this regional scheme, so that the subaerial eruptive manifestations are exclusively quaternary in the younger islands (La Palma and El Hierro) (Anguita et al. 2002; Carracedo 2011) (Fig. 8.1).

The consequence of this evolution is that the islands are emerged surface monoliths from depths of 3,000–4,000 m, by the superposition of lava flows and levels of intercalated pyroclasts, each of which is a separate block, except for

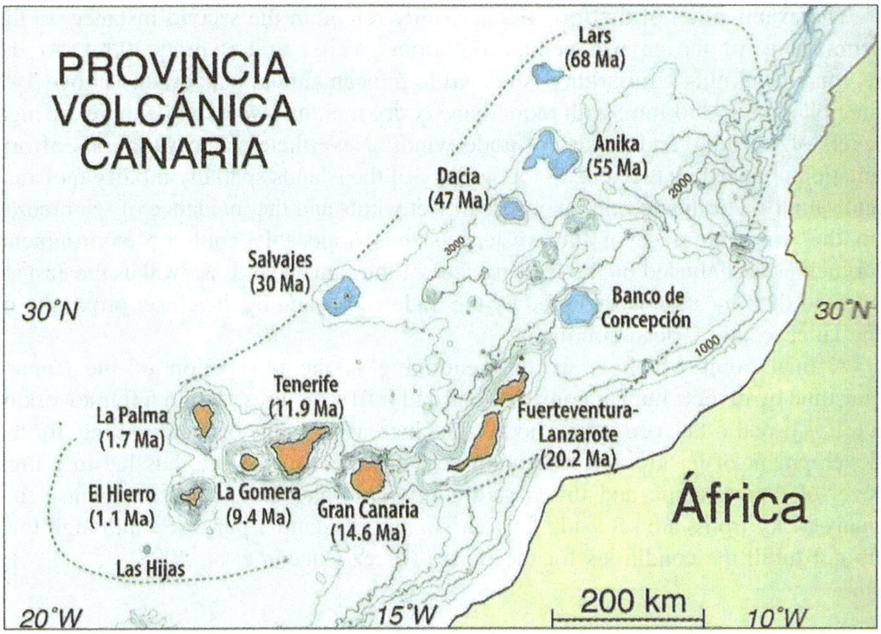

Fig. 8.1 Spatial location and components of the study area: emerged islands (ocher color) and seamounts (blue color). The numbers correspond to the ages of the oldest subaerial volcanism (*Source* Carracedo 2011)

Lanzarote and Fuerteventura which form a unitary set. The sea carves differentiated coasts in these blocks, although they share common traits because of their volcanic origin. The different temporal and spatial development of magmatism and the variable alteration of the resulting constructions show the differences the strong escarpment of ancient geological structures and the protagonism of the coastal scrublands in recent times (Yanes and Beltrán 2009). Thus, a more or less recti-linear coastline with low and high cliffs is a dominant feature in the igneous, sedimentary and volcanic rocks of the basal complex; the submarine substrate common to the archipelago, although at the coastal level is only recognizable in Fuerteventura and La Gomera. On the other hand, the basalts of the first subaerial eruptive phase form a coastline with large-sized cliffs which go back a long way from the sea occupying large areas in each island. The latter are the result of the intense and prolonged erosion of old dorsal massifs, which are complicated structures forming the basic components of the insular relief. The outline of the sea front is usually articulated, when the lavas of the second subaerial cycle reach the shore. This is defined by the alternation of stretches of low rocky coast between cliffs of varying height. The waves sculpt them in basalt, to which are added phonolites and trachytes in Tenerife and Gran Canaria. These materials are mostly associated with simple volcanic structures such as domes, cones and flows.

The examination of the forms under study refers, in the second instance, to the semi-aridity of the environmental conditions (Yanes and Beltrán 2009). At the regional level, this semi-aridity is related to a mean annual temperature above 19°, precipitation ≤ 350 mm, with more than six dry months a year, while there are high levels of sunshine and persistent trade winds. Nevertheless, the variations of orientation imposed by the general topography of the islands spatially modify moisture and salinity. For this reason, exposure to wet winds and the incidence of sea breezes on the windward face of the western islands noticeably cool the environment. Dryness is accentuated on the leeward face of the same islands as well as the eastern ones, as their location is sheltered by the trade winds and by the closer proximity of the latter to the African continent.

A final point which is worth mentioning is the occupation of the Canary coastline by diverse human constructions and infrastructures, within a framework of a tourism-real estate economic model. The increasing value of the coastline, for the development of leisure and recreation activities next to the sea, has led to a high level of demographic and urban saturation. This phenomenon is so intense that many rocky fronts are set aside for tourism and residential purposes, although they do not fulfill the conditions for these uses (Pérez-Chacón et al. 2007).

8.3 Morphological Characterization

The remarkable morphological diversity of the rocky coasts is a result of the features of the volcanic constructions and sedimentary accumulations which they form a part of. The insular contour includes cliffs of many different sizes, height, profile, inclination, lithology and structural layout of the rock, modeling and varying degrees of functionality. The steepness of slope, width, continuity, lithology and morphology of detail are aspects to contemplate in the abrasion platforms. For all the above reasons, the homogeneity that the rigidity of the forms themselves entails is more apparent than real. This occurs even though both categories co-exist in the same coastal stretch.

8.3.1 High and Low Cliffs in the Basal Complex

This term refers to the oldest but less frequent escarpments of the Canaries, because they are confined to a limited part of the north coast of La Gomera (Agulo-Vallehermoso) and the west coast of Fuerteventura (Los Molinos-Punta de Amanay). They are sculpted on the underwater basement of the archipelago and the almost uninterrupted erosion reveals their internal structure after they appear above the sea. The different composition of the insular substrate, and to a lesser extent its level and moment of emersion, are responsible for the differences in these types of cliffs. In Fuerteventura, the basal complex is composed of turbidite deposits, which

Fig. 8.2 Cliff in the basal complex of Fuerteventura, where rhythmic mesozoic sediments, crossed by dikes, give way to marine levels of sandstones and torrential formations (Puerto de La Peña, W of Fuerteventura) (Photo: Yanes, A.)

have layers of pillow lavas on top that are the first stage of the construction of the island from the ocean floor and give way to shallow water marine sediments (Anguita et al. 2002). The whole structure is crossed by intrusions of gabbros, pyroxenites, syenites and carbonatites, but above all by a profuse mesh of dikes. The same structure is so dense that in many points the host rock disappears, which is why the rock is highly resistant. However, marine erosion has carved vertical walls 15–20 m high, sometimes reaching up to 50 m (Fig. 8.2).

These walls are, in any case, low cliffs, because, according to Guilcher's altimetric criterion (1981), they do not reach 100 m which clearly distinguishes them from high, very high and mega-cliffs, as the height of the latter cliffs ranges between 100–250, 250–450 and 400–600 or more meters, respectively. The functionality of cliffs is another prominent feature, because the waves continue to model their base. The result is the appearance of small intertidal abrasion platforms and the exhumation of numerous dikes, which cause a certain irregularity in the coastal outline. Finally, the destruction and fossilization of the top of these cliffs by marine sandstones and glacis-cones with calcareous crusts are striking (Criado 1991).

In La Gomera, the sediments and lavas of the basal complex have a relatively lower compactness, since they appear as small finely chopped fragments between gabbros, pyroxenites and dikes. The latter accounts for 70–90% of the total rock mass (Anguita et al. 2002). The folds cut by fractures identified in the dikes reveal the intense deformation of this submarine group, while the erosion surface that cuts this group's culmination denotes a marked dismantling (Anguita et al. 2002). This process continues on the surface, where the surf has created subvertical escarpments of 100–200 m in height. The appearance of these high cliffs is generally one of a highly weather worn nature: on the one hand, because of the basal notch and open blow-holes at their base and the cavities and extensive flaking generated by the splashes of the waves between the dikes; and, on the other hand, as a result of the torrential incisions that open their way from the top. The existence of cordons of pebbles and fluvial-torrential deposits and/or of a slope at the foot of some of these walls turns them into stabilized cliffs (Arozena 1991; Yanes and Beltrán 2009).

8.3.2 High, Very High and Mega-Cliffs in Ancient and Dorsal Massifs

This group of cliffs is of great importance in the Canary Islands and are mainly located in the NW and S Lanzarote, SW of Fuerteventura, NW, W and SW of Gran Canaria, NW, N and NE of Tenerife, practically all La Gomera, N and S of La Palma and NW, N and E of El Hierro, where they stand out for their height, extension and orientation. In fact, they all have a difference in height of several hundred meters between their base and their summit, very noticeable in very high cliffs (250–450 m) but especially in the mega-cliffs (400–600 m) (Guilcher 1981). Their breadth is remarkable, because they extend almost without interruption along coastal stretches of several tens of kilometers. The 22 km of the Famara cliff (NW of Lanzarote), 11 km of Cofete (SW of Fuerteventura) and 8 km of Andén Verde (W of Gran Canaria) are examples of extensions of some of these cliffs (Yanes 2004). Finally, the adaptation to the structural axes that guide the spatial distribution of volcanism is manifest. The cliffs of Famara (NW of Lanzarote), Los Gigantes (NW of Tenerife) and Jedey-Las Indias (SW of La Palma) have, for example, a rectilinear arrangement, because they are inserted in edifices created from a single structural axis. The arched configuration of cliffs like those in Cofete (W of Fuerteventura), El Golfo and Las Playas (N and E of El Hierro) is a result of volcanic structures arising from the crossing of two or more of these axes (Yanes 2004, 2006; Yanes and Beltrán 2009).

These scarps also share structural and lithological similarities. This is due to their relationship with ancient massifs and dorsal, in the context of an eminently fisural volcanism, a high eruptive rate and a large volume of material emitted over a long period of time. Therefore, these rock walls are composed of the superposition of multiple lava flows of metric power and a horizontal and subhorizontal arrangement. Basalts are the predominant rocks in the composition of these cliffs, but phonolites and trachytes are also involved in the formation of some of the cliffs (Puerto Rico-Mogán, W of Gran Canaria and Anaga, NE of Tenerife). There are interspersed layers of accumulations of pyroclasts, cinder cones and baked soil between the lava flows; outcrops of hydrovolcanic tuff (Puerto Naos, La Palma, Hoya del Verodal, NW of El Hierro) and alluvial and colluvial deposits and consolidated dunes (Famara, W of Lanzarote) are occasionally found in certain escarpments. A mesh of vertical and subvertical dikes profusely cuts across this structure. Although these cliffs are ancient, the chronological time differences of volcanism mean some are myopliocene and others pleistocene.

The outline of the profile, the degree of functionality and the morphology define specificities in these cliffs. Thus, some have a steep slope, because they range from 60° to 70° to an almost absolute vertical. In large stretches of their development, they do not have a littoral platform, so they can be considered to be *plunging* type cliffs (Trenhaile 1987) (Andén Verde, W of Gran Canaria, Los Gigantes, NW of Tenerife, Garafía, N of La Palma) (Fig. 8.3).

The *notches*, *sea caves*, *blowholes*, stark dikes like breakwaters, *stacks* and reefs which the waves hit reflect the active nature of these escarpments and the retreat that has occurred. The ravines hanging in the front of some of the cliffs are perhaps the best evidence of the intensity of this retreat, since their base level is probably located 3–4 km from the present coast (Alojera, NW of La Gomera, Famara, NW of Lanzarote, Mogán, W of Gran Canaria) (Arozena 1991; Romero 2003). The importance of marine erosion, however, diminishes in sectors with small areas of destruction, narrow pebble beaches and alluvial and colluvial deposits protected by strong waves.

Fig. 8.3 The verticality highlights the size of numerous cliffs, as happens in the ancient volcanic massifs, the result of a continuous erosion on lava flows, levels of pyroclasts and intercalated baked soil (Garafía, N of La Palma) (Photo: Yanes, A.)

The physiognomy of other escarpments of this group is a result of strong slope dynamics, which has brought about powerful gravity, alluvial and colluvial deposits with an extensive lateral development. These deposits are detritical fans which support or imbricate calcareous crusts, beaches and/or consolidated quaternary dunes. They extend from the mid-slope of the escarpment like gentle sloping planes, which stand between the cliffs and the sea (Cofete, SW of Fuerteventura, Anaga, NE Tenerife, Hermigua-Agulo, N of La Gomera, Las Playas, E of El Hierro) (Fig. 1.4). In their advance towards the coast they usually cover wide abrasion platforms that are sometimes found at the foot of some escarpments (Valle Gran Rey, SW of La Gomera, Los Ajaches, S of Lanzarote) (Arozena 1991; Criado 1991; Romero 2003; Yanes 2006). The result is the non-functional character of these rock walls and the rupture of their profile, which ranges from a slope $\leq 30°$ in its lower part to a slope of 70°–90° in its upper stretch. Hence, the escarpments are of a *slope-over-wall* type (Trenhaile 1987).

The resumption of volcanism in specific enclaves of the old massifs and of some stretches in the dorsal ridges has led to inactivity in some cliffs (El Golfo and Hoya del Verodal, N and NW of El Hierro respectively, San Antonio-Teneguía and El Charco-Jedey, W of La Palma, Buenavista and Teno Bajo, NW of Tenerife, Jacomar, E of Fuerteventura). The lava flows emitted by the new volcanoes run through the ravines or the front of the littoral escarpments to the coast. There they form lava deltas attached to their base, from where the lava enters the sea (Yanes 2006; Yanes and Beltrán 2009; Carracedo 2011). They are rectangular, triangular or fan-shaped structural platforms which are practically flat, with an inclination of $\leq 5°$ and a variable surface of between less than one (Garachico, NW of Tenerife) and a little more than twenty square kilometers (El Golfo, N El Hierro). The waves create a new coastline in the edge of the lava deltas, since the primitive cliffs stayed back and evolve, then, by the dynamics of the slope (Yanes 2006) (Fig. 8.4). These primitive cliffs connect to these platforms by debris fans, so that their profile is, as in the previous case, *slope over wall* (Trenhaile 1987).

Fig. 8.4 The evolution of many large cliffs is marked by the presence of debris slopes (left image: Famara, NW of Lanzarote) and/or the provision of lava deltas at its foot (right image: Bay of Los Reyes, W of El Hierro) (Photos: Romero, C.)

8.3.3 High and Low Cliffs in Simple Volcanic Structures

These cliffs are found in all the islands and are linked to the erosion of pyroclastic cones, hydromagmatic structures, domes and lava flows. They are initially distinguished by their height, from a few meters to hundreds of meters, and by their length, of only a few hundred meters by the modest dimension of volcanic structures; also because of their functionality, since they are usually active escarpments with a vertical profile, because the flank of the volcanic structure that falls to the sea loses its normal dip. Their variable age should also be taken into account, since there are not only myopliocenes and pleistocenes but also holocenes.

The morphology is, however, the most notable aspect, because its modeling depends on the characteristics of each eruptive centre (Fig. 8.5). Numerous entrances and exits horizontally traverse the formations in pyroclastic volcanoes, depending on the different grain size and degree of welding of the aerial projection products (Montaña Abades and Montaña Roja, S of Tenerife, La Caldera, S of La Gomera, La Restinga, S of El Hierro). In hydromagmatic cones, cliffs are defined by the alternation of slits and protrusions in the hyaloclastites resulting from the interaction of magma with sea water. This preferentially affects the multiple lines of contact between the fine and well defined strata of these lavas (El Golfo, W of Lanzarote, Montaña Amarilla and Montaña Pelada, S of Tenerife). Finally, the escarpments in domes are uneven walls, which are especially interesting when the waves uncover their internal structure. The cliff of Los Órganos (N of La Gomera) is a magnificent example for its spectacular hexagonal prisms, one meter in diameters, which form the domelike structure. The sea continues to widen the joints between the hexagonal prisms.

In other cases, the sea remodels the flows poured out by the volcanoes. The high and low cliffs on the lava flows are the most common in the Canary Islands, with walls ranging from 5–15, 30–50 to 80–100 m high, depending on the power and number of stacked lava flows (Fig. 8.5). The basalt nature and the young geological age of many of them mean that the profile of the ledges tends to be regular and vertical, with a slope of 70°–90°. In those of a certain height, the slope can be $\leq 50°$ in its middle and upper sections, coinciding with small gravity slopes in the contact between the lava flows. In any case, the surf enlarges the columnar structure of the basalt and sculpts *notches*, *blowholes*, *natural arches*, *sea caves* and levels of destruction where loose rocks sometimes accumulate. All this shows the functionality of these escarpments where bees nests and *tafonis* are found on their front and top by the effect of the marine spray.

The potency, massiveness and less contrasted structure of the phonolitic and trachytic lava flows favour cliffs with a very uniform profile and morphology, although altered by caves of different depths. Fallen pyroclastic deposits and pyroclastic lava flows are an exception. Although the materials are welded well, their fragmentary origin predisposes them to erosion, so that the waves sculpt a very well developed notch. The upper rock wall is then in a position of pronounced overhang. Furthermore, there is the progression of vertical fissures, from degassing

Fig. 8.5 Cliffs in simple volcanic structures. Left image: cliff on the flank of the tuff cone of Montaña Pelada (El Médano, S of Tenerife), where erosion progresses from the parallel laminations forming the hyaloclastites (Photo: Yanes, A.). Right image: remodeling of recent basaltic castings and initial formation of an abrasion platform, after elimination of the level of surface debris slag (Los Hervideros, W of Lanzarote) (Photo: Mejías, J. L.)

and/or removal of lithic channels and, sometimes, the formation of abrasion surfaces that move the cliffs away from the sea.

8.3.4 Low Detritical Cliffs

Coastal modeling of the Canaries involves the existence of cliffs in ravine and slope accumulations, lahars, avalanches, beaches and quaternary dunes (Yanes 2016). They are preferentially located in the oldest parts of the islands, among which are enclaves in the NW of Lanzarote, W and S of Fuerteventura, NE and NW of Gran Canaria and Tenerife, N and W of La Palma and La Gomera and N and E of El Hierro. This is because the early initiation and cessation of eruption has contributed to the expulsion of powerful masses of debris into the sea, where they are usually joined to outcrops of old coastlines. The survival of the recent volcanism limits the development of low detritical cliffs, whose presence is more or less exceptional in other areas of the islands.

These sedimentary escarpments are defined by the contrasts in their height and size. In fact, the difference in height between the bases and peaks ranges from 1.5–3 to 50 m, although this can occasionally be 100 m (NE of Gran Canaria, W of La Palma). The length is less when the cliffs are formed in lahars, avalanches and marine outcrops, since they are small and fragmentary deposits. On the other hand, the sculpted formations, above all in slope accumulations, occupy hundreds of meters of the coastal front as a result of the evolution of the great escarpments of the old and dorsal massifs (Fig. 8.6).

Other features to be considered are the profile outline, which is generally *slope-over-wall* (Trenhaile 1987) and the greater or lesser complexity of its stratigraphic

Fig. 8.6 Detritical cliffs. Left image: cliff in fluviotorrencial deposits dislodged towards the sea, fruit of an active quaternary morphogenesis (Arguamul, NW of La Gomera) (Photo: Romero, C.). Right image: the heterometry and relatively reduced stratification and consolidation of these deposits have favored its retreat by marine action (Taganana, NE de Tenerife) (Photo: Yanes, A.)

sequence, which makes it possible to distinguish between simple and mixed detritical cliffs (Yanes 2016). The former are composed of a small number of deposits of the same origin. In some cases, these are detritical slopes and dejection cones with a predominance of angular and subangular heterometric elements, encompassed in a fine matrix (Cofete, SW of Fuerteventura, Bajamar, NE of Tenerife, Las Playas, E of El Hierro) (Fig. 8.6).

In other cases, the deposits are beach conglomerates and quaternary dunes of highly consolidated bioclastic sands, often of a *climbing dunes* type (Ugan, W of Fuerteventura, Tufía, E of Gran Canaria, La Playita, S of Tenerife). The latter are formed by the superposition or inter-stratifications of different sedimentary bodies. According to their origin, some mixed cliffs are formed by beaches and/or quaternary dunes arranged between either alluvial or colluvial deposits or they are covered by these deposits (Las Palmas de Gran Canaria, NE of Gran Canaria, Puntallana, E of La Gomera, Anaga, NE of Tenerife). Other escarpments are the result of the overlapping of lahar, avalanche and/or fluvial torrential deposits, among which are intercalated lava flows, pyroclastic clusters and/or old sea levels (Punta de Las Arenas, W of Gran Canaria, Punta del Guindaste, N of Tenerife) (Criado et al. 1998; Armas et al. 2001).

Detritical cliffs are ultimately characterized by stability. The marine dynamics only touches up the base during storms and very rough high tides, as the cliffs are protected by a beach or blocks which have become detached from their faces. In some cases, the lack of functionality lies in the anthropogenic reorganization of the coastal front (NE of Gran Canaria and Tenerife). Present erosion is reduced to a small *debris-flow* and torrential incisions that channel winter run-off.

8.3.5 The Contrasts of Abrasion Platforms

Abrasion platforms counteract the harsh character that cliffs impose on the whole coastline, especially if they are prominent and developed (Trenhaile 1987; Etienne 2007). The modification they introduce in the Canary coast is, however, relative, because only a few exceed a hundred meters of width and length and they have plenty of spatial continuity; on the other hand, there are many platforms which are fragmented and partially covered by beaches, dunes or old and/or recent detritical deposits (Yanes 2006).

Although there are numerous platforms associated with a sea level higher than the level today (enclaves at +5–8 and +10–50 m in the N of Tenerife and Gran Canaria, SW of Lanzarote and W of Fuerteventura, among others), most are intertidal. They are either horizontal abrasive platforms or have a gentle slope towards the sea. They usually extend to the foot of a cliff where they join the notch, especially if the waves act on the pyroclasts. This contact usually occurs by means of a ramp or a step of a few centimeters when the waves work on the hard nucleus of more or less recent lava flows. On the sea side, the edge of the platforms is, in some cases, a micro-cliff crowned by a small step and, in others, a level of subtidal destruction. On the other hand, the morphology of detail of its central sector is rich and varied, aside from the small topographical projections that are created by the waves in the contact between lava flows.

As in many other coasts, this morphology is controlled by the structure and age of the materials, availability of abrasives and wave energy (Trenhaile 1987; Etienne 2007). In fact, it is noteworthy that the platforms in old basaltic lava flows are usually smooth and polished from the action of an energetic sea (Vallehermoso and Alojera, N and W of La Gomera respectively); but if the basalt also has a prismatic disjunction, the traces of its truncation and the expansion of potholes and open dissolution pools in the fissures that mark where they make contact with each other (Barlovento and San Andrés and Sauce, NE of La Palma, Teno Bajo, NW of Tenerife). In sheltered areas, the irregularities of the recent basalt abrasion platforms are notorious, due to the partial elimination of the surface debris level of the lava flows. This fact emphasizes the toes caused by the cooling of the lavas, among which water pond, dissolution pools and *potholes* are found (Orchilla, SW of El Hierro, Güimar, E of Tenerife, Timanfaya, W of Lanzarote). Narrow but numerous parallel grooves alter the uniformity of these platforms in trachytic and phonolithic lavas. The opening and progressive widening of hollows formed in pyroclastic lava flows stand out on the surfaces, after the waves have eliminated many lithic fragments contained in the volcanic mass (coastline between Arico and Adeje, E and SW of Tenerife). Finally, the modeling of abrasion platforms originating from beach conglomerates and bioclastic sands from quaternary dunes is composed of small circular and oval cavities, which coalesce to create scalloped or lobulated depressions of a greater size. All of the above are open, usually, in the stratification planes of these sedimentary deposits (Arrieta-Orzola, E of Lanzarote, Las Canteras-El Confital, NE of Gran Canaria, El Médano-La Playita, S of Tenerife, Puntallana, E of La Gomera).

8.4 Biological Population and Anthropological Intervention

Singular organisms of defined spatial distribution populate the rocky insular coastline in spite of the limitations imposed by the steep slope, soil shortage, wind and sunlight. The cliffs exposed to the waves are colonized by the *Frankenio ericifoliae-Astydamietum latifoliae* association on the windward slope of the islands, while *Frankenio capitatae-Zygophylletum fontanesii* occupies the leeward side. The scarcity of flora and poor plant cover enable the existence of halophilous communities with individual specimens of prostrate and cushioned forms, which lose much of their aerial biomass in the dry months. The floral and plant composition and cover increase in the sedimentary escarpments, where the lack of substrate compaction and the accumulation of fine sediments allow the coexistence of *Argyranthemum frutecens*, *Tamarix canariensis*, *Euphorbia canariensis* and even *Phoenix canariensis* with halophytic species. The scenery in the escarpment fossilized by deposits from the slopes is undeniably outstanding, where furthermore the distancing of the sea and the greater height, in the western islands, favour the development of laurisilva (Yanes and Beltrán 2009). As for fauna, the most prominent feature is the noticeable presence of nesting seabirds (*Calonectrisierden diomedea*, *Puffinus assimilis*, *Bulweria bulwerii* and *Falco eleonorae*) and the existence of *Galloti* lizards on certain cliffs. Their value is unequivocal in the case of endemic species, some of which are also giant lacerates where these walls provide their last refuge (NE of La Gomera, NW of Tenerife, N of El Hierro) (Yanes and Beltrán 2009).

The algae, especially red algae (*Ceramiun*, *Corallina* and *Caulacanthus*), are predominant on many of the abrasion platforms, in the face of the swaying of the waves and the regular emersion and immersion caused by the tidal change. Dense clusters usually form on the solid rock where multiple species of invertebrates, crustaceans (*Chthamalus stellatus*) are found on the upper reaches of the mesolithic zone and molluscs (*Patella piperata* and *Osilinus atratus*) live alongside species with an infra-coastal affinity (*Aplysina aerophoba* and *Anemonia sucata Aiptasia*) (Fernández-Palacios and Martín 2001).

The rocky coastlines are not exempt from human action, despite the rigidity of the substrate, the verticality and the height of the cliffs. Although natural processes predominate, especially in areas of difficult access, the summit, the front and the base of many escarpments are the object of various forms of occupation. In relation to the above, agricultural practices and the growth of urbanization are forms of anthropogenic intervention (Pérez-Chacón et al. 2007).

The alteration of steep sloped coastal segments is linked, in principle, to the need to having suitable spaces available for agriculture near the coast. Thus, in fragmented islands with steep narrow gullied ravines, agricultural plots and terraces are not infrequent in detritical cliffs and in lava flows, whose preparation requires the mobilization of land of variable magnitudes. In the ancient and dorsal massifs, the use of lava deltas that advance towards the coast is noteworthy. The gentle topography and benignity of the climate facilitate agriculture, although it is necessary to have soil available from interior areas on the unaltered rock.

Urban development involves changes of a greater magnitude. It needs be taken into account, on the one hand, that the summit of several coastal escarpments is a place where population nuclei settle; and, on the other hand, that the construction of purpose built residential complexes on a slope is growing. On the other hand, overhead communication lines, the construction of port infrastructures and the expansion of the population nuclei, in the context of a constrained coast between rocky walls, result in the artificial enlargement of the maritime facade. Situations of instability appear in the areas surrounding these walls, with undesirable effects for the population and the territory.

8.5 Genesis and Evolution

8.5.1 Natural Marine Dynamics

The waves are the most effective marine agent in the coastal relief, because of the constancy of its work (Trenhaile 1987). In the Canary Islands, the usual system is characterized by the prevalence of the *sea* in relation to NE winds of 18–22 km/h, exponents of the incidence of trade winds on the coast. These do not favour a highly developed *swell*, with an estimated mean annual significant wave height of 1.5 m, a period of 4–8 s and a length of less than 100 m. From October to March, the *sea* cedes importance to the *swell* coming from the North Atlantic. Waves from the N and NW of more than 300 m in length, 2–3 m of significant wave height and a period of up to 18 s arrive in the Canaries, especially on the northern coasts. Their arrival sometimes coincides with winds of more than 80 km/h, with the possibility of triggering storms. During its development, the significant wave height can exceed 5 m and these waves increase the area of attack of the waves, which is appreciable due to the mesotidal character of the insular coastline (Yanes 2006).

The behavior of the waves is conditioned by the bathymetry, the topography of the sea bottoms and the degree of articulation of the coastal front. In this respect, the relatively young geological age of the archipelago and the fact that the islands are independent volcanic constructions do not contribute to them having, in general terms, an extensive and continuous littoral platform. If you add to this the recording of depths of 40–50 m and even 100 m less than 500 m from the breakwater area, the waves do not undergo much refraction and thus retain some of their initial energy. The irregular and contrasting underwater topography also increases the strength and height of the waves as they approach the shore. This fact is important when they are embedded in the spillways and lateral cooling walls of lava flows that extend under the water (W of Lanzarote, N of Tenerife, SW of La Palma and N of El Hierro). On the other hand, the entrances and exits that traverse many coastal scrublands favour the differential work of the swell. The importance of convergent refraction is manifest, then, where there is an exceptional advance of the lavas into the sea.

The combination of such factors means that the shaping of these coasts has been and is largely controlled by the hydraulic action of the waves. This can be clearly seen in the active cliffs and to a lesser extent in the stabilized ones, where their impact triggers high intensity momentary pressures, followed by a rapid decompression. The strength of this process is accentuated if there are abrasives thrown against the base of the cliffs or move the platforms extended at their bases. The alternation of weathering and drying cycles and the haloclasts enhance this mechanical work, although it is limited to the strip affected by the wetting of the waves and to the pools and crevices that cut across the rock.

The effectiveness of the waves is also dependent on the lithology, structure and arrangement of the volcanic products (Yanes 2006). The chemical composition, which is unstable because it is formed at great depths, and mineralogy of basalts means that the waves work quickly on this type of rock. The resistance of trachytes and phonolites is greater, since they enjoy stability as they come from shallower magmatic chambers (Carracedo 2011). Another factor to add to this is the variable progression of erosive processes depending on whether the rock is incoherent or coherent. This progression is considerable in pyroclasts, since its explosive origin determines its fragmentation, vacuolar texture and arrangement in successive layers of certain stratification. In this way, the retreat of the cliffs is accelerated when the base is made up of such aerial projection materials. The erosion is slowed, on the contrary, in lava flows, given the massive nature acquired by lava flows when they harden. For the aforementioned reason, the dismantling preferentially takes place starting from its lines of weakness: on the one hand, those lines resulting from the internal contrasts of the lava flows, because their base and roof are scoriaceous, because of the rapid cooling when in contact with the air, whereas their nucleus is compact. On the other hand, those lines appearing with the retraction of the solidifying lavas, which form a prismatic fissure in well developed columns, provided that the cooling has been slow and without any external disturbance (Carracedo 2011).

The competition of the waves increases, finally, when the lava flows overlap, with erosion preferentially acting in the fissures marking the contact between them. The fissures form horizontal discontinuities, of great significance in volcanic structures made up of the alternation of lava flows and pyroclastic levels (Yanes 2006). The coastal walls thus formed become unstable, bearing in mind that these discontinuities are also excellent sliding planes as a function of the porosity of the pyroclasts. These pyroclasts may be embedded with large amounts of water retained by the more impervious underlying lava flows (Anguita et al. 2002). On the other hand, the dikes that cross these materials create vertical discontinuities that increase the propensity of the cliffs to collapses and rock falls.

8.5.2 Natural Terrestrial Dynamics

Gravitational landslides and morphogenetic activity are also involved in the configuration of the insular coastlines. As in many volcanic archipelagos (Watts and Masson 1995; Mitchell 1998; Caniaux 2007; Carracedo et al. 2009; Carracedo 2011; Ramalho et al. 2013), these terrestrial mechanisms are part of the origin of mass movements whose deposits have been escarped by the sea.

The mobilization of large volumes of material is worth mentioning in the geological history of the Canary Islands, due to its decisive participation in the genesis of their relief. The mega landslides are of particular interest, because of their magnitude, as they took place in areas surrounding the old and dorsal massifs. The repetition of eruptive episodes, both numerous and close to one another in time, and the injection of a dense network of dikes have caused strong tensions in these extensive, high and abrupt structures, which have led to the lateral collapse of their flanks (Jandía, W Fuerteventura, La Aldea-Agaete, W of Gran Canaria, La Orotava and Güimar, N and E of Tenerife respectively, Cumbre Vieja, S of La Palma, El Golfo and Las Playas, N and E of El Hierro respectively, among others) (Watts and Masson 1995; Carracedo et al. 2009; Carracedo 2011). Large topographical depressions and walls open to the sea are example of this process, as well as hundreds and even thousands of cubic kilometers of detached rocks, which are dispersed in the archipelago's seabeds (Carracedo et al. 2009).

Alongside the mega landslides, other volcanoclastic deposits of smaller size are frequently found in oceanic volcanic islands, but with a rapid and massive emplacement (Ramalho et al. 2013), and these are an important source of sediments on the coast whose dynamics are still unknown. This is the case of fluvio-torrential and avalanche deposits from the detritical escarpments, linked to an active torrential morphogenesis and quaternary slope. The annual and interannual scarcity and irregularity of precipitation makes permanent runoff in the Canary Islands impossible, which is why these deposits are the result of the climatic and sea-level changes that occurred in that geological age (Arozena 1991; Criado et al. 1998; Armas et al. 2001; Romero 2003; Yanes 2006, 2016).

In this context, the detritical cliffs can be classified at least as *two-storied cliffs* (Trenhaile 1987), because morphoclimatic episodes of opposite signs alternate in their formation. Thus, in an initial stage of cold and dry climate, sporadic, brief and violent rains between phases of marked drought drag alluvial and colluvial materials from slopes largely devoid of vegetation. These materials overlap with soils and dunes near the coast, coinciding with a marine regression which causes the dismantling of the volcanic escarpments by subaerial erosion, with the consequent extension of a cover of scree at their feet. At a later stage, a warm and humid climate allows the recovery of the vegetation cover and a less irregular runoff, which cuts off the sedimentary accumulations on the coast. These environmental conditions coincide with a rise in sea level, which is due to the creation of beaches and cliffs from alluvial, colluvial and dune deposits.

8.5.3 Anthropogenic Dynamics

The rapid and profound transformations of the land uses that have recently taken place in the Canaries have changed the physiognomy and dynamics of many stretches of their rocky coastlines. The loss of importance, if not substitution, of marine processes due to subaerial processes is becoming more and more common on the coastlines, on islands where an economic model based on tourism has been implanted since the 1960s (Pérez-Chacón et al. 2007). The soft uses of the coast have nearly disappeared: from summer shellfish catching to salt and lime production, limited to a narrow strip between the shore and 50 m of altitude in a few enclaves where access to the sea is not too difficult (Sabaté 1993). In the face of a harmonious adaptation of such practices to an environment sensitive to impacts, the coast today is the object of inappropriate uses and/or intensities of use. Sun and beach tourism has turned it into a space of great commercial profitability. Consequently, the coast now plays an important part of the socio-economic dynamism of the islands.

The large scale litorization of the insular population to the coast has created a new landscape that has no relationship with the environment, since the readapting of the natural heritage to meet the demands of the tourist activity (S Gran Canaria and Tenerife) is constant. The conditioning of many cliffs is an example of this phenomenon, as residential and leisure requirements entail land clearings of considerable size, slope ruptures, rock falls and landslides, which are localized but frequent and of a noticeable magnitude. All these are increasingly recurring phenomena, especially in detritical cliffs due to the lower resistance of their materials (Yanes 2016). The destruction of buildings and the temporary disablement of coastal communication channels are increasing and with it the need for civil engineering works to contain these walls.

8.6 Final Considerations

The geoecological values that concur in the rocky coastlines of the Canary Islands are undoubted, as a function of their formal and functional diversity. The reason is that they are clear exponents, on the one hand, of the wide range of edifices created by the insular volcanism; and, on the other hand, the transformations of such edifices by interaction between their lithology, structure and age and the marine dynamics. Added to this is the possibility of being able to identify different stages in the progressive dismantling of what was formed by the volcanoes. Another point worth mentioning is the presence of several types of coast on the same stretch of coastline, which occurs when two or more of the abovementioned cliff patterns are associated. The position of old shores is in part due to specific mechanisms, consubstantial in the archipelago to the intermittent character of the eruptive activity; but also to the succession of contrasted environmental situations, such as those that have governed its quaternary evolution.

These singularities are influenced, in any case, by the impact generated by human activity. Its mark is such that the natural dynamics of these coastlines give way to the quantitative and qualitative changes introduced by their inappropriate use, in both modality and/or intensity. The consequent reduction, if not loss, of geomorphological, biological and landscape values, means that the conservation prospects of the rocky coastlines of the Canary Islands are not very encouraging.

References

Anguita, F., Márquez, A., Castiñeiras, P., & Hernán, F. (2002). *Los volcanes de Canarias. Guía geológica e itinerarios* (p. 222). Madrid: Editorial Rueda.

Armas, V., Romero, C., & García-Cacho, L. (2001). Significado morfológico de los depósitos volcanoclásticos y detríticos del Valle de La Orotava. Tenerife (Islas Canarias). In F. Manero (Ed.), *Espacio natural y dinámicas territoriales* (pp. 195–207). Valladolid: Edición de la Universidad de Valladolid.

Arozena, M. E. (1991). *Los paisajes naturales de La Gomera* (p. 346). Santa Cruz de Tenerife: Edición del Cabildo Insular de La Gomera.

Caniaux, G. (2007). Morphologie des littoraux aux Açores. In S. Etienne & R. Paris (Eds.), *Les littoraux volcaniques: une approche environnementale* (pp. 15–36). Clermont-Ferrand: Presses Universitaires Blaise-Pascal.

Carracedo, J. C., Pérez-Torrado, F., Paris, R., & Rodríguez Badiola, E. (2009). Megadeslizamientos en las Islas Canarias. *Enseñanzas de las Ciencias de la Tierra, 17*(1), 44–56.

Carracedo, J. C. (2011). *Geología de Canarias I (Origen, evolución, edad y volcanismo)* (p. 398). Madrid: Editorial Rueda.

Criado, C. (1991). *El relieve de Fuerteventura* (p. 318). Puerto del Rosario: Edición del Cabildo Insular de Fuerteventura.

Criado, C., Hansen, A., & Martín, A. (1998). Imbricación de procesos de vertiente, torrenciales y eólicos en el oeste de Gran Canaria: La génesis de La Punta de Las Arenas. In A. Gómez Ortiz (Ed.), *Investigaciones recientes de la geomorfología española* (pp. 357–366). Barcelona: Edición del Servei de Gestió y Evolució del Paisatge, Universidad de Barcelona.

Etienne, S. (2007). Les plates-formes d'érosion marine des littoraux volcaniques. In S. Etienne & R. Paris (Eds.), *Les littoraux volcaniques: une approche environnementale* (pp. 37–55). Clermont-Ferrand: Presses Universitaires. Blaise-Pascal.

Etienne, S., & Paris, R. (Eds.) (2007). Les littoraux volcaniques: une approche environnementale (p. 224). Clermont-Ferrand: Presses Universitaires. Blaise-Pascal

Fernández-Palacios, J. M., & Martín, J. L. (Eds.). (2001). *Naturaleza de las Islas Canarias. Ecología y conservación* (p. 474). Santa Cruz de Tenerife: Editorial Turquesa.

Guilcher, A. (1981). Tres grandes falaises et megafalaises d'iles macaronesiennes: Grande Canarie, Tenerife, Madere, Sao Miguel. *Livro Homenagem a O. Ribeiro. Vol. 1*, Lisbonne (pp. 357–366).

Mitchell, N. C. (1998). Characterizing the irregular coastlines of volcanic ocean islands. *Geomorphology, 23,* 1–14.

Pérez-Chacón, E., Hernández-Calvento, L., & Yanes, A. (2007). Transformaciones humanas y sus consecuencias sobre los litorales de las Islas Canarias. In S. Etienne & R. Paris (Eds.), *Les littoraux volcaniques: une approche environnementale.* Clermont-Ferrand: Presses Universitaires. Blaise-Pascal (pp. 173–191).

Ramalho, R., Quartau, R., Trenhaile, A., Mitchell, N., Woodroffe, C., & Avila, S. (2013). Coastal evolution on volcanic oceanic islands: A complex interplay between volcanism, erosion, sedimentation, sea-level change and biogenic production. *Earth-Science Reviews, 127,* 140–170.

Romero, C. (2003). *El relieve de Lanzarote* (p. 242). Arrecife de Lanzarote: Edición del Cabildo Insular de Lanzarote.

Sabaté, F. (1993). *Burgados, tomates, turistas y espacios protegidos.* Edición del Servicio de Publicaciones de la Caja de Ahorros de Canarias. Santa Cruz de Tenerife, 836 pp.

Trenhaile, A. (1987). *The geomorphology of rocks coasts* (p. 384). Oxford: Oxford University Press.

Watts, A. B., & Masson, D. G. (1995). A giant landslide in the north flank of Tenerife, Canary Islands. *Journal of Geophysical Research, 100,* 24487–24498.

Yanes, A. (2004). Los megaacantilados de Canarias: la diversidad dentro de una aparente homogeneidad. In R. Blanco, J. López Bedoya, & A. Pérez Alberti (Eds.), *Procesos geomorfológicos y evolución costera* (pp. 411–426). Santiago de Compostela: Edición del Servicio de Publicaciones de la Universidad de Santiago de Compostela.

Yanes, A. (2006). Caracterización morfológica de las costas de las Islas Canarias. In R. Pie & J. Vilanova (dirs), *Proyecto de directrices de ordenación del litoral de Canarias.* Edición de Equip BCpN. Barcelona, 30 pp.

Yanes, A., & Beltrán, E. (2009). *Acantilados con vegetación endémica de las costas macaronésicas. Bases ecológicas para la gestión de los tipos de hábitat de interés comunitario presentes en España (Directiva 92/43/CEE).* Ministerio de Medio Ambiente, Biosfera XXI Estudios Ambientales S.L. y Grupo Tragsa. Madrid, 96 pp.

Yanes, A. (2016). Caracterización y clasificación morfológica de los acantilados detríticos en Tenerife (Islas Canarias) e incidencia en ellos de los usos del suelo, en Tenerife. In M. E. Arozena & C. Romero (Eds.), *Temas y lugares. Homenaje a Eduardo Martínez de Pisón* (pp. 445–476). La Laguna: Edición del Servicio de Publicaciones Universidad de La Laguna.

Part II
Lineal Sandy Coasts and Beaches

Chapter 9
Beaches of Galicia

Ramón Blanco-Chao

9.1 Introduction

The coast of Galicia corresponds to the NW end of the Iberian Penninsula and extends from the estuary of the Eo River, in the north, to the mouth of the Miño River in the border with Portugal. This coast has distinctive characteristics compared with the rest of the Atlantic Iberian coasts given the importance of geological and structural controls that define the deeply indented coastline. Although is a coast exposed to a energetic storm and swell wave regime, at different scales there are great variations in the orientation of the coastline that causes a high variability in the exposition and the energetic regime.

Beaches represent about the 30% of the Galician coastline which is predominantly composed of rocky sectors. They show a great diversity of typologies and sizes, wave energy regimes and morphodynamics. There are a number of embayed and pocket beaches, many at the foot of cliffs meanwhile many of the longest beaches are associated to extensive lagoon complex or spits closing intertidal and estuarine systems, and dune systems are also common on many beaches. In total, and considering the emerged beach, about 876 beaches has been identified and mapped in the *Plan de Ordenación do Litoral* of the Galician Government, (www.xunta.es/litoral/web/index.php) but in the digital vector cartography available, more than 1,260 beach bodies were digitized, revealing the high fragmentation of the sedimentary bodies.

R. Blanco-Chao (✉)
Departamento de Geografía, Universidad de Santiago de Compostela,
15782 Santiago, Spain
e-mail: ramon.blanco@usc.es

J. A. Morales (ed.), *The Spanish Coastal Systems*,
https://doi.org/10.1007/978-3-319-93169-2_9

9.2 Geological Framework

The geology of Galicia is part of the Hesperian Massif of the NW of the Iberian Penninsula, where a variety of rock types and tectonic structures plays an important role in the location, shape and morphodynamic of beaches (Fig. 9.1). In the north

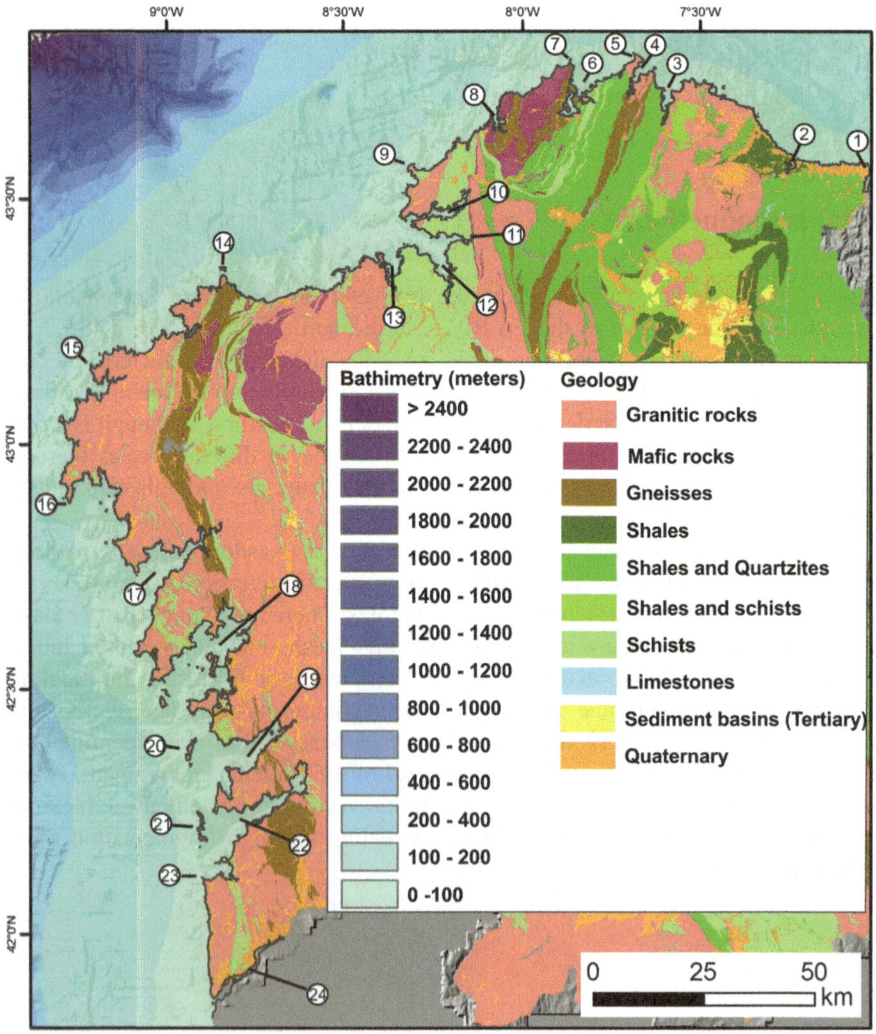

Fig. 9.1 Geological map of Galicia. The numbers correspond to places cited in the text: (1) Eo River, (2) Ria of Foz, (3) Ria of Viveiro, (4) Ria of Barqueiro, (5) Cape Estaca de Bares, (6) Ria of Ortigueira, (7) Cape Ortegal, (8) Ria of Cedeira, (9) Cape Prior, (10) Ria of Ferrol, (11) Ria of Ares, (12) Ria of Betanzos, (13) Ria of Coruña, (14) Cape San Adrián, (15) Cape Vilan, (16) Cape Finisterre, (17) Ria of Muros, (18) Ria of Arousa, (19) Ria of Pontevedra, (20) Ons Islands, (21) Cies Islands, (22) Ria of Vigo, (23) Cape Silleiro, (24) Miño River

coast, igneous rocks are less frequent, being dominated by Precambrian and Paleozoic metamorphic rock formations composed of shales and quartzites, and gneisses. In the NW corner between the Rias of Ortigueira and Cedeira, the basement is composed of a complex of mafic and ultramafic rocks. In the west coast Hercynian and Late Hercynian granitic rocks are the dominant rocks, together with formations of Precambrian micaschists and gneisses. The Hesperian Massif is affected by a dense network of faults and fractures running with directions NE-SW, NW-SE and N-S (Parga-Pondal 1969). This faults were reactivated during the Eocene and until at least the Early Quaternary (Pérez-Alberti 1991).

The lithological diversity and the tectonic structure is the main factor controlling the very indented coast of Galicia. Not only the great rias of the west coast, but along all the coast and at all scales, there is a succession of inlets and embayments, with some straight segments as the rocky southwest coast, or the cliffed coast at the west of the Eo river in the north coast. The coastal bays and inlets where beaches can develop are in many cases opened following weakness zones as fractures, contacts of different rocks or respond to differential erosion with embayments in weak rocks flanked by hard rocks.

9.3 Wave and Tidal Regime

The wave regime of the Galician coast can be characterized as a high energy regime, affected in winter and autumn by the circulation of storm tracks in the North East Atlantic. The coast is affected by swell waves from the fourth quadrant, with the most frequent waves (60%) between 1 and 4 m of Hs, and periods between 6 and 10 s arriving from the WNW, NW, NNW and N (Fig. 9.2). Waves from the NE and NNE are only present in the north and northwest coast (Buoys of Bares and Vilan) but totally absent in the southwest coast. Waves arriving from the NE and NNE are not generally high (<3 m) as they are often related with anticyclonic atmospheric situations. The wave regime has a marked seasonal behavior, with a marked high in winter, when waves with Hs higher than 4 m represent the 35%, a percentage that decrease in autumn and spring to 17.5 and 12% respectively, and becomes less than 2% in summer. Waves with Hs higher than 4 m are always long swell waves with periods longer than 10 s during all the year. A very similar distribution is recorded in Silleiro, in the southwest coast, with 31% of waves are higher than 4 m of Hs in winter, 12% in spring and 15% in autumn, and about 2.8% in summer, again corresponding to long swell with periods between 10 and 20 s.

Inside the south big *rias* the dominant wave regime respond to the propagation of open ocean waves, which are strongly affected by the presence of islands in front of the rias. This obstacles cause strong phenomena of wave diffraction and shadow zones of low energy that attenuate the wave energy. Added to that, fetch is too short inside the *rias* to develop high waves with local winds, and only some north and northeast winds can generate waves never higher than 2–3 m. Maximum waves of up 10–12 m high are recoded during fall and winter every year in the three buoys,

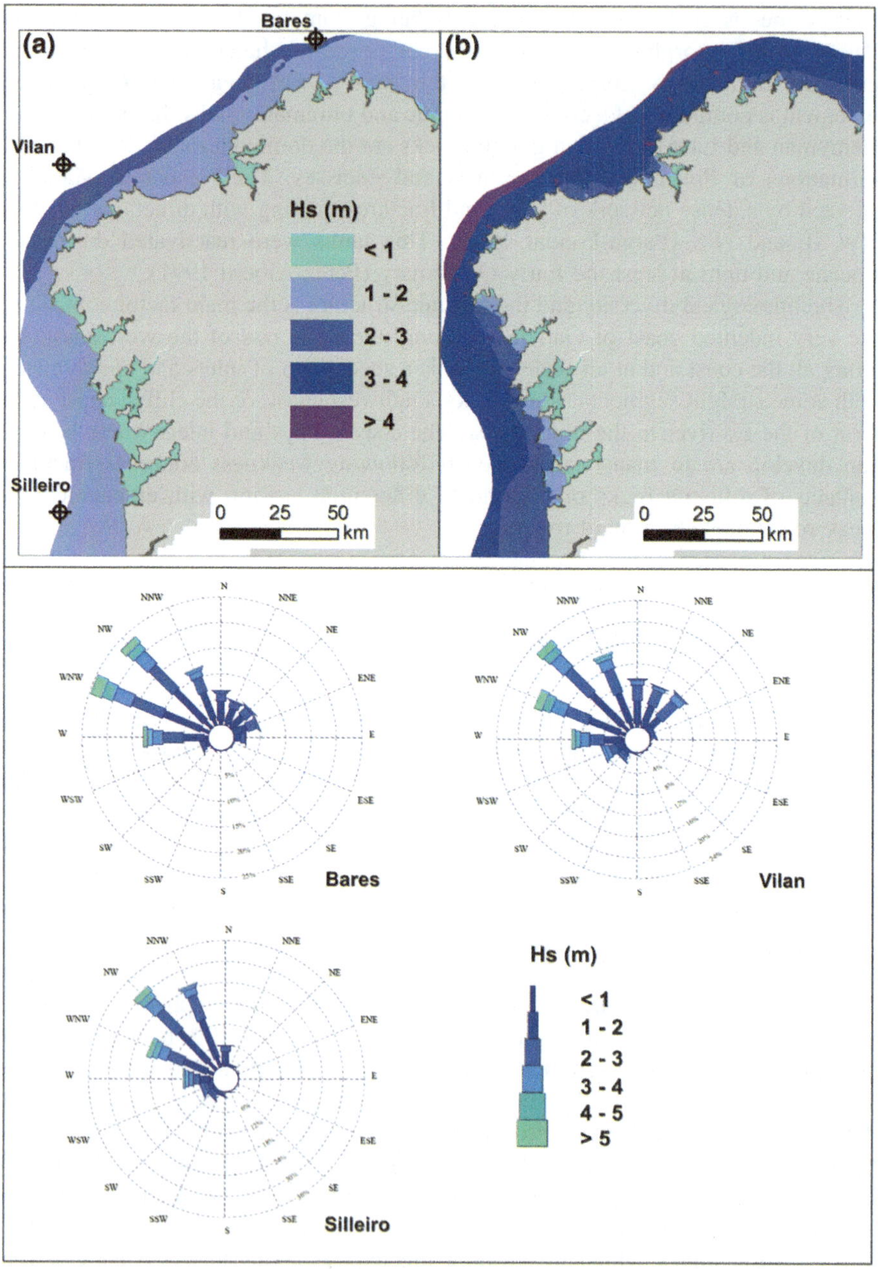

Fig. 9.2 Above, model of wave incidence: **a** wave height in the 50 percentile, **b** waves in the 90 percentile (Xunta de Galicia. Atlas de Ondas. www.meteogalicia.gal/modelos/atlas/atlasOndas.action?request_locale=es). Below, roses of wave component and height (Spanish Port Authority)

and although they represent a low percentage of the total wave record, their effects are important, both for beach morphodynamic and because the damages on coastal infrastructures.

Tidal regime is semidiurnal, with a mean spring range of 3.75 m. The tidal wave propagates northward with a maximum meteorogical tide in Vigo of 3.26 and 3.98 in A Coruña. Spring tidal range can exceed 4 m frequently under storm surge conditions.

9.4 Sediments

In many coasts of the world rivers are the major sources of sediments for beaches, but in Galicia the compartmentation of the coastline and the fact that the major rivers has their mouth in the bottom of the major inlets as the rias, inhibits the development of long coastal drifts and makes more important the local sources of sediment, and fluvial supplies are usually restricted to the innermost part of the rias (Bernárdez et al. 2012). The dominant component in the beaches along the entire coastline of Galicia is quartz, and the abundance of other minerals depends more on the lithological properties of the local geological context, the presence of rivers and the size of their catchment area and the exposition to waves. Sands are then predominantly siliciclastic, composed of quartz and feldspar derived from the coastal rock outcrops, with minor components of other minerals (Arribas et al. 2010), although the ratio of quartz/feldspar ratios is not only dependent on the source rocks and increase with the beach exposition (De Jong and Poortman 1970). Nevertheless quartz can be the major component of the total of the mineral fraction in the western coasts as both rock coastal sectors and the river catchments are predominantly granitic, although is common the formation of layers of heavy minerals as zircons or tourmalines. In the NW coast the abundance of heavy minerals, as garnets or olivines is much higher in the beaches just on the east flank of the mafic complex of Cape Ortegal, being scarce or absent on the beaches of the rias of Ortigueira, Barqueiro and Vivero (Bernárdez et al. 2012). In those coastal area dominated by low grade metamorphic rocks, as shales or phyllites, beaches are still mainly composed of quartz, but including rock fragments of sand size although usually in low proportions (Arribas et al. 2010).

The variations in the degree of exposition and the energetic regime of the beaches introduce a major factor of variation in the grain size. A complete range from fine sands to coarse gravel and clasts can be found in the beaches of Galicia, including boulder beaches. Considering only sand beaches, the grain size is dependent on the degree of exposition to waves, and is strongly influenced by the wave attenuation in the deep inlets, especially in the deeper ones as the *rias*. A study carried out in the Ria of Arousa found an increase in the grain size from the inner and less energetic sections of the Ria to the beaches in the outer and more exposed section, although a relative homogeneity of medium-coarse sands between 0.5 and 2 mm was detected (Rodríguez et al. 1987). In the Ria of Vigo, the Cies

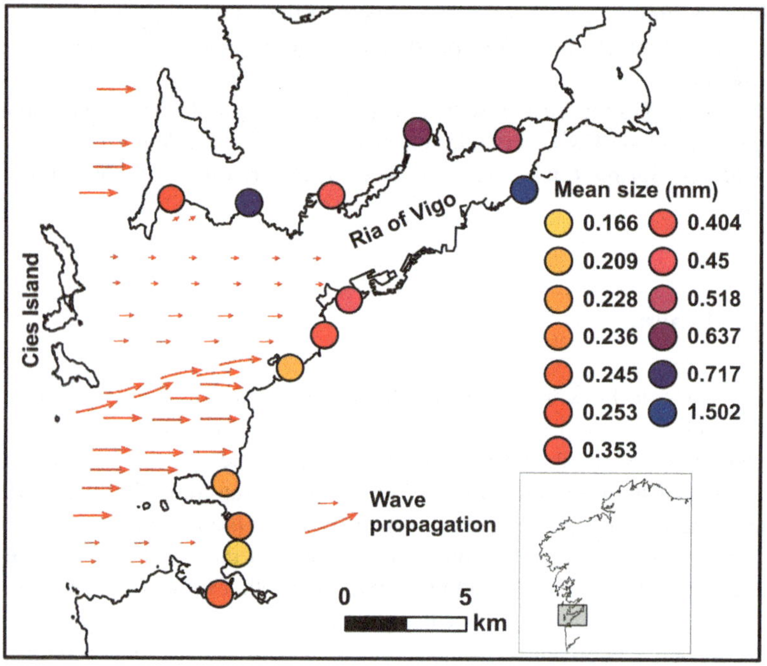

Fig. 9.3 Protective effect of the Cies Islands at the mouth of the Ria of Vigo and variation in grain size in the beaches. The wave propagation corresponds to a SW storm (Hs = 7 m, T = 16 s) (after Queralt et al. 2002)

Islands act as an obstacle for the wave propagation, creating a shadow zone and focusing the energy in the south more external beaches, which show a more dissipative profile and with finer sediments (0.1–0.5 mm) than in the north and inner sections (0.3–1.5 mm) (Queralt et al. 2002; Bernabeu et al. 2012) (Fig. 9.3). The different degree of exposure of the northern and southern margins of the *ria* and the progressive wave attenuation from the outer to the inner part is a good example of the relevance of the exposition factor in the indented coast. Other studies in the beaches of the Ria of Ares, revealed a predominance of well sorted fine sands (0.1–0.3 mm) (Asensio and Grajal 1982).

The carbonate content in the sediments in the Galician beaches is of biogenic origin, deriving from the shells of benthonic organisms and calcareous algae (Bernárdez et al. 2012; Vilas et al. 2005). The biogenic component show a high variability, with percentages almost negligible of less than 1% to be the major constituent with more than 75% (Flor et al. 2004). The percentage of the biogenic component was related with the degree of exposition of the beaches, with the maximum contents in the more exposed beaches and the higher supply of marine origin (Alejo et al. 2000; Rodríguez et al. 1987). The content in biogenic component in beach sediments has been also used as a marker of a modal eastward drift in the north coast and between Cape Vilan and Estaca de Bares, and southward in the

West Atlantic coast from Cape Vilan (Flor et al. 2004). The amount of bioclastic carbonate components is related with exposition, being concentrated in the coarser fractions of the sand beach, as was found in two exposed systems as Corrubedo and Louro, and in the other hand, the content of organic matter increases in the more sheltered beaches as San Francisco (Alejo et al. 2000). It is important to take in account that the biogenic content is strongly related with the biological productivity of the areas close to the beach. Beaches closed by rock points or linked to extensive intertidal rocky surfaces use to show a high amount of biogenic content.

9.5 Types and Characteristic of the Galician Beaches

One of the major factors introducing variability in the beach typology is the diversity of coastal energetic environments derived from the beach exposition to waves. As was cited above, the indented coastline makes difficult the occurrence of long longshore drifts, being common closed or semi-enclosed coastal sedimentary cells. The deeply indented coastline is the main factor in the dominance of small beaches. The 85% of the beaches have lengths of less of 500 m, 13% have lengths between 500 and 1,500 m and only six beaches have lengths above 1,500 m being the beach of Carnota the longest one with 5,600 m (Fig. 9.4).

Most of beaches fall in the intermediate typology of the classical definitions of Wright and Short (1984) and Masselink and Short (1993). The more exposed beaches usually show a mix profile with a low slope dissipative profile in the low tidal sector and a reflective sector with a higher slope in the high tide sections (Álvarez-Vázquez et al. 2003; Lorenzo et al. 2007), meanwhile in low energy beaches reflective profiles are dominant with reduced cross-shore transport as in the Cies Island. Occurrence of rhythmic forms as beach cusps are frequent in a wide range of the Galician beaches, and the presence of bars, arcuate, crescentic or oblique, characterizes the most exposed and energetic beaches (Costas et al. 2005). Queralt et al. (2002) found on the other hand a difference in the beach profiles in the Ria Vigo, with the most internal and less energetic beaches showing a composed profile with higher slopes and a coarser grain size than the most external, with finer sediments and more uniform and low slope profiles. The wave influence diminishes inside the deep inlets and *rias*, with a more important role of tides in the morphology of the beaches (Vilas et al. 2005) and can be a major factor in profile changes and cross-shore transport (Gonzalez-Villanueva et al. 2007).

The exposition to waves controls the occurrence of longshore sediment transport in the beaches. Many beaches are located inside inlets, or have their limits defined by salients of the coastline, forcing strong and persistent patterns of wave diffraction. The beaches in the north coast are protected from W and SW waves but exposed to NW waves. Open beaches show a net eastward drift, but in embayed beaches or those beaches located inside inlets or rias can show an eastward sedimentary drift caused by the strong wave diffraction as for example in Ria of Barqueiro (Delgado et al. 2002). On the other hand, in the south west coast, being

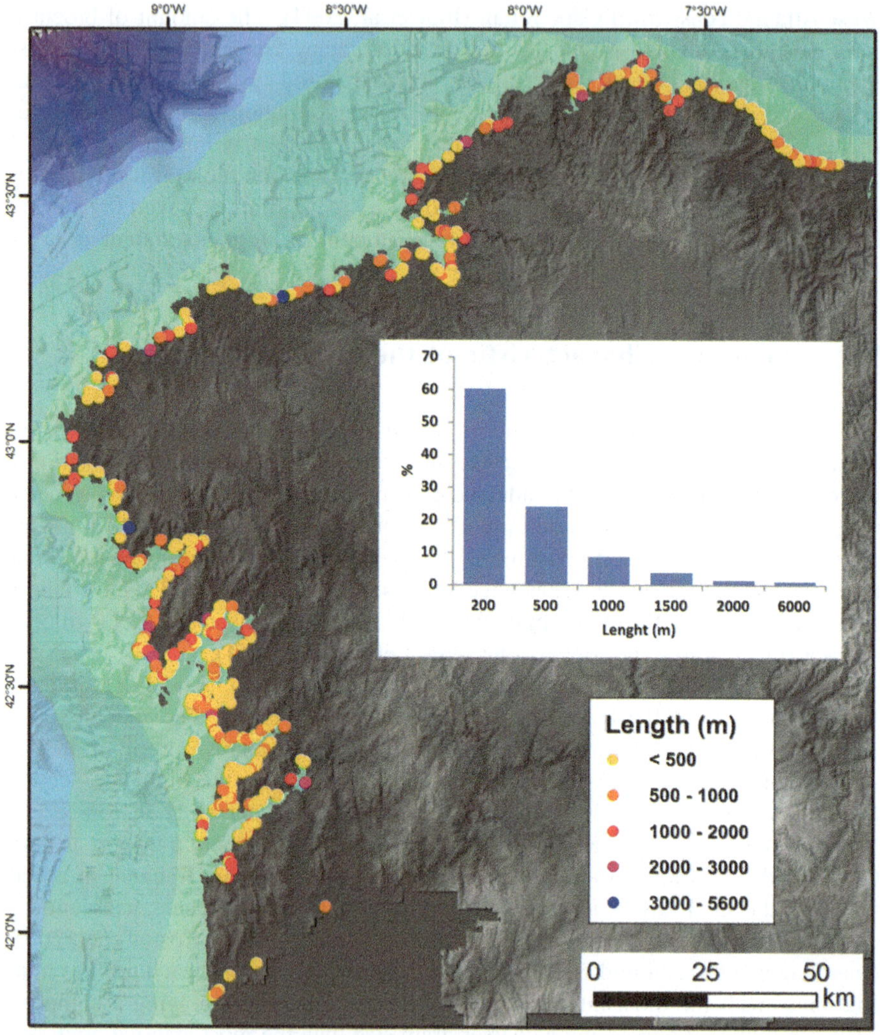

Fig. 9.4 Distribution of beaches along the coast of Galicia classified by their length (Xunta de Galicia: Plan de Ordenación do Litoral)

also dominant the NW waves, there are frequent storms with high waves from the W and SW. This variation in wave approach introduces changes in the longshore sediment transport that sometimes can be also affected by the presence of shoals in front of the beaches as was found in Louro Beach (Almécija et al. 2008), where the alongshore sediment transport is not sustained in one direction, but changes from southward to northward as approaching waves are from NW or SW (Fig. 9.5).

Coarse clastic beaches are present along the coast of Galicia, generally associated to high energy locations and to the occurrence of rocky coasts, cliffs or to

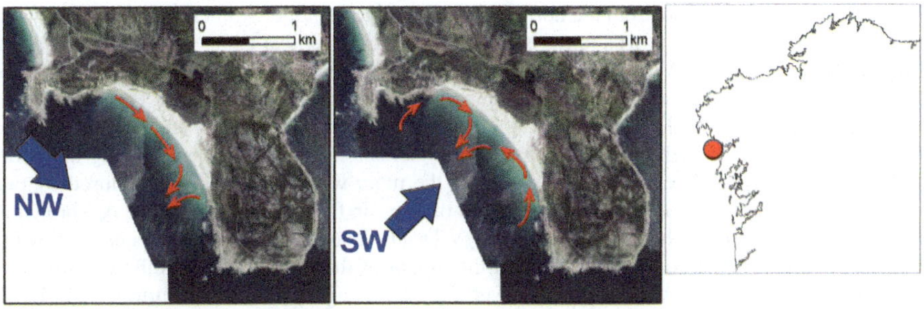

Fig. 9.5 Variations in the exposed beach of Louro at the west coast. The most common longshore transport is southward, but with SW storms the beach develop two cells forcing a northward transport in the south end of the beach (after Almécija et al. 2008)

ancient sedimentary deposits. The range of grain sizes includes gravels to boulder beaches, frequently found over rock platforms, which exerts a great influence in their morphodynamic processes. Gravel and clast beaches are present in coastal sectors with all types of rocks, although boulder beaches (sizes >25 cm in the long axis) are most frequent in granitic environments than in metamorphic rocks (Pérez-Alberti and Trenhaile 2015). The occurrence of gravel and clast beaches is normally linked to pocket and small beaches, where the main and frequently the only, source of sediment is the erosion of unstable cliffs with slope slumps, or the erosion of thick and extensive ancient continental sediments eroded with the Holocene rising sea level (Pérez-Alberti et al. 1997; Blanco-Chao et al. 2003, 2007). Gravel and clast beaches have a morphodynamic behavior characterized by typical high slope profiles, frequent beach crests and cross-shore classification of clast shapes. Boulder beaches respond only to very high energy waves able to move megaclasts (Pérez-Alberti and Trenhaile 2015), although longshore gradients can also be stablished in beaches with a wide range of particle size (Pérez-Alberti and Trenhaile 2015; Blanco-Chao et al. 2006).

A classification of beaches for Galicia has been proposed with several main types, attending to their morphology, space of accommodation and orientation to incoming waves (López Bedoya and Pérez-Alberti 2006; López-Bedoya and Freire-Boado 2013). Drift aligned: include those long beaches with near straight planform, in not very deep coastal entrances closed by rock points, usually exposed and energetic. Embayed: beaches in embayments with a concave planform, including the zeta-bay or log-spiral types, common both in protected areas inside the *rias* but also in exposed areas but protected by rock capes or salients that forces diffraction patterns. Longshore sediment transport can vary according wave approach although always conditioned by the rock limits of the cell. Pocket: beaches in protected sectors in relatively deep but not very wide embayments, closed by rock points and frequently backed by cliffs and continental slopes. Beaches on rock platforms: generally coarse clastic beaches over rock shore platforms, sometimes restricted to the high tide elevations and, in the case of coarser beaches as

Fig. 9.6 **a** A group of beaches in the north coasts forming an unique sedimentary cell. To the west there is a beach-dune system (**a1**), to the east the coast pass to vertical cliffs, with a cliff-foot beach (**a2**). **b** The double spit of Vilarrube, with a wide intertidal complex to the east and a smaller one to the west. Note the small island forming a tombolo exposed during low tides. **c** Pocket beach of Baleo, a system composed of two beaches, a sand beach in the first plane, and a smaller mixed beach with clast in the high tide elevations. Both systems remains isolated only during high tide. **d** A group of beaches in embayments opened in the more weathered schists and closed by two resistant granitic outcrops. **e** The log-spiral beach of Ares, in the north margin of the Artabro Gulf. **f** A group of pocket beaches in the high energy Death Coast. The inlets are opened following fractures and geological contacts. **g** A spit at the mouth of the estuary of the Anllons River. The hills to the north of the anchor point are covered by climbing dunes. **h** The Furnas-Sieira beach system. Two beach-dune systems forming a single, long sedimentary cell. The morphodynamic of the north beach is strongly controlled by the refraction of waves and the shadow zone created by the north rock coast. To the south, the beach pass to beaches on rock platforms, still backed by dunes. **i** Beach on a shore platform modelled on metamorphic rocks (Credits: **a**, **b**, **d**, **e**, **f**, **g** and **h**: 2016 © Digital Globe; **a1**, **a2**, **c** and **i**: Xunta de Galicia, POL)

boulders, can be located in coastal salients. River mouth beaches: are those beaches associated to rivers which mouth is in a relatively deep and narrow inlet, highly affected by the fluvial discharges, sometimes develop high berms that temporary enclose the river channel. Cuspate and tombolos: beaches responding to the typical geomorphological features that many times are related with wider sedimentary systems. Spits and double spits: spits are relatively frequent at the bottom of many inlets forming the closing bar of estuaries, or closing intertidal complex and anchored in rock points.

The longest and more complex beaches are those linked to coastal lagoons, intertidal and estuarine systems, but not associated to the major rivers. This type of beaches include single and double spits with extensive dunes, but also drift aligned beaches initially developed as spits, with fresh water bodies or humid areas at the back. The more exposed beaches, namely in the northwest coast have usually composed profiles, reflective at the high tide elevations and dissipative at the low tidal elevations, with longitudinal or crescentic bars. The indented coastline and the abundance of cliffs determine the occurrence of embayed and pocket beaches, with a number of cliff-foot beaches. Small rivers with mouths in deep inlets supplied the sediment for elongated beaches, which complexity is also high variable, sometimes with dune fields and marsh formations.

9.5.1 North-East Coast

The north coast exposition is mainly to NW swell waves and fully protected from SW waves. Between the Eo River and the Ria of Foz is straight compared with the most of the Galicia, and dominated by vertical cliffs, frequently with rock shore platforms at the foot. Cliff-foot beaches and beaches on rock platforms are the most common type, generally with wide dissipative profiles and with the morphodynamics strongly controlled by the underlying substrate. These geologically

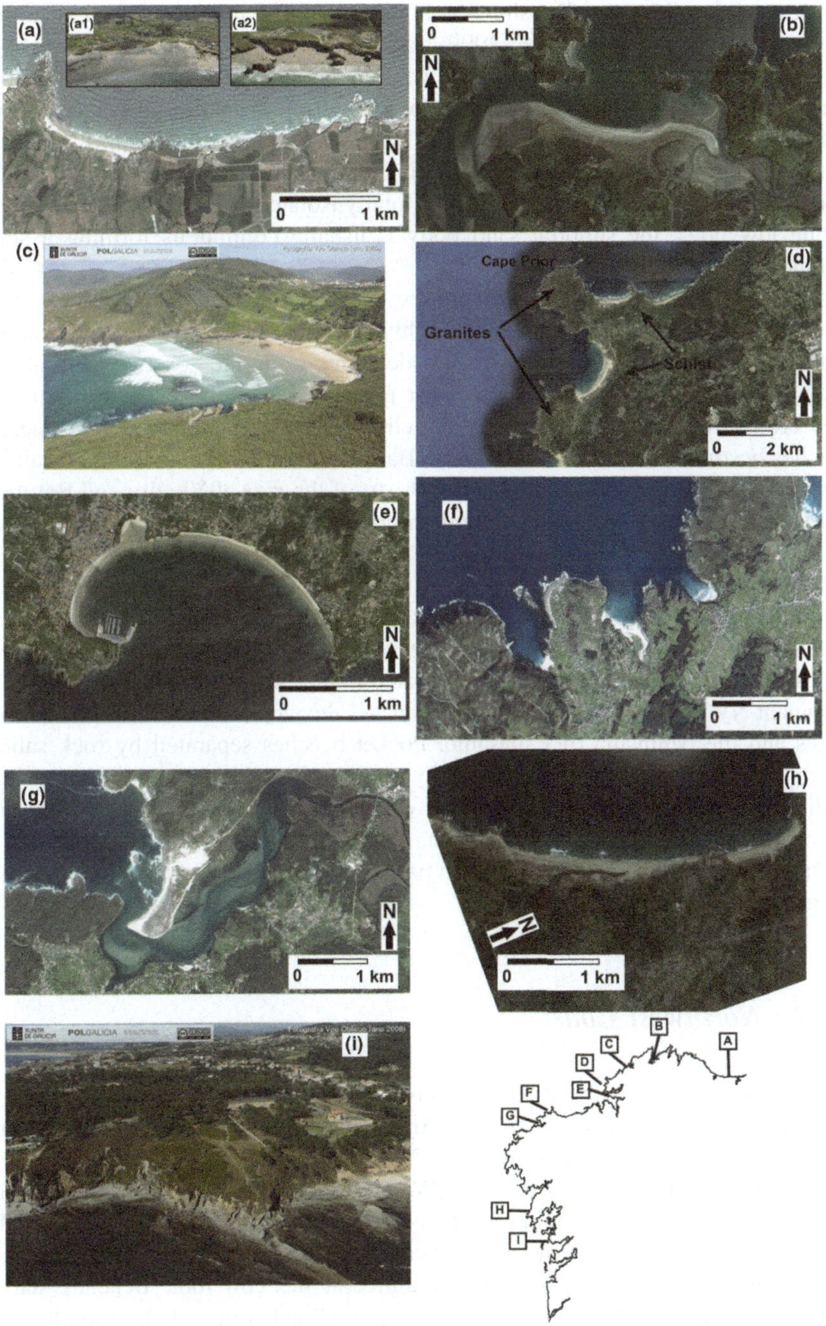
(a)
(a1)
(a2)
N
0 1 km
(b)
0 1 km
N
(c)
POLGALICIA
(d)
Cape Prior
Granites
Schist
N
0 2 km
(e)
N
0 1 km
(f)
N
0 1 km
(g)
N
0 1 km
(h)
N
0 1 km
(i)
POLGALICIA
A
B
C
D
E
F
G
H
I

constrained beaches usually show a higher mobility than those unconstrained beaches, (Jackson and Cooper 2009; Jackson et al. 2005), characterized by a marked reduction of sand volume and the exhumation of the underlying substrate, that leads to an increased control on the beach morphodynamics, as wave energy is conducted by the hard topography. On those sectors where the cliffs are absent beaches backed by dune systems appear, in coastal entrances closed by rock salient. Even though the emerged beaches can be totally isolated during high tide or even in spring low tides, the subtidal sand body is usually continuous forming a single sedimentary cell (Fig. 9.6a).

Beaches closing the Ria of Foz are the external complex of the estuary of the Masma River at the east margin and a highly artificialized beach at the west margin. Between the Ria of Foz and the Estaca de Bares Cape the coast changes its orientation and geology. Beaches become scarcer and generally reduced in length as they are restricted to small inlets in the cliffed coast. The higher development of beaches becomes in the Rias of Viveiro, Barqueiro and Ortigueira, three small *rias* with a NE-SW orientation. The innermost part of the *rias* of Viveiro and Barqueiro has estuarine complex closing beach-barriers, sometimes highly modified by harbors as in Viveiro. At the bottom of the Ria of Ortigueira the beach of Marouzos is a double spit closing a small intertidal system at the east and a wide estuarine complex in the west (Fig. 9.6b). In the flanks of the *rias*, beaches develop in small and usually not deep entrances, showing important differences in their degree of exposition, being more exposed those located in the east margin and oriented to the dominant NW waves. A set of cliff-foot beaches developed between Cape Estaca de Bares and the peninsula of Espasante. Pocket beaches separated by rock salients alternate with longer beaches frequently showing a great variability alongshore, with coarse grained beaches geologically constrained by shore platforms at the ends and sand beaches in the central section. The cliffs at the back of the beaches are unstable, with evidences of present and past slope movements which were the main sediment supply.

9.5.2 Northwest Coast

The coastline changes the direction between the cape Estaca de Bares and the Artabro gulf that groups the *rias* of Ares, Betanzos and Coruña. This coast is exposed to swell and storm waves from NW, WNW and W, but slightly protected from storms waves coming from the SW. The Capelada Range are coastal mountains up to 617 m high, forming a high cliffed coast where sand or gravel beaches are absent, being only present boulder beaches derived from the erosion of ancient Pleistocene deposits and from the instability at the cliff foot. Beaches start to develop in the Ria of Cedeira, with a very modified beach at the mouth of the Condomiñas river, and the double spit system of Vilarrube, which eastern end enclose a estuarine system.

Between The Ria of Cedeira and the cape Prior the coast is again dominated by cliffs, with pocket beaches at the foot (Fig. 9.6c), or beaches located at the mouth of rivers as the beach of Meira. The major beaches of Pantin and Frouxeira correspond to coastal lagoon systems, the first totally infilled, and both with dune systems at the back, including a climbing dune in Frouxeira, now totally stabilized by vegetation. Between the cliffs modelled in paragneisses and the granites of Cape Prior, the highly weathered schist formed an embayment with two beaches, backed by dune systems.

Between cap Prior and Cape Prioriño the coastline changes again to a north-south direction, with two beaches between 1.7 and 2.5 km of length, the northern in an embayment opened in the weathered schist, and the south, Doniños, in a embayment limited by fractures affecting granites, and closing a coastal lagoon (Fig. 9.6d). The mouth of the Ria of Ferrol is closed by two granitic areas that form a narrow channel, in some parts only 300 wide. This creates an extremely protected environment inside the ria, where the shores are today densely urbanized, namely by the town and harbor of Ferrol. Beaches are consequently scarce, and with a clear component of very fine sediments, especially the innermost ones. The Artabro Gulf is a wide embayment, totally protected from SW waves and only exposed to NW waves that are strongly attenuated by wave diffraction. At the bottom of which there are three entrances at the mouth of the rivers Eume, Mendo-Mandeo and Mero. Although called the *rias* of Ares, Betanzos and Coruña, two of them are really inlets where beach-dune systems close estuaries (Ares and Coruña) absent in the case of Betanzos. The most of the other beaches inside the Artabro gulf are small pocket beaches, with the exception of the Magdalena beach, a beach with a log-spiral or zeta bay planform (Asensio and Grajal 1982) (Fig. 9.6e). Other important beach is the Miño beach, a spit that enclose the mouth of the Baxoi river forming a marsh system. Many of these beaches have experienced important anthropic modifications, as sand nourishment, destruction of dunes and construction of promenades and walls at the back of beaches. In the west flank of the peninsula of A Coruña there is an urban beach, also highly modified and constantly subjected to sand nourishment. The exposition of the beaches to the east of A Coruña increases given the lack of coastal salient affecting the waves arriving from NW. The beaches are of high energy, with dune fields as Barrañán, or the long beach of Razo, another system closing a coastal lagoon that was hardly artificialized during the XXth century, including the regulation of the lagoon outlet.

From the cape San Adrian fronted by the Sisargas Island until Cape Finisterre extends the called Death Coast, because the number of shipwrecks recorded along centuries. The coastline changes slightly their general direction, being affected by high swells storms from the NW, but also exposed to the SW storms. This coastal sector is a cliffed coast, with a very irregular planform in which narrow and deep embayments were opened. The majority of beaches are small embayed and pocket beaches at the foot of cliffs (Fig. 9.6f), with the exceptions of the Barra beach, a spit in the estuary of the Anllóns River (Fig. 9.6g), and Traba beach at the front of a small coastal lagoon. The Trece beach is a small mixed system located in a coastal inlet closed by two rock headlands, with a extensive system of climbing dunes at the back and east flanks. In the wide embayment of the called *ria* of Camariñas, beaches are usually small, the longest one at the south of the estuary of the Grande River. Only small embayed or pocket beach appear until the coast between Cape Touriñan and Finisterre, with the highly energetic and exposed beaches of Nemiña, O Rostro and Mar de Fora.

9.5.3 West Coast

The west coast is where the differences in beach exposition are better defined, as is characterized by the four big *rias* of Muros, Arousa, Pontevedra and Vigo. In spite of their great dimensions, the fetch inside the *rias* are always short, and sea waves developed by local winds rarely exceed 2 m of significant height and always with short periods. This coast is in general exposed to the long swells from NW and W and to the SW storm waves, but the changes in the orientation of the coast introduce marked differences between the north and south margins of the *rias*, but not only. The beach of Langosteira, at the eastern protected flank of Cape Finisterre is much less energetic than the beaches located in the outer, exposed west flank. Meanwhile the small beaches located inside the deep embayment of Corcubion and Cee are protected from the most energetic waves, which are also attenuated by diffraction processes, the degree of exposition increase as we move to the south. The beach of Carnota is the longest one of Galicia, with close to 6 km in length. The beach has two well defined sectors, the north a type of small spit that encloses the marsh of Caldebarcos, and the south sector, elongated southward and with a intertidal and marsh system at the back. Both sectors are backed by dunes with an important aeolian activity dominated by the NE winds in summer. The coast sector between Carnota and the cape of Monte Louro is of high energy, with coarse clastic beaches located over granitic and very irregular rock platforms, except at the south with the two sand beaches of Ancoradoiro and Louro, the last another beach closing a coastal lagoon.

Inside the Ria of Muros, the orientation of the coastline and the diminishing wave energy from the outer to the inner sectors set a marked energy gradient reflected in the type of beaches. The north margin is exposed only to SW waves but protected from the most energetic waves from the NW and W. Waves entering in the Ria of Muros experienced a progressive attenuation as long swells are diffracted at the mouth of the *ria*. In the south margin of the *ria* of Muros, the energetic environment of the beaches are also controlled by the wave deformations inside the *ria*. But from the beach of Fonforrón, just at the south of the locality of Porto do Son, the coast runs NNE-SSW without any protection from the NW waves. The beaches of the exposed half south margin of the Ria of Muros are then energetic, with composed profiles and development of longitudinal or transverse bars. There are three shallow inlets with relatively long beaches (between 2 and 3 km), Baroña-Queiruga, Sieira-As Furnas and Espiñeirido. They are beaches totally exposed to the most energetic waves from the NW and W, but slightly protected from the SW that arrive to the beach after strong suffering a strong refraction. The complex of Baroña-Queiruga is composed of three beaches, the smallest one, the beach of Dique, in a small inlet with the mouth of a river. The three beaches form an only sedimentary cell, and they are isolated from each other only during high tides. A similar case is Sieira-As Furnas two beaches separated by a metamorphic cliffed point, but with a common subtidal sand prism (Fig. 9.6h). They are beaches with a modal composite profile, more reflective at the high tide elevations, commonly with beach cusps, and dissipative with longshore bars at the low tide elevations. To the north of this beach system there is a good example of a lithologically constrained beach, a beach on a metamorphic rock platform, in which present composition the biogenic component is dominant with up to 80–90%. Although separated by a low granitic point, the beach of Furnas-Sieira is not totally disconnected from the southern beach of Espiñeirido. On the rock point separating both beaches there are beaches on rock platforms, and the dune system is continuous at the back of the three beaches. The beach of Balieiros is more exposed and isolated, also backed with dunes.

Corrubedo is composed of a long beach (near 5 km), with a well developed dune system including a high dune of 15 m non stabilized by vegetation, a intertidal saltmarsh at the back and a small fresh water lagoon in the south end (Vilas et al. 1986). Inside the Ria of Arousa the waves propagated from open ocean enters forming two focus of diffraction caused by the Salvora Island, that are progressively attenuated traveling to the NE, especially to the north of the obstacles formed by the Arousa Island and the shoals and small islets to the west. The low energy inside the *ria* even allows to develop beaches allocated in salient as in As Sinas, in the NW of the Arousa Island. The beaches inside the *ria* are small in size, frequently modified by human interventions for touristic uses, including sand nourishment. In the outer south end of the *ria*, there is an intertidal complex protected from the open ocean waves by the isthmus of A Lanzada that enclosed the intertidal system after the XVIII century, forming now a sedimentary cell at the south of the Grove Penninsula.

Moving to the southern Ria of Pontevedra, beaches are still exposed, with disspative profiles and presence of low tide bars, although they are protected from the SW waves by the presence of the Ons Islands. Inside the Ria of Pontevedra, the schema of progressive wave attenuation to the innermost parts explains the transition from sand beaches more exposed in the outer sections to very fine sand and silt systems in the internal sections. In the south margin, the deep inlet of Aldán allocate a group of low energy beaches in very shallow entrances. The cliffed coast of A Vela is bare of beaches, until the south end. In the Ria of Vigo the Cies Islands exert a strong control in wave propagation inside the *ria*, forcing a net eastward propagation, highly attenuated by the archipelago and thus determining the variation in beach morphology, more exposed at the outer sections. At the south end of the *ria*, in a embayment, the beach of Panxón-América is separated by a rocky cliffed point from Ladeira Beach, that closes at the north end the intertidal system of A Ramallosa.

The last coastal sector before the Miño River, at the border with Portugal is a straight rock coast, where sand beaches are totally absent, with the exception of a small system in the inlet of Oia. The beaches along this shore are coarse-clastic, which the only source of sediment was essentially the erosion of Pleistocene sedimentary deposits during the Holocene sea-level rise, although it was probed that they are in part inherited from the last interglacial. There are only very small pocket beaches just to the north of the locality of A Guarda, and cuspate or salient beaches in the outer margin of the Miño River, fronting the best developed and extensive beach dune-system of Camiña and the beach of Moledo in the Portuguese coast.

9.6 Erosion Processes

The number of works about erosion in Galicia are still scarce, and given the explained variability of beach types and wave energy environments, there is quite difficult to extrapolate tendencies from single cases. The rate of the sea level rise in Galicia has been calculated in 2.21 and 2.91 mm/yr^{-1} in the tide gauges of Coruña and Vigo respectively during the period 1943–2001 (Marcos et al. 2005), but the effects an degree to which beaches of Galicia respond are questions that are still far of having an answer. The variability in wave regime, with a possible increase in the number of high energy storms will probably arise as an important factor in the future, as many Galician beaches have very limited spaces of accommodation and possibilities of landward migration. The coast of Galicia has been densely occupied since the second half of the XXth century, and the processes of urbanization experienced a remarkable increase in the last decades, with construction of marinas and harbors. Sand mining in dune systems was also a common and extended practice until the decades of 1970–1980.

The processes of beach erosion seem to be controlled by local factors, and human interference arise as an important cause, although not the only, of the most severe erosive processes (Alonso et al. 2000; Lorenzo et al. 2007; Vila-Concejo et al. 2002; Delgado et al. 2002). This is the case of three spit-beaches in the north *rias* of Barqueiro, Viveiro and Ortigueira, under very similar wave regime and sea-level changes, but with different recent evolution (Lorenzo et al. 2007). In Ortigueira the beach show an accretionary trend probably because the increased fluvial supply by human activities in the river catchments, but in Viveiro, where the coast was heavily artificialized the beach, today limited by the jettie of an artificial channel, remains stable because wave attenuation caused by the construction of a harbor. On the contrary, in Barqueiro the beach experiences a tilt with sustained erosion in the east end and a accumulative trend in the west end of the spit due to the westward drift (Lorenzo et al. 2007; Delgado et al. 2002). A similar tilt was detected in protected and small beaches after human interferences, as in Cies Island (Alejo et al. 2005; Costas et al. 2005). Beaches with strong human modifications, as the construction of hard sidewalks at the backshore, prevent beach retreat and lead to strong changes in beach morphodynamics and erosive tendencies as in the beach of Samil (Alejo et al. 2005).

9.7 Remarks

The great variety of beach types in Galicia derives first of the irregular and deeply indented of the coastline, controlled by geological and tectonic factors. This irregular planform introduces marked variation in the degree of exposition to incoming waves, which are characterized by high energy long fetch Atlantic swell and the pass of storms. The occurrence of inlets, embayments and *rias* of different size and shape also forces important phenomena of wave diffraction that frequently determines wave energy distribution and changes in the alongshore beach morphology. The supply of fluvial sediments is restricted to the innermost sections of the *rias* or bays, and except for those beach systems enclosing evolved lagoons or estuaries sediment sources for beaches are local, derived from the erosion of surrounding rock coasts, or from inherited sedimentary deposits. This coastal configuration inhibits the formation of long alongshore sediment drifts, and the most of the beaches forms closed sedimentary cells with any or little transfer to adjacent cells.

References

Alejo, I., Nombela, M. A., Francés, G., Alcántara-Carrió, J., Manso, F., & Vilas, F. (2000). El efecto de exposición en el patrón sedimentario del sector submarcal de tres playas en la Ría de Muros y Noia (NO de España). *Journal of Iberian Geology, 26,* 189–216.

Alejo, I., Costas, S., & Vila-Concejo, A. (2005). Littoral evolution as a response to human action: The case of two sedimentary systems in a Galician Ria. *Journal of Coastal Research, SI49,* 64–69.

Almécija, C., Alejo, I., & Pérez-Arlucea, M. (2008). Morfodinámica e hidrodinámica de una playa expuesta: ejemplo de la playa de Louro (Muros, NW Península Ibérica). *Geo-Temas, 10,* 487–490.

Alonso, A., Lorenzo, F., & Pagés, J. L. (2000). Dinámica litoral y erosión en la Ria de O Barqueiro: factores antrópicos y procesos naturales. *Geogaceta, 28,* 7–10.

Álvarez-Vázquez, R., Bernabeu, A. M., Nombela, M. A., Méndez, G., & Vilas, F. (2003). Evolución morfodinámica de la playa de Panxón (Galicia, NO de España). *Geogaceta, 33,* 19–22.

Arribas, J., Alonso, A., Pagés, J. L., & González-Acebrón, L. (2010). Holocene transgression recorded by sand composition in the mesotidal Galician coastline (NW Spain). *The Holocene, 20,* 375–393.

Asensio, I., & Grajal, M. (1982). Morfología litoral y sedimentación actual de la ría de Ares (La Coruña, Galicia, España). *Cuadernos do Laboratorio Xeolóxico de Laxe, 3,* 247–264.

Bernabeu, A. M., Lersundi-Kanpistegi, A. V., & Vilas, F. (2012). Gradation from oceanic to estuarine beaches in a ría environment: A case study in the Ría de Vigo. *Estuarine, Coastal and Shelf Science, 102,* 60–69.

Bernárdez, P., Prego, R., Giralt, S., Esteve, J., Caetano, M., Parra, S., et al. (2012). Geochemical and mineralogical characterization of surficial sediments from the Northern Rias: Implications for sediment provenance and impact of the source rocks. *Marine Geology, 291,* 63–72.

Bernárdez, P., Prego, R., Filgueiras, A. V., Ospina-Álvarez, N., Santos-Echeandía, J., Álvarez-Vázquez, M. A., et al. (2017). Lithogenic sources, composition and intra-annual

variability of suspended particulate matter supplied from rivers to the Northern Galician Rias (Bay of Biscay). *Journal of Sea Research*. https://doi.org/10.1016/j.seares.2017.05.006.
Blanco-Chao, R., Pérez-Alberti, A., Costa-Casais, M., & Valcárcel-Díaz, M. (2006). Abrasion processes in coarse-clastic beaches linked to rocky shore platforms. *Journal of Coastal Research, 48,* 21–28.
Blanco-Chao, R., Costa-Casais, M., Martínez-Cortizas, A., Pérez-Alberti, A., & Trenhaile, A. S. (2003). Evolution and inheritance of a rock coast: Western Galicia, northwestern Spain. *Earth Surface Processes and Landforms, 28,* 757–775.
Blanco-Chao, R., Pérez-Alberti, A., Trenhaile, A. S., Costa-Casais, M., & Valcárcel-Díaz, M. (2007). Shore platform abrasion in a para-periglacial environment, Galicia, northwestern Spain. *Geomorphology, 83,* 136–151.
Costas, S., Alejo, I., Vila-Concejo, A., & Nombela, M. A. (2005). Persistence of storm-induced morphology on a modal low-energy beach: A case study from NW-Iberian Peninsula. *Marine Geology, 224,* 43–56.
De Jong, J. D., & Poortman, H. H. (1970). Coastal sediments of the southeastern shores of the Ría de Arosa (Galicia, NW Spain). *Leidse Geologische Mededelingen, 37,* 147–167.
Delgado, I., Alcántara-Carrió, J., Alejo, I., & Louzao, M. (2002). Influence of hydrodynamics and sedimentary characteristics of Barqueiro Ria on Arealonga beach dynamics. *Journal of Coastal Research, 36,* 231–239.
Flor, G., Obeso, S., & Blanco, G. F. (2004). El componente bioclástico de playas arenosas de Galicia (NO de España). *Geogaceta, 36,* 143–146.
Gonzalez-Villanueva, R., Costas, S., Alejo, I., & Perez-Arlucea, M. (2007). Morphological changes forced by the tidal cycle in a low energy estuarine beach. *Journal of Coastal Research SI, 50,* 1010–1014.
Hernández-Vega, S., Castellanos, C., & Junoy, J. (2005). Cartography of the surface sediments of the Rıa de Foz (Lugo, Northwest Spain) after the construction of a jetty. *Thalassas, 21,* 83–92.
Jackson, D. W. T., & Cooper, J. A. G. (2009). Geological control on beach form: Accommodation space and contemporary dynamics. *Journal of Coastal Research, SI56*, 69–72.
Jackson, D. W. T., Cooper, J. A. G., & Del Rio, L. (2005). Geological control of beach morphodynamic state. *Marine Geology, 216,* 297–314.
López Bedoya, J., & Pérez-Alberti, A. (2006). Clasificación morfogenética de las playas de arena en Galicia como herramienta para abordar el uso sostenible de los complejos sedimentarios costeros. In A. Pérez-Alberti, J. López-Bedoya (Eds.), *Geomorfología y Territorio. Actas de la IX Reunión Nacional de Geomorfología* (pp. 843–859).
López-Bedoya, J., & Freire-Boado, M. A.: (2013). *Guia de boas prácticas e intervención en sistemas praia-duna.* Xunta de Galicia, Consellería de Medio Ambiente Territorio e Infraestructuras. Santiago. 341 p.
Lorenzo, F., Alonso, A., & Pagés, J. L. (2007). Erosion and accretion of beach and spit systems in Northwest Spain: A response to human activity. *Journal of Coastal Research, 23,* 834–845.
Marcos, M., Gomis, D., Monserrat, S., Álvarez-Fanjul, E., Pérez, B., & García-Lafuente, J. (2005). Consistency of long sea-level time series in the northern coast of Spain. *Journal of Geophysical Research, 110,* C03008.
Martínez, G. M., Pérez-Arlucea, M., González-Villanueva, R., & Campos, A. O. (2011). Anthropogenic influence on the Holocene sedimentation process along the Atlantic coast of Galicia (NW Iberian Peninsula). *Journal of Coastal Research, 64,* 1788–1792.
Masselink, G., & Short, A. D. (1993). The effect of tide range on beach morphodynamics: A conceptual model. *Journal of Coastal Research, 9,* 785–800.
Parga-Pondal, J. R. (1969). Sistemas de fracturación tardihercínicas delMacizo Hespérico. *Trabajos del Laboratorio Xeolóxico de Laxe, 37,* 3–15.
Pérez-Alberti, A. (1991). La geomorfologia de la Galicia Sudoriental (Problemas geomorfoló gicos de un Macizo Antiguo de la fachada Atlántica Ibérica: centrosudeste de Galicia). *Universidade de Santiago de Compostela*, Ph. D. thesis.
Pérez, I., Bernabeu, A., Rey, D., & Vilas, F. (2008). Formación del placer de la playa de Montalvo (Pontevedra). *Geogaceta, 44,* 231–234.

Pérez-Alberti, A., & Trenhaile, A. S. (2015). Clast mobility within boulder beaches over two winters in Galicia, northwestern Spain. *Geomorphology, 248,* 411–426.

Pérez-Alberti, A., Blanco-Chao, R., & Costa-Casais, M. (1997). La importancia de los balances de acumulación/erosión y fosilización/exhumación en la evolución geomorfológica de la costa de Galicia. *Dinámica Litoral-Interior. Actas XV Congreso de Geógrafos Españoles. Vol. I.* Cursos e Congresos da USC. n° 107. Serv. Public. e Interc. Científico da USC. AGE. pp. 205–219.

Perez-Arlucea, M., Mendez, G., Clemente, F., Nombela, M., Rubio, B., & Filgueira, M. (2005). Hydrology, sediment yield, erosion and sedimentation rates in the estuarine environment of the Ria de Vigo, Galicia, Spain. *Journal of Marine Systems, 54,* 209–226.

Queralt, S., Rodríguez-Fernández, N., Bernabeu, A. M., Rey, D., & Vilas, F. (2002). Características morfosedimentarias de las playas de la Ría de Vigo: influencia de las condiciones hidrodinámicas. *Geogaceta, 32,* 211–214.

Rey, D., Rubio, B., Bernabeu, A. M., & Vilas, F. (2004). Formation, exposure, and evolution of a high-latitude beachrock in the intertidal zone of the Corrubedo complex (Ria de Arousa, Galicia, NW Spain). *Sedimentary Geology, 169,* 93–105.

Rodríguez, M. D., Nombela, M. A., Vilas, F., & Rey, L. (1987). Estudio sedimentológico del litoral gallego: Relación entre la distribución granulométrica y el contenido en carbonatos biogénicos de las playas de las rías de Pontevedra y Arosa. *Cuadernos Laboratorio Xeolóxico de Laxe, 11,* 11–20.

Vilas, F., Arche, A., Ramos, A., Sopeña, A., Rey, L., & Nombela, M. A. (1986). El complejo de playa-lagoon de Corrubedo y sus submedios caracterı´sticos, Galicia, NW de España. *Acta Geologica Hispanica, 21–22,* 233–243.

Vila-Concejo, A., Alejo, I., & Vilas, F. (2002). Monitoring of a beach with a strong anthropogenic impact by means of topographic and bathymetric surveys, Samil Beach, Spain. *Shore and Beach, 70,* 3–10.

Vilas, F., Bernabeu, A. M., & Méndez, G. (2005). Sediment distribution pattern in the Rias Baixas (NW Spain): Main facies and hydrodynamic dependence. *Journal of Marine Systems, 54,* 261–276.

Wright, L. D., & Short, A. D. (1984). Morphodynamic variability of surf zones and beaches: A synthesis. *Marine Geology, 56,* 93–118.

Chapter 10
Beaches in Valencian Coast

Josep E. Pardo-Pascual and Eulàlia Sanjaume

10.1 Introduction

The beaches are the most abundant morphological element of the entire Valencian coast. Around 62% of the coast (except for the artificial areas), about 270 km, corresponds to beach segments, being by far the most usual type in the whole Gulf of Valencia and also throughout the southern part of the Valencian territory. Although there are substantial differences, it is clear that all of them present some common features that are determined by the existence of general similar conditions. The first one is the existence of a very small tidal range; the second would be the low energy of waves most of the time; the third would be that most of the beach sediments have has its primary origin in fluvial contributions; an finally the fourth element would be the human action that is very important and has determined the recent evolution of the Valencian beaches.

The current morphology of the beaches is related to the natural and/or anthropic factors affecting them. The beaches are, by definition, areas of accumulation; therefore it is evident that the availability of sediments is an essential factor. Also, another key factor is the energy with which the waves reach the shoreline. That will depend: on the characteristics of the incident waves; the morphology of the submarine beach profile; and the orientation of the coast with respect to the incident waves. In the Valencian coast, longitudinal wave-induced sediment transport takes

J. E. Pardo-Pascual (✉)
Geo-Environmental Cartography and Remote Sensing Group,
Department of Cartographic Engineering, Geodesy and Photogrammetry,
Universitat Politècnica de València, Camí de Vera s/n, Valencia 46022, Spain
e-mail: jepardo@cgf.upv.es

E. Sanjaume
Department of Geography, Universitat de València,
Avda. Blasco Ibáñez, 28, Valencia 46010, Spain
e-mail: Eulalia.Sanjaume@uv.es

J. A. Morales (ed.), *The Spanish Coastal Systems*,
https://doi.org/10.1007/978-3-319-93169-2_10

an essential role in many of its beaches. Human action, on another hand, will alter in a more or less local way, all these conditioning factors.

The alongshore transport plays its important role in each of the existing sedimentary cells along the Valencian coastline. The limits of these sedimentary cells are found where the littoral drift is zero. And this can happen due to both natural and human causes. The boundaries of coastal sedimentary cells are, basically, in those places where the depth next to the shoreline is high enough for waves to break directly on the own shoreline even during biggest storms. That happens, for instance, in the plunging cliffs, and also on the harbor jetties, if they reach a sufficient depth. From the sedimentary cells' perspective, three beach models could be differentiated: (i) beaches located in very small sedimentary cells, usually small bays or beach pockets closed between bedrock headlands where the plunging cliffs reach such depth that stops the sedimentary transport; (ii) long open beaches, inserted in large sedimentary cells that extend tens or hundreds of kilometres, along which the displacement of huge volumes of sand can be produced without any interruption; and (iii) an intermediate model, affecting beaches closed by bedrock headlands, but where the depth at the foot of the cliffs is not as important as preventing longitudinal transport, or where the cliff platform allows the transport.

Probably the characteristic that best defines the beaches is its great dynamism. The changes take place at different temporal and spatial scales. While all changes, at any scale, have high interest, those that occur in temporal ranges of a short or medium term (days, years, decades) are especially important from a social point of view since they affect human use of the beach as a resource. We must bear in mind that, currently, the beaches are the main tourist resource of a territory as the Valencian Community in which tourism concentrates 13.2% of regional GDP and 14.4% of employment (IMPACTUR 2015). For this reason, the alterations experienced by the beaches may have, in some cases, a social and economic impact of great importance. This explains, in turn, the large amounts of direct interventions that have made and are still being carried out in many of the Valencia beaches.

In this work, we intend to make a morphological characterization of the Valencian beaches. For this, the factors that determine the main differences between them, and their recent evolution (their changes on scales of weeks to decades) will be explained. Both natural and anthropic causes, which may explain the differences in the changes, will also be analyzed. With all this information, a general diagnosis about its current status will be proposed.

10.2 Main Factors that Affect Beaches

To understand the morphology and dynamics of the beaches it is necessary to take into consideration: the general structure of the coast, both emerged and submerged; the main characteristics of marine and meteorological agents; sediment supply; and human action.

10.2.1 General Relief Configuration

The main configuration of the Valencian coast is determined by the layout of the main geologic structures of the relief in the East of Iberian Peninsula: the Iberian Mountain Range on the northern area, and the Betic Mountain Range at the southern sector of the Valencian coast. The Iberian reliefs reach the Mediterranean through a series of faulted blocks that they are sinking near the coast, while the Betic reliefs reach directly the coast generating different types of cliffs (Sanjaume Saumell 1985; La Roca et al. 2005).

In the Valencian Oval, the bedrock reaches the shore only in a few segments: the Serra d'Irta a small mountain range parallel to the coast that stretches between Peníscola and Alcossebre (Fig. 10.1); as well as another small segment from Orpesa to Benicàssim, and the last one would be the Cullera Cape. It is important to note that within the Gulf of Valencia, two segments clearly differentiated according to their orientation would be distinguished: the northern part with a dominant direction NE-SW (or NNE-SSW), and the southern sector that follows the direction NW-SE (or NNW-SSE). The boundary between both is practically situated in the city of Valencia (Fig. 10.1).

If the artificial coastal areas (harbors and seawalls) will not count for the calculation, it is estimated that 78.8% of the Valencian Oval has beaches, and only 21.2% is the rocky coast. The latter are mainly located in the northern sector, where in addition to the segments previously cited, exist at the northernmost area, a sector of middle cliffs formed by Pleistocene fluvial deposits. They are dominant between the Ebro Delta and Vinaròs, but they are progressively disappearing southwards. For this reason, in the northern sector, there are mainly short beaches, inserted between middle cliffs. But later, moving southwards, the beaches tend to be much longer and they are only interrupted by artificial structures (harbors, groins, jetties, etc.). The average length of the Gulf of Valencia beaches is about 4 km. The plunging cliff of Sant Antoni Cape indicates the southern boundary of the Gulf of Valencia.

On the other hand, in the next structural segment (Sant Antoni-Les Hortes headlands) the cliffs are predominant. In fact, in this sector beaches represent only 28% of the coast (not artificial), while rocky coastal segments occupy 72%. Here the beaches are substantially shorter (1.1 km on average), although it is also possible found some long beaches, for instance, Benidorm (Llevant beach has 2.2 km, and Ponent beach, 3 km). Another example is the Sant Joan beach (6.4 km) located immediately north of the Les Hortes Cape. All these beaches have a very high tourist attraction. The southernmost part of the Valencian coast is again clearly dominated by beaches (68.8%), while only 32% is formed by the rocky coast. The average length of the beaches once again will be of the order of 4 km.

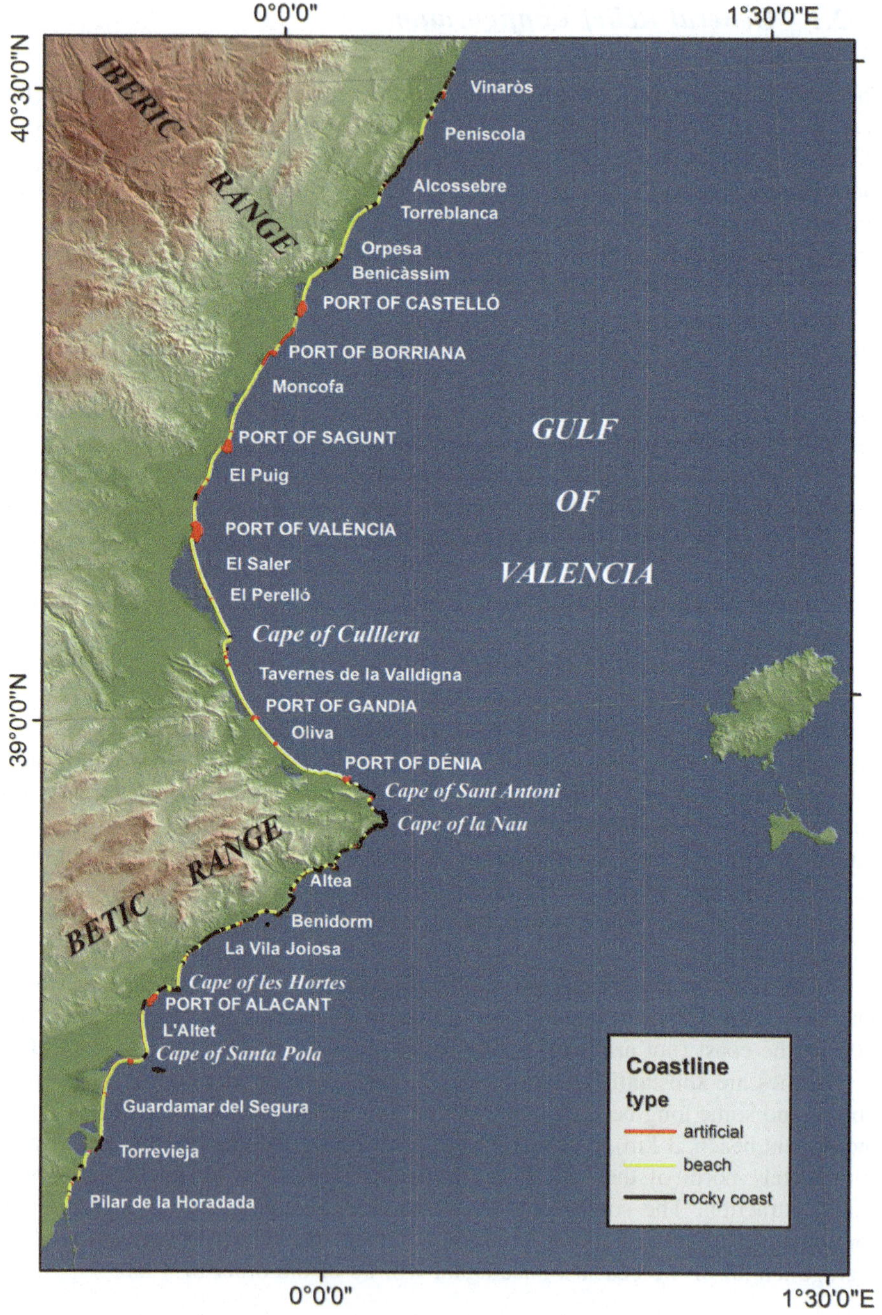

Fig. 10.1 Coastal types. The external edge of the harbor dikes and the seawalls are considered artificial coast. The rocky coast includes both cliffs and rocky low-beaches

10.2.2 Marine Dynamics

A very important factor to understand the behavior and morphology of Valencian beaches is its low tidal range. The analysis of the records of the historical series (1993–2013) from the tide gauge of the harbor of Valencia (REDMAR 2014) show that the average tide range is 0.24 m; the maximum range of astronomical tide reaches 0.39 m; and the maximum registered range (including both astronomical and meteorological factors) has been 1.32 m, in the analyzed period. It might be interesting to highlight these extreme values because they occur usually associated with conditions of very low-pressure and strong sea winds. These conditions will determine the over-elevation of the sea level during the time that these conditions occur, which can lead to flooding inland (behind the backshore), and also very significant morphological changes on the beach will be produced.

The general characteristics of the waves that affect the entire coast are deeply marked by the disposition and orientation of the own coast respect to the waves approach. The Valencian coast is located on the eastern side of the Iberian Peninsula, for that reason, they are to leeward of general mid-latitude zonal winds that usually flows from the west. Therefore, the only winds capable of generating waves are those of the first and second quadrants, especially those that arrive from NNE to SSE. Major waves are developed during storms associated with the passage of low-pressure systems from the west. The most frequent situation is the presence of an anticyclone in northern Europe and an area of low pressure located SW of the Iberian Peninsula. As a consequence, strong winds and waves of high energy come from NE and ENE to the Valencian coast (Pardo-Pascual 1991a). During storms, sometimes the waves are quite complex because the sea waves generate no too far away from the coast overlap the swell generates in the Gulf of Lyon, Gulf of Genoa or Gulf of Santa Eufemia (located on the Tyrrhenian side of the Calabria Peninsula). The energy of these waves will be more or less high depending on the fetch. In the northern part of the Valencian coast, the longest fetches are coming from Gulf of Genoa with more than 1,000 km with an orientation NE-ENE. On the contrary, in the southern part, the longest fetch coming from the E have almost 1,400 km.

The wave data recorded by the oceanographic buoys, (http://www.puertos.es/es-es/oceanografia/Paginas/portus.aspx) located in front of Valencia, (3.516°N, 0. 205E, 260 m of depth) and Alicante harbors (38,250°N, 0.41E, 52 m of depth), indicate that most of the waves have low energy. In Valencia, almost 24% of the waves have a significant height greater than 1.5 m. In Alicante, this height is reached only by the 14% of the waves. As regards the wave's period, in both buoys a clear predominance of short periods is detected, but in Valencia is more abundant than in Alicante.

Despite the clear prevalence of low energy wave conditions, there are times when the waves reach considerable heights. The larger storms come from NE or ENE and it is not uncommon for them to exceed 4 m of significant height. Last January 21th, 2017, was recorded the highest wave of the entire registered series in the buoy of Valencia. The wave had a significant height of 6.58 m; a maximum height of 11.13 m; and a peak period of more than 10 s. From the analysis of the records of the Valencia buoy, between 2005 and 2017, it is deduced that when the significant height exceed 4 m (0.43% of the total), most of the times the waves are coming from ENE (62%) and only 34% from NE, but if the waves are higher than 5 m, the all the waves are coming from NE. In Alicante, during the period 2007–2013 the waves with more energy are coming from ENE, but the most frequent are from E (Aragonés et al. 2016).

The angle of waves approach during the storms determines the way in which the longitudinal transport of sediments will be produced on the beaches. The longshore transport is more effective during the storms and will be faster and stronger on beaches oriented NE-SW or NNE-SSW, like happens in the northern part of Valencian Oval. The estimates of average annual potential transport of sediments southward made by Serra (1986) for the northern area (from Vinaròs to Valencia) range between 402,000 and 657,000 m^3/year. It is understood as a potential transport, because not always there is enough sediment to be transported. The scarcity of sand has been confirmed thanks to the real measurements on the beach itself. Thus, for example, next to the Castellón harbor, the real volume displaced was 100,000 m^3/year between 1950 and 2000. This magnitude has been decreasing during the recent years even more (HIDTMA 2002). The coastal orientation change from Valencia city until the southernmost part of the Gulf of Valencia to NW-SE. In this way, the littoral drift that has an annual net component in N-S direction will now be slower and less powerful since the wave trains do not need so much refraction to place the surf zone parallel to the shore but still moves solely to the south. However, from Oliva beaches to Dènia, the coast is almost parallel to the incident wave trains, so longitudinal transport is stopped, and only transverse transport will be produced.

The magnitude of longitudinal transport is substantially smaller in bays located in the structural sector. For instance, at the Ponent beach of Benidorm (Aragonés et al. 2015) the estimated potential values are 94,593 m^3/year, although the actual transport is substantially lower (13,222 m^3/year).

In the case of the open beaches existing south of Alicante, longitudinal transport is complex. Some works indicate that there is a tendency to the south, but assuming that this effect is not always clear and, many times, it will change according to the specific approach of the waves. This has been deduced from the dynamics of the beaches of Guardamar, located in the surroundings of the Segura river mouth, where there is erosion both north and south of it (Pagán et al. 2017). However, the heavy minerals analyzed from the almost 50 samples taken between l'Altet and Torrevieja tell another story. The results of this analysis indicate that the littoral drift has changed from the Pleistocene to the Holocene. According to the heavy minerals found in each sand sample, the sediments of l'Altet fossil dunes field where supplied by the ephemeral stream located north of this area. The Pleistocene beach barriers of the Elx lagoon have minerals carried out by Vinalopó River. Finally, the fossil dune field located south of Segura river mouth also have the same mineral composition than the samples taken from this river. So at that time, the littoral drift was N-S. On the other hand, the Holocene beach barrier of Elx lagoon has exactly the same heavy minerals than the Segura. The same occurs on the new dunes generated north of the Segura river mouth. At that place there are only a few fossil dunes, most of the dunes are Holocene's and the mineralogy shows that the source of sediment comes from the river. So looks like the littoral drift during the Holocene times has been S-N (Sanjaume 1985).

10.2.3 Sediment Supply

Sanjaume (1985) established that most of the sand of the Valencian beaches has been supplied—at least in the last instance—by fluvial emissaries that flow into our coast. Most of them act, still today, as a source of sediment supply, although some of the rivers have had different stages throughout history, such as the Segura River. The Vinalopó River (Fig. 10.2) is an atypical case since its influence was only during the Pleistocene and, at present, its incidence in the coastal accumulation is null.

Therefore, almost all Valencian beaches are formed by fluvial materials that have been redistributed by the littoral drift. There are, however, some places where there is a slight contribution from offshore submarine outcrops. On the contrary, there are some areas where self-supply predominates from fossil dunes and fossil beaches. The mineralogical analysis of the sand (Sanjaume1985) have allowed recognizing that the cliffs also act as a source of sediment supply in the structural stretches of the Betics mountain Range zone but, in the case, since there is no effective transport along the coast, its influence is much more spatially reduced. The middle cliffs at the northernmost part of the Valencian Oval also provide sediment to the local beaches, as evidenced by studies on coastal evolution (Pardo-Pascual 1991a).

We conclude with Sanjaume Saumell (1985), that there are two different sources of sediment supply. The first is more local (small ephemeral streams; marine nourishment due to the littoral drift; self-supply from fossil sediments; aeolian transport; and erosion or weathering of the cliffs bedrock). The second one is based on the contributions of the large rivers that transport sediments from their wide river basins. The type of supply varies from one area to another. Therefore, on the northern coast of the Valencia Oval, the main source of supply is the small local ephemeral streams that provide heterometric materials, given beaches with coarse sediments. There may also be some contributions from local middle cliffs, and from offshore. On the contrary, in the central and southern stretch of the Gulf of Valencia, the materials are of fluvial origin, mainly from Túria and Xúquer rivers, although there is also a certain aeolic contribution from the dunes. In the Betic structural sector, the sediments that move within each of the sedimentary cells located in this stretch come, mostly, from the weathering and erosion of the cliffs bedrock. Finally, in the southernmost sector there is a great variety of sources of supply (fluvial sediments coming from Segura or Vinalopó, marine with sediments transported by littoral drift or eroded from outcrops offshore, self-supply by erosion

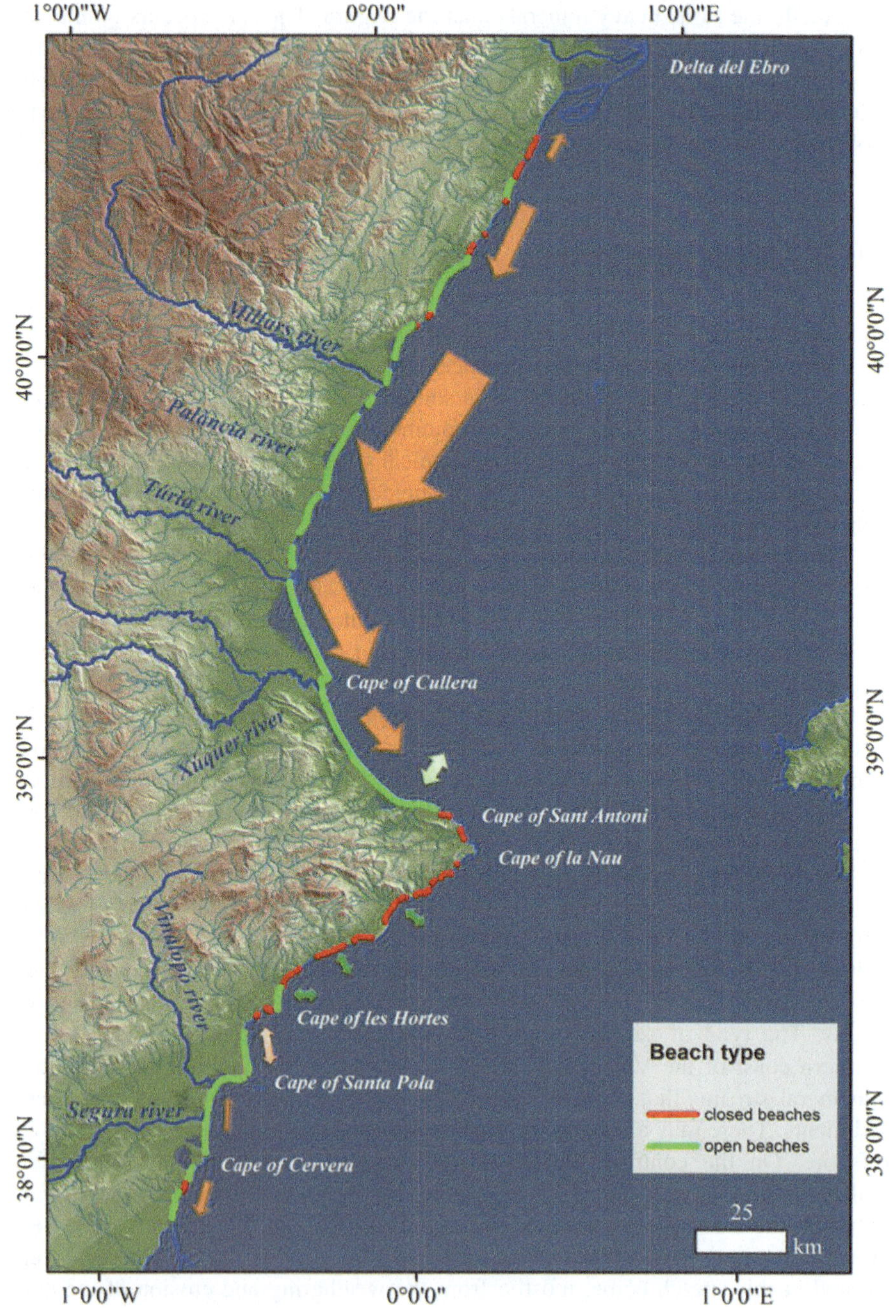
1°0'0"W
0°0'0"
1°0'0"E
40°0'0"N
39°0'0"N
38°0'0"N
Delta del Ebro
Millars river
Palància river
Túria river
Xúquer river
Vinalopó river
Segura river
Cape of Cullera
Cape of Sant Antoni
Cape of la Nau
Cape of les Hortes
Cape of Santa Pola
Cape of Cervera
Beach type
closed beaches
open beaches
25
km

◂**Fig. 10.2** The map shows the direction and magnitude (estimated) of the longitudinal sediment transport with an orange arrow; the places where there is an undefined drift, is represented by an orange arrow with two directions; those places where the perpendicular transport is much more significant the longitudinal one, the arrow is with a double direction and green colour. The type of beach is also indicated by colors: open beaches (green), closed beaches (red)

and weathering of fossil dunes, wind supply from current dunes to the beach, and weathering of rocky beaches).

10.3 Main Characteristics of the Valencian Beaches

Different classifications of the Valencian beaches could be done, based on the texture of the materials (http://www.mapama.gob.es/es/costas/temas/proteccioncosta/ecocartografias/default.aspx),[1] the dominant morphodynamic stage (Pardo-Pascual and Sanjaume 1995), but perhaps a first descriptive approach would be to differentiate those open beaches, basically continuous along several kilometers—while they may currently be interrupted by artificial structures such as harbors, groins or jetties—as opposed to those where there are natural obstacles and therefore, the beaches are closed. Figure 10.2 shows that practically the entire central and southern sectors have continuous and open beaches. These open beaches are the most common type on the Valencian coast (83%). The closed beaches are located in the northernmost of the studied area and also throughout the Betic structural sector. The length of these closed beaches is extremely variable, but on average they are about 800 m.

The texture of the material would be a second key feature of the beaches. If you look at the map in Fig. 10.3, you can clearly see the predominance of fine (47%) and medium-size sand (34%) on open beaches, while gravels are much more abundant in the closed beaches. Thus, a relationship between the type of beach-open-closed-and the dominant texture is appreciated, although a direct relationship between them cannot be established. In fact, the number of beaches with fine sand on closed beaches is exactly the same as gravel ones.

Sanjaume (1985) after a systematic sedimentological analysis of the material of the Valencian beaches, from Benicàssim to Torrevieja, distinguishes four large units:

(1) Coast of fluvial accumulation-marine washing. It is the one that we find in the whole northern part of the Gulf of Valencia. Genetically it is a coast of

[1]In this link you can find the eco-cartographic ordered by the Spanish Ministry of Agriculture, Fisheries and Environment from Alicante and Valencia (made in 2006 and 2007) and Castellón (made in 2009 and 2010). These studies include a bathymetrical survey (with isobaths every meter) from the shore until a depth of 40 m. In addition, a wide textural analysis campaign was made on the materials both emerged and submerged beach. All this information has been used as a basic source for the multiple maps that appear in this work.

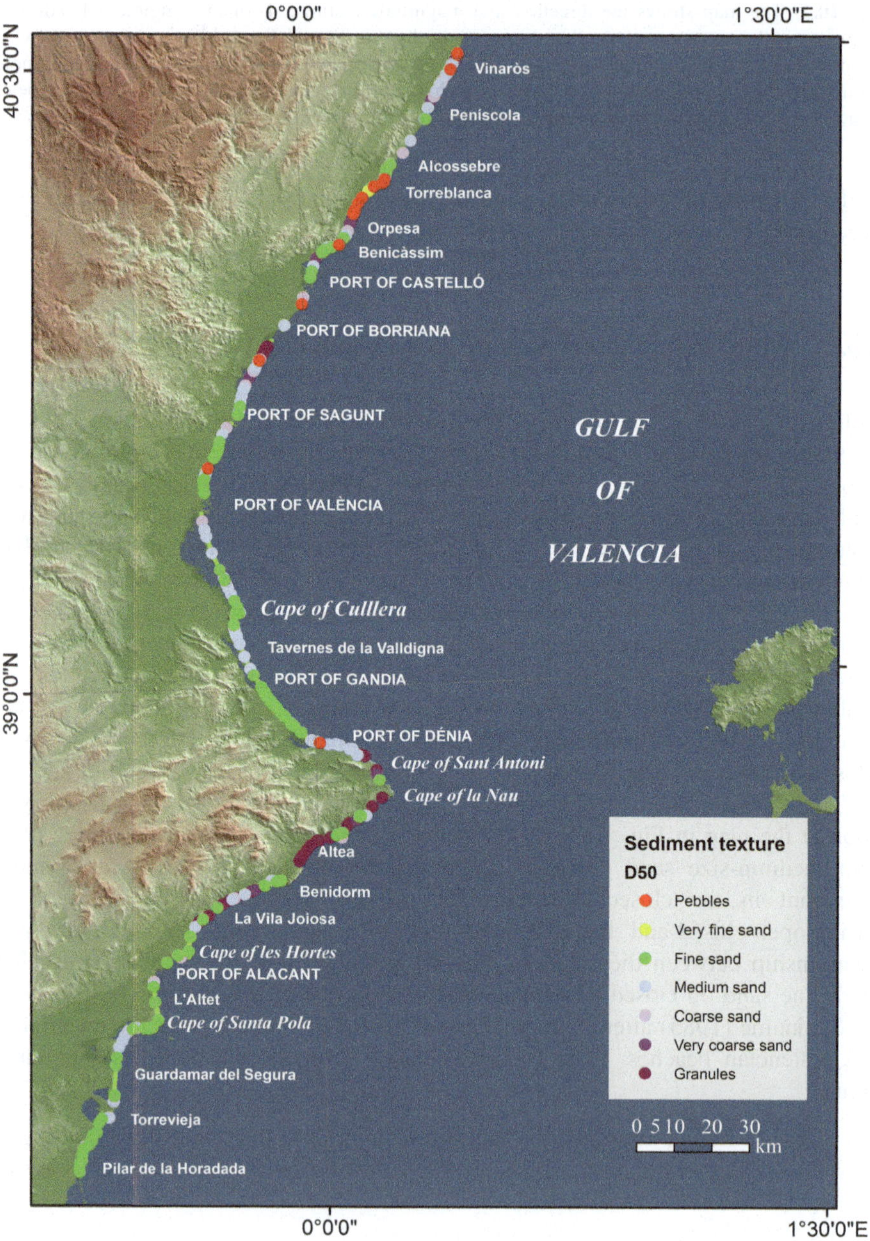

Fig. 10.3 Sediment size median in different beaches of Valencian coast. Original data used to create this map has been obtained from http://www.mapama.gob.es/es/costas/temas/proteccion-costa/ecocartografias/default.aspx

accumulation modelled by marine process from fluvial sediments supplied by ephemeral streams of a spasmodic regime. From sedimentological parameters, it is characterized by a constant presence of cobbles and pebbles; by the heterometry of materials, and by scarce classification of the sand fraction. The beach sediments do not have any of the characteristics of marine accumulation. These are materials of clear fluvial origin that are currently being washed by the waves, but they maintain their original peculiarities. In most of this sector, littoral drift suffers a scarcity of sediments, which intensifies the washing. The only sections where accumulation occurs, it is caused by the human action that modifies the natural conditions of the coastal system dynamics. For instance, the accumulations forced by groins in Benicàssim, as well as Castellón, Borriana, and Sagunt harbors.

(2) Coast of fluvial accumulation-marine accumulation. This sector extends from the Valencia harbor (including the forced accumulation on its northern side) to Dénia. The material is homometric, unimodal and well classified, which are typical from marine accumulation. Genetically, they continue to be beaches of fluvial accumulation, but their granulometric parameters have varied significantly with respect to the previous sector. The washing has completely disappeared and the accumulation exerted by the waves, due to the excess load of the littoral drift, is the dominant element. There is a complete absence of pebbles. These only appear during storm events, and survive only for a short period, because they are quickly buried by sandy contributions. Contrary to what happens in the previous sector, the alterations in the granulometric characteristics occur on the southern side of the artificial sediment traps. Since the erosion is increased in an induced way by the trap, consequently, the washing also increases. This is what happens south of Valencia harbor, south of jetty built at the Xúquer river mouth, and south of Gandia harbor (Fig. 10.3).

(3) Structural coast. It stretches between Dénia and Alicante. This sector is formed by closed beaches among rocky bedrocks. When the coast presents E-W orientation and in addition, the bays and coves are protected by rocky headlands (a clear example would be Benidorm beaches), their materials show typical characteristics of marine accumulation with homometric, unimodal and well classifies sediments. On the other hand, in NE-SW oriented sectors, exposed to the most powerful waves, sediments appear heterometric, little evolved and poorly classified. All these features would match with the peculiarities found in areas of fluvial accumulation with a predominance of washing. The only beach that does not follow the rules is Sant Joan beach, north of Les Hortes Cape, where accumulation is predominant even this beach has orientation N-S.

(4) Mixed coast: structural and marine accumulation. It encompasses the entire southern sector of the Valencian coast, from Alicante to Pilar de la Horadada. In this sector, a clear difference can be established between the areas in which the supply comes from a river (from the Segura River in this case), and those that present self-supply. The source for the supply is the erosion and weathering of Pleistocene fossil sediments (dunes or beaches). This is what happens on the beaches of l'Altet and Cape of Santa Pola (Figs. 10.1 and 10.5). In these

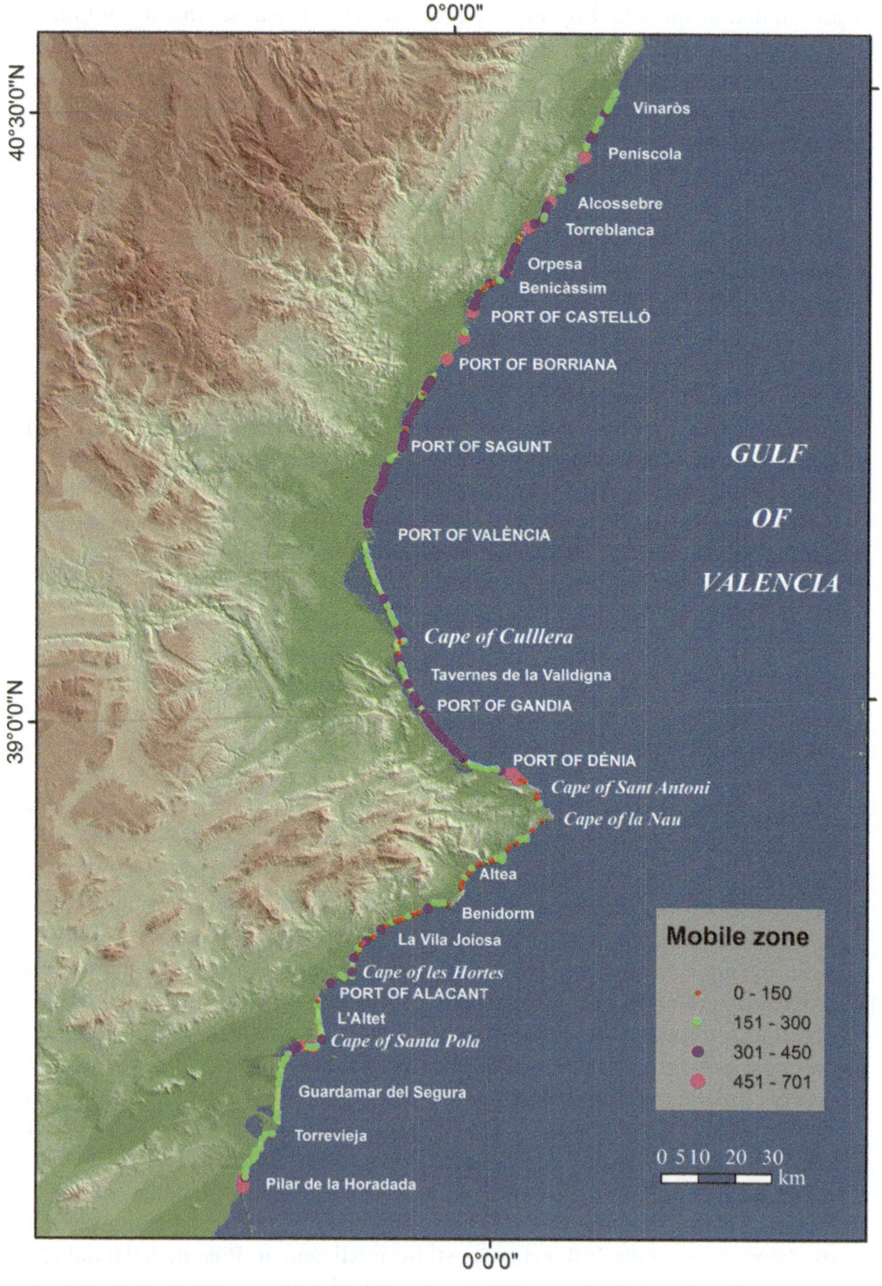

Fig. 10.4 Approximated closure depth distance to shore

Fig. 10.5 a Waterfront promenade makes the beach of Piles excessively narrow. Orthophotography taken from the National Plan of Airborne Orthophotography (PNOA) in May 2015. Subpixel coastline obtained from Sentinel 2 scenes (applying method of Pardo-Pascual et al. 2018), **b** handmade photo at the same place one week after the storms of December 2016 (with H_s of 4 m) and January 2017 (with H_s of 5.45 m). The waves have exposed the foundations of the promenade are in risk of destruction

beaches, the marine accumulation does not offer any doubt and is the predominant factor. However, the granulometric analysis indicates that the shore sediments are more evolved than the materials inside the dunes. That suggests that a recycling of old materials is being produced here at expense of which the current beach has been developed.

Another way to characterize the Valencian beaches is using the surf-scaling parameter that relates the wave dynamics with the offshore morphology (slope and existence or not of submarine bars). Using this parameter, Wright and Short (1983, 1984) distinguishes between: dissipative, reflective and intermediate beaches. The first ones have smooth profiles and a wide surf area; they are normally sandy beaches and the sediment transport is very high. They are optimal beaches for foredunes development (Short and Hesp 1982). In the reflective beaches, by the contrary, the nearshore slope is higher and the transported material is usually stored developing beach ridges with abundant beach cusps. The surf area is narrow and submarine bars are rarely formed (Wrigth and Short 1983). Potential transport is limited and gravel-size or small pebbles tend to be predominant; wind transport is virtually non-existent and therefore there are no dunes. The intermediate beaches incorporate elements of both dissipative and reflective beaches. In this type, the surf zone is quite complex, including crescentic submarine bars and rip currents systems (Short and Hesp 1982).

Pardo-Pascual and Sanjaume (1995) applied this methodology for the Valencian beaches. They work was based on the wave data available in that decade, as well as on more than 100 beach profiles carried out, in the Valencian low coasts, in the previous studies of Serra (1986), Esteban (1987), Pardo-Pascual (1991a, b), HIDRA (1992). The conclusion was that, as expected, the Valencian dissipative beaches were formed by fine-size sand, well classified, where the sediment supply was good enough. On the contrary, the intermediate and reflective beaches presented larger size and worse classification. The last ones usually occur in places with limited sediment supply. The dissipative beaches are, in fact, the most abundant, since they are located not only in almost the entire southern part of the Gulf of Valencia, but also on the sandy beaches situated south of les Hortes Cape, as well as on a few sandy beaches from the structural area, like Benidorm. However, in the last decades some dissipative beaches have become reflective due to anthropogenic causes. The scarcity of sediments determines an increase of the slope in the nearshore profile and the surf zone becomes narrower.

Another descriptive parameter of the Valencian beaches morphology is the position of closure depth in the submarine beach profile that means, the depth where the transport finishes. Aragonés et al. (2016), basing on many beach profiles registered along different beaches in the Valencia province, estimate the closure depth range between 5.13 and 7.86 m. Taking in account the Dean parameter of each one beach—obtained from the Ecocartogtraphy Study (http://www.mapama.gob.es/es/costas/temas/proteccion-costa/ecocartografias/default.aspx) and accepting that 5.13 m as closure depth, it has been calculated the offshore position where the cross shore transport finish in each one beach (Fig. 10.4). This map show as steeper beach profiles (in many cases depth closure is located close to 150) are dominating in Betic structural sector. Contrarily, in the Gulf of Valencia dominates smoother profiles, locating the closure depth to 300 or 450 m from shore. Beaches of southern end, and also El Saler beach and some others sited in the northern coast have intermediate slope, with closured depth ranging 150–300 m. Smoothest profiles are found in the supported beaches by port dikes or by natural obstacles, reaching its depth closure till 700 m from shore. Finally, the Valencian beaches could be classified according their morphology taking into account its texture, slope, as well as beach forms.

(i) *Sandy beaches*. The beaches are forms attached along its entire length to the outer edge of the coastal element. They can be found as a single element (bay beach, pocket beach), or they may be part of any other coastal accumulation forms as such as beach barriers, spits, tombolos, etc. Normally beaches present high mobility. They are wider or narrower depending on the characteristics of onshore and nearshore slopes, energy of the waves and sediment supply. In the Valencian coast, if the slopes are of very low gradient, the beach is dissipative, but if the slope is steep, then they are reflective. The main micro forms are: crescent marcs, and swash marcs. The middle-forms would be: sandy beach cusps, berms (that may be staggered showing how far different storms have come), an offshore there are submarine bars crescentic and longitudinal (this one usually is quite close to the shore). Both are very good elements to dissipate the energy of the waves. The longitudinal bar affects especially with sea waves and swells waves, while the crescentic bars are very effective during storms.

(ii) *Pebble beaches*. They are preferably located in the surroundings of the ephemeral streams, which due to its step it transport coarse materials to its mouth, after the waves wash the finest sediments. For this reason, the beaches located near the river mouth are formed with pebbles whose size it too coarse to be remobilized by the waves, that on the Valencian coast they almost always have very little energy. On the other hand, the water percolation from the wave's uprush among the pebbles also contributes to the stability of these sediments. In times of very strong storms, pebbles are moved up and stacked, giving rise to beach ridges and berms. But the beach cusps are perhaps the most characteristic of these beaches middle-forms. This type of beaches is common on the northern areas. They cover virtually the entire coast of Castellón province, and they are also possible to find in some areas of the southern coast as, for example, in Altea, La Vila Joiosa and el Campello beaches. Some decades ago the pebble beaches were much more abundant than at present. Most of these beaches have changed their texture due to recurrent sand discharges to carry out artificial beach nourishments.

(iii) *Rocky beaches*. This type of beaches is formed by conglomerates, cemented fossil beaches sand, and aeolianites (cemented fossil dune sand). Most of these materials are coming from the limestone lithology of the inland mountains. These rocky beaches normally present very interesting karst morphology, in which the dissolution is developing microforms of littoral karst. It is very interesting for the variety of its forms (metric, decimeter or centimeter scale). Although they also are some examples on the northern coast, the maximum representation occurs in the southernmost coast, and does not exist in the Gulf of Valencia beaches. The zones were the littoral karst is most spectacular would be: area between Torrevieja and cape Cervera, close of Santa Pola, the Albufereta of Alacant, Cape of les Hortes, l'Illeta dels Banyets del Campello, and Cap Negret in Altea, some stretches of Calp, the area between Cala Bassetes and Moraira, along most of Xàbia bay, and some beaches of Dénia (Sanjaume 1985). The morphology of the littoral karst undergoes modifications depending on the lithology and the profile slope, and basically depends on the frequency of humidity and drying cycles. Different kind of dissolution bowls in the most lover parts, as well as different types of lapiaz in the higher parts of the profiles are the most important characteristics of this littoral karst and rocky beaches (Sanjaume 1979).

10.4 Human Action

The human action is extremely important to understanding the current characteristics of the Valencian beaches and their evolution (Rosselló 1986; Pardo-Pascual 1991a; Sanjaume et al. 1996; Sanjaume and Pardo-Pascual 2005). These are very diverse actions, voluntary or involuntary, that may occur inside or outside of the coastal sedimentary cell and they have had or still have impacts which consequences have different temporal scale. Some impacts are local and can be noticed immediately, while some others are acting a regional spatial scale and they are a long-term impact. In this study the attention will be focused on the most significant actions.

10.4.1 Direct Actions on the Coastal System

The direct human actions carried out by the Spanish authorities on the Valencian coast have been the most important factor to explain the observed changes in the Valencian beaches. Based on the documentation about Actions on the Valencian Coast, Obiol (2003) indicates that during the period 1982–2002, on the Valencian coast were carried out: 104 actions of artificial beach nourishment; 60 building or repairing marine promenades; 54 interventions in groins or jetties, plus 13 on exempt groins; 16 interventions about urbanization works; 9 projects about dunes; and 28 actions on seawalls. These figures clearly prove that human actions are an essential factor to understand the morphology and evolution of the beaches. From all the previous actions, the most dangerous have been the rigid infrastructures, since they completely interrupt the longitudinal transport of sediments.

This sedimentary flow has remarkable effects the effect on the Gulf of Valencia and produces strong updrift over-accumulation and downdrift over-erosion around the artificial obstacle. In the sectors with excessive contributions (sand or gravel), there is a rapid filling of the profile that causes a progressive advance of the coastline. This effect is much more important as stronger is the coastal drift, as it happens mainly in the northern part of the Gulf of Valencia.

The construction of the port of Valencia began at the end of the 18th century, being expanded throughout the 19th, 20th up nowadays still with important extensions. The impact on the beaches was detected almost instantaneously. The over-accumulation recorded in the north has led to an advance of more than 1 km seaward along the shore so that part of the current city of Valencia is located on purely marine land from the middle of the 18th century (Pardo-Pascual 1997). In the south, the erosion effects have also been striking and even today, erosive processes are observed on the beaches of Saler (Sánchez-García et al. 2015). Throughout the twentieth century, the depth of the dikes of the port has exceeded the closure depth, so longshore transport has been completely interrupted. This has involved subdividing the littoral cell into two.

The ports of Gandia, Castellón, Borriana, and Sagunt were built from 1892, 1893, 1923 and 1953 respectively (Pardo-Pascual 1991a). These ports have split the original littoral cell into smaller cells, producing local lacks of sediments and consequent shoreline setbacks. In addition to the large ports, up to 11 medium and small fishing/recreational ports have been built in the Gulf of Valencia what makes 16 ports in this coastal segment. In the rest of the Valencian coast there are 23 other ports, mostly small ones, almost all of them since after the 70's of 20th century.

Groins for coastal defense are those with higher effects on the morphology and dynamics of the coast itself. All along the Valencian coast there are 159 groins, 141 of them at the Gulf of Valencia. Most of these dikes were built to slow down erosive processes. The objective was partially achieved, erosion slowed down punctually and only to a limited extent, but the sediment deficit migrated southwards. Other groins were introduced to create a beach, trying to improve the recreational use of the beaches or as a support point for further sand dumping.

Another model of coastal defense is seawalls built on the own shoreline. This defense is normally formed by breakwaters and, although prevents the retreatment of the coast, replaces the beach and the sand disappears. Some segments like that may be found at the south of the port of Castelló (4.3 km), Borriana (3.3 km) or Alicante (1.8 km) as well as top the north of the city of Valencia where 3 km long segment protects the northern motorway from coastal erosion.

Boardwalks, built near the shore or on the first line of dunes also have an impact on the shore. They prevent the sedimentary equilibrium of the beach profile (Sanjaume and Pardo-Pascual 2011) and generally cause a drop in the beach height. Consequently, this increases the risk of flooding in the nearby zones or even the destruction of the promenade itself. Piles beach, south of the port of Gandia (Fig. 10.5), is an example of this.

Since the 80s of the 20th century, a common technique for maintaining beaches is the artificial nourishment. Between 1982 and 2002, 100 km were nourished with

13.5 million m^3 of sand (Obiol 2003). The distribution of the sands has been unequal on the Valencian coast, the largest contributions have been made at the north of Sagunt. The most common inputs are sand (85%) although gravel has also been used. The origin of the sand is mainly marine (50%), but mixed, terrestrial or adjacent beaches have also been used. At the moment, artificial nourishment is a normal and accepted method despite its economic cost and environmental impact (extraction and dumping of sand). In addition, pouring sand does not guarantee its durability over time. In terms of environmental impact, the *Posidonia oceanica* is an important element for the dissipation of wave energy and the dumping of sand may damage them (Aragonés et al. 2016; González-Correa et al. 2008, Medina et al. 2001).

There are other actions carried out on the beach area, with clear implications on them but without any direct will to affect the beach. For example, the extraction of sand and gravel from the beach to agricultural and livestock exploitations. This practice-forbidden since the 1970s of 20th century—was a normal activity at least since the 19th century on the beaches near the city of Valencia (Pardo-Pascual 1991a). These actions produced, according to some technical reports, strong set-backs in normally cumulative beaches as the one next to the northern dike of the port of Valencia (Vilar 1931). Extractions at the coast south of the port could have reached 300,000 m^3 annually (Vilar 1934). However, quantifying these extractions is difficult because of the limited documentation.

The opposite action, involuntary dumping of materials, is less common but an important example must be noticed. Iron extraction was carried out in Sagunt. In this process there is a surplus material: slag. For much of the twentieth century, these slags were dumped south of the Sagunt port. This material, by its chemical composition, forged in contact with water and cemented the materials (Sanjaume 1985). This resulted in the cementation of the submerged beach materials up to about 6 km south of the dumping point (Pardo-Pascual 1991b).

10.4.2 Indirect Actions on the Coastal System

The human activities carried out in river basins affect the beaches indirectly. Changes in vegetation cover affect runoff and soil erosion, which ultimately constitute sediment inflows into the coastal system (Pardo-Pascual and Sanjaume 2001). The changes in the use of the watersheds' land in the 17th century had a decrease in sedimentary contributions to the coast by rivers (Mateu 1982). It has been throughout the 20th century, when Valencian rivers experienced bigger anthropic alterations which have led to a major hydro-sedimentary deficit (Segura-Beltrán and Sanchis-Ibor 2013; Sanchis-Ibor et al. 2017). One of the most recognized and studied impacts of dams are the retention of sediment that provokes river incision and an increase in coastal erosion (Petts 1984; Petts and Gurnell 2005). In the Mediterranean area, where water supplies are scarce and irregular, reservoirs have proliferated for the purposes of agricultural and urban supply, and flood control. These infrastructures have affected both the perennial rivers and those that have headwater flow. There are 54 reservoirs in the Xúquer river basin district, although the 27 main ones have a total water storage capacity coming to 3,300 hm^3; the largest are Alarcón, Contreras and Tous on the Xúquer River and Benagéber on the Túria River. There are 43 reservoirs in the Segura river basin district, almost all on this river, and 27 of them manage a volume of 1,141 hm^3.

The layout of these reservoirs on the Fig. 10.6 shows that over 75% of the river basins' area flowing to the Valencian coast is controlled by the reservoirs and they cannot provide sediments to the sea. In spite of having a National Strategy for River Restoration, the water authorities have not implemented corrective measures for these infrastructures. There is little data about the reservoirs' impact and it has only been evaluated in a few cases. Cobo (2008) examines some of the main reservoirs in the two river basin districts and produces figures about them that are summarised in Table 10.1. Total retention in the 11 reservoirs analysed on the Segura basin (from their construction until 1997) comes to 124.97 hm^3, whereas in the 15 reservoirs on the Xúquer, the Fig. 10. stands at 52.44 hm^3. The sediment volume retained overall is greater in the Segura river basin district than in the Xúquer's because its reservoirs are older and the conditions of the basin (higher aridity, scarce vegetation cover) result in higher erosion rates. The average loss ratio in the Xúquer river basin district comes to 12.3%. The most spectacular case is the Embarcaderos reservoir with a capacity decrease of 83.8%, followed a long way behind by Guadalest and Toba with 18.8 and 19.8% respectively. The annual loss ranges from 0.04% at Arenós to 2.70% at Embarcaderos. The average loss ratio in the Segura river basin district stands at 24.3% and the most affected reservoirs are Santomera and Valdeinfierno with a capacity decrease of 59.3 and 55.4%, respectively.

Gravel mining has a large number of effects on the riverbeds, including a change in slope that leads to a rise in the river's power and consequently, an increase in the erosion transmitted upstream and downstream from the mining point. The lack of sediment means that the flow has excessive energy ('hungry waters') which it offsets by eroding its own bed and banks in an attempt to load up with sediment (Kondolf 1997). This results in a sedimentary deficit that is also transmitted to the coastal system. Notwithstanding the many negative effects that this mining has on the rivers, the water authorities have not banned it completely and do not provide any information about the issue.

10.5 The Recent Evolution of the Valencian Beaches

As the beaches are always changing and evolving, the temporal behavior may be studied considering different temporal scales. Here we will focus on the short and medium term changes, i.e. those that occur during weeks or decades. These changes on the landform have also consequent effects on the beach as a socio-economic resource.

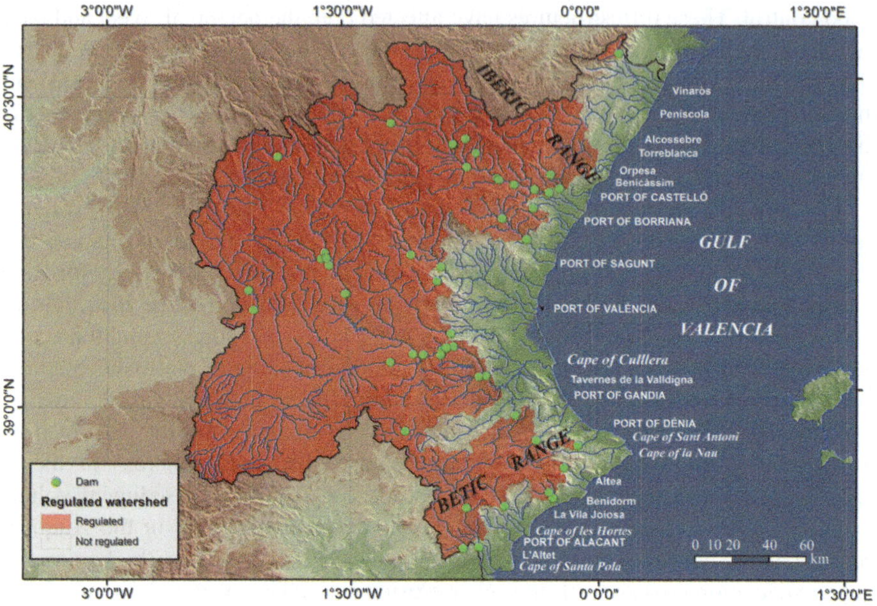

Fig. 10.6 River dams and reservoirs of the Xúquer river basin district. Red zone shows the regulated surface of different watersheds that flow to Valencian coast

Short-term changes, those that occur in a few days, perhaps a year, have relatively modest dimensions but, in many cases, have strong socio-economic impacts. Expectedly, the strongest changes will take place during storms. In the November 2001 storm, probably the most studied event on the Valencian coast, economic damage was officially estimated at 44.4 million euros (Obiol 2003), although the effects were not the same everywhere. Pardo-Pascual et al. (2014) analyzed the impact and coastal recovery of the southern part of the Gulf of Valencia (100 km) through 12 shoreline positions. These shorelines were extracted from Landsat 5 and 7 scenes using the methodology shown in Pardo-Pascual et al. (2012a). The first of the lines are taken on 8 November 2001, exactly before the two storms of 10 and 14 November 2001. The other 11 lines cover the next year. The results showed that both the impact of storms (magnitude of coastal retreat) and recovery were deeply related to the characteristics and conditions of each beach. Thus, beaches with steeper slopes—formed by coarser materials—receded considerably less than those formed by finer materials. However, the sandy beaches showed less retreat where the foredune existed than where dunes did not exist. That effect already has been noticed in many other places (Pye and Blott 2008). It was noted that, as other authors had already suggested (Haerens et al. 2011), the effects of the storm depended on the distance to the focus of the storm, as well as the orientation of the coast. The efficiency of alongshore transport (related to the waves approach, the orientation of the coastline and the existence of barriers that prevent that transport) had a major impact on the process of beach regeneration. In areas with limited

Table 10.1 Reservoir capacity, volume of retained sediments, total losses and annual losses. Estimated values for 1997. From Cobo (2008)

C.H. Segura	Initial capacity (hm^3)	Volume of retained sediments (hm^3)	Years	% losses	% annual losses
Dam					
Alfonso XIII	42.8	20.792	79	48.6	0.61
Anchuricas (Miller)	8	1.759	22	22	1
Argos	11.722	1.666	21	14.2	0.68
Cenajo	472	6.403	32	1.4	0.04
La Cierva	7.5	2.429	58	32.4	0.56
La Fuensanta	235	25.273	58	10.8	0.19
La Pedrera	235	27.901	68	11.9	0.17
Puentes	250	5.26	21	2.1	0.1
Santomera	31.56	18.726	101	59.3	0.59
Talave	10	0.91	8	9.1	1.14
Valdeinfierno	25	13.851	98	55.4	0.57
Total 1997		124.97		24.3	
C.H. Júcar					
Dam					
Alcora	2	0.075	18	3.8	0.21
Amadorio	16.55	0.723	31	4.4	0.14
Arenós	137.73	0.793	15	0.6	0.04
Arquillo S. Blas	22	0.965	28	4.4	0.16
Benageber	228	6.663	37	2.9	0.08
Beniarrés	30.835	3.831	20	12.4	0.62
Buseo	8	0.807	68	10.1	0.15
Contreras	872	19.595	19	2.2	0.12
Embarcaderos	9	7.539	31	83.8	2.7
Forata	39	1.664	31	4.3	0.14
Guadalest	16	3.008	24	18.8	0.78
La Toba	11	2.174	45	19.8	0.44
María Cristina	20	1.553	81	7.8	0.1
Regajo	7	0.326	20	4.7	0.23
Sichar	52	2.729	16	5.2	0.33
Total 1997		52.445		12.3	

transport, the recovery was very slow, spending sometimes several years to recover a position similar to that existing before the storm.

In the analysis of longer time scales—two or three decades—the influence of human impact and infrastructure disrupting longitudinal sediment transport is strikingly important. There are many works that show this (Sanjaume 1985;

Rosselló 1986; Pardo-Pascual 1991a; Sanjaume et al. 1996; López-García and Pardo-Pascual 1995; Brocal et al. 2005; Sanjaume and Pardo-Pascual 2005; Pardo-Pascual et al. 2012b; Sánchez-García et al. 2015). As an example, we can observe the impact of the jetties made for channeling the Xúquer river mouth, made between 1947 and 1956. These structures caused a strong sand accumulation on their northern side and a continuous retreat on the southern beaches. Small groins and seawalls were constructed during the seventies of the past century trying to stop the erosion processes (Fig. 10.7a). Thus, the problem of erosion was shifting towards the south. Figure 10.7 shows as 5.5 km south of river mouth, in La Goleta de Tavernes beach, at the end of the eighties the beach position were basically stable. However, during the nineties began a clear and sustained retreat of the shoreline. Due to rapid erosion process, residential houses initially built 60 m away from the shoreline during the eighties finished too much close to shore. In 2005, this beach had lost 50 meters wide. Given the strong residential and touristic usage of this beach, it became a socio-economic problem and several nourishment actions had to be made. Finally, in November 2008, a larger project was carried out with the "construction" of an artificial dune and the dumping of 252,000 m^3 of sand, what caused a substantial extension of the beach (up to 30 m). However (last part of Fig. 10.7c), the erosive trend continues after the sand dumping. After the January 2017 storm, the beach disappeared completely, the waves hit the houses originally built 60 m from the shoreline and the water even flooded the adjacent street, making emergency work necessary again.

The example of the La Goleta de Tavernes de la Valldigna beach is paradigmatic of the model of erosive problems of beaches when the alongshore transport that naturally nourishes the beach is interrupted and how this model of problem migrates downdrift. The coastal segment between Sagunt and Valencia has suffered the same erosive problems (Pardo-Pascual 1991b), as well as those located south of the port of Castelló such as Moncofa and the adjacent municipalities (Pardo-Pascual 1991a).

Artificial structures that interrupt sediment movement along the coast generate very marked erosion processes. It may happen even where the littoral drift is not so marked but there are well-defined longitudinal currents. Pagán et al. (2017) explain the impact that the construction of the channeling dikes at the mouth of the Segura River has on the adjacent beaches. In this zone, there are not clear drifts but strong currents may punctually move the sediments northward or southward. When it happens, it does not matter if moving to the north or to the south, the dikes trap the sediment, and the neighbor beaches left unsupplied of sediment, suffering local erosions. In fact, the rapid erosion suffered near the Segura river mouth has produced collapsing of some existing houses close to the shoreline.

The analysis of long sectors (Pardo-Pascual et al. 2013) shows the effect of interrupting longshore transport and also their uneven effect according to the dimension of the artificial structures and the coast orientation. Figure 10.8 shows the exchange rate (m/year) obtained from the linear adjustment of 106 shoreline positions obtained from Landsat images (5 and 7) recorded between 1984 and 2011 using the method proposed in Pardo-Pascual et al. (2012a). Three points may be

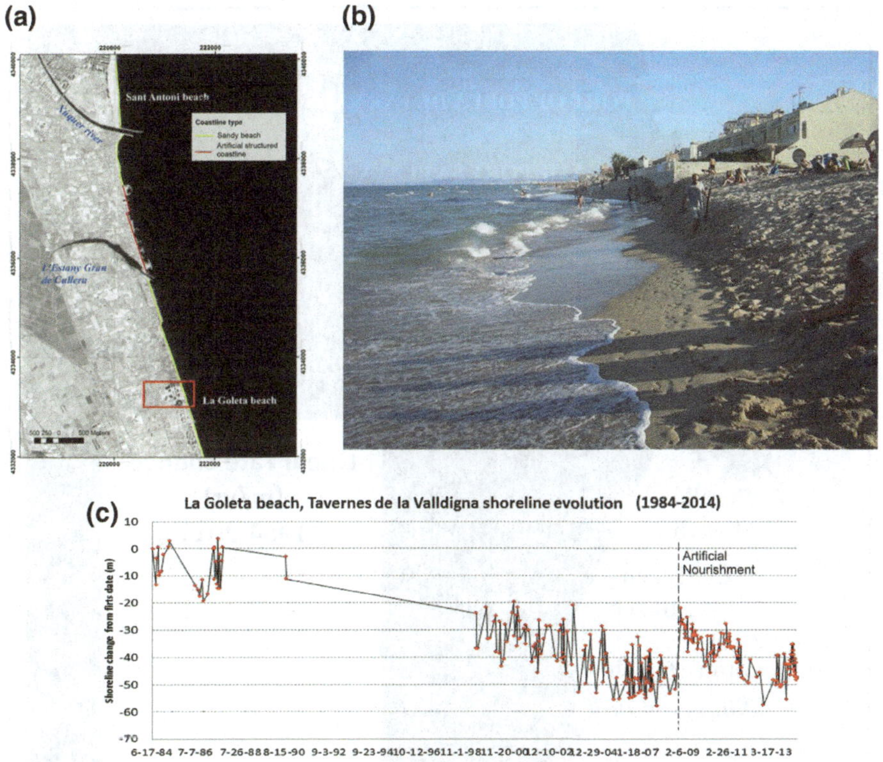

Fig. 10.7 **a** Indicates the position of La Goleta de Tavernes de la Valldigna beach in relation to the jetties at the Xúquer (Júcar) river mouth, in **b** is possible to see how was the beach in August 2008, the waves are coming into the houses, **c** shows the evolution of 278 shorelines between June 1984 and June 2014 of the segment marked with a red box in (**a**). Theses shorelines have been obtained from Landsat imagery applying the method proposed in Pardo-Pascual et al. (2012a)

analyzed: the harbor of Valencia, the Xuquer river mouth, and the Gandia harbor. At the northern side of all of them, an expansion of the beach can be appreciated. Southern from Valencia harbor and Xuquer river mouth the erosion is also clear. However, this erosion does not exist in the south of the Gandia harbor. It is because huge beach nourishments were done: 946,000 m^3 (between 1993 and 1995), and 371,000 m^3 (between 2002 and 2009), that have clearly mitigated the beach recession.

Figure 10.8 shows also that in the beaches of Oliva there is a light erosive trend. This fact was already observed by Sanjaume and Pardo-Pascual (2008) basing on the increment of sand size in these beaches between the seventies of last century and current situation. This erosive trend is the opposite of that observed during the last four centuries. In the XVI century, watchtowers were built to host military forces that protect the coast and were built on the beachfront to prevent pirate attacks. Nowadays, three of these watchtowers (in Oliva, Piles and Tavernes de la

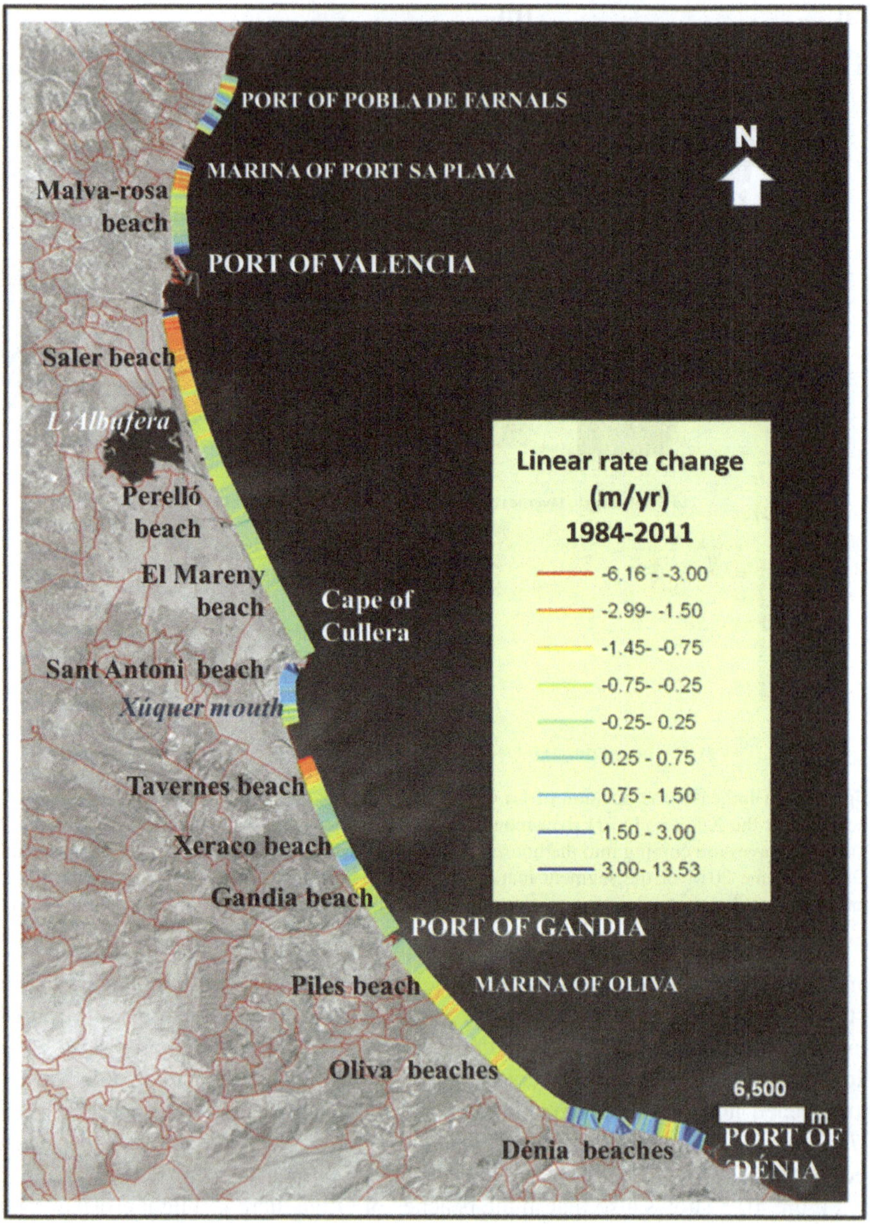

Fig. 10.8 The change rate of beaches in the period 1984–2011

Valldigna) remain stable but 390, 320 and 340 m (respectively) landwards. Therefore, from this cumulative scenario has changed to another erosive one, probably caused by a structural scarcity of sediments. Structural erosive signals also could be observed in other places. Just in the northern beach of Valencia, where the induced accumulation by the harbor, has shown a very strong growth of the beach even when large quantities of sand were extracted for agricultural and cattle raising, during the period 1977–2007, the shorelines remained stable, contrasting with previous rapid advancing situation. Close to 2007, the beach width increased again due to artificial sand nourishment and the implementation of modification of one the harbor dikes. The structural scarcity of sediments in many beaches has been caused partially by so many artificial obstacles to sediment transport, but evidence, as observed in Oliva, indicates a regional change of scenario, explained by the decrease of fluvial sediment supply due to the fluvial basins regularization.

10.6 Conclusions

The beaches cover most of the Valencian coast and they are an essential resource for the maintenance of the tourist industry. They are located mainly on the coast of the Gulf of Valencia and in the southern segment, south of Les Hortes Cape. In these two sectors, open and long beaches are predominant, contrasting with the smaller and pocket beaches found in the structural sector between the Cape of Sant Antoni and the cape of Les Hortes.

The current morphology and evolution of beaches are determined by the general structure of the relief and geological bedrock; the marine of waves climate; the sources of sediment supply to the beach; and by human action interferences. Most of the sand, gravel, and pebbles that are on the beaches originally were carried out from the rivers. However, the human action, especially the one produced for a century, is changing the rhythm and volume of sediments supplied to the sea. Although it does not produce an immediate effect, some evidence seems to indicate that the scarcity of sediments is causing a regional deficit of sediments inside the coastal sedimentary cells.

The origin of the waves during the storms explains the existence of a very efficient alongshore transport in the Valencian beaches. The artificial structures that interrupt the littoral drift cause imbalances in the availability of sediments, causing strong updrift accumulations of obstacles and serious erosion downdrift. This circumstance, as well as other direct human interventions on the coast (marine promenades, artificial sand nourishment, and dunes regeneration), has influenced the recent beeches evolution. Consequently, at present most of the Valencian beaches are deeply mediatized by direct human action.

The immediate future will also have to be mediated by human action, but the sing of the actions will necessarily have to be modified if we do not want to put at risk that natural resource (both natural and economic) which beaches are. It is essential to reopen the natural contribution of fluvial sediments or otherwise the beaches will be progressively and irremediably lost.

Acknowledgements The authors appreciate the financial support provided by the Spanish Ministry of Economy and Competitiveness in the framework of project CGL2015-69906-R. We are grateful to Dr. Francesca Segura by her help in section on human impact on fluvial systems and to Dr. Jaime Almonacid by his help during translation process.

References

Aragonés, L., García-Barba, J., García-Bleda, E., López, I., & Serra, J. C. (2015). Beach noudrishment impacto on *Posidonia oceánica*: Case study of Poniente beach (Benidorm, Spain). *Ocean Engieneering, 107,* 1–12.

Aragonés, L., Serra, J. C., Villacampa, Y., Saval, J. M., & Tinoco, H. (2016). New methodology for describing the equilibrium beach profile applied to the Valencia's beaches. *Geomorphology, 259,* 1–11.

Brocal, R., López García, M. J., & Pardo-Pascual, J. E. (2005). Análisis de la evolución de la línea de costa en el litoral valenciano mediante SIG e imágenes de satélite. El sector Tavernes-Cullera (1956–1999). *Geomorfologia litoral i quaternari* (pp. 307–324). Universitat de València.

Cobo, R. (2008). Los sedimentos de los embalses españoles. *Ingeniería del agua, 15*(4), 231–241.

Esteban, V. (1987). *Procesos litorales en las costas valencianas al sur del cabo de San Antonio.* Unpublished Doctoral Thesis ETS Ingenieros Caminos, Canales y Puertos, Universitat Politècnica de València, 4 vol.

González-Correa, J. M., Fernández Torquemada, Y., & Sánchez Lizaso, J. L. (2008). Long-term effect of beach replenishment on natural recovery of shallow *Posidonia oceanica* meadows. *Estuarine Coastal Shelf Science, 76,* 834–844.

Haerens, P., Bolle, A., Trouw, K., & Houthuys, R. (2011). Definition of storm thresholds for significant morphological change of the sandy beaches along the Belgian coastline. *Geomorphology, 143–144,* 104–117.

HIDRA. (1992). *Puerto de escala en Canet de Berenguer (Puerto Siles) y playa de Sagunt. Levantamiento topográfico y batimétrico (estado inicial).* COPUT, Generalitat Valenciana.

HIDTMA. (2002): *Proyecto de rehabilitación del frente litoral de Almazora.* Ministerio de Medio Ambiente, Dirección General de Costas.

IMPACTUR. (2015): *Comunitat Valenciana, Impactur 2015. Estudio sobre el impacto económico del turismo sobre la economía y el empleo.* Exceltur y Generalitat Valenciana, 44 pp. Available in http://www.exceltur.org/wp-content/uploads/2016/11/IMPACTUR-Comunitat-Valenciana-2015.pdf.

Kondolf, G. M. (1997). Hungry water: Effects of dams and gravel mining on river channels. *Environmental Management, 21*(4), 533–551.

La Roca, N., Sanjaume, E., & Gil, M. V. (2005). Principales características de los acantilados de la Nau, In Sanjaume E & Mateu JF (Eds.), *Geomorfología litoral i Quaternari.* Homenatge al profesor V. M. Rosselló i Verger, Universitat de València, València: 223–244.

López-García, M. J., & Pardo-Pascual, J. E. (1995). Avaluació dels canvis dels usos del sòl en un sector de litoral de Sagunt i Canet d'En Berenguer. *Braçal, 1995,* 465–477.

Mateu, J. (1982). *El norte del País Valenciano. Geomorfología litoral y prelitoral* (p. 286). València: Universitat de València.

Medina, J. R., Tintoré, J., & Duarte, C. M. (2001). Las praderas de Posidonia oceánica y la regeneración de playas. *Revista de Obras Públicas, 3*(409), 31–43.

Obiol, E. (2003). La regeneración de playas como factor clave del turismo valenciano. *Cuadernos de Geografía, 73/74,* 121–146.

Pagán, J. I., López, I., Aragonés, L., & Garcia-Barba, J. (2017). The effects of the anthropic actions on the Sandy beaches of Guardamar del Segura, Spain. *Science of Total Environment, 601–602,* 1364–1377.

Pardo-Pascual, J. E., Almonacid Caballer, J., & Palomar-Vázquez, J. M. (2013). Utilización de líneas de costa obtenidas de imágenes Landsat para caracterizar la evolución de amplios segmentos costeros. Aplicación al sur del golfo de Valencia, *Resúmenes de XII Jornadas Españolas de Puertos y Costas* (pp. 58–59). Autoridad Portuaria de Cartagena.

Pardo-Pascual, J. E. (1991a). *La erosión antrópica en el litoral valenciano*, Conselleria d'Obres Públiques, Urbanisme i Transport, Generalitat Valenciana, 240 p.

Pardo-Pascual, J. E. (1991b). Acción antrópica y evolución costera del sector Sagunt-Valencia, *XII Congreso Nacional de Geografía* (pp. 55–62). Valencia: AGE-Universitat de València.

Pardo-Pascual, J. E. (1997). El medi físic, *El port de València i el seu entorn urbà. El Grau i el Cabanyal-Canyamelar en la Història*, Ayuntamiento de Valencia (pp. 103–120).

Pardo-Pascual, J. E., Almonacid-Caballer, J., Mislata Cabo, R., & Roselló, R. (2012b). Caracterización de la evolución reciente de las playas de Sagunt y Canet d'En Berenguer. *Anales del Instituto de Estudios Saguntinos*, 55–79.

Pardo-Pascual, J. E., Almonacid-Caballer, J., Ruiz, L. A., & Palomar-Vázquez, J. (2012a). Automatic extraction of shorelines from Landsat TM and ETM multi-temporal images with subpixel precision. *Remote Sensing of Environment, 123*, 1–11.

Pardo-Pascual, J. E., Almonacid-Caballer, J., Ruiz, L. A., Palomar-Vázquez, J., & Rodrigo-Alemany, R. (2014). Evaluation of storm impact on sandy beaches of the Gulf of Valencia using Landsat imagery. *Geomorphology, 214*, 334–401.

Pardo-Pascual, J. E., Sánchez-García, E., Almonacid-Caballer, J., Palomar-Vázquez, J. M., Priego de los Santos, E., Fernández-Sarría, A., et al. (2018). Assessing the accuracy of automatically extracted shorelines on microtidal beaches from landsat 7, landsat 8 and sentinel-2 imagery. *Remote Sensing, 10*(2), 326.

Pardo-Pascual, J. E., & Sanjaume, E. (1995). Caracterización de las playas valencianas a partir del parámetro escalar de surf. In *Actas de las III Jornadas Españolas de Ingeniería de Costas y Puertos* (pp. 614–628). Universitat Politècnica de València.

Pardo-Pascual, J. E., & Sanjaume, E. (2001). Análisis multiescalar de la evolución costera. *Cuadernos de Geografía, 69*(70), 95–126.

Petts, G. E. (1984). *Impounded rivers: Perspectives for ecological management* (p. 326). Chichester: Wiley.

Petts, G. E., & Gurnell, A. M. (2005). Dams and geomorphology: Research progress and future directions. *Geomorphology, 71*(1), 27–47.

Pye, K., & Blott, S. J. (2008). Decadal-scale variation in dune erosion and accretion rates: An investigation of the significance of changing storm tide frequency and magnitude on the Sefton coast UK. *Geomorphology, 102*, 652–666.

REDMAR, Red de Mareógrafos de Puertos del Estado. (2014). *Resumen de parámetros relacionados con el nivel del mar y la marea que afectan a las condiciones de diseño y explotación portuaria. Puerto de Valencia.* Dirección Técnica Puertos del Estado, 33 p.

Rosselló, V. M. (1986). L'artificialització del litoral valencià. *Cuadernos de Geografía, 38*, 1–28.

Sánchez-García, E., Pardo-Pascual, J. E., Balaguer-Beser, A., & Almonacid-Caballer, J. (2015). Analysis of the shoreline position extracted from Landsat TM and ETM + imagery. In *ISPRS Archives: 36th International Symposium on Remote Sensing of Environment (Volume XL-7/W3)*, May 11–15, 2015, Berlin (pp. 991–998).

Sanchis-Ibor, C., Segura-Beltran, F., & Almonacid-Caballer, J. (2017). Channel forms recovery in an ephemeral river after gravel mining (Palancia River, Eastern Spain). *CATENA, 158*, 357–370.

Sanjaume, E. (1979). El karst marino en el litoral del norte de Alacant, *VI Coloquio de Geografía*, Palma de Mallorca (pp. 123–130).

Sanjaume, E., & Pardo-Pascual, J. E. (2005). Erosion by human impact on the Valencian coastline (E of Spain). *Journal of Coastal Research, SI 49*, 76–82.

Sanjaume, E., & Pardo-Pascual, J. E. (2008). Cambios de tendencias recientes en la evolución costera del golfo de Valencia: análisis espaciales y sedimentológicos In *Actas de las Jornadas Técnicas: Las Nuevas Técnicas de Información Geográfica al servicio de la gestión de zonas costeras: análisis de la evolución de playas y dunas* (pp. 1–9). Universitat Politècnica de València.

Sanjaume, E., & Pardo-Pascual, J. E. (2011). Degradación de sistemas dunaes. In E. Sanjaume & F. J. Gracia (Eds.), *Las dunas de España* (pp. 609–640). Cádiz: Sociedad Española de Geomorfología.

Sanjaume, E., Rosselló, V. M., Pardo-Pascual, J. E., Carmona, P., Segura, F., & López-García, M. J. (1996). Recent coastal changes in the Gulf of Valencia (Spain). *Zeistchrift für Geomorphologie, 102,* 95–118.

Sanjaume, E. (1985). *Las costas valencianas: sedimentología y morfología* (p. 505). Valencia: Universidad de Valencia, Sección de Geografía.

Segura-Beltrán, F., & Sanchis-Ibor, C. (2013). Assessment of channel changes in a Mediterranean ephemeral stream since the early twentieth century. The Rambla de Cervera, eastern Spain. *Geomorphology, 201,* 199–214.

Serra, J. C. (1986). *Procesos litorales en la costa de Castellón*. Unpublished doctoral thesis, ETS Ingenieros de Caminos, Canales y Puertos, Unviersitat Politècnica de València, 891 pp.

Short, A. D., & Hesp, P. (1982). Wave, beach and dune interactions in Southeastern Australia. *Marine Geology, 48,* 259–284.

Vilar, J. (1931). *Proyecto de defense de la playa frente a la caseta de amarre del cable de Baleares*. Unpublished document, Archive of Port of Valencia, March 1931.

Vilar, J. (1934). *Proyecto de defensa de la playa desde la desembocadura del río Turia al Perellonet*. Unpublished document, Archive of Port of Valencia, January 1934.

Wright, L. D., & Short, A. D. (1983). Morphodynamic of beaches and surf zones in Australia. In P. D. Komar (Ed.), *Handbook of coastal processes and erosion* (pp. 55–64).CRC Press.

Wright, L. D., & Short, A. D. (1984). Morphodynamic variability of surf zones and beaches: A synthesis. *Marine Geology, 56,* 93–118.

Chapter 11
Mediterranean Coastal Lagoons

Eulalia Sanjaume, Josep E. Pardo-Pascual and Francesca Segura-Beltran

Abstract On the eastern coast of the Iberian Peninsula—between the Ebro Delta and Cape Palos—the conditions during the Pleistocene, and especially after the Flandrian transgression, favoured the development of beach barrier systems and lagoons along large segments of the coast. The very small tidal range made connections between the sea and lagoons difficult, and this favoured sedimentation processes, which was often accelerated by human activity. Three very different sectors have been differentiated: the Gulf of Valencia, where the largest number of lagoons is found; the cliffed Betic structural sector between Cape Sant Antoni and Cape de les Hortes, in which there are just a few very small lagoons; and the southern sector in which Pleistocene formations occur on the surface or below the current sands. In the Gulf of Valencia there is practically a continuum of Holocene beach barrier-lagoon systems (post-Flandrian) that always start from a Quaternary alluvial structure. The structure and evolution of beach barrier-lagoon systems in this sector reveal the essential role of fluvial sediment supply to the system. The southern sector is formed by systems of a different nature, and in which the recent cumulative processes have played a much less important role, being so much more important the tectonic movements and Quaternary sea level changes. Human action is, in all cases, an essential factor in explaining the current landscapes and development of these ecosystems.

E. Sanjaume (✉) · F. Segura-Beltran
Department of Geography, Universitat de València, Avda. Blasco Ibáñez, 28, 46010 Valencia, Spain
e-mail: Eulalia.Sanjaume@uv.es

F. Segura-Beltran
e-mail: Francisca.Segura@uv.es

J. E. Pardo-Pascual
Geo-Environmental Cartography and Remote Sensing Group, Department of Cartographic Engineering, Geodesy and Photogrammetry, Universitat Politècnica de València, Camí de Vera s/n, 46022 Valencia, Spain
e-mail: jepardo@cgf.upv.es

J. A. Morales (ed.), *The Spanish Coastal Systems*,
https://doi.org/10.1007/978-3-319-93169-2_11

11.1 Introduction

Most of the Spanish Mediterranean coast is formed by beach barriers with active or relict lagoons. This structure is far from unique to the Mediterranean, although these systems occur here with two characteristics that distinguish them from those found in other parts of the world: namely, the small western Mediterranean tidal range and a high degree of anthropization. The average tidal range is just 20 cm. If meteorological factors are added, this average may reach 30 cm, with one metre being the maximum variability in annual sea level (Aragonés et al. 2016).

A common feature of all these wetlands is their high ecological value, given that they are home to specific flora and fauna, and are usually extremely productive spaces. For this reason, they are all protected by legislation. The environmental protection currently enjoyed by these coastal wetlands contrasts with the historical perception. For centuries, coastal wetlands were considered dangerous and unhealthy. Their difficult access, and above all, the presence of standing waters that facilitated the breeding of insects, meant they were considered a source of disease. For this reason, there were many attempts to drain the wetlands, and transform these landscapes into farmland.

The definition of these lagoons, as well as the origin of their beach barriers, has aroused controversies throughout the twentieth century. The lagoon/beach barrier systems appear at very different latitudes, and with differing conditions of wave energy, tidal range, climatic characteristics, and sea-level oscillations (raising, still standing, or falling). The development of barriers is controlled by four factors: (i) profile gradient; (ii) sediment supply; (iii) wave energy versus tide energy; and (iv) rates of sea level changes.

A gentle beach slope favours coastal sedimentary transport and so facilitates sedimentation. Prograding barriers will form during periods when the sea level remains stable under conditions of strong sedimentary contribution (Davis and FitzGerald 2004). During these moments, beach barriers continue prograding if there is excess supply. When the sediment contribution decreases the progradation stops. When a small sea level oscillation occurs, and/or a new increase in sediment supply occurs, a new beach barrier may be formed at a distance from the previous barrier. This is what has happened, for example, in the lagoons existing to south of the mouth of the River Xúquer (or Júcar) and in the Elx Lagoon (Fig. 11.1).

Most of the beach barriers on the Spanish Mediterranean coast developed during the Flandrian transgression, although along the Valencian coast there are also plenty of Pleistocene beach barriers adjacent to (or supporting) those of the Holocene. The fact that nearly all current beach barriers are basically Holocene does not mean that they have been homogeneously affected by changes in sea level. The existence or otherwise of subsidence—often related to neotectonic processes—has caused major differences between the various beach barrier-lagoon systems (including spit bars, and mid-bay bars) in the Mediterranean. Sometimes a developing spit may reach, at its seaward end, a structural element such as a cape, and then it becomes a beach barrier. This has happened in Valencia Lagoon. In the Mar Menor, we can see what

Fig. 11.1 Location of the beach barriers and wetlands

looks like a barrier island, although it is probably not a barrier island in the true sense.

The most striking feature that differentiates Mediterranean beach barriers from oceanic barriers is the size and dynamics of the inlets, due, above all, to the low energy levels in the tidal flows. The type of inlets that divides beach barriers in environments with high tidal ranges channels large flows of water into and out of the lagoon every day, which is why such inlets are usually wide and difficult for longshore transport to close. In the Mediterranean, with a negligible tidal range, there is practically no exchange of water and sediment (except during major storms when storm surges can cause the penetration of seawater). Conversely, when a lagoon receives a sudden rivers and ravines discharges, an over-elevation of the lagoon water level will cause drainage through the inlets. This lower level of activity explains why many inlets are closed by the actions of strong littoral drifts, and why human action is required to keep open connections between the sea and lagoon.

Lagoons have a variety of shapes, although the most common shape is elongated parallel of to the coast. Mediterranean lagoons are often tens of kilometres long although depending on the topographic configuration of the flooded area, they can adopt very different shapes (for example, the Valencia and Elx lagoons are round). Permanent lagoons are generally shallow, rarely reaching a meter in depth. In most cases, fresh, brackish, or salty water will appear only sporadically after periods of heavy rain. After such rainfalls, run-off contributions from ravines drain into the lagoons, as well as flooding waters from fans and adjacent floodplains (Segura-Beltran and Pardo-Pascual, in this book), and aquifer springs. Only during exceptional coastal storms does seawater enter most of the lagoons.

The anthropization of ancient lagoons is a particularly marked feature of the Mediterranean coast (Sanjaume and Pardo-Pascual 2000). Societies have long tried to take maximum advantage of these marshy spaces, and depending on the specific conditions of each lagoon, human intervention has acquired different objectives (mostly fishing, salt extraction, or farming).

The lagoon and marshy areas adjacent to beach barriers are clearly depositional environments in which sediments come from the immediate coast, from nearby continental areas, and from the organic waste generated by biocoenosis (that is, organisms that live in these ecosystems). The scarce coastal sediments are usually found near the beach since they will be relatively coarse sandy materials that require relatively large amounts of energy to be moved. In environments with high tidal ranges, the penetration of sediments through barrier inlets is noticeable (Carter 1988). However, the low tidal ranges of the Mediterranean mean that the methods by which materials can enter the wetlands are relatively few: either through water flows that occasionally penetrate the inlets; or by being dragged by the water or wind from the barrier; and very occasionally, during storms, where the barrier is sufficiently narrow and overwashing generates washover fans. Only in such cases can gravel-type sediments enter to lagoon.

Continental waters reach lagoons through surface and underground flows. Water from rivers and ravines may directly flow into the wetlands, or through adjacent floodplains during overflows in the fans (Segura-Beltran and Pardo-Pascual in this book). Given the hydrometeorological conditions of the Mediterranean, such contributions are sporadic but extremely important. Moreover, water flows from ravines can add considerable volumes of sediments that can fill the depression in which the wetland is located. The fluvial sediments that fill lagoons are fine, while the coarse fraction (sands and gravels) are deposited in the inner side of the wetland. Being an old base level, the profiles of ephemeral streams suffer a sharp break in slope when they reach lagoon, causing small fans prograding into wetlands. Thus, the size of the sediments decreases from the inner side of the wetland to the center of the lagoon. Lagoon sediments are therefore basically fine, and contain a high proportion of organic matter (peat sometimes forms) that creates clearly defined structures which are easily identifiable in core samples.

Groundwater contributions are necessary for maintaining permanent surface water. Being at sea level, the bottom of these lagoons often cut into the quaternary aquifer. This detrital aquifer is formed from the Quaternary fluvial sediementation that fill the coastal plains, and the aquifer is, in turn, fed by Mesozoic aquifers in the interior limestone reliefs. These are karstic aquifers, whose water table levels rise sharply following the sudden and intense rainfalls that are typical of the Mediterranean area. In these circumstances, springs provide large volumes of water into the lagoons and help ensure their persistence (Torreblanca and Valencia lagoons). In places where human action has carried out major drainage works, superficial water is only residual and sporadic.

This study focuses on the coastal wetlands found on the Spanish Mediterranean coast between the Ebro Delta and the Mar Menor. We intend to describe most of these landscapes and show the differences between them caused by a series of factors related to the local structure, coastal dynamics, sea-level, neotectonic processes, and human actions.

11.2 Beach Barriers and Coastal Wetlands

Along the 450 km of coastline that stretch between the Ebro Delta, to the north, and Cape Palos, to the south, there are up to 18 coastal wetlands of differing sizes and types. Figure 11.1 shows the 18 wetlands, the position of the beach barriers, and the local names, while Table 11.1 shows some essential numerical characteristics.

In most cases, the beach barriers and coastal lagoons are small—although some are large, such as Valencia Lagoon or the Mar Menor (Fig. 11.1). Together, they stretch 180 km in length, yet distribution is very uneven: 71% of the beach barrier coastline and nearly 54% of the wetlands studied are in the Gulf of Valencia (between the delta of Ebro and cape of Sant Antoni). The second area with numerous beach barriers is at the southern end of the study area, from the Cape de les Hortes to Cape Palos. In this sector, we find some 26% of the beach barrier coastlines, and 45.6% of the wetlands. In the cliffed coastal section between Cape Sant Antoni (marking the southern limit of the Gulf of Valencia) and Cape de les

Table 11.1 Main descriptive parameters of the studied wetlands

	Beach barrier characteristics			Wetlands characteristics	
	Offshore slope (degrees)	Barrier length (km)	Mean beach barrier width (m)	Surface (ha)	Width mean (m^2/m)
Albufera de Peníscola	0.474	4.7	145	105	223
Albufera de Torreblanca o Prat de Cabanes	0.421	10.2	68	1,061	1,043
Albufera d'Orpesa	0.718	1.9	100	55	283
Quadro de Castelló	0.387	7.5	181	980	1,314
Marjal de Nules	0.420	8.5	140	866	1,020
Marjal de Moncofa-Xilxes-Almenara	0.471	13.6	387	1,864	1,373
Marjal de Puçol-Alboraia	0.437	19.5	146	620	318
Albufera de València	0.621	29.5	616	22,300	7,113
Marjals de Tavernes Gandia	0.525	20.4	1,131	2,691	1,321
Albufera de Pego	0.628	7.8	1,256	1,290	1,644
Saladar de Xabia	1.347	3.2	152	117	369
Salines de Calp	1.218	1.4	192	40	294
Estany de l'Albir	0.956	1.2	111	24	193
Saladar d'Aigua Amarga	1.133	2.5	116	193	757
Salines de Santa Pola- El Fondo d'Elx (Albufera d'Elx)	0.328	7.9	246	5,390	6,363
Laguna de la Mata	0.758	5.0	646	996	2,000
Salinas de Torrevieja	0.999	5.4	561	2,366	4,366
Mar Menor	0.706	24.4	287	13,500	6,967

Hortes to the south, there are only a handful of very small beach barrier-lagoon systems that represent 3% of the beach barrier coastline and just 0.3% of the surface coastal wetlands.

These three geographical regions—the Gulf of Valencia; the cliffed section between Cape Sant Antoni and Cape de les Hortes; and the southernmost area from Cape de les Hortes to Cape Palos—establish three large models of coastal wetlands. The wetlands of the Gulf of Valencia (except for Valencia Lagoon and small lagoons in cliffed areas), have elongated shapes, with an average width that rarely exceeds a kilometre. The average wetland widths to the south are usually much greater (Table 11.1).

Another important feature is the offshore slope, which is a key factor in the development of a barrier. The average slope between the coastline and the −10 m isobath was estimated. The results of this parameter (Table 11.1) show a clear difference between the conditions found in the southern cliffed area and the other two geographical regions. In the Gulf of Valencia, the average slope is 0. 51° and slightly greater just opposite Valencia and Pego lagoons (around 0.6°); while those of the cliffed area reach 1.17°. The slopes in the southernmost area are in an intermediate position, although here, unlike the other two regions, the values vary greatly from one barrier to another: while the average slope of this region is 0.78°, that of the barrier that closes Elx Lagoon is 0.32°.

Another feature that distinguishes the three regions is the base on which the beach barrier sits. In the case of the Gulf of Valencia, all the beach barriers have proximal and distal supports. They usually start and end on an alluvial construction, usually fan-deltas, although sometimes on a flood plain (Segura-Beltran and Pardo-Pascual, in this book). Sometimes the distal end of the support is also a structural element (Cape Cullera). In the southernmost segment, the beach barriers usually sit on lithified quaternary materials that have undergone neotectonic elevation.

Finally, another differentiating feature in the three areas is the lagoons water characteristics: while in the Gulf of Valencia they are practically all freshwater, in the southernmost sector they are all brackish or saline. This can be related mainly to the climatic differences—the southernmost sector is much arid than the Gulf of Valencia—although anthropic action has helped increase the salinity to maintain the salt pans. The greater rainfall in the Gulf of Valencia causes greater volumes of continental water than in the southern zone, especially aquifer flows. For this reason, in most cases, the water has been transformed in historical times from saline to brackish to fresh. This ecological change has resulted in a radically different human exploitation throughout history: the areas flooded with freshwater have been used for farming (and fishing, to a lesser extent). In contrast, in those places with brackish or saline water, the main human uses have been the extraction of salt and fishing (as is the case of the Mar Menor).

Therefore, to describe the essential characteristics of the lagoon and beach barrier systems in the coastal segment analysed, we will divide the analysis into these three large geographic units.

11.3 Beach Barriers and Coastal Wetlands in the Gulf of Valencia

This geographical unit includes the lagoons between the Ebro delta and Cape Sant Antoni. The inclusion of these ten lagoon-beach barrier systems is based on their characteristics, but it has also been considered that the coastline of the Gulf of Valencia has behaved until recently, as a single coastal sedimentary cell. In this long coastal section, currently formed mainly of beaches, there is a littoral drift that transports sand, and even gravel, to the south. Throughout this sector the most efficient waves are those from the NE, or the ENE (Pardo-Pascual 1991). Therefore, because of the general orientation of the coast, longitudinal currents to the south are more frequent, and therefore, there is a net transport to the south. The frequency and speed of the longitudinal current to the south is greater in the northern part of the gulf (with a general NW-SE orientation) than in the southern part (NW-SE orientation), with the boundary between these two parts being near the city of Valencia. The annual net transport (or coastal drift) to the south practically ceases at Oliva, very near to the southern end of the Gulf of Valencia (with an almost EW orientation) and where the beach barrier that closes Pego Lagoon is located. Given this general transport to the south, it is to be expected that there will be a greater sedimentary contribution in the southern areas given that important volumes of sediment arriving from the north.

This sedimentary cell on the Valencia coast has been fragmented during the 20th century as a consequence of the construction of large seaports (Pardo-Pascual 1991; Sanjaume et al. 1996; Sanjaume and Pardo-Pascual 2005). These port structures have segmented the longitudinal transport of sediments, and so there have been strong accumulations and erosion updrift and downdrift around the ports. This has meant that some beach barriers located to the north of a port have significantly broadened their width because of the accumulation caused by these artificial works. This is the case of the Quadro de Castellon beach barrier located at north the port of Castellon. The opposite effect is produced to the south of these port structures, since the eroded sediments transported downdrift cannot be replenished because sediments are trapped to the north. Therefore, the width of these beach barriers clearly diminishes due to an accelerated beach retreat. To the south of the port of Castellon, the narrowness of the Nules Marsh beach barrier can be clearly seen. Another example is the northern part of the Puçol-Alboraia Marsh beach barrier—which has been eroded by the effect of the port of Sagunt.

The existence of a strong longitudinal transport of sediment is crucial to the development of beach barriers since the bulk of these sediments originate from the various rivers that flow into the sedimentary cell, as clearly demonstrated by Sanjaume (1985). The volume of sediments that each river contributes to the coastal system depends on the geology, climatic conditions, and above all, the size of its basin (Segura and Pardo-Pascual, this book). Some 61% of the lands that drain to Gulf of Valencia are drained by the River Xúquer, 18% by the River Turia, and 10% by the River Millars, leaving just 8.7% for the remaining rivers and ephemeral streams that reach the coast. There is, therefore, a considerable difference in the volume of sediments transported by the drift towards the south, depending on the position within the coastal sedimentary cell. Thus, in the northern part of the cell, the only fluvial systems that produce sediments are those that leave the small ravines that drain this sector and the small proportion coming from cliffs around the northern end of the Gulf of Valencia near Vinaròs and Benicarló. Therefore, the updrift from the mouth of the River Millars can only carry a small volume of sediment, since the fluvial courses of the sector only drain 4% of the surface that pours into the Gulf of Valencia. This partially explains why these beach barriers are so narrow (Table 11.1). The beach barriers formed by the downdrift from the mouths of the major rivers (the Turia and Xúquer) are much wider, the broadest part being just at the southern end of the sedimentary cell (Pego Lagoon). Indeed, in the two most southerly beach barriers (Tavernes-Gandia Marsh—Fig. 11.2—and Pego

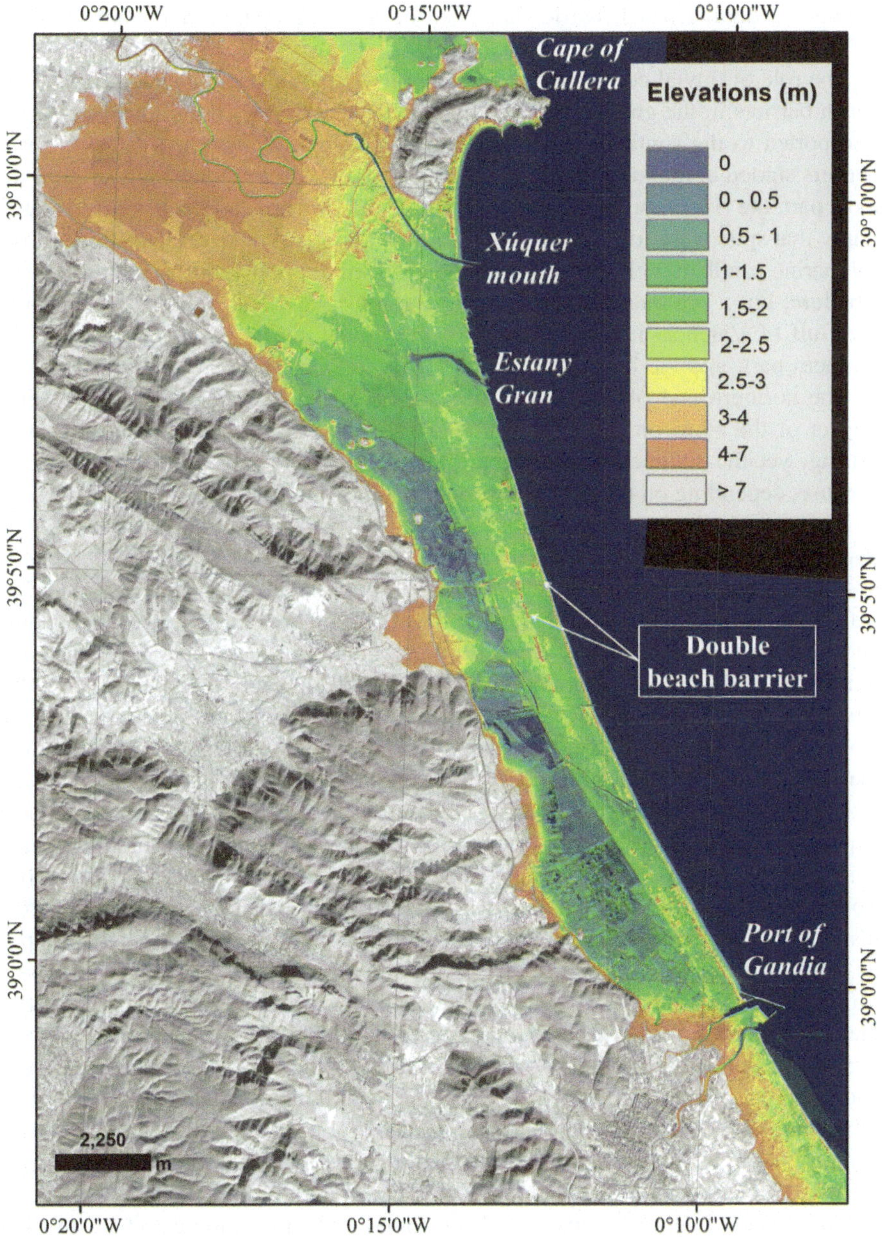

Fig. 11.2 Detailed topography of beach barrier and lagoons laying between Cullera and Gandia that discover the existence of two different beach barrier

Lagoon) a double beach barrier has been detected (Sanjaume and Pardo-Pascual 2003).

The role of fluvial contributions is crucial for explaining the development of the beach barriers in the gulf. These were mostly developed by the sedimentary excess transported to the south by the coastal drift. It is logical, then, that most of these barriers started at the mouths and grew progressively towards the south, and for the most part, the river mouths developed a fan-delta. This process does not necessarily imply that spits were formed, and it is possible that many barriers have arisen from submarine sandbanks. In some cases, the seaward support point is not an alluvial structure, but a pre-quaternary promontory (Orpesa and Valencia lagoons), but in the Gulf of Valencia it can be seen that the fluvial course always appears in the northern part, and the beach barrier starts from a Quaternary alluvial deposit.

The northern end of the Gulf of Valencia borders the Ebro delta. The limited impact of the Ebro on the coast immediately to the south may therefore be surprising, yet the sedimentary contribution can be considered almost zero. This is because, depending on the sea conditions, the sandy contributions of the River Ebro nowadays are distributed among each of its delta spits bars (Jiménez and Sánchez-Arcilla 1993; Guillén and Jiménez 1995; Valls et al. 2016). This reality has also been evidenced by the results of a mineralogical analysis made on Valencian beach sediments (Sanjaume 1985). There remains, however, a doubt about the contribution that the River Ebro potentially made to the Valencian coasts during the Holocene. The formation of the Ebro delta was a complex process that developed from various lobes during the Holocene (Somoza and Rodríguez-Santalla 2014; Cearreta et al. 2016). It is reasonable to suggest that since the Flandrian maximum, the process of delta formation has reduced the southward sedimentary contribution. An early sedimentary contribution by the Ebro could explain the sedimentary surplus needed for the development of the multiple beach barriers we find in the northern Valencian segment (the lagoons at Peníscola, Torreblanca, Orpesa, and Quadro de Castelló). These beach barriers are hardly justified by the scarce fluvial contribution of the local ravines. As the delta trapped its own sediments, the contributions towards the southern coasts decreased. Moreover, from the 1960s onwards, the construction of large reservoirs meant that the contribution of the River Ebro decreased drastically. This would explain regressive signs found on some beach barriers like the narrowness of the barriers, and the predominance of coarse sediments. It would also explain clear signs of an inland migration by the barrier at Torreblanca Lagoon, (Sanjaume et al. 1990). This explanation must be understood as a working hypothesis as there is insufficient reliable data to support it fully. On the sandy beach in the centre of the beach barrier of Torreblanca Lagoon, sands have been found with mineralogical characteristics typical of the Ebro basin (Segura et al. 1994), but these could also be the result of the destruction of the ancient Quaternary aeolianites submerged offshore.

Another key factor in understanding the evolution of existing lagoon-beach barrier systems is the position of sea level, determined by eustatic behaviour, and by local variations related to subsidence and/or neotectonic factors. All the beach barriers in the Gulf of Valencia are Holocene and have developed from the

Flandrian transgression. However, local subsidence is highly variable. As explained in the chapter on coastal river forms (Segura and Pardo-Pascual, this volume), in the Gulf of Valencia there is a clear difference between the northern sector (Ebro Delta—River Palancia) and the southern sector (River Palancia—Cape Sant Antoni). Indications of subsidence in the lagoons north of Sagunt are unclear; however, evidence of local subsidence in the lagoons of the centre and south of the Gulf of Valencia are well supported (Garay 1995). Thus, in the Puçol-Alboraia Marsh it has been possible to deduce the existence of a series of blocks that have followed a differential subsidence during the Quaternary (Pardo-Pascual et al. 1994; Segura et al. 1995). Clear subsidence processes have been reported on the plain where Valencia Lagoon sits (Sanjaume and Carmona 1995; Albarracín et al. 2013; Alcántara-Carrió et al. 2013), as well as in the southern coastal segment, between Cullera and Dénia. Subsidence is also very clear at Pego Marsh and where the Flandrian maximum, which is supposed to be 2 m above sea level, has been found at −2 m at 1.6 km inland from its current position on the coast (Viñals 1995). However, this does not mean that there has been no subsidence in the northern sector, or that subsidence occurred at a homogeneous speed in the southern sector.

Except for Valencia Lagoon, most of the lagoons along the Gulf are clogged with sediments and few permanently flooded areas remain. The filling of these ancient lagoons has been caused by the continental and marine processes described above, and, to a much greater extent, by anthropic action that, especially since the eighteenth century, has constantly sought to transform these lands. Mateu (2016) suggests that the impulse to drain the coastal wetlands is clearly related to the closure of the barrier inlets, which probably occurred during the second half of the seventeenth century and caused a continentalisation of the wetlands (Mateu et al. 1999). This closure of the natural inlets caused an expansion of the flooded areas within the wetlands (which were already largely filled with silt). In contrast, from the second half of the eighteenth century to the middle of the twentieth century, multiple human interventions were made to transform these spaces into farmland by draining the waters and digging artificial inlets. Determining the exact moment when the inlets were closed is not easy for the whole coast. In some cases, such as Valencia Lagoon, there is considerable documentation regarding the various actions taken (Sanchis 2001), but information is much less complete for the other areas.

Initially, it could be thought that closing the barrier inlets would alter the characteristics of the inland waters. However, there is evidence of changes in salinity (saline-brackish-fresh) prior to the closure of the inlets. In Valencia Lagoon, for example, the remains of freshwater fauna obtained from an several cores obtained from the current lagoon (and ^{14}C dated) show that the water was mostly fresh in 1110 ± 115 BP (Sanjaume et al. 1992). However, there is documentary evidence—including the existence salt pan operations next to one of the Valencia Lagoon inlets (Rosselló 1987)—that indicate the end of the seventeenth century as the moment when the transfer of water from the sea to the lagoon was stopped. At this historical moment three main factors combined: the Little Ice Age produced an increase in rainfall this caused flooding of the Xúquer and Turia rivers into the lagoon;

a the construction of a stable dike closing the old inlet for 30 years (1607–1637) which had been artificially regulated since the thirteenth century (Sanchis 2001); and the planting of rice fields alongside the Júcar river expanded toward the inlet and that contributed to the lagoon the surplus water and so favoured the growth of marsh vegetation (Sanchis Ibor 2016). All this suggests, as Rosselló and Sanchis (2016) correctly indicate, that the change of salinity was developed on a multi-secular scale, and human actions aimed at taking advantage of these environments could only take place once the lagoon had been transformed into freshwater.

Describing each of the existing lagoons on the coast of the Gulf of Valencia is beyond the objectives of this work, but it is worth focusing our attention on two different models: the Torreblanca Lagoon and Valencia Lagoon.

11.3.1 Torreblanca Lagoon (Albufera de Torreblanca)

The current lagoon covers an area of about 10 km^2 (Mateu 1977) and is closed by a 10 km long barrier that is supported by the alluvial fans of the River de les Coves to the north, and the River Xinxilla to the south. Its relatively large dimensions and some special characteristics have stimulated its study from various perspectives for decades (Rosselló 1969; Mateu 1977; Segura and Sanjaume 1986; Sanjaume et al. 1990; Segura et al. 1994; Dupré et al. 1994; Segura et al. 1995; Usera et al. 1990; Segura Beltran et al. 1997, Segura et al. 2005; Guillem et al. 2005; Ruiz and Carmona 2009; Carmona 2014; Segura et al. 2016; Carmona and Ruiz 2016; Mateu 2016). These works have focused mainly on: (i) the current shape and dynamics of the barrier, (ii) Quaternary evolution (fundamentally Holocene); (iii) possible changes in sea level or the coastline in relation to human occupation as indicated by archaeological remains; and (iv) natural and anthropic environmental changes in the lagoon.

The most distinguishing feature of this lagoon is undoubtedly its narrow beach barrier: in its widest sector it measures 125 m and at the narrowest point it measures just 7 m—almost all of which is occupied by a wide stoney beach ridge (Sanjaume et al. 1990). Numerous washover fans have formed over the stone barrier. The presence of the beach ridge, the abundance of washover fans, and the large size of stones that form the beachridge, make this beach barrier very distinct from the other Valencian barriers (Sanjaume et al. 1990; Segura, et al. 1995). Over some 10.5 km, three sectors of the barrier can be distinguished:

The northern part, between the prograding fan-delta of the River de les Coves and the Quarter Vell, is formed by a stoney beachridge (Fig. 11.3). Levels of peat have been found submerged in the sea near Torrenostra (already described by Rosselló 1969), indicating the recent retreat of the barrier (Fig. 11.4). An alignment of submerged eolianites was found in this sector at between −4 and −7 m in depth. The central sector is sandy and the beach is more developed than to the north. There are abundant remains of eolianites and Quaternary fossil beaches found among the

Fig. 11.3 Map of Torreblanca lagoon. It has been used a false colour Lansdat 8 scene acquired the 12-11-2017 as basemap. The black areas are flooded spaces. Marshy vegetation appears in red. It is interesting to observe the narrow beach barrier formed by very thick white cobbles

Fig. 11.4 Peat layers in the coastline indicate how this beach barrier is retrograding. Differences between shore and upridge size sediment could be observed

barrier boulders—and produced by the erosion of the submerged beaches. The mineralogy of these sands (Segura et al. 1994), suggests that either the sands come from the destruction of the submerged Pleistocene dunes, or they arrived before the Ebro delta became a barrier for southerly sedimentary transport.
The southern sector, which ends at the Torre de la Sal, has the widest barrier. Although boulders predominate, the distal sector finishes with surface outcrops of fossil dunes that form a small erosion platform submerged in the sea.

In addition to these longitudinal differences, there are also changes in the transversal direction: the diameter of the boulders is greater in the internal part (about 9 cm), than at the top (5–7 cm), or the external side (2.5–3 cm). This change in size is related to the energy of the waves that deposit the boulders.

It is difficult to establish the moment when the barriers started retrograding. From the sedimentary records of various surveys of this lagoon and barrier, it has been possible to sequence the evolution of the lagoon (Usera et al. 1990; Segura et al. 1997; Carmona and Ruíz 2014). Based on the above, Segura et al. (2016) indicate that, although various marine transgressions occurred over the past 15,000 years, only one Holocene lagoon has been located in its current location. If there was a precedent in the Pleistocene, it would have been located offshore. Regarding the space occupied by the wetland over the past 9000 years, we know from various surveys that the lagoon has moved over time. More sedimentary fill has been observed in the southern sector than in the northern sector, which could indicate greater subsidence in the south. Likewise, it has been observed that for practically 6000 years a humid environment has been maintained with low salinity levels in the water, as evidenced by foraminifera (Usera et al. 1990) and pollen records (Menéndez and Florschutz 1961; Dupré et al. 1994). These low levels of salinity, even shortly after the maximum Flandrian, are explained by the significant supply of water produced by the springs draining the aquifer.

In the southern part, there are dozens of old silos of between 0.5 and 1 m in depth carved into the outcrops of aeolianites and which archaeologists have estimated as being about 5000 years old (Guillem et al. 2005). On the alluvial fan of the Xinxilla river—an aggrading fan-delta (Segura and Pardo, this volume)—on which the southern part this barrier rests, some six hundred negative structures (holes, wells, silos, and buckets) were found. These structures have been dated to between 8000 and 7000 BP—although some are between 2000 and 2500 years old, and others date from historical times (Flors 2009). Here we can also find remains of urban structures from the Iberian period (2500 BP), some of which are now in the sea (Fernández 1990).

The location of these archaeological structures, especially the silos for storing cereals located on the fossil eolianites at Torre de la Sal, suggest that the coastline was much farther out and/or the sea level was lower. Likewise, the existence of probably urban settlements some 2000 or 2500 years ago on the fan of the River Xinxilla suggests that when the silos and buildings were made, the sea was not so close (Segura et al. 2005).

Already in historical times, there is evidence that, at least locally, salinity increased in some environments. Carmona and Ruíz (2014) point out that in core of the southern area, biological remains were found (dating from approximately 1650 BP) that suggest a brackish environment. Mateu (2016) also points to historical documentation indicating that some of the mouth was open to the sea until the mid-seventeenth century. This fact would imply the penetration, at least during storms, of seawater and a local increase in salinity.

It appears that this lagoon had a complex history. In its very early phase it was freshwater, then it was sometimes salty, before becoming fresh water again (and as it is today). Without denying the existence of subsidence blocks, the archaeological evidence and the results recorded by the various surveys suggest that the current structure could be explained by a change in the coastal sedimentary contribution from the Flandrian maximum to the present day. It is obvious that the area is currently depleted of sediments, although with the data that we currently have it is impossible to determine which factor explains this lack of supply. It is possible that in the initial phases of the development of this lagoon-beach barrier system, the area was partly fed by sediments from the Ebro. It is also possible that the River de les Coves, which is obviously now the main source of sediments, contributed a greater sedimentary load in earlier times. We cannot forget that the Late-Pleistocene fan-delta of the les Coves River was deposited offshore, and that it was subsequently destroyed by the sea (Segura-Beltrán and Pardo-Pascual, this volume).

A substantial decrease in fluvial contributions to the coastal system, and so a reduced transport to the south by the coastal drift, explains the systematic regression of the beach barrier, which would imply that populated areas (silos, buildings) were abandoned because of the increasing nearness of the sea. The coastal retreat and the progressive narrowing of the beach barrier caused, in turn, the inlet and widening of the mouth, facilitating the entry of seawater that would locally alter the nature of the lagoon. The structure of the beach barrier as a narrow beach ridge with clear signs of an inland migration can only be explained by the extraordinary sedimentary deficit that the sector has suffered in recent centuries.

The closure of the inlets at the end of the seventeenth century caused an increased flooding of the lagoon by the accumulated flow inside and this caused a significant increase in the problems of diseases such as malaria in the villages near the lagoon (Mateu Bellés 2016). From that moment, the wetland was seen as a health problem and drainage works began. Numerous channels were dug during the eighteenth and nineteenth centuries, and the sediments extracted by the excavation works were used to raise the level of the land between the channels, and convert the plots into farmland. The channelled water was carried towards the sea, and up to three artificial outlets were opened to empty the waters. A complete transformation was prevented by the high levels of underground water supply in the area, and so large flooded spaces remain. Indeed, many specialists consider that Torreblanca Lagoon is one of the best preserved coastal wetlands on the Valencian coast (Costa 1999, Gómez-Serrano et al. 2001). It was declared a natural park by the regional government in January 1989.

11.3.2 Valencia Lagoon (Albufera de Valencia)

The current Holocene beach barrier of Valencia Lagoon is about 30 km long and starts from the mouth of the River Turia and ends at the cliffs around Cape Cullera (Fig. 11.5). The barrier is the last of a series of beach barriers that have existed in the same place throughout the Quaternary (Rosselló and Sanchis 2016). The existence of Pleistocene beach barriers (Rey and Diaz del Rio 1983; Sanjaume and Carmona 1995) has been confirmed following a series of seismic profiles (Albarracín et al. 2013) and cores (Alcántara-Carrió et al. 2013). These authors describe a submerged fossilised beach barrier from the Tyrrhenian age that runs parallel to the current coastline and is about 1700 m wide and 70 km long (running between Valencia and Gandia). The roof of the rock structure is 10 m deep in the north and 40 m deep in the south (Alcántara-Carrió et al. 2013). The presence of Quaternary eolianites in the current beach barrier (Rosselló 1979) has been known for decades, and cemented sand samples from the Tyrrhenian age were also found in the lagoon at Pego (Fig. 11.1) (Viñals 1996; Viñals and Fumanal 1995) and this suggests a major subsidence of tectonic origin along the entire coastal segment. It is important to remember that the main Tyrrhenian sea level highstand was on average some 6–7 m above the present sea level (Pirazzoli 1991), so that its current position under the sea and at various depths reveals greater subsidence in the south than in the north.

The genesis of the Holocene beach barrier differs from the other ones, because at the beginning it was a spit developing several hooks. The closest one (proximal

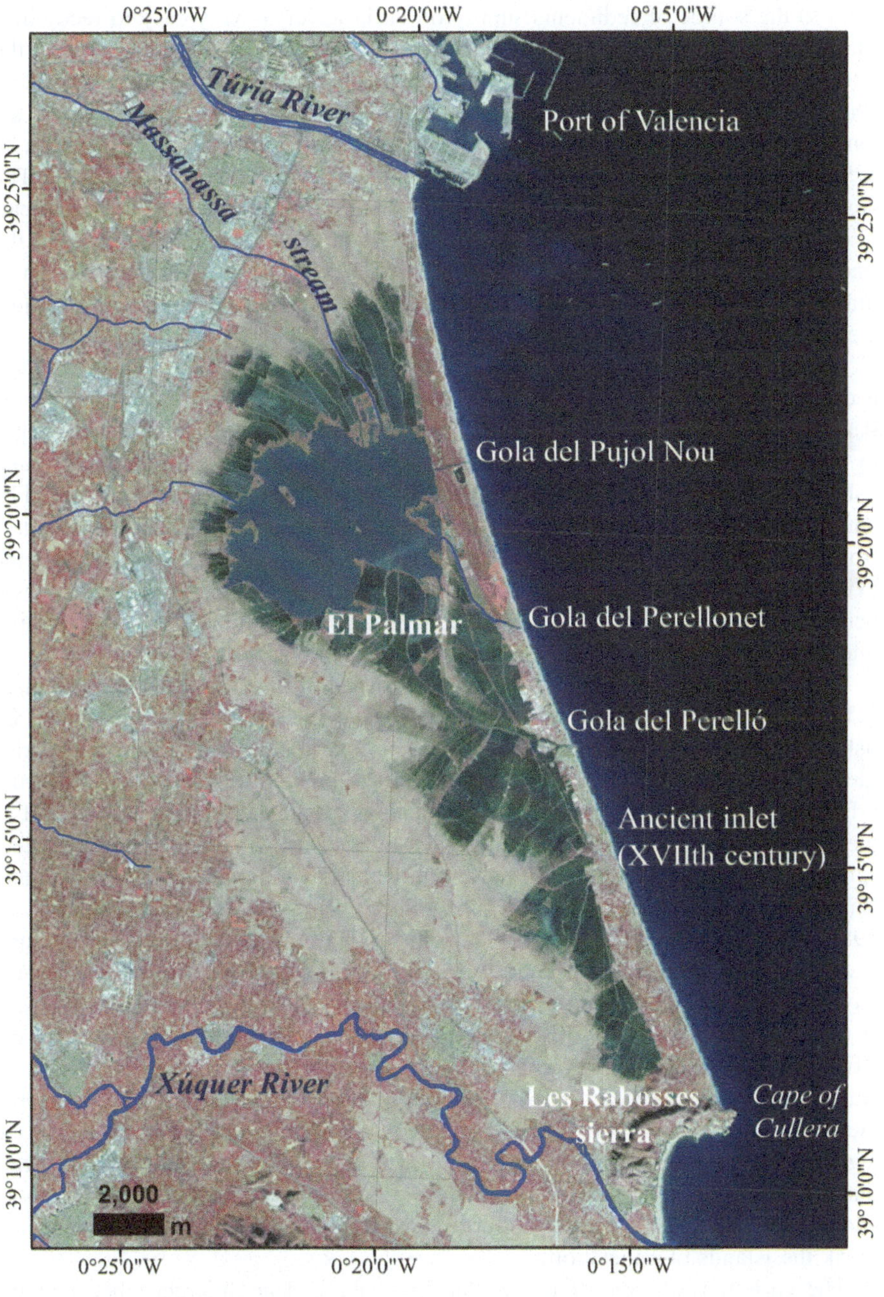

Fig. 11.5 Map of the Albufera de Valencia. The beach barrier—some hundreds of meters wide—lays between Túria river mouth and the mountains of Les Rabosses sierra. Here a Landsat false colour images has been used that allow to observe different environment into the wetland area: lake, marshy area—basically covered by rice crops that are partially flooded (green zone) and partially still dry (brown zone), and near the coastline, the beach barrier

side) to the source of sediments supply of Túria River, is very easy to recognize. Maybe because this hook has experienced less agricultural alterations since in the internal part of the second hook peduncle was installed the El Palmar settlement. The successive hooks towards the southern distal side are wider and wider, the human impact on this area has not allowed the peduncles identification, but can be recognized by the drainage channels drawing as well as the fields shape. The development of the spit ends when reach the sand trapped by the most external part of Cullera Cape. The inlets are showing the ends of these different peduncles of the free-tip spit (Sanjaume 1985). The origin of the beach barrier as is a succession of different beach barriers that start from the Túria and successively extend towards the east. The last beach barrier connects with Cape Cullera in the south.

The moment of the final closure of the beach barrier is difficult to determine, but it clearly occurred in historical times. The beach barrier currently has three inlets: Perelló, Perellonet, and Pujol (opened in 1953). All the inlets are regulated by gates to control the level of the lagoon, according to the needs of rice farmers and fishermen. Although some years ago it was thought that the inlet at Perelló could be natural, more recent studies suggest that both the Perelló and Perellonet inlets date from the 19th century (Rosselló 1995). The works of Sanchis (1998) indicate the existence of a large inlet in the southern zone that was open from the medieval period until the end of the seventeenth century and was regulated to protect fishing interests.

The evolution of the hydrological characteristics of the lagoon is closely related to the use made by people living nearby: it has changed from a salty lagoon used by fishermen (at least from the thirteenth century), to a freshwater lagoon used by rice farmers (eighteenth to twentieth centuries), to a hypereutrophic lagoon reflecting urban, industrial, and touristic growth (Sanchis et al. 2008) from the second half of the last century to the present. Studies during the last 20 years of the twentieth century confirmed a deterioration of lagoon waters (Soria et al. 1987; Vicente and Miracle 1992). After the lagoon and surroundings were declared a natural park (July 1986), a Sanitation Master Plan (1992) defined a wastewater treatment system; and sewage plants and water collectors were built around the lagoon. These measures have caused a substantial improvement in the quality of the water—which stabilised at the turn of the century (SDI 2011). However, the reductions of flows from the River Xúquer hinder the capacity of the ecosystem to assimilate the arrival of wastewaters (Sanchis et al. 2016). This loss of water quality has severely affected the ecosystems. This is most evident in the retreat of the marsh vegetation on the lagoon's banks and islands (*mates*). The growth of the communities of macrophytes that float on the surface has suppressed the wave action generated inside the lagoon, and this has made the wave action more powerful, and therefore able to erode and break the islands of vegetation.

The current landscape of the wetland reveals a clear difference between the marsh that covers an extension of 223 km^2 and the shallow lagoon (0.5–2 m) that covers some 28 km^2 and survives thanks to internal springs, the runoff from ditches and adjacent ravines, as well as flows from the rivers Turia and Xúquer after heavy rainfalls. The lagoon loses water by irrigation, evaporation, and through the inlets.

The rapid shrinking of the perimeter of the lagoon in recent centuries has attracted the interest of numerous researchers and generated public concern about the disappearance of the lagoon. During the second half of the twentieth century, several works described the problems that the lagoon faced. Most studies were based on theoretical models and many suggested that the sedimentation rates were so high that they would cause the total clogging of the lagoon in less than 200 years. García Labrandero (1959) indicated that the lagoon received 759,500 m^3/year of sediments and that it would be totally clogged by 2012! Alonso et al. (1974) estimated that the amount of sediment contributed by the irrigation ditches reached 160,000 m^3/year and filling would be complete by 2108. Dafauce (1975) considered that only fine sediments (silts and clays) settle in the lagoon and assuming that these represent only 40% of the total material, and that the volume of sediments reaching the lagoon is 275,000 m^3/year, then the lagoon would be totally filled by 2053. Mintegui et al. (1986) simulated erosion, transport, and sedimentation in the lagoon using various theoretical models of sediment production and estimated that the lagoon would be filled between 2066 and 2195.

Rossello (1976) proposed a different perspective and suggested that much of the contribution of the ravines would be deposited on the borders of the marshy area, while the deeper parts of the lagoon would probably remain stable, due to the processes of subsidence and/or compaction of accumulated sediments. Filling would therefore probably not occur. Sanjaume et al. (1992) used various surveys to establish the sedimentary evolution of the lagoon and historical sedimentation rates. Bearing in mind that the lagoon is a subsiding area, where the sediment is compact, and consequently, its volume is reduced, as well as the gradual disappearance of anthropic earthworks for more than a century, the conclusions were that the sedimentation rate was 0.47 mm/year between 2810 and 1110 BP, and 0.57 mm/year between 1110 BP and the present day—including all anthropic sedimentation (Sanjaume et al. 1992). These results suggest that the filling of Valencia Lagoon is not as imminent as predicted. Indeed, this work showed that only fine sediments were transported when floods reached the lagoon. Pebbles and gravel are not transported, and the small quantity of sand that arrives at the lagoon is deposited near the mouths of ravines, in the inner part of the marshy area.

11.4 Beach Barriers and Lagoons in the Betic Cliffed Coast

The coastal segment that stretches between the Cape Sant Antoni and the Cape de les Hortes is dominated mainly by wide stretches of cliffs and small beaches. The formation of beach barriers in this area is limited by the sharp slope of the offshore platform and by the small volume of sediment that moves through the area. Small beach barriers have only been produced in small bays. In this work, we have indicated the Saladar de Xàbia, l'Estany de l'Albir, and the Salinas de Calp

(Fig. 11.1). We could have introduced other old wetland areas such as l'Albufereta d'Alacant, or the Saladar de Sant Joan, but these former wetlands are now scarcely identifiable.

The old Xàbia lagoon sits in a structural Betic depression where Fumanal et al. (1993) detected a slight subsidence during the Pleistocene. The current beach barrier is 5 km long and up to 250 m wide and is attached to the cliffs of Cape Sant Antoni and Cape Sant Martí (thereby closing the morphostructural depression to the east). The beach barrier consists of Pleistocene calcarenites, while the absence of Holocene sedimentation is very noticeable (except for some constricted pebbles and gravels near the river mouth). The beach barrier is formed by the contributions from the River Gorgos (Sanjaume 1985), and looks like a midbay bar with a small sandy cove in the centre that interrupts the calcarenite alignments. Rosselló (1977) suggested that this cove could have been a mouth to the inner lagoon, but its current structure probably responds to tectonic and anthropogenic causes.

The Calpe wetland is a lagoon that has been transformed into a salt pan. It has been generated from the development of a double sand bar in the shadow area of the Penyal d'Ifac. This sector of the bay of Calp is Pleistocene (eolianites of that age are visible) and it is between 180 and 200 m wide. The northern area is Holocene although it may have underlying Pleistocene materials. The ancient lagoon has been used as a salt pan since Roman times (Sanjaume 1985).

11.5 Wetlands in the Southern Sector

The prominent role of neotectonic processes in the formation of the southern sector wetlands is their most distinguishing feature. The role of coastal dynamics is much less marked and the current structure reflects seawater flooding of depressions caused by subsidence and/or neotectonic processes that occurred during the Pliocene and continue to the present.

To the south of the Cape de les Hortes are a series of tectonic anticlinal and synclinal structures (Montenat et al. 1990), although some authors speak of intersect fractures (Rey et al. 1999) that have been active since the Pliocene (Garay 1995). These structures have shaped high reliefs—in anticlines or elevated blocks (Serra dels Colmenars, Domo de Santa Pola, Serra del Molar, Moncaio) and lagoons (Fig. 11.6) in the anticlines or sunken blocks (Saladar d'Aigua Amarga Lagoon, Elx Lagoon, Mata Lagoon, Torrevieja Lagoon). Tectonic activity during the Quaternary has produced a set of dislocated Tyrrhenian fossil beaches at different heights along the sector (Rosselló and Mateu 1978; Mateu and Cuerda 1978; Sanjaume and Gozálvez 1978; Goy et al. 1993). Moreover, these faults and folds are still active offshore (Maillard and Mauffret 2013). The beach barriers along this stretch are made of Pleistocene materials, as evidenced by the discovery of a huge outcrop of *Strombus bubonius* on the external part of Clot de Galvany.

To the south of the EW-running anticlinorum of Tabarca is the Torrevieja half-graben. Aligned with this is the San Javier sub-basin, a graben in the basement

Fig. 11.6 The map locates depressions where are the lagoons (Saladar d'Aigua Amarga, Clot de Galvany, Elx lagoon, salt pans of La Mata and Torrevieja) and also the hillocks (Serra dels Colmenars, L'Altet, Santa Pola dome, Serra del Molar, El Moncaio)

of the Murcia basin that is filled with 2000 m of Neogene-Quaternary sediments (Jiménez-Martínez et al. 2012). This sedimentary basin is fractured by a series of horst and grabens (Rodríguez Estella and Lillo 1992) in which Tortonian volcanic eruptions occurred—especially in its southern segment (Duggen et al. 2005). There are, in fact, multiple volcanic outcrops in the sector: five ones within the Mar Menor, and other ones at Cape Palos, and offshore near La Manga. There are also three volcanic islands in the Grosa Island group at depths of between 10 and 20 m, and other outcrops at greater depths (Acosta et al. 2013).

The structure and recent evolution of this group of wetlands varies greatly. The Saladar d' Aigua Amarga Lagoon, located south of Alicante, occupies one of the indicated tectonic depressions. It is interesting to note that the beach barrier that closes the lagoon contains current materials in its northern part, and Pleistocene calcarenite deposits to the south. This same type of calcarenite deposit encloses other depressions where small wetlands are found immediately to the south (such as Salar de la Sinieta and Clot de Galvany).

To the south of the Santa Pola dome is Elx Lagoon—one of the largest lagoons of the study area. It is currently formed of two clearly differentiated water bodies: the Santa Pola salt pan (by the sea) and the El Fondo d'Elx (to the west). The first is a seawater lagoon, while the latter is freshwater. This division of the original lagoon into two wetlands was produced by the growth of the fan-delta of the River Vinalopó (Segura-Beltran and Pardo-Pascual, in this book), and especially as a result of a series of drainage projects dating from the eighteenth century. The objective of these projects—as in so many other places on the Mediterranean coast—was to produce farmland. Previously, there was a lagoon known in classical texts as the Sinus Ilicitanus. The original lagoon—formed after the Flandrian maximum—extended 19 km from the coast to the village of San Isidro de Albatera (Blázquez and Usera 2005). Tent-Manclús (2013) reported that the maximum extension of the lagoon was reached in 5000 BP following subsidence in the basin, and that it reached the current position of the town of Orihuela (about 30 km from the coast). From this moment the surface of this great lagoon was diminished fundamentally by the action of two rivers that drain into it and formed a flood plain (River Segura), which filled the depression from west to east, and a delta-fan (River Vinalopó) that moved from north to south and eventually divided the lagoon.

It is a prograding lagoon and beach barrier system, in which up to three beach barriers have developed in parallel. The internal barriers are Pleistocene: the Tyrrhenian II are of the midbay bar type, and the Tyrrhenian III barrier only appears near Cape Santa Pola (although the rest may have disappeared under salt pans). The barriers were formed with materials from the River Vinalopó (the coastal drift had to then be N-S) in the Holocene. The barrier nearest the sea was generated with contributions from the River Segura, which indicates a net coastal drift in a S-N direction (Sanjaume 1985). The mineralogical data agrees that the paleogeography of the mouth of the River Segura was totally different from the current contents until relatively recently (Tent-Manclús 2013).

Las Salinas de Torrevieja and La Mata lagoons occupy the synclinal bottoms of successive depressions separated by an anticlinal threshold formed on calcareous materials (Lillo 1984). Both lagoons were separated from the sea by lines of dunes that Rosselló and Mateu (1978) date to the Riss glaciation—but were completed during the Würm (although during the Tyrrhenian II transgression, the Torrevieja lagoon would have flooded when sea levels rose above the current level). The sea connection is currently maintained through artificial canals built to facilitate operations in the salt pan (which currently draws water from the sea and a salt deposit 40 km inland at Monte Cabezo, near the village of Pinoso).

The Mar Menor is by far the largest of all the lagoons studied in this work (Fig. 11.7). It is a hypersaline lagoon that currently covers approximately 135 km^2 and has an average depth of 3.6 m and a maximum depth of 6 m. It occupies a depression that is closed to the sea by a 23 km long sandy beach barrier, known as La Manga (from Cape Palos to San Pedro del Pinatar). The barrier is segmented by five inlets that connect to the Mediterranean, most of them natural—although artificially maintained. One of the inlets is completely artificial (the Marchamalo which was opened at the end of the eighteenth century) (Lillo 1979). The most superficial composition of the Manga is sand (D 50 = 0.21 mm) although there are outcrops of volcanic rock and calcarenites (Sánchez-Badorrey and Jalón-Rojas 2015). These rocky outcrops have led some authors, such as Lillo (1979), to point out that La Manga is not strictly a beach barrier but a rocky structure that has been partially, or almost completely, covered by sandy sediments left by the sea. Certainly, when compared to the beach barriers in the Gulf of Valencia, the differences are noticeable (in the Mar Menor there is no nearby fluvial source of sediments) and it is believed that the supporting rocky substrate has been key to its formation. Acosta et al. (2013) found in the neighbouring platform at least eight relic beach barriers parallel to the current barrier. The northern part of this barrier island, just south of the Salinas de San Pedro del Pinatar is a spit without a defined peduncle, but with lines of sandy dunes curved by the refraction and diffraction of the waves.

The main geomorphological elements determining the ecology of the Mar Menor are: (1) the sandy barrier island (la Manga); (2) the inlets that determine the water inflow from the Mediterranean and, therefore, its hydrology and degree of confinement; (3) the islands and volcanic outcrops that provide a clear environmental diversity; (4) the ravines that provide runoff water from the farmland and mining areas that surround the lagoon; and (5) the marginal lagoons, now transformed into salt pans (Pérez-Ruzafa et al. 2005). The Mar Menor can be defined as a concentration basin in which evaporation clearly exceeds the water received from rainfall and runoff. The high annual water deficit (which ranges between 38 and 115 Hm^3 per year) is compensated by the entrance of Mediterranean waters due to differences in the level of the waters between the open sea and the lagoon (Arévalo 1988).

The surface of the lagoon has been reduced over time by natural processes and human actions. Currently, two of the marginal lagoons are used for salt extraction. Four other lagoons have been filled by sediment or drained for farming or construction. This process explains the reduction in the surface area of the lagoon: shrinking from 185 km^2 in 1868, to 172 in 1927, and from 138 in 1947 to 135 in

Fig. 11.7 Main lanscape elements in the Mar Menor

1969, with water depth also reducing (Lillo 1979; Pérez-Ruzafa et al. 1987). It has been estimated that because of extensive tree-felling in the surrounding areas in the sixteenth and seventeenth centuries, the annual sedimentation rates in the lagoon increased 0.3 to 3 mm (Pérez-Ruzafa et al. 1987). Dezileau et al. (2016) recently analysed a 4 m core sample taken from within the lagoon at 800 m from the sandy bar, and at 8500 m from the mouth of the nearest inlet. Sedimentological analyses were made of the macrofauna found and it was possible to date up to 18 samples. As a result, it has been possible to reconstruct the changes between the base of the core (dated at 6500 BP) to the present. With these results, a sedimentation rate for this extraction point of 0.6 mm per year was estimated. These values show that the sedimentary contribution is very limited and substantially lower than those found in the lagoons of the Gulf of Valencia (Sanjaume et al. 1992; Segura et al. 2005).

During the survey, Dezileau et al. (2016) located several moments of increase in the proportion of sand, and changes in fauna that reveal a greater marine influence. After the closure of the current lagoon, sedimentation of lagoonal deposits dominated between 6500 and 5000 BP, and the sandy barrier island was more than one kilometre offshore from its current position. As of that date, the barrier started moving inland to a position close to the current barrier, which would explain why from that moment, coarse-grain layers appeared more frequently in the extraction. At 2400 BP, fauna records indicate a change in the ecosystem as it becomes more isolated from the sea. According to (Dezileau et al. 2016) the lagoon changed from being leaky to becoming restricted and chocked (Kjerfve 1994). The lagoon accumulated fine sediments, and the intercalated layers of sand have been interpreted as overwash processes (Dezileau et al. 2016). The survey reveals up to eight high-energy events attributed to major storms, given the coincidence between the last sandy layer dated in the survey and documentary records of the largest storms (as evidenced in the fields of Cartagena). Specifically, in 1869 there is evidence of a change in the salinity of the lagoon, which favoured the penetration of new marine species (Navarro 1927) and affected the fisheries (Pérez-Ruzafa et al. 1991).

The fragility of the environmental balance in the Mar Menor is clear, and small changes in one of its essential elements can produce a substantial alteration in the whole system. The recent alterations associated with the intensification of farming activities in the surrounding fields (intensive fruit and vegetable farming) and urban growth (associated mainly with tourism) has especially affected the Manga and may be generating a substantial change in the ecological dynamics of this environment. Pérez-Ruzafa et al. (2005) highlight two human interventions that are producing a more significant change in the transformation of the lagoon: firstly, the expansion of the Estancio Channel in 1972 has produced an increase in the rate of water renewal, a decrease in salinity and extreme temperatures, and has enabled the entry of new colonising organisms (this process is known as lagoon mediterranization); secondly, modifications in the input of the water regime that are producing a chain of changes in water quality, benthic vegetation, and plankton.

11.6 Conclusions

On the eastern coast of the Iberian Peninsula—between the Ebro Delta and Cape Palos—the conditions during the Pleistocene, and especially after the Flandrian transgression, favoured the development of beach barrier systems and lagoons along large segments of the coast. The very small tidal range made connections between the sea and lagoons difficult, and this favoured sedimentation processes—which was often accelerated by human activity.

Three very different sectors have been differentiated: the Gulf of Valencia, where the largest number of lagoons is found; the cliffed Betic structural sector between Cape Sant Antoni and Cape de les Hortes, in which there are just a few very small lagoons; and the southern sector in which Pleistocene formations occur on the surface or below the current sands.

In the Gulf of Valencia there is practically a continuum of Holocene beach barrier-lagoon systems (post-Flandrian) that always start from a Quaternary alluvial structure—which indicates the main source of sedimentary supplies. These systems vary considerably depending on the sedimentary contributions received. There is a clear distinction between the northern part of the gulf where regressive beach barriers are common (the most striking example being the Torreblanca Lagoon), and the central and southern zones supplied by large river courses (Turia and Xúquer). The structure and evolution of beach barrier-lagoon systems in this sector reveal the essential role of sediment supply to the system.

The southern sector is formed by systems of a different nature, and in which the recent cumulative processes have played a much less important role (except for Elx Lagoon). Tectonic movements in this sector from the Pliocene to the present, and sea level changes, determined the current structure of beach barrier-lagoon systems.

Human action is, in all cases, an essential factor in explaining the current landscapes and development of the ecosystems. In the case of the lagoons of the Gulf of Valencia, the processes of closing the beach barriers and the changes in the characteristics of the waters (from brackish to fresh) determined the impulse of the local societies when economically taking advantage of these spaces. As a result, very important alterations and radical changes in appearance were produced. Such changes in the southernmost sector have been less radical throughout history, with many of the lagoons being used as salt pans. However, as the Mar Menor case shows, apparently small changes can produce substantial alterations in an ecosystem that is always delicately balanced.

Aknowledgements The authors appreciate the financial support provided by the Spanish Ministry of Economy and Competitiveness in the framework of projects CGL2013-46917-R, CGL2015-69906-R and CGL-2017-86839-C3-1-R.

References

Acosta, J., Fontan, A., Muñoz, A., Muñoz-Martín, A., Rivera, J., & Uchupi, E. (2013). The morpho-tectonic setting of the Southeast margin of Iberia and the adjacent oceanic Algero-Balearic Basin. *Marine and Petroleum Geology, 45,* 17–41.

Albarracín, S., Alcántara-Carrió, J., Barranco, A., Sánchez García, M. J., Fontán Bouzas, Á., & Rey Salgado, J. (2013). Seismic evidence for the preservation of several stacked Pleistocene coastal barrier/lagoon systems on the Gulf of Valencia continental shelf (western Mediterranean). *Geo-Marine Letters, 33*(2–3), 217–223.

Alcántara-Carrió, J., Albarracín, S., Montes, I. M., Flor Blanco, G., Fontán Bauzas, A., & Salgado, Rey. (2013). An indurated Pleistocene coastal barrier on the inner shelf of the Gulf of Valencia (western Mediterranean): Evidence for a prolonged relative sea-level stillstand. *Geo-Marine Letters, 33*(2–3), 209–216.

Alonso, J. J., Benet, J. M., & García Gómez, J. (1974). *Balance hídrico y estudio de aportes sólidos a la albufera de Valencia.* Unpublished document. Instituto de Hidrología y Medio Natural, Universidad Politécnica de Valencia, 209 pp.

Aragonés, L., Serra, J. C., Villacampa, Y., Saval, J. M., & Tinoco, H. (2016). New methodology for describing the equilibrium beach profile applied to the Valencia's beaches. *Geomorphology, 259,* 1–11.

Arévalo, L. (1988). El Mar Menor como sistema forzado por el Mediterráneo. Control hidráulico y agentes fuerza. *Boletín del instituto Español de Oceanografía, 5*(1), 63–95.

Blázquez, A. M., & Usera, J. (2005). Variaciones de la línea de costa en l'Albufera d'Elx (Alacant) durante el Cuaternario reciente. In E. Sanjaume & J.F. Mateu (Eds.), *Geomorfologia litoral i Quaternari: Homenatge al Dr. V. Rosselló* (pp. 59–72). Departament de Geografia, Universitat de València, València.

Carmona, P., & Ruíz, J. M. (2014). Erosión costera y cambio ambiental en el humedal de Cabanes-Torreblanca (Castelló). Datos para una gestión sostenible. *Boletín de la Asociación de Geógrafos Españoles, 66,* 161–180.

Carmona, P., & Ruiz, J. M. (2016). Registro geoarqueológico y cambio ambiental. La costa de Torre de la Sal y el humedal del Prat de Cabanes. In J. F. Mateu (coord.), *El territorio valenciano: Transformaciones ambientales y antrópicas* (pp. 107–110). València, Universitat de València.

Carter, R. W. G. (1988). *Coastal environments: An introduction to the physical, ecological, and cultural systems of coastlines.* AC Press, 617 pp.

Cearreta, A., Benito, X., Ibáñez, C., et al. (2016). Holocene palaeoenvironmental evolution of the Ebro Delta (Western Mediterranean Sea): Evidence for an early construction based on the benthic foraminiferal record. *The Holocene, 26*(9), 1438–1456.

Costa, M. (1999). *La vegetación y el paisaje en las tierras valencianas* (p. 167). Madrid: Rueda.

Dafauce, C. (1975). *La albufera de Valencia: un estudio piloto.* Madrid: Ministerio de Agricultura, (Monografias 4). Madrid, 127 pp.

Davis, Jr. R. A., & FitzGerald, D. M. (2004). *Beaches and coasts.* Blackwell Publication, 419 p.

Dezileau, L., Perez-Ruzafa, A., Degeai, P. H., Degeai, J Ph, Raji, O., Martinez, P. H., et al. (2016). Extreme storms during the last 6500 years from lagoonal sedimentary archives in the Mar Menor (SE Spain). *Climate of the Past, 12*(6), 1389–1400.

Duggen, S., Hoernle, K., van den Bogaard, P., & Garbe-Schönberg, D. (2005). Post-collisional transition from subduction-to intraplate-type magmatism in the westernmost Mediterranean: Evidence for continental-edge delamination of subcontinental lithosphere. *Journal of Petrology, 46*(6), 1155–1201.

Dupré, M., Obiol. R. P., & Roure, J. M. (1994). Análisis polínico del sondeo TU de la turbera de Torreblanca (Castellón, España). In *Trabajos de Palinología básica y aplicada: X Simposio de Palinología* (APLE, Valencia, septiembre 1994) Universitat de València, pp. 165–174.

Fernández, A. (1990). El yacimiento submarino de Torre la Sal (Ribera de Cabanes, Castellón): nuevas aportaciones. *Cuadernos de Prehistoria y Arqueología Castellóenses, 13,* 229–248.

Flors, E. (coord.) (2009). Torre la Sal (Ribera de Cabanes, Castellón) *Evolución del paisaje antrópico desde la Prehistoria hasta el Medioevo*. Monografías de Prehistoria y Arqueología Castellóense, Castellón, Diputación de Castellón, 606 pp.

Fumanal, M. P., Usera, J., Viñals, M. J., Mateu, G., Belloumini, G., Manfra, L., & Proszynska-Bordas, H. (1993). Evolución cuaternaria de la Bahía de Xàbia (Alacant). In: *Estudios sobre Cuaternario*, Universitat de València, València, pp. 17–26. ojo no se si es el Cuaternario del Pais Valenciano.

Garay, P. (1995). Marco geológico estructural y geotectónica. In: *El Cuaternario del País Valenciano*, Universitat de València, València, pp. 31–42.

García Labrandero, A. (1959). *Evolución y futuro de la Albufera de Valencia*. Valencia, unpublished document, 64 pp.

Gómez-Serrano, M. A., Mayoral, O., & Domingo, J. (2001). *Guía de la naturaleza de la costa de Castellón. Itinerarios para conocer su fauna, flora, paisajes e historia*. Antinea. Vinaroz, 206 pp.

Goy, J., Zazo, C., Bardaji, T., Somoza, L., Causse, C., & Hillaire-Marcel, C. (1993). Eléments d'une chronostratigraphie du Tyrrhénien des régions d'Alacant-Murcie, Sud-Est de l'Espagne. *Geodinamica Acta, 6*(2), 103–119.

Guillem, P. M., Martínez, R., Pérez, G., Jordá, G., Pérez Milian R., & Fernández López de Pablo J. (2005). El Prat de Cabanes (Cabanes, Castelló). Un jaciment prehistòric del III mil·lenni (aC). In E. Sanjaume & J. F. Mateu (Eds.), *Geomorfologia litoral i Quaternari. Homenatge al profesor M. Rosselló i Verger*, Universitat de València, València, pp. 195–202.

Guillén, J., & Jiménez, J. A. (1995). Processes behind the longshore variation of the sediment grain size in the Ebro Delta coast. *Journal of Coastal Research*, 205–218.

Jiménez, J. A., & Sánchez-Arcilla, A. (1993). Medium-term coastal response at the Ebro delta, Spain. *Marine Geology, 114*(1–2), 105–118.

Jiménez-Martínez, J., Candela, L., García-Arístegui, J. L., et al. (2012). A 3D geological model of Campo de Cartagena, SE Spain: Hydrogeological implications. *Geologica Acta: An International Earth Science Journal, 10*(1), 1–14.

Kjerfve, B. (1994). Coastal lagoons. In *Coastal lagoon processes* (Vol. 60, pp. 1–8). Elsevier oceanography series.

Lillo, M. J. (1979). Geomorfología litoral del Mar Menor y del Bajo Segura. *Papeles de Geografía, 8*, 9–55.

Lillo, M. J. (1984). Consideraciones paleogeográficas sobre el endorreísmo marginolitoral del sureste español (Provincias de Alicante y Murcia). *Limnetica, 1*(1), 86–95.

Maillard, A., & Mauffret, A. (2013). Structure and present-day compression in the offshore area between Alicante and Ibiza Island (Eastern Iberian Margin). *Tectonophysics, 591*, 116–130.

Mateu, J. F. (1977). El factor antrópico en la transformación del Prat d'Albalat. In *Medio físico, desarrollo regional y geografía: V Coloquio de Geografía*. Facultad de Filosofía y Letras de la Universidad de Granada, Sección de Geografía, Granada, 3 al 6 de octubre de 1977: 193–200.

Mateu, J. F., & Cuerda, J. (1978). Morfología cuaternaria de la costa entre l'Altet y el Cap de Santa Pola. *Cuadernos de Geografía, 23*, 63–82.

Mateu, J. F. (2016). El Prat de Cabanes-Torreblanca: cierre de la última bocana litoral, In J. F. Mateu (coord.), *El territorio valenciano. Transformaciones ambientales y antrópicas*, pp. 111–114.

Mateu, J. F., Sanchis, C., & Ferri, M. (1999). El golf de València durant els segles XVI i XVII. *Geoarqueologia i Quaternari litoral. Memorial María Pilar Fumanal* (pp. 367–374). València: Universitat de València.

Menéndez, J., & Florschutz, F. (1961). La concordancia entre la composición de la vegetación durante la segunda mitad del Holoceno en la costa de Levante (Castellón de la Plana) y en la costa W. De Mallorca. *Boletín de la Real Sociedad Española de Historia. Natural Sección Geológica*, 59(1–2), 97–100.

Mintegui, J. A., López Cadenas del Llano, F., López-Cuervo, S., Aguiló, J., Pérez-Soba, A., de Miguel, M., Garíca Bartual, M., Riesco, J. A., & Torrent, J. A. (1986). *La erosión en la sedimentación de la Albufera,* Valencia, Conselleria d'Obres Públiques i Urbanisme, Generalitat Valenciana, 108 pp.

Montenat, C., d'Estevou, P. O., & Coppier, G. (1990). Les bassins neogénes entre Alacant et Cartagena. In *Anonymous Doc. et Trav. IGAL* (Vol. 12, pp. 313–368).

Navarro, F. (1927). Observaciones sobre el Mar Menor (Murcia). *Notas y resúmenes del Instituto Español de Oceanografía, Ser.II, 16,* pp. 1–63.

Pardo-Pascual, J. E. (1991). *La erosión antrópica en el litoral valenciano* (p. 240). Urbanisme i Transports: Generalitat Valenciana. Conselleria d'Obres Públiques.

Pardo-Pascual, J. E., Segura, F., & Sanjaume, E. (1994). Evolución cuaternaria de la antigua albufera existente entre Puçol y Alboraia. *Cuadernos de Geografía, 59,* 63–86.

Pérez-Ruzafa, A., Marcos, C., & Gilabert, J. (2005). The ecology of the Mar Menor coastal lagoon: A fast-changing ecosystem under human pressure. In *Coastal lagoons: Ecosystem processes and modeling for sustainable use and development* (pp. 392–422). Boca Ratón, Florida: CRC Press.

Pérez-Ruzafa, A., Marcos, C., Pérez-Ruzafa, I. M. et al. (1987). Evolución de las características ambientales y de los poblamientos del Mar Menor (Murcia, SE de España). In: *Anales de Biología* (pp. 53–65).

Pérez-Ruzafa, A., Marcos-Diego, C., & Ros, J. D. (1991). Environmental and biological changes related to recent human activities in the Mar Menor (SE of Spain). *Marine Pollution Bulletin, 23,* 747–751.

Pirazzoli, P. A. (1991). *World atlas of Holocene sea-level changes* (Vol. 58, 291 pp). Elsevier Oceanography Series. Amsterdam: Elsevier.

Rey, J., & Diaz del Rio, V. (1983). Aspectos geológicos sobre la estructura poco profunda de la plataforma del Levante Español. In Castellví (Ed.) *Estudio oceanográfico de la plataforma continental* (pp. 53–74). Barcelona, Bustos.

Rey, J., Fernández-Salas, L. M., & Blázquez Morilla, A. M. (1999). Identificación de las unidades morfosedimentarias cuaternarias en la plataforma interna del litoral del País Valenciano: el rol de los factores morfoestructurales y eustáticos. *Geoarqueologia i quaternari litoral: memorial María Pilar Fumanal* (pp. 403–418). València: Facultat de Geografia i Història.

Rodríguez-Estrella, T., & Lillo, M. (1992). Geomorfología del Mar Menor y sectores litorales contiguos (Murcia-Alacant). *Estudios de Geomorfología en España,* 787–807.

Rosselló, V. M. (1977). Restos de marisma en el litoral de Xàbia (País Valenciano) transformaciones antrópicas. In *Medio físico, desarrollo regional y geografía*: V Coloquio de Geografía: Facultad de Filosofía y Letras de la Universidad de Granada, Sección de Geografía, Granada, 3 al 6 de octubre de 1977, 187–192.

Rosselló, V. M. (1976). Évolution recente de l'Albufera de Valenica et de ses environs. *Mediterranée, 4,* 10–30.

Rosselló, V. M. (1987). Les salines de l'Albufera: un enigma històric i una hipòtesi geogràfica. *Cuadernos de Geografía* (42), 113–132.

Rosselló, V. M., & Mateu, J. F. (1978). El litoral cuaternario de Santa Pola, consideraciones generales. *Cuadernos de Geografía,* (23), 1–18.

Rosselló, V. M. (1969). *El litoral valencià.* L'Estel, València, 171 pp.

Rosselló, V. M. (1995). *L'Albufera de València.* Publicacions de l'Abadia de Montserrat, Barcelona, 192 pp.

Rosselló, V. M. (1979). Una duna fósil Pleistocena en la restinga de la Albufera de Valencia. *Cuad Geogr, 25,* 111–126.

Rosselló, V. M. & Sanchis Ibor, C. (2016). La génesis de la Albufera de València. In J. F. Mateu (coord.), *El territorio valenciano. Transformaciones ambientales y antrópicas* (pp. 121–214).

Ruiz, J. M., & Carmona, P. (2009). Geomorfología y cambios ambientales desde el Holoceno Medio en el litoral de Torre la Sal (Castelló). Datos para su gestión sostenible, In E. Flors (coord.), Torre la Sal (Ribera de Cabanes, Castellón): *Evolución del paisaje antrópico desde la Prehistoria hasta el Medioevo,* 21–40.

Sánchez-Badorrey, E., & Jalón-Rojas, I. (2015). A new method for zoning of coastal barriers based on hydro-geomorphological and climate criteria. *International Journal of Environmental Research, 9*(1), 351–362.

Sanchis, C. (1998). *De la gola a les goles: canvi ambiental secular a l'Albufera de València* (p. 143). València: Fundació Bancaixa.

Sanchis, C. (2001). *Regadiu i canvi ambiental a l'Albufera de València* (p. 332). Collecció Oberta: Universitat de València. Centre Valencià d'Estudis del Reg.

Sanchis, C., Jegou, A., & Pech, P. (2008). L'albufera de València: Une lagune de médiance en médiance. *Geographie et cultures* (63), 1–14. https://gc.revues.org/1593. https://doi.org/10.4000/gc.1593.

Sanchis, C. (2016). Acción antrópica y cambio ambiental histórico en l'Albufera de València, in Mateu, J.F. (coord.): *El territorio valenciano. Transformaciones ambientales y antrópicas* (pp. 137–140). València: Universitat de València.

Sanchis, C., Benavent , J. M., & Pardo-Pascual, J. E. (2016). La eutrofización de l'Albufera de València y su impacto en la vegetación palustre. In Mateu, J. F. (coord.), *El territorio valenciano. Transformaciones ambientales y antrópicas* (pp. 141–144). València: Universitat de València.

Sanjaume, E., & Carmona, P. (1995) L'Albufera de Valencia: rasgos geomorfológicos y evolución cuaternaria. In *El Cuaternario del País Valenciano* (pp. 155–162). València: Universitat de València.

Sanjaume, E., & Pardo-Pascual, J. E. (2005). Erosion by human impact on the Valencian coastline (E of Spain). *Journal of Coastal Research*, 76–82.

Sanjaume, E., Rosselló V. M., Pardo-Pascual, J. E., Carmona, P., Segura-Beltrán, F. S., & López-García, M. J. (1996). Recent coastal changes in the Gulf of Valencia (Spain). *Zeitschrift fur Geomorphologie, Supplementband*, 95–118.

Sanjaume, E., Segura, F., López-García, M. J., & Pardo-Pascual, J. E. (1992). Recent sedimentation in Valencia Lagoon: Preliminary results. *Journal of Coastal Research*, 688–698.

Sanjaume, E., & Pardo-Pascual, J. E. (2000). Albuferas litorales: estado de la cuestión. In J. R. Andrés & F. J. Gracia (Eds.), *Geomorfología litoral. Procesos activos* (Vol. 8, pp. 115–140). Madrid: Istituto Geológico y Tecnominero de España.

Sanjaume, E. S. (1985). *Las costas valencianas: sedimentología y morfología* (p. 505). Sección de Geografía, Valencia: Universidad de Valencia.

Sanjaume, E., & Gozálvez, V. (1978). L'Albufera d'Elx y su litoral. *Cuadernos de Geografía, 23,* 83–105.

Sanjaume, E., & Pardo-Pascual, J. E. (2003). Características sedimentológicas y morfológicas de los espacios costeros de transición situados al sur de la desembocadura del Xúquer. *Cuadernos de Geografía, 73*(74), 183–206.

Sanjaume, E., Segura, F., & Pardo-Pascual, J. E. (1990). Procesos y formas de una restinga en retroceso: el caso de la albufera de Torreblanca, *I Reunión Nacional de Geomorfología*, 375–384, Sociedad Española de Geomorfología e Instituto de Estudios Turolenses, Teruel.

SDI. (2011). *Informe de seguimiento sobre la calidad de las aguas de l'Albufera de València.* Servici Devesa-Albufera, Ajuntament de València.

Segura-Beltran, F., Pardo-Pascual, J. E., & Sanjaume, E. (1997). Evolución cuaternaria de la albufera de Torreblanca. *Cuaternario y Geomorfología,* 11(1–2), 3–18.

Segura-Beltrán, F., Sanjaume, E., Pardo-Pascual, J. E., & Riquelme, J. (2005). Canvis del nivell del mar en l'evolució del Prat de Cabanes. In E. Sanjaume & F. Mateu (Eds.), *Geomorfologia litoral i Quaternari* (pp. 455–469). Homenatge al Dr. V. Rosselló. València: Universitat de València.

Segura, F., & Sanjaume, E. (1986). Análisis morfológico y sedimentológico de la llanura aluvial entre Orpesa y la Ribera de Cabanes. *Cuadernos de Geografía, 38,* 29–60.

Segura, F., Sanjaume, E., & Pardo-Pascual, J. E. (1994). *Albufera de Torreblanca. Evolución cuaternaria.* IVEI unpublished document, 86 pp.

Segura, F., Sanjaume, E., & Pardo-Pascual, J. E. (1995). Evolución cuaternaria de las albuferas del sector septentrional del Golfo de Valencia. In E. Sanjaume & F. Mateu (Eds.), *Geomorfologia litoral i Quaternari: Homenatge al Dr. V. M. Rosselló* (pp. 139–154). València: Universitat de València.

Segura, F., Sanjaume, E., & Pardo-Pascual, J. E. (2016). Evolución pleistocena de la Albufera de Torreblanca o Prat de Cabanes. In J. F. Mateu (coord.), *El territorio valenciano. Transformaciones ambientales y antrópicas* (99–102). València: Universitat de València.

Segura-Beltran, F., & Pardo-Pascual, J. E. (in this book) Fan deltas and floodplains in Valencian coastal plains.

Somoza, L., & Rodríguez-Santalla, I. (2014). Geology and geomorphological evolution of the Ebro River Delta. In *Landscapes and landforms of Spain* (pp. 213–227). Berlin: Springer.

Soria, J. M., Miracle, M. R., & Vicente, E. (1987). Aporte de nutrientes y eutrofización de la Albufera de Valencia. *Limnetica, 3*(2), 227–242.

Tent-Manclús, J. E. (2013). Cambio de la línea de costa en el Bajo Segura (Sur de Alacant) en los últimos 15.000 años. *Estudios Geográficos,* LXXIV (275), 683–702.

Usera, J., Robles, F., Martínez-López, F., et al. (1990). Fauna actual de gasterópodos y foraminíferos de la marjal de Torreblanca (Castellón). *Iberus, 9*(1–2), 515–526.

Valls, A., Segura, F., & Martínez-Clavel, B. (2016). El delta del Ebro a través de la cartografía histórica y la fotografía aérea: evolución morfológica de la llanura deltaica (1858–2012). In J. J. Durán, M. Montes, A. Robador, & A. Salazar (Eds.). *Comprendiendo el relieve: del pasado al futuro* (pp. 61–68). IGME. Actas de la XIV Reunión Nacional de Geomorfología Málaga.

Vicente, E., & Miracle, M. R. (1992). The coastal lagoon Albufera de Valencia: An ecosystem under stress. *Limnetica, 8*, 87–100.

Viñals, M. J. (1996). *El Marjal Pego-Oliva. Evolución geomorfológica* (p. 352). Valencia: Generalitat Valenciana Consellería de Agricultura y Medio Ambiente.

Viñals, M. J., & Fumanal, M. P. (1995). Quaternary development and evolution of the sedimentary environments in the Central Mediterranean Spanish coast. *Quaternary International, 29*, 119–128.

Viñals, M. J. (1995). Secuencias estratigráficas y evolución morfológica del extremo meridional del Golfo de Valencia (Cullera-Dénia), In *El Cuaternario del País Valenciano* (pp. 163–168). València: Universitat de València.

Chapter 12
Beach Systems of Balearic Islands: Nature, Distribution and Processes

Lluís Gómez-Pujol, Alejandro Orfila, Verónica Morales-Márquez, Montserrat Compa, Laura Pereda, Joan J. Fornós and Joaquin Tintoré

12.1 Introduction

The Balearic Islands (Western Mediterranean), including the subarchipelago of Cabrera, have 1,724 km of coastline, with 10% consisting of 867 predominately sandy beach systems (Fig. 12.1). These beaches occur along the coast of each island in temperate latitude (38 to 40°N) and are exposed to a microtidal

L. Gómez-Pujol (✉) · J. J. Fornós
Earth Sciences Research Group, Department of Biology,
Universitat de les Illes Balears, Cra. Valldemossa Km 7.5,
07122 Palma (Illes Balears), Spain
e-mail: lgomez-pujol@uib.es

J. J. Fornós
e-mail: joan.fornos@uib.es

A. Orfila · V. Morales-Márquez · L. Pereda · J. Tintoré
Mediterranean Institute for Advanced Studies, IMEDEA (CSIC-UIB),
Miquel Marquès 21, 07151 Esporles (Illes Balears), Spain
e-mail: aorfila@imedea.csic-uib.es

V. Morales-Márquez
e-mail: vmorales@imedea.csic-uib.es

L. Pereda
e-mail: lpereda@imedea.uib-csic.es

M. Compa
Instituto Español de Oceanografía, Centre Oceanogràfic de les Balears,
Moll de Ponent s/n, 07015 Palma (Illes Balears), Spain
e-mail: montse.compa@ieo.es

J. Tintoré
SOCIB, Balearic Islands Coastal Observing and Forecasting System
(MICINN-CAIB), Cra. Valldemossa km 7.1, 07121 Palma (Illes Balears), Spain
e-mail: jtintore@socib.es

J. A. Morales (ed.), *The Spanish Coastal Systems*,
https://doi.org/10.1007/978-3-319-93169-2_12

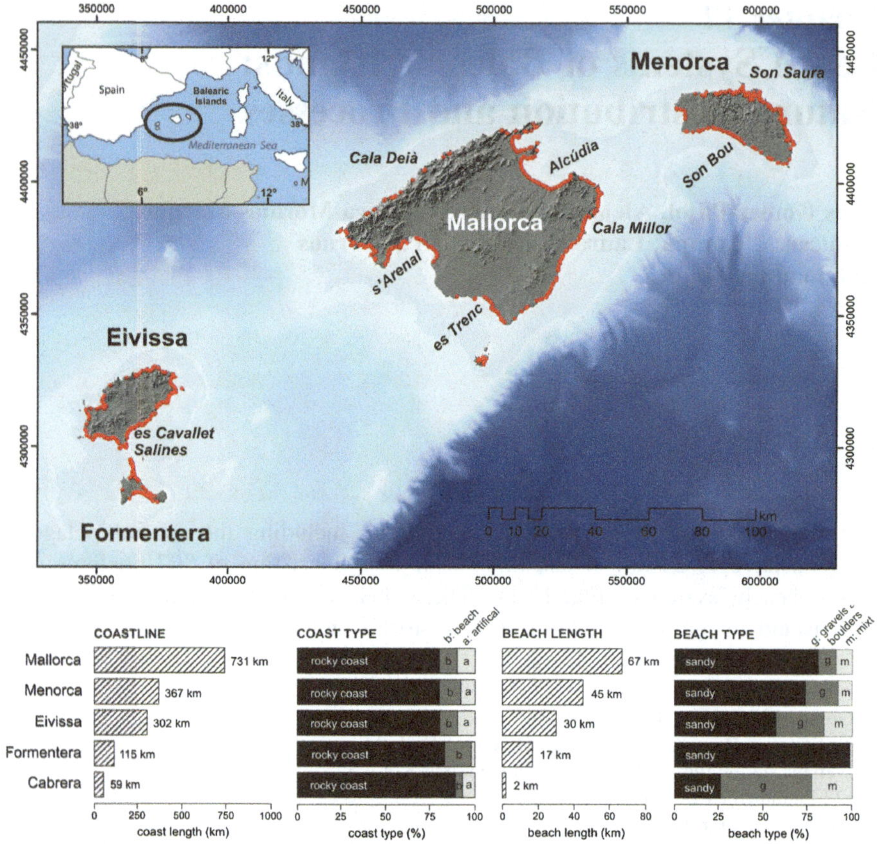

Fig. 12.1 Distribution of beaches (red dots) around the Balearic Islands (upper panel), coast type length and main features (lower panel)

environment. The absence of significant tides restricts beach morphodynamics to wave action and nearshore currents, especially during severe weather episodes. Waves rarely exceeds 8 m offshore; these values are considerably reduced nearshore where the maximum significant wave height is around 4 m during the most energetic sea storms that affect the Balearic Islands coasts at least once a year (Cañellas et al. 2010). Nevertheless there are substantial differences between islands and marine facades according to the wave exposition, beach sediment and geological framework. Therefore, the Balearic Islands' coast and its setting provide an ideal field site to explore the range of beach types and the control of wave, physiography and sediments conditions.

Despite coastal geomorphology having been the core of the geomorphological research in the Balearic Islands after the 2nd World War (i.e. Butzer 1962), beaches have not acquired a relevant role in literature since the second millennia (Gómez-Pujol and Pons 2007). The earliest studies of Balearic Islands beaches were concerned with reporting severe erosion caused by major storm events (Rosselló 1969, 1971). Twenty years later Jaume and Fornós (1992) initiated the characterization of the beach sediment along Mallorca and Servera and Martín (1996) started systematic studies on shoreline change due to the effects of a groyne construction in southern Mallorca. This group based at the University of the Balearic Islands concentrated their efforts on beach management issues and the relationship between foredune conservation and dry beach-backshore evolution (Rodríguez-Perea et al. 2000). However the first study of beach systems from a morphodynamic perspective commenced with Basterretxea et al. (2004) who described the wave, bar and beach profile evolution related to different wave forcing in a beach system fronted by the Mediterranean endemic seagrass *Posidonia oceanica*. By 2004 at the Mediterranean Institute for Advanced Studies, IMEDEA (CSIC-UIB), a multidisciplinary team incorporating biologists, coastal engineers, geomporhologists, and physical oceanographers commenced a systematic study of different beaches (Tintoré et al. 2005; Gómez-Pujol et al. 2007b; Orfila and Gómez-Pujol 2015) based on wave characterization, bathymetries and beach profiling surveys. At the same time Álvarez-Ellacuría et al. (2009) provided the first characterization of rips currents in the Balearic Islands, as well as the first operational systems for rips currents hazard prediction (Álvarez-Ellacuría et al. 2010). By 2007 the IMEDEA team developed a coastal videomonitoring system that operated temporally at Cala Millor (NE Mallorca) (Nieto et al. 2010) and finally in 2010, the Balearic Islands Coastal Observing and Forecasting System, SOCIB (MINECO-CAIB), initiated a long term beach survey programme combining coastal videomonitoring, beach survey and wave profiling at three beaches that is still in operation (Tintoré et al. 2013).

Considering this background, the aims of this chapter are (i) to report the full range of beach types that occur along the Balearic Islands coast, together with sediment and waves involved in their formation and dynamics; (ii) to describe the geographic distribution of these beaches along each island; and iii) to discuss some specific processes and agents in beach dynamics in the Balearic Islands.

12.2 The Balearic Islands Coastal Environment

12.2.1 Waves, Storms and Sea Climate

The Balearic Sea, the most western basin of the Mediterranean Sea, is a semi-enclosed and calm sea with a relatively moderate wave condition. The wave climate has already been identified to have, in general, a complex pattern as the result of the variability in the storm tracks; the complex orography and the relatively short fetch is controlled by the proximity of the continental Europe and the northern of Africa Maghreb (Cañellas et al. 2007; Ponce de Leon et al. 2012). The Balearic Islands are exposed to moderate winds that produce very low to moderate short-period waves

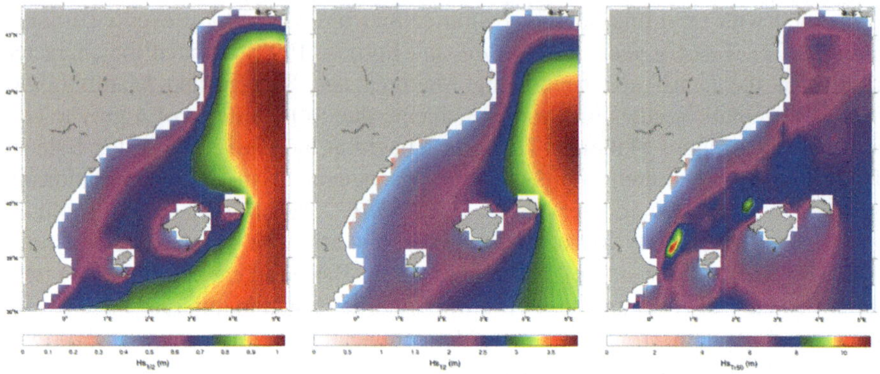

Fig. 12.2 Wave climate and extreme waves around the Balearic Islands. Mean significant wave, H_{50} (left panel), Annual extreme storm wave height, H_{12} (central panel) and 50 years return period wave height (right panel)

at the shore (H_s = 0.1–1 m; T = 3–6 s) (Fig. 12.2) (Gómez-Pujol et al. 2007b). There is a seasonal behaviour of the wave climate with maximum records occurring from December to February (average values of H_s > 1.1 m) and minimum values between June and August (average values of H_s < 0.6 m) (Cañellas 2010; Ponce de León et al. 2016). Regarding the sea storms (Fig. 12.2), the Balearic Sea is forced by northerly winds (Tramontane and Mistral) during the main part of the year, while the eastern part is generally modulated by a seasonal variability. Gale forced mistrals often develop over the Gulf of Geneva and the Gulf of Lion extending the effect over whole basin. Additionally, in the Pyrenees, the Iberian Peninsula, and the Alps, in the northeast area of the basin, there are decisive boundaries for the wind and pressure distribution (Orfila et al. 2005). All these factors explain why the northern quadrant of the islands is exposed to the most energetic environment and the 50-year return period for this area show values of significant wave height of around 11 m (Cañellas et al. 2007). This is also true for the eastern coast of the islands of Eivissa and Formentera that in the north and northeastern façades concentrate the winds and waves proceeding from north and mistral components, however here the 50-year return period show values of 9–10.5 m (Fig. 12.2). Otherwise in the southern part of the islands the 50-year return period is lower than 8 m. This is the result o the shadow effect of the islands over the intense north fetch produced by the storms (Ponce de León and Guedes Soares 2010). On the other hand, the significant wave height that is only exceeded during 12 h a year in the islands' northern façade is $H_{12} > 4$ m while the significant wave height in the southern façade is $H_{12} \sim 3$ m (Orfila and Gómez-Pujol 2015). Tides are almost negligible with a spring tidal range of less that 0.25 m, although changes in atmospheric pressure and wind stress can account for a considerable portion of sea level fluctuations (Gomis et al. 2012).

12.2.2 Sediments

The Balearic Islands' beaches are predominantly composed of medium to coarse moderately sorted marine biogenic carbonate sands, with some spatial variation controlled mainly by increasing contribution of lithoclasts and quartz along the north and northeastern coast of Mallorca, the northern coast of Menorca and most of the coast of Eivissa (Jaume and Fornós 1992; Gómez-Pujol et al. 2013c). Coarser sediments also appear in these regions, as well as coarse gravels and boulder beaches (Roig-Munar et al. 2012; Martorell 2014). There is a lack of studies on beach sediment properties at Eivissa and Formentera. Nevertheless based on the geological framework similarities between the Eivissa and the Northern ranges in Mallorca, and between Formentera and southeastern Mallorca and southern Menorca (Gómez-Pujol et al. 2013a), certain similarities can be envisaged. The average of the mean grain size from all the beaches of Menorca is 1.28 ± 0.81 phi, having the coarsest beach sediment a mean grain size of -0.16 ± 0.02 phi and the finest one, a mean grain size of 3.87 ± 0.04 phi (Gómez-Pujol et al. 2013c). In the case of Mallorca, the average of the mean grain size for all sandy beaches is 1.62 ± 0.64 phi ranging from 0.79 ± 0.65 phi to 2.34 ± 0.69 phi (Gómez-Pujol et al. 2007b). In Mallorca the 70.3% of sediment grains consists of fragments of crustose and free-living coralline algae (red algae), bryozoans, foraminifers, bivalves, gastropods and other skeletal carbonate grains. Lithoclasts, mainly calcareous, and quartz grains (29.7%) form subordinate components in comparison to skeletal carbonate grains (Jaume and Fornós 1992). This scenario is also true for the Menorcan beach sandy sediments; carbonate biogenic grains are the main component (78.4%) whereas lithoclasts, including quartz grains, sum 21.6% of the sediment (Gómez-Pujol et al. 2013c). The streams in the Balearic Islands, except for few beaches located in the mountain settings at the north coast of Mallorca and Menorca, and all around Eivissa, are ephemeral and exhibit a low sediment transport competence (Gelabert et al. 2005; Segura et al. 2007). They evidently do transport some quantities of sand- and mud-sized material during sporadic storms and floods (Estrany et al. 2009) but in beach sediment they remain masked by marine biogenic sands and lithoclasts elements detached from rocky cliffs. Jaume and Fornós (1992) and Fornós and Ahr (2006) show that biogenic carbonate particles are mainly produced in the *Posidonia oceanica* seagrass meadows adjacent to the beaches, as have been demonstrated for many beaches along the Western Mediterranean (De Falco et al. 2003; Gaglianone et al. 2017).

12.2.3 Geological Control

The Balearic Islands 867 beaches have an average length of 169 m (St. Dev. 359 m), ranging from 10 m to more than 5 km. This feature implies that physical boundaries in the form of headlands, rocks, reefs or beachrocks and minor islands

Fig. 12.3 Physiographical framework for selected beaches around the Balearic Islands. **a** Cala Varques, pocket beach in southeastern Mallorca; **b** Platja d'Alcúdia, an exposed beach-barrier in the northern bays of Mallorca; **c** Cala Tirant, a semi-enclosed beach exposed to northern waves in Menorca; **d** Son Bou, an exposed beach-barrier system fronting an lagoon in southern Menorca; **e** S'Arenal Gros de Portinatx (Eivissa) a semi-enclosed beach in northern Eivissa; **d** Es Trucadors and Platja de n'Adolf in Formentera, both are classical examples of exposed beaches

play a significant role in beach length and morphology along the coast (Fig. 12.3). The Balearic archipelago is located in the complex geological setting of the Western Mediterranean. Two tectonic phases affected the islands and configured its present geomorphological appearance. The first one was a compressive phase that lasted from the Palaeogene to the Middle Miocene; the second phase was an extensive one and occurred during the Upper Miocene generating a structure characterized by horsts and grabens that are bounded by Upper Miocene normal faults (Gelabert 1992). Succinctly and according to Gómez-Pujol et al. (2013a), Mallorca corresponds physiographically to a series of horsts and grabens that are not older than Langhian (16 Ma). The horsts consist of two subparallel mountain ranges orientated NE–SW, along with a series of small hills located in between. The grabens develop on the foreland of these ranges and are filled by sediments of Middle Miocene to Quaternary age, unaffected by tectonics and host the largest beaches often backed by lagoons. Menorca shows a much simpler physiographic morphology. The northern part of the island corresponds to a horst aligned W-E direction where the structured deposits (Silurian to Palaeogene) crop out, whereas the southern half is represented by an Upper Miocene carbonate platform. The island of Eivissa is composed of a series de thrust sheet (mainly of Middle Triassic to Middle Miocene carbonate deposits) formed during Alpine compression and trending NE–SW. In both islands the variability in rock strength, debility lines and differential erosion results in a large number of pocket beaches along the coastline. In the island of Formentera, only unfolded, upper Miocene and Pleistocene carbonate deposits are exposed to the surface. The impact of this geological control affects accommodation space (Gelabert et al. 2003); the beach length and the beach plan view shape or curvature. Accommodation space and physiographic controls is also relevant in diminishing wave attack on beaches by means of wave attenuation and refraction (Gómez-Pujol et al. 2007a). Very often the geological inheritance in the form of submerged beachrock on Plio-Quaternary dunes remnants protects the beach from storm incidence whereas in other cases it favours the formation of topographic rips (Álvarez-Ellacuría et al. 2009). Therefore, the typical Balearic beach is a relatively short and narrow beach that experiences attenuated waves at the shore and tend to have features such as rocks, reefs and islets or, very often, it appears at the bottom of a narrow wall-sided embayment. Therefore, geological inheritance is a major factor in shaping the boundaries of most Balearic beach systems and by reducing the breaker height it is a key actor in beach type.

12.2.4 Seagrass and Beach Dynamics

Mediterranean nearshore sandy and rocky bottoms are frequently colonized by the endemic reef-building seagrass *Posidonia oceanica* (L.) Delile, 1893, which is the most widespread seagrass of the basin (Pergent et al. 2012). *P. oceanica* plays an important role in many coastal processes, contributing to sediment deposition, attenuating currents and wave energy and stabilizing unconsolidated sediments (Vacchi et al. 2017). Infantes et al. (2012) suggest that at beaches fronted by *P. oceanica* seagrass, meadows can attenuate incident waves energy between 1.5 and 3.5% under incident waves with $0.5 \text{ m} \leq H_{rms} \leq 1.5 \text{ m}$ and $4 \text{ s} \leq T_p \leq 10 \text{ s}$. As a general trend, the greater the wave height and period, the greater is the attenuation per wavelength. Also, the shallower the water depth, the greater is the attenuation. This contributes to the little variability exhibited along the beach profile, mainly concentrated on the swash zone and in the small region (ca. 1.5 to 2 in depth) where the sand bar appears (Gómez-Pujol et al. 2011b). At some point the meadow is the exerting the equivalent role of a reef in a tropical beach, and for this reason most of the Balearic beaches can be envisaged as a perched beaches, since seagrass meadow attenuate and reduce the effect of waves.

P. oceanica loses leaves in autumn and leaf litter can be found mainly along sandy coasts forming wedge structures—leaf or seagrass berms—of few to several meters in thick that constitute one the major issues in western Mediterranean coastal management (Rodríguez-Perea et al. 2000; De Falco et al. 2008). Some authors pointed the importance of those *banquettes*—seagrass berms—for the protection of sandy beaches because they dissipate wave energy (Vacchi et al. 2017). Nevertheless Goméz-Pujol et al. (2013b), by means of a coastal videomonitoring study, indicate that seagrass berms are common beach features but they are not persistent though time and experience complex construction and destruction dynamics throughout the year. Seagrass berms, at least in semi-enclosed and open beaches, are continuously built up and destroyed and rarely persist before the arrival of new sea storms and for this reason the protective role of seagrass berms is underestimated against other more important contributions in beach protection such as the role of seagrass meadows in wave attenuations and as a physical barrier in nearshore sediment exchange.

12.2.5 Beach Risk Issues and Safety

The Balearic Islands are one of the most important "sun and beach" tourist destinations in Western Mediterranean. They receive around 13.6 millions of tourists each year. According to regional safety regulation and planning, there are 148 beaches that are in need of lifeguard services. Taking the data from the annual activity reports and statistics (2011–2013) summarized in Pereda et al. (2016) it was estimated that each year 20,157 incidents require of the lifeguard and sanitary services. It means that there are 316 incidences for each kilometre of beach. Around 95% of these incidences deal with jellyfish stings. If this kind of incidences is not considered in the analyses, then the ratios are much lower, and the numbers decrease to 960 incidences/year or 15 incidences per beach kilometre. Among them the 7% of the incidences correspond to drowning that means an average of 65 beach users drown each year rescued by lifeguard teams (Pereda et al. 2016). The number of fatal incidents (fatal drowning or other types of death occurred at beach) attends the 3% of the incidences. The current statistics does not allow separating the drowning related to rip currents from others; nevertheless according to lifeguard knowledge most of them are related to currents and wave agents (Pereda et al. 2016).

12.2.6 Man as a Coastal Agent

The Balearic Islands coast and their beaches have an iconic status in the tourism economy and in the inhabitants' culture and way of life. Tourism and recreational activities exert a significant pressure on the coast. It has been estimated that each ha of beach surface produces 16,537 €/month just in terms of recreational services (Riera et al. 2007). Therefore is not difficult envisaging the pressure on the coast that take the form in developments and boulevards that destroy coastal dune systems or build groins and sportive harbours along the coast (Pons and Rullan 2014). Garcia and Servera (2003) have identified massive construction on the coastal zone and nourishments as the major impacts from man as a coastal agent. By means of groins and sportive harbours, the sedimentary cellular system is altered and the sediments mass balance is unbalanced generating beach accretion at some points of the beach and erosion in the opposite ones as it has been demonstrated in s'Arenal —one of the largest beaches ins Mallorca—(Gómez-Pujol et al. 2011a). Whereas in coastal stations, such as Cala Millor in NE Mallorca, urbanised areas occupy foredunes and the dune field, disabling the sedimentary transfer between the nearshore and the subaerial beach. It contributes in modifying the beach geometry and enhancing the erosion processes because promenades and drain system reinforce the wave erosion (Tintoré et al. 2009). Coastal management practices have also a significant role in beach morphodynamics at Balearic Islands, especially in urban and overfrequented beaches. This issue is deeply considered in this volume by Roig-Munar and Rodríguez-Perea (2019).

12.3 Beach Types and Their Distribution

12.3.1 Regional Overview

The Balearic Islands have 1724 km of coast with 10% consisting of sandy beach systems (Balaguer et al. 2015), even though their distribution varies between islands (Fig. 12.4). Formentera is the island with the larger proportion of sandy coasts (15%) whereas in relative terms, both Cabrera and Mallorca exhibits the small percentage of sandy coasts (4 and 10% respectively). Despite of this, Mallorca exhibits 67 km of sandy beaches and hosts both the largest beaches and barriers among the archipelago's islands. In simple terms sandy beaches dominate the beach typology in all the islands, but in Mallorca, Menorca and Eivissa the contribution of gravel, boulder and mixed beaches exceeds 20% (Fig. 12.4). Eivissa and Formentera depart from the general schema. The first one because its structural arrangement results in many pocket beaches at the toe of high relief cliffs that generate the boulders and gravels that armour beaches (43%); whereas the second one is dominated by sandy beaches (99%) consisting of Upper Miocene and Quaternary rocks that do not generate large boulders and gravels when are dismantled being the procedente of sands from nearshore marine biogenous sediments.

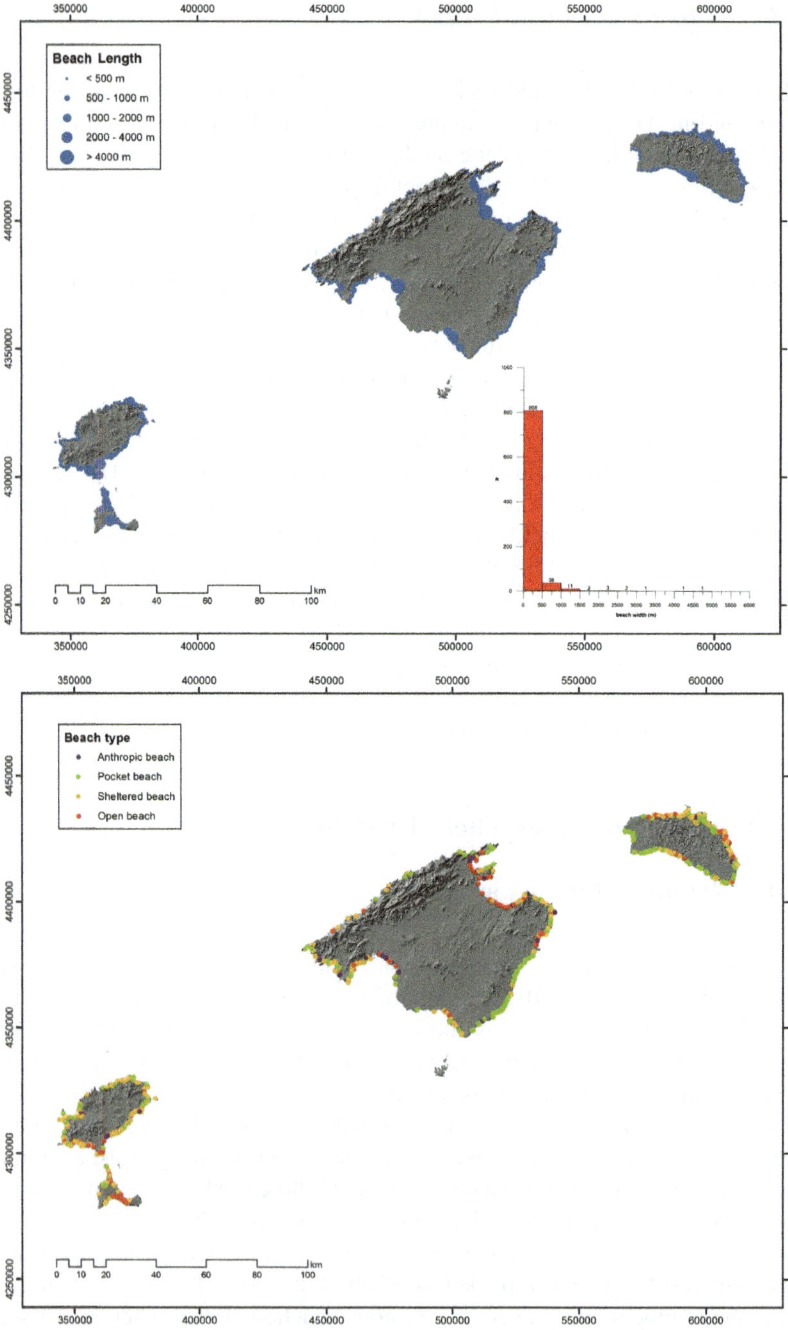

Fig. 12.4 Beach length (upper panel) and beach type (lower panel) along the Balearic Islands

So far, the average beach length for the Balearic Islands' beaches range from 5 m to 4.9 km, being the beach average length of 169 m (Gómez-Pujol et al. 2016). 93% of the beaches are narrow and have a beach length not larger that 500 m (Fig. 12.4). Just the 2.4% of the beaches exceed 1 km in length. Platja de Palma, the Alcúdia Bay beach System and Cala Millor are the largest beaches in Mallorca, Son Bou and Son Saura in Menorca and Platja d'en Bossa and Cavallet-Salines in Eivissa. The largest beaches appear closing the major coastal basins in each island or close related to structural accidents (Gómez-Pujol et al. 2007a, 2017a) very often in the form of beach-barriers fronting lagoons. On the other hand beaches below 500 m in length tend to appear sparsely along the shoreline and are related to medium to low-scale physiographical controls (i.e. dry-valley mouths or cliff toes).

12.3.2 Morphodynamic Classification

According to wave energy and sediments properties, beaches at Balearic Islands tend to fall in an intermediate-reflective state. Morphodynamic classification has been explored for the Mallorca beaches (Gómez-Pujol et al. 2007b) following morphodynamic model (Ω) from Wright and Short (1983). Nevertheless results can be extrapolated for the rest of the archipelago. Roughly three groups of beaches can be separated according to their modal morphodynamic state conditions and the physiographical context. From a physiographical point of view, beaches can be classified between enclosed beached (20%) that correspond with pocket beaches or beaches at the bottom of larger and narrow bays that reduce the action of waves; semi-enclosed beaches (42.4%) that relate with those beaches hosted in wither and narrow bays open to the wave action. Finally 27% of beaches fit with exposed and non-protected beaches to wave action. Against this background the truly reflective beaches correspond with enclosed and some sheltered beaches, which have a 75% of probability to fail below an Ω value of 1. Semi-enclosed beaches present modal conditions in the intermediate states although skewed to reflective states; Ω values of 2 and lower than 1 are between the 60 and 80% of modal states. This type of configuration relates with ridge-runnel and incipient transverse bars beach configuration. The larger semi-enclosed beaches and the exposed beaches have a 30% of probabilities to attend modal Ω values of 3 and 4, corresponding to intermediate morphodynamic states and showing longshore, crescentic and transverse bars.

12.4 Beach and Shoreline Evolution

12.4.1 Long-Term Evolution

Currently, there isn't a regional study addressing shoreline evolution at long-term scale. Nevertheless until recently there are several studies for specific locations that characterize shoreline evolution from 1956 to nowadays (i.e. Tintoré et al. 2009;

Gómez-Pujol et al. 2011a; Orfila and Gómez-Pujol 2015; Martín-Prieto et al. 2010, 2016). The entirety of these studies highlights the importance of human disturbance in beach system at long-term scales. For instance, according to Tintoré et al. (2009), in Cala Millor beach (north-eastern Mallorca) (Fig. 12.5), in 1956 the area of the emerged beach was of 44,260 m^2, corresponding to the original scenario when the beach, the lagoon and the field dunes in the southern part of the beach were not disturbed by urbanization and tourism development. In 1968 the beach surface increased significantly (+25,568 m^2) because the urban development affected the dune field and sand was removed and brought to the beach. Since then, aerial beach monitoring has indicated a continuous retreat associated to the sand redistribution and beach equilibrium adjustment. Since the 80s different nourishment works have been developed at Cala Millor despite there was not any problem of beach retreat and erosion. The nourishment had exclusively a tourism and leisure use motivation. In 2001 the beach experienced one of the most energetic storm inthe last 50 yrs in which the beach lost a significant amount of sediment. Since 2002, after the new nourishment, up until the 2014 the beach remained stable (Pilares et al. 2015). Other examples of human disturbance can be illustrated in Sa Rapita-Es Trenc (S Mallorca) (Martín-Prieto et al. 2016) and in s'Arenal (Palma Bay, W Mallorca) (Gómez-Pujol et al. 2011a). In both cases, the construction of sportive harbours has disrupted the sediment transport and plays a role as sediment trap. It results in sediment accumulation in some beach sectors and sediment erosion in others, but the averaged values indicate beach dynamic sedimentary equilibrium. On the other hand in the northern façades, which are subjected to the most energetic wave environments, long-term evolution indicates shoreline retreat just associated to natural conditions. This is case for Cala Deià (Fig. 12.5), a small and picturesque embayment in the Mallorcan northern range (Serra de Tramuntana). According to Gómez-Pujol et al. (2017b), Cala Deià during the period between 1956–2015 have experienced different cycles of erosion and accretion, being the sea storms of December 1980 and November 2001 the most damaging. During the 90s the beach achieved the maximum width because artificial sediment inputs, as well as a sea climate less energetic. From 2001 until now the beach is suffering a shoreline retreat because the effects of the 2001 extreme event, as well as, to an increase in the number of sea storms with wave significant heights larger than 2 m.

12.4.2 Medium-Term Evolution

Analysis of medium-term coastal changes reveals that the Balearic Islands coasts exhibit a great variety of shoreline trends (Fig. 12.6). The coastal change rates recorded between 2002 and 2012 are presented because since 2002, there have not been any significant beach nourishment work at the Balearic Islands and the human disturbance that has exerted a significant contribution in large-scale evolution can be discarded. Georectified aerial photographs spanning the period from 2002–2012 were compared in a GIS environment to calculate rates of shoreline change

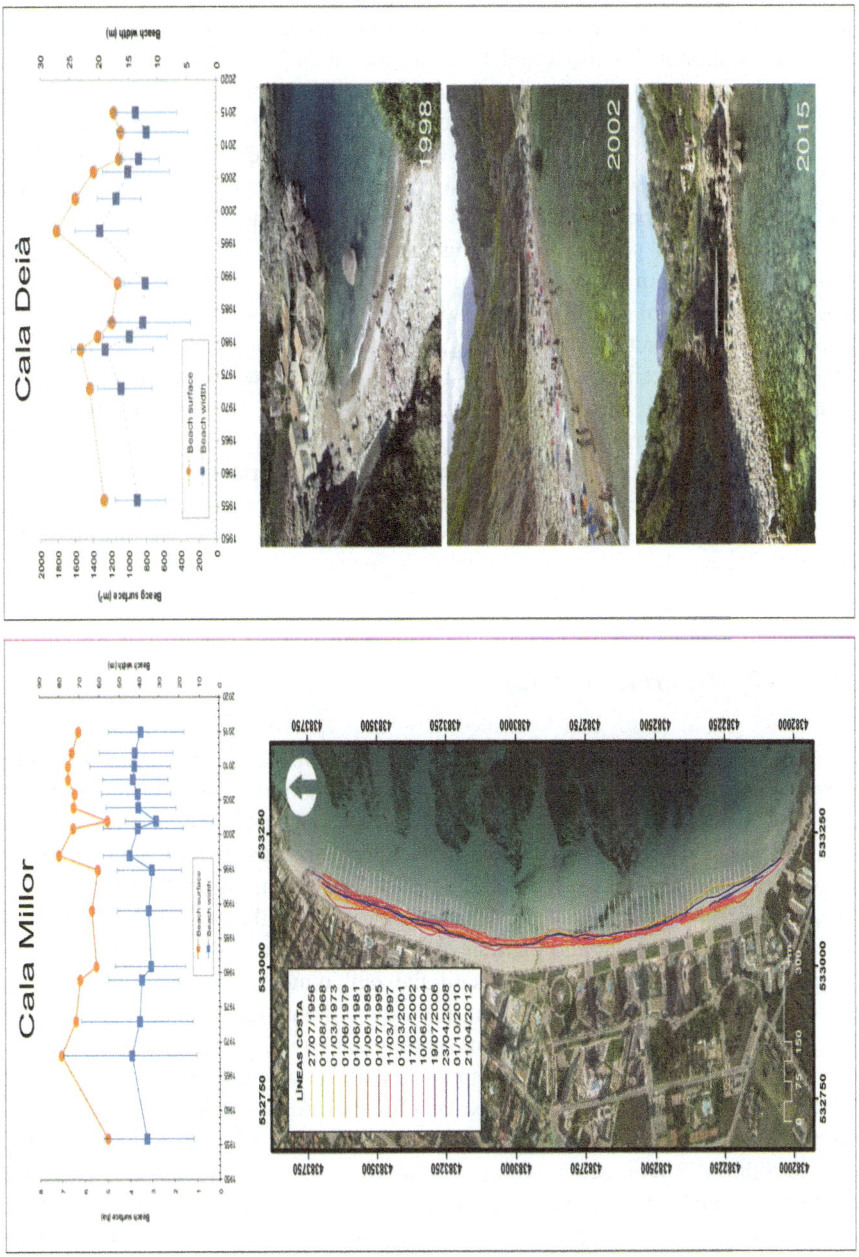

Fig. 12.5 Selected examples of long-term shoreline change along the Balearic Islands

(Gómez-Pujol et al. 2016) by means of linear regression erosion rate trends (Thieler et al. 2009). Results show that 79.7% of the beaches are stable and just 1.2% experience retreats rates larger than 1.5 m/yr and 3.8% experience accretion rates larger than 1.5 m/yr. Nevertheless there is a general and dominant beach stability scenario. In Mallorca, Menorca and Eivissa more than 70% of beaches are stable (shoreline change trend between −0.5 and 0.5 m/yr) whereas in Formentera the stable beaches account for less than the 60%. The number of eroding beaches (> −0.5 m/yr) changes between islands with Mallorca and Formentera having at least 20% of beaches with a negative sign, whereas Menorca and Eivissa attain around a 10%. All of the islands host between 9 to 18% of accreting beaches. There is not a clear spatial pattern in eroding, accreting or stable beach distribution. Nevertheless, except in the case of Formentera, northern and western exposures have a larger number of eroding beaches, whereas southern and eastern façades host a large number of accreting beaches (Fig. 12.6). On a broad sense, it can be noted that there are no differences in average shoreline change rates between beach types (Fig. 12.6). Except for the enclosed beaches in Mallorca, all the other types fall between −0.5 and 0.5 m/yr, indicating a stable beach dynamic behaviour. Exposed beaches in Formentera have been found to be the most variable falling above and below the average value in the stability interval. There are many erosive examples in the northwestern coast, as well as, many accretion beaches in the northeastern coast.

12.4.3 Short-Term Evolution

Seasonally Gómez-Pujol et al. (2011b) have demonstrated by means of EOF analyses of beach bathymetry and cross-shore profiling and sediment sampling that in carbonate beaches with negligible fluvial contribution, as is the case for most of the Balearic Islands beaches, significant morphology and sedimentary change needs for waves exceeding some intensity and duration (Hs > 1 m; 6 h). When waves do not exceed this threshold there is evidence of morphological change but constrained between the emerged beach and the rear of the bar and an attenuated beach rotation according to the size and dimensions of the embayment and wave energy flux variability. This leads to the conclusion that wave climate rather than wave forcing is the major control on sediment and morphological covariation in Balearic Islands beaches (Fig. 12.6). Additionally Álvarez-Ellacuría et al. (2011) identified spatial and temporal shoreline response to wave climate and identified two short-term temporal scales of interrelation between waves forcing ands shoreline position. They found that the averaged wave conditions for the four previous days correlate with the cross-shore change in shoreline displacement, whereas for the longshore sediment transport the best correlation corresponded to the average of the former seven days. This suggests that the beach, specifically the shoreline position, is acting in front of wave climate using two mechanisms: first, the beach is able to dissipate wave energy retrieving its position by adding sand to the submerged sand

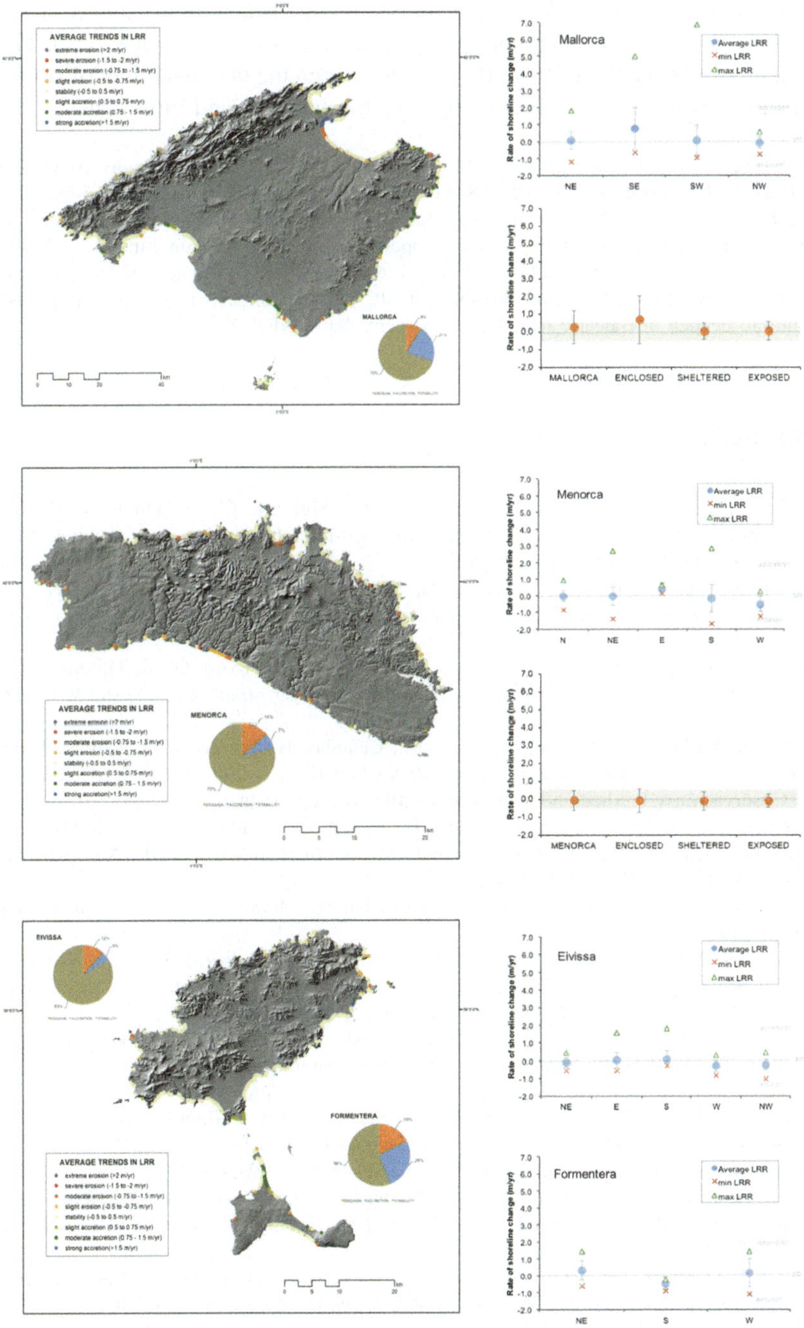

Fig. 12.6 Medium-term shoreline change rates (2002–2012) along the Balearic Islands. Lateral panels show the average, maximum and minimum shoreline change rates related to each island façade and to the beach type. Shadow areas indicate the rates understood as dynamic equilibrium interval

bar. When this mechanism is not sufficient, the beach redistributes the sand alongshore to better dissipate the energy from the waves and vice versa. Over a short-term time scale there is still the need to deepen the understanding of storm and groups of storms effects on beach recovery mechanisms and time scales.

Acknowledgements The authors would like to acknowledge the financial support from MINECO (AEI-FEDER, UE) projects CGL2013-48441-P, CGL2016-79246-P and CTM2015-66225-C2-2. MC and LP are supported by the regional Balearic Government FPI-CAIB co-financed by the European Social Fund (FSE, UE). VM is supported by FPU programm funded by Ministerio Educadión, Cultura y Deportes and the FSE (UE). Authors would like to acknowledge the support from SOCIB (MICINN-CAIB) and IMEDEA (CSIC-UIB) technicians that have contributed significantly to deepen in Balearic Island beach systems knowledge.

References

Álvarez-Ellacuría, A., Orfila, A., Gómez-Pujol, L., Simarro, G., & Obregón, N. (2011). Decoupling spatial and temporal patterns in short-term beach shoreline response to wave climate. *Geomorphology, 128,* 199–208.

Álvarez-Ellacuría, A., Orfila, A., Olabarrieta, M., Gómez-Pujol, L., Medina, R., & Tintoré, J. (2009). An alert system for beach hazard management in the Balearic Islands. *Coastal Management, 37,* 569–584.

Álvarez-Ellacuría, A., Orfila, A., Olabarrieta, M., Medina, R., Vizoso, G., & Tintoré, J. (2010). A nearshore wave and current operational forecasting system. *Journal of Coastal Research, 26,* 503–509.

Balaguer, P., Vallespir, J., Vizoso, M., Ruiz, M., Cañellas, B., Fornós, J., & Tintoré, J. (2015). *Atles de Sensibilitat Ambiental del Litoral de les Illes Balears. Edición 2015.* SOCIB, Sistema de Observación y Predicción Costero de las Illes Balears, Palma.

Basterretxea, G., Orfila, A., Jordi, A., Casas, B., Lynett, P., Duarte, C., et al. (2004). Seasonal dynamics of a microtidal pocket beach with *Posidonia oceanica* seabeds (Mallorca, Spain). *Journal of Coastal Research, 20,* 133–142.

Butzer, K. W. (1962). Coastal geomorphology of Mallorca. *Annals of the American Association Geographers, 52,* 191–212.

Cañellas, B. (2010). *Extreme Wave Height Events in the Balearic Sea: Characterization, Variability and Prediction* (Ph.D. thesis), Mediterranean Institute for Advanced Studies), IMEDA, University of the Balearic Islands, Palma, 84 pp.

Cañellas, B., Orfila, A., Méndez, F. J., Menéndez, M., Gómez-Pujol, L., & Tintoré, J. (2007). Application of a POT model to estimate the extreme significant wave height levels around the Balearic Sea (Western Mediterranean). *Jornal of Coastal Research, 50,* 329–333.

De Falco, G., Molinaroli, E., Baroli, M., & Bellacicco, S. (2003). Grain size and compositional trends of sediments from *Posidonia oceanica* meadows to beach shore, Sardinia, western Mediterranean. *Estuarine, Coastal and Shelf Science, 58,* 299–309.

De Falco, G., Simeone, S., & Baroli, M. (2008). Management of beach-cast *Posidonia oceanica* seagrass on the island of Sardinia (Italy, Western Mediterranean). *Journal of Coastal Research, 24,* 69–75.

Estrany, J. J., García, C., & Batalla, R. J. (2009). Groundwater control on the suspended sediment load in the Na Borges River, Mallorca, Spain. *Geomorphology, 106,* 292–303.

Fornós, J. J., & Ahr, W. M. (2006). Present-day temperate carbonates on a modern, low-energy, isolated ramp, the Balearic Platform, Spain. *Journal of Sediment Research, 67,* 364–373.

Gaglianone, G., Brandano, M., & Mateu-Vicens, G. (2017). The sedimentary facies of *Posidonia oceanica* seagrass meadows from the central Mediterranean sea. *Facies, 63,* 28.

Garcia, C., & Servera, J. (2003). Impacts of tourism development on water demand and beach degradation on the islands of Mallorca (Spain). *Geografiska Annaler, 85A*, 287–300.
Gelabert, B., Fornós, J. J., & Gómez-Pujol, L. (2003). Geomorphological characteristics and slope processes associated with different basins: Mallorca (Western Mediterranean). *Geomorphology, 52*, 253–267.
Gelabert, B., Fornós, J. J., Pardo, J. E., Rosselló, V. M., & Segura, F. S. (2005). Structurally controlled dranaige development in the south of Menorca (Western Mediterranean, Spain). *Geomorphology, 65*, 139–155.
Gelabert, B. (1992). An structural outline of the Serra de Tramuntana of Mallorca (Balearic Islands). *Tectonophysics, 203*, 167–183.
Gómez-Pujol, L., Balaguer, P., & Fornós, J. J. (2007). El litoral de Mallorca: síntesis geomórfica. In J. J. Fornós, J. Ginés & L. Gómez-Pujol (Eds.), *Geomorfología litoral: Migjorn y Llevant de Mallorca*. Palma: Societat d'Història Natural de Balears.
Gómez-Pujol, L., Compa, M., Orfila, A., Álvarez-Ellacuría, A., Balaguer, P., Roig-Munar, F. X., Fornos, J. J., & Tintoré, J. (2017) Las playas de Menorca: naturaleza y distribución. In L. Gómez-Pujol & G. X. Pons (Eds.), *Geomorfología litoral de Menorca: dinámica, evolución y prácticas de gestión*. Palma: Societat d'Història Natural de Balears.
Gómez-Pujol, L., Gelabert, B., Fornós, J. J., Pardo-Pascual, J. E., Rosselló, V. M., Segura, F. S., et al. (2013a). Structural control on the presence and character of calas: observations from Balearic Islands limestone rock coast macroforms. *Geomorphology, 194*, 1–15.
Gómez-Pujol, L., Orfila, A., Álvarez-Ellacuría, A., Cañellas, B., & Tintoré, J. (2011a). Evolución de la línea de costa en una playa urbana (s'Arenal, Mallorca): intervención antrópica y dinámica natural. In I. Montoya, I. Rodríguez, & M. J. Sánchez (Eds.), *Avances en Geomorfología Litoral*. Madrid: Universidad Juan Carlos I.
Gómez-Pujol, L., Orfila, A., Álvarez-Ellacuría, A., Terrados, J., & Tintoré, J. (2013b). Posidonia oceanica beach-cast-litter in Mediterranean beaches: a coastal videomonitoring study. *Jouranl of Coastal Research, 65*, 1768–1773.
Gómez-Pujol, L., Orfila, A., Álvarez-Ellacuría, A., & Tintoré, J. (2011b). Controls on sediment dynamics and medium-term morphological change in a barred microtidal beach (Cala Millor, Mallorca, Western Mediterranean). *Geomorphology, 132*, 87–98.
Gómez-Pujol, L., Orfila, A., Álvarez-Ellacuría, A., & Tintoré, J. (2017b). Evolución y dinámica de una playa encajada de bloques y gravas en la costa NW de Mallorca (Cala Deià, Mallorca). *Geotemas, 17*, 111–114.
Gómez-Pujol, L., Orfila, A., Cañellas, B., Álvarez-Ellacuría, A., Méndez, F. J., Medina, R., et al. (2007b). Morphodynamic classification of sandy beaches in low energetic marine environment. *Marine Geology, 242*, 235–246.
Gómez-Pujol, L., Pilares, A., Compa, M., Font, J., Álvarez-Ellacuría, A., & Tintoré, J. (2016). A regional assessment of the Balearic Islands sandy coast shoreline change, 2002–2012 (Western Mediterranean). In F. Dall'Orsso & A. Vila-Concejo (Eds.), *Book of Abstracts International Coastal Symposium ICS2016 Sydney, NSW, Australia 6th–11th March 2016*, School of Geosciences, The University of Sydney, Sydney.
Gómez-Pujol, L., & Pons, G. X. (2007). La geomorfología litoral de Mallorca cuarenta y cinco años después. In J. J. Fornós, J. Ginés, & L. Gómez-Pujol (Eds.), *Geomorfología litoral: Migjorn y Llevant de Mallorca*. Palma: Societat d'Història Natural de Balears.
Gómez-Pujol, L., Roig-Munar, F. X., Fornós, J. J., Balaguer, P., & Mateu, J. (2013c). Provenance-related characteristics of beach sediments around the island of Menorca, Balearic Island (western Mediterranean). *Geo-Marine Letters, 33*, 195–208.
Gomis, D., Tsimplis, M., Marcos, M., Fenoglio-Marc, L., Pérez, B., Raicich, F., Vilibic, I., Wöppelmann, G., & Montserrat, S. (2012). Mediterranean sea-Level variability and trends. In P. Lionello (Ed.), *The climate of Mediterranean region. From the past to the future*. Amsterdam: Elsevier.
Infantes, E., Orfila, A., Terrados, J., Luhar, M., & Nepf, H. (2012). Effect of a seagrass (*Posidonia oceanica*) meadow on wave propagation. *Marine Ecology Progress Series, 456*, 63–72.

Jaume, C., & Fornós, J. J. (1992). Composició i textura dels sediments de platja del litoral mallorquí. *Boll Soc Hist Nat Balears, 35,* 93–110.

Martín-Prieto, J. A., Roig-Munar, F. X., Rodríguez-Perea, A., Mir-Gual, M., Pons, G. X., & Gelabert, B. (2016). La erosión histórica de la playa de sa Ràpita (S. Mallorca). *Investment Geography, 66,* 135–154.

Martín-Prieto, J. A., Roig-Munar, F. X., Rodríguez-Perea, A., Pons, G. X. (2010). Evolució de la línia de costa de la platja de es Trenc. In J. Mayol, L. Muntaner & O. Rullán (Eds.), *Homenatge a Bartomeu Barceló Pons, Geògraf.* Palma: Lleonard Muntaner Editor.

Martorell, M. (2014). Morfogènesi i evolució dels codolars del Coll Baix, Sa Calobra i Cala Tuent. *Boll Soc Hist Nat Balears, 57,* 55–78.

Nieto, M. A., Garau, B., Balle, S., Simarro, G., Zarruk, G. A., Ortiz, A., et al. (2010). An open source, low cost video-based coastal monitoring system. *Earth Surface Processess Landforms, 35,* 1712–1719.

Orfila, A., Álvarez, A., Tintoré, J., Jordi, A., & Basterretxea, G. (2005). Climate teleconnections at monthly time scales in the Ligurian sea inferred from satellite data. *Progress in Oceanography, 66,* 157–170.

Orfila, A., Gómez-Pujol, L. (2015). *Estudio morfodinámico de Cala Deià.* Mediterranean Institute for Advanced Studies (IMEDEA, CSIC-UIB), Esporles, 40 pp.

Pereda, L., Gómez-Pujol, L., Tintoré, J., Pol, J., & De la Rosa, R. (2016). An examination of beach safety incidences and fatalities in the Balearic Islands (Western Mediterranean). In F. Dall'Orsso & A. Vila-Concejo (Eds.), *Book of abstracts international coastal symposium ICS2016 Sydney, NSW, Australia 6th–11th March 2016.* Sydney: School of Geosciences, The University of Sydney.

Pergent, G., Bazairi, H., Bianchi, C. N., Boudouresque, C. F., Buaia, M. C., Clabaut, P., et al. (2012). *Les herbiers de Magnoliophytes marines de Méditerranée. Résilence et contribution à l'atténuation des changements climatiques.* Gland: IUCN.

Pilares, A., Gómez-Pujol, L., Álvarez-Ellacuría, A., Orfila, A., & Tintoré, J. (2015). Cambios en la línea de costa de una playa urbana mediterránea a largo plazo: Cala Millor (Mallorca), 1956–2012. *Geotemas, 15,* 217–220.

Ponce de León, S., & Guedes Soares, C. (2010). The sheltering effect of the Balearic Islands in the hindcast wave field. *Journal of Ocean Engineering and Science, 37,* 603–610.

Ponce de León, S., Orfila, A., Gómez-Pujol, L., Renault, L., Vizoso, G., & Tintoré, J. (2012). Assessment of wind models around the Balearic Islands for operational wave forecast. *Applied Ocean Research, 34,* 1–9.

Ponce de León, S., Orfila, A., & Simarro, G. (2016). Wave energy in the Balearic Sea. Evolution from a 29 year spectral wave hindcast. *Renewable Energy, 85,* 1192–1200.

Pons, A., & Rullan, O. (2014). The expansion of urbanisation in the Balearic Islands (1956–2006). *Journal of Marine and Islands Cultures, 3,* 78–88.

Riera, A., Álvarez, M., Bestard, R., & Gómez-Pujol, L. (2007). El valor d'ús recreatiu de la badia de Santa Ponça (Calvià). *Cojuntura, 20,* 66–72.

Rodríguez-Perea, A., Servera, A., & Martín, J. A. (2000). *Alternatives a la dependència de les platges de les Balears de la regeneració artificial continuada: informe Metadona* (p. 108). Palma: Universitat de les Illes Balears.

Roig-Munar, F. X., Martín-Prieto, J. A., Rodríguez-Perea, A., Pons, G. X., & Mir-Gual, M. (2012). Morfogènesi i dinàmica de la platja d'Es Codolar (Eivissa, Balears). *Boll Soc Hist Nat Balears, 55,* 47–62.

Roig-Munar FX, Rodríguez-Perea A (2019) Coastal management in the Balearic Islands. This volume.

Rosselló, V. M. (1969). El litoral de Es Trenc (S de Mallorca). *Anales de la Universidad de Murcia, 27,* 223–245.

Rosselló, V. M. (1971). El oleaje y sus efectos geomórficos en las playas meridionales de Mallorca. *Cuadernos de geografía, 9,* 169–178.

Segura, F. S., Pardo, J. E., Rosselló, V. M., Fornós, J. J., & Gelabert, B. (2007). Morphometric indices of tectonic fluvial and karst processes in calcareous drainage basins, South Menorca Island, Spain. *Earth Surface Processess and Landforms, 32,* 1928–1946.

Servera, J., & Martín, J. A. (1996). Análisis y causas del retroceso de la línea de costa del arenal de sa Ràpita (Mallorca). In A. Grandal d'Anglade & J. Pagés (Eds.), *IV Reunión de Geomorfología.* A Coruña: Sociedad Española de Geomorfología.

Thieler, E. R., Himmelstoss, E. A., Zichichi, J. L., & Ergui, A. (2009). Digital shoreline analysis sytem (DSAS) version 4. US Geological Survey Open-file Report, 2081278.

Tintoré, J. (coord.) (2005). *Variabilidad y dinámica sedimentaria de las playas de Cala Millor y Cala Sant Vicenç. Alternativas de gestión basadas en el conocimiento científico* (Vol. 2, 688 pp.). Palma: Govern Illes Balears, IMEDEA (CSIC-UIB).

Tintoré, J., et al. (2013). SOCIB: The Balearic Islands coastal ocean observing and forecasting system responding to sciene, technology and society needs. *Marine Technology Society Journal, 47,* 1–17.

Tintoré, J., Medina, R., Gómez-Pujol, L., Orfila, A., & Vizoso, G. (2009). Integrated and interdisciplinary approach to coastal management. *Ocean Coastal Management, 52,* 493–505.

Vacchi, M., De Falco, G., Simeone, S., Montefalcone, M., Morri, C., Ferrari, M., et al. (2017). Biogeomorphology of the Mediterranean Posidonia oceanica seagrass meadows. *Earth Surface Processess and Landforms, 42,* 42–54.

Wright, L. D., & Short, A. D. (1983). Morphodynamics of beaches and surf zones in Australia. In P. D. Komar (Ed.), *Handbook of Coastal Processes and Erosion.* Boca Raton, FL: CRC Press.

Chapter 13
Lineal Sandy Coasts and Beaches of Málaga: Andalusian Mediterranean Coast

Gonzalo Malvarez and Fatima Navas

13.1 Introduction

The coast of Malaga is predominantly a low-lying beach and shore platform system exposed to the south and southeast. It is dominated by fluvial sediment input and subject to low energy Mediterranean wave energy conditions. This combined with a microtidal range (average 50 cm) represents the scenario for narrow and relatively steep beaches that are mainly composed by sandy material with sediment sizes of 1.5 mm or more and that are restricted to locations near river mouths and a narrow active nearshore zone.

The Province of Malaga has a coastline of approximate 158 km, of which only 20 are considered cliffs and/or platforms. Therefore, beaches are dominant and their characterisation highly representative of Malaga's shores.

The main sections that can be delineated to characterise the lineal sandy coasts of Malaga could use criteria such as exposure (orientation). The coast can be subdivided, and will be described in this chapter, according to exposure in south and East facing beaches (boxes A, B and C in Fig. 13.1). This characteristic is very relevant and influences morphodynamics in a variety of ways. Two further sections can be distinguished for its specific combination of intervening factors: The Bay of Malaga city (indicated D in Fig. 13.1), dominated by the filled estuaries of the Guadalhorce and Guadalmendina rivers and the Ensenada de Marbella (E in Fig. 13.1) with its significant cell circulation on the platform that provides the basis for the development of the substantial Cabopino dune system. Finally, a section is dedicated to

G. Malvarez (✉) · F. Navas
Coastal Environments Research Group, Universidad Pablo de Olavide, Seville, Spain
e-mail: gcmalgar@upo.es

F. Navas
e-mail: fnavcon@upo.es

J. A. Morales (ed.), *The Spanish Coastal Systems*,
https://doi.org/10.1007/978-3-319-93169-2_13

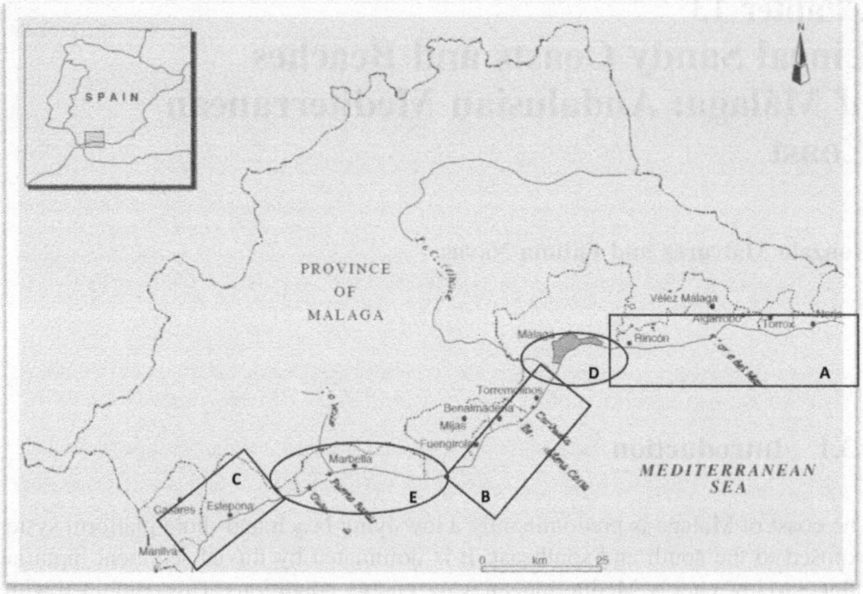

Fig. 13.1 Malaga Province and main locations. Municipalities along the coast are delineated. Boxes show south (**A**) eastern (**B** and **C**) facing shores and ovals depict the Bay of Malaga (**D**) and Ensenada de Marbella (**E**) respectively. Modified from Malvárez (1999)

description and characterisation of the artificial beach systems that are now found all along Malaga Province.

The climate in the Province varies from East to West: the Eastern reaches can be classified in terms of rainfall per annum as a semi-arid environment (below 300 mm yr^{-1}) whilst the West is very influenced by Atlantic fronts and the Föhn effects across the mountains facing the Strait of Gibraltar resulting in much greater annual precipitation rates in the hinterland (up to 800 mm yr^{-1}). The coast of Malaga Province is located at the westerly reaches of the Andalusian Mediterranean coast, considered part of the regional Alborán Sea. The continental shelf is dominated by a steep profile that extends along the first few kilometres seaward reaching depths coinciding with the accommodation space of the latest postglacial sea level rise which situated shorelines at 100 m depth worldwide.

13.1.1 Geological Set Up

The coastal system is, as is the hinterland and the continental platform, of significant geological complexity. The complex tectonic arrangement of the Beticas system here is dissected by current sea level and its coastal and nearshore formations and thus along the province some of the structural elements are linked to the internal and external Beticas system. The importance of this is that there are significant anomalies in facies which respond to the formations which are considered to be caused by folding faults, which introduces foreign newer materials to be present under older strata. This inversion of the stratigraphy also provides the grounds for extreme lithological variety, which is in turn reflected on beach and nearshore heteroclastic nature of materials. This heterogeneity is also mineralogical inducing a significantly variable morphosedimentary behaviour of sands once in the marine environment.

At a subregional scale, there are clearly variations caused by the presence in the hinterland of intrusions and great tectonic and contact metamorphism induced as the great peridotitic batholith of Ronda pushed the Jurassic and dolomitic limestones of the main mountain ridges (Tejeda to the East and Mijas and Blanca to the West). Along the eastern section (from Nerja to Malaga city; see Fig. 13.1) close to the rectilinear sandy coast there are outcrops of slates and phillites of the Maláguide complex (Silurian-Devonian) and schists from the Alpujárrides (Precambrian-Paleozoic). In between those formations the Plio-Quaternary infilled valleys, product of marine transgressions. One very significant event (in terms of transgression) resulted from the opening of the Strait of Gibraltar after the Messinian saline crisis, in the Pliocene, which turned the coast of Malaga into a succession of bays, including the Vélez, the Guadalhorce and the Fuengirola Rivers, as discussed in numerous works (Lhenaff 1977; Lario et al. 1995; Malvárez et al. 1998).

The Villafranquian crisis provokes an intense erosive phase that excavates some surfaces and fills others, resulting in frequent generation of travertines in continental zones (Lario et al. 1993); The proximity to the Atlantic explains that, within a generally semiarid climate, the proximity to a subhumid threshold provides a scenario for the predominance of erosion over the deposition and generation of glacis (Lhenaff 1977). The alluvial deposits cover levels of clay and some levels of Quaternary beaches in the interior of the infilled basins. More recently, during the rapid process of infilling of the relict estuaries a generally well preserved High Stand Flandrian shoreline exists (6000 BC) which represent the latest maximum after which all estuaries were relatively rapidly filled. More recently, the retreat of the sea due to a general trend to sea level drop after the Holocene maximum and the progressive deforestation come together generating the filling the estuary and smoothing the a narrow back beach and beach system.

The drainage network, of significance in this terrigenous sediment fed coastal system, is also diverse as lithologies change from limestones and marbles of the Tejeda, Almijara, Nieves, Blanca, Mijas and to ultrabasic rocks of Sierra Bermeja, Palmitera and the Ronda Peridotitic Complex. Limestones provide a well-developed karstic landscape and drainage network resulting from the many features present in the sierras (poljes, cavities, etc.) result in, both, a limited surficial run off and presence of abundant water springs which contribute to rivers in various parts of their basin increasing their erosive potential at many locations within the catchment.

The most hydrogeological significant features are those associated with the contacts between lithology of permeable (limestones) and impermeable nature (a variety of ultrabasic rocks mostly to the West of the province) that becomes a structural path for the main rivers. The western section of the province (from Malaga city to Manilva/Casares municipalities at the western edge with Cadiz) is dominated by the higher rainfall rates combined with drainage networks excavated into metamorphic and igneous rocks of the Sierra Bermeja and Genal valleys resulting in a very distinctive coastal setting where sediment production and distribution in the coastal cells are higher, which has been able to develop significant deposits on the shelf and even a large dune system at the eastern edge of the Ensenada de Marbella (Fig. 13.1).

13.1.2 General Environmental Settings

The main hydrodynamic agents affecting the coastline are wave-induced forces. The wave regime in this section of the Mediterranean Sea can be classified as of low energy and is affected particularly by storms that bring water level surges and greater waves (Fig. 13.2). The mean significant wave height is 1.0 m and mean zero crossing period is 5.0 s with typical directions of bimodal nature dominated by westerlies and easterlies nearly at 50% of times. These characteristics and the steep nearshore regions lead to beach types highly dependent on short term sediment supply. The beach morphology is adjusted to high frequency waves in the form of steep-narrow foreshores.

Fig. 13.2 Storm landing in Torre del Mar. Up to 7 m waves (deep water) are dissipated as the surf zone is expanded as a dissipative system

Tidal range is microtidal (<50 cm average astronomical tidal range). The effective fetch is limited to an average 500 km and only rarely do swell waves filter from the Atlantic Ocean. The morphology of the inner shelf is steep and narrow. Oceanic depths are reached within two kilometres from the coast in some sections. This results in a concentration of wave action on a narrow fringe of steep coastal shelf and sediment supply is mainly reworked fluvial sands and supply is episodic and concentrated in time around seasonal heavy rainfall.

Energy dissipation takes place on steep foreshores where off shore-directed sediment transport rapidly develops off-shore bars at the expense of beach face erosion. Excavation of the emerged beach may produce significant shoreline retreat. In reflective beach environments, during high-energy events (i.e. storms) higher rates of energy supply will translate into violent plunging or collapsing breaker types that will produce shoreline retreat and coarsening of beach material due to enhanced cross-shore sediment transport (Table 13.1).

One significant characteristic of morphodynamic evolution in Costa del Sol is the very limited development of cellular sedimentary circulation. Cell boundaries are extremely variable given the energy regimes of waves combined with the continental shelf morphology. This provides the basis for short crested, high frequency waves, caused by limited fetch and low energy achieved by local wind waves (Fig. 13.3). These waves are significantly steep and cannot generate bottom friction to initiate sediment transport in water depths beyond 3 m generally. These wave geometry (generally throcoidal) is highly reactive to changes in bathymetry and thus refraction is intense.

As tidal range is almost negligible wave action is constant on approximately the same water level. Despite the statistical presence of 70% calm wave fields, this is generally not observed on the coast as sea breeze is very active during the spring, summer and autumn periods. Additionally, the steep surf zones generate semi reflective scenarios and intense sediment transport in narrow mid surf zones. This implies strong-short lived drift and potential for rip currents.

Wave climate and coastal morphologies change significantly between nearby locations along the coast of Andalusia as local dynamics and mechanisms of

Table 13.1 Wave climate in Costa del Sol

	Deepwater significant wave height (Hs m)	Mean zero crossing point (TZ seconds)	Mean wave direction (degree)
Modal East	1.0	5.0	80
Storm East	2.5	7.0	80
Modal West	1.0	4.5	255
Storm West	2.5	7.0	255

Data from Puertos del Estado. Ministerio de Fomento

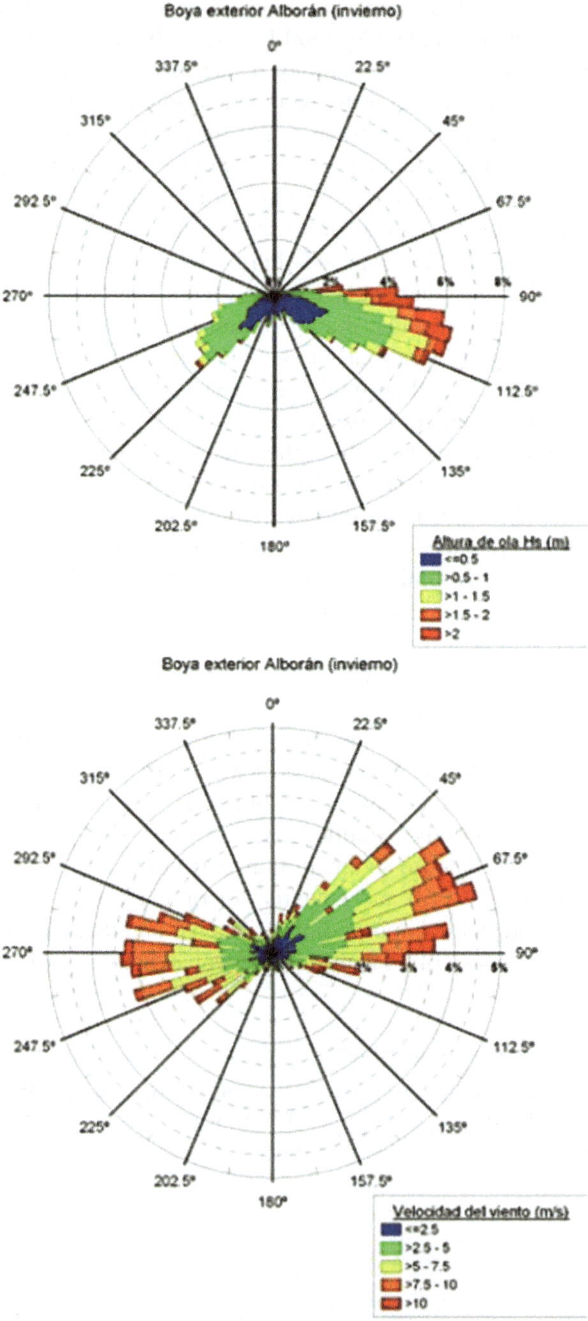

Fig. 13.3 Wave and wind roses for the Alboran Sea. Data from Ministry of environment. Diagram after Guisado-Pintado et al. (2015)

nearshore dynamics depend, among others, on the wave climate, the physical features of the cross-shore profile and the beach morphological profile (Malvárez and Cooper 2000).

Wave and wind approach is characterised by the alternation, almost 50%, of westerly and easterly winds, both in summer and winter seasons, with slightly dominance of easterly wave climate (112°-ESE) in summer time and SW (225°) during winter season. Westerly winds have their origin in the Atlantic Sea whereas easterly and south-easterly winds are originally from the Mediterranean Sea and could persist longer than 15 days in summer periods. In terms of high energy events, dominated by high frequency storm waves with periods of less than 7s and heights of over 5 m, cause water level surges and significant wave set up (Backstrom et al. 2008) and thus have a significant influence in the morphodynamic environments and coastal processes of Malaga's beach and nearshore systems.

Given that the Alborán Sea is adapted to low wave energy conditions morphodynamic processes induced by these extreme wave conditions are notably different to those observed under modal conditions. Nearshore dynamics are characterized by very low wave energy dissipation values and takes place rapidly along a short shallow-water section on the steep foreshore, almost over the shoreline (Fig. 13.4).

Fig. 13.4 La Carihuela, in Torremolinos, under moderate wave action showing a 2 bar configuration of the surf zone and short crested wind waves. *Source* Laboratorio Rediam. Service from Instituto Geográfico Nacional (PNOA)

13.2 The South-Facing Lineal Sandy Beaches of Malaga. Box A, Fig. 13.1

South facing beaches are present in the eastern sections of the province along the shorelines of Nerja, Torrox, Algarrobo, Velez-Málaga, Rincon de la Victoria (Fig. 13.1). The numerous beaches are linked to very localised swash aligned beach ridges with limited back beach development. Most beaches are semi-reflective exposing berms and scarps as wave energy fluctuates with the seasons. The sediment is entirely provided by during high torrential rainfall. These events result in large deposits of highly heterogeneous sands that are then producing poorly sorted beach materials. Sediment sizes in this southfacing eastern segment of the coast is close to coarse sands which introduces great percolation and steep profiles that rarely develop nearshore bars. Thus, the main wave breaking types are plunging and collapsing producing cross-shore sediment transport on the distal sections of the well developed deltas (Torrox, Algarrobo and Velez are the main ones).

The East-West balanced wave approach induces significant longshore transport despite the low energy wave regimes due to optimum refraction initiation from oblique wave incidence. This combined with the narrow steep surf zone forces dominant swash process to be significant because the surf zone is not activated under modal wave conditions. The surf zone is, however, very wide under storm conditions and longshore transport develops around deltas. These features represent the main accident that links the short term morphodynamic behavior with the main bodies of sand.

13.2.1 Beach-by Beach

The eastern most south facing segment, represented as box A in Fig. 13.1, includes the beaches at each sides of major deltas, like the Torrox and Velez river deltas. These deltas a recent fillings of the Pleistocene valleys and have been greatly affected by land use change in recent past and therefore the fresh sediment input can be noted in the beach sediments, mostly composed of dark, soil rich materials directly brought by surficial run off during peak precipitation periods. Thus, heteroclastic materials are dominant.

The El Morche and Peñoncillo beaches are swash aligned systems that are linked to the delta and sediment yields of the Torrox river. This rambla is capable of flash flood type sediment input and beaches are only operating with a limited capacity to remobilise long shore transported sands due to restricted littoral drift (Fig. 13.5).

Further West, at Torre del Mar, the prominent delta of the Velez river depicts an embayment to the East of the mouth that had been an end of circulatory cell sink for sands that circulated to and from the operational delta. The Velez river has a very dynamic recent evolution including episodes during the Holocene of extreme rapid filling of the previous estuary (Malvarez and Senciales 2003). During recent years the presence of the dam of La Viñuela has cut off sediment supply and thus the tip of the delta eroded significantly. Beaches at both flanks of the delta were relatively stable until the combined effects of the construction of the dam plus the cancellation of westward littoral drift along the surf zone induced by the construction of the marina La Caleta de Velez introduced a chronic erosive period. This was overcome by providing a beach nourishment scheme during the first implementation phases of the Ley de Costas of 1988. Figure 13.6 shows the relatively wide beach that fronts the coastal town of Torre del Mar. However, the dynamics of the delta have kept sediments moving on an infralittoral prism to the West of the current river mouth. Figure 13.7 illustrates the bathymetric configuration of the submerged delta which is conformed on the platform but not distal to the mouth. This geometric configuration, perhaps the result of net drift of sediments during storm wave action and flash floods from the Velez, can imply encroached erosion for the future for beaches at Torre del Mar. This is interpreted from the fact that as the nearshore becomes steeper near the marina of La Caleta del Velez and shallower to the West, wave dissipation will provide a constructive scenario to the West and the opposite to the Eastern section of the delta.

Fig. 13.5 Playas de El Morche and El Peñoncillo to West and East of the mouth of the Torrox, respectively. These beaches are an average of 60 m wide which is significant for these Mediterranean conditions. Image, Google Earth 2017

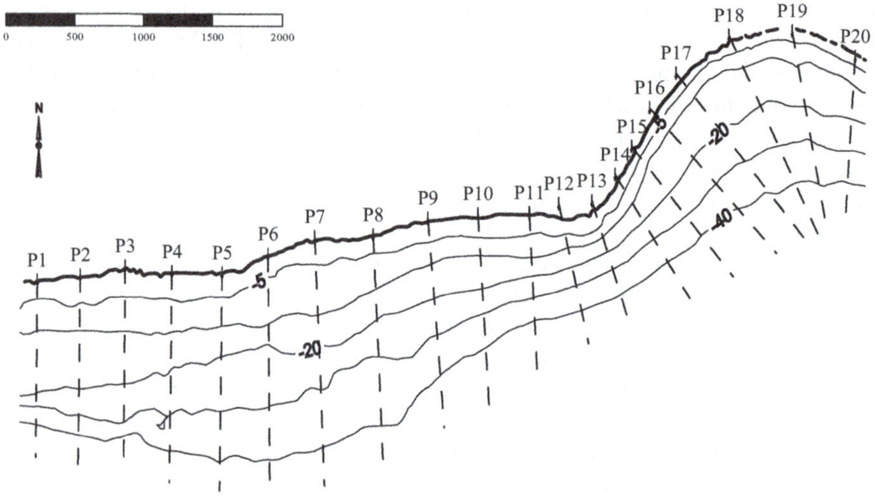

Fig. 13.6 To the East of the Velez delta, the beach suffered severe erosion caused by the construction of the Caleta de Velez marina which stopped westwards littoral drift. The current wide beach at Torre del Mar is the product of beach fill projects during the 1990s and 2000s

Fig. 13.7 The tip of the delta of the Velez and its submerged configuration to −50 m. Note the accumulation at P5 to P8 (at El Hornillo, in Fig. 13.8) to the West of the current river mouth at P13. Chart created by interpolation on data from bathymetric chart of IHM 1988

13.3 East Facing Lineal Sandy Beaches West of Malaga. Box B, Fig. 13.1

The group of beaches which face a general SE exposure facing the Levante winds directly are located immediately SW of Malaga city to the cape of Calaburras at the SW edge. Given the increase exposure to orthogonal approaching wave energy levels from a maximum possible fetch there are more dissipative environments in the surf zone here, some beaches are in intermediate-reflective state which is influenced by the slight sheltering around outcrops. Dissipative beaches as Carihuela, just North of Benalmadena, normally affected by spilling breakers, are observed which is linked to a constructive or accretion trends in their natural state but have turned further intermediate and reflective as development of the back beach has deactivated natural nourishment after storms. This section is influenced by the SW general direction of littoral drift and the significant supply from the Guadalhorce river when it was operational from a sedimentary stand point (i.e. prior to the construction of the three dam complex of the Guadalteba near Ardales). A rich variety of materials provides its composition to the marine environment in this stretch as metamorphic complexes of schists of the Montes de Malaga as well as the limestones of the Sierra de las Nieves at the head of the fluvial networks yielding in this section of the coast.

13.3.1 Beach-by Beach

The main beach in this stretch is La Carihuela, in Torremolinos which is a rectilinear beach which tends to dissipative behaviour but is sensitive to sediment supply from littoral drift from the North. The configuration is swash aligned and only in recent years (since the 1980s) after the construction of the large Puerto Marina, in Benalmádena, some accumulation of sand has been visible. The most significant element of La Carihuela is some signatures of rip currents introduced by the angle of incidence and steepness of the platform here which promotes the location of such currents when winds are prevailing from Levante. The beach is currently wider than it would be in natural conditions because La Carihuela was extensively renourished in the 1990s as a pilot implementation of the Ley de Costas of 1988. This project widened the dry beach to an average of 60 m by using off shore deposits. The quick adjustment of the nearshore introduced severe alterations to morphodynamics involving a departure from intermediate/dissipative modes of evolution to further intermediate/reflective ones. However, the beach continues stabilised despite the fact that urbanisation has encroached onto the back beach and has fixed the previous sedimentary deposits of the back beach by means of a significant promenade. Beach materials are directly linked to the supply of the Guadalrhorce and the extensive deposits that the mouth of the river develops at the Guardamar Aeolian planes, just south of the river mouth.

Given the recent alterations of the river input, it is significant how La Carihuela is subject to plumes of very fine sediments transported in suspended mode and deposited chaotically in the nearshore. This fraction is lost in the shelf and does not contribute to the finer fractions on the exposed beach. Figure 13.8 shows the complex behaviour of Carihuela beach under local winds from Levante and the construction of over 100 m wide surf zone with up to two sand bars. The rapid directional response of the short crested waves that are typical in these setting

Fig. 13.8 La Carihuela and Benalmadena beaches. The Puerto Marina harbour plays a central part in the evolution of these two beaches. Note the semi submerged breakwaters off shore of Benalmadena beach developing artificial tombolos. *Source* Google Earth (2017)

generates the severe indentations on the shoreline with potential for cusp development and localised rip currents. As far as circulatory cell dynamics is concerned the presence of Puerto Marina eliminates any possibility of sediment transfer under most conditions. Thus chronic erosion is affecting Benalmadena beaches just South of the marina given that no sediment supply is possible (there are no local streams or rivers either in that section).

After some 3 km of low cliffs and a limited shore platform developed in the interfase of the outcrops of schists at Los Boliches, the coast is again exposed to the Southeast at Fuengirola. Here a beach is the result of the confinement of the large river yielding in the area. The Fuengirola river occupies a filled in valley squeezed between the limestone and marble complexes of Sierra de Mijas and Sierra Blanca. The Sierra de las Nieves closes the head of the valley which is otherwise flat at the coastal plain providing a configuration that would have behaved as an open embayment during the Pleistocene. Now full of sediments the beach stretches for some 8 km with the River Fuengirola squeezed to the Southwest of the bay and two minor streams draining directly from the steep Sierra de Mijas. The materials are composed typically of a rich mix of sedimentary as well as metamorphic rocks from the nearby mountains. Little input appears to be coming from nearby coastal systems given that the shelf is steep and that two outcrops of metamorphic rock flank this section. Thus, the sands are clearer in colour at Fuengirola and sizes typically finer than further North.

Between the zones marked in boxes B and E in Fig. 13.1 is Cala de Mijas which is mentioned here in isolation from the rest because it represents an independent cell linked to the sediment inputs of the Arroyo Cala del Moral and a large deposit that is accumulated on the platform. This marine deposit seems the result of a combination of end of cell sediment transport coming from the Ensenada de Marbella (to the West) and the bay of Fuengirola (to the Northeast). The accumulation of sand on the nearshore has provided the basis for a semi dissipative morphodynamic environment

Fig. 13.9 Cala de Mijas. Left side is the 1956 aerial photograph and right is the Orthophotography from 2013 of the Instituto de Estadística y Cartografía de Andalucía. Note the significant erosion of the shoreline appreciated at the joint fo the photograhs. This can be considered correct given the accuracy that can be seen at the joint of the roads (indicated by arrow) as the old road was used to serve as one lane of the new motorway. Composition sourced via the Laboratorio Rediam

and the development of a *Cala* (which is not a geomorphological one). Although the beach at Cala de Mijas has been renourished recently, the beach material seems to be placed on the nearshore and its future evolution may be less erosive than in other sections. The materials could be locally sourced although the alteration of the topography of the nearshore could result in accentuated intermediate/reflective behaviour and a more prone to erosive morphodynamics (Fig. 13.9).

13.4 The Ensenada de Marbella. Box E, Fig. 13.1

The evidence that this configuration of the Ensenada is a very stable one is well documented from geologically recent episodes. These are deposits significantly more ancient than the usual postglacial forms on the coast of Malaga. Lhenaff (1977) highlighted a fossil beach of pebbles and sand at a height of +3 or 4 m. West of Cape Calaburras. It is in this deposit where warm fauna appears with *Strombus bubonius*. Remains of a marine deposit attached to the metamorphic substrate can be observed in this coastal stretch at Cabopino. It is a conglomerate of quartz edges and blocks of substrate with sandy matrix and fauna remains highlighting the presence of Tyrrenian indicators. The height of the deposit varies between +0.5 and 0.8 m. Among the fauna is found, we must highlight the presence of: (a) *Glycymeris violacescens*, calcareous algae and *Strombus bubonius*; and (b) *Acanthocardia tuberculata* and *Patella coerulea*. These species correspond to a deposit of very coastal facies.

In Torre de las Cañas we find two superimposed deposits separated by a karstified surface. In the lower, more ancient, we find fauna dated to the Tyrrenian-II. The upper, more modern, would correspond to Tyrrenian III. This same one is located further West towards Puerto Banús, associated with a cemented dune deposit, which has been described in other areas as characteristic of the TyrrenianIII. Towards the East the deposits do not present reference fauna of this epoch, although they display abundant fauna. The lack of associated dune deposits and the difference of the facies indicates it to be Tyrrenian-II in age (Lario et al. 1993).

In Torre Ladrones, a very cemented *lumaquin* conglomerate is found, on which a platform has been built at +0.5 to 0.7 m. This same deposit seems to be the one that in Torre de las Cañas where it is fossilised as a calcarenite karstified to ceiling. At this point, the deposit has a power of about 80 cm. reaching a height of +1.2 m. It is covered by another more recent deposit that reaches +1.8 m. The position and nature of this beachrock (see Fig. 13.11) may suggest that a frame is supporting the location and fixing of the contemporary dune field.

In terms of morphodynamics and processes affecting currently the Ensenada, the beaches found as the coastline faces South at the edge of the Ensenada de Marbella are characterised by plunging breakers, typical of semi-reflective beach types which break energetically at the shoreface due to the significant steepness of the beach with potentially erosive trends. Over all, surf zones and beaches at large become dissipative environments with broadened of the active zone under storm high energy events to the East of the Ensenada. There is a clear divide in terms of sediment transport and beach building (and ultimately dune building) around the centre of the Ensenada. This is clearly due to the change in exposure to prevailing wave approach and its effects on littoral drift. The other significant factor in this divide is the fact that hydrogeologically the domain of the *ramblas*, which dominated the Eastern coasts of Malaga are now definitely surpassed by a fluvial network that is supplying sediments all year long and operates on steep and well-developed basins direcly yielding to the coast and not infilling previously flooded valleys. Thus, the action of these river network is much more efficient in sediment yielding. The effects can be seen on the filling of the upper continental shelf and is particularly evident to the West of the Ensenada where the coupling of large riverine input and the submerged feature of the Placer de las Bovedas configure a large marine deposit in postglacial settings of sea level (see Fig. 13.10 at location opposite the Guadalmina river mouth) (Fig. 13.11).

These factors have contributed in the development of a dune system that can be characterised, in its natural setting, as a 12 km long by 400 m average coastal dune system including a sequence of parabolic and transgressive dunes that illustrates a combined history of high sediment input and active marine induced beach building during the last postglacial hemicycle. The significance of this system lies in its location and genetics. Most Atlantic facing coastal dune systems are associated to the period comprised between 18 K BP and the present when both glacial and riverine sediments have been placed in the active zone of current sea level to form

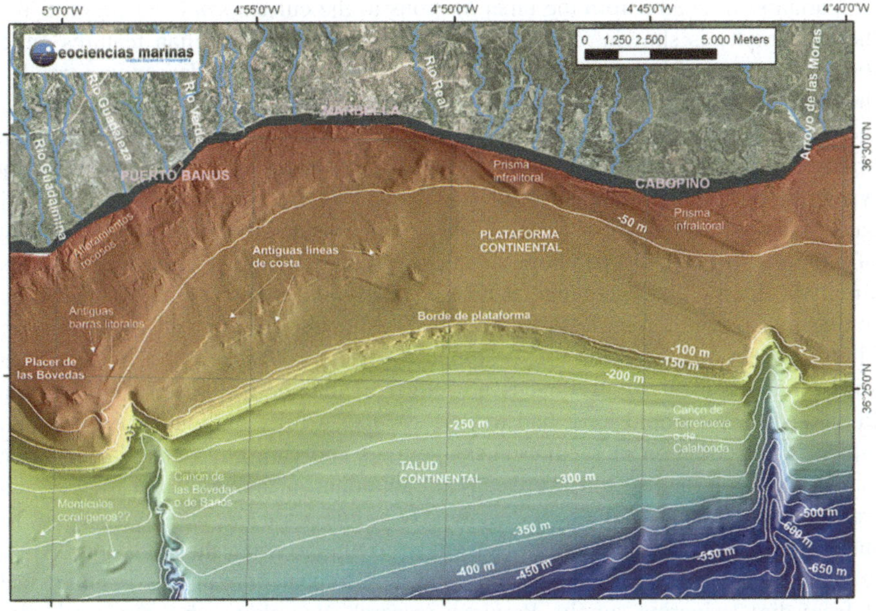

Fig. 13.10 The platform of the Ensenada de Marbella. Image from Dominguez and Malvárez (2015)

Fig. 13.11 Inserted in the active dune system, this karstified calcarenite is being exposed by shoreline erosion. West of Cabopino Marina. Photo: Coastal Environments Research Group (PAIDI 911)

beaches and dunes. However, Mediterranean coastal dunes fronted by steep continental shelves are commonly only present in association with river mouths and/or coastal lagoons. The dunes at Cabopino seems linked to a long term sedimentary and process cycle, which has generated a unique composite of Atlantic and Mediterranean coastal dune environment, representing one of the largest dune systems of the Spanish Mediterranean coast.

Variation in height from the inner sections to the current sea level suggests that the topography is showing a decreasing pattern typical of Aeolian features rather than a sequence of beach ridges that followed sea level in the descending trend of the last 6 k yr. Slip-face structures observed in situ on dune crests are interpreted as indications of the advance inland of the transgressive dunes at times when the large volumes of sediments created mobile dune trains. Figure 13.12 indicates the dune types along and across the system. The western section shows significant imbricated parabolic dunes whose development may be related to a process of reworking of previous structures of parabolic dunes, given the potential diminishing sediment supply scenario. The foredunes at this stage (1950s) were very developed illustrating that a continuous feature existed all along the 12 km system when combining current and degraded (i.e. modified) foredunes. This is a clear indication of substantial sediment supply that is no longer available. The beach supplying the initial sediments would have had to perform like a fully dissipative environment which is not common in Mediterranean settings and can only be explained by an active and continuous link between an abundant nearshore sand supply that transferred sediments of the necessary type (e.g. fine grain size and high sorting) for the transfer from beach to dune material to be so effective in a short timespan. These are characteristics that link the Cabopino system to nearshore processes and detach it from direct riverine supply. Previous research also shows that the continental shelf, down to depths of 50 m, is dominated by complex sediment transport related bedforms. Large deposits are storing sands and presumably promoting sorting in average to shallow water conditions behind the Placer de las Bovedas, 10–15 km to the West of the dune field. The availability of the large deposits of sands would have fed the older and most inner dune fields of the Cabopino system where large vegetated parabolic dunes existed on the inner sections of the system.

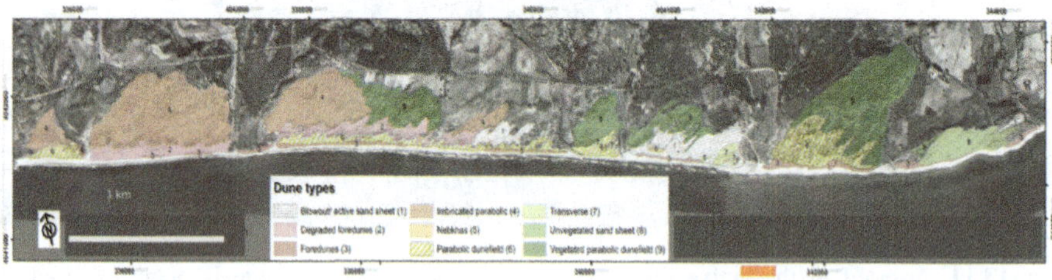

Fig. 13.12 Dune types as interpreted from 1956 aerial photographs. After Guisado-Pintado et al. (2016)

13.5 East Facing Beaches of Estepona, Casares and Manilva. Box C, Fig. 13.1

Finally, to the West of the town of Marbella and the Ensenada a new section is stretching to the edge of the Province of Malaga which includes the beaches of Estepona, Casares and Manilva. Beaches here show an exposure to the Levante again and their behaviour is dictated by stronger wave energies (as in Torremolinos) but also higher wind speeds as the topography of the mountains to the north promote a funnel effect that accelerates wind flows towards the Strait of Gibraltar. The mountain ranges north of Estepona are dominated by the great batholith of Ronda and Sierra Bermeja which is mostly peridotite. The significant alteration of the materials on the surface and the structural dominance of the geology promotes a hydrological network that provides direct feeding to beaches with significant forces due to increased precipitation regimes and the impermeable nature of lithologies promoting run off rather than infiltration as in limestones in other sections of the coast. Materials on the beaches is more heterogeneous given the steeper shorefaces and the increased wave energy coupled with river inputs that are commonly providing significant bed loads as well as suspended (see Fig. 13.13). Bed load is characterised by greater competences and the result is coarse and heterogeneous sands on beaches with a narrow surf zone that is not capable of sorting to higher degrees.

13.6 Artificial Shorelines in Malaga

Given the significant increase in interest from the tourism industry in the first consolidation years that followed the establishment of Costa del Sol as a global destination, during the 1960s beach front properties seek protection. This was granted and encouraged by Franco's government to protect the rising levels of investments and thus a number of sea walls were constructed which served also as the basis for promenades, infrastructure (sewerage and wiring projects) all of which was viewed as an advancement in the management of the eroding beaches (Fig. 13.14). However, as seawalls came under attack from the encroached sediment starved beach system, the erosiveness of the whole system became greater and soon, during the 1970s, sea walled promenades were "helped" by groynes and jetties. The solution appeared appropriate because the acute oblique angles of incident wave energy was proving that sea walls reflected energy erosion and problems alongshore to areas that were previously less unstable.

Sea walls seemed to have transformed the nearshore as waves approached the narrow surf zone on increased depth caused by undertoe of sea walls and general deepening of the nearshore generated as reflected waves coupled with incoming. Extending and enlarging sea walls was found to be counter producing the desired effects as erosion continued down drift. But promenades and sea walls were not

Fig. 13.13 Across the beach face materials are poorly sorted as nearshore processes are highly variable in energy and sediment sources varied. Note gravel coexisting with sands due to increased bedload transport from the river network

easily replaced because now they had become an essential part of the urban infrastructure. Additionally, sea walls had also introduced the issue of flooding from land sources as outlets of small streams had chocked with sediments or have directly been incorporated into the longshore infrastructure.

Keeping most of the sea walls and their infrastructure groynes now were set to stop the power full loss of sands that was affecting the narrowed down system. South facing beaches were affected mostly by the strong oblique wave approach and related littoral drift that was taking sediment away. Therefore, during the 1970s

Fig. 13.14 Malagueta beach in Malaga after effects of sea-wall. Beach is gone. Photograph from Archivo Histórico Provincial de Málaga

a frequent image was that of extensive groyne fields emplaced along the waterfront of Pedregalejo, in Malaga, Benalmádena and Marbella/Banús beaches (Fig. 13.15). All of these groyne fields include a T shape that was designed to stop longshore sediment loss as well as onshore wave attack. The abundance of marina constructing during the late 1970s and early 1980 (from Torre del Mar to Estepona over 12 marinas were constructed) all fought this issue of "excessive" littoral drift. The solution in most cases came in the shape of further groyne fields that intended to stop siltation at entrances. However, engineers continued to hesitate in guessing the directional component of littoral drift, as demonstrated by the opposing directions of entrances of Fuengirola and Cabopino marinas, among others). This lack of understanding at a subregional scale and speedy action at project or local scale was not just an issue of Malaga province but a common issue during these decades of intense intervention on coastal systems (Cooper and Pilkey 2004).

Severe storms during the later years of the 1980s brought about the reality that it seemed that intervening with the natural system had gone too far. At that moment, sediment input had been severely reduced due to massive intervention on river drainage networks (to promote water management for the flourishing tourism industry) and beaches were all showing erosive trends. Management could not stay and watch how the main asset generating this economic "take off" was being damaged by nature. Thus, as the 90s started the brand new Ley de Costas of 1988 was put to action in the most complex scenario that there was in the whole of Spain: Costa del Sol in Malaga. The favoured coastal protection technique in the Ley de Costas (and most environmentally friendly) was beach nourishment or beach fill. The principles of beach nourishment imply that the sediment that should feed the beach must be off shore and that the sediment should be pumped back into position by replacing the slow natural process of beach building. However, as mentioned before, the strong bi-directional component of wind generated waves and related drift soon proved that nourishment was going to be a problem as marinas silted and choked and beaches up drift loss their sands within the first storm. The constant

Fig. 13.15 Frequent layout of hammerhead groyne fields in Malaga's beaches in 1970s. Photo of Marbella in 1974. *Source* Ayuntamiento de Marbella Archivo

60 metre wide beach that was mostly constructed in 1990 to 1993 disappeared rather quickly and the shoreline was transformed for good as new volumes of sand entering the system and began to drift. Although most was lost due to inadequate sand sizes and shapes, the starvation was generalised. This way, from the mid 90s various sediment sources were attempted (once dredging of offshore was deemed too expensive and inaccessible). Lack of understanding of beach morphodynamics in the coast of Malaga brought about strange situations where sand replacements came from sources where abundant talk and silt was present (e.g. from Guadalhorce river mouth near Malaga) to nourish beaches elsewhere along the province's more developed beaches. Sediments with high content on dust from limestones consolidated on the shoreface soon after providing a totally unnatural profile that responded chaotically to wave action.

In most cases beaches became totally full of crushed shells and were not particularly liked by tourists who frequently were injured by the sharp nature of the artificially fed sediments. In all, in 1992 alone, over 7 million cubic metres of sand were pumped or transported onto the beaches (some 27 km stretch) and the problem of sand starvation and erosion continues. Each year, as storm hit the East facing sections of the coast of Malaga, severe erosion takes its toll on infrastructure and beach width. Meanwhile, south facing sections end up receiving vast quantities of sediment that either pulses onto sediment sinks (like Cabopino) or rush away into depths beyond wave base and feed the extensive canyons of the continental shelf.

The issue of littoral drift along the Costa del Sol is one of the main issues in coastal stabilization. As elsewhere, (Cooper and Pilkey 2004) drift is not well understood at a regional level and the bidirectional wave climate severely influences the functioning of groyne fields and jetties at marinas. The port of Fuengirola, constructed in the 1970s on an existing mooring jetty provides a good illustration. Port engineers chose opposite directions for net longshore drift in the 1950s and 1970s designs, but none was fully successful and sedimentation in the marina has been a constant issue for navigation and shoreline erosion downdrift (both East and West).

A new approach to protection was introduced with the implementation of the Ley de Costas of 1988; this was aimed at starting a new era in soft coastal protection that superseded the hard solutions of the past. It involved the widespread application of beach nourishment. Beach nourishment, however, turned out to be unsuccessful in terms of the longevity of the beach fill and the associated economic implications. Projection of outcomes of nourishment projects are very much dependent upon understanding local morphodynamic conditions, and particularly the processes operating when modal conditions are exceeded (during storms). Experiences in Malaga show that the application of beach fill has not been entirely satisfactory partly because of difficulties in solving the fundamental problem of sediment starvation in the coastal system. The former riverine sources of sediment have been much reduced by dam building and the sediment stored in coastal dunes has been lost under urban development. Offshore sediment deposits were used in beach nourishment schemes and reasonably resilient beaches were built, but only because the material that was pumped onto the beach was so coarse or had such a high component of limestone that it became highly consolidated and partially cemented.

In other cases, off-shore material was excessively shelly, making it unpleasant for recreational purposes. The first attempts to nourish Marbella's beaches failed partly through use of sands drawn from the mouth of the Guadalhorce River. The upper surface of these sands rapidly cemented, creating an abrasive surface that was unsuitable for recreational use. In contrast, sediments utilized for nourishment of Malaga's beach had ideal characteristics, although it was still prone to sand loss during storms. The failure to achieve a reasonable life span is strongly influenced by the distortion that beach nourishment introduces into the littoral system. The sudden addition of large volumes of beach material creates a significant shift in the hydrodynamics of shoaling waves, well beyond the expectations of the design phase. This seems to have been the case on beaches such as Marbella and Malaga, where replenishment had to be repeated soon after the first fill. The reason for this could be the entrainment of large volumes of sediment in a littoral cell whose dynamics have been distorted into a more effective transport machine. Over seven million m^3 of sand were pumped onto 27 km of beaches along the Costa del Sol in 1992 alone. Continued annual expenditure on beach nourishment is necessary to maintain the beach resource and stabilize the shoreline.

References

Backstrom, J. T., Jackson, D. W. T., Cooper, J. A. G., & Malvárez, G. (2008). Storm-driven shoreface morphodynamics on a low-wave-energy delta: The role of nearshore topography and shoreline orientation. *Journal of Coastal Research, 24,* 1379–1387.

Cooper, J. A. G., & Pilkey, O. H. (2004). Longshore drift: trapped in an expected universe. *Journal of Sedimentary Research, 74,* 599–606.

Dominguez, L. M., & Malvarez, G. C. (2015). La Plataforma Continental. In N. Malvarez & P. Guisado (Eds.), *Geomorfología de la Ensenada de Marbella.* ISBN: 978-84-606-8671-2. 80 pp.

Guisado-Pintado, E., Malvarez, G. C., & Ojeda, J. (2015). Procesos y dinámica costera en Ensenada de Marbella: evolución morfosedimentaria reciente. In *Revista Geotemas* (Vol. 15).

Guisado-Pintado, E., Malvarez, G. C., Jackson, D., & Navas, F. (2016). Inconsistencies in coastal dune genesis and development in the western Mediterranean Cabopino Dune system, southern Spain. Abstract for EGU General Assembly. EPSC2016–13715.

Lario, J., Zazo, C., Somoza, L., Goy, J. L., Hoyos, M., Silva, P. G., et al. (1993). Los episodios marinos cuaternarios de la costa de Málaga (España). *Revista de la Sociedad Geológica de España, 6,* 41–46.

Lario, J., Zazo, C., Dabrio, C. J., Somoza, L., Goy, J. L., Bardají, T., et al. (1995). Record of recent Holocene sediments inputs on spit bars and deltas of south Spain. Holocene cycles: Climates, sea levels and sedimentation. *Journal of Coastal Research (SI), 17,* 241–245.

Lhenaff, R. (1977). *Recherches geomorphologiques sur les Cordillères Bétiques Centro-Occidentales (Espagne), Thése* (p. 713). Lille: Université de Lille III.

Malvárez, G., Lario, J., Zazo, C., Goy, J. L., & Luque, L. (1998). Evolución de la Costa de Málaga durante el Pleistoceno Superior y Holoceno, y morfodinámica actual de los sistemas litorales. In Gómez Ortíz, et al. (Eds.), *Itinerarios Geomorfológicos por Andalucia Oriental:Trabajo de Campo de la V Reunión Nacional de Geomorfología* (pp. 9–27). Barcelona: Universitat de Barcelona.

Malvárez, G. (1999). Procesos morfodinámicos litorales en la Costa del Sol. En J. M. Senciales & E. Ferre (Eds.), *Elementos de los Paisajes de la Provincia de Málaga Studia Malacitana* (Vol. 11, pp. 169–228). Málaga: Servicio de Publicaciones de la Univ. de Málaga.

Malvárez, G., & Cooper, J. A. G. (2000). A whole surf zone numerical modelling approach to the characterisation of nearshore environments and beaches. *Journal of Coastal Research, 16*(3), 808–823.

Malvárez, G., & Senciales, J. M. (2003). La desembocadura del río Vélez (provincia de Málaga, España) evolución reciente de un delta de comportamiento Mediterráneo. *Cuaternario y Geomorfología, 17*(1–2), 47–61.

Chapter 14
Beaches of Cadiz

Laura Del Río, Javier Benavente, F. Javier Gracia, Giorgio Anfuso, María Aranda, Juan B. Montes, María Puig, Lara Talavera and Theocharis A. Plomaritis

14.1 Introduction

Cadiz province has a 230 km long coast and is situated in the Andalusia Region, Southwest Spain (Fig. 15.1). It principally faces the Atlantic Ocean, i.e. from the Guadalquivir River mouth to the North, to Punta Europa in the South and, secondarily, the Mediterranean Sea, i.e. from Punta Europa to Punta Chullera, both

L. Del Río (✉) · J. Benavente · F. J. Gracia · G. Anfuso
M. Aranda · J. B. Montes · M. Puig · L. Talavera
Department of Earth Sciences, Faculty of Marine and Environmental Sciences, University of Cadiz, 11510 Puerto Real, Cadiz, Spain
e-mail: laura.delrio@uca.es

J. Benavente
e-mail: javier.benavente@uca.es

F. J. Gracia
e-mail: javier.gracia@uca.es

G. Anfuso
e-mail: giorgio.anfuso@uca.es

M. Aranda
e-mail: maria.aranda@uca.es

J. B. Montes
e-mail: juan.montes@uca.es

M. Puig
e-mail: maria.puig@uca.es

L. Talavera
e-mail: lara.talavera@uca.es

T. A. Plomaritis
CIMA, University of Algarve, Campus of Gambelas, 8005-135 Faro, Portugal
e-mail: haris.plomaritis@uca.es; tplomaritis@ualg.pt

J. A. Morales (ed.), *The Spanish Coastal Systems*,
https://doi.org/10.1007/978-3-319-93169-2_14

sectors being divided by the Strait of Gibraltar (Fig. 15.1). The Mediterranean side of the Cadiz Province occupies the western part of the Alborán Sea meanwhile the Atlantic side occupies the central-southern part of the Gulf of Cadiz. The Gulf of Cadiz is an arcuate physiographic unit extending from Cape San Vicente (in SW Portugal) to Tangier (in Morocco), with a crescent-shaped continental shelf bounded by the bathymetric contour of 100 m depth (García-Lafuente et al. 2006). Along the Atlantic investigated areas the continental shelf is about 30–40 km in width, while it is virtually absent in the Strait of Gibraltar and Mediterranean areas.

The coastline shows different orientations: the 155 km-long straight Atlantic sector is broadly NW–SE oriented, the 50 km long Strait of Gibraltar area is broadly SW-NE oriented, and the 25 km long rectilinear Mediterranean sector is NNE-SSW oriented. Two large bays must be mentioned (Fig. 15.1): the Bay of Cadiz is located on the Atlantic coast and is constituted by a sheltered and relatively shallow environment, while the Bay of Algeciras, in the Strait of Gibraltar, shows higher depths due to the aforementioned absence of continental shelf.

From the geomorphological point of view, the coast is characterized by a great diversity of landforms and environments including low coastal areas consisting of beaches and dunes, at places developed on sand spits and barriers (the most important ones being Valdelagrana, Sancti-Petri and Barbate spits) and protecting saltmarshes. Rocky coasts, i.e. cliffed sectors and rocky shore platforms, are also present, particularly in the southern sector. In this respect, beaches constitute the dominant coastal typology. Along the Atlantic sector, they are usually long, exposed and drift-aligned; they are mainly composed by medium and fine golden sand consisting of quartz (c. 70%), feldspars (18%), carbonates (10%) and heavy minerals (2%) (Anfuso and Gracia 2005). Near the Strait of Gibraltar there are also small, sheltered swash-aligned beaches forming pocket beaches of different dimensions. They are composed by coarse to fine sand of golden quartz or dark litharenitic deposits. Mediterranean beaches usually are long, drift-aligned and composed by dark litharenitic deposits (Bello-Smith et al. 2011).

Small foredunes and developed dune ridges border many sandy beaches along the Atlantic sector, especially in correspondence with sand spits and barriers, e.g. at Valdelagrana and Sancti-Petri (Gracia et al. 2006). Very high and mobile dune systems are observed in the Southern part of the Atlantic investigated sector, namely at Bolonia and Los Lances embayments. In the Bay of Algeciras dunes are only observed in Palmones sandspit, because of the high level of human occupation in the rest of the Bay and the shelter role played by the surrounding relieves. Along the Mediterranean sector, remnant dunes and small secondary ridges are observed only at limited areas (e.g. La Atunara beach).

Extensive saltmarshes are present in the Bay of Cadiz and around Barbate River mouth, while minor systems appear in sheltered and estuarine areas protected by sand spits in the Bay of Algeciras (e.g. Palmones River) and the Mediterranean coast (Guadiaro River).

Soft cliffs, essentially composed by unconsolidated Pliocene and Quaternary deposits (mainly sands, clays and conglomerates) are observed mainly between Chipiona and El Puerto de Santa María, while at Cape Roche and Conil they are composed of cemented detritic deposits, namely Middle Miocene to Plio-Pleistocene conglomerates and sandstones, with numerous lateral changes of facies to sands, clays and marls (Del Río et al. 2009). Higher relief areas appear South of Cape Trafalgar, with hard cliffs composed by Miocene calcareous sandstones at Barbate, while rocky headlands towards the Strait of Gibraltar are composed by Miocene sandstones and in some cases Quaternary conglomerates (Del Río et al. 2009). No cliffed areas appear in the Mediterranean sector. Rocky shore platforms are mainly observed between Chipiona and Rota, between Cadiz and Cape Trafalgar, and at the Strait of Gibraltar.

Regarding the hydrodynamic conditions, tides are semidiurnal and the tidal range considerably varies along the investigated area, from mesotidal in the North Atlantic coast to microtidal in the Bay of Algeciras and the Mediterranean coast, with a rapid decrease in tidal range between Cape Trafalgar and Punta Europa. In particular, mean spring tidal ranges (MSTR) show values of 2.96 m at Cadiz, 2.30 m at Barbate, 1.22 m at Tarifa and 1 m at Algeciras (Fig. 14.1).

The area is affected by westerly (blowing from II and IV quadrants) and easterly (from I and II quadrants) winds. Westerly winds (*Poniente*) are related to Atlantic low-pressure systems and approach the Gulf of Cadiz from NNW (c. 11% of annual occurrence), NW and W (c. 9% each), SW, WSW and NNW (c. 6% each). In the Mediterranean sector they approach from W (c. 17% of annual occurrence), WSW (c. 11%), WNW (8%) and SW (4%). Easterly winds (*Levante*) blow from the Mediterranean Sea and hence greatly affect the Mediterranean sector of the study area, where they approach from E (c. 24% of annual occurrence), NE (c. 12%) and ESE (c. 4%) with maximum velocities above 80 km/h. Levante winds are channelized through the Strait of Gibraltar acquiring strong velocities and spreading northward and southward (Dorman et al. 1995), greatly affecting the south part of Cadiz Province, where they approach from E (c. 15% of annual occurrence) and ESE (c. 14%).

According to the previously described wind patterns and coastal orientation, westerly winds give rise to both swell and sea waves along the Atlantic area, approaching from W (c. 50% of annual occurrence), WSW (c. 13%) and WNW (c. 9%, Fig. 14.2a) with wave heights lower than 1 m during fair weather conditions and higher than 3 m during storms (Rangel-Buitrago and Anfuso 2011a, b; Del Río et al. 2012; Anfuso et al. 2016). Because of coastal orientation, westerly waves have a limited influence in the Mediterranean sector, approaching the coast with low wave heights from WSW (c. 23%) and W (c. 19%, Fig. 14.2b).

Conversely, easterly winds greatly affect the investigated Mediterranean coast, giving rise to swell and sea waves that approach from E (c. 46% of annual occurrence, Fig. 14.2b) with significant wave heights lower than 0.75 m (75% of cases). In the Atlantic area, easterly waves acquire a limited importance and they only generate sea waves approaching from SSE with a moderate wave height and annual occurrence (c. 18%, Fig. 14.2a).

Along the Atlantic sector dominant longshore drift flows south-eastward, although local and temporal reversals, related to waves approaching from the South,

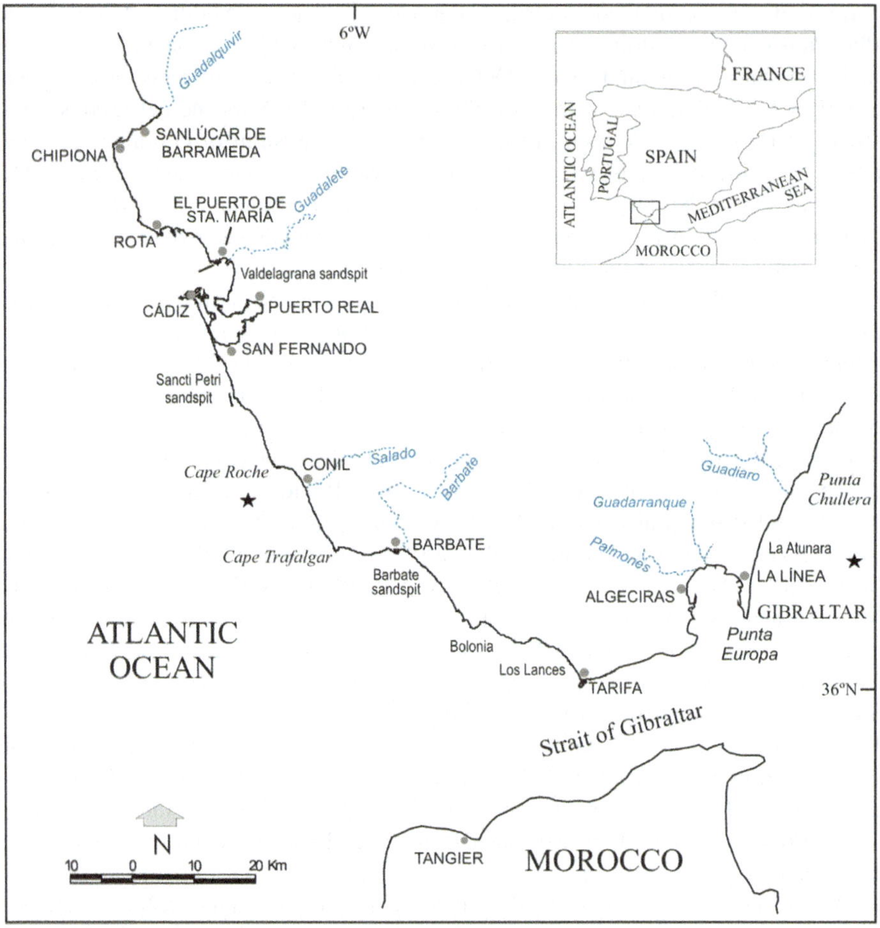

Fig. 14.1 Location map indicating the position of the two wave modelling points belonging to the SIMAR network (Spanish National Ports Authority)

are also observed at certain places (Anfuso et al. 2008). Along the Mediterranean sector the main longshore drifts flows south-westward.

14.2 Origin and Historical Evolution

The Atlantic coast of Cadiz presents a broadly rectilinear outline structurally controlled by NW-SE fractures, interrupted by several bays and embayments related to the Quaternary activity of faults: Cadiz, Barbate and Algeciras bays are the most important ones, forming tectonic grabens with recent subsiding trends

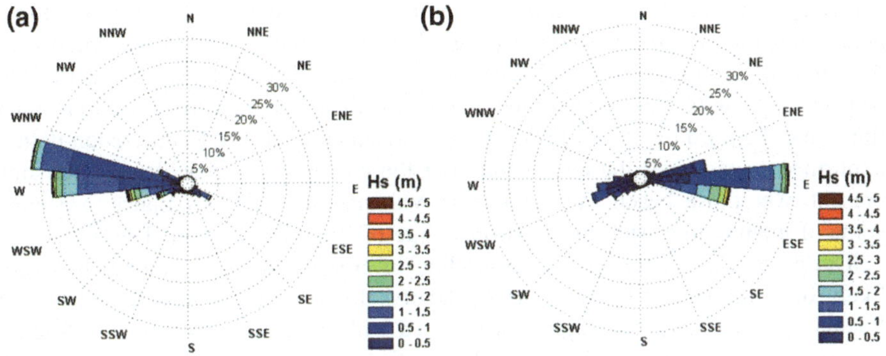

Fig. 14.2 Wave roses for the period 1958–2001 in: **a** the Atlantic sector, SIMAR network point 1055045; **b** the Mediterranean sector, SIMAR network point 2022074. Data provided by Puertos del Estado (National Ports Authority)

(Goy et al. 1995; Gracia et al. 1999, 2008). During Quaternary times those subsiding blocks and bays collected the surface drainage of broad areas, leading to the differentiation of the main rivers of the region: Guadalete River (whose mouth is located in the Bay of Cadiz), Barbate River (on the Bay of Barbate) and Palmones River (located on the Bay of Algeciras) (Fig. 14.1). Other minor rivers are located on secondary embayments (Salado River at Rota, Conilete River at Conil, Valle River at Bolonia Bay and Jara River at Los Lances Bay). All these places receive a continuous sediment supply through watercourses and are at the same time sheltered from wave action. As a consequence, important sedimentary systems formed once during periods of sea level stillstand or slightly lowering trend.

In fact, the present distribution and development of beaches along the Cadiz coast is a consequence of the generation of such coastal sedimentary systems during the last eustatic hemicycle. According to Zazo et al. (2008), the postglacial sea level rise occurred in southern Spain in two phases: one with a rapid rise until 6500 cal BP and a later decelerating or even stable one with minor, metric oscillations. As a consequence, six coastal prograding sedimentary episodes were formed (Zazo 2006; Zazo et al. 2008): H1 (7400–6000 yr BP); H2 (5400–4200 yr BP); H3 (4000–3000 yr BP); H4 (2700–1900 yr BP); H5 (1900–1300 yr BP) and H6 (last 500 yr). Since the development of coastal sedimentary bodies not only depends on the sea level trends but also on other local or subregional factors (like local subsidence, sediment supply, recent tectonic movements, etc.), the number of prograding episodes varies from one sedimentary system to another even within the same physiographic unit.

The best preserved historical Holocene records of coastal evolution along this coast are those existing at Valdelagrana spit barrier, inside the Bay of Cadiz, where a set of over 20 different prograding episodes can be found (Rodríguez-Polo et al. 2009; Del Río et al. 2015a), with complex inter-relationships. This system generated over 3000 years ago. Very energetic events, like strong storms and especially tsunamis, as well as human activity, have played a key role in the evolution of this area (Alonso et al. 2015). Another prograding sedimentary system is located near the Strait of Gibraltar, at Los Lances Bay (Tarifa), where at least three parallel beach ridges can be clearly identified, the intermediate one being ascribed to Roman times and the third one to a Medieval age (Gracia et al. 2004). Radiocarbon dating confirmed previous estimations made after the analysis of archaeological remains present over those beach ridges, both at Valdelagrana and Los Lances systems, some of them dating back to the Bronze Age (Gómez Ponce et al. 1997).

During the last centuries the prograding or eroding trend of beaches has very likely been influenced by two sets of factors: the first one is represented by the fluctuating wave energetic regime, associated with large-scale atmospheric forcing affecting the frequency and magnitude of storms in the region (Plomaritis et al. 2015), and/or with the occurrence of tsunamis, which has been repeatedly documented in the area (Alonso et al. 2015). The second one is related to the human occupation and use of the fluvial catchments draining to the coast. Deforestation practices, held during Roman, Medieval and 19th century times, favoured the increase of sediment supply to the coast, while hydrological regulation and dam construction during the 20th century has drastically reduced sediment input to the coast and triggered erosional trends in many beaches located near the river mouths (Del Río et al. 2015b).

14.3 Morphodynamic Characterization

In order to classify morphodynamic behaviour of beaches in Cadiz province, a total of 23 beach profiles on the Atlantic coast and 4 beach profiles in the Mediterranean coast were analysed, including their respective sediment characteristics. For the morphodynamic characterisation of these areas, a range of widely used parameters were employed: Surf Scaling parameter (Guza and Inman 1975), Surf Similarity parameter (Battjes 1974) and dimensionless fall velocity or Dean parameter (Gourlay 1968; Dean 1973). Finally, in order to apply the classification by Masselink and Short (1993), relative tidal range (RTR) was determined.

Nearly all the studied profiles were classified among the various intermediate types proposed by Wright and Short (1984) or Masselink and Short (1993) (Anfuso et al. 2006; Benavente et al. 2015b). The only exceptions observed, at one extreme of the range, were profiles of the Mediterranean coast, mainly La Atunara and Sotogrande beaches, which presented typically reflective morphology (Fig. 14.3). At the other end is Valdelagrana beach, which presents a clearly dissipative type morphology (Fig. 14.3) without presence of sand bars and a small temporal variability, as indicated by the classifications of Wright and Short (1984) and Wright et al. (1987).

The collation of the results of numerous monitoring programmes carried out over the last 20 years has allowed the determination of the morphodynamic behaviour of beaches in the study area. Table 14.1 shows the results of the parameters used for

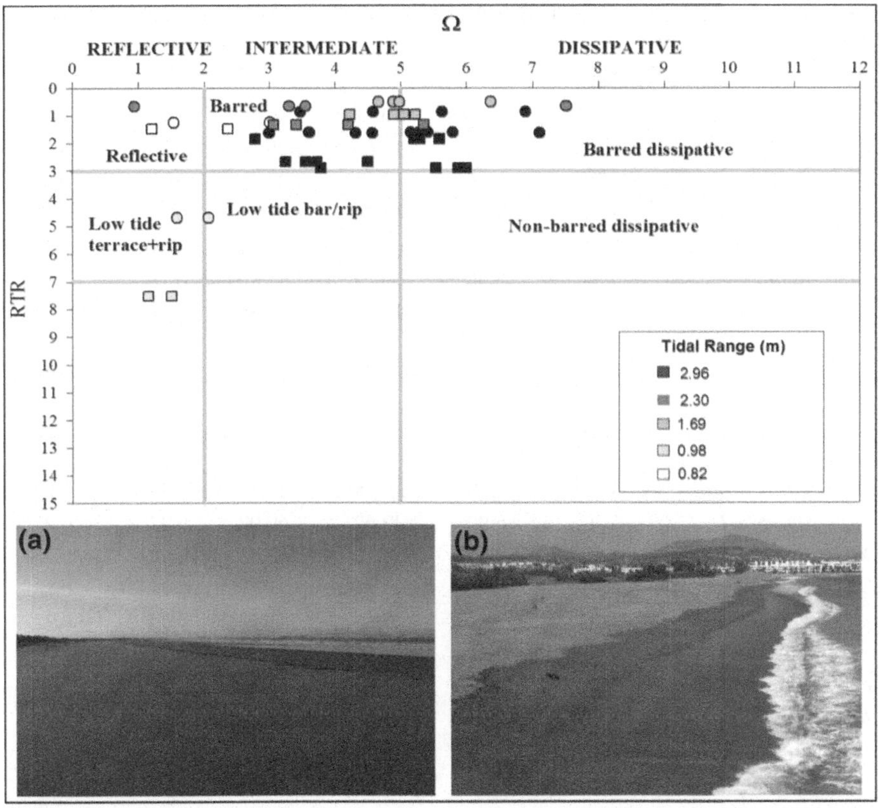

Fig. 14.3 Classification of beach profiles along the study area in winter (circles) and summer (squares) according to the Masselink and Short (1993) scheme. Darker colours correspond to beaches in the North sector of the Atlantic coast, and lighter colours correspond to beaches in the Bay of Algeciras and Mediterranean coast. Photo A: Valdelagrana dissipative profile. Photo B: Sotogrande reflective profile

calculation of the entire set of morphodynamic variables for both summer and winter months. The tidal range selected corresponded to the MSTR for each of the areas, without seasonal variations. As can be observed, according to both Scaling and Surf Similarity parameters, most of the studied profiles are classified as intermediate, i.e. Surf Scaling values between −1 and 30 and Surf Similarity values between 0.4 and 2 (Wright and Short 1984), with a dissipative behaviour during winter conditions, i.e. Surf Scaling parameter higher than 30 and Surf Similarity parameter lower than 0.4 (Wright and Short 1984) (Table 14.1). This is clearer around the Bay of Cadiz, where Surf Scaling parameter accurately describes the behaviour of these beaches, which is clearly intermediate in the exposed areas and more dissipative in the inner bay. In the case of the clearly dissipative profile (profile 18, Valdelagrana), both parameters presented limited variability, its morphology being defined as dissipative even during summer (fair weather) conditions.

Table 14.1 Morphodynamic parameters calculated on different beach profiles along the coast of Cadiz province

Nr	Name	Dean		Surf scaling		Surf similarity	
		Summer	Winter	Summer	Winter	Summer	Winter
11	*Regla*	5.54	3.61	11.60	58.56	0.57	0.24
12	*Tres Piedras*	3.79	5.15	45.38	33.68	0.29	0.31
13	*Aguadulce*	5.99	5.80	41.24	35.21	0.30	0.31
14	*Rota Norte*	5.87	5.42	7.86	27.71	0.69	0.35
15	*Rota*	5.54	5.28	6.19	41.89	0.78	0.28
16	*Vistahermosa V3*	3.24	4.32	10.78	42.38	0.58	0.28
17	*Vistahermosa V5*	4.50	3.00	7.33	17.42	0.70	0.44
18	*Valdelagrana VL6*	3.55	7.11	50.99	72.17	0.27	0.22
21	*La Victoria*	3.73	4.58	19.46	46.80	0.43	0.27
22	*Barrosa 1*	5.30	6.90	32.62	69.99	0.32	0.22
23	*Barrosa 2*	5.59	4.59	18.12	42.05	0.43	0.28
24	*Torre Puerco*	2.78	5.63	21.34	41.39	0.40	0.28
25	*Palmar*	5.20	3.47	29.45	41.36	0.34	0.28
27	*Caños*	3.41	3.31	1.90	5.77	1.32	0.75
28	*Barbate*	5.36	7.51	13.90	47.24	0.49	0.26
29	*Retín*	4.21	0.94	20.36	3.22	0.40	1.01
30	*Zahara*	3.07	3.55	5.57	13.58	0.77	0.49
31	*Bolonia*	4.89	6.37	5.26	30.76	0.79	0.33
32	*Punta Paloma*	5.05	4.89	9.01	2.92	0.61	1.06
33	*Valdevaqueros*	5.22	4.67	1.79	2.44	1.36	1.16
34	*Los Lances*	4.23	4.98	4.70	21.14	0.84	0.39
35	*Rinconcillo (S)*	1.51	2.07	4.01	4.62	0.86	0.77
36	*Rinconcillo (N)*	1.15	1.59	3.17	2.44	0.97	1.05
37	*La Atunara*	2.36	3.01	9.07	12.69	0.70	0.71
38	*Sotogrande*	1.20	1.54	4.80	8.12	0.97	0.88

Towards the South, profiles are still classified as intermediate according to the above parameters although with a more "reflective" character, despite their morphology being definitely closer to that of typically reflective profiles or presenting swash bars and/or rhythmic forms (Benavente et al. 2015b).

Despite of the above result, some of the areas presented a clear reflective profile morphology, mainly in the Mediterranean area, but the aforementioned parameters did not adequately show this character. These profiles nonetheless presented reflective morphology in both summer and winter months with high beach front slopes (Montes et al. 2017).

When using the Dean parameter (Table 14.1) the obtained values fall completely within the range of reflective domain for the case of the Mediterranean beaches. In this respect, some of the Atlantic beaches located in the central sector appear in the intermediate zone but close to the reflective one, due to the influence of the coarsest sediments found in this area. These beaches, already distant from the main rivers, exhibit overall coarser grain size that has a strong influence on the beach classification (Benavente et al. 2015b). Exceptions to this are located in the proximity of the mouth of smaller water courses like Barbate River.

In order to avoid the previously mentioned discrepancies it seems necessary to introduce the tidal range as an additional variable in the morphodynamic classification of these beaches. In Fig. 14.3, the classification of the profiles studied is shown for both winter conditions as well as summer ones, based on the model proposed by Masselink and Short (1993).

According to this, the classification of Mediterranean beach profiles matches their morphological characteristics, displaying reflective behaviour in both summer and winter (Fig. 14.3). Their low variability, which can be observed on a morphological and morphodynamic level, is due to the way they behave regarding erosion, namely by means of a parallel retreat with a very low change in beach slope. On the contrary, the morphology of Atlantic profiles, mainly in the central area, showed a pivoting behaviour between summer and winter periods.

In general, as the tidal range increases, profiles tend toward morphologies with mesoforms, whether in the form of swash bars (intermediate profile with bars), low tide bars or low tide terrace (Benavente et al. 2015b). Even in nearly microtidal areas, close to Strait of Gibraltar, a low tidal range seems to favour the appearance of intertidal bars. In general, almost all profiles would be classified amongst the morphological types defined as wave-dominated (Masselink and Short 1993; Short 2006), having low tidal influence, as their RTR is lower than 3. Such beaches normally appear in microtidal coasts (Short 1999), this being in agreement with the findings for the Mediterranean coast only.

On the other hand, these behaviours are strongly influenced by the sediment source areas, where the Atlantic coast of Cadiz is fed by the Guadalquivir and Guadalete rivers. These river basins are highly regulated and drain fine-grained marls, sandstones and shales. The grain size variability is not especially relevant in Masselink and Short (1993) classification, being considered less significant that H_b variations. Nevertheless, in areas of low energy where H_b variations are small, spatial and temporal changes in grain size become more relevant (Benavente et al. 2015b).

In this regard, the proximity to the Strait of Gibraltar implies an increased distance from main river sediment supply points. Therefore, beach sediments become generally coarser, coming from smaller river catchments with immature sediment, as well as from the erosion of coastal headlands and from offshore marine sources. The same occurs on the Mediterranean coast, where there are no important rivers to directly influence sediment supply beyond that of beaches located right at the river mouth.

The presence of typically reflective morphologies, such as the low tide terrace, or clearly reflective profiles, such as those near the Strait of Gibraltar or those of the Mediterranean coast, would be explained by the easier water percolation and drainage that occur in the intertidal zone due to the coarser sediment (Masselink and Short 1993; Turner 1993, 1995). These morphological types would therefore be highly controlled by the type of sediment, although agreement between the two variables (grain size and beach slope) is not perfect, due to the variability of the hydrodynamic conditions (waves and tides). In any case, it is clear that the type of sediment available, which in turn depends on the type and proximity of sediment source, controls the final beach morphology.

Regarding the relationship between morphodynamic characterization and medium-term evolution, according to Wright et al. (1985, 1987), profiles with an erosive trend should be more reflective. However, this is only seen in profile 29 (Retín) (Table 14.1), due to winter months leaving the profile eroded and composed of much coarser sediments. Conversely, profiles such as 11 (Regla), 14 (Rota Norte) or 17 (Vistahermosa V5) (Table 14.1), which are located at the end of a morphological cell (Anfuso et al. 2008) and hence show accretion trends, nonetheless present more "reflective" trends than adjacent areas. For such scenarios low energy prevents sediment-accumulating profiles from exporting sediment seawards, thereby maintaining a dissipative profile (Wright et al. 1979), with sand stored on the upper parts of the profile.

Nevertheless, for the Gulf of Cadiz coast it has been demonstrated how the existing erosive trend results in beaches with typically dissipative morphology and behaviour that have little capacity for change (Benavente et al. 2002a, b). Consequently, the response is the opposite of what could be expected, that is, the profiles with long-term erosive trends usually have a dissipative morphology that does not change with the modifications of the incident energy.

In spite of the poor variation in values obtained for the classification of Masselink and Short (1993), morphologies presented variability. This is clearly shown by the behaviour of the Surf Scaling and Surf Similarity parameters (Table 14.1); these generally reveal an increase in "reflectivity" towards the Strait of Gibraltar, although modulated by the boundary conditions mentioned above (such as the presence of small river mouths).

14.4 Recent Trends and Current Behaviour

14.4.1 Recent Evolution Trends

Beaches in Cadiz coast represent a major socio-economic resource and they provide significant ecosystem services. For this reason, the analysis of shoreline trends is of prime interest from both the scientific and applied points of view.

According to several authors (Anfuso et al. 2007; Martínez et al. 2001; Del Río et al. 2013, 2015; Puig et al. 2014; Benavente et al. 2015a, among others) and based on the analysis of aerial photographs and orthophotos dating back to 1956, most part of Cadiz coast has shown a relatively stable behaviour over the last decades, although there are numerous sectors with a clearly erosive trend. Most of the eroding beaches are located in the Northern half of the province, while stable or even accreting trends prevail in the Southern coast (Fig. 14.4). The great spatial and temporal variability of shoreline changes observed in the study zone is related to the aforementioned heterogeneity of the coast, as well as to the diversity of factors contributing to erosion processes on each coastal sector.

The Northermost part of Cadiz coast, between Sanlúcar de Barrameda and Rota, is the one most severely affected by coastal erosion, and most beaches in this area show a clear retreating trend. Shoreline recession is continuous and especially significant along the area between Aguadulce beach (Chipiona) and Punta Candor (Rota) (Fig. 14.4), with a mean erosion rate over the last decades of 0.7 m/yr that reaches up to 1.6 m/yr at some points. Beach and dunes at Punta Candor have been

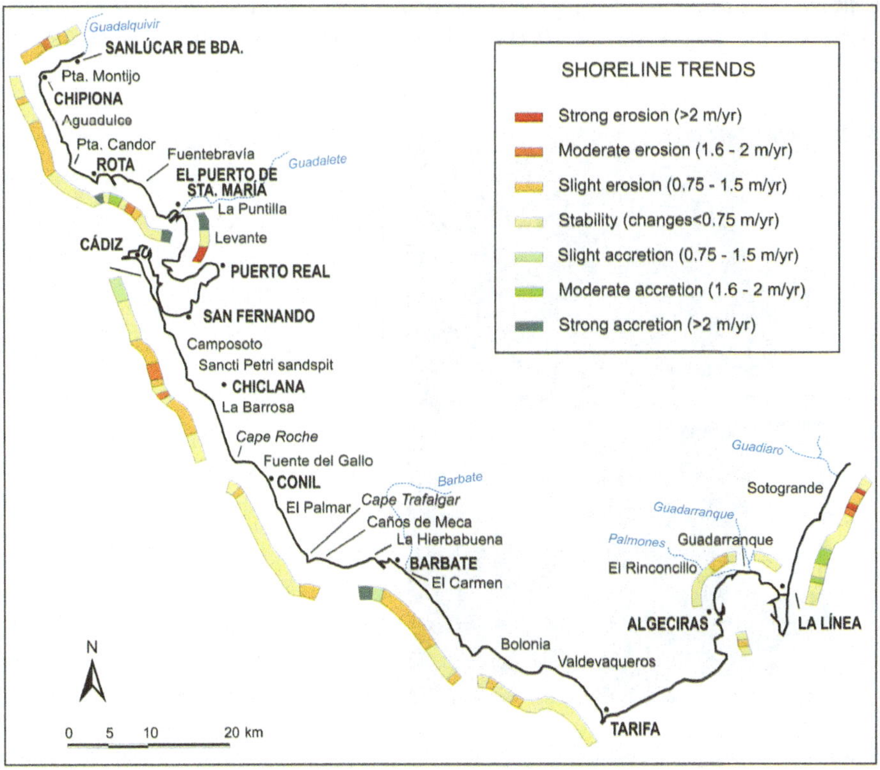

Fig. 14.4 Shoreline changes recorded over the last 60 years (1956–2016) in beaches and dunes along the coast of Cadiz

retreating at a roughly constant rate since the 1950s (Fig. 14.5a), mainly due to nearshore bathymetry (i.e. subtidal rocky shore platforms) and coastal orientation triggering the continuous action of erosion processes in this area (Anfuso et al. 2007; Del Río et al. 2013). Also the coast around Punta Montijo shows particularly severe recession, with an average retreat rate of 1.4 m/yr.

Around the Bay of Cadiz there are also some areas where significant beach erosion has been observed in the last decades. For instance, in Fuentebravía beach (El Puerto de Santa María) (Fig. 14.4) a continuous recession has been recorded over the period 1956–2008, despite repeated nourishment works having been carried out since the 1990s (Benavente et al. 2006a; Del Río et al. 2013). In this case, beach erosion can mainly be attributed to sediment deficit caused by interruption of longshore sediment transport by the groins of NATO Base at Rota, located immediately updrift Fuentebravía beach, which have moreover led to the need for shoreline armouring, thus in turn exacerbating the erosion problem (Cooper et al. 2009). A few kilometres South, the most remarkable erosional hotspot in Cadiz coast can be found at Punta de los Saboneses, the southernmost end of Levante beach (Fig. 14.4) in Valdelagrana spit barrier (El Puerto de Santa María). Here average beach and dune retreat rate for the last decades was −7.7 m/yr, with extreme rates of up to −11 m/yr at some points (Del Río et al. 2015a) (Fig. 14.5b), involving the loss of nearly 60 Ha of beach, dune and salt marsh areas along a

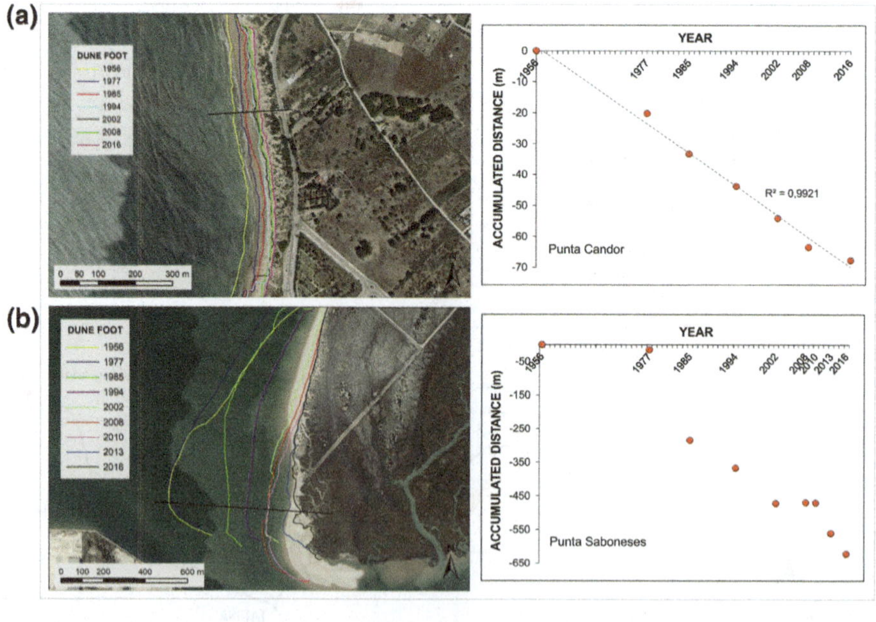

Fig. 14.5 Examples of contrasting temporal distribution of shoreline erosion; black lines correspond to transects represented in the graphs on the right (Background: 2016 orthophotographs ©PNOA). **a** Constant erosion rates at Punta Candor. **b** Time-changing erosion rates at southern Levante beach

1.2 km-long coastal trait. Such extreme erosion is mainly a consequence of the different phases of construction and lengthening of the jetties at Guadalete river mouth (located immediately updrift of Valdelagrana spit barrier), which since the 1970s disrupted the log-spiral equilibrium planform of the system by shifting the upcoast control point (Martínez et al. 2001); for this reason, erosion rates in this area have changed accordingly over time (Fig. 14.5b). In fact, the central sector of the spit has remained relatively stable in the last decades, while the northern sector has shown an accreting trend corresponding to the aforementioned changes in the equilibrium planform shape. Immediately updrift of the jetties, La Puntilla beach experienced massive accumulation of sand and shoreline advance at an average rate of 4.2 m/yr (Fig. 14.4), constituting the most rapidly accreting area along the whole Cadiz coast (Del Río et al. 2013). However, the beach has now reached equilibrium and shoreline position has remained stable over the last 10 years.

In the outer sector of the Bay of Cadiz there are both eroding and stable or accreting beaches. General shoreline stability has prevailed over the last decades between Cadiz urban beach and the north of Sancti Petri sandspit (San Fernando), although artificial beach nourishments in the former could be masking its naturally erosive trend (Del Río et al. 2013) and contribute to prolonged stability in downdrift beaches (Lentz and Hapke 2011). The most significant retreat has been recorded along Sancti Petri sandspit, with a mean recession rate of nearly 1 m/yr in Camposoto beach that reaches 1.4 m/yr at some points (Puig et al. 2014) (Fig. 14.4). One of the reasons for this erosion is nearshore morphology, as the presence of a discontinuous reef could trigger wave energy concentration in this area (Puig et al. 2016).

Further South, prevailing erosive trends appear in northern La Barrosa beach (Chiclana), although recent artificial replenishments have been masking this natural trend. Part of this recession can be attributed to urban development on the back-beach, as dune destruction in the 1970s and 1980s eliminated sediment buffer and altered cross-shore sediment budget (Del Río et al. 2013). Beaches between the southern sector of La Barrosa and Cape Roche have shown stability over the last decades, as recorded also in most beaches around Conil de la Frontera. Nevertheless, slight erosion appears at Fuente del Gallo beach and southern El Palmar beach, with average recession rates around 0.5 m/yr (Del Río et al. 2013) (Fig. 14.4).

South of Cape Trafalgar, shoreline trends are quite irregular, partly due to the compartmentalisation of the coast into embayments separated by headlands with distinctive behaviour. The most severely eroding area is Caños de Meca, with an average beach and dune retreat rate of 1 m/yr that reaches 1.8 m/yr at some points (Fig. 14.4). This can be partly attributed to the presence and shape of submarine reefs offshore Cape Trafalgar, which partially block longshore drift, triggering sediment deficit at Caños de Meca (Del Río et al. 2013). Furthermore, backbeach artificialisation at the Eastern end of the beach led to the destruction of a wide dune system in the 1970s, thus increasing vulnerability to erosion, and the rip-rap revetment built to protect the seafront houses has made matters worse, as it increases wave reflection and prevents cross-shore adaptation of beach profile to wave conditions (Trenhaile 1997). Significant shoreline retreat of up to −1.1 m/yr occurs also at the eastern end of El Carmen beach (Barbate), which is located on Barbate sandspit; here beach rotation generates accretion in the western zone and erosion at the distal end of the spit. This pattern could be related to sediment deficit due to massive sand accumulation updrift, in La Hierbabuena beach, where shoreline advance has been continuous over the last decades at an average rate of 2.5 m/yr (Del Río et al. 2013) (Fig. 14.4).

Beaches in the southernmost sector of the Atlantic coast can be considered relatively stable, except for some small areas (Fig. 14.4). The central part of Bolonia beach has eroded at a mean rate of 0.9 m/yr, while beach and dune recession at the western end of Valdevaqueros embayment is over 1 m/yr but on a very local scale, related to exposure to energetic Easterly waves (Del Río et al. 2013).

Recent shoreline evolution in the Bay of Algeciras has been mainly linked to the extensive industrial and harbour development occurred in this area from the 1960s onwards, with the building of numerous infrastructure on the coastline (jetties, groins, seawalls, land reclamation, etc.) (Manno et al. 2016). The most significant changes have been recorded in El Rinconcillo beach (Algeciras). The southern sector of the beach has experienced accretion related to the shelter provided by the massive port of Algeciras, while beach and dunes at its northern end (Palmones sand spit) have retreated at an average rate close to −0.5 m/yr (Montes et al. 2016) (Fig. 14.4), related to changes in longshore drift due to the aforementioned coastal infrastructure and to the regulation of Palmones river (Vallejo et al. 2000). The same reasons led to significant erosion on beaches located in the north and northeastern sector of the Bay (such as Guadarranque and Puente Mayorga beaches) between the 1970s and 2000, but at present they have been stabilized by engineering interventions, and their sheltered location at the bottom of the Bay prevents further erosion.

Finally, most beaches in the Mediterranean sector of Cadiz coast show a prevailing stability over the last decades. However, there is a remarkable erosional hotspot around the mouth of Guadiaro River (San Roque) (Fig. 14.4), with an average recession rate of −3.1 m/yr at Sotogrande beach (Del Río et al. 2015b). At certain points in this sector, shoreline retreat since the mid-1970s reaches 160 m, probably due to a great extent to the building of Sotogrande marina. Here the disappearance of the beach in front of a developed sector has recently led to the construction of a rip-rap revetment and two groins in order to prevent damage to the high-income houses at the seafront.

Besides the aforementioned factors responsible for recent shoreline evolution on each sector of Cadiz coast, it is important to note that the northern and central areas of the province are greatly affected by the regulation of Guadalquivir, Guadalete and Barbate river basins, where a number of dams were built specially between the 1960s and 1970s. Fluvial sediments get trapped in the reservoirs, hence causing sediment deficit in the coastal zone and shoreline erosion (Komar 2000). The southernmost beaches in the Atlantic coast are less influenced by this impact, due to their distance to the above rivers, the particular hydrodynamic regime in the area and the lower intensity and sediment load of longshore drift in this area (Del Río et al. 2013). Nevertheless, river regulation of minor basins (Palmones, Guadarranque, Guadiaro) also affects sediment budget of beaches within the Bay of Algeciras and in the Mediterranean coastal sector.

14.4.2 *Present-Day Coastal Hazards*

As stated in the previous section, along the coast of Cadiz there are several coastal hazard hotspots, mostly related to extreme meteorological events and short-term trends, namely coastal erosion and flooding. The frequency and severity of the risks derived from these hazards have increased over the last decades, mainly due to the increased anthropogenic pressure and urban development on the coast (Benavente et al. 2015b).

Important beach erosion problems appear in the NW sector of the study zone, around Chipiona and Rota. Beaches in this area are backed by low, soft cliffs, thus preventing flood risks, but short- and medium-term erosion have led to frequent investments in beach nourishment (Gomez-Pina et al. 2006). In this respect, during mild winter seasons beaches can even show accretionary trends, while high-energy storm periods trigger strong erosive processes that can cause significant damage (Anfuso and Del Río 2003; Anfuso and Benavente 2006). This has been mainly observed during winters characterized by very negative values of the NAO index, such as 1995–1996 and 2009–2010. These results are in accordance with the modulation of the storm wave energy and associated surge levels by the NAO index as stated by Plomaritis et al. (2015). In general, the medium-term erosive trend in this area is related to the decrease in sediment supply from Guadalquivir river (Anfuso et al. 1999), except in beaches located North (i.e. updrift) of jetties and groins.

Towards the South, beaches in the northern sector of El Puerto de Santa María are also remarkable from the perspective of coastal hazards. As mentioned above, Fuentebravía is a dissipative beach without significant morphodynamic changes due to its clearly erosive trend (Benavente et al. 2002a) both in the short and medium terms, which has led to numerous interventions (Muñoz-Pérez et al. 2001). To the South, Vistahermosa beach shows a seasonal behaviour, so there can be local erosion problems due to high-energy storm events but beach recovery generally occurs in the short-medium term scale (Benavente et al. 2000; Puig et al. 2016). This can be attributed to the location of the beach at the end of a sediment transport cell, thus providing an adequate sediment supply (Benavente and Gracia 2003; Anfuso et al. 2008).

The central and southern sectors of the Bay of Cadiz constitute a low-lying environment where besides coastal erosion, storm-induced coastal flooding can also be highlighted as a significant hazard. The high population of this area contributes to increase the risk, as occurs in the densely developed seafront of Cadiz city, where urban beaches as La Victoria frequently experience damage to infrastructure as a consequence of erosion and flooding generated by winter storms, particularly when storm events coincide with spring tides (Fig. 14.6a–c) (Del Río et al. 2012).

The nearby Valdelagrana spit barrier presents an extreme retreating trend in its southern area (see previous section), which is also observed over the short term, with a slow evolution typical of dissipative profiles with limited ability to change (Benavente et al. 2002a; Del Río et al. 2015a); here most erosion occurs during high-energy storms (Rangel-Buitrago and Anfuso 2011a, b). Regarding flood hazard, Valdelagrana is also one of the main hotspots in the study area due to its low elevation (less than 5 m above MSL) and to the aforementioned erosion experienced by the low and discontinuous foredunes separating the beach from the salt marsh area. This way, flood episodes have been described in this zone, as well as formation and reactivation of washover deposits during high-energy events (Fig. 14.7) with return periods lower than 10 years, despite the dissipative morphology of the beach and its relatively sheltered location (Benavente et al. 2006b). Loss of habitats occurs as a consequence of this, but economic damage is limited due to the natural character of the central and southern sectors of the spit barrier. In fact, the developed northern sector is protected from wave action by the jetty at Guadalete River mouth, which also contributes to its accretionary trend (see Fig. 14.4).

Fig. 14.6 Examples of storm impacts at urban beaches along Cadiz coast. **a** Damage to coastal protections at La Victoria beach (Cadiz) in April 2015. **b** Seafront overtopping by storm surge at La Victoria beach (Cadiz) in January 2010. Note that dry beach width in this area is commonly 40 m. **c** Flooding of beach facilities at Cortadura beach (Cadiz) in March 2010. **d** Damage to seafront house at El Rinconcillo beach (Algeciras) in March 2017

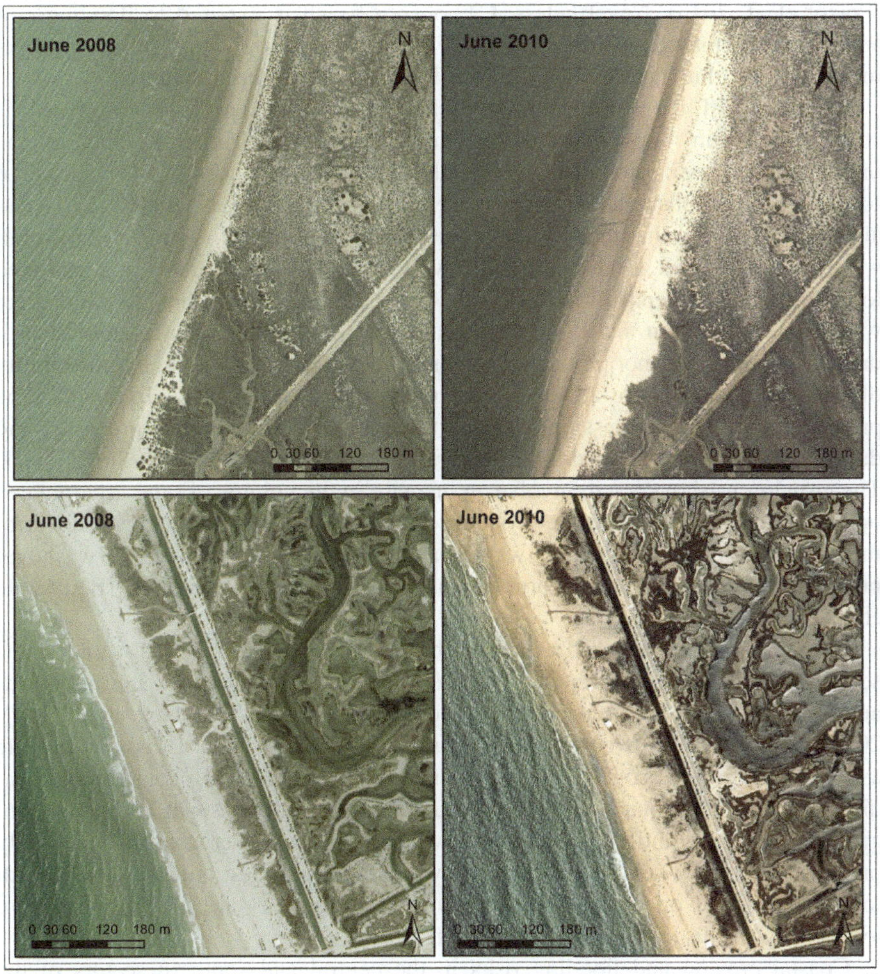

Fig. 14.7 Development of washover deposits related to the 2009/2010 winter storm season in the southern area of Valdelagrana sandspit (upper images) and in the central area of Camposoto beach in Sancti Petri sandspit (lower images). Background orthophotographs ©PNOA

Similar circumstances can be found at Camposoto beach in Sancti Petri sandspit, an exposed, seminatural area in the southern limit of the Bay of Cadiz. As was previously mentioned, this is an intermediate-dissipative beach with a clearly erosive trend in the medium term. Short-term behaviour is also characterised by significant erosion episodes that have triggered the need for several beach nourishments (Muñoz-Pérez et al. 2001, 2014) and foredune recovery interventions. Despite beach and dune system having been severely eroded during high-energy storms over the last decades (Benavente et al. 2002b, 2013), beach recovery occurs during fair weather periods, with an alternation between summer and winter profiles

(Poizot et al. 2013). Nevertheless, Camposoto beach is exposed to flooding episodes that generate overwash processes (Fig. 14.7) and reactivate existing washover morphologies, thus determining a general weakening of the foredune.

In this regard, the main cause of washover generation in Camposoto beach is structural erosion (supported by signs indicating coastal recession, such as the outcropping of fossil salt marsh layers in the foreshore), followed by the presence of foredune depressions related to high wave conditions (Benavente et al. 2013). It has also been observed that these features are related to a lack of sediment during fair weather periods that hinders dune recovery, thus related to the aforementioned long-term erosion trend. In fact, calculation of minimum wave heights needed for washover reactivation and flooding reveals that the storm events capable of producing this effect are not particularly intense, thus indicating the high vulnerability of Camposoto beach to storm impacts (Benavente et al. 2013). It can thus be predicted that system recession will lead to an increase in overwash processes and dune breaching over the medium term, with the formation of wide washover areas. The growing washovers will involve an increase in the volume of sand accumulated on them, thus generating further system retreat. This way, a continuous rollover of the system is to be expected, as well as an increase in the relevance of washover processes even during modal storms mainly due to the lack of sediment (Benavente et al. 2013).

A few kilometres South, La Barrosa beach (see Fig. 14.4) has also undergone several nourishment interventions related to sediment deficit and the enormous anthropogenic pressure generated by tourist activity (Muñoz-Pérez et al. 2001, 2014). This beach showed a clearly seasonal behaviour, with a dissipative state associated to erosive conditions and an intermediate one during constructive conditions. This behaviour could reflect beach stability but seasonal variations progressively reduced until a certain time in which a dissipative erosive state definitely prevails, when new small replenishment works are needed in order to avoid beach loss and damage on coastal infrastructure (Benavente et al. 2006a). Backshore topography in this area prevents coastal flooding.

Coastal erosion and flooding risks in the central and southern sectors of the Atlantic side of Cadiz are reduced, due to the lower dependence on fluvial sediment supply and the limited urban pressure on the coast. Nonetheless, local erosion problems appear at points such as Caños de Meca, where beach loss due to dune destruction for building purposes (see previous section) leads to the need for periodic artificial nourishments.

Anthropogenic pressure on the coast is higher in the Mediterranean side of the study area, from the industrial, maritime transport and tourist sectors. The Bay of Algeciras has been deeply modified over the last decades, and present-day erosion and flood risks are higher in the most developed zones. In this respect, strong Levante storms (see Sect. 14.1) are responsible for beach erosion in this area, but flood risk is lower because Levante winds are mostly associated to high atmospheric pressures, so storm surge is not particularly important in this case (Del Río et al. 2017). Levante storms cause significant damage to coastal infrastructure, both in the western coast of the Bay and in the exposed Mediterranean beaches located

North of Gibraltar. For instance, between February and April 2017 a series of high-energy Levante events caused severe erosion and damage to diverse infrastructure located along this coast. Several houses were seriously damaged at El Rinconcillo beach, water pipelines and sewage drains were left unprotected on the beach and affected by wave action (Fig. 14.6d), and 3 m high escarpments were cut on Palmones dunes. In the exposed sector, at La Atunara and Sotogrande beaches some sections of the seafront and rip-rap revetments were damaged and had to be replaced.

To sum up, the most significant risks along Cadiz coast occur in the areas with the highest anthropogenic pressure (namely the Bay of Cadiz and the Bay of Algeciras). Coastal erosion impacts are particularly important in those sectors where sediment supply is linked to river mouths, while coastal flooding events appear mostly in low-lying areas exposed to wave action.

14.5 Concluding Remarks and Future Perspectives

The coast of Cadiz province constitutes a complex system with a wide variety of hydrodynamic and geological-geomorphological conditions. Besides, the combination of large natural systems and high-value socioeconomic activity at the coast makes it a unique environment for multidisciplinary research on coastal processes and risks. In this respect, there are several topics in which research questions still remain open and are presently being investigated.

In terms of morphodynamic behaviour, the short- and medium-term evolution of the coast, as well as its short-term response to storm events, have been intensively studied as described in the previous sections. However, the interannual variability of shoreline behaviour is still focus of research due to its interest for medium- and long-term coastal stability. For instance, the post-storm recovery (relaxation time) after extreme events and the sediment mobility over the inner continental shelf are of large importance for the estimation of the resilience of a coastal system against changes in sediment supply or wave regime (Vousdoukas et al. 2012; Houser et al. 2015).

Regarding the role of sediment supply in beach stability, it is clear that many beaches in the coast of Cadiz depend directly on fluvial sediment input, yet the alongshore distance from the river mouths over which this influence extends is not known. In this respect, the sediment deficit caused by river damming has been detected on beach volumetric evolution in some cases, but in other areas there are additional factors masking these effects (Del Río et al. 2013). One of these factors might be the presence of cliffs on soft materials backing the beach, and in fact the role played by cliff-beach interaction in the sedimentary equilibrium of sandy coasts is one of the main open research lines in this area (Rendón et al. 2009).

Concerning the impact of wave energy on coastal response, it has been clearly demonstrated that beaches in the study area show a seasonal behaviour directly linked to seasonal changes in wave regime (e.g. Anfuso and Gracia 2005;

Benavente et al. 2015b). In this regard, no significant patterns are observed in the NAO index over the last decades (Plomaritis et al. 2015), but in some exposed beaches the average erosion trends are increasing, as well as related phenomena such as overwash and rollover processes in sandy barriers. New methods are being applied for investigating these events in the study zone, including UAS surveying (Talavera et al. 2017), and the relationship between coastal changes and recently proposed atmospheric indices will be explored in this respect (Castelle et al. 2017).

The role of boundary conditions such as tidal range and geological framework, which greatly influence e.g. grain size, is of great importance in modulating beach behaviour in Cadiz coast (Benavente et al. 2015b). For instance, beach susceptibility to suffer damage by storms is partly determined by the ability of beaches to change their morphodynamic state through the year, as this involves the availability of enough sand to seasonally adapt their profiles to the new energetic conditions (Benavente et al. 2002a). However, no clear patterns have been found relating tidal range and beach stability in the study zone. In areas with submerged or intertidal rocky shore platforms, beaches tend to be protected from modal waves, but they become less seasonal (due to the geological control on beach profile changes) and this may involve a higher vulnerability against storm waves (Anfuso et al. 2008; Del Río et al. 2013).

As for the influence of human activity on the evolution and present state of Cadiz beaches, there are numerous examples of negative impacts of coastal development and defence structures, especially between the 1970s and the 2000s (e.g. Del Río et al. 2013; Benavente et al. 2015a). Nevertheless, some human interventions have also had positive impacts on beach stability, such as the dune preservation works extensively performed at many points along Cadiz coast, and the periodic nourishments carried out in many urban beaches, where they are particularly necessary due to the (human-induced) disappearance of the dune-beach exchange and the reduction in natural sediment supply (Del Río et al. 2013).

Finally, the coast of Cadiz represents an ideal area for the evaluation of the consequences of climate change, given the exposure to storm-related hazards and the environmental and socioeconomic value of the coastal zone. On a global scale, it has been stated that due to sea level rise the frequency of coastal flooding episodes will be doubled (Vitousek et al. 2017). On a regional scale, the assessment of climate-related coastal hazards such as erosion, overwash and flooding is currently being performed along the coast of Cadiz for present-day conditions and for various climate change scenarios, in order to evaluate current risks and also how these will be behave in the future under different sea level and storminess conditions. According to official estimates, wave height in the Gulf of Cadiz is predicted to decrease, thus involving a reduction in shoreline erosion by longshore transport and in dune recession (IH Cantabria 2014). However, other studies show a statistically significant increase in the number of storms (particularly the most severe ones) over the last century in the Gulf of Cadiz (Ribera et al. 2011). If this situation continues in the future, an increase in storm-induced coastal erosion should be expected. Moreover, sea level rise is expected to increase around 7 cm by 2040 (IH Cantabria 2014), which would also influence coastal erosion processes, particularly in the lowest elevation areas. In this context, future increase in population affected by coastal hazards along the study area triggers the need for investigating adequate prevention, adaptation and mitigation strategies in various sectors of the coastline.

References

Alonso, C., Gracia, F. J., Rodríguez-Polo, S., & Martín-Puertas, C. (2015). El registro de eventos energéticos marinos en la Bahía de Cadiz durante épocas históricas. In J. Rodríguez-Vidal (Ed.), *Eventos energéticos marinos históricos y ocupación costera en el Golfo de Cadiz*. Rev C & G 29 (1–2) (pp. 95–117).

Anfuso, G., Achab, M., Cultrone, G., & López-Aguayo, F. (1999). Utilidad de la distribución de minerales pesados y de los análisis granulométricos en el estudio de la dinámica litoral: aplicación al tramo costero entre Sanlúcar de Barrameda y Rota (Cadiz, suroeste de la península Ibérica). *Bol Inst Esp Oceanogr, 15*(1–4), 243–250.

Anfuso, G., & Benavente, J. (2006). Approximation to the classification and morphodynamic behaviour of the beaches between Chipiona and Rota (Cadiz, SW Spain). *Cien Mar, 32*(4), 605–616.

Anfuso, G., Benavente, J., Gracia, F. J., & Del Río, L. (2006). Morphodynamic characterization of Cadiz beaches (SW Spain). *Journal of Coastal Research, 48,* 8–15.

Anfuso, G., Benavente, J., Gracia, F. J., & Del Río, L. (2008). An approximation to short-term evolution and sediment transport pathways along the littoral of Cadiz Bay (SW Spain). *Environment Geology, 56,* 69–79.

Anfuso, G., & Del Río, L. (2003). Cuantificación de las variaciones volumétricas y evolución del litoral entre Chipiona y Rota (Cadiz) durante el periodo 1996–1998. *Rev C&G, 17*(3–4), 17–27.

Anfuso, G., Domínguez, L., & Gracia, F. J. (2007). Short and medium-term evolution of a coastal sector in Cadiz SW Spain. *CATENA, 70*(2), 229–242.

Anfuso, G., & Gracia, F. J. (2005). Morphodynamic characteristics and short term evolution of a coastal sector in SW Spain: Implications for coastal erosion management. *Journal of Coastal Research, 21*(6), 1139–1153.

Anfuso, G., Rangel-Buitrago, N., Cortés-Useche, C., Iglesias-Castillo, B., & Gracia, F. J. (2016). Characterization of storm events along the Gulf of Cadiz (eastern central Atlantic Ocean). *International Journal of Climatology, 36,* 3690–3707.

Battjes, J. A. (1974). Surf similarity. In *Proceedings 14th Coastal Engineering Conference* (pp. 446–480). ASCE.

Bello-Smith, A., Cerasuolo, G., Perales, J. A., & Anfuso, G. (2011). Environmental sensitivity maps: The north coast of Gibraltar Strait example. *Journal of Coastal Research, 64,* 875–879.

Benavente, J., Anfuso, G., Del Río, L., Gracia, F. J., & Reyes, J. L. (2006a). Evolutive trends of nourished beaches in SW Spain. *Journal of Coastal Research, 39,* 765–769.

Benavente, J., Del Río, L., Anfuso, G., Gracia, F. J., & Reyes, J. L. (2002a). Utility of morphodynamic characterization in the prediction of beach damage by storms. *Journal of Coastal Research, 36,* 56–64.

Benavente, J., Del Río, L., & Gracia, F. J. (2015a). Riesgos de erosión costera en el litoral de Cadiz: problemática actual y perspectivas futuras. In J. Rodríguez-Vidal & M. C. Núñez-Lozano (Eds.), *El litoral de Andalucía* (pp. 65–91). Norma y naturaleza: Serv Publ Universidad de Huelva.

Benavente, J., Del Río, L., Gracia, F. J., & Martínez-del-Pozo, J. A. (2006b). Coastal flooding hazard related to storms and coastal evolution in Valdelagrana spit (Cadiz Bay Natural Park, SW Spain). *Continental Shelf Research, 26*(9), 1061–1076.

Benavente, J., Del Río, L., Plomaritis, T. A., & Menapace, W. (2013). Impact of coastal storms in a sandy barrier (Sancti Petri, Spain). *Journal of Coastal Research, 65,* 666–671.

Benavente, J., & Gracia, F. J. (2003). Modelización del transporte sedimentario longitudinal en playas de poca energía. Playa de Vistahermosa (Bahía de Cadiz). *Geogaceta, 33,* 7–10.

Benavente, J., Gracia, F. J., Del Río, L., Anfuso, G., & Rodríguez-Ramírez, A. (2015b). Morphodynamic characterization of the Spanish beaches of the Gulf of Cadiz. *Boletin Geologico y Minero, 126*(2–3), 409–426.

Benavente, J., Gracia, F. J., & López-Aguayo, F. (2000). Empirical model of morphodynamic beachface behaviour for low-energy mesotidal environments. *Marine Geology, 167,* 375–390.

Benavente, J., Martínez-del-Pozo, J. A., Gracia, F. J., Reyes, J. L., & Del Río, L. (2002b). Procesos de desbordamiento en la flecha litoral de Sancti-Petri (Bahía de Cadiz): riesgos asociados. In E. Serrano & A. García de Celis (Eds.), *Estudios Recientes en Geomorfología* (pp. 23–32). Valladolid: Univ Valladolid.

Cantabria, I. H. (2014). *Website of the viewer of C3E Project (Climate Change in the Spanish Coast).* www.c3e.ihcantabria.com.

Castelle, B., Dodet, G., Masselink, G., & Scott, T. (2017). A new climate index controlling winter wave activity along the Atlantic coast of Europe: The West Europe pressure anomaly. *Geophysical Research Letters, 44,* 1384–1392.

Cooper, J. A. G., Anfuso, G., & Del Río, L. (2009). Bad beach management: European perspectives. *Geological Society of America Series, 460,* 167–179.

Dean, R. G. (1973). Heuristic models of sand transport in the surf zone. In *Proceedings of the Conference on Engineering Dynamics in the Surf Zone* (pp. 208–214). Sydney, NSW.

Del Río, L., Benavente, J., Gracia, F. J., Alonso, C., & Rodríguez-Polo, S. (2015a). Anthropogenic influence on spit dynamics at various timescales: Case study in the Bay of Cadiz (Spain). In G. Randazzo, D. W. T. Jackson & J. A. G. Cooper (Eds.), *Sand and gravel spits* (pp. 123–128). Springer.

Del Río, L., Benavente, J., Gracia, F. J., & Chica, J. A. (2015b). Riesgos de erosión en la costa de Cadiz: gestión actual y perspectivas futuras. *Geo-Temas, 15,* 149–152.

Del Río, L., Gracia, F. J., & Benavente, J. (2009). Mass movements and cliff retreat along the SW Spanish coast. *Journal of Coastal Research, 56,* 717–721.

Del Río, L., Gracia, F. J., & Benavente, J. (2013). Shoreline change patterns in sandy coasts: A case study in SW Spain. *Geomorphology, 196,* 252–266.

Del Río, L., Montes, J. B., Plomaritis, T. A., Martinelli, A. B. M., & Benavente, J. (2017). Aplicación de modelos empíricos para el análisis de la inundación por temporales en playas con diferentes regímenes de marea y oleaje. *Geo-Temas, 16,* 131–134.

Del Río, L., Plomaritis, T. A., Benavente, J., Valladares, M., & Ribera, P. (2012). Establishing storm thresholds for the Spanish Gulf of Cadiz coast. *Geomorphology, 143–144,* 13–23.

Dorman, C. E., Beardsley, R. C., & Limeburner, R. (1995). Winds in the strait of Gibraltar. *Quarterly Journal of the Royal Meteorological Society, 121,* 1903–1921.

García-Lafuente, J., Delgado, J., Criado-Aldeanueva, F., Bruno, M., Del Río, J., & Vargas, J. M. (2006). Water mass circulation on the continental shelf of the Gulf of Cadiz. *Deep Sea Research, 2*(53), 1182–1197.

Gómez Ponce, C., Borja, F., Lagóstena, L., López, J. J., Ruiz, J. A., & Borja, C. (1997). Primeras fases de la evolución de la flecha litoral de Valdela-grana (El Puerto de Santa María, Cadiz). Datos arqueológicos. In J. Ro-dríguez-Vidal (Ed.), *Cuaternario ibérico* (pp. 165–167). AEQUA: Huelva.

Gómez-Pina, G., Fages, L., Ramírez, J. L., Muñoz-Pérez, J. J., & Enríquez, J. (2006). A critical review of beach restoration projects in the northern coast of Cadiz (Spain) after thirteen years. *Coastal Engineering Procceding, 1*(30), 4167–4178.

Gourlay, M. R. (1968). *Beach and dune erosion tests.* Delft Hydraulics Lab Report M935/M936.

Goy, J. L., Zazo, C., Silva, P. G., Lario, J., Bardají, T., & Somoza, L. (1995). *Evaluación geomorfológica del comportamiento neotectónico del Estrecho de Gibraltar (Zona Norte) durante el Cuaternario* (pp. 51–69). SECEG, Madrid: IV Coloquio Internacional sobre el enlace fijo del Estrecho de Gibraltar.

Gracia, F. J., Alonso, C., Martínez-del-Pozo, J. A., Del Río, L., Anfuso, G., & Benavente, J. (2004). Evolución geomorfológica holocena e histórica de la Ensenada de Los Lances (Tarifa, Cadiz). In R. Blanco, J. López, & A. Pérez-Alberti (Eds.), *Procesos geomorfológicos y evolución costera* (pp. 255–269). Santiago de Compostela: Univ.

Gracia, F. J., Del Río, L., Alonso, C., Benavente, J., & Anfuso, G. (2006). Historical evolution and present state of the coastal dune systems in the Atlantic coast of Cadiz (SW Spain): Palaeoclimatic and environmental implications. *Journal Coastal Research, 48,* 55–63.

Gracia, F. J., Rodríguez-Vidal, J., Belluomini, G., Cáceres, L. M., Benavente, J., & Alonso, C. (2008). Diapiric uplift of an MIS 3 marine deposit in SW Spain. Implications in Late Pleistocene sea level reconstruction and palaeogeography of the Strait of Gibraltar. *Quaternary Science Reviews, 27*(23–24), 2219–2231.

Gracia, F. J., Rodríguez-Vidal, J., Benavente, J., Cáceres, L., & López-Aguayo, F. (1999). Tectónica cuaternaria en la Bahía de Cadiz. In L. Pallí & C. Roqué (Eds.), *Avances en el estudio del Cuaternario español* (pp. 67–74). University of Girona.

Guza, R. T., & Inman, D. L. (1975). Edge waves and beach cusps. *Journal of Geophysical Research, 80,* 2997–3012.

Houser, C., Wernette, P., Rentschlar, E., Jones, H., Hammond, B., & Trimble, S. (2015). Post-storm beach and dune recovery: Implications for barrier island resilience. *Geomorphology, 234,* 54–63.

Komar, P. D. (2000). Coastal erosion—Underlying factors and human impacts. *Shore and Beach, 68*(1), 3–16.

Lentz, E. E., & Hapke, C. J. (2011). Geologic framework influences on the geomorphology of an anthropogenically modified barrier island: Assessment of dune/beach changes at Fire Island, New York. *Geomorphology, 126,* 82–96.

Manno, G., Anfuso, G., Messina, E., Williams, A. T., Suffo, M., & Liguori, V. (2016). Decadal evolution of coastline armouring along the Mediterranean Andalusia littoral (South of Spain). *Ocean Coast Manage, 124,* 84–99.

Martínez, J. A., Anfuso, G., & Gracia, F. J. (2001). Recent evolution of a tidal delta in Cadiz Bay (SW Spain) due to human interventions. In E. Özhan (Ed.), *Proceedings of MEDCOAST'01* (Vol. 3, pp. 1425–1433). Hammamet, Tunisia.

Masselink, G., & Short, A. D. (1993). The effect of tide range on beach morphodynamics and morphology: A conceptual beach model. *Journal Coastal Research, 9*(3), 785–800.

Montes, J., Benavente, J., & Del Río, L. (2016). Caracterización preliminar y evolución reciente de las playas de la Bahía de Algeciras (Cadiz). In J. J. Durán, M. Montes, A. Robador, & A. Salazar (Eds.), *Comprendiendo el relieve: del pasado al futuro* (pp. 539–545). Madrid: IGME.

Montes, J., Del Río, L., & Benavente, J. (2017). Caracterización morfodinámica y cambios recientes en las playas mediterráneas de la costa de Cadiz. *Geo-Temas, 16,* 63–66.

Muñoz-Pérez, J. J., de San, López, Román, B., Gutiérrez-Mas, J. M., Moreno, L., & Cuena, G. J. (2001). Cost of beach maintenance in the Gulf of Cadiz (SW Spain). *Coast English, 42*(2), 143–153.

Muñoz-Pérez, J. J., Román-Sierra, J., Navarro-Pons, M., Neves, M. G., & Del Campo, J. M. (2014), Comments on "Confirmation of beach accretion by grain-size trend analysis: Camposoto beach, Cadiz, SW Spain" by E Poizot et al (2013) Geo-Marine Letters 33(4). *Geo-Marine Letters, 34*(1): 75–78.

Plomaritis, T. A., Benavente, J., Laiz, I., & Del Río, L. (2015). Variability in storm climate along the Gulf of Cadiz: The role of large scale atmospheric forcing and implications to coastal hazards. *Climate Dynamics, 45,* 2499–2514.

Poizot, E., Anfuso, G., Méar, Y., & Bellido, C. (2013). Confirmation of beach accretion by grain-size trend analysis: Camposoto beach, Cadiz, SW Spain. *Geo-Mar Letters, 33*(4), 263–272.

Puig, M., Del Río, L., Plomaritis, T. A., & Benavente, J. (2014). Influence of storms on coastal retreat in SW Spain. *Journal Coastal Research, 70,* 193–198.

Puig, M., Del Río, L., Plomaritis, T. A., & Benavente, J. (2016). Contribution of storms to shoreline changes in mesotidal dissipative beaches: case study in the Gulf of Cadiz (SW Spain). *Natural Hazards and Earth System Sciences, 16,* 2543–2557.

Rangel-Buitrago, N., & Anfuso, G. (2011a). Coastal storm characterization and morphological impacts on sandy coasts. *Earth Surface Processes and Landforms, 36*(15), 1997–2010.

Rangel-Buitrago, N., & Anfuso, G. (2011b). Morphological changes at Levante Beach (Cadiz, SW Spain) associated with storm events during the 2009–2010 winter season. *Journal of Coastal Research, 64,* 1886–1890.

Rendón, J. J., Gracia, F. J., & Del Río, L. (2009). Los procesos de erosión hídrica en el retroceso erosivo de acantilados sobre rocas blandas en la provincia de Cadiz. In A. Romero, F. Belmonte, F. Alonso, & F. López (Eds.), *Advances in Studies on Desertification* (pp. 387–390). Murcia: Univ Murcia.

Ribera, P., Gallego, D., Peña-Ortiz, C., Del Río, L., Plomaritis, T. A., & Benavente, J. (2011). Reconstruction of Atlantic historical winter coastal storms in the Spanish coasts of the Gulf of Cadiz, 1929–2005. *Natural Hazards and Earth System Sciences, 11,* 1–8.

Rodríguez-Polo, S., Gracia, F. J., Benavente, J., & Del Río, L. (2009). Geometry and recent evolution of the Holocene beach ridges of the Valdelagrana littoral spit (Cadiz Bay, SW Spain). *Journal of Coastal Research, 56,* 20–23.

Short, A. D. (1999). Wave-dominated beaches. In A. D. Short (Ed.), *Beach and Shoreface Morphodynamics* (pp. 173–203). Chichester: John Wiley and Sons.

Short, A. D. (2006). Australian beach systems—Nature and distribution. *Journal of Coastal Research, 22*(1), 11–27.

Talavera, L., Del Río, L., Benavente, J., Barbero, L., & López-Ramírez, J. A. (2017). Eficiencia del uso de drones en la evaluación de cambios costeros: primeros resultados en la playa de Camposoto (San Fernando, SO de España). *Geo-Temas, 16,* 127–130.

Trenhaile, A. S. (1997). *Coastal dynamics and landforms* (p. 382). Oxford: Clarendon Press.

Turner, I. (1993). Water table outcropping on macro-tidal beaches: A simulation model. *Marine Geology, 115,* 227–238.

Turner, I. (1995). Simulating transported the influence of groundwater seepage on sediment by the sweep of the swash zone across macro-tidal beaches. *Marine Geology, 125,* 153–174.

Vallejo, I., Sánchez, E., & Ojeda, J. (2000). *Transformaciones territoriales y de la dinámica litoral en la Bahía de Algeciras: Aplicaciones de los SIG y la Teledetección* (pp. 148–163). Tecnologías geográficas para el desarrollo sostenible: Univ Alcalá.

Vitousek, S., Barnard, P. L., Fletcher, C. H., Frazer, N., Erikson, L., & Storlazzi, C. D. (2017). Doubling of coastal flooding frequency within decades due to sea-level rise. *Scientific Reports, 7,* 1399.

Vousdoukas, M. I., Almeida, L. P. M., & Ferreira, Ó. (2012). Beach erosion and recovery during consecutive storms at a steep-sloping, meso-tidal beach. *Earth Surface Processes and Landforms, 37,* 583–593.

Wright, L. D., Chappell, J., Thom, B. G., Bradshow, M. P., & Cowell, P. (1979). Morphodynamics of reflective and dissipative beach and inshore systems: Southeastern Australia. *Marine Geology, 32,* 105–140.

Wright, L. D., & Short, A. D. (1984). Morphodynamic variability of surf zones and beaches: A synthesis. *Marine Geology, 56,* 93–118.

Wright, L. D., Short, A. D., & Green, M. O. (1985). Short term changes in the morphodynamic states of beaches and surf zones: An empirical predictive model. *Marine Geology, 62,* 339–364.

Wright, L. D., Short, A. D., Boon, J. D., Hayden, B., Kimball, S., & List, J. H. (1987). The morphodynamic effects of incident wave groupiness and tide range on an energetic beach. *Marine Geology, 74*(1–2), 1–20.

Zazo, C. (2006). Cambio climático y nivel del mar: la Península Ibérica en el contexto global. *Rev C&G, 20*(3–4), 115–130.

Zazo, C., Dabrio, C. J., Goy, J. L., Lario, J., Cabero, A., Silva, P. G., et al. (2008). The coastal archives of the last 15 ka in the Atlantic-Mediterranean Spanish linkage area: Sea level and climate changes. *Quaternary International, 181,* 72–87.

Chapter 15
Beaches of Huelva

Juan A. Morales, Antonio Rodríguez-Ramírez and Mouncef Sedrati

15.1 Introduction

The Huelva Coast forms the littoral sector of the SW Iberian Peninsula which extends from the mouth of the Guadiana River, on the border with Portugal, to the mouth of the Guadalquivir River. From a physiographic point of view, the Huelva Coast can be considered a linear, low and sandy coast formed by long beaches, although it could be divided in different sectors according to its topographic configuration and distribution of sedimentary environments. Using these criteria, two sectors are distinguished. The first, in form of beaches developed in front of ancient barrier islands systems, the second constituted by beaches attached on the front of a Pleistocene palaeo-cliff system (Fig. 15.1). All this extension of beaches is only interrupted by three estuarine mouths and two tidal inlets. From a dynamic point of view, the coastline is divided in 5 cells, separated by these channels. The total longitude of continuous beaches is 110 km, corresponding 74 km to old barrier islands or spits and 36 km to cliff fronts.

J. A. Morales (✉) · A. Rodríguez-Ramírez
Department of Earth Sciences, University of Huelva, Huelva, Spain
e-mail: jmorales@uhu.es

A. Rodríguez-Ramírez
e-mail: arodri@uhu.es

M. Sedrati
Géosciences Océan, UMR CNRS 6538, Université Bretagne Sud, Campus de Tohannic, 56017 Vannes, France
e-mail: mouncef.sedrati@univ-ubs.fr

J. A. Morales (ed.), *The Spanish Coastal Systems*,
https://doi.org/10.1007/978-3-319-93169-2_15

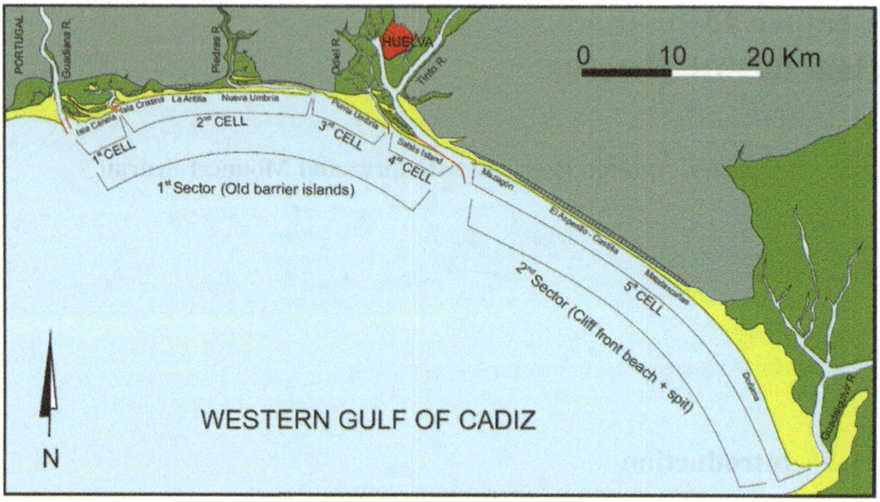

Fig. 15.1 Huelva beaches, indicating the distinguished coastal dynamic cells

15.2 Dynamic Factors

15.2.1 Winds and Waves

The dominant wind blows from the southwest (22.5% of time); northwest (18.5%), southeast (14.0%) and northeast (12.0%) winds are also significant (Borrego 1992; Morales et al. 2004). The provenance of these winds is directly related with the position of high and low pressure centers in the north-Atlantic area.

Taking into account the direction of this coastal segment, the winds from southwest and southeast are responsible for most of the waves, so the dominant wave provenance is from the southwest. Waves have a significant wave height ($H_{1/3}$) minor than 0.5 m for 75% of the time, approaching the coast from N 35–40°E (MOPU 1991; Borrego 1992; Morales et al. 2004). Waves higher than 1.5 m are associated with storms from the Atlantic Ocean or the Gibraltar Strait. The influence of these storms is more important during the winter. A mean of 15 storms a year can affect this coast (7.9% of total time), with mean significant height of 3.80 m and extreme waves reaching 6 m (Morales et al. 2014a).

The coastline orientation, oblique to the main wave direction, induces a littoral drift from west to east. The potential transportation values of this drift diminish from 3.0×10^5 Hm^3/year in the western sector with a W–E direction (CEEPYC 1990) to 1.9×10^5 m^3/year in the eastern area with a WNW-ESE direction (Cuena 1991). Its continuity is interrupted by the refraction of the waves, which produces an inversion in the direction of this on the eastern side of the mouths of the main rivers (Morales et al. 2001, 2004). This fact gives rise to compartmentalisation of the sand transport along the coast in the form of independent cells bounded by estuary mouths (Fig. 15.1).

15.2.2 Tides

The tide in the Huelva Coast presents a progressive displacement in an east-to-west sense, from the Gibraltar Strait to the southern Portuguese coast. The tidal regime has a mesotidal character including a slight semidiurnal inequality and cycles of longer amplitude (Borrego 1992). The mean tidal range in this sector is 2.10 m, ranging from 0.70 m in an Extreme Equinox Neap tide to 3.85 m in an Extreme Equinox Spring tide. The magnitude of associated tidal currents in the open beaches does not exceed 0.20 m/s.

15.2.3 Sediment Availability

Two rivers (Guadiana and Guadalquivir) are attributed to be the main sediment source of the littoral of Huelva, with a mean discharge of 144 and 185 m^3/s, respectively (Vanney 1970). In the Guadiana case, the sediment accumulation rates in the outer estuary suggest an annual supply of 0.5 Hm^3 for the last 6000 years, with a rate suspension/bedload near to 1/4 (Morales 1993). Nevertheless, in the present moment the annual balance of sand supply to the sea is only 352.59 m^3 (Morales et al. 2014b). This sediment can be transported longshore eastwards by the littoral drift.

In the case of the Guadalquivir River, the rate suspension/bedload remains unstudied but some authors suggest that the suspended load is higher than the bedload. The suspended matter supplied by the Guadalquivir was estimated for the last decade to be 71,087.25 m^3/year (Contreras and Polo 2010). This big amount of suspended matter generates a patch of muds just at the mouth of the river but not contribute to supply the coast westwards.

The rivers located in the central coast of Huelva (Piedras, Tinto and Odiel) have very limited flows, with a scarce importance in the sedimentary dynamics of the adjacent open coast.

The east sector of the Huelva Coast is also supplied by the erosion of a long cliff system incised on early Pleistocene sandy materials during the late Pleistocene and early Holocene. It is necessary to remark that the Huelva Coast is the only place in Spain where a tertiary basin is directly in contact with the sea. The materials of this basin can be easily dismantled and reworked by marine processes, so this sector of the coast became a well-supplied littoral from a sedimentary point of view.

15.2.4 Human Actions

The first remarkable human action was done just on the river supplies. The sediment transport capacity of the five rivers has decreased since 1960, with the construction of 65 dams regulating over 75% of the water flow (Ojeda 1988). Is easy to deduce that a big part of the sediment supplied by the rivers in the past is retained in these dams and not arrives now to the coast, generating a sedimentary deficit. In the Guadiana River, is estimated that the present bedload supplied to the coast is near a 20% of the potential bedload due to this cause (Morales et al. 2014a).

On the other hand, this coast is heavily altered with modern jetties built to stabilize the river-mouth estuarine channels. These dams modified artificially the wave refraction layout and contribute to the compartmentalisation of the coast into independent cells, giving rise to considerable modifications to the erosional/sedimentary dynamics of the adjacent sectors and causing either total (Vila Real do Santo Antonio and Huelva) or partial (Isla Cristina and Punta Umbria) obstacles to the sedimentary transition. Consequently, five sedimentary dynamic cells may be delimited in this human-altered area. Three of the four cell limits are directly related with the effects of the main jetties (Ojeda and Vallejo 1995). Each one of these cells includes two different areas: (a) affected by erosional processes (downdrift the jetty) and (b) affected by progradational processes (updrift the jetty).

The replenishment of the erosive coastal tracks is a constant human process of these beaches. Periodically, the erosive areas of the beach are supplied with sand extracted in areas located below the base level of the swell. Replenishment works have been carried out for the last 15 years along the entire coast although in the whole Coast of Huelva. In general regeneration projects have an average life of around 6–7 years, after which new regeneration work is necessary.

The occupation of the beaches on a massive scale, with illegal constructions and the lack of control of access to these beaches have modified the evolution of sand bars with either partial or total destruction of forms and accentuation of the deflation areas.

15.3 Morphodynamic Characterisation

According to Wright and Short (1984), the concept of morphodynamic states to imply the complete assemblages of depositional forms and coupled hydrodynamic process signatures, including waves, tides and wind-induced currents. According to their dynamic and morphological characteristics, exposed sandy beaches can be classified into several morphodynamic types (Wright and Short 1984; Short 1999), from the "dissipative state" to the "reflective extremes". Dissipative beaches are wide with low slopes and composed of finer sediment. The shoaling and surf zone are wide with waves tend to break far from the intertidal zone and dissipate force progressively along wide surf zones, characterized by spilling breakers. Reflective beaches present high slopes and coarse sand. The shoaling and surfing area is narrow or they have no surf zone. The waves break brusquely on the intertidal zone.

Beach profiles and planforms are directly related to sediment characteristics and coastal hydrodynamic parameters, therefore gaining knowledge on these relationships between morphology and dynamics can help in predicting beach behaviour.

For this reason, numerous attempts have been made to suggest parameters and indexes aimed at classifying beaches according to their morphodynamic state. The first models were mainly based on wave and sediment characteristics (Dean and Maurmeyer 1983), the influence of tide on wave hydrodynamics and beach slope (Wright et al. 1985). However, other variables may also influence the morphodynamic state of beaches, mainly linked to local and regional structural setting (Benavente et al. 2016).

15.3.1 Long-Term Morphology and Behavior

The morphodynamic characteristics of the beaches located along the Huelva shores was analyzed by Benavente et al. (2015), from a beach survey program carried out for four years (2000–2004). Surveying of profiles, sediment samples were taken from the foreshore, as well as different hydrodynamic data were carried out. Beach morphology and behavior presents notable differences between sinter and summer. These differences were characterized through the use of diverse parameters and indexes from the literature: surf scaling parameter (Guza and Inman 1975); surf similarity parameter (Battjes 1974); dimensionless grain fall parameter (Gourlay 1968), with the classifications proposed by Wright et al. (1985) and Masselink (1994); and relative tidal range (Masselink and Short 1993).

Acording to Surf Scaling Parameter (ε), the Huelva beaches present during the winter intermediate profiles very close to be reflective, with values between 1 and 15. Nevertheless, in summer, beaches are clearly dissipative, with values between 15 and 60. According to Surf Similarity Parameter (ξ) the winter values are situated between 0.25 and 0.55, in the limit between dissipative and intermediate profiles, while the summer profiles are intermediate, more dissipative to the east, with values between 0.5 and 1.6. In general the Huelva beach winter profiles showing morphologies closer to the "low tide terrace" to "low tide bar" type, according to the classification of Masselink and Short (1993). The summer profiles showing morphologies of "barred" to "barred dissipative" type. However, according to Benavente et al. (2015), the result obtained after the application of the models of classical classification has not been satisfactory because these models don′t integrated waves, tidal influences and sediment grain size.

The presence of reflective morphologies in Huelva coast, according to ε distributions, is therefore likely to be attributed to a higher infiltration of water in the intertidal zone due to the coarser sediment (Masselink and Short 1993). These profiles would thus be strongly controlled by sediment characteristics, which in turn, depend on the distance from, and type of, sediment source. Other deviations from the predicted morphodynamic behavior are related to the increase in disipativeness close to river mouths and littoral barriers associated, which occurs in some points in the study area. This characteristic is strongly controlled by the sediment source areas.

As commented in the previous chapter, the Huelva coast is the best sandy supplied coastal transect of the Iberian Peninsula. In different cells, the sand comes from different sources. The western coast of Huelva (1st sector) receives supplies from the Guadiana River and from the Portuguese coast. Thus, the sediment is coarser and presents a large content in bioclasts. The cell located to the SE (2nd sector) is supplied from the cliff erosion of Neogene formations of the Guadalquivir Basin in Huelva, consisting mainly of sands with a finer character. In consequence, the western beaches would have more dissipative profiles than the eastern ones. Nevertheless, the observations are totally contrary, being the eastern beaches more reflective than those located in the western cells. The cause in these changes would be interpreted as a result of the amount of sand feeding more than properly the grain size.

Considering these peculiarities, it can be deduced that changes in wave dimensions, tidal range and sediment properties have a great influence in the morphodynamics of these beaches, as example of a low-energy coastline where spatial and temporal variations of Hb are generally less than 1 m. Applying the criteria of Bernabeu et al. (2003), which gives greater importance to the tidal range and grain size of sediment, the Huelva beaches can be characterized as "tide-modified beaches" (Short 2006), hence an intermediate type between which is tide-dominated and wave-dominated ones.

15.3.2 Short-Term Dynamics and Evolution

As is described, the most part of time the Huelva display ridge and runnel systems. Beaches with intertidal bars usually occur systematically along low step sandy shores subject to waves of short fetch and tidal schemes which can be varied. Mesotidal beaches similar to the described in this chapter were studied by Mulrennan (1992) or Houser and Greenwood (2005). The genesis of these bars and their long-term (seasonal and interannual) mobility was previously well studied, being the waves, through the processes of swash, surfing and shoaling, the most important dynamic factor (Kroon and Masselink 2002), whereas longitudinal tidal currents or those enforced by lateral winds are responsible of their longitudinal mobility (Sedrati 2006; Sedrati and Anthony 2007). Nevertheless, the short-term dynamic (daily) is scarcely studied.

Some experiments in the easternmost cell of the Huelva beaches studied this daily dynamic through a rise of topographic profiles accompanied by non-directional wave gauge data and currents measured using a currentmeter-holograph *Aquadopp* (Sedrati and Morales 2017).

Initially, the beach presented a dissipative profile, with an intertidal bar located in the limit with the subtidal zone. During the first three days of calm weather conditions (Hs near 0.7 m) the beach experienced the development of an intertidal bar that migrates landwards rapidly. Highlight the presence of rip channels separating longitudinally different bars. Step to higher energetic conditions occurred during the two following days (Hs near to 1.1 m) led to an important landwards migration of the bar and a longitudinal displacement of the rip channels. During these two days the bar was flattened with an infilling of the runnel located in the upper area of the foreshore. An event of erosion was registered the seventh day of measurements with waves of Hs near 1.5 m. During the twelve hours of this event, the bar was completely eroded. The following four days the conditions were intermediate (Hs near to 1.0 m) and contributed to a planar growing of the beach without development of bars.

This beach behavior suggests a sensible and immediate response of the beach to the wave conditions where a bar can be created, attached to the backshore and destroyed in only two weeks. So, the long term dynamic is the synthesis of all the movements and the consequent sedimentary balance occurred in all these short-term dynamic changes.

15.4 Huelva Beaches

As commented in the introduction, the coast is compartmented in five dynamic cells. In each one of these cells long beaches are developed (Fig. 15.2). The dynamic and evolution of all of these beaches are explained in the following chapters.

15.4.1 First Cell: Isla Canela Beach

Isla Canela beach extends between the Guadiana River Mouth and the Carreras Inlet in the frontal area of a barrier island chain located along of the delta formed by this river (see Chap. 24 of this book). It is a higly dynamic beach, which experienced many changes since the construction of regulation jetties in both channels limiting the cell (Guadiana and Carreras). The coastal evolution in this area was studied by Garel et al. (2014) based on a time series of 16 ortho-rectified vertical photographs ranging from 1956 to 2011. A coastal vulnerability along Isla Canela beach had been observed by previous papers, related to the changes in shoals of the existing frontal ebb-tidal delta and attributed to the reduction in both fluvial exports by dams and longshore transport by jetties (Garel and Ferreira 2011; Gonzalez et al. 2001).

Before the jetties, bypassing was a discontinuous process achieved through bank breaching and attachment of the shoals to the coast (Morales 1993, 1997). Presently, the jetties at the mouth of the Guadiana produce a divergent wave refraction system, creating a compartment of the predominant longshore transport. Other effect created by the jetties (1977) was the breaching and disconnection of the shoals from the Portuguese margin. Since then, the shoals are in a continuous process of approaching to the beach front (Fig. 15.3). In 2000 some of these shoals

Fig. 15.2 Pictures of different Huelva beaches located in the distinguished coastal dynamic cells. **a** Isla Canela Beach. **b** Nueva Umbría Beach in the front of El Rompido Spit. **c** Punta Umbría Beach. **d** Cheniers of Saltes Islands. **e** Mazagón Beach. **f** Castilla Beach in the front of El Asperillo Cliff. **g** Matalascañas Beach. **h** Doñana Beach in the front of an extensive dune system

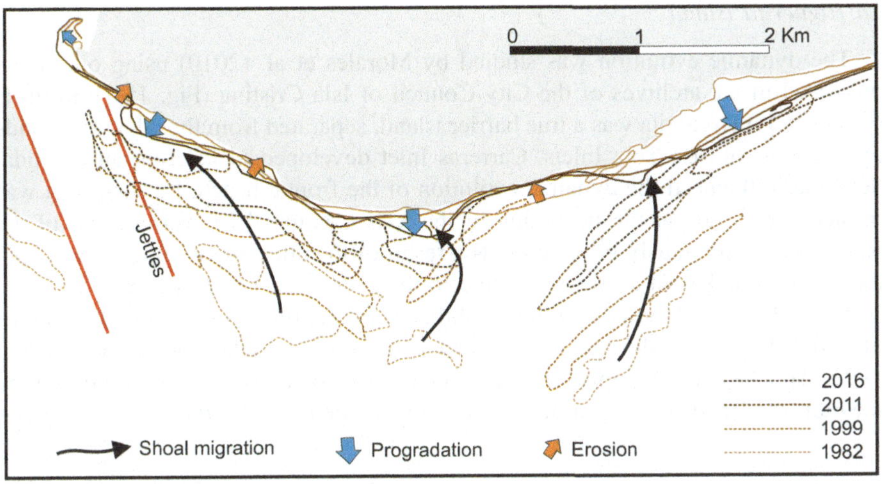

Fig. 15.3 Decadal evolution of the coastline along Isla Canela Beach related with the shoal migration after the breaching caused by the jetties

started to attach the coastline. Locally, shoal attachment produces a convex beach (Fig. 15.2a) due to convergent longshore transport promoted by wave refraction over the shoal, when the shoal is relatively close to the shore (Garel et al. 2015). On the contrary, the accretion segment is bordered by retreating areas created by divergence in the wave trains. So, the beach is presently highly compartmented, alternating progradational and erosional areas.

The main part of the touristic buildings is located just in this highly dynamic area. The erosion problem is especially obvious at the central part of the beach, where tourism-driven anthropogenic pressures has motivated the implementation of various coastal defense measures from the mid 90s.

To the east, the presence of another jetty limiting the Carreras mouth inhibits the bypassing of sand to the next cell. The sand started to be retained by the jetty transforming this beach segment in a high-rate depositional area.

15.4.2 Second Cell: Isla Cristina-La Antilla-Nueva Umbría Beaches

The coastal stretch comprished between the mouths of the Carreras Creek and the Piedras Estuary is currently a straight beach evolved from several barrier islands. Built the western most of them (La Higuetita Island) in the framework of the Guadiana River delta and the rest as a island string in the interfluve Guadiana-Piedras having continuity on the ancient Gato Island, located at the mouth of the Río Piedras.

La Higuerita Island

The dynamic evolution was studied by Morales et al. (2010) using old charts preserved in the archives of the City Council of Isla Cristina (Fig. 15.4). In these old charts La Higuerita was a true barrier island, separated from the adjacent islands by Carreras and La Tuta Inlets. Carreras Inlet developed in its front an ebb-tidal delta that influenced the dynamic evolution of the frontal beach. La Tuta inlet was infilled and closed at the beginning of the XIX Century. The western part of the island was continuously prograding as a result of the attachment of sand bars from the ebb-tidal deltas in a process similar to those described in Isla Canela Beach.

In 1974, two jetties were built to stabilize Carreras Inlet. Since then, the sand of the old delta was transformed into sand bars and attached to the front of the beach in 1992. The beach developed in the front of the island is mainly composed by medium sand and presents a dissipative low slopped profile. Presently, the beach

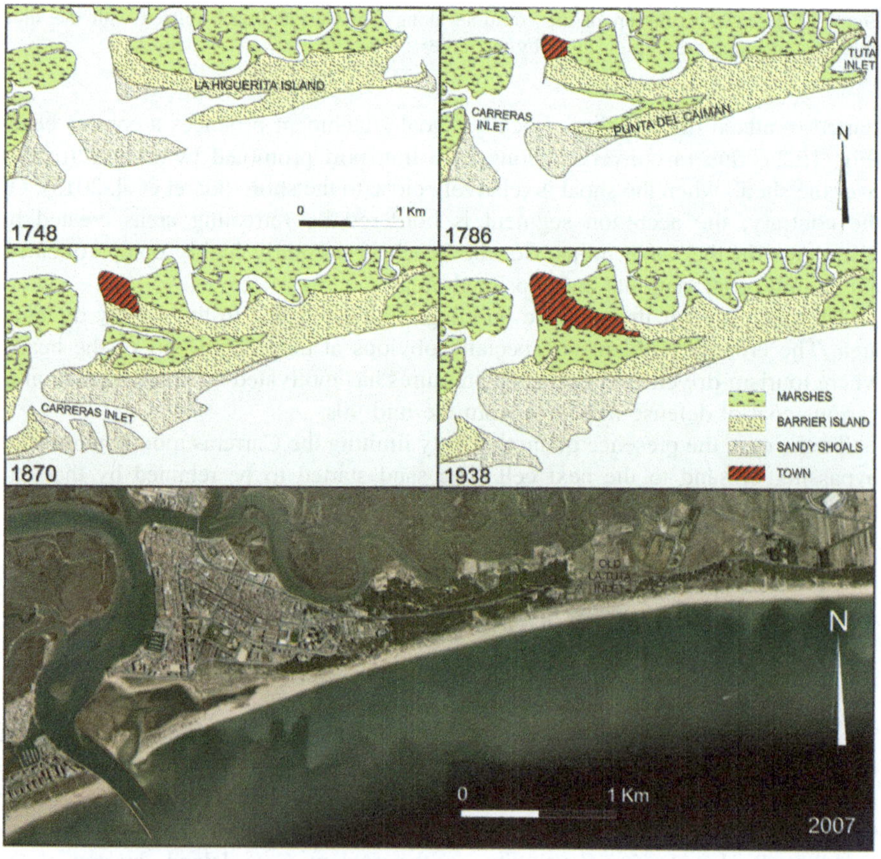

Fig. 15.4 Recent evolution of Isla Cristina Beach in front of La Higuerita Island. Modified from Morales et al. (2010)

front is under an erosional process, motivated by the sand deficit caused by the presence of the jetties that inhibit the bypass of sand from the previous cell.

La Antilla Beach

Old charts show this coastal stretch as a string of barrier islands that would represent the continuation towards the East of La Higuerita Island (Fig. 15.5). The mainland limit of the coastal lagoon located in the backbarrier area is a palaeo-cliff raised over Plio-Pleistocene materials, representing the higher reliefs of the area.

Aerial photograph of 1956 (Fig. 15.5b) shows this coastal track as a straight coast developing beaches with a well feeded dune system. In the back barrier area a salt marsh was developed. This morphology and environments indicate that the inlets that separated the old islands so as the coastal lagoon located behind the barriers were already infilled in 1956. Presently the entire system barrier island-lagoon is totally urbanized Fig. 15.5c).

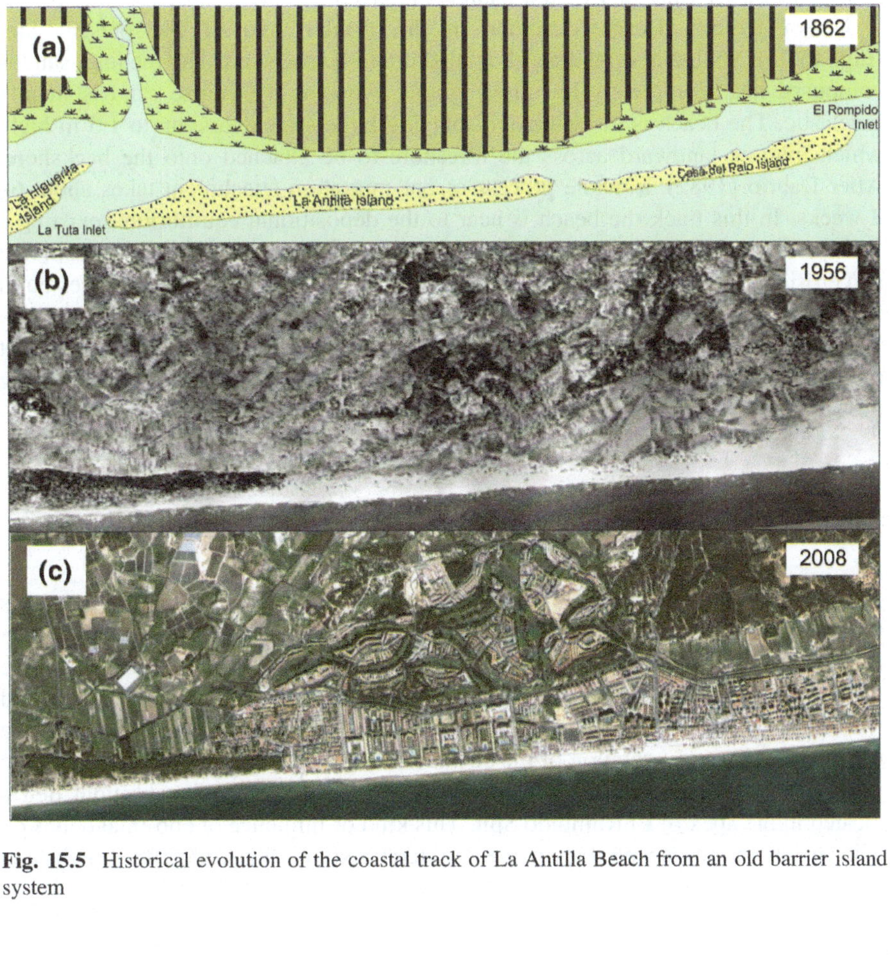

Fig. 15.5 Historical evolution of the coastal track of La Antilla Beach from an old barrier island system

The beach located in the frontal area is very similar to Isla Cristina, a dissipative beach with sand bars composed by medium sand. The beach is presently under an erosive balance, menacing the first line of buildings, which are directly attacked during severe storms. In some occasions, the tidal black muds deposited in the back barrier lagoon appear in the frontal beach after severe storms as prove of the coastline retrogradation. That forced the coastal authorities to replenish periodically the beach front.

Nueva Umbría

The spit that closes the Piedras River Estuary (named El Rompido Spit) has been studied from different approaches by various authors during the last decades. Dabrio (1989), Zazo et al. (1994) or Ojeda and Vallejo (1995) studied the sedimentary dynamics of the spit, but are Borrego et al. (1993) who described the evolution from a previous barrier island. Dabrio (1982) centered the attention in the beach dynamics and its sediments. A most recent paper (Morales et al. 2001) asseverated that the spit was formed by the longitudinal growth of a previous barrier island named Levante or El Gato Island (Fig. 15.6).

Nueva Umbría Beach develops in the frontal area of the present Spit (Fig. 15.2b). Sediment on the beach is medium to coarse sand. The medium sand is composed near completely by quartz grains, whereas the coarse sand is mainly bioclastic. The beach is dissipative, displaying large swash bars, up to 1.5 m high, which migrate landward across the foreshore to be attached onto the backshore. After Dabrio (1982), a whole process of bar migration-attachment takes about 6–8 weeks. In this track the beach is near to the depositional equilibrium, existing a west-to-east longshore bypassing and seasonal alternances between depositional and erosional stages. The migrational bars occurred during depositional stages build a sedimentary sequence of landward dipped sets of cross stratificated sands.

In the apex of the spit the bars acquire a curved shape to be adapted to the inlet morphology by entering waves. This part has a positive sedimentary balance doing the spit grow in longitude by attachment of successive bars.

15.4.3 Third Cell: Punta Umbría Beaches

Coastal spit of Punta Umbria is one of the elements closing the estuary of the rivers Odiel and Tinto. This spit is formed by the accretion of curved bars and beach ridges. The spit evolution, as the previously described in El Rompido, is an example of apical initial accretion in West-to-East direction, but in this case was coupled with a process of subsequent progradation which marks a migration of the coastline towards the sea and a rotation of the costline direction. The arrival of sand from the previous cell is intermittent and significantly influenced by the ebb-tidal deltas located in the apex of El Rompido Spit. This kind of influence of ebb-tidal deltas on adjacent shorelines is well documented in literature (e.g. Fitzgerald 1984; Hicks and

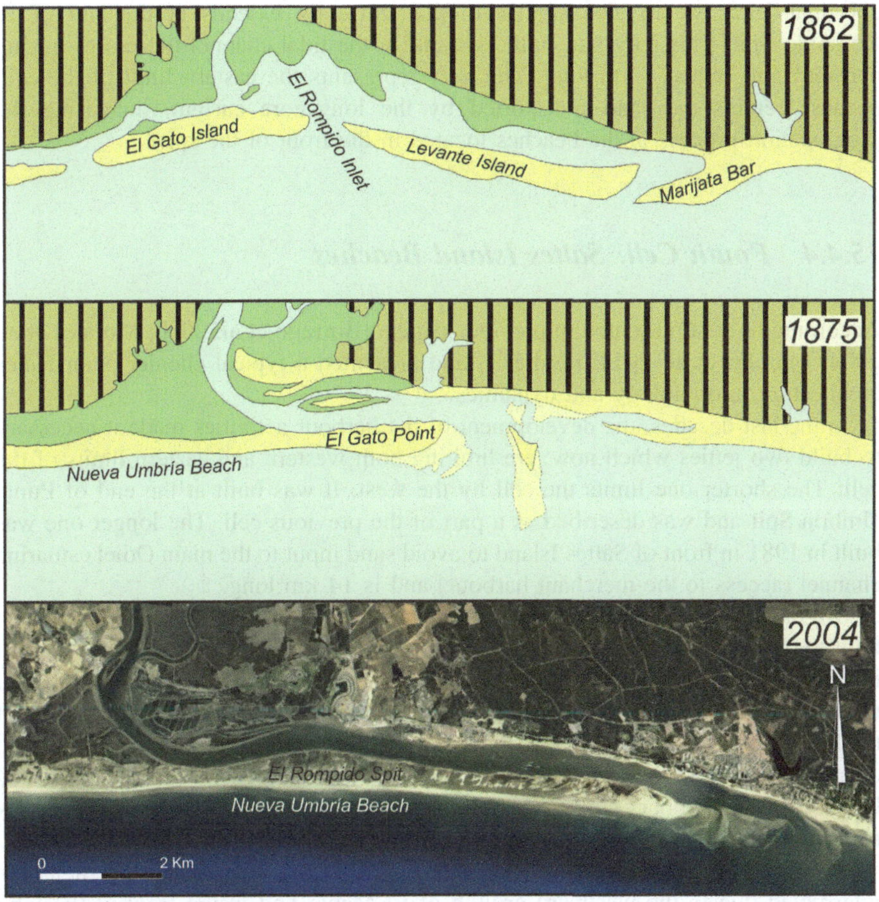

Fig. 15.6 Historical evolution of el Rompido Spit and Nueva Umbría Beach from an old barrier island. Modified from Morales et al. (2001)

Hume 1996; Hicks et al. 1999; Fitzgerald et al. 2002; Van Heteren et al. 2006) and was described in this coast in the case of the Guadiana mouth (Garel et al. 2015).

Beach sediment corresponds to well sorted medium sand. Textural and compositional maturity of the sediment is high and the grains are mostly subrounded and composed by quartz with some shell fragments. The beach presents a dissipative profile with a normal dynamic of migrating ridge-and-runnels. During storms the beach experiments erosional periods, but the normal balance is depositional, including occasions of high energy with a depositional regime, where the sediment accumulated in the beach is coarse sand or gravel composed mainly by bioclastic clasts.

Presently, in the apex of Punta Umbría Spit there is a rock jetty built in 1984 (Fig. 15.2c). Before the construction of this dam the spit developed an ebb-tidal-delta system also similar to that currently extends at the end of El Rompido Spit. This jetty was built to stabilize the tidal channel (access to fishing harbour) and is just 1 km long. The jetty represents the eastern limit of the cell, because retains the sand transported by the longshore current, increasing the progradation process in the beaches located in the front of the spit.

15.4.4 Fouth Cell: Saltes Island Beaches

According to data presented in previous papers (Borrego et al. 2000; Morales et al. 2004; Morales et al. 2014b) Saltés Island presented a typical chenier plain facies architecture, morphology and dynamics.

In the last decades, the development of the harbour activities made it necessary to build two jetties which now, are limiting both western and eastern limits of the cell: The shorter one limits the cell by the west. It was built at the end of Punta Umbria Spit and was described as a part of the previous cell. The longer one was built in 1981 in front of Saltés Island to avoid sand input to the main Odiel estuarine channel (access to the merchant harbour) and is 14 km long.

Migration and build-up of coarse storm barriers (Cheniers) is the process which built the coastline in this cell during storm periods previous to the construction of the jetties. It consists of the accumulation of coarser material in the highest areas of tidal systems. Normally the material comes from the dismantling of the shoals of the Punta Umbría ebb-tidal deltas and takes on the form of littoral bars built on top of muddy facies. In the centre of Saltes Island is located a chenier plain constituted by successive sand bars built up on a tidal plain and separated at present by bodies of salt marshland (Fig. 15.2d).

The process of chenier build-up remained active up until 1994, when it was interrupted due to the westward growth of El Manto spit in the front of the long jetty. This spit closed access to the chenier plain by the waves. The growth of this spit was induced by the construction of the jetty of Punta Umbría which modified the wave refraction patterns in this section of the coast.

In the front of this spit was developed a sandy beach, which is in a dynamic equilibrium, since it do not received inputs of sand from the previous cell due to the presence of the western jetty. Under these conditions, the beach sand is in continuous rework, typical of a dissipative beach with bars.

15.4.5 Fifth Cell, Erosional Track: Mazagón Beach

Mazagón Beach was historically considered a typical protected beach located in the back area of a complex of ebb-tidal deltas at the mouth of the Odiel-Tinto estuary. Today, this dynamics has been modified by different human constructions, including the long jetty of Saltes Island and a marina dock at the west of the proper beach (Fig. 15.2e).

Previous papers distinguished two dynamic areas with different accretionary tendencies (Ballesta et al. 1998; Morales et al. 2004): The first of these areas, located in the eastern sector of the beach, is clearly erosional, although the retrogradation rate increased since 1987. The second sector is located to the west, limiting easterly the present marina (Fig. 15.7). This coastal sector changed from being erosive before the construction of the Saltés jetty, to having a high progradational rate after its construction; another clear increase in this rate was also observed after the construction of the marina dock (1991–1993).

In the eastern erosive area, fair weather waves occurred during anticyclonic conditions induced a ridge and runnel dynamics, but long term studies estimated some permanent erosion affect this area, with a negative balance of 30,000 m^3/year in a total surface of 42,000 m^2. It represents a loss sand of 0.71 m^3 per m^2 per year (Muñoz Pérez et al. 2001; Del Río et al. 2002). During periods of storms intense wave events destroy the backshore and some trends of aeolian dunes. A good example was the succession of storms occurred during the winter of 1995/96, that generated a sedimentary deficit of 1.25 m^3 per m^2 in only two months (Ballesta et al. 1998).

The western progradational area has experienced a very high progradational rate since the building of the jetties (Morales et al. 2004), but especially after the building of the marina dock. During the period from December 1995 to February 1996, when the storms occurred, a slight coastline setback was observed. However, during the fair weather months, a quick landward migration of sandy bars was observed, not only contributing to a recovering of the initial profile but also increasing progradation. A mean seawards migration of the coastline of more than 65 m/year was observed before 1998. After this date, this ratio decreased for two reasons: (a) because the beachline surpassed the front of the dock and (b) because an artificial bypassing of sand to the erosive area was initiated by the coastal authorities.

The cause of the presence of these two different dynamic sectors is attributed to the influence of the jetty of Saltes. The construction of this jetty introduced very significant modifications in the beach dynamics. On one hand, the jetty is a physical barrier that inhibits the bypassing of sand supplied by the littoral drift from the western cell, generating a sand deficit eastwards (Del Río et al. 2002). On the other hand, the wave refraction model was extremely modified, so a wave divergence zone appeared just in front of the urbanized zone and reversing the sense of littoral drift that started to circulate westwards in the accumulative area (Morales et al. 2004). The construction of a marina dock on the western beach meant a barrier to the sand bypassing. As a consequence, the sand started to be accumulated in the east side of the dock.

Presently, the beach maintenance represents an annual cost of near Euros 100,000 for artificial bypassing of sand between these two areas.

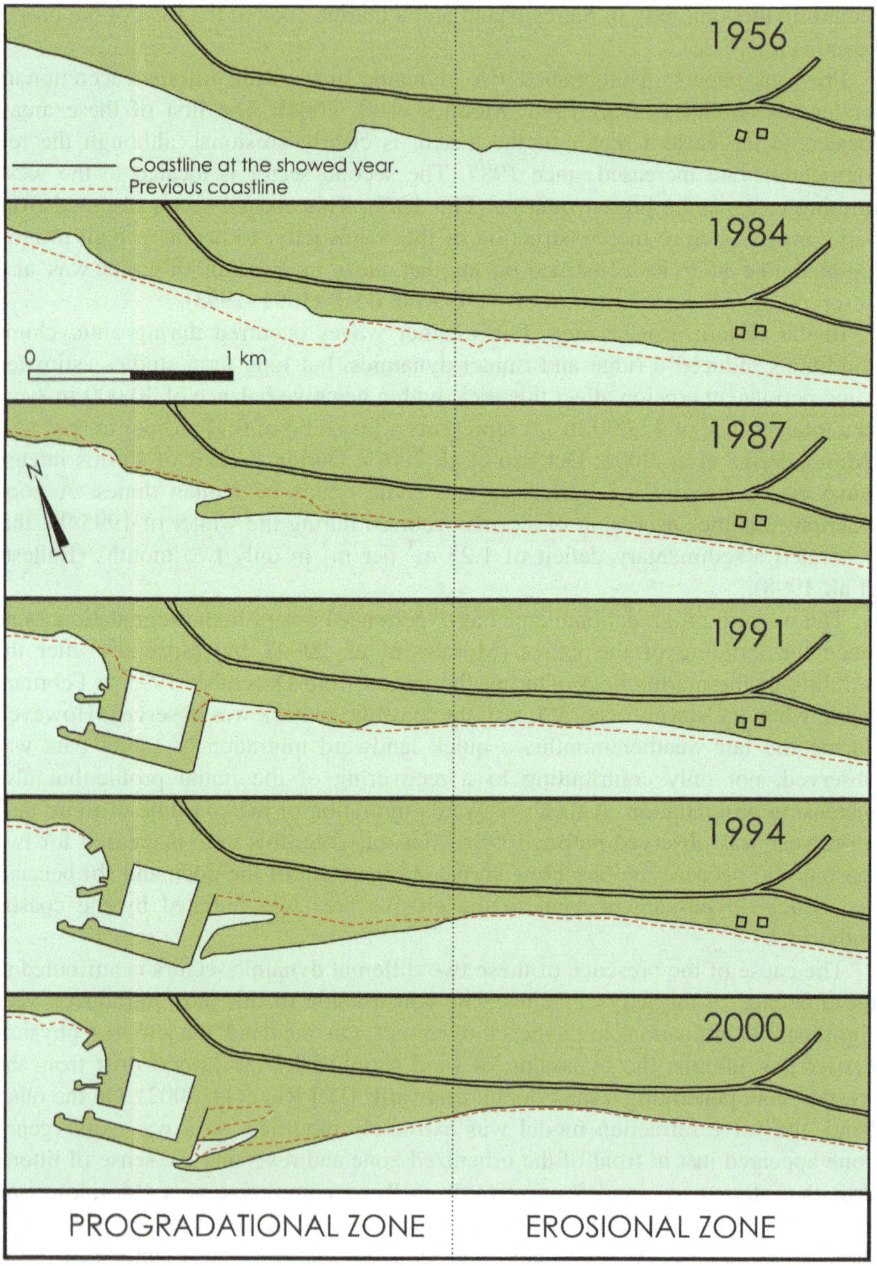

Fig. 15.7 Recent evolution of Mazagon Beach comparing between the two distinguished dynamic areas. Modified from Morales et al. (2004)

15.4.6 Fifth Cell, Bypassing Track: El Asperillo-Castilla Beaches

Between the tourist towns of Mazagón and Matalascañas is one of the most significant and peculiar coastal stretches of the Gulf of Cádiz, that extends along about 25 km of coastline. The dynamic processes that develop in this coastal section are of great importance for the rest of the littoral, and especially for the understanding the coastal morphodynamics that takes place towards the mouth of the Guadalquivir River, where is located one of the most important natural protected areas in the world: Doñana National Park.

Matalascañas and Mazagón were built during the Spanish tourist development of the decade of the sixties, before that, this coast was completely depopulated between the mouth of the Tinto-Odiel and the Guadalquivir Rivers. Just for its protection, given the continuous landing of pirates, a series of towers of surveillance (Almenara towers) were built between the end of the 16th and the beginning of the 17th centuries.

This coastal section experiments a continuous erosion and retreat, visible in the dynamics of the Asperillo cliff (Fig. 15.2f). This cliff is the most spectacular morphology of this coastal sector. It´s a sandy cliff of about 10–15 m high, on which is a spectacular dunes system of more than 100 m high. This cliff loses altitude towards the SE, buried by the mobile dune systems of Doñana spit.

To establish the rates of coastal retrogradation-erosion of this sector, the precise location of different historical constructions were studied by Menanteau (1979). The different Almenara towers built along this coastal stretch are magnificent points of reference for establishing their spatio-temporal evolution. The sector of greater erosion since the 16th century has been the central sector, where is located the Asperillo tower which ruins are visible during very low spring tides. This tower was built twice, the last one in the middle of the 18th century. Menanteau (1979) estimated an erosive retreat of about 1.2 m/year for the Asperillo tower sector, 0.4 m/y for the Arroyo del Oro sector and 0.7 m/year for La Higuera sector.

The slope processes associated with the cliff retreat are of great importance for the contribution of sediments to the beach. Two types of processes can be identified (Rodríguez Ramírez 1998). The first, dominant in the western sector, is linked to the circulation of runoff water, and the result is a morphology of badlands and large landslides that give the cliff a more irregular and discontinuous appearance. The second it is a more gravitational process in which materials slide or fall to the foot of the cliff. This process is dominant where the cliff has a more active profile due to marine erosion (waves), greater accumulation of dunes in the upper part (Asperillo dunes) and greater infiltration due to greater thickness of the dune systems. These characteristics dominate in the eastern section. The result is a cliff with a more rectilinear and continuous profile.

Once the sediments arrive at the foot of the cliff, the marine dynamics erode the deposits and redistribute it along the coast by the west to the east drift current. The result is a sandy beach of scarce slope, attached to the Asperillo sandy cliff

(Fig. 15.2f). These beaches can be classified as intermediate and mesotidal, with low tide bars, in summer, and somewhat more reflective in winter, although it depends on the incidence of winter storms (Fig. 15.8). The profile of the beaches attached to the cliff remains more or less stable over time, but subject to migration to the east of erosive-sedimentary lobes (Fig. 15.9), like a sedimentary transit zone towards Doñana spit.

Fig. 15.8 Panoramic view of the coastal sector of Asperillo. **a** 7/3/2012: winter profile with a longshore bar, **b** 29/10/2012: summer profile

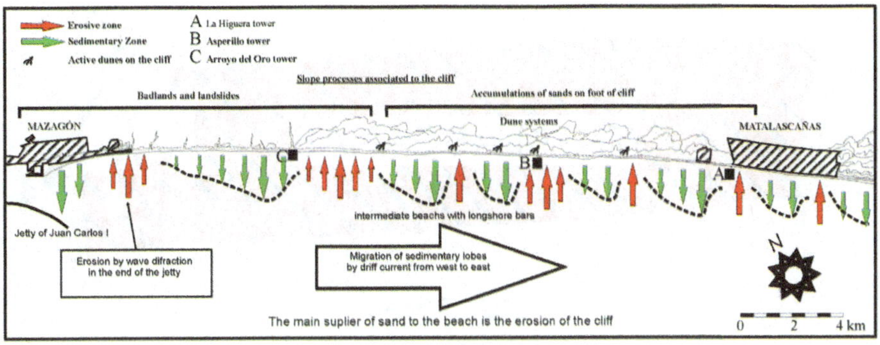

Fig. 15.9 Geomorphological outline of the Asperillo coast

15.4.7 Fifth Cell, Urban Track: Matalascañas Beach

The urbanization of Matalascañas extends along 5 km of coastline. It has a sandy beach of scarce slope attached to a sandy cliff about 10 m high, which loses height as we move towards the East, being buried by the mobile dune systems of Doñana. It is precisely these beaches, together with the proximity of Doñana N. P., the true interest of this enclave, making it a tourist attraction amount of enormous economic interest for the region.

This coastal sector, from the geological point of view, is clearly erosive. Proof of this is the existence of the coastal cliff as well as the different buildings eroded by the sea, including an Almenara tower of the 16th century (Fig. 15.2g). To the natural processes were superimposed, from the decade of the seventies, the construction of several groins and urban complexes on the beach, which led to the alteration of the natural coastal dynamics. From the construction of urbanization, in the decade of the sixties, the problems of sand loss on the beaches and retreat of the cliffs is continuous. In 1979, this problem was mitigated by the construction of 20 groins perpendicular to the beach, of a height of about 3 m and spaced approximately every 200 m.

A topographic survey of a series of cross-shore topographic profiles at Matalascañas beach highlights complex three-dimensional and highly dynamic morphologies commonly involving two distinct sandbar systems (Fig. 15.10).

To the northwest of the Almenara tower, the beach exhibits a *Transverse Bar and Rip* morphology according to the bar beaches classification (see Wright and Short 1984; Masselink and Short 1993) (Fig. 15.10b, P1). Along the urbanized area of Matlascañas beach (the zone with a seawall and 20 groins), the intertidal beach shows the presence of non-pronounced bar-trough systems with double to triple bars with a mean wavelength of 200 m and the bars ranged in width from 40 to 50 m (Fig. 15.10b, P2). To the southeast of the urbanized area (out of the protected zone by groins), three intertidal bar-trough systems situated around the mid-tide are more pronounced and ranged in width from 20 to 50 m (Fig. 15.10b, P3). Troughs

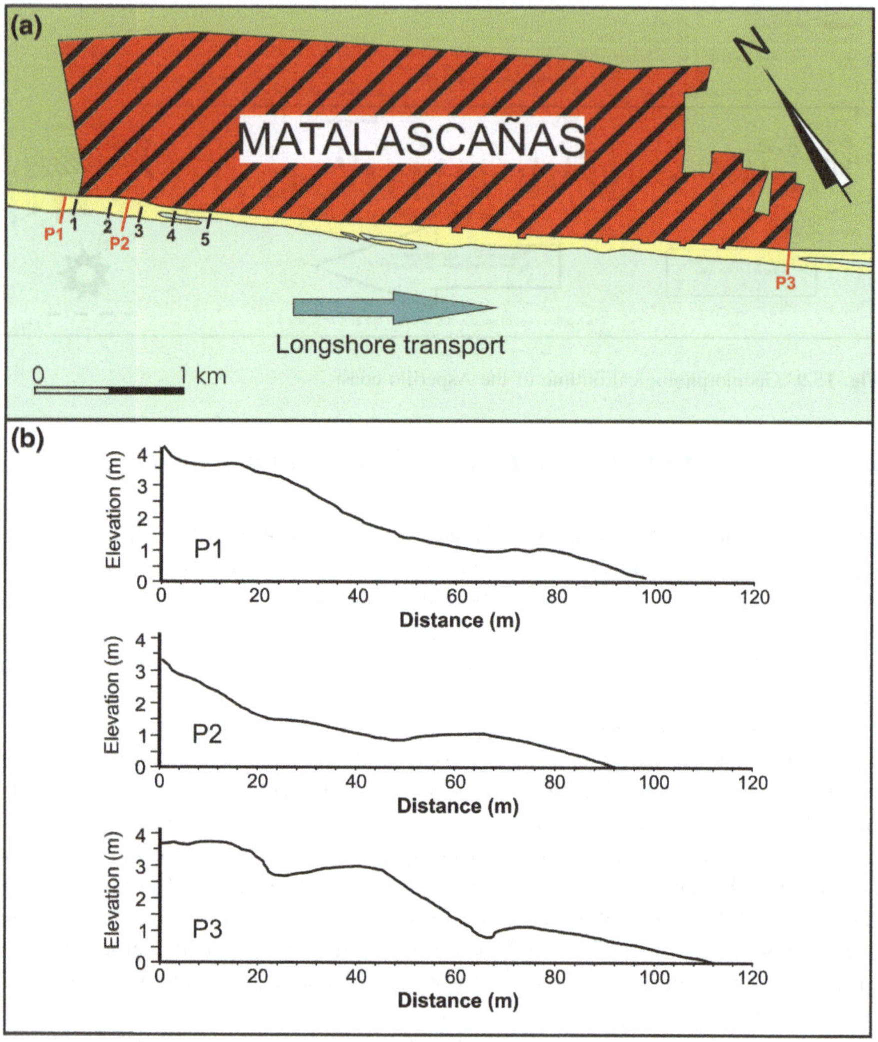

Fig. 15.10 **a** Location of three surveyed topographic profiles at Matalascañas beach. P1 is the beach profile representing the non urbanized zone, P2 represent a beach profile located between two groins and P3 present a beach profile to the south of the urbanized area. 1, 2, 3, 4, 5 indicates the position of the emerged groins of Matalascanas beach. **b** Profiles representing the mean morphology of the three sectors of Matalascañas beach

on this zone are larger and much more subdued. It's seems to be clear that the 20 implemented groins are in the origin of the contrasting morphological behavior of the intertidal zone beach. Mid-term photographic and beach profiles topographic surveys (One profile by sector) highlight the persistence of this morphological configuration (Three contrasting bars behavior) between the protected and non-protected zones at Matalascañas beach under low energy conditions. However, during high energy conditions the bars may be completely flattened. Sedrati and Morales (2017a, b) have suggested that bar-trough systems morphodynamics at this beach is drived by wave-processes. To the north of Almenara tower, the longshore bars migration is much related to the wave littoral drift while across-shore weak landward migration of the upper bar is related to swash processes during spring tide. To the south of Almenara Tower, alongshore sediment transport is partially perturbed and interrupted by groins, thus this deprives the urbanized beach area from longshore bars migration process. In fact, bar-trough systems situated between the groins highlights across-shore mobility (seaward or landward) depending on wave and wind-driving currents conditions (Sedrati and Morales 2017a). To the south of the urbanized area, the bar-trough systems highlights both across-shore and alongshore mobility.

The protective role of the 20-implemented groins in Matalascañas beach is questionable. Sedrati and Morales (2017a, b), pointed out that cross-shore vs longshore bar mobility may even be mitigated by the presence of the groins, which favor onshore than alongshore bar migration. These degraded groins whose stones are dislocated, appear to be a hindrance to the natural hydrodynamic processes and sediment transport than to participate in the protection and the rehabilitation of the urbanized beach.

15.4.8 Fifth Cell, Depositional Track: Doñana Beach

Doñana beach is situated between Matalascañas and the Guadalquivir River mouth, and extends along Doñana spit in a system with high progradation rates. It is the least disturbed coastal system of the Huelva coast because it is inside of the Doñana National Park. The main supply of sediment comes from the erosion of the Asperillo cliff, transported longshore by the littoral drift towards the east. The slope of the beaches decreases towards the mouth of the Guadalquivir where, in addition, the progradation rates are higher. In the entire track there is a rotation of 30° in the coastline direction from N128°E to N158°E, being finally the beach parallel to the arrival of the main wave trains. In the first track the beach is dissipative-intermediate with a longshore bar and runnel beaches consist of a shore parallel bar separated from the beach by a deep trough that progressively progresses towards the foreshore (Fig. 15.2h). In the area near the mouth, the beaches are very dissipative with a wide surf zone including shore bars, and a low-sloping and wide beach face consisting of medium-fine sand.

There is a close relationship between coastal dynamics and maritime climate. The storms cause a high variability in the beach, with differential effects over successive sectors of the same coastal track with different hydrodynamic features. These events increase the coastal erosion and move sand rapidly offshore, while lower energetic conditions may cause gradual beach accretion (Komar 1976). The record of storms in Doñana coast during the last 4 decades(1956–1996) were analyzed by Rodríguez-Ramírez et al. (2003), delimiting the effects in beaches and spits, and compared with a statistical study of the dominant winds and other climatic-oceanographic features, in order to establish the periodicity of these events. The results obtained show that these storms periodicity are related with the NAO (North Atlantic Oscillation) with a periodicity of 9–10 years of high storms. Each high-energy period induces the creation of new beach-ridges in littoral spits in the post storm period (40 to 150 days). The number of these sedimentary beds coincides closely with the storm periods included in each interval studied by aerial photographs, whereas the intermediate swales are characteristic of fair weather conditions.

The coastal evolution of the storm deposits as a part of the Guadalquivir estuary was studied by Rodríguez Ramírez (1998) and Rodríguez Ramírez et al. (1996, 2008, 2014, 2015, 2016) from the analysis of cores and geomorphological study of ortho-rectified vertical photographs (see Chap. 22 of this book), evidencing its intense dynamics during the recent Holocene.

The dynamics of Guadalquivir ebb-tidal delta is also very intense. The sandy shoals of this tidal delta are in a continuous process of approaching to the beach front, covering the rocky shoals of Pleistocene formations. This fact caused numerous shipwrecks, especially between the XVI and XVII centuries, during the period of West Indies Fleet of Spain, between the port of Seville and South America. Today, even the entrance channel must be continuously dredged to favour the entry of ships into the port of Seville.

«Anse perdido en esta barra i salida del puerto de los mejores navíos de España, i de los más ricos que della salían para las flotas de las Indias, i ahogándose grande número de la gente que en ellas estaba embarcada, sin que por maravilla salga ninguna flota sin dexar perdido.»

«Thus the best ships of Spain and the richest, they left for the fleets of the Indies were lost in this bar entering or exiting the port, drowning a large number of people who were embarked on them, without leaving a fleet without no losses.»

Agustin de Horozco (1598).

Are just these approaching bars that finally prograde the spit in a double sense: On a hand apically displacing the estuarine channel to the East, on the other hand, frontally getting the rotation of the coast.

References

Ballesta, M., Morales, J. A. y., & Acosta, E. A. (1998). Efecto erosivo de 10s temporales del invierno 1995–1996 sobre la Playa de Mazag6n (Huelva, SO Espaiia): Influencia de las construcciones costeras. *Revista Soc. Geol. España, 11*(3–4), 285–296.

Battjes, J. A. (1974). Surf similarity. In *Proceedings 14th Coastal Engineering Conference* (pp. 446–480). ASCE.

Benavente, J., Gracia, J., Del Río, L., Anfuso, G., & Rodríguez-Ramírez, A. (2016). Morphodynamic characterization of the Spanish beaches of the Gulf of Cadiz. *Boletin Geológico y Minero, 126*(2), 409–426.

Bernabeu, A. M., Medina, R., & Vidal, C. (2003). A morphological model of the beach profile integrating wave and tidal influences. *Marine Geology, 197,* 95–116.

Borrego, J. (1992). *Sedimentología del estuario del río Odiel (Huelva, S.O. España)* (Ph. D. Thesis). University of Sevilla, 315 pp.

Borrego, J., Morales, J. A., & Gil, N. (2000). Facies Deposicionales y Evolucion Reciente de una Llanura de Cheniers en la Desembocadura de la Ría de Huelva (S. O. España). Revista de la Soc. *Geol. de España, 13,* 405–416.

Borrego, J., Morales, J. A., & Pendón, J. G. (1993). Holocene filling of an estuarine lagoon along the mesotidal coast of Huelva: The Piedras River mouth, southwestern Spain. *Journal of Coastal Research, 9,* 242–254.

CEEPYC. (1990). Centro de Estudios y Experimentación de Puertos y Costas "Ramón Iribarren". Plan de estudio de la dinámica litoral de la Provincia de Huelva. Informe. Dirección General de Puertos y Costas, Madrid, 37p.

Contreras, E., & Polo, M. J. (2010). Propuesta metodológica para diagnosticar y pronosticar las consecuencias de las actuaciones humanas en el Estuario del Guadalquivir. Vol. 2: aportes desde las cuencas vertientes. Autoridad Portuaria de Sevilla. Informe inédito. 123 pp.

Cuena, G. J. (1991). *Proyecto de regeneración de las playas de Isla Cristina* (p. 100). M.O.P.T.: Servicio de Costas, Informe inédito.

Dabrio, C. J. (1982). Sedimentary structures generated on the foreshore by migrating ridge and runnel systems on microtidal and mesotidal coast on S. *Spain. Sedim. Geol., 32,* 141–151.

Dabrio, C. J. (1989). Playas e Islas barrera-lagoon. In A. Arche (Ed.), *Sedimentología* (Vol. 1, pp. 349–394)., Serv Madrid: Publ. C.S.I.C.

Dean, R. G., & Maurmeyer, E. M. (1983). Models for beach profile response. In P. D. Komar (Ed.), *CRC Handbook of coastal processes and erosion* (pp. 151–165). Boca Raton, Florida: CRC Press.

Del Río, L., Benavente, J., Gracia, F. J., Anfuso, G., Martínez del Pozo, J. A., Domínguez, L., et al. (2002). The quantification of coastal erosion processes in the South Atlantic Spanish Coast: Methodology and preliminary results. In *Littoral 2002: The changing coast* (pp. 383–390).

Garel, E., Sousa, C., & Ferreira, Ó. (2015). Sand bypass and updrift beach evolution after jetty construction at an ebb-tidal delta. *Estuarine, Coastal and Shelf Science, 167,* 4–13.

FitzGerald, D. M. (1984). Interactions between the ebb-tidal delta and landward shoreline: Price Inlet, South Carolina. *Journal of Sedimentary Petrology, 54,* 1303–1318.

FitzGerald, D. M., Buynevich, I. V., Davis, R. A., & Fenster, M. S. (2002). New England tidal inlets with special reference to riverine-associated inlet systems. *Geomorphology, 48,* 179–208.

Garel, E., & Ferreira, Ó. (2011). Effects of the Alqueva Dam on Sediment Fluxes at the Mouth of the Guadiana Estuary. *Journal of Coastal Research, 64,* 1505–1509.

Garel, E., Sousa, C., Ferreira, Ó., & Morales, J. A. (2014). Decadal morphological response of an ebb-tidal delta and down-drift beach to artificial breaching and inlet stabilization. *Geomorphology, 216,* 13–25.

Gonzalez, R., Dias, J. M. A., & Ferreira, Ó. (2001). Recent rapid evolution of the Guadiana Estuary mouth (southwestern Iberian Peninsula). *Journal of Coastal Research Proceedings of International Coastal Symposium, 2000,* 516–527.

Gourlay, M. R. (1968). *Beach and dune erosion tests. Delft hydraulics laboratory* (Report, M935/ M936).

Guza, R. T., & Inman, D. L. (1975). Edge waves and beach cusps. *Journal of Geophysical Research, 80,* 2997–3012.

Hicks, D. M., & Hume, T. M. (1996). Morphology and size of ebb-tidal deltas at natural inlets on open-sea and pocket bay coasts, North Island, New Zealand. *Journal of Coastal Research, 12,* 47–63.

Hicks, M. D., Hume, T. M., Swales, A., & Green, M. O. (1999). Magnitudes, spatial extent, time scales and causes of shoreline changes adjacent to an ebb–tidal delta, Katikati Inlet, New Zealand. *Journal of Coastal Research, 15*, 220–240.

Horozco, A. de. (1598). Historia de la ciudad de Cádiz.

Houser, C., & Greenwood, B. (2005). Profile response of a lacustrine multiple barred nearshore to a sequence of storm events. *Geomorphology, 69*, 118–137.

Komar, P. D. (1976). *Beach processes and sedimentation* (p. 429). New Jersey: Prentice-Hall Inc.

Kroon, A., & Masselink, G. (2002). Morphodynamics of intertidal bar morphology on a macrotidal beach under low-energy wave conditions, North Lincolnshire, England. *Marine Geology, 190*, 591–608.

Masselink, G. (1994). *Morphodynamics of microtidal beaches. Examples from Central Queensland, Australia* (PhD Thesis). Sydney University, Sydney, Australia.

Masselink, G., & Short, A. D. (1993). The effect of tide range on beach morphodynamics and morphology: A conceptual beach model. *Journal of Coastal Research, 9*(3), 785–800.

Menanteau, L. (1979). Les Marismas du Guadalquivir. Exemple de transformation d'un paysage alluvial au cours du Quaternaire récent. These 3e cycle, 154 p. Univ. Paris-Sorbonne.

M.O.P.U. (1991). Estudio sobre el oleaje para la primera fase del estudio de evolución de la playa de Castilla (Huelva). Dirección General de Puertos y Costas. Informe inédito, 33p.

Morales, J. A. (1993). *Sedimentología del estuario del río Guadiana (S.O. España-Portugal)* (Ph D Thesis). Servicio Publics Univ Huelva. 300 pp.

Morales, J. A. (1997). Evolution and facies architecture of the mesotidal Guadiana River delta (S. W. Spain, Portugal). *Marine Geology, 138*, 127–148.

Morales, J. A., Borrego, J., & Ballesta, M. (2004). Influence of harbour constructions on morphosedimentary changes in the Tinto-Odiel estuary mouth (southwest Spain). *Environmental Geology, 46*, 151–164.

Morales, J. A., Borrego, J., & Davis, R. A., Jr. (2014a). A new mechanism for chenier development and a facies model of the Saltés Island chenier plain (SW Spain). *Geomorphology, 204*, 265–276.

Morales, J. A., Borrego, J., Jiménez, I., Monterde, J. R., & Gil, N. (2001). Morphostratigraphy of an ebb-tidal delta system associated with a large spit in the Piedras Estuary mouth (Huelva Coast, Southwestern Spain). *Marine Geology, 172*, 225–241.

Morales, J. A., Cantano, M., & Rodríguez-Ramírez, A. (2010). Relación entre la evolución geomorfológica y el uso humano de las zonas costeras: Ejemplo de Isla Cristina (Huelva, SO España). *Geogaceta, 48*, 15–18.

Morales, J. A., Sedrati, M., Cantano, M., Rodríguez-Ramírez, A., Delgado, I., Lozano, C., & Flor-Blanco, G. (2014a). From the continent to the coast: the bedload transport across the lower sector of the Guadiana River Mouth (Spain-Portugal). *Geomorphologie: Relief, Processus, environnement, 20*(3), 251–260.

Mulrennan, M. E. (1992). Ridge and runnel beach morphodynamics: An example from the Central East Coast of Ireland. *Journal of Coastal Research, 8*, 906–918.

Muñoz-Pérez, J. J., de San, López, Román-Blanco, B., Gutiérrez-Mas, J. M., Moreno, J., & Cuena, G. (2001). Cost of beach maintenance in the Gulf of Cadiz (SW Spain). *Coastal Engineering, 42*, 143–153.

Ojeda, J. (1988). *Aplicaciones de la teledetección espacial a la dinámica litoral (Huelva), Geomorfología y Ordenación del Territorio* (P. D Thesis). Univ. Sevilla, 411 pp.

Ojeda, J., & Vallejo, I. (1995). La Flecha de el Rompido: análisis morfométrico y modelos de evolución durante el periodo 1943–1991. *Rev. Soc. Geol. España, 8*, 229–238.

Rodríguez-Ramírez, A. (1998). *Geomorfología del Parque Nacional de Doñana y su entorno* (p. 146). Madrid: Organismo Autónomo Parques Nacionales del Ministerio de Medio Ambiente.

Rodríguez-Ramírez, A., Flores-Hurtado, E., Contreras, C., Villarías-Robles, J. J. R., Jiménez-Moreno, G., Pérez-Asensio, J. N., et al. (2014). The role of neotectonics in the sedimentary infilling and geomorphological evolution of the Guadalquivir estuary (Gulf of Cadiz, SW Spain) during the Holocene. *Geomorphology, 219*, 126–140.

Rodríguez-Ramírez, A., Pérez-Asensio, J. N., Santos, A., Jiménez-Moreno, G., Villarías- Robles, J. J. R., Mayoral, E., et al. (2015). Atlantic extreme wave events during the last four millennia in the Guadalquivir estuary, SW Spain. *Quaternary Research, 83,* 24–40.

Rodríguez-Ramírez, A., Rodríguez Vidal, J., Caceres, L., Clemente, L., Belluomini, G., Manfra, L., et al. (1996). Recent coastal evolution of the Doñana National Park (S.Spain). *Quaternary Science Reviews, 15,* 803–809.

Rodríguez-Ramírez, A., Ruiz, A., Cáceres, L., Rodríguez-Vidal, J., Pino, R., & Muñoz, J. M. (2003). Analysis of the recent storm record in the southwestern Spanish Coast: Implications for littoral management. *The Sciense of the Total Environment, 303,* 189–201.

Rodríguez-Ramírez, A., Villarías-Robles, J. J. R., Pérez-Asensio, J. N., Santos, A., Morales, J. A., Celestino-Pérez, S., et al. (2016). Geomorphological record of extreme wave events during Roman times in the Guadalquivir estuary (Gulf of Cádiz, SW Spain): An archaeological and paleogeographical approach. *Geomorphology, 261,* 103–118.

Rodríguez-Ramírez, A., & Yañez, C. M. (2008). Formation of chenier plain of the Doñana marshland (SW Spain): Observations and geomorphic model. *Marine Geology, 254,* 187–196.

Sedrati, M. (2006). *Morphodynamique transversale et longitudinale de plages à barres intertidales en domaine macrotidal et en condi- tions de forte agitation: Baie de Wissant.* Nord de la France: Thèse de Doctorat, Université du Littoral Côte d'Opale, Dunkerque.

Sedrati, M., & Anthony, E. J. (2007). Storm-Generated morphological change and longshore sand transport in the intertidal zone of a multi-barred macrotidal beach. *Marine Geology, 244,* 209–229.

Sedrati, M., & Morales, J. A. (2017a). Efecto de los espigones transversales en la dinámica de las playas disipativas con barras: Ejemplo de Matalascañas, Huelva. *Geo-Temas, 17,* 239–242.

Sedrati, M., & Morales, J. A. (2017b). On the role impermeable groins on barred beach morphodynamics: Example of Matalascanas beach, Spain. *Geophysical Research Abstracts, 19,* EGU2017–14160.

Short, A. D. (1999). *Handbook of beach and shoreface morphodyanmics* (p. 379). West Sussex, UK: Wiley.

Short, A. D. (2006). Australian beach systems-nature and distribution. *Journal of Coastal Research, 22*(1), 11–27.

Vanney, J. R. (1970). *L'hydrologie du Bas Guadalquivir* (p. 176). Madrid: Publ. Dpt. Geogr. Aplic. C.S.I.C.

van Heteren, S., Oost, A. P., van der Spek, A. J. F., & Elias, E. P. L. (2006). Island-terminus evolution related to changing ebb-tidal-delta configuration: Texel, The Netherlands. *Marine Geology, 235,* 19–33.

Wright, L. D., & Short, A. D. (1984). Morphodynamic variability of surf zones and beaches: A synthesis. *Marine Geology, 56,* 93–118.

Wright, L. D., Short, A. D., & Green, M. O. (1985). Short-termchanges in themorphodynamic states of beaches and surf zones: An empirical predictive model. *Marine Geology, 62,* 339–364.

Zazo, C., Goy, J. L., Somoza, L., Dabrio, C. J., Belluomini, G., Improta, S., et al. (1994). Holocene secuence of sea-level fluctuation in relation to climatic trends in the Atlantic-Mediterranean linkage coast. *Journal of Coastal Research, 10*(4), 933–945.

Chapter 16
Classification and Characteristics of Beaches at Tenerife and Gran Canaria Islands

Ignacio Alonso, Mariona Casamayor, María José Sánchez García and Isabel Montoya-Montes

16.1 Introduction

16.1.1 Geological Context

Canary archipelago is formed by seven main islands and several islets that extends for nearly 500 km across the eastern Atlantic, between latitudes 27°N and 30°N, lying its eastern edge only 100 km from the NW African coast (Fig. 16.1).

The Canary Islands have a volcanic origin. They are the result of a hot spot acting in a passive continental margin on a very slow-moving tectonic plate (Carracedo et al. 1998). Submarine volcanic activity begun around 35 m.y. in Fuerteventura (Cantagrel et al. 1993) and it has continued until recent times, since the last eruption took place only few years ago at El Hierro island (Rivera et al. 2014; Oglialoro et al. 2017).

This large period of magmatism has not been continuous. Periods of volcanic activity have been followed by relaxation periods during which erosion took place, and huge amounts of sediments were eroded through a very dense network of gullies to the coast (Menéndez et al. 2008). In five of the seven islands (Lanzarote, Gran Canaria, Tenerife, La Palma and El Hierro) there are evidences of volcanic

I. Alonso (✉) · M. Casamayor · M. J. Sánchez García · I. Montoya-Montes
Instituto de Oceanografía y Cambio Global IOCAG, Universidad de Las Palmas de Gran Canaria ULPGC, Las Palmas, Spain
e-mail: ignacio.alonso.bilbao@ulpgc.es

M. Casamayor
e-mail: moia4@hotmail.com

M. J. Sánchez García
e-mail: mariajose.sanchez@ulpgc.es

I. Montoya-Montes
e-mail: isabel.montoya@ulpgc.es

J. A. Morales (ed.), *The Spanish Coastal Systems*,
https://doi.org/10.1007/978-3-319-93169-2_16

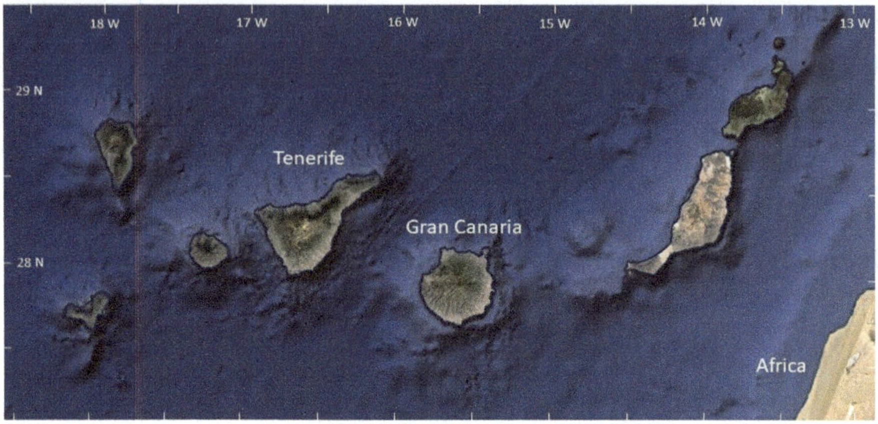

Fig. 16.1 Location of the Canary Islands. Tenerife and Gran Canaria Islands are located in the center of the archipelago. Eastern islands are older while those located westward are much younger. Image modified from Google Earth

activity in the last five centuries, and the lava flows resulting from this activity have also reached the coast.

Other geological factor that has contributed to shape the different insular volcanic edifices, both the emerged and the submerged parts, are the landslides. These avalanches are the result of gravitational instabilities on these very high volcanic edifices, since the original stratovolcanoes could be higher than 7000 m measured from the ocean bottom (Carracedo et al. 1998; Urgeles et al. 1997; Masson et al. 2002).

Because of all the aforementioned geological factors, as well as to the effect of waves that steadily contribute to erode the northern coast of the islands, the shelf presents very strong differences between some islands and the others. In this context, Lanzarote and Fuerteventura present quite large shelves, as opposite to Tenerife, La Palma and El Hierro were the shelf is practically nonexistent. There are also significant differences between different coastal stretches in certain islands: in Gran Canaria the shelf is very narrow at the NE while it is quite large in the rest of the island, or in Gomera where the shelf is much wider at the north than at the south. Age of the islands, recent volcanism and wave incidence are the key factors to understand these differences (Maestro-González et al. 2005).

Criado et al. (2011) point out that as the littoral shelf becomes wider, there are more opportunities for marine organisms such as moluscs, equinoids, artropods, foraminiphera and corallinaceous algae to find out the adequate habitat to grow. These organisms are the key ingredient of marine sedimentary inputs, which are transported by waves from the nearshore to the foreshore. Therefore, as far as the shelf in the Canaries is narrow, the amount of marine sedimentary inputs is scarce in long stretches of the coastline.

The other possible source of sediments to the coast is terrestrial, which is related either to cliff erosion or to fluvial runoff. Criado et al. (2011) state that both possible sources represent a very small contribution of sediments to the coast. This is explained by the slow response of volcanic materials to wave action and to the weak precipitation in most parts of the archipelago. Probably the only exception is located at Tazacorte beach (west coast of La Palma Island). This particular beach receives the runoff from the Caldera de Taburiente, located in the central part of the island. The very large catchment area, the height difference and the episodic but extremely heavy rains determine that the beach receives from time to time large amounts of sediments. Resulting from these inputs the beach is clearly prograding and the nearby harbor has siltation problems (Marrero et al. 2017). Because of the limited inputs of sediments to the coast, particularly in areas with narrow shelf, beaches in the islands are normally short and naturally limited by rocky headlands.

16.1.2 Socio-economic Context

The Canaries have an area of just over 7500 km^2 and a population over 2.1 million people. This population is very irregularly distributed, since 83% of the total population concentrates in Tenerife and Gran Canaria islands, which have 895.000 and 843.000 inhabitants respectively (ISTAC 2018).

More than 40% of the islands surface has been declared natural protected areas, where the population is very scarce. Therefore, the pressure of population on the territory is quite large, with values of 475 inhabitants/km^2 for the whole archipelago, which rise up to 850 and 950 inhabitants/km^2 for Tenerife and Gran Canaria respectively. This population pressure is not homogeneously distributed over the territory, since the largest towns and infrastructures in each island are located along the coast. Therefore, most of the coastline is built up and there are more roads per km^2 than in any other European islands (Moreno Gil 2003).

An additional factor related to this pressure is the main economic activity in the islands: tourism accounts for at least 50% of the GDP of the islands (Garín-Muñoz 2006), though according to Parra-López and Baum (2004) tourism represents around 83% of the GDP. In fact, Canary Islands received more than 16.7 million visitors in 2017, distributed in the following way: 6.2 million in Tenerife, 4.6 million in Gran Canaria and 5.9 million in the rest of the islands (ISTAC 2018). More than 96% of these visitors come to the islands for holydays, and most of them stay at tourist resorts located within few hundreds of meters from the coastline (Morales and Santana 1993). Therefore, large touristic complexes have been built along the islands coast, and it explains what Moreno Gil (2003) has called "concrete tourism".

The Canaries have become a global player in tourism due to a number of factors. Climate conditions such as low rainfall and relative isolation have become fundamental pillars of the tourism development scheme that has been put into action on the islands in recent decades. Another invaluable regional resource that has also contributed is the magnificent natural conditions of the islands, among which the beaches and coastal dunes are a key aspect for any tourist destination based on sun, sand and surf (García-Rodríguez et al. 2016).

Both the native population and the huge number of visitors are mostly concentrated in the coastal zone. Therefore, related activities are also located close to the shoreline. They include not only urban and tourist developments, but all sort of infrastructures. Among them marinas, ports, desalinization and sewage treatment plants, artificial beaches and all kind of roads and parking lots can be cited. Therefore, population, mass tourism and related activities generate a big pressure on the coastal zone. This pressure originates different impacts along the coast, whose effects have been pointed out by several authors (Alonso et al. 2002; Hernández-Calvento et al. 2005; García-Romero et al. 2016; Ferrer-Valero et al. 2017).

16.2 Beaches at Tenerife and Gran Canaria Islands

16.2.1 Criteria Used for Beach Classification

A good approximation to the beaches in the whole archipelago has been performed considering the beaches in the two most populated islands: Tenerife and Gran Canaria. To characterize the beaches along these two islands, we have made a thorough search of the whole coastline along both islands. Beaches have been classified in four groups following two different criteria: Firstly according to their nature (natural or artificial), and secondly the natural ones have been divided according to the dominant sediments: sandy beaches, mixed beaches and pebble-cobble beaches. Table 16.1 summarizes the main characteristics of each of the resulting groups.

Table 16.1 Main characteristics of the four types of beaches considered in this work

Beach type	Description
Sandy beach	Wave-built foreshore accumulation of dominant sand size sediments, usually composed of both emerged and submerged bodies
Mixed beach	Wave-built accumulation of sand to boulders. These materials can be either randomly distributed all year round or following a seasonal pattern: pebbles and cobbles dominant size across the whole foreshore during winter, while at summer time the lower foreshore is mostly composed by sand
Pebble-cobble beach	Wave-built accumulations of pebbles and cobbles dominant grain size, usually forming berms and supratidal ridges. These materials do not change along the year
Artificial beach	Beach that have been created by man-made structures (groins, detached breakwaters, etc.) and/or by artificial inputs of sand. In some cases there were natural beaches at that particular spot, but they have been altered to be larger and/or wider (e.g. Alcaravaneras, La Laja, Tauro and Meloneras beaches in Gran Canaria Island)

In this search we have considered only beaches larger than 100 m. Using different orthophotos we have measured the length, maximum width and minimum width at each particular beach. To avoid problems related to tide, width has always been measured from the high tide mark to the back of the beach. This upper limit can be either artificial (normally roads, pedestrian promenades or agriculture fields) or natural (cliffs, dunes or natural lowlands). In the latter case the upper limit of the beach was selected following either a geomorphologic (cliff toe, first foredunes) or a sedimentologic criteria (where the dominant beach material was observed). The dominant grain size was obtained from scientific papers, technical reports, in situ observations and analysis of photographs available in the web for each of the different spots.

16.2.2 Beaches at Tenerife Island

There are 93 beaches all around Tenerife Island, but there is no clear pattern in the distribution of the different types of beaches along the coast (Fig. 16.2). Nevertheless, sandy beaches are mostly concentrated in three areas: N coast (between beach numbers 15–25), W coast (numbers 32–40) and SE coast (between beaches 62–77). Artificial beaches are mostly found along three areas: at the SW (between beaches 42–52) along Las Américas area, and in the north coast at Puerto de la Cruz (beaches 20–21). In both areas the concentration of tourist resorts is very high, and therefore there is a big touristic pressure for sandy beaches. The third sector is along the E-NE coast, where four artificial beaches are found (beaches 80, 86–88). In this case they are mostly designed for the native population, since along this area there are big villages and towns (Güímar, Candelaria, Radazul and Santa Cruz de Tenerife). Therefore, the demand for sandy beaches as recreation areas is not only due to tourism but also to local residents.

Regarding the coarse-grain beaches, including both the mixed beaches and the pebble-cobble ones, they are found all along the island. However, they are mostly located at the three corners of Tenerife Island: along Anaga at the NE, Teno at the NW and Las Galletas at the S. These corners correspond to the distal part of the three-armed rift system (Carracedo et al. 2007; Geyer and Martí 2010). The geological configuration along these areas, characterized by basaltic lava flows cut by a dense network of dykes, total absence of shelf and very short and steep gullies, do not favor the presence of beaches. The few beaches present in these areas are very narrow and dominated by coarse grained materials. Figure 16.3 shows several examples of the beaches at Tenerife Island.

Table 16.2 shows the main characteristics associated to all those beaches. In some cases there are several beaches located very close to each other and responding all of them to the same name. They are called sub-beaches. In case of artificial beaches, these sub-beaches are normally separated by groins, while in natural areas are separated by headlands. Length of each one of them is shown in

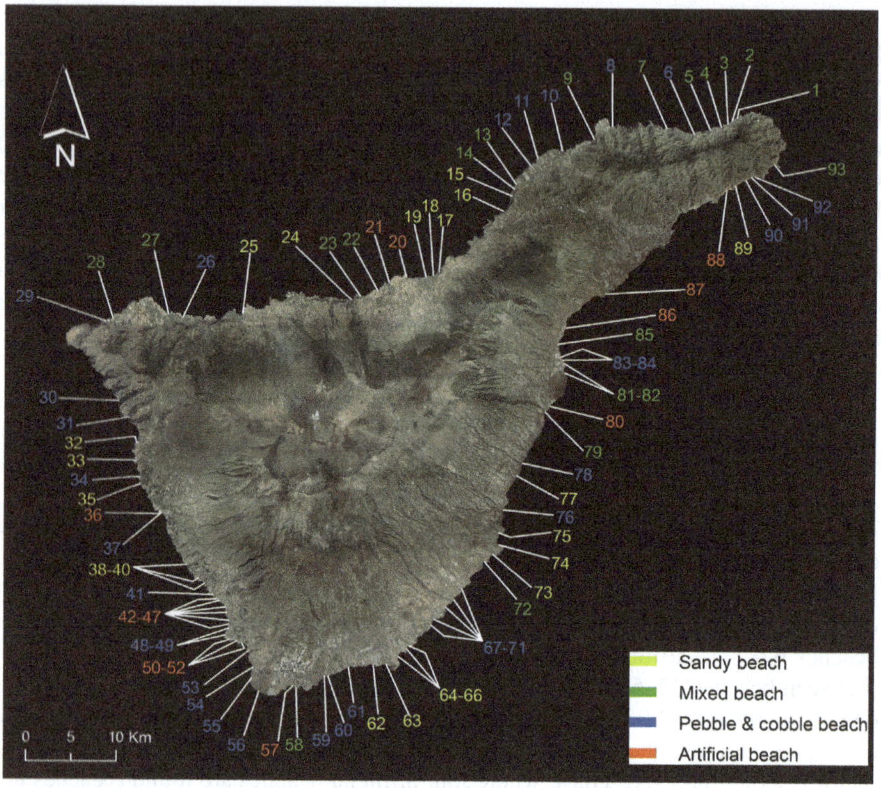

Fig. 16.2 Location and classification of all the beaches at Tenerife Island. Note that beaches can be found all around the island except at the NE and NW ends on the island, as well as in the city area (between beaches 87 and 88). The colors key indicates the type of beach considered

the length column of Table 16.2 (e.g. Playa Jardín beach, number 21, includes three sub-beaches of 280, 130 and 260 m length).

16.2.3 Beaches at Gran Canaria Island

Beaches at Gran Canaria Island are present all along the coastline, but beach concentration is denser in the S and E coast. Beach distribution (Fig. 16.4) follows a clearer pattern than at Tenerife. Sandy and artificial beaches are mostly located in two sectors along the coastline: at the S-SW (between beach numbers 22–41) and at the E-NE (between beach numbers 53–62 and 1–2), while both mixed and pebble-cobble beaches are dominant along the N-W coast (beach numbers 3–21) and at the SE (beaches 42–52).

Fig. 16.3 Photographs of several beaches in Tenerife Island. **a**, **b** Typical sandy beach in the northern coast backed by high cliffs, so that debris are common in the upper part of the beach (La Garañona, number 16). **c** Three artificial beaches in the SW coast. Two of them limited by groins and the one on the right is within the harbor area (El Camisón, Las Vistas and Los Cristianos beaches, 50–52 from left to right). **d** Artificial beach of Las Teresitas (number 88), quite large and wide because is the closest beach to the main town in the island. **e** Cobble beach of Playa Honda (number 48). It is quite narrow and a pedestrian promenade occupies the upper limit, which is quite common in tourist areas

Table 16.2 Main characteristics of the beaches at Tenerife Island

#	Beach name	Beach type	Length (m)	Width (m)
1	El Draguillo	2	130	0–10
2	Unknown name	2	180	10
3	Benijo	2	270	15
4	Almáciga	2	290	15
5	Roque de las Bodegas	2	350	0–20
6	Tachero	3	200	0–10
7	Tamadite	2	210	10–20
8	Punta del Hidalgo	3	580	0–10
9	El Arenal	2	260	0–10
10	Jover	3	120	0–20
11	La Barranquera	3	130	0–10
12	Chamorro	3	150	20

(continued)

Table 16.2 (continued)

#	Beach name	Beach type	Length (m)	Width (m)
13	El Pris	2	160	10
14	San Juan	2	150	0–10
15	La Arena	1	210	10–35
16	La Garañona	1	1100	0–10
17	El Ancón	1	210 + 140	0–20
18	Los Patos	1	115 + 650	0–40
19	Bollullo	1	210 + 100	0–35
20	Martiánez	4	400	0–25
21	Playa Jardín	4	280 + 130 + 260	10–60
22	Los Roques	2	340	15
23	de Castro	2	120	15
24	El Socorro	1	250	15–60
25	San Marcos	1	200	5–30
26	Caleta de Interián	3	235	5–20
27	Agua Dulce	2	115 + 95	10–20
28	La Arena	2	100	0–20
29	El Fraile	3	250	0–15
30	Masca	3	200	0–10
31	Barranco Seco	3	500	0–15
32	Los Gigantes	1	120	0–20
33	La Arena	1	160	40
34	Punta Blanca	3	120	15
35	La Jaquita	1	45 + 45 + 120	0–30
36	San Juan Puerto	4	60 + 290	5–30
37	San Juan	3	200	5–25
38	Callao Salvaje	1	120	15–50
39	El Puertito	1	160	0–5
40	Diego Hernández	1	70 + 75	0–10
41	La Enramada	3	360	0–30
42	Sin nombre	4	365	0–15
43	El Duque	4	400	15–55
44	Fañabé	4	920	5–55
45	La Pinta	4	195	20–35
46	Las Américas	4	140	15–50
47	Troya	4	420	5–70
48	Playa Honda	3	450	0–20
49	El Guincho	3	120	0–5
50	El Camisón	4	360	10–20
51	Las Vistas	4	880	25–75
52	Los Cristianos	4	360	30–85
53	Los Callaos	3	220	5–10
54	Palm-Mar	3	160	5–10

(continued)

Table 16.2 (continued)

#	Beach name	Beach type	Length (m)	Width (m)
55	Malpaís de Rasca	3	140	5–20
56	Faro de Malpaís de Rasca	3	140	10–20
57	Las Galletas Puerto	4	400	0–35
58	Las Galletas	2	430	0–10
59	Colmenares	3	330	5–10
60	El Barranco	3	360	0–15
61	San Blas	3	190	45
62	Los Perros	1	170	25
63	La Tejita	1	950	20–110
64	El Médano	1	330 + 220 + 75 + 35 + 100	10–60
65	El Cabezo	1	70 + 70	15–30
66	La Jaquita	1	70 + 130	10–60
67	Unknown name	3	480	0–50
68	El Medio	3	100 + 195 + 190	0–45
69	Los Tarajales	3	220	15
70	Mareta del Río	3	860	5–15
71	Callao Hondo	3	120 + 165	7–30
72	Ensenada de Abades	2	160 + 20 + 45 + 30	5–10
73	Abades	1	40 + 100 + 95	20–65
74	Punta de Abona	1	145	10–35
75	El Porís	1	60 + 90	0–20
76	El Sombrerito	3	210 + 60 + 120	10–15
77	El Abrigo	1	50 + 130	0–30
78	Barranco de Herques	3	225	15
79	Las Bajas	2	660	5–10
80	Puertito de Güimar	4	730	0–55
81	El Socorro	2	815	0–15
82	La Restinga	2	255	10
83	Lima	3	420	0–10
84	La Viuda	3	120 + 45	0–20
85	Candelaria	2	320	0–5
86	Punta Larga	4	600 + 270 + 260	0–30
87	La Nea	4	290	5–35
88	Las Teresitas	4	1230	40–75
89	Las Gaviotas	1	110	0–5
90	El Burro	3	140	0–5
91	El Llano	3	300	5
92	Igueste	3	380	5–10
93	Antequera	2	330	5

More than one value in length indicates that there are several beaches very close to each other and responding to the same name. These are called "sub-beaches" (see text). Width is measured from high tide mark. Two values refer to the wider and narrower parts of the beach. Beach type description: sandy beach (1), mixed beach (2), pebbles-cobbles beach (3), artificial beach (4)

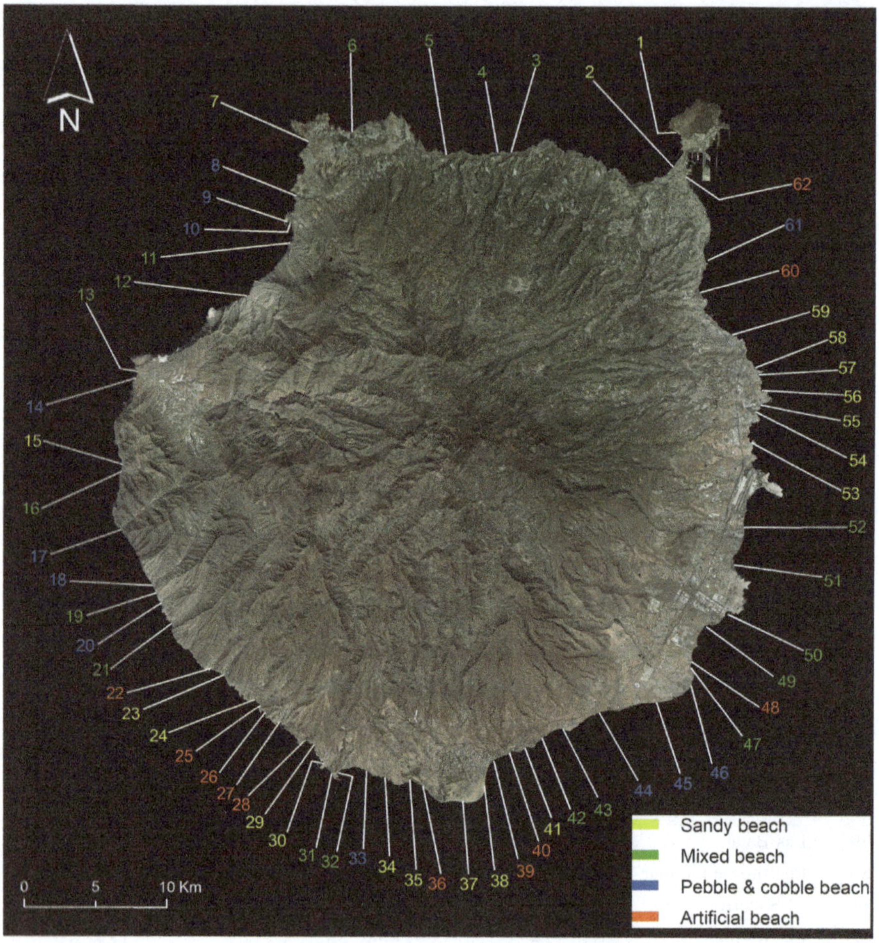

Fig. 16.4 Location and classification of the different beaches at Gran Canaria Island. Note that at the northern and northwestern coast the amount of beaches is much lower than at the rest of the island. The colors key indicates the type of beach considered

In the former case, beaches 3–21 are located along the northern and western coast. All along this area, waves are higher and therefore they break against the shoreline with more energy. As a result, sandy beaches are very unusual. There are only three exceptions along this area: Las Canteras, Sardina and Guguy Chico (beaches number 2, 7 and 15 respectively). Las Canteras and Sardina beaches are located in small bays where sandy sediments are the dominant material in the seafloor (MAGRAMA 2002). Both of them are protected from incident waves by the coastline configuration as well as by a natural offshore rocky bar in the case of Las Canteras beach and a nearby groin at Sardina beach. These specific conditions allow the existence of sandy beaches at Gran Canaria northern coast.

Fig. 16.5 In situ and aerial view of several beaches at Gran Canaria. **a** Guguy Chico beach (number 15), a magnificent sandy beach in the western coast. Note the dune attached to the cliff. **b** Example of mixed beach, with sand in the lower foreshore and pebbles in the upper part of the profile (photograph taken at low tide, Vargas beach, number 51). **c** Aerial view of El Veril beach (number 39), an artificial sandy beach formed by three sub-beaches, separated by curved groins. **d** Aerial view of El Cura, Tauro and Amadores beaches (numbers 24–26 respectively from top to bottom). Note the different color of the sand, which indicates the different source of the material: El Cura is a natural beach, Tauro has been nourished with sand form the Sahara Desert, and Amadores has sand form a submerged sand bank rich in corallinaceous algae, which explains the white colour

Guguy Chico is a different case. This facing west beach is completely open to the ocean with a very high cliff at the back. The very wide sandy shelf and the existence of a dune attached to the cliff are key aspects to understand why this beach is sandy all year round (Fig. 16.5).

The other group of coarse grain beaches is located at the SE coast (beaches 42–52 in Fig. 16.3). They are associated to the mouth of three of the largest gullies in the island (Guayadeque, Arinaga and Tirajana gullies), which according to Menéndez et al. (2008) have evacuated the amount of nearly 22 km^3 of material to the coast. The deposition of such huge amount of material along the coast has formed three fan deltas that can be clearly identified from aerial photography. Even though the eroded material covers a wide range of particle size, pebbles and cobbles are the dominant fraction along the coastline, since waves and coastal currents wash away the finer fractions. Table 16.3 shows the main characteristics of all those beaches.

16.2.4 Comparison Between Beaches in Both Islands

Considering the values shown in Tables 16.2 and 16.3, it has been possible to obtain the total length of beaches in both islands, as well as the average length and width corresponding to each of the different beach types. Average length has been obtained both for the beaches, as well as for the associated "sub-beaches" listed in Tables 16.2 and 16.3. The average area is also shown in Table 16.4. It has been obtained after multiplying the average length by the average width. Since the width

Table 16.3 Idem Table 16.2 for Gran Canaria Island beaches

#	Beach name	Beach type	Length (m)	Width (m)
1	El Confital	1	440	0–15
2	Las Canteras	1	2830	0–70
3	El Altillo	2	400	0–15
4	San Andrés	2	810	0–15
5	San Felipe	2	200	20
6	Bocabarranco	2	190	50
7	Sardina	1	75 + 40	0–25
8	Juncal	3	120	25
9	Puerto de Las Nieves	3	215	15
10	Agaete	3	170	0–20
11	Guayedra	2	140	35
12	El Risco	2	390	25
13	La Aldea	2	105 + 80	20
14	El Charco	3	360	15
15	Guguy Chico	1	550	0–30
16	Guguy Grande	2	210	10–25
17	Tasartico	3	460	0–20
18	Tasarte	3	700	0–25
19	El Ambar	2	275	0–20
20	Los Secos	3	140	15
21	Veneguera	2	370	15–30
22	Mogán	4	215	15–55
23	Taurito	1	220	0–60
24	El Cura	1	250	0–25
25	Tauro	4	190	15
26	Amadores	4	490	30
27	Puerto Rico	4	280	40–80
28	Anfi del Mar	4	200	25–50
29	Patalavaca	1	90 + 125	5–30
30	Marañuelas	1	270	10

(continued)

Table 16.3 (continued)

#	Beach name	Beach type	Length (m)	Width (m)
31	Cementera	2	380	15
32	El Pajar	2	75 + 140	10–25
33	Triana	3	220 + 145 + 115	15
34	Montaña Arena	1	240	10
35	Pasito Blanco	1	180	5–40
36	Meloneras	4	270	45
37	Maspalomas	1	2710	10–80
38	El Inglés	1	2700	70
39	El Veril	4	110 + 280 + 330	0–70
40	Las Burras	4	330	10–90
41	San Agustín	1	195 + 400	0–80
42	El Aguila	2	400	15
43	Tarajalillo	2	230	10
44	Juncalillo del Sur	3	2950	10
45	El Matorral	3	1060 + 750	10
46	Tenefé	3	2270	10
47	Pozo Izquierdo I	2	360	10
48	Pozo Izquierdo II	4	290	10
49	Bahía de Formas	2	500	10–30
50	Arinaga	2	320	0–10
51	Vargas	2	1200	15
52	El Burrero	2	400	10–20
53	Aguadulce	1	100	15
54	Salinetas	1	270	25
55	Melenara	1	310	15–85
56	El Hombre	1	230	15–80
57	La Garita	1	230	20–65
58	San Borondón	1	190	50
59	Bocabarranco	1	330	60
60	La Laja	4	1260	0–40
61	San Cristóbal	3	270	0–15
62	Alcaravaneras	4	440	0–95

was measured from the high tide line until the upper limit of the beach, the average area corresponds to the area landward from the foreshore, what is mostly used for sunbathing. This is what Komar (1998) relates to as the backshore, or what Short (1999) defines as subaerial beach.

At Tenerife Island there are 93 beaches larger than 100 m and this number expands to 124 when considering the sub-beaches. These beaches cover a total

Table 16.4 Total and average values for the different beach types and for both considered islands

Island	Beach type	Total length (m)	Beaches		Sub-beaches		Average width (m)	Average area (ha)
			Number	Average length (m)	Number	Average length (m)		
Tenerife	1	7140	23	310	39	183	19	0.34
	2	5835	20	292	24	243	9	0.23
	3	9505	33	288	39	244	12	0.29
	4	9240	17	544	22	420	35	1.47
	Total	31,720	93	341	124	256	17	0.44
Gran Canaria	1	14,025	20	701	23	610	31	1.92
	2	7195	19	379	21	343	18	0.61
	3	9945	12	829	15	663	13	0.86
	4	3635	11	330	13	280	35	0.98
	Total	34,800	62	561	72	483	24	1.17

Beach type: sandy beach (1), mixed beach (2), pebbles-cobbles beach (3), artificial beach (4)

length of nearly 32 km, which represents 9.3% of the 342 km of coastline. The four types of beaches are well represented, both in number and total length, but there is a clear difference in size between natural and artificial beaches. While the first ones have average length of roughly 300 m and widths ranging from 9 to 19 m, the average values for artificial beaches reach 550 m length and 35 m width. Consequently, the average area of natural beaches is only 0.28 ha while the artificial ones are more than 5 times that size, expanding up to 1.47 ha.

The lower values of width in natural beaches (both sandy, mixed and pebble beaches) is mostly due to the very narrow shelf, so that marine sediments are very scarce. On the contrary, artificial beaches are much wider thanks to a sort of coastal protection structures that creates a shelter to waves action, as well as thanks to the artificial nourishment carried out in most of them.

At Gran Canaria Island the situation is slightly different. There are 62 beaches larger than 100 m (72 considering sub-beaches), but they extend over nearly 35 km. Since the island perimeter is 256 km, it indicates that beaches represent 13.6% of the island perimeter. The most abundant ones are the natural sandy beaches, which cover 14 km, followed by the pebble and cobble beaches that extends over nearly 10 km. Beaches in these two categories are the largest ones, since they have 700 and 830 m average length respectively. On the contrary, mixed and artificial beaches are much shorter, since they extend for only roughly 350 m. Regarding beach amplitude, artificial and natural sandy beaches are the widest with average values of 31–35 m, while pebble-cobble and mixed beaches are much narrower, since the average width is only 13 and 18 m respectively.

Natural sandy beaches are by far the largest type of beaches, with an average area of nearly 2 ha. Artificial beaches are half that size and both mixed and pebble-cobble beaches are slightly smaller, but much larger in any case than those at Tenerife Island.

The lower number and size of artificial beaches in Gran Canaria compared to Tenerife is explained by the much larger extension of natural sandy beaches (14 km at Gran Canaria versus 7.1 km at Tenerife). In both islands the pressure derived both from tourism and local residents is very high, which demand adequate infrastructures, and sandy beaches are a key infrastructure for recreational purposes. Therefore, administrations have had to create more artificial beaches at Tenerife Island to compensate the lower number of natural sandy beaches. As stated by Cooper and Alonso (2006), nourishment is justified on the basis of creating recreational space.

In Gran Canaria there are a good number of natural sandy beaches, some of them quite large (Las Canteras, Maspalomas and El Inglés beaches, each one of them around 2.8 km length) and with adequate services for users. On the contrary, there is only one natural sandy beach at Tenerife larger than 1 km (La Garañona, beach number 16 in Table 16.2). Even though this beach is spectacular (Fig. 16.3a, b), it is not heavily used because a number of factors: it is very narrow at high tide; it is backed by a high cliff, so that during many hours it is in the shade; it can be dangerous because of big waves and debris falling from the cliff and it is not easily accessible to people.

16.3 Case Studies

16.3.1 Movement of Pebbles and Cobbles at San Felipe Beach (Gran Canaria)

San Felipe beach is a mixed beach, 200 m long, located in the northern cost of Gran Canaria Island (number 5 in Fig. 16.4 and Table 16.3). It is a perfect example of how these beaches evolve both in the medium and in the short term. It follows a clear seasonal pattern. In summer time, the lower foreshore is fully covered by sand, with pebbles only in the upper foreshore and the backshore. The general appearance is very similar to that of Fig. 16.5b. On the contrary, during winter, the beach is exposed to higher waves, and the increase in wave energy results in the offshore movement of the sand and the onshore movement of part of the pebbles, which creates a higher winter berm (Casamayor et al. 2017).

Nevertheless, the beach behavior in the shorter term, and particularly during the stormy period, shows that the coarser particles move not only in the cross-shore direction, but mostly in the longshore one. To follow the movement of individual pebbles and cobbles on the beach, Casamayor et al. (2015) used passive integrated transponders in 198 particles, according to the methodology proposed by Allan et al. (2006). Tracer weight ranges between 82 and 2837 g with an average of 450.6 g.

The beach was divided in six sectors considering both the alongshore and the cross-shore directions: East, center and West in the alongshore direction, and upper and lower foreshore in the cross-shore direction. Particles used as tracers were displaced on the beach on March 14, 2013, with nearly the same number of particles in each sector (Fig. 16.6). The distribution of the pebbles on the beach was randomly, so that in each sector there was approximately the same number of lighter and heavier particles.

Location of tracers on the beach was measured three times in less than one month after their initial location on the beach. Therefore, the movement followed by these particles during this period (from March 14 until April 10, 2013) can be separated in three shorter episodes: First one took place from March 14 until March 19. During this episode wave height was very small and the tagged particles showed slight movements. Most of them remained in their original sector, but a small number of pebbles initially located in the upper foreshore fell down to the lower foreshore.

Second episode took place between March 19 and April 2. During this period, a stormy event with waves around 3 m significant wave height hit the area. These NNW waves pulled mostly the lower foreshore, and therefore particles located along this area were detected quite far eastwards from their original location.

During third episode (lasting between April 2 and April 10) another stormy event took place in the area. In this case significant wave heights were also around 3 meters and offshore direction was from the NNW, but this event lasted for nearly 60 h while the previous one lasted for only 36 h. The resulting situation was a new longshore transport towards the east of most particles located in the lower foreshore. In addition, some particles previously located in the upper foreshore showed an eastward movement, which indicates that they fell from the upper berm and then were also pulled eastwards by the waves.

Wave data shown in Fig. 16.6f corresponds to SIMAR point 4035011, located approximately 3 km offshore San Felipe beach. They are the outcome of WAM and WaveWatch coupled models (Puertos del Estado 2015) and therefore they were not directly measured on the field and have not been propagated to the coast. Nevertheless, this experiment shows that pebbles and cobbles move on the beach depending on wave conditions, so that higher wave height determines that pebbles move longer distances in the longshore direction, and this transport takes mostly place in the lower foreshore.

16.3.2 Medium Term Sedimentary Balance at Las Canteras Beach

Las Canteras beach is an urban beach, since it is located adjacent to Las Palmas de Gran Canaria, the largest city in the Canaries. This is a sandy beach nearly 3 km long located in the northeastern cost of Gran Canaria Island (number 2 in Fig. 16.4 and Table 16.3). Boundary conditions are the key aspect when talking about beach

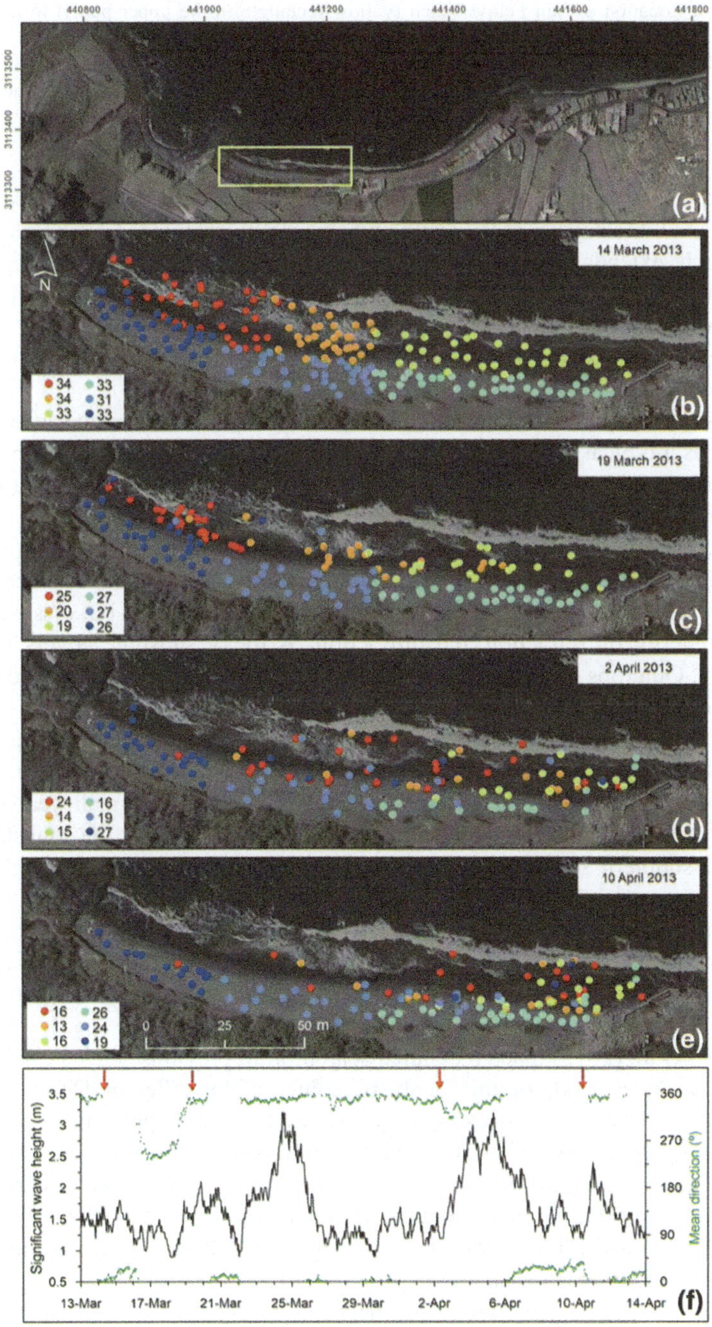

440800
441000
441200
441400
441600
441800
3113500
3113400
3113300
(a)
14 March 2013
N
34
34
33
33
31
33
(b)
19 March 2013
25
20
19
27
27
26
(c)
2 April 2013
24
14
15
16
19
27
(d)
10 April 2013
16
13
16
26
24
19
0
25
50 m
(e)
Significant wave height (m)
Mean direction (°)
3.5
3
2.5
2
1.5
1
0.5
360
270
180
90
0
13-Mar
17-Mar
21-Mar
25-Mar
29-Mar
2-Apr
6-Apr
10-Apr
14-Apr
(f)

◂**Fig. 16.6** **a** Location of San Felipe beach (yellow rectangle in the upper panel) in the northern coast of Gran Canaria. **b** Original location of the traces (colour dots) in the six defined sectors (see text). **c**, **d**, **e** Tracers location in the subsequent surveys. **f** Evolution of wave height and wave direction measured offshore during the experiment. The red arrows refer to the dates of determining the tracers location on the beach

morphodynamic at this beach. The beach is bounded by a rocky headland at the north, a small breakwater at the south and a partially fragmented offshore sandstone bar, which extends parallel to the shoreline 200–300 m offshore (Fig. 16.7a). According to Alonso and Vilas (1994), the boundary conditions determine the existence of three different sectors along the beach, which resemble to the beach states defined by the well-known beach morphodynamic model proposed by Wright and Short (1983).

The southern part of the beach is completely exposed to incident waves and it behaves as a typical dissipative beach: mild foreshore slope, very occasional beach cusps and an offshore sand bar in winter that migrates onshore, forming a subaerial sand bar during late spring and summer.

The central sector of the beach is partially protected from wave energy by the two main fragments of the offshore rocky bar, but the opening between them is large and deep enough for waves to come in without breaking, but dissipating part of their energy by refraction and diffraction. A rocky substrate covers the lower foreshore. Outcrops of this substrate may be either exposed or fully buried by sand, depending on the amount of sand on the area.

The northern sector is the most sheltered one due to the presence of the offshore bar as well as by the general shoreline configuration. This area is much more reflective, with a steeper beach slope and regular beach cusps all year round. It shows a natural trend to accumulate sediments transported by the longshore drift from the southern sector. This trend has been observed for many years (Alonso 1993; Alonso et al. 2000), and different administrations have carried out along the years, topographic measurements and digital terrain models (DTM) of the area, to determine the necessity to remove part of this accumulated material. Because of it, in 2003 and 2009 different amounts of sand were extracted to nourish other beaches in Gran Canaria.

Alonso et al. (2015) made a detailed search of all previous studies carried out in the beach. As a result of this analysis, some of the different DTMs were not comparable to the others because they followed different procedures in the data acquisition. Only four of these studies were considered perfectly comparable between them. These are the studies carried out by Bolaños and Santana (1997), Medina et al. (2006), Gómez Llorca (2012), and Alonso et al. (2015). All of them included a detailed DTM of the beach face and the subaerial beach at the northern sector of Las Canteras beach.

Figure 16.7 shows the comparison between consecutive DTMs and Table 16.5 shows the values (in m^3) obtained after subtracting one DTM from the following one. To make a right interpretation of the values shown in Table 16.5, it is necessary to take into consideration several elements.

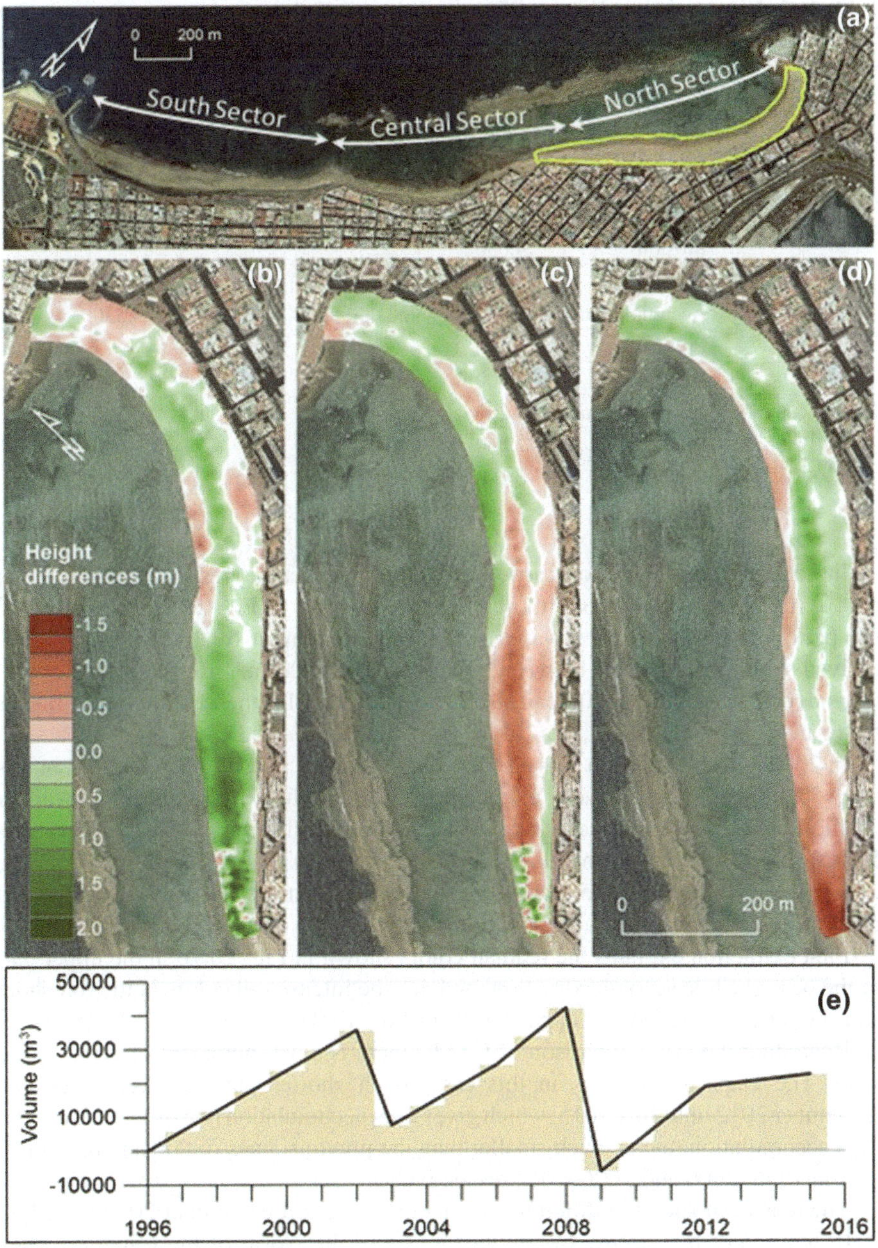

Fig. 16.7 **a** Aerial view of Las Canteras beach, showing the three sectors that form the beach. Note the presence of the offshore bar. The yellow polygon corresponds to the area measured in the different DTMs. **b** Difference between DTMs carried out in August 1996 and August 2006. **c** Difference between DTMs carried out in August 2006 and November 2012. **d** Difference between DTMs carried out in November 2012 and July 2015. **e** Evolution of the amount of sediments at the beach considering the accumulation rates and the two extractions carried out in 2003 and 2009

Table 16.5 Volume changes between DTMs carried out by Bolaños and Santana (August 1996), Medina et al. (August 2006), Demarcación de Costas (November 2012) and Alonso et al. (July 2015). Volume of material removed in each extraction, elapsed period between DTMs and accumulation rate are also shown

Period	Change	Volumen change (m^3)	Extraction (m^3)	Time interval (years)	Accumulation rate (m^3/year)
August 1996–August 2006	Accumulation	29.450	34.000	10	5.940
	Erosion	4.050			
	Difference	25.400			
August 2006–November 2012	Accumulation	12.350	56.400	6.3	8.330
	Erosion	16.250			
	Difference	−3.900			
November 2012–July 2015	Accumulation	15.260		2.67	1.200
	Erosion	12.060			
	Difference	3.200			

In 2003 the administration carried out a sand extraction of sand from the upper part of the beach profile along most of the area shown in Fig. 16.7b. Therefore, in the period 1996–2006 there was a sand accumulation of 59.400 m^3 (25.400 m^3 derived from the balance between DTMs and 34.000 m^3 from the extraction). Considering a period of 10 years between both DTMs, the accumulation rate is 5.940 m^3/year.

In 2009 a new extraction was also carried out in this part of the beach, but in this case 56.400 m^3 of sand were extracted both from the foreshore and the upper profile. Therefore, the net accumulation of sediments in the period 2006–2012 was 52.500 m^3 (−3.900 m^3 from the balance between MDTs and 56.400 m^3 from the extraction). In this case the elapsed period is 6.3 years (from August 2006 until November 2012), and the accumulation rate increases up to 8.330 m^3/year.

First extraction explains the reddish colors shown in Fig. 16.7b in the upper part of the beach, while second extraction explains the intense red colors in the foreshore and the reddish ones in the upper beach in Fig. 16.7c.

Regarding the last comparison of DTMs, there is a net increment of 3.200 m^3 of sand. The elapsed period is in this case much shorter, only 2.7 years between November 2012 until July 2015, which gives a net accumulation rate of 1.200 m^3/year. This accumulation rate is much smaller than the previous ones (6.000 and 8.300 m^3/year in 1996–2006 and 2006–2012 respectively).

The reason of such big difference is that in the period November 2012 until July 2015 three very big stormy events took place in the North of the Canaries. They happened in February 2014, November 2014 and February 2015. Significant wave heights larger than 5.3 m were recorded in each one of them (Alonso et al. 2015). Wave heights were recorded at Gran Canaria wave buoy located offshore in the North coast of Gran Canaria. Propagation of such swell to the coast could perfectly give values of maximum breaking wave heights around 9 m. These three events not only lasted each one of them for many hours, but what it is really unusual, is that the three of them took place just in the period of twelve months. The first of these events was particularly strong, and its effects in the north of Spain were reported by Flor et al. (2015). Las Canteras beach could not recover from the erosive effect of this event because few months later a new storm took place.

The sedimentary balance shown in Fig. 16.7d reflects the effect of those storms. The red colors represent a loss of nearly 12.000 m^3 of sand. The eroded area shows a shape like a wedge, being more intense towards the South and particularly in the lower foreshore. Both elements (shape and location) reflect that the loss of sand is related to erosive processes. Therefore, the low accumulation rates obtained for the period 2012–15 are related with the mentioned stormy events.

Figure 16.7e shows the evolution of the volume of sediments in this northern sector of Las Canteras beach in nearly two decades relative to the situation measured in August 1996. This figure has been drawn considering that the amount of sand is cero in August 1996, and that the accumulation rates mentioned in previous paragraphs keep constant during the periods between consecutive DTMs. It shows a steady increment of sand between 1996–2003 at the rate of 6.000 m^3/year. A sharp decrease in 2003 of 34.000 m^3 represents the first extraction. Since then the accumulation processes continues until 2006, when the accumulation rate increases at 8.300 m^3/year. In 2009 the second extraction took place, and the volume of material was reduced to values even lower than those of 1996. Nevertheless, the beach quickly recovered and by November 2012 there was a new excess of nearly 20.000 m^3 of sand. Since then the accumulation rate is much smaller due to the high energy events recorded in the area. In July 2015 the amount of material is approximately 23.000 m^3 more than the amount existing in 1996.

This sector of the beach presents a natural trend to gain sediments even under very adverse circumstances as it has happened in 2014 and 2015. Nevertheless, the future evolution is uncertain, since the accumulation rate depends on wave climate. The administration should consider the convenience of making new DTMs in order to determine if new extractions may be necessary in the near future.

References

Allan, J. C., Hart, R., & Tranquili, J. V. (2006). The use of Passive Integrated Transponder (PIT) tags to trace cobble transport in a mixed sand-and-gravel beach on the high-energy Oregon coast USA. *Marine Geology, 232*, 63–86.

Alonso, I. (1993). *Procesos sedimentarios en la playa de Las Canteras (Gran Canaria)* (PhD Thesis, Univ Las Plamas de Gran Canaria), 333 pp.

Alonso, I., Alcántara-Carrió, J., & Cabrera, L. (2002). Tourist resorts and their impact on beach erosion at Sotavento beaches Fuerteventura Spain. *Journal of Coastal Research SI, 36*, 1–7.

Alonso, I., Sánchez-García, M. J., Rodríguez-Valido, S., & Casamayor, M. (2015). *Estudio del cálculo del balance sedimentario en la playa de Las Canteras* (Technical report). Ayuntamiento de Las Palmas de Gran Canaria, 72 pp.

Alonso, I., & Vilas, F. (1994). The influence of boundary conditions on beach zonation. In A. S. Arcilla, M. J. F. Stive, & C. N. Kraus (Eds.), *Proc coastal dynamics* (pp. 417–431). ASCE.

Alonso, I., Vilas, F., & Alcántara-Carrió, J. (2000). Importancia de la escala temporal en estudios de dinámica litoral. In J. R. de Andrés & F. J. Gracia (Eds.), *Geomorfología litoral Procesos activos* (pp. 31–43). UCA SEG and ITGE.

Bolaños, A., & Santana, S. (1997). *Batimetría y topografía de la playa de Las Canteras* (Academic report). Universidad de Las Palmas de Gran Canaria.

Cantagrel, J. M., Fuster, J. M., Pin, C., Renaud, U., & Ibarrola, E. (1993). Age Miocene inferieur des carbonatites de Fuerteventure (23 Ma:U-Pb zircon) et le magmatisme précoce d´une ile océanique (Iles Canaries). *Comptes Rendus de l'Academie des Sciences de Paris, 316,* 1147–1153.

Carracedo, J. C., Day, S., Guillou, H., Rodríguez Badiola, E., Canas, J. A., & Pérez Torrado, F. J. (1998). Hotspot volcanism close to a passive continental margin: The Canary Islands. *Geological Magazine, 135,* 591–604.

Carracedo, J. C., Rodríguez Badiola, E., Guillou, H., Paterne, M., Scaillet, S., Pérez Torrado, F. J., et al. (2007). Eruptive and structural history of Teide Volcano and rift zones of Tenerife Canary Islands. *Bulletin of the Geological Society of America, 119*(9–10), 1027–1051.

Casamayor, M., Alonso, I., Cabrera, J., Rodríguez, S., & Sánchez-García, M. J. (2015). Long term recovery rates obtained using RFID technology at a mixed beach. *Geologica Acta, 13*(2), 85–96.

Casamayor, M., Alonso, I., Sánchez, M. J., Montoya-Montes, I., & Rodríguez, S. (2017). Procesos de acumulación y erosión en playas mixtas en relación con el oleaje. *Geo-Temas, 17,* 83–86.

Cooper, J. A. G., & Alonso, I. (2006). Natural and anthropic coasts: Challenges for coastal management in Spain. *Journal of Coastal Research, SI 48,* 1–7.

Criado, C., Yanes, A., Hernández, L., & Alonso, I. (2011). Origen y formación de los depósitos eólicos en Canarias In E. Sanjaume & F. J. Gracia (Eds.), *Las dunas en España* (pp. 447–465). SEG.

Ferrer-Valero, N., Hernández-Calvento, L., & Hernández-Cordero, A. I. (2017). Human impacts quantification on the costal landforms of Gran Canaria Islands (Canary Islands). *Geomorphology, 286,* 58–67.

Flor, G., Flor-Blanco, G., Flores-Soriano, C., Alcántara-Carrió, J., & Montoya-Montes, I. (2015). Efectos de los temporales de invierno de 2014 sobre la costa asturiana. *Geo-Temas, 15,* 17–20.

García-Rodríguez, J.-L., García-Rodríguez, F. J., & Castilla-Gutiérrez, C. (2016). Human heritage and sustainable development on arid islands: The case of the Eastern Canary Islands. *Island Studies Journal, 11*(1), 113–130.

García-Romero, L., Hernández-Cordero, A. I., Fernández-Cabrera, E., Peña-Alonso, C., Hernández-Calvento, L., & Pérez-Chacón, E. (2016). Urban-touristic impacts on the aeolian sedimentary systems of the Canary Islands: Conflict between development and conservation. *Island Studies Journal, 11*(1), 91–112.

Garín-Muñoz, E. (2006). Inbound international tourism to Canary Islands: A dynamic panel data model. *Tourism Management, 27,* 281–291.

Geyer, A., & Martí, J. (2010). The distribution of basaltic volcanism on Tenerife Canary Islands: Implications on the origin and dynamics of the rift systems. *Tectonophysics, 483,* 310–326.

Gómez Llorca, L. M. (2012). *Levantamiento topográfico de la playa de Las Canteras*. Technical report. Demarcación de Costas de Canarias (MAGRAMA).

Hernández-Calvento, L., Ruiz, P., Pérez-Chacón, E., Alonso, I., Alcántara, J., & Suárez, C. (2005). Transformaciones ambientales en ecosistemas dunares producidas por el desarrollo turístico: el ejemplo de Maspalomas (Gran Canaria Islas Canarias). In E. Sanjaume & J. F. Mateu (Eds.), *Geomorfología litoral y Quaternari* (pp. 203–214).

ISTAC. (2018). Instituto Canario de Estadística, Gobierno de Canarias. http://www.gobiernodecanarias.org/istac/.

Komar, P. D. (1998). *Beach processes and sedimentation* (2nd ed., 544 pp). Prentice Hall.

Maestro-González, A., Medialdea-Cela, T., Llave-Barranco, E., Somoza-Losada, L., & León-Buendía, R. (2005). El margen continental de las Islas Canarias. In A. Martín Serrano (Ed.), *Mapa geomorfológico de España y del margen continental a escala 1:1000000* (pp. 229–232). IGME.

MAGRAMA. (2002). *Estudio ecocartográfico de la zona norte del litoral de Gran Canaria* (Technical report). Secretaría de Estado de Aguas y Costas Dirección General de Costas.

Marrero, N., Montoya, I., & Alonso, I. (2017). Evolución de la línea de costa de Tazacorte entre 1964 y 2015 (La Palma Islas Canarias). *Geo-Temas, 17,* 247–250.

Masson, D. G., Watts, A. B., Gee, M. J. R., Urgeles, R., Mitchel, N. C., Le Bas, T. P., et al. (2002). Slope failures in the flanks of the western Canary islands. *Earth Science Reviews, 57,* 1–35.

Medina, R., Bastón, S., Cánovas, V., Torres, A., Luque, A., Alonso, I., et al. (2006). *Estudio Integral de la Playa de Las Canteras* (Technical report). Dirección General de Costas, 301 pp + anejos.

Menéndez, I., Silva, P. G., Martín-Betancor, M., Pérez-Torrado, F. J., Guillou, H., & Scaillet, S. (2008). Fluvial dissection isostatic uplift and geomorphological evolution of volcanic islands (Gran Canaria Canary Islands Spain). *Geomorphology, 102*(1), 189–203.

Morales, G. M., & Santana, A. S. (1993). Procesos de construcción y transformación del espacio litoral grancanario inducidos por el fenómeno turístico. *Eria, 1993,* 225–246.

Moreno Gil, S. (2003). Tourism development in the Canary Islands. *Annals of Tourism Research, 30*(3), 744–747.

Oglialoro, E., Frezzotti, M., Ferrando, S., Tiraboschi, C., Principe, C., Groppelli, G., et al. (2017). Lithospheric magma dynamics beneath the El Hierro Volcano Canary Islands: Insights from fluid inclusions. *Bulletin of Volcanology, 79,* 70.

Parra-López, D. E., & Baum, T. (2004). An analysis of supply-side relationships in small island destinations: The role of tour operators travel agencies and tourism transport in the Canary Islands. *Tourism and Hospitality Planning & Development, 1*(3), 201–218.

Puertos del Estado. (2015). *Conjunto de datos SIMAR* (Technical report). Available at http://www.calipsopuertoses//BD/informes/INT_8pdf.

Rivera, J., Hermida, N., Arrese, B., Gonzalez-Aller, D., Sánchez de Lamadrid, J. L., Gutierrez de la Flor, D., et al. (2014). Bathymetry of a new-born submarine volcano: El Hierro Canary Islands. *Journal of Maps, 10*(1), 82–89.

Short, A. D. (1999). Beaches. In A. D. Short (Ed.), *Handbook of beach and nearshore morphodynamics* (pp. 3–20). Willey.

Urgeles, R., Canals, M., Baraza, J., Alonso, B., & Masson, D. G. (1997). The most recent mega landslides on the Canary Islands: El Golfo Debris Avalanche and the Canary Debris Flow west El Hierro Island. *Journal Geophysical Research, 104,* 25331–25348.

Wright, L. D., & Short, A. D. (1983). Morphodynamics of beaches and surf zone in Australia. In P. D. Komar (Ed.) *CRC Handbook in coastal processes and erosion* (pp. 35–64). CRC Press.

Part III
Fluvio Marine Systems

Chapter 17
The Galician Rías. NW Coast of Spain

Federico Vilas, Ana Bernabéu, Belén Rubio and Daniel Rey

17.1 Introduction

The rías on the coast of Galicia make up a series of deep sea inlets along the entire 1720 km long coastline. In plant view, their morphology is cone or funnel-shaped and features two distinct sectors: the first makes up the larger cone, which corresponds to the ría itself, and the second is the smaller cone corresponding to the river channel which discharges into the headwaters. Rías can be classified as drowned river valleys that were flooded by seawater after the last transgression, with rivers at their headwaters, all of which bring about conditions more typically seen in estuaries. They are also associated with a complex variety of sedimentary environments, such as bays, intertidal plains, deltas, marshlands, beaches and sand dunes.

The orientation of the rías is structurally controlled by a sub-continental basement fracture system related to the break-up of Gondwanaland. From a geological point of view, Galician rías trend north-eastward, nearly at right angles to the main north trending Palaeozoic structural trends. Its coast is characterized by an irregular morphology tectonically controlled by three main fault systems: NE-SW, ENE-WSW and N-S trending faults. These fracture systems control the orientation

---

F. Vilas · A. Bernabéu · B. Rubio · D. Rey (✉)
Grupo GEOMA. Dpto. Geociencias Marinas y O.T. Universidad de Vigo,
36310 Vigo (Pontevedra), Spain
e-mail: danirey@uvigo.es

F. Vilas
e-mail: fvilas@uvigo.es

A. Bernabéu
e-mail: bernabeu@uvigo.es

B. Rubio
e-mail: brubio@uvigo.es

J. A. Morales (ed.), *The Spanish Coastal Systems*,
https://doi.org/10.1007/978-3-319-93169-2_17

of the main longitudinal axis of most of the major rías, particularly those facing the Atlantic Ocean.

In the scientific literature, the terms estuary and ría have always been used interchangeably, and as a result many of the studies carried out on estuaries were similarly applied to rías, especially with regard to the sea and river water mixing processes (Boyd et al. 1992; Dalrymple et al. 1992). Nevertheless, today we know that rías are in fact different and that they have a more complex arrangement of environments, dominated by processes related to wave energy, in which estuarine areas can be clearly distinguished (Vilas et al. 2010).

Etymologically, the word "estuary" is derived from the Latin "aestuarium", meaning marshland or canal, which in turn is derived from the word "aestus", which means "tide". Therefore the term is applied to any coastal environment where the tide is particularly important. As a result, in the second half of the 14th century, the terms "ría" and "estuary" were used interchangeably and both referred to the part of the river at the sea outlet. There is a great variety of definitions, depending on the different fields which include the study of estuaries. Many are contradictory, due to the varying knowledge of researchers, or due to the specific characteristics of those bodies of water partially enclosed in a coastal area, which at the same time receive fluvial inputs. The criterion for the definition put forward by Pritchard (1952, 1967) is salinity, whereas Fairbridge (1980) takes into account how far the sea reaches into a river valley. Both definitions are complementary and they are also valid in so far as determining the upstream limits of salinity penetration, as well as of tidal penetration, and both work well for those estuaries linked to river mouths. Other definitions try to adapt to the different peculiarities that are present in different coastal inlets, such as coastal lagoons, which are only occasionally connected to the sea (Day 1980) or including a biological component (Perillo 1995). There are other varying definitions for the term 'estuary', depending on whether they emphasize physical, eustatic, tectonic or sedimentological criteria. Taking into account the latter, Dalrymple et al. (1992) establishes differences in the upper and lower limits of estuaries with regard to definitions based on salinity. He defines an estuary as "the seaward portion of a drowned valley system which receives sediment from both fluvial and marine sources and which contains facies influenced by tide, wave and fluvial processes. The estuary is considered to extent from the landward limit of tidal facies at its head to the seaward limit of coastal facies at its mouth". From a geomorphological point of view, the term "ría" is used by Perillo (1995), albeit to refer to a type of estuary and only in reference to those rías in which an area may be influenced by estuarine processes.

The term "incised valleys" has also been used in studies into the sedimentary fill of these environments (Dalrymple et al. 1994), given that these are areas where it is possible to identify the transgressive (TST) and the highstand sequence (HST), which, in general, fills the resulting incision during one lowstand-transgressive-highstand sequence.

As occurs in estuaries, the "rías" along the Galician coast are influenced by mechanisms such as variations of the tidal range (Davies 1964), the degree of energy dominating in different sectors and also the fluvial discharge into the headwaters (Vilas and Nombela 1985; Vilas et al. 2005). An understanding of these environments makes it possible not only to characterize a specific environment, but also to interpret the various facies preserved in the fossil record. Furthermore, these are areas of high biological productivity and therefore subject to various economic and human activities; as a result, any information regarding their sedimentology and hydrodynamics is of great interest, due to variations in the sedimentary bottoms and their influence on the distribution of species.

17.2 The Term Ría

The term "ría" has a relatively old origin, according to the revision by Méndez and Rey (2000) and Méndez and Vilas (2005). It is found in the 1495 edition of a Spanish-Latin vocabulary by Elio A. de Nebrija, denoting a "river port, ostium fluminis". In 1780, the Royal Academy of the Spanish Language generalised its use to refer to a geographical area with a characteristic topography or morphology, defined as "the part of the river at the sea outlet". Later, Von Richthofen (1986) adopted the term "ría" to designate a type of coast characterized by the existence of a valley occupied by the sea, taking as an example the Galician rías. Numerous works appear once the term "ría" is introduced in the scientific literature. It is worth highlighting the cartographic works by Schurtz (1902), Scheu (1913), Torre Enciso (1958), geomorphological and tectonic works by Carlé (1947, 1949, 1950) and Nonn (1966), who establishes a morphological classification of the rías of the coast of Galicia, as well as the works by Pannekoek (1966a, b, 1970) in which he attributes the main characteristics of the relief of the rías coastline to the Hercynian faults reactivated during the Tertiary.

Other authors consider that the term "ría" should be restricted to specific coasts of the Iberian Peninsula coastline and other areas of high relief such as Brittany in France, Devon and Cornwall in the United Kingdom, Korea, Southeast China and the South of Patagonia in Argentina, among others (Castaing and Guilcher 1995; Goudie 2018). In his review of the term "estuary", Perillo (1995) also establishes a similar definition, although he associates the term "ría" to an estuary developed in a river valley on a high relief coast, giving as an example the rías on the Galician coast, in contrast to the estuaries occupying low relief coasts, such as the Thames or the Gironde. However, there is a dilemma regarding this assessment, given that it contemplates the existence of estuarine circulation in both cases; in the case of the Galician rías, although they feature similar hydrodynamic processes, typical estuarine circulation is only present in the inner most area of the ría, namely where the river flows into the ría (Ruiz-Villarreal et al. 2002; Souto et al. 2003; Piedracoba et al. 2005; Vilas et al. 2005). Besides, the characteristics and distribution of sediment in rías (Rubio et al. 2001; Rey et al. 2005; Vilas et al. 2005) also display significant differences with respect to the facies models described for estuaries. Furthermore, several works carried out on the Galician coast to establish the genesis and evolution of the north-western sector of the peninsula (Vidal-Romaní 1996;

Twidale and Vidal-Romaní 1994; Vilas et al. 1995, 1996, 1999b; Pagés-Valcarlos 2000, among others), as well as the evolution of the Galician coastline since the last glacial maximum (García-Gil et al. 2002; Vilas et al. 2005; Muñoz-Sobrino et al. 2016; Gonzalez-Villanueva et al. 2015, among others) make it possible to consider the term ría in a broader sense than a simple geomorphological context, as considered by Evans and Prego (2003). Recently, the Spanish Royal Academy of Exact, Physical and Natural Sciences (RAC), in its Scientific and Technical-Geological Lexicon (Vera et al. 2013), defines the term "ría" as: "A flooded river valley, developed on a rocky coast, submerged as a result of a relative sea level rise. In plan view, it is shaped like a funnel with deep entrances. A cross-section view reveals a trough-like shape and a flat bottom, except nearest the head, where it has a V shape. Most of the surface presents marine conditions, with swell that controls the distribution of sediments (sands in the shallowest areas and lutite in the deepest). Actual estuarine processes are only present at the mouths of the rivers flowing into the ría."

Types of Rías

Taking into account only the most significant geomorphological features present in the rías on the Galician coast, Nonn (1966) establishes three types of ría (Fig. 17.1). The first corresponds to the lowest area of a drowned river system, where the river is responsible for the amplitude of the ría and often for the course. Nevertheless, alteration processes of the rock along the banks may also be involved or small-scale tectonic processes. It is on the Cantabrian shore of the North coast of Galicia where the best examples can be found (rías of Ortigueira, Barqueiro, Foz, Ribadeo) although it is also possible to identify partially similar examples on the northwest coast (rías of Cedeira, Ferrol, Ares and Betanzos, Laxe and Camariñas). The second type is characterized by the preponderance of tectonics and, above all, when it is impossible to justify the size of the rías based on their main rivers. Consequently, the hydrographic systems of the river Verdugo-Oitavén corresponding to the Ría of Vigo, that of the river Lérez in the Pontevedra Ría and of the river Tambre in the Muros and Noia Ría, according to Nonn (1966), demonstrate that the present-day volume of water does not justify the dimensions of these rías; as a result he concludes that tectonics (by means of subsidence, uplift and rotation of the tectonic blocks of the emerged part) are responsible for the configuration of this sector, where the rías are known as Rías Baixas. The third type corresponds to drowned basins altered during the Tertiary; these rías are reached by rivers of a certain importance, whose courses are developed during periods of low sea level. They feature a characteristic "globular or amoeboid" shape (divided into branches, when seen from above) and exemplified on the Galician coast by the rías of A Coruña and Arousa.

Despite all of the above, it is also possible to interpret the existence of rías with mixed characteristics, given that these different types are determined by the prevalence of certain mechanisms over others in order to establish this classification.

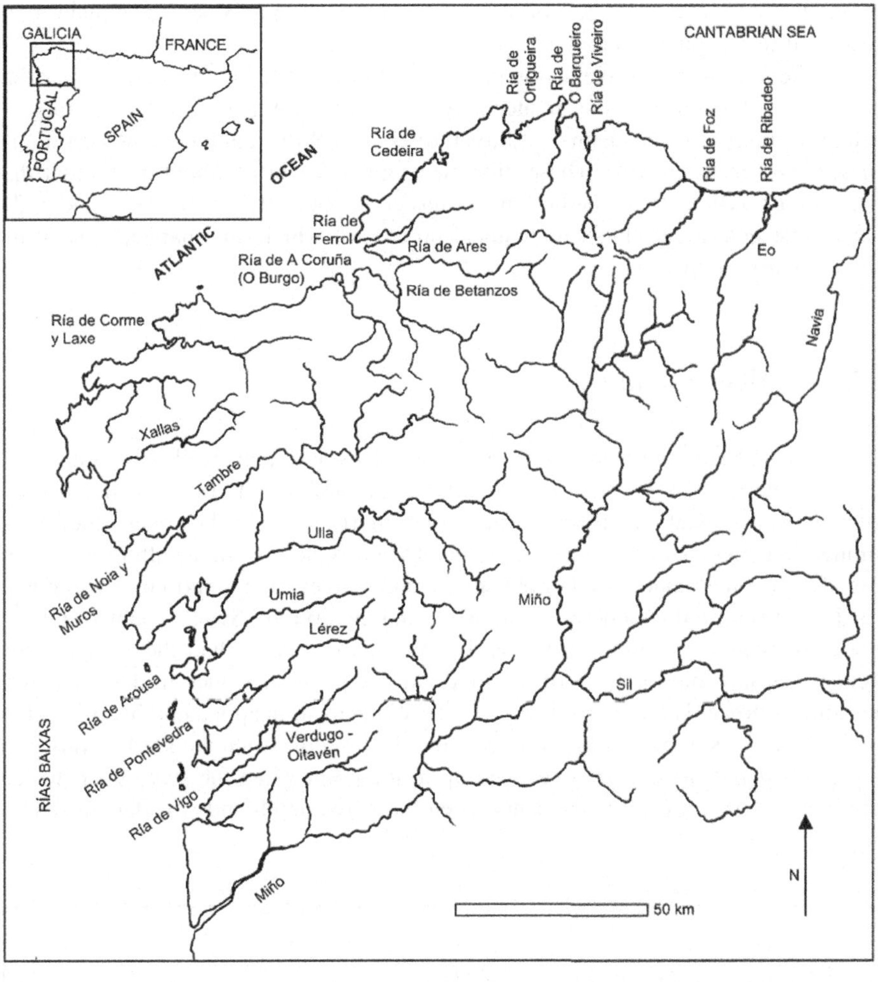

Fig. 17.1 Map of Galicia showing the rías along the coastline and the distribution of their hydrographic network (modified from Méndez and Vilas 2005)

17.3 Physical Processes

The movements of water masses and their variability in space and time are controlled by the balance between the tides, which in the coast of Galicia varies from 2 to 3 on the Atlantic shore, and up to 4 in the cantabrian coast area.

The mixing of fresh and seawater is a fundamental aspect in estuaries and rías, given that salinity gradually increases from the river towards the sea. In estuaries, this process may take place over a distance of dozens or even hundreds of kilometres, as occurs in the Gironde estuary (where it reaches 65 km) or in the Gambia River estuary (up to 200 km). In spite of the fact rías can extend more than 30 km in length, for example the Ría of Vigo, these water mixing processes remain more restricted to the inner sectors.

Unlike in estuaries, wave action in rías constitutes one of the main processes that control the dynamics of the greater part of this environment. On the contrary, in estuaries, forcing mechanisms are fluvial discharge at the head and daily variations in sea level at the mouth. These differences are what bring about a different distribution of sediments on the bottoms (Vilas et al. 2005; Rey et al. 2005) and at the same time determine physical-chemical processes, which are characteristic in the fluvial-marine transition.

17.3.1 Wave Action

In the open sea, there are two types of waves: sea waves and swell. The first, which are characterised by isolated and irregular crests and a wide range of heights and periods, can appear in the inner areas of estuaries or rías; however, due to the limited dimensions of these areas, the height and period of the resulting waves are low. On the other hand, swell-type waves are able to exert an important effect due to the interaction produced between the wave and the bottom. Swell-type waves reach the coast with well defined fronts and wavelengths greater than their height, thus generating an orbital motion of the fluid particles, which reduce in diameter from the surface towards the bottom (Fig. 17.2). Whereas in deep waters this movement of fluid particles diminishes before reaching the bottom, when the swell approaches the coast, depth (h) is slowly reduced up to the point when the swell starts to feel the bottom (base level). This occurs when the wave depth and wavelength ratio is

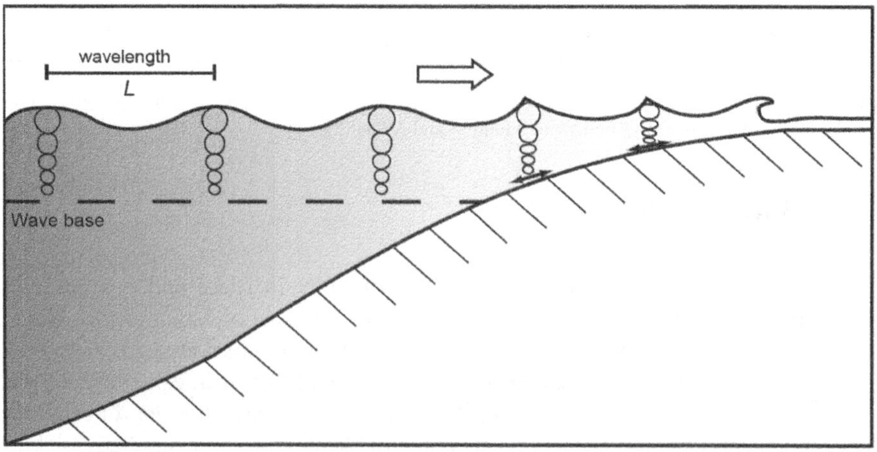

Fig. 17.2 Swell transition from deep to intermediate waters, showing the orbital movement of fluid particles as they approach the coast (modified from Vilas et al. 2010)

equal to or lower than 0.5. Under these circumstances, at depth, the circular orbit changes to elliptic and, at the bottom, the motion is limited to a back and forth motion. Swell energy increases as it moves towards the coast, and although in the case of estuaries it reaches its maximum at the mouth, in the case of rías, where the mouth is completely open and between 40 and 60 m in depth, swell is able to propagate without abatement almost up to the most internal areas, establishing different trajectories depending on the existing bathymetry.

17.3.2 Tides

Any tidal wave that reaches the coast and enters a semi-confined body of water in a coastal inlet, ría or estuary, generates a series of resounding effects which will modify the characteristics of the latter. These effects will be conditioned by two main parameters: the tidal prism and the cross-section of the estuary or ría basin. (The tidal prism can be defined as the volume of water between high tide and low tide that flows into and out of the estuary during the tidal cycle.) The variation of the area of said section together with the tidal prism will determine variations in the range of the tide and in the speed of the residual current in the basin of the estuary or ría. As the area of the basin diminishes due to the narrowing of the inner section nearer the head, the amplitude of the tide will increase and the current will intensify. This effect is offset by the gradual decrease in depth towards the head, which heightens the effect of the friction between the water column and the bottom. The behaviour of tidal waves in rías, given that the water column depth is much greater than the tidal range, corresponds to the hyposynchronic propagation model described by Le Floch (1961). On the other hand, in the inner most part it corresponds to the hyposynchronic type, because friction with the bottom and the margins of the basin bring about a reduction in the tidal parameters towards the river. Tidal processes in this area determine the mobilisation of sediments and their effect is often recognisable in sedimentary deposits. Similarly to meso and microtidal estuaries (Davies 1964), cyclical changes in the ebb and flow of the tidal currents may favour the resuspension of sediment from the bottom and the formation of the so-called turbidity maximum (Uncles and Stephens 1989). Recent works regarding the tidal effects on the external and central areas of rías (Rey et al. 2005; Vilas et al. 2005) have brought to light its limited influence on the transport of sediments and post-sedimentary processes. It is only in the inner most part, close to the mouths of the rivers, that the effect of the tide is clearly recognizable in the sedimentary deposits (Nombela et al. 1995).

17.3.3 Fluvial Discharges

The intensity and relative significance of fluvial currents decrease within the marine bodies of water that occupy the basins of the rías, as occurs in an estuary in the direction of the sea. This is due to the fact that the hydraulic gradient decreases in the area close to the river mouth. To this regard, the flushing capacity of fresh water or the outlet velocity in an estuary is a quantifiable parameter, which is based on the relationship between the annual average fluvial discharge and the transversal section of the area at the transition point of fresh and saline water (Gibbs 1977). The value of this parameter makes it possible to determine the landward intrusion limit of saline water and, consequently, the transition segment of the sedimentary sub-environments of the estuary.

Regardless of the discharge volume, fluvial currents are more significant in estuaries than in rías; they condition the longitudinal and transversal salinity gradients and therefore the density gradients, which in turn control estuarine circulation. They are also a source of sedimentary deposits.

17.3.4 Water Mixing and Estuarine Circulation

The interaction of fresh water and sea water in rías begins at the mouths of the river that flow into the rías. As a result of the difference in density between these two masses of water, a residual circulation is established, referred to as "estuarine circulation" (Dyer 1997), through which both water masses separate according to their different density: the denser (sea) waters below fresh river waters, as can be observed in estuaries. The velocity of the current is what determines the degree of the mixture between both, generating diffusion processes between them in the case of slow currents and more effective mixes for stronger currents. This variation in salinity not only generates a density gradient able to act as a driving force behind estuarine circulation; it also favours a series of physico-chemical flocculation processes and influences the formation of the turbidity maximum and the entrapment of fine particles from the middle of the ría up to the estuarine sector at the head of the ría.

In spite of the fact that the mixing and circulation processes in estuaries and rías are very similar, the differences that exist between the morphology of rías and that of estuaries bring about differences in the dynamic agents that dominate estuarine circulation. The mouths of estuaries are generally no deeper than 10 m, as a result only the tidal regime enables the entry of ocean water and the flow of fresh water towards the exterior predominates. The depth of rías, on the other hand, can be more than 50 m at their mouths, thus making it easy for a free and constant exchange of water mass and energy with the continental shelf (Souto et al. 2003).

17.4 Biogeochemical Processes

The variation in salinity from the river to the sea in rías, the same as in estuaries, is not only going to generate a density gradient that can act as a driving force behind estuarine circulation, but it also conditions a series of processes (flocculation, pelletisation, early diagenesis and methanogenesis) which are of interest in so far as recognizing the different sedimentary facies of these environments.

Flocculation by salts is a process which takes place whenever Van der Waals forces of attraction are involved. These forces are not particularly strong; nevertheless, their intensity varies inversely to the square of the distance between two particles of clay, and they eventually become important when said particles come close enough to each other. In fresh water, contrary to what happens in salt water, flocculation does not take place because clay particles are negatively charged and, when they come into close proximity, they repel each other because of their like charges. On the other hand, salt water contains free cations, which consequently interact with the negatively charged clay particles, thus reducing repulsion. In these conditions, the Van der Waals forces are able to overcome the repulsion forces. The formation of floccules will take place when the particles come into close enough proximity to each other.

The cohesive nature of the muddy sediments, which predominate in the central area of the rías and at the heads of estuaries, suggests that most of the sediment flocculates, except in extremely energetic conditions (Kranck and Milligan 1992). At present, it is known that particle flocculation is not an exclusively physico-chemical process, but rather a dynamically active process that can be easily modified by changes in hydrodynamic conditions (Manning and Bass 2006). Therefore, flocculation is also the function of the mechanisms that bring particles into contact, namely Brownian motion, turbulence (McCave 1985) and differential sedimentation (Manning and Dyer 1999) and of the mechanisms that bind them, which are salinity and the organic matter content (Van Leussen 1988). In some cases, the concentration can become so high that fluid muds are generated (an aqueous concentration of fine matter greater than 10 g/l). When tidal currents slow down, the sediment in suspension is deposited, creating a layer close to the bottom with a high concentration. If the dense suspension remains stationary, the basal part of the layer is consolidated in such a way that when the currents move faster again, they are not able to erode the material. The abundance and thickness of these muddy layers is directly related to the longitudinal variation of the matter in suspension. In the fluvial area, the concentration is low and the muddy layers are thin. Towards the sea, these layers are thicker and more abundant, due to the increase in the concentration of the matter in suspension; they reach their maximum just below the maximum turbidity peak.

As a result of the aforementioned mechanisms, the cohesive fine sediments can form a great range of aggregates known as floccules (Burban et al. 1989). Although they are less dense than the particles they are composed of, they sediment faster. As the floccules increase in size, their effective density generally decreases, but their sedimentation velocity increases. Wave or tidal energy is often able to hinder the sedimentation of floccules; nevertheless, most of them sediment in the fluvio-marine transition.

In the central and inner parts of rías, in the same way as at the heads of estuaries, the muddy deposits are also made up of biodeposits, faecal pellets and pseudo faeces. Organic matter plays an essential role in the formation of these, by means of agglomeration or aggregation processes by filter feeders; they are then added to the flocculation processes and can form particles large enough to sediment the bottom. Most of the particles between 1 and 5 μm are ingested and after passing through the digestive tract, they are compressed and expelled into the water as compacted faecal pellets, the sizes of which vary in length from 50 to 300 μm. Some of the material is rejected before ingestion and expelled back into the water, constituting what are known as pseudofaeces. Many filter feeders, which inhabit these environments, such as copepods, tunicates, mussels and oysters, among others, transform the organic matter in suspension into pellets, the sedimentation velocities of which are much greater than that of the individual constituents. Many of the clay-organic pellets end up sedimented in areas of high velocity currents, zones where loose clays and floccules would not be deposited. The pellets are usually very resistant because the fine grain particles are very compacted and bound by mucus, on the other hand, the pseudofaeces are packed more loosely. These are very common processes in estuaries and in the central and inner areas of the Galician rías, where the cultivation of mussels on rafts contributed to the formation of the same (Rubio et al. 2001; León et al. 2004).

There are other biogeochemical processes which also play a key role in the recognition of different sedimentological facies in these transition environments, namely mineralogical and geochemical transformations that take place during early diagenesis. These are fundamentally controlled by the oxidation of the organic matter. This oxidation, or remineralisation, consumes the oxygen present in the interstitial sediment pore water. The size of these pores and their intercommunication diminishes with regard to the grain-size of the sediment. As a result, the renewal of oxygen in these muddy sediments takes place at a much lower rate than in sandy sediments, resulting in the generation of anoxic areas of high organic matter content.

Early diagenesis is very intense in muds with high organic matter content, which are typical in estuaries (Turner and Millward 2002; Bush et al. 2004) and rías (Rubio et al. 2001, 2011; Rey et al. 2005). In the latter, the deepening of the redox boundaries varies from inner to outer sector of the ría (Fig. 17.3). The oxic zone expands as it gets deeper toward the outer ría, in a similar way as the suboxic, anoxic and methanic zones. This spatial trend can be explained, either by a progressive change in the hydrodynamic conditions along the ría or by the different origin (terrestrial or marine) of the organic matter and their aging in the water column (Andrade et al. 2011) In these type of sediments, oxygen is depleted extremely quickly generally just scarce centimetres from the surface. Under these conditions, the sediment progressively acquires suboxic conditions, which eventually become anoxic as the oxygen disappears completely. However, the oxidation

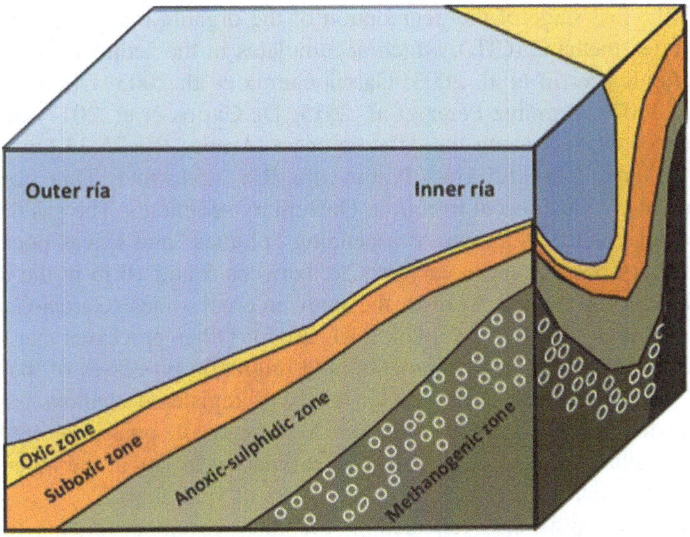

Fig. 17.3 Block diagram illustrating the deepening of the redox boundaries from inner to outer ría (modified from Rubio et al. 2011)

of organic matter continues with the use of alternative oxidants, with the intervention of bacteria, according to the general sequence described by Froelich et al. (1979), depending on their energy production. Therefore, the compounds used during the final stages of early diagenesis are sulphate and CO_2 dissolved in sea water. In so far as the first, the resulting product is sulphuric acid (H_2S); this compound reacts with the iron released during the dissolution of Fe oxides and oxyhydroxides and produces pyrite with a generally framboidal morphology. This mineral is not generated directly, but rather after a series of stages in which different iron sulfur-precursors intervene. First of all, monosulfides precipitate and are responsible for the blackish colour of these sediments. Afterwards, these evolve into greigite, pyrrhotite or mackinawite. These intermediaries are interesting in so far as the magnetic properties of the sediment, given that greigite and pyrrhotite exhibit a magnetic behaviour slightly less intense than that of magnetite. The results of these studies in the rías along the Atlantic coast have demonstrated that the concentration of magnetic minerals decreases rapidly with depth, until a very stable background value (Emiroglu et al. 2004; Mohamed 2006). This behaviour is related to the diagenetic dissolution of the detrital magnetic iron oxides and oxyhydroxides, with high dissolution rates of the same and a predominance of pyrites and greigites in the inner most areas of the rías (Rey et al. 2005), as well as Fe oxyhydroxides, oxides and silicates towards the exterior zone. These authors have determined half lives for the dissolution of detrital oxides such as magnetite of between 4 and 900 years in rías, depending on the type and content of organic carbon in the sediment (Emiroglu et al. 2004; Mohamed 2006).

During the last stage of the degradation of the organic matter, the reduction of CO_2 generates methane (CH_4), which accumulates in the sediments and can even form gas fields (Ferrín et al. 2003; García-García et al. 2003; Durán et al. 2007; Diez et al. 2007a; Ramírez-Pérez et al. 2015; De Carlos et al. 2017) of a considerable scale in the rías along the Atlantic coast (Arousa Ría 25.33 km^2, Vigo Ría 12.6 km^2, Muros Ría 11.5 km^2, Pontevedra Ría 4.54 km^2). This gas tends to accumulate in the most recent fine grain Quaternary sediments. The gas that escapes into the water column is known as ascending "plumes" and leaves pockmarks on the sedimentary bottom; these vary in size, between 5 and 10 m in the outer most areas and between 20 and 50 m in the more internal zones (García-García et al. 1999; Durán et al. 2007; García-Gil et al. 2015). Other processes can also bring about facies control on the accumulation and migration processes of the gas (Diez et al. 2007b; Vilas et al. 1999a). The levels of registered shallow gas and gas seepage, in the case of the rías along the Atlantic coast, present a similar spatial arrangement, characterized by a greater accumulation of gas in the inner and axial areas of the rías and of seepage along the edges of said areas.

Anthropic action is also reflected in the most recent sediments of rías and estuaries (Álvarez-Iglesias et al. 2006). On the one hand, an increase has been detected in the rate of sedimentation over recent years (Diz et al. 2002; Mohamed 2006; Álvarez-Iglesias et al. 2007), and on the other, most of the polluting heavy metals adhere to the muddy sediment particles (Rubio et al. 2000; Álvarez-Iglesias et al. 2003), which establishes an efficient process for their elimination from the water column. In the sedimentary record, the generation of anoxic conditions also has other important implications from an environmental point of view. When it comes to reducing or minimising the toxicity of trace metals, not only the formation of sulphurs (such as pyrite) in anoxic sediments, but also complexation by organic matter are both very important (Álvarez-Iglesias and Rubio 2008, 2012). However, when anoxic sediments oxidise, metals are released by sulfurs into the adjacent water column or into interstitial water and/or redistributed in other geochemical phases of the sediment; these metals can also interact with the benthonic fauna through their incorporation into tissue.

17.5 Zonation and Sedimentology

From a sedimentological and stratigraphic point of view, rías exhibit distinctive features of a unique sedimentation environment which is different from estuaries. Since the last Holocene transgression, the sedimentary filling found in the estuaries of these basins reveals a progressive transition towards the current conditions present in the rías (Fig. 17.4). Although the physical and biochemical processes acting in the rías are, to a large extent, similar to those present in different types of estuaries, thanks to their interaction with the existing geomorphological characteristics, it is possible to establish clearly differentiating elements. The most investigated in this regard (see references in Méndez and Vilas 2005) are the second

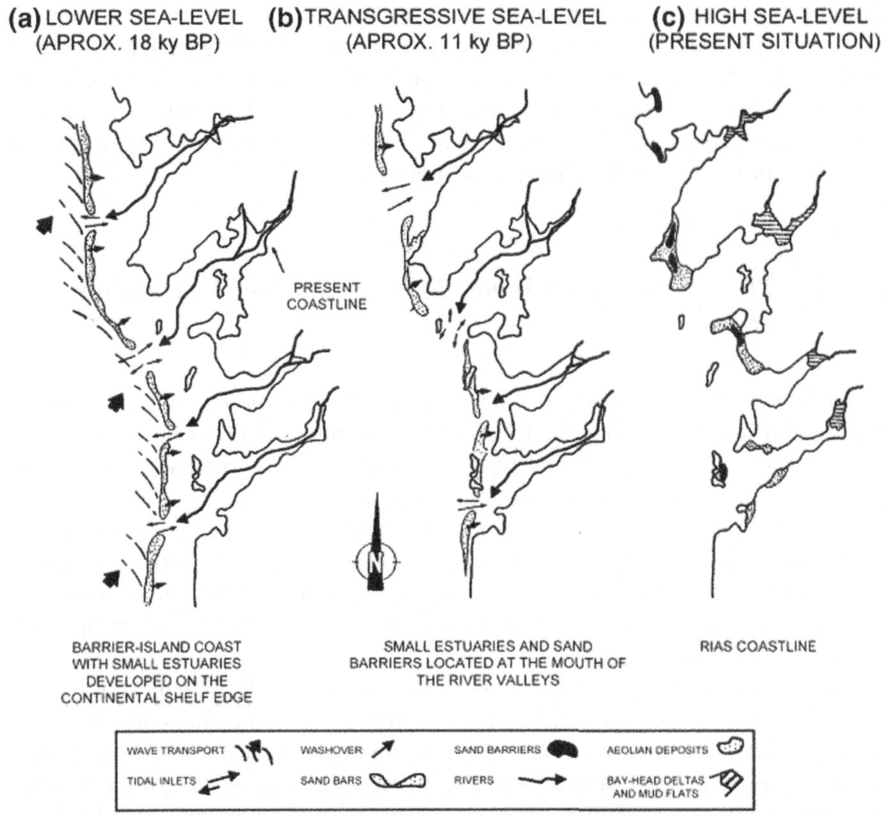

Fig. 17.4 Theoretical diagram showing the evolution of the coastline of the Rías Baixas in Galicia since the last sea-level transgression. Regression of the coastline since: **a** the Last Glacial Maximum (18 ky BP), **b** position during the Younger Dryas (11 ky BP) and **c** current location (modified from Vilas et al. 2010)

type, described by Nonn (1966) as those in which tectonic processes are responsible for their configuration. This type of ría matches those located along the Atlantic margin, known as the Rías Baixas (North to South: Ría of Muros-Noia, Arousa, Pontevedra and Vigo). This section and the following will pay greater attention to this part of the coast than to the more Northern rías or those along the coast of the Cantabrian Sea.

Seen from above, the Rías Baixas have a funnel-shaped morphology in which there are two distinguishable sectors; a larger cone corresponding to the ría itself, where the salinity of the waters is marine, and a smaller cone, which corresponds to the main fluvial channel discharging into the headwaters. Their orientation is approximately parallel to their longitudinal axis (e.g. the Tambre River in the Ría of Muros, the River Ulla in Ría of Arousa). Their surface areas vary between 140 km^2 (Ría of Pontevedra) and 330 km^2 (Ría of Arousa) and their length ranges from

20 km in the Ría of Pontevedra to 33 km in the Ría of Vigo; they are approximately between 8–12 km wide at the outer-most points and 1–3 km in the internal sector. Depths vary between 50–60 m at the outer-most points and 5–10 m in the inner-most areas. Geomorphologically, they are large valleys with a trough-like cross-section profile and flat bottoms with a slight slope towards the sea. All of them, except the Muros Ría, have islands at their entrances.

17.5.1 The Movement of Water Masses and the Distribution of Energy

In these rías, the movements of water masses have a mesotidal regime with a semidiurnal cycle (Alonso et al. 1993) and a maximum bottom velocity of around 15 cm/s (Souto et al. 2003), as well as oceanic coastal circulation (especially phenomena such as seasonal upwelling and downwelling). In winter, there is intense swell and significant maximum wave heights (H_{smax}) of around 8 m (Rey et al. 2005; Varela et al. 2005; Vilas et al. 2005); fluvial water flow is relatively unimportant and drainage basins are significantly small (<2500 km^2 surface area) and there are low average discharge rates (Ríos et al. 1992; Pazos et al. 2000; Pérez-Arlucea et al. 2005).

The general pattern of water mass circulation observable in the rías is very similar to what has been established for estuaries by several authors, not only in so far as salinity, but also in terms of water mixing (Pritchard 1967). Regarding hydrodynamics, it is possible to identify the main features of the typical estuarine circulation, with a considerable change to inverse estuarine or anti-estuarine during seasonal downwelling episodes (Varela et al. 2005). Although these circulation processes means that they tend to be associated to estuaries, their geomorphological characteristics, wide but deep "trough-shaped" valleys, contrary to what happens in estuaries, means that the effect of the propagation of tidal waves is limited within the whole area. This is due to the fact that the relationship between the tidal prism and the cross-section of the rías, and therefore the relationship between the convergence and friction of water masses with the bottom, which is so important in estuaries, is only relevant for the distribution of facies in the estuarine sector at the head of the ría, and in the Rías Baixas this area only corresponds to 20% of the depositional surface. As a result, in the remaining 80%, at the moment, typically estuarine facies distributions are not developed; sedimentary inputs are therefore depending on the distribution and transport capacity of the swell (Vilas et al. 2005, 2010). This variation of energy and its corresponding facies distribution (Fig. 17.5) is in contrast with that described by Dalrymple et al. (1992) for estuaries. Estuary zoning is characterized by high energy distribution at both extremes but low in the central area. In the internal part (proximal or head) and in the external part (distal or mouth), the energy of the river on the one hand and the wave action on the other, lead to coarse sedimentation being transported towards the central area. It is in the latter where the energy is at its lowest and where the marine and fluvial influence are balanced out and fine sediments are deposited.

The distribution of energy in rías (Fig. 17.5a) is characterized by high swell energy initiated in open sea. It reaches the external part (distal or mouth) and moves progressively through the external and internal sectors of the ría, until the head of the ría or estuarine zone, where the marine and fluvial influences are balanced out. The fluvial influence is limited to this area, where the rivers discharging into these bodies of water lack sufficient energy to redistribute the coarser sediments that they deposit. The maximum tidal reach takes place at the limit between the inner ría area

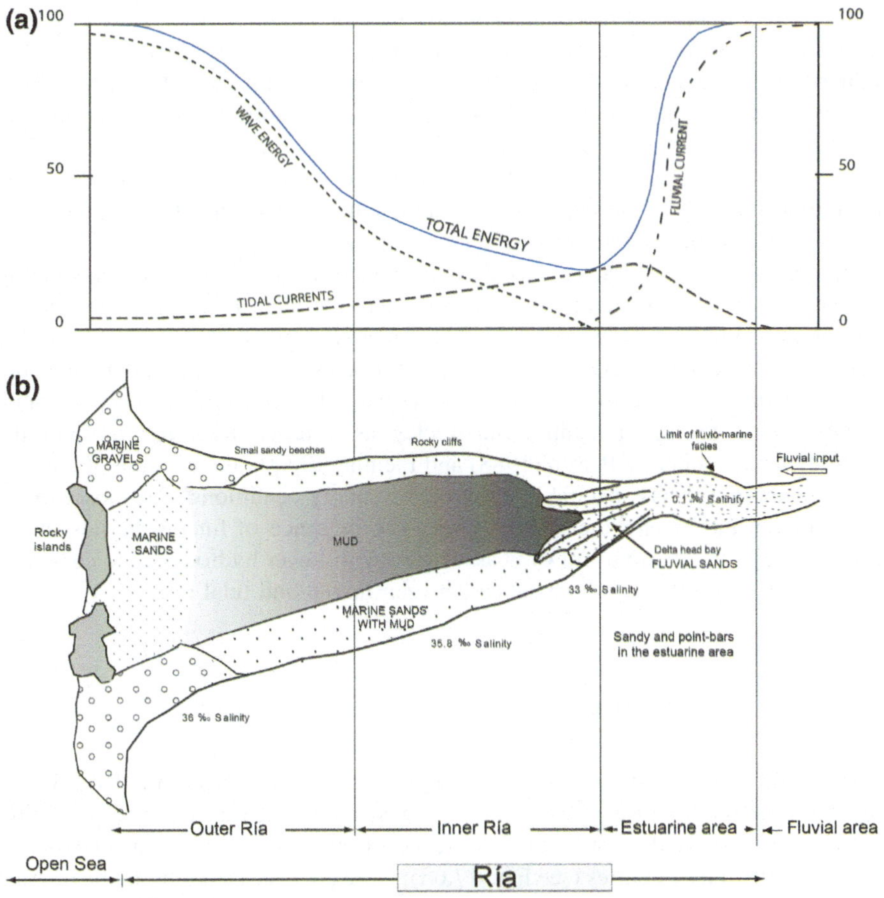

Fig. 17.5 Energy variation and facies distribution in the Rías (modified from Vilas et al. 2010): **a** Energy Variation: high wave energy at the mouth and in the external sector of the ría, progressively decreasing towards the internal sector of the ría. Fluvial energy of little importance in the estuarine zone at the head of the ría. **b** Facies Distribution: coarse marine sands in the distal part, passing to muddy sands and sandy muds, with high bioclastic content, in the outer and inner ría sectors. Estuarine mud with sand lenses and fluvial sands in the fluvial-tidal boundary

and the actual estuarine area, where it is possible to observe processes such as maximum turbidity and fine sediment accumulation due to changes in salinity and also flocculation. It is in this area, where swell energy is diminished, that the circulation can be classified as typically estuarine, as described for tide-dominated estuaries (see Fig. 17.6 II). The internal sector of the ría is dominated by waves and its action gradually increases toward the sea and it is here where the energy of swell-type waves is able to prevail over the tidal energy. The external area of the ría is clearly dominated by swell and eventually by the action of local processes, such as coastal outcropping. In this latter sector, with depths at the entrance reaching even more than 50 m, the distribution of energy towards the internal sector varies depending on the bathymetry; it is generally lower along the longitudinal axis due to the greater depth and more intensive towards the margins. Analyses carried out using numerical models establish that the swell in the internal part of the Ría of Pontevedra is concentrated along well-defined entry corridors (see Fig. 17.7); these vary depending on the swell conditions in open sea, mainly the wave period. The greater the period, the transition between deep and intermediate waters takes place at a greater depth (see Fig. 17.2). Therefore, the action of the swell will strengthen the turbulence and the mixing of waters and maintain the sediment in suspension in the exterior and internal sector of the ría.

In general terms, rías are exposed environments with an important tendency to preserve sporadic high-energy events, and where the main force, from a sedimentary point of view, is the wave energy. This fact is demonstrated in Fig. 17.8, that compares the textural trend of superficial sediments in the Vigo and Pontevedra Rías with that of an estuary, within the ternary diagram proposed by Fleming (2000), using the relationship sand/silt/clay as a basis. Keeping in mind the hydrodynamic model by Pejrup (1988) and Fleming (2000), the location of the data in this diagram reflects specific hydrodynamic energy conditions. The closer to the silt extreme, the greater the energy level. The presence of finer materials in the estuary (clay content up to 60%) is associated with lower hydrodynamic energy in the system, controlled by the balance between fluvial and tidal inputs.

17.5.2 Facies Distribution

The general distribution pattern of the sediments on the bottoms of the Rías Baixas reveals a longitudinal trend oriented lengthwise along their axes (see Fig. 17.5b), with wave-dominated facies, but an absence of features that are characteristic of those described for estuaries (see Fig. 17.6 I); these are restricted only to the internal area of the head (Nombela et al. 1995). This tendency is well represented for the partially open rías, due to the presence of islands at the mouth, but it can present slight variations if the entrance is completely open, as is the case in the Ría of Muros.

The finer sediments (silt and clay) extend from the internal part of the ría, occupying mainly the central basin (Vigo, Pontevedra and Arousa Rías) or sheltered areas on the northern shore (Muros Ría); they reach the external sector of the

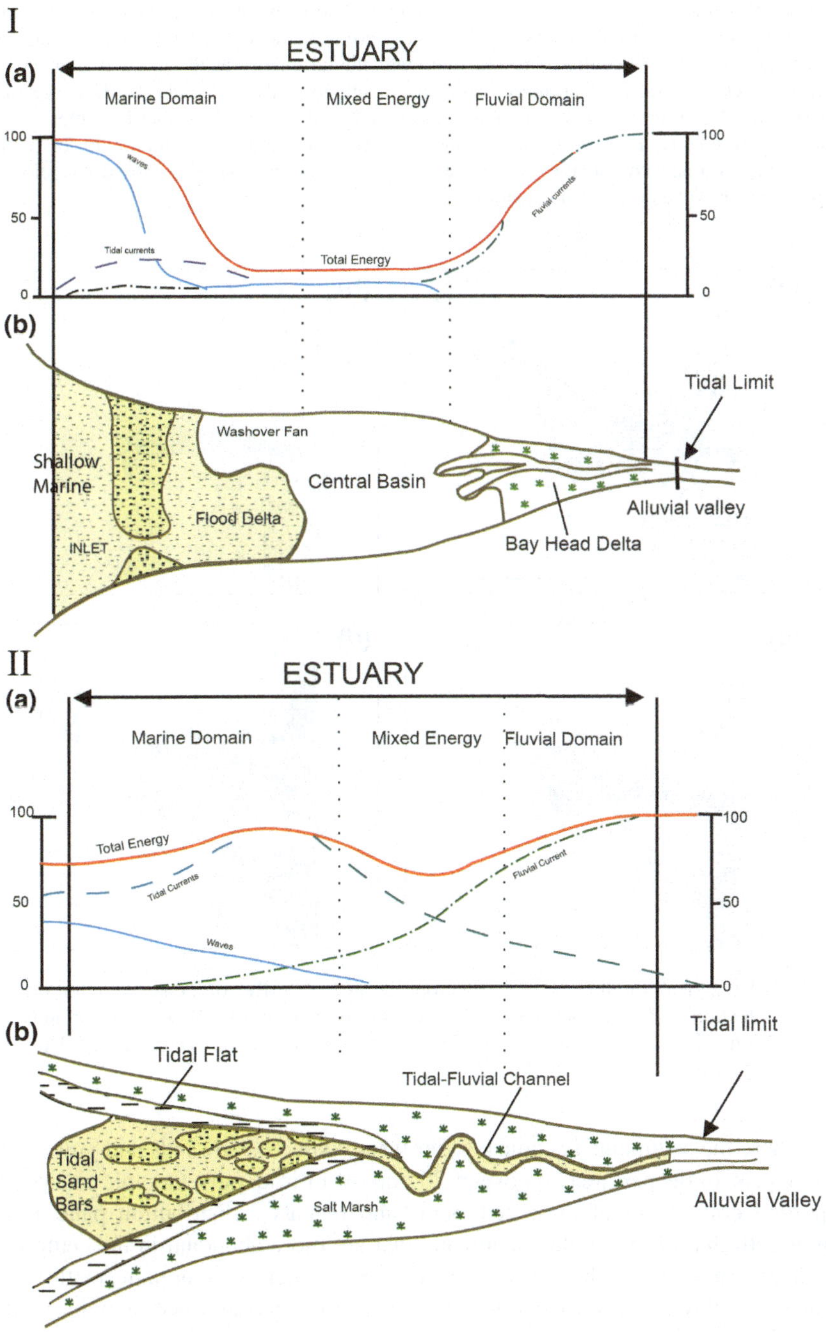
I
ESTUARY
(a)
Marine Domain
Mixed Energy
Fluvial Domain
100
50
0
Waves
Tidal currents
Total Energy
Fluvial currents
(b)
Washover Fan
Shallow Marine
Central Basin
Flood Delta
INLET
Tidal Limit
Alluvial valley
Bay Head Delta
II
ESTUARY
(a)
Marine Domain
Mixed Energy
Fluvial Domain
100
50
0
Total Energy
Tidal Currents
Fluvial Current
Waves
(b)
Tidal limit
Tidal Flat
Tidal-Fluvial Channel
Tidal Sand Bars
Salt Marsh
Alluvial Valley

◂**Fig. 17.6** Dynamic processes and major morphological components along the estuary (modified from Dalrymple et al. 1992). (I) Wave dominated estuaries: **a** high energy at the mouth, very distinct minimum in the central part and important fluvial energy at the head. **b** Characteristic "coarse-fine-coarse" facies distribution from the mouth towards the head. (II) Tide dominated estuaries: **a** tidal current energy dominating the mouth and longitudinal tidal bars breaking any existing wave energy. Decreasing fluvial energy downstream and energy minimum less defined than in wave dominated estuaries. **b** Characteristic "coarse-fine-coarse" facies distribution less defined than in wave dominated estuaries

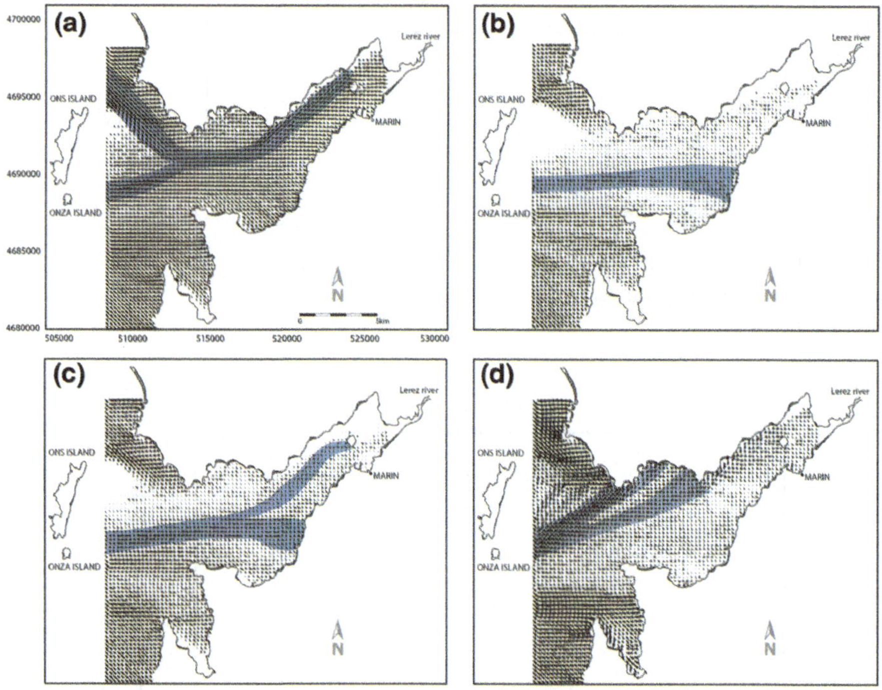

Fig. 17.7 Numerical simulation of wave propagation in the Ría of Pontevedra for the most frequent direction and different significant wave height (Ho) and peak period conditions (Tp): **a** Ho = 2.5 m, Tp = 14 s; **b** Ho = 3 m, Tp = 10.5 s; **c** Ho = 4 m, Tp = 12 s; **d** Ho = 6 m, Tp = 18 s (modified from Rey et al. 2005)

rías, where the sediments become coarser (sandy-silt, coarse sands and gravels) and only appear in deeper areas (between 20 and 50 m in depth). Organic matter is an important component of these sediments and reveals a distribution pattern coinciding with that of the muddy sediments that are more abundant in the central axis and in internal areas. This association is a consequence of organic matter being attracted to clay-like sediments, to which it is adsorbed and occasionally agglutinated, thus speeding up its decomposition. The concentration of organic matter, for example, in the sediments of the Ría of Vigo reaches very high values, between 2 and 10% (Vilas et al. 1995), especially if compared to deep-sea sediments, which

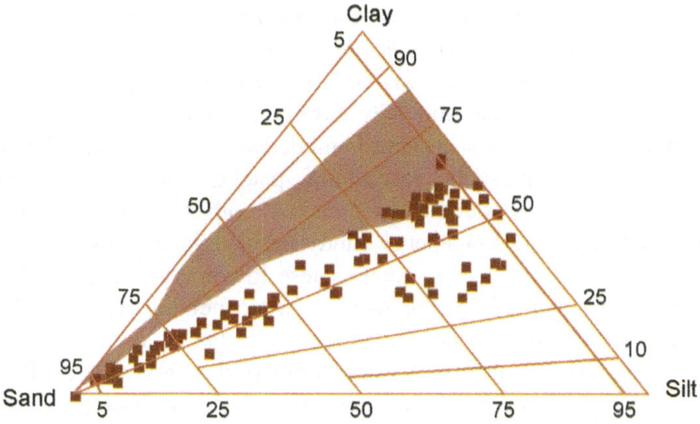

Fig. 17.8 Ternary diagram proposed by Fleming (2000) to assess the energy level of different sedimentation environments, based on the distribution of facies. The grey band represents estuary data (Dyfi Estuary, in Wales, UK). The dots correspond to sediment samples collected in the Rías of Vigo and Pontevedra (Galicia, NW Spain) (modified from Vilas et al. 2005)

are generally below 0.5%. This abundance is a result of the increased biological production in these rías, strengthened by the influence of coastal outcropping, which introduces nutrients into the rías and consequently promotes photosynthesis.

The predominant fauna in the rías is largely of marine origin. Only in the innermost areas, which are influenced by river contributions, is it possible to detect the presence of different communities than those found in saline waters. The influence of marine waters combined with contributions from the rivers, and the complex topography and bathymetry, determine the presence of a great sedimentary and environmental variability which has important biological consequences. Within a ría, we may find practically the whole range of coastal sediments, both within the intertidal and the subtidal zone (Vilas 2007).

In the intertidal mud flats areas, communities of species with a greater or lesser resistance to variations in salinity are abundant. These species are frequently sustained by the presence of marine phanerogams (*Zostera marina* and *Z. Noltii*), involving a highly interesting environment as they are particularly favourable zones for numerous infauna and epifauna species, spawning areas, juvenile refuges, sources of oxygen and organic matter.

Mud bottoms appearing in the inner areas and central areas of the rías are occupied by species, which may also be found in the open waters in wide sectors of the continental shelf. These include communities fundamentally dominated by surface and subsurface deposit feeder species. The predominating groups are polychaeta, such as Capitellidae (*Notomastus*, *Heteromastus*), the Maldanidae (*Euclymene*), crustacea and molluscs. Molluscs are a further dominant group within the muddy subtidal area, and there is an important presence of bivalves (*Nucula turgida*, *Abra alba*, *Pandora albida*, *Thysira flexuosa*, etc.). Small-sized crustaceans are also particularly numerous, in almost all cases, the Amphipoda (*Ampelisca*, *Gammarous*) or Thainadaceae (*Apsweudes latreilli*).

The most exposed areas to the waves are characterised by sandy bottoms near the coast and shallow depths and they exhibit continuity with the beaches and gradually change to silty muds in central and deeper areas. In more sheltered areas, sands appear more dispersedly, resembling patches, between finer sediments. Moving inwards from the outer-most area, on both sides of the coastline that outlines the rías, there are low-energy sandy beaches located in areas partially or completely sheltered from ocean waves. These low-energy beaches characterized by calm conditions and minimal non-storm wave heights (Hsig −0.25 m) have received specific attention by several authors (Costas et al. 2005; Vila-Concejo et al. 2010; Bernabeu et al. 2012) since some studies revealed a different morphological behaviour from the open-ocean beaches (Short 2006).

These sandy bottoms are also densely populated by different organisms, both in the subtidal and intertidal environments. The type of community depends on the type of sands on which they settle: fine, medium or coarse grained, although as a general rule, the three types are dominated by polychaetae, molluscs, crustaceans and echinoderms, with an important presence of filter feeder species and surface deposit feeders. Although a large number of the species cited for silty bottoms may be present in sandy areas, (particularly where this is coarse-grained), certain species, which may be cited as characteristic of muddy bottoms, also appear here, such as molluscs of the family Cardiaceae (*Cerastoderma edule, Acanthocardia* spp., etc.) Veneraceae (*Venus striatula*, *Venerupis* spp.) and Tellinacea (*Tellina tenuis*, *T. fabula*).

Gravel appears at the mouths of the rías, which are controlled by swell. These sands and gravels are mainly made up of calcareous and siliceous bioclastic fragments, as well as minerals of siliciclastic origin, such as quartz, potasic feldspar, sodic feldspar, muscovite and biotite. Their abundance is greater towards the banks and in the more exterior areas, which are more exposed to swell action. In these areas, where there are mainly very coarse bioclastic fragments, the $CaCO_3$ content reaches values above 90% (Vilas et al. 2005). These fragments mainly comprise shell remains, although occasionally, patches of calcareous algae of the genus Lithothamnium are detected.

17.6 Evolution and Stratigraphic Interpretations

Rías constitute real sediment traps, where transgressive (TST) and highstand (HST) systems tracts are well represented (Durán 2005; Durán et al. 2007; Diez et al. 2007a); studies carried out confirm the tectonic-eustatic character of the Rías Baixas and their quaternary filling. The lowstand systems tract (LST) is not registered inside the rías due to the subaerial exposure of the entire continental shelf (Ferrín 2005), but it can remain in the deepest areas, between rocky outcroppings or distal parts of the continental shelf and/or on the edge of the slope. By means of high-resolution seismic studies, several authors (Acosta 1982, 1984; Acosta and Herranz 1984; Herranz and Acosta 1984; Rey 1993; García-Gil et al. 1999, 2000; García-García et al. 2005; Durán et al. 2007; Diez et al. 2007a, b; Martínez-Carreño et al. 2017a) have offered a geometric vision and evolution of the sedimentary bodies; these are supported by the calculation of the thickness of the quaternary sediments over the basement of the rías, by the interpretation of sedimentary sequences and by the identification of paleoreliefs, particularly since the last glacial maximum (18,000 years BP). Consequently, the presence of granitic and metamorphic basements has been identified in the different rías and an important paleorelief has developed over these; over the latter, in turn, there are several units (Fig. 17.9). This studies and those realised by a multiproxy approach (Andrade et al. 2014) make it possible to obtain an approximate vision of the correlation between the evolution of the different stages of the rías filling and the relative changes in sea level documented in the Rías Baixas since the last glacial maximum. The oldest sedimentary filling is represented by a basal unit, which has been attributed to Pleistocene sediments of fluvial origin (Hinz 1970; Rey 1993; Durán 2005). The upper limit of the strata in this unit is interrupted by an important erosion surface generated during the subaerial exposure of the rías associated to the fall in sea level in the last glacial maximum, between 20 and 18 ky BP. It is estimated that at this moment, the relative sea level was around 120–130 m below the current sea level (Hanebuth et al. 2000; Dias et al. 1987, 2000; Rodrigues et al. 1991, 2000) and close to the edge of the current continental shelf. This unit is construed as estuarine successions developed during these periods of low sea level, prior to the last glacial maximum, equivalent to those described by Dalrymple et al. (1992) and Allen and Posamentier (1993). Over this last unit, there are several units that were deposited during different sea positions within the last eustatic cycle; they demonstrate a progressive variation in facies, not only vertically, but also horizontally, towards the internal part of the heads of the rías. It is even possible to define transitions from fluvial to estuarine facies, moving from the external sector towards the internal area of the rías, in a progressively rising sea. The presence of pockmarks and acoustic blankings, observed in many of the seismic records of the mid/internal sector of the rías, are proof of the accumulations of methane of microbial origin that are present in the sedimentary filling (Martínez-Carreño et al. 2017b). The origin of this gas seems to be related to the presence of the abundant organic matter contained in the intertidal mud flats and in the channels of the estuarine environments, which ended up sealed off by overlying units during the last transgression. Seismic records often reveal disorganized and chaotic facies in the deepest units, which turn into subparallel facies with more tubular geometries towards the surface (see Fig. 17.9); these represent a reduction of the energy gradient determined by the physiography of the basin and the reduction of the confinement of the sedimentation zone as the filling takes place (Diez 2007a, b). It is from this moment on that the V-shaped cross-section profile changes to a flat bottom "trough-shape", characteristic of the cross-section of rías.

There is scarce information regarding this type of environment, but in general terms, the ideal facies model could be considered the type described by Dalrymple

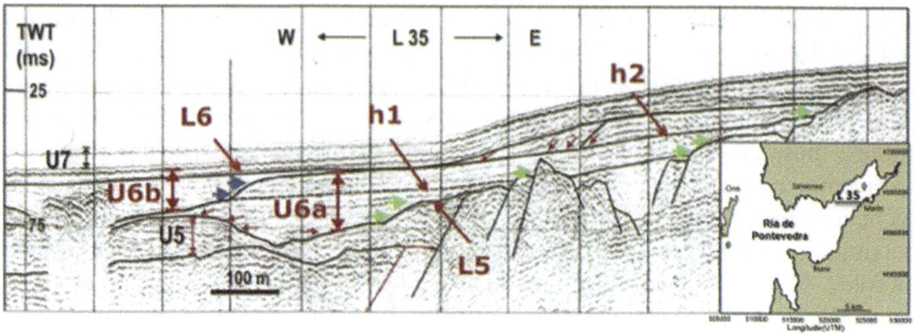

Fig. 17.9 Stratigraphic architecture of a seismic reflection profile (L-35) of the Ría of Pontevedra. The paleorelief (L5) represents the erosion surface originating on the basement or previous units, during the last glacial maximum (LGM) 18 ky ago, over which unit U5 is deposited followed by units U6 a, b. Its geometry is interpreted as being due to the global eustatic fall 11 ky ago (Younger Dryas), which brings about the partial erosion of the previous materials and it is over this paleorelief (L6) where the most recent unit (U7) is located, with the sea in progressive ascent. In the TST (U5 and U6) and in the HST (U7) sediments, numerous shallow accumulations of gas have been located (modified from Durán et al. 2007)

et al. (1994) as incised-valley systems, in which the fluvial incision has been produced over a former tectonic valley feature. Thus, over one complete sea-level cycle (sea-level fall to subsequent highstand) two main sedimentary sequences can be differentiated: the ría central basin and the estuarine zone.

The ideal sequence generated in the central basin of the rías is represented by an erosion surface underlying fluvial and estuarine-fluvial facies at the base, passing on to sandy/silt facies and/or silty/clay facies at the top of the sequence, depending on the variation of energy due to swell propagation (see Fig. 17.6) from the mouth to the more interior parts of the central basin of the ría. Due to these variations, the resulting differences that come about generate fining-upward sequences. These deposits can reach thicknesses of over 20 or 30 m because there is scarce erosive capacity, as the orbital movement of the swell does not reach the bottom. As mentioned above (see Fig. 17.2), whereas sediment is kept in suspension above the wave base level, below this the sedimentation of the finest sediments (silt and clay) is constant.

In the estuarine zone, the ideal sequence generated is similar to the one described by Dalrymple et al. (1992) for the bay-head delta of estuaries. In this sequence, as the estuary evolves and turns into a delta, there is a movement of facies towards the interior sector of the ría basin and an expansion of sandy bars interdigitated with intertidal mud flats and/or marshes with the development of tidal channels (Roy et al. 1980; Harris 1988). As a result, the fluvial and estuarine-fluvial facies at the base pass on to fluvial facies at the top of the sequence, generating a symmetric (fining upwards–coarsening upwards) sequence.

In the fossil record, the deposits of these types of environments, rías or estuaries, are not easy to interpret, given that they are often not formed by a single and simple filling process. They have been reworked periodically, occupying the same spaces during several transgressions, and therefore any outcropping is very complex; neither is it easy to distinguish proximal from distal facies, which is essential for the vast majority of sedimentary interpretations, as suggested by Dalrymple and Choi (2007).

References

Acosta, J. (1982). Apantallamientos acústicos en la Ría de Muros y Noya y el margen continental de Cádiz. *Boletín del Instituto Español de Oceanografía, 2,* 125–149.

Acosta, J. (1984). Occurrence of acoustic masking in sediments in two areas of the continental shelf of Spain: Ria de Muros (NW) and Gulf of Cadiz (SW). *Marine Geology, 58,* 427–434.

Acosta, J., & Herranz, P. (1984). Contribución al conocimiento del cuaternario marino en la Ría de Muros y Noya. *Thalassas, 2,* 13–21.

Allen, G. P., & Posamentier, H. W. (1993). Sequence stratigraphy and facies model of an incised valley fill: The Gironde Estuary, France. *Journal of Sedimentary Research, 63,* 378–391.

Alonso, J., Garcia, M. J., & Diaz Del Rio, G. (1993). Current measurements on the Galician shelf. Wind and tide effects. *Boletin del Instituto Espanol de Oceanografia, 9,* 323–342.

Álvarez-Iglesias, P., Quintana, B., Rubio, B., & Pérez-Arlucea, M. (2007). Sedimentation rates and trace metal input history in intertidal sediments from San Simón Bay (Ría de Vigo, NW Spain) derived from 210Pb and 137Cs chronology. *Journal of Environmental Radioactivity, 98,* 229–250.

Álvarez-Iglesias, P., & Rubio, B. (2008). The degree of trace metal pyritization in subtidal sediments of a mariculture area: Application to the assessment of toxic risk. *Marine Pollution Bulletin, 56,* 973–983.

Álvarez-Iglesias, P., & Rubio, B. (2012). Early diagenesis of organic-matter-rich sediments in a ría environment: Organic matter sources, pyrites morphology and limitation of pyritization at depth. *Estuarine, Coastal and Shelf Science, 100,* 113–123.

Álvarez-Iglesias, P., Rubio, B., & Pérez-Arlucea, M. (2006). Reliability of subtidal sediments as "geochemical recorders" of pollution input: San Simón Bay (Ría de Vigo, NW Spain). *Estuarine, Coastal and Shelf Science, 70,* 507–521.

Álvarez-Iglesias, P., Rubio, B., & Vilas, F. (2003). Pollution in intertidal sediments of San Simón Bay (Inner Ria de Vigo, NW of Spain): Total heavy metal concentrations and speciation. *Marine Pollution Bulletin, 46,* 491–503.

Andrade, A., Bubio, B., Rey, D., Álvarez-Iglesias, P., Bernabeu, A. M., & Vilas, F. (2011). Paleoclimatic changes in the northwestern Iberia peninsula during the last 3000 years as inferred from diagenetic proxies in the sedimentary record of the Ría de Muros. *Climate Research, 48,* 247–259.

Andrade, A., Rubio, B., Rey, D., Álvarez-Iglesias, P., Bernabeu, A. M., & Fedi, M. (2014). Environmental changes at the inner sector of the Ría de Muros (NW Spain). *Estuarine, Coastal and Shelf Science, 136,* 91–100.

Bernabeu, A. M., Lersundi-Kampistegui, A. V., & Vilas, F. (2012). Gradation from oceanic to estuarine beaches in a ría environment: A case study in the Ría de Vigo. *Estuarine, Coastal and Shelf Science,* 60–69.

Boyd, R., Dalrymple, R., & Zaitlin, B. A. (1992). Classification of clastic coastal depositional environments. *Sedimentary Geology, 80,* 139–150.

Burban, P. Y., Lick, W., & Lick, J. (1989). The flocculation of fine-grained sediments in estuarine waters. *Journal Geophysical Research, 94,* 8323–8330.

Bush, R. T., McGrath, R., & Sullivan, L. A. (2004). Occurrence of marcasite in an organic-rich Holocene estuarine mud. *Australian Journal of Soil Research, 42,* 617–621.

Carlé, W. (1947). Las Rías Bajas Gallegas en versión de J. Gómez Llarena. *Estudios Geográficos, 35,* 323–330.

Carlé, C. W. (1949). Las rías bajas gallegas. *Estudios Geográficos, 35,* 320–330.

Carlé, W. (1950). Resultado de investigaciones geológicas en las formaciones antiguas de Galicia. In *Publicaciones extranjeras sobre Geología de España, tomo V* (pp. 61–90). Madrid.

Castaing, P., & Guilcher, A. (1995). Geomorphology and sedimentology of rias. In G. M. E. Perillo (Ed.), *Geomorphology and sedimentology of estuaries* (pp. 69–111). Amsterdam: Elsevier.

Costas, S., Alejo, I., Vila-Concejo, A., & Nombela, M. A. (2005). Persistence of storm-induced morphology on a modal low-energy beach: A case study from NW-Iberian Peninsula. *Marine Geology, 224,* 43–56.

Dalrymple, R. W., Boyd, R., & Zaitlin, B. A. (1994). *Incised-valley systems: Origin and sedimentary sequences.* SEPM Special Publication, No 51.

Dalrymple, R. W., & Choi, K. (2007). Morphologic and facies trends through the fluvial–marine transition in tide-dominated depositional systems: A schematic framework for environmental and sequence-stratigraphic interpretation. *Earth Science Reviews, 81,* 135–174.

Dalrymple, R. W., Zaitlin, B. A., & Boyd, R. (1992). Estuarine facies models: Conceptual basis and stratigraphic implications. *Journal of Sedimentary Research, 62,* 1130–1146.

Davies, J. L. (1964). A morphogenetic approach to world shorelines. *Zeitschrift fur Geomorphologie, 8,* 127–142.

Day, J. H. (1980). What is an estuary? *South African Journal of Science, 76,* 15–17.

De Carlos, A., Martinez-Carreño, N., Barros-García, D., Luis, J. R., & García-Gil, S. (2017). Geochemical and microbial context of the gassy sediments in the Ría de Vigo (NW coast of Spain). *Marine Geology, 385,* 1–12.

Dias, J. M. A., Boski, T., Rodrigues, A., & Magalhaes, F. (2000). Coast line evolution in Portugal since the Last Glacial Maximum until present—A synthesis. *Marine Geology, 170,* 177–186.

Dias, J. M. A., Rodrigues, A., & Magalhaes, F. (1987). Evoluçao da linha de costa em Portugal, desde o último máximo glaciario: síntesis dos cohecimentos. *Estudos do Quaternário, 1,* 53–66.

Diez, R., García-Gil, S., Durán, R., & Vilas, F. (2007a). Gas accumulations and their association with particle size distribution patterns in the ría de arousa seabed (Galicia, NW, Spain): An application of discriminant analysis. *Geo Marine Letters, 27*(2–4), 89–102.

Diez, R., García-Gil, S., Durán, R., & Vilas, F. (2007b). Gas charge sediments in the ría de Arousa: Short to long term fluctuations. *Estuarine, Coastal Shelf and Science, 71*(3–4), 467–479.

Diz, P., Francés, G., Pelejero, C., Grimalt, J. O., & Vilas, F. (2002). The last 3000 years in the Ría de Vigo (NW Iberian Margin): Climatic and hydrographic signals. *The Holocene, 12,* 459–468.

Durán, R. (2005). *Estratigrafía sísmica desde el Último Máximo Glacial en la Ría de Pontevedra (NO España)* (Tesis Doctoral). Pbl. Universidad de Vigo.

Durán, R., García-Gil, S., Diez, R., & Vilas, F. (2007). Stratigraphic framework of gas accumulations in the Ría de Pontevedra (NW Spain). *Geo-Marine Letters, 27,* 77–88.

Dyer, K. R. (1997). *Estuaries: A physical introduction* (2nd ed., p. 195). Chichester: Willey.

Emiroglu, S., Rey, D., & Petersen, N. (2004). Magnetic properties of sediment in the Ría de Arousa (Spain): Dissolution of iron oxides and formation of iron sulphides. *Physics and Chemistry of the Earth, 29,* 947–959.

Evans, G., & Prego, R. (2003). Rías, estuaries and incised valleys: Is a ria an estuary? *Marine Geology, 196,* 171–175.

Fairbridge, R. H. (1980). The estuary: Its definition and geodynamic cycle. In E. Olausson & I. Cato (Eds.), *Chemistry and biogeochemistry of estuaries* (pp. 1–16).

Ferrín, A. (2005). Cenozoic seismic stratigraphy of the SW Galician continental shelf. *Comparative study with the Canterbury shelf (SE New Zealand) during the Quaternary* (Ph. D. Thesis). Pbl. Universidad de Vigo.

Ferrín, A., Durán, R., Diez, R., García-Gil, S., & Vilas, F. (2003). Shallow gas features in the Galician Rías Baixas (NW Spain). *Geo-Marine Letters, 23,* 207–214.

Fleming, B. W. (2000). A revised textural classification of gravel-free muddy sediments on the basis of ternary diagrams. *Continental Shelf Research, 20,* 1125–1137.

Froelich, P. N., Klinkhammer, G. P., Bender, M. L., Luedtke, N. A., Heath, G. R., Cullen, D., et al. (1979). Early oxidation of organic matter in pelagic sediments of the eastern ecuatorial Atlantic: Suboxic diagenesis. *Geochimica et Cosmochimica Acta, 43,* 1075–1090.

García-García, A., García-Gil, S., & Vilas, F. (1999). A seeping seafloor in a Ría environment: Ría de Vigo (NW Spain). *Environmental Geology, 38,* 296–300.

García-García, A., García-Gil, S., & Vilas, F. (2003). Monitoring the Spanish gas fields in the Ría de Vigo (1991–2001). *Geo-Marine Letters, 23,* 200–206.

García-García, A., García-Gil, S., & Vilas, F. (2005). Quaternary evolution of the ría de Vigo, Spain. *Marine Geology, 220,* 153–179.

García-Gil, S., Cartelle, V., De Bls, E., De Carlos, A., Diez, R., Durán, R., et al. (2015). Shallow gas in the Iberian continental margin. *Boletín Geológico y Minero, 126*(2–3), 575–608.

García-Gil, S., Duran, R., & Vilas, F. (2000). Side Scan Sonar image and geologic interpretation of the Ría de Pontevedra seafloor (Galicia, Spain). *Scientia Marina, 64,* 393–402.

García-Gil, S., Vilas, F., & García-García, A. (2002). Shallow gas features in incised-valley fills (Ría de Vigo, NW Spain). *Continental Shelf Research, 22,* 2303–2315.

García-Gil, S., Vilas, F., García-García, A., & Durán, R. (1999). Holocene storm delta in incised-valley fill sediments of Ría de Pontevedra, NW Spain. *EOS, Transactions, 80*(46), 559.

Gibbs, R. (1977). Suspended matter transport and the turbidity maximum. In C. Officer (Ed.), *Estuaries, geophysics and the environment* (pp. 104–109). National Academy Science Publication.

Gonzalez-Villanueva, R., Pérez-Arlucea, M., & Costas, S. (2015). 8000 years of environmental evolution of barrier-lagoon systems emplaced in coastal embayments (NW Iberia). *The Holocene, 25*(11), 1786–1801.

Goudie, A. (2018). Rias: Global distribution and causes. *Earth-Science Reviews, 177,* 425–435.

Hanebuth, T., Stattegger, K., & Grootes, P. M. (2000). Rapid flooding of the Sunda Shelf: A late-glacial sea-level record. *Science, 288,* 1033–1035.

Harris, P. T. (1988). Large-scale bedforms as indicators of mutually evasive sand transport and the sequential infilling of wide-mouthed estuaries. *Sedimentary Geology, 57,* 273–298.

Herranz, P., & Acosta, J. (1984). Estudio geofísico de la Ría de Muros y Noya. *Boletín del Instituto Español de Oceanografía, 1,* 48–78.

Hinz, K. (1970). Seismic reflection measurements with a pneumatic sound in the Ría de Arosa (Mw. Spain). *Leidse Geologische Mededelingen, 37,* 169–184.

Kranck, K., & Milligan, T. G. (1992). Characteristics of suspended particles at an 11-hour anchor station in San Francisco Bay, California. *Journal of Geophysical Research, 97,* 11373–11382.

Le Floch, P. (1961). *Propagation de la maree dans l''Estuaire de la Seine et en Seine Maritime.* These, Univ. Paris, 507 pp.

León, I., Méndez, G., & Rubio, B. (2004). Geochemical phases of Fe and degree of pyritization in sediments from Ria de Pontevedra (NW Spain): Implications of mussel raft culture. *Ciencias Marinas, 30,* 585–602.

Manning, A. J., & Bass, S. J. (2006). Variability in cohesive sediment settling fluxes: Observations under different estuarine tidal conditions. *Marine Geology, 235,* 177–192.

Manning, A. J., & Dyer, K. R. (1999). A laboratory examination of floc characteristics with regard to turbulent shearing. *Marine Geology, 160,* 147–170.

Martínez-Carreño, N., García-Gil, S., & Cartelle, V. (2017a). An unusual Holocene fan-shaped subaqueous prograding body at back of the Cies Islands ridge (Ría de Vigo, NW Spain): Geomorphology, facies and stratigraphic architecture. *Marine Geology, 385,* 13–26.

Martínez-Carreño, N., García-Gil, S., Cartelle, V., Ramírez-Pérez, A. M., & Insua, T. L. (2017b). Geochemical and geological factors controlling the spatial distribution of sulfate-methane transition zone in the Ría de Vigo (NW Spain). *Continental Shelf Research, 140,* 47–59.

McCave, I. N. (1985). Mechanics of deposition of fine-grained sediments from nepheloid layers. *Geo-Marine Letters, 4,* 243–245.

Méndez, G., & Rey, D. (2000). Perspectiva histórica del conocimiento geológico de las rías gallegas. *Journal Iberian Geology, 26,* 21–44.

Méndez, G., & Vilas, F. (2005). Geological antecedents of the Rias Baixas (Galicia, northwest Iberian Peninsula). *Journal of Marine Systems, 54,* 195–207.

Mohamed, K. (2006). *Influencia climática, diagenética y antropogénica sobre las propiedades magnéticas y geoquímicas de los sedimentos marinos cuaternarios del noroeste de la Península Ibérica* (Tesis Doctoral). Universidad de Vigo Pbl., 298 pp.

Muñoz-Sobrino, C., García-Moreiras, I., Martínez-Carreño, N., Cartelle, V., Insua, T., Ferreiro-Da Costa, J., et al. (2016). Reconstruction of the environmental history of a coastal insular system using shallow marine records: The last three millennia of the Cies Islands (Ría de Vigo, NW Iberia). *Boreas, 45*(4), 729–753.

Nombela, M. A., Vilas, F., & Evans, G. (1995). Sedimentation in the mesotidal Rías Bajas of Galicia, (NW Spain): Ensenada de San Simón, inner Ría de Vigo. In B. W. Flemming & A. Bartholoma (Eds.), *Tidal signatures in modern and ancients sediments* (Vol. 24, pp. 133–149). Spec. Publ. IAS.

Nonn, H. (1966). Les régions cotières de la Galice (Espagne). *Étude géomorphologique.* Publications de la Faculté des Lettres de l'Université de Strasbourg, Paris.

Pagés-Valcarlos, J. L. (2000). Origen y evolución geomorfológica de las rías atlánticas de Galicia. *Revista de la Sociedad Geológica de España, 13*(3–4), 393–403.

Pannekoek, A. J. (1966a). The geomorphology of the surroundings of the Ría de Arosa (Galicia, NW Spain). *Leidse Geologische Mededelingen, 37,* 7–32.

Pannekoek, A. J. (1966b). The Ría Problem. The role of antecedence, deep weathering, and Pleistocene slop-wash in the formation of the West-Galician rías. *Tijdschr. K Ned Aardr. Gemotsch., 83,* 289–297.

Pannekoek, A. J. (1970). Additional geomorphological data on the ria area of western Galicia (Spain). *Leidse Geologische Mededelingen, 37,* 185–194.

Pazos, O., Nombela, M. A., & Vilas, F. (2000). Continental contribution of suspended sediment to an estuary: Ría de Vigo. *Scientia Marina, 64,* 295–302.

Pejrup, M. (1988). The triangular diagram used for classification of estuarine sediments: A new approach. In *Tide-influenced sedimentary environments and facies* (pp. 289–300).

Pérez-Arlucea, M., Mendez, G., Clemente, F., Nombela, M., Rubio, B., & Filgueira, M. (2005). Hydrology, sediment yield, erosion and sedimentation rates in the estuarine environment of the Ria de Vigo, Galicia, Spain. *Journal of Marine Systems, 54,* 209–226.

Perillo, G. M. E. (1995). Geomorphology and sedimentology of estuaries. An introduction. In *Developments in sedimentology* (Vol. 53, pp. 1–16). Elsevier.

Piedracoba, S., Souto, C., Gilcoto, M., & Pardo, P. C. (2005). Hydrography and dynamics of the Ría de Ribadeo (NW Spain), a wave driven estuary. *Estuarine, Coastal and Shelf Science, 65,* 726–738.

Pritchard, D. W. (1952). Salinity distribution and circulation in the Chesapeake Bay estuarine system. *Journal of Marine Research, 11,* 106–123.

Pritchard, D. W. (1967). What is an estuary: Physical viewpoint. In G. H. Lauff (Ed.), *Estuaries* (Vol. 83, pp. 3–5). AAAS Washington, DC. Pub.

Ramírez-Pérez, A. M., de Blas, E., & García-Gil, S. (2015). Redox processes in pore water of anoxic sediments with shallow gas. *Science of the Total Environment, 538*(15), 317–326.

Rey, J. (1993). *Mapa fisiográfico submarino de la Ría de Arosa, (E 1:50.000).* Publicaciones IEO, 15 pp.

Rey, D., Mohamed, K. J., Bernabeu, A., Rubio, B., & Vilas, F. (2005). Early diagenesis of magnetic minerals in marine transitional environments: Geochemical signatures of hydrodynamic forcing. *Marine Geology, 215,* 215–236.

Ríos, A., Nombela, M. A., Perez, F. F., Roson, G., & Fraga, F. (1992). Calculation of runoff to an estuary Ria de Vigo. *Scientia Marina, 56,* 29–33.

Rodrigues, A., Magalhaes, F., & Dias, J. A. (1991). Evolution of the North Portuguese coast in the last 18.000 years. *Quaternary International, 9,* 67–74.

Roy, P. S., Thom, B. G., & Wright, L. D. (1980). Holocene sequences on an embayed high-energy coast: An evolutionary model. *Sedimentary Geology, 26,* 1–19.

Rubio, B., Álvarez-Iglesias, P., Bernabeu, A. M., León, I., Mohamed, K. J., Rey, D., et al. (2011). Factors controlling the incorporation of trace metals to coastal marine sediments: Cases of study in the Galician Rías Baixas (NW Spain). In J. Ibukun Agboola (Ed.), *Relevant perspectives in global environmental change*. InTech, www.intechopen.com.

Rubio, B., Nombela, M. A., & Vilas, F. (2000). Geochemistry of major and trace elements in sediments of the Ria de Vigo (NW Spain): An assessment of metal pollution. *Marine Pollution Bulletin, 40,* 968–980.

Rubio, B., Pye, K., Rae, J., & Rey, D. (2001). Sedimentological characteristics, heavy metal distribution and magnetic properties in subtidal sediments, Ria de Pontevedra, NW Spain. *Sedimentology, 48,* 1277–1296.

Ruiz-Villarreal, M., Montero, P., Taboada, J. J., Prego, R., Leitão, P. C., Pérez-Villar, V. (2002). Hydrodynamic model study of the Ria de Pontevedra under estuarine conditions. *Estuarine, Coastal and Shelf Science, 54,* 101–113.

Scheu, E. (1913). Die Rias von Galicien. ihr Werden und Vergehen. Z. *Gesteine Erdkunde.* Berlin.

Schurtz, H. (1902). An der Riasküste Galiciens. In *Deutsche Geographischer Bläter* (Vol. 25, pp. 50–74). Bremen.

Short, A. D. (2006). Australian beach systems—Nature and distribution. *Journal of Coastal Research, 22*(1), 11–27.

Souto, C., Gil-Coto, M., Fariña-Busto, L., & Pérez, F. F. (2003). Modeling the residual circulation of a coastal embayment affected by wind-driven upwelling: Circulation of the Ría de Vigo (NW Spain). *Journal Geophysical Research, 108*(C11), 3340–3356.

Torre Enciso, E. (1958). Estado actual del conocimiento de las rías gallegas. In *Homaxe a R. Otero Pedrayo* (pp. 237–250). Ed. Galaxia, Vigo.

Turner, A., & Millward, G. E. (2002). Suspended particles: Their role in Estuarine biogeochemical cycles. *Estuarine, Coastal and Shelf Science, 55,* 857–883.

Twidale, C. R., & Vidal-Romaní, J. R. (1994). The Pangea inheritance. *Cuad. Lab. Xeol. Laxe, 19,* 1–45.

Uncles, R. J., & Stephens, J. A. (1989). Distributions of suspended sediment at high water in a macrotidal estuary. *Journal Geophysical Research, 94,* 14395–14405.

Van Leussen, W. (1988). Aggregation of particles, settling velocity of mud flocs: A review. In J. Dronkers & W. van Leussen (Eds.), *Physical Processes of Estuaries* (pp. 347–403). Berlín: Springer.

Varela, R. A., Rosón, G., Herrera, J. L., Torres-Lopez, S., & Fernández-Romero, A. (2005). A general view of the hydrographic and dynamical patterns of the Rias Baixas adjacent sea area. *Journal of Marine Systems, 54,* 97–113.

Vera et al., (Vilas, F.). (2013). Real Academia de Ciencias Exactas, Físicas y Naturales de España. El termino Ría. In *RAC, Vocabulario Científico y Técnico-Geología* (VCTRAC.es).

Vidal-Romaní, J. R. (1996). Geomorfología de Galicia. In *Galicia. Geografía. Tomo* (Vol. XVII, pp. 36–63). Hércules de Ediciones. A Coruña.

Vila-Concejo, A., Hughes, M. G., Short, A. D., & Ranasinghe, R. (2010). Estuarine shoreline processes in a dynamic low-energy system. *Ocean Dynamics, 60*(2), 285–298.

Vilas, F. (2007). Rías and tidal-sea estuaries. Encyclopedia of life support systems. In *UNESCO-EOLSS PUBLISHERS* (*Coastal zone and estuaries: estuarine systems,* 2.6.3.1.).

Vilas, F., Bernabeu, A. M., & Mendez, G. (2005). Sediment distribution pattern in the Rias Baixas (NW Spain): Main facies and hydrodynamic dependence. *Journal of Marine Systems, 54,* 261–276.

Vilas, F., Bernabeu, A., Rubio, B., & Rey, D. (2010). Estuarios, rías y Llanuras intermareales. In A. Arche (Ed.), *Sedimentología. Del proceso físico a la cuenca sedimentaria* (pp. 619–673). Madrid: CSIC.

Vilas, F., García-Gil, S., & García-García, A. (1999a). Shallow gas in incised-valley fills of the Ría de Vigo, NW Spain. *EOS, Transactions, 46*(80), 557.

Vilas, F., García-Gil, S., García-Gil, E., Nombela, M. A., Alejo, I., Francés, G., et al. (1999b). *Cartografía de Sedimentos Submarinos: Ría de Arousa* (Mapa 1:50.000). Xunta de Galicia. Consellería de Pesca, Marisqueo y Acuicultura (Ed.) Santiago de Compostela, 32 pp.

Vilas, F., García-Gil, S., García-Gil, E., Nombela, M. A., Alejo, I., Rubio, B., et al. (1996). *Cartografía de sedimentos submarinos. Ría de Pontevedra* (Mapa 1: 50000 y memoria). Xunta de Galicia. Consellería de Pesca, Marisqueo y Acuicultura (Ed.) Santiago de Compostela, 40 pp.

Vilas, F., & Nombela, M. A. (1985). Las zonas estuarinas de las costas de Galicia y sus medios asociados, NO de la Península Ibérica. *Thalassas, 3,* 7–15.

Vilas, F., Nombela, M. A., García-Gil, E., García-Gil, S., Alejo, I., Rubio, B., et al. (1995). *Cartografía de Sedimentos Submarinos de las Rías de Vigo y Pontevedra* (5 Mapas). Xunta de Galicia, Consellería de Pesca, Marisqueo y Acuicultura (Ed.) Santiago de Compostela, 32 pp.

Von Richthofen, F. (1986). *Führer für Forrschungsreisende* (310 pp). Oppenheim, Berlin.

Chapter 18
Cantabrian Estuaries

Germán Flor-Blanco and Germán Flor

18.1 Introduction

Asturias and Cantabria are two Spanish regions (southern Bay of Biscay, NW Iberian Peninsula) with a W-E cliff coast along 422 km. It contains more than 50 estuaries with different features and stages of evolution, including the Rías in Galicia. They have very varied sizes and lengths (Fig. 18.1) as well as sediment types; they are linked to short rivers and, in general, high and low waters allow during high spring tides that mixing waters achieve variable lengths of 5–15 km.

According to morphogenetic criteria (Perillo 1995) every Cantabrian estuaries are *primary*, that has been almost completely infilled by sediment, generally on a bedrock valley; tides are the most important dynamic agent of marine origin determining from the tide range the distribution of sand bodies in the mouth (Hayes 1975) with a mesotidal character in this region. Rias can be included, according to Vilas (2002), which are more exposed to incoming waves and the hydrodynamic processes, represented by tidal currents and fluvial discharges, and have a scarce or deficient sedimentary fill. Along the coasts, all the possible states of evolution of estuaries and rías can be deduced.

These Cantabrian estuaries are developed since the Upper Pleistocene according to a global Holocene transgression well recorded in the old estuary of Gijón-Piles (Flor and Lharti 2008) until the Flandrian maximum sea level.

The studied estuaries (Fig. 18.1) are mixed systems between marine and fluvial influenced and some of them are rock-bounded controlled (Allen 1993) along a coast where many old erosion marine levels (named "rasas") were developed due to

G. Flor-Blanco (✉) · G. Flor
Department of Geology, University of Oviedo, 33005 Oviedo, Spain
e-mail: gfb@geol.uniovi.es

G. Flor
e-mail: gflor@geol.uniovi.es

J. A. Morales (ed.), *The Spanish Coastal Systems*,
https://doi.org/10.1007/978-3-319-93169-2_18

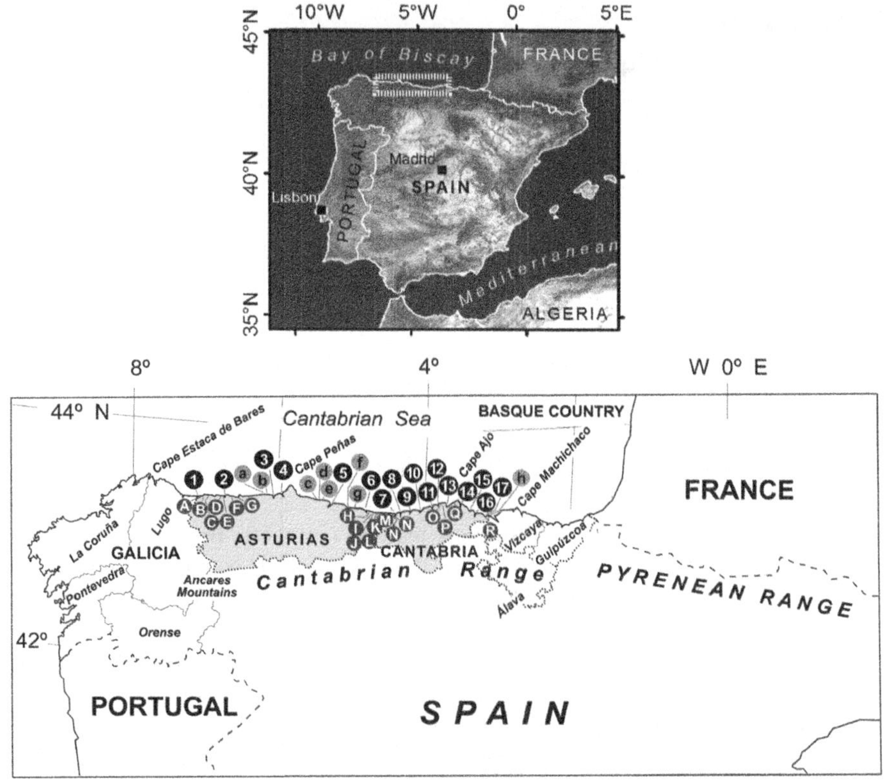

Fig. 18.1 Situation of the largest estuaries of the Cantabrian coast. (1) Eo, (2) Navia, (3) Nalón, (4) Avilés, (5) Villaviciosa, (6) Ribadesella, (7) Tina Mayor, (8) Tina Menor, (9) San Vicente de la Barquera, (10) La Rabia, (11) San Martín de la Arena, (12) Mogro, (13) Santander, (14) Ajo, (15) Cabo Quejo, (16) Asón, and (17) Oriñón. Other minor mentioned estuaries are: (A) Sarello, (B) Anguileiro, (C) Porcía, (D) Viavélez, (E) Ortiguera, (F) Puerto de Vega, (G) Esva, (H) Aguamía, (I) Cuevas de Mar, (J) Niembro, (K) Poo, (L) Llanes, (M) Purón, (N) Marimuerto, (Ñ) La Franca, (O) San Juan de la Canal, (P) La Maruca, (Q) Galizano, and (R) Mioño. Other small inactive estuaries are linked to functional river mouths: (a) Esqueiro, (b) Uncín, (c) La Ñora, (d) España, (e) Merón, (f) Libardón, (g) Espasa, and (h) Brazomar

uplift since the Late Miocene (Flor and Flor-Blanco 2014a; Domínguez-Cuesta et al. 2015).

All estuaries are trumpet-shaped in plan, incised valley systems (Posamentier and Vail 1988), sometimes with sinuous axes, generally with a sandy confining barrier: Navia, Barayo, Nalón, Villaviciosa (Figs. 18.1 and 18.2b), San Vicente de la Barquera, San Martín de la Arena, Mogro (Figs. 18.1 and 18.2c), Santander, Asón, and Oriñon; and less frequent gravel or mixed barrier: Porcía, Esva, Ribadesella, Tina Mayor (Figs. 18.1 and 18.2e), and Mioño or only rocky inlet: Eo (Figs. 18.1 and 18.2a), Tina Menor (Figs. 18.1 and 18.2e), and Niembro. Regarding to the geometry, the Cantabrian estuaries are mesotidal estuaries confined by sand barriers (Hayes 1975; Pritchard 1967; Fairbridge 1980) but without an ebb tidal delta because Cantabrian Sea is characterized by a relative higher tidal range and these type of morphologies are smaller or simply not are formed. It is important to highlight that the mouth of the estuaries of Navia, Nalón, Avilés, Villaviciosa, San Vicente de la Barquera and San Martín de la Arena has been channeled by jetties (Flor-Blanco et al. 2015a) changing the morphodynamic and sedimentary features and in other cases the dune systems have been anthropized eliminating a great part or all dunes by urbanization, roads, car parks and promenades.

Mainly, these estuaries are drained by a unique trunk river but there are some exceptions leading to the appearance of two jointly estuaries, really two estuarine subsystems, that share a certain area with its own river one of which is a higher hierarchy (Table 18.1) and a typical morphology of wave-dominated estuaries in a partial state of infilling with many morphosedimentary units.

There are some examples in Asturias and Cantabria coasts as the Eo estuary; the outer one is developed as a wave-dominated estuary without sedimentary barrier and the inner estuary of Eo (Flor and Flor-Blanco 2011) is characterized by a segment (3.5 km long) with tidal sand bars parallel to the axis estuary (Figs. 18.1 and 18.2a) where rocky margins are narrowed (Flor 1995); so, the latter one reproduces the characteristics of a macrotidal estuary according to Hayes (1975). Also the estuary of Avilés in Asturias (López Peláez and Flor 2006; López Peláez 2016) is the result of the sedimentary infill of one principal valley and another small tributary. The estuary of San Vicente de la Barquera comprises one more entire estuary linked to a coastal river (El Escudo) or subsystem of Rubín and other western muddier branch (subsystem of Pombo) drained by the coastal stream of Gandarilla (Flor-Blanco 2007; Flor-Blanco et al. 2015c).

Other similar examples are also the small estuary of La Rabia is represented by two subsystems linked to coastal streams (Flor-Blanco et al. 2012). In the case of the bay of Santander (Table 18.1) is constituted by a broad subsystem drained by several streams and other eastern subsystem of Cubas of lesser dimension, linked to the mountain river of Miera (Flor and Flor-Blanco 2014b).

There are other singular estuaries, generally of small sizes where the dominant fluvial or marine influence allows the emergence of well-developed upper marshes or they build a large confining barrier, almost 50% of the total estuarine surface, where the dune field is very well represented, i.e. the estuary of Barayo. Other estuaries have been developed in small rock bounded fractures or sinkholes, and nowadays are almost infilled as Guadamía, Poo and Río Purón in Asturias and Galizano and Mioño in Cantabria (Fig. 18.1). Moreover, in Asturias has been developed an exceptional example; it is a small estuary in a sinkhole (with an oval floor, 120 m long, and 45 m width) connected to the waterfront through a karstic fissure that replaces the inlet (estuary of Marimuerto; Fig. 18.2c). Some small estuaries with a natural deficit of sediments were dredged for port uses: Viavélez, Puerto de Vega and Llanes (Fig. 18.1).

From a morphological and dynamic point of view, these estuaries are included in wave-dominated (WDE) types (Dalrymple et al. 1992) but widely represented in all

(a)
(b)
(c)
(d)
(e)
(f)

◂**Fig. 18.2** Different examples of Cantabrian estuaries. **a** Eo, with two estuarine subsystems and a central flood-tidal delta and several sand shoal including spill-over lobes in the inner Eo. **b** The estuary of Villaviciosa is confined by two jetties since 1925. It develops a flood-tidal delta in a break of the right dike and its marshes have been filling by sand. **c** The small estuary of Viavélez with a sinuous plant which was almost totally dredged. **d** The singular and unique Marimuerto estuary in scientific bibliography, formed in a calcareous sinkhole and connecting with the sea, 30 m seaward, by a hole. **e** Mogro is an estuarine system with a great sand spit culminated with parabolic dunes and a broad climbing vegetated dune field. **f** Tina Mayor (West) and Tina Menor (East) are rock-bounded estuaries, the first one, the outer sector is predominantly filled with gravel fractions and the second one by sands

estuaries, only in the mouth because the tidal currents and fluvial discharges are most important to the rest of the estuary.

This study is focused to join the different studies about Asturias and Cantabria estuaries from the point of view of morphodynamic and sedimentation behavior. Fortunately, in these regions, there are many examples with particular features that can understand the evolution of these sensitive systems to any natural or anthropogenic changes.

18.2 Regional Setting

This coastal region is under the influence of the temperate humid Atlantic climate with average rainfalls oscillate between 800–900 mm year^{-1} and 2,200 mm year^{-1}; river floods occur principally in spring and autumn and low discharge are typical during summer and sometimes until winter.

The winds come from SW, NW and E and NE. Prevailing winds are southwesterly in winter and autumn and northeasterly, upwelling favorable, in summer and spring (Lavín et al. 2004). However, along the Cantabrian coasts, main directions of wave seas with low significant highs (<1.5 m) are from WNW and NNW and frequencies of 7.9 and 7.6%, while the main direction of wave swells (Hs > 1.5 m) is NW with frequency of 23.5%, and the significant storm number: H1/3 > 4 m (Lechuga et al. 2012). The heavy waves mean that coastal sand transport tends to move in an eastwards direction. To sum up, NW-SE, NE-SW and W winds are relationship with the dune fields of sand spits barriers of the estuaries mouth (Flor and Flor-Blanco 2014c) mainly in the formation of the foredunes, longitudinal dunes and some blowouts. In some cases, there are sand sheets formed by south winds in the inner areas of estuaries but with low representation.

Tides are semidiurnal with an average range of about 2.47 m (obtained from the tide gauge at the Gijón Port representative of the year 2010) about 72% in a year, but tides vary between macrotidal (4.75 m) and microtidal (1.0 m). This coast can be defined as "low mesotidal" during neap tides and "high mesotidal" during spring tides (Flor-Blanco et al. 2015b). Moreover, tide waves are gradually dissipated

Table 18.1 Distribution of the greater estuaries of Asturias and Cantabria in four morphological zones, including surfaces, the most important morphosedimentary and dynamic units and sedimentary distribution

ESTUARY	estuarine subsystem	MOUTH COMPLEX: mouth bar	MOUTH COMPLEX: confining barrier – exposed beach	MOUTH COMPLEX: confining barrier – aeolian dune field	MOUTH COMPLEX: inlet	MOUTH COMPLEX: TOTAL SURFACE ha	BAY: main channel	BAY: flood-tidal delta	BAY: secondary channel	BAY: spill-over lobes	BAY: sand flats	BAY: estuarine beaches	BAY: estuarine dune field	BAY: TOTAL SURFACE ha	TIDAL FLATS: main channel	TIDAL FLATS: mud flats	TIDAL FLATS: marshes	TIDAL FLATS: tidal creeks	TIDAL FLATS: TOTAL SURFACE ha	UPPER CHANNEL: main channel	UPPER CHANNEL: fluvio-tidal plains	UPPER CHANNEL: TOTAL SURFACE ha
EO	outer	○			○	5.95	○	○	○	○		○●		121.60		◌	◌	◍	144.85	○●	3.40	4.53
	inner									○		○●		195.60	○	◌	◌	◍	260.00	○●	96.97	129.20
NAVIA		○	○	○	○ [built dikes]	30.50	○ [built dikes]							34.63	○	◌	◌	◍	261.60	●	24.79	44.19
NALÓN		○	○	○	○ [built dikes]	102.90	○ [built dikes]				○	○	○	98.18	○	◌	◌	◍	419.61	●	112.63	141.26
AVILÉS	Avilés	[fully removed]	○	○	○ [built dikes]	471.07	○ [built dikes]				○	○	○	86.84	○				440.80			235.90
	Llodero						○				○	○	○	23.02	○	◌	◌	◍	44.49	○●		10.14
VILLAVICIOSA		○	○	○	○ [built dikes]	77.50	○ [built dikes]	○	○	○	○		○	100.72	●	◌	◌	◍	528.75	●	167.14	172.90
RIBADESELLA		○	●		●	40.48	● [built dikes]	●	○			●		23.53	●	◌	◌	◍	319.77	●	31.60	46.06
TINA MAYOR			●		○	6.55	●					○●		15.92	○●	◌	◌	◍	128.00	●	36.45	67.13
TINA MENOR						3.95	○			○	○	○●	○	39.12	○				108.83	●	20.31	32.67
SAN VICENTE DE LA BARQUERA	Rubín		○	○	○ [built dikes]	36.10						○	○	72.59	○				307.32	●	4.41	9.45
	Pombo						○	○	○	○	○	○			○	◌	◌	◍	83.43		48.66	49.55
LA RABIA	Rioturbio		○	○	○	23.51	○				○	○		13.32	○	◌	◌	◍	40.12	○	19.89	34.90
	El Capitán														○	◌	◌	◍	31.67	○	14.39	15.52
SAN MARTÍN DE LA ARENA		○	○	○	○ [built dikes]	15,55	○ [built dikes]					○		63.48	○	◌	◌	◍	309.94	○●	386.74	390.16
MOGRO		○	○	○	○	118.95	○			○	○	○	○	133.50	○	◌	◌	◍	144.30	●	45.11	59.33
SANTANDER	Santander	[fully removed]	○	○	○	100.18	○				○	○		1,475.64	○	◌	◌	◍	1,715.73	○	48.84	203.65
	Cubas				○		○	○	○		○			69.44	○	◌	◌	◍	210.22	●	30.99	56.20
AJO						9.15	○			○	○	○	○	62.51	○	◌	◌	◍	59.15	○	7.32	15.05
CABO QUEJO			○	○	○	24.17				○	○	○		15.58	○	◌	◌	◍	80.35			
ASÓN		○	○	○	○	770.72	○				○	○		246.60	○	◌	◌	◍	2,212.55	○●	220.45	267.72
ORIÑÓN			○	○	○	68.10	○ [built dikes]				○	○		20.74	○		◌	◍	74.82	●	87.70	90.78

○ sandy bottoms ● gravelly bottoms ◌ muddy and fine sandy bottoms ◍ complex bottoms: mud, sands and shell debris
[light grey] fully removed [dark grey] built dikes

Modified from Flor-Blanco and Flor (2012)

upstream until the estuary head, according to the hyposynchronic model (Le Floch 1961), but some large estuary can sometimes play as a hypersynchronic estuary, i.e. in the estuaries of Ribadesella (Flor and Camblor 1989), San Vicente de la Barquera and Santander.

18.3 Morphological and Sedimentary Features

The types of sedimentary rocks, structural disposal and faults have determined geometry in plant of the valley, previously excavated during the last low sea-level stand: straight, arcuate and sinuous when river process embedded meanders, as well as the amplitude.

The morphological reality of the mesotidal estuaries along the Cantabrian coast, even Atlantic estuaries, suggests its differentiation in four longitudinal geomorphological zones, according to Flor criteria (1995), and subsequent works (Flor-Blanco 2007; Flor-Blanco et al. 2015b). From the mouth to the inner area, the distribution of zones is the next: *Mouth complex*, *Bay*, *Tidal flats* including the marshes, and *Upper channel.*

This differentiation is a consequence of the transition of energy level and types of sedimentation from the mouth of the estuary, where the coastal agents prevail, to the innermost limit (estuarine head) where the river discharges are the main dynamic agent, as it is shown (Fig. 18.3) in conjunction with the morphological zonation.

Each zone is composed of a variety of other minor morpho-sedimentary and dynamic units, some of them are substantially representative, but others appear in a unique way. The sedimentary fractions are a determining factor in the development of certain geometries but so are the river valley shape and how narrow the confining estuary is. For instance, the main channel, the mud flats, marshes and tidal creeks are the widely developed units in the Cantabrian estuaries. Moreover, the presence of the flood-tidal deltas, spill-over lobes, estuarine aeolian dunes and beaches is relatively significant. Nevertheless, the gravel barriers are less frequent.

Normally, during low tides, the sedimentary bottoms are fully emerged except drainage channels of greater hierarchy.

Regarding average main grain along of an estuary, it decreases from the barrier to the inner tidal flats, and upper channel is represented by the coarsest sizes due to the direct flood river discharges (Fig. 18.3).

From the morphological point of view, the Cantabrian estuaries could divided as follows:

(a) **Mouth complex**, normally are represented by a sandy barrier with a dune field that in many cases can developed one or more foredunes and other minor dune morphologies. These exposed beaches can be dissipative or reflective as Navia, but estuaries linked to a trunk river mostly draining gravel fractions and being

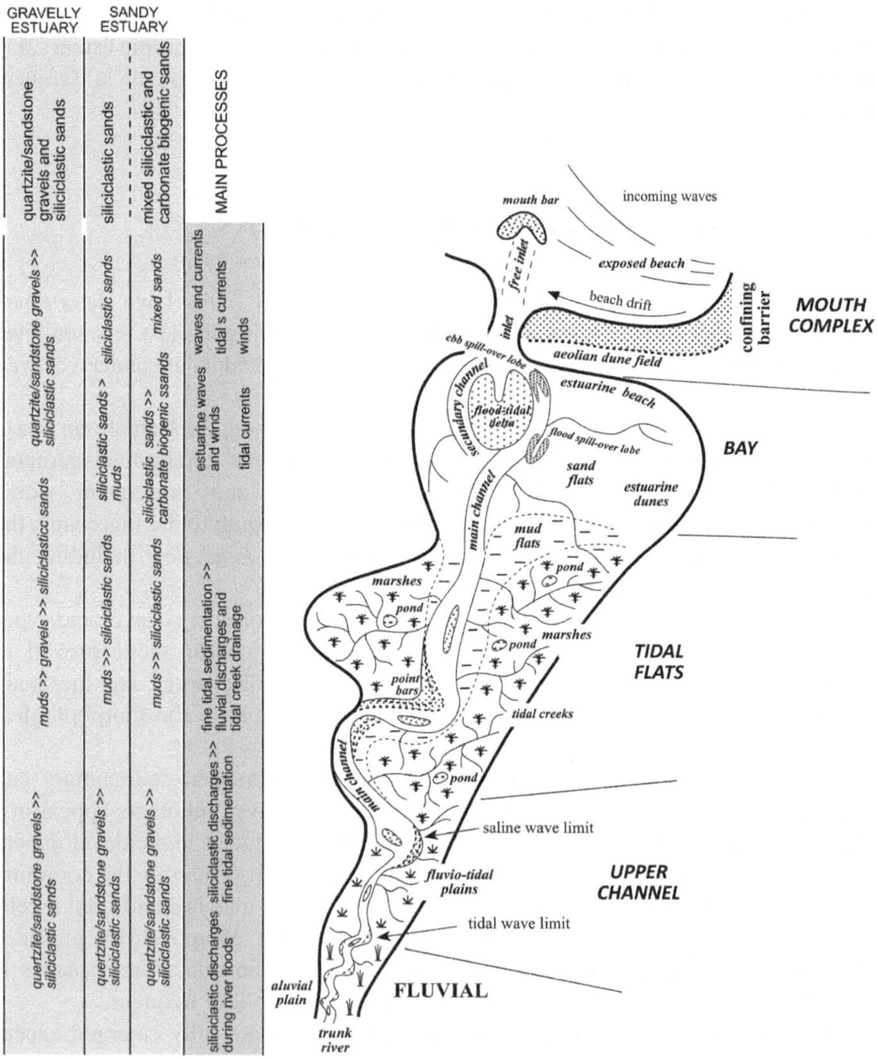

Fig. 18.3 Morpho-sedimentary zonation along a mesotidal confining estuary developed in a fluvial valley. Main morpho-sedimentary and dynamic units as well as some large bedforms are included. Average grain-sizes and mineralogical composition allow characterize the sedimentological transitions and processes from the mouth to the inner limit, distinguishing between sandy and gravelly estuaries

able to form gravel barriers, for instance as Porcía, Esva, Ribadesella and Tina Mayor (Fig. 18.1).

Highlight than inlets are narrow and in some cases, channels were considerably lengthened with lateral jetties (Flor-Blanco et al. 2015a). Seaward main channel can be a free inlet (Navia, Avilés, Villaviciosa, San Vicente de la Barquera and

Cuchía) and sometimes connect directly with a mouth bar, a permanent submerged shoal, generally arched with the convexity towards the sea as Villaviciosa (Flor et al. 2015). Wave storms remove this structure even allowing its temporal disappearance or, in the non-common case of Tina Menor, waves and tidal currents replaced this morphology by an extensive erosive depression or scour (Flor-Blanco et al. 2015b).

(b) **Bay** is the wide outer mostly sandy or gravelly area where the main channel is the strongest energetic unit. Lateral and central bars are very characteristic with fusiform geometries where sand waves, local megaripples, and a wide variety of superimposed small bedforms are generated. Other smaller channels are stabilized in certain estuaries (Villaviciosa, Tina Menor, estuarine subsystem of Cubas in Santander). Filtering bivalves are present as i.e. *Ensis siliqua, Cerastoderma edule, Phollas dactylus*.

Some shoal and bars are developed and in few estuaries stable flood-tidal delta have formed (Eo, Villaviciosa, San Vicente de la Barquera and the subsystem of Cubas of Santander bay), or several spill-over lobes (Eo, Tina Menor, Mogro, Ajo) and in others elongated sand flats if rock bounded prevents the formation of larger structures (i.e.: Villaviciosa, Tina Menor, La Rabia, Ajo). Wave and current ripples cover the entire sandy surface and channel banks. In many parts of these units are developed sand bubbles escaping and are bioturbated by the worm *Arenicola marina*, and others as *Nereis diversicolor, Owenia fusiformis, Diopatra napolitana*; the burrowing ghost shrimp *Callianasa subterranea*, echinid *Echinocardium cordatum*, etc. (Flor and Flor-Blanco 2014d).

Estuarine beaches, irregularly distributed on the upper limits but always very narrow, are developed on the margins, as well associated to small aeolian estuarine dunes formed by southern winds (Villaviciosa, San Vicente de la Barquera and Asón).

(c) **Tidal flats and marshes** are the broadest estuarine zone area, traversed longitudinally by the sinuous main channel (Fig. 18.3). Consequently, sandy and mixed sandy and muddy point bars are developed as well as erosive margins on the muddy flats. Active mud sedimentation is restricted to narrow belts near the channels. In the lower and saline areas, the bottoms are colonized with algal vegetation (*Enteromorpha* ssp.), and as height is gained, generally landward, other halophytic species stratifying constitute a marsh according to the estuarine zonation from marine to supratidal marshes, simplified with the plant genus: *Zostera, Spartina, Sarcocornia, Suaeda, Halimione, Juncus, Phragmites*. Consequently, they develop a very high productivity.

Many tidal creeks drain these muddy surfaces whose low slope favors their sinuous path. Some abandoned creeks can constitute ponds where gastropods are found in very high densities.

(d) **Upper channel** So called this zone because it is the functional main channel most of the time, in most cases with a meandering layout in plan. It is dominated by the river discharges with the tidal wave influence travelling upstream until its dissipation. During river floods, it can be overflowed by affecting the fluvial-tidal plains which are narrow and can receive sand sheets or gravel lobes. Tidal creeks are scarce disappearing upstream. The gravelly bottoms are dominant generating many bars constituted by siliciclastic clasts that vary in shape from rounded to angular.

18.4 Dynamic Behavior

The rocky coast of the Cantabrian Sea is characterized by an eastward coastal current that is the main factor of the net sediment transport from rivers, cliffs and platform deposits. On the other hand, the tides are very important for understanding the dynamics and sedimentary distribution in these estuaries, assuming a mesotidal range detailed above.

The salinity distribution along the water column during a tidal cycle shows partially the current dynamic behavior inner the estuary. Mixing saline and fresh waters are very variable in each estuary and over the year, depending on the spring and neap tidal ranges and the low and flood fluvial discharges. Spring and average tidal ranges and low and average water discharges allow mixing waters according to a vertically mixed model, and salt-wedge model under the remaining conditions.

Taking into account the average numerical values (flow ratio) of Simmons (1955) and regarding to the mixed saline and fresh waters, the studied estuaries of Eo, Villaviciosa and San Vicente de la Barquera (Fig. 18.4) are vertically homogeneous. Nalón, Ribadesella and Tina Menor (Fig. 18.4) are partially stratified, and only Tina Mayor is a salt wedge estuary (Fig. 18.4).

Figure 18.4 shows an estuarine classification, which joins different values as the average tidal range, fluvial discharge (QF) and tidal flood (QT). With this triangular model, each estuary could be classified in relation to the type of salinity stratification, and can be indirectly related to the geomorphological areas of the mouth, including important morpho-sedimentary units. Some estuaries have been studied front the point of view of dynamic currents, salinity, sedimentation etc., So, three estuaries have shown with different features between them (Flor-Blanco 2007). Tina mayor is drained by Cares (tributary) and Deva rivers with a broad watershed and the fluvial influenced are widely developed along the estuary. However, Tina Menor is a mixed estuary, between tidal and fluvial influenced, with an important river, the Nansa, but less than Cares-Deva. In the other hand, San Vicente de la Barquera is drained by the stream of Gandarilla and the small coastal river of El Escudo, with small fluvial basins and consequently little river flow, being the tidal prism much more voluminous than the fluvial discharges.

With this different estuarine behavior, the triangular model allows to classifying any estuary according to these parameters, and can be extrapolated to most of the Atlantic estuaries.

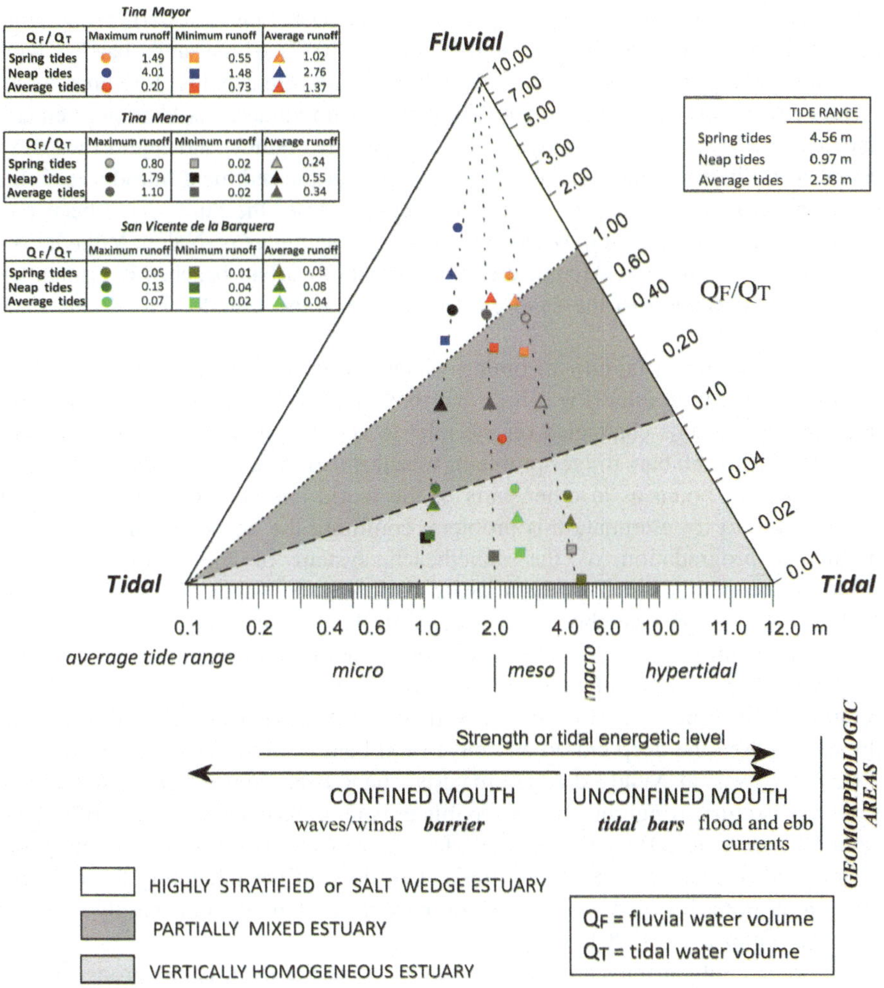

Fig. 18.4 Triangular diagram of fluvial-tidal range controls that shows the variations in water mixing, QF/QT (modified flow ratio limits of Simmons 1955), in the estuaries of Tina Menor, Tina Menor and San Vicente de la Barquera (Flor-Blanco 2007; Flor-Blanco et al. 2015b). The estuarine confined (barrier) and rock bounded confined mouths are included because they are related with tidal strength and water mixing

Although despite of the Cantabrian estuaries are classified as wave-dominated estuaries criteria (Dalrymple et al. 1992), this wave dynamic is only determinant in the mouth, mainly in the sandy dune-beach barrier and locally in the inner estuarine beaches. First and four quadrants are the main directions of the incoming waves, being the NW waves the strongest. Since 2006, some high wave storms have produced different changes in the mouth of the estuaries. The destructive effects are visible in surge events as in February and March of 2014 when the strongest surge event caused some damages in European Atlantic coast (Flor et al. 2014; Masselink et al. 2016; Castelle et al. 2017; Flor-Blanco et al. 2017). Some sandy barriers have critically damaged, promoting the retreatment of the dune fields, foredunes completely eroded, crevasse of sediment landward the foredunes, sea-front promenades and marinas damages. To mitigate the erosion, the policy and authorities have established different management but a great part of the sediment introduced in the sea front promenades has sent to garbage dumps. Also, the dunes have been preserved by plant colonization, wooden walkways, fencing areas (Nalón, Villaviciosa, Somo, San Martín de la Arena), dredgings inputs to urban beaches in Santander, defense accumulation to the sand eroded as Navia and Laredo (Flor-Blanco and Flor 2016).

In the outer part (mouth), incoming waves refract to induce the beach drift of sediment toward the inlet (Fig. 18.2), but they can break over the outer mouth bar and the ebb currents channeled on the inlet to stabilize that shallow bar. In some estuaries, these ebb bars trigger problems to sail if there is a marina or harbour inner the estuary. In Spain as in other parts of the world has been common the jetties construction to try eliminate this problem, confining the artificially inlet and promoting a progradation of the dune/beach system of the confining barrier (Flor-Blanco et al. 2015a). In Asturias, the estuaries of Navia, Nalón, Avilés, and Villaviciosa, and in Cantabria: San Vicente de la Barquera and San Martín de la Arena are examples of heavy changes in the estuarine mouths. In other cases, the dredging of these kind of ebb bars, but also bottoms of the main channel, has promoted different scenarios. In the estuary of Villaviciosa, the sediment was dumped bellow the deep closed and have been beneficial to the system (Flor et al. 2015), however, in Avilés, dredged materials have been dumped out of the system since last century, triggering an irreversible erosion in beach/dune barrier of Salinas (Flor-Blanco et al. 2013). The estuary of Urdaibai (Vizcaya, Basque Country), out of the studied area, was intensive dredged in the ebb bar, changing the dynamic conditions along few years and the major part of the dune system sharply retreated (Monge-Ganuzas et al. 2013).

Flood and ebb currents are well developed along the inlet, being strongest from high to low tides the first ones, and ebb tides are more important during low tides when the estuary is emptying the tidal prism and the river water. A strong jet-like current favors the presence of a longer free inlet seaward, that is more important during river floodings and spring tides (van Lancker et al. 2004), such as in Navia. In all estuaries, the Coriolis effect starts its influence, and, for example in estuaries with high fluvial currents or great rain and flood periods, is the unique part of the estuary where these effects are effective and can be deduced by the salinity.

The dynamics development in the inner estuary is different along it and the Coriolis effect is the main factor to understanding the dynamic and sedimentary distribution in Cantabrian estuaries and Rias. This effect has been detected in other estuaries around the world as Lawrence estuary (Prandle and Crookshank 1974), Oregon (Boggs and Jones 1976), Chesapeake Bay (Valle-Levinson and Atkinson 1999) or Kennebec estuary (Fenster et al. 2001). On the other hand, some studies

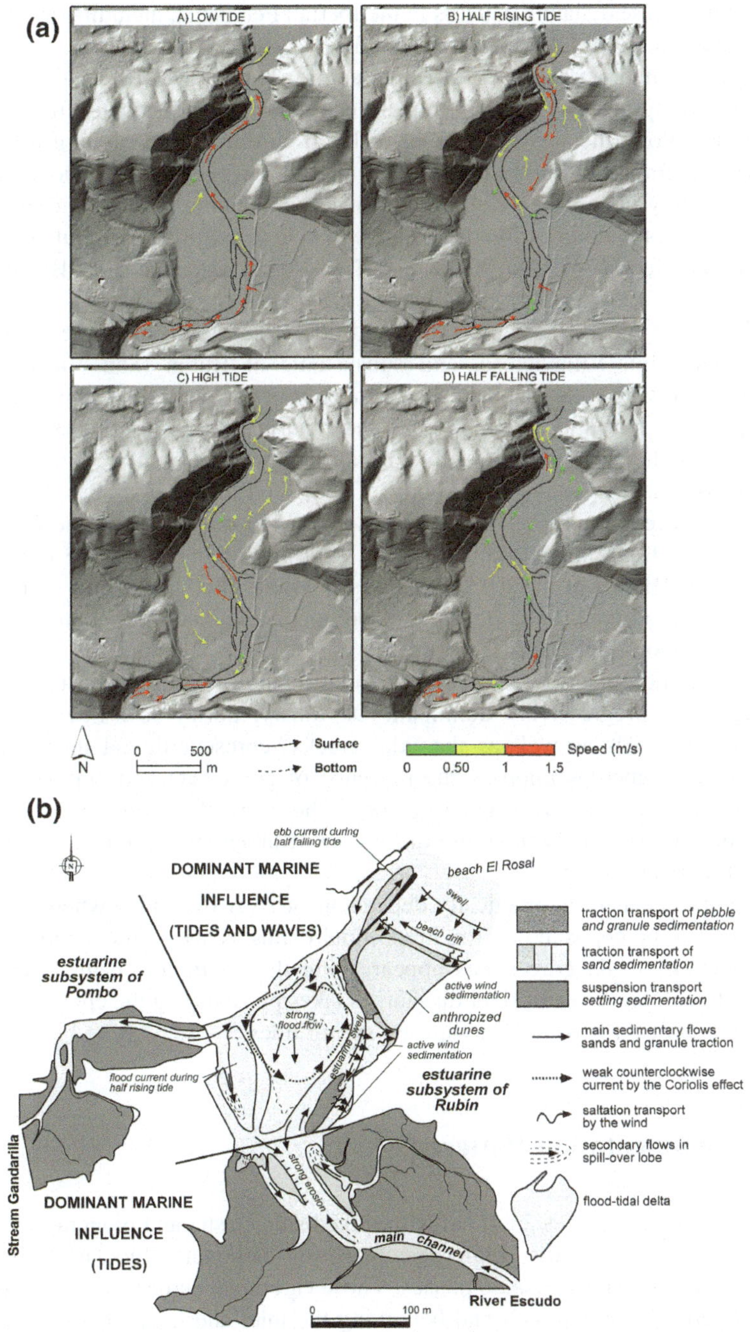

Fig. 18.5 Schemes of representative currents recorded in the water column during a tidal cycle and the importance of Coriolis effect in Tina Menor (**a**) and in the outer area of the Rubín subsystem of San Vicente de la Barquera where dynamic and sedimentary processes are modeled (Flor-Blanco et al. 2015b) (**b**). Modified from Flor-Blanco et al. (2015c)

have verified these dynamic patterns in the Iberian Peninsula, including the Galician Rías (Dale et al. 2006; Álvarez et al. 2006).

Into the Cantabrian estuaries, the circulation in a tidal cycle is relatively simple with the development of flood currents during the rising marine water from low tide to high tide along the western margin and ebb currents during falling tide in the opposite margin, according to the Coriolis effect. However, the rock bounding and the morphology of some estuaries can change the dynamic patterns and during the half rising tide, the flood is diverted to the east side (Fig. 18.5a) without developing the Coriolis effect as in Tina Mayor and Tina Menor estuaries (Flor-Blanco et al. 2015b).

If the bay is broad, the ramp of a flood-tidal delta is linked to the inlet of the mouth: outer Eo, Villaviciosa, and subsystems of Rubín (San Vicente de la Barquera) and Cubas (Santander), but if the bay is narrow this is substituted by a large spill-over lobe/s (Tina Menor and Ajo). During half rising tide until an hour before the high tide, approximately, the strong flood current from the ramp is radially dissipated over the main body of the flood-tidal delta. A surficial weak levogirous current can be activate during spring high tides, clear in the Eo estuary (Flor et al. 1992), Villaviciosa (Flor et al. 1996) and San Vicente de la Barquera estuary (Flor-Blanco et al. 2015c). This is possible after the strong current activating the flood-tidal delta that is smoothing reworked the contours of the heart-shape flood-tidal delta (Fig. 18.5b).

Along the tidal flat zone, flood and ebb currents decrease, but upstream fluvial discharges are progressively dominant. Obviously, during low tides, main and secondary channels as well as the tidal creeks represent fluvial discharge, and meandering geometries improve the presence of point bars and opposite erosive margins. Long-term sands migrate upstream the channel bottoms of tidal creeks. During high tides, mainly in spring ones, finer sedimentation dominate in the mud flats and marshes.

The upper channel is practically dependent of river dynamics where overwash sediments can be deposited on the fluvio-tidal plains as sheet and/or lobes of sand and/or gravels, while gravel bars appear along the confining channel. The most important geological role of river discharges takes place during the episodes of river floods and better if it coincides with the spring tides.

18.5 Sedimentation Patterns

Most Cantabrian estuaries are filled with mixed siliciclastic and biogenic sediments (Fig. 18.3) with a relative high percentage of carbonate (bioclasts), probably inherited from previous eustatic phases. Those carbonate percents of sand decrease upper stream from the mouth and bay along the inlet and main channel including sand shoals, estuarine dunes and beaches, and tidal flats and marshes devoid of all. A few estuaries are only filled with siliciclastic sands and/or gravels supplied by the trunk rivers of Navia, Nalón, Sella, Deva, Nansa, and Besaya (Table 18.1); others siliciclastic fractions are supplied by rivers linked to the small estuaries of Porcía, Barayo, Esva, Bedón, Ajo, and Mioño (Table 18.1).

Coal and carbonaceous sands and silts come from mining activity in the watershed which are common in the Central Asturian Coal Basin (supplied by the Nalón River) due to the historical inputs to the Nalón estuary (Flor et al. 1998) and this sediment has been transported eastward to input in several beaches at least up to a distance of several kilometers. This is one of the most evident examples of the anthropization processes in a coast under the sedimentary point of view. Also, the Nalón estuary has received inputs from two Hg mining areas with a long history of As and Hg contamination, which have been identified as As and Hg 'hotspots' (Fernández-Martínez et al. 2005; Higueras et al. 2014). However, only a few researches have been performed concerning these elements in the Nalón estuary. Recently, Garcia-Ordiales et al. (2015, 2016, 2017) showed with As and Hg geochemical profiles of saltmarsh sediment cores and channel boreholes a preliminary geochemical background values for both elements; moreover, they inferred the important enrichment in both elements due to mining contributions.

In addition to the mining activities in the Nalón river, other changes were produced in the mouth of the estuary after the jetties construction for channelization of the inlet (Flor et al. 2015); promoting sedimentary changes in neighboring beaches dune systems as Los Quebrantos and Bayas (just to the E of the mouth).

In coastal segments of Asturias where longitudinal sand drift is deficient or there is a sediment by-passing, away from large supplying rivers, small estuaries linked to coastal streams (until 1.5 km length and narrow banks less than 275 m). Most of the small estuaries can be completely filled by sediments, varying from those that have a dominance of gravels: Porcía, and Mioño (Fig. 18.1) to mixed ones, generally of gravels and sands, with an incompletely filling; if a port is located inside (Viavélez, Puerto de Vega and Llanes) must be periodically dredged (Flor and Flor-Blanco 2008). Other estuaries are partially filled by sand and/or gravel in the mouth and mud and gravel in the inner side, without marshes. Generally, they are straight if slope is high and gravel sedimentation and scarce sands are very thin, prevailing rocky bottoms. Also, they can be sinuous in plant, exhibiting a pattern of meander carved in a rocky bottom (Viavélez and Puerto de Vega in Asturias).

In the karstic coast of eastern Asturias, other small estuaries were completely filled by mixed gravel and sands also linked to coastal streams. The irregular sandy estuary of Niembro occupies an old blind valley (1.0 km length) where more siliciclastic sands were introduced from a quartzite quarry located in the upper part of the watershed. The small estuary of Poo (0.85 km length), constituted by some sinkholes connected a coastal stream and open to the sea, has been filled with fine siliciclastic sands. The removed estuary of Llanes (0.80 km length) developed some small mud flats and marshes in its inner limit but nowadays is completely anthropized. Elongated estuaries as Guadamía (0.70 km) and Purón (0.85 km) have mouth widths of 50 and 120 m, narrowing to a few meters, and are filled only with siliciclastic sand fractions; their morphology is similar to the "calas" in Balearic Islands (Butzer 1962; Rosselló 2005; Gómez-Pujol et al. 2013). The singular estuary of Marimuerto (Fig. 18.2c), so far the only one cited in the world, has been filled in a karst hollow linking with the coastline through a fissure of 100 m, which represents the inlet. It has an oval plant with the elongated major axis N-S in 100 m and the maximum width is 45 m, and the maximum width being 45 m; moreover, sandy bay, mainly represented by a tidal-flood delta, and marshy tidal flats are well represented.

In Cantabria, also heavy metals concentrations of Fe, Mn, Cu, Pb, Zn, Ni and PAH's were detected in the Bay of Santander in sediment profiles; they began increasing in the 1910s, and anthropogenic inputs have continued up to now (Viguri et al. 2007). Also in the estuary of San Martín de la Arena, analyzing Al, Zn, Pb, Cd, Cu, Ni, Cr and As) in surface and core sediments (Irabien et al. 2008). Subtidal sediments at the upper layers were less polluted than the intertidal sediments, and levels of metal pollution were: Zn (heavily to extremely); Pb (heavily); Hg (moderately to heavily); Cu (moderately); Cd (moderately); Cr (moderately), and Ni (unpolluted to moderately) (Bárcena et al. 2017).

Regarding the small estuaries in Cantabria (Fig. 18.1), San Juan de la Canal (0.65 km length) and La Maruca with a length of 1.2 km, are partially filled with mixed biogenic sands but both develop some sparse muddy facies and marshes in the upper limit. The estuary of Galizano (0.75 km length) is mainly filled by sand, containing beach deposits and small aeolian dune fields (foredunes and a climbing dune) and sandy marshes in the inner areas.

Many Cantabrian estuaries are undergoing migration processes of sandy volumes from the mouth complex towards the outer bay, and mainly along the western sides due to the Coriolis effect. A good example is the bay of the La Rabia estuary (Cantabria), which gradually receives sands from the exposed beach where aeolian dune front is sharply retreating and all beach is losing sandy volume since the last 10 years. A similar trend has been observed in the Villaviciosa estuary (Fig. 18.2b), where sandy intrusion has been produced on western side of the bay. These filled processes are common in high sea level rise scenario and have been studied in few estuaries in the Cantabrian Sea (Bruschi et al. 2013).

Other effects by sea level rise and climatic changes are being the concurrence of surge events promoting the erosion of the dune fields in the confining area of all estuaries. The strong wave storms during second half of the twenty century and, evidenced in the winter of 2014 (Martínez Cedrún 2008; Flores-Soriano 2015; Borghero 2015; Flor-Blanco et al. 2017) have generated washover sheets that covered the topmost aeolian dunes of the confining barrier in some of them, advancing towards the internal dune field, the retreatment of all dune fields with maximums up 35 m, and vertical slopes until between 2 and 8 m. This sediment has been incorporated to the beach or the shoreface and in some instances, since the last surge event, the natural dynamics have changed the morphologies of the beaches and even, smaller dunes are forming (Flor-Blanco and Flor 2016).

Finally, several small estuaries were inactivated from this Flandrian transgression, currently incorporated as part of the mouths of rivers: Esqueiro, Uncín, La Ñora, España, Merón, Libardón, Espasa, and Brazomar (Fig. 18.1) or have a limited dynamic and sedimentation since half rising to half falling tides: Esva (Villegas et al. 1992).

18.6 Managements

After World War I, large although irregularly investments were made in ports. In several estuaries the lower main channels and inlets were channeled and lengthened with jetties seaward, triggering a migration of the beach/dunes system with the formation of a new aeolian dune field (Navia, Nalón, Villaviciosa, San Vicente de la Barquera), as well as the creation of new port docks and marinas. Most of the ports located inside estuaries were built in sandy bays and a minority in the tidal flats. Obviously, these ports need maintenance dredging which results in a loss of sedimentary volume and the retreat of the dune field of the confining barrier (Flor and Flor-Blanco 2005; Flor et al. 2006) as happening in Santander by dredging or Laredo after port construction (Flor-Blanco and Flor 2016).

Estuaries of Avilés and Santander were intensively dredged to develop their commercial ports. The loss of sedimentary volumes, which products were dumped on the continental shelf, has caused the retreat of the dune field belonging to the confining barrier: Salinas-El Espartal and Somo-Loredo (Flor and Flor-Blanco 2014b; de Sanjosé et al. 2016), respectively.

Wide natural marshes were reclaimed during the Middle Ages and even in the late nineteenth and early twentieth centuries (Rivas and Cendrero 1991) with greater emphasis on the estuaries of Navia, Nalón, Avilés, Villaviciosa, Ribadesella, Tina Mayor, San Martín de la Arena, Mogro, Santander, and Oriñón. Large natural areas were lost to be transformed into urban and industrial spaces, roads, railways, airports, etc., but also as agricultural land and pastures. Since 21th century approximately, parts of these reclaimed marshes have been recuperated, destroying the dikes and it has been allowed the transformation the soils due to the decreasing frequency of tidal flooding (Santín et al. 2007). Also, in some cases, the recover reclaimed wetlands have developed to pedogenetic horizons with differences between more or less flooding frequency, taking to account that some features differ depends on factors as biota an land use (Fernández et al. 2010).

The estuaries of Navia, Avilés and Santander are strongly altered by dredging, channelization and mainly reclamation which were specially developed since the mid-twentieth century. Marsh reclamation began in the 17th century, and was particularly intense from the second half of the 19th century. From old maps since 17th century and aerial photographs since 1945, the morphology evolution has been calculated by some authors. According to López-Peláez (2016) in some Asturian estuaries, Flor-Blanco (2007) in western Cantabrian estuaries and Cendrero and Díaz de Terán (1977) and Cendrero et al. (1981) in Santander estuary (Cantabria), the human activity in the major part of the estuary surfaces have changed irreversibly.

On the other hand, in Eo estuary (Asturias), where the construction of a port in its western margin, the circulation changed; so, a slowly and continuous migration started. The flood-tidal delta moved to the SE side of the bay filling and narrowing the secondary channel, join to the marshes filling of La Linera, affecting to the aquaculture industry (Flor and Flor-Blanco 2011; Prumm and Iglesias 2016) with a ratio of 5 m/year. Nowadays, the channel must be continuously dredging. Similar patterns show in San Vicente de la Barquera estuary (Cantabria), where the construction of mouth jetties in the second half of the last century, has changed the currents and the sedimentation with a slow migration of a flood-tidal delta to the SE (Flor-Blanco et al. 2015b).

For these reasons, policy managements must be clear, trying to recover the natural scenarios and it is convenient to distinguish the activities which are harmful to the environment mainly in the bay and the marshes, the most threatened parts of the Cantabrian estuaries, and taking account the future sea level rise and their consequences.

References

Allen, J. R. L. (1993). An introduction to estuarine lithosomes and their controls. *Sedimentary Revue, 1,* 123–138.

Álvarez, I., Varela, M., Prego, R., & Gómez-Gesteira, M. (2006). Miño River dams discharge on neighbor Galician Rias Baixas (NW Iberian Peninsula): Hydrological, chemical and biological changes in water column. *Estuarine, Coastal and Shelf Science, 70*(1), 52–62.

Bárcena, J. F., Claramunt, I., García-Alba, J., Pérez, M. L., & García, A. (2017). A 130 year record of pollution in the Suances estuary (southern Bay of Biscay): Implications for environmental management. *Marine Pollution Bulletin, 56,* 1727–1729. A method to assess the evolution and recovery of heavy metal pollution in estuarine sediments: Past history, present situation and future perspectives. *Marine Pollution Bulletin, 124,* 421–434.

Boggs, S., Jr., & Jones, C. A. (1976). Seasonal reversal of flood-tide dominant sediment transport in a small Oregon estuary. *Geological Society of America Bulletin, 87,* 419–426.

Borghero, C. (2015). *Study of morphological evolution of dune fields in Cantabria (N. Spain) during the Anthropocene* (Master thesis). Uppsala University.

Bruschi, V. M., Bonachea, J., Remondo, J., Gómez-Arozamena, J., Rivas, V., Méndez, G., et al. (2013). Analysis of geomorphic systems response to natural and human drivers in northern Spain: Implications of global geomorphic change. *Geomorphology, 196,* 267–279.

Butzer, K. W. (1962). Coastal geomorphology of Mallorca. *Annals of the Association of American Geographers, 52,* 191–212.

Castelle, B., Bujan, S., Ferreira, S., & Dodet, G. (2017). Foredune morphological changes and beach recovery from the extreme 2013/2014 winter at a high-energy sandy coast. *Marine Geology, 385,* 41–55.

Cendrero, A., de Terán, J. R., & Salinas, J. M. (1981). Environmental-economic evaluation of the filling and reclamation process in the Bay of Santander, Spain. *Environmental Geology, 3*(6), 325–336.

Cendrero, A., & Díaz de Terán, J. R. (1977). Caracterización cuantitativa del desarrollo histórico del relleno de la Bahía de Santander; un proceso natural activado por el hombre. *Revista de Obras Públicas, 124*(3150), 797–808.

Dale, A., Gomez-Gesteira, M., Prego, R., & Alvarez, I. (2006). Hydrographic and atmospheric analysis of an autumnal upwelling event in the Ria of Vigo (NW Iberian Peninsula). *Estuarine, Coastal and Shelf Science, 68*(3), 529–537.

Dalrymple, R. W., Zaitlin, B. A., & Boyd, R. (1992). Estuarine facies models: Conceptual basis and stratigraphic implications. *Journal of Sedimentary Petrology, 62,* 1130–1146.

de Sanjosé, J. J., Serrano, E., Berenguer, F., González-Trueba, J. J., Gómez-Lende, M., González-García, M., et al. (2016). Evolución histórica y actual de la línea de costa en la playa de Somo (Cantabria), mediante el empleo de la fotogrametría aérea y escáner láser terrestre. *Cuaternario y Geomorfología, 30*(1, 2), 119–130.

Domínguez-Cuesta, M. J., Jiménez-Sánchez, M., González-Fernández, J. A., Quintana, L., Flor, G., & Flor-Blanco, G. (2015). GIS as a tool to detect flat erosional surfaces in coastal areas: A case study in North Spain. *Geologica Acta, 13*(2), 97–106.

Fairbridge, R. H. (1980). The estuary: Its definition and geo-dynamic cycle. In E. Olausson & I. Cato (Eds.), *Chemistry and biochemistry of estuaries* (pp. 1–16).

Fenster, M. S., FitzGerald, D. M., Kelley, J. T., Belknap, D. F., Buynevich, I., & Dickson, S. M. (2001). Net ebb sediment transport in a rock-bound, mesotidal estuary during spring-freshet conditions: Kennebec River estuary, Maine, U.S.A. *GSA Bulletin, 113*, 1522–1531.

Fernández, S., Santín, C., Marquínez, J., & Álvarez, M. A. (2010). Saltmarsh soil evolution after land reclamation in Atlantic estuaries (Bay of Biscay, North coast of Spain). *Geomorphology, 114*(4), 497–507.

Fernández-Martínez, R., Loredo, J., Ordóñez, A., & Rucandio, M. I. (2005). Distribution and mobility of mercury in soils from an old mining area in Mieres, Asturias (Spain). *Science of the Total Environment, 346*(1), 200–212.

Flor, G. (1995). Clasificación y caracterización de los estuarios asturianos. In O. Cendrero & I. Olaso (Eds.), *Proceed. IV Coloquio Internacional sobre Oceanografía del Golfo de Vizcaya* (pp. 133–141). Santander.

Flor, G., Busto, J. D., & Flor-Blanco, G. (2006). Morphological and sedimentary patterns of ports of the Asturian Region (NW Spain). *Journal of Coastal Research*, 35–40.

Flor, G., Ceñal, R. C., González, M. S., & Ortega, M. I. (1998). Aspectos morfológicos, dinámicos y sedimentológicos del estuario del Nalón (Asturias, noroeste de España). *Trabajos de Geología, 20*(20), 3–39.

Flor, G., & Camblor, C. (1989). Características dinámicas y sedimentológicas del estuario de Ribadesella (Asturias, N de España). *Trabajos de Geología, 18* Univ. Oviedo, 13–36.

Flor, G., Fernández Pérez, L. A., Menéndez Fidalgo, R., Martínez Cueto-Felgueroso, E., & Rodríguez-Casero, G. (1996). Dynamics and sedimentation of the mesotidal estuary of Villaviciosa (Asturias, Northern Spain). *Revista Sociedad Geológica de España, 9*(3–4), 205–225.

Flor, G., Fernández-Pérez, L. A., & Cabrera-Ceñal, R. (1992). Características dinámicas y sus relaciones sedimentarias en la Ría del Eo (Galicia-Asturias, Noroeste de España). *Revista de la Sociedad Geológica de España, 5*, 7–25.

Flor, G., & Flor-Blanco, G. (2005). An introduction to the erosion and sedimentation problems in the coastal regions of Asturias and Cantabria (NW Spain) and its implications on environmental management. *Journal of Coastal Research*, 58–63.

Flor G., & Flor-Blanco, G. (2008). Estuarios de arroyos costeros del occidente de Asturias. *Territoris, 7*, 135–143.

Flor, G., & Flor-Blanco, G. (2011). *La influencia humana sobre las playas y estuarios de Galicia oriental y Asturias occidental*. Geolodía 11. Sociedad Geológica de España, 45 pp. file:///C:/Users/Usuario/Desktop/GEOLODIA%202011.pdf.

Flor-Blanco, G., & Flor, G. (2012). Clasificación geomorfológica y características de los estuarios cantábricos (NW de España). In: A. González-Díez et al. (Eds.), *Avances de la Geomorfología en España* 2010-2012. 315–318.

Flor-Blanco, G., Flor, G., Martínez-Cedrún, P., & Bruschi, V. (2012). La costa occidental de Cantabria (Santanderestuario de Tina Menor). In: A González-Díez, G. Flor-Blanco, J. R. Díaz de Terán Mira (Eds.), *Guía de excursions de campo. XII Reunión Nacional de Geomorfología*. Santander 2010-2012. 25–44.

Flor, G., & Flor-Blanco, G. (2014a). Raised beaches in the Cantabrian coast. In F. Gutiérrez & M. Gutiérrez (Eds.), *Landscapes and landforms of Spain* (pp. 239–248). Netherlands: Springer.

Flor, G., & Flor-Blanco, G. (2014b). Subsistemas estuarinos de la bahía de Santander (Cantabria, N de España). In S. Schnabel & A. Gómez Gutiérrez (Eds.), *Proceed. XIII Reunión Nacional de Geomorfología, Cáceres* (pp. 568–571), Univ. Extremadura.

Flor, G., & Flor-Blanco, G. (2014c). Componentes de viento generadores de morfologías y campos de dunas costeras en Asturias (no de España). *Cuaternario y Geomorfología, 28*(3–4), 47–68.

Flor, G., & Flor-Blanco, G. (2014d). Actividades de campo en estuarios como recurso educativo. *Enseñanza de las Ciencias de la Tierra, 22*(2), 157.

Flor, G., Flor-Blanco, G., & Flores-Soriano, C. (2014). Cambios ambientales por los temporales de invierno de 2014 en la costa asturiana (NO de España). *Trabajos de Geología, 34* Univ. Oviedo, 97–123.

Flor, G., Flor-Blanco, G., & Rey, J. (2015). Dynamics and morpho-sedimentary interactions in the lower mesotidal estuary of Villaviciosa (NW Spain): A management proposal. *Geologica Acta, 13*(2), 107–121.

Flor, G., & Lharti, S. (2008). Estratigrafía y sedimentología del recubrimiento costero de la ciudad de Gijón (Asturias). *Trabajos de Geología, 28* Univ. Oviedo, 137–157.

Flor-Blanco, G. (2007). *Características morfosedimentarias y dinámicas y evolución ambiental de los estuarios de Tina Mayor, Tina Menor y San Vicente de la Barquera (Costa Occidental de Cantabria)* (PhD Thesis (unpublished)). Dpt. Geology. Univ. Oviedo.

Flor-Blanco, G., Alcántara-Carrió, J., Flor, G., Flores-Soriano, C., García-Ordiales, E., & Pando, L. (2017). Alteraciones en la costa asturiana debido a los efectos catastróficos de oleaje de 2014. *Geo-Temas, 17,* 207–210.

Flor-Blanco, G., & Flor, G. (2016). Management of dune fields on the coasts of Asturias, Cantabria and the Basque Country (Cantabrian Sea, NW Iberian Peninsula). In F. X. Roig-Munar (Ed.), *Restoration and management of dune systems. Case studies* (pp. 35–70). Publisher: project "Life Pletera" (LIFE13 NAT/ES/001001).

Flor-Blanco, G., Flor, G., Morales, J. A., & Pando, L. (2015a). Hydrodynamic controls of morpho-sedimentary evolution in a rockbounded mesotidal estuary. Tina Menor (N Spain). *Journal of Iberian Geology, 41*(3), 315–332.

Flor-Blanco, G., Flor, G., & Pando, L. (2013). Evolution of the Salinas-El Espartal and Xagó beach/dune systems in north-western Spain over recent decades: Evidence for responses to natural processes and anthropogenic interventions. *Geo-Marine Letters, 33*(2–3), 143–157.

Flor-Blanco, G., Flor, G., Pando, L., & Abanades, J. (2015b). Morphodynamics, sedimentary and anthropogenic influences in the San Vicente de la Barquera estuary (North coast of Spain). *Geologica Acta, 13*(4), 279–295.

Flor-Blanco, G., Pando, L., Morales, J. A., & Flor, G. (2015c). Evolution of beach-dune fields systems following the construction of jetties in estuarine mouths (Cantabrian coast, NW Spain). *Environmental Earth Sciences, 73*(3), 1317–1330.

Flores-Soriano, C. (2015). *Evolución de los campos dunares de Asturias desde mediados del siglo XX hasta 2014* (Master thesis). University of Oviedo.

Garcia-Ordiales, E., Cienfuegos, P., Roqueñí, N., Covelli, S., Flor-Blanco, G., Fontolan, G., et al. (2017). Historical accumulation of potentially toxic trace elements resulting from mining activities in estuarine saltmarshes sediments of the Asturias coastline (northern Spain). *Environmental Sciences and Pollution Research.*

Garcia-Ordiales, E., Flor-Blanco, G., Cienfuegos, P., Rumayor, M., Covelli, S., Roqueñí, N., et al. (2016). Determinación de valores geoquímicos de fondo para la evaluación del enriquecimiento por metales y metaloides en los sedimentos estuarinos del Río Nalón (Norte de España). *Geo-Temas, 16*(1), 547–550.

Garcia-Ordiales, E., Loredo, J., Cienfuegos, P., Covelli, S., Flor-Blanco, G., Fontolan, G., et al. (2015). Metales pesados y metaloides en sedimentos de las Marismas del Estuario del río Nalón (Norte de España). *LNEG Comunicações Geológicas, 102*(1), 69–72.

Gómez-Pujol, L., Gelabert, B., Fornós, J. J., Pardo-Pascual, J. E., Rosselló, V. M., Segura, F. S., et al. (2013). Structural control on the presence and character of calas: Observations from Balearic Islands limestone rock coast macroforms. *Geomorphology, 194,* 1–15.

Hayes, M. O. (1975). Morphology of sand accumulation in estuaries: An introduction to symposium. In L. E. Cronin (Ed.), *Estuarine Research, II, Geology and Engineering,* 3–22 (Academic Press, Inc.).

Higueras, P., Fernández-Martínez, R., Esbrí, J. M., Rucandio, I., Loredo, J., Ordóñez, A., et al. (2014). Mercury soil pollution in Spain: A review. In *Environment, energy and climate change* (Vol. I, pp. 135–158). Springer Int Pub.

Irabien, M. J., Cearreta, A., Leorri, E., & Viguri, J. R. (2008). A 130 year record of pollution in the Suances estuary (southern Bay of Biscay): Implications for environmental management. *Marine Pollution Bulletin, 56,* 1727–1729.

Lavín, A., Valdés, L., Sánchez, F., Abaunza, P., Forest, A., Boucher, J., Lazure, P., et al. (2004). The Bay of Biscay: The encountering of the ocean and the shelf (18b,E). In A. R. Robinson & K. H. Brink (Eds.), *The sea* (Vol. 4, Chapter 24, pp. 933–1001).

Le Floch, P. (1961). *Propagation de la marée dans l'estuaire de la Seine et en Seine Maritime* (Thése Doctorat d`Ètat). Univ. Paris, 507 p.

Lechuga, A., de la Peña, J. M., Antón, A. I., & Díez, G. F. (2012). Behavior of the beaches on the North of Spain is global warming involved? *Coastal Engineering Proceedings*, 10 p. https://doi.org/10.9753/icce.v33.sediment.74.

López-Peláez, J. (2016). *Evolución morfosedimentaria e histórica de los estuarios asturianos de Navia, Nalón, Avilés, Villaviciosa y Ribadesella* (PhD Thesis (unpublished)). Dpt. Geology. Univ. Oviedo.

López Peláez, J., & Flor, G. (2006). Cambios morfológicos por canalización de la desembocadura en los estuaries asturianos de Navia, Nalón y Villaviciosa. *Actas de la IX Reunión Nacional de Geomorfología*, 419–431.

Martínez Cedrún, P. (2008). *Caracterización morfológica y sedimentológica de los campos dunares de Cantabria. Evolución ambiental* (Phd Thesis (unpublished)). University of Oviedo.

Masselink, G., Scott, T., Poate, T., Russell, P., Davidson, M., & Conley, D. (2016). The extreme 2013/2014 winter storms: Hydrodynamic forcing and coastal response along the southwest coast of England. *Earth Surface Processes and Landforms, 41*(3), 378–391.

Monge-Ganuzas, M., Cearreta, A., & Evans, G. (2013). Morphodynamic consequences of dredging and dumping activities along the lower Oka estuary (Urdaibai Biosphere Reserve, southeastern Bay of Biscay, Spain). *Ocean and Coastal Management, 77,* 40–49.

Perillo, G. M. E. (1995). Definitions and geomorphologic classifications of estuaries. In G. M. E. Perillo (Ed.), *Geomorphology and Sedimentology of Estuaries. Developments in Sedimentology, 53,* 17–47. Amsterdam: Elsevier.

Posamentier, H. W., & Vail, P. R. (1988). Eustatic control son clastic deposition II: Sequence and systems tracts models. In C. K. Wilgus, B. S. Hastings, C. G. Sr. Kendall, H. W. Posamentier, C. A. Ross, J. C. Wagooner (Eds.), *Sea level changes: An integrated approach. Society of economic paleontologists and mineralogists special publications, 42,* 125–154.

Prandle, D., & Crookshank, N. L. (1974). Numerical model of St Lawrenece River Estuary. *J. Hydraulics Div., ASCE, Vol. 100 (HY4). Proceed. Paper 10472,* 517–529.

Pritchard, D. W. (1967). What is an estuary? Physical point of view. In G. H. Lauff (Ed.), *Estuaries* (Vol. 83, pp. 3–5). Am. Assoc. Adv. Sci. Pub.

Prumm, M., & Iglesias, G. (2016). Impacts of port development on estuarine morphodynamics: Ribadeo (Spain). *Ocean & Coastal Management, 130,* 58–72.

Rivas, V., & Cendrero, A. (1991). Use of natural and artificial accretion on the north coast of Spain: Historical trends and assessment of some environmental and economic consequences. *Journal of Coastal Research, 7*(2), 491–507.

Rosselló, V. M. (2005). Cala, una mesoforma litoral: concepte, models i aproximació morfomètrica. *Cuadernos de Geografia, 77,* 1–18.

Santín, C., Otero, X. L., Fernández, S., González-Pérez, M., & Álvarez, M. Á. (2007). Variations of organic carbon stock in reclaimed estuarine soils (Villaviciosa estuary, NW Spain). *Science of the Total Environment, 378*(1), 138–142.

Simmons, H. B. (1955). Some effects of upland discharge on estuarine hydraulics. *Proceedings of the American Society Civil Engineer, 81* (792-1-729/20).

Valle-Levinson, A., & Atkinson, L. P. (1999). Spatial gradients in the flow over an estuarine channel. *Estuaries and Coasts, 22*(2), 179–193.

van Lancker, V., Lanckneus, J., Hearn, S., Hoeckstra, P, Levoy, F., Miles, J., et al. (2004) Coastal and nearshore morphology, bedforms and sediment transport pathways at Teignmouth (UK). *Continental Shelf Research, 24*, 1171–1202.

Viguri, J. R., Irabien, M. J., Yusta, I., Soto, J., Gómez, J., Rodriguez, P., et al. (2007). Physico-chemical and toxicological characterization of the historic estuarine sediments: A multidisciplinary approach. *Environment International, 33*, 436–444.

Vilas, F. (2002). Rias and tidal-sea estuaries. In *Knowledge for sustainable development. The Encyclopedia of life support systems* (Vol. 2, theme 11.6.3, pp. 799–829).

Villegas, M. L., Garrido, F., González, M. T., & Toledo, M. M. (1992). Características físico-químicas de la desembocadura del río Esva como respuesta a la acción fluvial y mareal. *Boletín Ciencias Naturales R.I.D.E.A., 42*, 39–50.

Chapter 19
Estuaries of the Basque Coast

Manu Monge-Ganuzas, Alejandro Cearreta, María Jesús Irabién and Ane García-Artola

19.1 General Characteristics of the Basque Estuaries

The Basque coast at the southeastern Bay of Biscay is mainly the result of the Quaternary modelling related to variations in sea level. It is an exposed coast, east-west oriented, of about 150 Km in length with steep cliffs mostly made of calcareous rocks—between 20 and 200 m of altitude—formed by a folded and faulted series of Cretaceous and Paleogene materials and abrasion platforms located at the foot of these rocky cliffs. They are occasionally interrupted by estuaries and pocket beaches developed during the Holocene (Fig. 19.1).

The Basque coast is located within the middle latitudes of the eastern North Atlantic Ocean and the annual mean temperature is >10 °C. Climate is moderate, oceanic, with mild winters and temperate summers, and rainy, with over 1500 mm of precipitation each year. Therefore, a Köppen's classification of *Cfb* climate (marine west coast-mild) can be assigned to the area. The average relative humidity

M. Monge-Ganuzas (✉)
Service of Urdaibai Biosphere Reserve, Directorate of Land Planning, Biodiversity and Climate Change, Basque Government, Madariaga tower. San Bartolome, 48350 Busturia, Spain
e-mail: manu-monge@euskadi.eus

A. Cearreta · M. J. Irabién · A. García-Artola
Departamento de Estratigrafía Y Paleontología, Facultad de Ciencia Y Tecnología, Universidad Del País Vasco UPV/EHU, Apartado 644, 48080 Bilbao, Spain
e-mail: alejandro.cearreta@ehu.eus

M. J. Irabién
e-mail: mariajesus.irabien@ehu.eus

A. García-Artola
e-mail: ane.garcia@ehu.eus

J. A. Morales (ed.), *The Spanish Coastal Systems*,
https://doi.org/10.1007/978-3-319-93169-2_19

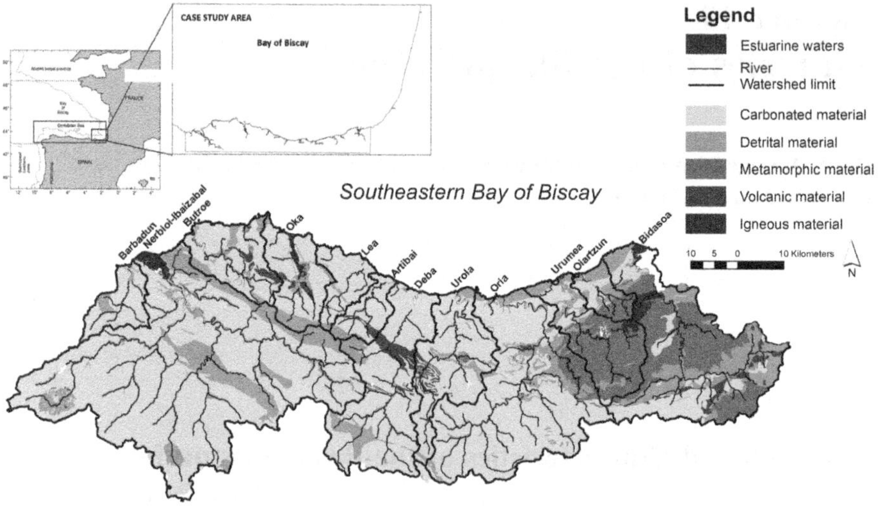

Fig. 19.1 Location and delimitation of the watersheds and main rivers of the Basque estuaries superimposed on the lithological map

during the year is 78%, and the winds blow predominantly from the northwest and south, with 27.5% of calms.

Due to its orientation and position, the Basque coast is exposed to strong waves (mean Hs: 1.5–2 m; mean Tp: 14 s) with large fetches that extend to distances of more than 1500 km. These swell-type waves coming from the fourth quadrant—270°/360°—are the most common in the area (25%). At high and low tides, wave action represents the dominant dynamic process at the estuarine mouths and causes sand transport from the adjacent beaches located to the east of the inlets. The tidal wave is semi-diurnal in character. The mean tidal range is of approximately 1.5 m at neaps and 4 m at springs, while the maximum annual tidal range is over 4.5 m, during equinoctial tides in March and September. Thus, the region is defined as mesotidal.

Generally, the surface sediments of the Basque estuaries consist of semi-consolidated (mud) and non-cohesive (sand) materials, the first carried by rivers and the latter introduced from the adjoining platform by waves and tides (Churruca 1897). Provided that the tidal current velocities are sufficiently high, landwards sediment transport will take place as a consequence of the asymmetry of the tidal wave propagating inside the estuaries (Al-Ragum et al. 2014). Moreover, as the highest velocities occur during the flood, when this phase is long enough, net transport will take place in the upwards direction and trend to fill up the estuary, reducing its depth.

Along the Basque coast there are 12 main rivers with relatively small basin areas, ranging from 93 Km^2—Oiartzun catchment—to 1820 Km^2—Nerbioi-Ibaizabal catchment—with a total surface of 3785.82 Km^2 that constitutes 0.82% of the drainage basin area of the Iberian Peninsula (Fig. 19.1; Table 19.1).

Table 19.1 Main characteristics of all major rivers draining the Basque Country and discharging into the Cantabrian Sea (for locations see Fig. 19.1)

Catchment	Area (Km^2)[a]	Maximum height (m)[a]	River length (Km)[a]	Estuarine length (Km)[a]	Mean fluvial Q (m^3/s)[b]	Sediment supply (t $year^{-1}$)[c]
Barbadun	134	580	26.89	4.53	2.9	–
Nerbioi-Ibaizabal	1820	1000	58.33	22.6	35.56	61,300
Butroe	236	600	36.58	8.53	4.73	9040
Oka	219	625	14.39	12.22	3.6	7160
Lea	128	550	23.54	2.87	1.8	–
Artibai	110	600	23.06	5.27	2.5	1990
Deba	554	750	60.33	6.67	14.08	29,300
Urola	349	680	58.11	7.74	7.98	10,850
Oria	908	749	66.44	11.35	25.66	69,900
Urumea	302	800	47.05	11.74	11.13	10,600
Oiartzun	93	500	14.44	5.37	2.67	1830
Bidasoa	751	780	66.00	15.81	27.19	14,300

Data have been obtained from: [a]AVA/URA (2017), [b]Uriarte (1998), and [c]Prego et al. (2008)

The Basque rivers correspond either to the ‘upland’—500 to 1000 m—or the ‘mountainous’ classification—>1000 m—of Milliman and Syvistki (1992), depending on the criterion adopted. These rivers are characterised by their small length and the steepness of their water courses. Their final parts become estuaries under the marine tidal influence with a variable size: in the Nerbioi-Ibaizabal case it has a length of 22.6 km while in the case of the Lea only 2.87 km are influenced by the sea. The southern limit of the catchments of the Butroe, Lea, Oka and Artibai rivers are strongly conditioned by the Bilbao-Alsasua fault and the axial syncline existing in the west of the area—see Fig. 3.3 of Vera (2004)—while the others have their origin in the reliefs located further south around the border with the Mediterranean watershed.

On the other hand, the rivers Barbadun, Nerbioi-Ibaizabal, Butroe, Oka, Lea, Artibai, Deba and Urola, to the west, mainly drain carbonate and siliciclastic materials and secondarily basaltic volcanic rocks, while the rivers Oria, Urumea, Oiartzun and Bidasoa, located to the east, run mainly on metamorphic and igneous materials and secondarily on sedimentary rocks. This circumstance determines the nature of the sediments that export to the estuaries and continental shelf. Although river discharges to the coast have resulted in extensive research on a global scale (Dyer 1986), it seems that it is necessary to conduct more research on suspended sediment fluxes for small rivers of the world (Maneux et al. 1999) and specifically of the Iberian rivers (Meybeck et al. 2003). Similarly, the contribution from Basque rivers is poorly known and understood (Uriarte et al. 2004). However, in the case of watershed discharge to the Basque coast, there have been some quantitative approaches (Uriarte 1998; Prego et al. 2008), although more work is to be done.

It seems that all these rivers contribute, at least, with a total of 216,270 t year^{-1} of suspended sediments to their estuaries (Prego et al. 2008). The calculated sediment supply of each river by these authors varies from 1830 t year^{-1}—Oiartzun catchment—to 69,900 t year^{-1}—Oria catchment—in virtue of its geological and spatial characteristics, as well as of the existing land uses and the climatology, among other factors (Table 19.1).

This chapter follows geographically from the west to the east the main Basque estuaries, and its purpose is firstly to compile and summarise the existing information about their origin and Holocene natural development and, secondly, to describe their Anthropocene environmental evolution during recent decades.

19.2 Holocene Development and Anthropocene Evolution of the Basque Estuaries

Overview and Methods

Basque estuaries, as all others in the world, are ephemeral environments (Dyer 1995). Since the last climate change (11,500 year ago) previous Pleistocene river valleys were drowned and rapidly converted into estuaries. Thus, since the Holocene, these sedimentary environments present an estuarine circulation and, consequently, are being infilled with sediments of marine (quartz sand, bioclasts and lithoclasts), estuarine (bioclastic) and continental (quartz sand, mud and silt) origin. In addition to the record of their natural evolution, these materials have preserved the fingerprint of the human activities developed in the surrounding area, which in some estuaries have led to a severe deterioration of their original characteristics (Belzunce et al. 2004; Cearreta et al. 2004; Rivas and Cendrero 1992). The analysis of 43 boreholes (Holocene) and 28 small cores (Anthropocene) drilled along the Basque estuaries during the last 30 years has allowed the reconstruction of their sedimentary depositional structure and the natural evolution processes related mainly to the variations of sea level during the last 8500 year, together with their anthropogenic impact as registered in their most recent geological record (Fig. 19.2; Table 19.2).

A rotary drill of 10 cm in diameter was used to drill the boreholes. In all cases, the borehole arrived at the pre-Holocene substrate. Samples for analysis were taken at approximately 20–50 cm intervals. On the other hand, cores were obtained by the insertion of PVC tubes (12.5 cm diameter) into the sediment. Samples for analysis in this case were taken at approximately 1–2 cm intervals. The integrated high-resolution microfaunal-geochemical-sedimentological approach to study the sedimentary successions of these boreholes and cores together with radiocarbon and $^{210}Pb/^{137}Cs$ dating performed have allowed the environmental interpretation of the sedimentary record of the Basque estuaries at different temporal scales (Holocene and Anthropocene).

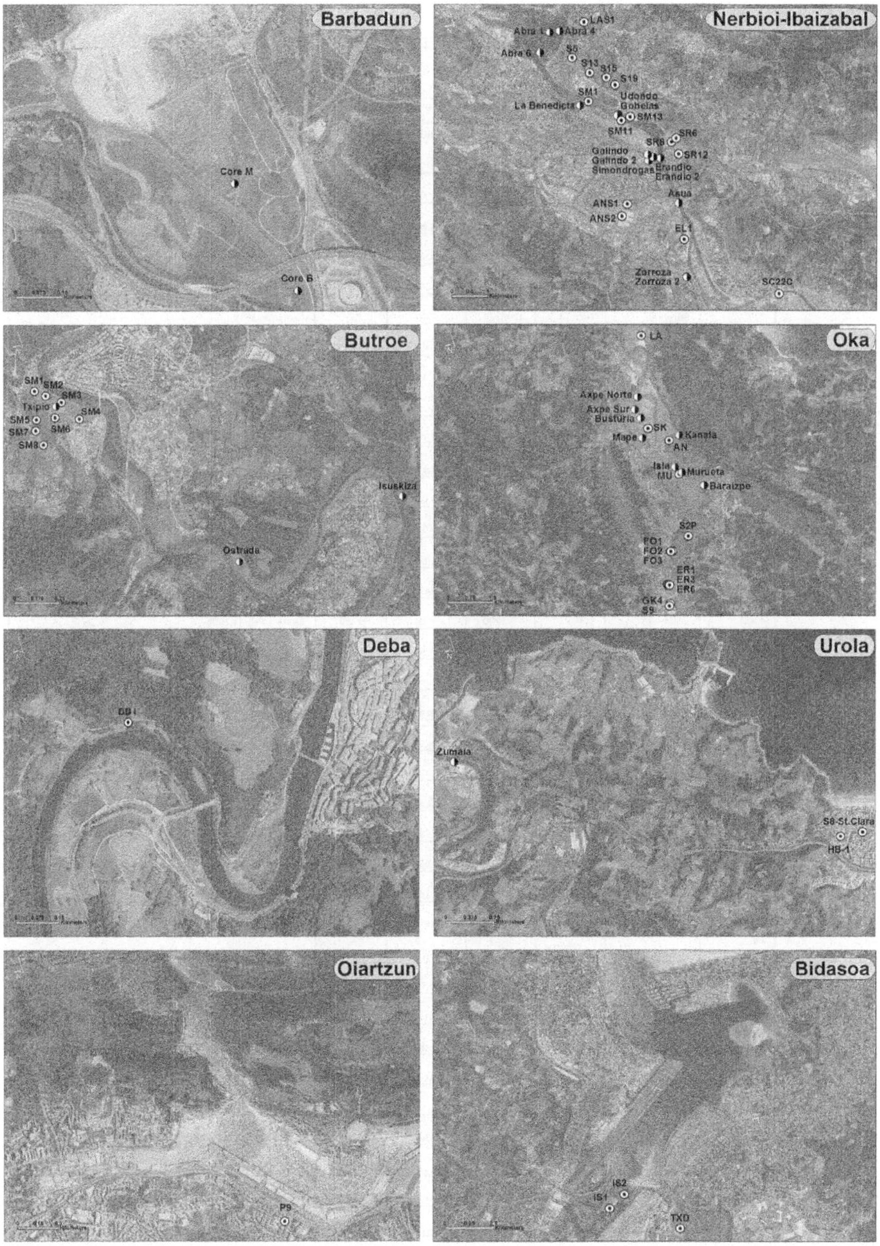

Fig. 19.2 Location of the Holocene boreholes (represented by a dot) and Anthropocene cores (represented by a semi-circle) drilled along the Basque estuaries

Table 19.2 List of coordinates and length of the boreholes (in white) and cores (in grey) drilled along the Basque estuaries with reference to the publications which contain their original information

Estuary	ID	Type	X UTM ETRS89	Y UTM ETRS89	Z (m) ordnance datum	Length (cm)	References
Barbadun	Core B	Core	490903	4798557	intertidal	100	Cearreta et al., 2008
	Core M	Core	490700	4798900	4.50	50	Cearreta et al.; 2008, Leorri and Cearreta, 2009
Nerbioi-Ibaizabal	SC22C	Borehole	505012	4790710	3.30	2050	Cearreta, 1998
	EL1	Borehole	502374	4792220	3.90	2700	Cearreta, 1992; Cearreta 1993; Cearreta 1998
	SR8	Borehole	502012	4794935	3.70	2900	Cearreta, 1998
	S15	Borehole	500210	4796724	4.90	3800	Cearreta, 1998
	SR12	Borehole	502215	4794590	5.43	2730	Leorri and Cearreta, 2004
	SR6	Borehole	502143	4795036	5.42	2650	Leorri and Cearreta, 2004
	SM13	Borehole	500869	4795632	6.77	4150	Leorri and Cearreta, 2004
	SM11	Borehole	500626	4795532	7.19	1730	Leorri and Cearreta, 2004
	SM1	Borehole	499714	4796063	7.19	2300	Leorri and Cearreta, 2004
	S19	Borehole	500451	4796518	6.02	1750	Leorri and Cearreta, 2004
	S13	Borehole	499744	4796852	5.23	1850	Leorri and Cearreta, 2004
	S5	Borehole	499259	4797273	5.83	3400	Leorri and Cearreta, 2004
	LAS1	Borehole	499585	4798264	10.73	2740	Leorri and Cearreta, 2004
	ANS1	Borehole	500795	4793200	9.30	1250	Cearreta et al., 2006
	ANS2	Borehole	500651	4792863	7.50	1310	Cearreta et al., 2006
	Erandio	Core	501701	4794479	intertidal	50	Cearreta et al., 2000; Cundy et al., 2003
	Zorroza	Core	502440	4791182	intertidal	50	Cearreta et al., 2000
	Udondo	Core	500547	4795696	intertidal	50	Cearreta et al., 2000
	Zorroza2	Core	502450	4791171	0.429	600	Cearreta et al., 2002b
	Asua	Core	502217	4793227	0.644	600	Cearreta et al., 2002b
	Erandio2	Core	501706	4794474	0.716	800	Cearreta et al., 2002b
	Galindo	Core	501377	4794399	1.905	1000	Cearreta et al., 2002b
	Simondrogas	Core	501507	4794496	1.395	1000	Cearreta et al., 2002b
	Gobelas	Core	500518	4795670	0.755	800	Cearreta et al., 2002b
	La Benedicta	Core	499479	4795952	intertidal	20	López, 2016
	Galindo2	Core	501354	4794571	intertidal	20	López, 2016
	Abra1	Core	498626	4797977	subtidal	64	Cearreta et al., 2017
	Abra4	Core	498912	4798014	subtidal	56	González-Lanchas, 2017; Cearreta et al. 2017
	Abra6	Core	498361	4797416	subtidal	50	González-Lanchas, 2017
Butroe	SM1	Borehole	503829	4805627	3.33	1210	Cearreta and García-Fernández, 2015
	SM2	Borehole	503916	4805591	3.56	2140	Cearreta and García-Fernández, 2015
	SM3	Borehole	504037	4805543	2.58	2485	Cearreta and García-Fernández, 2015
	SM4	Borehole	504175	4805416	3.88	1470	Cearreta and García-Fernández, 2015
	SM5	Borehole	503845	4805411	3.58	1710	Cearreta and García-Fernández, 2015
	SM6	Borehole	503988	4805426	2.94	1610	Cearreta and García-Fernández, 2015
	SM7	Borehole	503839	4805327	4.58	1365	Cearreta and García-Fernández, 2015
	SM8	Borehole	503900	4805217	5.40	1055	Cearreta and García-Fernández, 2015
	Ostrada	Core	505412	4804331	3.391	50	Cearreta et al., 2002a
	Txipio	Core	503996	4805513	2.507	30	Cearreta et al., 2002a
	Isuskiza	Core	506655	4804824	3.110	50	García-Artola et al., 2011; Leorri et al., 2013; García-Artola et al., 2017

(continued)

(continued)

Oka	FO1	Borehole	526417	4797809	3.76	387	Irabien et al., 2015
	FO2	Borehole	526485	4797831	2.80	551	Irabien et al., 2015
	FO3	Borehole	526426	4797815	3.77	580	Irabien et al., 2015
	LA	Borehole	525429	4805410	4.70	4975	García-Artola et al., 2015
	SK	Borehole	525660	4802138	5.29	1750	García-Artola et al., 2015
	AN	Borehole	526363	4801716	3.51	2920	García-Artola et al., 2015
	MU	Borehole	526703	4800534	4.13	3010	García-Artola et al., 2015
	S2P	Borehole	527020	4798362	4.68	3860	Hernández, 2013; García-Artola et al., 2015
	ER1	Borehole	526320	4796643	5.49	1420	Cearreta et al., 2006; García-Artola et al., 2015
	ER3	Borehole	526360	4796627	6.39	1330	Cearreta et al., 2006; García-Artola et al., 2015
	ER6	Borehole	526387	4796622	6.38	1380	Cearreta et al., 2006; García-Artola et al., 2015
	GK4	Borehole	526414	4795971	7.20	1310	García-Artola et al., 2015
	S9	Borehole	526379	4795889	7.73	1260	García-Artola et al., 2015
	Busturia	Core	525421	4802512	3.722	37	Cearreta et al., 2013
	Baraizpe	Core	527558	4800145	3.659	24	Cearreta et al., 2013
	Mape	Core	525473	4801804	3.35	50	Leorri et al., 2013
	Kanala	Core	526703	4801901	4.055	38	Leorri et al., 2013; Leorri et al., 2014
	Isla	Core	526567	4800781	3.599	44	Cearreta et al., 2013; García-Artola et al., 2017
	Axpe Sur	Core	525218	4802813	4.054	50	Leorri et al., 2014
	Axpe Norte	Core	525313	4803260	3.98	50	Irabien et al., 2015
	Murueta	Core	526803	4800597	4.128	50	García-Artola et al., 2015
Deba	DB1	Borehole	551532	4793661	5.53	2940	Hernández, 2013
Oiartzun	P9	Borehole	587839	4796700	4.05	1200	Irabien et al., 2015
Urola	S8-St. Clara	Borehole	567139	4792831	5.91	1600	Mendicoa, 2011
	HB-1	Borehole	566797	4792804	6.18	900	Mendicoa, 2011
	Zumaia	Core	560838	4793715	intertidal	47	Goffard, 2015
Bidasoa	IS1	Borehole	597864	4800225	3.40	1500	Cearreta, 1992;Cearreta 1994
	IS2	Borehole	598013	4800376	3.30	3100	Cearreta, 1992; Cearreta, 1993; Cearreta, 1994
	TXD	Borehole	598585	4800009	3.73	2500	Irabien et al., 2012

The general characteristics of the benthic foraminiferal assemblages contained in the Holocene and Anthropocene sedimentary sequences of the Basque estuaries are described by Cearreta and Murray (1996) and Leorri and Cearreta (2004) and some methodological limits and suggestions about their correct interpretation are stated by Cearreta and Murray (2000). From the comparison of the modern distribution and abundance of the foraminiferal assemblages from this coastal zone, their microfaunal content is divided between autochthonous species, those that live and reproduce within the estuary, and allochthonous taxa, Those that have been transported from the adjacent marine shelf into the estuary (Cearreta 1988). These foraminiferal assemblages are organized into sedimentary tracts. Each of these sedimentary tracts is composed of a set of assemblages according to its location within the estuaries. They normally appear separated by continuous stratigraphic surfaces that constitute good correlation markers.

19.2.1 Barbadun Estuary

Cearreta et al. (2008) and Leorri and Cearreta (2009) analysed two cores extracted at the Barbadun estuary mudflat and saltmarsh areas in 2001 and 2003 respectively. Both cores B and M were collected from a middle estuarine location (Fig. 19.2; Table 19.2).

On the one hand, geochemical profiles of Core B show low metal and PAH values throughout the whole sedimentary record, in contrast with results obtained from other moderately polluted nearby estuaries (Cearreta et al. 2002b) where low metal contents, characteristic of a pre-industrial time, are followed upwards by increased values associated to anthropogenic effluents during the 20th century. The sediment and its microfaunal content indicate the initial presence of a sandy intertidal flat developed under normal-salinity conditions in the middle estuary which existed until 1914 CE. Then, these sediments were followed by an increase in brackish estuarine conditions until 1951 CE when a low marsh extended over the area during a short period of time. These environments were developed before the installation on the original estuarine domains of an oil refinery in 1970 CE. The core included evidences of reconstruction of the main estuarine tidal channel and the excavation and dumping of sediments during this human-induced severe physical alteration. Finally, modern muddy environmental conditions became established since 1997 CE.

On the other hand, Core M sediments can be divided into two well-defined environments. Firstly, a sandy intertidal flat under normal-salinity conditions equivalent to the one identified in Core B, was followed after 1914 CE by a vegetated saltmarsh environment, slightly enriched in heavy metals, that continues up to the present.

The recent environmental evolution of the Barbadun estuary is not related to persistent historical pollution but rather to the extensive occupation of the original estuary after 1970 CE by the previously mentioned oil refinery. This transformation included modification of the original main tidal channel and an impoverishment of the general environment with the almost complete loss of the estuary and its various sub-environments.

19.2.2 Nerbioi-Ibaizabal Estuary

A total of 9 boreholes were bored during 1988–1990 period at the Nerbioi-Ibaizabal estuary (SC22C, EL1, SR8, S15, SR12, SR6, S19, S13 and S5). In 1996 two supplementary boreholes (ANS1 and ANS2) were completed. As a final point, in 1998 four extra boreholes (SM13, SM11, SM1 and LAS1) were drilled (Fig. 19.2; Table 19.2).

The main species and established the diagnostic assemblages that characterize the estuarine benthic foraminiferal record were described by Cearreta (1992, 1993,

1998), Leorri and Cearreta (2004) and Cearreta et al. (2006). *Ammonia tepida* and *Haynesina germanica* composed the basic indigenous assemblage. In brackish environments additionally these two are accompanied by *Elphidium oceanense*, *Elphidium excavatum* and *Elphidium williamsoni*; and by *Bolivina britannica*, *Bolivina pseudoplicata* and *Quinqueloculina seminula* in nearshore marine environments. In the lower and middle estuary boreholes the samples showed the saltmarsh species *Entzia macrescens*, *Trochammina inflata* and *Arenoparrella Mexicana*. However, these are never abundant. At the upper estuary low marsh assemblages were barely found. *Lobatula lobatula* together with *Rosalina irregularis*, *Rosalina anomala*, *Gaudryina rudis*, *Massilina secans*, *Textularia truncata*, *Quinqueloculina lata*, *Miliolinella subrotunda* and *Elphidium crispum* dominated the allochthonous component. As these robust foraminiferal tests have been transported estuary inwards, a changing degree of marine influence along the sedimentary record can be interpreted from their variable proportions at the borehole sections.

Based on geotechnical studies performed by the Bilbao Port Authority during the period 1879–1995 it is possible to state that the average thickness of the Holocene sequence in this estuary varies extremely. This variation ranges from around 10 m in the upper estuary, 20 m in the middle and 30 m in the lower. Moreover, it can see a basal sandy gravel package of possible fluviatile origin along the central axis of the estuary that corresponds with the original pre-industrial estuary. It is also interesting that the basement depth varies significantly within the same estuarine area, even among very closely located boreholes. The sediments that compose the estuarine record range from fluviatile to supratidal materials, through brackish estuarine and near marine.

On the whole, the analysis of the sedimentological data bares fining-upward sequences along the boreholes that range from gravels at the base, sand in the middle and mud at the top. The sedimentary record has been characterized by micropaleontological evidences as follows: basal coarse materials almost barren of foraminifera; middle sandy sediments with abundant, diverse and near-marine foraminiferal assemblages; and upper muddy deposits with abundant, low-diversity and brackish foraminiferal assemblages. It can state by the combination of geological interpretation and radiocarbon dating that basal sediments present Lateglacial-lowermost Holocene fluvial character, followed upwards by lower and middle Holocene transgressive deposits, with an upper Holocene regressive sequence in the lead. SR6, SR12, SR8, SM11, SM13 and SM1 boreholes drilled in the middle estuarine area represent clearly this general sedimentary sequence. However, depending on their palaeolocation their sedimentary and microfaunal aspects differ. On the one hand, increasing muddy sediments and absence of open marine elements are distinctive in the upper estuary and on the other hand, the lower estuary increasing sandy sediments and reworked basal fluviatile materials are typical. This sedimentary sequence was triggered by the last Pleistocene glacio-eustatic sea-level fall. So, it can be affirmed that this interval could represent a eustatic cycle at almost fourth-order scale. Even though the are some sequence lacks, all the characteristic elements of depositional sequences including a basal sequence boundary (SB), a lowstand systems tract (LST), a transgressive systems tract (TST), a transgressive surface (TS), a maximum flooding surface (MFS) and a highstand systems tract (HST) are shown by this sedimentary record. Moreover, a tidal ravinement surface was identified in addition to the surfaces that define the systems tracts.

The boreholes were interpreted following this depositional sequences model. In there also were defined foraminiferal assemblage zones (FAZs). Figure 19.3a has been built based on foraminiferal assemblages present along the boreholes of the right bank of the estuary (Fig. 19.2; Table 19.2). In this figure the sequence-stratigraphic interpretation of the Nerbioi-Ibaizabal estuary fill is schematically shown. The sequence boundary (SB) has been defined by the location of unconsolidated Quaternary gravels and sands over the Cretaceous basement. A deposit of fluviatile gravels and coarse sands that corresponds in the upper and middle estuarine areas to the basal FAZ overlaps this boundary (Fig. 19.3a). These fluvial deposits are almost barren of foraminifera. Consequently, they have been interpreted as the lowstand systems tract (LST). However, due to the nonexistence of carbonaceous material this LST has not been radiocarbon dated. But, as its stratigraphic position underlies lower Holocene estuarine deposits is possible to insinuate that they correspond to a Lateglacial-lowermost Holocene time. From the sedimentary analysis carried out on these sediments in the lower estuarine area clearly can state that these have been re-worked and included in the following systems tract during the Holocene marine transgression (Fig. 19.3a). A later accumulation of estuarine sediments with copious foraminifera over fluviatile deposits noticeably indicates a marine transgression within the earlier fluvial valley. Different FAZs that incorporate the transgressive systems tract (TST) bounded by various stratigraphic surfaces (Fig. 19.3a), were accumulated during this interval, all of them showing increasing numbers of allochthonous foraminifera and, thus, indicating a rising in open marine influence as transgression advanced with time.

At the bottom of these estuarine deposits disconnecting them from the coarse LST material can be defined a transgressive surface (TS). As the transgression continued, sediments from the inner shelf, containing dominant allochthonous foraminifera, that augmented near-marine conditions within the estuary superimposed the estuarine sediments. This surface situated between the marine and the brackish conditions has been interpreted as an erosional surface called tidal ravinement surface (TRS) (Allen and Posamentier 1993). The radiocarbon dates obtained in this TST oscillate from 8520 to 1685 year cal BP. In spite of this, some of these were gotten from re-worked materials [i.e. boreholes SM11, SR8 and S15 (see Fig. 19.2; Table 19.2)]. It seems that is clear that dates obtained from estuaries under high-energy conditions should be discussed with carefulness, specifically while the study is based on boreholes taken near to the estuary entrance which normally is under more energetic conditions.

It can state that sea level reached almost its present location at around 3000 year cal BP by various curves from the Bay of Biscay (Pirazzoli 1991). The reversal from eustatic transgression to relative regression, with deposition of estuarine sediments analogous to those of the underlying TST but containing

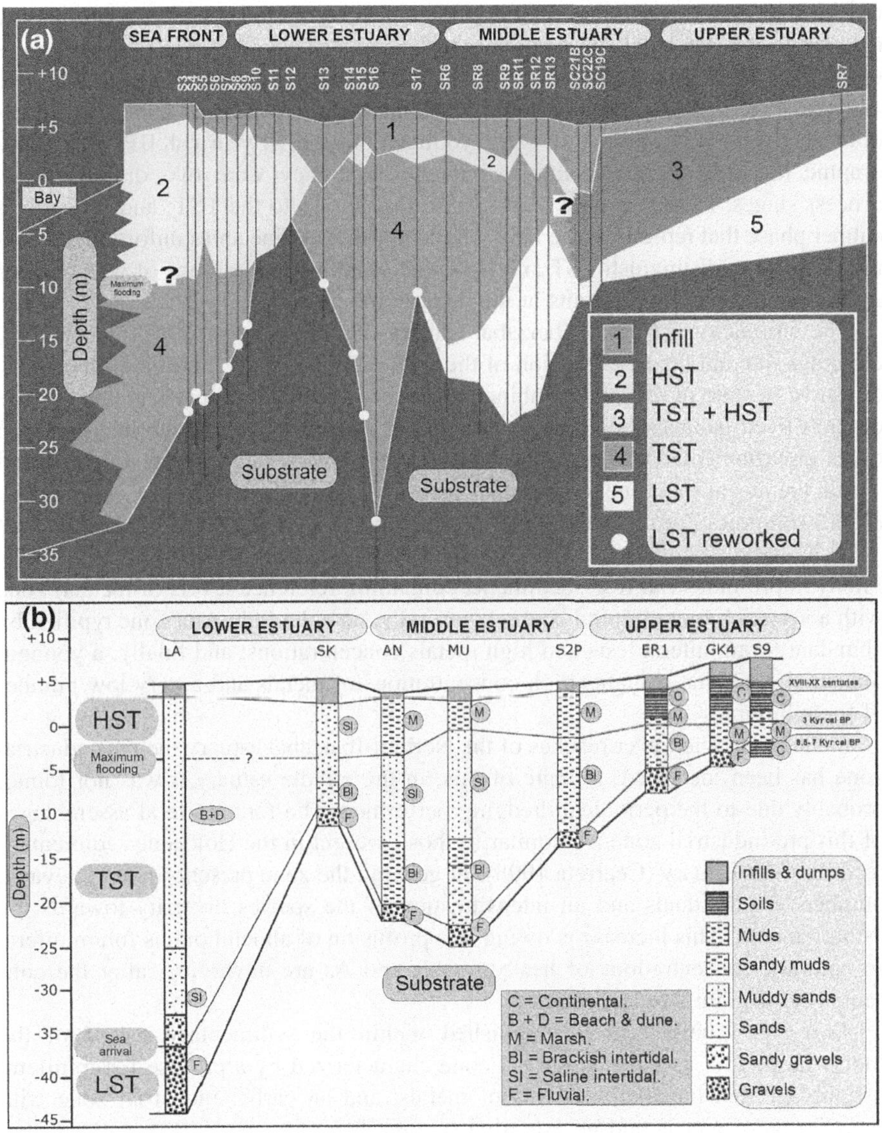

Fig. 19.3 **a** Sequence-stratigraphic interpretation of the Nerbioi-Ibaizabal estuary Holocene deposits. Vertical lines indicate borehole locations. LST: Lowstand Systems Tract; TST: Transgressive Systems Tract; HST: Highstand Systems Tract. Depth appears with reference to local ordnance datum (modified from Leorri and Cearreta 2004). **b** Sequence-stratigraphic interpretation, lithology and foraminifera-based palaeoenvironmental interpretation of the Oka estuary Holocene deposits. Isochronous lines are derived from calibrated radiometric dating (modified from Cearreta and Monge-Ganuzas 2013). Depth appears with reference to local ordnance datum

brackish and shallower foraminiferal markers can be considered as the sedimentary reaction to this sea level calming. This highstand systems tract (HST) incorporates the highest FAZs in the boreholes (Fig. 19.3a) while the surface between the HST and the TST signifies the maximum flooding surface (MFS). Radiocarbon dates obtained from HST sediments were younger than 2810 year cal BP. The topographic location of dated samples indicates that they were two different infill process stages. Firstly, a quick early phase that agrees to the TST; and secondly a calmer phase that represents the HST. In the upper estuarine areas unfortunately it is not possible to distinguish TST and HST due to the nonappearance of well-defined mid-Holocene marine elements at the estuary mouth.

The chronicle of Nerbioi-Ibaizabal estuary has been powerfully influenced by the industrial and urban expansion of the city of Bilbao since the mid-19th century onwards. In spite of this, the combined analysis of benthic foraminifera and a range of short lived isotopes (including ^{210}Pb and ^{137}Cs) have allowed the interpretation of its environmental history. Thus, three different zones have been differentiated within the recent estuarine sedimentary record based on the results of Cearreta et al. (2000) (Erandio, Zorroza and Udondo cores) and Cearreta et al. (2002a) (Zorroza2, Asua, Erandio2, Galindo, Simondrogas and Gobelas cores) (Fig. 19.2; Table 19.2). Firstly, a pre-industrial (e.g., sediments containing reference levels of metals) zone with a very rich foraminiferal content; secondly, an elder industrial zone typified by abundant foraminiferal tests and high metals concentrations; and finally, a younger industrial zone that contains high concentrations of metals and a very low number of foraminifera.

In the upper and lower reaches of the Nerbioi-Ibaizabal estuary the pre-industrial zone has been identified. In spite of this, in the middle estuary it was not found, probably due to the periodical dredging performed. The foraminiferal assemblages of this pre-industrial zone are similar to those present in the Holocene sedimentary record of this estuary (Cearreta 1998). In general, the zone presents a very elevated number of individuals and an intensification of the species diversity towards the estuary mouth. This increase is owing to a profusion of allochthonous foraminifera. In contrast, concentrations of heavy metals and As are unvarying, after the consequence of grain size is corrected.

Two zones have been distinguished within the sedimentary record of the industrial period: an elder industrial zone characterized by a profuse foraminiferal assemblage and remarkable levels of metals, and an earlier industrial zone with great levels of metals and no estuarine foraminifera.

In the cores where the pre-industrial zone appears, the foraminiferal assemblages are analogous in terms of species supremacy. However, the elder industrial assemblage contains a low number of individuals and species in comparison to the pre-industrial assemblage. The dating of these materials have been impossible (see Cundy et al. 2003) because the relatively unpredictable distribution of short-lived radionuclides (Cearreta et al. 2000). Consequently, it was also impossible to determine if both zones are continuous in time or, in contrast, separated by time lags induced by the dredging carried out since the 20th century.

The earlier industrial zone along the estuary in contrast, presents a sheet deeply contaminated with variable thickness. The highest values of pollution have been found in the middle estuary area where dredging have been more intense. This zone fits with the second part of the industrial development dated around 1950–2000 CE, when metropolitan and industrial disposals was highest. These flows provoked minimum oxygen levels, high carbon-based matter and as well metal concentrations. As a consequence, under those conditions the estuarine fauna disappeared completely and this situation have remained throughout many years. This environmentally negative situation have been reinforces by the high concentrations of heavy metals and As founded in the examined cores. The content of metals in these cores varies greatly with profundity. Probably, this fact is due to changes in the anthropogenic enters, early diagenetic responses and physical reworking of deposits. Unfortunately, there are not enough detailed historic records to characterize properly the temporal anthropogenic disposals tendencies. However, the vertical spreading of metals in the cores advises that concentrations of Zn, Pb and Cu (and possibly As) intensified earlier in the estuary and that Cr and Ni were disposed later. Luckily, the passing of environmental strategies, the implementation of waste-treatment techniques and the cessation of industrial units during the 1990s lead to a certain environmental recuperation of this estuary (Cajaraville et al. 2016; García-Barcina et al. 2006; Leorri et al. 2008a). This is corroborated by the identification of a less contaminated surficial layer in the channel area (cores La Benedicta and Galindo analysed by López 2016) and in the Abra bay (cores Abra1 and Abra4 analysed by Cearreta et al. 2017 and cores Abra4 and Abra6 studied by González-Lanchas 2017) (Fig. 19.2; Table 19.2).

19.2.3 Butroe Estuary

Eight boreholes (SM1-SM8) were studied in the Butroe estuary by Cearreta and García-Fernández (2015) in order to reconstruct its environmental transformation process during the Holocene. Previously mentioned multi-proxy analyses were performed on these materials. Based on the obtained results and by comparison with the evidences from other Basque estuaries, the Holocene environmental evolution of its sedimentary environments is interpreted as the result of the previously described relative sea-level rise occurred during the Holocene. Sea level exhibits a progressive rise until 3000 year cal BP, followed by a stabilization since then until human reclamation of saltmarsh areas during the second part of the 19th century. Altogether, lower estuarine deposition is interpreted as the infilling of a former meander of the river Butroe that currently compressed estuarine and inland materials. During the low sea-level conditions in the Lateglacial, sedimentation in this area was represented by fluvial gravels without foraminifera. As marine transgression was progressing an estuarine sedimentation took place included within the TST presenting muddy-sandy sediments of brackish and continental origin in the lower estuary. Finally, HST deposited from 3000 year cal BP up to the human occupation of the original saltmarshes in the 19th century, represents intertidal conditions, progressively with more brackish and marginal characteristics under conditions of stable sea level.

After the analysis of the Ostrada and Txipio cores, (Fig. 19.2; Table 19.2), Cearreta et al. (2002b) noticed that topographically higher foraminiferal assemblages were dominated by agglutinated species (*E. macrescens* and *T. inflata*) even though all samples contained a fluctuating content of hyaline tests, and lower zones were governed by calcareous hyaline species (*A. tepida, H. germanica, E. williamsoni* and *E. oceanense*) but the assemblages all the time displayed an significant agglutinated content. This configuration is suitable for both saltmarshes even if the calcareous constituent in the more raised assemblages augmented significantly from the upper to the lower estuary. This circumstance gives the impression that saltmarsh foraminiferal dissemination in the Butroe estuary is controlled by a mixture of both altitude on top of mean sea level (degree of uncovering) and salinity.

García-Artola et al. (2011, 2017) and Leorri et al. (2013) described the environmental regeneration process of the formerly reclaimed Isuskiza saltmarsh as a consequence of tidal water inundation (Fig. 19.2; Table 19.2). In the basal 5 cm of the core foraminifera were away. This interval likely signifies the anthropogenic accumulation established during the reclamation epoch that happened in this area until around the 1960s. The upper 38 cm contained very low amounts of foraminiferal tests. The calcareous hyaline species *A. tepida* and *E. oceanense*, plus the agglutinated species *E. macrescens* and *T. inflata* were principal in those samples. The quantity of species was very little. This middle interval was placed during the restoration process, from the cultivated ground to the restored salt marsh, as revealed by the upward-increasing amount of tests. The well-defined supremacy of calcareous species and the temperate sand content imply that these accumulations were placed in a low height environment for instance a tidal flat (Cearreta 1988). In the top 7 cm, foraminiferal tests were temperately profuse. Agglutinated forms were more notable than calcareous hyaline types. *Entzia macrescens*, *T. inflata*, and *Miliammnina fusca* were prevailing. Species quantity was modest but higher than in the preceding interval. This interval with steadied foraminiferal content symbolizes the contemporary restored saltmarsh environment.

19.2.4 Oka Estuary

With the purpose of examine the ecological evolution of the estuary throughout the Holocene, 13 boreholes (FO1, FO2, FO3, LA, SK, AN, MU, S2P, ER1, ER3, ER6, GK4 and S9) were studied by Cearreta et al. (2006), Cearreta and Monge-Ganuzas (2013), Hernández (2013), Irabien et al. (2015) and García-Artola et al. (2015).

Previous studies of the Holocene infill in the Oka estuary based on microfossils were carried out by Pascual and Rodríguez-Lázaro (1996) and Pascual et al. (1998, 1999, 2000, 2001, 2002), although none of their cores reached the bedrock. Consequently, initiation of the marine transgression in this estuary could not be determined in these publications. They described saltmarsh initiation at around 4600 year BP, in contrast with the regional establishment of saltmarsh formation at about 3000 cal year BP (Leorri and Cearreta 2004; Leorri et al. 2012), and observed environmental evolution from an estuarine into a higher-energy environment that provoked a major erosion during a transgressive pulse at around 3400–3200 year BP. A final transgressive pulse was described by Pascual and Rodríguez-Lázaro (2006) when marine foraminifera invaded the saltmarsh platform around 1900 year BP.

The foraminiferal assemblages allowed the rebuilding of relative Holocene sea level (Leorri et al. 2012). Cearreta and Monge-Ganuzas (2013) found three systems tracts typified by definite foraminiferal assemblages in relation to their position inside the upper, middle or lower estuary, and divided by a uninterrupted stratigraphic plane that allows correlation (Fig. 19.3b). As it has been described previously for the Nerbioi-Ibaizabal estuary, for the period of low sea-level states sedimentary accumulations (lowstand systems tract) were distinguished by fluvial gravels and coarse sands not including foraminifera and other marine fauna. As the marine transgression took place, coarse deposits were well-looked-after by superimposing estuarine sediments progressing in the direction of the mainland (transgressive systems tract). The lowest segment of this systems tract is characterized by enormous volumes of marine deposits in the lower estuary, a combination of marine and brackish deposits in the middle estuary and brackish materials in the upper estuary, and the top segment (seaward) by open sea supplies (such as abundant allochthonous foraminiferal tests) until a maximum flooding surface was achieved at its upper boundary. As a final point, the highstand systems tract, was placed at some point in the late Holocene (since 3000 year cal BP until the saltmarsh reclamation at the commencement of the 18th century) and exemplified by intertidal and supratidal brackish situations in the inner estuary.

Leorri et al. (2012) designated descriptive samples of diverse estuarine environments and topographic altitudes from these boreholes gathered in the Oka estuary (simultaneously with samples from the near Nerbioi-Ibaizabal and Deba estuaries) to create sea-level index points (SLIPs) and rebuild the relative sea-level curve over the last 10,000 years in northern Spain. This curve exhibits two main stages: (1) quick relative sea-level rise from −27 m at ca 10,000 year cal BP to −5 m at ca 7000 year cal BP, at between 9 and 12 mm year^{-1}; and (2) a quite slow sea-level rise since ca 7000 year cal BP at between 0.3 and 0.7 mm year^{-1} (seemingly steadied about 3000 year cal BP) until contemporary anthropogenic degrees were reached at the commencement of the 20th century.

The Murueta core (Fig. 19.4a) analyzed by García-Artola et al. (2015) describes the natural evolution of a saltmarsh area in the Oka estuary at least during the time span between 1957 and 2008 CE. This core contains exclusively agglutinated tests (*E. macrescens* and *T. inflata*) are the dominant taxa. Species number is extremely little all over the core. The assemblages in the Murueta tiny core are unchanging with profundity and can be associated to a unceasing high saltmarsh environment (Cearreta et al. 2002a; Leorri et al. 2008a), permanently overhead mean highest high water (Leorri et al. 2010). Sand content is extremely little all over the

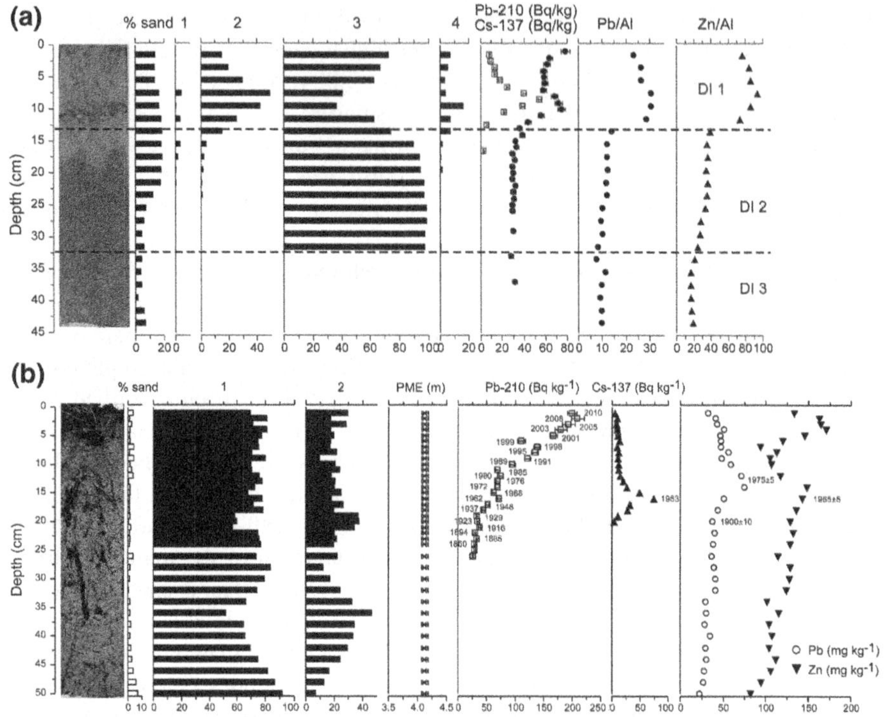

Fig. 19.4 **a** Core photograph, sand content (%), main foraminiferal species (1: *A. mexicana*; 2: *S. moniliforme*; 3: *E. macrescens*; 4: *T. inflata*), total ^{210}Pb (dots) and ^{137}Cs (squares) concentrations (Bq kg^{-1}), and Al-normalized Pb and Zn distributions with depth (cm) in the Isla saltmarsh core (Oka estuary). Defined depth intervals (DIs) are also shown (taken from Cearreta et al. 2013). **b** Core photograph, sand content (%), main foraminiferal species (1: *E. macrescens*; 2: *T. inflata*) (%), palaeomarsh elevation (PME, m), ^{210}Pb and ^{137}Cs activities (Bq kg^{-1}) and total Pb and Zn distribution (mg kg^{-1}) with depth (cm) in the Murueta saltmarsh short core (Oka estuary). Ages based on the CRS model, the ^{137}Cs peak and total Pb and Zn are shown (taken from García-Artola et al. 2015)

sedimentary register in reaction to the lesser dynamism of these zones with a fewer tidal control.

The supremacy of fine-grained deposits maintains the palaeoenvironmental explanation of this zone. Subsequently the Murueta core has been decoded as a continued high saltmarsh environment, the palaeomarsh altitude (PME) persisted stable through time. The ^{210}Pb profile exhibited a typical exponential decay with profundity, which implies minimum sediment combination or interruption. This, in turn, insinuates that this short core can provide a consistent chronology based on ^{210}Pb. This chronology has been provided by means of the CRS model and reinforced by a clear ^{137}Cs activity peak at 16 cm profundity that can be appointed to 1963 CE. Chronology is further maintained by total Pb and Zn profiles that indicate two concentration peaks in 1975 CE (at 12 cm depth) and 1965 CE (at 16 cm depth), correspondingly (Leorri et al. 2008b, 2014; Leorri and Cearreta 2009). The total Pb concentration profile presents a clear difference from pre-industrial values at 20 cm depth that has been linked with the 1900s (Leorri et al. 2014). The rebuilt relative sea-level curve was achieved by detracting the above described PME from the contemporary altitude of each sample and drawn against the age, achieving the resultant relative sea-level disparity curve for the final 150 years (Fig. 19.5). Utilizing a midpoint linear regression, a sea-level rise rate of 1.7 ± 0.2 mm $year^{-1}$ was found for the 20th century (error computed at the 95% confidence). The rate achieved here is equal to former regional investigations that fluctuate from 2.0 ± 0.3 mm $year^{-1}$ for the epoch 1884–1994 CE (Leorri et al. 2008b) and 2.0 mm $year^{-1}$ for the 20th century (García-Artola et al. 2009) to 1.9 ± 0.3 mm $year^{-1}$ since 1923 CE (Leorri and Cearreta 2009). In spite of this, the sea-level rise rate achieved based on the local transfer function is inferior and adjusts the vertical error of the earlier reconstructions that has been diminished from ±7.8 cm stated by Leorri et al. (2008b) and ±5.3 cm described by García-Artola et al. (2009) to the present ±3.5 cm. Equal 20th century speeding up have been established based on analogous foraminifera-based reconstructions obtained in SW Europe (Rossi et al. 2011) and the Atlantic coastline of North America (Gehrels et al. 2005; Kemp et al. 2009). Kemp et al. (2011) described a more sudden speeding up at the commencement of the preceding century than other authors for the West Atlantic shoreline, whereas Long et al. (2014) attributed it to a local sea-level signal. In order to confirm this data, the geological reconstruction was contrasted with the nearest tide-gauge record placed at Santander (Fig. 19.5), and in spite of the variances in the time period involved by both reconstructions, they depict similar tendencies.

The Santander tide-gauge record offers a rate of 2.08 ± 0.33 mm $year^{-1}$ for the epoch 1943–2004 CE (Chust et al. 2009), a number marginally greater than the geological reconstruction. However if we limit the study of the geological record to a similar time period (1937–2004 CE) we achieve a rate of 2.1 mm $year^{-1}$, alike to the rate offered by the Santander tide-gauge record. The Brest tide-gauge record supplies a similar tendency, even if it shows a lower rate (1.4 mm $year^{-1}$) for the 20th century (Wöppelmann et al. 2007). These incongruities could be clarified by diverse vertical land movements (e.g., Leorri et al. 2012, 2013) or water masses relocation resulting from atmospheric (Marcos and Tsimplis 2008) and steric influences (Tsimplis et al. 2011). Simultaneously, these dissimilarities convincingly insinuate the necessity to implement local reconstructions since sea-level variations happen at diverse time and spatial scales. We can deduce that sea-level rise rate in the Oka estuary during the 20th century was 3–6 times greater than the reconstructed rate for the previous 7000 years. No such rates have been obtained during the upper Holocene, even if the analytical resolution of the boreholes and the short core is very dissimilar.

On the other hand, the benthic foraminiferal content of the Isla core (Figs. 19.2 and 19.4b; Table 19.2) studied by Cearreta et al. (2013) was very high, except for the basal 12 cm. Foraminiferal abundance even though variable, presented a general upward enhance in the core. Eleven diverse species were discovered with

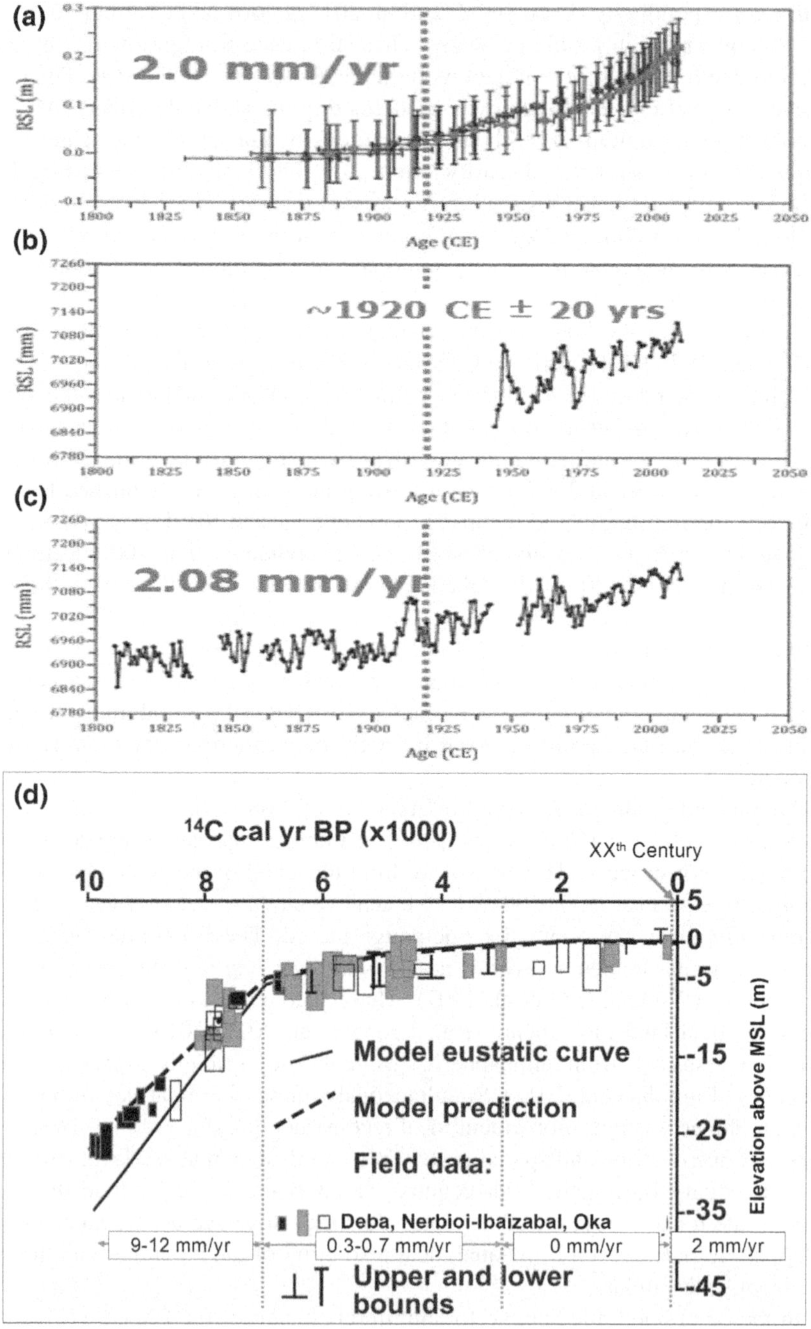
(a)
2.0 mm/yr
RSL (m)
Age (CE)
(b)
~1920 CE ± 20 yrs
RSL (mm)
Age (CE)
(c)
2.08 mm/yr
RSL (mm)
Age (CE)
(d)
^{14}C cal yr BP (x1000)
XXth Century
Elevation above MSL (m)
Model eustatic curve
Model prediction
Field data:
Deba, Nerbioi-Ibaizabal, Oka
9-12 mm/yr
0.3-0.7 mm/yr
0 mm/yr
2 mm/yr
Upper and lower bounds

◂**Fig. 19.5** **a** Relative sea-level curve from the Murueta saltmarsh (Oka estuary) based on foraminiferal reconstructions. The normal and dashed lines represent sea-level trends derived from Anthropocene data and late Holocene data respectively (see text for discussion); **b** Annual relative sea-level values recorded at the Santander tide gauge (Spain); **c** Annual relative sea-level values provided by the Brest tide gauge (France) (modified from García-Artola et al. 2015); **d** A-plot of Holocene sea-level index points from the Nerbioi-Ibaizabal, Oka and Deba estuaries, showing calibrated age against depth relative to present mean sea level. Grey boxes represent both vertical and age errors. Dashed line represents model predictions based on the following values: lithospheric thickness (71 km), upper mantle viscosity (0.3×10^{21} Pa s) and lower mantle viscosity (50×10^{21} Pa s). Black line represents model eustatic curve for the last 12,000 year. Sea-level rise velocity for each section of the curve is also indicated (modified from Leorri et al. 2012)

E. macrescens as dominant species all over. As it has been explained before for the Isuskiza core of the Butroe estuary three different profundity intervals were differentiated here. The bottommost 12 cm was described by the non-appearance of foraminifera and an extremely low sand content. Historic records and the concentration profiles of ^{210}Pb and ^{137}Cs indicate this zone as an anthropogenic accumulation established during the reclamation period. The subsequent 19 cm contained slight numbers of foraminiferal tests, little species numbers and cumulative upwards sand contents. The only dominant species was agglutinated *E. macrescens*. *Scherochorella moniliforme* was a subordinate species. Using historical images and the sediment build-up rates derived from the short-lived radionuclides profiles, this interval is explained as being accumulated during the restoration process of the formerly reclaimed cultivated land. The upper 13 cm was extremely controlled by agglutinated foraminifera. *Entzia macrescens*, *S. moniliforme* and *T. inflata* were the key species. *Arenoparrella mexicana* was a subordinate species. The sand content augmented in contrast with preceding intervals. The foraminiferal concentration displayed a high mean number of individuals. The mean number of species was modest. This upmost interval signifies the contemporary restored saltmarsh environment.

In addition, the occurrence of well-built saltmarsh waterways over an earlier agrarian use in the Isla area by 1957 CE is patent on aerial pictures, telling environmental restoration by this time in development. Subsequently, if cultivated soil signified by the basal interval is earlier to 1950s and the ^{137}Cs maximum abundance distinguished at 8 cm depth in the upper interval would exemplify the 1963 CE peak, a very fast sedimentation rate can be assumed during the restoration progression (about 18 mm year^{-1}), in conjunction with accretion of cumulative amounts of sand during the intermediate interval. The sedimentation speed diminished harshly during the enlargement of the restored saltmarsh environment (around 1.7 mm year^{-1}). Basal (anthropogenic deposit) and intermediate (quickly restoring saltmarsh during the 1950s–1960s) intervals are described by little concentrations of metals, while the upper interval (regenerated saltmarsh) exhibits augmented values. Even though this model insinuates that sediments from the intermediate interval could symbolize pre-industrial supplies deposited before substantial anthropogenic records, historic data denote that metal processing industrial units have been operating in this estuarine area since the commencement of 1910s. Thus, this augmentation is more possible related to the above stated reduction in sediment accretion, given that elevated sedimentation rates may "dilute" contaminant inputs due to combination with coarse sedimentary elements or non-polluted materials (Valette-Silver 1993).

19.2.5 Deba Estuary

Borehole BD1 was studied micropalaeontologically by Hernández (2013) in the Deba estuary (Fig. 19.2; Table 19.2) and revealed an environmental evolution during the Holocene similar to other Basque estuaries described previously. The basal foraminiferal assemblage (10,150–9130 year cal BP) has been interpreted as corresponding with an intertidal muddy environment influenced by the sea. This unit is followed by a brackish intertidal environment (9280–8000 year cal BP) with relatively little marine influence. The following assemblage continues corresponding to a brackish intertidal environment (7170–6810 year cal BP) but under more restricted conditions than previously. Further up, there is a unit without foraminifera deposited in an environment under fresh water influence. Finally, the foraminiferal assemblage situated on top of the sequence corresponds to a brackish intertidal environment dominated by only one species. This fact implies a very restricted environment without open sea influence.

In summary, a transgressive sequence produced by the Holocene sea-level rise can be observed also in this estuary. However, this middle part of the Deba estuary did not register saltmarsh areas at the top of the sedimentary sequence as occurred in the Oka or Nerbioi-Ibaizabal estuaries. The environments in this estuary present intertidal characteristics with an important marine influence at the beginning, and progressively became more fresh-water influenced.

19.2.6 Urola Estuary

A core drilled at Zumaia and studied by Goffard (2016) (Figs. 19.1 and 19.2; Table 19.2) shows a clear anthropogenic impact over the estuary due to industrial activities. The sedimentary record of this core is characterized by sandy and muddy sediments whereas microfossil content decreased upwards as the estuarine pollution increased. Pollutants consist mainly of heavy metals that were supplied by industries along the watershed and, consequently, into the estuary since 1960s. During the more industrial interval (between 1960s and 1980s) the pollution of this estuary was highest and the presence of foraminifera in the estuarine sediments was nearly null. Since 1980s the pollutant content decreased significantly and the benthic foraminiferal assemblages recovered again their normal abundance. This was an indicator of the improvement in the environmental conditions of this transitional ecosystem induced by some restoration measures and the recent reduction of the industrial activities.

19.2.7 Oiartzun Estuary

Borehole P9 drilled in the Oiartzun estuary (Fig. 19.2; Table 19.2) and analysed by Irabien et al. (2015) reflects the impact of Roman mining on this estuary. The core is composed by fluctuating gravels, sandy muds, and muds hidden by a thick layer of anthropogenic infilling supplies. Concentrations of Pb from 7.65 to −0.85 m depth continue low and steady. These values are in suitable conformity with those suggested for pre-industrial supplies from other estuaries of the nearby area (Legorburu et al. 1989; Cearreta et al. 2000), comparable to background data in the area (Rodríguez et al. 2006; Cearreta et al. 2013). Higher than −0.85 m profundity, between 3380–3100 and 930–680 year cal BP, there is a five-fold augmentation in Pb concentrations. As the Oiartzun river empties mineralized supplies of Palaeozoic age (Fig. 19.1) before running into the estuary, this spread is expected to be associated to the beginning of upstream galena use during Roman times (nearby 1900 year cal BP) and the resulting discharge and transference of Pb- enhanced particles. Vertical dissemination of Pb in this borehole is reasonably analogous to that found formerly by Irabien et al. (2012) neighbouring to the age-old Roman harbour of Oiasso (contemporary Irun, Bidasoa estuary) (Fig. 19.1), where post-Roman deposits were contaminated as a result of the intensive use of regional rocks. Furthermore, in both estuaries greatest contents of this metal exist several centuries later Roman settlement, but previous to the Industrial Revolution. Assumed that no noteworthy industrial actions are recognized for this past period, these augmentations could be explicated by the remobilization of formerly contaminated materials from unknown waste loads during flood episodes. In that circumstance, mineral use not only exerted an acute ecological impact at local magnitude during Roman times (Irabien et al. 2012), but additionally left a remnant of contamination for the subsequent centuries.

Similarly, Cearreta et al. (2005) performed a research with the key goal to make available geological gears for the environmental understanding of the Roman neighbourhood of Forua (Oka estuary, Fig. 19.1). The three boreholes pierced next to the archaeological site (FO1, FO2 and FO4) contained grey muds with some stratums of sandy and conglomeratic sediments and plant and bioclastic rests, covered by a brown soil (1.10–1.50 m thick). Radiocarbon dating of a bivalve shell from the bottom of the lengthiest borehole (FO4) shown an age of 6210–5970 year cal BP. Foraminiferal assemblages corroborated that both sedimentary evidences from Oiartzun and Forua signified a representative series of estuarine infilling, with the previous revealing more profuse inputs of fluviatile sediments. This could be associated to the amplified average water stream (4.8 $m^3 s^{-1}$) of the

Oiartzun river when matched to other regional tributaries with greater watersheds, which is probable to be a result of the high precipitation recorded in its catchment (González et al. 2012).

Contents of Pb in all materials from Forua persisted low. Highest concentrations were a little higher than those established in pre-Roman materials from Oiartzun and Irun (Irabien et al. 2012). Conversely, this metal presented a moderately suitable correlation with Al, a conservative element that has been often employed as a proxy for grain-size (Cearreta et al. 2002b; Alvarez-Iglesias et al. 2007). Thus, this noticeable augmentation was more probable to reveal a granulometric influence than a meaningful anthropogenic contribution. Preceding investigations have proven the omnipresent characteristics of the Roman Pb peak in peat—swamps, ponds and ice cores (see Renberg et al. 2001). Nevertheless, the regional significance of its environmental effect is more uncertain. As presumed, and nevertheless the moderately short gap existing between the Oiartzun and Oka estuaries (70 km by the shortest possible route, Fig. 19.1), no noteworthy clue of the grave impact triggered by Roman quarrying activities in the previous was identified in the last.

19.2.8 Bidasoa Estuary

The environmental evolution of the Bidasoa estuary (Figs. 19.1 and 19.2; Table 19.2) was studied by means of three boreholes: IS1 (Cearreta 1992, 1994), IS2 (Cearreta 1992, 1993, 1994) and TXD (Irabien et al. 2012).

The sedimentary records of IS1 and IS2 boreholes show the typical foraminiferal assemblages of the Bay of Biscay previously described in this chapter. Both boreholes present a succession of estuarine environments that indicate progressive changes in depth and salinity due to the Holocene transgression. In the case of IS1, its basal unit is dated at 7810 year BP and its intermediate unit shows an age of 2740 year BP. The IS2 borehole presents two episodes of maximum marine influence in its sedimentary record. The first is represented at the beginning by its basal unit and passes to a shallowing sequence dated around 6630–6590 year BP. Later, the second marine influence is characterized by a high increase of allochthonous tests. Finally, a new shallowing of the environment is observed and a saltmarsh environment established.

The sedimentary record of both boreholes is variable in the vertical but also laterally (500 m). While in IS1 there is a unique marine flooding pulse and a later shallowing registered, in IS2 two marine flooding episodes were identified. This impossibility to correlate two nearby boreholes of the same marginal environment was already stated by Murray and Hawkins (1976) and Culver and Banner (1978). It is necessary to take into account that their records are non-continuous, probably with numerous sedimentation units eroded along the temporal development of this estuary. For instance, according to the radiocarbon dates, in IS1 4 m of sedimentary record are registered during 40 years while in IS2 also 4 m of sediment were settled during around 5000 years. This huge difference in the deposition ratio of both boreholes is a fact that must be precisely taken into account for the interpretation of these highly dynamic estuarine environments. However, if we compare the sedimentary record of these two boreholes with the results provided by other studies made in the nearby estuaries is possible to state that they fit very well with the general model proposed regionally for the Basque estuaries mainly based on the evidences collected at the Nerbioi-Ibaizabal and Oka estuaries.

On the other hand, the TXD borehole represents a standard estuarine infilling sequence, presenting a development on the way to more confined environmental states from a sandy unprotected estuarine environment to a fine-grained high marsh environment gradually more inaccessible from the estuarine waters. An archaeological method is enhanced to the explanation of this estuarine successions with the intention of explain the environmental influence caused by Roman withdrawal actions upon the Bidasoa estuary. The geological evidence displays high contents of Pb, undoubtedly higher than those endorsed currently for tolerable sediment quality. Isotope components corroborate use of local galenas as the main supply for this metal. Results from TXD show that Pb contamination spread moreover to the nearby estuarine area. The commencement of Roman removal activities is verified by a noticeable spread in Pb concentrations. In spite of this, the greatest value emerges delayed in time (afterwards 660 CE). Even though this enhancement could be associated to the upstream remobilization of formerly contaminated sediments, a short time ago achieved archaeological evidences indicate that occurrences of previously non-documented quarrying events could have taken place in the nearby area following Roman times. Average values and variety of concentrations of heavy metals in the materials gathered from the Roman harbour of Irun (Santiago Street) (see Irabien et al. 2012, Table 3) equalled with the background values suggested for the Basque estuaries by Rodríguez et al. (2006) exhibit that the contents of Ni and Cr persist near to local background, while scarce samples reveal discreetly augmented levels of Zn and Cu. In spite of this, all the samples look as if they have been really enhanced in Pb. Smallest contents of this metal appear in the deepest stratigraphic levels (see Irabien et al. 2012, Table 2) which is expected to signify the initial phases of the building of the Roman harbours (dated 70–95 CE). These ages were achieved by dendrochronological investigation of the timber beams and the typology of the archaeological ceramic compendium related (Urteaga 2005). Equally, materials from the upper stratigraphic levels show clearly much greater values. Archaeological data (Urteaga 2005) point out that they were built up throughout the 2nd century and the first years of the 3rd century, as soon as human actions in the Roman polis accomplished the highest greatness. It is remarkable that highest values are even greater than levels identified by Sáiz-Salinas et al. (1996) in the contemporary Bidasoa estuary, being analogous to those discovered during contemporary times in other Basque streams and estuaries harshly contaminated by up-to-date withdrawal and industrial activities (Irabien et al. 2008; Leorri et al. 2008a; Sánchez et al. 1994, 1997).

19.3 Conclusions

The Basque estuaries represent drowned river valleys that were firstly flooded by the sea due to a climate-change-induced sea-level rise 8500 years ago, and subsequently, infilled with a general shallowing sequence. Basque estuaries developed from an initial fluvial environment followed by a firstly marine and secondly brackish depositional system.

The rising velocity of sea level during the period between 10,000 and 7000 year cal BP was around 9–12 mm $year^{-1}$ whereas from 7000 to 3000 year cal BP the sea-level rise decreased to values of 0.3–0.7 mm $year^{-1}$. At about 3000 year cal BP sea level mostly established at its present position until the 20th century increase when sea level started to rise at around 2 mm $year^{-1}$.

In the same way, the sedimentary signal of human activities and their environmental consequences in the context of the current climate change and accelerated sea-level rise are being registered in the estuarine sedimentary record.

In detail, Holocene evolution of the different Basque estuaries present distinctive features as the result of the coastal contour, catchment morphology, meteorological parameters, erosion/sedimentation rates (upstream and downstream), differences in the wave regime and effect at their littoral, and distribution of the tidal wave, among others.

Any natural or human-induced change on these characteristics of the Basque estuaries may cause different sedimentary, morphological and environmental changes at the adjacent coastal areas. A discussion about this topic will be presented in Chap. 33 and this issue should be studied in the next future from a multi- and transdisciplinary approach. In fact, human activities in the Basque estuaries have provoked sometimes critical environmental situations. Some of them have been irreversible and others, fortunately, could be reversed with time and due to changes in human behaviour. These are the cases, for instance, of the Oiartzun and Bidasoa estuaries during Roman ages or the Barbadun, Butroe and Nerbioi-Ibaizabal estuaries during recent times.

Oka estuary, the only World Biosphere Reserve of the Basque Country, was declared recently as Global Geosite (in 2016), and it can be considered as a quasi-natural laboratory for Geosciences. This estuary has been the main source of regional information about current sea-level rise and its sedimentary and environmental consequences. Although it seems almost pristine, several anthropogenic changes have occurred from its formation until present time, such as an early extensive reclamation of saltmarshes, recent development of public infrastructures (like ports, railway or roads), and modern dredging and dumping of sand in the lower estuary.

Other estuaries like Deba and Urola, both located within the Basque Coast Global Geopark declared in 2015, have been less studied and further work should be done in order to describe and characterize adequately their natural sedimentary record and the increasing human influence on them.

During more than 25 years, different boreholes and cores have been drilled and studied along the Basque estuaries by the Harea-Coastal Geology research group of the University of the Basque Country UPV/EHU (www.ehu.eus/en/web/harea-geologialitoral/home). In fact, the study of the microfossil content (mainly benthic foraminifera) of the Basque estuaries, together with other sedimentological, geochemical or archaeological approaches, has become the main source of geological information in order to reconstruct their Holocene and Anthropocene development and the main drivers behind their environmental evolution. However, much additional knowledge about these variable coastal environments should be gathered in the next years and decades using a multidisciplinary geological approach.

Acknowledgements This research was sponsored by the projects Harea-Coastal Geology Group (Basque Government, IT976-16) and ANTROPICOSTA (MINECO, CGL2013-41083-P). It embodies Contribution 42 of the Geo-Q Zentroa Research Unit (Joaquín Gómez de Llarena Laboratory).

References

Allen, G. P., & Posamentier, H. W. (1993). Sequence stratigraphy and facies model of an incised Valley fill: The Gironde estuary, France. *Journal of Sedimentary Petrology, 63,* 378–391.

Al-Ragum, A., Monge-Ganuzas, M., Amos, C. L., Cearreta, A., Townend, I., & Manca, E. (2014). An evaluation of the Rouse theory for sand transport in the Oka estuary, Spain. *Continental Shelf Research, 78,* 39–50.

Alvarez-Iglesias, P., Quintana, B., Rubio, B., & Pérez-Arlucea, M. (2007). Sedimentation rates and trace metal input history in intertidal sediments from San Simón Bay (Ría de Vigo, NW Spain) derived from ^{210}Pb and ^{137}Cs chronology. *Journal of Environmental Radioactivity, 98,* 229–250.

AVA/URA. (2017). http://www.uragentzia.euskadi.eus/u81-0002/es.

Belzunce, M. J., Solaun, O., González Oreja, J. A., Millás, E., & Pérez, V. (2004). Contaminants in sediments. In A. Borja & M. B. Collins (Eds.), *Oceanography and marine environment of the Basque Country*. Elsevier Oceanography Series (Vol. 70, pp. 283–315).

Cajaraville, M. P., Orive, E., Villate, F., Laza-Martínez, A., Uriarte, I., Garmendia, L., et al. (2016). Health status of the Bilbao estuary: A review of data from a multidisciplinary approach. *Estuarine, Coastal and Shelf Science, 179,* 124–134.

Cearreta, A. (1988). Distribution and ecology of benthic foraminifera in the Santoña estuary, Spain. *Revista Española de Paleontología, 3,* 23–38.

Cearreta, A. (1992). Ecoestratigrafía (foraminíferos bentónicos) del relleno estuarino holoceno en el litoral vasco (Golfo de Bizkaia). *Rivista Italiana di Paleontologia e Stratigrafia, 98,* 243–266.

Cearreta, A. (1993). Paleoenvironmental interpretation of Holocene coastal sequences in the southern Bay of Biscay. *Geologische Rundschau, 82,* 234–240.

Cearreta, A. (1994). Análisis micropaleontológico e interpretación paleoecológica del relleno sedimentario holoceno en el estuario del Bidasoa (Golfo de Bizkaia). *Geobios, 27,* 271–283.

Cearreta, A. (1998). Holocene sea-level change in Bilbao estuary (North Spain): Foraminiferal evidence. *Micropaleontology, 44,* 265–276.

Cearreta, A., Alday, M., Irabien, M. J., Etxebarria, N., & Soto, J. (2008). Modern conditions and recent environmental development of the Muskiz estuary: Historical disturbance by the largest oil refinery of Spain. *Journal of Iberian Geology, 34,* 191–213.

Cearreta, A., García-Artola, A., Leorri, E., Irabien, M. J., & Masque, P. (2013). Recent environmental evolution of regenerated salt marshes in the southern Bay of Biscay: Anthropogenic evidences in their sedimentary record. *Journal of Marine Systems, 109–110,* 203–212.

Cearreta, A., & García-Fernández, Z. (2015). Holocene sedimentary infilling of the Txipio salt marsh (Plentzia-Barrika, Biscay): Record of environmental changes in the Butroe estuary during the last 8,500 years. *Geogaceta, 57,* 111–114.

Cearreta, A., Irabien, M. J., Gómez, J., Kortabitarte, I., & González-Lanchas, A. (2017). The Anthropocene geological record in the Abra of Bilbao: Evidences of its natural and human history. *Geogaceta, 61,* 11–14.

Cearreta, A., Irabien, M. J., Leorri, E., Yusta, I., Croudace, I. W., & Cundy, A. B. (2000). Record of anthropogenic impacts on the Bilbao Estuary, N. Spain: Geochemical and microfaunal evidence. *Estuarine, Coastal and Shelf Science, 50,* 571–592.

Cearreta, A., Irabien, M. J., Leorri, E., Yusta, I., Quintanilla, A., & Zabaleta, A. (2002a). Environmental transformation of the Bilbao estuary, N. Spain: Microfaunal and geochemical proxies in the recent sedimentary record. *Marine Pollution Bulletin, 44,* 487–503.

Cearreta, A., Irabien, M. J., & Pascual, A. (2004). Human activities along the Basque coast during the last two centuries: Geological perspective of recent anthropogenic impact on the coast and its environmental consequences. In A. Borja & M. B. Collins (Eds.), *Oceanography and marine environment of the Basque Country. Elsevier oceanography series* (Vol. 70, pp. 28–50).

Cearreta, A., Irabien, M. J., Ulibarri, I., Yusta, I., Croudace, I. W., & Cundy, A. B. (2002b). A Recent salt marsh development and natural regeneration of reclaimed areas in the Plentzia estuary, N. Spain. *Estuarine, Coastal and Shelf Science, 54,* 863–886.

Cearreta, A., Irabien, M. J., Uriarte, E., & Monge-Ganuzas, M. (2005). Herramientas Geológicas Para la Interpretación Paleoambiental del Yacimiento Romano en Forua (Reserva de la Biosfera de Urdaibai). Unpublished Technical Report, 31 pp.

Cearreta, A., Leorri, E., & Santos, L. (2006). The stratigraphic significance of reworked Quaternary microfossils in the Holocene of the Bilbao estuary (northern Spain). *Revista Española de Micropaleontología, 38,* 207–217.

Cearreta, A., & Monge-Ganuzas, M. (2013). Evolución paleoambiental del estuario del Oka (Reserva de la Biosfera de Urdaibai, Vizcaya): respuesta al ascenso del nivel marino durante el Holoceno. *Geo-temas, 14,* 163–166.

Cearreta, A., & Murray, J. W. (1996). Holocene paleoenvironmental and relative sea-level changes in the Santoña estuary, Spain. *Journal of Foraminiferal Research, 26,* 289–299.

Cearreta, A., & Murray, J. W. (2000). AMS 14C dating of Holocene estuarine deposits: Consequences of high-energy and reworked foraminifera. *The Holocene, 10,* 155–159.

Churruca, E. (1897). Estado y progreso de las obras del puerto y ría de Bilbao. *Revista de Obras Públicas, 44*(II), (1157) 574–580.

Chust, G., Borja, A., Liria, P., Galparsoro, I., Marcos, M., Caballero, A., et al. (2009). Human impacts overwhelm the effects of sea-level rise on Basque coastal habitats (N Spain) between 1954 and 2004. *Estuarine, Coastal and Shelf Science, 84,* 453–462.

Culver, S. J., & Banner, F. T. (1978). Foraminiferal assemblages as Flandrian paleoenvironmental indicators. *Paleogeography, Paleoclimatology, Paleoecology, 24,* 53–72.

Cundy, A. B., Croudace, I. W., Cearreta, A., & Irabien, M. J. (2003). Reconstructing historical trends in metal input in heavily-disturbed, contaminated estuaries: Studies from Bilbao, Southampton Water and Sicily. *Applied Geochemistry, 18,* 311–325.

Dyer, K. R. (1986). *Coastal and estuarine sediments dynamics* (p. 358). Chichester: Wiley.

Dyer, K. R. (1995). Sediment transport processes in estuaries. Geomorphology and sedimentology of estuaries. *Developments in Sedimentology, 53,* 423–471 (Elsevier Science).

García-Artola, A., Cearreta, A., & Irabien M. J. (2017). Recent agricultural occupation and environmental regeneration of salt marshes in Northern Spain. In C. W. Finkl & C. Makowski (Eds.), *Coastal wetlands: Alteration and remediation. Coastal research library series* (Vol. 21, pp. 47–79). Springer.

García-Artola, A., Cearreta, A., & Leorri, E. (2015). Relative sea-level changes in the Basque coast (northern Spain, Bay of Biscay) during the Holocene and Anthropocene: The Urdaibai estuary case. *Quaternary International, 364,* 172–180.

García-Artola, A., Cearreta, A., Leorri, E., & Irabien, M. J. (2011). Saltmarsh environmental regeneration from anthropogenic reclaimed coastal areas: Interest of their study under current sea-level rise scenario. *Geogaceta, 50,* 161–164.

García-Artola, A., Cearreta, A., Leorri, E., Irabien, M. J., & Blake, W. H. (2009). Las marismas costeras como archivos geológicos de las variaciones recientes del nivel marino. *Geogaceta, 47,* 109–112.

Garcia-Barcina, J. M., González-Oreja, J. A., & de la Sota, A. (2006). Assessing the improvement of the Bilbao estuary water quality in response yo pollution abatement measures. *Water Research, 40,* 951–960.

Gehrels, W. R., Kirby, J. R., Prokoph, A., Newnham, R. M., Achterberg, E. P., Evans, H., et al. (2005). Onset of recent rapid sea-level rise in the western Atlantic Ocean. *Quaternary Science Reviews, 24,* 2083–2100.

Goffard, A. (2016). Registro del impacto humano en el estuario del Urola (Geoparque de la Costa Vasca) durante el Antropoceno. *CKQ, 6,* 43–60.

González, M., Blasco, J., Del Campo, A., Espino, M., Ferrer, L., Fontán, A., et al. (2012). Un método de corrección de flujo explícito en elementos finitos para turbidez y tiempo de residencia de las aguas portuarias. aplicación al puerto de Pasaia. *Revista de Investigación Marina, 19,* 632–644.

González-Lanchas, A. (2017). Reconstrucción paleoambiental del impacto humano en el Abra de Bilbao durante los últimos 200 años. *CKQ Quaternary Studies, 7,* 61–73.

Hernández, S. (2013). Evolución ambiental de los estuarios vascos (Urdaibai y Deba) durante el Holoceno como consecuencia del ascenso en el nivel marino. *CKQ Quaternary Studies, 3,* 51–62.

Irabien, M. J., Cearreta, A., Leorri, E., Gómez, J., & Viguri, J. (2008). A 130-year record of pollution in the Suances estuary (southern Bay of Biscay): Implications for environmental management. *Marine Pollution Bulletin, 56,* 1719–1727.

Irabien, M. J., Cearreta, A., & Urteaga, M. (2012). Historical signature of Roman mining activities in the Bidasoa estuary (Basque Country, northern Spain): An integrated micropalaeontological, geochemical and archaeological approach. *Journal of Archaeological Science, 39,* 2361–2370.

Irabien, M. J., García-Artola, A., Cearreta, A., & Leorri, E. (2015). Chemostratigraphic and lithostratigraphic signatures of the Anthropocene in estuarine areas from the eastern Cantabrian coast (N. Spain). *Quaternary International, 364,* 196–205.

Kemp, A. C., Horton, B. P., Culver, S. J., Corbett, D. R., van de Plassche, O., Gehrels, W. R., et al. (2009). Timing and magnitude of recent accelerated sea-level rise (North Carolina, United States). *Geology, 37,* 1035–1038.

Kemp, A. C., Horton, B. P., Donnelly, J. P., Mann, M. E., Vermeer, M., & Rahmstorf, S. (2011). Climate related sea-level variations over the past two millennia. *Proceedings of the National Academy of Sciences, 108,* 11017–11022.

Legorburu, I., Ramos, A., & Sola, M. (1989). Heavy metals in coastal sediments in Guipuzcoa (Spain). *Toxicological and Environmental Chemistry, 23,* 129–134.

Leorri, E., & Cearreta, A. (2004). Holocene environmental development of the Bilbao estuary, northern Spain: Sequence stratigraphy and foraminiferal interpretation. *Marine Micropaleontology, 51,* 75–94.

Leorri, E., & Cearreta, A. (2009). Quantitative assessment of the salinity gradient within the estuarine systems in the southern Bay of Biscay using benthic foraminifera. *Continental Shelf Research, 29,* 1226–1239.

Leorri, E., Cearreta, A., García-Artola, A., Irabien, M. J., & Blake, W. H. (2013). Relative sea-level rise in the Basque coast (N Spain): Different environmental consequences on the coastal area. *Ocean and Coastal Management, 77,* 3–13.

Leorri, E., Cearreta, A., & Horton, B. P. (2008a). A foraminifera-based transfer function as a tool for sea-level reconstructions in the southern Bay of Biscay. *Geobios, 41,* 787–797.

Leorri, E., Cearreta, A., Irabien, M. J., & Yusta, I. (2008b). Geochemical and microfaunal proxies to assess environmental quality conditions during the recovery process of a heavily polluted estuary: The Bilbao estuary case (N. Spain). *Science of the Total Environment, 396,* 12–27.

Leorri, E., Cearreta, A., & Milne, G. (2012). Field observations and modelling of Holocene sea-level changes in the southern Bay of Biscay: Implication for understanding current rates of relative sea-level change and vertical land motion along the Atlantic coast of SW Europe. *Quaternary Science Reviews, 42,* 59–73.

Leorri, E., Gehrels, W. R., Horton, B. P., Fatela, F., & Cearreta, A. (2010). Distribution of foraminifera in salt marshes along the Atlantic coast of SW Europe: Tools to reconstruct past sea-level variations. *Quaternary International, 221,* 104–115.

Leorri, E., Mitra, S., Irabien, M. J., Zimmerman, A. R., Blake, W. H., & Cearreta, A. (2014). A 700 year record of combustion-derived pollution in northern Spain: Tools to identify the Holocene/Anthropocene transition in coastal environments. *Science of the Total Environment, 470–471,* 240–247.

Long, A. J., Barlow, N. L. M., Gehrels, W. R., Saher, M. H., Woodworth, P. L., Scaife, R. G., et al. (2014). Contrasting records of sea-level change in the eastern and western North Atlantic during the last 300 years. *Earth and Planetary Science Letters, 388,* 110–122.

López, I. (2016). Evolución ambiental reciente de los sedimentos de la Ría de Bilbao: evidencias micropaleontológicas y geoquímicas del proceso de recuperación. *CKQ Quaternary Studies, 6,* 75–87.

Maneux, E., Dumas, J., Clément, O., Etcheber, H., Charritton, X., Etchart, J., et al. (1999). Assessment of suspended matter input into the oceans by small mountainous coastal rivers: The case of the Bay of Biscay. *Earth and Planetary Science Letters, 329,* 413–420.

Marcos, M., & Tsimplis, M. N. (2008). Coastal sea level trends in Southern Europe. *Geophysical Journal International, 175,* 70–82.

Mendicoa, J. (2011). El relleno sedimentario holoceno de la depresión de Zarautz: respuesta al ascenso eustático del nivel marino. *CKQ Quaternary Studies, 1,* 115–131.

Meybeck, M., Laroche, L., Dürr, H. H., & Syvitski, J. P. M. (2003). Global variability of daily total suspended solids and their fluxes in rivers. *Global and Planetary Change, 39,* 65–93.

Milliman, J. D., & Syvistki, J. P. M. (1992). Geomorphic/tectonic control of sediment discharge to the ocean: The importance of small mountainous rivers. *Journal of Geology, 100,* 525–544.

Murray, J. W., & Hawkins, A. B. (1976). Sediment transport in the Severn estuary during the past 8.000-9.000 years. *Journal of the Geological Society of London, 132,* 385–398.

Pascual, A., & Rodríguez-Lázaro, J. (1996). Micropaleontología en materiales holocenos de la ría de Gernika (Golfo de Bizkaia). *Geogaceta, 20,* 209–212.

Pascual, A., & Rodríguez-Lázaro, J. (2006). Marsh development and sea level changes in the Gernika Estuary (southern Bay of Biscay): Foraminifers as tidal indicators. *Scientia Marina, 70,* 101–117.

Pascual, A., Rodríguez Lázaro, J., Weber, O., & Jouanneau, J. (2001). Cambios paleogeográficos durante el Holoceno en las marismas de Forua, Reserva de la Biosfera de Urdaibai. *Geogaceta, 30,* 191–194.

Pascual, A., Rodríguez-Lázaro, J., Weber, O., & Jouanneau, J. M. (2002). Late-Holocene pollution in the Gernika estuary (southern Bay of Biscay) evidenced by the study of Foraminifera and Ostracoda. *Hydrobiologia, 475*(476), 477–491.

Pascual, A., Weber, O., Caballero, F., Rodríguez Lázaro, J., & Jouanneau, J. (1999). Análisis integral (micropaleontología y sedimentología) de un sondeo supramareal Holoceno en la ría de Gernika (Golfo de Vizcaya). *Geogaceta, 28,* 113–116.

Pascual, A., Weber, O., Rodriguez-Lázaro, J., & Jouanneau, J.-M. (2000). Evolución sedimentaria del pólder de Anbeko (Reserva de Urdaibai, Golfo de Vizcaya). *Geogaceta, 26,* 75–78.

Pascual, A., Weber, O., Rodriguez-Lázaro, J., Jouanneau, J.-M., & Puyos, M. (1998). Le comblement de la ria de Gernika (Golfe de Gascogne) a l'Holocene Tereminal. *Oceanologica Acta, 21,* 263–269.

Pirazzoli, P. A. (1991). *World atlas of holocene sea-level changes. Elsevier science series* (Vol. 58, 299 p). London.

Prego, R., Boi, P., & Cobelo-García, P. (2008). The contribution of total suspended solids to the Bay of Biscay by Cantabrian Rivers (northern coast of the Iberian Peninsula). *Journal of Marine Systems, 72,* 342–349.

Renberg, I., Bindler, R., & Brännvall, M. L. (2001). Using the historical atmospheric lead deposition record as a chronological marker in sediment deposits in Europe. *The Holocene, 11,* 511–516.

Rivas, V., & Cendrero, A. (1992). Análisis histórico de la evolución superficial de los estuarios del País Vasco. *Lurralde, 15,* 199–227.

Rodríguez, J. G., Tueros, I., Borja, A., Belzunce, M. J., Franco, J., Solaun, O., et al. (2006). Maximum likehood mixture estimation to determine metal background values in estuarine and coastal sediments within the European Framework Directive. *Science of the Total Environment, 370,* 278–293.

Rossi, V., Horton, B. P., Corbett, D. R., Leorri, E., Perez-Belmonte, L., & Douglas, B. C. (2011). The application of foraminifera to reconstruct the rate of 20th century sea level rise, Morbihan Golfe, Brittany, France. *Quaternary Research, 75,* 24–35.

Sáiz-Salinas, J. I., Ruiz, J. M., & Francés-Zubillaga, G. (1996). Heavy metal levels in intertidal sediments and biota from the Bidasoa estuary. *Marine Pollution Bulletin, 32,* 69–71.

Sánchez, J., Marino, N., Vaquero, M. C., Ansorena, J., & Legórburu, I. (1997). Metal pollution of old lead-zinc mines in Urumea river valley (Basque Country, Spain). Soil, biota and sediments. *Water, Air, and Soil Pollution, 107,* 303–319.

Sánchez, J., Vaquero, M. C., & Legorburu, I. (1994). Metal pollution from old lead-zinc mine works: Biota and sediment from the Oiartzun Valley. *Environmental Technology, 15,* 1069–1076.

Tsimplis, M., Spada, G., Marcos, M., & Flemming, N. (2011). Multi-decadal sea level trends and land movements in the Mediterranean Sea with estimates of factors perturbing tide gauge data and cumulative uncertainties. *Global and Planetary Change, 76,* 63–76.

Uriarte, A. (1998). *Sediment dynamics on the inner continental shelf of the Basque Country (N. Spain)* (Unpublished PhD thesis). Faculty of Science Oceanography. University of Southampton, 302 p.

Uriarte, A., Collins, M., Cearreta, A., Bald, J., & Evans, G. (2004). Sediment supply, transport and deposition: Contemporary and Late Quaternary evolution. In A. Borja & M. B. Collins (Eds.), *Oceanography and marine environment of the Basque Country. Elsevier oceanography series* (Vol. 70, pp. 97–131).

Urteaga, M. (2005). El Puerto romano de Irún (Guipúzcoa). In M. Urteaga & M. J. Noain (Eds.), *Mar Exterior: el occidente atlántico en época romana, Actas del Congreso Internacional, Pisa 2003* (pp. 85–106). Escuela Española de Historia y Arqueología-CSIC, Rome.

Valette-Silver, N. (1993). The use of sediment cores to reconstruct historical trends in contamination of estuarine and coastal sediments. *Estuaries, 16,* 577–588.

Vera, J. A. (Ed.). (2004). *Geología de España* (p. 890). Madrid: SGE-IGME.

Wöppelmann, G., Martin Miguez, B., Bouin, M.-N., & Altamimi, Z. (2007). Geocentric sea-level trend estimates from GPS analyses at relevant tide gauges worldwide. *Global and Planetary Change, 57,* 396–406.

Chapter 20
The Ebro River Delta

Inmaculada Rodríguez-Santalla and Luis Somoza

20.1 Geographical and Geological Setting

The delta of the Ebro River is located on the Spain Mediterranean coast, in Tarragona province, about 170 km southwest of Barcelona (Fig. 20.1). It is one of the largest deltas of the Mediterranean, began to develop when the last post-glacial eustatic sea level rise was attained (Maldonado 1972). The present extension for the emerged area is 325 km^2 and represents only 15% of 2,171 km^2 of the whole delta. The length of the emerged coast is approximately 50 km, and the altitude is 4–5 m above sea level (Serra 1997). The morphology of the deltas depends of the effects of tides, waves and currents (Galloway 1975). The Ebro Delta is a microtidal environment delta dominated currently by wave action. The river sediment input has been reduced drastically due to the construction of dams, especially Mequinenza, Flix, Ribarroja (Fig. 20.2), during the first half of the 20th century, changing the position of the Ebro Delta inside the Galloway diagram from river-wave domain to wave domain, increasing greatly the landforms associated by wave storms.

Despite having a large legal framework of protection (Rodríguez et al. 2010), they are some effects which present risk for the stability of the system. In fact, the absence of contribution of the river due to control of uses of fluvial channel is one of them (irrigation, dams and generation hydropower production). Floods and river overflowing encourage vertical accretion and compensate the natural subsidence, avoiding the effects associated to sea-level rise, intensified by global warming (Ibáñez et al. 2010). Dams interrupt the continuity of the sediments transport

I. Rodríguez-Santalla (✉)
Universidad Rey Juan Carlos, C/ Tulipán s/n. Móstoles, 28933 Madrid, Spain
e-mail: inmaculada.rodriguez@urjc.es

L. Somoza
Marine Geological Mapping Dv., Geological Survey of Spain (IGME), Madrid, Spain
e-mail: l.somoza@igme.es

J. A. Morales (ed.), *The Spanish Coastal Systems*,
https://doi.org/10.1007/978-3-319-93169-2_20

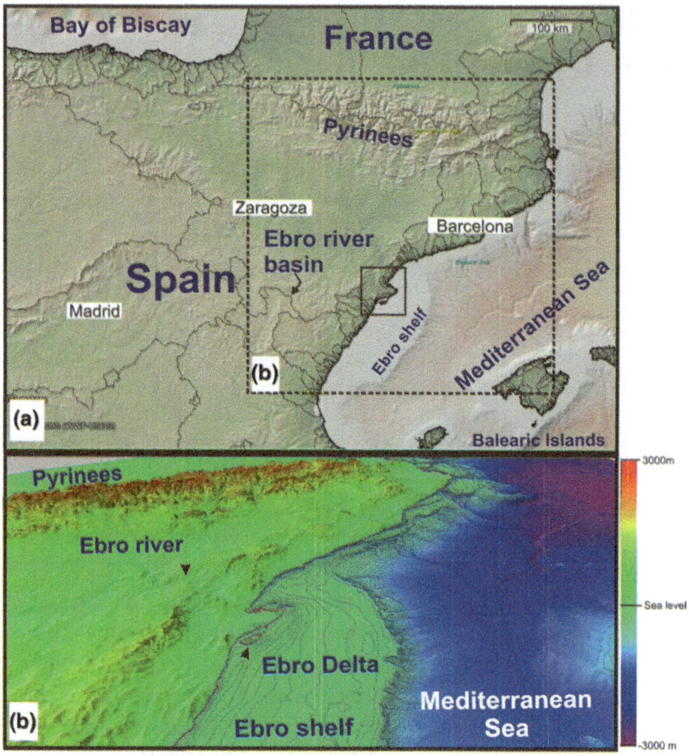

Fig. 20.1 **a** Northeastern sector of the Iberian Peninsula showing the location of the Ebro Delta. The boundaries of the Ebro River basin are also shown. **b** Shaded relief DTM of the emerged and submerged region of the Ebro Delta (Somoza and Rodríguez-Santalla 2014)

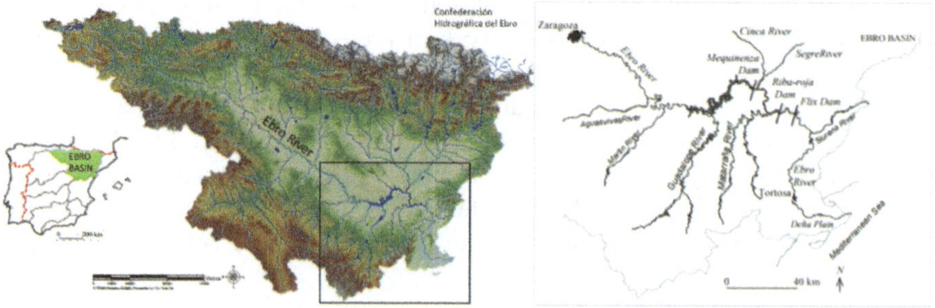

Fig. 20.2 Location of dams in lower Ebro River (modified from Vericat 2005)

causing morphological changes downstream, as well as the deltaic and coastal ecosystems (Kondolf 1997), in addition to enabling salt wedge intrusion which modifies the capacity of solids transport and decreasing the thickness of the freshwater layer, preventing their use for agricultural holdings. Guillén and

Palanques (1992) and Movellán (2004) have studied the effects of saline wedge in the Ebro Delta in detail.

The Ebro delta represents an area of great ecological values (salt marshes, lakes, springs, etc.) that is combined with a considerable local economic activity relates with rice crops, yields 98% of the production in Catalonia and is the third in the European market (Fatoric and Chelleri 2012).

Ebro River is one of the Spanish rivers presenting greater number of dams in his river basin, which is regulated in 97%. As indicated above, the dams of Mequineza (operative from 1966 and with a capacity of 1,534 hm^3) and Ribarroja (1969, 207 hm^3) are the main responsible of hydrological changes downstream, because they retain practically 99% of the sediments transported by the river (Dolz et al. 1997), and this has produced a very negative impact on the Ebro delta. According Vericat (2005) downstream from the dam of Flix (1948, 11 hm^3), the last before the mouth and 100 km of it, the total load of the Ebro river represents 3% of what was transported at the beginning of the 20th century in the absence of dams. Tena et al. (2011) indicate that the Tortosa hydrological station registers 1% of flow what was moving at the beginning of the 20th century. On the other hand, forestry plantation between 1960 and 1970, have also contributed to reduce material inputs of the river (Maldonado 1972). These situations produce the subjection of the Ebro delta to the waves and currents, favoring its erosion and subsidence.

20.2 Morphology of the Delta Coast

The different configurations of Ebro delta from its origins are well documented in Maldonado (1972), Somoza et al. (1998), Somoza and Rodríguez-Santalla (2014). The current geometry of the Ebro Delta is related to various processes acting directly on it. Firstly, the retention of the sediment flux by dams built along the river and the increase in water consumption, for irrigation especially. Secondly, marine processes that continuously reshape the delta and alter its configuration (Serra et al. 2012). The current morphological configuration is established by two hemideltas separated by the river Ebro (Fig. 20.3a). It is formed by the delta front composed of the present river mouth (Fig. 20.3c), and two large spit bars that partially close two adjacent lagoons: El Fangar spit bar to the NW (Fig. 20.3b) and Los Alfaques spit bar to the SW (Fig. 20.3d). The latter spit is joined to the rest of the delta by the Trabucador bar, approximately 250 m wide and 6 km long (Fig. 20.4d). Tortosa Cape, in Ebro River mouth, is where marine agents are more effectives due its position with regard to Eastern waves, the most energetic. This way, a current is generated transporting the sediments eroded in the mouth and the reduced amount of material that the Ebro River provides, toward North and South, nourishing El Fangar and La Banya spits (Fig. 20.3a). The mouth shows a very flat submerged

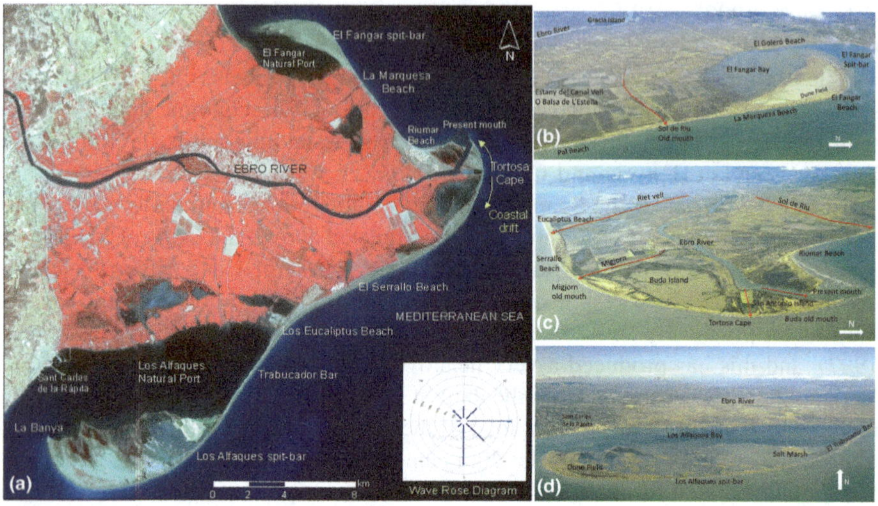

Fig. 20.3 **a** Present-day morphology of the Ebro Delta and Wave Rose diagram; **b** El Fangar spit bar; **c** Mouth of Ebro River: to the left, the abandoned Buda delta. To the right, the present deltaic lobe at the mouth of the Ebro River; **d** Los Alfaques spit bar, linked to the main delta by the El Trabucador bar. *Images Source* Carol (1998)

profile with a low slope, which favours the development of bars, resulting in a wetland (El Garxal), and in whose front develops the dunes field of the Riumar beach (Fig. 20.3c).

The mouth has changed its geometry since the second half of 20th Century due to strong erosion. Ramírez-Cuesta et al. (2016) analysed the changes in the different environments that develop in the mouth, showing that an extensive active eolic mantle has been accumulated over the old beach ridges that are deposited in the mouth. This demonstrates the importance of aeolian processes in the whole Ebro Delta.

El Fangar spit is a sand bar with 2 km long with a maximum width of 1.4 km at its centre, where an active dune field is developed. Los Alfaques spit is bigger than El Fangar, with foredunes developed on the coast. Even within the same deltaic system, both dune fields show different geomorphological characteristics, due to differences on the conditions in which the modellers agents (wind and wave) operate and especially, to coastal orientation (Rodríguez-Santalla et al. 2017). More information on the morphological characteristics and evolution of both dune field may be found in Rodríguez et al. (2009), Serra et al. (2012), Rodríguez-Santalla et al. (2015), Barrio-Parra et al. (2017), Barrio-Parra and Rodríguez-Santalla (2016).

20.3 Geological Structure of the Delta

The sedimentary succession of the Ebro Delta is underlain by 500–2,500 m of Plio-Quaternary deposits, which in turn unconformably overlie Miocene deposits (IGME 1987). The basement consists mostly of Lower Jurassic limestone and dolomites and Upper Cretaceous marls and bioclastic limestone (e.g. Maldonado 1975). These Mesozoic formations are cross-cut by a dense network of normal faults with prevailing N70E and N110E strikes, corresponding to the structural grains of the Catalan Coastal Chain and the Iberian Chain, respectively. These faults bound concealed structural highs and lows parallel to the present coast. The top of the Miocene sediments beneath the delta is marked by a strong reflection that

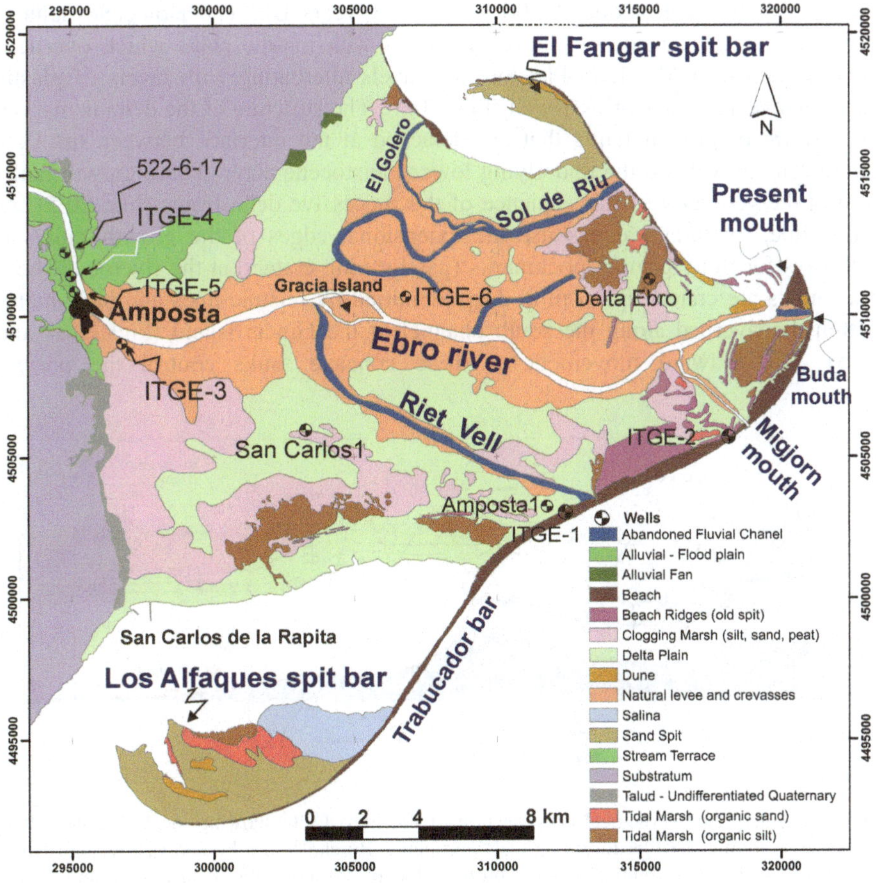

Fig. 20.4 Map of the Ebro Delta showing the different morphosedimentary domains. Location of boreholes used to construct the cross-section in Fig. 20.5 is also shown (Somoza and Rodríguez-Santalla 2014)

corresponds to a regional erosional surface (unconformity) developed during the desiccation of the Mediterranean in the Messinian salinity crisis. The Messinian unconformity is overlain by the Plio-Quaternary Ebro Group, which comprises the Pliocene Lower Ebro Clays that grade upwards and shorewards into the Plio-Pleistocene Ebro Sandstones. These sandstones contain also interbedded shallow marine conglomerates and marly clays. The Ebro Sandstones and the underlying Ebro Clays forms a huge landward-thinning wedge beneath the delta plain.

Figure 20.4 shows a geomorphological map with the location of boreholes drilled to investigate the internal structure of the delta (ITGE 1996). The thickness of the Holocene deposits of the delta ranges from 18 m on the landward side to 51 m at the delta front (Fig. 20.5). The maximum thickness of delta sediments occurs near the mouth of the present-day Ebro Delta (Maldonado 1972). Offshore radiocarbon ages (Díaz et al. 1990) indicate that deposition of the prodelta on the shelf began between about 10,000 and 11,000 years BP. The Holocene delta is composed of a sequence of alternating sands with marine clays which overlie a thick sequence of Mid-Late Pleistocene gravels alternating with layers of plastic clays as revealed San Carlos 1 well (Fig. 20.5). The structure of the delta is marked by a series of growth faults that are detached at the interface between the Late Pleistocene gravel and the underlying lower Pleistocene clays. These growth faults control the geometry of the sequence of the successive delta fronts (Fig. 20.5). At same time, overpressure gas use the extensional edges of the growth faults as pathways for fluid forming pockmarks (gas-forming craters) at the riverbed. These pockmarks are craters up 16 m in deep identified by seismic lines and bathymetry within the riverbed along the south channel of the Gracia Island (Figs. 20.4 and 20.5). The seaward movement of these concave faults control the present

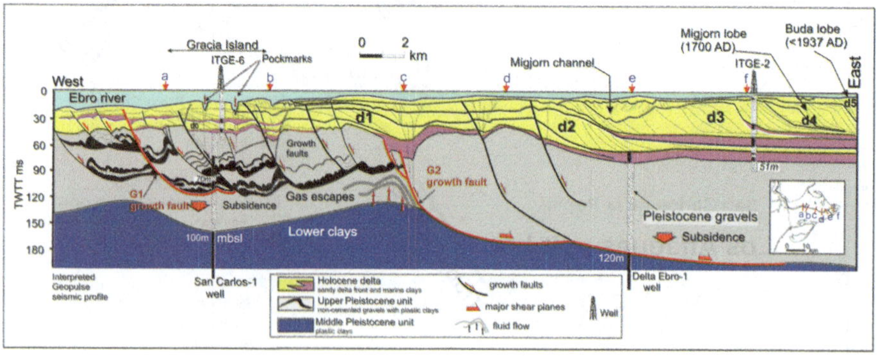

Fig. 20.5 Interpreted geological cross-section of the Ebro Delta showing the Holocene deltaic sequence overlying the Pleistocene gravels and clays. Boreholes are located in Fig. 20.4. Inset shows tracks of high-resolution seismic profiles (Geopulse) along the main river channel from Amposta to present mouth (modified from Somoza et al. 1998)

subsidence of the delta affecting mostly to the inner side (e.g. segments *a* to *c* in Fig. 20.5). Thus, the highly meandering shape of the Sol de Riu channel and its landward current segment (Fig. 20.4) is probably controlled by the depressions associated to growth faults near Gracia Island (Fig. 20.5).

20.4 Evolution of the Delta During the Last 7,000 Year

The onset of deltas around the world margins are mostly related to the last maximum flooding by sea level reached at ca. 7,000 year and the later retreat (e.g. Stanley and Warne 1994). In the Ebro delta, the foraminiferal record obtained from the Carlet borehole support no evidence of marine flooding older than 7,500 year (Benito et al. 2015).

The Holocene deposits of the delta has been interpreted as a depositional sequence being composed of a transgressive system track (TST), formed mainly of a basal mollusc-shell lag and marine gray or black clays overlapping Pleistocene gravels, and a highstand system tract (HST) (Somoza et al. 1998). The top of the maximum flooding surface (MFS) separating the TST from the HST was dated at 7,860–6,900 year BP near Amposta city (in the inner border of Delta (Arasa 1994). The HST, which overlies the MFS, includes a total of five progradational units ("d" in Fig. 20.6): basal unit d0 made of bioclastic coarse sand containing marine mollusks and deposited during the TST before 7,000 year BP; units d1 and d2 composed of coarse and medium sands, accumulated above the MFS and dated between 6,150 and 3,600 year BP. These three lower units are considered as

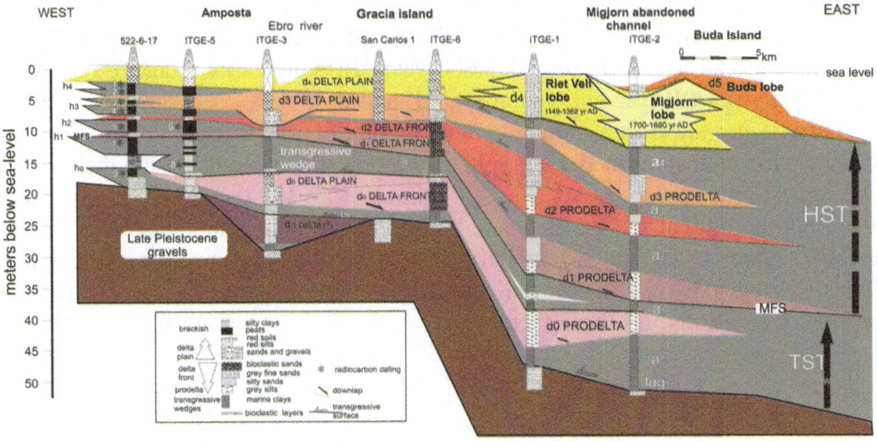

Fig. 20.6 Correlation of wells along the delta showing five progradational sand unit interbedded with agradational marine clays composing the HST

delta-front and nearshore deposits; and finally, units d3 and d4 characterized by sands with scattered pebbles and silty sands are defined as delta-plain deposits younger than 2,700 year BP.

20.5 Holocene Coastline Evolution

The morphological evolution of the delta during the Holocene is the result of the successive accumulation of deltaic lobes s at the mouth of the river. These morpho-stratigraphic units advance radially seawards from an avulsion point across the lowest lateral zones of the delta-plain (Fig. 20.6).

Based on borehole logs and radiocarbon ages from peat deposits, a tentative chronology for the formation of the present morphology of the Holocene Ebro Delta has been proposed (Somoza et al. 1998, Fig. 6). This chronology is based on the assumption that the main deltaic constructions (d1, d2, d3, d4 and d5) were formed during periods of sea level stability (stillstands) or slight falls, whereas peat units recorded in drillholes and deposited in inland lagoons are related to flooding during periods of slight sea-level rise.

Periods of slight fall in sea-level and/or major river floodings seem to have an important role in switching the location of deltaic constructions (Fairbridge 1988). In the case of the Ebro Delta, the main periods of delta-lobe switching seem to have a recurrence of thousands of years whereas the frequency of the morphological modifications of the mouth is of the order of hundreds of years. The correlation between peat levels and deltaic lobes ages with radiocarbon data frequency of peat layer in deltas show that the main growth stages of deltaic lobes are intercalated with stages of flooding which form agradational peat layers (Fig. 20.6). Therefore, five Holocene stages of growth of deltaic lobes are differentiated in the Ebro Delta and allow to construct a relative sea level curve (Fig. 20.7): d1 (6,150–5,350 years BP conventional ages), d2 (4,400–3,600 years BP conventional ages), d3 (2,910 and 2,700 years BP conventional ages), d4, and d5.

These stages are closely similar to those reported for Holocene periods of intense progradation which formed long spit bars along the Atlantic South Iberian and Western Mediterranean coasts (Fig. 20.7, Zazo et al. 1994). Therefore, these data point out that global sea-level/climate high-frequency fluctuations have driven the growth periods of the Ebro Delta during the Holocene. In contrast, historic deltaic lobes (d4 and d5) may have been affected by anthropogenic factors. In the next section, we will discuss these factors.

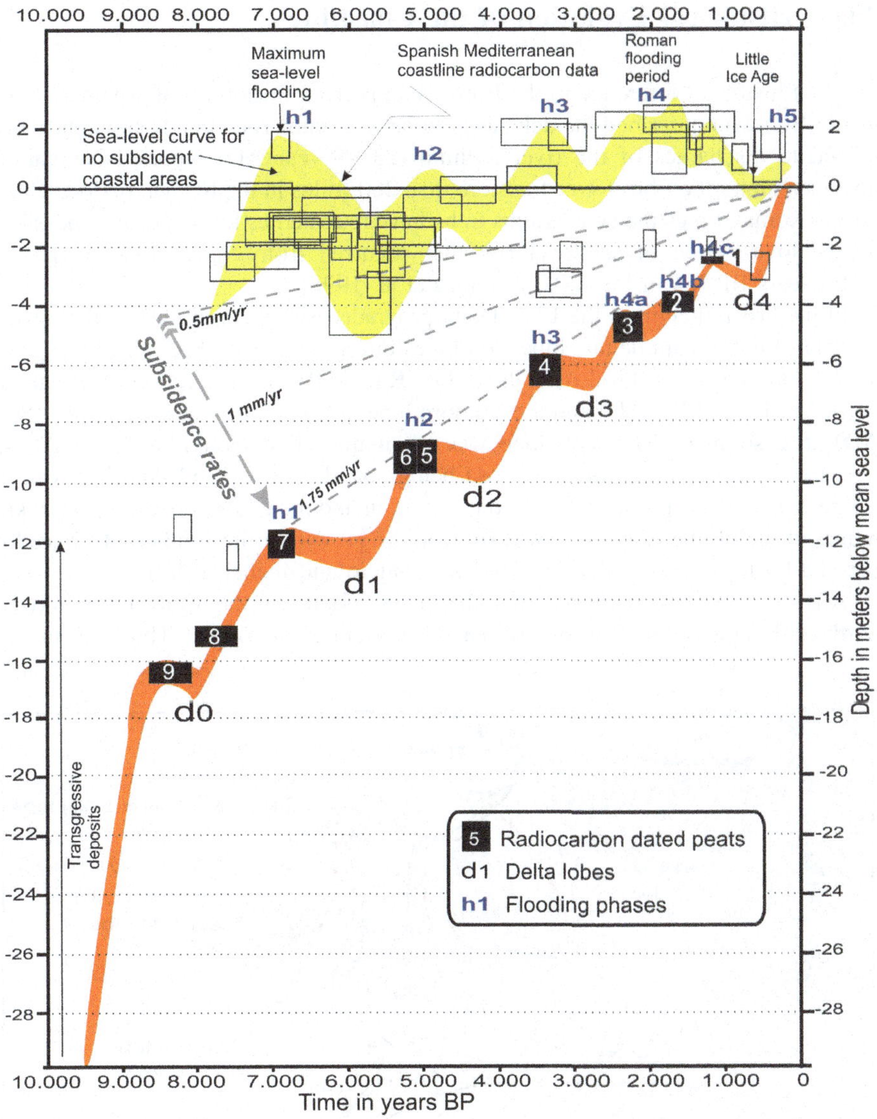

Fig. 20.7 Relative sea-level curve for the Ebro Delta based on the correlation between peat levels and deltaic lobes (Somoza et al. 1998, orange line) and radiocarbon data from Spanish Mediterranean coastline (Zazo et al. 1994, yellow line). The five stages of deltaic growth (d1–d5 in Fig. 20.8) are associated with relative sea-level falls whereas peat layers are associated with relative flooding stages (h1–h5). The maximum Holocene sea-level flooding corresponds to h1 stage at ca. 7,000 year BP. Linear subsidence trend are also shown (dashed grey lines)

20.6 Historical Evolution of the Coastline

The first historical reference to the Ebro Delta is from Posidonio of Apamea (born in 135 BC) who described the delta as a "zone with numerous floods apparently not related to discharges of the river". Plinio (23–79 year AD) reported navigation upstream along the Ebro River for more than 260 miles in the Roman period. Thus, during Roman times the coast seems to be have retreated as far as Tortosa, possibly related to a period of sea-level rise that occurred from ca. 2,700 years BP to 1,100 years BP (h4 in Fig. 20.7, Somoza et al. 1998).

In the last millennium the Ebro Delta prograded significantly due to the development of three main deltaic lobes (d4 lobes in Fig. 20.8): the Riet Vell lobe (d4a) at the south (ca. 1149–1362 AD), the Sol-de-Riu (or Riet de Zaida) lobe (d4b) at the north (ca. 1575–1700 AD) and the Migjorn lobe (d4c) at the centre (ca. 1700–1800 AD). The southern Riet Vell lobe was the main active mouth in 1149 AD, as reported by the conquerors of Tortosa being abandoned in 1362 AD. It has been suggested that the partial destruction of this abandoned lobe provoked an 8 km retreat of the old headland that will sourced the growth of the southern Trabucador bar and La Banya/Los Alfaques spit bar (Canicio and Ibáñez 1999).

After this period, it seems that the main mouth of the Ebro Delta moved northward to the Sol de Riu (or also named as Riet de Zaida) lobe. Thus, this mouth

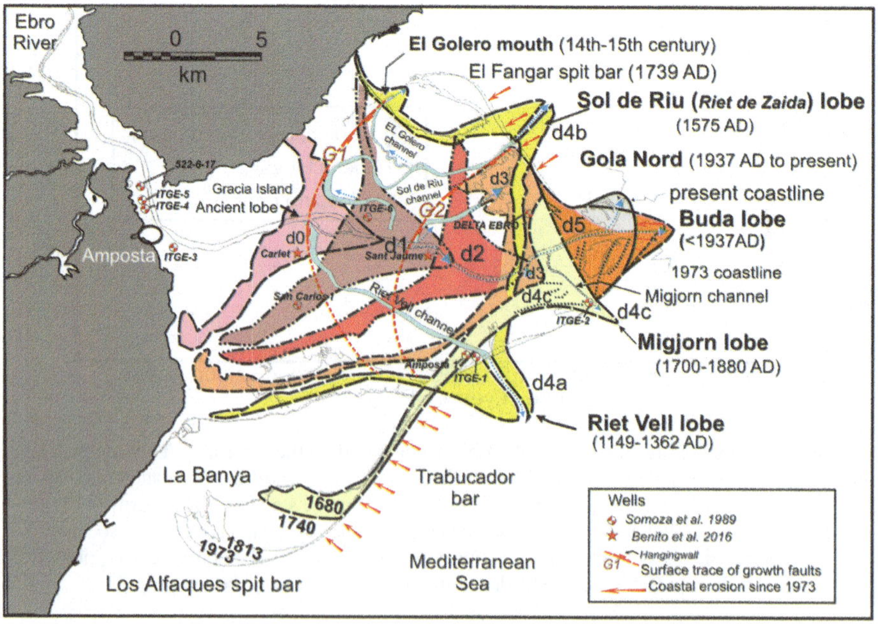

Fig. 20.8 Proposed schematic evolution of the coastline of the Ebro Delta during the Holocene based on boreholes, seismic lines, morphology and historic charts (modified from Somoza and Rodríguez-Santalla 2014)

was active at least in 1575 AD, as reported during the construction of the disappeared "Tower of the Sol-de Riu", built up to prevent the entrance of pirates to the Amposta harbour. This lobe was separated from the old Riet Vell lobe by large palaeo-bay known as Port Fangós (Benito et al. 2015). The Sol-de-Riu lobe was result of the avulsion of the channel from the Gracia Island forming a highly meandering channel in the north (Fig. 20.8). The area of channel meandering coincide with the area of maximum subsidence related to the movement of main listric growth faults (G1) identified in the high-resolution seismic profiles at the landward side of the Gracia Island and forming gas escapes structure in the river bed (G1 in seismic section, Fig. 20.5). The trace of this growth fault extends from the El Golero mouth in the north to the Trabucador bar in the south (G1 in Fig. 20.8) (Maestro et al. 2002). This depression forced a highly meandering channel in the north finishing into the El Golero mouth from the 14th to 15th centuries. This meandering channel is constituted by two segments where the channel flowed landwards. We suggest that this anomalous pathway may have been forced by the formation of a depression at the hanging wall of the G1 growth fault (Fig. 20.5). Moreover, the final segment of the channel follows the trace of the G1 to form the El Golero mouth (Fig. 20.8). Therefore, we suggest also that the depression generated by the subsidence at the hanging wall of the growth fault G1 may be also responsible of the formation of a large palaeobay-lagoonal system named as Port Fangós at this time (Benito et al. 2015). The interaction between subsurface growth faults and river switching is also identified previously. Therefore, eastwards Gracia Island, the river switched with a northeast direction following the trace of the subsurface G2 growth fault. This switching of the river is correlated with an old Sol de Riu lobe formed by the d3 progradational stage (ca 2,910–2,700 year BP) before Roman flooding stage (Fig. 20.8). The breach of this highly sinuous meander forced a new channel that generate the Sol de Riu lobe, active at least to 1575 AD and being the main entrance to Amposta harbour at this time. Probably El Golero mouth (in the north was contemporaneous with the Riet Vell lobe in the south, but once breached the meandering channel and forming a shorter and with more hydraulic gradient to sea, provoked a quick decay of the old Riet Vell main channel (Benito et al. 2015).

The next detailed navigation maps from 1733 AD and 1749 AD show the complete filling of the previous paleobay generated landwards of the northern Sol de Riu and southern Riet Vell deltaic lobes from 1149 AD to, at least, 1575 AD (Benito et al. 2015).

Between the beginning of the 16th Century and the middle of the 17th Century, the sharp increase in the construction of ships, following the discovery of America, led to a great deforestation in this sector of the Iberian Peninsula. This is considered to be one of the main factors responsible for a new expansion of the delta, as it induced a great increase in the sediment supply feeding the delta mouth. The development of the Migjorn lobe, active around 1700 AD (Fig. 20.8), coincides with this period of deforestation. The Migjorn channel was result of a new river switching vent that took place at a location called La Cava closed to the Gracia Island as consequence of a man-made in the outer levee of a pronounced meander

(Ribas 1996; Benito et al. 2015). Modern maps show a rapid progradation of the central Migjorn lobe until 1880 AD due to the infill of the shallow inner palaeobay of Port Fangós (Ibáñez et al. 1997).

The major advance of the Migjorn deltaic lobe might be also related to a small sea-level fall between 0.5 and 1 m below the present mean sea level (MSL), closely related to the neoglacial "Little Ice Age" event, between 1600 and 1800 AD (Fig. 20.7). It has been proposed that a relative sea-level rise event following this period might have been the cause for the growth of major spit bars during the 17th and 18th centuries, that are observed in the current morphology of the Ebro Delta. Thus, the southern Los Alfaques spit bar and Trabucador bar were formed since at least in 1680 AD by littoral drift of sediment sourced from the destruction of the Riet Vell lobe. On the other hand, the northern El Fangar spit bar was constructed with sediments eroded from the destruction of the Sol-de-Riu lobe (Fig. 20.8). The most recent deltaic growth period (d5) corresponds to the Buda lobe (active till 1950 AD) and the present active mouth.

20.7 Coastal Changes in the 20th Century

As has been mentioned previously, the evolution of the Ebro Delta in the last decades is associated to construction of dams started at the beginning of the 20th Century (the first in 1913), but most were built in the period 1940–1975. Before massive dam construction, maximum floods reached 20,000 $m^3\ s^{-1}$, whereas minimum flows of about 50 $m^3\ s^{-1}$ were common in summer (Maldonado 1972). The construction of the dams has entailed almost the suppression of river flooding over the delta plain and a drastic reduction of sediment feeding the mouth. The last major floods in the lower Ebro took place in 1907 and 1937, with peaks of around 23,000 $m^3\ s^{-1}$ in Tortosa (40 km upstream from the mouth). During the recent decades, only small floods (2,000–3,000 $m^3\ s^{-1}$) have occurred from time to time. These last floods were probably the cause of the present configuration of the Ebro Delta mouth, where the river discharge point was changed northward. Somoza and Rodríguez-Santalla (2014) show a sequence of images that points out to that the switch from the Buda delta lobe to the present-day one took place between 1950 and 1956. The opening of Gola Nord caused the modification of delta coast, especially Tortosa Cape and both spits: Delta front is the zone with greater erosion by the ravages of east waves and, conversely, the north and south spits record high deposition and rapid progradation of the shoreline. According with Rodríguez et al. (2010), this spatial pattern results from the combined effect wave and wind action. The longitudinal sediment drift from the old eastern mouth has NW and SW components along the flanks of the delta, and the main winds, coming from the NW, influence the current coastal design and the distinctive configuration and evolutionary behavior of the two hemi-deltas. Figure 20.9 illustrates the areas of the coast affected by net retreat and accretion processes. The mouth area of the delta front is the zone with greater erosion, near 2.8 km of mean retreat after almost

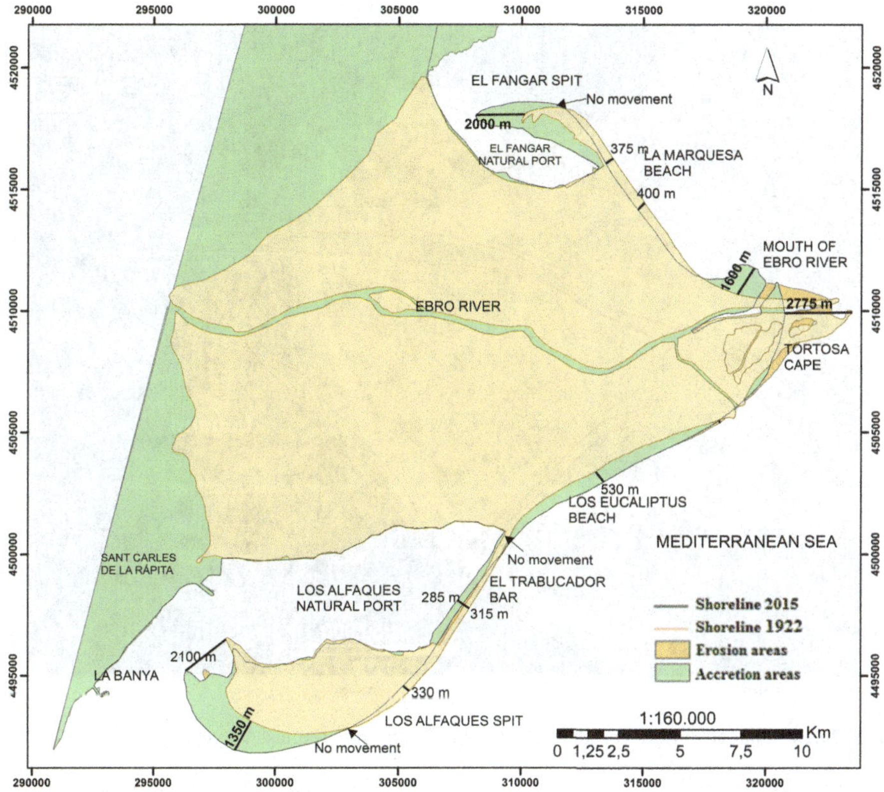

Fig. 20.9 Distribution of erosion and sedimentation areas for the period 1922–2015 (modified from Rodríguez 1999, 2010)

ninety years, reaching approximately 30 m/year. Figure 20.10 shows how the retreat not was lineal. The shoreline at Tortosa Cape recorded an extremely rapid retreat, 1,500 m between 1957 and 1984, which supposes a mean rate of more than 55 m/year; and between 1984 and 2015 the deltaic front continued its retreating trend reaching at a mean rate of 19 m/year. Now, the velocity of the retreat has decreased, despite remain the erosion in the mouth Rodríguez (1999).

The areas with more coastal accretion are the both spits. The El Fangar spit, located in the hemidelta north, is a 6 km long sand spit length with a maximum width of 1.4 km in the middle part, which spreads to the north-west forming a bay (Fig. 20.11). The inner coast of the bay presents a smooth slope easily overrun by seawater during meteorological tide episodic (Rodríguez et al. 2003). It has a landscape of dunes, type barchans, practically all along the outside of the spit (Figs. 20.4 and 20.11), which represents the most extensive formation of mobile dunes of the delta. The development and evolution of dunes field has been widely researched by Sánchez (2008), Rodríguez et al. (2009), Serra et al. (2012), Barrio-Parra and Rodríguez-Santalla (2016), Barrio-Parra et al. (2017),

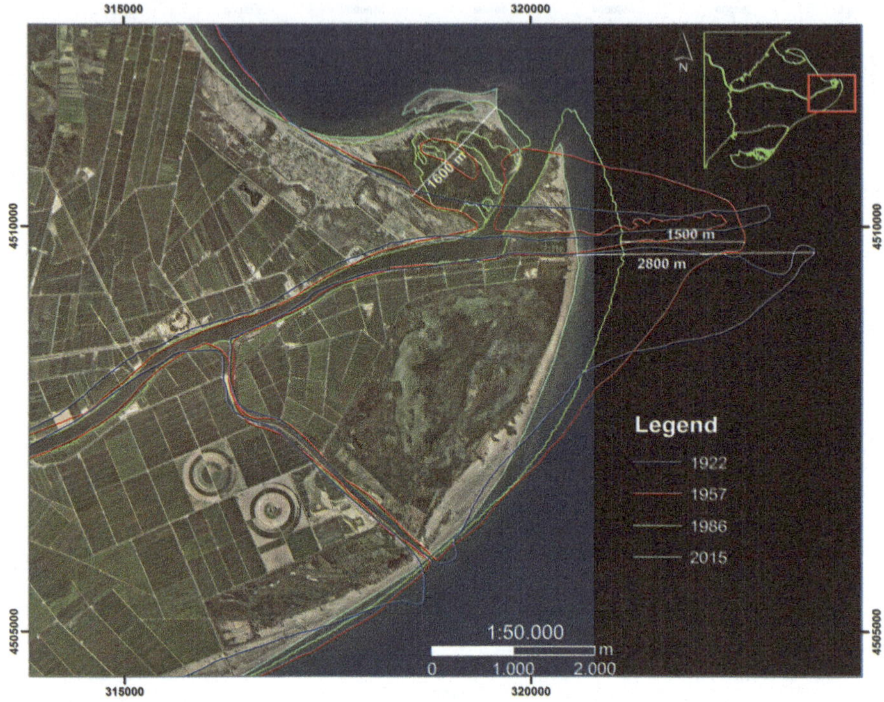

Fig. 20.10 Changes in the coastline of the Ebro Delta over the last 90 years

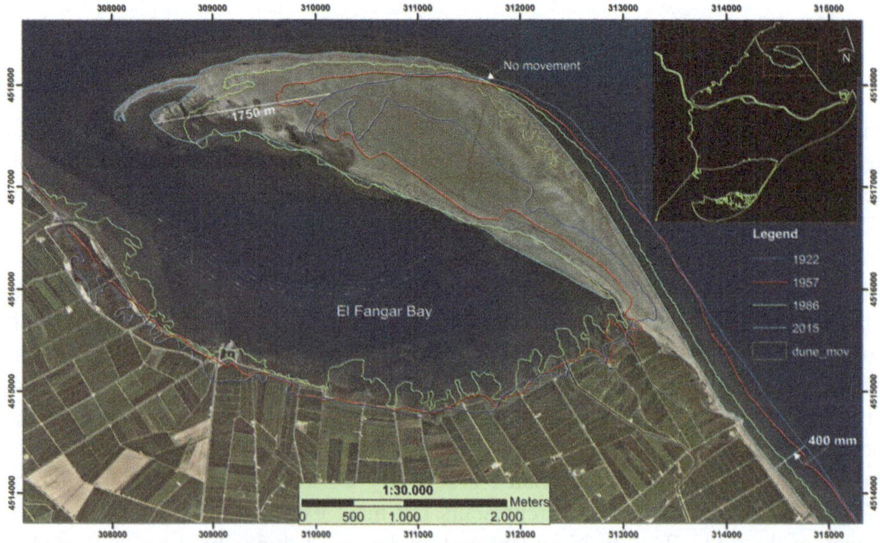

Fig. 20.11 Evolution of El Fangar spit over orthopho of 2015 of Institut Cartogràfic de Catalunya

Rodríguez-Santalla et al. (2017). The Fangar spit has been developed as a sequential addition of sandbars with hook shape, from sediments originating by the erosion of the delta front. As a general trend, the spit evolution shows an important accretion on its end, and it turns towards the continental coast closing progressively the bay.

Figure 20.11 shows the evolution of El Fangar spit. The outer coast shows an important backward movement of the outer shoreline of Fangar spit in La Marquesa and Pal beaches. This trend decreases towards the north until a null-point (no accumulation and no erosion), where changes from a regressive movement to an accretion of sediments, which is especially predominant at the tip of the bar. The increase of the tip has been 1750 m between 1922 and 2015, reaching a velocity of 19 m/year. From 1957 to 1986 the spit has experienced an increase greater than 900 m which represents an increase speed near 32 m/year; and between 1986 and 2015 the increase has been greater than 300 m, reaching a rate near 11 m/year. The accretion in second period may seem to be greater than expected, according with Rodríguez et al. (2011). They indicate that a breakwater in front of the a Restaurant located at the base of El Fangar spit, and built due to the consequences of the severe storms in 1997, has been causing the rupture of the dunes and, secondly, the acceleration of the erosion by increase the reflectivity of the coast. The sediments are being eroded in this area since then, and transported and deposited in the tip of El Fangar.

Los Alfaques spit, together with the Fangar spit, it is one of the main areas for the deposition of eroded material in the delta. Like El Fangar, has a western orientation (Fig. 20.12), and is formed by sediments accretion from the erosion of delta front, and transported by longshore currents. The different positions of coastline can be distinguishing through the beach ridges, which have been dated for Rodríguez-Santalla et al. (2015) through OSL technique and whose results are consistent with earlier reconstructions of Maldonado (1972), Canicio and Ibáñez (1999) and Somoza et al. (1998). All of them suggest that in XVII century Los Alfaques has already begun its formation. Like El Fangar spit, Los Alfaques spit has dune fields. In its formation, they begin being barchan dunes associated to every growth ridge, but as they are taking inner positions in the spit, they are colonized by vegetation turning into foredunes.

Los Alfaques spit has two areas with different behaviour: half north is an erosive area to the point where the movement is null, and from which begins the accretion area (Fig. 20.12). The mean retreat of 355 m between 1922–2015 (3.8 m/year), in the area next to El Trabucador bar, and accretion it reaches up to 1,540 m (16.5 m/year). La Banya tip had moved over 2,000 m westward in same period. A more detailed description is given in Rodríguez (1999, 2005), Sánchez-Rodríguez et al. (2014).

Los Alfaques are joined to delta body by El Trabucador Bar which is a narrow sandy formation, about 250 m wide and 6 km long. It has also been formed by annexation of sand bars with material from the erosion of the delta front transported by the effects of longshore drift. At certain times and under heavy storm conditions the bar is overwashed, It has sometimes even been broken (Sánchez-Arcilla et al. 1997;

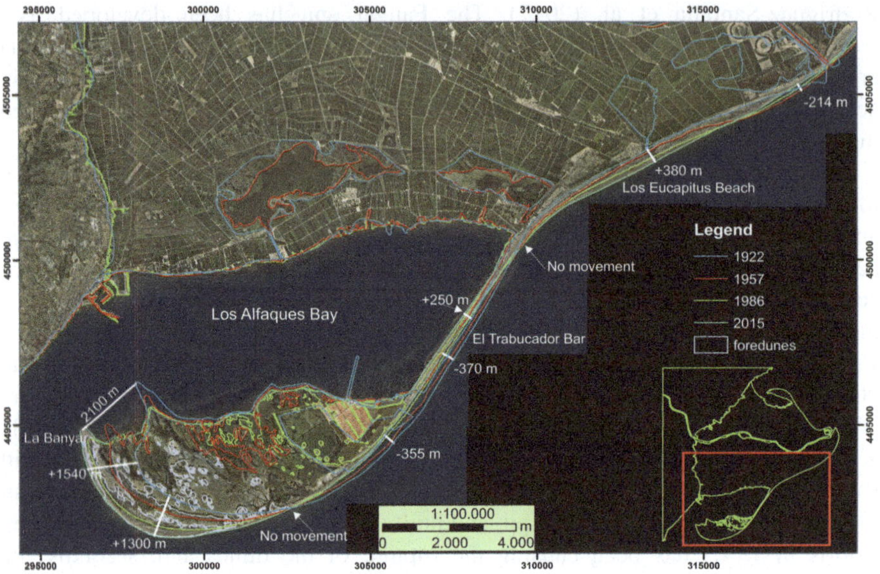

Fig. 20.12 Evolution of Los Alfaques spit, El Trabucador bar and Los Eucaliptus Beach over orthopho of 2015 of Institut Cartogràfic de Catalunya

Rodríguez et al. 2010), introducing sediments inside the bay and form washover fans, which causes the spit to migrate towards the coast. As the principle of the bar retains its width, this pattern of erosion in the outer coast and slight sedimentation in the inner, cause the turning of 4° towards land, value slightly higher the estimated by Rodríguez (1999). In the Fig. 20.12 it can be appreciated that the mean accretion in inner coast is 250 m between 1922 and 2015 (2.6 m/year), while in outer coast the erosion reaches 374 m (4 m/year) in the same period, indicating a gradual reductions of the El Trabucador bar width, increasing his vulnerability to waves storm.

20.8 Threats and Forecast of the Delta Evolution

According to Somoza and Rodríguez-Santalla (2014), the deltas are ephemeral morphologies built up as response to the sensitive equilibrium established between fluvial sediment supply, global sea level, wave-induced energy and regional subsidence. Currently, effects of climate change can significantly accelerate the process. As seen above, Ebro Delta has an important sediments deficit which do not allow the progradation, and therefore, thereby favouring the effects of subsidence.

The reduction in water flow allows a quasi-permanent wedge of marine water (Rodríguez et al. 2010). This salt wedge alters and reduces the depth of the fresh

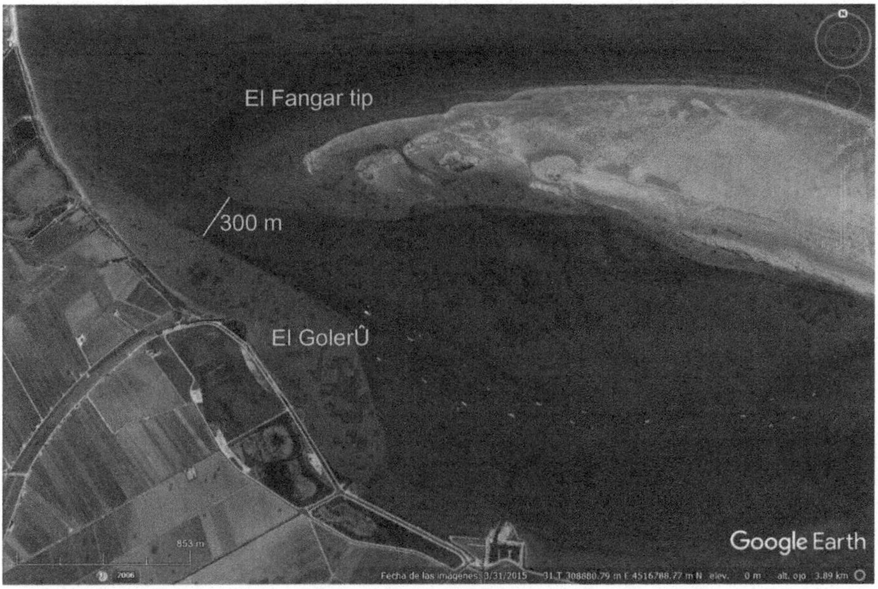

Fig. 20.13 Detailed image that show the submerged coast in El Fangar Bay

water layer, making it more difficult for the delta's agricultural operations to make use of it. This effects, are very significant in Ebro delta because all delta plain is covered by agriculture.

The current major threat is the erosion of the delta front. That, so as the pattern of longshore sediments transport along the coast, represents the main factor capable of modifying the coastal morphologies of the delta. One of effects observed, that concerns to fishermen, is the high growth of the El Fangar tip (Fig. 20.13). At the same time, El Goleró beach, just opposite the point of the sand spit, also is prograding to the inside of the bay and could lead to the closure of the bay, if the hydrodynamic conditions governing the system allow it (Rodríguez et al. 2011).

The erosion problem in Ebro Delta has been addressed from a variety of perspectives and different actuations have been undertaken to mitigate and reduce the impact. Several proposals have been made as to how to recover the delta ecosystem and its natural dynamics. A review on the studies and measures proposed and adopted may be found in Rodríguez (2005) and Rodríguez et al. (2010, 2011).

In the medium and long term, the other threat is that, in the case of deltas, the progressive subsidence must be added to the sea-level rise in a scenario of global warming. Average subsidence rate in the Ebro Delta has been estimated at 1.75 mm/year during the last 7,000 years (Fig. 20.14 modified from Somoza et al. 1998). Ibáñez et al. (1997) considered a subsidence of 2 mm/year for the last 0.3 k year and recent subsidence to be 1–3.2 mm/year. Recent research combining subsidence and sea-level rise data on the Ebro Delta coast estimate a variable

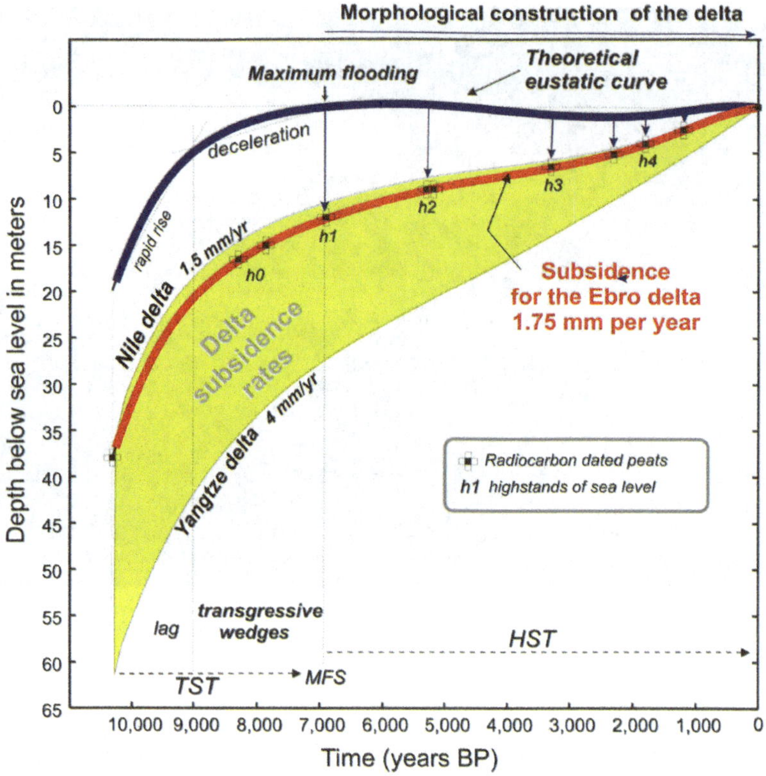

Fig. 20.14 An average subsidence rate of 1.75 mm/year has been calculated for the Ebro Delta on the basis of the depth of dated buried peat deposits. The rates estimated for other worldwide deltas are also shown

relative sea-level rise (RSLR) ranging between 2 and 6 mm/year (Jiménez et al. 1997), and from 4 to 56 mm/year. This subsidence rate that falls within the range of other deltas (1–4 mm/year) is mainly caused by compaction of deltaic sediments, degassing of peats and growth faults developed at the base of deltaic sediments. Anyway, the fact of that subsidence is partly related to the recent movement of the hanging walls of the growth faults this make to vary the values within the delta (Fig. 20.5).

Somoza and Rodríguez-Santalla (2014) show in Fig. 20.15 how a potential sea-level rise will affect to the present morphology of the delta (although the model presents in Fig. 20.15 does not consider adaptation processes to sea-level rise which may reduce potentially floodable areas). Flooding of areas that presently are only 0.75 m above sea level would reduce dramatically the extension of the delta producing a great retreat and influencing mainly the rice fields. The only areas in the delta that would remain emerged would be those formed by the growth of natural levees created by historical floodings of Ebro River that do not occur nowadays due

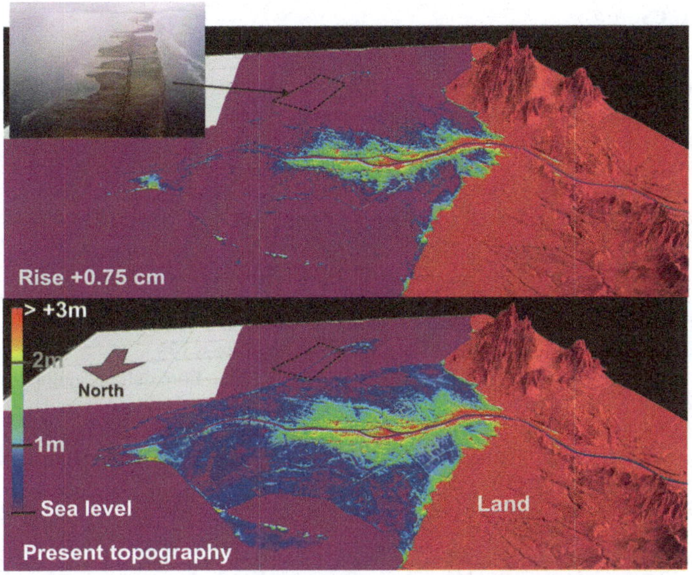

Fig. 20.15 Model of the morphology Ebro Delta after a potential relative sea level rise of 0.75 m associated with global warming and local subsidence. The digital model, with a spatial resolution of 5 m, has been constructed with data from the IGN (Instituto Geográfico Nacional) (Somoza and Rodríguez Santalla 2014)

to the construction of dams. Similar analysis can be found in Sánchez-Arcilla et al. (2008).

Policies aimed at reducing the apparent levels of risk often employ costly engineering solutions that may be inherently unsustainable (Tessler et al. 2015). Somoza and Rodríguez-Santalla (2014) offer some recommendations of IPCC to reduce the impact of climatic changes in coastal areas: retreat, protection, and adaptation. In this sense, the IRTA (Instituto de Investigación y Tecnología Agroalimentarias of Generalitat of Catalonia) is running a research project (2014–2018) financed by LIFE Program of UE, which is oriented to the restorations of the sediments flux through the Ebro River channels, from last dam of the river, to cause the accretion vertical in delta plain, as a measure against sea-level rise.

The coastal deltas are the zones with high ecological, geomorphologic and economic values. At the same time, they are the most vulnerable in relation to climate change effects. Tessler et al. (2015) find that although risks are distributed across all levels of economic development, wealthy countries effectively limit their present-day threat by gross domestic product-enabled infrastructure and coastal defense investments. The future of the deltas is conditioned by the socioeconomic capability to prepare for and mitigate exposure to hazardous conditions, which implies maintain the monitoring of the coastal processes both the short and medium term in order to design adequate measures aimed at reducing and alleviating their effects and consequences on the natural.

References

Arasa, A. (1994). *Estratigrafia i sedimentologia dels materials Plio-Quaternaris del Baix-Ebre i sectors adjacents* (PhD). University of Barcelona, 447 pp.

Barrio-Parra, F., & Rodríguez-Santalla, I. (2016). Cellular automata to understand the behaviour of beach-dune systems: Application to El Fangar Spit active dune system (Ebro delta, Spain). *Computers & Geosciences, 93,* 55–62.

Barrio-Parra, F., Rodríguez-Santalla, I., Pires de Matos, R., & Ribeiro, M. (2017). A modeling approach to assess the key factors in the evolution of coastal systems: The Ebro North Hemidelta case. *Estuaries and Coasts, 40*(3), 758–772.

Benito, X., Trobajo, R., Ibáñez, C., Cearreta, A., & Brunet, M. (2015). Benthic foraminifera as indicators of habitat change in anthropogenically impacted coastal wetlands of the Ebro Delta (NE Iberian Peninsula). *Marine Pollution Bulletin, 101,* 163–173.

Canicio, A., & Ibáñez, C. (1999). The Holocene evolution of the Ebre delta, Catalonia, Spain. *Acta Geographica Sinica, 54,* 462–469.

Carol J. (1998). *Aeroguía del litoral de Cataluña.* Barcelona: Planeta, 272 pp.

Díaz, J. I., Nelson, C. H., Barber, J. H., & Giro, S. (1990). Late Pleistocene and Holocene sedimentary facies on the Ebro continental shelf. *Marine Geology, 95,* 333–352.

Dolz, J., Gómez, M., & Nieto, J. (1997). El Ebro en el delta. *Revista de Obras Públicas, 3368,* 7–14.

Fairbridge, R. W. (1988). Mississippi delta-lobe switching during Holocene eustatic fluctuations. *American Association of Petroleum Geologists Bulletin, 72,* 183–184.

Fatoric, S., & Chelleri, L. (2012). Vulnerability to the effects of climate change and adaptation: The case of the Spanish Ebro Delta. *Ocean & Coastal Management, 60,* 1–10.

Galloway, W. E. (1975). Process framework for describing the morphologic and stratigraphic evolution of deltaic depositional systems. In M. L. Broussard (Ed.), *Deltas, models for exploration* (pp. 87–98). Houston Geological Society.

Guillén, J., & Palanques, A. (1992). Sediment dynamics and hydrodynamics in the lower course of a river highly regulated by dams: The Ebro River. *Sedimentology, 39,* 567–579.

Ibáñez, C., Canicio, A., Day, J. W., et al. (1997). Morphologic development, relative sea level rise and sustainable Management of water and sediment in the Ebre Delta, Spain. *Journal of Coastal Conservation, 3,* 191–202.

Ibáñez, C., Sharpe, P. J., Day, J. W., Day, J. N., & Prat, N. (2010). Vertical accretion and relative sea level rise in the Ebro Delta Wetlands (Catalonia, Spain). *Wetlands, 30,* 979–988.

IGME. (1987). *Contribución de la exploración petrolífera al conocimiento de la geología de España.* Madrid: Publicaciones Instituto Geológico y Minero de España.

ITGE. (1996). Estudio geológico del Delta del Ebro. Proyecto para la evaluación de la tasa de subsidencia actual. Informes técnicos Instituto Tecnológico y Geominero de España, Madrid.

Jiménez, J. A., Sánchez-Arcilla, A., Valdemoro, H., Gracia, V., & Nieto, F. (1997). Processes reshaping the Ebro delta. *Marine Geology, 144,* 59–79.

Kondolf, G. M. (1997). Hungry water: Effects of dams and gravel mining on river channels. *Environmental Management, 21–4,* 533–551.

Maestro, A., Barnolas, A., Somoza, L., et al. (2002). Geometry and structure associated to gas-charged sediments and recent growth faults in the Ebro Delta (Spain). *Marine Geology, 186,* 351–368.

Maldonado, A. (1972). El delta del Ebro: estudio sedimentológico y estratigráfico. Boletín de Estratigrafía 1, Universidad de Barcelona.

Maldonado, A. (1975). Sedimentation, stratigraphy, and development of the Ebro Delta, Spain. In M. L. Broussard (Ed.), *Delta models for exploration* (pp. 311–338). Houston, Texas: Houston Geological Society.

Movellán, E. (2004). *Modelado de la cuña salina y del flujo de nutrientes en el tramo estuarino del río Ebro* (Ph D). Universidad de Barcelona. Barcelona, 232 pp.

Ramírez-Cuesta, J. M., Rodríguez-Santalla, I., Gracia, F. J., Sánchez-García, M. J., & Barrio-Parra, F. (2016). Analysis of recent geomorphological changes occurred in the Ebro River mouth by using changes detection techniques (period 1957-2013). *Applied Geography, 75,* 12–27.

Ribas, X. (1996). A propósito de La Cava. *Informatiu del Museu del Montsià, 39,* 5–6.

Rodríguez, I. (1999). *Evolución geomorfológica del Delta del Ebro y prognosis de su Evolución* (Ph D). Departamento de Geografía, Universidad de Alcalá de Henares.

Rodríguez, I. (2005). Shoreline management guide. Case studies Ebro delta. Technical Report. European Program EUROSION. http://copranet.projects.eucc-d.de/files/000155_EUROSION_Ebro_delta.pdf.

Rodríguez, I., Galofré, J., & Montoya, F. (2003). El Fangar spit evolution. Coastal Engineering VI: Computer Modelling and Experimental Measurements of Seas and Coastal Regions (pp. 419–425). WIT Press, UK.

Rodríguez, I., Sánchez, M. J., Montoya, I., Gómez, D., Martín, T., & Serra, J. (2009). Internal structure of the aeolian sand dunes of El Fangar spit, Ebro Delta (Tarragona, Spain). *Geomorphology, 104*(3–4), 238–252.

Rodríguez, I., Serra, J., Montoya, I., & Sánchez, M. J. (2010). The Ebro Delta: From its origin to present uncertainty. In *River deltas types structures & ecology* (pp. 161–171). Nova Science Publishers Inc.

Rodríguez, I., Serra, J., Montoya, I., & Sánchez, M. J. (2011). El Delta del Ebro. Características dinámicas y ambientales y propuestas para su protección. In I. Montoya, M. J. Sánchez, & I. Rodríguez (Eds.), El litoral Tarraconense. JMC Ofimática S.L, Barcelona.

Rodríguez-Santalla, I., Gómez, D., Sánchez, M. J., Montoya-Montes, I., Martín, T., Martín-Velázquez, S., et al. (2017). Comparación de la dinámica dunar entre las formaciones situadas en los hemideltas norte y sur del río Ebro. *Geo-Temas, 17,* 287–290.

Rodríguez-Santalla, I., Gracia, F. J., Pérez-Morales, C., Serra, J., Medialdea, A., & Barrio, F. (2015). Análisis de los cambios geomorfológicos ocurridos en la flecha de los Alfaques (Delta del Ebro, Tarragona). *Geo-Temas, 15,* 73–76.

Sánchez, M. J. (2008). *Evolución y análisis morfodinámico del campo dunar de La Flecha del Fangar (Delta del Ebro)* (PhD). Universidad Rey Juan Carlos, Madrid, 285 pp.

Sánchez-Arcilla, A., Jiménez, J. A., Gelonch, G., & Nieto, J. (1997). El problema erosivo en el Delta del Ebro. *Revista de Obras Públicas, 3*(368), 23–32.

Sánchez-Arcilla, A., Jiménez, J. A., Valdemoro, H., & Gracia, V. (2008). Implications of climatic change on Spanish Mediterranean low-lying coasts: The Ebro Delta case. *Journal of Coast Research, 24*(2), 306–316.

Sánchez-Rodríguez, V., Rodríguez-Santalla, I., & Barrio-Parra, F. (2014). Evolución de la flecha de Los Alfaques (delta del Ebro) por medio de LiDAR. In *Avances de la Geomorfología en España 2012-2014.* Susanne Schnabel y Álvaro Gómez Gutiérrez Ed, 607 pp.

Serra, J. (1997). El sistema sedimentario del Delta del Ebro. *Revista de Obras Públicas, 3,* 15–22.

Serra, J., Rodríguez, I., Sánchez, M. J., & Montoya, I. (2012). Delta del Ebro: papel del sistema dunar frente a la regresión deltaica (actuaciones y medidas paliativas). In A. Rodríguez-Perea, X. Roig, G. X. Pons, & J. A. Martín (Eds.), La gestión integrada de playas y dunas: experiencias en Latinoamérica y Europa (Vol. 19, pp. 237–242). Mon. Soc. Hist Nat Balears.

Somoza, L., Barnolas, A., Arasa, A., Maestro, A., Rees, J. G., & Hernandez-Molina, F. J. (1998). Architectural stacking patterns of the Ebro delta controlled by Holocene high-frequency eustatic fluctuations, delta-lobe switching and subsidence processes. *Sedimentary Geology, 117,* 11–32.

Somoza, L., & Rodríguez-Santalla, I. (2014). Geology and geomorphological evolution of the Ebro River Delta. In F. Gutiérrez & M. Gutiérrez (Eds.), *Landscapes and landforms of Spain* (pp. 213–227). Berlin: Springer.

Stanley, D. J., & Warne, A. G. (1994). Worldwide initiation of Holocene marine deltas by deceleration of sea-level rise. *Science, 265,* 228–231.

Tena, A., Batalla, J. R., Vericat, D., & López, J. A. (2011). Suspended sediment dynamics in a large regulated river over a 10-year period. *Geomorphology, 125,* 73–84.

Tessler, Z. D., Vörösmarty, C. J., Grossberg, M., Gladkova, I., Aizenman, H., Syvitski, J. P. M., et al. (2015). Profiling risk and sustainability in coastal deltas of the world. *Science, 349,* 638–643.

Vericat, D. (2005). *Sediment transport and processes in a highly regulated river. The lower Ebro, NE Iberian Peninsula* (PhD Thesis). Departamento de Medio Ambiente y Ciencias del Suelo, Universidad de Lérida.

Zazo, C., Goy, J. L., Somoza, L., Dabrio, C. J., Belluomini, G., Improta, S., et al. (1994). Holocene sequence of sea-level fluctuations in relation to climatic trends in the Atlantic-Mediterranean linkage coast. *Journal of Coastal Research, 10,* 933–945.

Chapter 21
Fan Deltas and Floodplains in Valencian Coastal Plains

Francesca Segura-Beltran and Josep E. Pardo-Pascual

21.1 The Relief: Large Structural Units

The València region consists of two great mountain alignments formed during the Alpine orogeny—to the north, the Iberian System (NW-SE), to the south, the Baetic System (NE-SW)—and a central sector which is a transition area between the two. In the northern Iberian sector, carbonated Mesozoic materials of marine origin predominate, which were folded into anticlines and synclines during the paroxysmal phase of the Alpine orogeny during the Oligocene period. During the Miocene, a series of folds transversal to the direction of the mountain system were formed (Garay 1995). At the end of the Tertiary and the beginning of the Quaternary periods, two new distensive phases took place, which fractured the reliefs about 10–30 km from the coast to form a stepped series of horsts and graben going NE-SW down towards the sea. During the Quaternary, the rivers eroded the horsts and filled up the coastal plains.

The central area of València region is a transition between the Iberian and the Baetic mountain chains. The Caroig platform is a sub-tabular relief formed by limestone from the Jurassic and Cretaceous periods and located about 40 km from the coast. In the eastern part of this platform, there is a wide trench with a base of tertiary materials fill up by quaternary fluvial sediments carried by the Xúquer and Túria rivers (Fig. 21.1).

F. Segura-Beltran (✉)
Department of Geography, Universitat de València,
Avda. Blasco Ibáñez, 28, 46010 Valencia, Spain
e-mail: francisca.segura@uv.es

J. E. Pardo-Pascual
Geo-Environmental Cartography and Remote Sensing Group,
Department of Cartographic Engineering, Geodesy and Photogrammetry,
Universitat Politècnica de València. Camí de Vera, 46022 València, Spain
e-mail: jepardo@cgf.upv.es

J. A. Morales (ed.), *The Spanish Coastal Systems*,
https://doi.org/10.1007/978-3-319-93169-2_21

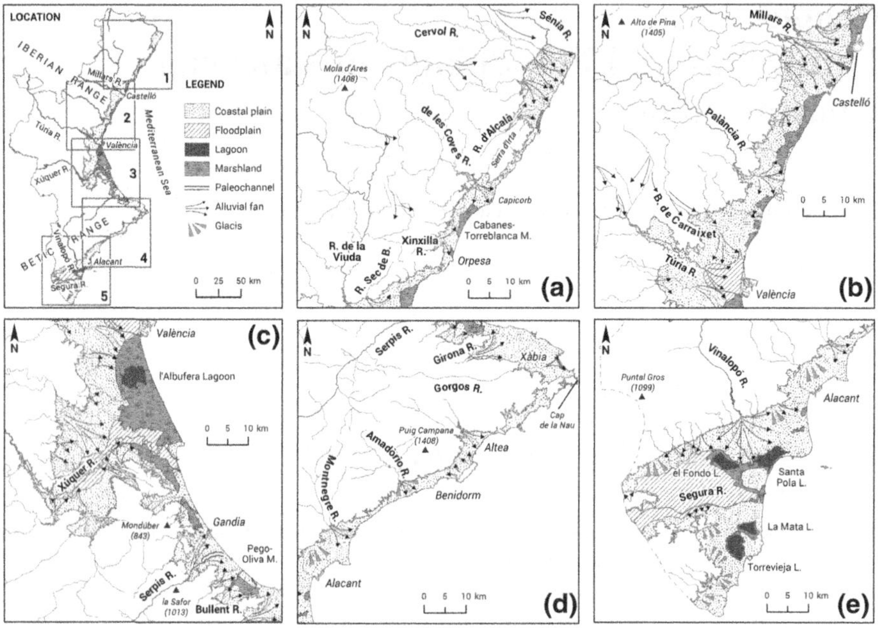

Fig. 21.1 Valencian coastal plains: the study area and the most important fluvial forms

The southern area belongs to the Baetic System and was folded during the first half of the Miocene. The northern part belongs to the External Prebaetic and consists of numerous folds running in an ENE-WSW direction. The reliefs are anticlines formed by Jurassic and Cretaceous limestone materials and the valleys are synclines formed by Miocene marls. Unlike the Iberian System, the reliefs reach the coast to form cliffs. To the SW is the Sub-Baetic area formed by small isolated Jurassic and Cretaceous reliefs. Further south, the Baetic area is made up of chains formed by Triassic dolomites. Finally, in the southern part of the Valencian region there is the Vega del Segura, the continuation of the Guadalentín trench which is home to a wide plain made up of Miocene deposits and Quaternary fluvial inputs carried by the Segura and Vinalopó rivers (Garay 1995).

Although there are small outcrops of siliceous sandstone and Triassic gypsum, carbonate materials (limestone, dolomite and marl) predominate in the Valencian mountains which explains their widespread and more or less accelerated karst dissolution processes. By contrast, coastal plains are formed by detritic materials of fluvial or marine origin. Thus, València region is formed by reliefs very close to the sea with a coastal plain of varying dimensions at its foot. In the northern part, this is only about 10 km but in the centre it extends to about 30 km and then disappears when the Baetic reliefs reach the sea where it is restricted to small plains between the capes. Finally, in the southern sector, the Lower Segura plain is a depression over 20 km long.

These plains are filled with Quaternary fan and floodplain deposits alternating with coastal lagoon/marsh deposits (Sanjaume 1985). These deposits are sometimes affected by neotectonic phenomena. In the central sector, between Sagunt and Dènia, the marine platform also has steps formed by faults running parallel to the coast. The major subsidence in this sector—over 200 m of sedimentary thickness—is related to neotectonics, and the same is also to be found on the Alacant coast where it has been responsible for the uplifting of some fossil beaches and the subsidence of some blocks. Furthermore, in this southern sector, from the Pliocene, NE flexures are produced, crossed with others in a N120°E direction, which forms a series of undulations with bulges (Cap de Santa Pola and Serra del Molar) and subsidence zones occupied by the lagoons of the Mata, Torrevieja, Fondó d'Elx and Santa Pola salt lake (Fig. 21.1d).

Finally, subsidence also affects the Vega del Segura. This structure would have been occupied by the sea during the Late Pliocene, but the uplifting of the reliefs mentioned above caused a regression that moved the sea to a position close to its present one (Goy et al. 1990) (Fig. 21.1d).

21.2 Characteristics of Valencian Rivers

From the hydrological point of view, there are two types of Valencian Rivers: perennial and ephemeral. Perennial streams channels fed by groundwater during dry periods are usually large (hundreds of km in length and thousands of km^2 in basin) and flow through non-Mediterranean climatic zones. They usually have gentle slopes and meandering courses, carry fine sediment and create wide floodplains.

Perennial rivers have river regimes that take on the characteristics of the climates they pass through. At their headwaters, they often have a fluvio-nival regime, which evolves to Mediterranean in their middle and lower basins. They cause flooding with abundant discharge, and if brought about by heavy Mediterranean-type rain (convective or cold low pressure) they have sharply peaked hydrographs with very high maximum discharges. The most important from N to S in the study area are the Túria, Xúquer and Segura rivers.

The ephemeral rivers (locally called ramblas, rius secs or barrancs) are similar to wadis and only flow after heavy and intense Mediterranean rains. They are usually short (less than 100 km in length) with small basins measuring only a few hundred km^2 developed under a Mediterranean climate. In some cases, these ephemeral streams may have headwater flow when fed by local aquifers (when they are called rius), but this flow disappears downstream due to transmission losses. The reliefs' arrangement so close to the coast leads to courses with steep slopes, carrying coarse sediment (gravel, pebbles or blocks). The high sedimentary load generates rivers with a braided or wandering path. When the channels reach the coastal plains, the slope changes abruptly and the channels lose confinement (Segura 1990).

In addition, when passing from the calcareous rocks, dominant of the nearby reliefs, to the detritic materials of the plains, they lose part of their flow due to transmission losses. The conjunction of these factors results in a reduction in their velocity and the consequent deposition of materials forming alluvial fans, some of which reach the coast as fan deltas. When the basins are smaller, they form diminutive torrential cones at the mountain fronts.

As a consequence of heavy Mediterranean rains, the ephemeral rivers generate flash floods with sharply peaked hydrographs, very high maximum discharges, vertical rising limbs and lag time of less than one hour in many cases (Segura 1990; Camarasa and Segura 2001). Despite their small size, they cover most of the territory studied. The most important from N to S are Riu de la Sénia, Cervol, Rambla de Cervera, Riu de les Coves, Riu Millars, Riu Palància, Carraixet, Riu Serpis, Gorgos, Riu d'Algar, Amadorio, Vinalopó and Montnegre (Fig. 21.1).

21.3 The Coastal Plains: Interaction Between Fluvial Deposits and Lagoon/Marshes

The thickness of the Plio-Quaternary deposits of the coastal plains depends on several factors, including the tectonic structure of the trenches, subsidence, neotectonics, fluvial sedimentary inputs and eustatic changes. The combination of all these factors has created complex coastal plains where fluvial, marine and transitional processes coexist. The fluvial sediments are deposited by rivers and ephemeral streams which form wide floodplains and fan deltas. Resting over the distal parts of these forms, beach barriers have emerged closing former Holocene marine gulfs which have turned into lagoons. Over time, these spaces have been silted up and turned into marshes. Thus Valèncian's coastal plain is an alternating sequence of fans, fan deltas and floodplains with coastal lagoons and/or marshes (Sanjaume 1985). The current lagoons/marshes were formed in the last Flandrian transgression when the sea level rose above its present level. During the maximum Flandrian transgression large marine gulfs were formed which were closed off by beach barriers in the following regressive period. The fluvial network was adapted to this base level. As a result, current rivers with smaller basins continue to flow into the inner side of the marshes, while rivers with larger basins reach the sea. Figure 21.1 shows the sequence of these fluvial forms and the lagoons that mark the coastal plains. Tectonics and neotectonics have fragmented these plains and the continental shelf into blocks which had differentiated behaviours during the Quaternary period. Stability, uplift and subsidence have determined the distribution and sedimentary thickness of the fluvial deposits and lagoons. Broadly speaking, in the northern sector stability or slight subsidence predominated. By contrast, the Valencian continental shelf between Sagunt and Cap de la Nau sank significantly in the Pleistocene and the subsidence rate declined progressively during the Holocene. Garay (1995) refers marked subsidence near the coast where Quaternary sediment thickness over 200 m has been detected. There was an uplift of blocks in the south-west of the province of Alacant as far as the Mar Menor, suggesting a different dynamic (Rey et al. 1999), although some blocks in the Segura basin indicate major subsidence. The coastal environments are also subject to continuous displacement on the continental shelf. Studies carried out by some researchers show that the river inputs form prograding bodies on the shelf, albeit with differential deposition. In general, the shelf is very wide in the Gulf of València and has very thick deposits, whereas from Cap de Sant Antoni, the shelf and sedimentary thickness decrease due to the Baetic sector's morphostructure. Fluvial deposits corresponding to current and relict pro-deltas have been identified right across the shelf (Rey et al. 1999).

21.4 River Deposits on the Coastal Plain

The arrangement and characteristics of the different sedimentary forms on the coastal plains vary significantly depending on the type of river, basin dimensions, tectonics, neotectonics, base-level changes, etc. (Bardají et al. 1990; Silva et al. 2003; Viseras et al. 2003). These variables result in two types of deposits on the Valencian coastal plains: alluvial fans/fan deltas and floodplains. By contrast, in the raised southern areas, the rivers cut through ancient Plio-Quaternary glacis.

21.4.1 Origin and Types of Alluvial Fans and Fans Deltas

As noted above, the characteristics of the Valencian coastal plains lead to the formation of alluvial fans and fan deltas. These forms have been described in the local literature as alluvial fans (Mateu 1982; Segura 1990; Viseras et al. 2003) although some of them discharge into the sea or the lagoons and therefore can also be considered as fan deltas. The terminological discussion conducted in the literature also appears in this case (McPherson et al. 1987; Nemec and Steel, 1988; Dijk et al. 2012). Fan deltas represent interaction between heavily sediment-laden alluvial-fan systems and marine or lacustrine processes (Nemec and Steel 1988). It is difficult to use the term fan delta in the strict sense in these coastal plains. The particle size is large and fluvial processes are dominant in the study area due to its low wave energy and negligible tidal range. Furthermore, prograding of various levels of fans throughout the Quaternary means that only Holocene build-ups have a marine influence today. Consequently, the terms alluvial fan and fan delta are used interchangeably in this paper.

These fluvial forms can be classified in different ways:

(a) By their **geometry and depositional facies**. According to Nemec and Steel (1988), there are a range of forms in the Valencian alluvial plains (Table 21.1):

Tab. 21.1 Physical characteristics of perennial rivers and ephemeral streams of Valencian coastal plains

	Drainage basin area (km^2)	Length (km)	Runoff (m^3/s)	River planform	Depositional form	Fan delta style/ floodplain style
Riu de la Sénia	196.7	49	1.15	Braided	Fan-delta complex	Prograding
Riu Cervol	343.1	59	Ephemeral stream	Braided	Fan-delta complex	Prograding
Rambla de Cervera	358.5	59	Ephemeral stream	Braided	Fan-delta complex	Prograding
Riu de les Coves	505.1	20	Ephemeral stream	Braided	Fan-delta	Prograding
Riu Millars	4028.2	156	9.06	Braided	Fan-delta	Prograding
Riu Palància	911.2	85	2.2	Braided	Fan-delta	Aggrading
Riu Túria	6393.6	280	15	Braided	Braidplain delta	Prograding
Riu Xúquer	21578.5	497.5	49.22	Meandering	Flood plain	Convex/ aggrading
Riu Serpis o d'Alcoi	752.8	74.5	3.3	Braided	Fan-delta	Aggrading
Riu Girona	117.7	38.6	Semiperennial/ ephemeral stream	Wandering	Fan-delta	Prograding
Riu Gorgos	283.2	39	semiperennial/ ephemeral	Braided	Fan-delta	Aggrading
Riu d'Algar	216.2	12.2	0.9	Braided	Fan-delta	Aggrading
Riu Amadorio	205.2	28.4	0.1	Braided	Glacis incised/fan delta	Erosion
Riu Montnegre	486.6	39.8	0.27	Braided	Fan-delta	Prograding
Riu Vinalopó	1691.7	81.2	0.37	Braided	Alluvial fan	Aggrading
Riu Segura	14,925	325	5.16	Meandering	Flood plain	Convex/ aggrading

1. Alluvial fans are semi-conical or triangular-shaped (in plan view) accumulations at the mouths of valleys, associated with active tectonic scarps. They develop when a stream emerges from a confined region onto a plain. Examples would be the alluvial fan of the Magre River, tributary of the Xuquer River, near of its mouth (Fig. 21.1c).

2. A fan delta is a coastal prism of sediments formed by an alluvial-fan system and deposited, mainly or entirely subaqueously, at the interface between the active fan and a standing body of water (marine or lacustrine). Most of the examples analysed in this paper are of this type. A fan delta is the actual delta of an alluvial fan, while a delta is formed by a river. There are many transitional forms between these forms.
3. A fan-delta complex consists of coalescent or vertically stacked fan-delta lobes created by an alluvial fan or fans (Nemec and Steel 1988).
4. A braidplain delta is the delta of a braidplain that shows no direct upstream transition into 'proximal' alluvial-fan deposits. This could be the case of the Túria River floodplain (Fig. 21.1b).

(b) **By fan style**

Throughout the Quaternary, there were different stages of sedimentation and incision, which have led to terraces and alluvial fans/fan deltas. Terraces in the Valencian coastal plains can be correlated with different levels of fans/fan deltas at the mountain front. Harvey (2012) suggests that regional aggradation of alluvial fans and river terraces appears to have coincided with Pleistocene glacial phases and periods of dissection with interglacials. By contrast, the Holocene coincides with an aggrading phase that is associated with anthropic action.

There are two types of alluvial fans and terraces based on their arrangement: prograding and overlapping or aggrading (Fig. 21.2). The first is a type of fan-head trenching and distal aggradation and has been called in different ways by other authors (telescoping fans, Nemec and Steel 1988; prograding fans, Harvey 2012). Prograding can be caused by several factors s: (a) a high sedimentary load (Harvey 2012); (b) lowering of the sea level; or (c) little subsidence. Aggrading fan (Nemec and Steel 1988; Harvey 2012) presents different levels of overlapped Quaternary fans. The cause is usually faulting or downwarping persistence and subsidence is fairly uniform, so the fan is very thick at its apex (Nemec et al. 1988).

(c) **By their spatial location**

The Valencian coastal plains' fan deltas may drain directly into the sea or the lagoons. The former are bigger and generated by streams with larger basins. The latter border the lagoons on their inner side since this was the base level of these rivers during the Flandrian transgression. As they have small basins, the low sediment supply leads to small progradation over the wetland, although they are not able to completely fill the marshes.

From the point of view of sedimentation, fan deltas are formed by fluvial processes, and some cases of levels of erosive coast (highstands) and intercalated beaches can only be found in their distal part, especially on the Alacant coastline (Bardají et al. 1990).

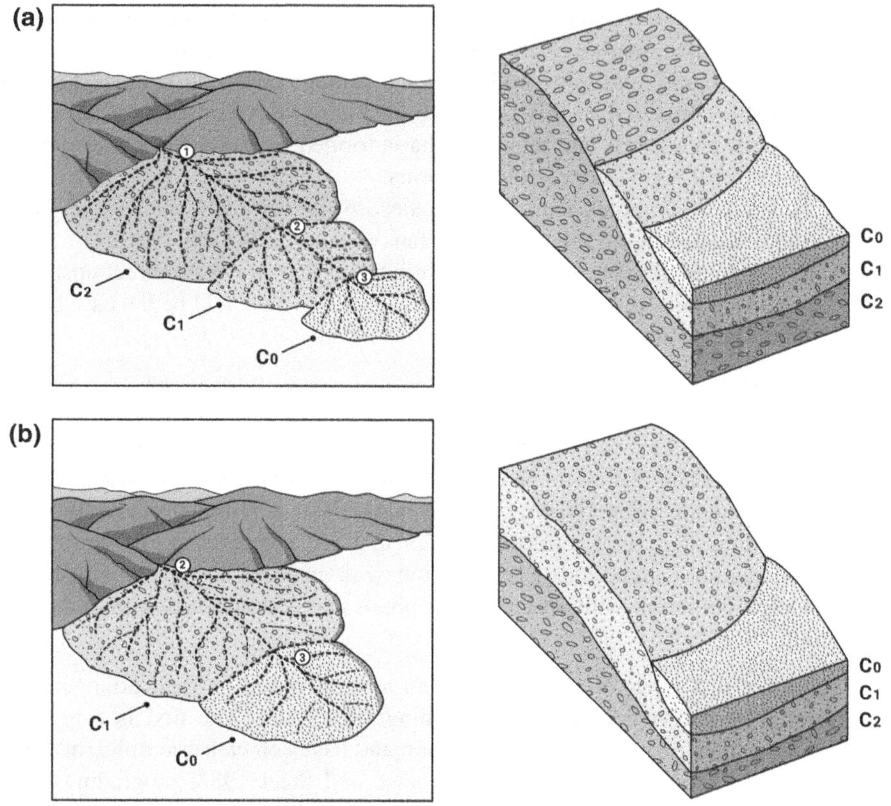

Fig. 21.2 Alluvian fan delta models. C_2: Early and Middle Pleistocene fan; C_1: Late Pleistocene fan; C_0: Holocene fan. **a** Prograding fan model. On the surface, there is the encrusted fan of the Early-Middle Pleistocene and in a prograding succession, the Late Pleistocene and the Holocene (both not encrusted). **b** Aggrading fan delta model. The Late Pleistocene fan (not crusted) appears on the surface and the Holocene fan on the distal part. If the river is incised, the crusted fan of the Eaely Pleistocene appears on the riverbed

21.4.2 Floodplains

The largest perennial rivers are meandering and develop major convex floodplains with a channel that is raised and bordered by levees. Vertical accretion and fine materials (silts and clays) predominate. Their forms include dykes or levees, lateral flood basins, yazoos, point bars, abandoned meanders, and crevasse splays. They are low-energy plains, whose specific stream power at bank full stage is generally lower than 10 Wm^{-2} (Charlton 2008). This group includes the Xúquer and Segura floodplains (Figs. 21.1, 21.7 and 21.8).

21.4.3 Glacis

To the south of Cap de la Nau, coinciding with a sector where precipitation decreases and aridity increases, the drainage network blurred and there are some sectors where it has been disorganised by neotectonics. In these areas, there are large glacis with a gentle slope of between 3° and 8°. They are often stepped and dissected by fluvial courses and have several levels of terraces. They frequently have crusted surfaces on which there are sometimes flat-bed ravines. Their origin is complex and the lateral shift from glacis to fluvial terraces is frequent (Fig. 21.1d, e). The age of these forms has been widely discussed. The older ones are considered to be from the Late Pliocene or Early Pleistocene, although several subsequent levels occur. As for their origin, Cuenca and Walker (1985) associate the Plio-Pleistocene glacis (2.6–2.8 mya) with the existence of lacustrine zones fed by currents that descended from the slopes in a warm climate with strong contrasts of precipitation. In addition, there are several levels dating between 39,000 (Late Pleistocene) and the Holocene with different climatic oscillations (Rosselló 1978).

21.5 Geographical Distribution of the Main Coastal Alluvial Forms

There are a number of sectors on the Valencian coastal plains with similar fluvial forms. From north to south, they are as follows:

21.5.1 Sector Ebro Delta—Serra D'Irta

This is a sector formed by horst and graben crossed by ephemeral streams which build large alluvial fans. There are two subareas:

(a) Vinaròs-Benicarló-Peníscola plain (Fig. 21.3). This is a tectonic trench running between the Ebro Delta and the Peníscola tombolo. It is filled by a fan delta complex formed by the Sénia, Cervol, Rambla de Cervera and Rambla d'Alcalà rivers. The first three form a thick alluvial bajada formed by a prograding sequence of three levels of alluvial fans. The oldest, from the Early-Middle Pleistocene, have been deposited at the front of the reliefs and they are formed by a conglomerate of cobbles, pebbles and gravel with a clay-silt matrix and covered with a calcareous crust. The Late Pleistocene fans have finer materials (gravel, sand, silt and clay) not crusted. Finally, near the coastline, the Holocene fans are made up of fine materials (silt, sand and clay). They have been eroded by the sea in their distal part leaving a coastline of medium-height cliffs. This erosion is directly related to the lack of coastal nourishment

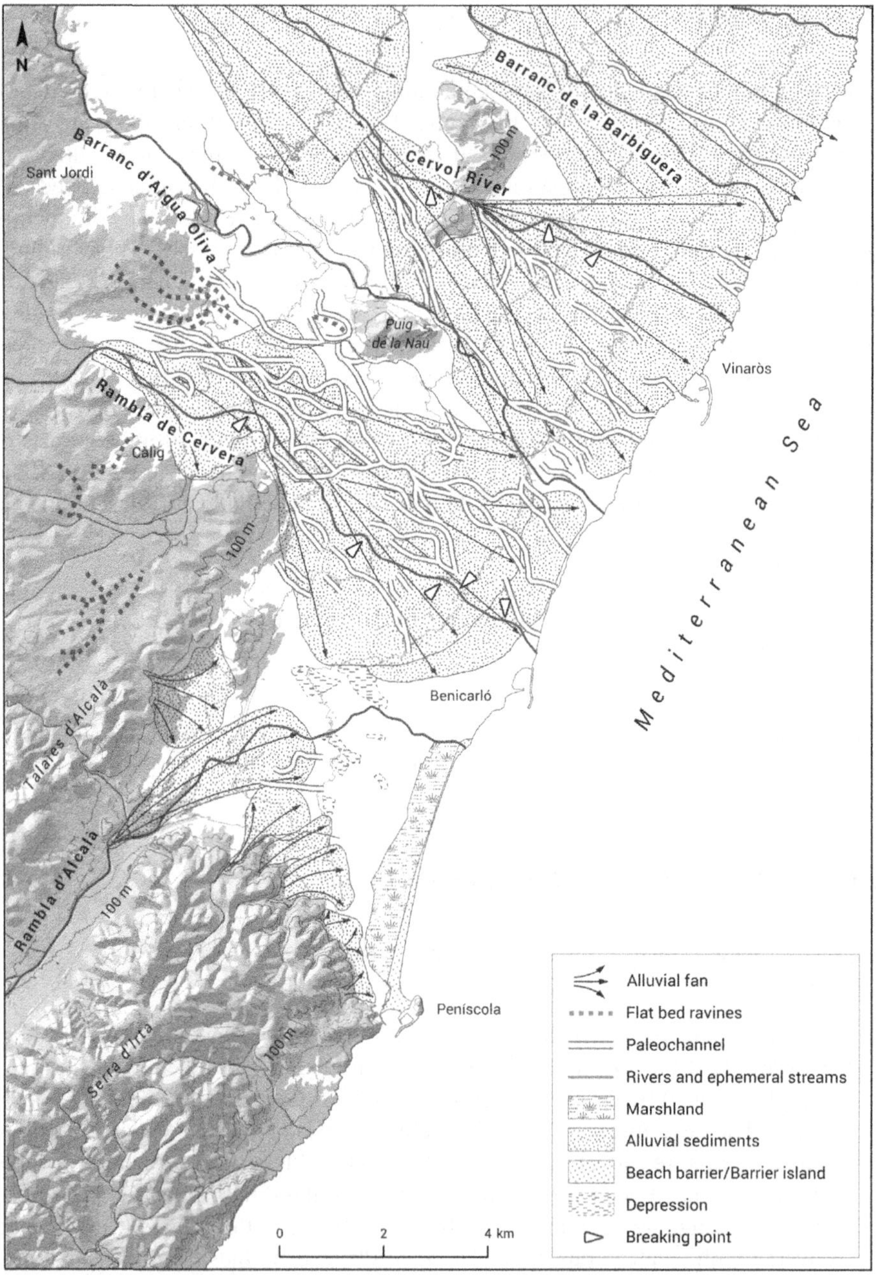

Fig. 21.3 Vinaròs-Benicarló coastal plain. A fan delta complex formed by the Riu de la Sénia, Riu Cervol, Rambla de Cervera and Rambla d'Alcalà agradding fan deltas. In the southern part, the Peníscola lagoon is formed. Barranc d'Aigua Oliva and Barranc de la Barbiguera are interfan rivers

generated by the growth of the Ebro Delta, which has retained the sediments supplied by this river (Sanjaume et al. 1996; Pardo Pascual and Sanjaume Saumell 2001). The southern part of the Vinaròs-Benicarló plain is formed by an Early-Middle Pleistocene fan, deposited by the Rambla d' Alcalà. This form lies perpendicular to the previous fans given the NNE-SSW route of this ephemeral stream (Fig. 21.3). When it meets the Rambla de Cervera fans, the Rambla d'Alcalà turns through a 90° angle, which means that the Late Pleistocene deposits run parallel to those of the northern ephemeral streams. Holocene materials are deposited near the mouth. On them rests the beach barrier that closes the small Peníscola lagoon, now silted up. There is a semi-endorheic space between the Rambla d'Alcalà and Rambla de Cervera fans (Segura 1990; 1996).

(b) The Serra d'Irta is a horst, which reaches the coast where it forms a cliff. At the foot of this cliff, there are small fan deltas whose distal part has been eroded by the sea to create a low cliff on a coastline broken up by small *calas* (Fig. 21.1a).

21.5.2 Sector Orpesa-Torreblanca Plain

This is a trench formed during the Miocene distension and compartmentalised in several blocks. The thickness of the Plio-Quaternary sediments ranges between 20 and 150 m. A number of studies see greater subsidence in the southern sector than in the northern one (Fig. 21.1a) (Segura et al. 1997, 2005).

In the northern part, the Riu de les Coves generates a prograding fan delta whose distal end forms the coastal protuberance of Capicorb (Fig. 21.4). It has a crusted surface formed by pebbles and blocks that is attributed on a regional basis to the Early-Middle Pleistocene. The Late Pleistocene level, formed by loose materials with an abundant matrix, appears in the form of terraces, suggesting that the fan delta was deposited offshore and has been destroyed. The Holocene sediments are located south of the current mouth, forming a small protuberance that could be associated with some perhaps very recent diffluence. It is thus a prograding fan delta (Segura 1990) which was larger in the past and has since been eroded by the sea, probably due to the lack of sediment input found throughout this northern Valencian sector because of the Ebro Delta's growth.

The complex alluvial fan formed by the Riu Xinxilla-Barranc de la Font del Campello-Barranc dels Llorenços (Fig. 21.1a) in the southern part of the plain is smaller than the previous one. It supports the Torreblanca and the Orpesa lagoon beach barriers to the north and south respectively. It is an aggrading fan delta whose growth has divided a continuous lagoon developed between Torreblanca and Orpesa into two parts. Finally, the inner side of the Torreblanca lagoon is framed by small prograding fan deltas formed by streams that descend from the nearby reliefs (Fig. 21.1a) (Segura et al. 1997, 2005).

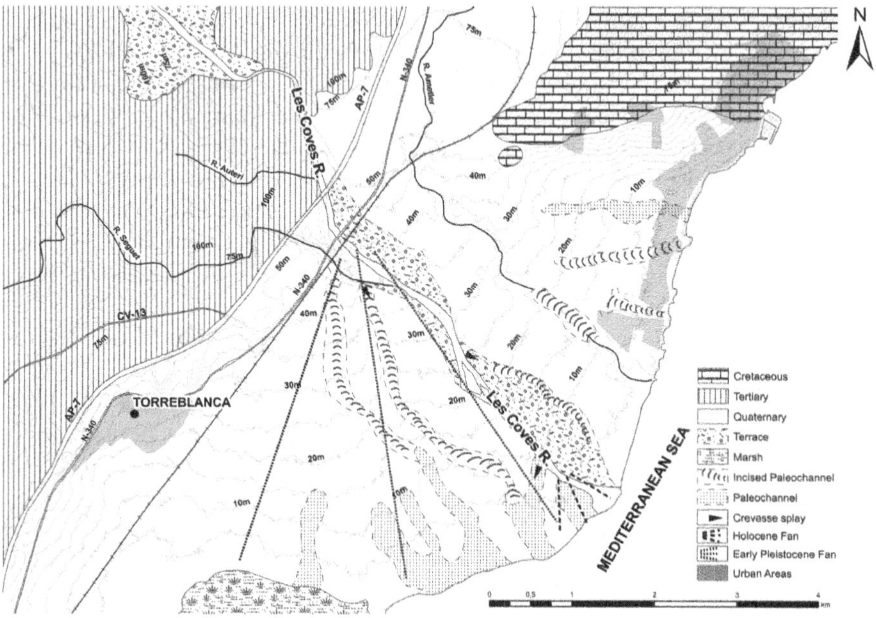

Fig. 21.4 Les Coves River fan delta. The prograding fan forms the Capicorb protuberance in the coastline. Several paleochannels show the channel mobility throughout the Quaternary period

21.5.3 Sector Benicàssim—Castelló

The coastal plain has a morphogenetic scheme similar to the rest of the Gulf of València: at the foot of the reliefs, a few kilometres from the coast, an alluvial bajada starts in Benicàssim and ends with Riu Millars-Rambla de la Viuda fan-delta (Fig. 21.5). A long, narrow beach barrier grew between the two protuberances that closed off a lagoon, now turned into a marsh. It is a fan-delta complex formed by the coalescence of the fans growing at the foot of the Massís del Desert de les Palmes. It is wider in Benicàssim, where several fan deltas flow into the sea filling the marsh. It narrows considerably on the inner side of the marsh, where the smaller ravines end and the wetland widens. In the southern sector, the Riu Sec de Borriol flows into the coastal plain after draining a small interior trench. The convex topography of the alluvial fan of the Riu Millars-Rambla de la Viuda forces the Riu Sec de Borriol to make a 90° turn and build its own fan by encroaching on the Prat del Quadro marsh. Although the fan progrades in the wetland, the natural channel disappears at this point. Construction of an artificial channel in the second half of the 20th century enabled this river to flow into the sea (Segura 2001).

The fan-delta complex formed by the Riu Millars-Rambla de la Viuda is prograding and in its advance has managed to fill the marsh, which at this point is interrupted. With a very complex structure, which has been caused by the mobility of both channels throughout the Quaternary, its deposits contain several levels of

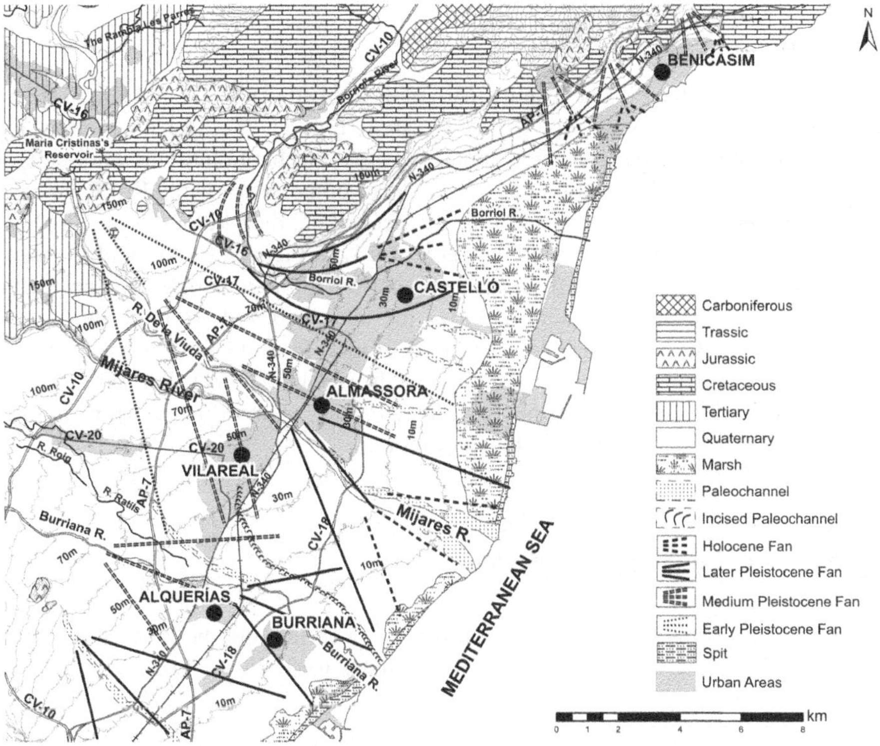

Fig. 21.5 The prograding fan delta of the Millars River and Rambla de la Viuda coalesces with the others created by the Riu Sec de Borriol and Riu Sec de Borriana, forming a protuberance that limits and interrupts a continuous space of marshes

fans (dated on a regional basis) (Fig. 21.5). The oldest is represented by fragments of Tertiary deposits, which may have been the result of the first avulsions of the Millars river. In the Early-Middle Pleistocene, two fans were formed represented by a powerful conglomerate with a calcareous crust on top on which sheet floods can be seen. At this time the Millars and the Rambla de la Viuda did not end together and some authors suggest the latter drained into an interior trench (Mateu 1982). Its mouth in the coastal plain would be later than this period. Downstream of the confluence of the two rivers, there is a new intersection point where a new Late Pleistocene fan starts. It is formed by loose materials and with a smaller size than the oldest level. Also downstream (Almassora town) there is a Holocene fan delta formed by loose materials with a silt-clayey matrix which extends towards the coast. It should be noted that there are several paleochannels, hanging over the current channel, which drain fan runoff during torrential rains and cause flooding, independently of the main channel. The Millars-Rambla de la Viuda fan coalesces in the south with another fan formed by the Riu Sec de Borriana and its numerous tributaries that have been interpreted as ancient paleochannels of the Millars River (Segura 2006) (Fig. 21.5).

21.5.4 Sector Castelló-Sagunt

A beach barrier extended between the Millars-Rambla de la Viuda and Riu Palància fans closing off an uninterrupted marshy area (Fig. 21.1b). The alluvial bajada becomes smaller and there are only small fans that discharge into the inner side of the marsh. The most important depositional form in the sector is the Riu Palància's aggrading fan delta which has an almost perfect conical morphology (Fig. 21.6). At the front of the mountain, there is the apex of a Late Pleistocene fan with deposits similar to those of other fans from the same period. Numerous paleochannels have been identified on this fan hanging above the main bed with different degrees of incision. During flooding periods, they collect runoff from the fan and take it to the coast or in some cases to the surrounding marshes (Segura 1991). This fan overlaps another crusted fan formed in the Early Pleistocene, exposed by erosion into the bed. A Holocene fan starts near the river mouth.

21.5.5 Sector Sagunt-València

A beach barrier has formed, resting on the Palància fan and the Túria floodplain, which closes off a narrow uninterrupted marshy area (Fig. 21.1b). The alluvial

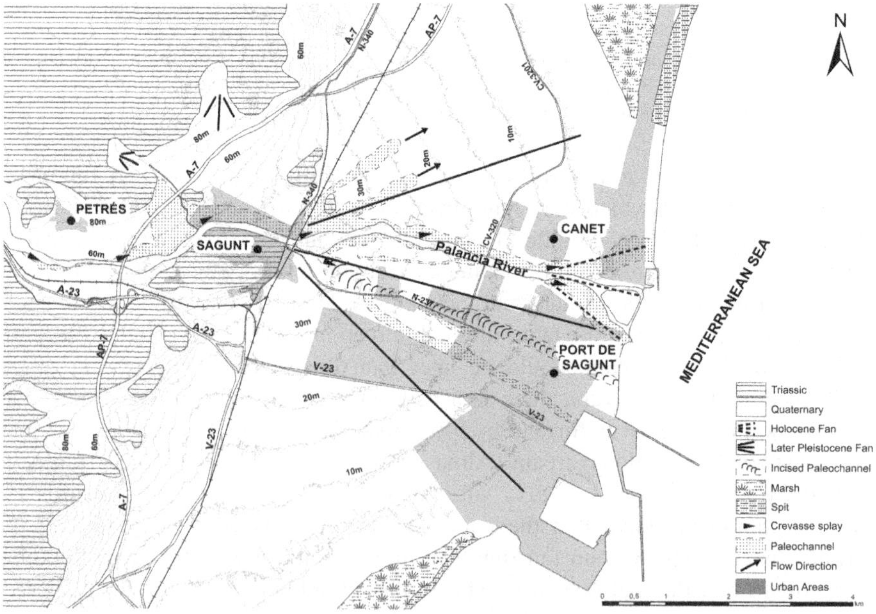

Fig. 21.6 The aggrading fan of the Palancia River interrupts a marshy area. It is crossed by numerous paleochannels that drain surface water to the sea and lateral marshes during periods of flooding

bajada that borders the inner side of this area is formed by small fan deltas that begin in the adjoining reliefs. The most important forms include the Carraixet's fan delta which, when it reached the plain, formed a fan delta dated to the Late Pleistocene. The Carraixet's inputs narrowed the marsh (Fig. 21.7). The interaction of natural and anthropic processes has been exceptionally active in this sector. In fact, the Carraixet's final course is probably of human origin to avoid the repeated flooding which this horticultural sector to the north of the city of València would otherwise experience.

21.5.6 Sector València-Cullera

This sector is made up of the inputs from the two great Valencian rivers: the Túria and the Xúquer. The first forms a braidplain delta where a Pleistocene alluvial fan with a 0.45% slope can be clearly identified, formed by red silt and clays with calcareous nodules and calcretes. About 6 km from the coast, Holocene deposits bury the middle and distal sector of the Pleistocene fan, forming an alluvial plain with a 0.2% slope (Fig. 21.7). Residual marshes are located on its sides and on the distal part of this floodplain is the beach barrier that closes the València lagoon, which extends to Cap de Cullera.

The city of València was founded on a terrace on the right bank in 138 BC in Roman times. At this time, the coastline was deep inland (Carmona 1990; Carmona and Ruiz 2011). On the plain, in addition to the main channel, several paleochannels have been identified which come into operation during floods. There are two on the right bank: (a) the Mercat paleochannel, which borders the Arab wall, and (b) the Rambla de Predicadors, which skirts an ancient river island. On the left bank, there are two more: the Camí del Grau and la Rambla (Carmona and Ruiz 2011). There are two more paleochannels on the right bank in the river mouth area which take water flows to the València lagoon and two more that carry them towards the coast marshes (Carmona and Ruiz 2014; Portugués-Mollá et al. 2016). There are also several lacustrine areas in this floodplain. The València lagoon, which runs from the Túria's mouth to Cap de Cullera, is still functional today but there are also others which have been silted up by the progradation of the Túria (Carmona and Ruiz 2011).

Based on the sedimentological, stratigraphic and archaeological data, Carmona and Ruiz (2011) has determined the Holocene evolution of the sector. The Flandrian transgression penetrated towards the interior of the continent and the pre-Holocene valleys formed shallow and possibly branched inlets in a shallow coastal lagoon environment. During the late Holocene, the progradation of the coastal sandy ridges and the aggradation of the floodplain took place, caused by an increase of the sediment supplied. The floodplain evolution included several growth periods with different sedimentary inputs: (a) High magnitude flood events (2800 BP to the 3rd century BC); (b) Slow riverbed aggradation and progressive levee formation in Roman times (from the 2nd century BC to the 5th century AD);

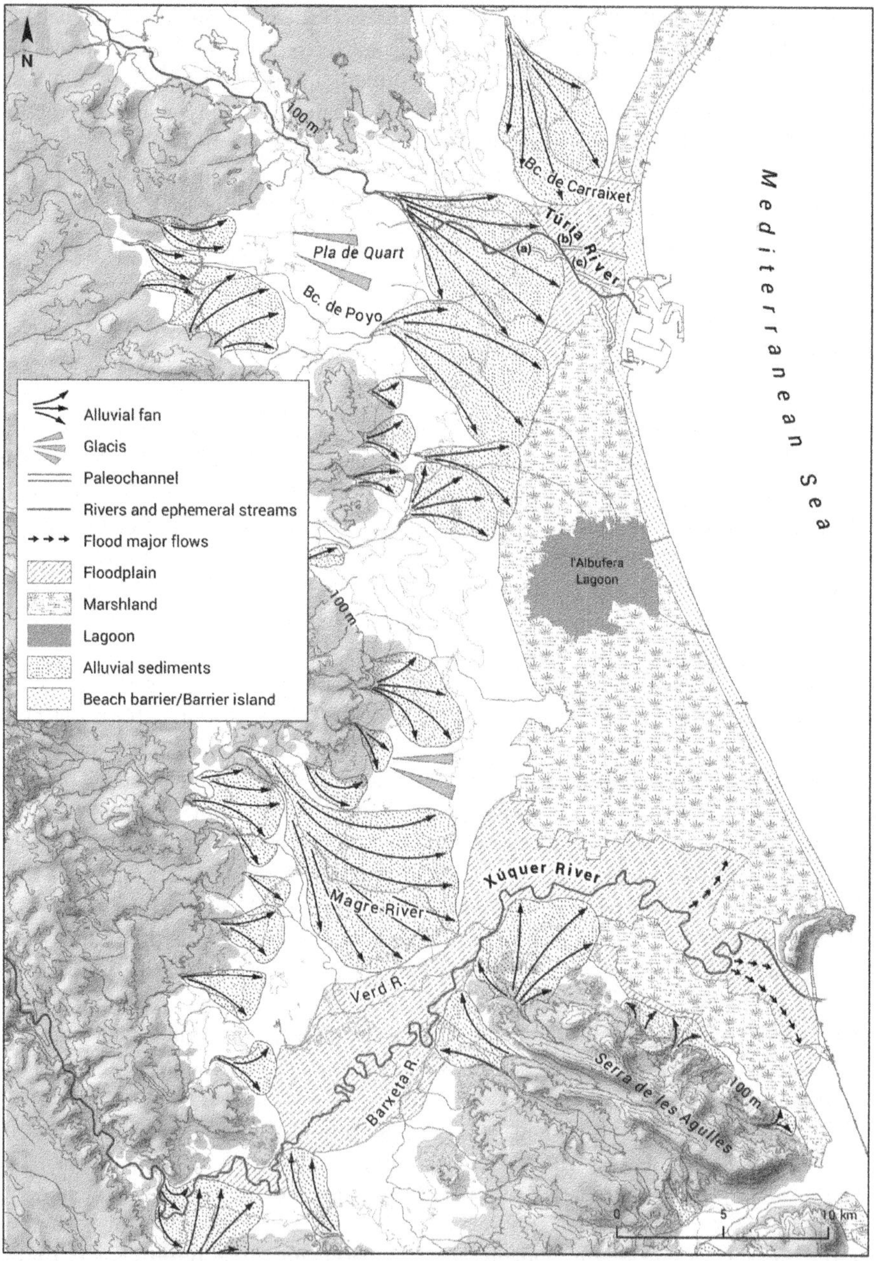

Fig. 21.7 The central floodplains formed by the Turia and Xúquer Rivers, are separated by the Albufera lagoon and marshy area. During flood events, both rivers bring flow to the sea and to the adjacent lagoons (the València lagoon and others located the north and south of both rivers). **a** Mercat and Predicadors paleochannels, **b** Braç de Roca paleochannel, **c** Braç del Grau paleochannel. With information from Ruiz, Carmona and Mateu (2006) and Carmona and Ruiz (2011)

(c) Swampy environmental processes (6–9th centuries); and (d) Second phase of high-magnitude floods (11th century). River floods in the Roman and Islamic periods are interpreted as a consequence of anthropic activity (deforestation) associated with the expansion of crops. The formation of swamps in Visigothic times (6–9th centuries) is believed to be due to the opposite effect, i.e. to the recovery of vegetation resulting from a decline in agriculture (Carmona 2016).

The current arrangement of the main channel and the paleochannels generates a particular dynamic during extraordinary flooding: part of the flow drains into the sea, but another part flows into the lateral marshes, especially towards the València lagoon, where the river built an interior delta during the Holocene (Rosselló 1972; Carmona and Ruiz 2014).

21.5.7 Sector Cullera-Pego

The most important river form in the sector is the Xúquer floodplain. It has a slope of 0.1%, with a concave or convex morphology depending on the sectors. The river is meandering, presenting some historical mobility (Mateu 2000; Ruiz, 2002) and is bordered by dikes or levees, raised between 3 and 6 m above the lateral flood basins, where fluvial marshes have grown. Near its mouth, these floodplain marshes join to the coastal ones to form a large wetland that floods during high magnitude flood events (Fig. 21.7). It is a Holocene floodplain, embedded between Pleistocene terraces and alluvial fans that constrain the plain in some sectors (Magre and minor alluvial fans). In the lateral flood basins, overflows are organised in the two yazoos, the Riu Verd (left bank) and the Riu Barxeta (right bank) (Mateu 2000).

An important feature of the river is that it has overflow branches that are activated during extraordinary floods (Fig. 21.7). Part of the water flows through the current channel, but another part flows through the Barranc del Duc, a branch that collects the overflow and steers it towards the sea and the marshes on the right bank. On the left, the overflowing water is guided towards the Albufera marshes by another secondary branch (La Roca and Carmona 1983, Carmona and Ruiz 2014). The construction of the deltaic plain has been studied by Ruiz (2002), Ruiz and Carmona (2005), Ruiz et al. (2006). During the Flandrian transgression, the coastline was several hundred metres further back from where it is today and a beach barrier isolated a wide lagoon area that extended between the Túria and the Xúquer. An inner delta which developed between 6000 and 3000 BP drained into this lagoon, as happened in the north with the Túria. Between 3000 BP and the Middle Ages the lower basin's meander train and floodplain were built over the previous deltaic environments and there was significant Islamic settlement in these areas at the end of the Middle Ages. As a consequence of this fluvial activity, there has been strong vertical and horizontal accretion in the floodplain altering the topography. This has led to the depopulation of some historic urban centres located in lateral marshy areas (Mateu 2000). In adition, there is a narrow marsh to the south of the Xúquer's mouth running as far as Gandia (Fig. 21.1c), which experienced significant progradation throughout the Holocene, as evidenced by the existence of a twin beach barrier. However, on the inner side of this area there are hardly any significant river inputs (Sanjaume and Pardo-Pascual 2003). The Gandia plain is a triangular saging structure formed in the transition area between the Iberian system and the Baetic ranges. The trench was filled in the Quaternary with inputs from the Riu Serpis and other minor channels. The Riu Serpis discharges in the north of the plain but two more southern paleochannels can be observed in the fan which suggests that the river has migrated northwards. It is an aggrading fan delta since surface deposits typical of the Middle Pleistocene are not found there (Segura and Carmona1999). At 1.5 km from the coast, the Serpis fan has a step, delimited by the 5 m isohipse, which has been interpreted as a Flandrian beach (Viñals 1996).

21.5.8 Sector Pego-Dénia

In the Pego sector, tectonic forces formed a triangular trench with a Mesozoic substrate featuring several blocks that descend in steps towards the coast, flanked by raised blocks to the north, south and west (Ballesteros et al. 2009). There are several aggrading fan deltas at the front of the reliefs on the edges of the triangle and the Pego-Oliva marsh at its distal part (Fig. 21.1c). Throughout the Quaternary, the major subsidence in the sector and the numerous transgressions and regressions led to the formation of an aggrading fan delta created by the Riu Bullent. In the sedimentary complex, continental facies alternate with different sedimentary environments. Continental sedimentation is due to the growth of the alluvial fans between Marine Isotopic Stage (MIS) 11 and MIS 7 whereas between MIS 15 and MIS 12, transition environments predominated (brackish marsh-lagoon periods); subsequently a lagoon system took shape between MIS5 and MIS 1 which has lasted until today (Dupré et al. 1988; Viñals 1996; Torres et al. 2014).

To the south of Pego there is a new triangular trench filled by Tertiary and Quaternary sediments from the Riu Girona and other smaller streams that form a fan delta complex of prograding alluvial fans (Fig. 21.1d). The internal stratigraphy of these systems reveals an 'ageing' sequence. In the inner part of this depression is the apex of the Early Pleistocene fan formed by the Riu Girona. The Middle and Late Pleistocene fans are prograding to the sea. The latter has several lobes that suggest that the river migrated laterally during this period, forming a small Holocene fan delta (Segura 2009; Segura-Beltrán et al. 2016).

21.5.9 Alluvial Plains of the Baetic Mountain Ranges

When the Baetic ranges reach the sea, they form a seafront of cliffs. These mountainous alignments, consisting mainly of Mesozoic limestone and marls, are interrupted by an extensive network of orthogonal faults that compartmentalise the relief in small trenches filled with Plio-Quaternary sediments. There are several fluvial plains formed by the Gorgos, Algar, Amadorio and Montnegre rivers in this sector (Fig. 21.1d).

The Bay of Xàbia is a triangular trench between the spurs formed by Cap de Sant Antoni and Cap de la Nau. The Riu Gorgos flows into it, forming an alluvial fan in the Late Pleistocene embedded between three levels of terraces from the Middle Pleistocene. Two of these terraces are elevated by the neotectonic activity of the faults that compartmentalise the sector in blocks (Blázquez 1999). The thickness of fluvial sediments varies in the floodplain in line with the differing subsidence of the blocks, which varied throughout the Quaternary. Xàbia as a whole has much higher subsidence rates than the northernmost sectors of the València coastal plains, although the rate decreases from the Tyrrhenian stage onwards (Viñals and Fumanal 1995).

To the south of Cap de la Nau the coast forms a cliff broken up by numerous capes and bays resulting from recent fractures. Subsidence in this area during the Early Pleistocene led the coastal front to retreat. This disorganised the drainage network flowing into the sea by increasing its slope towards the E. As a result, starting in the Middle Pleistocene, the drainage network was hierarchised by headward erosion and became embedded in a Villafranchian glacis, with thick calcareous crusts and the few Quaternary deposits covering the valley (Blázquez 1999).

Further south, between Altea and Benidorm, there is the shift from the external Prebaetic system to the Subbaetic or internal Prebaetic system. A complex network of fractures forms a large number of individual blocks, and this, combined with the presence of Keuper deposits, greatly complicates the relief arrangement. The diapiric outcrops distort the deposits and neotectonic movements become prominent. The area's geomorphology is characterised by the development of glacis and alluvial fans, which start in the nearby reliefs tilted towards the sea. In their distal part, they connect with narrow and very short systems of terraces which the rivers run between.

The Altea coastal plain would be a good example of this geomorphological structure, although in this case it would be necessary to add the halokinetic-movements due to the plasticity of the Trias that have lifted up some parts of the plain. The Riu Algar's fluvial deposits barely form a narrow fan, which must have been larger during the Pleistocene period since they extend under the sea on the shelf (Fig. 21.1d) (Blázquez 1999). The Riu Amadorio reaches the coastal plain constricted by flysch deposits. It traces a wide alluvial fan with some diffluence. Further to the south, the Riu Montnegre and some small ravines form a prograding fan delta complex with Pleistocene deposits on the surface and a small protuberance at the mouth of the Montnegre formed by Holocene deposits (Fig. 21.1d).

21.5.10 Sector Segura-Vinalopó Plain

This consists of a subsidence structure delimited to the north by a strike-slip fault (Crevillent fault, the continuation of the Cadiz-Alicante fault system, CF) and to the south by the Segura fault (ENE-WSW) (BSF) where the river flows. In addition, the San Miguel and Torrevieja Faults cross the Lower Segura valley, reaching the Torrevieja and La Mata lagoons (Fig. 21.8). The complex tectonics of the sector have left a set of sagged and elevated blocks. The subsidence axes are: (a) Elx-N of Santa Pola, (b) Lower Segura and (c) Orihuela-San Miguel de Salinas, which hosts the Fondó d'Elx and the Santa Pola, La Mata and Torrevieja lagoons. Offsetting this subsidence are three raised areas: Serra dels Colmenars, Serra del Molar and Serra de Santa Pola. Some authors have interpreted these structures as anticlines and synclines (García and Martínez 2006) while others consider them horst and graben (Blázquez 2001).

The evolution of the sector is complex and has been interpreted in a number of ways, although all authors recognise the importance of neotectonics (Alfaro 1995; Soria et al. 2001), which have affected continental and littoral Pleistocene and Holocene sediments. The pre-Quaternary substrate is of marine origin from the Mio-Pliocene epoch and is formed by calcarenites, marls, silts and clays that emerge in the Crevillent, Colmenar, Molar and Santa Pola mountain ranges, although the latter is a reef platform. The neo-Quaternary fill is deposited on an irregular basement characterised by the existence of the graben and horsts mentioned above. From the Tortonian age to the present, tectonic movements have led to a number of uplift and subsidence stages with differential behaviour among the blocks (Soria et al. 2001).

The sector's current morphological configuration consists of the Segura alluvial plain, the Vinalopó fan and the Fondó d'Elx and Santa Pola lagoons, nowadays separated but which in other times during the Quaternary formed a single lake (Fig. 21.8). During the Late Pliocene the sea occupied this entire sector. The coastline was at the foot of the reliefs of the Sierra de Crevillent to the north, Macizo del Segura and Sierra de Carrascoy to the west and the reliefs of the Sierra de Cartagena to the south. During the Early Pleistocene, the sea began to retreat to the southeast as a result of the uplifting of the Serra de Crevillent, leaving a lagoon area inland which grew significantly and is separated from the sea by a system of beach barriers. At the same time, alluvial fan systems were generated at the front of the reliefs. In the transition from the Early to the Middle Pleistocene, the Elx Depression subsided and the Lower Segura basin formed as a result of the coming together of several strike-slip faults and the reactivation of others. These movements restructured the corridor creating a marine environment in the river's current mouth. During the Middle Pleistocene, subsidence continued in the Lower Segura

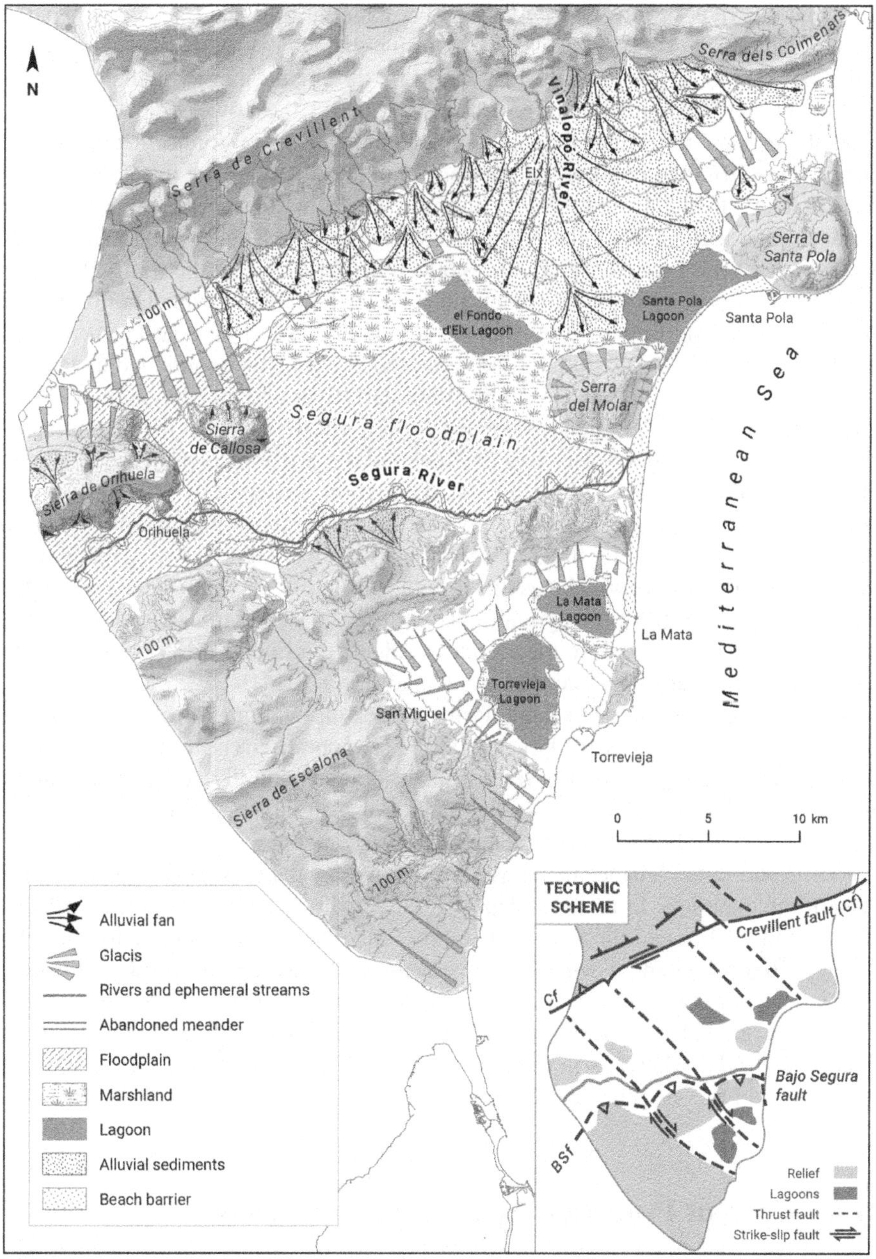

Fig. 21.8 The Segura floodplain and the Vinalopó aggrading fan is a complex space formed by sunken and elevated blocks created by neotectonics. The sunken areas are occupied by lagoons and salt lakes, separated by small relieves. A wide marine gulf formed during the tertiary period turned into a large lake (Sinus Ilicitanus) and was filled by alluvial deposits of the Segura River and the overlapping fan of the Vinalopo River. With information of Ferrer and Blázquez (1999), Blázquez (2001), Viseras et al. (2003)

basin and this river provides a large amount of material from the Baetic reliefs in the form of fan deltas (Goy et al. 1990). During the Holocene, between the years 8,500 BP to 8,200 BP (6,500 BC to 6,200 BC), the small estuaries at the mouths of the Segura and Vinalopó silted up. Around 4,200 BC, the Lower Segura plain was flooded, forming the Sinus Ilicitanus, a lagoon with a sea connection. It spread along 19 km from the coast to the interior, limited to the N by the distal deposits of the Vinalopó, to the E and NW by the alluvial formations descending from the Serra de Crevillent and to the S by the Moncayo and Rojales mountain ranges (Blázquez, 2001). From 4,000 BC to 3,000 BC subsidence in the Sinus Ilicitanus lengthened the lagoon to its maximum point, near the city of Orihuela (3,000 BC). In the last period starting in 3,000 BC, the lagoon dried up. Two deltas were formed within the Sinus Ilicitanus: to the southwest, the Segura River continued advancing from west to east, while to the north the Vinalopó River and the alluvial fans on the edge of the Serra de Crevillent prograded from north to south. At the beginning of our era (1 AD) the lagoon was very shallow and the Segura River provided a large amount of sediment so it started to cross the lagoon protected by natural dikes. In just a few hundred years, its mouth moved from the centre of the lagoon (1 AD) to Guardamar (500 AD). By the Late Middle Ages the lagoon was already very narrow according to historical chronicles, and starting in the 18th century when this area began to be settled for agricultural purposes it was completely drained; the remains of the Elx lagoon are the current wetlands the Fondó d'Elx-Crevillent and the salt lake of Santa Pola (Tent-Manclús and Soria 2014).

The Vinalopó, on the other hand, forms a thick alluvial fan with an approximate radius of 8 km, an area of 40 km^2 and an 8.5% slope (Ferrer and Blázquez 1999; Blázquez 2001) (Fig. 21.8). The river loses its course naturally and currently flows into the Assarb de Dalt that crosses the Santa Pola salt lake before draining into the Mediterranean. It forms an aggradational fan delta on whose surface there are Holocene and Late Pleistocene deposits which overlap the crusted Early and Middle Pleistocene materials. In its distal deposits, the Vinalopó fan separates the wetlands of the Fondó d'Elx and the Salinas de Santa Pola which were joined until the 19th century (Sinus Ilicitanus) when improvement works were carried out (Sanjaume 1985). The thickness of sedimentation of the Vinalopó fan, is indicative of this generalised subsidence. In addition, the disappearance of the Vinalopó river channel and other channels in this area as a result of the lack of slope in its final stretch, would be one more indicator of the tectonic subsidence affecting the zone.

21.6 Current Processes: Ephemeral Fan Deltas

Since fan deltas are generated by ephemeral rivers, the absence of discharge during a good part of the year generates special processes in the fluvial mouths (Fig. 21.9). The waves and coastal drift close the mouths by forming bars of pebbles and gravels and remain in place until a flood occurs which can break through them. This happens only rarely, as the rivers discharge is temporary and spatially discontinuous. When the event is high magnitude it breaks the bar, and if the sedimentary load is abundant it is able to form a coarse-grained delta that emerges on the surface in a variety of forms, as happened when the Riu de les Coves flood in March 2015. As can be seen in Fig. 21.9, the diffraction of the waves and the double drift generated a micro-delta with two spit bars, which are destroyed by the waves over time. The processes indicate that it is a delta dominated by the waves given that the continental input is ephemeral. The balance of these processes is irregular and it forms a protuberance only when the fluvial input is high and the fan delta progrades as in this case. Despite this ephemeral character, research carried out using seismic, sedimentary and bathymetric data on the continental shelf has detected the presence of two types of submerged fans at the mouths of the main rivers (Les Coves, Millars, Palància, Túria, Gorgos, Algar, Amadorio, Montnegre, Vinalopó and Segura).

One group is formed by relict fans oriented towards the north. The other group is made up of the most recent fans and are south-facing (Rey et al. 1999). Likewise, Alcántara-Carrió et al. (2013) observed incised channels in the continental shelf that they attribute to the Rivers Túria, Xúquer and Serpis and believe formed at a time when the sea level fell in the last glacial period, which suggests the fan deltas and floodplains grew in response to a drop in the sea level.

21.7 Discussion and Conclusions: The Relationship Between Fan Deltas and the Coastline

Basin dimensions, sediment input, tectonics and neotectonics in the study area all determine the size, style and characteristics of fluvial forms. There are complex fan deltas between the Ebro Delta and the Palància fan formed by rivers and ephemeral streams with small basins where several alluvial fans come together to form alluvial bajadas that silt up the coastal plains. It is a sector with little subsidence, which is why prograding fans dominate. Between the Palància fan and Cap de la Nau, an obvious subsidence sector, the plain widens. Consequently, the Palància River forms an aggrading fan whereas the larger rivers (Xúquer and Túria) generate large and complex alluvial plains. Towards the south, the Baetic cliff sector (Cap de la Nau-Santa Pola) is highly compartmentalised by tectonics and affected by neotectonics. It is a complex area where cliffs alternate with small trenches, and alluvial fans or glacis develop. It is common to find raised or sunken blocks which distort the sedimentation and fluvial forms. Finally, the Santa Pola-Segura floodplain sector consists of a wide plain crossed by the other great river, the Segura, with highly complex tectonics and subsidence resulting from extremely significant neotectonics that have conditioned the morphology of the alluvial deposits and sedimentary rhythm.

An analysis of the fan deltas shows the importance of subsidence due to neotectonics in the style of these forms. Prograding fan deltas dominate the area from

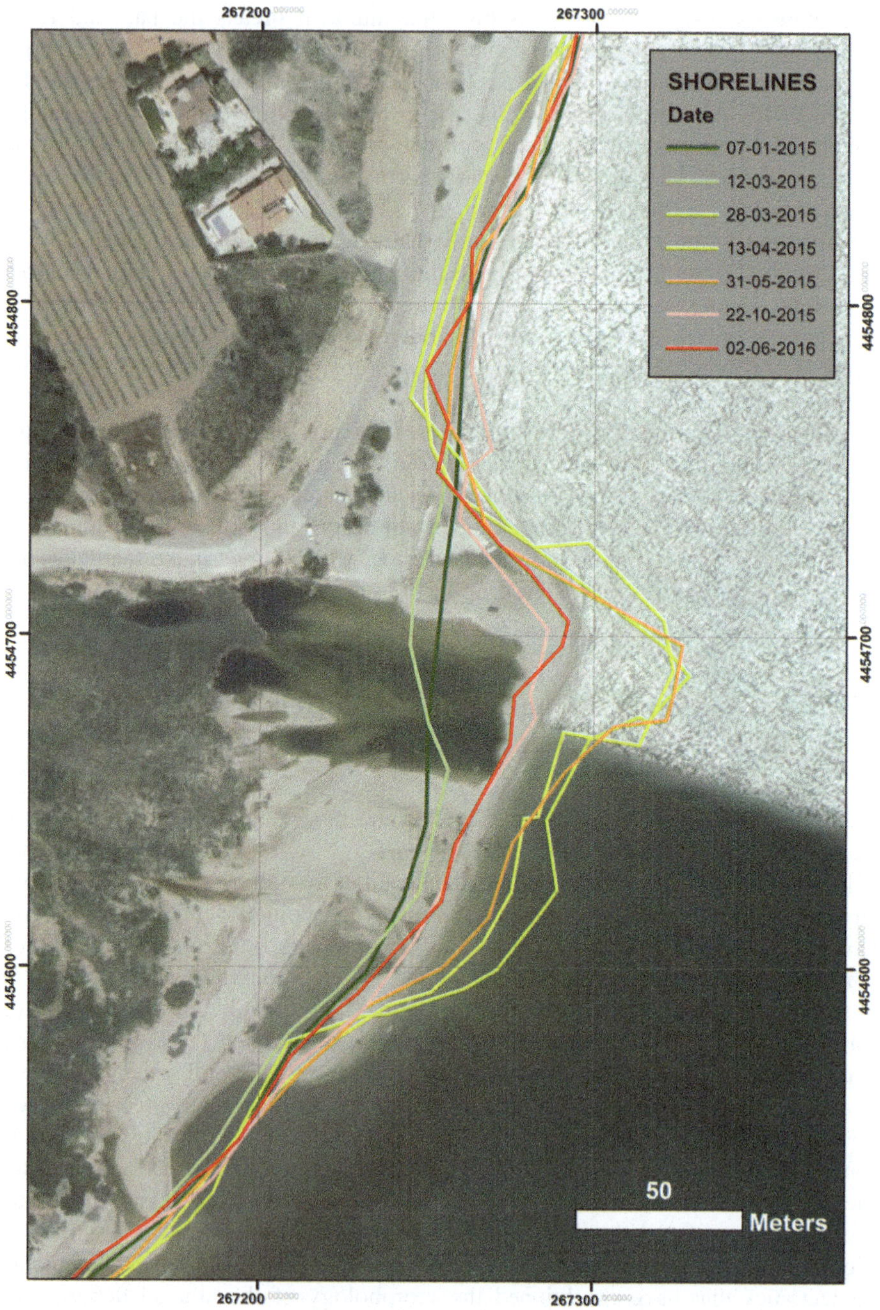

Fig. 21.9 The ephemeral delta of the riu de les Coves is formed when the floods provide plenty of fluvial sediments. The diffraction of the waves and a double drift form a small delta with two spit bars, which are destroyed by waves

the Ebro Delta to the Palancia, suggesting little subsidence or even a certain degree of stability in the sector. By contrast, the Palancia's alluvial fan is aggrading which coincides with the start of major subsidence that also affects the Túria and Xúquer floodplains. Similarly, in the Santa Pola-Segura floodplain sector, the Vinalopó fan is of an aggradational type and the Segura valley shows signs of a strong subsidence.

On the other hand, the Baetic sector has very active neotectonics (with areas of strong subsidence and others with elevation that have generated misalignments in the drainage network), soft lithologies and a semi-arid climate that have boosted the formation of glacis and flat-bottomed ephemeral streams. In this area, erosion predominates over accumulation processes. It should also be noted that the prograding fan deltas that drain into the sea form protuberances along the coastline (R. de les Coves, Millars, Riu Girona), while aggradational fan deltas are associated with a rectilinear coastline. When they flow into a lagoon, the former tend to fill the humid space, being able to cross it, while the latter barely fill the wetland.

In addition, the dynamics in the formation of the submerged part of the fan delta are quite complex. Given the wave and tidal regime, as well as the ephemeral regime of sediment supply, an exposed accumulation is formed at the time of the flood, which disappears after a few days. Hence, the current layout of the coastline and the shape and silting up of the lagoons making up València's coastal alluvial plains are obviously dependent on fluvial processes and forms. While the rivers with larger basins with their high sedimentary load are able to prograde, silt up the lagoons and shape the coastline, the small ones' low sedimentary load means they do not silt up the marshes and only shape their inner side without reaching the coast.

From of point of view of sediment supply, the coastline in the northern sector has less sedimentary input because the streams are ephemeral and have small basins. By contrast, in the central sector, the Xúquer and Túria rivers are perennial, with basins that provide greater sedimentary input and therefore the coasts are stable or prograding. In the Baetic sector, there are hardly any fluvial sedimentary inputs due to the compartmentalisation of the relief and the predominance of cliffs. Finally, the most southerly sector again has input from a perennial river with a large basin and sedimentary input again rises.

Acknowledgements The authors appreciate the financial support provided by the Spanish Ministry of Economy and Competitiveness in the framework of projects CGL2013-44917-R, CGL2015-69906-R and CGL-2017-86839-C3-1-R.

References

Alcántara-Carrió, J., Albarracín, S., Montoya, I., et al. (2013). An indurated Pleistocene coastal barrier on the inner shelf of the Gulf of València (western Mediterranean): Evidence for a prolonged relative sea-level stillstand. *Geo-Marine Letters, 33*(2), 209–216. https://doi.org/10.1007/s00367-012-0316-9.

Alfaro P. (1995). Geotectónica en la cuenca del Bajo Segura (Extremo oriental de la Cordillera Bética). Tesis doctoral. Dpto. de Ciencias de la Tierra y del Medio ambiente. Universidad de Alicante, Alicante, 218 pp.

Ballesteros, B. J., Domínguez, J. A., Díaz-Losada, E., García, O., et al. (2009). Zonas húmedas mediterráneas y acuíferos asociados. Condicionantes hidrogeológicos del Marjal de Pego-Oliva (Alicante-València). *Boletín Geológico y Minero, 120*(3), 459–478.

Bardají, T., Dabrio, C. J., Goy, J. L., Somoza, L., & Zazo, C. (1990). Pleistocene fan deltas in southerastern Iberian península: sedimentary controls and sea-level changes. *Spec Pubs int Ass Sediment, 10,* 129–151.

Blázquez, A. M. (1999). Geomorfología continental y submarina del espacio costero entre Dènia y Benidorm (Alicante): notas bibliográficas. *Investigaciones geográficas, no, 22*(1999), 111–128.

Blázquez, A. M. (2001). *L'Albufera d'Elx: Evolución cuaternaria y reconstrucción paleoambiental a partir del estudio de los foraminíferos fósiles* (p. 562). Dpt. De Geografia: Universitat de València.

Camarasa, A. M., & Segura, F. S. (2001). Flood events in Mediterranean ephemeral streams (ramblas) in València region. *Spain. Catena, 45*(3), 229–249.

Carmona, P. (1990). Evolución Holocena de la llanura costera del R. *Cuaternario y Geomorfología, 4*(990), 69–81.

Carmona, P. (2016). Progradación Fluvial y Cambios en la línea de Costa en época Histórica en el Golfo de València (España). In Leveau et al. (Eds.), *Environmental Reconstruction in Mediterranean Landscape Archaeology* (p. 169).

Carmona, P., & Ruiz, J. M. (2011). Historical morphogenesis of the Turia River coastal flood plain in the Mediterranean littoral of Spain. *CATENA, 86*(3), 139–149.

Carmona, P., & Ruiz, J. M. (2014). Procesos geomorfológicos en llanos de inundación y lagos costeros mediterráneos. El cambio ambiental histórico en la Albufera de València (España). *Cuaternario y Geomorfología, 28*(3–4), 95–106.

Cuenca, A., & Walker, M. J. (1985). Consideraciones generales sobre el Cuaternario continental en Alicante y Murcia. *Cuadernos de Geografia, 36,* 21–32.

Charlton, R. (2008). *Fundamentals of fluvial geomorphology* (p. 224). London; New York: Routledge.

Dijk, M. V., Kleinhans, M. G., Postma, G., et al. (2012). Contrasting morphodynamics in alluvial fans and fan deltas: Effect of the downstream boundary. *Sedimentology, 59*(7), 2125.

Dupré, M., Fumanal, M. P., Sanjaume, E., et al. (1988). Quaternary evolution of the Pego coastal lagoon (Southern València, Spain). *Palaeogeography, Palaeoclimatology, Palaeoecology, 68* (2–4), 291–299.

Ferrer, C., & Blázquez, A. M. (1999). El abanico del Vinalopó a lo largo del Holoceno superior: una aproximación geoarqueológica. *Cuadernos de Geografia, 65–66,* 347–358.

Garay, P. (1995). Marco geológico estructural y neotectónica del País Valenciano. *En El Cuaternario del País Valenciano*, Universitat de València-AEQUA, 31–42.

García, J., & Martínez, J. J. (2006). Caracterización sísmica del anticlinorio del bajo segura (Alicante): fallas del Bajo segura, torrevieja y san miguel de salinas. *Geogaceta, 40,* 19–22.

Goy, J. L., Zazo, C., Somoza, L., et al. (1990). Evolución paleogeográfica de la depresión de Elche-Cuenca del Bajo Segura (España) durante el pleistoceno. *Estudios Geológicos, 46*(3–4), 237–244.

Harvey, A. M. (2012). The coupling status of alluvial fans and debris cones: A review and synthesis. *Earth Surf Process Landforms, 37*(1), 64–76.

La Roca, N., & Carmona, P. (1983). Fotointerpretación de la Ribera del Xúquer después de la inundación de octubre de 1982. *Cuadernos de Geografía, 32–33,* 121–134.

Mateu, J. F. (1982). *El Norte del País Valenciano: geomorfología litoral y prelitoral* (p. 286). València: Universidad de València.

Mateu, J. F. (2000). El contexto geomorfológico en las inundaciones de la Ribera del Xúquer. *Cuadernos de Geografia, 67–68,* 241–259.

McPherson, J. G., Shanmugam, G., & Moiola, R. J. (1987). Fan-deltas and braid deltas: Varieties of coarse-grained deltas. *Geological Society of America Bulletin, 99*(3), 331–340.

Nemec, W., Steel, R. J. (1988). What is a fan delta and how do we recognize it. *Fan Deltas: Sedimentology and tectonic settings* 3–13.

Pardo, J. E., & Sanjaume, E. (2001). Análisis multiescalar de la evolución costera. *Cuadernos de Geografia, 69*(70), 95–126.

Portugués-Mollá, I., Bonache-Felici, X., Mateu-Bellés, J. F., & Marco-Segura, J. (2016). A GIS-based model for the analysis of an urban flash flood and its hydro-geomorphic response. The València event of 1957. *Journal of Hydrology, 541,* 582–596.

Rey, J., Fernández, L.M., Blázquez, A.M. (1999). Identificación de las unidades morfosedimentarias cuaternarias en la plataforma interna del litoral del País Valenciano: el rol de los factores morfoestructurales y eustáticos. In P. Fumanal (Ed.), *Geoarqueologia i Quaternari litoral. Memorial María Pilar Fumanal* (pp. 403–418). Universitat de València, València, ojo no estann citats 93 i 96.

Ruiz, J. M. (2002). Hidrogeomorfologa del llano de inundacin del Xúquer. Tesis doctoral inédita. Facultad de Geografa e Histria, Universitat de València (201 pp). Departamento de Geografía, Universitat de València.

Ruiz, J. M., Carmona, P. (2005). La llanura deltaica de los ríos Júcar y Turia y la Albufera de Valencia. *Geomorfologia i Quaternari litoral, Homenatge al Dr. V. Rosselló* 399–419.

Ruiz, J. M., Carmona, P., Mateu, J. F. (2006) El marco físico del regadío de la Ribera. En Hermosilla J (dir): Las Riberas del Xúquer: paisaje y patrimonio. Agua, territorio y cultura del regadío en la Ribera (pp. 22–44). Confederación Hidrográfica del Xúquer. València.

Rosselló, V. M. (1972). Los ríos Júcar y Turia en la génesis de Albufera de Valencia. *Cuadernos de geografía, 11,* 7–25.

Rosselló, V. M. (1978). Llanos y piedemontes: un dominio subárido. *López Gómez A, Rosselló i Verger, VM (1978) Geografía de la provincia de Alicante* (pp. 37–52). Alicante: Diputación Provincial de Alicante.

Sanjaume, E. (1985). *Las costas valencianas: sedimentología y morfología* (p. 505). Sección de Geografía, València: Universidad de València.

Sanjaume, E., Rosselló, V. M., Pardo-Pascual, J. E., Carmona, P., Segura, P., Lopez, M. J. (1996). Recent coastal changes in the Gulf of València (Spain). *Zeitschrift fur Geomorphologie Supplement Band* 95–118.

Sanjaume, E., & Pardo-Pascual, J. E. (2003). Características sedimentológicas y morfológicas de los espacios costeros de transición situados al sur de la desembocadura del Xúquer. *Cuadernos de Geografía (València), 73–74,* 183–205.

Segura, F. S. (1990). *Las ramblas valencianas: algunos aspectos de hidrología, geomorfología y sedimentología* (p. 229). València: Universitat de València.

Segura, F. S. (1991). Geomorfología fluvial y trazado de mapas de riesgo de inundación: el cono aluvial del Palancia. In *Sociedad y territorio. XII Congreso Nacional de Geografía* (pp. 221–227).

Segura, F. S. (1996). La geomorfología y las inundaciones en la plana de Vinarós-Benicarló. *Cuadernos de Geografia, 59,* 87–120.

Segura, F. S. (2001). Evolución urbana e inundaciones en Castelló. *Cuadernos de Geografia, 60* (70), 253–278.

Segura, F. S. (2006). Las inundaciones de la Plana de Castelló. *Cuadernos de Geografia, 79,* 75–100.

Segura, F. S. (2009). Geomorfología, inundaciones y alteración antrópica del espacio inundable: el caso del riu Girona (Alacant, octubre de 2007). *Boletín de la AGEN, 49,* 83–103.

Segura, F., Pardo, J. E., & Sanjaume, E. (1997). Evolución cuaternaria de la Albufera de Torreblanca. *Cuaternario y Geomorfología, 11*(1–2), 3–18.

Segura, F., & Carmona, P. (1999). Las inundaciones en la plana de Gandía: formas y procesos. *Cuadernos de Geografía, 65–66,* 45–60.

Segura, F., Sanjaume, E., Pardo-Pascual, J. E., et al. (2005). *Canvis del nivell del mar en l'evolució del Prat de Cabanes* (pp. 455–469). Geomorfologia litoral i Quaternari. Homenatge a V.Mª. Rosselló i Verger: Universitat de València.

Segura-Beltrán, F., Sanchis-Ibor, C., Morales-Hernández, M., et al. (2016). Using post-flood surveys and geomorphologic mapping to evaluate hydrological and hydraulic models: The flash flood of the Girona River (Spain) in 2007. *Journal of Hydrology, 541*, 310–329.

Silva, P. G., Goy, J. L., Zazo, C., et al. (2003). Fault-generated mountain fronts in southeast Spain: Geomorphologic assessment of tectonic and seismic activity. *Geomorphology, 50*(1), 203–225.

Soria, J. M., Alfaro, P., Fernández, J., et al. (2001). Quantitative subsidence-uplift analysis of the Bajo Segura Basin (eastern Betic Cordillera, Spain): tectonic control on the stratigraphic architecture. *Sediment Geol, 140*(3), 271–289.

Tent-Manclús, J. E., & Soria, J. M. (2014). Formación y desecación del Sinus ilicitanus (Sur de Alicante) en los últimos 15.000 años. *Geogaceta, 55*, 35–38.

Torres, T., Ortiz, J. E., Martín-Sánchez, D., Arribas, I., Moreno, L., Ballesteros, B., et al. (2014). The long Pleistocene record from the Pego-Oliva marshland (Alicante-València, Spain). *Geological Society, London, Special Publications, 388*(1), 429–452.

Viñals, M. J. (1996). El marjal Pego Oliva evolución geomorfológica. *Generalitat Valenciana Conselleria de Agricultura y Medio Ambiente*. València.

Viñals, M. J., & Fumanal, M. P. (1995). Quaternary development and evolution of the sedimentary environments in the Central Mediterranean Spanish coast. *Quaternary International, 29*, 119–128.

Viseras, C., Calvache, M. L., Soria, J. M., et al. (2003). Differential features of alluvial fans controlled by tectonic or eustatic accommodation space. Examples from the Betic Cordillera, Spain. *Geomorphology, 50*(1), 181–202.

Chapter 22
The Guadalquivir Estuary: Spits and Marshes

Antonio Rodríguez-Ramírez, Juan J. R. Villarías-Robles, José Noel Pérez-Asensio and Sebastián Celestino-Pérez

22.1 Introduction

The estuary of the Guadalquivir River is the largest in the Gulf of Cadiz, covering an area of some 185,000 ha. It enshrines Doñana National Park, one of the most important protected areas in Spain and possibly the best known internationally; in 1994 UNESCO distinguished it as 'World Heritage Site.' The park, established in 1969, consists of a restricted zone and a buffer protection area, extending over some 108,087 ha. It presents a great variety of ecosystems and biodiversity that is unique in the world; it is the habitat, for instance, of the Spanish imperial eagle and the Iberian lynx, listed as Endangered Species on the IUCN Red List.

From a geological standpoint, the estuary represents the culmination of the marine filling of the Cenozoic Guadalquivir Basin, constrained between the

A. Rodríguez-Ramírez (✉)
Departamento de Ciencias de la Tierra, Universidad de Huelva,
Avenida 3 de Marzo s/n, 21007 Huelva, Spain
e-mail: arodri@uhu.es

J. J. R. Villarías-Robles
Instituto de Lengua, Literatura y Antropología, Centro de Ciencias
Humanas y Sociales (CCHS), Consejo Superior de Investigaciones
Científicas (CSIC), Calle Albasanz 26-28, 28037 Madrid, Spain
e-mail: juanjose.villarias@cchs.csic.es

J. N. Pérez-Asensio
Grup de Recerca de Geociències Marines, Departament d'Estratigrafia,
Paleontologia i Geociències Marines, Universitat de Barcelona,
Carrer Martí i Franquès s/n, 08028 Barcelona, Spain
e-mail: Noel.PerezAsensio@unige.ch

S. Celestino-Pérez
Instituto de Arqueología de Mérida, Consejo Superior de Investigaciones
Científicas (CSIC)-Junta de Extremadura, Plaza de España 15, 06800 Mérida, Spain
e-mail: scelestino@iam.csic.es

J. A. Morales (ed.), *The Spanish Coastal Systems*,
https://doi.org/10.1007/978-3-319-93169-2_22

Variscan Massif—the northern passive margin of the basin—and the Baetic Cordillera, which is the southern collisional Alpine orogenic belt. Beneath the sediments of the marshland, Plio-Quaternary deposits as much as 400 m thick unconformably cover Late Miocene-Early Pliocene blue clays, sands and silts (Salvany et al. 2011).

The present-day configuration of the estuary is the result of the post-glacial transgression of the Atlantic Ocean, starting ca. 15,000 years, that developed during the latest Pleistocene-Holocene up until some 5,500 years ago, when the level of the sea stabilised (Melières 1974). The lower Guadalquivir valley was transformed into a wide estuary as the interfluves turned into pronounced headlands. Marine and fluvial dynamics, dependent upon climate and tectonics, thus shaped the present landscape, which features extensive dune systems, marshes and spits, as well as erosion of the headlands (cliff formation).

Over the past few decades the Holocene evolution of the estuary has been the subject of a number of studies. Significant results have been a chronology of events of progradation and erosion (Zazo et al. 1994; Rodríguez-Ramírez et al. 1996; Dabrio et al. 1999), an analysis of a long core (Zazo et al. 1999; Lario et al. 2001; Pozo et al. 2010), research on the period between the Late Pliocene and the Quaternary—without concentrating on the evolution during the Holocene, however (Salvany et al. 2011)—and a multidisciplinary, detailed analysis of cores, geomorphic patterns and historical data that has yielded new evidence on the sedimentary infilling and geomorphic evolution of the Guadalquivir estuary during the Holocene. Results from the latter project have provided valuable information with which to model the coastal evolution of the Gulf of Cadiz in the Late Holocene

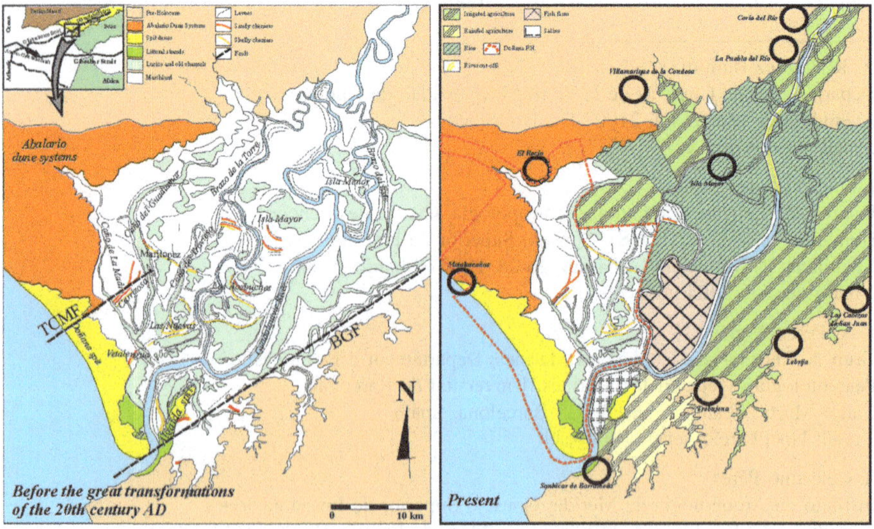

Fig. 22.1 Geomorphological elements of Guadalquivir estuary and current transformation of territory

(Ruiz et al. 2004, 2005; Rodríguez-Ramírez and Yáñez 2008; Rodríguez-Ramírez et al. 2014, 2015, 2016) (Fig. 22.1).

22.2 Geographical and Geological Setting

The Guadalquivir estuary has a central position on the Iberian side of the Gulf. One of the largest wetlands in Europe, it comprises two morphogenic systems: the littoral (the dunes and spits) and the estuarine system (the marshland). Vegetation in the dunes and spits is dominated by *Pinus pinea*, *Juniperus* sp. and *Cistus* sp. In the marshland, the soils are saline; vegetation is dominated by *Arthrocnemum macrostachyum* and *Suadea vera*, with *Scirpus* sp. in the most flooded areas.

Hydrodynamics is controlled by the fluvial regime, the tidal inflow, the action of the waves and the drift currents as well as the climate. The estuary has a typically Mediterranean climate—with Atlantic influence. Summers are hot and dry, while winters are relatively mild and wet. The mean annual temperature is 16.7 °C and precipitation is 537 mm, according to measurements taken at Doñana Palace Weather Station. SW winds prevail: they are felt in 22.5% of the days in the yearly cycle, as concluded by Instituto Nacional de Meteorología for the City of Huelva, 1960–1990 (Rodríguez-Ramírez et al. 2003). As they come from the Ocean, SW winds are commonly associated with storm events (Rodríguez-Ramírez 1998).

The Guadalquivir is the main river in the estuary. It drains the Spanish southwest and produces most of the fluvial sediments. The mean annual discharge is 164 m^3/s, even though winter spates can easily exceed 5000 m^3/s (Vanney 1970). The highest runoff (>1000 m^3/s) takes place from January to February, with fluvial current velocities up to 1 m/s (Vanney 1970; Menanteau 1981).

The tidal inflow in the Gulf exhibits a progressive displacement in an East-to-West direction, from the Strait of Gibraltar to the southern Portuguese coast. The maximum tidal range observed at the Guadalquivir river mouth is 3.86 m, the average range being some 2 m (source: mareographer of Bonanza: period from 1997 to 2003. Data of Spain's Departamento de Clima Marítimo of Organismo Autónomo Puertos del Estado, OAPE: http://www.puertos.es/es-es/oceanografia). The coastline can be described as mesotidal, with a slight semidiurnal inequality.

The action of the waves depends directly upon the prevailing SW winds. In the wintertime, Atlantic cyclones are common; they give rise to strong SW winds which generate 'sea-type' waves more than 6–8 m high (Hsmax; data of Spain's Departamento de Clima Marítimo of Organismo Autónomo Puertos del Estado, OAPE: http://www.puertos.es/es-es/oceanografia). These waves cause significant erosion in the littoral zone (Rodríguez-Ramírez et al. 2003), yet they represent only around 3–5% of the total waves in a year; in general, the wave regime in the Gulf is a medium-to-low energy one, with waves usually less than 0.6 m high (data of Departamento de Clima Marítimo: http://www.puertos.es/es-es/oceanografia). Most wave fronts approach the coast obliquely and induce littoral currents that transport sand from the Portuguese coast to Spanish nearshore areas. The magnitude of this drift ranges from 180 × 103 hm^3/year (Cuena 1991) to 300 × 103 hm^3/year (CEEPYC 1990).

22.3 The Spits

The littoral of the Gulf presents a series of broad zones under tidal influence associated with the mouths of the main rivers. These zones show the growth of littoral spits and barrier-islands that tend to seal the river mouths from the Ocean. Such developments are the product of an intense longshore current that moves along the coast from west to east and depends on the oblique incoming direction of the SW winds. A number of physical and geological parameters govern the growth and geometry of a spit: the wave regime, the sea-level oscillations, climate, tectonics, the geological setting and ultimately human activity.

The chief broad zone in the Gulf is, again, the Guadalquivir estuary. From a geomorphological standpoint, one of the most notable features in this estuary is the existence of two large spit systems, Doñana and Algaida (Rodríguez-Ramírez et al. 1996) (Fig. 22.1). The former is the most extensive, lying on the right bank of the estuary after growing towards the E and SE. It has a length of about 20 km and a maximum width of 5 km. It is partly covered today by the active dunes of two dune systems (Rodríguez-Ramírez et al. 2016): the first extends from the shoreline towards the core of the spit, displaying formations that range from foredunes to transversal bodies that can be as high as 35 m at Cerro de Los Ánsares, while the second system spreads below such formations and extend further inland, consisting of rather blurred parabolic dunes no more than 10 m high (Fig. 22.2).

The Doñana spit system also comprehends a number of littoral strands which represent intense coastal progradation from west to east. These strands appear grouped into two sets, one separated from the other by a soft sedimentary break. The strands of the first set slightly curve towards the west by ancient longshore currents. Radiocarbon dates obtained from the oldest strand indicate a beginning of this progradation around the first century before the Christian era (CE) (Rodríguez-Ramírez et al. 2016). The second, more recent set of strands, subsequent to an erosional surface, exhibits a general development that includes ridges and swales progressively curving towards the east. The beginning of this second development is estimated at around the 6th century CE. The erosional surface that marks the morphological anomaly, clearly separating the two progradation processes, can be dated to the 2nd or 3rd century CE (Rodríguez-Ramírez et al. 2016). The most recent formations in the spit, corresponding to the past 200–300 years, are covered by dunes.

The Algaida spit system stands on the left bank of the estuary, after a growth towards the NNE favoured by the drift current moving in the stretch of coast between the town of Chipiona and that of Sanlúcar de Barrameda in the same direction. In this second spit system, dune systems are less conspicuous than in the Doñana spit: Algaida presents a number of erosional scarps that are superimposed

Fig. 22.2 Dune systems of Doñana spits

on top of one another (Rodríguez-Ramírez et al. 1996) and break the geomorphological continuity of previous formations that resulted from progradation. More than one progradation set can be identified, owing much to the intense fluvial dynamics of the Guadalquivir River. There was a time when Algaida was an isle in the estuary and, therefore, was embraced by two channels of the river (Rodríguez-Ramírez et al. 1996). A complex tidal delta would cause the riverbeds to migrate cyclically, in a way similar to the dynamics which animates most other estuaries in the Gulf (Morales et al. 2006).

22.4 The Marshland

Behind the spit systems of Doñana and Algaida there extends a large freshwater marshland of 140,000 ha, which is the end product of the infilling of a large marine gulf or paleo-estuary that existed in the area some 5,500 years ago (Fig. 22.3). The infilling took place as the growth of the littoral formations isolated the paleo-estuary from the sea. There is, in effect, a direct correlation between the littoral formations and the estuarine ones. The modelling of the infilling was shaped by both the fluvial sediments—contributed by the Guadalquivir and convergent rivers, especially the Guadiamar—and the tidal inflow. In a low-energy environment facilitated by the growth of large littoral spits, the intense dynamics of the rivers in the marine paleo-estuary brought about the formation of a finger delta there (Rodríguez-Ramírez 1998). Extensive wetlands of great environmental interest also emerged as a result, the

Fig. 22.3 Dune systems of Doñana spit and the marshland on the horizon

most important—for its size and diversity—being Doñana National Park: it is the last tract of relatively undisturbed marshland in this estuarine system, thanks to which a detailed geomorphological study has been possible over the years.

22.4.1 The Holocene Sedimentary Sequence in the Marshland

Analytical core-drilling in the marshland to depths as far down as 18 m has revealed a series of Holocene deposits (marshland, dunes and beaches) resting on pre-Holocene formations of sands and clays that indicate dunes and marshes (Rodríguez-Ramírez et al. 2014, 2015). Northwest of the marshland, a number of different semi-stable dune systems—known as 'the Abalario Dune Systems'—move in NW and W directions. South and east, these systems gradually fossilize underneath the marshland and the Doñana spit. Laterally towards the inner, central sector of the paleo-estuary, the ancient dune formation interdigitates with the pre-Holocene marshland, the facies of which suggests either an emerged environment or an environment in transition—i.e. subject to periodic tidal flooding as well as freshwater inputs (Rodríguez-Ramírez et al. 2015).

The earliest sedimentary evidence of the progressive Holocene sea-level rise in the centre of the paleo-estuary is the presence of sandy silts with abundant transported marine macro- and microfauna (the bivalve *Glycimeris* sp., the benthic foraminifer *Triloculina trigonula* and other species) at the bottom of the sequence; the sand and the marine fauna decrease gradually towards the top. The sequence consists mostly of a facies of clayey silts in which the dominant macro-fauna is the estuarine bivalve *Tellina tenui*. In the lower and middle parts of this facies, the micro-paleontological analysis, based on benthic foraminifera, revealed an estuarine *Ammonia tepida* assemblage, which included *Elphidium translucens* and *E. granosum* as associated taxa; in the upper part, the most abundant macro-fauna includes the estuarine bivalves *Cerastoderma* sp. and *Ostrea* sp., as most abundant taxa, while the foraminifera group appeared dominated by *H. germanica*, with *A. tepida* as accompanying species. The top of the sequence registers the prevalence of silty clay and terrestrial, freshwater gastropods (*Cochlicella* sp., *Helix* sp., *Melanopsis* sp.), with stems and roots at the very top. A number of decimetre-scale layers with a higher sand content and transported marine fauna, related to high-energy events (see Sect. 22.5), interrupt at intervals the clay sediments. Being conditioned by the fault of the lower Guadalquivir River, the thickness of the Holocene sedimentary sequence increases progressively towards the south (11 m in S11, 28 m in PN) to >60 m in the southern margin of the basin.

22.4.2 The Geomorphology of the Marshland

The intense fluvial and marine dynamics has produced diverse morphologies in the marshland. A geomorphological analysis (Rodríguez-Ramírez and Yáñez 2008) made it possible to recognize the clay-rich fluvial levees flanking the river and its former courses as the most characteristic feature of the formation (Fig. 22.4). These levees have a variable width (300–2000 m) and great length (5–10 km), reaching heights of 0.5–1.5 m above the adjacent inter-levee marshes. While they cause the marshland to be topographically as high as 2–3 m above sea level, the elevation gradually diminishes towards the south, down to 0.5–1 m near the mouth of the Guadalquivir River, the slope being on the order of 0.004%. The levees are part of the deltaic system which has been filling the paleo-estuary over the past five millennia, advancing in two main directions: north–south and northeast–southwest. The evolution of the fluvial dynamics during the Middle and Late Holocene turned many of the old channels of the delta (e.g. Travieso, Madre de las Marismas, Guadiamar) into flood basins. Currently, the only active riverbed is that of the Guadalquivir River as known today; up to a few decades ago, the river had two other beds running across the marshland: Brazo de La Torre and Brazo del Este, both annulled by human-made cut-offs. The levees of Guadalquivir River, higher than the rest of the surrounding plain, delimit to the rest of the marshland plain like an endoreic basins, its importance is such that in the local jargon it is called the "Montaña del Río", "River Mountain".

The inter-levee marshes represent the topographically lower zones in such a plain; these marshes, therefore, are the zones in the plain that remain flooded longer than other zones: from December to May, being dry the rest of the year. Flooding is

Fig. 22.4 Old channel of Travieso (Caño Travieso) and levee (pacil de los Almajos Dulces)

of fluvial and pluvial origin; there is no longer any significant tidal influence in the plain. Sub-areas in the inter-levee zones are the so-called '*lucios*,' where flooding lasts the longest; they have soils covered with salt scabs in the summertime.

22.4.3 *The Chenier Plain*

Overlying the clayey infilling of the marshland are beach ridges of sandy and shelly deposits with a littoral strand morphology (Figs. 22.1, 22.5 and 22.6). These complex deposits are known as 'cheniers' (Rodríguez-Ramírez and Yáñez 2008). They have been widely used for reconstructing paleo-environmental and coastal evolutions, as they register the location of ancient shorelines. Their sedimentary record potentially stores past coastal processes, shoreline changes, sea-level fluctuations, co-seismic land-level changes and even climatic and paleo-oceanographic changes (Augustinus 1989). They are found singly or in multiple sets of ridges that run roughly parallel to the shoreline, thus making up a chenier plain; the ridges form sequentially, recording the successive positions of an advancing shoreline (Hoyt 1969).

Otvos and Price (1979) defined chenier plains as net progradational coastal plains that feature two or more parallel to subparallel stranded ridges (cheniers) of coarse sediment (sand, gravel or shell) overlying and separated by finer-grained littoral sediments or marshy mudflat deposits. This definition has been further refined, to the effect that today chenier plains are understood as those that exhibit two or more chenier complexes, each of which, in turn, consisting of two or more beach ridges, cheniers or spits (McBride et al. 2007). These ridges can be further classified in relation to the movement of the shoreline, as Curray (1964) has

Fig. 22.5 Sandy chenier of Vetalengua

Fig. 22.6 Shelly chenier of Las Nuevas

proposed, distinguishing between transgressive (landward migration), regressive (seaward migration) and laterally-accreted ridges. A transgressive ridge surrounded by littoral mud deposits is also known as a chenier; a regressive ridge is known as a beach ridge; a laterally accreted ridge is known as a spit or a recurved spit (McBride et al. 2007).

Usually the evolution of cheniers and chenier plains is strongly influenced by episodic sedimentation. Chenier ridges form by means of punctuated deposition of coarse sediments accessed from the shelf and foreshore by wave action (Augustinus 1989). By contrast, rates of foreshore mudflat progradation are controlled by a variable supply of fine sediments. Delta switching is a classic cause of chenier-plain development (Gould and McFarlan 1959). However, many factors, such as storm impacts, sea-level fluctuations, climate oscillations, longshore currents and tidal inlet dynamics are known to be important (McBride et al. 2007); in addition, factors that facilitate development in one area may not have the same influence in another (Augustinus 1989).

The large sandy and shelly ridges in the Guadalquivir estuary rest on the clay sediments supplied by the Guadalquivir River (Rodríguez-Ramírez et al. 1996). These formations have similar characteristics to those of the cheniers and chenier plains studied in other coastal zones of the world, yet represent an important regional site for deciphering chenier evolution and coastal history in the Gulf of Cadiz during the Holocene. They have been the object of conscientious research since 1990 (Rodríguez-Ramírez et al. 1996, 1998, 2008; Rodríguez-Ramírez and Yáñez 2008). Their peculiarity lies in their being protected from marine dynamics by large littoral spits; normally, however, coastal zones comprehending chenier plains are open to the sea (Augustinus 1989), and in this open environment the effects of both marine dynamics and wind-induced setups, especially during storms, are more pronounced.

The chenier complex in the Guadalquivir estuary has been studied by means of historical aerial photographs and extensive, detailed fieldwork aimed at geomorphological identification and mapping and a radiocarbon-based chronology of processes. As remarked above, this chenier complex consists of a series of sandy and shelly formations with a littoral strand morphology. The sandy formations are distally attached to the Doñana spit and oriented SW–NE (Figs. 22.1 and 22.5). Two main systems can be recognized in them: Carrizosa-Marilópez, to the north, and Vetalengua-Las Nuevas, to the south. Secondary formations (e.g. Los Acebuches) lie in between. Agricultural transformations have erased any trace of a likely extension of these sandy formations further north. The Carrizosa-Marilópez system includes at least three correlative sandy ridges, each 50–100 m wide, with elevations oscillating between 3.00 and 2.25 m in the case of the two oldest ridges and averaging 1.50 m the newest one. The Vetalengua system comprises two 50 m-wide ridges, with an elevation of 2.15–2.25 m. The dominant lithology in these sandy cheniers presents a sequence of fine to medium yellow quartz-rich sands (65–69%), with very good sorting. The thickness of the deposit is variable, 0.3–2.50 m, forming a wedge shape that progressively thins landwards. The internal structures show subparallel lamination that dips seawards and is interrupted by an erosive level that includes abundant, highly reworked and transported mollusc remains—many of which of marine origin (*Glycymeris* sp., *Chlamys* sp.), although with abundant presence of estuarine forms (*Cerastoderma* sp.)—and dispersed pebbles and an assortment of small pottery shards from the Neolithic period and the Copper Age, also highly reworked and transported.

The shelly formations are on the south-western banks of the fluvial levees and on the sandy littoral strands (Figs. 22.1 and 22.6). These formations have beach-ridge morphology, with narrow, small ridges and landward-dipping crests. They extend as far as 10 km in length, their width ranging from 5 to 30 m. In Las Nuevas they display multiple sets of ridges as evidence of the continuous advance of the ancient shoreline. Topographically, the highest ridges are between 1.75 and 1.95 m high; the less developed ones, between 1.30 and 1.50 m and partly buried by modern clay deposits. These formations are basically made up of lenticular lumachel accumulations of invertebrate shells 0.25–1.5 m thick, mostly of *Cerastoderma* sp. and *Ostrea* sp. There are also numerous bioclasts (30–40% dry weight), within a sparse silty–clayey matrix that becomes more developed upwards (48%), there mixing in with soil development and abundant roots. The sand fraction ranges from 30% at the base to 14% at the top. The internal structure is characterized by gentle landward-dipping lamination, interrupted by large-scale cross-bedded sets, with small amounts of mud at the base. This cross lamination progressively disappears landwards, giving way to a fine bed some 10 m long. The sequence is interrupted by several erosive surfaces which show a number of overlying beds that exhibit the same characteristics, thus highlighting vertical aggradation.

The chronology of the chenier plain rests upon Radiometric-Standard and AMS methods of C14 assayed on shells, calibrated and corrected for the reservoir effect with the parameter suggested for the Gulf. The sandy systems of Carrizosa-Marilópez yielded an age of c. 4,000 cal. year BP; those of Vetalengua-Las Nuevas produced an age of c. 2,000 cal. year BP (Rodríguez-Ramírez et al. 2015, 2016).

Both sandy systems, Carrizosa-Marilópez and Vetalengua-Las Nuevas, can be considered as laterally-accreted sandy ridges (spits), with a direction of progradation that is perpendicular to the direction of the main coastal barriers and a lateral support that is provided by earlier marshland deposits. The shelly ridges, on the other hand, are the classical cheniers, made of transgressive and regressive ridges. The transgressive ridges are the main chenier formations in the estuary, as they show the greatest geomorphological prominence in the landscape. They formed when the progradation of muddy formations in the estuary was interrupted by the erosion caused by high-energy events, which gave way to wave reworking and ridge development, in turn followed by mudflat progradation—as explained by Hoyt in a seminal paper (1969). Thus the transgressive chenier implies strong erosive processes with intense shoreline retrogradation. In the regressive mode, by contrast, the chenier itself is reworked by waves while new sedimentation episodes are generated seawards, without muddy progradation. Thus the regressive chenier implies minor erosive processes with shoreline progradation. These formations have characteristics which resemble those of wave-built beach ridges (Otvos 2000).

22.5 Extreme Wave Events

The estuary presents substantial erosive and sedimentary evidence of its having been hit by extreme wave events (EWEs) throughout its geological history. Tsunamis and coastal storm surges are two of the most dangerous of these events and yet the most common in coastal locations (Morton et al. 2011). They are high-energy events which result in the deposition of sedimentary beds that have certain specific sedimentological and paleontological characteristics; however, differentiation in this evidence between coastal storm surges and tsunamis as the cause thereof is difficult, as both types of EWEs generate very similar deposits (Fujiwara et al. 2000).

Geomorphological and sedimentological features as the products of EWEs are well known for the coasts of SW Iberia; they have been attributed to either storm surges or tsunamis, or both (Lario et al. 2010). At present, violent storms occur in the Gulf on a cycle regulated by the North Atlantic Oscillation (NAO; periodicity of $\sim$6 year) and solar irradiation (sunspot cycles) (periodicity of $\sim$11 year) (Rodríguez-Ramírez et al. 2003). Concerning tsunamis, SW Iberia is a low-probability tsunamigenic area (Reicherter 2001). The epicentres of the corresponding earthquakes have commonly been placed at some 200 km southwest off Cape Sâo Vicente, near the Gorringe Bank (Martínez Solares et al. 1979); current analyses, however, point to movements along the Azores-Gibraltar Fault or along minor faults associated with it; e.g., the Marques de Pombal Fault (Terrinha et al. 2003). Still other likely epicentres can be posited in connection with movements of faults that are even closer to the coasts of the Gulf of Cadiz (Silva et al. 2005). Historically, the most recent tsunami occurred on November 1, 1755. It is the notorious 'Lisbon earthquake'; the large waves impacted on extensive coastal areas of the Iberian Peninsula and Morocco (Gracia et al. 2006), their oscillation reaching as far as the south coast of England (Foster et al. 1991).

The geological record of the Guadalquivir estuary offers a wide spectrum of geomorphological and sedimentary developments in connection with EWEs: wash-over fans, paleo-cliffs or erosional scarps, crevasse splays, coarse gravel deposits and sedimentary lags of sand and shells (Lario et al. 2001; Pozo et al. 2010; Rodríguez-Ramírez et al. 2015, 2016). Spits, levees and cheniers are affected by both crevasse splays and wash-over fans in the coarse of EWEs; human settlements are affected by EWEs as well, of course. Cheniers, however, cannot be considered in and of themselves as straightforward, unambiguous evidence of either tsunamis or storm surges, or both (Rodríguez-Ramírez and Yáñez 2008). The noticeable presence of marine fauna and pebbles in chenier deposits, several kilometres from their zone of origin, does point to the relevance of high energy events (such as tsunamis or storm surges) as suppliers of detrital material (sand and shells) upstream the estuary. Yet it is not possible to determine precisely, from a geological perspective of the formations in the estuary, whether detrital supplies come from storms or tsunamis; a number of successive winter storms may give rise to the advance of detrital material of considerable size upstream the estuary in several consecutive phases (Kortekaas 2002). The Guadalquivir chenier plain is the result of the reworking and subsequent deposit of the residual lag of sands, shells, pebbles and archaeological remains that entered several kilometres far into the estuary by the energy of events of various kinds (storms, tsunamis, stream floods) in different ages, ending up accumulating in longshore bars or ridges (the cheniers). Thus cheniers cannot by themselves be considered as direct, obvious evidence for differentiating one kind of event from the other.

Analysis of long and short cores has revealed evidence of a number of events of higher marine influence that affected the estuary in the Late Holocene. Lario et al. (2001) have recognised a rapid episode of coarser sediment input which may be related to high energy events occurred around 2400 cal year BP. Pozo et al. (2010) and Zazo et al. (1999) have reported on clear signs of an event c. 4000 cal year BP. More recently, Rodríguez-Ramírez et al. (2015 and 2016), on the basis of radiocarbon dates obtained with the AMS method on mollusc shells retrieved from deep and shallow deposits of sand, have identified four majors events: A ($\approx$4000 cal year BP), the most intense and cataclysmic, as it transformed dramatically the geography of the estuary; B ($\approx$3550 cal year BP); C ($\approx$3150 cal year BP); and D ($\approx$1700–1600 cal year BP). As a general mark, these four events produced a very characteristic facies which consists of a sand layer containing a massive accumulation of shells (both articulated and disarticulated bivalves) and shell fragments in a sandy–muddy matrix with gravel and lithoclasts, on an erosive base. The shells and shell fragments belong to macro- and micro-faunal species of diverse environments: mostly species native to an open marine environment, the rest indicatory of a protected, low-energy one. The subsequent estuarine dynamics caused these remains to accumulate in sandy and shelly cheniers, as did the archaeological debris.

22.6 Neo-Tectonics of the Estuary

The Mediterranean basin presents numerous examples of neo-tectonic activity that conditions the morphodynamic evolution of the coastlines, specifically vertical displacements related to plate collision (Mastronuzzi and Sansò 2012). Comparable scenarios have been identified along the Pacific coasts of North and South America. The Guadalquivir estuary finds itself rather close to a contact zone between the tectonic plates of Eurasia and Africa, as it lies right across the northern boundary of the African plate and the Azores-Gibraltar fault line. The present geological structure of the Gulf of Cadiz is the result of the European–African plate convergence motion, which is a dextral strike-slip along the Azores–Gibraltar plate boundary (Medialdea et al. 2009) (Fig. 22.1). Westward drift and collision of the northern African and southern Iberian margins in the Early–Middle Miocene brought about the radial emplacement of huge allochthonous masses (the so-called 'Olistostrome Unit') in the Guadalquivir Basin (Iberian foreland) and the Gulf of

Cadiz (Torelli et al. 1997). This emplacement favoured an active subduction zone in the estuary.

The multidisciplinary study of cores drilled in the estuary has revealed a great deal of information about developments conditioned by neo-tectonic activity (Rodríguez-Ramírez et al. 2014). In combination with other parameters, neo-tectonics caused changes in sedimentation rates, resulted in new geomorphological features (spits, dunes, cheniers, marshes, levees, alluvial soils), brought about sea-level oscillations and facilitated certain dynamic processes (Rodríguez-Ramírez et al. 2014). Because such effects point to neo-tectonics as a critical factor in the evolution of the entire Atlantic–Mediterranean linkage coast after the transgressive maximum of the Atlantic Ocean, it must be so considered for present and future understandings of this evolution.

In the structure of the Plio-Quaternary deposits of the lower Guadalquivir basin, Armijo et al. (1977), Viguier (1977) discovered two sets of normal faults linked to an assumed Pliocene extensional phase: (1) a main set of N-S-oriented faults which includes the lower Guadalquivir and Guadiamar-Matalascañas faults and (2) a subordinate system of E-W-oriented faults. Studies by Goy et al. (1994), Salvany and Custodio (1995), Zazo et al. (2005) have described comparable E-W-oriented faults shaping the more recent sedimentation in the Pleistocene. Furthermore, studies by Flores-Hurtado (1993), Salvany (2004) have recognized NW-SE- and NE-SW-oriented systems of alignments and a regional tilting of the basin towards the SSE as the dominant structural features affecting the sedimentation. More recently, Rodríguez-Ramírez et al. (2014) have identified a series of geomorphological anomalies in the marshland, such as peculiar drainages, rectilinear alignments marked by the hydrographical network, subdued morphologies, erosive forms and even patches of vegetation caused by weaknesses and changes of humidity in the ground. All of these anomalies present SW-NE and NW-SE directions, which define the northern and southern boundaries of the marshland area.

Analysis of the deep cores has revealed a progressive sinking of the lower Guadalquivir basin towards the south, along the alignment known as 'The Torre Carbonero-Marilópez Fault' (TCMF) (Rodríguez-Ramírez et al. 2014) (Fig. 22.1). The alignment appears intersected at intervals by other lines orthogonal to it, following a NW-SE direction, the most visible being the faults 'Madre de las Marismas' (MMF) and 'Marilópez' (MF). The sinking facilitated over-wash and the impact of tsunamis, including the impact of these EWEs on human settlements of the Neolithic period and the Copper Age. The Carrizosa-Vetalarena littoral strands cross the marshland area by following the TCMF alignment, thus making a rather clear geomorphological and paleo-geographical boundary.

In addition, the progressive subsidence of the lower basin conditioned sedimentation in the estuary (Fig. 22.7). The analysis has unveiled a first stage of sediment infilling and tilting of the zone south of the TCMF alignment between 4000 and 2000 years BP, which resulted in retro-aggradational sedimentary systems nurtured by relative sea-level rise as well as progressive sinking of the ground surface. The oldest Holocene formations on the surface date from 4706–4071 cal. year BP (the Carrizosa strands), which stand north of the TCMF line; south of this fault, by contrast,

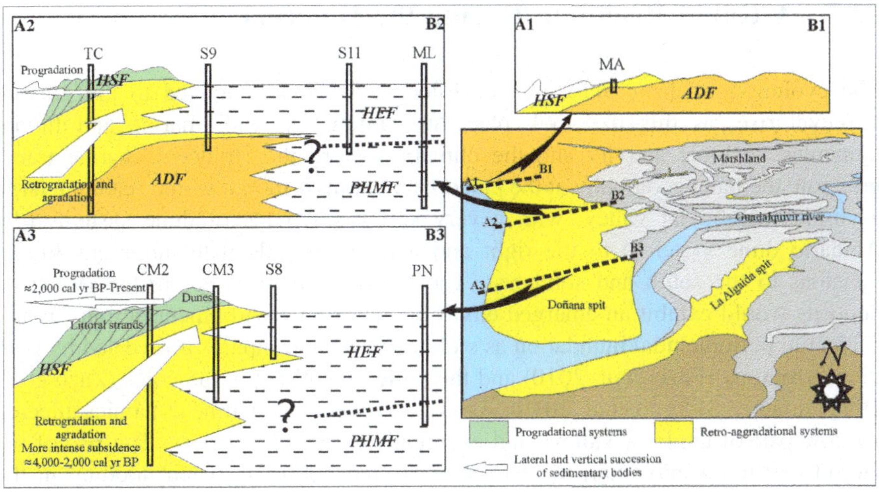

Fig. 22.7 Lateral and vertical succession of sedimentary bodies in the Guadalquivir palaeoestuary and cores analysed (ADF: Abalario Dune Formations, HSF: Holocene Spit Formations, HEF: Holocene Estuarine Formations, PHMF: Pre-Holocene Marshland Formations) (modificated from Rodríguez-Ramírez et al. 2014)

the visible formations date to c. 2000 cal. year BP or are more recent. Although the Doñana spit is the largest in the Gulf of Cadiz, the exposed beach–barrier systems in the spit are about 2000 years of age at the oldest. In effect, the geomorphic, stratigraphic and chronological analysis of the most recent phases in the growth of the spit, as well as in the formation of chenier systems in the estuary, showed the formation of progradational systems in a context of relative cessation of the sinking processes—at least south of the TCMF alignment—in the past 2000 years or so. As subsidence had been intense there between approximately 4000 and 2000 cal. year BP, the date 2000 cal. year BP represents a chronological transition between a retrograding, aggradational system—conditioned by the subsidence-prompted sea-level rise—and a system that progrades in connection with a relatively stable sea level.

Consequently, the growth phases of the coastal barrier prior to 2000 cal. year BP lie buried several metres under the ground surface, fossilized by later formations, c. 2000 cal. year BP old or younger. Such buried deposits may contain archaeological remains from prehistoric and early historic times. Although archaeologists excavating in the Doñana spit system since the 1920s have found no traces of human settlement older than the Roman Imperial period, they have failed to dig in much further down.

22.7 A Rapid Evolution During the Holocene

The evolution of the estuary in the Holocene is directly related to the gradual sea-level rise, as this rise took place in conjunction with marine and fluvial dynamics, tectonic activity and the climate. Before the Holocene marine transgression, the paleogeography at the paleo-Guadalquivir mouth would feature a large asymmetrical fluvial valley, with a very soft relief made of dune systems (the Abalario dune systems) to the right and a relief of hills with numerous fluvial incisions to the south and southeast. The paleo flood plain before the Holocene deposits would exhibit an emerged environment or an environment in transition—subject to periodic tidal inundation as well as freshwater inputs, as revealed by both the microfauna (Pozo et al. 2010) and the pollen identified in the cores (Zazo et al. 1999; Jiménez-Moreno et al. 2015). As the sea level rose, marine dynamics affected the low part of the river valleys and eventually formed an estuary. The rise of the Ocean reached a maximum in this sector around 5000 BP, when flooding in the valley and the erosion of inter-fluvial areas would be connected. The effect of the sea-level rise over the headlands was the generation of large cliffs, with significant coastal erosion developing in such features. By studying the changing position of a number of 16th- and 17th-century watchtowers in the area vis-à-vis the shoreline, a rate of cliff recession of 0.7–0.8 m/year on average was calculated for the La Higuera watchtower, at the beach of Matalascañas. To get some idea of the approximate coastal recession accrued there over longer time spans, one can apply such a rate to the past 2500 and 5000 years, obtaining as a result that the shoreline may have receded as much as 1800 and 3600 m, respectively, since the mid Holocene (Rodríguez-Ramírez et al. 2014). Back in that period, therefore, the landscape featured a shoreline that extended far seawards than its present position. Pollen from today's coastal southwestern Iberia for the period of $\approx$10,020 $\pm$ 50 to 5380 $\pm$ 50 ^{14}C year BP suggests a wet, temperate climate at the time (Santos et al. 2003).

The first morpho-sedimentary units of the Holocene in the Doñana coast took the form of estuarine barriers and marshland at river mouths. Favoured by the low, soft relief to the right of the valley, the Holocene sedimentation would have spread gradually northwards from there. Paleontological evidence suggests a lagoon environment under considerable marine influence (Fig. 22.8a). In the outer margins of the estuary, the earliest Holocene deposits formed spits which enclosed part of the estuary and, consequently, enabled sedimentation therein of finer materials. In a context of relative sea-level rise, the progressive subsidence of the ground surface facilitated the formation of a retro-aggradational sedimentary system, showing high sediment thickness.

By 4000 cal. year BP, the southernmost sector of the Abalario dune systems would have submerged by subsidence. This process was conditioned by a number of orthogonal fault systems (following NW-SE, NE-SW and E-W directions) which outline individual blocks that control swamped areas and elevated areas. The continental area would then be gradually covered by subsequent estuarine and

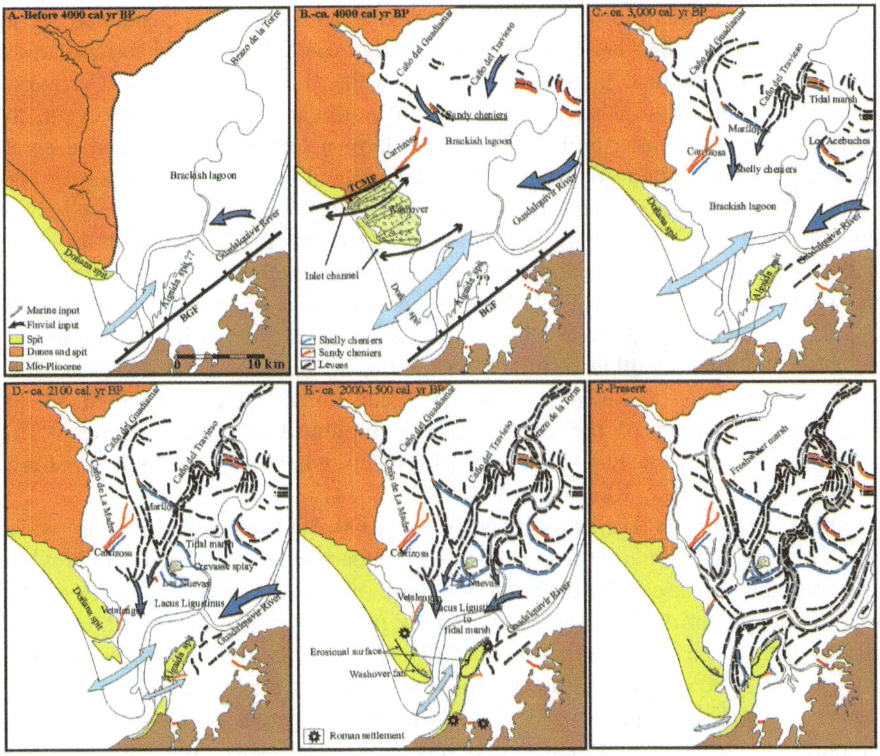

Fig. 22.8 Paleogeographical approach of the Guadalquivir estuary during the last 5000 years (modificated from Rodríguez-Ramírez et al. 2014, 2015, 2016)

coastal developments in the course of the Holocene, burying and fossilizing the dune formations of El Abalario. The macrofaunal and foraminifera assemblages from this period indicate a brackish lagoon, with tidal flows that would introduce marine fauna towards the inner lagoon areas (Rodríguez-Ramírez et al. 2014). It was a period of notable marine transgression into the estuary and high marine influence in it, which increased its size and enlarged its communication with the open sea.

Such a gradual process of marine input and sedimentary infilling was then interrupted by a series of events of higher marine influence that occurred from c. 4000 to 3000 cal year BP. The largest was the first (Event A), around 4000 cal year BP (Fig. 22.8b). It produced a sand layer that contains a massive accumulation of shell, with predominance of open marine species. This layer sits erosively upon the aeolian dune formation of El Abalario, extending both laterally and longitudinally over many kilometres in the estuary. It may have been unknowingly detected still farther inland, near the northernmost boundary of the original estuary—the earliest outlet of the Guadalquivir River after the Holocene marine maximum—, interrupting there for a long time the formation of a delta (Arteaga et al. 1995). The deposit exhibits morphostratigraphic traits of a wash-over fan generated by an over-wash process that broke the existing sandy coastal barrier and formed a tidal channel, which in turn favoured, after the event, the growth of the Carrizosa–Vetalarena sandy littoral strands towards the northeast. Also after the event, the barrier morphology was probably dominated by a low-relief ridge and a swale topography that invited greater marine influence and the formation of sandy cheniers inside the estuary. The sequence of sedimentary facies in the analysed cores, the extensive area in the paleo-estuary affected by the event, and the great paleo-geographical changes that the event unleashed, including a large overwash of the southernmost section of El Abalario, are all clear signs of a tsunamigenic process of a large magnitude (Fig. 22.8a). Such a violent marine incursion may have affected the Neolithic and Copper Age human settlements that existed at the time in the area, to judge by the artefacts found widely scattered in the paleo-estuary. Some of the debris later accumulated in sandy cheniers. This tsunamigenic process has been recognized elsewhere in the Gulf of Cadiz (Camacho et al. 2017; Koster and Reicherter 2014).

Other high-energy events followed in the 4000–3000 cal year BP span, which was a phase of great marine dominance in the estuary when compared with previous and subsequent phases. The magnitude of these other events was apparently smaller than that of Event A, however. Event B took place ~3550 cal year BP, leaving a record that can be correlated, at a regional level, with that of an earthquake which occurred ~3600 cal year BP in the Southwest Portuguese Margin (Vizcaíno et al. 2006). Event C took place ~3150 cal year BP, impacting on an extensive geographic area in the estuary and terminating a Middle Bronze Age settlement near the mouth of the Guadalquivir River. The record of this third event might be correlated with that of an event—likely a tsunami—recognized in the Tinto–Odiel (Morales et al. 2008) and Guadalete estuaries (Lario et al. 1995), also in the Gulf of Cadiz.

Though occurring in a 1000-year phase of great marine influence in the estuary, the three EWEs punctuated a gradual process of isolation of the area from the ocean, which produced deposits in the estuary of ever finer facies with ever less allochthonous elements. Indeed, analysis of the sandy littoral strands, which contain shells from different environments, indicate an open estuary—not restricted by sandy barriers—up to c. 3000 cal year BP or even later (Fig. 22.8c). The oldest evidence of emerged sedimentary units in the estuary is the sandy littoral strands of Carrizosa–Vetalarena, which might also represent the highlight of the maximum sea-level rise in the Gulf of Cadiz during the Holocene. The Guadalquivir chenier plain, as remarked above, resulted from the reworking and subsequent sedimentation of the residual lag of sand, shells, pebbles and archaeological remains that had been carried far into paleo-estuary by the energy of events of different sorts (storms, tsunamis, stream floods), ending up accumulating in longshore bars or ridges (the cheniers).

The transition from an open-estuary scenario to a semi-closed one culminated ~3000 cal year BP, as a result of both the acceleration of coastal progradation and the growth of the sandy barriers that flanked the mouths of the river

(Fig. 22.8c). This gradual process left a record of successive littoral strands of the shelly chenier plain (Rodríguez-Ramírez and Yáñez 2008). The shelly littoral strands suggest a partial closure of the estuary; the cores contain scant input of sand as well as remains of fauna from more restricted environments. After ~3000 cal year BP, the estuary started to find itself increasingly confined. Inputs of transported marine fauna decreased dramatically while both macro- and micro-fauna from sheltered environments became paramount. The fluvial sedimentary inputs in an ever more restricted environment combined with the growth of the sandy barriers to bring about a number of silty–clayey formations in the estuary.

Infilling of the estuary with such formations must have accelerated still further in the course of the 1st millennium BCE, when the area entered historical times and for which archaeological data and paleo-geographical information from written testimonies supplement geological findings and inferences. In such early historical period, the sedimentary effects of agriculture-, mining- and metallurgy-related deforestation north of the lower Guadalquivir basin multiplied in an environment increasingly isolated from the Ocean. These economic activities developed significantly in the periods of the kingdom of Tartessus (7th–6th centuries BCE) and the republic of Rome (3rd–1st century BCE). The Roman authors Mela (1987: 8) and R. F. Avienus (in Gavala y Laborde 1959: Apéndice, xviii) inform of the presence in the estuary of a coastal lagoon from about the 6th century BCE—if not earlier—to the 1st century AD. Avienus called it *Lacus Ligustinus* (Fig. 22.8d). Feeding on the Guadalquivir, the Guadiamar and other convergent rivers, the lagoon emptied into the Ocean by means of two outlets. The information is upheld in part by the Greek geographer and ethnologist Strabo of Amasia, who wrote (1966: 45) that the river *Baetis*, nowadays the Guadalquivir River, had two mouths rather than one. In the first centuries of the Christian era, coastal progradation as well as infilling of the estuary must have proceeded even more rapidly as the sedimentary impact of deforestation was compounded by the slowing down or cessation of subsidence processes (Rodríguez-Ramírez et al. 2014).

Geological data substantiate the information provided by Mela, Avienus and Strabo. There is geomorphic, sedimentary and chronological evidence that two outlet channels emerged from a brackish coastal lagoon north of the present-day mouth of the Guadalquivir River and flowed towards the Ocean, one channel being larger than the other (Fig. 22.8d). The larger, or 'Vetalengua,' channel spanned the distance between the Doñana spit system and the present-day Algaida spit system; the other channel run between the latter system and the hills that rise east of Sanlúcar de Barrameda. In between both channels lay Algaida, which was therefore an isle rather than a spit at the time; Avienus called it *Cartare Insula*, 'the isle of Cartare.' Periodic migration or transformation of riverbeds in a wide estuary is characteristic of deltaic dynamics, like those at work in the Guadalquivir estuary since the highest stand reached by the Ocean in the Holocene. Together with the relict channel outlets, the very morphology of today's Algaida spit system is a clear instance of such dynamics, as they have generated in this system erosional scarps or paleocliffs besides transforming an isle back into a peninsula.

During the first centuries of the Christian era, successive littoral strands formed in both areas, Algaida and Doñana, which resulted in the closing of the eastern as well as the western channel outlet (Fig. 22.8e). Inside the estuary, the fluvial levees and the chenier of Las Nuevas grew larger; in effect, this shelly chenier is the clearest sign that the estuary had become progressively isolated from the sea as the growth of the sandy coastal barriers brought ever less marine sedimentation into the old basin (Rodríguez-Ramírez and Yáñez 2008). The Doñana spit system grew some 5 km towards the SE while, south of the isle of Algaida, a tombolo started to form which eventually connected the mainland to the former isle, turning it back into a spit system. Such processes of coastal progradation and sedimentary infilling of the estuary caused some Roman settlements living off marine resources to be abandoned; e.g., Tesorillo, northeast of the Algaida spit, and Ebora, northeast of Sanlúcar de Barrameda. Coastal progradation from the 1st century BCE to the 2nd or 3rd century CE would be the local manifestation in the Guadalquivir estuary of what Zazo et al. (2008) have defined as 'Progradation Phase H_5' for the Gulf of Cadiz.

As coastal progradation developed, storm surges took place. These events generated wash-over fans in the Doñana spit in the 1st century BCE; earlier, in the 4th–3rd centuries BCE, they had generated crevasse-splay formations in the levees of the inner side of the estuary (Rodríguez-Ramírez et al. 2016). In the 2nd or 3rd century CE, 'Progradation Phase H_5' in the estuary was interrupted by a EWE, leaving as geomorphological and sedimentary manifestations: (1) a significant erosive scar on the oldest littoral strands in the Doñana spit, (2) wash-over fans and (3) a striking sedimentary lag in the estuary that contains abundant shells (Rodríguez-Ramírez et al. 2016). Furthermore, accumulation in the chenier of Las Nuevas crested between the 2nd and the 4th century CE, especially in the 3rd century CE—a development which can be explained as the result of the reworking and subsequent build-up of the basal residual lag in the estuary that the EWE had originated.

Following this EWE, progradation of the spits and sedimentary infilling of the estuary resumed. In the spits, noticeable sets of ridges and swales emerged, with substantial aeolian developments which have continued up to the present. The onset of this new progradation phase would be the same as that of 'Progradation Phase H_6' for the Gulf of Cadiz as defined by Zazo et al. (2008). The western, or remaining, outlet channel of the Guadalquivir River narrowed considerably as it migrated towards the south, away from the Roman settlement by Cerro del Trigo, on the Doñana spit. At this site, archaeologists have recognised evidence of an economic decline in the 5th century CE and an abandonment of the settlement in the 6th century CE (Campos et al. 2002), in all likelihood because of the gradual confinement of the estuary. This confinement affected the shelly cheniers of Vetalengua and Las Nuevas, where the marks of the progressive advance of the fluvial levees can be identified. The evolution of the riverbed and the gradual infilling of the estuary would turn *Lacus Ligustinus* into a tidal marsh and subsequently into a freshwater marsh (Fig. 22.8e).

22.8 The Anthropogenic Degradation

The rapid sedimentary infilling of the estuary over the past two millennia might have been the effect of agricultural activity and deforestation practiced on a scale larger than before (Fig. 22.1). Moreover, there is the progressive degradation in recent decades of the fluvial-tidal dynamics in the estuary, as a result of river channelling, the construction of numerous dams in the hydrographical basin and other initiatives encroaching upon the river (Rodríguez-Ramírez et al. 2008).

The dams—as many as 113 in the hydrographical basin—have reduced the river flow dramatically, noted especially on the occasion of the large winter floods. Because of these floods, the flow of the river used to exceed 10,000 m^3/s, the mean annual discharge being 164 m^3/s. Since 1945, according to Menanteau (1981), the mean winter flow has decreased 3,000 m^3/s, from 5,000 to 2,000 m^3/s. The hydrographical basin covers an area of 56,978 km^2 and has a length of 657 km, the reservoir capacity being 8,782 hm^3. In the 1960s the dams caused the extinction of species, such as sturgeon (*Acipenser* sp.).

The Guadalquivir is the only navigable river in Spain (Fig. 22.9). In the past decades, new river cut-offs have eliminated most meanders and some of the river channels while the artificial bunds have isolated large portions of the marshland from the river and tidal floods (Fig. 22.1). These anthropogenic transformations have altered the tidal wave, to the extent of confining it mostly to the lower course of the river and increasing its range and speed (Vanney 1970); there the speed now attains 1.35 m/s at high tides and 0.45 m/s at low tides, with acceleration in the rectilinear sectors of the cut-offs that can reach 48 km/h in the lower section (Vanney 1970). The first of the river cut-offs took place in 1755; the last in 1972. This hydrodynamic alteration even affects Doñana National Park, where the fluvial-tidal contributions are very scarce; it causes serious problems to the fauna every year. There are other threats besides. Intensive agriculture is practiced in the marshland outside the Park, rice being the main crop. Salt-work factories and fish hatcheries have proliferated in the outlet area. Regular dredging of the river—especially at its mouth, where sandy lowlands multiply—is carried out to facilitate navigation of large ships.

Fig. 22.9 A ship crossing the Guadalquivir marshland

Today, longshore drift is becoming stronger and sedimentation is increasing at the end of the Doñana spit. Beach ridges are forming with a NE orientation, encroaching into the Guadalquivir riverbed with a progradation rate of 375 m from 1956 to 1996 (9 m/year) and a cross-sectional diminution of the outlet (Rodríguez-Ramírez et al. 2003). In contrast to other estuaries in the Huelva coast, no jetties have been built in the mouth of the Guadalquivir River; instead, there has been growing urbanization on the left margin of the mouth, where Sanlúcar de Barrameda stands. It is precisely in this area where the highest rates of retrogradation occur.

References

Armijo, R., Benkhelil, J., Bousquet, J. C., Estévez, A., Guiraud, R., Montenat, C., et al. (1977). Chapitre III, Les résultats de l'analyse structurale en Espagne. Groupe de recherche néotectonique de l'Arc de Gibraltar. L'histoire tectonique récent (Tortonien à Quaternaire) de l'Arc de Gibraltar et des bordures de la mer d'Alboran. *Bulletin de la Société Géologique de France, 7–19,* 575–614.

Arteaga, O., Schulz, H. D., & Roos, A. M. (1995). El problema del 'Lacus ligustinus': Investigaciones geo-arqueológicas en torno a las marismas del Bajo Guadalquivir. In D. Ruiz Mata (Ed.), *Tartessos 25 años después, 1968–1993: Actas del Congreso Conmemorativo del V Symposium Internacional de Prehistoria Peninsular* (pp. 99–133). Jerez de la Frontera, Spain: Ayuntamiento de Jerez de la Frontera.

Augustinus, P. G. E. F. (1989). Cheniers and chenier plains: A general introduction. *Marine Geology, 90*(4), 219–229.

Camacho, S., Bosky, T., Moura, D., Scott, D., Connor, S., & Pereira, L. (2017). Paleo-environmental evolution of the Guadiana Estuary, Portugal, during the Holocene: A modern foraminifera analog approach. *The Holocene, 27*(2), 197–235.

Campos, J. M., Gómez-Toscano, F., Vidal, N. O., Pérez, J. A., & Gómez, C. (1999). La factoría romana de El Cerro del Trigo (Doñana, Almonte, Huelva). *Anuario Arqueológico de Andalucía, III*(1), 330–348.

CEEPYC, Centro de Estudios y Experimentación de Puertos y Costas 'Ramón Iribarren'. (1990). *Plan de estudio de la dinámica litoral de la provincia de Huelva. Informe*. Dirección General de Puertos y Costas. Madrid, 37 pp.

Cuena, G. J. (1991). *Proyecto de regeneración de las playas de Isla Cristina*. Informe del Servicio de Costas. Unpublished report. Ministerio de Obras Públicas y Turismo de España. Madrid, 100 pp.

Curray, J. R. (1964). Transgression and Regression. In R. C. Miller (Ed.), *Papers in Marine Geology* (pp. 175–203). Amsterdam: Elsevier.

Dabrio, C. J., Zazo, C., Lario, J., Goy, J. L., Sierro, F. J., Borja, F., et al. (1999). Sequence stratigraphy of Holocene incised-valley fills and coastal evolution in the Gulf of Cadiz (southern Spain). *Geologie en Mijnbouw, 77,* 263–281.

Flores-Hurtado, E. (1993). *Tectónica reciente en el margen ibérico suroccidental* (p. 458). Huelva: Tesis doctoral, Universidad de Huelva (Spain).

Foster, I. D. L., Albon, A. J., Bardell, K. M., Fletcher, J. L., Jardine, T. C., Mothers, R. J., et al. (1991). High-energy coastal sedimentary deposits: An evaluation of depositional processes in Southwest England. *Earth Surface Processes and Landforms, 16,* 341–356.

Fujiwara, O., Masuda, F., Sakai, T., Irizuki, T., & Fuse, K. (2000). Tsunami deposits in Holocene bay mud in Southern Kanto region, Pacific coast of central Japan. *Sedimentary Geology, 135* (1–4), 219–223.

Gavala y Laborde, J. (1959). *La geología de la costa y la bahía de Cádiz y el poema 'Ora Maritima'*. Madrid: de Avieno. IGME.

Gould, H. R., & McFarlan, E., Jr. (1959). Geologic history of the Chenier plain, Southwestern Louisiana. *Gulf Coast Association of Geological Societies Transactions, 9,* 261–270.

Goy, J. L., Zazo, C., Dabrio, C. J., & Lario, J. (1994). Fault-controlled shifting shorelines in the Gulf of Cadiz since 20 Ky BP. In: *1st symposium on the Iberian Atlantic Continental Margin, Abstract*, Vol. 24. Lisbon.

Gracia, F. J., Alonso, C., Benavente, J., Anfuso, G., & Del Río, L. (2006). The different coastal records of the 1755 tsunami waves along the South Atlantic Spanish coast. *Zeitschrift für Geomorphologie Supplement, 146,* 195–220.

Hoyt, J. H. (1969). Chenier versus barrier, genetic and stratigraphic distinction. *Bulletin of the American Association of Petroleum Geologists, 53,* 299–306.

Jiménez-Moreno, G., Rodríguez-Ramírez, A., Pérez-Asensio, J. N., Carrión, J. S., López-Sáez, J. A., Villarías-Robles, J. J. R., et al. (2015). Impact of late-Holocene aridification trend, climate variability and geodynamic control on the environment from a coastal area in SW Spain. *The Holocene, 25*(4), 607–617.

Kortekaas, S. (2002). *Tsunamis, storms and earthquakes: Distinguishing coastal flooding events*. (Ph.D. Thesis, Coventry University (UK). Coventry, 171 pp.

Koster, B., & Reicherter, K. (2014). Sedimentological and geophysical properties of a ca. 4000 year-old tsunami deposit in Southern Spain. *Sedimentary Geology, 314,* 1–16.

Lario, J., Zazo, C., Dabrio, C. J., Somoza, L., Goy, J. L., Bardají, T., et al. (1995). Record of recent Holocene sediment input on spit bars and deltas of South Spain. *Journal of Coastal Research, 17,* 241–245.

Lario, J., Zazo, C., Plater, A. J., Goy, J. L., Dabrio, C. J., Borja, F., et al. (2001). Particle size and magnetic properties of Holocene estuarine deposits from the Doñana National Park (SW Iberia): Evidence of gradual and abrupt coastal sedimentation. *Zeitschrift für Geomorphologie, 45,* 33–54

Lario, J., Luque, L., Zazo, C., Goy, J. L., Spencer, C., Cabero, A., et al. (2010). Tsunami versus storm surge deposits: A review of the sedimentological and geomorphological records of extreme wave events (EWE) during the Holocene in the Gulf of Cadiz, Spain. *Zeitschrift für Geomorphologie, 54*(3), 301–316.

Martínez Solares, J. M., López Arroyo, A., & Mezcua, J. (1979). Isoseismal map of the 1755 Lisbon earthquake obtained from Spanish data. *Tectonophysics, 53,* 301–313.

Mastronuzzi, G., & Sansò, P. (2012). The role of strong earthquakes and tsunami in the late Holocene Evolution of the Fortore River coastal plain (Apulia, Italy): A synthesis. *Geomorphology, 138*(1), 89–99.

McBride, R. A., Taylor, M. J., & Byrnes, M. R. (2007). Coastal morphodynamics and Chenier-Plain evolution in Southwestern Lousiana, USA: A geomorphic model. *Geomorphology, 88*(3–4), 367–422.

Medialdea, T., Somoza, L., Pinheiro, L. M., Fernández-Puga, M. C., Vázquez, J. T., León, R., et al. (2009). Tectonics and mud volcano development in the Gulf of Cádiz. *Marine Geology, 261*(1–4), 48–63.

Mela, P. (1987). De chorograhia. In V. Bejarano (Ed.), *Fontes Hispaniae Antiquae VII*. Barcelona: Instituto de Arqueología y Prehistoria.

Melières, F. (1974). *Recherche sur la dynamique sédimentaire du Golfe de Cadix (Espagne)*. Ph.D. Thesis, Université de Paris VI. Paris, 325 pp.

Menanteau, L. (1981). Les Marismas du Guadalquivir: Exemple de transformation d'un paysage alluvial au cours du Quaternaire récent. Ph.D. Thesis, Université de Paris-Sorbonne. Paris, 252 pp.

Morales, J. A., Cantano, M., Rodríguez-Ramírez, A., & Martín, R. (2006). Mapping geomorphology and active processes on the coast of Huelva (SW Spain). *Journal of Coastal Research, 48,* 89–99.

Morales, J. A., Borrego, J., San Miguel, E. G., López-González, N., & Carro, B. (2008). Sedimentary record of recent tsunamis in the Huelva Estuary (Southwestern Spain). *Quaternary Science Reviews, 27*(7–8), 734–746.

Morton, R. A., Gelfenbaum, G., Buckley, M. L., & Richmond, B. M. (2011). Geological effects and implications of the 2010 tsunami along the central coast of Chile. *Sedimentary Geology, 242,* 34–51.

Otvos, E. G., Jr. (2000). Beach ridges—Definitions and significance. *Geomorphology, 32*(1–2), 83–108.

Otvos, E. G., Jr., & Price, W. A. (1979). Problems of chenier genesis and terminology—An overview. *Marine Geology, 31*(3–4), 251–263.

Pozo, M., Ruiz, F., Carretero, M. I., Rodríguez-Vidal, J., Cáceres, L. M., Abad, M., et al. (2010). Mineralogical assemblages, geochemistry and fossil associations of Pleistocene-Holocene complex siliciclastic deposits from the Southwestern Doñana National Park (SW Spain): A palaeoenvironmental approach. *Sedimentary Geology, 225,* 1–18.

Reicherter, K. R. (2001). Paleoseismologic advances in the Granada Basin (Betic Cordilleras, Southern Spain). *Acta Geologica Hispanica, 36,* 267–281.

Rodríguez-Ramírez, A. (1998). *Geomorfología del Parque Nacional de Doñana y su Entorno* (p. 146). Madrid: Organismo Autónomo Parques Nacionales, Ministerio de Medio Ambiente.

Rodríguez-Ramírez, A., & Yáñez, C. M. (2008). Formation of chenier plain of the Doñana marshland (SW Spain): Observations and geomorphic model. *Marine Geology, 254*(3–4), 187–196.

Rodríguez-Ramírez, A., Rodríguez-Vidal, J., Cáceres, L., Clemente, L., Belluomini, G., Manfra, L., et al. (1996). Recent coastal evolution of Doñana National Park (Southern Spain). *Quaternary Science Reviews, 15,* 803–809.

Rodríguez-Ramírez, A., Ruiz, F., Cáceres, L. M., Rodríguez-Vidal, J., Pino, R., & Muñoz, J. M. (2003). Analysis of the recent storm record in the South-western Spain coast: Implications for littoral management. *Science of the Total Environment, 303,* 189–201.

Rodríguez-Ramírez, A., Morales, J. A., Delgado, I., & Cantano, M. (2008). The impact of man on the morphodynamics of the Huelva Coast (SW Spain). *Journal of Iberian Geology, 34,* 73–108.

Rodríguez-Ramírez, A., Flores-Hurtado, E., Contreras, C., Villarías-Robles, J. J. R., Jiménez-Moreno, G., Pérez-Asensio, J. N., et al. (2014). The role of neo-tectonics in the sedimentary infilling and geomorphological evolution of the Guadalquivir Estuary (Gulf of Cadiz, SW Spain) during the Holocene. *Geomorphology, 219,* 126–140.

Rodríguez-Ramírez, A., Pérez-Asensio, J. N., Santos, A., Jiménez-Moreno, G., Villarías-Robles, J. J. R., Mayoral, E., et al. (2015). Atlantic extreme wave events during the last four millennia in the Guadalquivir Estuary. *SW Spain. Quaternary Research, 83*(1), 24–40.

Rodríguez-Ramírez, A., Villarías-Robles, J. J. R., Pérez-Asensio, J. N., Santos, A., Morales, J. A., Celestino, S., et al. (2016). Geomorphological record of extreme wave events during roman times in the Guadalquivir Estuary (Gulf of Cádiz, SW Spain): An archaeological and paleogeographical approach. *Geomorphology, 261,* 103–118.

Ruiz, F., Rodríguez-Ramírez, A., Cáceres, L. M., Rodríguez-Vidal, J., Carretero, M. I., Clemente, L., et al. (2004). Late Holocene evolution of the Southwestern Doñana National Park (Guadalquivir Estuary, SW Spain): A multivariate approach. *Palaeogeography, Palaeoclimatology, Palaeoecology, 204,* 47–64.

Ruiz, F., Rodríguez-Ramírez, A., Cáceres, L. M., Rodríguez-Vidal, J., Carretero, M. I., Abad, M., et al. (2005). Evidence of high-energy events in the geological record: Mid-Holocene evolution of the Southwestern Doñana National Park (SW Spain). *Palaeogeography, Palaeoclimatology, Palaeoecology, 229,* 212–229.

Salvany, J. M. (2004). Tilting neotectonics of the Guadiamar drainage basin, SW Spain. *Earth Surface Processes and Landforms, 29,* 145–160.

Salvany, J. M., & Custodio, E. (1995). Características lito-estratigráficas de los depósitos plio-cuaternarios del bajo Guadalquivir en el área de Doñana: Implicaciones hidrogeológicas. *Revista de la Sociedad Geológica de España, 8,* 21–31.

Salvany, J. M., Larrasoaña, J. C., Mediavilla, C., & Rebollo, A. (2011). Chronology and tctonosedimentary evolution of the Upper Pliocene to Quaternary deposits of the lower Guadalquivir foreland basin, SW Spain. *Sedimentary Geology, 241,* 22–39.

Santos, L., Sánchez-Goñi, M. F., Freitas, M. C., & Andrade, C. (2003). Climatic and Environmental Changes in the Santo André Coastal Area (SW Portugal) during the Last 15,000 years. In M. B. Ruiz Zapata, et al. (Eds.), *Quaternary climatic changes and environmental crises in the mediterranean region* (pp. 175–179). Alcalá de Henares, Spain: Ediciones de la Universidad de Alcalá de Henares.

Silva, P. G., Borja, F., Zazo, C., Goy, J. L., Bardají, T., De Luque, L., et al. (2005). Archaeoseismic record at the ancient Roman City of Baelo Claudia (Cadiz, South Spain). *Tectonophysics, 408,* 129–146.

Stzabo of Amasia. (1966). *Géographie, tome II (Livres III et IV): Texte établie et traduit par François Laserre*. Paris: Les Belles Lettres.

Terrinha, P., Pinheiro, L. M., Henriet, J. P., Matias, L., Ivanov, M. K., Monteiro, J. H., et al. (2003). Tsunamigenic–seismogenic structures, neotectonics, sedimentary processes and slope instability on the Southwest Portuguese margin. *Marine Geology, 195*(1–4), 55–73.

Torelli, L., Sartori, R., & Zitellini, N. (1997). The giant chaotic body in the Atlantic Ocean off Gibraltar: New results from a deep seismic reflection survey. *Marine and Petroleum Geology, 14,* 125–138.

Vanney, J. R. (1970). *L'hydrologie du Bas Guadalquivir* (p. 176). Madrid: Editorial CSIC.

Viguier, C. (1977). Les grands traits de la tectonique du basin néogène du Bas Guadalquivir. *Boletín Geológico y Minero, 88,* 39–44.

Vizcaíno, A., Gràcia, E., Pallàs, R., Garcia-Orellana, J., Escutia, C., Casas, D., et al. (2006). Sedimentology, physical properties and ages of mass-transport deposits associated to the Marquês de Pombal Fault, Southwest Portuguese Margin. *Norwegian Journal of Geology, 86,* 177–186.

Zazo, C., Goy, J. L., Somoza, L., Dabrio, C. J., Belluomini, G., Improta, S., et al. (1994). Holocene sequence of relative sea level highstand–lowstand in relation to the climatic trends in the Atlantic-Mediterranean linkage coast: Forecast for future coastal changes and hazards. *Journal of Coastal Research, 10*(4), 933–945.

Zazo, C., Dabrio, C. J., González, A., Sierro, F., Yll, E. I., Goy, J. L., et al. (1999). The record of the latter glacial and interglacial periods in the Guadalquivir marshlands (Mari López Drilling, SW Spain). *Geogaceta, 26,* 119–122.

Zazo, C., Mercier, N., Silva, P. G., Dabrio, C. J., Goy, J. L., Roquero, E., et al. (2005). Landscape evolution and geodynamic controls in the Gulf of Cadiz (Huelva Coast, SW Spain) during the Late Quaternary. *Geomorphology, 68*(3–4), 269–290.

Zazo, C., Dabrio, C. J., Goy, J. L., Lario, J., Cabero, A., Silva, P. G., et al. (2008). The coastal archives of the last 15 ka in the Atlantic-Mediterranean Spanish linkage area: Sea level and climate changes. *Quaternary International, 181*(1), 72–87.

Chapter 23
Estuaries of the Huelva Coast: Odiel and Tinto Estuaries (SW Spain)

Berta Carro, José Borrego and Juan A. Morales

Abstract The estuary of the Odiel and Tinto rivers is located in the southwest of the Iberian Peninsula on the western Gulf of Cadiz (Atlantic Ocean). The Huelva estuary is constituted by the common mouth of the Odiel and Tinto Rivers in a "Y" shape oriented in a N-S direction across 35 km long incised valley. This estuary was generated after the Flandrian Transgression (Holocene), which signified the marine inundation of the main fluvial valleys incised by the rivers during the last Pleistocene lowstand. It extends along the south-western coastal margin of the Guadalquivir sedimentary basin that was incised on Cenozoic non-consolidated sediments during the late Pleistocene and early Holocene when sea level was located up to 100 m below the present position. This estuary is presently completely filled with sediments and has started to prograde to build a delta. The fluvial basin of both rivers is seriously affected by acid mine drainage, so this estuary have a induces important changes in the chemical characteristics of the water, suspended particulate matter (SPM) and sediments.

23.1 Introduction

Located in the southwest of the Iberian Peninsula on the Atlantic coast north of the Gulf of Cadiz (Fig. 23.1a), the estuary of the Odiel and Tinto rivers, also known as the Ria of Huelva, extends along the southwestern coastal margin of the Guadalquivir sedimentary basin. The estuary was incised into unconsolidated Cenozoic sediments during the late Pleistocene and early Holocene, when the sea level was up to 100 m lower than its present position. The sea began to flood the

B. Carro (✉) · J. Borrego · J. A. Morales
Department of Earth Sciences, University of Huelva, Huelva, Spain
e-mail: berta.carro@dgeo.uhu.es

J. Borrego
e-mail: borrego@uhu.es

J. A. Morales
e-mail: jmorales@uhu.es

J. A. Morales (ed.), *The Spanish Coastal Systems*,
https://doi.org/10.1007/978-3-319-93169-2_23

fluvial valley formed during this lowstand period in the course of the Holocene transgression. It was approximately 8,700 years BP that the lowest unit of the Holocene estuarine infilling commenced deposition (Borrego et al. 1999), but the estuary is now entirely infilled with sediment and has started to prograde to form a delta.

From a physiographic perspective, we can define this estuarine system as a bar-built estuary (Borrego et al. 1995), following the Fairbridge (1980) criteria, or a highly evolved wave-dominated estuary according to the criteria set forth by Dalrymple et al. (1992). The interior of this coastal environment is made up of wide tidal flats and salt marshes which developed on top of estuarine accretionary bodies of fluvial-tidal origin (Borrego et al. 1999). The mouth of this estuarine system consists of the following three barriers (Fig. 23.1b) separated by the Punta Umbría and Padre Santo channels:

(a) Punta Umbría spit to the west.
(b) Saltés Island, a complex chenier plain system constituted by shelly ridges running subparallel to the coastline.
(c) Torre Arenillas counterspit, which has formed on the eastern margin and joins directly onto a Pleistocene cliff.

The fluvial basin of the Tinto and Odiel rivers lies, to a great extent, on top of Palaeozoic materials and, more specifically, on a volcano-sedimentary complex (VSC), containing some of the largest metallic ore deposits in Europe (Fig. 23.1a), which have been mined for at least the last 4,500 years (Leblanc et al. 2000). Natural changes to these sulphide deposits, in conjunction with mining activity has led to the secular pollution of the Odiel and Tinto rivers, whose waters now contain very high concentrations of heavy metals and have extremely low pH values—less than 3 (Grande et al. 2000). The combination of acidic waters from mines, industrial effluents, and seawater has played a decisive role in the development of the chemical composition of the water in the estuary.

This estuary is therefore seriously affected by acid mine drainage, giving rise to a twofold mixing process: the combination of saline and acid water. However, additionally, while the fluvial water reveals extremely low pH values and high concentrations of dissolved metals, the seawater has a pH between 7 and 8.2, resulting in a mixing process between acid river water and neutral or slightly alkaline seawater. These peculiarities lead to significant changes in the chemical characteristics of the water, suspended particulate matter (SPM) and sediments (Elbaz-Poulichet et al. 1999; Grande et al. 2000; Borrego et al. 2002; Achterberg et al. 2003).

The hydrochemical properties of the water and the hydrodynamic characteristics of this system, together with its spatial and time variations, allow the Ria of Huelva to be divided into three different zones (Carro 2002; López-González 2002):

Tinto river mixing zone. This is the zone under direct influence from the Tinto river input (Fig. 23.1b) with a pronounced longitudinal gradient, water pH values of between 2.5 and 7.0 and chlorinity measuring between 1.7 and 19.5 g L^{-1}. Large volumes of heavy metals supplied by the river Tinto make the concentrations of

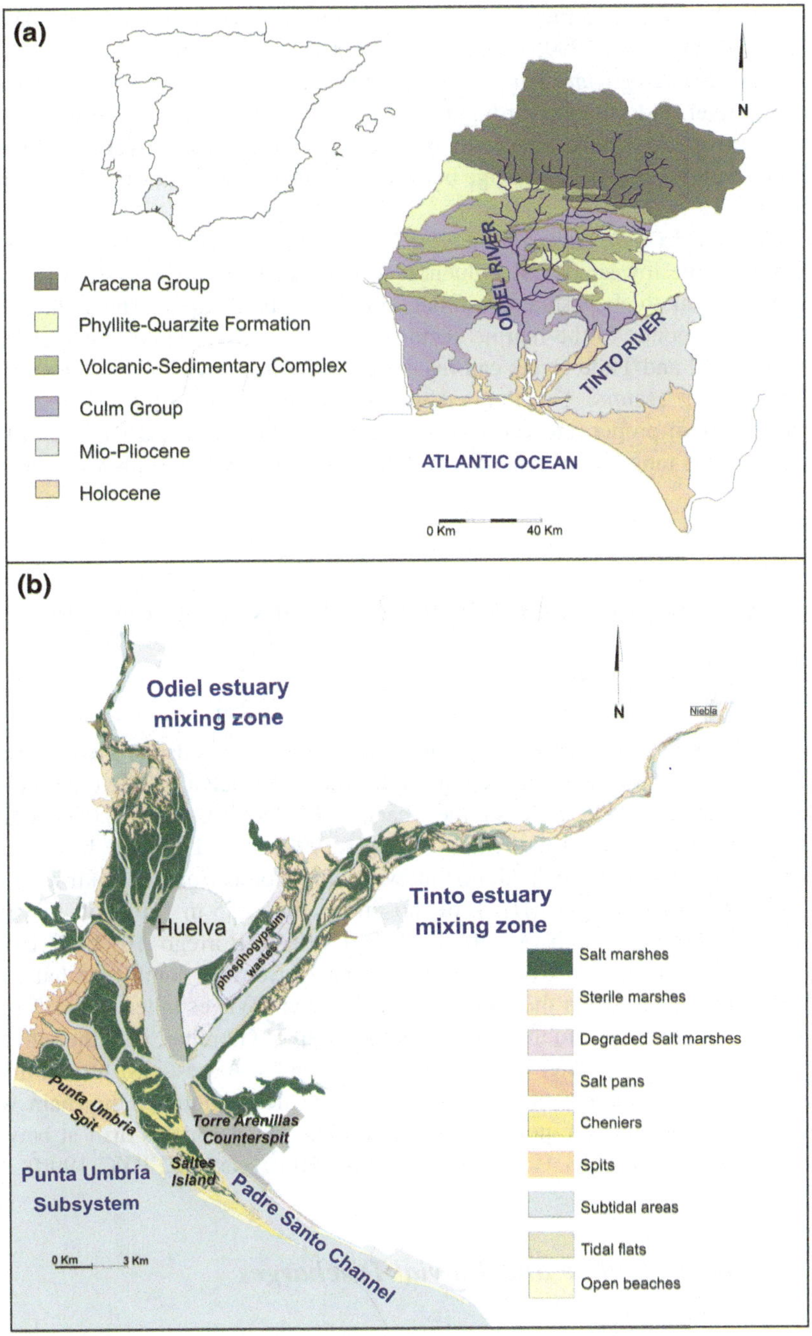

Fig. 23.1 **a** Geological setting of the Odiel and Tinto river basin. **b** Cartography of sedimentary environments

heavy metals dissolved in the water in this sector the highest in the Ria of Huelva (Elbaz-Poulichet et al. 1999; Grande et al. 2000).

Odiel river mixing zone. This sector of the estuary sees the mixing of water from the Odiel river and seawater brought in by the tide (Fig. 23.1b). The water properties reveal a pronounced longitudinal gradient, as in the Tinto river mixing zone, with the result that the pH can range between 3.5 and 8.0 and chlorinity between 2.6 and 21.0 g L^{-1}.

Punta Umbría Subsystem and Padre Santo Channel. This is the sector of the estuary with the lowest fluvial influence. The Punta Umbría subsystem is a tidal system adjacent to the estuary without any direct tidal input. The Padre Santo channel corresponds to the marine sector of the estuary, so while punctual floods from the Odiel and Tinto rivers can reach this area, they have already been diluted in their mixing sectors. The Punta Umbría system is laterally connected to the estuarine system proper and can receive a certain input of mixed waters. In these areas, water pH ranges between 6.5 and 8.2, while chlorinity presents values greater than 17.0 g L^{-1}.

23.2 Hydrodynamic and Hydrochemical Characteristics

23.2.1 Wave Regime

To understand the dynamics of this coastal sector, we must consider the importance of factors such as littoral agents acting on the coast of Huelva. While the dominant wind is southwesterly (22.5% of the year), NW (18.5%), NE (12.0%) and SE (14.0%) winds are also significant (Borrego 1992). Wind regime has a direct bearing on wave regime, so the dominant wave direction is from the southwest with a significant wave height ($H_{1/3}$) of no more than 0.5 m 75% of the time, approaching the coast at N 35–40° E (MOPU 1991; Borrego 1992). It is only during Atlantic storms and when there are seawaves from the southwest that waves taller than one metre reach the coast (7% of the time). Waves taller than 1.5 m are associated with storms from the Atlantic or the Strait of Gibraltar, with the influence of these storms being greater during the winter months. A west-east littoral drift is induced by the orientation of the coastline, which lies oblique to the main wave direction. We calculate the potential transportation values for this drift at between 1.9×10^5 m^3/year (Cuena 1991) and 3.0×10^5 Hm^3/year (CEEPYC 1990).

23.2.2 Tidal Regime and Fluvial Discharges

The Ria of Huelva exhibits a strong tidal influence, which governs the salt-induced mixing processes of large masses of water. As it lies on a mixed-energy mesotidal coast, with a mean tidal range of 2.10 m, and slight diurnal inequality, a range of

3.06 m is reached during spring tides and 1.70 m during neap tides. (Borrego et al. 1995). The tidal wave moves along the estuary following a hypersynchronous model, with the tidal range increasing landwards and reaching 4 m in the inner part during astronomic spring tides (Borrego et al. 2002). This oscillation in the duration of the tidal wave leads to zonation along the length of the intertidal sector with each of the various zones being separated by different critical tide levels (e.g. Doty 1949; Borrego et al. 1993, 1994).

The river Tinto (Fig. 23.1) rises in the Rio Tinto mining area, flows 100 km to its mouth into the estuary, and drains a basin of 720 km^2. The Odiel river has its source in the Sierra de Aracena mountain range and is 140 km long with a basin covering 2,300 km^2 (Fig. 23.1). Both rivers are torrential in nature and flow through a climatic zone with extremely irregular rainfall (Sáinz et al. 2004). Fluvial discharge into the estuary can be characterised as markedly seasonal and is very irregular from one year to the next. The mean inflow is normally less than 10 $m^3\ s^{-1}$ although in the event of substantial floods, inflow can exceed 400 $m^3\ s^{-1}$ (Borrego et al. 1994). At this latitude, the climate is characterised as having a short, mild winter when most of the annual rainfall occurs, and warm, dry summers (Cánovas et al. 2007). Freshwater inflow into the interior of the estuary from the Tinto and Odiel rivers, both of which are heavily affected by acid mine drainage, reveals significant seasonal and annual variation. Between 1960 and 1996, the average monthly inflow from both rivers was 49.8 Hm^3, with an average annual inflow of 598 Hm^3. The pronounced seasonality of this inflow is caused by a rainy season lasting from October to March, with an inflow which can at times reach monthly averages of 100 Hm^3 and a dry period (between May and September) with mean monthly volumes of less than 5 Hm^3 (Borrego 1992).

Changes in water volumes due to the river water regime inflow and the tidal prism result in a variety of mixing models in the estuary. In the rainy season, when there is a mean flow of 21 $m^3\ s^{-1}$ for any tidal situation, the mixing conditions within the estuary can be defined as "*partially stratified*" (according to the criteria set forth by Simmons 1955). On the other hand, in the dry season, when the volume of flow is less than 6 $m^3\ s^{-1}$, they can be described as a "*good mixing*" (Borrego 1992).

23.2.3 *Hydrochemical Zonation*

The hydrochemical properties of estuary systems are typically defined by the mixing processes between masses of fresh water from rivers and salt water from the sea. A typical salt-induced dilution process can be observed in the mouths of the Odiel and Tinto rivers; however, this system is peculiar due to the nature of the fluvial contributions to both rivers from their drainage basins. Acid mine drainage (AMD) affects both rivers and, consequently, the entire system with all the characteristics associated with processes of this type, e.g. acid waters with a pH lower than 3 containing high concentrations of heavy metals (Carro et al. 2011; Nelson and Lamothe 1993; Sáinz et al. 2004; España et al. 2005).

The hydrogeochemical characteristics of an estuary where acid river water and sea water mix allow the interaction of two geochemical processes to be defined: there is a typical process of salt-induced mixing and also a process involving the neutralisation of acid water from AMD. The product resulting from the convergence of these two mixing processes will not only affect specifically how the heavy metals and nutrients that form the system behave but will also allow us to develop a hydrochemical zonation of the estuary. How these zones are defined depends on the pH and chlorinity of the water, as well as on the concentrations of heavy metals in their dissolved and particulate phases.

The three zones are: (Fig. 23.2a)

Zone 1. This is a sector with pH values ranging between 2.4 and 4.5, where the chlorinity of the water is no more than 3 g L^{-1}. Here, the onset of the neutralisation process results in the precipitation of particulate sulphates, which scavenge a large portion of the concentration of metals transported by the acid water in a dissolved phase.

In the absence of salt-induced mixing, this neutralisation process leads to the precipitation of sulphates and therefore the rapid removal of dissolved S and C. The precipitated sulphates scavenge a large portion of the concentration of certain heavy metals carried by the water in their dissolved phase (e.g. Cu, Zn), as these sulphated salts act as temporary storage for heavy metals and hydrogen atoms (acidity). Along with this process, there might also be a process of oxidation of Fe^{2+} to Fe^{3+}, which produces water and consumes acidity in the following reaction (Younger et al. 2002):

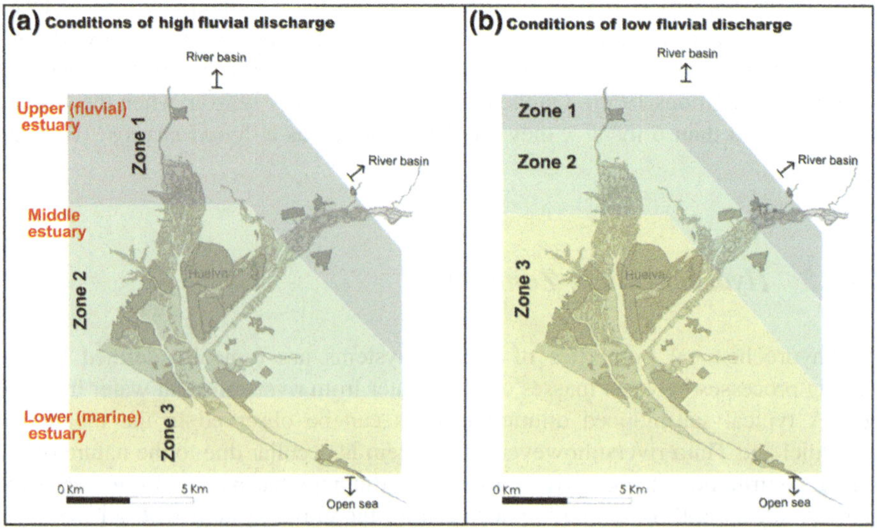

Fig. 23.2 Location and extent of zonation of the estuarine system with seasonal variations depending on the volume of fluvial discharge: **a** high fluvial discharge and **b** low fluvial discharge

$$4Fe_2^+ + O_2 + 4H^+ \rightarrow 4Fe_3^+ + 2H_2O$$

We must consider Fe_2^+ the dominant species of Fe in the inputs of fluvial water that reach the estuary (Cánovas et al. 2007).

The neutralisation process caused by the mixing of water masses is not associated with weathering of carbonates or aluminium silicates in their mineral phases; charts show that there is no link between the concentration of elements such as dissolved Ca, Na, Mg or K and pH values. At the same time, this process shows no effect resulting from the dilution of sea water with pH values higher than 7, since this would typically lead to a rise in the concentration of dissolved marine elements such as Cl, Na, Ca and S.

Zone 2. In this zone, pH and chlorinity are directly related (ranging between 4.5 and 7.5, and 3 and 15 g L^{-1}, respectively). This shows that the neutralisation process is induced by the dilution of the river water from Zone 1 with salt water, introducing dissolved (typically marine) elements into the mixture. The same behaviour that shows a link between pH values and chlorinity also suggests that other dissolved elements, such as Na, Ca, K, Mg and S, are contributed by dissolved salts in sea water.

During this mixing process by dilution, a marked fall is observed in the concentrations of certain heavy metals dissolved in river water which comes from acid mine drainage. This is true for Cu and Zn, which revealed lower concentrations in the dissolved phase, dropping from 1500 to 600 μg L^{-1} and, from 4000 to 1000 μg L^{-1} respectively. At the same time, a portion of these dissolved metals is absorbed as solid particles that go to make up the SPM, resulting in a progressive rise in the concentration of these metals in this phase. This process has also been observed in the laboratory (Achterberg et al. 2003; España et al. 2006).

Zone 3. This zone is characterised by slightly basic pH water (between 7.5 and 8.2), where the salt-induced mixing process typical of estuarine systems occurs. Here, the concentration of certain dissolved elements such as Cu or U increases as a result of remineralisation caused by saline shock, a process which sees some of the concentration of the chemical element in its particulate phase re-dissolved and transferred to the dissolved phase.

The average increase in the dissolved concentrations of Cl, Na and S and the distribution of the concentration of heavy metals dissolved in water both depend on the chemical element in question. While certain elements such as dissolved Mn or Zn experience a marked drop in concentration as a result of the dilution of estuarine water with open sea water, other elements such as dissolved Cu or U increase in concentration.

These three zones move over the hydrological year due to seasonal chemical variations induced by the presence or absence of discharges. Thus, during periods of low fluvial inputs, the tidal influence area moves upstream causing the mixing zone to be pushed northwards (Fig. 23.2b).

23.3 Sedimentary Characteristics

The Huelva estuary—essentially the shared mouth of the rivers Odiel and Tinto in the shape of a Y running north to south across a 35 km-long incised valley—was formed after the Flandrian Transgression (Holocene), when sea water flooded the main river valleys incised during the last Pleistocene lowstand. The Huelva estuarine system is presently in an advanced state of clogging as a result of the sea level stand over the last 4,500 years (Rodríguez Ramírez et al. 1996) and an elevated accumulation rate during the Holocene (Dabrio et al. 2000).

The estuaries of the Odiel and Tinto rivers display a morphodynamic longitudinal division in 3 domains (Fig. 23.2a).

Upper (fluvial) estuary. A braided channel system separating salt-marsh deposits characterises the innermost environment. The morphological framework is governed by tidal control in the deeper channels where silt and clay are deposited. The fluvial component is sand, with coarse sand and gravel patches accumulating mainly in the distributary channels. The river exerts strong control of the sedimentation in this domain; this is both from a dynamic point of view and also by the volume of sediments. The sediment supplied by these rivers is very similar and consists of red sands and yellow silts with a high iron content.

Middle estuary. The inner domain of both estuaries is a sector with mixed energy between river and tide, although tides dominate over the weak fluvial action. The tidal regime is the main control factor on the sedimentary dynamics of the central domain of the estuaries. The velocities and range of the tidal currents directly control the conditions of sedimentation in the intertidal band and the estuarine channels. From a physiographical point of view, this domain is characterised by salt-marsh bodies with a fusiform morphology. Various distributary channels create elongated salt-marsh bodies where the vegetation is dominated by halophytes which can tolerate variable salinity. Water mixing (salinity and pH) takes place in this domain, and the main sedimentary processes are flocculation and tidal reworking of fluvial sands supplied by the river floods.

Lower (marine) estuary. Three sandy barriers separated by an intermediate salt-marsh body can be seen in a cross-section of the outer zone. Saltés Island lies in the middle of the estuary between the deeper Padre Santo Channel, which is the main passage to the Huelva harbour, and the narrower Punta Umbría Channel. The tidal effect is very pronounced in both the channels and the protected intertidal zones, while the swell action is considerable in the more exposed intertidal areas. Sedimentation is controlled by the linked action of tides and waves. Waves from the SW are responsible for transporting the sand supplied to grow the spits that partially close the estuary.

In addition, a vertical division of sub-environments throughout the estuary is controlled by tidal exposure and submersion levels. In terms of this work, three sub-environments are considered: subtidal (channels and estuarine accretion bodies), non-vegetated intertidal (active channel margins and tidal flats) and vegetated intertidal (salt marshes).

23.3.1 Sedimentary Environments

Seven sedimentary environments have been identified and characterised according to the relationships between processes and sediments (Borrego et al. 2000), originating facies associations, and the sequences in each that allow them to be recognised in the stratigraphic record (Fig. 23.1b).

1. *Estuarine accretion bodies*: these bodies constitute the primary infilling of the subtidal zones in the estuary in the marine, central and inner domains (Frey and Howard 1986). Sedimentary bodies in the central and inner domains usually present a muddy lithology with low sand content and are normally very bioturbated by the activity of annelids (*Arenicolides ecaudata*), bivalves (*Scrobicularia plana*, *Crassostrea angulata* and *Mytilus galloprovincialis*) and gastropods (*Cymbium olla*, *Murex brandaris* and *Cerithium rupestre*).
2. *Estuarine channels*: the estuarine channels of the Tinto and Odiel rivers connect both fluvial channels with the outer estuary. The facies consists of homogeneous, red, medium-coarse sand deposits that lack bioturbation. The sand in shallow areas is fine-medium with a clay-silt matrix and is bioturbated. It also contains a large number of shell fragments because of the mixture of fluvial sands and marine sediment from the intense tidal reworking of the sediment in the channel environment. This tidal bedding of interbedded sands and muds can be seen in channel sequences.
3. *Tidal creeks*: channels with exclusive tidal transport develop in the inner and central zones of the estuaries. Bioturbated organic muds containing many shells and shell fragments are widespread in the tidal creek system. It is possible that these shells were the main component of the sediment in channels as the result of lag erosional processes. Sand may often be absent due to the remoteness of the source of the fluvial or marine sand supply.
4. *Tidal flats*: these correspond to intertidal sectors of areas with a minor stage of infilling. In the past, these environments developed occupying wide intertidal areas, but today the environment is reduced to some locations in the marine zone. Presently, they are topographically and stratigraphically to be found above the swash platforms of the Punta Umbría subsystem. In the sedimentary record, their sediment is fine-to-medium sand with an organic-rich muddy matrix (3–7 wt%). The internal structures are observed to be horizontal laminations with a significant degree of bioturbation (20–60%), which is mainly produced by bivalve (*Cerastoderma edule* and *Chamelea gallina*) and annelid (*Arenicolides ecaudata* and *Nereis diversicolor*) activity. Also common are marine sea grasses (*Zostera noltii* and *Zostera marina*).
5. *Active channel margins*: this is the environment that presently occupies the intertidal area attached to the tidal channels, which evolved from the previous tidal flats. The main facies consist of muds and sandy muds with high organic matter content. These muds reveal thick shell layers intercalated with finer layers. Scattered muddy sand intercalations which range in thickness from a few millimetres to a number of centimetres and muds (wavy tidal bedding) are found

in the neighbouring estuarine channels. These intercalations are caused by changes in the environment energy resulting from the alternation of spring and neap tides.

6. *Salt marshes*: salt marshes have developed on the tidal flats and channel margins, caused by tidal accretion. The dominant sediments are organic-rich, medium-to-fine, muddy sand or sandy mud with 5–10% organic carbon content. The sequence ranges from horizontal parallel lamination to massive bedding. Characteristics include intense bioturbation by halophyte roots (*Sarcocornia perennis*, *Salicornia ramosissima* and *Spartina maritima*) and annelid and crustacean activity at the base of the sequence.
7. *Open beaches*: the exposed beach is a sub-environment generated on the front of the barriers closing the estuary (spits). This is a dissipative beach which develops bars that migrate across the nearshore. The sediment comprises medium-to-coarse, moderately well-sorted quartz sand, with a large number of decimetre-scale, normal-graded coarse shell beds (storm beds). The internal structure reveals planar cross-bedding which dips landward with underlying beds that dip seaward. This sequence corresponds to beach accretion due to the migration of ridge and runnel systems. Today there is another beach which has developed in the frontal area of a long jetty that protects the entrance to the Padre Santo Channel.
8. *Ebb-tidal deltas*: found at the outer (marine) part of the system, this is a sandy platform stretching to the southeast. In the past, a wide system of platforms developed in both (the Padre Santo and Punta Umbría) channels. Today, however, it is only present in the Punta Umbría inlet, as a result of the intense human alteration of the mouth of the Padre Santo channel. Swash platforms developed in the intertidal portion of a system of sub-parallel ebb-tidal deltas. Additionally, sand flats separated the main ebb channels, with wave action inducing the landward migration of curved swash bars. Low energy bars and linguoid megaripple trends developed on the surface of the swash platforms due to waves and ebb currents. The identified facies are made up of moderately well sorted medium quartzitic sand, with metric-scale herringbone-trough cross-bedding.
9. *Cheniers*: chenier ridges are long sandy bodies which form on the intertidal flats (or low supratidal zone) and run parallel to the coastline approximately 2 m above Mean Neap High Water level. The sand which comprises the sediment is mainly well sorted and medium to very coarse. Compositionally, the most abundant components are quartz grains and marine shell fragments. The internal structures consist of planar cross-strata that dip seaward in the lower part of the sequence, while the dip decreases in the upper part and can become horizontal or even dip slightly landward. The sets may display a large number of erosional and reactivation surfaces formed by storm events.

23.3.2 Holocene Sedimentary Sequence

Holocene infilling of the Odiel and Tinto Estuary is made up of a sequence of six lithological units on top of a basal body of Pleistocene fluvial conglomerate and dominated by muddy and, to a lesser extent, sandy facies. These units are: (Fig. 23.3).

Lower muds (Estuarine Unit E1): between 10 and 13 m thick, composed mainly of muds (clayey silts)—black or dark grey in colour corresponding to high organic-carbon content.

Lower sandy muds (Estuarine Unit E2): the dominant lithology corresponds to very fine sands with a muddy matrix and a characteristic dark grey colour similar to E1.

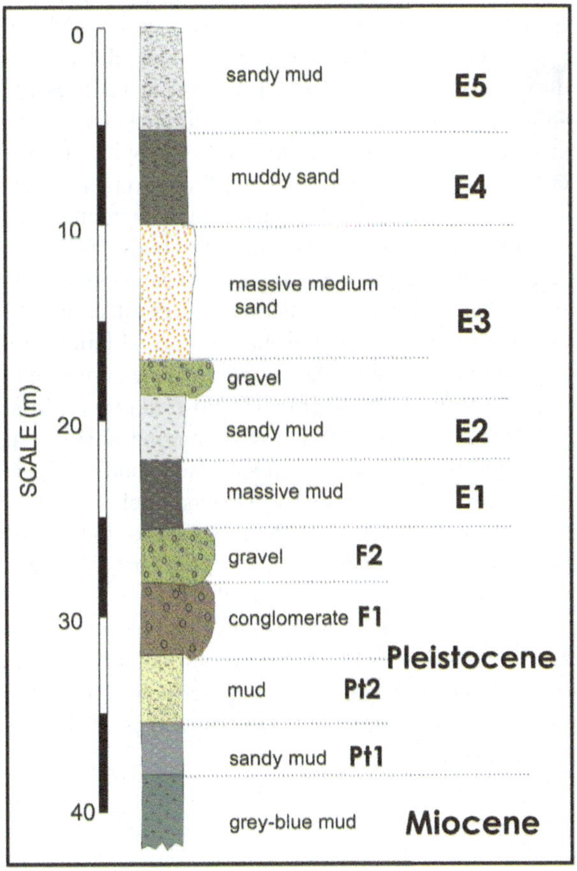

Fig. 23.3 Type of column where the stratigraphic units described are observed

Intermediate sands (Estuarine Unit E3): the characteristic facies of this unit is clean, very well sorted fine-to-medium sand which is beige in colour.
Upper sandy muds (Estuarine Unit E4): very fine, poorly sorted sands with a muddy matrix—greyish in colour.
Upper muds (Estuarine Unit E5): black, very badly sorted clayey silts which are completely disrupted by bioturbation.
Surficial rooty muds (Estuarine Unit E6): a thin layer measuring no more than 0.5 m in thickness at the top of the sequence consisting of brown silty clays with a visible parallel lamination interrupted by plant bioturbation.
Surficial sands (Waves –O– and wind –V– generated units): Discontinuous bodies of sands and gravelly sands appear, closing the estuary on the top of the sequence and corresponding to facies generated by waves and winds.

The overall thickness of the complete sequence reaches more than 30 m at its deepest. While the lower three units represent a transgressive sequence deposited under sea-level rise conditions, the upper part of the sequence is a regressive (progradational) sequence which was sedimented under a sea level that was more or less stable. All the units present intercalated successive shell layers deposited during high energy events. The sand in E3 presents a clear channel shape, corresponding to the position of the original estuarine channel, located under the current city and its harbour complex and occupying a position different from the present channel track. Whereas an aggrading geometry is dominant in the innermost areas, in the outer areas, the presence of a topographic gradient induced the development of units prograding to the deeper zones. The hanging wall of faults would be responsible for the topographic gradients that conditioned the geometry of the sedimentary units, as well as the present course of the estuarine channels. This geometry is due to the presence of normal faults which were active in recent times.

The O units are bodies of limited size composed of medium to very thick sand, even up to the size of gravel, whose dominant lithology is quartz grains together with the bioclastic fragments of marine shells, especially Glycymeris variabilis. Arranged on the body of upper sandy muds (E4) or even on the upper mud body (E5), these are large sandy barriers, such as the Punta Umbría Arrow (O1), the Saltés Island cheniers and Punta del Sebo (O2) and the Torre Arenillas Arrow (O3) that currently close the estuary. They are interpreted as being the result of wave action on the emerged areas in the marine sector of the estuary.

On all these sandy bodies described above, others composed of fine to very fine sand develop, which are also bioturbated by roots and have a higher content of organic matter, giving them a darker color, at least on the surface. This body is the most recent formation and is interpreted as an edaphised and plant-bioturbated body of wind origin.

23.3.3 A Peculiar Closing System: Facies Model of Saltés Island Chenier Plain

Saltés Island is composed by several **shelly-sandy ridges** disposed in a sub-parallel way to the open coast. There are wide salt-marsh bodies and tidal flats separating successive ridges (Fig. 23.1).

In the open side, two complex systems of ebb-tidal deltas extend by the mouths of both, Punta Umbría and Padre Santo channels. Padre Santo inlet displayed the widest of these systems consisting of six main ebb channels separated by a number of levees. Those outside Punta Umbría channel exerted an influence on the dynamics of the chenier plain due to the wave refraction scheme across them to arrive to the tidal flat front.

A singular type of evolutionary model is observed in the ridge system on Saltés Island. An architecture scheme and geomorphological study of its development reveal a model of typical chenier plain facies that migrated onto the tidal flat, describing a succession of cheniers. However, in contrast to the accepted models of chenier evolution, this migration did not only occurs during winter storms; independently, they mainly evolve during high spring tides that cover the ridges. This process consists in a migration of sand bars in a way very close to dissipative beaches, except that in this case bars migrate over the sandy muds of a tidal flat.

In geomorphology, cheniers are sandy ridges produced by storms acting on muddy sediments along open coasts, with systems of these ridges typically running parallel to the coastline. They are usually found on muds sedimented in upper tidal flats or supratidal areas. It is universally accepted that there are two types of cheniers, depending on the origin of the coarse sediments that make up the ridge as well as the relative significance of the transversal and longitudinal wave transport. Storms are the main transport agent, causing a typical facies model of a basal body of muds superimposed by the sandy or shelly chenier body.

The evidence of the architecture observed under the sandy crests of Saltés Island describes a typical chenier plain; however, its development is not as a succession of classic sandy ridges or barrier islands (Morales et al. 2013). The surficial distribution of environments in terms of critical tidal levels and their elevation relative to wave action (Fig. 23.4) are the determining factors for the facies architecture of this coastal sector, displaying a classical chenier plain architectural scheme like those described in literature for microtidal and mesotidal coasts. The processes which gave rise to this chenier system point to an unusual form of development because it differs from other models (Morales et al. 2013).

Morphostratigraphic characteristics suggest that Saltés Island has functioned as a chenier plain for at least the last 200 years (Borrego et al. 2000). The architectural disposition of facies (Fig. 23.4) demonstrates that the sequences in every the ridges are similar to the described in the literature for stormy cheniers. The origin of the entire architectural facies model of Saltés Island can be explained if during the last 3,200 years acted the same processes observed over the last 200.

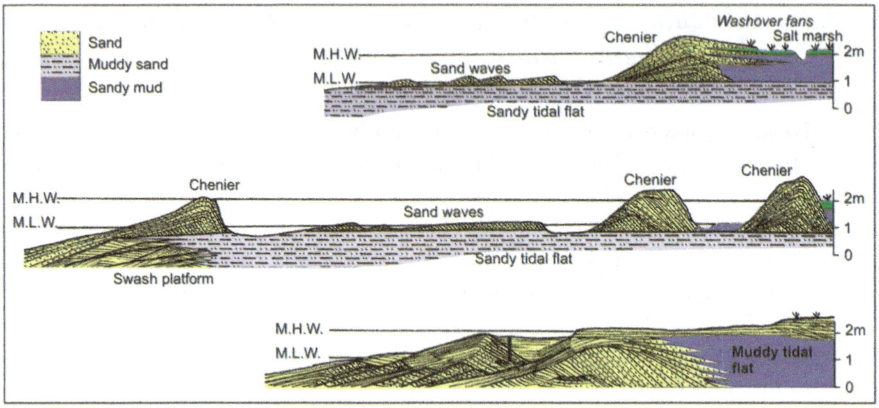

Fig. 23.4 Depositional facies scheme architecture (the horizontal scale varies in this perspective) indicating topographic location in relation to certain tide levels: MHW = Mean high water, MLW = Mean low water. Modified from Morales et al. (2013)

The tops of the cheniers are currently exposed to the air and is colonised by upper plants. Also wind action select the finer grains to form incipient foredunes. These tops are only submerged and affected by waves during extreme high tides or storms, which can produce washovers. Muddy sediments fill the back-chenier tidal flats reaching the supratidal areas and forming salt marshes.

23.4 Holocene Evolution

The architecture system of sedimentary bodies at the mouth of the rivers Odiel and Tinto suggests that the basal units of the estuary were deposited in a fluvial system which developed during the sea-level lowstand that occurred during marine isotope stage (MIS) 2. This low sea level induced a fluvial incision in previous Pleistocene units deposited during MIS 3 and the beginning of MIS 2.

The first three properly Holocene estuarine units were deposited between 11,000 and 5,000 year BP, presented onlap upper limits and corresponded to a transgressive sequence deposited during the Flandrian sea-level rise.

Sandy barriers that probably protected this transgressive estuarine system have not been observed, which suggests that they were not preserved or perhaps were preserved outside our study area.

The deposition of estuarine unit E3 occurred around 5,000 year BP and represents the invasion of the estuary by sands from the open coastal area, which were deposited with a clearly channelled morphology. The track of this old estuarine channel differs from that which exists today. The sea level when this E3 unit was deposited was some metres lower than at present.

Upper estuarine units provide a regressive sequence, deposited with a more or less stable sea level (although with slight oscillations). These units represent the establishment of a network of tidal drainage and almost complete estuarine infilling.

The sandy bodies partially enclosing the estuary began to grow until the upper estuarine units reached a high degree of estuarine clogging. The presence of these wave-generated bodies constrains the system, conditioning hydrodynamic processes that contribute to the final filling of the estuary. Thus, since the growth of these sandy barriers, the interior of the estuary has become tide-dominated. Almost complete filling of the innermost part of the estuary meant the start of a process of progradation that began a number of decades ago.

Units that fill the estuary are affected by neo-tectonic activity in the context of postorogenic lift from the margin of a foreland due to extensional cortical thinning of the Betic Range.

Studies state that the Holocene infilling of the estuary began around 9,060 year BP (Lario et al. 2002), when it was flooded during the last Flandrian transgression, leading to a continuous rise in sea level over a very long time. During this period, sedimentation in the centre of the estuary consisted of alternating sand and mud estuarine accretion bodies with a clear marine influence shown by its biological associations. This transgressive phase stopped just 6,500 years ago when the rate of rise decreased dramatically before reaching its present level. Following this stabilisation, sandy barriers formed in the mouth of the estuary, and tidal flats and marshes developed in the interior. From a physiographical point of view, during this second phase, the estuary acquired a morphology closer to an anastomosing system, with a large number of tidal channels separated by salt-marsh bodies. The transgressive period ended around 4,500 year BP, when the sea was positioned in its present level. Sandy barriers formed later using the sand of the estuarine body E3. Then the estuarine mouth was closed, and tidal flats and marshes subsequently developed in the interior.

Thus, we can divide the estuarine evolution in three stages according with the nature of the sedimentary bodies: (a) 9,060–4,500 year BP: the estuary start to be infilled with fluvial silts and authocthonous-flocculated muds under transgressive conditions; (b) 4,500–3,200 year BP: after the sea-level stabilisation marine sands arrive to the estuary to be mixed to fluvial sands eroded from Neogene formations. (c) 3,200 years BP to the present: the high rate of sedimentation experienced during the previous stage joint the partial closure of the estuary by barriers. Then acidic fluvial inputs from Tinto and Odiel rivers dominated the estuary. Authochtonous sedimentation became the main source of sediments (Fig. 23.5).

Human influence can also be seen on the natural dynamics because of the implementation of coastal protection measures in the outer estuary. Between 1975 and 1980, the long Huelva bank was constructed at the entrance of the estuary to prevent silting of the Padre Santo Channel, a dredged ship channel. Another bank was built near Punta Umbría to protect the fishing harbour.

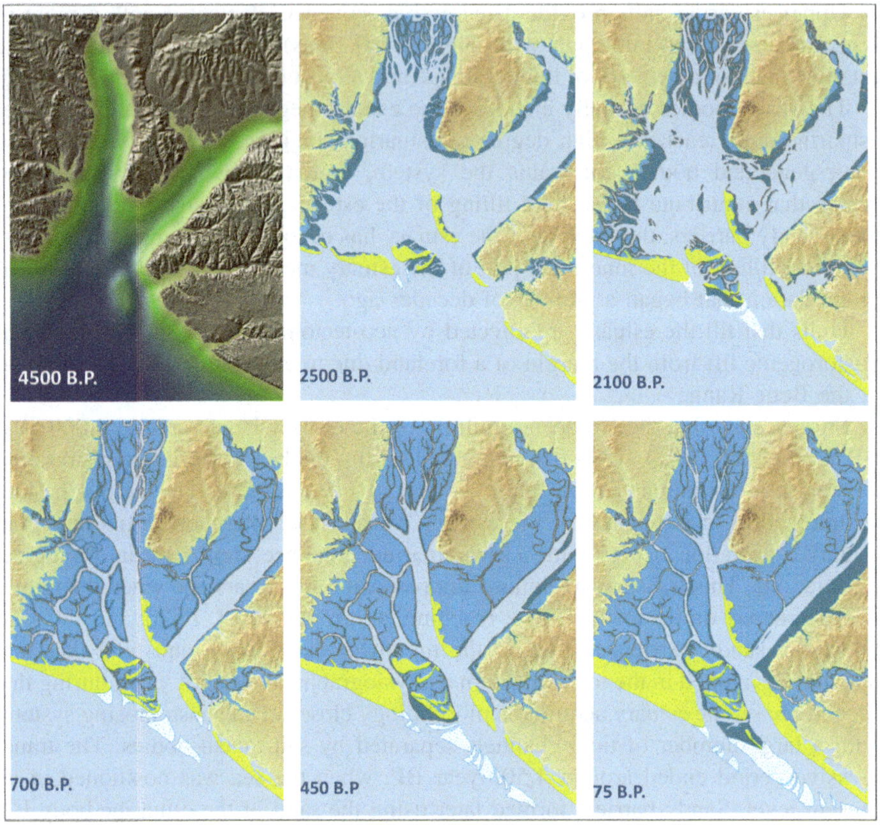

Fig. 23.5 Holocene evolution of the Odiel and Tinto estuary

23.5 Effects of AMD and Industrial Activity on the Estuarine System

Tinto and Odiel rivers drain one of the most important metallogenic area of Europe, the Iberian Pyrite Belt. The massive sulphide deposits of this district are the largest in the world with abundant base metal sulphides (Fe, Zn, Cu and Pb) and associated trace metals (Cd, As, Tl, Sn and Hg), including significant amounts of Au and Ag. These ores were exploited since prehistoric times, at least since 6,000 years ago. The mining history of this district can be divided into successive phases, separated by period of low activity. In a first phase, Cu deposits were exploited extensively during the chalcolithic age (6,000–4,500 year BP). In a second period (3,100–2,500 year BP) Ag and Au were mined by Tartessian people to commerce with Phoenician and Greek people. A third intensive stage, initiated when the arrival of Roman people (2,000–1,800 year BP), but were subsequently abandoned by the Visigoths (1,600–1,300 year BP) and Arabs (1,300–500 year BP) (Freijero and

Rothenberg 1981; Ferrero 1988). Pyrite production peaked between 1875 and 1930 with important mines like Riotinto, Tharsis and Sotiel. Over the last 150 years, the mining activity was intense, exploiting pyrite and other polimineral concentrates. All of these activities caused a significant environmental impact, including large areas covered by wastes and ashes which are presently under erosion and lixiviation processes. Today, only a small number of mines are still working (Riotinto, Aguas Teñidas, Sotiel and Magdalena). Due to the action of waters on the mine and waste areas favoured secularly AMD processes doing these rivers extremely polluted. Their waters are consequently very acidic, showing pH values lower than 3 and containing big amounts of dissolved and suspended toxic metals which are discharged into the Estuary (Davis et al. 2000; Elbaz-Poulichet et al. 1999, 2000; Grande et al. 2000; Borrego et al. 2002).

The pollution of this estuary, affected by acidic waters since the XIX Century, increased between 1966 and 1985 due to the entrance of acidic waters coming from industries located in its margins. These industries produced fertilizers, foundries, petroleum-derived products and other chemicals and the residual waters were thrown directly to the estuary. One by-product formed in the chemical process of fertilizer production is phosphogypsum, which is composed mainly of gypsum ($CaSO_4{-}_2H_2O$) but also contains small quantities of trace elements, rare earth elements and fluorine (Arocena et al. 1995). These factories process over 2×10^6 tonnes of phosphate rock a year, generating approximately 3×10^6 tonnes of phosphogypsum which were stockpiled in the intertidal margin (Bolivar et al. 2000). Phosphogypsum stacks cover 12 km^2 of the salt marshes in the Tinto estuary, with intense radioactive impact on this marsh-land (Bolivar et al. 2002). Occasionally, these stacks were affected by erosion processes occurring during the rainy months and extreme tidal fluxes. Effluents associated with phosphogypsum introduce large quantities of P, As, U, Th and ^{226}Ra into the estuary (Ruiz et al. 1998; Elbaz-Poulichet et al. 2000; Grande et al. 2000; Alcaraz Peregrina and Martınez-Aguirre 2001; Borrego et al. 2002). Nevertheless, since 1985, this zone has come under a corrective plan for the control of industrial waste disposal.

This mixture of industrial waste, river water and water from the open sea that enters the estuary during tidal floods creates a complex system of internal circulation and patterns of biochemical behaviour that have not been studied sufficiently. This biochemical behaviour is the origin of continuous dissolution/precipitation processes controlled by volumes of tidal water, the inflow of river water and industrial waste. There are two reasons for this contamination: (1) contributions of acid river water and sediments with high concentrations of Fe_2O_3, Cu, Zn, Pb and Ba in the estuary; (2) industrial effluent contributions with large amounts of P_2O_5, As, Hg and U. Tidal streams redistribute these elements throughout the system and deposit them in moments of low energy together with fine sediments.

Three main factors control directly the distribution of pollution in sediments: (a) the volumes of fluvial river inputs, (b) tidal currents and (c) location of effluents into the estuary. The combined action of these factors controls spatial variations in the geochemical composition of estuarine sediments. The singularity of this estuary is the presence of a pH-induced water mixing process apart the typical salt-induced

Table 23.1 Average concentrations of trace metals in dissolved phase (DP) and suspended matter (SM) in each domain in the estuarine system

	Fe		Co		Ni		Cu		Zn		Cd		Pb		U	
	DP (mg L^{-1})	SM (mg L^{-1})	DP (μg L^{-1})	SM (mg L^{-1})	DP (μg L^{-1})	SM (mg L^{-1})	DP (μg L^{-1})	SM (mg L^{-1})	DP (μg L^{-1})	SM (mg L^{-1})	DP (μg L^{-1})	SM (mg L^{-1})	DP (μg L^{-1})	SM (mg L^{-1})	DP (μg L^{-1})	SM (mg L^{-1})
Zone 1 upper estuary	82	359381.9	300.5	3.1	95.3	4.1	9833	5424.6	23040	1198.1	123.6	0.6	101.4	1169.0	7064	42.5
Zone 2 middle estuary	0.9	48878.2	36.19	48.2	17.4	16.57	790.3	5064.5	949.2	854.9	5328	112.4	1336	1019.4	0.5	0.5
Zone 3 lower estuary	0.3	14817.5	5.3	15.9	5.2	70.4	1307	503.8	202	608.5	1.1	3.37	3.7	284.5	2.1	0.7

wedge. So, the metals, which were previously dissolved in the acid environment, precipitate as the pH increases upon arrival in the tidal influence area. This additional process controls specially the behaviour of heavy metals and nutrients and the pass of heavy metals from the dissolved to the suspended particulate phases to be finally sedimented.

When fluvial water (characterised by high concentrations of sulphates, Fe and other trace metals, among them Cu, Zn and Cd) reaches the system, the acid neutralisation starts if pH values increase to 4. Then, precipitation of sulphated salts is induced, resulting in a decrease in the concentration of sulphates and heavy metals in the dissolved phase. Subsequently, the progressive influence of tidal water volumes results in a rapid increasing in conductivity and pH values, reaching finally typical marine values near the mouth. In this context, Fe and other trace metals present in the dissolved phase show a non-conservative behaviour. These metals pass to the suspended fraction when the pH value reaches values between 5 and 7.5. This process is observed as an increase in the concentration of Fe and metals in the suspended matter and an increase in the concentration of solid particles present in the water. The effect of these processes on the concentration of heavy metals in the system and between the dissolved and particulate phases are shown in Table 23.1:

Neverthelessthe water in the fluvial estuary continues showing metals concentrations higher than those typically found in sea water. For that, this area is considered one of the main sources of metals in the Gulf of Cadiz.

Finally, the suspended matter is physically deposited during the hydrodynamic tidal settlings. Surficial sediments contain high toxic metal concentrations (in mg/kg

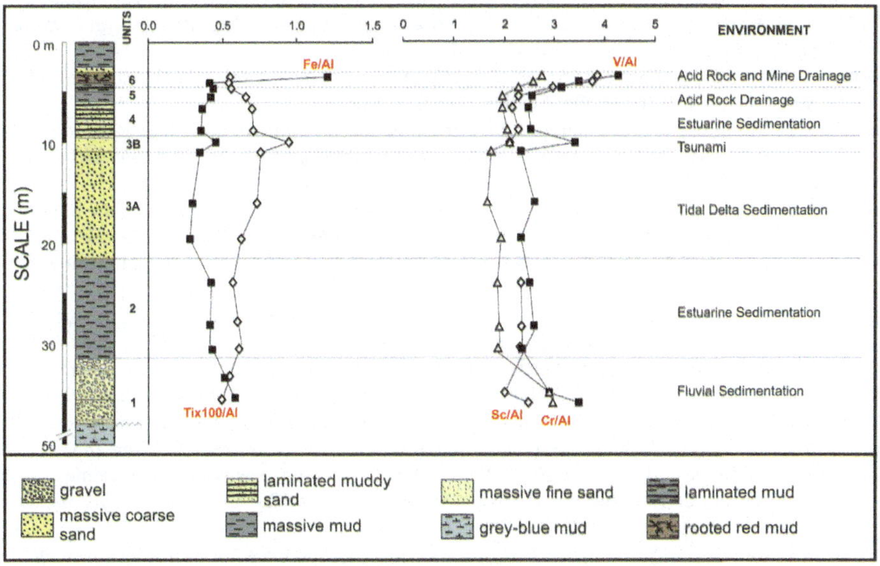

Fig. 23.6 Stratigraphic sequence studied by Borrego et al. (1999) in the Odiel central estuary showing the vertical variation of Al_2O_3-normalised Ti, Fe, Sc, Cr, and V

dry wt, fraction ~63 mm): 4500–6300 Cu; 2600–5100 Zn; 1900–3400 As; 1700–10 400 Pb; 200–700 Mn; 9–39 Cd; and 15–49 Hg, according to data supplied by Cabrera et al. (1992), Nelson and Lamothe (1993).

These metal concentrations were varying along the geologic history of the estuary, being preserved in the sedimentary record. The geochemical evolution of sedimentary infilling was studied by Borrego et al. (1999). In this study, centered in the central basin of the Odiel Estuary, both, provenance of sediment and environmental conditions were determined using the heavy metal concentrations in sediments for a period of 10,000 years. This work emphasizes that the increasing of metals in the sedimentary units occurs in a parallel way to the historical mining periods described above (Fig. 23.6).

References

Achterberg, E. P., Herzl, V. M. C., Braungardt, C. B., & Millward, G. E. (2003). Metal behaviour in an estuary polluted by acid mine drainage: the role of particulate matter. *Environmental Pollution, 121,* 283–292.

Alcaraz Peregrina, J. M., & Martínez-Aguirre, A. (2001). Natural radioactivity in groundwaters around a fertilizer factory complex in South of Spain. *Applied Radiation and Isotopes, 55,* 419–423.

Arocena, J. M., Rutherford, P. M., & Dudas, M. J. (1995). Heterogeneous of trace elements and fluorine in phosphogypsum by-product. *Science of the Total Environment, 162,* 149–160.

Bolivar, J. P., Garcia-Tenorio, R., & Vaca, F. (2000). Radioecological study of an estuarine system located in the south of Spain. *Water Research, 34,* 2941–2950.

Bolivar, J. P., García-Tenorio, R., Mas, J. L., & Vaca, F. (2002). Radioactive impact in sediments from an estuarine system affected by industrial wastes releases. *Environment International, 27*(8), 639–645.

Borrego, J. (1992). *Sedimentología del estuario del río Odiel (Huelva, S.O. España)* (Ph.D. Thesis). Universidad de Sevilla, Sevilla, Spain, 315 pp.

Borrego, J., Morales, J. A., & Pendon, J. G. (1993). Holocene filling of an estuarine lagoon along the mesotidal coast of Huelva: The Piedras river-mouth, southwestern Spain. *Journal Coastal Research, 9,* 242–254.

Borrego, J., Morales, J. A., & Pendón, J. G. (1994). Holocene estuarine facies along the mesotidal coast of Huelva, south-western Spain. In W. F. Flemmin & A. Bartholoma (Eds.), *Tidal signatures in modern and ancient sediments*. International Association of Sedimentologists, Special Publication 24, pp. 151–170.

Borrego, J., Morales, J. A., & Pendón, J. G. (1995). Holocene estuarine facies along the mesotidal coast of Huelva, south-western Spain. In B.W. Flemming & Bartholomä A (Eds.), Tidal signatures in modern and ancient sediments. *Spec Publ Int Ass Sedimentology 24*, pp. 151–170.

Borrego, J., Ruiz, F., Gonzalez-Regalalo, M. L., Pendon, J. G., & Morales, J. A. (1999). The Holocene transgression into the estuarine basin of the Odiel River Mouth (Cadiz Gulf, SW Spain): Lithology and faunal assemblages. *Quaternary Science Review, 18,* 769–788.

Borrego, J., Morales, J. A., & Gil, N. (2000). Evolución sedimentaria reciente de la desembocadura de la Ría de Huelva (Suroeste de España). *Revista de la Sociedad Geológica de España, 13*(3–4), 405–416.

Borrego, J., Morales, J. A., de la Torre, M. L., & Grande, J. A. (2002). Geochemical characteristic of heavy metal pollution in surface sediments of the Tinto and Odiel river estuary (southwestern Spain). *Environmental Geology, 41,* 785–796.

Cabrera, F., Conde, B., & Flores, V. (1992). Heavy metals in the surface sediments of the tidal river Tinto (SW Spain). *Fresenius Environmental Bulletin, 1,* 400–405.
Cánovas, C. R., Olías, M., Nieto, J. M., Sarmiento, A. M., & Cerón, J. C. (2007). Hydrogeochemical characteristics of the Tinto and Odiel Rivers (SW Spain). Factors controlling metal contents. *Science of the Total Environment, 373,* 363–382.
Carro, B. (2002). *Sedimentación reciente en el estuario del Río Tinto* (p. 137). Msc: Huelva University.
Carro, B., López-González, N., Grande, J. A., Gómez, T., de la Torre, M. L., & Valente, T. (2011). Impact of acid mine drainage on the hydrochemical characteristics of the Tinto-Odiel estuary (SW Spain). *Journal Iberian Geology, 37*(1), 87–96.
CEEPYC, Centro de Estudios y Experimentación de Puertos y Costas 'Ramón Iribarren'. (1990). *Plan de estudio de la dinámica litoral de la provincia de Huelva.* Informe. Dirección General de Puertos y Costas. Technical report. Madrid, Spain.
Cuena, G. J. (1991). *Proyecto de regeneración de las playas de Isla Cristina.* Huelva, Spain: Memorias del Ministerio de Obras Públicas y Transporte. Technical report.
Dabrio, C. J., Zazo, C., Goy, J. L., Sierro, F. J., Borja, F., Lario, J., et al. (2000). Depositional history of estuarine infill during the last postglacial transgression (Gulf of Cadiz, Southern Spain). *Marine Geology, 162,* 381–404.
Dalrymple, R. W., Zaitlin, B. A., & Boyd, R. (1992). Estuarine facies models: Conceptual basis and stratigraphical implications. *Journal of Sedimentary Petrology, 62,* 1130–1146.
Davis, R. A., Jr., Welty, A. T., Borrego, J., Morales, J. A., Pendon, J. G., & Ryan, J. G. (2000). Rio Tinto estuary (Spain): 5000 years of pollution. *Environmental Geology, 39,* 1107–1116.
Doty, M. S. (1949). Critical tide factors that are correlated with the vertical distribution of marine algae and other organisms along the Pacific Coast. *Ecol., 27,* 315–328.
Elbaz-Poulichet, F., Braungardt, C. B., Achterberg, E. P., Morley, N. H., & Cossa, D. (1999). A synthesis of the results of TOROS and CANIGO projects on metal contamination in the Tinto–Odiel rivers (southern Spain) and the Gulf of Cadiz. In: J. M. Pacyna, Kremer, H., Pirrone, N., Barthel, K. (Eds.), *Socioeconomic aspects of fluxes of chemicals into marine environment* (pp. 101–109). European Commission, Project Report of the Energy, Environment and Sustainable Development 19089.
Elbaz-Poulichet, F., Dupuy, C., Cruzado, A., Velasquez, Z., Achterberg, E. P., & Braungardt, C. B. (2000). Influence of sorption processes by iron oxides and algae fixation on arsenic and phosphate cycle in an acidic estuary (Tinto River, Spain). *Water Research, 34,* 3222–3230.
España, J. S., Pamo, E. L., Santofimia, E., Aduvire, O., Reyes, J., & Barettino, D. (2005). Acid mine drainage in the Iberian Pyrite Belt (Odiel river watershed Huelva SW Spain): Geochemistry mineralogy and environmental implications. *Applied Geochemistry, 20,* 1320–1356.
España, J. S., Pamo López, E., Santofimia, E., Reyes, J., & Matin, J. A. (2006). The removal of dissolved metals by hydroxysulphate precipitates during oxidation and neutralizations of acid mine waters Iberian Pyrite Belt. *Aquatic Geochemistry, 12,* 269–298.
Fairbridge, R. W. (1980). The estuary: Its definition and geodynamic cycle. In E. Olausson & I. Cato (Eds.), *Chemistry and biogeochemistry of estuaries.* John Wiley: New York, pp. 1–35.
Ferrero, M. D. (1988). Los conflictos de Febrero de 1888 en Riotinto: Distintas versiones de los hechos. *Huelva Hist, 2,* 603–623.
Freijero, B., & Rothenberg. (1981). *Studies in ancient mining and metallurgy in South-West Spain: Explorations and excavations in the Province of Huelva.* London: Institute of Archeology.
Frey, R. W., & Howard, J. D. (1986). Mesotidal esturiane sequence: a perspective from the Georgia bight. *Journal of Sedimentary Petrology, 56,* 911–924.
Grande, J. A., Borrego, J., & Morales, J. A. (2000). A study of heavy metal pollution in the Tinto-Odiel estuary in southwestern Spain using factor analysis. *Environental Geoogy, 39,* 1095–1101.
Lario, J., Zazo, C., Goy, J. L., Dabrio, C. J., Borja, F., Silva, P. G., et al. (2002). Changes in sedimentation trends in SW Iberia Holocene estuaries (Spain). *Quaternary International,* 93e94, 171e176.

Leblanc, M., Morales, J. A., Borrego, J., & Elbaz-Poulichet, E. (2000). 4,500-years-old mining pollution in southwestern Spain: longterm implications for modern mining pollution. *Economic Geology, 95,* 655–662.
López-González, N. (2002). *Sedimentación reciente en el estuario del Río Odiel* (p. 137). Msc: Huelva University.
Morales, J. A., Borrego, J., & Davis, R. A., Jr. (2013). A new mechanism for chenier development and a facies model of the Saltés Island chenier plain (SW Spain). *Geomorphology, 204,* 265–276.
Nelson, C. H., & Lamothe, P. J. (1993). Heavy metal anomalies in the Tinto and Odiel River and estuary system, Spain. *Estuaries, 16,* 496–511.
Rodríguez Ramírez, A., Rodríguez Vidal, J., Gracia, F. J., Cáceres, L. M., Guerrero, V., & Cantano, M. (1996). Post-Fandrian eustatic curve in Cadiz Gulf (SW Spain). State of the art. Mediterranean and Black Sea Shorelines. *INQ°A Newsletter, 18,* 7–12.
Ruiz, F., Gonzalez-Regalado, M. L., Borrego, J., Morales, J. A., Pendon, J. G., & Muñoz, J. M. (1998). Stratigraphic sequence, elemental concentrations and heavy metal pollution in Holocene sediments from the Tinto-Odiel estuary, south-western Spain. *Environmental Geology, 34,* 270–278.
Sáinz, A., Grande, J. A., & de la Torre, M. L. (2004). Characterisation of heavy metal discharge into the Ria of Huelva. *Environment International, 30,* 557–566.
Simmons, H. B. (1995). Some effects of unpland discharge on esturine hydraulics. *Proc ASCE, 81,* 1–20.
Younger, P., Banwart, S. A., & Hedin, R. S. (2002). *Mine water* (p. 442). Hydrology pollution remediation: Kluwer Academic Publishers.

Chapter 24
The Guadiana River Delta

Juan A. Morales and Erwan Garel

Abstract The Guadiana Estuary is a good example of rock bounded estuary which consists of a single narrow estuarine channel with a meandering morphology imposed by faults systems affecting the hard geology of the substrate. Only along the last kilometers of the estuarine channel, the valley opens when Cenozoic Guadalquivir Basin formations appear. In this area, the Guadiana develops a prograding coastal system constituted by successive sandy barriers separated by salt marshes which configure a wave dominated delta. This progradation is possible thanks an interaction of the coastal agents which enhanced the silting in addition to a good availability of sediments. This chapter explains an explanation of the dynamic functioning of the open coastal environment, so as the resulting facies model.

24.1 Introduction

The Guadiana Estuary (Fig. 24.1) is a 80 km long estuary at the Portugal–Spain southern border, constituted by a single and narrow (50–700 m wide) channel (Lobo et al. 2004). It belongs to the type of "rock-bound estuaries", i.e. long and narrow systems incised in bedrock valleys with large basin drainage areas (Fitzgerald et al. 2000a; Garel et al. 2009). Its meandering morphology is imposed by faults systems affecting the hard geology of the substrate (Morales 1997; Morales et al. 2006). The estuary can be separated into 3 sectors with distinct ecological and hydrological characteristics (Morales 1993; Chícharo et al. 2001): the marine estuary (from the mouth to ~10 km) which is strongly influenced by seawater, the central estuary (from ~10 to 20 km) with brackish water and,

J. A. Morales (✉)
Department of Earth Sciences, University of Huelva, 21007 Huelva, Spain
e-mail: jmorales@uhu.es

E. Garel
Centre for Marine and Environmental Research (CIMA), University of Algarve, Faro, Portugal
e-mail: egarel@ualg.pt

J. A. Morales (ed.), *The Spanish Coastal Systems*,
https://doi.org/10.1007/978-3-319-93169-2_24

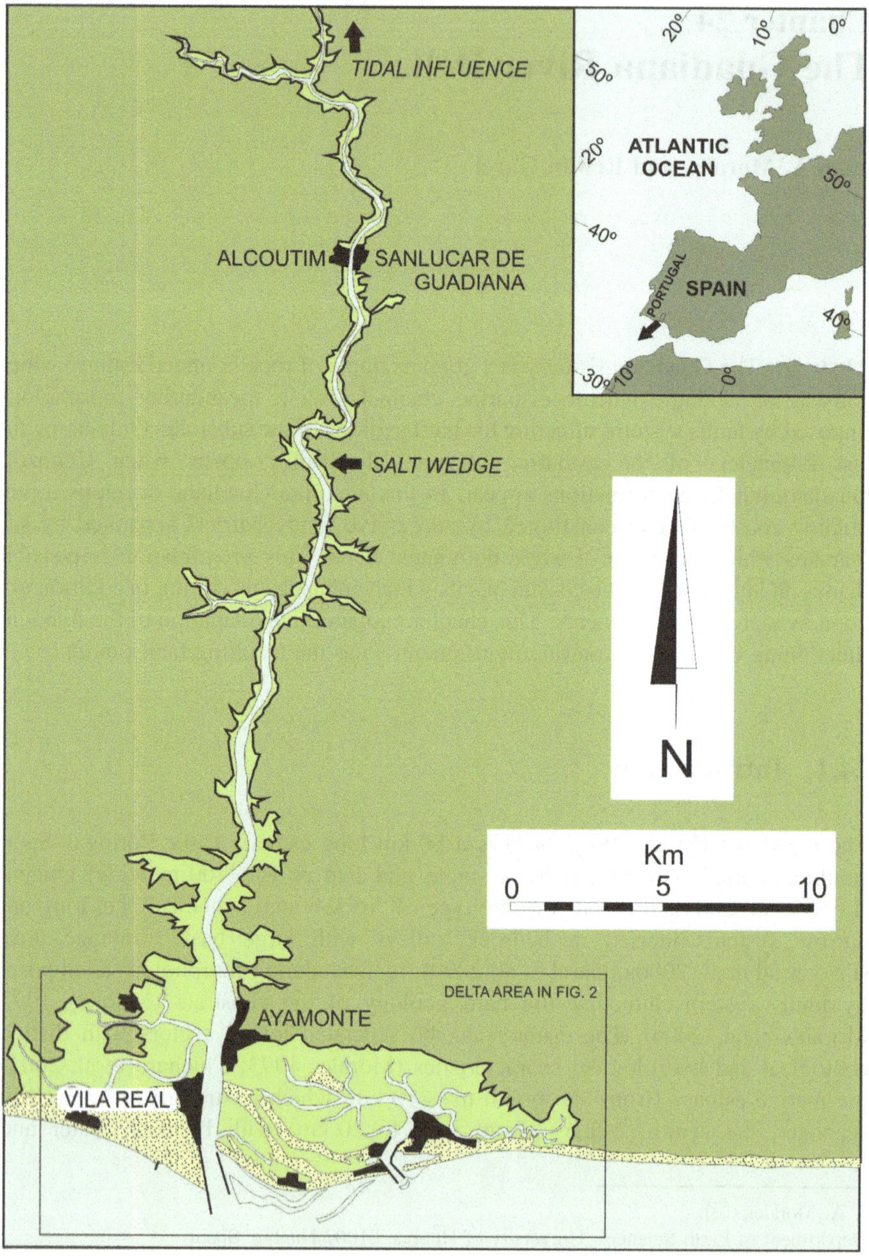

Fig. 24.1 Location of the Guadiana River Delta in the southern border between Spain and Portugal

upstream, the fluvial estuary which is filled up with freshwater. The fluvial and central sectors of the estuary have a clear meandering morphology, but these meanders have been superimposed by the hard lithology and some faults existing in the bedrock. The transversal section of the channel across the meanders presents a pool and a lateral tidal bar system. The straight tracks of the channel show a rectangular symmetric section displaying an average depth, without pool nor bar (Morales 1997; Morales et al. 2006).

However, when the estuary enters in the Mio-Pliocene formations of the Guadalquivir Basin, along the 7 km of the marine estuary, erosional processes could excavate a wider bay (Morales 1993). Infilling this bay, the system developed a deltaic system. The fluvial delta evolved during the Holocene period by means of prograding chains of sandy barriers, which developed tidal flats and salt marshes in their back barrier area. The whole system acquired a wave-dominated delta morphology (Morales 1997).

This morphological development represents a typical example of a mature estuarine system in advanced stage of sediment infilling (Morales 1997). This evolution was mainly due to an interaction of the coastal agents which enhanced the silting in addition to large sediment availability.

As the Coast of Huelva, the Guadiana delta has a mesotidal regime, with a mean tidal amplitude of 2.1 m, but oscillating between 3 m during spring tides and 1 m during neap tides. During the propagation into the estuary, tidal range varies little along 60 km of tidal influence (generally less than 10% of the amplitude at the mouth), except along the later 15 km where significant damping occurs as a result of bed shallowing (Morales et al. 1994; Garel 2017a). Waves display a mean significant height between 0.5 m (Morales et al. 1994) and 1.0 m (Costa et al. 2001) coming mainly from the southwest, producing a dominantly eastward longshore transport.

24.2 Physiography and Sedimentary Environments

The physiography of the Guadiana system is entirely conditioned by the inherited nature and morphology of the substrate. In the inner estuarine area the presence of Carboniferous Slates and Quartzites conditioned the fluvial incision during the Pleistocene. In this erosive process the presence of tectonic structures like faults exert a strong control of the channel track (Oliveira 1990). This nature of the substrate inhibited the river generating a wide estuary with the typical funnel-shape morphology, limiting its inner volume and its availability to maintain a large amount of sediments in this inner estuary. As a consequence, the sediment started quickly to be deposited in the open area generating a deltaic zone in its front.

All this sediment arrived to the open area developing two different physiographic features: (1) a wide, barrier island/salt marshes prograding complex and (2) an ebb-tidal delta.

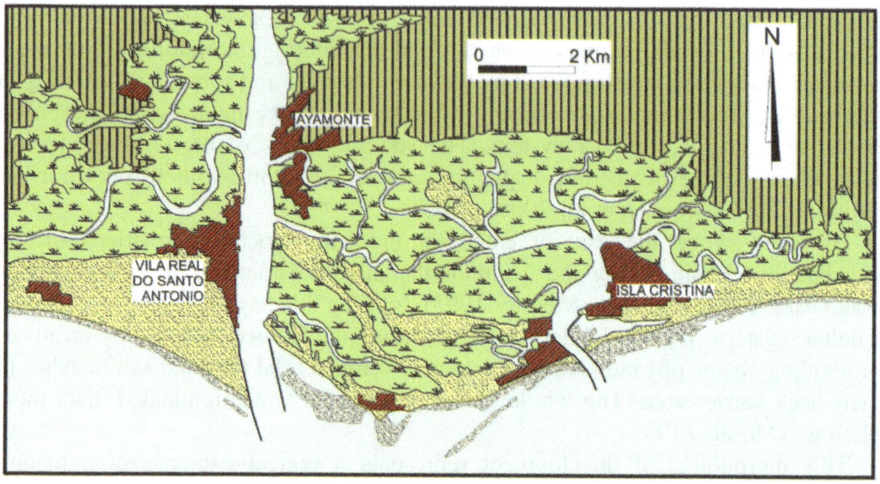

Fig. 24.2 Tidal network and environments in the Guadiana River Delta. In yellow the sandy barriers and shoals, in green the salt marshes, in red the towns

24.2.1 The Wave-Dominated Delta Plain: Spits, Barrier Islands and Salt Marshes

There are fundamental morphologic differences between the two margins of the main channel. The location updrift of the western (Portuguese) margin, allowed the waves to build a littoral spit that experienced a transversal progradation. This sandy area results from the accumulation of littoral drift that originates from the erosion of a wide system of cliffs at west. On the outer spit side, a slight-sloped and dissipative foreshore is highly accumulative. This high rate of sedimentation by addition of swash bars produced a longitudinal and transversal accretion of the spit. In this spit large aeolian dunes up to 20 m high are developed on the sandy beach facies. On the backbarrier area, a large tidal flat and salt-marsh area was developed. This tidal area directly feed from the main estuarine channel.

The eastern area (Spanish side), located downdrift, is constituted by a progradational succession of barrier islands separated by wide salt-marsh areas. In this side of the delta plain, the tidal network drains the salt-marshes through a distributary inlet directly to the open sea (Fig. 24.2).

The presence and distribution of sedimentary environments and facies present in the delta plain are determined by: (1) action or absence of waves; (2) proximity from fluvial sediment supply (open coast or estuarine channel); and (3) tide-related exposure and submersion rate. In these environments are included: (a) in the open area: barrier islands, littoral spits, inlets, ebb tidal deltas, beaches, dunes and washover fans; and (b) in the back-barrier area: bypassing estuarine channel, lateral tidal bars, lagoons, creeks, active intertidal margins, tidal flats and salt marshes.

24.2.2 The Delta Front (Ebb-Tidal Delta)

The sandy sediment deposited in front of the main estuarine channel is continually reworked by tide and wave activity. This rework generates shoals that act as swash platforms by continuous bar attachment. The presence of these shoals modifies the angle of wave approaching the foreshore (Fig. 24.2). The main ebb channel of the delta is directly connected to the estuarine channel. The interaction between wave action, tidal currents, river discharge and sediment supply produces large morphological changes at seasonal to decadal time scales (Garel et al. 2014). The fast growth and the mobility of sand bars over the swash platform, associated to migration and shallowing of the inlet channel, made it necessary to stabilize the mouth artificially by jetties (Garel et al. 2015).

24.3 Dynamic Evolution

24.3.1 Long Term Evolution

The present morphology of the Guadiana delta plain was built by sedimentary infilling of the primitive bay existing in the southernmost area of the system by sediments coming from the river and the littoral drif. Since the sea level stabilization occurred after the Flandrian transgression, 2.49×10^9 m^3 of sediments have been deposited extended on an area of 73.5 km^2. That means an average accumulation rate of 5.0×10^5 m^3/year (Morales 1997). This volume of sediment was the responsible of the coastal evolution during the Holocene, inducing a strong progradation of frontal bars of the ebb-tidal delta and the final development of a wave-dominated delta with a wide delta plain.

Paleogeographic information obtained from dated cores and surficial sediments (Morales 1997, Delgado et al. 2012) allowed interpreting the oldest sediment infilling and progradation of this bay (Fig. 24.3). This progradation occurred in a different way on either sides of the estuary mouth. On the western side (Portugal) a transverse growth of a long spit (Monte Gordo Beach) took place. At the same time on the eastern side (Spain), the progradation occurred in an intermittent way forming successive barrier islands from active frontal sand bars. These differences are related with the sense of movement of the sand supply caused by the west-to-east littoral transport. The cartography of these old barrier islands existing on the Spanish delta plain demonstrates an evolution from a morphology typical of a mixed-energy coastline to a wave-dominated coast (according with the criteria by Hayes 1979). An explanation for this morphologic change would be related with a progressive lost of tidal energy caused by a decreasing tidal prism on this side (Morales 1997). Also an increasing of the wave energy arrived to the coastal front would be possible if the primary barrier island displayed in their front a large shallow shelf dissipating the wave energy by surf before it arrived to the first beach

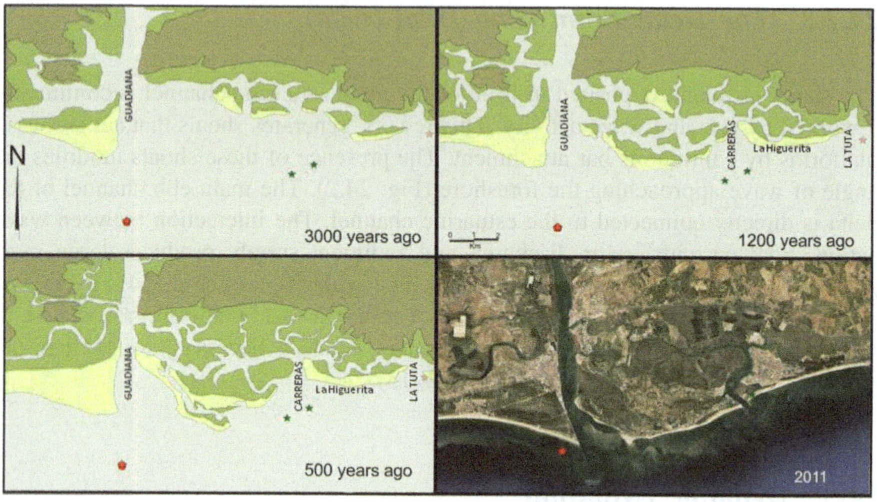

Fig. 24.3 First stages of progradation in the deltaic area compared to the present aerial picture

line. The extension of this shelf decreased with the deltaic progradation, so increasing the wave energy applied on the further barrier islands. This same process was described by Davis et al. (1989) in the Keys that close Sarasota Bay in the Gulf Coast of Florida.

A succession of nautical charts since 1755 allowed a detailed reconstruction of the historical evolution during the last 200 years (Fig. 24.4). These charts show the growing of the most recent barrier islands and spits adjacent to the main estuarine channel. In the open front, the tidal currents associated with the wave regime resulted in the development of an asymmetrical ebb-tidal delta. From a morphological point of view we can distinguish: (a) A very large western shoal (named the O'Brill bank) which grew updrift as an extension of the Monte Gordo spit; and (b) a broad shoal located eastward on the downdrift side of the inlet and (c) a wide frontal lobe.

In fair weather conditions, the predominant southwestern swell refracts and becomes larger toward the east on the western shelf transporting sediment north-eastward across the shoal (Fig. 24.4, 1793 situation). New sand coming from the west is supplied by the littoral longshore transport, contributing to the growing of the Portuguese spit and its associated shoal. Consequently, the main channel, connected with the estuarine channel, migrates landwards. Nevertheless, the platform is in relative morphodynamic equilibrium because ebb currents at spring tides potentially enhanced by river discharge during fluvial floods transport the sandy sediment in the contrary sense counteracting the wave-induced onshore transport over the long term.

Special situations occurred during high energy conditions (fluvial discharges, storms and even tsunamis). These events may breach the western shoal opening a new channel in N-S direction closer to the estuary mouth, and separating the shoal from Monte Gordo Beach (e.g. Fig. 24.4, 1838). Breaching is favored when part of the inlet becomes too shallow and is not efficient enough (hydraulically speaking) to

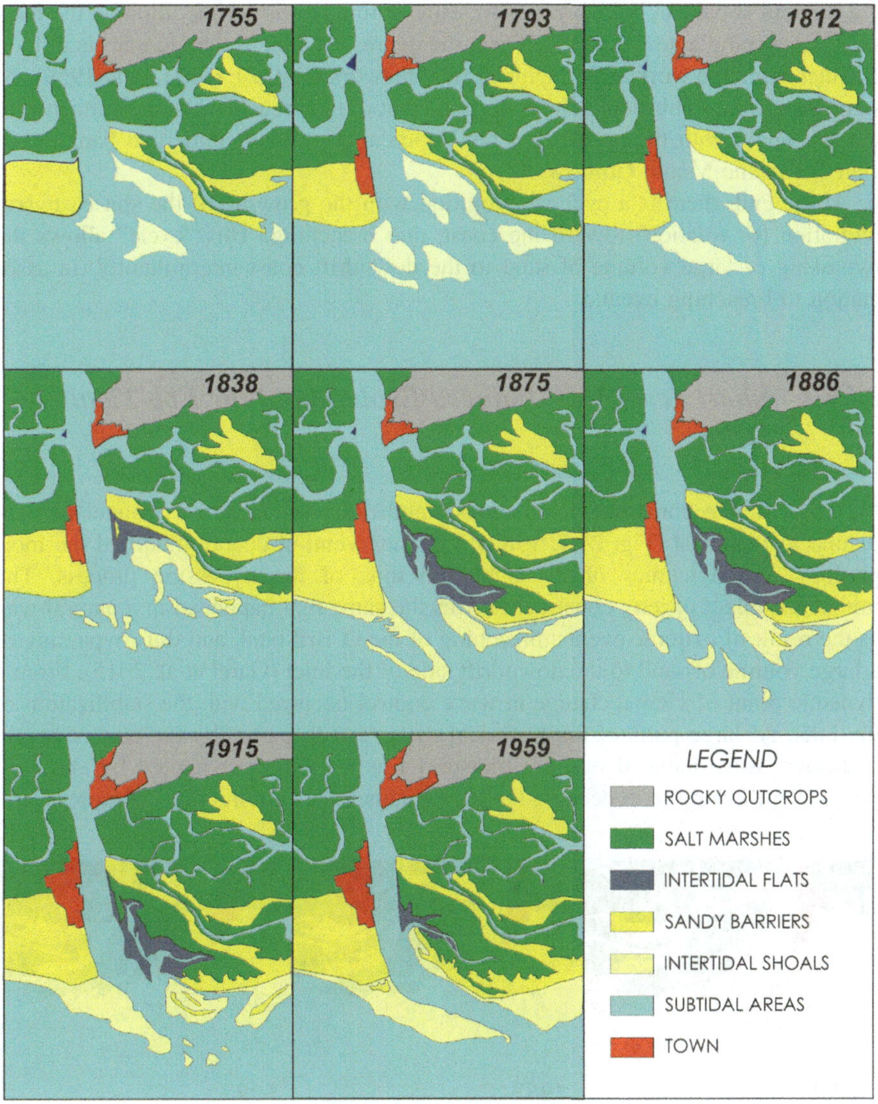

Fig. 24.4 Natural evolution of the deltaic area in the last 250 years

flush out the tidal prism on ebb tides. After the breaching, the main channel became transversal to the coastline. Then the abandoned main ebb channel acquires the dynamics of a marginal flood channel. Swash bar migration landwards on the shoal becames quicker under these conditions because the ebb tide ceases to act on them. So, the entire platform can migrate until the secondary flood channel loses its functionality (Fig. 24.4, 1875). The process implies in fact the bypassing of a large volume of sand to the downdrift part of the delta. Just then, the attachment of the old western shoal can build a new spit, establishing restricted condition on the back side, developing a new tidal flat. Then, the depressed area of the old marginal flood channel begins to work as an incipient tidal network (Fig. 24.4, 1915 and 1959). At same time, a new ebb-tidal delta is born in front of the breached channel, attached to the spit apex, with the subsequent development of a new submerged bank, associated with the Monte Gordo spit.

At the end, there is a cyclical evolution with the growing of the spit eastward, migration (or rotation) toward the coast, and breaching. This "cycle" allows the bypassing of large volume of sand to the downdrift coast intermittently (in association to breaching events).

24.3.2 *Short Term Man-Induced Evolution of the Ebb Tidal Delta*

In 1972–74, the construction of jetties stabilized artificially the channel perpendicular to the coast (Fig. 24.5, 1982). This intervention deeply modified the morphology and dynamics of the inlet and thus of the bypassing process. The immediate effect of jetty construction was the same that happened in a natural way as the result of extreme events: breaching of the O'Bril bank and thus bypassing of a large volume of sand to the downdrift side of the inlet (Garel et al. 2015). From a dynamic point of view a change in wave control occurred with the stabilization of the inlet. At large parts of the historical delta (and in particular the eastern swash platform), wave-induced onshore transport was no longer countered by ebb tidal flows over the long term because the inlet channel does not migrate anymore (Garel

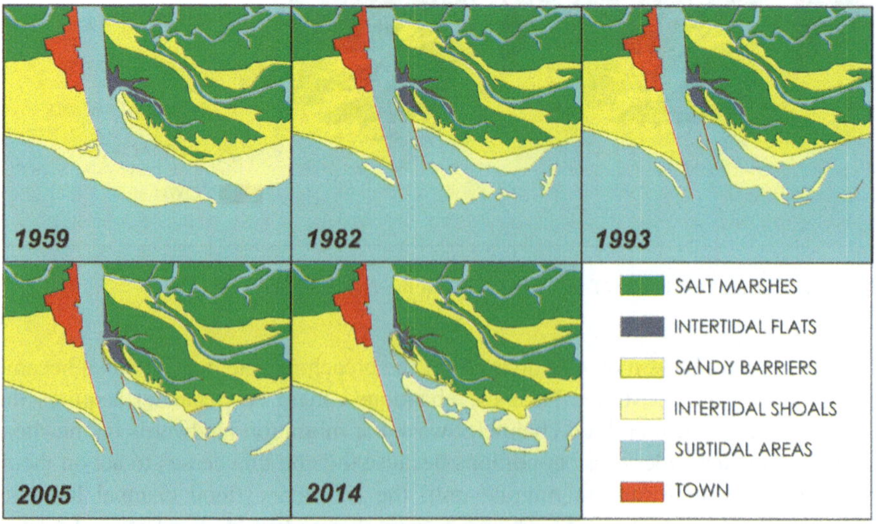

Fig. 24.5 Man-induced evolution of the deltaic area in the last 50 years

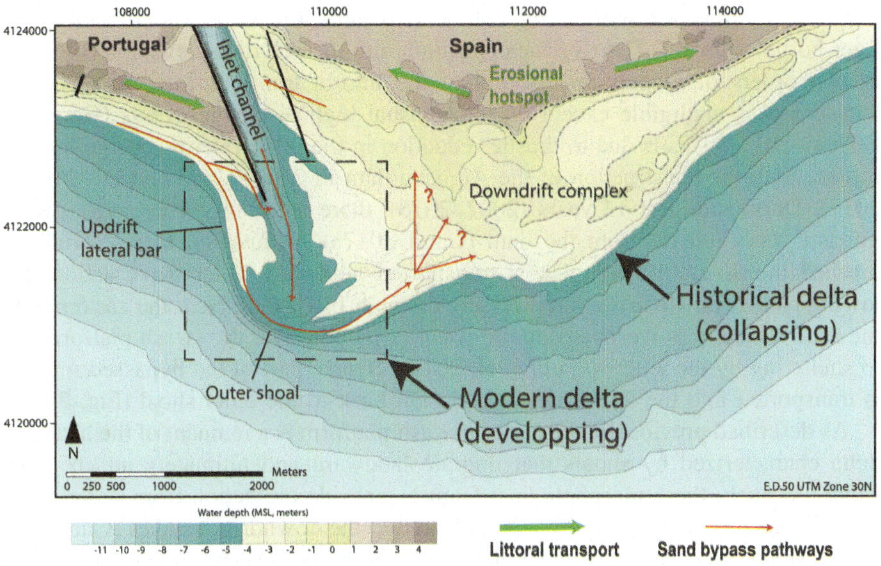

Fig. 24.6 Topo-bathymetry of the Guadiana Estuary mouth area from 2014 with indication of the main morphological elements of the ebb delta. The main transport pathways from the updrift beach to the downdrift area are indicated with a thin red line. The wide green arrows indicate the direction of the littoral transport: eastward along the updrift beach; divergent along the downdrift beach producing an erosional hotspot at the central part of the island

et al. 2014). Such collapsing part of the delta is characterized by intense erosion due to wave action (Fig. 24.5, 1993). This erosion is well-evidenced with the landward migration of shoals over the swash platform at a migration rate of the order of 10–100 m/year (Garel et al. 2014).

In a parallel way, the confinement of the flow between the jetties resulted in the development of a narrow and more symmetrical modern ebb delta in the seaward stream of the stabilized estuarine ebb jet (Garel et al. 2015). This modern ebb delta is characterized by an updrift bar and outer shoal (or "ebb shoal proper" following Kraus (2000) terminology) which developed relatively rapidly (few years) due to a large contribution of local sand eroded from the O'Bril bank (Garel 2017b). The downdrift area consists of the eroding broad complex including landward migrating shoals which are relict of the historical delta (Garel et al. 2014). Recent bathymetric studies show that the modern delta is presently (2017) still growing and migrating offshore at a rate of 7.5 m/year (Garel 2017b).

Sand deposited at the new delta is remobilized by waves mainly during storms. Bypassing is presently a relatively continuous but slow process that contrasts with the intermittent (cyclical) bypassing of large volume of sand associated to breaching of the historical delta. The morphological elements of the modern delta constituted by the updrift lateral bar and outer shoal delineate a major pathway for sand transport from the updrift coast to the downdrift complex (Fig. 24.6). Probably, some updrift material also passes the western jetty and reaches the inlet channel, from where it is transported by ebb jets towards the outer shoal, along with river-borne sediment. The riverine contribution to the ebb delta development is one to two orders of magnitude lower than the littoral transport rate and generally considered as negligible except during unusual high discharge events (Garel and Ferreira 2011). This is due to drastic reduction in the freshwater discharge into the estuary with the construction of the Alqueva dam in 2002 (Garel and D'Alimonte 2017). During the last 14 years (2002–2016), there were only 4 episodes of significant water release from the dam (Garel 2017a). Ultimately, the material that reached the downdrift complex is transported landward under wave action, following pathways which are not precisely defined (Fig. 24.6). Near the eastern jetty, the sand transport is westward due to wave refraction over the swash platform and to sheltering by the jetties (Garel et al. 2014). Hence, part of the bypassed material is transported into the inlet channel and then back to the outer shoal (Fig. 24.6).

As described previously, the eastern swash platform is a remnant of the historical delta characterized by shoals that migrate landward and ultimately attach to the shore. As such, these morphological features provide protection from wave action and episodically deliver large volume of sand to the downdrift beaches (Garel et al. 2014). In the vicinity of the shoal attachment point, both accretion and erosion are generally observed over the short-term (years). Erosion, due to wave refraction over the bar, is transient and restricted to the portion of the beach behind (or facing) the shoal (for details about the process, see Kana et al. 1999). However, at a longer time scale, shoal attachment results in large beach accretion—at least locally—which is for example much greater in magnitude than that resulting from beach nourishments operations implemented at these beaches (see Garel et al. 2014).

With the individualisation of several very large shoals since jetty construction (some of them are still in the process of attachment), the downdrift coast is characterized by overall accretion at a decadal time scale. Only the central part of the downdrift barrier island (i.e., the apex of Isla Canela) is submitted to long-term erosion due to the lack of (sheltering) shoal on its shoreface (Fig. 24.6). This erosional hotspot results from the temporally constant divergent littoral transport which is produced by the refracted waves thus providing sand towards the adjacent parts of the beach.

At the Guadiana ebb delta, the downdrift effect (i.e., large accretion) of the jetty construction is therefore opposite to the typical situation where jetties inhibit sediment bypassing resulting usually in downdrift erosion. Similar accretion has been observed at the updrift beach, due to the cross-shore transport (as shoal) of large volume of local sand issued from the erosion of the O'Bril bank rather than littoral transport as generally predicted (see Garel et al. 2015).

It is important to note that the local sediment source will drain out in the next decades with the complete collapse of the historical delta. The large decrease of this sediment supply will most probably lead to large erosion of the downdrift coast, several decades after jetty construction.

This shows that the knowledge of system morphodynamics at adequate temporal and spatial scales is essential to correctly plan management actions. In particular,

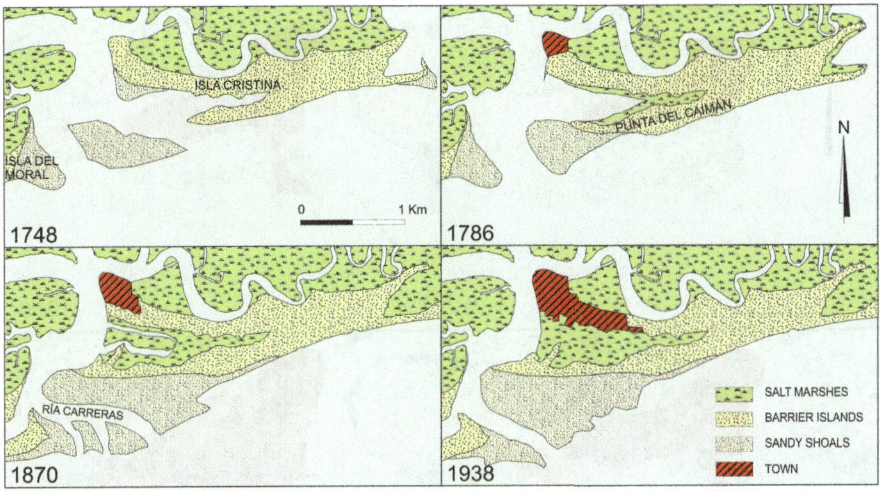

Fig. 24.7 First stages of evolution in the eastern deltaic area

the modality for sand bypassing at ebb deltas should be determined to properly assess the potential impacts of engineering structures.

24.3.3 Short-Term Evolution in the Eastern Deltaic Area

On the easternmost part of the system, downdrift the Carreras Inlet, the progradation did not build successive barrier islands, but contributed to grow La Higuerita Island as a drumstick-type barrier (Fig. 24.7). The intermittent arrival of sand from from the ebb-tidal delta formed in front of the Carreras Inlet, small spits were attached to the original barrier island. In the opposite boundary of the island was located "La Tuta" inlet which was closed by waves before 1850. This closure was caused by a loss of energy of tidal currents associated with a decreasing tidal prism (Fitzgerald et al. 2000a, b). After the closure, washover phenomena and coastal erosion problems continued in this area with a less developed foredune. A constant migration of Carreras Inlet and changes in dimensions and morphology of the ebb-delta gave rise made the harbor managers to build two jetties to stabilize the channel. The modern evolution of this island under the human influence is similar to that described by Bettencourt (1988) and Pilkey et al. (1989) in the Algarve barrier islands. These hard structures modified the wave refraction scheme, contributing to the complete rework of the ebb-tidal delta to grow a new spit attached in the island apex (Fig. 24.8).

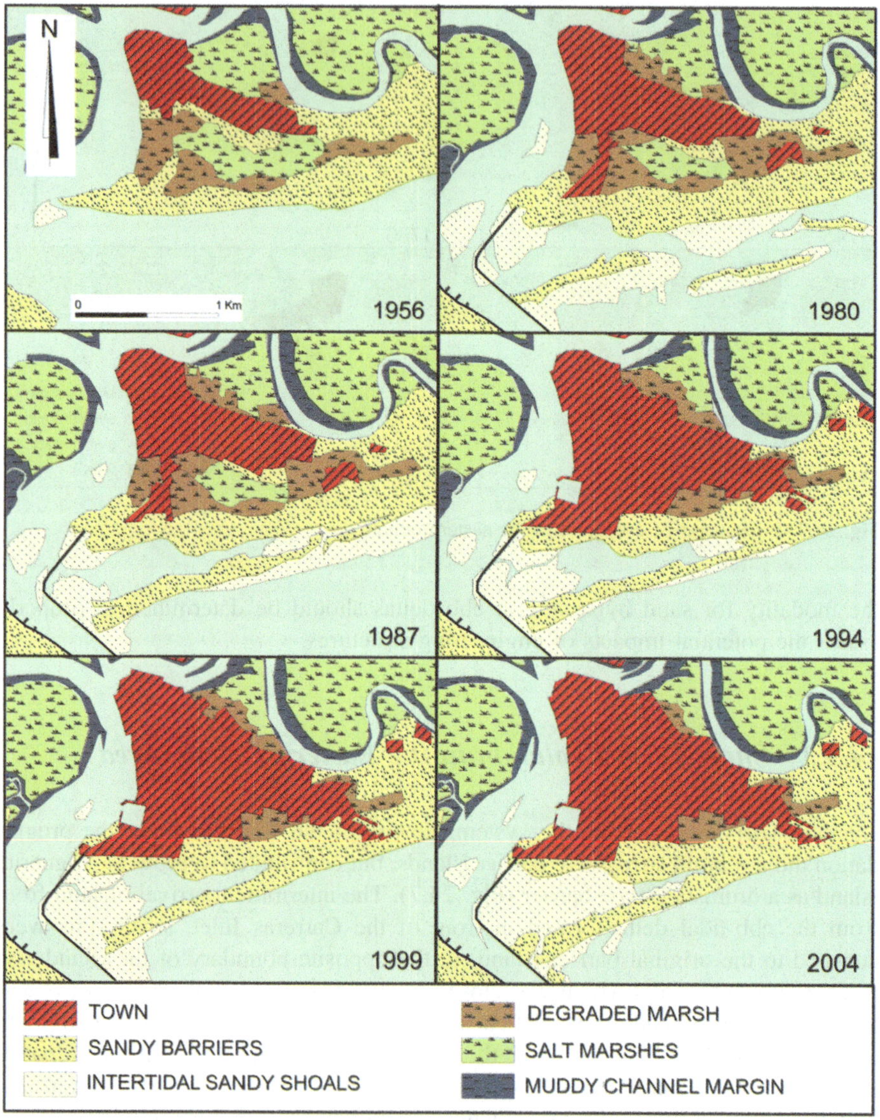

Fig. 24.8 Man-induced evolution in the eastern margin of Carreras Inlet

24.4 Sediment Distribution and Facies Architecture

24.4.1 Sandy Facies

Sandy facies are the main component of some of the present environments as: estuarine channel, submerged delta (shoals), beaches, Aeolian dunes and washover fans.

Estuarine channel facies corresponds to homogeneous, red, medium-coarse sand deposits that lack bioturbation. In marginal sectors or during minor tidal currents, fine-medium bioturbated sand with a clayey silt matrix is typical. In both cases, the sand is composed by shell fragments and lithoclasts because an intense tidal reworking occurred in this channel mix sands of fluvial and marine origins. Facies sequences display tidal alternations of interbedded sands and mudy sands.

Ebb-tidal delta facies (swash platforms, levees or shoals) are mainly constituted by medium to coarse, well-sorted sands and composed by shell fragments. The appearing sedimentary structures are related with the migrations of bedforms of variable scales (planar and trough crossbeddings with abundant reactivation surfaces and scours). The total sequence reaches 1.5 m thick.

Beach facies consists of medium, well-sorted clean sands without any bioturbation, typical of dissipative conditions with bar development. The characteristical vertical sequence of sedimentary structures consists of a base constituted by slaghtly seaward sloped parallel lamination. On these facies a set of ripple laminated fine sands is developed and on them, a 0.5 m thick set of landward sloped cross beds is normally deposited. The sequence culminates by another seaward sloped parallel laminated set. Shell levels deposited during winter storms are also interbedded.

Aeolian dune facies are composed by fine to very fine sands with a complex through crossbedding and abundant collapse structures. Their internal structure is strongly disturbed by plants.

24.4.2 Muddy Facies

The facies are mainly muddy in the tide-dominated environments, which are those protected from waves by the sandy barriers: Lagoons, subtidal creeks, tidal flats, intertidal channel margins and salt marshes.

Lagoon facies appear in the subtidal zone of the back barriers in the initial stages of evolution. These facies are usually constituted by bioturbated muds, with a variable sand percentage according to the proximity of sandy sources (estuarine channel or washover fans).

Tidal creeks have a variety of facies. In the base of the channels a lag of shells and fragments is visible. This lag can be covered by sand, but the sandy component may be often absent in creeks located far from the source of supply. Less active channels have usually filled by bioturbated muds.

Tidal flats facies are normally constituted by bimodal muddy sands. The coarser mode is comprised between coarse and fine sand (1.000–0.125 mm) and is dominant in these facies. The second one is the silty matrix. The organic carbon content is about 10% of total sediment mass. These available nutrients induce a high activity annelids and crustaceans causing bioturbation. On washover fans or in areas around tidal inlets muddy sands can occasionally be covered by layers of non-bioturbated clean medium sands.

Active channel margins present different facies according with the place where these are located. Intertidal margins of the estuarine channel (close to the sandy source) present non-bioturbated fine and very-fine, grey sand with some muddy matrix. Intertidal margins of creeks present facies consisting of poorly sorted muddy sand with a entire sediment finer than 0.125 mm forming a fining-upwards sequence. These facies are usually strongly bioturbated. Both, proportion of finer sizes and degree of bioturbation increase inwards. The innermost channel margins facies are only composed by monotonous heavily bioturbated muds that can alternate with thick layers of shells. In the Carreras Inlet channel margins thin, millimetric to centimetric alternations of muds, sandy muds and muddy sands (wavy tidal bedding) appear. These laminations are originated by oscillating energy between weather seasons. Occasional centimetric peat levels are formed on these margins after the decomposition of green algae accumulations.

Salt marsh facies are mainly integrated by rooted muds with a muddy matrix finer than 0.063 mm. Locally the sediment may content a fine sand fraction if the marsh is located close to sandy barriers or main channels. Storms may also generate clean sand layers. A continuous aggradation process raises the altitude of the marsh above the high spring tidal level, so only few tides a year can inundate it. Evaporation causes then an almost total loss of the vegetation. These unvegetated areas have been defined as "sterile marshes" by Borrego et al. (1993) although they should be called "sterile flats" because without vegetation they would not be called marshes. S*terile flat facies* are composed by silty clays with parallel lamination and a high salt content, never a thicker than a decimeter. Root bioturbation is then absent but desiccation cracks are common. This facies is near to those generated in salt pans but the main differences are topographic position and surficial extension.

24.4.3 Facies Architecture

The information from field observations, mapping and vibracores was processed to elaborate three stratigraphic sections (Fig. 24.9). The first section (profile I–I') was designed transversally to the eastern Portuguese margin of the estuarine channel, crossing the sand barrier and the salt marsh to arrive the bedrock. The second profile (II–II') crosses the eastern Spanish side of the delta, to the west of Carreras inlet. This second profile cuts some barrier islands and intermediate tidal flats-salt

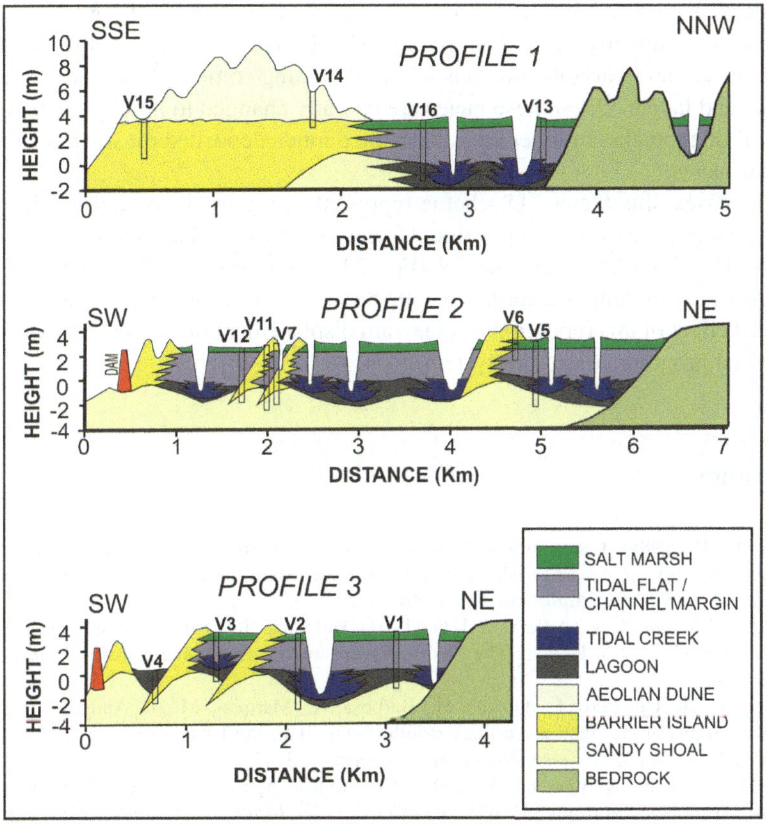

Fig. 24.9 Depositional facies architecture of the Guadiana River Delta. Modified from Morales (1997)

marsh bodies. The last section (profile III–III') is located to the east of the inlet and cuts the western margin of La Higuerita across three of their successive curved spits so as a the tidal marsh body to reach the mainland.

All, the three profiles show vertical and transversal facies changes. These changes are mainly controlled by the process responsible of the sediment deposition and the sedimentary environment in which it takes place. The base of the three profiles, which appears in the lower part of every core are the clean shelly sands corresponding with the ebb-tidal shoals. On the barrier fronts only clean sands with the sequence of structures typical of beaches are deposited above the shoal facies. In the central barrier very fine sands of dune facies culminate the sequence. In the back-barrier areas, muddy lagoonal and tidal flat facies overlap the shoal facies. These facies are generated in areas protected from the wave action by the own barrier. Intercalated with the tidal facies in these back-barriers metric layers of clean sand corresponding with washover fans are also present.

In cores obtained from the salt marsh, muddy facies typical of lagoons and tidal flats are also directly deposited on the shoal facies. These muddy lithofacies sequences contain variable amounts of sand, grading sometimes as sandy mud and muddy sand layers. Often these facies are laterally changed to muddy or shelly sand layers of tidal creeks. In inner areas the rooted muds deposited on salt marsh are the surficial facies.

In all cases, this facies 3D scheme represents a regressive model. Each vertical sequence is a succession of depositional facies that shows a shallowing-aggradational system. The described facies model (Fig. 24.9) is longitudinally repetitive on the eastern estuary margin in accordance with the successive generation of sandy barriers. The final effect of this repetition is a seaward displacement of the coastline that imply the coastal progradation of the system.

References

Bettencourt, P. (1988). Contribution of the sedimentological study for the understanding of the recent evolution of a barrier island system (Algarve, South of Portugal). *Bulletin Institut de Géologie du Bassin d'Aquitaine, 44,* 81–96.

Borrego, J., Morales, J. A., & Pendon, J. G. (1993). Holocene filling of an estuarine lagoon along the mesotidal coast of Huelva: The Piedras River mouth, southwestern Spain. *J Coast Res, 9,* 242–254.

Chícharo, M. A., Chícharo, L., Galvão, H., Barbosa, A., Marques, M. H., Andrade, J. P., et al. (2001). Status of the Guadiana estuary (south Portugal) during 1996–1998: An ecohydrological approach. *Aquatic Ecosystem Health Management, 4,* 1–17.

Costa, M., Silva, R., & Vitorino, J. (2001). Contribução para o estudo do clima de agitação marítima na costa portuguesa. In: *Proceedings of 2as Jornadas Portuguesas de Engenharia Costeira e Portuária*. International Navigation Association PIANC, Sines, Portugal.

Davis, R. A., Jr., Knowles, S. C., & Bland, M. J. (1989). Role of hurricanes in the Holocene stratigraphy of estuaries: Examples from the Gulf Coast of Florida. *Journal of Sedimentary Petrology, 59,* 1052–1061.

Delgado, J., Boski, T., Nieto, J. M., Pereira, L., Moura, D., & Gomes, A., et al. (2012). Sea-level rise and anthropogenic activities recorded in the late Pleistocene/Holocene sedimentary infill of the Guadiana Estuary (SW Iberia). *Quaternary Science Reviews, 33,* 121–141.

Fitzgerald, D. M., Buynevich, I. V., Fenster, M. S., & McKinlay, P. A. (2000a). Sandy dynamics at the mouth of a rock-bound, tide-dominated estuary. *Sedimentary Geology, 131,* 25–49.

FitzGerald, D. M., Kraus, N. C., & Hands, E. B. (2000b). *Natural mechanisms of sediment bypassing at tidal inlets* (pp. 9): US Army Corps of Engineers.

Garel, E. (2017a). Present dynamics of the Guadiana estuary. In: D. Moura, A. Gomes, I. Mendes, J. Anílbal (Eds.), Guadiana River estuary—investigating the past, present and future, pp. 15–37. 1st edition. University of Algarve. Faro, ISBN 978-989-8859-18-1. http://hdl.handle.net/10400.1/9887.

Garel, E. (2017b). Efficient dredging strategy for channel maintenance of the Guadiana ebb-delta. *Proc of Coastal Dynamics 2017*, Helsingor, Denmark, paper 232, 12 p.

Garel, E., & D'Alimonte, D. (2017). Continuous river discharge monitoring with bottom-mounted current profilers at narrow tidal estuaries. *Continental Shelf Research, 133,* 1–12. https://doi.org/10.1016/j.csr.2016.12.001.

Garel, E., & Ferreira, Ó. (2011). Effects of the Alqueva Dam on sediment fluxes at the mouth of the Guadiana estuary. *Journal of Coastal Research, 64,* 1505–1509.

Garel, E., Pinto, L., Santos, A., & Ferreira, Ó. (2009). Tidal and river discharge forcing upon water and sediment circulation at a rock-bound estuary (Guadiana estuary, Portugal). *Estuarine, Coastal and Shelf Science, 84*, 269–281.

Garel, E., Sousa, C., Ferreira, Ó., & Morales, J. A. (2014). Decadal morphological response of an ebb-tidal delta and down-drift beach to artificial breaching and inlet stabilisation. *Geomorphology, 216*, 13–25. https://doi.org/10.1016/j.geomorph.2014.03.031.

Garel, E., Sousa, C., & Ferreira, Ó. (2015). Sand bypass and updrift beach evolution after jetty construction at an ebb-tidal delta. *Estuarine, Coastal and Shelf Science, 167*, 4–13. https://doi.org/10.1016/j.ecss.2015.05.044.

Hayes, M. O. (1979). Barrier island morphology as a function of tidal and wave regime. In S. P. Leatherman (Ed.), *Barrier Islands* (pp. 1–29). New York, NY: Academic Press.

Kana, T. W., Hayter, E. J., & Work, P. A. (1999). Mesoscale sediment transport at Southeastern US tidal inlets: Conceptual model applicable to mixed energy settings. *Journal of Coastal Research, 15*(2), 303–313.

Kraus, N. C. (2000). Reservoir model of ebb-tidal delta evolution and sand bypassing. *Journal of Waterway, Port, Coastal, and Ocean Engineering, 126*, 305–313.

Lobo, F. J., Plaza, F., González, R., Dias, J. M. A., Kapsimalis, V., Mendes, I., et al. (2004). Estimations of bedload sediment transport in the Guadiana estuary (SW Iberian Peninsula) during low river discharge periods. *Journal of Coastal Research, 41*, 12–26.

Morales, J. A. (1993). *Sedimentología del Estuario del Guadiana (S.W. España-Portugal)* (Ph.D. Thesis). University of Sevilla, Sevilla, Spain, 300 pp.

Morales, J. A. (1997). Evolution and facies architecture of the mesotidal Guadiana River delta (S. W. Spain–Portugal). *Marine Geology, 138*, 127–148.

Morales, J. A., Pendon, J. G., & Borrego, J. (1994). Origen y evolución de flechas litorales recientes en la desembocadura del estuario mesomareal del rio Guadiana (Huelva, SW España). *Revista de la Sociedad Geologica de Espana, 7*, 155–167.

Morales, J. A., Delgado, I., & Gutiérrez-Mas, J. M. (2006). Sedimentary characterisation of bed types along the Guadiana Estuary (SW Europe) before the construction of the Alqueva Dam. *Estuarine, Coastal and Shelf Science, 70*, 117–131.

Oliveira, J. T. (1990). The south Portuguese zone. Stratigraphy and synsedimentary tectonism. In R. D. Dallmeyer & E. Martínez García (Eds.), *Pre-mesozoic geology of Iberia* (pp. 334–347). Berlin: Springer-Verlag.

Pilkey, O. H., Neal, W. J., Monteiro, J. H., & Dias, J. M. A. (1989). Algarve barrier islands: A noneoastal-plain system in Portugal. *Journal of Coastal Research, 5*, 239–261.

Part IV
Coastal Dunes

Chapter 25
Aeolian Dune Fields in the Coasts of Asturias and Cantabria (Spain, Nw Iberian Peninsula)

Germán Flor, Germán Flor-Blanco, Patricio Martínez Cedrún, Carmen Flores-Soriano and Cecilia Borghero

25.1 Introduction and Regional Settings

Aeolian dune fields are relatively well represented along the cliff coast of Asturias and Cantabria (NW Iberian Peninsula) (Fig. 25.1; Tables 25.1 and 25.2). Siliciclastic sands are supplied by large rivers to their estuaries and coastal upwelling and nutrients from some estuaries contribute with the formation of carbonate sands, but inherited and currently generating bioclastic sands are important in several stretches; easterly winds produce upwelling in this coast and shelf mainly in spring and summer (Lavín et al. 2004). Surplus sands were preferably drifting eastward, allowing the formation of sandy beaches and dune fields. In confining estuarine barriers, dune fields are the broadest, and on the nearest embayed beaches, these aeolian morphosedimentary transitional environments are best developed.

With the exception of the steep cliff edge, the coastal relief is smooth, bordered by several erosion surfaces of continental and marine origin staggered and irregularly distributed down to 280 m in height (Flor and Flor-Blanco 2014a; Domínguez-Cuesta et al. 2015; Díaz-Díaz et al. 2016). The mountains lay parallel to the coast, with a discontinuous range at a variable distance from less than 1.0 to

---

G. Flor (✉) · G. Flor-Blanco · C. Flores-Soriano
Department of Geology, University of Oviedo, 33005 Oviedo, Spain
e-mail: gflor@geol.uniovi.es

G. Flor-Blanco
e-mail: gfb@geol.uniovi.es

P. M. Cedrún
Department of Earth Sciences and Condensed Matter Physics, University of Cantabria, Santander, Spain
e-mail: patricio.martinez@unican.es

C. Borghero
Department of Earth Sciences, University of Uppsala, Uppsala, Sweden
e-mail: cecilia.borghero@gmail.com

J. A. Morales (ed.), *The Spanish Coastal Systems*,
https://doi.org/10.1007/978-3-319-93169-2_25

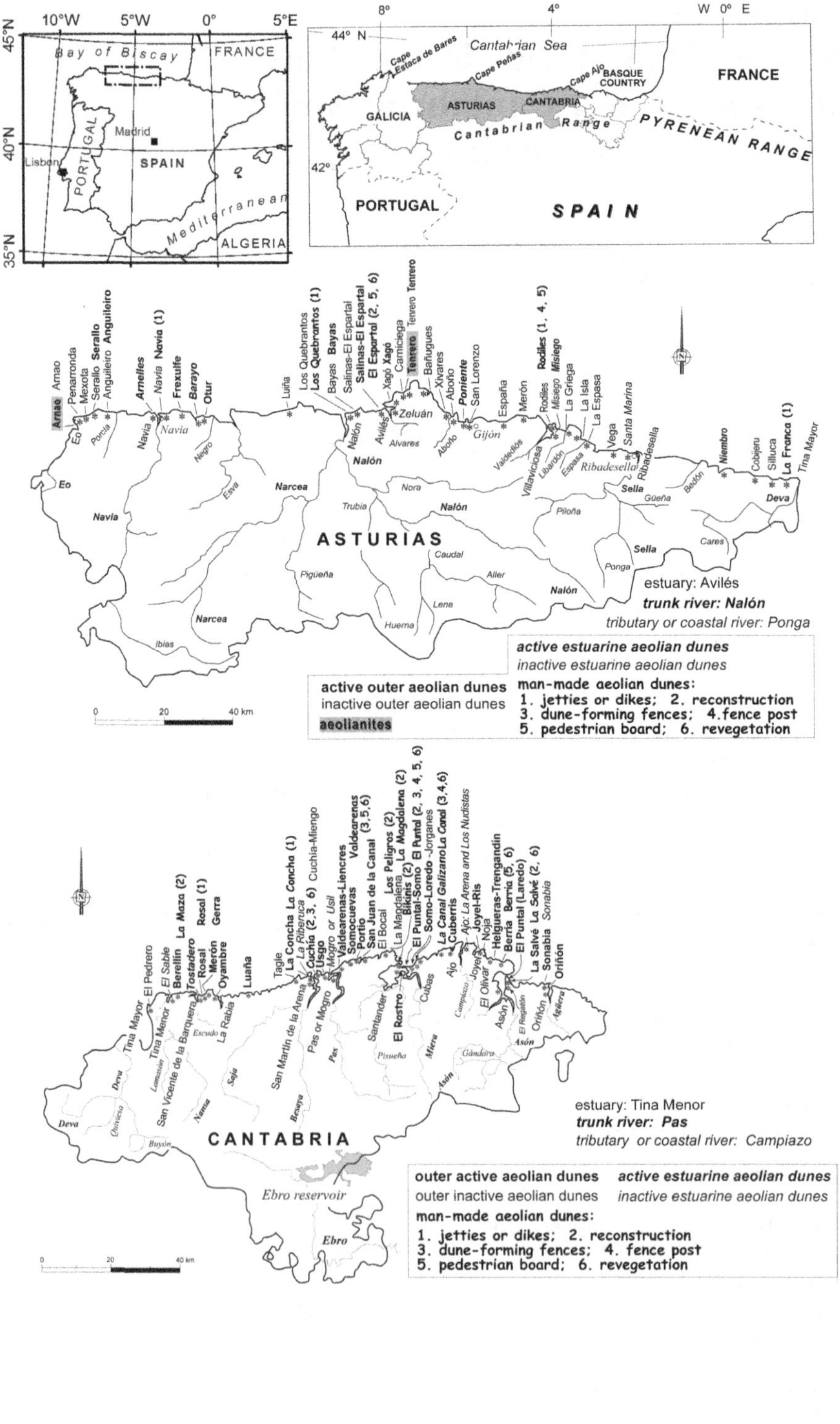
10°W
5°W
0°
5°E
45°N
40°N
35°N
Bay of Biscay
FRANCE
Madrid
SPAIN
PORTUGAL
Lisbon
Mediterranean
ALGERIA
8°
4°
W 0° E
44° N
42°
Cantabrian Sea
Cape Estaca de Bares
Cape Peñas
Cape Ajo
BASQUE COUNTRY
FRANCE
GALICIA
ASTURIAS
CANTABRIA
Cantabrian Range
PYRENEAN RANGE
PORTUGAL
SPAIN
Arnao
Penarronda
Mexota
Serallo
Anguileiro
Arnelles
Navia
Navia (1)
Frexulfe
Barayo
Otur
Luiña
Los Quebrantos
Los Quebrantos (1)
Bayas
Salinas-El Espartal
El Espartal (2, 5, 6)
Xagó
Carniciega
Tenrero
Bañugues
Xivares
Aboño
Poniente
San Lorenzo
España
Merón
Rodiles
Rodiles (1, 4, 5)
Misiego
La Griega
La Isla
La Espasa
Vega
Santa Marina
Ribadesella
Niembro
Cobijeru
Silluca
La Franca (1)
Tina Mayor
Zeluán
Gijón
Eo
Porcía
Negro
Esva
Narcea
Nalón
Avilés
Alvares
Nora
Trubia
Villaviciosa
Libardón
Sella
Güeña
Bedón
Deva
Piloña
Cares
Ponga
ASTURIAS
Caudal
Pigüeña
Aller
Lena
Huerna
Ibias
estuary: Avilés
trunk river: Nalón
tributary or coastal river: Ponga
active estuarine aeolian dunes
inactive estuarine aeolian dunes
active outer aeolian dunes
inactive outer aeolian dunes
aeolianites
man-made aeolian dunes:
1. jetties or dikes; 2. reconstruction
3. dune-forming fences; 4. fence post
5. pedestrian board; 6. revegetation
0 20 40 km
El Pedrero
El Sable
Berellín
Tostadero
Rosal
Merón
Oyambre
La Maza (2)
Rosal (1)
Gerra
Luaña
Tagle
La Concha
La Concha (1)
La Riberuca
Cuchía (2, 3, 6)
Cuchia-Miengo
Usgo
Mogro or Usil
Valdearenas-Liencres
Somocuevas
Valdearenas (3, 5, 6)
Portio
San Juan de la Canal
El Bocal
Los Peligros (2)
La Magdalena
La Magdalena (2)
Bikinis (2)
El Puntal-Somo
El Puntal (2, 3, 4, 5, 6)
Somo-Loredo-Jorganes
La Canal Galizano
La Canal (3, 4, 6)
Cuberris
Ajo: La Arena and Los Nudistas
Joyel-Ris
Noja
Helgueras-Trengandín
Berria
Berria (5, 6)
El Puntal (Laredo)
La Salvé
La Salvé (2, 6)
Sonabia
Oriñón
Tina Mayor
Tina Menor
San Vicente de la Barquera
Escudo
La Rabia
San Martín de la Arena
Pas or Mogro
Santander
El Rostro
Cubas
Ajo
Joyel
Campiazo
El Olivar
Asón
El Regatón
Oriñón
Agüera
Deva
Lamasón
Quiviesa
Nansa
Saja
Besaya
Pas
Pisueña
Miera
Gándara
Buyón
CANTABRIA
Ebro reservoir
Ebro
estuary: Tina Menor
trunk river: Pas
tributary or coastal river: Campiazo
outer active aeolian dunes
outer inactive aeolian dunes
active estuarine aeolian dunes
inactive estuarine aeolian dunes
man-made aeolian dunes:
1. jetties or dikes; 2. reconstruction
3. dune-forming fences; 4. fence post
5. pedestrian board; 6. revegetation
0 20 40 km

◂**Fig. 25.1** Situation of Asturian and Cantabrian aeolian dune fields in northwestern Spain (Cantabrian Sea, NW Iberian Peninsula), including the interesting bioclastic dunes partially cemented (aeolianites). They are characterized by active or inactive processes, as well as those due to the human management

6.0 km. The catchment higher limit of the Cantabrian Range is located between 40 and 20 km to the coastline, so rivers are relatively short with steep gradients (2,000 m in 30–70 km) and deep incisions (Gutiérrez-Elorza et al. 2002); river basins are relatively small, so most of the 10 Cantabrian river basins have less than 1,000 km^2 of surface, Deva, Navia, Nervión, Saja and Sella rivers range between 1,000 and 2,000 km^2; the Nalón basin is the exception with a surface of 4,866 km^2 (Prego and Vergara 1998).

The coastline is extremely rugged consisting of successive capes or points and embayments shaped by the differential erosion of waves, as well as a large number of mountain and coastal river mouths. From a sedimentary point of view, Cantabrian coast and the adjacent continental shelf can be considered undernourished (Fernández-Valdés 1997) due to the scarce fluvial sandy contributions; a fundamental consequence is the majority formation of embayed beaches on a rocky coast some of them filled with boulders and megaclasts, including sands and cobbles and exposed bedrocks.

The coastal climate is mild and humid due to the influence of the Atlantic Ocean at mid-latitude 42–43°N (Carballas et al. 2016); summer coastal temperatures average around 25 °C and spring and autumn are mild. The rains are quite abundant around 900/1200 mm per year on the coast, exceeding 2000 mm in mountainous areas. Consequently, due to humidity and plant colonization the active dune belt, including the transgressive units, is narrow about an average of 20–40 m.

Tides are semidiurnal and mesotidal with typical spring-neap cycle tidal ranges varying between about 4.8 m (extreme spring) and 0.7 m (extreme neap), according to Gómez et al. (2014), in central Asturias. Tidal range slightly increases to Cantabria, but exceptionally the estuaries of Santander (5.43 m) and San Vicente de la Barquera can be ocasionally macrotidal and hypersynchronous during spring tides (Flor-Blanco 2007; Flor-Blanco et al. 2015a), reaching upper ranges.

Finally, the waves, mainly in surge events, are an important factor to understanding the evolution of Atlantic dunes. In general, prevailing swell waves come from the NW, followed by N and NE, with an annual average significant height in the average regime about 2.50 m and the average period is about 5 s. The significant storm wave height in the scalar regime for a confidence of 90% is approximately 18 m, while the peak period is 17 s (Díaz Sánchez 2014).

Table 25.1 Responsible rivers and estuaries linked to the aeolian dune fields, characteristics and main dune typologies, occupied surfaces and main winds that generated them in the Asturian coast. Modified from Flor et al. (2011). Main dune typologies are described from the outer to the inner area of each aeolian dune field. Grey arrows represent the sand transfer from a trunk river, generally, to the E, as the main supplier to generate beaches and associated dune fields

TRUNK RIVER	ESTUARY	AEOLIAN DUNE FIELD	MORPHO-SEDIMENTARY UNIT	CHARACTERISTICS	MAIN DUNE TIPOLOGIES	TOTAL SURFACE (m²)	WIND DIRECTIONS
EO	EO	Arnao	exposed beach	inactive vegetated	aeolianites tabular cliff-top	148,520	SW
		Penarronda	exposed beach	embryo inactive vegetated	embryo foredune low mounds tabular cliff-top	39,902	NW and NE NW and NE SW
		Mexota	exposed beach	inactive vegetated	climbing dune tabular cliff-top	13,856	NE
		Sarello	exposed beach	embryo inactive vegetated	foredune low mounds climbing dune	9,462	NW and NE
		Anguileiro	exposed beach	embryo inactive vegetated	foredune climbing dune climbing dune	3,535	NW and NE
NAVIA	NAVIA	Navia	estuarine barrier estuarine dunes	embryo active vegetated inactive vegetated	tabular dune low foredunes low mounds tabular dune	112,980	NW
		Frejulfe	exposed beach estuarine barrier	embryo active vegetated inactive vegetated	foredune irregular field climbing dunes	36,520	NW and NE
		Barayo	estuarine barrier	embryo active vegetated inactive vegetated	tabular dune foredune foredune and tongue-shaped dunes	70,469	NW and NE
		Otur	exposed beach	embryo active vegetated inactive vegetated	tabular dune embryo foredune irregular field	14,239	NW and NE
ESQUEIRO	LUIÑA	San Pedro de Luiña	exposed beach	anthropized	low foredune and tabular dune	54,418	NE and NW
NALÓN	NALÓN	Los Quebrantos	estuarine barrier	embryo active vegetated anthropized	tabular dunes tabular dunes and low foredunes foredunes	527,042	NW and NE
		Bayas western	exposed beach	embryo active vegetated inactive vegetated	tabular dune tabular dune foredunes	92,995	NW
		Bayas eastern	exposed beach	embryo active vegetated inactive vegetated	tabular dunes tabular dunes foredunes, tongue-shaped dunes and climbing dunes	113,677	NW and W
	AVILÉS	Salinas (Avilés)	estuarine barrier	embryo inactive vegetated anthropized recovered	tongue-shaped dunes foredunes foredunes low tabular dunes	3,240 849	NW and W
		Llodero (Avilés)	estuarine dunes	inactive vegetated anthropized	foredunes foredunes	65,702	NW
		Xagó	exposed beach	embryo active vegetated inactive vegetated anthropized	tabular dune foredune and tongue-shaped dunes foredunes, tongue-shaped dunes and climbing dunes foredunes	421,118	NW, W and NE
		Aguilera or Carniciega (Verdicio)	exposed beach	inactive vegetated	tabular and climbing dunes	14,642	NW
		Tenrero (Verdicio)	exposed beach	embryo active vegetated inactive vegetated anthropized	tabular and climbing dunes climbing and cliff-top dunes tongue-shaped dunes longitudinal dunes (aeolianites)	140,518	NW and NE
ABOÑO	ABOÑO	Aboño	estuarine barrier	destroyed	foredune and irregular field	62,117	NW
PILES	GIJÓN	San Lorenzo (Gijón)	estuarine barrier	destroyed	foredunes climbing dunes	931,573	NW and NE
LINARES	VILLAVICIOSA	Rodiles (Villaviciosa)	estuarine barrier	embryo inactive vegetateda anthropized	tabular dunes foredunes foredunes and climbing dunes	253,633	NW
		Misiego (Villaviciosa)	estuarine dunes	embryo inactive vegetated anthropized	foredunes foredunes	55,844	NW and SW
LIBARDÓN	COLUNGA	La Griega	exposed beach	inactive vegetated	tabular dunes	40,782	NW
	LA ISLA	La Isla	sheltered beach	inactive vegetated	tabular dunes	195,079	NW
ACEBO	VEGA	Vega	exposed beach	embryo inactive vegetated	foredunes irregular field	40,782	NW
SELLA	RIBADESELLA	Santa Marina	estuarine barrier	destroyed	foredunes irregular field	195.079	NW

RIVERS: SELLA: mountain river; *ACEBO: coastal river or stream* ***WINDS:*** NW: prevailing; NE: secundary **VEGA** inactive full filled estuary

Table 25.2 Responsible rivers and estuaries linked to the aeolian dune fields, characteristic and main dune typologies, occupied surfaces and main winds that generate them in the Asturian coast. Modified from Flor et al. 2011). Main dune typologies are described from the outer to the inner area of each aeolian dune field. Grey arrows represent the sand transfer from a trunk river, generally, to the E, as the main sediment supplier to generate beaches and associated dune fields

TRUNK RIVER	ESTUARY	AEOLIAN DUNE FIELD	MORPHO-SEDIMENTARY UNIT	CHARACTERISTICS	MAIN DUNE TIPOLOGIES	TOTAL SURFACE (m²)	WIND DIRECTIONS
CARES-DEVA	TINA MAYOR	El Pedrero	estuarine barrier	embryo inactive vegetated	tabular dune tabular dune and incipient foredune	13,633	NW and SW
NANSA	TINA MENOR	El Sable	inner estuarine barrier	embryo inactive vegetated	tabular dune incipient foredune	6,424	NW
ESCUDO GANDARILLA	SAN VICENTE DE LA BARQUERA	El Rosal Tostadero	estuarine barrier estuarine beach	embryo active vegetated inactive vegetated anthropized	tabular dune irregular mounds irregular mounds irregular mounds	161,923	NW and NE
		Merón	exposed beach	active vegetated inactive vegetated	tabular dune, foredune irregular mounds climbing dune	24,855	NW
TURBIO El CAPITÁN	LA RABIA	Oyambre	estuarine barrier	inactive vegetated anthropized	foredune and tongue-shaped dunes irregular mounds	104,116	NW
		Luaña	exposed beach	embryo inactive vegetated anthropized	tabular dune and gentle foredune gentle foredune irregular mounds	13,049	NW
		Tagle	exposed beach	embryo destroyed	incipient foredune climbing and cliff-top dunes	32,494	NW and W
SAJA-BESAYA	SAN MARTÍN DE LA ARENA	La Concha	estuarine barrier	embryo inactive vegetated destroyed	tabular dune and foredune irregular mounds irregular mounds	101,718	NW and NE
		La Ribera	esturine beach	embryo inactive vegetated	tabular dune tabular dune	15,792	NW and NE
		Cuchia outer	exposed beach	artificially recovered	foredunes	669,965	NW
		Cuchia inner	exposed beach	destroyed	longitudinal dunes		W
PAS	MOGRO	Liencres	estuarine barrier	embryo inactive vegetated anthropized natural recovered	foredune and gentle mounds parabolic dunes climbing dunes tongue-shaped and climbing dunes tongue-shaped dunes	1,457 243	NW and SW
		Mogro or Usil	estuarine beach	inactive vegetated destroyed	tabular dune tabular dune	39,641	NW
		La Unquera	estuarine beach	destroyed	tabular dune	28,100	NW
		El Sardinero	exposed beach	destroyed	foredunes and tabular dune	84,330	NE
MINA CUBÓN BÓO MIERA	SANTANDER + CUBAS	Puntal de Somo	estuarine barrier	embryo inactive vegetated	tabular dune foredunes, tongue-shaped dunes, irregular mounds and saucer & bowl blowouts	308,149	NW and SSW
		Somo-Loredo	exposed beach	embryo active vegetated inactive vegetated destroyed	tabular dune foredunes and tabular dune irregular mounds irregular mounds	739,041	NW
		Jorganes	exposed beach	destroyed	tabular cliff-top dune	1,127 289	NW and W
LA COLINA		Galizano	beach - estuary	inactive vegetated recovered	foredunes and tabular dune tabular dune	8,041	NW and NE
CUBERRIS		Cuberris	exposed beach	inactive vegetated anthropized	tabular dune climbing dunes	49,451	NW
CAMPIAZO	AJO	La Arena and Los Nudistas	estuarine beach	active vegetated inactive vegetated anthropized	foredunes foredunes and climbing dunes tabular dune	36,972	NW
karstic springs	CABO QUEJO	Joyel - Ris - Noja	exposed beach	embryo active vegetated inactive vegetated destroyed	foredunes foredunes and climbing dunes irregular mounds and climbing dunes saucer blowouts irregular mounds	1,060 158	NW
ASÓN		Trengandin	exposed beach	embryo active vegetated inactive vegetated destroyed	tabular dune foredunes irregular mounds irregular mounds	284,789	NW and NE
		Berria	exposed beach	active vegetated inactive vegetated anthropized destroyed	tabular dunes foredunes irregular mounds irregular mounds	479,561	NW and NE
	ASÓN	El Olivar	estuarine beach	destroyed	climbing dunes	689,213	NE
		El Regatón	estuarine beach	inactive vegetated	foredunes	29,784	NW
		El Puntal + La Salvé	estuarine barrier	embryo active vegetated inactive vegetated anthropized destroyed	tabular dunes foredunes and tongue-shaped dunes irregular mounds and saucer & bowl blowouts irregular mounds irregular mounds	5,694 975	NE and NW
AGÜERA		Sonabia	exposed beach	active vegetated inactive vegetated	foredunes & bowl blowouts mixed dunes (longitudinal + barchans) transverse dune + saucer blowouts	96,774	NW
	ORIÑÓN	Oriñón	estuarine barrier	embryo inactive vegetated destroyed	tabular dune and foredunes foredunes irregular mounds	301,738	NW and NE

25.2 Winds

Coastal dunes are the result of the vegetation, while is not essential, where humidity is high, and the physical processes as a coevolution of topography and vegetation (Durán and Moore 2013), but the main physical agent acting on dune fields is the wind (Chapman 1990). It is well known that the shape of dunes depends on wind regimes and sand availability, particularly in the sand deserts (Wasson and Hyde 1983). In many cases, a simple correlation between wind directions and dune geometry can be developed (Flor and Flor-Blanco 2014b).

The main winds come from the WSW and SW (Fig. 25.2), typical during autumn and winter, which are very dry and hot ("Foehn Effect"), and strong (maximum gusts in Santander), but they do not affect the stability of the Cantabrian dune fields. W (humid) and NE (relatively dry) components are strong, generating small elongated dune geometries (longitudinal dunes) some of them superimposed to foredune (Flor and Flor-Blanco 2014b). The weak and moderate NW wind humid component, with a low frequency, and maximum frequency and intensity of these winds occurs during the winter.

Southwestern winds only are responsible to generate de inactive dune field of Arnao cliff-top tabular dune (Flor and Flor-Blanco 2014b), the estuarine dune field of El Tostadero (San Vicente de la Barquera estuary), the inner eastern part of Liencres dunes as inactive climbing dunes (Martínez Cedrún and Flor 2004), as well as the climbing dunes of La Magdalena, and Suaces in Noja (Flor et al. 2006).

NW winds are more important to generate foredunes, which are the most frequent dune morphology but also generates parabolic dunes in Liencres, a unique dune typology in the Cantabrian Sea; also the singular dune field of Sonabia originated from NW winds. However, NE winds are most frequent under anticyclonic conditions as in summer and can generate irregular dunes in the western side of some embayed beaches, as Berria, or climbing and cliff-top dunes in Mexota (Flor and Flor-Blanco 2014b), half northern segment of El Puntal-La Salvé spit, and Oriñón.

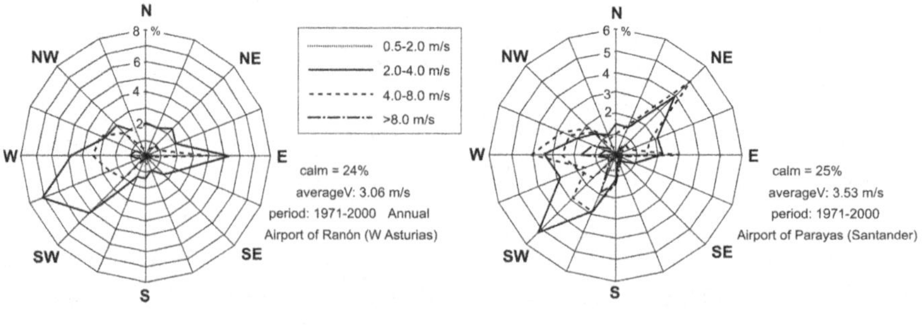

Fig. 25.2 The yearly average of wind rose diagrams with frequency and speed distributions (1971–2000) of representative meteorological stations in central coasts of Asturias (airport of Asturias) and Cantabria (airport of Santander)

25.3 Dune Geometries

Basically, aeolian dune fields are present under three modalities (Figs. 25.1 and 25.3; Tables 25.1 and 25.2):

- Natural dunes are mainly foredunes, tongue-shaped dunes, and climbing dunes; also blowouts, parabolic, cliff-top, and longitudinal dunes are represented. They are generated in the backsides of the beaches along the whole coast and this type of dune assume most of the existing dune fields in the Cantabrian Sea,

Fig. 25.3 **a** Bayas, on the eastern overlap to the foredune, few intrusive vegetated tongue-shaped dunes (Google Maps. 2017). **b** Since 80s, an embryo tabular dune (light green) landward the upper beach was generated due to the dumping of dredged sands in the neighboring western estuary of Avilés (2017). **c** Transgressive washover sand sheets migrating landward in the dune field of Navia (2017). **d** Eroded foredune (2016) is retreating quickly during the last decade (Barayo). **e** Isolated tongue-shaped dune (roll-over dune) in the Salinas-El Espartal, induced by trampling from the left side (2015). **f** Set of free tongue-shaped dunes (1990) migrating landward (El Puntal-La Salvé)

specifically in Asturias and Cantabria; these are best represented and considered mostly as exposed aeolian dunes as the associated beach supplying the sand without human influences.

- Less important and scarce are the *estuarine dunes*, as foredunes and embryo dunes, that are linked to estuarine beaches around the estuarine sandy bays, reaching reduced dimensions (i.e. Avilés, Villaviciosa, San Vicente de la Barquera, Pas and Asón) and generally have been generated by south winds. There are examples as the estuarine dunes in the estuary of Navia, also, in the subsystem of Llodero (Avilés), and several aeolian patches coexist in the estuarine bay of Villaviciosa (Bornizal dunes), some colonized by halophytic vegetation as *Halimione portulacoides* (Flor-Blanco and Flor 2009). In some cases have been generated by dredging feedbacks, such as inner San Vicente de la Barquera bay (Flor-Blanco 2007; Flor-Blanco et al. 2015a).
- Induced artificially dunes could have generated indirectly by human constructions or dredgings and inputs managements (Martínez et al. 2013). These morphologies are widely represented by foredunes with a fast seaward progradation. In other cases, the policy application has favored the new formation with sand feedbacks and planting, restoring old dunes eroded or triggering new geometries.

The orientation of the coast cliff segment in which the embayed beach is installed is fundamental to locate the aeolian dune fields and the dune typology; so, coastal section have main W-O trends followed by NE-SW and NW-SE, consequently, the components of wind of the first and fourth quadrant have the greater importance in these coasts to deflate the beaches and build their dune fields. The presence of cliff slopes allows the formation of climbing, cliff-top and falling (scarce) dunes as different geometries, mainly tabular sand sheets, lobes, hummocky topography, etc. Wind direction and geometry (crests and troughs, and dune contour) are satisfactorily correlating in dune fields of Asturias (Flor and Flor-Blanco 2014b).

In the aeolian dune fields of the Cantabrian coast (Fig. 25.1), basically two main geometries of free dunes are developed: transverse dunes, as foredunes (established, according Hesp 2002) (Fig. 25.3a–d; Tables 25.1 and 25.2), and longitudinal dunes (linear dunes), preferably tongue-like or tongue-shaped dunes (Flor 1986) (Fig. 25.3e, f; Tables 25.1 and 25.2); these are superimposed from the lee-side of foredunes with moderate to strong and humid W winds; the maximum wind gusts on the coast are due to W winds (Fernández García and Rasilla 1992). Other dune fields deposited by W winds are climbing dunes (Tables 25.1 and 25.2), and the mined and the destroyed large longitudinal ones of Cuchía in Cantabria (Flor 1980; Martínez Cedrún 2008).

There are different types of dunes widely localized along the Cantabric coast. The foredunes are the best-represented typologies in this coast but a variety of other geometries, even small shapeless or irregular dunes, can be identified (Table 25.1), as tabular dunes or incipient foredunes (embryo) (Hesp 2002), tongue-shaped dunes, mixed barchans and longitudinal dunes, blowouts (some units developing sedimentary rims), longitudinal dunes, climbing and cliff-top dunes as tabular and lobular geometries (Flor et al. 2011; Flor and Flor-Blanco 2014b); parabolic dunes, only on Liencres (Martínez Cedrún and Flor 2004, 2008).

Along the Cantabric coast, there are dune morphologies less common and hardly ever developed. A singular dune field is on Sonabia (Cantabria), constituted by a trellis complex of longitudinal and barchanoid dunes, generated on a slope of a few degrees (Flor and Martínez Cedrún 1991). Erosive or mixed dunes as blowouts are identified in the oldest dune fields (eastern area of Salinas-El Espartal, climbing dune of Liencres and Noja) and inactive dune fields (Somo, La Salvé and Sonabia). Saucer blowouts (Cooper 1958; Hesp 2002) and bowl blowouts (Hesp 2011) are scarce, practically nonexistent in Asturias and present in the dune fields of El Puntal-Somo and El Puntal-La Salvé spits, Noja (Martínez Cedrún 2008) and Sonabia (Flor and Martínez Cedrún 1991) (Table 25.2).

Except for the active dune belt in contact with the upper beach, many irregular geometries can be assigned to hummocky low mounds in the broad inner fields, fully vegetated. According to Goldstein et al. (2017) that boundary dune is referred as hummocky foredunes.

With respect to the vegetation, which shows the evolution degree of the dune, the marram grass (*Ammophila arenaria*) is the dominant vegetal species on temperate coasts, colonizing the active foredunes on most Cantabrian dunes, largely responsible for the great dimension (<7 m in height) and irregular geometry of this frontal dune. It is very efficient in catching and stabilizing the aeolian sand coming from the adjacent beach (Hart et al. 2012). Other pioneer species are the sand-couch grass (*Elymus fractus*) and other plants as (*Calystegia soldanella*, *Euphorbia paralias*, *Eryngium maritimum*, *Cakile maritima*, *Lagurus ovatus*, *Pancratium maritimum*), that are stable along the embryo foredune. Some dune fields contain a great variety of native plants far away of the beach, as *Smilax aspera*, *Ulex europaeus*, *Abutus unedo* y *Rhamnus alaternus*.

25.4 Sedimentary Features

The river basins supplied large volumes of siliciclastic sands building the coastal submerged prism linked to estuaries that are confined by barriers developing beach and associated dune fields, (Fig. 25.1; Tables 25.1 and 25.2), but available surplus sand allowed the formation of other beaches and aeolian dunes down current of the coastal drift eastward: Eo-Tapia, Navia-Otur, Nalón-Xagó, Villaviciosa-Vega, Deva + Nansa-Tagle, etc. (Fig. 25.1; Tables 25.1 and 25.2).

If there is an excess of sedimentary volume and an amount of sand available as well onshore winds, between the most important variety of factors, aeolian dune fields can grow. So, great dune fields were built near the siliciclastic sand supply by mountain rivers and it can be evidenced in the rivers of Navia, Nalón, Sella, Deva and Nansa and their estuaries. The other major component is biogenic carbonate, generated by processes of coastal upwellings and as a result of the nutrients supplied by rivers and estuaries, that filled other large estuaries (i.e. Eo, Santander and

Asón). After the sea-level maximum in the Middle Holocene, great dune fields of the Cantabrian coast prograded seaward.

Many dune fields have been characterized according to granulometric features (Folk and Ward 1957) and mineralogical composition (biogenic carbonates versus siliciclastic components). These studies allow establishing in part the origin and the coastal sedimentary dynamics until constituting the beach/dune systems. Also calculating the average values (Martínez Cedrún et al. 2014) is possible to discriminate sand beaches and their associated dune fields.

The transverse transitions between the beach and associated dune field follow the general tendencies, as the beaches have average larger grain sizes, sortings are worse and asymmetries are negative, containing high biogenic carbonate percentages. The dune sands show finer grain sizes, best sorting, positive asymmetries and lower carbonate percentages, due to the selective remobilization of the finest grains from the beach by wind (Flor 1981a, b; Flor and Martínez Cedrún 1991; Flor and Flor-Blanco 2009; Martínez Cedrún et al. 2014). In the most of the cases, the differences of the values of the granulometric parameters and the carbonate content versus silica between beach and dunes are very subtle.

There are many works in desert dunes (i.e. Ahlbrandt 1979; Lancaster 1986; Livingstone and Warren 1996) about the surficial granulometric patterns, differentiating some dune geometries like the crest, trough, stoss and lee sides, downwind trough, interdune area, etc. However, in coastal dunes, there are few contributions including the compositional mineralogy, highlighting those in tongue-shaped dunes (Flor 1986), blowouts (Flor 1984; Martínez Cedrún 2008), and foredunes (Flor 1984), etc. These interesting studies were detailed by Flor (1981b) to distinguish between transverse (foredune) and longitudinal dunes (tongue-shaped dunes) in the dune field of Xagó (Table 25.3). In the interdune areas between tongue-shaped dunes, where wind is strongly channeled, the average sizes are coarser than in the crests. Sands of foredunes are coarser along the crests and finer sizes in the downwind trough; the windward side contains variable sizes and lee side sands are middle and fine.

25.5 Dune Development

Most of the Cantabria and Asturias dune fields were formed since the mid-Holocene (Flandrian transgression) once the sea level fell and the prograding processes seaward were favored. It is essential to know the evolutionary patterns from this early stage to those that have occurred since the middle of the 20th century as a result of man's intervention through dredging of the estuaries to which they belong.

Table 25.3 Main granulometric parameters and biogenic carbonate percent which discriminate different areas of longitudinal (tongue-shaped dunes) deposited by western winds and transverse (foredune) dunes due to NW winds in the set of beaches/dunes of Xagó (n = number of samples). Modified from Flor (1981a)

	Zones	Mean size ϕ	Sorting ϕ	Skewness	Kurtosis	Biogenic carbonates %
Beach	Upper beach (n = 12)	Variable	Well and very well	Negative trend	Meso and leptokurtosis	Maximum
Tongue-shaped dunes	Crest (n = 8)	Fine and very fine	Very well	Positive	Platy and mesokurtosis	Middle
	Normal flank (n = 5)	Fine and very fine	Very well	Slightly positive	Platy and mesokurtosis	Middle
	Avalanche flank (n = 5)	Fine and very fine	Very and grea telly very well	Positive	Platykurtosis	Middle and minimum
	Interdune area (n = 6)	Middle	Very well	Positive	Platy and mesokurtosis	Maximum
	Lee side (n = 8)	Fine and very fine	Greatelly very well	Positive	Platykurtosis	Middle and minimum
Foredunes	Stoss side (n = 7)	Variable	Well and very well	Variable	Meso and leptokurtosis	Variable
	Crest (n = 16)	Middle	Well and very well	Variable	Meso and leptokurtosis	Maximum
	Lee side (n = 5)	Middle and fine	Well and very well	Positive	Leptokurtic trend	Middle
	Trough downwind (n = 10)	Fine and very fine	Very well	Variable	Mesokurtosis	Variable

25.5.1 Since Mid-Holocene

Most authors agree that most of the current dunes were generated in the middle Holocene as sea level dropped, triggering the progradation of beach/dune systems (Orford et al. 2003, Aagaard et al. 2007, Pye and Tsoar 2009, Anthony et al. 2010), but the human activity was also documented (Roskin et al. 2015).

Based on estuarine sediments of Cantabrian coast from boreholes and short core, relative sea-level rise trends were reconstructed. Since 8,500 until 7,000 cal BP sea-level rise of 9e12 mm year^{-1} from −21 to −5 m, and 0.3–0.7 mm year^{-1} since 7,000 cal BP until 20th century; during the 20th century the sea-level rise is 1.7 ± 0.2 mm year^{-1} during the (García-Artola et al. 2014).

In the absence of absolute dating, some approach was tried in the largest and most conserved aeolian dune fields of Asturias: Bayas, Salinas-El Espartal and Xagó, fed by the sandy contributions of the watershed of the river Nalón (Flor et al. 1998). In the last aeolian dune field of Xagó was supported by GPR studies (Flor-Blanco et al. 2016) and are a contribution to the understanding of the evolution of many Atlantic dune fields. In these dune fields, three dune belts were identified along the backshore (inner, middle and outer dune fields), being the inner one the oldest and less represented, as climbing dunes (Bayas and Xagó) or a large ridge (Salinas-El Espartal), related to a generalized process of seaward progradation.

The evolution of the Xagó complex (Fig. 25.4) could be extrapolated to the other two fields in Asturias with similar developing, previously mentioned. Their formation began during the high sea-level in the middle Holocene (Flandrian transgression), that is well represented by some solifluction deposits and culminating beach sediments (sands and/or gravels) along this coast (Flor and Flor-Blanco 2014b), constituting terraces at heights of 4.5–5.0 m, regarding the zero topographic level. Initially, a beach was built at the base of the cliff (Fig. 25.4a).

According to Flor-Blanco et al. (2016), plentiful sedimentary contributions from the Nalón river was following by sea level falls, allowing the progradation of the aeolian dunes/beach seaward (Fig. 25.4b, c). A climbing dune was the first dune sedimentation (Fig. 25.4b) and later a large dune ridge (Fig. 25.4c). Outcrops of dune trenches in the Salinas-El Espartal dune field were deduced by GPR radargrams and suggest that an interruption of the sand sedimentation occurred, probably due to a sea-level rise which reached a lower height, with the consequent erosion (Fig. 25.4d) and the formation of longitudinal dunes, oblique to the foredune. In the next stage, a new sea-level fall allowed a new sea level stabilization and a new foredune was formed (Fig. 25.4f). On the other hand, sea-level rise induces dune erosion in the long-term evolution as well as storm waves (de Winter and Ruessink 2017) in short time, resulting in transgressive dunes (Thom 1978; Giannini et al. 2007; Pye and Tsoar 2009; Hesp 2013).

Other examples in the Cantabrian coast that were built only in two phases due to seaward progradation; so dune fields extend parallel to the water line: inner (older) and outer attached to the active beach. The internal field of Tenrero is partially cemented due to the high biogenic carbonate percent, and has been occupied by a colony of bungalows keeping a good part of the dune morphology; probably, it can be correlated with the aeolianites of Górliz (Biscay), dated with ^{14}C by Cearreta (1993) in 6.020–5.710 ± 50 years BP. The widest dune complex has been developed in Liencres with two inner and outer belts are characterized by parabolic dunes and an inner broad field of climbing dunes (Martínez Cedrún and Flor 2004).

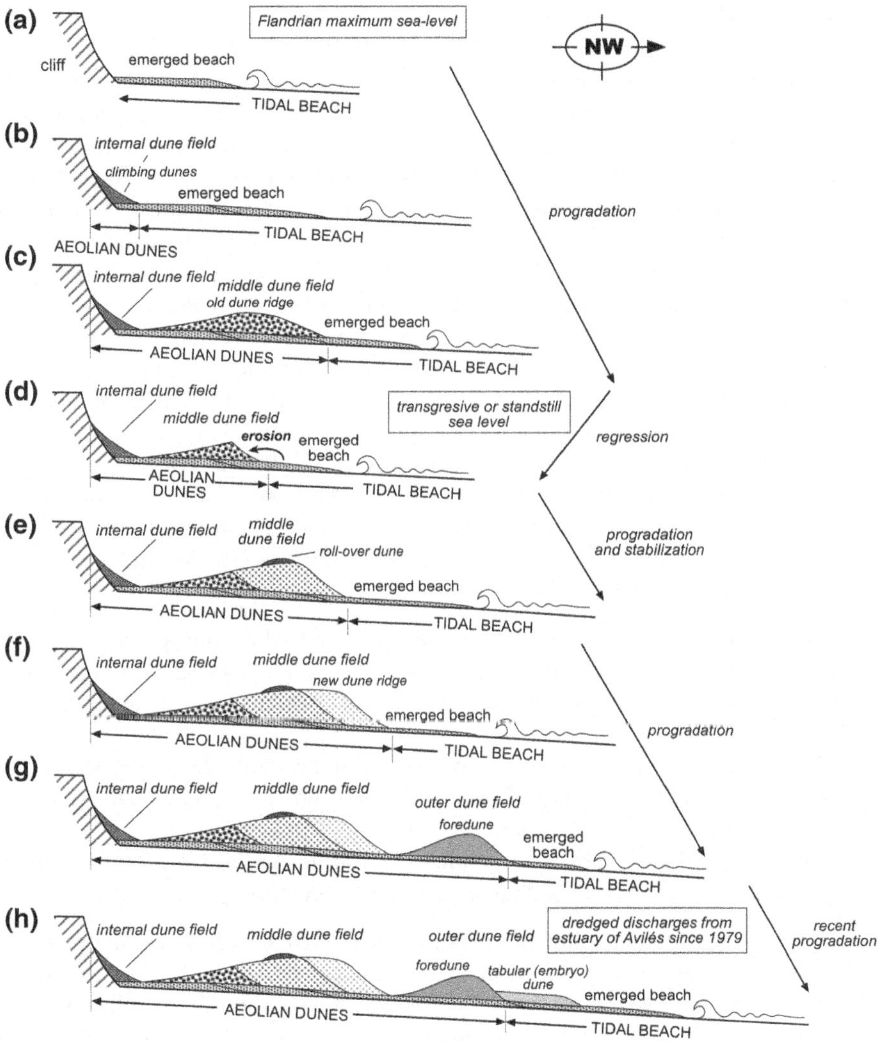

Fig. 25.4 Simplified evolution according transverse profiles of the aeolian dune field of Xagó (Asturias), since the Flandrian transgression. Modified from Flor-Blanco et al. (2016)

25.5.2 Anthropocene

Since the end of the 19th century, dredging, reclamation of marshes and artificial sea margins in estuaries have been carried out for the development of some Cantabrian ports (Rivas and Cendrero 1991), mainly in Asturias (Flor et al. 2006; López Peláez 2015). The consequence has been manifested in the erosion (retreat) of the dune front in the fields associated with the confining barriers and the morphodynamic and sedimentary behavior in the outer part of the estuary.

Throughout the first half of the twentieth century, some estuaries were canalized, the jetties were enlarged seaward a few hundred meters with a new longer tidal inlet, resulting the formation of new dune fields (induced dunes) due to a rapid progradation of the beach/dune system of the confining barrier (Flor-Blanco et al. 2015b). This process was materialized in a few decades in Navia, Nalón and Villaviciosa estuaries, as well as more narrowly in Avilés, all in Asturias, and in San Vicente de la Barquera and Cuchía in Cantabria. This process has been produced throughout the first half of the 20th century since 1916–2015.

In other cases, the estuarine intensive dredging in Avilés and Santander has promoted a great retreatment in the aeolian dunes belonging to the confining barriers (Salinas-El Espartal and Somo, respectively) and neighboring beaches as Bayas in Asturias, and in the case of the spit of Somo an elongation towards the interior of the estuary (Hart et al. 2012, Sanjosé et al. 2016).

Using old vertical photos of 1945 or 1956 (US Air Force), other later photo sets and orthophotos supplied by the Spanish IGN, Google Earth and Bing Maps, as well historic maps, dune evolution can be established. So, three models of dune evolution are recorded.

(1) The first one is represented by rapid changes due to the anthropic intervention, generally in estuarine mouths where a port was built and elongation of jetties in the inlet promoted new aeolian dune fields, generally of great magnitude (Fig. 25.5)
(2) The second model is represented by natural dune fields deducting two alternating processes of progradation and two erosion phases deduced by Flores-Soriano (2015), Borghero (2015).
(3) The last model corresponds to Aguilera or Carniciega, Mexota, and Oyambre which are small dune fields that are constant retreatment.

In spite of the different behaviour between the groups, they have similar patterns. This is due to they have been affected, to a greater or lesser degree, by similar natural and anthropogenic agents. They have been exposed to anthropogenic interventions such as the intensive dredging activities, river canalization, sand quarrying, urban development, trampling of the outer dune areas, etc. On the other hand, among natural factors, stand out level rise, between 2.0 and 4.3 mm/year, in the northern coast of Spain (Lorenzo et al. 2007) and storms, which are becoming more energetic and frequent. The fact that could be observed in winter of 2014.

25.5.2.1 Asturian Dune Fields

Dune field evolution from 1957 to 1983 is different between the three groups. In the first group, it is possible to observe several behaviours: some dune fields experienced a fast progradation after the construction of jetties along one or both sides of

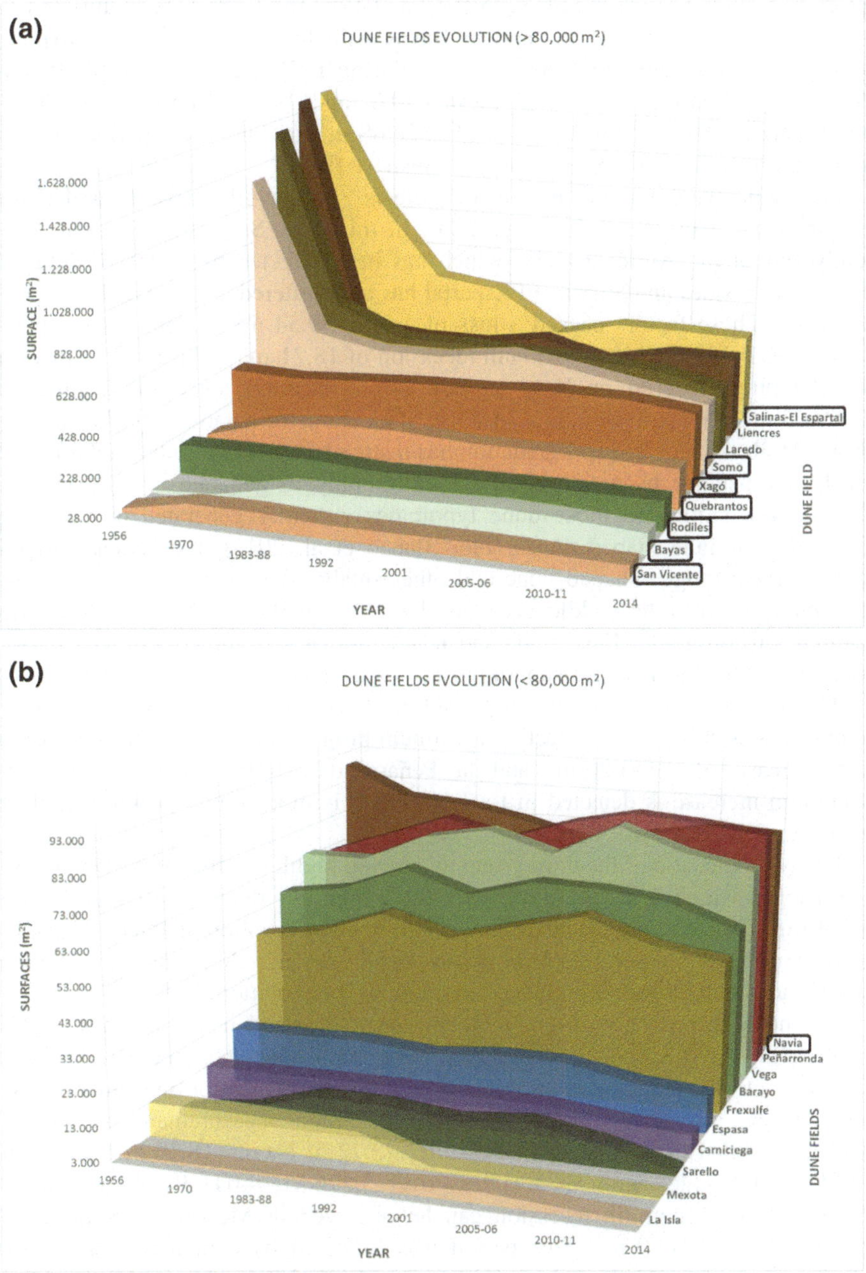

Fig. 25.5 Evolution area of dune fields **a** more than 80,000 m^2 **b** less than 80,000 m^2. In boxes, highlights the dune fields anthropized

the river mouth, as occurred between 1957 and 1983 in Quebrantos (Asturias) with an increase of 126,406.42 m^2 and in Bayas with 89,829.71 m^2. The maximum progradation in Quebrantos was 141.39 m, during 1957 and 1970, and 57.05 m in Bayas (neighbouring dune field) during 1970 and 1983 (Diego-Cavada 2014). Nevertheless, Navia (Asturias) lost 11,226.08 m^2 in the same period, with a maximum recession of 58.91 m in 1957 respect 1970.

For understanding the man-made effects, Salinas-El Espartal and Xagó (Asturias) are clear examples (Fig. 25.5). In this case, Salinas-El Espartal is the sandy spit of the Avilés estuary, which has been affected by intensive dredging operations besides and Salinas-El Espartal has also suffered a strong anthropogenic influence. These factors caused a loss of 1,007,759.54 m^2 in Salinas-El Espartal during 1957–1983, whit a maximum recession of 18.21 m (1970–1983). However, the dumping material at Xagó caused a progradation since the eighties of 7,080.08 m^2, with a maximum of 11.43 m in the dune front between 1970 and 1983 (Figs. 25.3b, 25.4h). This progradation has restored a new dune ridge and this is an ideal circumstance by a voluminous sandy supply to generate a roll-over dune (Fig. 25.4e), similar to those dune typologies currently generated in Salinas-El Espartal as tongue-shaped dunes (Flor-Blanco et al. 2013). Extrapolated to this stage, a great tongue-shaped dune and other smaller ones generated, overrunning the trough between the middle and inner dune field of Bayas. The large dune ridge continues to prograde (Fig. 25.4f) and it is culminates in the eastern side by low foredunes (Flor 1981a), and superimposed worm dunes (Flor-Blanco et al. 2016).

The behaviour between 1957 and 1983 of the second group in Asturias is more homogeneous (Fig. 25.5), detecting maximum progradations of 62.36 m in Sarello and increase of 8,444.20 m^2 and in Peñarronda 51.39 m and 14,244.01 m^2. Minimum increase is detected in La Isla, which is more stable in this period (an increase of 12.95 m^2).

Lastly, in the case of the third group this period highlights for an erosive process: Carniciega, which lost 21.10 m^2 (2.32 m along the dune front) and Mexota 608.19 m^2 (0.4 m along the dune front), both being relic climbing dunes, formed in a context of paleo-environment of a coast with greater sedimentary contributions and without any anthropic influence throughout its evolution.

All these groups have in common one erosive period in the nineties of 20th century (Fig. 25.5), produced mainly by strong anthropogenic interventions with a common activity related to the occupation and disappearance for trails, marinas, sea promenades, etc. In Asturias, maximum recession measurement is observed between 1983–1992, 22.92 m in Frexulfe, 22.72 m Navia, 18.19 m Salinas and 14.98 m in Rodiles with an area loss over 5,876.51; 4,041.39; 48,658.76 and 21,598.29 m^2. Minimum recession can be observed in Mexota, 1.28 m, with a global loss of 724.89 m^2. This period was followed by a natural progradation. There is some dune fields in the period of time 1992–2001, whose surface increase with a maximum progradation of: 23.27 m (6,585.79 m^2) Peñarronda, 20.35 m (1,695. m^2) La Isla, 14.16 m (8,742.96 m^2) Rodiles. Minimun progradation in the dune front is observed in Barayo, with only 1.53 m and 5,388.56 m^2 and in Otur an increase of 1,836.02 m^2 and frontal progradation of 2.49 m.

Finally, a new erosive period started in 2006–2011, caused mainly by natural processes (Fig. 25.5). Maximum recession is observed between 2001 and 2006 in Salinas 4.50 m, with a decrease of 77,341.79 m^2, in Barayo 4.13 m (507,42 m^2) and 2.03 m in Carniciega, with a decrease of 395.97 m^2. By comparison, some dune fields continued to grow during this period with a recovery the area of: 22,460.73 m^2 Xagó; 10,391.24 m^2 Vega; 7,687.26 m^2 Rodiles. In 2011 occurred a general recession respect 2006 with maximum erosion in: Vega 17.55 m (7,204.8 m^2), Frexulfe 17.14 m (7,271.53 m^2) and Sarello 15.85 m (3,510.13 m^2).

The most destructive wave storms detected in the Cantabrian coast were in February 1st and 2nd and in March 3rd and 4th with waves of 11 m heights (Flor et al. 2014). For this reason, the maximum recessions were measured between 2014 and 2011 ortophotographs (Fig. 25.5). The most damage has been in the dune field of Sarello, which lost 5,620.24 m^2 and had a recession in dune front of 40.61 m; Otur 8,606.56 m^2 (35.04 m); Bayas 19,081.69 m^2 (19.70 m); Quebrantos 8,491.18 m^2 (16 m); Barayo 7,309.92 m^2 (17.99 m). Furthermore, Anguileiro dune field disappeared completely with a loss of 425.60 m^2. Moreover, the temporal evolution of the areas since the first records to 2014 points out marked net losses of sand in various systems. This is observed in large aeolian dune systems: Navia (15,065.44 m^2), Barayo (1,225.62 m^2) y Salinas (1,216,046.28 m^2) but also in small dune fields: Mexota (9,544.11 m^2), Sarello (1,140.86 m^2), Otur (1,982.18 m^2) and Espasa (5,591.60 m^2). Nevertheless, there are also some dune systems which have prograded significantly as Quebrantos (95,019.29 m^2), Bayas (9,577.59 m^2) and Xagó (16,980.90 m^2).

25.5.2.2 Cantabrian Dune Fields

In Cantabria, San Vicente de la Barquera is the best example of the firs group with an increasing of the dune field area since jetties construction (1944) over 64,000 m^2 in 20 years and a maximum progradation of 274.55 m (Flor-Blanco et al. 2015a). From this moment, parts of the back old dune field was occupied by trail, carparks and a camping but the front dune continue increasing until 2010, when the great storms of 2011 and mainly in 2014 promoting continuously recession, quantified with a maximum of 7.35 m (Flor-Blanco et al. 2015b). However, Cuchía also was affected by a jetties construction but it evolution has been related with mining and destroying of the backward dune field, losing a third part of it surface (Flor-Blanco et al. 2015b).

On the other side, the Somo, in Santander estuary, is the dune field which has suffered most of the human pressure. The dune retreat in this segment has been driven by the intensive dredging activities on the Santander estuary inlet for the development of the port, especially during the 80s (Fig. 25.5).

Some dune fields of the second group, the third is not represented in Cantabria, experienced intense recession of several meters, most likely after storm, as it happened in the period 2005–2007, as well as 2010–2014, most recent and more widespread (Fig. 25.5). Since 1956 recession rates are quite variable over time but in general dunes have been eroding at average rate of 0.5 m/year, despite periods of expansion and grow have been reported. In the period 1956–2014 the dune field external limits have been retroceded of approximately 80 m in Somo (average rate 1 m/year), about 40 m in Liencres (average rate 0.7 m/year), 16 m in Oyambre (average rate 0.2 m/year) and 17 m in Laredo (average rate 0.2 m/year).

About the areas, the dune fields show an average decrease of 25% since 1956 (Fig. 25.5). The natural environment suffered mostly in the period 1956–1988, when urban expansion occurred. The natural dune field in Oyambre in 1956 was 9,166 m^2, whereas in 2014 was 3,852 m^2; in Liencres (Natural Park-Coastal Reserve) the natural area in 1956 was 168,848 m^2 whereas in 2014 it was 445,521 m^2; in the case of Laredo the dune field in 1956 was 1,565,579 m^2, but it has been decreasing since two decades, with the increasing of erosion processes due the construction of a port in 2011 (Fig. 25.6a). Storm and surge events of 2014 reduced the dune field area to 362,318 m^2. The case of Somo is different because the human pressure has been higher by dredging in the Santander estuary since 70 s, favouring a difference of 1,349,497 m^2 in 1956 which was only 37,493 m^2 in 2014.

The rest of Cantabrian dune fields without anthropogenic influence as Galizano, Cuberris, La Arena, Joyel-Ris, Trengandín, Berria, Sonabia and Oriñón, developed a progradation since 1956 until 1988, except La Arena and Trengandín and different behaviour in the first decade of the 21th century, so mainly with retreatment trend. Nevertheless, since 2010 the most part of dune fields shortened except Trengandín, that is protected from NW wave storms. As of the rest of the Cantabrian coast, storm and surge events of 2014 winter, caused great damage in all front dunes.

After that destructive event, detected in some parts of the Atlantic Coast (Flor et al. 2014; Masselink et al. 2016; Castelle et al. 2017; Flor-Blanco et al. 2017) it was possible to observe different erosive features which were similar to erosive models according to Flor et al. (2014). On the one hand, dune fields with minor erosion show the migration of sediments towards the land and minor erosion in the dune front. This processes consisted of sand intrusion as washover fans or as aeolian sand sheets. Otherwise, in dune fields with major erosion rates, the dune front has been total or partially eroded; climbing dunes and the largest dune field developed large scarps. In coastal systems where a stream flows into, it has observed a sand intrusion upper stream or the output of coarse fluvial fraction can be produced in the intertidal beach building a granule fan.

The lost sand in most dune fields affected by the storms in 2014 was deposited in the backshore and the foreshore of the beach, improving the dune field aspect and recovering any dune morphologies. An evidence of this is the formation of new tabular dunes, incipient foredunes and small and ephemeral echo dunes, or the most widespread regulation of the escarpment, formed by the shock of the swell violence, building a new small climbing dune.

25.6 Management

Several aeolian dune fields are destroyed directly or damaged in Asturias and Cantabria due to: urbanization (Salinas-El Espartal, inner Rodiles, San Lorenzo, Santa Marina, La Concha, playa Segunda, Noja, La Salvé), industry (Salinas-El Espartal, Aboño), mining (Xagó, Cuchía, Liencres), road construction and seafronts

◂**Fig. 25.6** **a** Large and width sandy spit (Puntal-La Salvé) of the estuary of Asón (port of Laredo in the right side) retreating foredune and strongly urbanized since second part of the 20th century (2015, Bing Maps). **b** The spit barrier of El Puntal-Somo (1988) is crossed by many washover channels. Restoration works were made in corridors: fences, revegetation, soft protected enclosures, etc. **c** Large aeolian dune field of Liencres (estuary of Pas); parabolic dunes are unique in Cantabrian coast (2016, Google Earth). **d** In Salinas-El Espartal (Asturias), where built artificial dune field (large and asymmetric) was developed with an outer active foredune, including some tongue-shaped dunes. **e** Sinuous fences in El Puntal-Somo dune field. **f** Isolated areas along the seaward margin of the dune field (wood posts and nylon ropes), where the *Ammophila arenaria* plant is natural colonizing the protected active dunes (Rodiles aeolian dune/beach system) in 2015; it is complemented with some pedestrian boards that access to the beach. Dune fields to protect: **g** Climbing lobe narrowing windward (in the background) that occupy a low trough (inner dune field) of Xagó (Asturias). **h** Aeolian dune complex constituted by an orthogonal set of longitudinal and barchan dunes in Sonabia (Cantabria)

(Salinas-El Espartal, San Lorenzo in Gijón, Ribadesella, La Concha in Suances, La Salvé in Laredo).

Nourishing sand beaches was applied in 2005 in Salinas-El Espartal, braking temporarily the dune erosion and the defense of the seafront. Within the set of beaches of Bikinis-La Magdalena-Los Peligros (Santander) since 1990, allowed the indirect creation of aeolian dune patches (climbing dunes), something wider in the first one.

Some dune problems and different dune restoration techniques were applied in the Spanish coast (Gómez-Pina et al. 2002); these authors present the cases of the Somo (Fig. 25.6b) and Liencres spits (Fig. 25.6c), and Cuchía but they were more widely developed by Martínez Cedrún (2008), including other dune fields of Cantabria. Many treaties on dune management have been edited (Kidd 2001; Pye et al. 2007; de Vega et al. 2007; Nordstrom 2008; Martínez et al. 2013, Swann et al. 2015). A detailed and extensive article about the management of the main involved coastal dune fields of Asturias, Cantabria and Basque Country was published by Flor-Blanco and Flor (2016), now emphasizing the construction of wide artificial foredunes in the El Puntal-La Salvé beach (Fig. 25.6a).

Interventions to improve some dune fields of Asturias and Cantabria are not numerous, but there are significant examples, implemented in the decades of the 90 s until the beginning of the 21th century (Tables 25.1 and 25.2; Fig. 25.6).

The unique example where a new foredune was created (60,000 m^2) using beach sand dumpings (2002) was in the eastern area of Salinas-El Espartal (Fig. 25.6d). Moreover, woody walkways elevated over the dune were located according to a longitudinal (promenade) and several perpendicular configurations (access to the beach or to a site of panoramic sight) with the implementation of planting native vegetation, mainly *Ammophila arenaria*. Also, in the neighboring dune field of Xagó, some walkways were supported on the surface of the old destroyed and mined areas.

Another case is in El Puntal-La Salvé (confining spit of the Asón estuary); an artificial foredune was built after the strong wave storms of winter 2014

(Flor-Blanco and Flor 2016). Sand was obtained from dredging of the shallow bottoms of the outer mouth bar and submerged beach (Fig. 25.6a).

Fencing techniques were applied in Liencres to increase the sedimentation in several sand bottoms without vegetation was unsuccessful, including a few number of elevated walkways; inner climbing dunes (dark green) were vegetated with pines since 1950. In this same dune field, isolated areas were planted with this marram grass but other native species were later included improving the vegetal biodiversity. Another method is the application of geotextile sheets that only was applied locally in exposed areas of washover channels crossing the dune field in the Somo spit (Fig. 25.6b) in 1993. Sinuous or zig-zag configurations, but parallel to the shore, are built; on the top of geotextiles, a 0.5 or 1.0 m of sand was deposited and up fences were implanted (Fig. 25.6e). Cordoning off areas with timber cylindrical wood piles spaced and joined simply by a solid braided nylon rope is a very positive technique. This technic has induced a relative fast progradation of the dune front in the upper beach of Rodiles (estuary of Villaviciosa, Asturias) during the last 16 years (Fig. 25.6f), well studied by Flor-Blanco et al. (2015b), Flor et al. (2015b).

Indirectly, dumping dredged sands or beach nourishment triggered new dune morphologies. The first case took place in Xagó, where an embryo tabular dune was generated (Fig. 25.3b), after developing to a new foredune, even vegetated by pioneers. However, intentional nourishment in Peligros-La Magdalena-Bikinis beaches (Santander), has promoted small dunes, mostly as climbing dunes.

Some dune fields are protected by regional authorities through conservation designa-tions: Natural Monuments of Peñarronda, Frexulfe, Isla of Deva and beach of Bayas, El Espartal (small area of 0.056 km^2), Barayo (Partial Natural Reserve) and Villaviciosa estuaries. In Cantabria, SCIs (Sites of Community Importance) of Rías occidentales (western estuaries), dunes of Oyambre (Natural Park), dunes of El Puntal-Somo (Natural Park), Liencres dune field (Fig. 25.6c), climbing lobe of the inner belt of Xagó (Fig. 25.6g) and all dune field of Sonabia (Fig. 25.6h) must be declared as some figures of protection in the future.

To sum up, all dune fields of Asturias and Cantabria have been suffered a natural recession due sea level rise and the continuous surge event, increasingly more often. Policy, laws and management must go hand in hand for try to preserve in the best possible way this type of habitats of great importance for the coast.

References

Aagaard, T., Orford, J. D., & Murray, A. S. (2007). Environmental controls on coastal dune formation; Skallingen Spit, Denmark. *Geomorphology, 83,* 29–47.

Ahlbrandt, T. S. (1979). Textural parameters of aeolian deposits. In E.D. McKee (Ed.), A study of global sand seas: U.S. Geological survey professional paper (Vol. 1052, pp. 21–51).

Anthony, E. J., Mrani-Alaoui, M., & Héquette, A. (2010). Shoreface sand supply and mid-to late Holocene aeolian dune formation on the storm-dominated macrotidal coast of the southern North Sea. *Marine Geology, 276,* 100–104.

Borghero, C. (2015). Study of morphological evolution of dune felds in Cantabria (N. Spain) during the Anthropocene.

Carballas, T., Rodríguez-Rastrero, M., Artieda, O., Gummuzio, J., Díaz-Raviña, M., & Martín, Á. (2016). Soils of the temperate humid zone. In J. F. Gallardo (Ed.), *The soils of Spain* (pp. 49–144). New York: Springer.

Castelle, B., Bujan, S., Ferreira, S., & Dodet, G. (2017). Foredune morphological changes and beach recovery from the extreme 2013/2014 winter at a high-energy sandy coast. *Marine Geology, 385,* 41–55.

Cearreta, A. (1993). Palaeoenvironmental interpretation of Holocene coastal sequences in the southern Bay of Biscay. *Geologische Rundschau, 82,* 234–240.

Chapman, D. M. (1990). Aeolian sand transport-an optimized model. *Earth Surface Processes and Landforms, 15,* 751–760.

Cooper, W. S. (1958). Coastal sand dunes of Oregon and Washington. Geological Society of America 72.

de Vega, Ley, Seoane, C., Gallego Fernández, J. B., & Vidal Pascual, C. (2007). *Manual de restauración de dunas costeras* (p. 251). Madrid: Dirección General de Sostenibilidad de la Costa y del Mar. Magrama.

de Winter, R. C., & Ruessink, B. G. (2017). Sensitivity analysis of climate change impacts on dune erosion: Case study for the Dutch Holland coast. *Clim Chan, 141,* 685–701.

Díaz Sánchez, R. (2014). *Predicción de remonte extremo en playas: aplicación a los sectores litorales de la costa española* (Ph.D. Thesis). E.T.S. Ingenieros de Caminos, Canales y Puertos. Universidad Politécnica de Madrid. 236 p. http://oa.upm.es/32923/1/Roberto_Diaz_Sanchez.pdf.

Díaz-Díaz, L. M., Flor-Blanco, G., López-Fernández, C., Luis, P. (2016). Analysis and interpretation of marine/continental terraces in the central coast of Asturias (NW Spain. In *Proceedings of EGU General Assembly Conference Abstracts* (Vol. 18, pp. 12499).

Diego-Cavada, J. M. (2014). *Sedimentología, morfología y evolución del campo dunar de Bayas (Asturias)* (M.Sc. Thesis). Department of Geology. University of Oviedo.

Domínguez-Cuesta, M. J., Jiménez-Sánchez, M., González-Fernández, J.A., Quintana, L., Flor, G. & Flor-Blanco, G. (2015). GIS as a tool to detect flat erosional surfaces in coastal areas: a case study in North Spain. *Geologica Acta, 13*(2), 97–106. http://dx.doi.org/10.1344/GeologicaActa2015.13.2.2

Durán, O., & Moore, L. J. (2013). Vegetation controls on the maximum size of coastal dunes. *Proceedings of the National Academy of Sciences, 110*(43), 17217–17222.

Fernández García, F., & Rasilla, D. (1992). El viento en la Cornisa Cantábrica: avance de un estudio sobre los temporales del S. *Cuadernos de sección. historia, 20,* 271–295.

Fernández-Valdés, J. M. (1997). *Morfología y sedimentación de la plataforma continental interna del área central de Asturias. Gijón, España* (Ph.D. Thesis, unpublished). Department of Marine Science and Technology. University of Oviedo. 232 p.

Flor, G. (1980). Las dunas costeras de Cantabria: valores singulares geológicos. *Actas I Reunión de Geología Ambiental y Ordenación del Territorio*. Santander. 22 p.

Flor, G. (1981a). Las dunas eólicas costeras de la playa de Xagó (Asturias). *Trabajos de geología, 11,* 61–71.

Flor, G. (1981b). Los parámetros texturales en las playas y dunas arenosas del borde occidental del Cabo Peñas (Xagó y Verdicio, Asturias). *Boletín de la Real Sociedad Española de Historia Natural. Sección Geológica, 79,* 89–102.

Flor, G. (1984). Estudio sedimentológico y morfológico de una duna costera "blowout" (Cabo Frouxeira, La Coruña). *Trabajos de Geología, 14,* 161–174.

Flor, G. (1986). Sedimentología de una duna lingüiforme en la playa de Xagó (Asturias). In *Actas del IX Congreso Nacional de Sedimentología* (Vol. I, pp. 317–328). University of Salamanca, Diputación Provincial de Salamanca.

Flor, G., & Flor-Blanco, G. (2009). Sedimentología de los depósitos arenosos superficiales de playa de Vega (Concejo de Ribadesella, Asturias). *Rev Soc Geol España, 22,* 105–121.

Flor, G., & Flor-Blanco, G. (2014a). Raised beaches in the Cantabrian Coast. In F. Gutiérrez & M. Gutiérrez (Eds.), *Lands Landf Spain 20* (pp. 239–248). Dordrecht: Springer.

Flor, G., & Flor-Blanco, G. (2014b). Componentes de viento generadores de morfologías y campos de dunas costeras en Asturias (no de España). *Cuat y Geom, 28*(3–4), 47–68.

Flor, G., & Martínez Cedrún, P. (1991). Características morfológicas y sedimentológicas de las dunas eólicas de Sonabia (zona oriental de Cantabria, NO de España). *Est Geol, 47,* 317–337.

Flor, G., Cabrera-Ceñal, R., Suárez, M., & Iglesias Ortega, M. A. (1998). Aspectos morfológicos, dinámicos y sedimentológicos del estuario del Nalón (Asturias, noroeste de España). *Trab Geol* (Vol. 20, pp. 3–37). University of Oviedo.

Flor, G., Carriedo Veci, J., & Martínez Cedrún, P. (2006). Aspectos morfológicos y sedimentarios de los campos dunares de la costa de Noja (Cantabria). In A. Pérez Alberti, J. López Bedoya (Eds.), *Reunión Nacional de Geomorfología* (pp. 361–373). University of Santiago de Compostela.

Flor, G., Martínez Cedrún, P., & Flor-Blanco, G. (2011). Campos dunares de Asturias, Cantabria y País Vasco. In E. Sanjaume, F. J. Gracia Prieto (Eds.), *Dunas en España* (Vol. 5, pp. 127–159). Sociedad Española de Geomorfología.

Flor, G., Flor-Blanco, G., & Flores-Soriano, C. (2014). Cambios ambientales por los temporales de invierno de 2014 en la costa asturiana (NO de España). In *Trabajos de geología* (Vol. 34, pp. 97–123). University of Oviedo.

Flor-Blanco, G. (2007). *Características morfosedimentarias y dinámicas y evolución ambiental de los estuarios de Tina Mayor, Tina Menor y San Vicente de la Barquera (costa occidental de Cantabria)* (Ph.D. Thesis, unpublished). Department of Geology. University of Oviedo.

Flor-Blanco, G., & Flor, G. (2009). Aspectos geomorfológicos del tramo inferior del estuario de Villaviciosa (Asturias) en relación con su evolución histórica. *Revista de la Sociedad Geológica de España, 22*(1–2), 123–136.

Flor-Blanco, G., & Flor, G. (2016). Management of dune fields on the coasts of Asturias, Cantabria and the Basque Country (Cantabrian Sea, NW Iberian Peninsula). In F.X. Roig Munar (Ed.), *Restoration and management of dune systems. Case studies* (pp. 35–70). Edition: Càtedra *d'Ecosistemes* Litorals Mediterranis, Parc Natural del Montgrí, les Illes Medes i el Baix Ter. Museu de la Mediterrània, Publisher: project "Life Pletera" (LIFE13 NAT/ES/001001).

Flor-Blanco, G., Flor, G., & Pando, L. (2013). Evolution of the Salinas-El Espartal and Xagó beach/dune systems in north-western Spain over recent decades: evidence for responses to natural processes and anthropogenic interventions. *Geo-Mar Letters, 33*(2–3), 143–157.

Flor-Blanco, G., Flor, G., Pando, L., & Abanades, J. (2015a). Morphodynamics, sedimentary and anthropogenic influences in the San Vicente de la Barquera estuary (North coast of Spain). *Geologica Acta, 13*(4), 279–295.

Flor-Blanco, G., Pando, L., Morales, J. A., & Flor, G. (2015b). Evolution of beach-dune fields systems following the construction of jetties in estuarine mouths (Cantabrian coast, NW Spain). *Environmental Earth Sciences, 73,* 1317–1330.

Flor-Blanco, G., Alcántara-Carrió, J., Flores-Soriano, C., García-Ordiales, E., & Pando, L. (2017). Alteraciones en la costa asturiana debido a los eventos catastróficos de oleaje de 2014. *Geotemas (Madrid), 17,* 207–210.

Flores-Soriano, C. (2015). *Evolución de los campos dunares de Asturias desde mediados del siglo XX hasta 2014 (107 p)* (Master Thesis). Department of Geology. University of Oviedo.

Folk, R. L., & Ward, W. C. (1957). Brazos River bar: a study in the significance of grain size parameters. *Journal of Sedimentary Petrology, 27,* 3–26.

García-Artola, A., Cearreta, A., & Leorri, E. (2014). Relative sea-level changes in the Basque coast (Northern Spain, Bay of Biscay) during the holocene and anthropocene: the Urdaibai estuary case. *Quaternary International, 364,* 172–180.

Giannini, P. C. F., Sawakuchi, A. O., Martinho, C. T., & Tatumi, S. H. (2007). Eolian depositional episodes controlled by Late Quaternary relative sea level changes on the Imbituba-Laguna coast (southern Brazil). *Marine Geology, 237,* 143–168.

Goldstein, E. B., Moore, L. J., & Durán Vinent, O. (2017). Vegetation controls on maximum coastal foredune 'hummockiness' and annealing time. *Earth Surf Dynam Discuss, 5*, 417–427.

Gómez, A. G., Bárcena, J. F., Juanes, J. A., Ondiviela, B., & Sámano, M. L. (2014). Transport time scales as physical descriptors to characterize heavily modified water bodies near ports in coastal zones. *J Env Manag, 136*, 76–84.

Gómez-Pina, G., Muñoz-Pérez, J. J., Ramírez, J. L., & Ley, C. (2002). Sand dune management problems and techniques, Spain. *J Coast Res SI, 36*, 325–332.

Gutiérrez-Elorza, M., García-Ruiz, J. M., Goy, J. L., Gracia, F. J., Gutiérrez-Santolalla, F., Martí, C., Martín-Serrano, Á., Pérez-González, A., Zazo, C. (2002). In W. Gibbons, T. Moreno (Eds.), *The geology of Spain* (Vol. 14, pp. 335–357). The Geological Society. Quaternary.

Hart, A. T., Hilton, M. J., Wakes, S. J., & Dickinson, J. M. (2012). The impact of *Ammophila arenaria* foredune development on downwind aerodynamics and parabolic dune development. *Journal of Coastal Research, 28*(1), 112–120.

Hesp, P. (2002). Foredunes and blowouts: initiation, geomorphology and dynamics. *Geomorphology, 48*(1–3), 245–268.

Hesp, P. A. (2011). Dune Coasts. In E. Wolanski & D. S. McLusky (Eds.), *Treatise on estuarine and coastal science 3* (pp. 193–221). Waltham: Academic Press.

Hesp, P. A. (2013). Conceptual models of the evolution of transgressive dune field systems. *Geomorphology, 199*(1), 138–149.

Kidd, R. (2001). *Coastal dune management: A manual of coastal dune management and rehabilitation techniques* (p. 96). Coastal Unit, Newcastle: NSW Department of Land and Water Conservation.

Lancaster, N. (1986). Grain-size characteristics of linear dunes in the southwestern Kalahari. *Journal of Sedimentary Petrology, 56*(3), 395–400.

Lavin, A., Valdes, L., Sanchez, F., Abaunza, P., Forest, A., Boucher, J., Lazure, P., Jegou, A. -M. (2004). The Bay of Biscay: The encountering of the ocean and the shelf (18b, E). In A. R. Robinson, K. H. Brink (Eds.), *The Sea* 14, 24: 933–1001.

Livingstone, I., & Warren, A. (1996). *Aeolian geomorphology: An introduction* (p. 211). Harlow: Longman.

Lorenzo, F., Alonso, A. & Pagés, J.L. (2007). Erosion and accretion of beach and spit systems in Northwest Spain:A response to human activity. *Journal of Coastal Research, 23*(4), 834–845.

López Peláez, J. (2015). *Evolución morfosedimentaria e histórica de los estuarios asturianos de Navia, Nalón, Avilés, Villaviciosa y Ribadesella* (Ph.D. Thesis, unpublished). Department of Geology. University of Oviedo. 307 p.

Martínez, MªL, Gallego-Fernández, J. B., & Hesp, P. A. (Eds.). (2013). *Restoration of coastal dunes* (p. 347). Springer Series on Environmental Management: Springer.

Martínez Cedrún, P. (2008). *Caracterización morfológica y sedimentológica de los campos dunares costeros de Cantabria. Evolución ambiental* (Ph.D. Thesis, unpublished). Department of Geology. University of Oviedo. 816 p.

Martínez Cedrún, P., Flor, G. (2004). Geocantabria II. Excursión costera Liencres-San Vicente de la Barquera. *Geocantabria. Itinerarios Geológicos. Itinerarios Didácticos del XI Simposio sobre la Enseñanza de la Geología* (pp. 26–47). Gobierno de Cantabria.

Martínez Cedrún, P., & Flor, G. (2008). Rasgos morfológicos y sedimentarios del campo dunar de Liencres (Cantabria). In J. Benavente & F. J. Gracia Prieto (Eds.), *Trabajos de Geomorfología en España, 2006–2008* (pp. 275–278). Actas X Reunión Nacional de Geomorfología: Universidad de Cádiz.

Martínez Cedrún, P., Flor, G., Flor-Blanco, G., & Maroto González, G. (2014). Relaciones texturales y composición mineralógica de los sistemas de playa/dunas en una costa rocosa: caso de Cantabria (NO de España). *Rev Soc Geol España, 27*(2), 13–27.

Masselink, G., Scott, T., Poate, T., Russell, P., Davidson, M., & Conley, D. (2016). The extreme 2013/2014 winter storms: hydrodynamic forcing and coastal response along the southwest coast of England. *Earth Surface Processes and Landforms, 41*(3), 378–391.

Nordstrom, K. F. (2008). *Beach and dune restoration* (p. 187). Cambridge: Cambridge University Press.

Orford, J. D., Murdy, J. M., & Wintle, A. G. (2003). Prograded Holocene beach ridges with superimposed dunes in north-east Ireland: mechanisms and timescales of fine and coarse beach sediment decoupling and deposition. *Marine Geology, 194,* 47–64.

Prego, R., & Vergara, J. (1998). Nutrient fluxes to the Bay of Biscay from Cantabrian rivers (Spain). *Oceanologica Acta, 21*(2), 271–278.

Pye, K., & Tsoar, H. (2009). *Aeolian sand and sand dunes* (p. 458). New York. p: Springer.

Pye, K., Saye, S., & Blott, S. (2007). *Sand dune processes and management for flood and coastal defence. Part 4: Techniques for sand dune management.* R&D Technical Report FD1392/TR. Joint Defra/EA Flood and Coastal Erosion Risk Management R&D Programme.

Rivas, M. V., & Cendrero, A. (1991). Use of natural and artificial accretion on the North Coast of Spain: Historical trends and assessment of environmental and economic consequences. *Journal of Coastal Research, 7*(2), 491–507.

Roskin, J., Silvan, D., Shtienberg, G., Roskin, E., Porat, N., & Bookman, R. (2015). Natural and human controls of the Holocene evolution of the beach, aeolian sand and dunes of Caesarea (Israel). *Aeolian Research, 19*(A), 65–85.

Sanjosé, J. D., Serrano, E., Berenguer, F., González-Trueba, J. J., Gómez-Lende, M., González-García, M., et al. (2016). Evolución histórica y actual de la línea de costa en la playa de Somo (Cantabria), mediante el empleo de la fotogrametría aérea y escáner láser terrestre. *Cuaternario y Geomorfología, 30*(1–2), 119–130.

Swann, C., Brodie, K., & Spore, N. (2015). *Coastal foredunes: Identifying coastal, aeolian, and management interactions driving morphologic state change.* ERDC/CHL TR-15-17. Coastal Inlets Research Program. U.S. Army Corps of Engineers. 56 p.

Thom, B. G. (1978). Coastal sand deposition in southest Australia during the Holocene. In J. L. Davies & M. A. S. Williams (Eds.), *Landform evolution in Australasia* (pp. 197–214). Canberra: ANU Press.

Wasson, R., & Hyde, R. (1983). Factors determining desert dune types. *Nature, 303,* 337–339.

Chapter 26
Coastal Dunes in the Ebro Delta

María José Sánchez-García, Isabel Montoya-Montes, Mariona Casamayor, Ignacio Alonso and Inmaculada Rodríguez-Santalla

26.1 Introduction

The Ebro Delta is one of the most significant coastal sedimentary formations in Spain, not only for being the largest and most important delta in the country, but also for presenting several active dune fields on it. In addition, other ecosystems such as salt marshes and lagoons contribute to conform a characteristic and unique habitat.

The importance of all these elements prompted to declare and protect the zone as Natural Park in 1983, extending the protected area in 1986 to the southern hemi-delta. Besides, different areas of the Ebro Delta are also under other protection figures, which enhance the role of the whole delta in safeguarding the biodiversity of the species as well as the good state of the ecosystems (Biosphere Reserve, Site of Community Importance and Special Protection Area, among others) (Fig. 26.1).

The Ebro delta is basically a plain that does not exceed 4–5 m above sea level, and whose emerged surface represents an approximate area of 325 km^2 (Rodríguez

M. J. Sánchez-García (✉) · I. Montoya-Montes · M. Casamayor · I. Alonso
Instituto de Oceanografía y Cambio Global, IOCAG, Universidad de Las Palmas de Gran Canaria, ULPGC, Campus de Tafira, 35017 Las Palmas de Gran Canaria, Spain
e-mail: mariajose.sanchez@ulpgc.es

I. Montoya-Montes
e-mail: isabel.montoya@ulpgc.es

M. Casamayor
e-mail: mariona.casamayor101@alu.ulpgc.es

I. Alonso
e-mail: ignacio.alonso.bilbao@ulpgc.es

I. Rodríguez-Santalla
Dpto. de Biología y Geología, ESCET, Universidad Rey Juan Carlos,
C/Tulipán s/n, 28933 Móstoles, Madrid, Spain
e-mail: inmaculada.rodriguez@urjc.es

J. A. Morales (ed.), *The Spanish Coastal Systems*,
https://doi.org/10.1007/978-3-319-93169-2_26

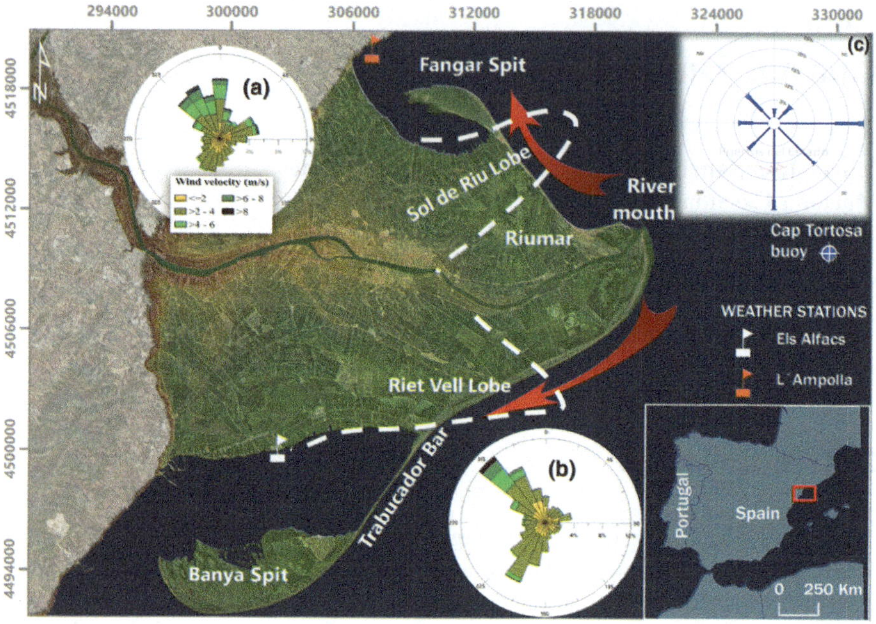

Fig. 26.1 Ebro Delta and associated dune field location. Wind charts represent hourly wind data recorded during the period 1997–2007 at Fangar (**a**) and Els Alfacs (**b**) weather stations. In **c** is represented significant wave height (Hs) recorded during the period 2004–2017, data from Puertos del Estado. Dashed lines represent old deltaic lobes described in Somoza et al. (1998). Red arrows symbolise littoral transport in both hemideltas. Orthophoto from Instituto Geográfico Nacional (2011).

1999), while the submerged surface (delta front and prodelta) has an extension about 2,172 km^2 (Maldonado 1972). The maximum sediment thickness reaches 60 m in the most distal areas of the deltaic plain and decreases both landward and seaward (ITGE 1996).

Originally Ebro Delta was considered a microtidal delta dominated by fluvial regime and waves (Galloway 1975), but the fluvial effect has greatly decreased as the river flow has become fully regulated due to the construction of dams. Therefore, nowadays it is a delta mainly dominated by waves and they are morphological changes responsible at the Ebro Delta coast (Jiménez and Sánchez-Arcilla 1993; Guillén and Palanques 1997). The lack of sediments in the lower course of the river, consequence of the sediment retention by many dams, has generated the erosion at the apex of the delta, from where sediments are moved both northward and southward (Guillén and Palanques 1992). Resulting from this longshore transport, sediment accumulates at the ends of the N and S spits, which are clearly prograding. These prograding areas compensate the erosion that takes place at the river mouth, and therefore there is no net loss nor gain of the extension of delta (Rodríguez 1999).

In the last decade of the 19th century, dune fields developed along the entire coastline of the delta (Mallada 1889), but the drastic reduction of the sedimentary inputs from the river, in conjunction with storms that hit the area, have caused the decrease of the extension of the dune bodies. Nowadays, they are located only at the north and south ends of the delta, as well as in isolated patches along the coastline. The main dune fields are found at Fangar Spit, Riumar beach, Trabucador Bar and Banya Spit (Fig. 26.1). In addition, small nebkhas can develop along the rest of the coast associated with winds from the NNW, but being quickly destroyed after easterly waves.

26.2 Regional Setting

26.2.1 Wind

In general, prevailing winds at Ebro Delta blow from two main directions: NNW and SSW, but some winds from the ENE are also present in the area, though with lower intensity and frequency (Bolaños et al. 2009). Nevertheless, there are several meteorological stations in the delta and on adjacent beaches, from which Cateura Sabrí et al. (2004) described the wind climate in Ebro area. These authors detect a different wind behaviour depending on the area of the delta in which the station is located suggesting that these differences are produced by large-scale orographic effects. Those differences are shown in the wind chart obtained from wind data recorded at Els Alfacs and L'Ampolla weather stations (Fig. 26.1a, b). At El Alfacs there are two dominant winds directions (SSW and NW), while at L'Ampolla station the wind shows a multidirectional spectrum, but the stronger winds are those from NNW. Since the wind characteristics differs from the northern to the southern side of the delta, the coastal dune behaviour will also be different.

Regarding the strongest winds, northern ones can reach gusts up to 150 km/h. Between 1997 and 2007, wind gusts higher than 100 km/h from NNW directions were recorded every year (Sánchez García 2008). However, Bolaños et al. (2009) state that large velocities have been recorded from eastern winds in agreement with storm conditions associated with cyclonic activity over the NW Mediterranean.

Wind climate shows a strong seasonal pattern. During fall and winter there is a clear predominance of northern and north-western winds, while in spring prevailing winds are from the east and during summer dominant ones are from the south.

Sánchez García (2008) analysed hourly wind velocities from 1998 to 2007 (almost 500,000 data). She found that only 1.1% of the wind records were lower than 0.3 m/s (known as *calm* in Beaufor wind force scale). The rest of data were distributed between 0.3 and 23.7 m/s. Maximum wind gusts recorded in the area reach up to 39 m/s (140 km/h), values that were recorded during a windstorm that took place in December 1998.

Wind has a double effect on the dune fields. On one hand dunes can grow, evolve, and be moved as consequence of the wind effect, but on the other hand, wind can also generate a significant loss of sediment, which is moved by aeolian transport to the sea. Eastern winds generate waves with great height and energy. When these waves reach the coast cause an intense erosion along the beach that can eventually affect the dune fields. Both northern and north-western winds have a great negative effect on dunes located on the northern hemidelta: sediments from the vegetated dunes located along the inner coast are blown to the sea under strong northern winds, and mobile dunes that form the outer coast are pulled towards the beach under north-western winds.

26.2.2 Tides

The effect of the astronomical tides can be considered negligible in the delta environment due to its microtidal nature, since tidal range varies between 20 and 25 cm (Jiménez 1996).

Nevertheless, the role of meteorological tides is much more important than astronomical tides and it has strong morphodynamical implications both on the beaches and in coastal dunes. This effect is particularly important when sea level rise due to meteorological tides is coupled with storm waves. Two of the most aggressive storms registered in the last decades happened in October 1990 and October 2003. In both cases they resulted from a combination between surprisingly high easterly waves and sea level elevations (Jiménez et al. 1991).

Sea level elevations can be associated with situations of low pressure (Rodríguez 1999) or episodes of strong landward winds (Jiménez 1996; Curcó 2003). During these storm surges, sea level can reach up to 1 m above the mean level, with the highest probability of occurrence between October–December and February–May (Curcó 2003).

26.2.3 Waves

Both waves and wave-induced currents are strongly influenced by the wind regime, both also by the effective fetch. At the Ebro Delta, the wave climate is characterised by a predominance of NW conditions (which coincides with the predominance of NW winds) with also significant E and S storms (Bolaños et al. 2009; Ràfols et al. 2017). However, E and S winds, acting over a larger fetch, produce the highest and more energetic waves that reach the delta. These waves have a strong damaging effect on the littoral. The frequency of waves from E is greater than the rest of directions, and it coincides with the most energetic waves that reaches the coast. The most vulnerable areas to the storms are Marquesa Beach (close to Fangar Spit southeast area), eastern end of Ebro Delta coast, Trabucador Bar (Jiménez et al. 2011).

Bolaños et al. (2009) present the distribution of significant wave height (Hs) and mean period (Tz) for the four XIOM wave buoys distributed along the Catalan coast. Wave data recorded at Tortosa wave buoy, just in front of Ebro river mouth, presents a wider distribution for larger periods with Hs ranging from 1 to 5 m for 7 s mean period.

26.2.4 Littoral Transport

The scheme of the net longshore transport along the coast is determined by waves that impact in the delta front. Waves from E generate two different transport directions depending on the area of the delta: northwards in the northern hemidelta and southwards in the southern hemidelta (Fig. 26.1). This transport is responsible for the greatest erosion that is concentrated on the delta apex (river mouth and Buda Island), while the largest accretions occur at both ends of the north and south spits (Jiménez and Sánchez-Arcilla 1993).

Several authors have calculated sediment transport rates in the area (CEDEX 1996; Jiménez 1996; Barrio-Parra et al. 2013, 2017) in order to know the future evolution of this sedimentary system. All of them coincide in the described bidirectional pattern, but the transport rates obtained are substantially different between them. The determination of reliable transport rates is a key factor to know the amount of sediment that will be available for its remobilization and incorporation into the dune system. In this context, Barrio-Parra et al. (2017) have recently developed a model that allows computing transport rates considering the wind-induced longshore currents and the beach-dune sediment exchange due to aeolian processes.

Even though the described general pattern in the alongshore sediment transport, Ribeiro et al. (2012) have measured longshore drift towards the south along Fangar Spit, in the northern hemidelta.

26.3 Background

The first references about aeolian deposits in Ebro Delta are descriptive works that mention the location and extension of the dune deposits.

In the physical description of geology of the province of Tarragona, Mallada (1889) mentions the presence of dune ridges along south hemidelta. Extension covered by these aeolian deposits ranges between 500 and 3000 m width, with heights between 50 and 80 cm depending on the dominant winds. This author describes the dynamics of the outer dune ridge in the following way: it is destroyed in the winter by storms, but in summer it is aligned normal or oblique to the coastline.

In 1958, Duboul-Razavet refers to dune formations in Trabucador bar and Fangar spit as prograding coastal ridges. He also points out the importance of the wind and the orientation of the coast in the morphology and development of the delta.

Maldonado (1972) includes a more detailed description of the sedimentary bodies in the delta. He relates the formation of the dune ridge between Fangar and Riumar with winds of NNE by the active transfer of sediments along the coast by the wind. Although some years later, those wind directions were adjusted on the north hemidelta where coast direction run into strongest wind direction. This wind is less important in the southern hemidelta, because on one hand the same delta prevents the formation of waves and on the other the wind that crosses deltaic plain does not find loose sand to transport since salt-marshes and crops retain the sediments. For this reason, on the southern hemidelta dune ridges are not formed except at Banya Spit. Maldonado (1972) carried out the first cartography of the Ebro Delta, which includes the most outstanding dune fields.

In the last decade of 20th century the number and quality of papers dealing with different aspects of the delta geomorphology significantly increased. Research studies were carried out on coastline retreat and coastal dynamics (Jiménez et al. 1997; Montori Blanch 2005), sedimentary balance (Guillén 1992), long-term evolution (Jiménez 1996; Rodríguez 1999), sediment discharges by the river (Guillén 1992), effects of sea level rise (Jiménez and Sánchez-Arcilla 1997a) and so forth. Most of them highlight the role of aeolian transport in the general dynamics of Ebro Delta.

Guillén (1992) made a first approximation to the quantification of wind transport in the delta by applying the Bagnold equation (Bagnold 1941). He estimated the total volume of sand transported by the wind in the dune field areas in about 100,000 m^3/year. The direction of resultant transport was towards the SE, so that the winds responsible for most of the aeolian transport were those from the NW. Guillén (1992) highlighted the dune fields of Fangar Spit due to the greater availability of sediments and the orientation of the coast, parallel to the prevailing NW winds.

Legoff (1993) carried out an approximation to the net wind transport at the dune fields located in Fangar Spit. With that purpose, he studied the evolution of the dune field between 1957 and 1991. Based on aerial photographs, he mapped the dune field and obtained the area at different times. He estimated the volume of the dune field by assuming a constant dune height in the entire period. He concluded that Fangar dune field was growing towards the NW. Based on data from Legoff (1993); Jiménez (1996) estimated an average storage rate of material in form of dunes of 1.3 m^3/m/year

CEDEX (1996) defined three coastal directions to calculate the aeolian transport: one was along the northern hemidelta (Fangar Spit) and two in the southern hemidelta (Trabucador Bar and Banya Spit). They concluded that there were two main aeolian transport directions in the delta: one towards the NW with rates of nearly 3,000 kg/m/year and another one towards the SE of more than 48,000 kg/m/year.

The use of sediment traps to determine the gross wind transport in dune fields at Fangar Spit was carried out by Serra et al. (1997). They performed two field experiments using Leatherman traps in different wind conditions to characterise the wind transport in a maximum range of winds velocities both in Fangar Spit and Trabucador Bar. They estimated the aeolian transport in the north hemidelta about 10,000 m^3/year towards the SE, assuming a dune field width of 250 m.

In spite of the differences among the aeolian transport quantification made in previous studies, all of them agreed that the most important transport direction is toward SE.

26.4 Delta Evolution and Dunes Formation

The evolution of the dune fields is directly related to the evolution of the whole delta, as well as to the wind direction and wind speed in the area. According to Guillén and Palanques (1997), a very fast growth of the Ebro Delta took place during XIV and XV centuries because of the intense deforestation inland. Nowadays, the delta front is suffering strong erosion due to the dramatic sediment cut down. The resultant coastline retreat is directly related to the building of Mequinenza and Ribaroja dams at the end of 1960. From that date, the delta front is not prograding anymore but it is being reshaped (Jiménez 1996). The river mouth is under a high erosion rate in contrast with the accretion at the end of both spits. A retreat larger than 2 km has been measured at the river mouth, while in the period between 1956 and 2006, the spits are growing to the west more than 600 meters in Banya Spit and almost 1.1 km in Fangar Spit.

Reconstruction of the delta evolution between 1600 and 1900 was performed by Maldonado (1972) from old charts. On the other hand, Somoza et al. (1998) established the delta evolution from drill holes in the delta plain, high-resolution seismic profiles, surface electrical conductivity, sea level changes and historical data. Those works greatly contributed to establish the periods of delta lobes formation, the coastline reshaping, the growth of the spits, and the origin of the dune fields (Fig. 26.1). Riet Vell lobe was the first river mouth and was active from approximately 1000–1300 AD. After that, river mouth migrated to the NW developing a new lobe called Sol de Riu by Somoza et al. (1998) and Riet the Zaida by Cearreta et al. (2016). This new lobe lead to the formation of Fangar Spit due to its erosion about 1700 AD. In between both lobes, first field dunes of Banya Spit were formed from the sediment eroded from the lobe and transported by longshore drift to the west. River mouth dunes and Riumar dunes were formed after Banya and Fangar field dunes considering that the present river mouth is the last lobe developed in the delta.

26.5 Coastal Dune Fields

According to Klijn (1990), the adequate combination of the following elements can determine the presence or the absence of coastal dunes, as well their morphology. These elements are wind, waves, sediment and vegetation. All of them can be found at Ebro Delta.

Dune fields distribution in the Ebro Delta is directly related to the coast orientation and the direction of the stronger winds, which are those from 315 °N (Maldonado 1972; Serra et al. 1997). There are two sediment sources to feed the coastal dunes at the Ebro Delta: on one hand, sediments coming from the delta front that are drifted along the coastline by waves and currents. They are deposited mainly in the spits ends that are currently growing and providing space for sand drying and available sediment for wind transport. On the other hand relict materials resulting from the erosion of ancient delta lobes.

Dune fields area in the Ebro Delta is decreasing due to the reduction in sediment inputs to the coast, and to the heavy erosion shown at the nearby coastline. For this reason, they are adjusting their accommodation space according to coastal behaviour.

In certain areas, the landward area of the dune field was occupied for different purposes (residential units, agriculture, etc.), so that the dunes are bounded between this anthropic areas and the coastline. In the case where the coastline is under erosion, dunes have not the possibility of moving landward due to the human occupation. An example of the influence of this anthropic occupation is the loss of the dune ridge in the north hemidelta. In 1957 there was a continuous dune ridge from Fangar Spit to Riumar, with only one interruption at the mouth of a channel (dune field digitalised based on 1957 orthofoto from ICGC). Most part of this dune ridge was lost due to a storm events sequence (Jiménez et al. 2012). The only areas in which there are nowadays some dunes are those with enough beach width to allow for coastline and dune retreat, which is the case of the dunes at Fangar Spit (Fig. 26.2). Land uses that prevent from landward migration are mostly related with agriculture (rice crops, irrigation channels and roads), but also with buildings.

26.5.1 Fangar Dune Field

Fangar dune field had been studied thoroughly including short and medium term dune evolution (Sánchez García 2008), aeolian transport (Guillén 1992; CEDEX 1996; Universidad de Barcelona 1997; Sánchez García 2008), stratigraphy using ground penetrating radar (Rodríguez Santalla et al. 2009, Gómez Ortíz et al. 2009) and dune modelling (Barrio-Parra and Rodriguez-Santalla 2014). In contrast with the scarce studies carried out in other Ebro Delta coastal dunes.

Fangar Spit is characterised by a wide and large sandy plain, partially covered by dunes both in the outer and in the inner coasts (Fig. 26.2). There is a big difference

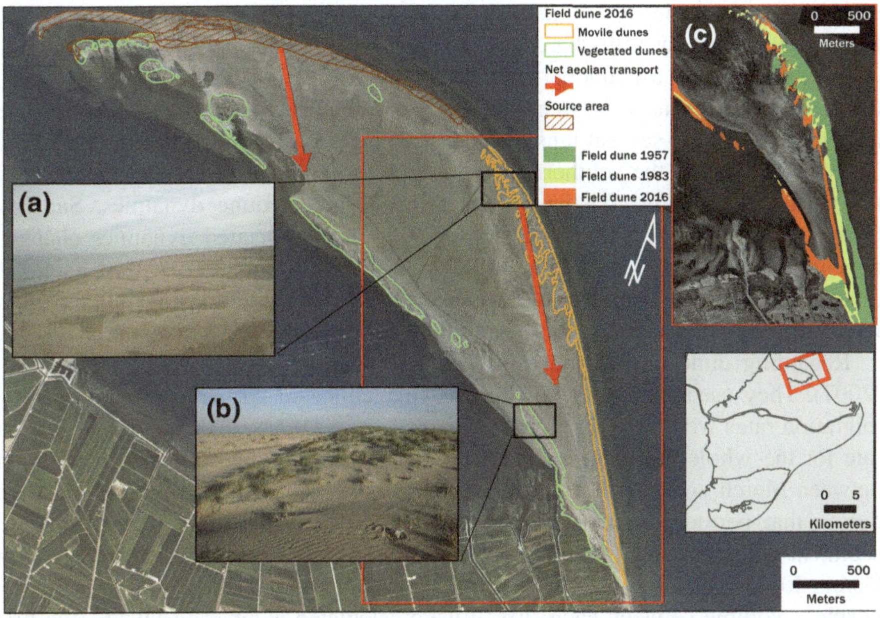

Fig. 26.2 Coastal dune fields at Fangar Spit. Figures **a** and **b** display an example of inner and external coastal dune fields. **c** Dune fields extension in 1957, 1983 and 2016 is plotted over a 1957 orthophoto (Institut Cartogràfic i Geològic de Catalunya (ICGC))

between the dune fields located along the outer and the inner coasts of Fangar Spit. In the outer coast, facing Mediterranean Sea, we can find the only active dune field in the whole delta, while in the inner coast, there are isolated patches of vegetated dunes. The area source of sediments in both cases is the north coast of the spit. According to Serra et al. (1997), only the third part of the northern coast feeds the dune field, while the rest feeds the floodplain.

Outer Field Dune

Mobile dune field expands parallel to the coastline all along the outer coast of the spit. It reaches 4 km in length, and its width ranges from 50 to 250 m. The surface covered by dunes is about 1,000,000 m^2, and the volume of sediments involved is 200,000 m^3 approximately. Sánchez et al. (2007) classified dune morphology attending to the dunes medium height and the dunes shape; then the field can be divided in four different areas named zone 1, 2, 3 and 4. The first one has generally isolated small barchan with an average height from 1 to 2 m. Dunes in zone 2 appear forming barchanoids ridges and they average height varies from 2 to 3 m (Fig. 26.2a). Pye and Tsoar (1990) related these morphological changes to an increase in the sediment supply. Dunes height rise up to five metres in zone 3 and their morphology oscillates between barchanoid and linear dunes. Interdune areas are bigger here than in the other zones. In zone 4, dune height suffers a decrease reaching an elevation of 2.7 m. Dune surface turn round isolated barchan like in zone 1 but in this case this barchans are shoreline aligned.

Dune dynamics are under different processes depending on their location. Considering the distance between the dunes and the beach, at the north of the dune field this distance is several tens of meters, while it is much narrower at the south. Therefore, dunes in the southern area are more vulnerable to storms than those in the north. Changes in coastline orientation also affect dune dynamics. Sánchez García (2008) observed that dune morphology and associated dynamics changes depending on the coast orientation relative to the wind direction. A third factor is dune location inside the dune field, since northern dunes are under higher wind intensities than dunes at the end of the dune field (Rodríguez Santalla et al. 2009).

Regarding dune movement, migration rates were measured by Sánchez et al. (2007). They used six DEM built from DGPS collected throughout one year, and migration rates were estimated between consecutive DEMs. The average migration rate for the whole period was 100 m/year. The highest migration rate was found between March and April, which was associated to a windy storm that took place during that period. Comparison between DEMs showed a net positive trend in sediment volume, indicative of an accumulative pattern in this dune field (Sánchez García 2008). The net transport direction was SE, in agreement with other authors. Displace volume between each survey were calculated comparing DEMs (the first and the last of the entire period) the results were obtained for each zone mentioned before. Combining migration rates and displace volume, the four zones were characterised: Zone 1 shows a bigger volume of gained sediments, due to the small size of the dunes in this area. It is also a sediments contribution source for the whole dune field. In zone 2 the height and complexity of the dunes increase, which implies a drop of the moved sand volume. Dunes in zone 3 are much higher and thus, its sand movement is slower if we relate it to the rest of the field. Zone 4. Although the morphological and dimensional characteristics of its dunes are similar to those in zone 1, these ones have a less percentage of gained sediment because they are more exposed to the wave action.

The evolution of Fangar dune field was considered under two time scales: short-term and medium-term. Short-term studies include dune migration, dune morphology net aeolian transport calculation and surface dune evolution; all these parameters were measured in Sánchez García (2008). Also, medium-term field dune evolution was established thorough orthophotos comparison. In general, mobile dune fields almost change in surface but it is easy to observe dune movement landward caused by coastal erosion (Fig. 26.2c). Almost two third parts of dune field are not only moving thorough SE but to W because coastline is under high erosion (Fig. 26.2c). Since aerial photograph are available (1927) only the northern part of the dune field is still in the same position although the surface had been changing.

Ground penetrating radar had shown relationships between internal structure reflectors and dune activity: (i) high dune activity in small dunes with overlapping radar sequences; (ii) larger dunes (up to 5 m height) exhibiting internal structure with low dipping angle; (iii) small dunes with low activity present partially overlapping elongated radar sequences defined by sub horizontal reflections (Gómez Ortíz et al. 2009).

Inner Field Dune

Parallel to the inner coast of the spit there are several patches of vegetated dunes. Its origin is associated to the beginnings of the aquaculture activities in the area. A sandy ridge was built between 1971 and 1983 to protect these activities from the northern winds. In fact, this dune field was not present in 1956 orthophoto, but in 1983 there is a slight and sparsely vegetated sand hill (Fig. 26.2c). It was built along the coastline with sand both from the inner and the outer coasts. Years later, a new fence was also built adjacent to a new aquaculture area. It was made of sand and organic fragments from the cleaning of the shellfish culture areas.

This vegetated dune field is increasing in volume and surface because it reduces wind intensity and acts as a sediment trap. Nebkas and foredunes dominate dune morphology. Maximum dune height reaches up to 4.5 m and dune field surface is about 400,000 m^2, according to a Digital Elevation Model based on LiDAR data corresponding to 2011.

26.5.2 Riumar Dune Field

The origin and evolution of Riumar dune field is related to changes in the river mouth and urban development in the nearby area, in contrast with the rest of dune fields in the delta.

The Ebro river mouth area is evolving very quickly (Fig. 26.3). The coastline extracted from the aerial photograph taken in between 1945 and 1946 is approximately 1,200 m inland compared to the coastline in 2016 (Fig. 26.2a, b). In fact, from 1945 to 2016 the north river mouth has grown more than 680,000 m^2 in surface, allowing the formation of a new dune area, which reaches an extension of 440,000 m^2. NW winds carry sand from the beach and deposit it in the dune field, being this sedimentary transfer unique in the north hemidelta (Barrio-Parra et al. 2017). LiDAR data from ICGC in 2011 showed that maximum dune height was higher than 6 m, being the highest dunes located close to the lagoon.

Although it has been created a new dune field on the eastern part, western part of the dune field shows another behaviour. Comparison between Fig. 26.3a and b show that the houses in 2016 are built over the dune field. Urban development is an important process affecting dune evolution and destruction because of several reasons. First of all, the houses are blocking sand transport to the South-East, this fact implies a reduction in the area available for dune movement, sand enters into the urbanization and periodic maintenance works have to be undertaken to remove it from the streets. In addition, dune morphology has changed since 1946, since embryo dunes or barchan dunes have evolved into foredunes. The reason could be related to the home building effect in wind speed and direction. Different managing strategies are being developed by the administration. Some of them are the building

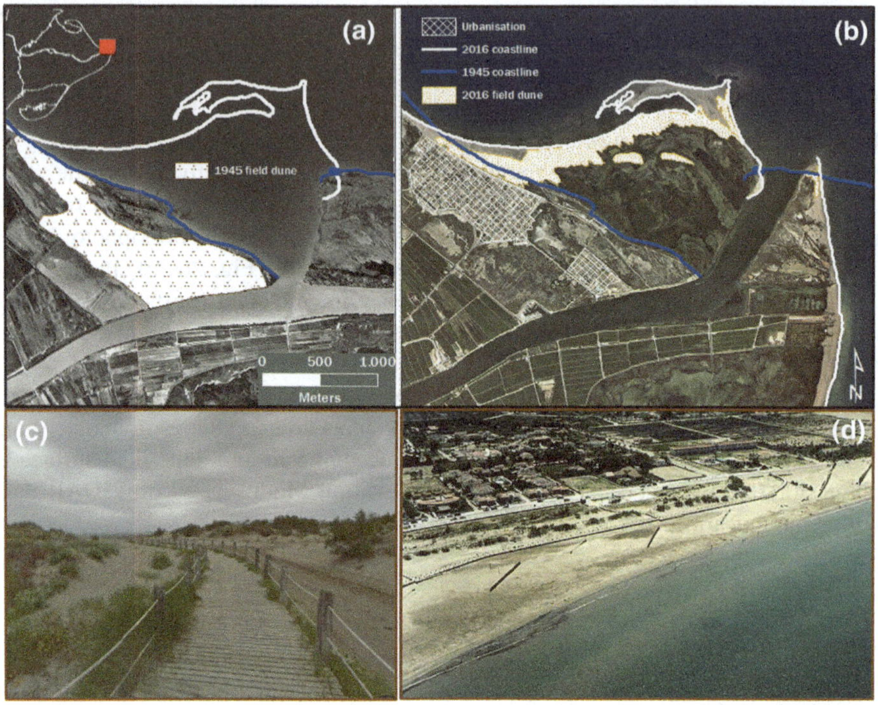

Fig. 26.3 **a** 1946 orphothoto showing the coastline in 2016. **b** 2016 orthophoto, showing the new dune field that has recently developed. **c**, **d** Examples of some of the management actions carried out in this area: wooden walkway over the dunes (**c**) and protection fence around vegetated dunes and sand traps (**d**)

of aeolian fences designed to prevent the sand blowing into the urban area (Fig. 26.3d), the building of wood paths over the dunes to prevent dune destabilisation due to trampling (Fig. 26.3c), and other related to environmental education issues (installation of information boards into the field dune and viewpoints).

26.5.3 Trabucador Dune Field

The dunes at Trabucador Bar are under extreme conditions due to wind and waves action. This is the most vulnerable area along the Ebro Delta coast (Sánchez-Arcilla and Jiménez 1994) The development of stable dunes is difficult because of the different alignment of the coastline relative to the prevailing winds. As a result of this situation, sand is blown from the dunes to the sea, where it becomes incorporated in the marine dynamic (Maldonado 1972; Serra et al. 1997) (Fig. 26.4). Despite of that, inner coast accommodates several dune vegetated patches which are destroyed under easterly wave storms. Regarding waves, easterly storms can generate waves that can overtop the bar, and the dunes are washed off (Jiménez and Sánchez-Arcilla 1997b). An example of these situations took place in October 1991, when waves generated a washover and the bar broke. To regenerate the bar, central administration built an artificial dune 2 m height all along the bar (Montoya Font and Galofré i Saumell 1997). Since then, Trabucador Bar has broken and the artificial dune ridge has been destroyed several times by easterly storms. It has been artificially regenerated one time after the other. Following photographs show the situation before and after the storm in 1991 (Fig. 26.4c, d).

Trabucador bar presents two different morphologies. Along the inner coast foredunes are the dominant morphology, while on external coast barchan dunes are the prevailing form. The latest may be washed away and built up during the same year depending on wave conditions. Considering the influence of the coastline evolution, Trabucador Bar is retreating as a consequence of erosion after easterly storms. This retreat implies the landward movement of the dunes (Fig. 26.4)

26.5.4 Banya Dune Field

Dune fields at Banya Spit cover an area of nearly 23 Ha, where dunes height ranges from 1 to 4 m. These dunes are disposed along curved dune ridges parallel between them and to the shoreline. The formation of these dune ridges is related to the progradation of the spit (Fig. 26.5).

As it has been previously described, Banya spit grows from the joining of littoral bars to the shoreline (Fig. 26.5c), so that there is a huge amount of sediments to be transported by the wind and incorporated to the dunes. Although the dominant and stronger winds are from the North, the southern hemidelta is also under the influence of SW winds (Fig. 26.1b). From February until October moderate winds from the SSW are determinant to pull sand onshore, and the new dunes become stabilized by vegetation.

Just as Trabucador Bar, main dune morphologies are foredunes and barchan dunes. Depending on the foredune location they are going to present more or less field fragmentation, for instance, dunes far from coastline are rounded while dunes close to shoreline (in the growing spit direction) present space continuity. All the dune areas are vegetated, but the vegetation density and mature ecosystems are only present in the inner dunes (Fig. 26.5a, c).

26.6 Threats

Main threats to the dune fields of Ebro Delta are the effects derived from climate change, the decrease in sedimentary contributions by the river and coastal erosion, and all of them are interrelated.

Fig. 26.4 Trabucador bar location and associated dune field. Central image shows the 2011 DEM, where red colours are linked with coastal dunes in both sides. **a** Photograph of the field dune in the inner coast. **b** Photograph of the field dune in the inner coast. **c**, **d** Photographs of Trabucador bar after and before the storm occurred in October 1991

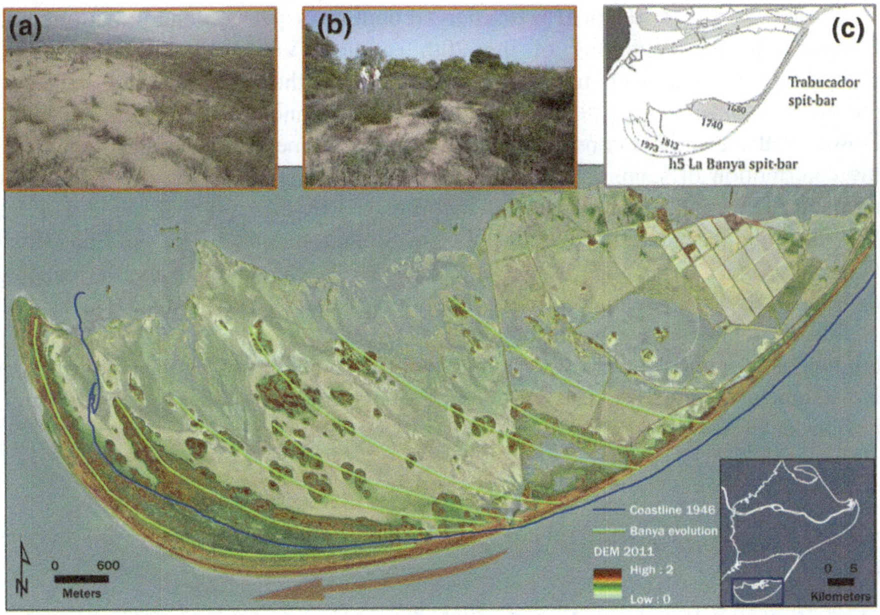

Fig. 26.5 2011 DEM of Banya Spit over 2016 orthophoto. Reddish colours indicate higher dune height. **a**, **b** different dune morphologies present in Banya Spit. **c** Evolution of Banya Spit from Somoza et al. (1998). Brown arrow symbolise littoral transport. Green lines are digitalized from 2016 orthophoto

26.6.1 Coastal Erosion

The most threatened dune fields by costal erosion are those found in Trabucador Bar, river mouth and the central part of North hemidelta, due to these areas are currently in retreat. The dune fields along the coastline avoid coastal erosion moving towards land, as long as they have enough space to do so, as it is the case in the dunes at Fangar spit. However, when there is not space for their movement the response of the dunes is the disappearance as it happens in the dune field of Riumar.

26.6.2 River Sediment Discharges

The sediment load in suspension has been decreasing since the beginning of the 20th century up to the present due to the high water regulation of the Ebro River (Nelson 1990; Jiménez et al. 1990; Guillén and Palanques 1992; Vericat and Batalla 2006). However, the reforestation carried out between 1960 and 1970 has also had a negative effect on the availability of sediment to river transport (Maldonado 1972).

Nowadays, 3% of sediment is discharged both in suspension and in bottom load compared to the beginning of the 20th century (Vericat and Batalla 2006). Considering the sediments transportable by the river, the sands are those that form the beaches and are deposited in the dunes (Jiménez and Sánchez-Arcilla (1997a). However, these authors considered that of all the sediment discharged by the river, the contribution of sands is insignificant. They also estimated the minimum flow necessary to produce the transport of 200 μm sand by the river and found that this flow is rarely exceeded throughout the year. So that, regulations of the Ebro River flow cause that coarse material is trapped in the various dams upstream, reaching the delta only the finer material, essentially consisting of clay and silts.

Some initiatives have been attempted to recover sediment trapped in the dams but there are important obstacles such as sediment contamination, the way to return the sediment to the river, and the amount of sediment needed to reach the equilibrium in Ebro Delta coast.

26.6.3 Climate Change Effects

There is an overall concern about the delta future. In order to make an estimation of delta evolution many models and predictions had been built up (Sánchez-Arcilla et al. 2008; Fatorić and Chelleri 2012; Alvarado-Aguilar and Jiménez 2007; Caspani 2014; Sayol and Marcos 2018). Sánchez-Arcilla et al. (2008) classified relevant climatological variables related to Ebro Delta dynamics, that could be affected by climate changes and divided them in two groups: those linked with delta formation and those related with delta reduction (Table 26.1).

Those authors pointed out the main threats facing Ebro Delta and its dunes: increasing frequency and intensity of storms and the elevation of the mean sea level. Although it exists an agreement identifying main processes that could be intensified because of climate change, the quantification of the sea level rise or storminess is an important goal that has to be solved. Another factor usually not taken into account in the modelling, is erosion and redistribution of sediment that is going to be different from the current one (Sayol and Marcos 2018).

According to the report presented by the IPCC in 2013, for a RCP 8.5 scenario, the global sea level rises between 0.52 and 0.98 m by 2100, estimated values are higher than the estimated values in IPCC 2007 report (Meehl et al. 2007). A relative

Table 26.1 Variables implied in Ebro Delta evolution which could be affected by climate change (from Sánchez-Arcilla et al. 2008)

Delta process	Climatological variable
Formation	Rainfall rate
	Desertification (catchment area)
Reduction	Sea level rise
	Storm frequency
	Wave characteristics

rise of sea level of 0.5 m without associated sedimentary response would suppose the disappearance of around 50% of Ebro Delta (Cendrero et al. 2005). The rise in sea level will cause that dunes, which are more vulnerable to storms, increase its exposition to the effect of waves.

Alvarado-Aguilar et al. (2012) considered that it would be relatively easy to decrease the potential flood areas in the northern hemidelta by ensuring control points in the channels using the infrastructure existing.

If the relative rise in sea level is added to the subsidence of the delta, the situation will worse as it is a markedly subsiding area. Direct estimates of subsidence rate made by ITGE (1996) obtained an approximate subsidence value for the delta of 0.20 cm/year.

Regarding to storms events, dune fields that are most vulnerable to storms are those of Fangar Spit and Trabucador Bar, since they are located at a lower distance from the shoreline.

Although various possible scenarios are predicted in the face of climate change, there is still great uncertainty due to the complexity of functioning of the climate system and its inclusion in predictive models (Zazo 2006). But recently Sayol and Marcos (2018) estimated impact of local sea level rise approximation using novel methodology and extreme surges and waves in coastal areas under climate change scenarios. They concluded that extreme surges and waves increase substantially the flood risk with respect to mean sea level, since extreme events contribute about 20% into different scenarios.

Acknowledgements Part of this work was carried out under the project "Cuantificación y Contribución del Transporte Eólico en los Procesos Dinámicos y Ambientales en el Delta del Ebro. Aplicación a su Gestión Integrada y a la Conservación de los Ambientes Marginales" funded by the Ministry of Science and Technology of Spanish State.

References

Alvarado-Aguilar, D., & Jiménez, J. A. (2007). A pseudo-dynamic approach to coastal flood hazard mapping. In *International Symposium on GIS and Computer Mapping for Coastal Zone Management* (pp. 1–10). Spain: Santander.

Alvarado-Aguilar, D., Jiménez, J. A., & Nicholls, R. J. (2012). Flood hazard and damage assessment in the Ebro Delta (NW Mediterranean) to relative sea level rise. *Natural Hazards, 62,* 1301–1321.

Bagnold, R. (1941). *The physics of blown sand and desert dunes* (p. 256). London, UK: Methuen.

Barrio-Parra, F., & Rodriguez-Santalla, I. (2014). A free cellular model of dune dynamics: Application to El Fangar spit dune system (Ebro Delta, Spain). *Computers & Geosciences, 62,* 187–197.

Barrio-Parra, F., Ribeiro, M., Bosnic, I., Taborda, R., Rodriguez-Santalla, I., & Cascalho, J. (2013) Sediment budget at Fangar Spit (Ebro Delta, Spain): Matching short-term observations with long-term evolution. In *6th SCACR—International Short Course/Conference on Applied Coastal Research* (pp. 1–10), Lisbon, (Portugal).

Barrio-Parra, F., Rodríguez-Santalla, I., Taborda, R., & Ribeiro, M. (2017). A modelling approach to assess the key factors in the evolution of coastal systems: The Ebro North hemidelta case. *Estuaries and Coasts, 40,* 758–772.

Bolaños, R., Jorda, G., Cateura, J., Lopez, J., Puigdefabregas, J., Gomez, J., et al. (2009). The XIOM: 20 years of a regional coastal observatory in the Spanish Catalan coast. *Journal of Marine Systems, 77,* 237–260.

Caspani, D. (2014). *DeltaLab: Designing delta ecosistems: The Ebro Delta* (M.Sc. Thesis Landscape Architecture). Wageningen University, Netherlands, 87 pp.

Cateura Sabrí, J., Sánchez-Arcilla, A., & Bolaños Sánchez, R. (2004). Clima de viento en el Delta del ebro. Relación con el estado de la mar. In J. C. García Cordón, C. Diego Liaño, P. Fernández de Arróyabe Hernáez, C. Garmendia Pedraja, & D. Casilla Álvarez (Eds.), *El Clima entre el Mar y la Montaña* (Vol. 4, pp. 463–472). Asociación Española de Climatología y Universidad de Cantabria, Serie A.

Cearreta, A., Benito, X., Ibáñez, C., Trobajo, R., & Giosan, L. (2016). Holocene palaeoenvironmental evolution of the Ebro Delta (Western Mediterranean Sea): Evidence for an early construction based on the benthic foraminiferal record. *The Holocene, 26*(9), 1438–1456.

CEDEX. (1996). *Estudio de la dinámica litoral del Delta del Ebro y prognosis de su evolución.* Technical report, Madrid, Spain.

Cendrero, A., Sánchez-Arcilla, A., & Zazo, C. (2005). Impactos sobre las zonas costeras. In M. de Medio Ambiente (Ed.), *Evaluación preliminar de los impactos en España por efecto del cambio climático* (pp. 469–524), Madrid, Spain.

Curcó, A. (2003). *Estudis sobre el poblament vegetal del delta de l'Ebre: flora, vegetació i ecologia de comunitats vegetals* (Ph.D. thesis). Universidad de Barcelona, Barcelona, Spain.

Duboul-Razavet, C. (1958). Le regime des courants superficielles aux bords des cotes du delta de l'Ebre. *Bulletin Comite Central D'oceanographie et d'Etude des cotes, X, 7,* 392–406.

Fatorić, S., & Chelleri, L. (2012). Vulnerability to the effects of climate change and adaptation: The case of the Spanish Ebro Delta. *Ocean and Coastal Management, 60,* 1–10. https://doi.org/10.1016/j.ocecoaman.2011.12.015.

Galloway, W. E. (1975). Process framework for describing the morphologic and stratigraphic evolution of deltaic depositional systems. In M. Lou Broussard (Ed.), *Deltas, Models for exploration* (pp. 87–98). Houston, USA: Houston Geological Society.

Gómez Ortíz, D., Martín-Crespo, T., Rodríguez, I., Sánchez, M. J., & Montoya, I. (2009). The internal structure of modern barchan dunes of the Ebro River Delta (Spain) from ground penetrating radar. *Journal of Applied Geophysics, 68,* 159–170.

Guillén, J. (1992). *Dinámica y balance sedimentario en los ambientes fluvial y litoral del Delta del Ebro* (Ph.D. thesis). Universidad Politécnica de Cataluña, Barcelona, Spain, 580 pp.

Guillén, J., & Palanques, A. (1992). Sediment dynamics and hydrodynamics in the coger course of a river regulated by dams: The Ebro river. *Sedimentology, 39*(4), 567–579.

Guillén, J., & Palanques, A. (1997). A historical perspective of the morphological evolution in the lower Ebro river. *Environmental Geology, 30,* 174–180.

IPCC. (2007). Resumen para Responsables de Políticas. In M. L. Parry, O. F. Canziani, J. P. Palutikof, P. J. van der Linden, & C. E. Hanson (Eds.), *Cambio Climático 2007: Impactos y Vulnerabilidad. Contribución del Grupo de Trabajo II al Cuarto Informe de Evaluación del IPCC.* Cambridge University Press, Cambridge, United Kingdom, 21 pp.

IPCC. (2013). Climate change 2013: The physical science basis. In T. F. Stocker, D. Qin, G. K. Plattner, M. Tignor, S. K. Allen, J. Boschung, A. Nauels, Y. Xia, V. Bex, P. M. Midgley (Eds.), *Contribution of Working Group I to the Fifth Assessment Report of the Intergovernmental Panel on Climate Change.* Cambridge University Press, Cambridge, United Kingdom and New York, NY, USA, 1535 pp. https://doi.org/10.1017/cbo9781107415324.

ITGE. (1996). *Estudio geológico del Delta del Ebro. Proyecto para la evaluación de la tasa de subsidencia actual.* Technical report, Madrid, Spain.

Jiménez, J. A. (1996). *Evolución costera en el Delta del Ebro. Un proceso a diferentes escalas de tiempo y espacio* (Ph.D. thesis) Universidad Politécnica de Cataluña, Barcelona, Spain, 274 pp.

Jiménez, J., García, M. A., & Sánchez-Arcilla, A. (1990). *Análisis y propuestas para estabilizar el Delta del Ebro: Estimación de transporte de sedimento en el río Ebro*. Contribución a la evolución costera. Technical report LT-2/t. Dirección General de Puertos y Costas, Generalitat de Cataluña.

Jiménez, J., & Sánchez-Arcilla, A. (1993). Medium-term coastal response at the Ebro delta, Spain. *Marine Geology, 114,* 105–118.

Jiménez, J., & Sánchez-Arcilla, A. (1997a). Physical impacts of climatic change on deltaic coastal systems (II): driving terms. *Climatic Change, 35,* 95–118.

Jiménez, J., & Sánchez-Arcilla, A. (1997b). El problema erosivo en el Delta del Ebro. *Revista de Obras Públicas, 3368,* 23–32.

Jiménez, J., Sánchez-Arcilla, A., & García, M. A. (1991). Erosión de la barra del Trabucador durante la tormenta de octubre de 1990. *Revista de Obras Públicas, 3299,* 23–30.

Jiménez, J. A., Sánchez-Arcilla, A., Valdemoro, H. I., Gracia, V., & Nieto, F. (1997). Processes reshaping the Ebro delta. *Marine Geology, 144*(1), 59–79.

Jiménez, J. A., Valdemoro, H. I., Bosom, E., Gracia, V. (2011). Storm-induced coastal hazards in the Ebro Delta (NW Mediterranean). In *The Proceedings of the Coastal Sediments 2011* (pp. 1332–1345).

Jiménez, J. A., Sancho-García, A., Bosom, E., Valdemoro, H. I., & Guillén, J. (2012). Storm-induced damages along the Catalan coast (NW Mediterranean) during the period 1958–2008. *Geomorphology, 143–144,* 24–33. https://doi.org/10.1016/j.geomorph.2011.07.034.

Klijn, J. A. (1990). Dune forming factors in a geographical context. In T. W. Bakker, P. D. Jungerius, & J. A. Klijn (Eds.), *Dunes of the European coasts* (pp. 1–13). Cremlingen, Germany: Catena Verlag.

Legoff, M. (1993). *Etude du champ de dunes du Delta de l'Ebre*. Departamento de Geología Dinámica, Geofísica y Paleontología, Universidad de Barcelona, 52 pp.

Maldonado, A. (1972). El delta del Ebro. Estudio sedimentológico y estratigráfico. In *Boletín de Estratigrafía*, Vol. 1, 486 pp.

Mallada, L. (1889). Reconocimiento geográfico y geológico de la Provincia de Tarragona. In J. B. Macelwane (Ed.), *Boletín de la Comisión del Mapa Geológico de España* (Vol. 16). Madrid, Spain, 175 pp.

Meehl, G. A., Stocker, T. F., Collins, W. D., Friedlingstein, P., Gaye, A. T., & Gregory, J. M., et al. (2007). Global climate projections. In S. Solomon, D. Qin, M. Manning, Z. Chen, M. Marquis, K. B. Averyt, M. Tignor, & H. L. Miller (Eds.), *Climate change 2007: The physical science basis. contribution of working group I to the fourth assessment report of the intergovernmental panel on climate change*. Cambridge University Press.

Montori Blanch, C. (2005). *Sistema de drenaje de playas. Investigación y desarrollo* (TPh.D. thesis) Universidad de Barcelona, Barcelona, Spain, 377 pp.

Montoya Font, F. J., & Galofré i Saumell, J. (1997). La particularidad sobre la actuación física en la costa: caso del Delta del Ebro. *Revista de Obras Públicas, 3368,* 33–45.

Nelson, C. H. (1990). Post Messinian deposition rates and estimated river loads in the Ebro sedimentary system. *Marine Geology, 95,* 395–418.

Pye, K., & Tsoar, H. (1990). *Aeolian sand and sand dunes*. London: Unwin Hyman. (458 pp).

Ràfols, L., Pallares, E., Espino, M., Grifoll, M., Sánchez-Arcilla, A., Bravo, M., et al. (2017). Wind-wave characterization in a wind-jet region: The Ebro Delta case. *Journal of Marine Science and Engineering, 5*(1), 12. https://doi.org/10.3390/jmse5010012.

Ribeiro, M., Barrio, F., Taborda, R., Cascalho, J., Bosnic, I., Rodríguez, I., Sánchez, M. J. (2012). Longshore drift: experimental and empirical predictors. In *The example of the Ebro Delta. Actas Das 2as Jornadas de Engenharia Hidrográfica. Instituto Hidrográfico*, Lisbon, Portugal, pp. 271–274.

Rodríguez, I. (1999). *Evolución geomorfológica del Delta del Ebro y prognosis de su evolución* (Ph.D. thesis). Universidad de Alcalá de Henares, 161 pp.

Rodríguez Santalla, I., Sánchez García, M. J., Montoya Montes, I., Gómez Ortiz, D., Martín Crespo, T., & Serra Raventos, J. (2009). Internal structure of the aeolian sand dunes of El Fangar spit, Ebro Delta (Tarragona, Spain). *Geomorphology, 104*(3–4), 238–252.

Sánchez García, M. J. (2008). *Evolución y análisis morfodinámico del campo dunar de la Flecha del Fangar (Delta del Ebro)* (Ph.D. thesis). Universidad Rey Juan Carlos, 185 pp.

Sánchez, M. J., Rodríguez, I., & Montoya, I. (2007). Short term coastal dune evolution of Fangar Spit (Ebro Delta, Spain). In *International Conference on Management and Restoration of coastal dunes* (pp. 39–40). Spain: Santander.

Sánchez-Arcilla, A., Jiménez, J. A., Valdemoro, H. I., & Gracia, V. (2008). Implications of climatic change on spanish mediterranean low-lying coasts: The Ebro Delta case. *Journal of Coastal Research, 24*(2), 306–316.

Sánchez-Arcilla, A., & Jiménez, J. A. (1994). Breaching in a wave-dominated barrier spit: The trabucador bar (north-eastern spanish coast). *Earth Surface Processes and Landforms, 19,* 483–498.

Sayol, J. M., & Marcos, M. (2018). Assessing flood risk under sea level rise and extreme sea levels scenarios. Application to the Ebro Delta (Spain). *Journal of Geophysical Research: Oceans, 123,* 794–811. https://doi.org/10.1002/2017JC013355.

Serra, J., Riera, G., Argullos, J., & Parente-Maia, L. (1997). El transporte eólico en el Delta del Ebro: Evaluación y contribución al modelado litoral. *Boletín Geológico y Minero, 108*(4), 477–485.

Somoza, L., Barnolas, A., Arasa, A., Maestro, A., Rees, J. G., & Hernández-Molina, F. J. (1998). Architectural stacking patterns of the Ebro delta controlled by Holocene high-frequency eustatic fluctuations delta-lobe switching and subsidence processes. *Sedimentary Geology, 117,* 11–32.

Universidad de Barcelona. (1997). *Estudio de la situación actual y evolutiva de las formaciones eólicas litorales del Delta del Ebro.* Technical report. Facultad de Geología, Universidad de Barcelona.

Vericat, D., & Batalla, R. J. (2006). Balance de sedimentos en el tramo bajo del Ebro. *Revista Cuaternario y Geología, 20*(1–2), 79–90.

Zazo, C. (2006). Cambio climático y nivel del mar: la Península Ibérica en el contexto global. *Revista Cuaternario & Geomorfología, 20*(3–4), 115–130.

Chapter 27
Littoral Dunes on Valencia Coast

Eulalia Sanjaume and Josep E. Pardo-Pascual

27.1 Introduction

Along the studied area, the sediment supply from perennial and ephemeral rivers is the main nourishment of the beaches, and consequently of dunes as well. The littoral dunes need for their development very strong nourishment with a large excess of sediment supply, winds relatively strong coming from the sea area, grain sediment according to the wind speed and very low atmospheric humidity. For that reasons, is not strange that the most important dune fields are related to de sediments carried out from the perennial rivers belonging to our studied coast, such us: Túria and Segura. The sediments from the Xuquer River have been used to fill up his alluvial plain, the two prograding beach barriers located south of his river mouth and some small dunes, but not any dune field as large as the related with Túria and Segura ones. Even though fossil dunes (emerged or submerged) are possible to find more or less everywhere, we may also study the emerged ones. The fossil dunes appear, for example, at the distal part of de Torreblanca beach barrier, Almenara, Perellonet, and they are very common in the southern part, from Xàbia to Torrevieja. But the most special are the fossil climbing dunes attached to the Serra Gelada cliff. The Holocene dunes that we find on our coastline offer different types: foredunes, dune ridges, shadow dunes, hummocks, small barchans, transverse, reverse, parabolic and active dunes. Most of these dunes are located on the three main dune fields (Saler, L'Altet, and Guardamar). The foredunes were probably all along the coast in the past, but now we only find some fragments of the sandy

E. Sanjaume (✉)
Departamento de Geografía, Universitat de València, València, Spain
e-mail: eulalia.sanjaume@uv.es

J. E. Pardo-Pascual
Departamento de Ingeniería Cartográfica, Geodesia y Fotogrametría,
Universitat Politècnica de València, València, Spain
e-mail: jepardo@cgf.upv.es

J. A. Morales (ed.), *The Spanish Coastal Systems*,
https://doi.org/10.1007/978-3-319-93169-2_27

beaches belonging to the southern part of the Valencia Oval. At the present time, most of the Holocene dunes have succumbed under the construction of buildings and houses too close to the shore of the beach, roads, parking lots, and marine promenades. Regarding the dune fields, each of them presents a different state of conservation. The Saler dune field has experienced an important human action and a big part of its dunes have been destroyed. L'Altet dunes have been very lucky because they are just in front of the airport runways and, obviously, any type of building constructions that can hinder the aircraft takeoffs and landings were allowed. This is the only area where fossil Pleistocene dunes coexist with current active dunes. Finally, the dunes of Guardamar are well preserved, although they were colonized with pine trees at the beginning of the 20th century to prevent that the village of Guardamar was invaded by the active sand dunes.

27.2 Pleistocene Dunes

The dunes quickly react to environmental changes, primarily to variations in the sediment supply and sea level changes. There is some confusion regarding temporary adscription of dunes, although, on many times, is based on the different dating systems used. According to some opinions, like Fornós et al. (2009) among the others, the dunes are generated at the time of sea level negative pulsations. Conversely, some other authors point out that the dunes are built during transgressive periods (Riquelme and Blazquez 2001), or during periods of sea level still standing. In this case, dunes would be developed over the pre-existing ones, if those ones are easy to erode since, in this way, there is a new sediment supply for the aeolian transport (Nordstrom et al. 1990).

If we apply logic, which does not always occur in Nature that is rather chaotic, the marine regression seems to be the best time for the creation of dunes because the beach would grow in width as the sea goes down, which would mean a huge increase of sediments able to be transported by wind. These dunes would go prograding as the shore was moving away. Later on, during the next transgressive period quite a lot of the preexisting dunes would be eroded, by the waves of the new raising sea level, but at the same time dunes located on the upper part of the beach without erosion will start the weathering process of cementation, especially in our areas were limestones are predominant. This process requires warm and relatively dry climate, but with enough humidity to allow the growth of vegetation. It is possible to recognize the existence of fixing vegetation on the fossil dunes thanks to the tubes concretions than were coating each root of the plant (root-concretions). At the present time, the Holocene dunes, with a slight sea level rise, are being eroded and not generate new dunes. For that reason, we think that the main dune generation is at sea level negative pulsations, but their transformation in aeolianites would be during the transgressive periods (Sanjaume and Pardo-Pascual 2011a).

The comparison of the sedimentological and mineralogical (heavy minerals) characteristics between the Holocene and Pleistocene dunes of the Alicante coast shows the self-supply of sandy materials in the Holocene dunes (Sanjaume 1983, 1985a). The Pleistocene aeolianites are especially abundant on the Alicante coast, where tectonic uplift movements have recently occurred. There are fewer outcrops because of the predominance of the subsidence in the northern part of the Valencia Oval. However, some Pleistocene fossil dunes have been found (Fig. 27.1a) in the beach barrier of Torreblanca lagoon (Segura et al. 1990, 2005). Small fragments have been found in Benicassim (Castelló), as well as loose debris, not on-site, at the beach barrier south of Millars river mouth. Grater extension presents the calcarenite located in the inner part of the beach barrier of Estanys de Almenara (Castelló) old lagoon (Sanjaume 1985b). In the middle and southern part of Valencia Oval, there are new outcrops. Some remains are small, such as we found in the beach barrier of Puçol-Alboraia old lagoon (Segura et al. 1997; Pardo et al. 1996). The Penyeta del Moro fossil dune found at the beach barrier of Valencia lagoon, close to Perellonet village (Rosselló 1979), is quite big and the heavy minerals of its sandy material are identical to those of the Holocene of beach barrier sands (Sanjaume 1985a). The dunes located in the innermost beach barrier of Tavernes lagoon (Ruíz and Carmona 2005) perhaps are covering oldest dunes, as it seems to suggest its morphology and texture (Sanjaume and Pardo 2003).

We must get to Denia (Alicante) to find again Pleistocene dunes. From here they are very abundant in virtually all the rest of the Alicante coastline. In Xábia, the Pleistocene dunes reach a huge development along the beach barrier which closed the old lagoon. About these aeolianites has carved an interesting littoral karst in which are three areas clearly defined with the morphologies typical of each of them, as described in the works of Sanjaume (1979, 1985a).

Further South the Pleistocene dunes reappear as climbing dunes that are attached to the external parts of the small creeks (calas) existing between Moraira and Calp as, for example, Cala Bassetes, Cala Fustera, etc. (Sanjaume 1979; Riquelme and Blazquez 2001; Riquelme 2005). One of the largest and more important outcrops is located on the Serra Gelada cliff, where the climbing fossil dunes (Fig. 27.1b) reach over 150 m of height. Sometimes they have widths exceeding 200 m and its estimated total volume is about 15,750,000 m^3 (Chapapría and Rosselló 1996). The aeolianites come back immediately to the South of Alicante city and stretching to San Pedro del Pinatar. In the area of L'Altet-Carabassí fossil, dunes are at the base of the current dunes (Sanjaume 1985a), and they are the source of supply for the current dunes that form the dune field of L'Altet, as we will see later in this paper. These dunes are prolonged by the Quaternary coastline of Santa Pola (Rosselló and Mateu 1978; Cuerda and Sanjaume 1978; Sanjaume and Gozálvez 1978; Gozálvez and Rosselló 1978). South of Moncaio they are located preferably under newer dunes, although there are in the immediate beach areas and/or where wind erosion has fleshed them.

Further South the Pleistocene dunes are mainly under the Holocene dune field of Guardamar that goes on until the drainage channel of La Mata salt pans. The fossil dunes probably extended much further South, but have disappeared under the hundreds of houses built to feed the coastal urbanizations existing until Torrevieja. Fortunately, a small part of this dune field in La Mata (Fig. 27.1c) has been saved.

Fig. 27.1 **a** Fossil dune in Torre la Sal beach, located at the southern end of the Torreblanca lagoon's beach barrier. **b** Panoramic view of climbing fossil dunes of Serra Gelada cliff. **c** Fossil dune located at the Local Natural Park of Molino del Agua belonging to La Mata (Alicante)

In this way, we can see parabolic Pleistocene dunes, blowouts, as well as mixed forms, thanks to the fact that in 2006 it was declared a Local Natural Area by the City Hall of La Mata. This area can be crossed by wood footbridges making several circuits so that people can admire these formations.

The dunes have experienced strong anthropic pressure. Some have been used as a quarry for masonry stones since Roman times, for the construction of churches, watchtowers, or ornaments (Xàbia, Moraira, for example), others were excavated to make pools to conserve fish that would later be used to obtain "garum" (a very strong sauce very appreciated by the Romans). Examples of these pools can be seen in Banys de la Reina of Calp. Another type of use was the extraction of millstones that we found in Illeta dels Banyets del Campello (Sanjaume 1985a).

27.3 Foredunes

The Holocene dunes are predominant along the southern part of the Valencia Oval. Sometimes those dunes are covering the Pleistocene ones, or they have been destroyed by human impact. The loss of the dunes supposes a serious problem for the beach. It has been demonstrated that the dunes are essential to maintaining the beach sedimentary balance (Sanjaume and Pardo 1991b). However, we can still discuss the main characteristics of the remains of foredunes found. Later we will study the main dune fields of our coast.

There is currently about 60 km of foredunes, though does not occur continuously, but with many interruptions due to the constructions made in first-line of the beach (buildings and marine promenades, mainly). Most of these dunes have just in front embryo dunes, shadow dunes, small barchans and hummocks in the southern coastline of Valencia Oval (Sanjaume et al. 2011).

The old maps showed the existence of two alignments of transverse dunes, more or less parallel to the shoreline along 150 km, from the surroundings to Castelló city to Denia (Sanjaume and Pardo 1992). Now we just found remains of foredunes, with different degrees of degradation on the beaches of Canet, El Saler, Tavernes-Xeraco, the northern part of Gandia and, especially, along with the municipality of Oliva.

Most of the foredunes are transversal, arranged parallel to the shore, but in the central and southern part of Valencia Gulf transverse dunes of reverse type can be found due to the seasonal alternation between E and W winds. In Oliva (Valencia) the foredunes were developed enough to generate a second alignment that remained until the middle of the 1970s when disappeared in many places under the buildings. The dunes of Canet beach, located in the northern part of the Palancia river mouth, deserve special attention. Despite not having a source of nourishment as abundant as other areas, it has developed a small dune zone (about 34 ha), with a very precarious balance. There are low-rise dunes with irregular morphology (embryo, shadow dunes, and hummocks). They had a strong alteration by the continuous movement of people crossing then trying to reach the shore.

However, we want to emphasize that this is one of the few points of the Valencian coast (another maybe could be Xeraco) where the Administration has had the good sense to built the marine promenade just behind the dunes saving them from destruction and, thereby, ensuring the dynamic equilibrium of the beach and a better defense of the beach against the problems of accelerated erosion (Sanjaume and Pardo 2005).

27.4 Devesa del Saler Dune Field

The most important dune fields are located close to de main perennial rivers (Túria and Segura). Their contributions have historically been an excellent source of supplies not only to generate their own alluvial plain, but their surpluses have been of such magnitude that they managed to develop large dune fields.

The Devesa del Saler is located in the northern part of the Valencia lagoon beach barrier. This area starts about 6 km from the old river mouth of Túria and of Valencia City. It has a maximum width of about 1000 m and stretches 10.5 km South until the Perellonet inlet. The beach is dissipative. In its nearshore, there are two submarine bars. The outer bar is parallel to the coast, while the internal, closer the shoreline is a crescentic bar. The crescentic bar wings connect to the shore where generating a lot of rhythmic sinuosities. In the backshore small storm berms, sandy beach cusps and wind ripples would be the most abundant microforms.

This dune field is quite special because it was almost completely destroyed. The external part was completely destroyed by heavy machinery and the central depression was filled with the sand of the external dunes. The inner part did not suffer as much damage, most of them have remains, although influenced by infrastructure works. After such a terrible loss we have been able to compensate the disaster since the external area has become in a natural laboratory to analyze the spontaneous dune regeneration as well as the dunes artificially induced, for more than 40 years (Sanjaume and Pardo-Pascual 2011b).

27.4.1 Morphology Prior to Destruction

The Holocene dune field, before their leveling, had two large dune groups with slightly contrasting characteristics in terms of height, extent, and morphology, separated by a wide longitudinal depression of variable width and temporary pounding with water from groundwater level (Sanjaume 1974).

(A) Inner Group

We can only guess the morphology of this part by topographic maps made in the works prior to their destruction (Fig. 27.2). This area was formed by transverse dunes that were mostly oriented oblique with respect to the current shore. They were dissected dunes, which could be barchans in origin, but later became parabolic dunes due to NW prevailing winds. However, in the majority of cases, the morphologies were much more complex, probably by the juxtaposition of various simple shape dunes. Ahead of these dunes also existed lower transverse dunes

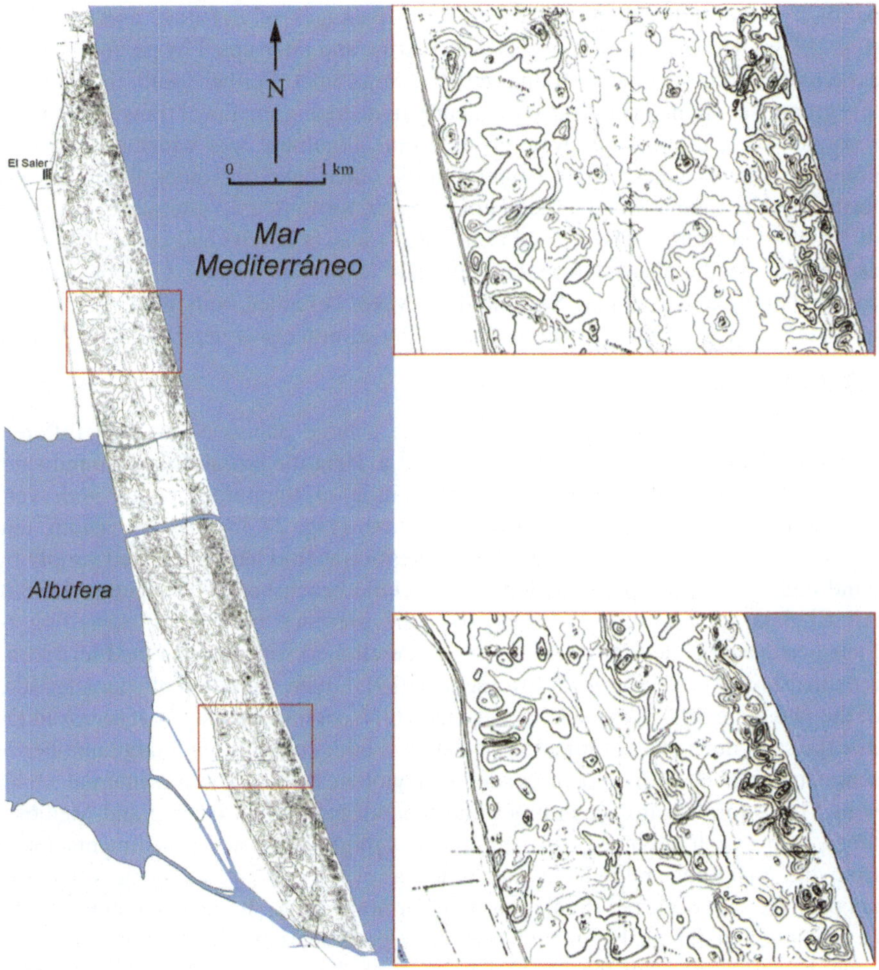

Fig. 27.2 Devesa del Saler dune field topography prior to the destruction in earlier 1960s

locates parallel to the shore. This would indicate a change in prevailing winds, becoming the eastern winds responsible for these new formations. Most of the dunes have relatively few lengths and were pretty wide with a rather chaotic distribution. The maximum heights (6.5 m) were found in the vicinity of the Albufera lagoon. Deflation depressions had also different orientations (oblique, perpendicular and parallel to the shore. Greater alteration of its heavy minerals suggested that the dunes of this internal group were older than the external ones (Sanjaume 1985a). All these dunes are colonized by pine trees and scrubs.

(B) Outer Group

This area was irreversibly altered. According to the primitive topography, this area was formed by the foredunes and a series of sub-alignments of the transverse dunes, parallel to the shore, dissected by frequent passages of deflation and blowouts (Fig. 27.3a). However, small swales very narrow and interrupted by parabolic dunes and complex dunes could be found until the Pujol inlet. Further south, the dunes of this group were transformed into two clearly defined alignments of transverse dunes, separated by a fairly wide swale. The second alignment was wider and at some points, the crest exceeded 5 m height. On the other hand, the outer alignment was formed by the foredune and a second dune ridge, forming a narrow subset with a lot of deflation corridors. The second dune, also transverse type had crests that could exceed, in specific places 6.6 m, but generally had lower heights. Close of the golf course, belonging to Parador Nacional Luís Vives del Saler, both alignments joined by an impressive parabolic dune surrounded by several blowouts (Sanjaume 1985a).

(C) Central Depression

Between the two large dunes, sets stretched a large depressed area, of different width, which ran the central road of La Devesa. Here the proximity of groundwater and drainage difficulties caused the installation of temporary ponds with very typical brackish and hygrophilous vegetation (Fig. 27.3b). By its width and extension does not seem likely that this depression would have generated simply by wind deflation but by progradation of the whole beach barrier (Sanjaume 1985a). All El Saler dunes were Holocene. But the presence of a small upwelling of Pleistocene aeolianite (Penyeta del Moro at Perellonet) suggests that the Pleistocene remains should not be far away (Rosselló 1979). Given the different characteristics of the two dune sets as well as the presence of the central depression, it seems likely that the sand dunes were formed in two stages, both contributing a large number of sediments, related to historical episodes of violent floods. The old internal sector would have formed in Roman times and the external one in Islamic and Medieval times. Depression remained between both sets might be related to sediments deficit from the Visigothic period. At that time, as evidenced by sedimentary records, major floods not occurred (Sanjaume and Carmona 1995). In those moments would have enough sediment budget for the beach progradation, but there would be no enough sand surplus for the dunes development. The overabundance of sediments would begin to occur in Islamic times favoring the formation of the outer dunes set.

27.4.2 Dunes Devastation and Restoration

Unlike what was happening with the sand dunes of northern and southern sectors where dunes were destroyed for agricultural use, El Saler dune field remained without major changes until the second half of the 20th century, perhaps because of the history of la Devesa del Saler. Until 1865 was owned by the Crown and was used as a hunting reserve, which managed to keep its natural characteristics. It was later owned by the State until 1927 when the Valencia City Council bought the Albufera lagoon and the Devesa of Saler. In October 1962 the Mayor of Valencia decides that this area, so close to Valencia and used for Valencian people during the

Fig. 27.3 **a** Foredune at La Punta (Saler). Foredune and shadow dunes. **b** Central depression regenerated by the works of the Technical Bureau Devesa-Albufera. **c** Foredune recreated artificially, middle part of Devesa del Saler

weekends for picnics and make the Sunday's "paella" under pine trees, could be sold to make a huge urbanization and get lots of money. In December of the same year, the Mayor's office gives free of charge to Housing Ministry to build the "Parador Nacional Luín Vives" and the immediate golf course. The subsequent transformations, although they supposed a quite big vegetation alteration, did not have too much significance for the dunes morphology.

In 1965 the Housing Ministry approved the Plan of Management of the Saler, saying yes to the private development project against the dunes. In March 1967 it was decided what companies would do the works of urbanization. In 1970 heavy machinery started to destroy the dunes and in 1973 the external set was swept completely, but a lot of buildings were built as well as a section of the planned maritime promenade, many asphalted areas for parking lots, internal roads, etc. On the other hand, the internal set was widely disturbed by the construction of some houses, parking lots, pathways, drainage channels, water pipes and some other infrastructure works. As a consequence of these works, lots of dunes and pine trees (30%, in 1973) were lost. Some were cut down to provide for the infrastructure, whereas others have slowly died from the harmful effects of salinity, having lost the foredunes protection (Sanjaume 1988). When the democracy was restored in Spain in 1979, the newly-elected City Council responded to popular demand and financed the necessary research in an attempt to save and repair as much as possible of this harmed area. As soon as they assumed the administrative control, decided to invest about 12,000,000 euros (equivalence of the money spent on the old currency "pesetas") between 1979 and 1987. Most of this funding was dedicated to the repurchase of land parcels sold to private individuals, in order to stop the construction project. The City Council commissioned a multidisciplinary group of experts the feasibility studies of the Saler dunes regeneration. The recommendations of the first studies (González et al. 1981) divided the whole Davesa into three areas: (i) **Filter zone** (northern part), including the most destroyed area, impossible to recover immediately, and that would be used to accommodate most of the beach visitors; (ii) **Park zone** (middle part), had very few buildings and consequently higher ecological value. For that reason car driving was prevented, leaving access only to pedestrian traffic. In this area later on were develop, with more or less success, some projects for artificial (bulldozing sand) dune recreation or semi-artificially accelerating sand accumulation using permeable fences, and as soon as the accumulation reached a critical stage planting autochthonous vegetation (Sanjaume 1988); (iii) **Unaltered zone** (southern part). The area was completely swept, the dunes completely destroyed, but no more impacts (Fig. 27.4). For that reason, we decided to do absolutely nothing, and let this area called La Punta del Perellonet to recreate by them-self, like a natural laboratory to follow its natural evolution. From 1979 to 1985 la Punta remained opened and people and vehicles were able to cross it to get to the shore. Finally, to protect its ecology, la Punta was fenced off. The whole zone of Saler was declared a Natural Protected Environment by the Valencia City Council in 1983. In 1986 the Albufera lagoon and its beach barrier were declared Natural Park (Pardo et al. 2005a).

The Technical Bureau Devesa-Albufera has been working recreating dunes in a semi-artificial way all over the Devesa, except in La Punta until the present day. In addition, the State Administration (Ministry of Public Works, at that time) has also regenerated some dunes in the most degraded areas using artificial methods (Benavent et al. 2004, Sanchis Ibor and Vizcaino 2016) (Fig. 27.3c).

Fig. 27.4 Image of la Punta del Perellonet immediately after being flattened

During the last decade the remains of the planned marine promenade have been eliminated, as well as parking lots and roads have been removed, so the asphalt has disappeared completely. For this reason, the Devesa del Saler looks again less anthropogenic.

27.4.3 Evolution of La Punta del Perellonet

After its devastation, the area of La Punta del Perellonet (28 ha) has become in a natural laboratory to research geomorphologic processes of dune formation and, especially useful to check the natural behavior of previous surplus beach sediments, but with erosion problems in recent decades (Sanjaume and Pardo 1991a). For the reason, a detailed monitoring of the area has been done, since its devastation until some years ago, through the implementation of different topographic surveys and field observations. For more than 40 years, we have followed the evolution of this area and the recreated dunes. La Punta can probably be one of the most studied

Table 27.1 Summary table of the different available topographic surveys of La Punta

Topographic survey	Source	Method	Altimetric data	References
1980	Aerial photographs 1:3,000	Photogrammetric survey	Contour lines map 1:1,000	Gironés Sempere and López Galera (1997)
1989	Topographic profiles transverse to the shore	Automatic interpolation	Profiles each 200 m	Sanjaume and Pardo (1991a)
1994	Aerial photographs 1:5,000	Photogrammetric survey	Contour lines map 1:1,000	Gironés Sempere and López Galera (1997)
2001	Aerial photographs	Photogrammetric survey	Contour lines map 1:2,000 and dimension points	Carytop for Valencia City Council
2003	Dimension points	GPS-RTK	More than 50,000 dimension points	Fons and Sánchez (2004), Pardo et al. (2008)
2005	Dimension points	LiDAR	More than 180,000	Gracia Martínez (2008), Pardo et al. (2008)
2008	Dimension points	LiDAR	More than 180,000	Gracia Martínez (2008), Pardo et al. (2008)

areas along the Mediterranean coast. Our research has allowed us to know as indirect factors affect the development of the dunes. Currently, there are seven topographic surveys, performed with different techniques and, therefore, with different levels of precision (Table 27.1), which has allowed to characterize and quantify the changes occurred the last decades.

The availability of data of different origin and credibility of each measurement levels are different. Although it is always complex to establish what the original situation was, in this case, fortunately, there are aerial photographs (Fig. 27.4) that demonstrate the complete dunes devastation, leaving a completely flat surface. On the other hand, when we were making the topographic profiles surveys of 1989 measurement, levels of dirt (even with plastic remains) were found in quite a lot of dunes that marked the zero level after the devastation (Fig. 27.5a). Based on these data, the first quantification between 1873 and 1889 was made. In 1997 new topographic surveys were done, using photogrammetric techniques corresponding to the situation of 1980 and 1994, checking that the sedimentary increase had substantially accelerated compared to the data of 1989.

The work of Pardo et al. (2005b) described in detail the methodology applied to calculate volumes in each date until 2001. GPS-RTK was used in the 2003 topographic surveys and was evident that the level of accuracy improved substantially as well as the ability to observe details. Subsequent data have been obtained from two surveys carried out by LiDAR techniques, with one data per square meter and altimetric accuracy about 15 cm. The first was held in May 2005 and the second in December 2007. In these cases, to make the calculations were necessary to do editing work to remove the records corresponding to areas with vegetation (Gracia Martínez 2008; Pardo et al. 2008).

27.4.4 Basic Keys of Dune Restoration Process at la Punta

After la Punta artificial flattening in 1973, this sector has evolved creating a new natural dune morphology, but substantially different from the existing morphology prior to their destruction. Two groups of basic processes have been involved in its evolution. Both have been taking place, with greater or lesser intensity, throughout these years.

The first process would be part of the sediment mobilization from the innermost part of la Punta to the beach. This would facilitate the foredunes development. Although this process was much more exaggerated immediately after the devastation due to the lack of vegetation, the process has continued, albeit with less intensity, until a few years ago. This model of change (Sanjaume and Pardo 1991a) was mainly responsible for the initial feedback from the dune system. Later the unequal distribution of both vegetation and incipient dune morphology determined that the topographic configuration was more and more complex, as is possible to see from the comparison of 1980, 1994 and 2001 topographies (Fig. 27.5b).

The second process was the arrival to la Punta of new sediment coming from the relocation of sands belonging to the beach nourishment made in the beaches located much more to the north. This could explain the impressive increase in the volume of sediment that has been produced, above all, throughout the nineties of the 20th century.

Three phases can be distinguished in the evolution of regenerated dunes of la Punta:

Initial Development Phase

In 1980 an incipient foredune, parallel to the shore, was beginning to stand out. It was relatively wide, but not too high, fed by the western prevailing wind. It was also appreciated the development of very wide blowouts in the border area of the golf course, as well as one swale behind the foredune. In the most southern area, close to Perellonet inlet there were two crests more the 3 m height. In 1989, the dune morphology was more complex, slightly increasing the volume of sediments

Fig. 27.5 **a** Level of dirt (detritus) found in the new dunes of la Punta, indicating the level zero after the prior dunes devastation. **b** Dune field of l' Altet. Example of Holocene dunes overlapping fossil dunes. **c** Carabassí area with active parabolic Holocene dune

in the area. Unlike the original dunes now the foredune has transverse morphology, but it is the reverse type because the winds acting are bidirectional E-W (Sanjaume and Pardo 1991a).

Accelerated Growth Phase

Five years later, in 1994, the increase of the dunes was very significant. The foredune was completely developed. It was narrow with a lot of transversal corridors, dividing it into several compartments. Most of the foredune reached more than 4 m height and, even, some specific crests reach 6 m height. In addition, a second alignment was built behind the swale. Later on, this longitudinal swale was transformed into a string of blowouts with different sizes and shapes. All this growing phase showed a very positive sedimentary balance. The sand self-sufficiency of the first phase practically had turned to strong growth since la Punta began to withhold part of the sand transport. The littoral drift increased the sand transport because was carrying part of the sand used during the beach nourishment of northern beaches. Between 1989 and 1994 the volume of accumulated material was 79,062 m^3. A turning point occurred in the amount of sand supply, the 3,526 m^3/year of sand between 1973 and 1989, changed to 15,812 m^3/year of sand between 1980 and 1994 (Pardo et al. 2005a). In 2001, it is noted that the pre-existing dunes were still increasing not just in a few places like in the past. Now, most of the first alignment, which merges the foredune and the rear dune, has experienced a strong increase in height, giving a group with dune crest between 5 and 6.5 m height (Fig. 27.6). Transverse corridors cutting the longitudinal dunes and blowouts are abundant. In addition, there are a lot of embryo dunes aligned just in front of the foredune, indicating progradation by sedimentary surplus. All of this suggests that la Punta had very strong nourishment from 1994 to 2001. To understand the paradox of sandy surplus in a sedimentary cell with negative deficit of sand supply (Sanjaume and Pardo 2008) must take into account the different phases of artificial beach nourishment made by the State Administration to regenerate the beaches located North of la Punta where have been poured 820,000 m^3 of sand (Pardo et al. 2005a).

A Slowdown of Growth and New Self-sufficiency Phase

Between 9th and 16th November 2001, the Valencian coast (as well as much of the Spanish Mediterranean coast) was whipped by large marine storms consecutive throughout the same week. In these storms, the buoy off the Valencia harbor recorded waves exceeding 4 m significant height and periods of more than 10 s. Consequently, much of the foredunes were eroded. This would explain the significant decline of sediment volume on la Punta between 2001 and 2003 (Sanjaume and Pardo-Pascual 2011b).

Further changes are apparently minor, even though they have a slight negative trend. The analysis by using GPS-RTK and LiDAR, indicate that there is a certain clearing in the internal area of the beach, as well as from the blowouts bottoms. The sand coming from this source has been increasing the foredunes height. This fact confirms the initial theory presented by Sanjaume and Pardo (1991a), based on less

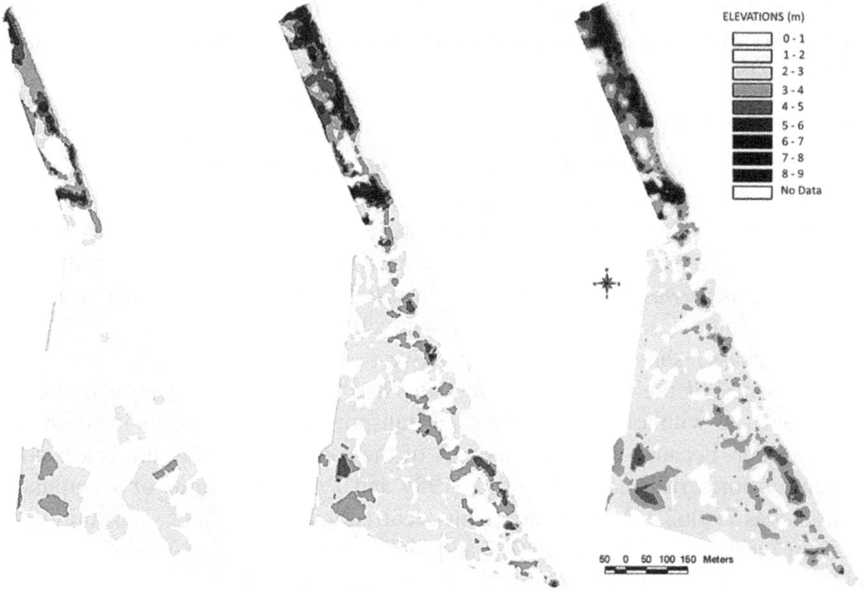

Fig. 27.6 La Punta dunes evolution comparing topographies of different years (1980, 1994 and 2001)

reliable data, that the process of dune morphology regeneration was possible thanks to the eroded sand in the internal zones of la Punta.

All the studies conducted in the Saler have allowed us to know that sand dunes of new recreation are much more inland than the old dunes surviving the devastation. That shows the recession experienced by the beaches as a result of accelerated erosion by sediment deficit. If we compare the distance between the shore and the golf course dune (which was not destroyed) and the distance between the new dunes of la Punta and the shore we will see that they are between 15 and 20 m more inward. This confirms the strong recessing that is having our coastline.

27.5 L'Altet Dunes

This dune field of 170 ha is pretty special and unique on the Valencian coast because fossil Pleistocene dunes are together with active Holocene dunes. As a result of this, most of the dunes have a complex morphology. They are quite old dunes that have experienced a long period of erosion and weathering. Holocene dunes were fed by sand coming from the old ones and many times the new dunes are adapted to pre-existing morphology or sometimes appear completely new forms.

The northern part of this dune field has experienced the characteristic urban pressure of the Alicante coast, but the closeness of the airport airstrips prevented the destruction of a large area which is of great interest. Fortunately, it was declared Local Natural Park in 2005 and probably in the future, it will be one of the few untouched dune areas along the Valencian coast. This dune field should be really called Altet-Carabassí, since it covers from Urbanova (located few kilometers South of Alicante city, to the foothills of Carabassí, stretching 5.5 km. Currently, the urban pressure of Los Arenales del Sol has stolen 53 ha to the original surface (170 ha).

It is actually a huge fossil dunes field whose erosion generates sand that supplies to small sand seas fixed by vegetation, zones of active dunes, deflation depressions, etc., alternating with outcrops of aeolianites completely bare of vegetation, as well as small sand cords that have been migrating inland and they become escarpments in the innermost part of this field. There are also deflation corridors, shallow swales, and blowouts of different types. All the area presents a rather complex morphology and a vegetation cover very scarce.

According to the analysis of heavy minerals found in the aeolianites, it seems they were fed with sediments from Barranc de Les Ovelles, located south of Alicante harbor. On the other hand, the heavy minerals of the active dunes present a composition almost identical the aeolianites. The only difference is a remarkable increase of epidotes and altered minerals which presupposes that the sands have experienced more than one sedimentary cycle (Sanjaume 1985a). Therefore, due to the great mineralogical similarity is clear that recent dunes have been generated by erosion of fossil dunes. The vegetation cover (Fig. 27.5b) consists of different psammophile species, such as *Crucianella maritima, Ammophila arenaria, Othantus maritimus, Lotus creticus y Thymalaea hirsuta* (Seva et al. 1989).

This dune field is currently divided by "Los Arenales del Sol" urbanization, which has destroyed most of the dunes close to the beach. But the huge fossil dunes are preserved, for the time being, in the innermost part. The northern area called l'Altet dunes occupies the greatest extent and is divided into two sub-areas (northern and southern l'Altet). Between Los Arenales del Sol and the foothills of Carabassí, where is located the Gran Alacant urbanization, we find the Carabassí dunes (Fig. 27.5c).

Some years ago, when did not exist the Digital Elevation Model (DEM), and the aerial photos did not good resolutions neither appropriate scale, the only thing that we found was the existence of two alignments separated by a quite shallow swale as well as many blowouts with different orientations in the l'Altet dune field. At present, the new techniques allow a much more detailed interpretation.

Thanks to the Digital Elevation Model (DEM) it is possible to observe that the height of sedimentary deposits increases zonally, forming stripes parallel to the coastline. Morphological differences allow, in turn, divide this section into two sub-areas, whose boundary would be marked by the road going from L'Altet village to the beach (the dashed black line of Fig. 27.7).

Northern sub-area of l'Altet: presents a foredune quite short (less than 2 m height) that is parallel to the shoreline. The foredune has some erosion corridors (more by trampling than by wind erosion). For that reason, the foredune looks quite compartmentalized. In some areas, there are embryonic dunes, just in front of the foredune. Immediately behind the foredune, start the earliest forms of the huge Pleistocene dune field. Between both systems, there is a very narrow swale, parallel to the foredunes, through which flows a path (Fig. 27.7). The Pleistocene dune complex presents two alignments, separated by a depressed area, relatively large, interrupted by dunes of neoformation. Many of these dunes maintain their original forms, although the sand detached from them by weathering is easily carried out by wind and can modify the primitive Pleistocene forms.

The circular base dunes, sometimes very eroded have crests (more than 15 m height) located in what would be the tip of the cone. There are also remains of a parabolic dune, much shorter. Its blow out has been created by winds from SE. Further to the South, following this same alignment there is a transverse dune more or less parallel to the shoreline, generated also by winds from E, and another conic dune, so eroded than is impossible to recognize its original morphology (Fig. 27.7). Some blowouts are at its NNE flank. All these dunes have heights between 4 and 9 m. Later, a new depression is found, carved on the own aeolianite or sometimes like a swale excavated in the sands coming from aeolianite erosion. All this depressed area separates the alignment describes above and the most internal alignment, whose maximum levels are between 15 and 20 m high. The internal alignment is marked by the discontinuity of their forms, which are, in addition, wider than the previous ones. There are with elliptical base and more or less pyramidal shape; a huge parabolic dune with his neck oriented E–W; transverse dunes; and many eroded ones without clear morphology, but most probably they were transverse. There are some blowouts shaped saucer or trough (Sanjaume and Pardo-Pascual 2011a).

Southern sub-area of l'Altet: is much more complex. The two alignments perfectly clear in the previous sub-area are mixed in this sector, so that the overall appearance is much more chaotic (Fig. 27.7). Also, presents characteristics quite different from the northern sub-area. The Pleistocene dunes are eroded much more so, sometimes, it is difficult to characterize their morphology. In this section, the foredune presents newly active dunes and scraps of fossil dunes. There are a lot of blowouts of different sizes and shapes.

Some are so small that they are even located at the top of the dune crest. Regarding the forms, there are different types: saucer, trough, triangular and arrowhead-shaped. There are also some escarpments carved or in the Pleistocene aeolianite or in incipient active parabolic dunes. In the southernmost part small ridges (quite short, but very long) more or less parallel to the shoreline, are staggering toward the highest central area and they after precipitate abruptly western wards. In this side, the escarpments are of great magnitude. It seems that the eastern wind erodes the pre-existing dunes and generates the ridges that are migrating until they precipitate on the roads or parking lots. On this internal part, the blowouts are also abundant. In the middle part, wherein the former sub-area was the second alignment and the longitudinal depression; we find here fossil dunes cut by corridors, deflation depressions, and active dunes created with fossil dunes eroded sands.

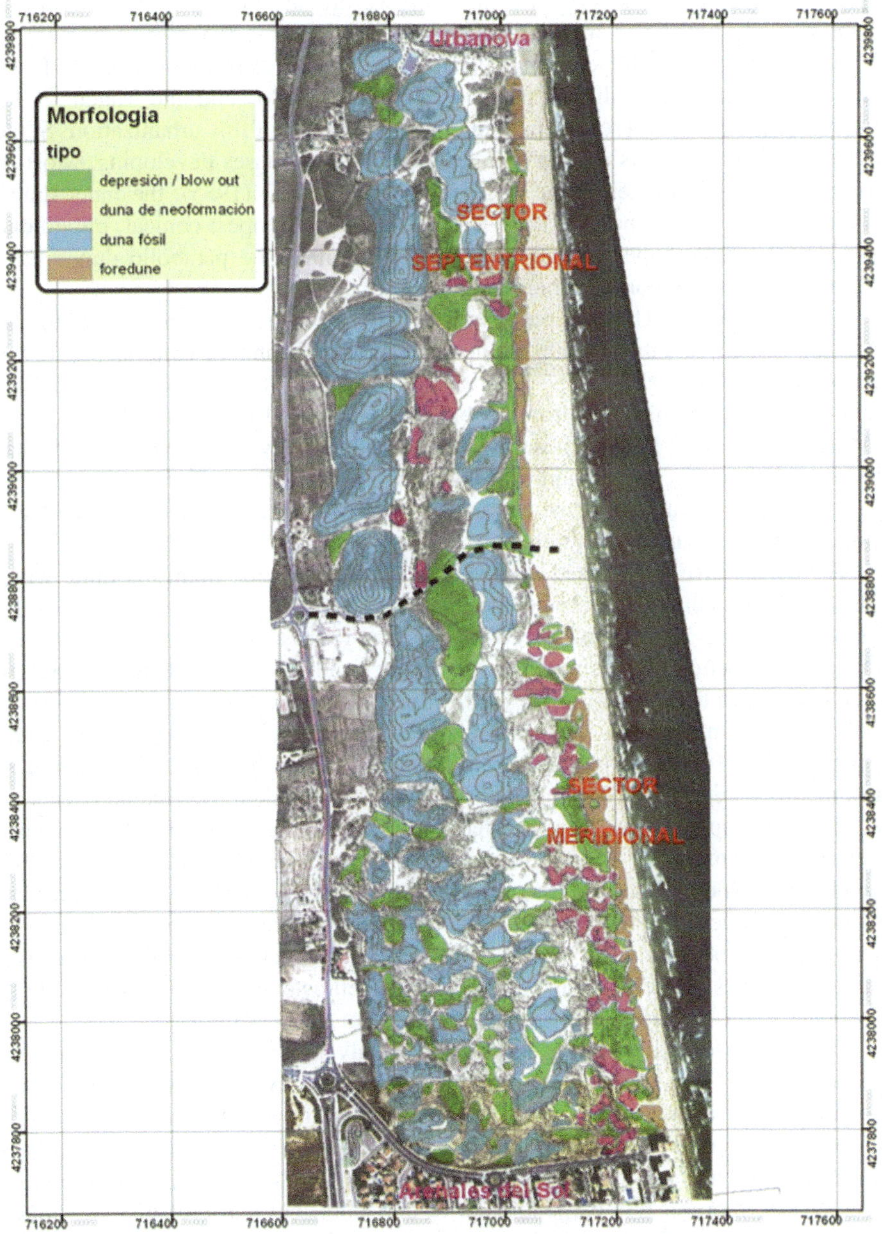

Fig. 27.7 Geomorphological map of the l'Altet dune field. It shows the contour lines with equidistance of 2.5 m. An orthophotograph of 2008 is the base of the map. The dashed black line is separating the area into two sub-sectors

Dunes located behind Los Arenales del Sol urbanization: The dune field has been virtually razed and transformed into buildings. This has meant the loss of 53 ha of dunes, both Holocene and Pleistocene, as well as some outcrops of fossil beaches, which were studied at the end of the 1970s (Rosselló and Mateu 1978). Fossil dunes are still preserved in some parcels no built of this urbanization. Some others are being used as a quarry to get materials for houses developing. Most of these dunes are extremely large. According to Gozálvez (1985), the inner dunes could exceed 30 m in height. They have very diverse shapes: conical, pyramidal, and dorsal whale-shaped, dome-shaped, or parabolic. The parabolic dunes have formed blowouts of considerable dimensions.

Carabassí Dunes: This section is located between the southernmost buildings of Arenales del Sol and the foothills of the Carabassí. Including the external part of the depression of Clot de Galvany, where was found one of the Pleistocene beaches with the highest number of Strombus bubonius per m^2 of the entire Alicante coast (Mateu and Cuerda 1978). These dunes have a relatively small extension (59 ha) because this area is much more narrow than the other ones. The dunes are constrained by the urbanization; the entrance of Clot de Galvany depression; Carabassí slopes buttresses; and the shoreline.

The dunes have a dispersed disposition. They are not too high, but quite wide. They present a chaotic appearance because the Holocene dunes are overlaying the Pleistocene ones. The fact of the co-existence of fossil and Holocene active dunes makes this section very similar to the l'Altet dunes. The aeolianites outgrow practically until the shore. They are scraps more or less scattered and their height is very scarce. On top and around of these dunes, there are many narrow paths generated by people walking and crossing them to reach the shore. The paths have lost incipient vegetation, so the naked sand may be reworked by wind and generate a lot of blowouts. Some blowouts have quite big dimensions. They are mainly saucer type. Among the fossil dunes, some are parabolic as a consequence of NE winds, and some others are round like dome-shaped dune, but very low at present, due to they have been strongly eroded, and never got more height by the scarcity of sediments, or because most of the dune is buried under the current beach. The Holocene dunes are active transverse dunes as well as embryos in their front, both generate by the easterly wind, overlying on top of the fossil dunes.

The section located just in front of Clot de Galvany depression exit, the erosion predominates. So no too many accumulation forms exist. Blowouts and small deflation depressions are abundant. However, there are active dune accumulations on both sides created by sand coming from fossil dunes. The Cot de Galvany is quite special. Now is a wetland, attached to Santa Pola dome, long and narrow at present day, more or less perpendicular to the shore, but that probably was much larger and this wetland could connect with the northern wetlands as Saladar d'Aigua Amarga, close to Alicante city. All these wetlands probable were a lagoon closed by a Pleistocene barrier. The barrier was nourished by the ephemeral stream Barranc de Les Ovelles as heavy minerals demonstrated. The barrier was getting wider and wider until was attached to Santa Pola dome. The Pleistocene dunes sure have had more than 20 m high (the same as the dunes at Guardamar dune field), and later have been eroded and lost a lot of sediments. These dunes are created during Pleistocene sea-level regression as show the dunes that are possible to find on top of the Pleistocene beach Tirrenian II with a lot of Strombus bubonius. In the southernmost part of this area, very close to the Carabassí slopes, the Pleistocene dunes are divided by a narrow swale that in some places looks like was carved in the aeolianites substrate. The fossil dunes are staggering between 5 and 15 m in height. The same are doing some blowouts and embryonic dunes. Most of the blowouts are here again saucer type. The fossil dunes have an oval base and mainly dome-shaped and parabolic forms.

The Altet-Carabassí dune field, due to dry climate conditions and vegetation scarcity, has a big sand budget. Here the mobility of sediments is much bigger than in any other of analyzed dune fields along the Valencian coast. In the Carabassí, the area with sand ready to be moved with moderate wind energy has been estimated that covers 18.38 ha. But with a strong wind, the area will come bigger (33.95 ha). That represents almost 62% of the dune field surface. On the other hand, in the l'Altet dune field, the zone with sand potentially able to be transported by moderate energy wind (10.68 ha) may change and increased with the strong wind until 42.9 ha. That means 68% of the area of the dune field. The main reasons for this mobility are lack of vegetation ant the trampling of people and vehicles. The scarcity of vegetation, in turn, is motivated by a shortage of rainfall (less than 200 mm/year; strong evaporation; and the aeolianites hardness, which prevent the roots penetration. The degradation of these dunes presents, therefore, natural reasons inherent to space itself, as well as the use of that space in areas where the human activity has been so strong.

27.6 Guardamar Dune Field

The dune field of Guardamar del Segura is the most extensive of the complete Valencian coast. Although strictly speaking it would cover only the municipality of Guardamar beaches, in the broad sense it would start north of La Marina village and would finish at Torrevieja. The Segura river mouth divides the dune field into two parts clearly differentiated by the heavy minerals composition, its height, dunes orientation respect to the shoreline, and the own dunes morphology (Sanjaume 1985a). The dunes cover an area of approximately 700 ha. There was a huge transgressive dune field with precipitation dunes in the southern part. This field has presented a good degree of conservation, much better than any other one along the Valencian coast. Nevertheless, the urban population pressure of Guardamar and some residential areas in its internal limits are threatening to these dunes which were active until the beginning of the 20th century. In the innermost part dunes clearly exceed 20 m, even more in the southern part where they almost reached 30 m height. The mobility of the dunes was very high since they were moving at a rate of 3–8 m/year. This mobility was a danger to the population. In its advance, they were transformed into precipitation dunes, who manage to bury an Arab

Mosque, a lot of farmlands and at the end; they were threatening to precipitate on the first houses of the Guardamar town. To solve this problem in 1906 was carried out a campaign of dunes fixation by mass planting of pine trees (*Pinus pinea*, *Pinus halepensis*) and palm trees (*Phoenix dactylifera*) (Sanjaume and Pardo 1992; Seva et al. 1989). This dune field presents a great complexity for its own evolution, from the upper Pleistocene to the present day. The orientation of these dunes suggests that there have been different stages of generation and that, in many cases, some overlap with others. On the other hand, the mineralogy of heavy minerals analyzed (Sanjaume 1985a) also corroborated such assumptions.

27.6.1 Northern Zone of Guardamar Dune Field

It extends between the beaches of La Marina and the Segura river mouth. The Pleistocene beach barrier of the Elx lagoon stretches to the southern part of Salinas del Pinet (Pinet salt pans). In this section, there are few outcrops that preserved dune morphology, although they are considered the base of the present coast. In the surroundings of La Marina, the dunes have already acquired enough entity, so the width of the dune field is extended, as well as the number of dune alignments, reaching its maximum width at the Segura river mouth. The morphology of these dunes is known in detail thanks to the available Digital Elevation Model (DEM). Unfortunately, in some sections, the DEM does not cover the complete dune field but is enough to be able to imagine how the behavior of the different dune ridges is. In the northernmost part, there are three alignments more or less parallel to the shore, while close to the Segura river mouth channel there are seven. External and intermediate dunes can be followed along the Elx lagoon (Sanjaume and Gozálvez 1978). The dunes of the area are transverse. The internal ridges are the widest and they have a lot of blowouts generated by easterly wind. Wind from the East is also responsible for parabolic dunes which are attached side by side.

Sometimes the blowouts have acquired such dimensions that the parabolic dune horns have been isolated, forming small ridges perpendicular to the beach. They are making a rectangular pattern by the superposition of different blowouts generations (Fig. 27.8). Only in the southernmost part of this zone appear parabolic dunes generated by wind from NE, and that will be the most frequent in the southern zone of Guardamar dune field. From the first alignment until the last one the swales are completely chaotic due to the presence of these detached horns and blowouts and deflated areas. The only large swale is located between the foredune and the first alignment.

The swale maximum width is close the river mouth channel. All over this area is possible to find blowouts of many types, but mainly trough and sauce-shaped. The foredune, that in some places reaches more than 7 m, has a great continuity. In some sectors, the beach is very wide and large number of pyramidal dune embryos or shadow dunes is found. Normally a lot of these embryos are attached to aeolianites remains. In the wider stretches are also generated small active parabolic

Fig. 27.8 Digital Elevation Model (DEM) of the northern part of Guardamar dune field. The scale can be recognized by the grid (200 m)

dunes. They are located between the foredune and embryonic dunes. Orientation parallel to the shore, its low height and mineralogical composition, very similar to the sediments that the Segura rives is carrying out still today, suggest that the dunes of this zone are Holocene and that they do not overlie large-scale Pleistocene dune buildings (Sanjaume 1985a), although the base is still the aeolianite found immediately South of Alicante city.

27.6.2 Southern Zone of Guardamar Dune Field

This sector of Guardamar dune field extends between The Segura river mouth and the channel of La Mata salt pans, although probably remain, but with the predominance of Pleistocene dunes, until Torrevieja surroundings. In addition, Guardamar town divides this southern zone into two tranches.

The first tranche (Fig. 27.9) extends between Segura river mouth and Moncaio foothills, having Guardamar just behind the dunes. This huge part of this dune field was an important transgressive dune field with a big amount of precipitation dunes in the innermost part. The town of Guardamar was built on top of fossil dunes, and some of them still can be recognized at present time. South of the city farmlands and touristic urbanizations have been destroying dunes and for that, the dune field is extremely narrow at that place. The second tranche starts at the southern slopes of Moncaio and finished at La Mata salt pans channel. In both areas, the dune field shows a big predominance of huge parabolic dunes.

In the first segment of this field, the parabolic dunes are extremely large, so big than de blowouts are transformed in deflation basins, and the horns or wings of these big parabolic forms in the aerial photographs looks like oblique alignments of transverse dunes, but they are not. Thanks to DEM is possible to know the real original form and was amazing to realize that really they were large parabolic dunes generated by NE wind (Sanjaume and Pardo-Pascual 2011a). From Segura river mouth, the newest dunes are advancing through the older parabolic dunes deflation basins. New contributed material overlaps with dunes already formed before. Thereby reaching the maximum height in the SE part of the town, where these precipitation dunes reach 30 m.

Dune morphology is always the same: parabolic, but in some places, the drawing surface is complicated since former blowouts are filled with contributions from the advance of new parabolic dunes.

The height, therefore, will rise as much as the contributions are overlapping. In this way, in an NE-SW transect, the first parabolic dunes have heights between 2.5 and 5 m; towards the inner part reaches 7 m; later 10 m, although a large volume of these dunes is located between 15 and 20 m. Finally, in the innermost part and quite far away from the river mouth (which has been the main sediment supply for those dunes) the parabolic dunes reach almost 30 m height.

The foredune is a transverse dune with a lot of small blowouts, Between the foredune and the parabolic ones, there is a quite wide swale. In the second part of

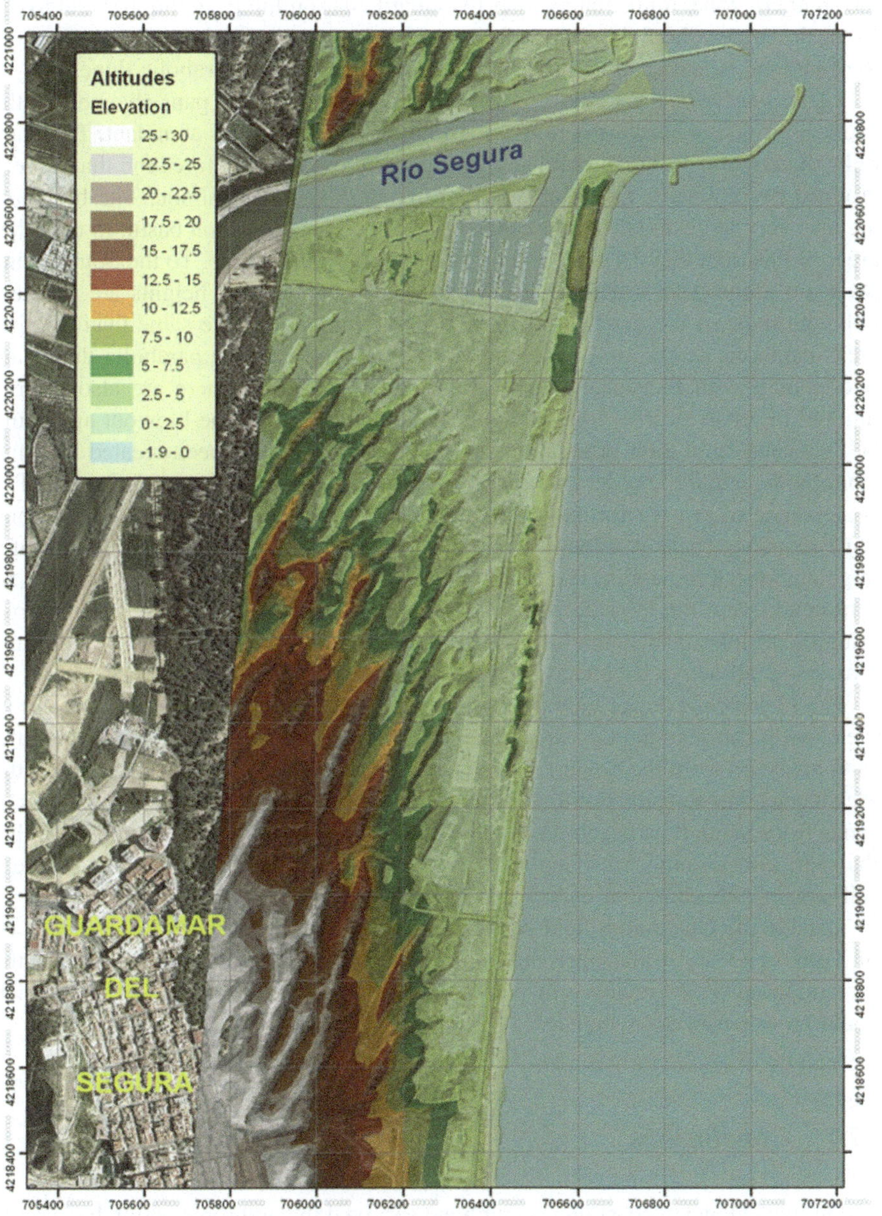

Fig. 27.9 The central zone of Guardamar dune field, extending between the Segura river mouth and the town of Guardamar del Segura

the southern zone, immediately after the Guardamar buildings, going parallel to the Moncaio slopes, the field almost disappear is very narrow and there only survives the foredune, some shadow dunes and small barchans and parabolic dunes.

The second tranche is going from Moncaio to La Mata salt pans channel. At the Moncaio foot, the beach is getting wider, but the dunes are less abundant. There are only shadow dunes, small barchans, and parabolic. After the foredune that is very cropped by blowouts, the inner dunes are predominantly parabolic. The blowouts are also very large and by coalescence finally become a deflation basin. In the other areas of this dune field, swales and deflation basins were scarce. To the South of the Moncaio, dune field widens again. Maximum of about 17 m heights are reached, although most of the dunes are kept between 5 and 12 m. The same than in other areas parabolic dunes are dominant. Many times these active parabolic dunes are overlapping with Pleistocene fossil dunes. In this section, there are also a large number of erosive forms like blowouts ancient and recent. The blowout age could be evaluated from their orientation since the Holocene have been created by wind from the E, while the older ones derived from the erosive action of NE wind. The coalescence of erosive forms generates, in many cases, deflation corridors among dune remains. There is a longitudinal swale very chaotic for the overlapping of a large number of blowouts. The foredune is quite wide, in some places more than 5–7 m height, and the lee side is precipitating to the next vegetated swale and covering older dunes. This area has also a lot of shadow dunes attaches to fossil dune remains small active backhands and parabolic.

The Pleistocene dunes sand, that represents the base of this southernmost part of the Guardamar dune field, have a heavy mineral composition almost identical to the sand analyzed from the Segura river. That confirms that the littoral drift during the Pleistocene was towards South. Now is mainly going North because the analysis of sands from Santa Pola beaches also have the same heavy minerals composition that the sediments coming from Segura river. That means that the present dunes, had a self-supply, the same that occurs at l'Altet.

Outside the Guardamar field itself, the Saladaret beach and Los Locos beach (E of Torrevieja) and Cala Ferris had embryonic dunes. On the other hand, on the eastern shore of Torrevieja salt pans, there is kind of dunes made of silt, which could be categorized as lunettes.

27.7 Conclusions

Along the Valencian coast dunes that have survived the anthropic action have quite distinct characteristics, especially in relation to the dune fields: in El Saler all dunes are Holocene and more than half of them were completely destroyed by human action; in l'Altet, fossil dunes of large size are predominant and the active dunes generated by self-supply are overlapping the older ones creating a very chaotic morphology; and finally in Guardamar fossil dunes are predominantly parabolic dunes (from NE) and the recent materials are superimposed on pre-existing formations, although north of the Segura river mouth there is predominance of small parabolic (from E) over transverse dunes with blowouts in also different direction may confirm that they are newer. This change on the wind regime and as a consequence of the direction of the littoral drift has been also demonstrated by the heavy minerals compositions in the sand of both areas. All the foredunes are transverse, except the new foredune of la Punta which is reverse. In the few areas where there is a surplus of sediments, is possible to find a lot of embryonic dunes such as shadow dunes, small barchans, and small parabolic. The relative scarcity of fossil dunes in the Valencia Oval could be explained by the predominance of subsidence. On the other hand, the wealth of Pleistocene dunes along the Alicante coast would be linked to processes or emersion. All of the dunes have been and still are subject to great anthropogenic pressure (Sanjaume and Pardo-Pascual 2011c). The dunes have been destroyed just to give free space to build houses, roads, parking lots, marine promenades, etc., without taking into account the dunes are the best beaches defenders because they contribute to the beach balance giving or withholding sediments.

Acknowledgements The authors appreciate the financial support provided by the Spanish Ministry of Economy and Competitiveness in the framework of project CGL2015-69906-R.

References

Benavent, J. M., Collado, F., Martí, R. M., Quintana, A., Sánchez, A., & Vizcaíno, A. (2004). *La restauración de las dunas litorales de la Devesa de l'Albufera de Valencia* (p. 65). Valencia: Ajuntament de València.

Chapapría, E., & Rosselló, V. M. (1996). Formes terrestres, dipòsits eòlics i marins de la Serra Gelada. *Cuadernos de Geografía, 60,* 223–242.

Cuerda, J., & Sanjaume, E. (1978). Datos paleontológicos y sedimentológicos del litoral Pleistoceno de Santa Pola. *Cuadernos de Geografía, 23,* 19–96.

Fons, D., & Sánchez, S. (2004). *Levantamiento GPS de las dunas de la Punta (El Saler).* Proyecto Final de Carrera, ETSI Geodésica, Cartográfica y Topográfica, Valencia, 141 pp.

Fornós, J. J., Clemmensen, L. B., Gómez-Pujol, I., & Murray, A. S. (2009). Late Pleistocene carbonate aeolianites on Mallorca, Western Mediterranean: A luminescence chronology. *Quaternary Science Reviews, 28,* 2697–2709.

Gironés Sempere, M. J., López Galera, N. (1997). *Estudio evolutivo del campo de dunas del Saler (zona de la Punta) entre los años 1980 y 1994.* Proyecto Final de Carrera. ETSI Geodésica, Cartográfica y Topográfica, Valencia, 2 vol.

González, V., de Felipe, G., Guara, M., Mira, A., Morell, I., Navarro, V., et al. (1981). *Estudios Previos para la Ordenación del Monte de la Devesa del Saler* (p. 93). Valencia: Ayuntamiento de Valencia.

Gozálvez, V. (1985). Precisiones sobre los depósitos cuaternarios en la antigua albufera del Saladar de Alicante. In *Pleistoceno y Geomorfología Litoral. Homenaje a Juan Cuerda.* Valencia: Universidad de Valencia.

Gozálvez, V., & Rosselló, V. M. (1978). La Serra del Molar y sus yacimientos pleistocenos. *Cuadernos de Geografía, 23,* 107–126.

Gracia Martínez, G. (2008). *Evaluación de las nuevas fuentes de información geográfica para la realización de estudios de evolución costera y dunar (Propuesta para el empleo de datos*

LIDAR). Proyecto Final de Carrera. ETSI Geodésica, Cartográfica y Topográfica, Valencia, 151 pp.

Mateu, J. F., & Cuerda, J. (1978). Morfología cuaternaria de la costa entre l'Altet y el Cap de Santa Pola. *Cuadernos de Geografía, 23,* 63–82.

Nordstrom, K. L., Psuty, N., & Carter, B. (1990). *Coastal dunes* (p. 392). Chichester: Wiley.

Pardo, J. E., Palomar, J. M., Sanjaume, E., & Recio, J. (2005b). Morphometrical analysis of the geomorphological changes of the Devesa del Saler disturbed dune field (Valencia, Spain): 1965–2001. In *Abstracts vol. 6th International Conference onGeomorphology inregions of environmental contrasts*, Zaragoza.

Pardo, J. E., Ruíz, L. A., Palomar, J. M., Calaf, X., Colmenar, G. R., Almonacid, J., & Gracia, G. (2008). Teledetección, GPS y Lidar: Nuevas técnicas de análisis y evolución de la línea de costa y de los espacios playa duna. In J. E. Pardo, L. A. Ruíz (Eds.), *Actas de las Jornadas Técnicas: Las nuevas técnicas de Información Geográfica al servicio de la gestión de zonas costeras: análisis de la evolución de playas y dunas.* Valencia: Universidad Politécnica de Valencia, Publicación electrónica. ISBN 978-84-8363-336-6.

Pardo, J. E., Sanjaume, E., Gironés, M. J., & López, M. N. (2005a). Análisis del proceso de regeneración natural de las dunas de la Punta del Perellonet entre 1973 y 2001. In E. Sanjaume & J. F. Mateu (Eds.), *Geomorfologia Litoral i Quaternari. Homenatge A V.M. Rosselló I Verger.* Valencia: Servei de Publicacions de la Universitat de València.

Pardo, J. E., Segura, F., & Sanjaume, E. (1996). Evolución cuaternaria de la antigua albufera existente entre Puçol y Alboraia. *Cuadernos de Geografía, 86,* 63–86.

Riquelme, J. (2005). Las eolianitas oleistocenas entre entre la Punta de Moraira y la Pnya d'Ifac (litoral valenciano). In E. Sanjaume & J. F. Mateu (Eds.), *Geomorfologia Litoral i Quaternari. Homenatge A V.M. Rosselló I Verger.* Valencia: Servei de Publicacions de la Universitat de València.

Riquelme, J., & Blazquez, A. M. (2001). Caracterización de las eolianitas cuaternarias entre Cap Blanc y el Panyal d'Ifac (litoral valenciano). *Cuadernos de Geografía, 69,* 164–186.

Rosselló, V. M. (1979). Una duna fósil Pleistocena en la restinga de la Albufera de Valencia. *Cuadernos de Geografía, 25,* 111–126.

Rosselló, V. M., & Mateu, J. F. (1978). El litoral cuaternario de Santa Pola, consideraciones generales. *Cuadernos de Geografía, 23,* 1–18.

Ruíz, J. M., & Carmona, P. (2005). La llanura deltaica de los ríos Júcar y Turia y la Albufera de València. In E. Sanjaume & J. F. Mateu (Eds.), *Geomorfologia Litoral i Quaternari. Homenatge A V.M. Rosselló I Verger.* Valencia: Servei de Publicacions de la Universitat de València.

Sanchis Ibor, C., & Vizcaíno, A. (2016). Las comunidades vegetales de l'Albufera de València. In J. F. Mateu (Ed.), *El territorio valenciano.* Transformaciones ambientáles y antrópicas.

Sanjaume, E. (1974). El cordón litoral de la Albufera de Valencia: Estudio sedimentológico.). *Cuadernos de Geografía, 14,* 61–96.

Sanjaume, E. (1979). Karst marino en el litoral del norte de Alacant. In *VI Coloquio de Geografía.* Palma de Mallorca.

Sanjaume, E. (1983). Aspectos sedimentológicos de las dunas pleistocenas de la costa alicantina. In A. Obrador (Ed.), *Actas X Congreso Nacional de Sedimentología. Menorca'83.* Mahón: GES.

Sanjaume, E. (1985a). *Las costas Valencianas. Sedimentología y Morfología Litoral.* Valencia: Servei Publicacions Universitat de València, 505 pp.

Sanjaume, E. (1985b). La calcoarenita de la restinga dels Estanys d'Almenara. In *Pleistoceno y Geomorfología Litoral. Homenaje a Juan Cuerda.* Valencia: Universidad de Valencia.

Sanjaume, E. (1988). The dunes of Saler, Valencia, Spain. *Journal of Coastal Research SI, 3,* 63–69.

Sanjaume, C., & Carmona, P. (1995). L'Albufera de Valencia: rasgos geomorfológicos y evolución cuaternaria. In *El Cuaternario del País Valenciano.* Valencia: Universitat de València.

Sanjaume, E., & Gozálvez, V. (1978). L'Albufera d'Elx y su litoral. *Cuadernos de Geografía, 23*, 83–106.
Sanjaume, E., Gracia, F. J., & Flor, G. (2011). Introducción a la geomorfología de sistemas dunares. In E. Sanjaume & F. J. Gracia (Eds.), *Las dunas de España*. Cádiz: Sociedad Española de Geomorfología.
Sanjaume, E., & Pardo, J. E. (1991a). Dune Regeneration on a Previously Destroyed Dune Field, Devesa del Saler, Valencia, Spain. *Zeitschrift für Gepmorfologie, 81*, 125–134.
Sanjaume, E., & Pardo, J. E. (1991b). The possible influence of sea-level rise on the precarious dunes of Devesa del Saler Beach, Valencia, Spain. *Landscape Ecology, 6*(2), 57–64.
Sanjaume, E., & Pardo, J. E. (1992). The dunes of the Valencian Coast (Spain): Past and present. In R. W. G. Carter, T. G. E. Curtis, & M. J. Dheehy-Skeffington (Eds.), *Coastal Dunes*. Rotterdam: Balkema.
Sanjaume, E., & Pardo, J. E. (2003). Características sedimentológicas y morfológicas de los espacios costeros de transición situados al sur de la desembocadura del Xúquer. *Cuadernos de Geografía, 73*(74), 183–206.
Sanjaume, E., & Pardo, J. E. (2005). Erosion by human impact on the Valencian coastline. *Journal of Coastal Research, 49*, 76–82.
Sanjaume, E., & Pardo, J. E. (2008). Cambios de tendencias recientes en la evolución costera del golfo de Valencia: análisis espacilaes y sedimentológicos. In J. E. Pardo, L. A. Ruíz (Eds.), *Actas de las Jornadas Técnicas: Las nuevas técnicas de Información Geográfica al servicio de la gestión de zonas costeras: análisis de la evolución de playas y dunas*. Valencia: Universidad Politécnica de Valencia, Publicación electrónica. ISBN 978-84-8363-336-6.
Sanjaume, E., & Pardo-Pascual, J. E. (2011a). Las dunas de las costas valencianas. In E. Sanjaume & F. J. Gracia (Eds.), *Las dunas de España*. Cádiz: Sociedad Española de Geomorfología.
Sanjaume, E., & Pardo-Pascual, J. E. (2011b). Las dunas de la Devesa del Saler. In E. Sanjaume & F. J. Gracia (Eds.), *Las dunas de España*. Cádiz: Sociedad Española de Geomorfología.
Sanjaume, E., & Pardo-Pascual, J. E. (2011c). Degradación de sistemas dunares. In E. Sanjaume & J. F. Gracia (Eds.), *Las dunas de España*. Sociedad Espanola de Geomorfología. Cádiz.
Segura, F., Pardo, J. E., & Sanjaume, E. (1990). Procesos y formas en una restinga en retroceso: el caso de la albufera de Torreblanca. In *Actas 1ª Reunión Nacional de Geomorfología*. Teruel: Instituto de Estudios Turolenses.
Segura, F., Pardo, J. E., & Sanjaume, E. (1997). Evolución cuaternaria de la Albufera de Torreblanca. *Cuaternario y Geomorfología, 11*(1–2), 3–18.
Segura, F., Sanjaume, E., Pardo, J. E., & Riquelme, J. (2005). Canvis del nivel del mar en l'evolució del Prat de Cabanes. In E. Sanjaume & J. F. Mateu (Eds.), *Geomorfologia Litoral i Quaternari. Homenatge A V.M. Rosselló i Verger*. Valencia: Servei de Publicacions de la Universitat de València.
Seva, E., Gutierrez, A., Martín, J., & Escarré, A. (1989). Plantas psamófilas. In A. Escarré, J. Martín, & E. Seva (Eds.), *Estudios sobre el medio y la biocenosis en los arenales costeros de la provincia de Alicante*. Alicante: Instituto de Estusios Alicantinos.

Chapter 28
Dunes in the Gibraltar Strait Realm

F. Javier Gracia, Laura Del Río, María Aranda, Giorgio Anfuso, Lara Talavera, Juan B. Montes and Javier Benavente

28.1 Introduction

The central northern side of the Strait of Gibraltar, greatly controlled by recent faults (Gracia et al. 2005), is constituted by a NE-SW rectilinear portion of 27 km length that extends from Tarifa to the Rock of Gibraltar. The Strait of Gibraltar was formed during the Lower Pliocene after the Messinian Salinity Crisis that almost dried up the Mediterranean Basin, as a consequence of fluvial erosion and tectonic activity that connected the Atlantic and Mediterranean (Blanc 2002). The Strait is an erosive corridor inset over N-S oriented Miocene sandstone units belonging to the Betic Ranges, which are cut forming a rocky coast with steep slopes and cliffs. The coast around the Strait, and especially its Atlantic side (Gulf of Cádiz), presents

F. Javier Gracia (✉) · L. Del Río · M. Aranda · G. Anfuso · L. Talavera
J. B. Montes · J. Benavente
Dpto. Ciencias de la Tierra, Facultad de Ciencias del Mar y Ambientales,
Universidad de Cádiz, 11510 Puerto Real, Spain
e-mail: Javier.gracia@uca.es

L. Del Río
e-mail: laura.delrio@uca.es

M. Aranda
e-mail: maria.aranda@uca.es

G. Anfuso
e-mail: Giorgio.anfuso@uca.es

L. Talavera
e-mail: lara.talavera@uca.es

J. B. Montes
e-mail: juan.montes@uca.es

J. Benavente
e-mail: Javier.benavente@uca.es

J. A. Morales (ed.), *The Spanish Coastal Systems*,
https://doi.org/10.1007/978-3-319-93169-2_28

an outstanding development of coastal dune systems, associated with favorable environmental conditions existing in the zone: strong winds, wide beaches with fine sands fed by important rivers, prevalence of medium to low energy waves, a significant tidal range (around 3 m), and a moderate to disperse human occupation of the coastal fringe. Tidal range progressively diminishes towards the Strait (around 1 m) to reach typical microtidal values (under 0.5 m) in the Mediterranean side, where beaches are much narrower and have coarser sediments (Gracia et al. 2011). On January 2003 all this coastal zone was declared as a protected area by the Andalusian regional government, with the figure of Natural Park.

Coastal relieves in the zone are formed by low mountains of the Betic Ranges and the Gibraltar Tectonic Arc, reaching the coast and forming low to medium escarpments (mostly lower than 10 m). While the Atlantic shoreline side shows an important structural control following a NW-SE direction, differential movements between tectonic blocks near the Strait produced a series of small structurally-controlled embayments generated by differential erosion between Miocene sandstones and Cretaceous marls and clays (Barbate, Bolonia, Valdevaqueros, Algeciras; Fig. 28.1). In the Mediterranean side a long and rectilinear NNE-SSW coast develops to the South of the Guadiaro River mouth, forming the isthmus of the Gibraltar tombolo, which closes the Algeciras Bay by its Eastern side.

Climate in the zone is of Mediterranean type, although the Atlantic exposure in its Western side increases atmospheric humidity and precipitations, bringing milder temperatures. In any case, the dry, hot season usually lasts longer than the humid, temperate winter. Annual rainfall is about 600 mm (Sánchez 1988) and concentrates mainly in autumn-winter. The hot and dry summer conditions are imposed by the strengthening of the Azores High over central Europe. Under these circumstances, dry and strong winds blowing from the East (*Levante* winds) prevail in the Strait. When such air masses go through the Strait, they are accelerated due to funnel effect, reaching very high speeds, with occasional gusts over 110 km/h. Once in the Atlantic side of the Strait, this wind diverges and widens laterally, blowing along the SW Iberian coast following a SE-NW direction (Fig. 28.1). Levante winds present an annual frequency of 19.6%, much higher during drought periods, and an annual average speed of 27.8 km/h (Sánchez 1988).

During the winter season, the Azores High becomes weaker, allowing low pressure systems and fronts to reach the Southern Iberian Peninsula. Under these circumstances Atlantic humid winds prevail in the Strait (*Poniente* winds), blowing from W and WNW (Fig. 28.2), with a 12.8% occurrence and an average velocity of 15.8 km/h (Sánchez 1988). Levante and Poniente winds represent more than 50% of the total winds recorded in the zone. Other minor winds in the Strait blow from SW and WSW, although most of them, with a Westerly component, are usually associated with breezes or with meso-Atlantic cyclonic events.

Climate and winds in the Strait of Gibraltar are conditioned by the North Atlantic Oscillation (NAO). During years with low or negative values in the NAO index, atmospheric pressure diminishes in the subtropical zone and the Azores High weakens. Under these conditions the Atlantic low pressure systems reach more

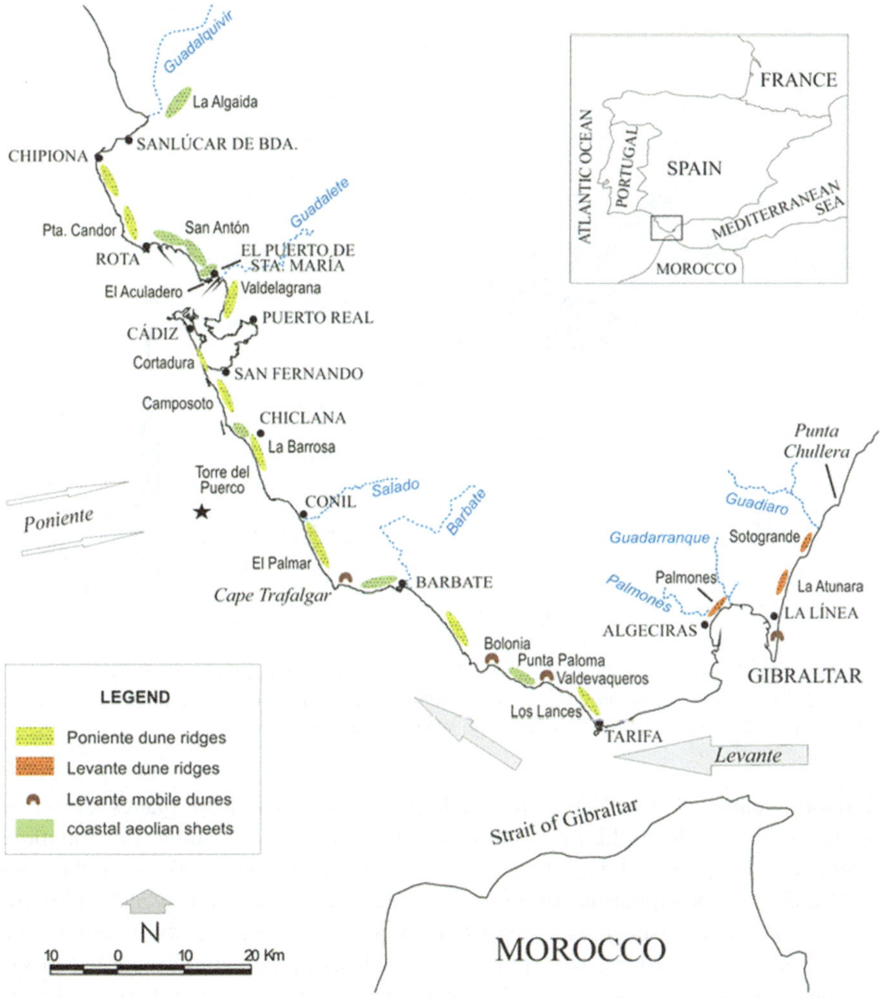

Fig. 28.1 Location of the main dune systems in the Iberian side of the Strait of Gibraltar. The black star indicates the position of the SIMAR point used in Fig. 28.2

easily the Iberian Peninsula and the Gulf of Cádiz, causing intense precipitations and even coastal flooding (Plomaritis et al. 2015). On the contrary, during years with a positive phase of the NAO, atmospheric pressure increases above the Azores islands and decreases over Iceland, so high pressure conditions become stronger in the Gulf of Cádiz and Easterly (Levante) winds prevail in the Strait (Hurrell 1995). This situation sometimes blocks the normal atmospheric circulation in the zone, with strong Easterly winds prevailing for long periods. During the 20th century and especially in the last decades, a certain positive trend is being recorded in the NAO index in the South Iberian Peninsula (Hurrell 1995). In the last decades, the

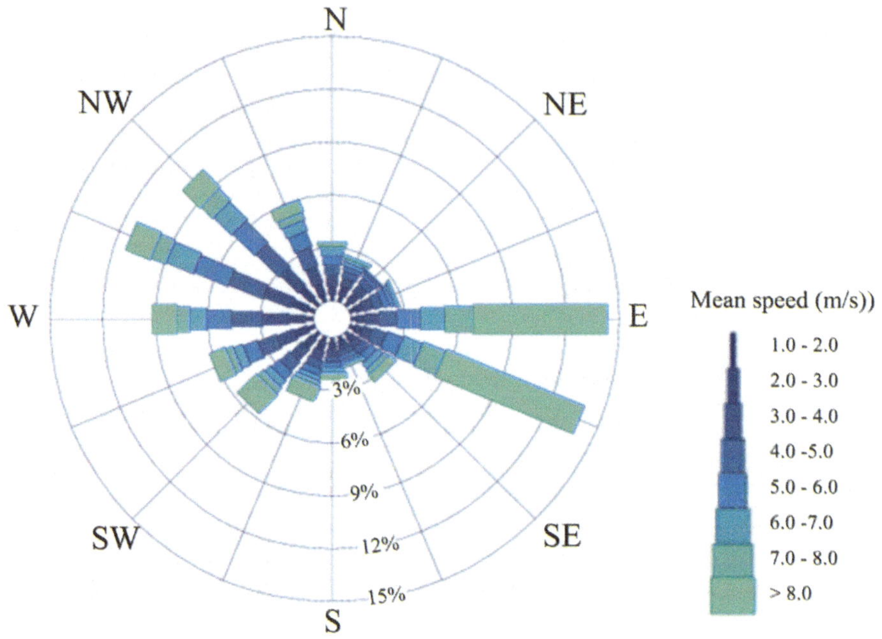

Fig. 28.2 Wind rose obtained in the SIMAR point located about 8 km from the coast (see Fig. 28.1 for location), considering data from 2005 to 2016 with an hourly sampling (Puertos del Estado, Spanish Ministry of the Environment; www.puertos.es)

Northern centre of the NAO (Icelandic low) has moved to places closer to Scandinavia, which could be a cause of the decrease of winter precipitations recorded in this period (Trigo et al. 2004). Goodess and Jones (2002) compared the NAO index and precipitations during winter seasons for the last 40 years and found a good correlation, especially for the Gulf of Cádiz and the nearby Guadalquivir River Basin (Aberg 2005). Millennial-scale analysis of vegetation recorded in salt marshes and lakes in South and Western Spain suggests a trend of northward migration of the westerlies associated with higher persistency of positive NAO index, leading to less winter rainfall (Jiménez-Moreno et al. 2015). A direct consequence of this tendency would be a higher intensity and frequency of Levante winds in the Strait of Gibraltar.

With regards to the dune generation potential, Levante winds are, in theory, the most important ones due to their low humidity and high speed. However, winds only affect directly the Mediterranean side of the Strait, where beaches are composed by coarse sediments supplied by short, mountainous, and highly fluctuating rivers. Such available sediments, together with the microtidal conditions, limit Levante wind action in forming dunes. In the Atlantic side, Levante winds blow roughly parallel to the shoreline, a situation which prevents the generation of big, continuous dune ridges. Only some isolated, high and mobile dunes can be found at the lee side of the small embayments located near the Strait (Fig. 28.1). In contrast,

Poniente winds blow at a roughly perpendicular direction to the shoreline in the Atlantic side, but air masses are very humid and then aeolian transport is much more limited. These Westerly winds generate dune ridges (Fig. 28.1) with a certain lateral continuity, but made of modest dunes due to a much more important vegetal cover and very low to null mobility (Gracia et al. 2006).

The sediment supply necessary for dune development is given mainly by rivers. The Atlantic side of the Strait includes two important estuaries, linked to Guadalquivir and Guadalete rivers, while the Mediterranean side presents one significant fluvial mouth related to the Guadiaro River. Along the very Strait coast, where Betic mountains meet the sea, between Tarifa and Gibraltar, water courses are very small and their sediment supply is almost negligible. Prevailing longshore drift on both sides flows towards the South, favouring the distribution of sediments from the main river mouths, along beaches. However, the hydrological regulation of the main fluvial basins (mainly Guadalquivir, Guadalete, Barbate and Guadiaro rivers) has drastically reduced fluvial sand supply to the coast in the last 50 years, leading to sediment deficit in many beaches and the related negative consequences in the dune systems connected to them (Del Río et al. 2008).

28.2 Origin and Historical Evolution of Coastal Dunes

The oldest aeolian deposits in the Strait of Gibraltar coast date back to Pleistocene times. They appear in the form of poorly cemented eolianites, due to the siliceous nature of the geological substratum, which significantly reduces their preservation potential and makes them highly vulnerable to erosion processes. Apart from the well-known Late Pleistocene dune deposits outcropping at El Aculadero (Puerto de Santa María), dated in 63 ky BP by Santonja and Pérez González (2010), the most important eolianites in the Atlantic side of the Strait appear at Cape Trafalgar. They are composed of about 15 m of quartz-rich sands moderately cemented and well laminated, plenty of rhizoliths. The slope and geometry of laminae (Fig. 28.3a) suggests that the dune complex was originated by Easterly, Levante winds.

Fig. 28.3 Pleistocene eolianites formed by Levante winds. **a** Cape Trafalgar; **b** Gibraltar Rock

The deposit fossilizes a beachrock developed around the present sea level and dated in cal 107 ± 2 ky by Zazo et al. (1999), which indicates that the sedimentary complex was formed during the MIS 5.

On the contrary, the carbonate composition of the Gibraltar Rock has favoured the cementation of a number of Quaternary coastal deposits of different nature, including aeolian sands. Dunes and aeolian deposits in Gibraltar appear both inside coastal caves and forming sand ramps and cliff-front accumulations, always on the eastern side of the Rock, where they locally form sandy cliffs more than 10 m high (Fig. 28.3b). Climbing dunes develop along that rocky face, covering previous forms and deposits. Their sedimentological and palaeoecological characteristics and geochronology were studied in detail by Rodríguez-Vidal et al. (2010) and Cáceres et al. (2013). The main aeolian episodes in the Rock are associated with MIS 5 and 3. The base of the deposit gives an OSL age of 130 ± 15 ky, while the upper levels of the sand cliff date to 95 ± 9 ky. The top of the accumulation connects to an extended sand ramp which enters into Ibex Cave, a rock shelter located at 300 m a.s.l., whose deposits correspond to MIS 3.

The majority of the original aeolian deposits in Gibraltar have been eroded due to a prolonged coastal retreat of the eastern side of the Rock during the Late Quaternary. Cáceres et al. (2013) and Rodríguez-Vidal et al. (2013) reconstructed the environmental conditions which made possible the development of this important aeolian system. According to these authors, between 125 and 30 ky BP, during sea-level fall and lowstand, the Rock of Gibraltar was bounded on its eastern and SE side by a wide shore platform, of up to 5 km wide at times, covered by dunes and associated seasonal wetlands. The landscape would have been composed by an open savannah-type parkland with a mosaic of shrubs where Levante winds piled up aeolian sands against the rocky cliffs. The post-glacial sea level rise drowned this platform, which at present lies about 20 m under sea level.

After the rapid post-glacial sea level rise, during Holocene and historical times several episodes of aeolian deposition have been recognized along this coast, very probably related to climatic and eustatic fluctuations. Sea-level maximum, about 2–3 m above the present sea level, was achieved in the region between 5300 and 6500 year BP (Zazo et al. 1994; Gracia and Benavente 2000). From then on, different spit-barrier systems enclosed estuaries, while beach-ridge systems were developed in open coast beaches. The evolution of all those sedimentary systems and their eustatic signification was analysed by different authors, a synthesis of which can be found in Dabrio et al. (2000). Associated with this evolution, aeolian systems recorded different episodes of expansion and retraction. Borja et al. (1999) made a synthesis of the main historical aeolian phases recorded in the Gulf of Cádiz and defined three main systems: D1 (2700–2000 cal BP), D2 (14th to 17th centuries AD) and D3 (17th century).

After the 8th century B.C. many Phoenician settlements were abandoned and subsequently covered by thick aeolian deposits, possibly triggered by deforestation practices for cereal harvesting. Some poorly cemented grey inactive dunes, fixed by vegetation, outcropping in the Bay of Cádiz (east of San Fernando, limiting the inner side of salt marshes) and to the South (between La Barrosa and Torre del Puerco beaches) could be ascribed to this aeolian episode.

A slight climatic transition to more temperate and humid conditions occurred in the region around the 5th century AD, favouring the development of organic soils upon previous aeolian deposits. This soil level is coeval with new human settlements, even on open areas, which lasted the whole Roman period (between 2000 and 700 cal. BP; Borja et al. 1999). Later on, by the end of the Medieval period, this coastal region was once more affected by the development of wide aeolian sheets generated by Levante winds. Many Roman and Early Middle Age settlements placed on the western sides of the bays were covered by aeolian sediments (Alonso et al. 2004). This aeolian phase probably lasted until the 17th century. During the Little Ice Age (17th to 19th centuries) humid Atlantic winds (Poniente) prevailed, which together with higher precipitations and temperature decrease, favoured a new phase of pedogenesis. Finally, during the 20th century the new climatic trends towards warming brought about a renewed prevalence of Levante winds, which reactivated mobile dunes all along this coast, a situation which is still active at present. The alternation between humid periods, responsible for the edaphic development and human settlement, and dry periods with Levante winds and aeolian deposition, can be recognized at different stratigraphic sections along the Atlantic side of the Strait of Gibraltar, like the one illustrated by Gracia et al. (2006, 2011) in a quarry made on the old town of Cádiz city.

28.3 Present Day Distribution, Morphology and Dynamics of Dune Systems

Taking into account the dynamic and morphological characteristics of the coast around the Strait, three main sectors can be distinguished: Atlantic-N, Atlantic-S and Mediterranean. The first one includes the low coast extending from the Guadalquivir River mouth to Cape Trafalgar (Fig. 28.1). This NW-SE oriented coast is exposed to Poniente humid winds and is characterized by wide and long sandy beaches, fed by the main fluvial channels arriving to the coast in this zone, Guadalquivir and Guadalete rivers. The important sediment supply has favoured the development of wide dune systems all along this coast, most of them artificially fixed by pine trees: La Algaida, Punta Candor, San Antón, Valdelagrana, Cortadura, La Barrosa and El Palmar. The prevailing orientation of this coast facilitates the development of longitudinal dune ridges, mostly covered by vegetation (Fig. 28.4a).

The coast between Chipiona and Rota presents interesting examples of dunes, some of them reaching more than 10 m high, forming ridges with notable continuity. The artificial regulation of fluvial discharge in the Guadalquivir River basin during the last decades has produced an important reduction in the sediment supply to this coastal stretch, leading to erosion (Del Río et al. 2002, 2013). Negative

Fig. 28.4 Some examples of coastal dunes near the Strait of Gibraltar. **a** Vegetated dunes generated by Poniente winds near Tarifa; **b** vertical escarpments on dunes near Rota village; **c** mobile dune generated by Levante winds south of San Fernando; **d** big mobile "Levante" dune advancing upon a pine forest at Valdevaqueros Bay

sediment budget in many of these beaches makes energetic storm waves reach the dune base, excavating outstanding vertical escarpments (Fig. 28.4b). Punta Candor, near Rota village, presents one of the most important and interesting dune systems of this coast, coexisting both mobile dunes (advancing at about 1 m/year in average) and well preserved fixed aeolian deposits; it constitutes a protected area due to its natural values, mainly vegetation cover.

To the South the aeolian system continues until the Guadalete River mouth (Fig. 28.1). Some historical settlements related to aeolian sediments indicate a complex relationship between human occupation and dune dynamics during the Late Holocene, with short periods of intense aeolian activity, lasting between 200 and 400 years, alternating with other longer periods of edaphic stability and vegetation expansion (Borja et al. 1997).

Inside the Bay of Cádiz, low although continuous dune ridges generated by Poniente winds can be found on the northern embayment, at Valdelagrana beach, and in the outer, barrier between Cádiz and San Fernando cities (see Fig. 28.1 for location). Valdelagrana beach-barrier and dune complex, with a total length of

7.2 km and an average width of 1.5 km, is one of the most outstanding sedimentary coastal systems in the Gulf of Cádiz. It is composed by 20 different prograding episodes, the oldest one dated in 3500 year BP (Zazo et al. 1994). The ulterior morphological evolution is a consequence of complex relationships among sea level fluctuations, climate changes, Guadalete River sedimentary supply, very energetic-catastrophic historical events (storms and tsunamis) and human interventions (Alonso et al. 2014; Del Río et al. 2002, 2015a). At present the system consists of a northern urbanized sector, almost 2 km long, and a southern natural one, protected under the figure of Natural Park. This beach-barrier is very sensitive to human activities (Rizzo et al. 2017), especially those related to the geometry and location of the Guadalete River mouth, and several episodes of intense erosion and flooding have been recorded over the last decades, leading to the progressive erosion and retreat of the foredune ridges (Benavente et al. 2006; Rodríguez Polo et al. 2010).

The second sector, South of Cape Trafalgar, is characterized by several bays and embayments, broadly E-W oriented, and more exposed to the strong Levante winds. As mentioned above, these winds mainly blow with a SE-NW direction (Fig. 28.2), producing aeolian transport of sand along the beaches of the bays and deposition in the lee (western) sides. As a consequence, important dunes form in all those favorable sites, a process also present at given points of the Bay of Cádiz (Fig. 28.4c). High mobile dunes linked to Levante winds develop at Cape Trafalgar, Bolonia Bay and Valdevaqueros embayment (Fig. 28.1). The dimensions and height of the dunes increase from W to E, parallel to the increasing intensity of easterly winds. Gibraltar Rock lacks this type of active dune due to the present-day absence of sandy beaches in its Mediterranean side. The most outstanding dune of this south-Atlantic sector, and one of the biggest dunes in the Iberian Peninsula, is Valdevaqueros dune complex (Fig. 28.4d). It reaches more than 35 m in height and presents a high mobility, leading to periodical interruptions of a nearby road (Gracia et al. 2013).

Bolonia and Valdevaqueros dunes are located on the Punta Paloma aeolian complex. Historically, this zone was occupied by a sand corridor with transversal and climbing dunes which connected both the Bolonia and Valdevaqueros embayments (Gracia et al. 2013). The width of this second embayment is of approximately 2 km, and the Valle River mouth is located in the centre of the bay. This river was an open estuary in the beginning of the past century, but nowadays is a small fluvial channel with reduced water discharge disconnected from the open sea by a closing sand barrier. During the mid-past century, the construction of different military infrastructures in Punta Paloma area, along with their associated access roads, derived in the implementation of coastal actions that aimed to reduce the dune growth and protect these facilities. Thus, artificial alongshore sandy mounds acting as natural trenches were built. However, opposite to what was expected, the strong Levante winds mobilized the sediment into the military zone. In addition, due to the wave and longshore drift action at this area, the sandy mounds located at the western part of the embayment notably increased the pattern of sediment transport, causing the progressive filling of sediment into the Valle

River mouth and its final closure. The resulting sand barrier has gradually increased the supply of sand that arrives at Valdevaqueros dune in the last decades, forced by Levante winds (Gómez-Pina et al. 2003; Bello-Millán et al. 2016).

In recent years, Valdevaqueros dune has been subject of different studies (Muñoz-Pérez et al. 2009; Navarro et al. 2011, 2013, 2015). The dynamics of this dune has several components. On one hand, the Westerly swell waves and long-shore drift supply the sediment necessary to maintain the Valdevaqueros beach in equilibrium, and the rising wind erodes and transports the sand present on the dry beach towards the base of the dune and upwards (Bello-Millán et al. 2016). On the other hand, the easterly winds are intensified in this area by the funnel effect of the Strait, causing the dune landward migration at a rate of 17.5 m/year, and sand accretion at an annual rate of up to 140 m^3/m. As a result, Valdevaqueros dune grows and moves westwards, with ridges oriented perpendicular to the dominant windstream (Muñoz-Pérez et al. 2009).

Close to the Strait of Gibraltar a very interesting dune system appears at Los Lances Bay, near Tarifa village. It forms a set of historical parallel ridges hosting a complete sequence of typical coastal dunes developed on a prograding coast; from sea to land: embryo dune patches, foredune ridge, dune slacks, low dunes covered with grass, and finally relict dunes covered by a pine forest. Relationships between relict ridges and archaeological remnants made Gracia et al. (2004) consider a probable genesis during Roman times, with an important progradation phase during the Middle Ages.

Finally, the Mediterranean coastal side of the Strait ends at Punta Chullera, following a NNE-SSW direction. This section is totally exposed to Levante winds and this, in combination with the high urban pressures and the limited sediment supply, condition the natural dynamics in this area. Two different dune systems can be distinguished in this sector, both with a similar orientation. The first one is located inside the Algeciras Bay, and the other one between La Atunara and Sotogrande beaches. The system located inside the bay (Palmones dunes) consists of fully vegetated, continuous, and longitudinal dunes of 3 m height, reaching 5 m at some points. During the most energetic storm conditions, waves reach the dune foot and generate escarpments. The second system, open to the Mediterranean, is characterized by continuous dune ridges sparsely vegetated and low dune crests that typically reach 2 m high.

28.4 Dune Degradation and Management

Coastal dunes play an important role in the sedimentary stability of beaches, constituting a natural barrier against coastal storms. In addition, their ecological and landscape values make them attractive ecosystems from the educational and recreational perspectives. As a consequence, dune management and conservation is a major environmental priority, both from the national and international points of view (Bonnet 1989). However, as in other coastal zones in the world (Hoozemans et al. 1993), many dunes around the Strait of Gibraltar coast are seriously threatened due to three different factors: the narrow strip of land they occupy, which makes them highly vulnerable; the anthropogenic pressures to which they are subject, including increasing human settlements; and the associated changes in land uses. From the 7800 km of Spanish coastlines, dunes occupy around 40%, and approximately only 45% of the dune systems remain natural (Ley et al. 2007). Dunes are also especially vulnerable to consequences derived from sea level rise and climate change, a situation which is expected to worsen due to the ongoing human action (De Winter and Ruessink 2017).

The Spanish Law 42/2007 on Natural Heritage and Biodiversity establishes the legal framework for conservation, sustainable use, enhancement and restoration of the national natural heritage, combining the preservation of biodiversity and geodiversity, the latter including geological Spanish contexts of global importance like the Iberian Peninsula coastal lowlands (Gracia 2008). At the same time, the European Council Directive 92/43/EEC on the conservation of natural habitats constitutes the main reference of Spanish nature conservation, as it ensures the preservation of a wide range of rare, threatened or endemic animal and plant species. The directive includes 116 different EU habitats of interest, from which 11 are dune systems (García de Lomas et al. 2011). Coastal dunes of the Atlantic side of the Strait of Gibraltar are amongst the most complete dune systems existing in Spain. This way, dune systems such as Punta Candor, San Antón, Valdelagrana or El Palmar include a number of EU protected habitats like the following (Table 28.1):

Table 28.1 Main EU-protected habitats present on the coastal dunes of the Strait of Gibraltar. Modified from García de Lomas et al. (2011)

Dune system	2110	2120	2130	2150	2190	2230	2250	2260	2270
Chipiona-Punta Candor	*	*			*		*		*
San Antón-Valdelagrana	*	*	*						*
Camposoto	*	*			*				*
La Barrosa-Torre del Puerco	*	*		*			*	*	*
El Palmar	*	*							
Trafalgar	*	*	*		*	*	*	*	*
Zahara	*	*			*		*		*
Bolonia	*	*					*		*
Valdevaqueros	*	*	*		*	*	*	*	*
Los Lances	*	*	*		*	*	*	*	*
Sotogrande-La Atunara	*	*		*		*		*	
Palmones	*	*							

2110 Embryonic shifting dunes
2120 Shifting dunes along the shoreline with *Ammophila arenaria* (white dunes, foredune)
2130 Fixed dunes with herbaceous vegetation (grey dunes)
2150 Eu-Atlantic decalcified fixed dunes (*Calluno-Ulicetea*)
2190 Humid dune slacks
2230 *Malcomietalia* dune grasslands
2250 Dune juniper thickets (*Juniperus* spp.)
2260 Dune scleorophyllous scrubs (*Cisto-Lavenduletalia*)
2270 Wooded dunes with *Pinus pinea* and/or *Pinus pinaster*

Morphological and ecological alterations on dune systems lead to dune weakening and disappearance, not only interpreted as a loss of habitat and support for organisms, but also as a loss of natural protection against forcing agents such as energetic waves during storms, which produce damage to human properties and infrastructure. For this reason, the main altering threats have to be diagnosed in order to diminish or stop disturbances. There is an important set of pressures/hazards capable of altering dune structure and functions and, in consequence, their area (Nordstrom 2008; Lithgow et al. 2013).

Among the natural processes affecting sediment equilibrium on coastal systems, climate variability is one of the most important one, since changes in temperature or precipitation lead to changes in the weathering processes and rates, which affect the production of sediments and their transport rates. The stability and growth of beaches in the Strait of Gibraltar are strongly dependent on the sediment yield produced by physical-chemical weathering, slope and fluvial processes on the surrounding sandstone relieves (Gracia et al. 2008). On the other hand, climatic changes also determine sea level oscillations that can completely eliminate dune systems in the long term, partially modify them or, on the contrary, create new ones. Certain types of climatic changes can increase the frequency and magnitude of marine storms. In these cases, foredunes are the most exposed areas to suffer erosion, together with the inner dune areas, if erosive trends progress. Specifically, the Cadiz coast is directly exposed to strong winter storms acting on the beach-dune systems, leading to berm-dune erosion and overwash (Rangel-Buitrago and Anfuso 2011). The most exposed coastal zones in the central and northern sectors of the Atlantic side of the Strait are affected by persistent shoreline retreat (Del Río et al. 2013, 2015b), especially on places where bathymetry favours the concentration of wave energy producing intense erosion during storm episodes. Overwash processes cause dune ridge breaching (Fig. 28.5a). Multiple overwash events, possibly related to edge wave action during storms, can give rise to multiple fragmentation, ultimately leading to the generation of flame structures (Fig. 28.5b), a morphology indicative of severe dune erosion (Carter 1988).

At present, a vast number of human activities modify the coastal sediment supply and, therefore, have an impact on dune systems. Many beaches and dunes of this coast record medium to long-term erosion, with retreat values of up to 1 m/year at some places, like north of Punta Candor dune field (Domínguez et al. 2005; Del

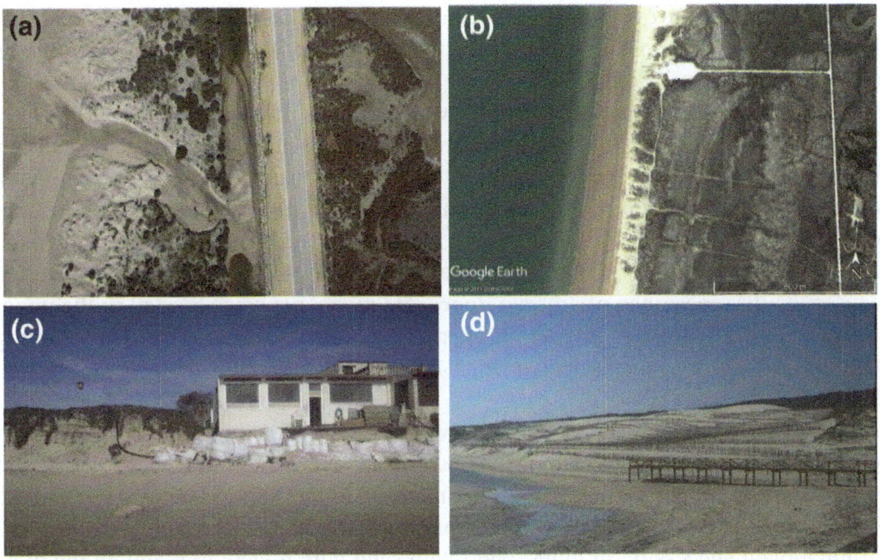

Fig. 28.5 Examples of dune degradation in the Cádiz coast. **a** Vertical aerial view of a washover structure (channel and fan) cutting dune ridges at Camposoto Beach, south of San Fernando (sea to the left). Vertical aerial photo taken with an UAV (Unmanned Aerial Vehicle) in February 2017. **b** Flame structures as a consequence of intense erosion of dune ridges at Valdelagrana beach-barrier (photo taken in June 2010). **c** Invasive species (*Carpobrotus* spp.) and urban structures on Palmones dunes (Algeciras) affected by erosion. **d** Stoss side of Valdevaqueros dune covered by osier fences

Río et al. 2013, 2015b). On the other hand, coastal engineering structures normal to the shoreline interrupt the prevailing longshore transport, promoting beach and dune erosion in downdrift areas. That is the case of Valdelagrana beach-barrier in the Atlantic side (Fig. 28.5b), or Sotogrande beach and dunes in the Mediterranean side of the Strait.

Another factor to consider is the historically small or absent social awareness regarding coastal dunes. Traditionally, dunes were considered as unproductive spaces or unhealthy environments, and frequently their elimination was justified on these grounds. The agricultural activities occupied the internal strips of the dune systems and even the foredune ridges in some cases, like at El Palmar or Bolonia beaches. Sand extraction for constructive purposes or for the elaboration of glass was a historical practice at the San Antón dunes (see location in Fig. 28.1). In the last decades, however, the greatest impacts on dune systems have been linked to urban development (Fig. 28.5c), which introduces rigid elements within the zone of sedimentary exchange between the beach and the dunes, causes strong destabilizations of the dune dynamics, increases fragmentation or directly destroys the foredune ridges. The presently applicable Coastal Law in Spain (Shores Act, 1988 modified in 2013) specifically prohibits the modification or destruction of dunes, whatever the purpose.

The introduction of exotic plants causes the displacement of native taxa and transforms the composition and structure of the existing dune plant formations (Fig. 28.5c). In many cases, the changes are not only qualitative (variation of species) but also quantitative (density of vegetation). Similarly, the vegetation cover is substantially reduced as a consequence of the indiscriminate access of walkers, bikers, equestrians, and even motorized vehicles. In these circumstances, significant changes in the evolution of the dune dynamics occur, for instance the mobilization of a greater volume of sand that triggers the genesis and development of blowouts. During the tourist season the mechanical cleaning of beaches and dunes often causes the disappearance of embryo dunes and pioneer species located in the upper beach. To mitigate this impact, it is recommended to apply manual cleaning methods which are more selective than mechanical ones (García de Lomas et al. 2011).

The main consequences of these activities are the breaking of plant continuity, changes in dune morphodynamic characteristics (height, slope, width, etc.), and sand compaction. Moreover, there is an increase in the volume of waste generated by humans. In such cases it is necessary to implement measures to restore the original plant succession in dunes and retain sand with a view to maintain the dune geometry (Hesp 2002).

Plant restoration needs to eliminate the exotic species threating the wildlife and morphosedimentary equilibrium of the dune. Perhaps the most problematic exotic species in the coastal dunes of the Strait of Gibraltar is the ice plant (*Carpobrotus* spp.), very widespread along the Iberian coast. Between 2008 and 2013, the Environment Department of the regional Andalusian government developed a management plan in the Tarifa Island in order to eliminate this exotic species, since it was displacing the endemic *Limonium emarginatum*, declared as highly vulnerable. The works consisted in the removal of 556 Tons of the exotic plant and the recovery of circa 2 Ha of previously invaded area (García de Lomas et al. 2014).

In very dynamic dunes related to Levante winds, where human activities or infrastructures are threatened by sand burial, interventions are focused on slowing down or directly stopping dune mobility. Attempts to fix very active dunes include a number of varied, and not always successful, methods. In recent years the Spanish Ministry of Environment carried out several measures for sand dune stabilization in Valdevaqueros dune due to its intense mobility: deployment of physical structures (e.g. wooden and osier fences along the dune stoss side, Fig. 28.5d), planting of vegetation to retain sand, or even the mechanical extraction of the sand that frequently invades the nearby road (Gracia et al. 2013). Nevertheless, these actions are not sufficient to cope with the management issues of the dune. Plantations are not effective or at least not in the most mobile fringes, and the entire system of wooden fences is often easily and rapidly covered by sand due to the intense aeolian activity in the area.

Dune reforestation was a common measure in the last decades of the 20th century, especially on mobile dune systems (Barbate, Bolonia-Valdevaqueros, or Los Lances) and on those systems close to densely populated areas, for instance Rota and Puerto de Santa María (Fig. 28.6). Pine trees were used for this purpose,

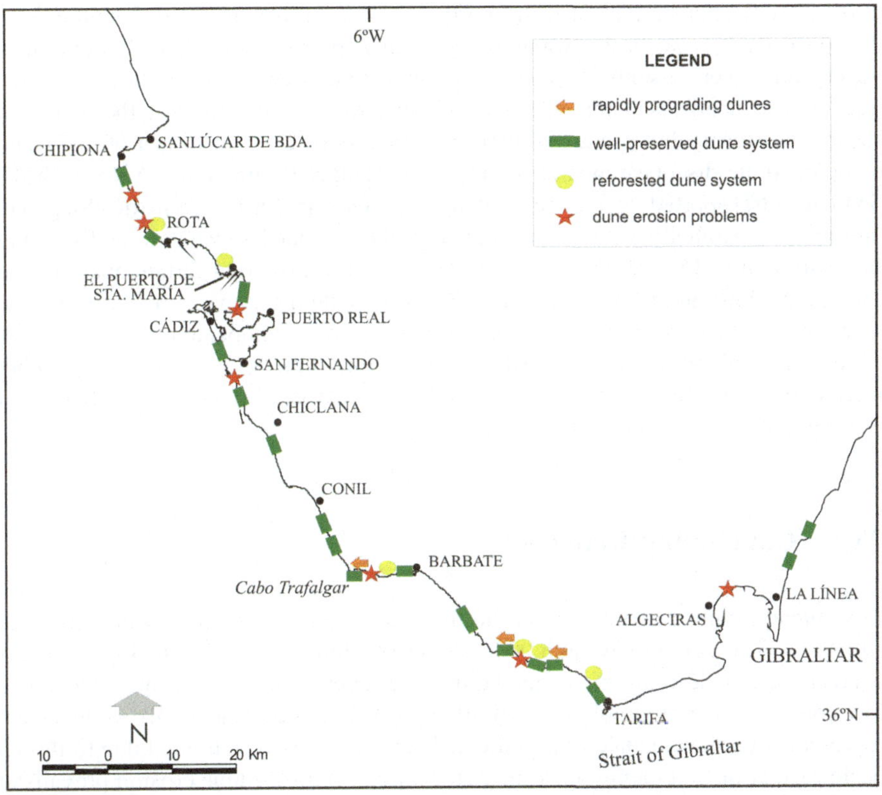

Fig. 28.6 Present environmental state of the coastal dunes around the Strait of Gibraltar in its Iberian side

in order to maintain the natural plant succession typical of these Mediterranean habitats.

In other cases, dune fixation aims to preserve a certain sand volume enough to facilitate plant colonization and habitat restoration. Such measures are applied to eroding dunes, especially at places where dunes are inserted within an ecological space of interest (natural parks or other protected coastal areas). At Valdelagrana beach-barrier, the central dunes recorded erosion and retreat during the last decades (Rodríguez Polo et al. 2010). The main problems are related to human impacts because beachgoers tend to walk or sunbath over the dunes, especially in the northernmost area, where dune ridges present a greater height. Vegetation is essentially composed by marram grass and, secondarily, by different invasive (i.e. broom, "Cytisus White Lion") and native Mediterranean bushes.

In 2007 the Andalusian region started a plan for active coastal management which included the recovery of coastal dunes in zones subject to severe erosion. High fences were emplaced in the central Valdelagrana beach to prevent beach users from accessing dunes. The program succeeded until heavy storms during the

2009–2010 winter partially (in the northern) and completely (in the central part) destroyed the fences. In the Southern part, dunes present lower heights but denser vegetation cover, essentially composed by native Mediterranean bushes in the transition to nearby salt marshes. Specifically, dunes failed against the action of successive storms that produced dune foot retreats of c. 25 m in just 60 days and complete dune disappearance at some points (Rangel-Buitrago and Anfuso 2011). Sand was transported several tens of meters landward in the form of elongated, narrow, and parallel washover fans covering the saltmarsh vegetation and forming flame structures (Fig. 28.5b). Strong easterly winds transported certain quantities of sand to the dune and formed small foredunes. Despite the above, fences are still the best solution to restore dunes, especially if they are accompanied by beach nourishment for enlarging the dry beach and protecting dunes (Ley et al. 2007). Other interventions in the coastal dunes around the Strait consist in the installation of wooden pathways/wooden piles.

28.5 Concluding Remarks

In summary, dune systems around the Strait of Gibraltar present a wide variety of situations. The northern Atlantic side, between Chipiona and Conil (Fig. 28.6), is characterized by low and fragmented dune ridges generated by Poniente winds, on a low coast with dense population and urban pressure. Sediment supply to the coast, represented by the Guadalquivir and Guadalete rivers, is very limited due to fluvial basin regulation and damming. Under such circumstances, dune erosion prevails at many places and restoration measures have been applied to different retreating dune systems (e.g. Rota, Valdelagrana or Conil).

The southern Atlantic side, closer to the Strait, is affected by strong Levante winds, which create very dynamic, mobile, and prograding dunes. Coastal relieves are characterized by hills and low mountains of the Betic Ranges, and less regulated rivers are flowing to the coast, some of them maintaining their sediment supply. Human occupation is low and urban areas are sparse, often dedicated to military purposes and activities which pose a much lesser impact on coastal dunes. As a consequence, environmental health of coastal dunes in this sector is satisfactory (Fig. 28.6), especially those included in the Strait of Gibraltar Natural Park, where more effective protection measures are applied.

Finally, the Mediterranean side is characterized by a dense population and intense human interventions along the coast of the Algeciras Bay, leading to erosion problems in the Palmones dune system (Fig. 28.6). Small isolated dune fields between Gibraltar and Sotogrande hardly survive despite the significant reduction in the sediment supply of the highly regulated Guadiaro River, the coarser nature of beach sediments in this sector or the longshore drift blockage produced by the marina at Sotogrande.

Climate change predictions for the next decades (Jiménez-Moreno et al. 2015) suggest an increase in the frequency and intensity of Levante winds in the Strait.

This situation would favour the reinforcement and growth of the Levante-derived dunes around the Strait, and perhaps recovery of the small dunes existing in the Mediterranean side. Dune dynamics on the highly mobile Trafalgar, Bolonia and Valdevaqueros dunes would surely require more frequent interventions. In contrast, Poniente-derived dune ridges are expected to experience little changes.

References

Aberg, U. (2005). *Recent berm ridge development inside a mesotidal estuary: The Guadalquivir River mouth case* (Unpublished Degree Dissertation). University of Uppsala, 62 pp.

Alonso, C., Gracia, F. J., & Benavente, J. (2004). Las marismas, alfares y salinas como indicadores para la restitución paleotopográfica de la Bahía de Cádiz durante la antigüedad. In *XVI Encuentros de Historia y Arqueología* (pp. 263–287). Ayuntamiento de San Fernando: Exmo.

Alonso, C., Gracia, F. J., & Rodríguez-Polo, S. (2014). Modelo de evolución histórica de la flecha-barrera de Valdelagrana (Bahía de Cádiz). In S. Schnabel & A. Gómez (Eds.), *Avances de la Geomorfología en España 2012–2014* (pp. 584–587). Cáceres: XIII National Geomorphological Meeting.

Bello-Millán, F. J., Somoano, M., Clavero, M., Gómez-Pina, G., & Losada, M. A. (2016). El sistema dunar de Valdevaqueros: evolución histórica y alternativas de gestión. *RIBAGUA - Revista Iberoamericana del Agua, 3,* 46–55.

Benavente, J., Del Río, L., Gracia, F. J., & Martínez, J. A. (2006). Coastal flooding hazard related to storms and coastal evolution in Valdelagrana spit (Cádiz Bay Natural Park, SW Spain). *Continental Shelf Research, 26,* 1061–1076.

Blanc, P. L. (2002). The opening of the Plio-Quaternary Gibraltar Strait: Assessing the size fo a cataclysm. *Geodinamica Acta, 15,* 303–317.

Bonnet, J. (1989). Aspects of conservation and management of the sand-dune areas in Spain. In F. Van der Meulen, et al. (Eds.), *Perspectives in coastal dune management: Towards a dynamic approach* (pp. 269–275). Hague: SPB Academic Publishing.

Borja, F., Gutiérrez, J. M., & Díaz del Olmo, F. (1997). Fases de dunas durante el Holocene Reciente en el litoral de El Puerto de Santa María (Cádiz). *Geogaceta, 21,* 39–42.

Borja, F., Zazo, C., Dabrio, C. J., Díaz del Olmo, F., Goy, J. L., & Lario, J. (1999). Holocene aeolian phases and human settlements along the Atlantic coast of Southern Spain. *The Holocene, 9*(33), 333–339.

Cáceres, L. M., Rodríguez-Vidal, J., Abad, M., Ruiz, F., Finlayson, C., Finlayson, G., et al. (2013). Modelo de construcción de dunas topográficas pleistocenas en la costa oriental de Gibraltar. In R. Baena, et al. (Eds.), *VIII Reunión de Cuaternario Ibérico* (pp. 155–158). La Rinconada, Sevilla: AEQUA.

Carter, R. (1988). *Coastal environments.* Academic Press, 617 pp.

Dabrio, C. J., Zazo, C., Goy, J. L., Sierro, F. J., Borja, F., Lario, J., et al. (2000). Depositional history of estuarine infill during the last postglacial transgression (Gulf of Cadiz, Southern Spain). *Marine Geology, 162,* 381–404.

De Winter, R. C., & Ruessink, B. G. (2017). Sensitivity analysis of climate change impacts on dune erosion: case study for the Dutch Holland coast. *Climatic Change, 141,* 685–701.

Del Río, L., Benavente, J., Gracia, F. J., Alonso, C., & Rodríguez-Polo, S. (2015a). Anthropogenic influence on spit dynamics at various timescales: Case study in the Bay of Cadiz (Spain). In G. Randazzo, et al. (Eds.), *Sand and Gravel Spits* (pp. 123–138). Dordrecht: Springer. (Coastal Research Library 12).

Del Río, L., Benavente, J., Gracia, F. J., Alonso, C., Rodríguez-Vidal, J., & Anfuso, G. (2008). Los espacios protegidos del litoral gaditano. In F. J. Gracia (Ed.), *Geomorfología de los espacios naturales protegidos de la provincia de Cádiz* (pp. 103–195). S.E.G., University of Cádiz.

Del Río, L., Benavente, J., Gracia, F. J., Anfuso, G., Martínez-del-Pozo, J. A., & Domínguez, L., et al. (2002). The quantification of coastal erosion processes in the South Atlantic Spanish coast: Methodology and preliminary results. In *Proceedings of 6th International Conference on LITTORAL'2002*, Eurocoast, Porto, Vol. 2, pp. 383–390.

Del Río, L., Benavente, J., Gracia, F. J., & Chica, J. A. (2015b). Riesgos de erosión en la costa de Cádiz: gestión actual y perspectivas futuras. In G. Malvárez, et al. (Eds.), *VIII Jornadas de Geomorfología Litoral*. Geotemas, Vol. 15, pp. 149–152.

Del Río, L., Gracia, F. J., & Benavente, J. (2013). Shoreline change patterns in sandy coasts. A case study in SW Spain. *Geomorphology, 196*, 252–266.

Domínguez, L., Anfuso, G., & Gracia, F. J. (2005). Vulnerability assessment of a retreating coast in SW Spain. *Environmental Geology, 47*, 1037–1044.

García de Lomas, J., Dana, E. D., Gimeno, D., García-Morilla, J., & Ceballos, G. (2014). Control de uña de león (Carpobrotus spp.; Aizoaceae) en la Isla de Tarifa. *Revista de la Sociedad Gaditana de Historia Natural, 8*, 31–41.

García de Lomas, J., Gracia, F. J., & García, C. M. (2011). Las dunas como hábitats de interés comunitario. Problemas de conservación. In E. Sanjaume & F. J. Gracia (Eds.), *Las dunas en España* (pp. 585–606). Cádiz: Spanish Geomorphological Society.

Gómez-Pina, G., Ramírez, J. L., Muñoz, J. J., & Ley, C. (2003). Valdevaqueros, historia de una duna. In M. A. Losada & M. C. Castillo (Eds.), *VII Jornadas Españolas de Ingeniería de Costas y Puertos* (pp. 1–11). Almería: Fundación para el Fomento de la Ingeniería del Agua.

Goodess, C. M., & Jones, P. D. (2002). Links between circulation and changes in the characteristics of iberian rainfall. *International Journal of Climatology, 22*, 1593–1615.

Gracia, F. J. (2008). Costas bajas de la Península Ibérica. In A. García-Cortés, et al. (Eds.), *Contextos geológicos españoles. Una aproximación al patrimonio geológico español de relevancia internacional* (pp. 192–199). Madrid: Instituto Geológico y Minero de España.

Gracia, F. J., Alonso, C., Martínez-del-Pozo, J. A., Del Río, L., Anfuso, G., & Benavente, J. (2004). Evolución geomorfológica holocena e histórica de la Ensenada de Los Lances (Tarifa, Cádiz). In R. Blanco (Eds.), *Procesos geomorfológicos y evolución costera* (pp. 255–269). University of Santiago de Compostela.

Gracia, F. J., & Benavente, J. (2000). Geomorfología de la costa atlántica gaditana. In J. R. De Andrés & F. J. Gracia (Eds.), *Geomorfología Litoral. Procesos activos* (pp. 235–255). Cádiz: Monogr. S.E.G. 7, ITGE, Serv. Publ. Univ.

Gracia, F. J., Benavente, J., Alonso, C., Del Río, L., Abarca, J. M., García de Lomas, J., & Anfuso, G. (2011). Las dunas del litoral gaditano. In E. Sanjaume & F. J. Gracia (Eds.), *Las dunas en España* (pp. 359–384). Cádiz: Spanish Geomorphological Society.

Gracia, F. J., Del Río, L., Alonso, C., Benavente, J., & Anfuso, G. (2006). Historical evolution and present state of the coastal dune systems in the Atlantic coast of Cádiz (SW Spain): Palaeoclimatic and environmental implications. *Journal of Coastal Research, 48*, 55–63.

Gracia, F. J., Nuño, T., Del Río, L., & Benavente, J. (2013). Valdevaqueros, Cádiz: La difícil gestión de una gran duna móvil. *Enseñanza de las Ciencias de la Tierra, 21*(3), 332–335.

Gracia, F. J., Ojeda, F., & Macías, A. (2008). El Parque Natural de Los Alcornocales. In F. J. Gracia (Ed.), *Geomorfología de los espacios naturales protegidos de la provincia de Cádiz* (pp. 57–82). S.E.G., University of Cádiz.

Gracia, F. J., Rodríguez-Vidal, J., & Cáceres, L. M. (2005). Structural geomorphology and coastal neotectonics of the Gibraltar Strait. In F. J. Gracia (Eds.), *Geomorphology of the South-Atlantic Spanish Coast* (pp. 59–64). Zaragoza: Sixth International Conference on Geomorphology, Field trip guide A-4.

Hesp, P. (2002). Foredunes and blowouts: initiation, geomorphology and dynamics. *Geomorphology, 48*, 245–268.

Hoozemans, F. M. J., Marchand, M., & Pennekamp, H. A. (1993). *A global vulnerability analysis: Vulnerability assessment for population, coastal wetlands and rice production on a global scale* (2nd ed., p. 184). Delft: Delft Hydraulics.

Hurrell, J. W. (1995). Decadal trends in the North Atlantic Oscillation regional temperatures and precipitation. *Science, 269,* 676–679.
Jiménez-Moreno, G., Rodríguez-Ramírez, A., Pérez-Asensio, J. N., Carrión, J. S., López-Sáez, J. A., Villarías-Robles, J. J. R., et al. (2015). Impact of late-Holocene aridification trend, climate variability and geodynamic control on the environment from a coastal area in SW Spain. *The Holocene, 25*(4), 607–617.
Ley, C., Gallego-Frenández, J. B., & Vidal, C. (2007). *Manual de restauración de dunas costeras.* Madrid: Spanish Ministry of Environment. (258 pp).
Lithgow, D., Martínez, M. L., Gallego-Fernández, J. B., Hesp, P. A., Flores, P., Gachuz, S., et al. (2013). Linking restoration ecology with coastal dune restoration. *Geomorphology, 199,* 214–224.
Muñoz-Pérez, J. J., Navarro, M., Román-Sierra, J., Tejedor, B., Rodríguez, I., & Gómez-Pina, G. (2009). Long-term evolution of a transgressive migrating dune using reconstruction of the EOF method. *Geomorphology, 112,* 167–177.
Navarro, M., Muñoz-Pérez, J. J., Román-Sierra, J., Ruiz-Cañavate, A., & Gómez-Pina, G. (2013). Análisis de la capacidad real de transporte eólico en el sistema dunar de Valdevaqueros (Tarifa). *VII Jornadas de Geomorfología Litoral. Geotemas, 14,* 91–94.
Navarro, M., Muñoz-Pérez, J. J., Román-Sierra, J., Ruiz-Cañavate, A., & Gómez-Pina, G. (2015). Characterization of wind-blown sediment transport with height in a highly mobile dune (SW Spain). *Geol Acta, 13*(2), 155–166.
Navarro, M., Muñoz-Pérez, J. J., Román-Sierra, J., Tsoar, H., Rodríguez, I., & Gómez-Pina, G. (2011). Assessment of highly active dune mobility in the medium, short and very short term. *Geomorphology, 129,* 14–28.
Nordstrom, K. F. (2008). *Beach and dune restoration.* Cambridge: Cambridge University Press. (200 pp).
Plomaritis, T. A., Benavente, J., Laiz, I., & Del Río, L. (2015). Variability in storm climate along the Gulf of Cadiz: the role of large scale atmospheric forcing and implications to coastal hazards. *Climate Dynamics, 45,* 2499–2514.
Rangel-Buitrago, N., & Anfuso, G. (2011). Morphological changes at Levante Beach (Cadiz, SW Spain) associated with storm events during the 2009–2010 winter season. *Journal of Coastal Research, 57,* 1886–1890.
Rizzo, A., Aucelli, P. P. C., Gracia, F. J., & Anfuso, G. (2017). A novelty coastal susceptibility assessment method: Application to Valdelagrana area (SW Spain). *Journal of Coastal Conservation.* https://doi.org/10.1007/s11852-017-0552-2.
Rodríguez Polo, S., Gracia, F. J., & Del Río, L. (2010). Retroceso costero en la flecha de Valdelagrana, El Puerto de Santa María (Cádiz). In X. Úbeda, et al. (Eds.), *Avances de la geomorfología en España, 2008–2010* (pp. 75–78). Solsona: XI Nat Geomorph Meeting, University of Lleida, Centre Tecnològic Forestal de Catalunya.
Rodríguez-Vidal, J., Cáceres, L. M., Ruiz, F., Abad, M., Finlayson, C., Finlayson, G., et al. (2010). Las arenas eólicas del Pleistoceno Superior en Gibraltar. *Geogaceta, 48,* 27–30.
Rodríguez-Vidal, J., Finlayson, G., Finlayson, C., Negro, J. J., Cáceres, L. M., Fa, D. A., et al. (2013). Undrowning a lost world—The Marine Isotope Stage 3 landscape of Gibraltar. *Geomorphology, 203,* 105–114.
Sánchez, J. L. (1988). *Estudio climatológico de la Bahía de Cádiz.* Instituto Hidrográfico de la Marina, Cádiz. Spanish Ministery of Defense, Madrid. 150 pp.
Santonja, M., & Pérez González, A. (2010). Precisiones en torno a la edad y la industria lítica de El Aculadero (El Puerto de Santa María). *Cuaternario y Arqueología. Libro homenaje a Francisco Giles Pacheco.* Diputación de Cádiz, Asociación Profesional del Patrimonio Histórico-Arqueológico de Cádiz, pp. 19–26.
Trigo, R. M., Pozo, D., Osborn, T. J., Castro, Y., Gámiz, S., & Esteban, M. J. (2004). North Atlantic Oscillation influence of precipitation, river flow and water resources in the Iberian Peninsula. *International Journal of Climatology, 24,* 925–944.

Zazo, C., Goy, J. L., Somoza, L., Dabrio, C. J., Belluomini, G., Improta, S., et al. (1994). Holocene sequence of sea-level fluctuations in relation to climatic trends in the Atlantic-Mediterranean linkage coast. *Journal of Coastal Research, 10,* 933–945.

Zazo, C., Silva, P. G., Goy, J. L., Hillaire-Marcel, C., Ghaleb, B., Lario, J., et al. (1999). Coastal uplift in continental collision plate boundaries: data from the Last Interglacial marine terraces of the Gibraltar Strait area (South Spain). *Tectonophysics, 301,* 95–109.

Chapter 29
The Active Dune System of Doñana National Park

Ismael Vallejo and José Ojeda

29.1 Introduction

Along the last 50 km of Huelva's east coast and covering more than 430 km^2, the aeolian sand sheet of El Abalario-Doñana constitutes the biggest example of an aeolian complex on the whole coast of Spain. In it different sequences or episodes of aeolian formations can be discerned, extending in time from the Late Glacial/ Holocene up to the present (Borja and Díaz del Olmo 1996), encompassing areas of entirely stabilised dunes and others where active dunes prevail, including outstanding examples of mobile dunes. This chapter is dedicated to the latter episode of active dunes and more specifically to the part thereof situated inside Doñana National Park (Fig. 29.1).

The complex hence labelled the Doñana Active Dune System presents a number of specificities that make it a unique case on both national and European level. Among them its surface area must be mentioned, altogether covering more than 60 km^2, while the total volume of sand sediments has been estimated in 2,137 Hm^3 by mean of DEM analysys (Vallejo 2007). Especially notable is the mobile nature of its dune formations, all the more so if we take into account the bioclimatic zone in which they are located, where vegetation usually gains over wind action by fixing dunes in place. Conversely, in this case wind and vegetation emerge as the basic agents of a very rich morphological and ecological variety, in an equilibrium that still allows remarkable advances of the dune fronts.

I. Vallejo (✉) · J. Ojeda
GIS and COAST Research Group, Department of Physical Geography and Regional Geographic Analysis, University of Seville, Calle María de Padilla s/n, 41004 Seville, Spain
e-mail: ivallejo@us.es

J. Ojeda
e-mail: zujar@us.es

J. A. Morales (ed.), *The Spanish Coastal Systems*,
https://doi.org/10.1007/978-3-319-93169-2_29

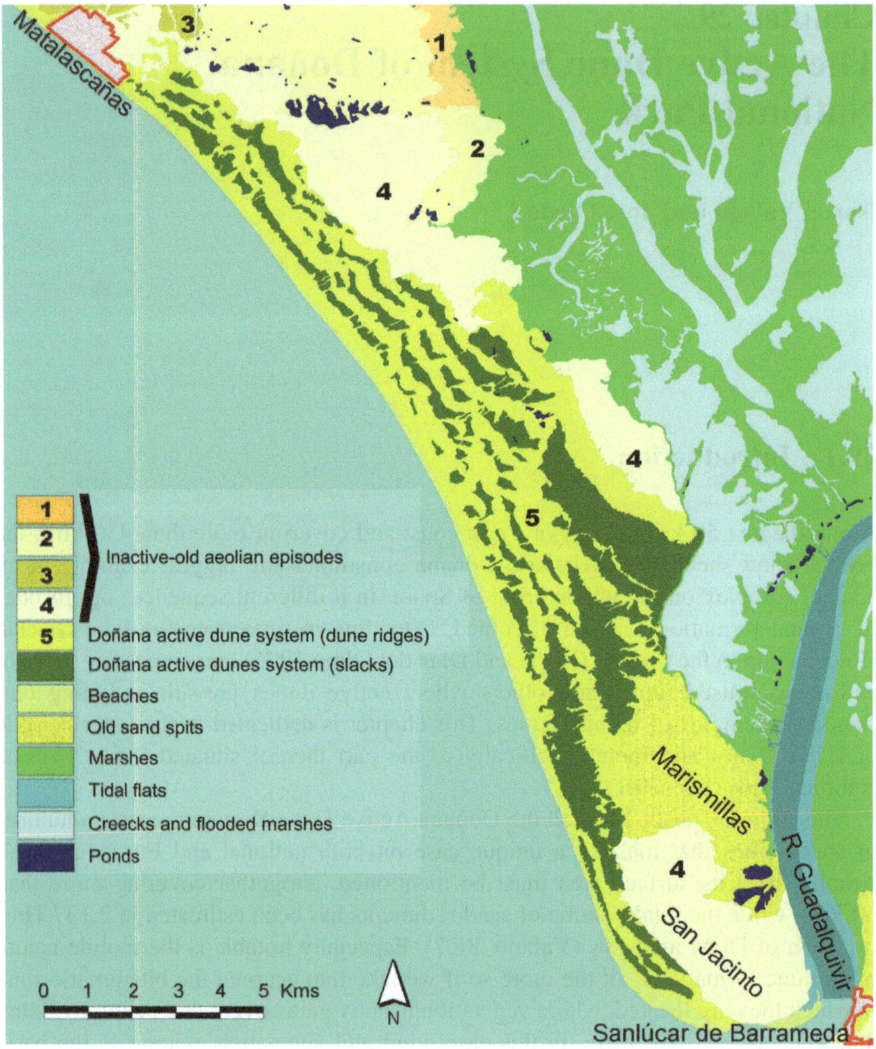

Fig. 29.1 Location of the Doñana active dune system in the context of the El Abalario-Doñana littoral aeolian sand sheet

At national level, only the examples in the Canary Islands (see corresponding chapter in this volume), in very different bioclimatic surroundings, present cases of large complexes of mobile dunes, although vegetation tends to play a less relevant role in them. Finally, given the general context of alteration and destruction along Spanish coasts, the Doñana dunes' state of preservation can be considered a truly exceptional case which should be linked to its inclusion since the late 1960s in a protected area of maximum international importance. Yet despite this, we should never consider that the system concerned is foreign to human intervention, as the dunes' own active nature may be related to multiple actions that have taken place in the none too remote past (Granados 1987; Borja et al. 1999).

Despite these singularities, widely recognised at national and international level, the relative lack of studies that have focused on this dune group is surprising. Its treatment has mostly been limited to including it as a particular unit in the various sector divisions concerning, at the very least, the ensemble of Doñana National Park; standing out in numbers are those with an ecological focus (Valverde 1958; Allier et al. 1974; Clemente et al. 1997; Montes et al. 1998); a few present a basically geomorphologic approach, with either a more quaternary focus (Borja and Díaz del Olmo 1996) or as studies on current formations (Vanney and Menanteau 1979; Rodríguez-Ramírez 1998).

The work carried out by the team led by Professor García-Novo (Torres Martínez 1975; García-Novo et al. 1975) must also be mentioned; besides establishing sectors and units it conducted the first research on the Doñana dunes' mobility. Despite the passage of time and the technical limitations conditioning those valuable studies, until very recently there were no other references concerning the advance of those formations, even though, as has been commented, it is one of the aspects that makes this dune system most remarkable (Vallejo 2007).

Based mainly on the last work cited, in this chapter we make a first analysis of the dynamic context of the Doñana coast before going on to describe the main dune system units and their most characteristic elements; finally, the chapter closes with a study on key developmental aspects of the dune formations in the last half-century.

29.2 Littoral Dynamics and Wind Action on the Doñana Coast

This section makes mention of those aspects of littoral dynamics that exclusively affect the sector of the Doñana National Park coast. In this regard, owing to its particular influence on the processes of the dune complexes' formation and development, special attention has been paid to coastline trends and to the area's aeolian potential.

29.2.1 The Progressive Trend of the Beaches

The first of the conditions explaining this impressive Doñana dune complex is the existence of a significant and continuous source of sediments in this case comprising 26 km of beach with a clear prograding tendency between the locality of Matalascañas and the mouth of the Guadalquivir. This continuous growth, as indicated in the chapter devoted to the beaches of Huelva in this same volume, is the result of receiving contributions from a clear drift toward the east which

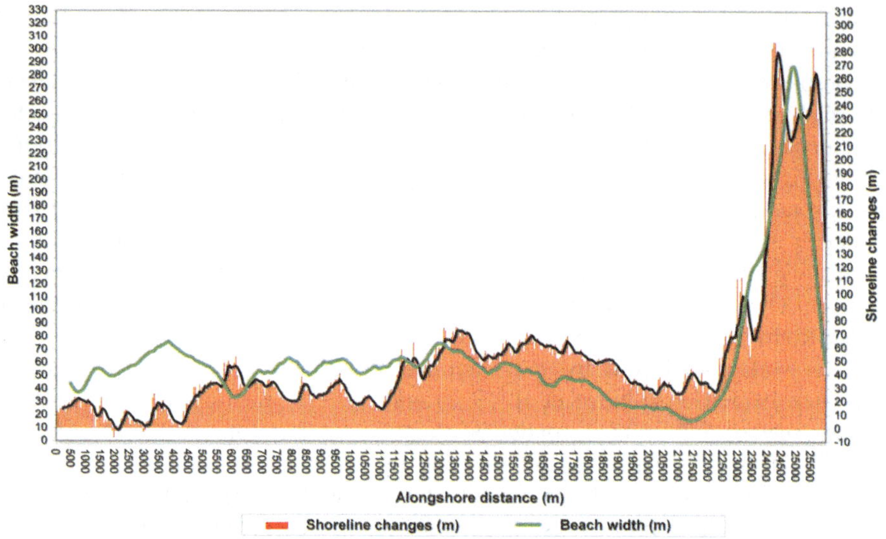

Fig. 29.2 Width of the back-shore in 2001 and coastline changes between 1956 and 2001 on the Doñana coast (measurements taken for transects delineated every 50 m)

dominates the entire Huelva coast. The prograding nature of Doñana's beaches can be clearly observed via the historical evidence, such as the current interior position of certain watchtowers in the area (Carbonero, Zalabar y and Jacinto) built in the 16th century as surveillance points very close to the shoreline. This prograding behaviour of the beaches remains active at present, as seen in more current data obtained by comparing coastlines depicted in orthophotographs. Based on two of them, corresponding to 1956 and 2011, Fig. 29.2 enables estimation of a mean current high beach width of more than 60 m, while overall changes between both coastlines shows a clear positive balance, with almost constant progression from Matalascañas to the mouth of the Guadalquivir.

Added to these favourable conditions of the Doñana beaches are a further two essential aspects per their consideration as sources of aeolian sediments. In first place, a suitable grain-size characterisation of aeolian potential parameters, as medium-sized sands are involved, modally measuring around 0.25 mm, and in second place a generally NNW-SSE orientation which assures maximum effectiveness of winds, which as shall be seen proceed mainly from the SW.

29.2.2 Aeolian Potential

The climatic features of the Doñana environment coincide with those of an oceanic Mediterranean Atlantic coast climate type (Pita López 2003); this would be characterised by average annual precipitation of between 500 and 600 mm, mostly

Table 29.1 Aeolian drift potential in Doñana

	DP	RDP	RDP/DP	RDD
Year	4715.9	3848.4	0.82	41.4
Jan	871.7	733.4	0.84	37.1
Feb	326.2	262.8	0.81	49
Mar	440.8	380.4	0.86	33.7
Apr	579.3	455.4	0.79	54.5
May	309	250.4	0.81	48.4
June	127.1	113.3	0.89	70.1
July	59	49.4	0.84	61.7
Aug	60.9	58.7	0.96	76
Sept	160.2	136.1	0.85	47.2
Oct	377.2	311.2	0.82	29
Nov	579.3	486.1	0.84	41.3
Dec	825.2	677.5	0.82	31.6

Monthly drift potential in kg/m/month; Yearly drift potential in kg/m/year
DP drift potential; *RDP* resultant drift potential; *RDD* resultant drift direction; *RDP/DP* directional variability index

concentrated in winter, a typically Mediterranean summer dry season and milder temperatures due to the oceanic influence, in this case is manifested by the inexistence of monthly averages below 10°.

The data on wind intensity and direction are from the El Arenosillo weather station. Translated into transport capacity (Fryberger and Dean 1979), this data shows a clear predominance of winds from the third quadrant and thus a clearly unimodal transport system in the NE direction which substantially agrees with the overall positioning of the dune fronts (NW-SE). For its part, the annual net rate would stand at around 3,848.4 kg/m/year, with the seasonal contribution of the winter months from November to January standing out and secondarily the spring months of March and April (Table 29.1).

29.3 Basic Units of the Doñana Active Dune System

In the different approaches to this dune complex the description of a relatively simple general configuration stands out, responding to a succession of dune trains or ridges parallel to the coastline (transverse ridges), separated from each other more or less clearly by a series of inter-dune depressions or slacks which are locally called *corrales*. These same approaches usually coincide when it comes to indicating the existence of two large complexes which would be established by their

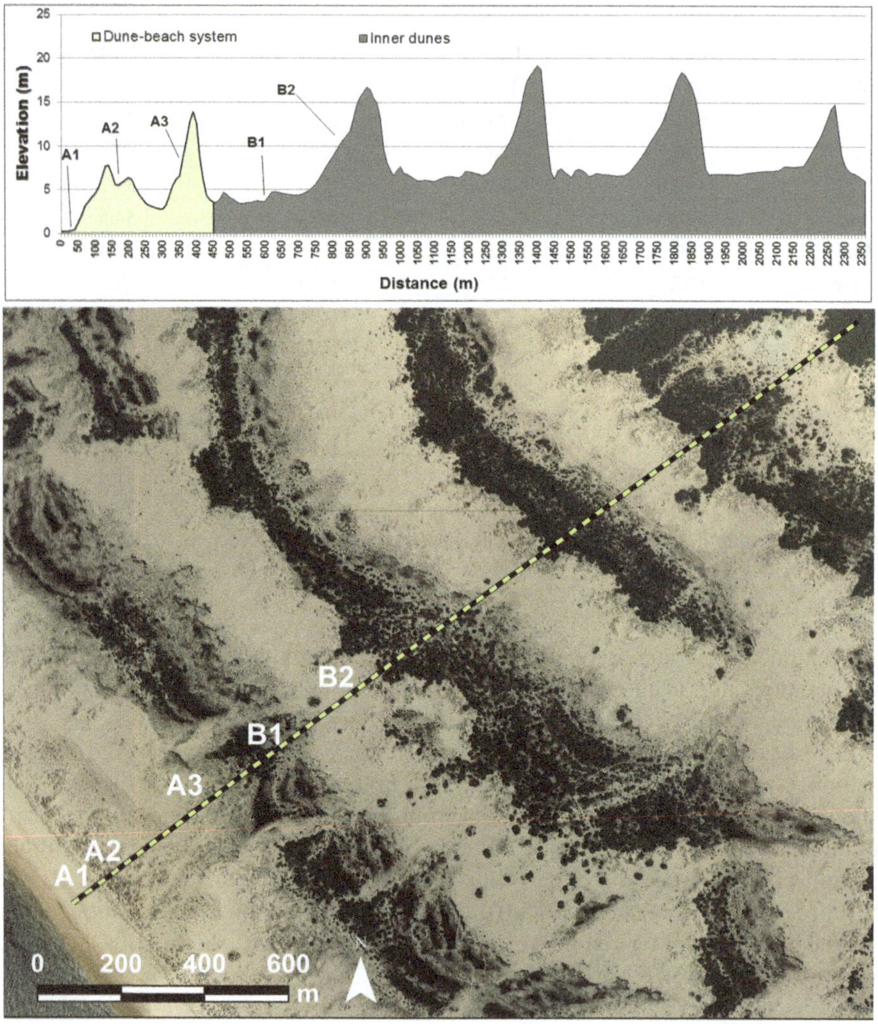

Fig. 29.3 Sector arrangement of the Doñana active dune system (A—coastal complex: A1—beach; A2—primary dune; A3—secondary dune. B—interior complex: B1—inter-dune trough or slack; B2—interior dunes)

proximity to the coastline, thus speaking, for example, of a 'train of littoral dunes' and some 'internal trains' (García-Novo et al. 1975). The sector arrangement indicated below is based on previous works (Montes et al. 1998; Vallejo 2007) and is shown in Fig. 29.3.

29.3.1 The Coastal Complex (Beach/Dune System)

This group encompasses both the beach and a first band of dunes extending around 350 m into the interior on average. Although with exceptions along the coast, which will be presented later, this complex can be described as a succession of three parallel bands described as follows:

- The first comprises the beach (Fig. 29.3, A1), which usually presents a series of incipient or embryonic dune formations in its back-shore. These are normally ephemeral elements that can disappear during storms; they present in their typical forms of ridges, ramps or terraces (Hesp 2002). Given their location and the marine influence, there is scant colonisation by vegetation, limited to scattered groups of different pioneer species;
- In second place, (Fig. 29.3, A2) the established foredune (Hesp 2002) or primary dune (Psuty 2004) is positioned, with much lateral continuity along the entire coast, though with a very broad range of configurations and morphologies (unitary ridge, parallel ridges, group of hummocks, etc.). The most characteristic features of this unit are constituted by a dissymmetric profile that is steeper toward the beach and less so toward the interior, and by very intense plant colonisation dominated by marram grass (*Ammophila arenaria*). The terminology of established foredune derives from this second characteristic, as that intense colonisation would correspond to a vocation of more permanence and stability. At the same time, the term primary dune is used, alluding to its active sedimentary connection with the beach, which makes it the first stage or phase of sediment incorporation in the dune system;
- Lastly, the evolved coastal dune or secondary dune (Psuty 2004) represents a more advanced phase in which an old primary dune has undergone migration toward the interior and is now separated from the current primary dune by an incipient slack. The secondary dune often presents in the form of a more or less continuous transverse ridge or as a succession of various depositional lobes associated to blowouts. In all cases it is usually a dune with a dissymmetric profile, gentle on its upwind side and very steep on its downwind side which defines as a typical front of sub-vertical advance. The area of separation between the primary and secondary dune results from an intense process of aeolian deflation and may be considered an incipient slack; in certain areas this intense deflation exposes the water table, which facilitates the colonisation of such spaces by different hydrophilic species (*Juncus acutus*). In some sections this intermediate corridor presents much lateral continuity, separating the two primary and secondary dune units very clearly; in other cases, however, it appears interrupted by different residual elements, either in the form of groups of hummocky dunes or as longitudinal forms that mostly correspond to lateral o erosional walls of blowouts. In all these cases the protection offered by the primary dune vis-à-vis the marine influence facilitates the appearance of other woody species such as *Armeria pungens*, *Artemisia christmifolia* or *Helychrisum picardii* and, with a nature more limited to certain sections,

junipers (*Juniperus oxicedrus*). In the latter case, the sections of coastal dune where this species appears are quite singular and may represent the only vestiges of the surroundings' original dune formation.

Per this general outline, along the Doñana coast's 26 km a variety of configurations in the beach/dune system can be found, whose classification and systemisation enables distinguishing up to four differentiated types depending on morphometric parameters (height and width) and plant density, in the foredune's case, while for the beach the data used concerns behaviour (coastline change rates from 1956 to the present) as well as back-shore width; these parameters were obtained for 516 transects perpendicular to the coastline, based on which cluster analysis was conducted (Vallejo et al. 2006; García and Vallejo 2012). The four types obtained are shown in Fig. 29.4, inserted in a functional model that modifies the one proposed by Psuty (2004) for beach/dune systems.

Departing from the north end of the beach, for a length of 6 km type 1 is associated to beaches which present the lowest growth rates of the whole group, with averages always below 1 m/year, and back-shore widths of around 60 m. The primary dune is presented as a ridge about 80 m wide with scant vegetation (10–15%) and frequent blowouts in the form of narrow passages connected to the beach. Behind this primary dune a space of about 200 m opens, occupied by a deflation surface in which the water table may emerge, whereby the presence of significant plant coverage is recorded therein; this depressed zone is followed by a transverse ridge of mobile dunes.

Type 2 is installed in the following 10 km, in which the beaches gradually present more significant advance rates while their width also increases. In the primary dune the width increases (130 m) along with height and plant density (20–30%), diminishing the succession of blowouts that open the way to the beach. This

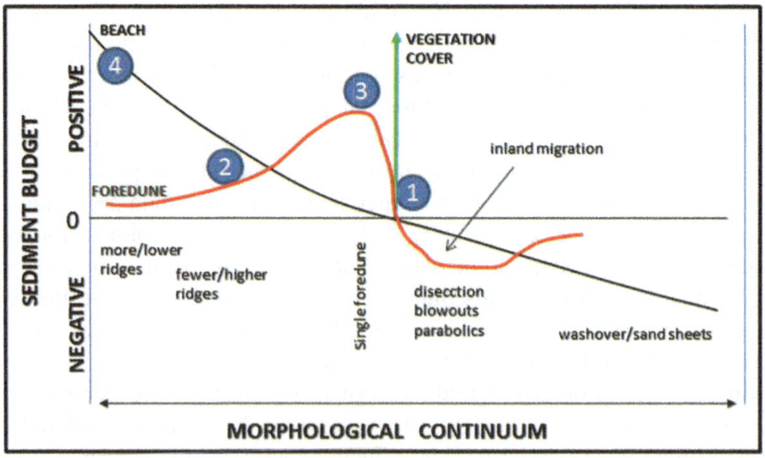

Fig. 29.4 Basic configuration of beach-dune system along Doñana coast, inserted in the Psuty model (Psuty 2004)

first dune ridge is followed by a wide space of about 300 m, in which the dune formations appear more disordered than in the previous case; in it can be found transverse dunes, though dunes also develop in internal blowouts and hummocks disconnected from the beach whose lateral walls often remain as residual longitudinal forms that break the predominant transversal succession of this coastal dune.

Type 3 appears before the river mouth sector is reached, in a section where the beach advance rates undergo a significant decline; the reduction experienced by the high beach is quite notable, with widths that represent minimums on the entire Doñana coast, averaging 45 m. With respect to the primary dune, what stands out most is the presence of a single well-developed ridge achieving the greatest heights of this first coastal dune (around 9 m on average) and also presenting very high percentages of plant coverage (70%), with abundant shrub species in the more protected areas. From this first ridge of well-developed dunes extensive deflation surfaces often open, usually also colonised by vegetation, followed by different dune forms that may be transverse mobile ridges or large hummocky dunes generated by the presence of large vegetation (*Juniperus oxicedrus*).

Finally, type 4 agrees very well with Psuty's original model and is situated at the far end of the Doñana spit. The coastline's most spectacular advances occur in this sector, with average annual rates of around five metres; naturally, this sector corresponds to large back-shore widths (average 90 m). These beach characteristics condition the appearance of extensive dune formations arranged as many successive ridges with very high plant coverage values, though they are unable to develop significant heights owing to their rapid disconnection from the feed source represented by the beach when a new ridge accordingly appears on the back-shore. Hence, as indicated when GPR techniques were applied (Morales et al. 2017), the dune ridges do not migrate toward the interior but rather maintain their position after being colonised by vegetation while new ridges appear as the beach grows.

29.3.2 The Interior Complex

With some exception along the coast, the secondary dune ends in the form of an advance front on a first and well developed inter-dune depression or slack (see Fig. 29.3, elements A3 and B1). Based on this contact the complex of interior dunes begins, which would comprise a succession of slacks and ridges of transverse dunes whose number varies along the system with a maximum of six ridges over a distance of more than 4 km toward the interior. These dune trains or ridges present a markedly mobile nature and consequently appear with almost no plant coverage. The morphology of these ridges is that typical of transverse dune formations, where the gentlest slopes correspond to the upwind side, rarely attaining 10°, while on the downwind side a typical advance slip-face appears with short and steeper slopes, often of more than 30° (Fig. 29.5).

Although the lateral continuity of these ridges is quite significant, their planar arrangement is far from rectilinear; rather, it often corresponds to a model in which

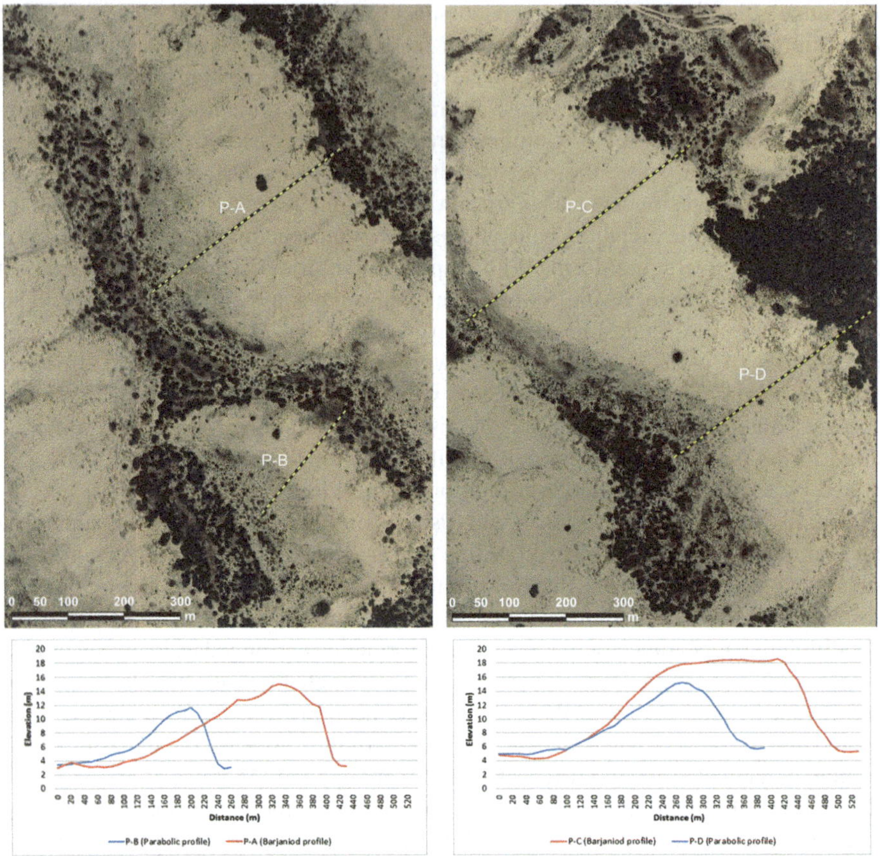

Fig. 29.5 Plan and profile of barchanoid and parabolic sections along transverse dune ridges

barchanoid and parabolic sections succeed each other, endowing the fronts with a clear wavy course and which are ultimately indicative of differences in advance speed in favour of the former. This greater mobility is clearly reflected in the profiles corresponding to each section, more extensive in the barchanoid ones and more pointed in the parabolic ones (Fig. 29.5). In certain cases, some of those sections are able to isolate themselves from the ridges they integrated, giving rise to individualised barchanoid or parabolic dunes, something much more frequent in the latter case (Fig. 29.5, P-B).

Given this dominant transversal pattern, the presence of different longitudinally-arranged dune formations, i.e. aligned parallel to wind direction, must be mentioned. The longitudinal forms in Doñana are of two types, already recognised and briefly described in the aforementioned pioneering works (García-Novo et al. 1975).

The first type of longitudinal dune is the most well known and is associated to vegetation fixing in place the lateral sectors or less active arms of parabolic dunes. Those arms are usually isolated between two transverse ridges, while the central sector to which they were joined progresses at a faster pace. They therefore appear on many occasions as residual morphologies, occasionally breaking the lateral continuity of the slacks where they remain and eventually end fixed by vegetation. The second type, instead of an isolated dune, presents as a series of parallel formations perpendicularly positioned on the transverse ridge itself. Its presence is particularly evident in the trains of more internal dunes, as well as in others located in areas further south; ultimately, in both cases its appearance is apparently associated to dune ridges with less transgressive activity. Thus, as a transverse ridge is almost stagnated, wind erosion opens a series of corridors or grooves that cross the dune from tail to front separated each other by positive longitudinal morphologies; these morphologies often present widths of under 80 m and heights rarely over 3 m, and they usually appear lightly colonized by vegetation (Fig. 29.6).

With respect to the inter-dune depression or slacks, they would occupy a surface area of 23.5 km^2 and are clearly distinguished by their depressed topography vis-à-vis the dune formations that surround them and the abundant vegetation installed on them. Depending on water table depth they are differentiated into wet and dry slacks, with a predominance of "*monte blanco*" and "*monte negro*" (white scrub and black scrub), respectively, and in both cases with a very substantial presence of

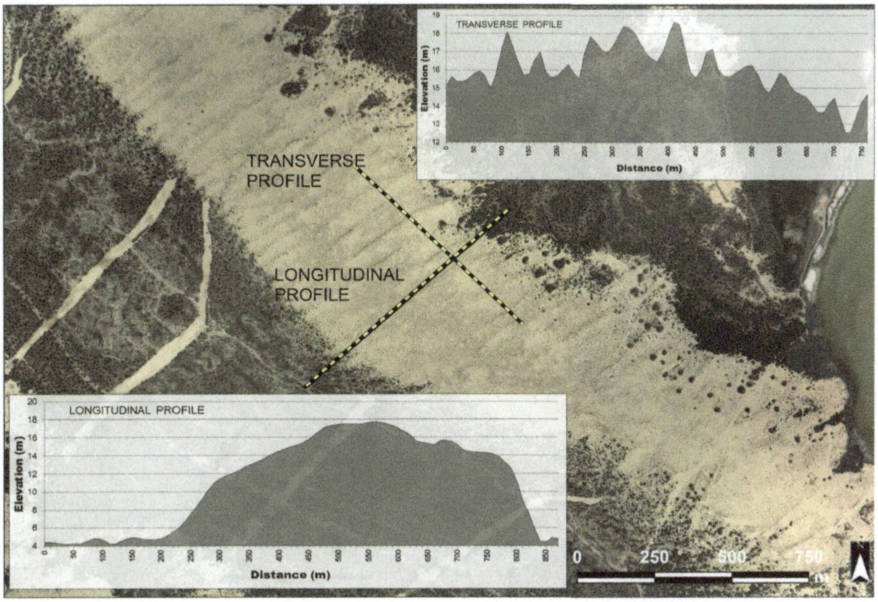

Fig. 29.6 Plan and profile of longitudinal dunes on a transverse ridge

pines (*Pinus Pinea*) introduced by man since the 18th century (García-Novo et al. 2007).

The formation of these extensive depressions is due to the widespread action of aeolian deflation on the tails of the different dune trains. This process acts by lowering the dune's topography in this sector of the tail, which ends up bringing the water table toward the surface, resulting in damper sands whose compacting largely prevents wind erosion while facilitating the introduction of vegetation. A slack thus grows at the expense of withdrawal of a dune train on its upwind side, while its gradual disappearance occurs due to the coverage it will be subject to by the advance of the next dune front downwind.

In the framework of these inter-dune depressions two characteristic dune formations can be observed which present a residual nature. The first consists of a series of ridges parallel to each other and with respect to the tail of the preceding dune; those ridges are usually not more than a metre high and have been labelled, in Doñana and in other dune fields, 'counter-dunes', 'worms', "gegenwalle" or residual dune ridges (García-Novo et al. 1975; Levin et al. 2009; Hesp et al. 2011) and their formation is still under debate. Such counter-dunes represent marks of successive positions of the tail of a transverse dune train in the process of advancing. In Doñana, their formation has been associated to periods of less wind activity in which vegetation is able to colonise that sector of the dune tail owing to the water table's proximity; in the next activity period deflation will begin behind this fixed ridge, which would remain as a residual formation (Fig. 29.7 Left).

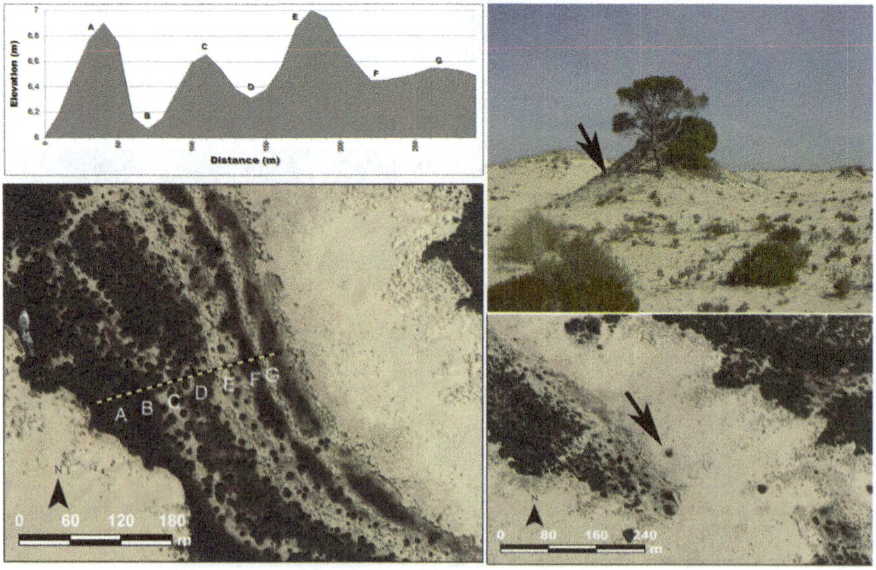

Fig. 29.7 Residual dune forms integrated in slacks: counter-dunes or worms (Left) and pyramidal dunes (Right)

The second residual formation consists of a type of dune whose morphology is pyramidal or conic, presenting in an isolated manner after the passage of a transverse ridge. This formation, described in other contexts and labelled a remnant knob (Hesp and Thom 1990) is formed by the resistance to aeolian deflation presented by certain transverse dune sectors thanks to the presence of significant examples of trees (pines) or shrubs (needle- and scale-leaf junipers). Thus, in the first phase of its formation this dune type situates on the downwind side of the transverse dune itself, progressively integrating the slack as the dune advances (Fig. 29.7 Right).

29.4 Evolving Dune Systems Patterns in the Last 50 Years

The mobile nature of the Doñana dunes makes their monitoring especially interesting, owing to their specific description and the impact the detected changes and trends may have as indicators of more significant environmental changes. Nevertheless, as stated in the introduction, studies on this particularity were limited to measurements made in the late 1970s by the team led by Professor García-Novo; while reiterating the value of those studies, there is no doubt that with respect to dune advance measurements various limitations must be taken into account, whereby comparisons between those measurements and those derived from other more recent work will not be made (Vallejo 2007).

In those latest studies three orthophotographs corresponding to the years 1956, 1977 and 2001 were used, along with the three digital elevation models (DEMs) used to produce them based on the respective photogrammetric flights. The main results obtained are shown below for the coastal and interior complexes.

29.4.1 The Coastal Complex

The evolution of the coastal complex is naturally closely linked to that of the beach itself; it is thus important to take up again what was previously said about the prograding nature of the Doñana coast (Fig. 29.2), where after analysis of coastline changes in recent years a mean growth of more than 50 m has been verified; however, that average includes major differences between sectors that can be considered stable or slow-growth, above all in the first sectors of the littoral barrier of Doñana and others at its far end, in which the coast has advanced at very significant rates.

Bearing in mind these two ends that were represented in foredune types 1 and 4, this complex's evolution also corresponds to those two fundamental models. In the case of the initial sections, the coastline's moderate advance first implies an enlargement of the first dune formations. However, as discerned in Fig. 29.8 Left, that enlargement entails major modifications in dune dynamics leading to more intense aeolian deflation after the initial obstacle represented by the first dune front

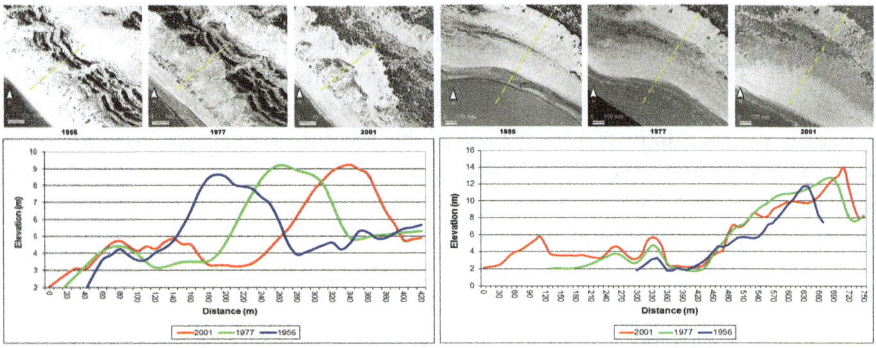

Fig. 29.8 Evolution of two coastal foredune sectors (types 1 and 4 in Fig. 29.4)

(primary dune); this enables the formation of blowouts or the generation of marked deflation surfaces that in turn serve to feed a second dune front (secondary dune). This second formation would therefore amount to a transitional unit between the coastal and interior complexes and can be considered the initial phase of a new transverse ridge that will end up joining the succession of interior dune ridges.

In the case of sectors where the coast presents accelerated growth (Fig. 29.8 Right), the continuous development of new dune fronts enables the previous ones to be quickly colonised by vegetation, as they become relatively isolated from the sedimentary source represented by the beach. This fact also implies scant development of the succeeding dune formations.

29.4.2 The Interior Complex

Analysis of the evolutionary aspects of the interior complex is grounded on three different approaches: a first one that shows the changes undergone by the dune and slack surfaces at the different reference dates, as well as the interchanges occurring between both units in the two periods that those dates define (1956–1977 and 1977–2001); a second centred on estimating dune front advance rates during those same periods; and a third which for those same intervals studies the topographical and volumetric changes that take place in the dune ridges.

With respect to the first question, the surfaces that dunes and slacks represent in each of the years studied show opposite trends. Thus, a clear trend of diminishment of the dunes' extent is detected, as it goes from 3,997 km^2 in 1956, to 3,523 km^2 in 1977 and, finally, 3,523 in 2001. In the case of slacks, their surfaces naturally get bigger, growing from 1,712 to 1,957 and 2,354 in 2001.

In terms of interchange between units in the two reference periods, the previous process is associated to a decrease in the second period (1977–2001) of the surfaces that shift from slacks to dunes (advance of dune fronts on the slacks) and to a

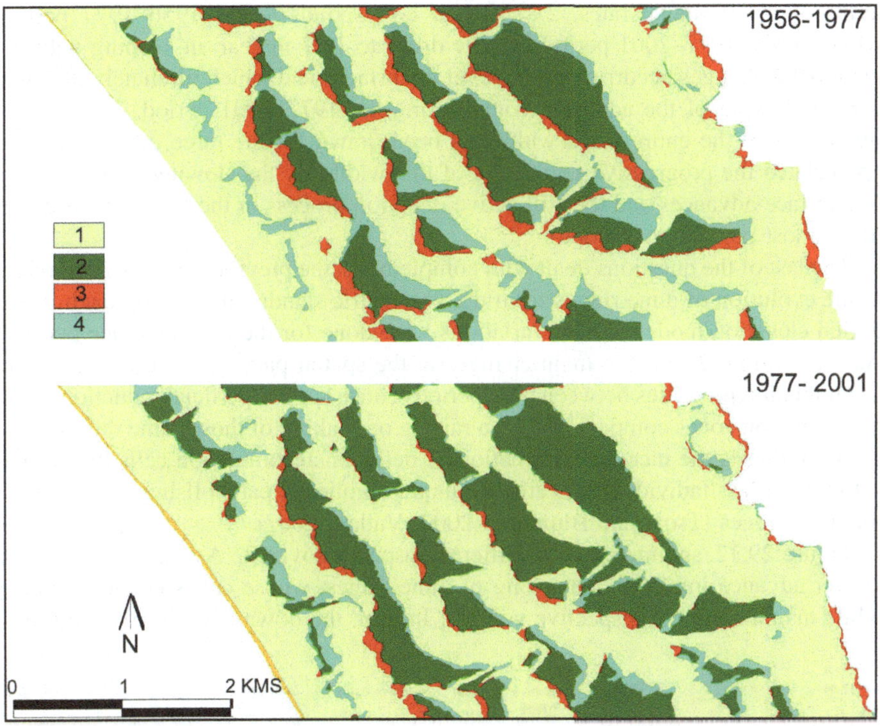

Fig. 29.9 Interchange between surfaces of dunes and slacks in a sector of active dunes: *1* No change (dune-dune); *2* No change (slack-slack); *3* Dune advance (slack-dune); *4* New slack (dune-slack)

certain maintaining of those that involve the transformation of dunes into slacks (withdrawal of a dune ridge's tail and the area's colonisation by vegetation) (Fig. 29.9).

The second question centres on calculating the dune fronts' advance rates, ascertained from their shifting positions as indicated across the three reference dates (Fig. 29.10). In this case, movement was measured at 1,250 points throughout the dune system, with activity sampling in the two reference periods (Fig. 29.10). In

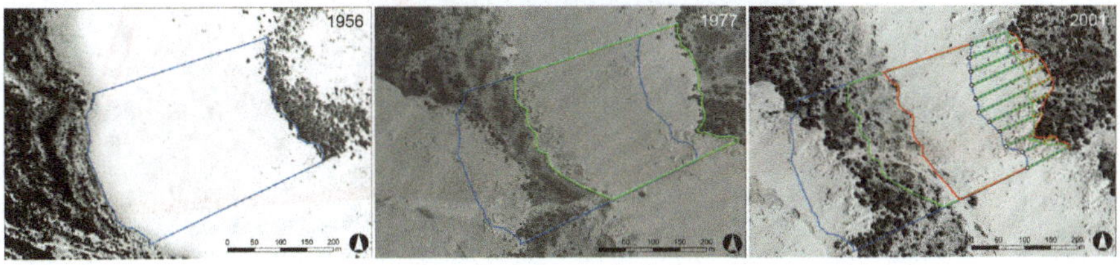

Fig. 29.10 Example of calculation of dune front advance rates

terms of mean annual rates, 2.37 m/year corresponds to the 1956–1977 period, while for the 1977–2001 period the rate drops to 1.27 m/year, in keeping with the mentioned changes occurring at the level of surfaces and which ultimately denote a certain slowing of the advance fronts during the 1977–2001 period. In the same figure, where the entire ridge width has been drawn for the three dates, it can be appreciated the progressive narrowing of this width, by the slowing down process of slip-face advances and the still active deflation process in the tail areas of dunes (downmost part of stoss slope).

The last of the questions dealt with complements the previous analyses and refers to the evolution of dune ridges from the volumetric standpoint. To approach it two digital elevation model (DEM) variations were done for the two reference periods. As seen in Fig. 29.11, the main changes in the spatial pattern of the deflation and accumulation processes between both periods consist of a superficial reduction of the accumulation zones compensated by a raising or peaking of those same dunes. Also worth noting is the increased formation of deflation/accumulation cells that would act more or less individually to create a larger number of parabolising sectors along the dune ridges (Tsoar and Blumberg 2002; Vallejo 2007).

Figure 29.12 summarises the different aspects covered. As can be seen, the slower advance implies a narrowing of dune ridges, whose crests become sharper while maintaining the respective volume. In turn, the downwind profile acquires a

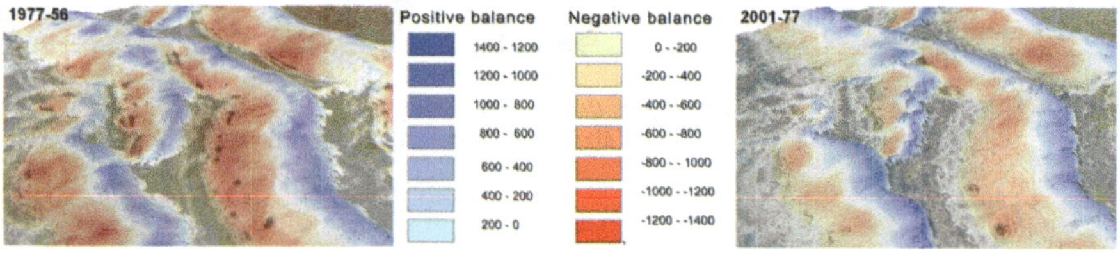

Fig. 29.11 Volumetric balance (m^3) in a sector of active dunes

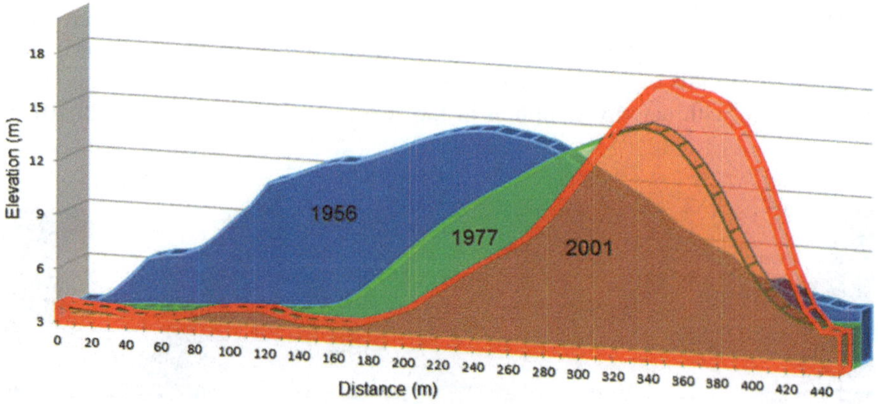

Fig. 29.12 Profiles showing the evolution of a dune ridge

more convex outline, indicative of more deflation. This same figure serves to illustrate an evident trend toward gradual deceleration of the advance fronts. This facts enables two working hypotheses of very distinct significance to be noted (Vallejo 2007). The first would entail cyclic dune advance behaviour that would still require a return to the situation of the 1956-type profiles, whence the cycle would begin once again. The second, on the contrary, holds that the deceleration is a prelude to potential stabilisation of the dunes, in the same line as what other authors have indicated for similar Mediterranean littoral dune environments (Tsoar and Blumberg 2002; Levin and Ben-Dor 2004).

In these cases, the reasons put forward to explain this stabilisation include widespread expansion of vegetation associated to the suppression of different human-induced activities that limited its development (livestock farming, grazing), which would imply a decline in sediment supply rates, resulting in turn from the application of environmental protection policies in dune environments. To verify just how far this model may be reproducing in Doñana is a working line of major interest for coming years, maximum when it becomes evident that its declaration as a national park in 1969 is a milestone in the culmination of a process of gradual abandonment of practices such as uprooting and cutting of vegetation (for construction and heating), livestock farming or artisanal fishing (Ojeda Rivera 1987), all of which hindered the spread of vegetation.

References

Allier, C., Gonzalez Bernaldez, F., & Ramirez-Diaz, L. (1974). *Mapa ecológico 1/10000 de la Reserva Biológica de Doñana.* Consejo Superior de Investigaciones Científicas Sevilla.

Borja, F., & Díaz del Olmo, F. (1996). Manto eólico litoral (MEL) del Abalario (Huelva España). Episodios morfogenéticos posteriores al 22000 BP. In A. Pérez Alberti, et al. (Eds.), *Dinámica y Evolución de Medios Cuaternarios.* Santiago de Compostela: Xunta de Galicia.

Borja, F., Zazo, C., Dabrio, C. J., Díaz del Olmo, F., Goy, J. L., & Lario, J. (1999). Holocene aeolian phases and human settlements along the Atlantic coast of southern Spain. *Holocene, 9,* 333–339.

Clemente, L., Siljeström, P., & Rodríguez, A. (1997). Relación suelo-geomorfología en el Parque Nacional de Doñana. *Cuaternario y Geomorfología, 11*(1–2), 33–42.

Fryberger, S. G., & Dean, G. (1979). Mechanisms for the formation of aeolian sand seas. In E. D. McKee (Ed.), *A study of global sand seas* (Vol. 1052). United States Geological Survey Professional Paper.

García, D., & Vallejo, I. (2012). Caracterización y modelos de funcionamiento del sistema playa-duna en el Parque Nacional de Doñana (Huelva, España). In AGE_CSIC (Ed.), *XV Congreso Nacional de Tecnologías de la Información Geográfica.* Madrid: AGE-CSIC.

García-Novo, F., Martín Vicente, A., & Toja Santillana, J. (2007). *La frontera de Doñana.* Sevilla: Universidad de Sevilla, 317 pp.

García-Novo, F., Ramírez, L., & Torres, A. (1975). El sistema de dunas de Doñana. *Naturalia Hispánica, 5,* 1–56.

Granados, M. (1987). *Transformaciones históricas de los ecosistemas del Parque Nacional de Doñana* (Doctoral thesis). Universidad de Sevilla, Sevilla, 485 pp.

Hesp, P. A. (2002). Foredunes and blowouts initiation geomorphology and dynamics. *Geomorphology, 48,* 245–268.

Hesp, P. A., & Thom, B. G. (1990). Geomorphology and evolution of transgressive dunefields. In K. F. Nordstrom, et al. (Eds.), *Coastal dunes form and processes*. Chichester: Wiley.
Hesp, P. A., et al. (2011). Transgressive dunefi eld landforms and vegetation associations, Doña Juana, Veracruz, Mexico. *Earth Surface Processes and Landforms, 36,* 285–295.
Levin, N., & Ben-Dor, E. (2004). Monitoring sand dune stabilization along the coastal dunes of Ashdod-Nizanim Israel 1945–1999. *Journal of Arid Environments, 58*(3), 335–355.
Levin, N., et al. (2009). Modelling the formation of residual dune ridges behind barchan dunes in North-east Brazil. *Sedimentology, 56,* 1623–1641.
Montes, C., Borja, F., Bravo, M. A., & Moreira, J. M. (1998). *Doñana Una Aproximación Ecosistémica* (p. 311). Sevilla: Consejería de Medio Ambiente Junta de Andalucía.
Morales, J. A., et al. (2017). Estructura interna de los cordones frontales de dunas eólicas en la flecha litoral de Doñana mediante el uso del georradar. *GeoTemas, 17,* 271–274.
Ojeda Rivera, J. F. (1987). *Organización del territorio en Doñana y su entorno próximo (Almonte) Siglos XVIII-XX* (p. 459). Madrid: ICONA.
Pita López, M. F. (2003). El clima de Andalucía. In A. Lopez Ontiveros (Ed.), *Geografía de Andalucía.* Barcelona: Ariel.
Psuty, N. P. (2004). The coastal foredune A morphological basis for regional coastal dune development. In M. L. Martinez, et al. (Eds.), *Coastal dunes ecology and conservation*. Berlin: Springer.
Rodríguez-Ramírez, A. (1998). *Geomorfología del Parque Nacional de Doñana y su Entorno* (p. 146). Madrid: Ministerio de Medio Ambiente.
Torres Martínez, A. (1975). *Estudio ecológico cuantitativo de los sistemas de dunas y marismas de la Reserva Biológica de Doñana* (Doctoral thesis). Universidad de Sevilla, Sevilla, 303 pp.
Tsoar, H., & Blumberg, D. G. (2002). Formation of parabolic dunes from barchan and transverse dunes along Israel's Mediterranean coast. *Earth Surface Processes and Landforms, 27,* 1147–1161.
Vallejo, I. (2007). *Caracterización geomorfológica y análisis de la evolución reciente del sistema de dunas activas del PN de Doñana (1956–2001)* (Doctoral thesis). Universidad de Sevilla, 468 pp.
Vallejo, I., et al. (2006). Characterization and classification of the beach-dune system of the Doñana National Park, Southern Spain. *Journal of Coastal Research, 48,* 112–117.
Valverde, J. A. (1958). An ecological sketch of the Coto Doñana. *British Birds, 51,* 1–23.
Vanney, J. R., & Menanteau, L. (1979). Tipes de reliefs littoraux et dunaires en Basse-Andalousie (De Huelva à l'embouchure du Guadalquivir). *Mélanges de la Casa Velázquez, 15,* 5–52.

Chapter 30
Aeolian Sedimentary Systems of the Canary Islands

Antonio I. Hernández-Cordero, Carolina Peña-Alonso, Luis Hernández-Calvento, Nicolás Ferrer-Valero, Aarón M. Santana-Cordero, Leví García-Romero and Emma Pérez-Chacón Espino

30.1 Introduction

The aeolian sedimentary systems of the Canary Islands have certain natural and socio-economic characteristics that make them differentiated environments, both in the context of the dune fields of Spain and Europe. This is a consequence of their location in intraplate hot spot volcanic islands, their climatic conditions, the high pressure exerted on them by society and the lack of adequate management measures. The relationship between their elements and the continuous change to which they are exposed make them environments with high complexity and fragility. In recent decades, some research lines have been opened around these systems, based on geomorphological, biogeographic, historical, cultural, landscape and

A. I. Hernández-Cordero (✉) · C. Peña-Alonso · L. Hernández-Calvento · N. Ferrer-Valero · A. M. Santana-Cordero · L. García-Romero · E. P.-C. Espino
Grupo de Geografía Física y Medio Ambiente, Instituto de Oceanografía y Cambio Global, IOCAG, Universidad de Las Palmas de Gran Canaria, ULPGC, Parque Científico-Tecnológico de Taliarte, Calle Miramar, 121, 35214 Telde, Las Palmas, Spain
e-mail: hernandez.cordero@ulpgc.es

C. Peña-Alonso
e-mail: cpena@becarios.ulpgc.es

L. Hernández-Calvento
e-mail: luis.hernandez.calvento@ulpgc.es

N. Ferrer-Valero
e-mail: nicolas.fvg@ulpgc.es

A. M. Santana-Cordero
e-mail: aaron.santana@ulpgc.es

L. García-Romero
e-mail: levi.garcia@ulpgc.es

E. P.-C. Espino
e-mail: emma.perez-chacon@ulpgc.es

J. A. Morales (ed.), *The Spanish Coastal Systems*,
https://doi.org/10.1007/978-3-319-93169-2_30

socio-ecological analysis. These new approaches have allowed to expand the knowledge about the diversity and complexity of the processes that occur in these systems, as well as to obtain information regarding their management. This chapter, structured in four sections, aims to explain the origin and nature of these systems, their particular characteristics and processes, the interactions of human activities and their consequences. Finally, a series of recommendations for their management are proposed.

30.2 Origin and Distribution of the Aeolian Sedimentary Systems of the Canary Islands

The Canary Islands conform a volcanic archipelago, composed of seven islands, a minor island and five islets. It is located in the Atlantic Ocean, between 27° and 30° North latitude and 13° and 19° West longitude. The beginning of the formation of the archipelago dates from at least 20.6 million years BP (Carracedo et al. 1998), with the formation of the central volcanic shield of Fuerteventura. From this moment, the main volcanic activity has followed a general displacement towards the west, forming new islands, until generating an E-O chain where the typical succession of a hot spot model is recognized (McDougall 1971).

The location of the archipelago on an old oceanic crust (165–176 m.a.), rigid and of slow displacement, determines the practical absence of tectonic subsidence in the Canary Islands (Carracedo et al. 1998). This results in periods of subaerial survival of atypical amplitude in oceanic islands. For this reason, although the Canary coasts are predominantly rocky (dominate cliffs and coastal platforms) and the beaches are only on 17% of their perimeter (≈1,550 km) (Criado et al. 2011), aeolian coastal sedimentary systems are developed on them. This happens especially in the eastern islands, older and, therefore, exposed for more time to erosive processes. In some cases, the aeolian sedimentary systems extend along wide coastal stretches, eventually encapsulating an island (La Graciosa) or parts of others (Fuerteventura or Lanzarote) (Hernández Calvento et al. 2017). From this perspective, the aeolian sedimentary systems of the Canary Islands are geographical elements of great singularity.

The main factors that favor the development of these systems, under natural conditions, are: (i) the availability of sandy sediments in the seabed and their mobility along the coasts; (ii) the possibility of sediment accumulation on beaches; and (iii) the presence of favorable climatic and topographic conditions for aeolian transport.

With respect to the first factors, the natural sources of sediments in the Canary Islands are mainly terrestrial and contributions of marine organisms (bioclasts). The first ones come mainly from the erosion of rocky coasts and paleodunes and from transport of ravines. As for the latter, the current productivity of the seabed of the archipelago is limited, especially when compared with other geological periods

(Meco et al. 2007, 2008). The availability of sandy sediments is determined by the development of insular platforms, by forming shallow sloping bottoms capable of favoring the processes of production and accumulation of carbonate and mineral sediments.

On the other hand, the mobility of the sediments along the coasts and the possibility of accumulating in the form of beaches, is closely linked to the nature of the islands. In the older islands, with longer exposure to erosive agents, the interferences are less, facilitating the coastal transport of sediments (Criado et al. 2011). As for the topographic conditions of the coasts, in these mountainous islands it is only possible the development of the aeolian transport when low coasts are developed by tectonic, volcanic or erosive processes. Finally, the current climate conditions of the Canary Islands favor the mobility of the sand, as will be seen later.

The combination of the factors described determines that the abundance of aeolian sedimentary systems presents strong contrasts between islands, distributed according to a clear gradient E-W (Fig. 30.1). On the coasts of the western islands (El Hierro, La Palma, La Gomera) these systems are very scarce, being more abundant towards the central islands (Tenerife and Gran Canaria) and eastern ones (Fuerteventura, Lanzarote and La Graciosa). This gradient is observed both in absolute terms, of surface area, and in relative terms, with respect to the coast length of each island (Fig. 30.1).

The aeolian sedimentary systems of the western Canary Islands are located in small coastal enclaves exposed to the prevailing winds of NNE, in the context of a

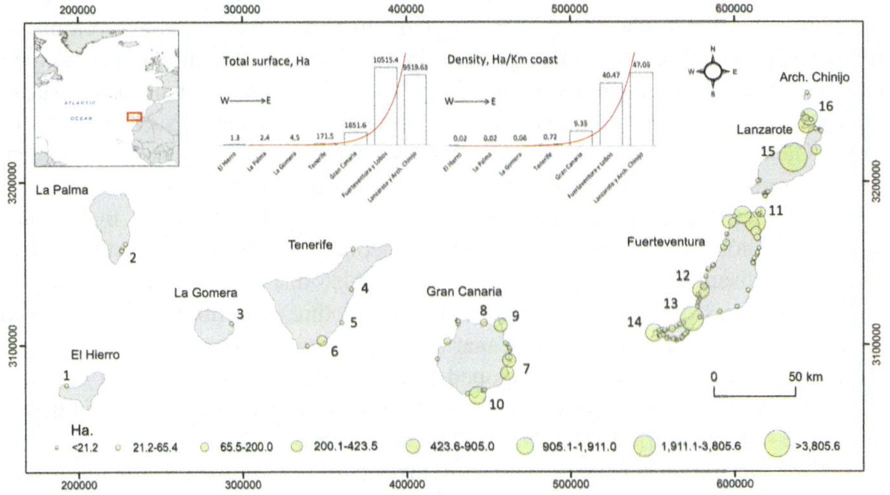

Fig. 30.1 Potential distribution of the aeolian sedimentary systems in the Canary Islands (including those that have disappeared due to human causes), indicating their surfaces, ranges, and numerical identification of the systems mentioned in the text: *1* Arenas Blancas; *2* southeast coast of La Palma; *3* Puntallana; *4* Güimar; *5* Punta de Abona; *6* El Médano-La Tejita; *7* Tufia, Gando, Arinaga; *8* Bañaderos; *9* Guanarteme; *10* Maspalomas; *11* Corralejo; *12* Las Salinas; *13* Punta de Jandía; *14* Istmo de Jandía; *15* El Jable; *16* La Graciosa

general scarcity of sandy sediment on the coasts. The Arenas Blancas system, in El Hierro, forms a singular deposit of organogenic sand, disconnected from the marine source of sediments, which advances about 150 m in a northwest direction on a lava plain generated by recent emissions. It forms a low volume sand sheet on which a small field of nebkhas develops. In La Palma, inputs of dark sands (lithoclastic) can be observed in three enclaves of the southeast coast, where they develop aeolian sedimentary systems of between 1 and 2 ha of surface and scarce volume. The lava plains they cover were originated by recent emission centers. The erosion of the scoriaceous surfaces generates beaches of terrigenous sand (of dark colors), from which these systems extend southwestward. Meanwhile, La Gomera has no recent volcanic activity and has a strong cliff perimeter. The low coastal areas are limited to ravines' mouths with flat-bottomed and lobes of gravitational movements. In one of them, Puntallana, is located the only aeolian sedimentary system documented on this island. It is a quaternary fossil system of clear bioclastic sands that conserves a residual activity by erosion and remobilization of the consolidated marine levels (Morales et al. 2001).

The active systems of Tenerife develop on the flat surfaces of the eastern volcanic platforms. Apart from small fields of nebkhas, one extinct on the coast of Güimar and another preserved in Punta de Abona, the system of greatest extension is formed in El Medano-La Tejita. It is a deposit with more than 100 ha which develops NE-SW, mainly composed by cemented biogenic sands, although with significant residual activity. It is based on recent emissions of basaltic flows. In Gran Canaria, aeolian sedimentary systems are located on almost 40 km of coastline. However, these systems have disappeared from approximately 45% of the coastal perimeter where they were located before the urban-tourist development (Ferrer-Valero et al. 2017). Although they thrive in diverse structural contexts, their natural development has preferentially taken place in the eastern half of the island, which is flatter and more exposed to prevailing winds. Systems such as those of Tufía, Gando and Arinaga, form old inputs of marine sands, mostly consolidated, that advance towards the SSW on lava platforms originated in recent cycles of volcanic reactivation. The coastal platforms of the north, originated by tectonic upheavals, host the formation of small aeolian sedimentary systems, like Bañaderos, disappeared by urban development. Also disappeared, the Guanarteme dune system was developed on a sedimentary structure in tombolo, now occupied by the city of Las Palmas de Gran Canaria. Maspalomas is the only one example in the Canary Islands of a system developed on a fan delta. It is also the most vigorous transgressive dune field in the archipelago, with dunes that exceeded 20 m in height in the pre-touristic period (Hernández Calvento 2006).

In the eastern (older) islands, these systems become a dominant feature of coastal environments. Thanks to the abundance of sandy beaches, predominantly organic, low coastal reliefs and constant winds, a great variety of aeolian sedimentary systems has developed, in number, size and processes. This includes a great heterogeneity of ages and degrees of consolidation, given rise by the formation of successive systems, depending on climatic crises, from the Miocene (Meco et al. 2008). The high marine biological productivity during the hot and humid periods would have generated a lot of sediments. When these sands were exposed to the winds, with the descent of the sea level in the coldest and driest periods, the massive transport of sands towards the inner islands was favored. Extensive dune fields of pre-Quaternary reached elevations of around 400 m.o.s.l. in inland areas of Fuerteventura (Meco et al. 2008).

The extensive basaltic plains of northern Fuerteventura and Lobos, originated in eruptions of the Upper-Pleistocene and Holocene, host large systems that penetrate several kilometers to the south, such as Corralejo, El Cotillo and Majanicho. With the exception of the transgressive dune field of Corralejo, the rest of them are formed by sand sheets of variable volume and nebkhas fields. Along the entire western and southern coast, the extensive tectonic plains associated with the Pliocene marine abrasion platform (IGME 2004), host numerous aeolian sedimentary systems, most of them in a fossil state, although with different degrees of remobilization. Highlights include Las Salinas, Punta de Jandía and Istmo de Jandía. The latter one is, by extension, the second largest in the archipelago. In Lanzarote and La Graciosa, the total extension of coastal dunes and aeolianite is, in relative terms, greater than in Fuerteventura, although it is concentrated in a smaller number of systems. With approximately 8,000 potential hectares (one part has disappeared under urban and agricultural developments, or sand extractions), El Jable is the largest aeolian sedimentary system in the Canary Islands. In this, the sands input from the northwest, through a wide inlet, and crossed the island toward SSE over Quaternary lava plains. The aeolian sedimentary activity is progressively smaller towards the south (Cabrera Vega 2010). The rest of the systems of the islands of Lanzarote and the islets of the Chinijo Archipelago are also developed on extensive lava platforms, the result of a Quaternary volcanic reactivation. Apart from some small and medium-sized systems in the north and northeast of Lanzarote, the most remarkable ones are those of the island of La Graciosa. These are sand sheets and nebkhas fields of variable volume that crossed the island from end to end (8 km), conditioned by Quaternary volcanic reliefs.

The aeolian sedimentary systems in the Canary Islands currently occupy 18,915.48 ha, which is equivalent to 2.54% of the surface (Table 30.1). They are present in all the major islands, as well as in the small island of La Graciosa and in the islets of Alegranza and Lobos.

30.3 Characteristics and Natural Dynamics of the Current Aeolian Sedimentary Systems of the Canary Islands

30.3.1 Aridity: A Key Factor

The aeolian sedimentary systems of the Canary Islands are located mainly on the coast, a factor that conditions their climatic characteristics. The data from the meteorological stations of Maspalomas and La Graciosa (Servicio Hidráulico de

Table 30.1 Surface area occupied by the aeolian sedimentary systems in the Canary Islands

Island	Transgressive dune sheets	Transgressive dunefields	Fossil dunes	Total (ha)	%
Fuerteventura	1,093.69	1,837.19	7,488.88	10,419.76	55.09
Lanzarote	6,286.88	0.00	161.19	6,448.07	34.09
La Graciosa	1,274.75	0.00	8.01	1,282.76	6.78
Gran Canaria	0.91	340.93	201.37	543.21	2.87
Tenerife	103.21	0.00	0.00	103.12	0.55
Lobos	99.20	0.00	0.00	99.20	0.52
Alegranza	0.00	0.00	9.59	9.59	0.05
La Gomera	1.06	0.00	4.22	5.28	0.03
La Palma	2.40	0.00	0.00	2.40	0.01
El Hierro	2.00	0.00	0.00	2.00	0.01
Total (ha)	8,864.10	2,178.12	7,873.26	18,915.48	100
%	46.86	11.52	41.62	100	–

Las Palmas y Agencia Estatal de Meteorología: *Hydraulic Service of Las Palmas and State Meteorological Agency*), representative of the general climatic conditions in the coasts, show that these systems are exposed to high temperatures and scarce rainfall. The average annual temperature is 21 °C in Maspalomas and 19.7 °C in La Graciosa, while the average annual rainfall ranges between 81 mm in Maspalomas and 110 mm in La Graciosa. Both aeolian sedimentary systems present an arid climate (Fig. 30.2), feature that is constant throughout all the months of the year.

The wind is another key factor in the sedimentary dynamics, since it conditions the greater or lesser mobility of the sediment and the formation of aeolian landforms. In the case of Maspalomas, the winds present two main directions: the westerly winds, which account for 19% of the annual frequency, and the E-ENE winds, which represent 15%. The highest speeds are reached in the months of autumn and winter, with an average of 4.5 m/s, predominating the E and NE winds (Máyer Suárez et al. 2012). The effective winds, which produce aeolian sediment transport (>5.1 m/s), suppose, in the case of Maspalomas, 24% of the total annual frequency, concentrated in the winter months (in some months they represent 40% of the records). They come from the E (34%), the W (24%) and the ENE (18%) (Pérez-Chacón et al. 2007; Máyer Suárez et al. 2012, Fig. 30.2).

The most frequent wind components in La Graciosa are NE (27%), NNE (21%) and ENE (17%), being in the summer months when higher speeds are reached, with an average of 4.7 m/s (Pérez-Chacón et al. 2012). Effective winds represent 26% of the total annual frequency, with NE being the dominant direction (32%), followed by ENE (27%) and NNE (15%) (Fig. 30.2).

As a consequence, the aeolian sedimentary transport in Maspalomas is produced from E-NE to W-SW (Hernández Calvento 2006, Máyer Suárez et al. 2012), while in La Graciosa it is from NE to SW (Pérez-Chacón et al. 2012). Although the local

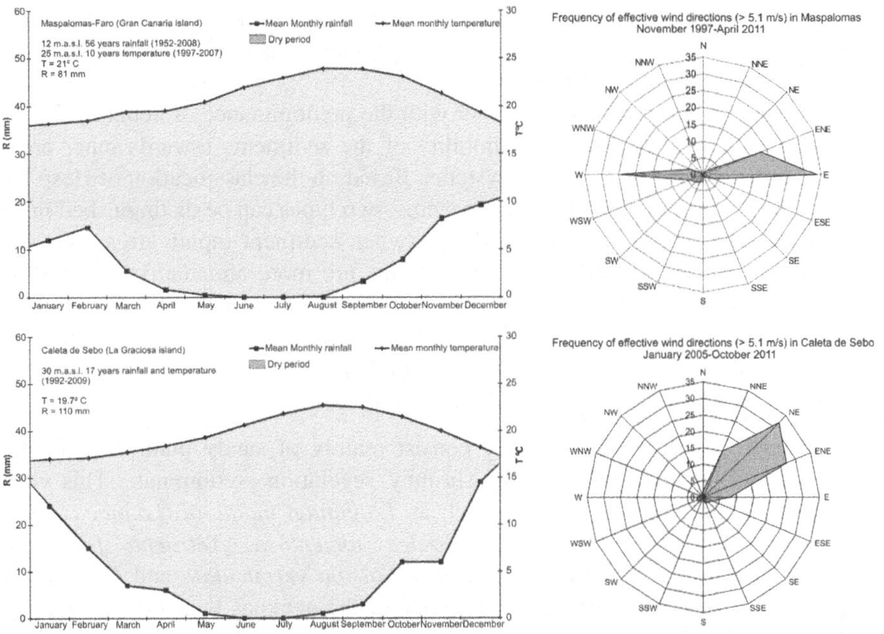

Fig. 30.2 Ombrothermal diagrams and wind roses of aeolian sedimentary systems of Canary Islands. Modified from Hernández Calvento (2006), Pérez-Chacón et al. (2010, 2012), Máyer Suárez et al. (2012) and Hernández-Cordero et al. (2015c)

relief can induce local variations, these transport directions are representative of the totality of the aeolian sedimentary systems of the Canary Islands.

30.3.1.1 Geo-ecological Characteristics and Types of Aeolian Sedimentary Systems

The arid climate and the constant and intense winds (NE trade winds) limit the development of the vegetation, by hydric stress, and condition the geomorphological processes. The scarcity of rainfall favors the aeolian transport throughout the year and the concentration of the salts, contributed by the marine spray, in the substrate. Consequently, plant density is generally low, except in certain areas where water availability increases and/or where sediment transport is reduced. For these reasons, halophilous and xerophilous shrubs predominate, most of them also present in the northwest coast of Africa, such as *Traganum moquinii*, *Salsola vermiculata*, *Polycarpaea nivea* and *Launaea arborescens*. On the other hand, the presence of grasses and rhizomatous or stoloniferous herbaceous species, frequent in the temperate dunes, is scarce (Santos Guerra 1993; Hernández-Cordero et al. 2012, 2015a). In the Canary Islands, the scarce development of the vegetation and the predominance of shrubs have as a consequence that the first dunes generated by the plants (embryonic dunes) do not evolve into dune ridges. The nebkhas thus become characteristic landforms of both the primary stages of the dune formation, as well as the more advanced ones.

The low vegetation density, together with the predominance of nebkhas and the climatic characteristics, favor the mobility of the sediments towards inner areas, being therefore transgressive dune systems. Based on the classification of Hesp and Walker (2013) for transgressive dune systems, two types can be distinguished in the Canary Islands: transgressive dune sheets (when sediment inputs are scarce) and transgressive dunefields (when sediment inputs are more abundant) (Table 30.1). Although they are not the subject of this work, there are also fossil dune systems, some of which present active aeolian sedimentary processes due to the erosion of eolianites and the re-mobilization and accumulation of sands in the form of sand sheets, nebkhas and falling dunes (Alonso et al. 2011).

- *Transgressive dune sheets*: They consist mainly of sandy plains where sand sheets and nebkhas generated by shrubby vegetation predominate. This vegetation is represented by species such as *Traganum moquinii*, *Launaea arborescens*, *Atriplex glauca*, *Chenoleoides tomentosa*, *Tetraena fontanesii*, *Polycarpaea nivea*, *Euphorbia paralias*, *Salsola vermiculata* and *Ononis hesperia*, among others (Fig. 30.3). The size of the nebkhas is associated with the height of the plants that generates them, ranging from a few centimeters to 5 m in height. Apart from the nebkhas, in some sand sheets there are climbing dunes, falling dunes and precipitation ridges. The low volume of sediments in these systems prevents the formation of more complex free dunes such as barchan dunes and barjanoid ridges. However, they can cover considerable extensions and cross the islands from one part to another, giving rise to input and output systems. The input ones are developed in the windward coasts, whereas the output systems are located in the leeward coasts.
  According to Hesp and Walker (2013) these systems can constitute initial stages of transgressive dune fields. Also can represent later phases, induced by natural or anthropogenic reduction of sand supplies. However, their origin can be in the re-stabilization of a previously destabilized system, such as El Jable in Lanzarote (Cabrera Vega 2010; Alonso et al. 2011).
  In the Canary Islands it is common to observe transgressive dune sheets that currently lack marine sediment inputs. Also located in the islands are systems where aeolian sedimentary processes depend on feedback mechanisms, where the water erosion of the deposits located inland, produces the transport of sands, by rills, to the sea, from where they can be reintroduced into the beach-dune systems (Pérez-Chacón et al. 2010).
  The smaller volume of sand facilitates plant colonization, developing different herbaceous and/or shrub plant communities depending on the distance to the coast, the mobility and volume of sediments and the type of substrate (sand or mixture of sand and volcanic materials). Considering the degree of mobility of

Fig. 30.3 Top, active sand sheet and nebkhas field with halophilous shrubs, as *Salsola vermiculata*, *Polycarpaea nivea* and *Traganum moquinii* (Lambra, La Graciosa); bottom, stabilized sand sheet and nebkhas field with grasses and xerophilous shrubs, as *Cenchrus ciliaris* and *Launaea aborescens* (El Jable, Lanzarote)

the sand, it is possible to distinguish between active and stabilized systems (Fig. 30.3):

Active sand sheets and nebkhas fields: the main landforms are mobile sand sheets with ripples, nebkhas and shadow dunes (Pérez-Chacón et al. 2010; García-Romero et al. 2016, Fig. 30.3). On the backshore, ephemeral nebkhas formed by seasonal species such as *Cakile maritima*. In the second band of vegetation develops a scrub dominated by halophilous and psammophilous small shrubs and perennial herbs, such as *Polycarpaea nivea*, *Cyperus capitatus* and *Euphorbia paralias* (phytosociological association *Euphorbio paraliae-Cyperetum capitati*), which induce the formation of small stable nebkhas. Next to this band appears a scrub of *Traganum moquinii* (phytosociological association *Traganetum moquinii*) that forms nebkhas up to 5 m high (Pérez-Chacón et al. 2010; Hernández-Cordero et al. 2015a). The proportion and distribution of these plant communities depend on the specific environmental conditions of each system.

Stabilized sand sheets and nebkhas fields: as the wind weakens inland or sediment inputs are reduced, the mobility of the sand decreases, increasing the density of vegetation. In these cases, sand sheets stabilized with nebkhas (Fig. 30.3) and/or nebkhas dismantled by water and wind erosion predominate. In the latter case, if the sand is not restored by new sedimentary inputs, the underlying substrate may emerge (Pérez-Chacón et al. 2010). Under these environmental conditions, therophytes are abundant and psammophilous herbaceous communities plant are developed, dominated mainly by *Ononis tournefortii* and *Cyperus capitatus* (phytosociological association *Ononido tournefortii-Cyperetum capitati*), which usually develop on greater sand deposits. There are also scrubs of xerophilous and halophilous shrubs that, depending on local environmental conditions, are mainly dominated by *Launaea arborescens*, *Salsola vermiculata*, *Chenoleoides tomentosa*, *Ononis hesperia*, *Lotus spp.* and/or *Polycarpaea nivea* (phytosociological associations *Chenoleoideo tomentosae-Salsoletum vermiculatae*, *Chenoleoideo tomentosae-Suaedetum mollis*, *Launaeo arborescentis-Schizogynetum sericeae*, *Cenchro ciliaris-Launaeetum arborescentis* and *Launaeo arborescentis-Ononidetum hesperiae*) (Pérez-Chacón et al. 2010; Hernández-Cordero et al. 2015a, Fig. 30.3).

- *Transgressive dune fields*: They are represented by the dune fields of Corralejo (Fuerteventura) and Maspalomas (Gran Canaria) (Table 30.1), systems with a large mobile sedimentary volume. The sedimentary dynamic of both systems responds to a cyclical model of sediment input and output (Hernández Calvento 2006; Malvárez et al. 2013). In Maspalomas, sediments input by the East coast (Hernández Calvento 2006) and in Corralejo, along the N-NE coast (Criado 1987), and return to the sea through the S and SE respectively. Subsequently, they are returned by marine dynamics to the input areas (Hernández Calvento 2006; Malvárez et al. 2013). In this process, part of the sediment falls to depths greater than 20 m, which prevents its reintroduction (Hernández Calvento 2006) and causes sediment deficit (Hernández-Calvento et al. 2014).

  Transgressive dunefields show a greater variety and complexity of aeolian landforms, as free dunes (sand sheets, barchan dunes and barchanoid ridges), dunes conditioned by vegetation (eco dunes, parabolic shape dunes, nebkhas, shadow dunes, stabilized dunes and precipitation ridges), erosive landforms (blowouts, slacks and deflation surfaces) or dunes conditioned by relief (eco, climbing, cliff top and falling dunes) (Criado 1987; Hernández Calvento 2006;

Fig. 30.4 **a** Foredune with nebkhas formed by *Traganum moquinii.* Behind it shadow dunes and parabolic shape dunes are observed (Maspalomas, Gran Canaria); **b** mobile dunes in Maspalomas, where barchanoid ridges and slacks are observed (the vegetation located in the slack is formed by a scrub of *Launaea arborescens* and individuals of *Tamarix canariensis*); **c** mobile dunes, where barchan dunes and barchanoid ridges without vegetation and sand sheets with halophilous and xerophilous shrubs as *Salsola vermiculata* and *Launaea arborescens* are observed (Corralejo, Fuerteventura); **d** semi-stabilized dunes with incipient colonization of *Cyperus capitatus* (in forefront); stabilized dunes with seasonal psammophilous herbaceous vegetation of *Cyperus capitatus* and *Ononis tournefortii* and xerophilous scrub of *Launaea arborescens* (in the middle); and mobile dunes and semi-stabilized dunes without vegetation or with sparse individuals of *Tamarix canariensis* (in the distance) (Maspalomas)

Criado et al. 2007; Alonso et al. 2011; Hernández-Cordero et al. 2015c; García-Romero et al. 2017, Fig. 30.4).

While Corralejo is located on a lava platform, Maspalomas sits on a fan delta that determines the existence of shallow water (Hernández Cordero et al. 2006; Hernández-Cordero et al. 2015c) which conditions its geo-ecological characteristics. In Maspalomas wet slacks are developed, absent in Corralejo, which favors the existence of hygrophilous plant species, such as *Tamarix canariensis*, *Cyperus laevigatus* and *Juncus acutus* (Hernández-Cordero et al. 2015c).

The mobility of the dunes in inner areas, together with the aridity, condition the distribution of the vegetation. Plant communities do not present a well-defined spatial arrangement in bands parallel to the coastline. This is due to the fact that the coastal-interior environmental gradient is altered by constant changes produced by sand burial in habitats potentially colonized by plants (Hesp 1991; Hernández-Cordero et al. 2015c). Therefore, plant communities are not linearly associated with different stages of plant succession according to the distance to the coast. This does occur in other dune systems, where pioneer species are generally grasses and herbaceous plants with rhizomes or stolons, while the most advanced species, as the age of the dunes increases, are shrubs and trees (Olson 1958; Chapman 1964). In this sense, the tendency to increase the biomass and the stratification of plant communities inland occurs mainly in dune systems with significant rainfall, which enables the development of vegetation and greater geomorphological stability (Hesp 1991), contrary to what happens in arid regions, such as the Canary Islands coasts. In the transgressive dunefields of these islands, the vegetation presents a mosaic zonation determined by aeolian sedimentary activity and the type of landforms, according to the existence of dunes or slacks. The herbaceous, shrub and arboreal plant communities are distributed indistinctly throughout the whole dune system (Hernández-Cordero et al. 2015c).

There are two clearly differentiated zones in these transgressive dunefields (Hernández-Cordero et al. 2015c): active dunes and stabilized dunes. A transition zone of semi-stabilized dunes, with intermediate characteristics, is also developed (Fig. 30.4).

Active dunes: they present two well differentiated geomorphological and ecological units: the foredune (Fig. 30.4a) and the mobile dunes (Fig. 30.4b, c).

The **foredune** develops in the areas sediment inputs and it is conformed by the first dunes formed on the backshore by plant colonization. They can constitute dune ridges (Hesp 2002) or nebkhas. The latter appear with a balance of positive sand, moderate wind energy and scarce or ineffective vegetation cover (Pye 1990), as in the Canary Islands. The role played by the *Traganum moquinii* shrub as a generator of the foredune is very significant in the dune fields of the Canary Islands, similar to that of *Ammophila arenaria* in the rest of the European and Spanish dunes (Sunding 1972; Hernández-Cordero et al. 2015a, Fig. 30.4a). This species acts as a pioneer and it is also present in the most advanced stages of plant succession (Hernández-Cordero et al. 2012), which differs from the usual geo-ecological succession of the foredune, where the pioneer species are substituted by others, as dune age and distance to the coast increases (Hesp 2002; Hesp and Walker 2013). The prevalence of shrubs and a low plant density in the foredune of the aeolian sedimentary systems of the Canary Islands determine a morphology in hummocks (Hernández-Cordero et al. 2012, Fig. 30.4a).

The studies carried out in Maspalomas indicate that *Traganum moquinii* requires significant water inputs from the phreatic level and relatively sweet water (Viera Pérez 2015), so it colonization begins on slacks, as well as on other areas where access to water table is possible, preferably close to the sea (Hernández-Cordero et al. 2015b, c). The sand burial stimulates the growth of this species, favoring the formation of nebkhas (Hernández-Cordero et al. 2015b; Viera Pérez 2015). These nebkhas slow the sedimentary transport into the system, forming shadow dunes. The entrance of a dune between nebkhas, and its contact with the shadow dunes, can generate parabolic shape dunes that mark the transition to mobile dunes (Hernández-Cordero et al. 2012; Hernández-Calvento et al. 2014; Viera Pérez 2015, Fig. 30.4a).

While in the foredune of Maspalomas the community of *Traganum moquinii* is monospecific, in Corralejo there are other plant species, such as *Euphorbia paralias*, *Cyperus capitatus* and *Cakile maritima* (Fernández Galván et al. 1982; Hernández-Cordero et al. 2015c; Peña-Alonso 2015), due to a lower sand input.

Mobile dunes are characterized by the predominance of free dunes (sand sheets, barchan dunes and barchanoid ridges), along with slacks (Fig. 30.4b, c). The average displacement rate of mobile dunes ranges between 5.8 m/year in Corralejo (Jiménez et al. 2006a) and 8 m/year in Maspalomas, reaching 30 m/year (Pérez-Chacón et al. 2007). The dunes have maximum heights that range between 14 m in Maspalomas (Hernández-Cordero et al. 2015c) and 17 m in Corralejo (Hidtma-Iberinsa 2005). The high mobility rates, together with the large volume of sediments, make plant colonization difficult. In Maspalomas, the critical threshold for the rate of displacement is around 12–13 m/year, after which the vegetation gradually disappears, to do it definitively after 20 m/year (Hernández Cordero et al. 2015). However, there are also zones without vegetation in areas where rates are very low, which seems to be related to the absence of slacks, which are fundamental for the distribution of vegetation in mobile dunes (Hernández-Cordero et al. 2015b, c)

Hernández-Cordero et al. (2015b) determined that plant colonization of mobile dunes in Maspalomas begins in the dry and wet slacks (Fig. 30.4b). Plant species have two patterns of plant colonization against the movement of the dunes: vertical and horizontal. The horizontal colonization pattern, developed by the herbaceous community of *Cyperus laevigatus* and the thicket of *Launaea arborescens*, consists in the progressive distancing, in the direction of the advance of the slip face of the dune, of a new generations of plants. Its survival depends on the displacement of the dune rate and its proximity to the slip face. The vertical colonization pattern, developed mainly by the small tree *Tamarix canariensis*, consists in its growth in height. Its survival depends on its height when the dune and the plant interact: if the plant equals or exceeds the height of the dune, then it will survive; otherwise, it will die buried. When two surviving specimens are close enough, they can transform the barchanoid ridges into parabolic dunes. This reduces the displacement of the dune rate and facilitates the colonization of new plants in the slacks localized to leeward. *Traganum moquinii* can develop both colonization patterns (horizontal and vertical), due to its positive reaction to sand burial and its location in areas where the dunes reach different heights. In this sense, it is located in wet slacks close to the coast, as well as in the output areas of the sediments to sea (Criado 1987, Hernández-Cordero et al. 2015c).

In Maspalomas, the main pioneer species in the direct colonization of the dunes is *Cyperus capitatus* (Hernández-Cordero et al. 2015c; Hernández-Cordero et al. 2017, Fig. 30.4d), associated with the phytosociological association *Ononido tournefortii-Cyperetum capitati*.

Also in Corralejo most of the vegetation is located in the slacks, although the dunes have a greater plant colonization than in Maspalomas, acting as pioneer species some psammophilous species, such as *Euphorbia paralias* and *Cyperus capitatus* (phytosociological association *Euphorbio paraliae-Cyperetum capitati*). This seems to be related to the lower displacement rates of the dunes in this system or to the stabilization processes due to the reduction of sand inputs.

Stabilized dunes: when the dunes are totally colonized by the vegetation, the aeolian sand transport is reduced. In this cases, the dunes become sandy accumulations with a topography formed by high and depressed areas that coincide with the dunes and the slacks of the pre-existing mobile dunes (Fig. 30.4d). In Maspalomas and Corralejo, the processes of stabilization of mobile dunes have been originated mainly by human activities (Hernández Calvento 2006; Alonso et al. 2011; Málvarez et al. 2013; García-Romero et al. 2016; Hernández-Cordero et al. 2017).

Stabilized dunes can also develop naturally due to the progressive remoteness of sediment sources, the increase of fine sediments and plant colonization (Fernández Galván et al. 1982; Criado 1987). In the stabilized dunes, herbaceous, shrubby, and, in the case of Maspalomas, arboreal communities are developed (Fig. 30.4d). In the dunes with a greater accumulation of sand and coinciding with the top of old barchanoid ridges and barchan dunes, therofitic meadows of *Cyperus capitatus* and *Ononis tournefortii* (phytosociological association *Ononido tournefortii-Cyperetum capitati*) are established. As the process of plant succession progresses, it becomes a thicket of *Launaea arborescens* (phytosociological association *Launaeo arborescentis-Schizogynetum sericeae*), in the case of Maspalomas. In Corralejo, scrubs of *Salsola vermiculata* (phytosociological association *Chenoleoideo tomentosae-Salsoletum vermiculatae*), *Polycarpaea nivea* or *Ononis hesperia* (phytosociological association *Launaeo arborescentis-Ononidetum hesperiae*) are also developed, depending on local environmental conditions. The existence of wet slacks in Maspalomas favors the development of small forests of *Tamarix canariensis* (phytosociological association *Atriplici ifniensis-Tamaricetum canariensis*) and herbaceous communities of *Juncus acutus* (phytosociological association *Scirpo globiferi-Juncetum acuti*).

30.4 Land Uses and Their Consequences

The recent territorial transformation that took place in the Canary Islands, derived from the change of the socio-economic model, has had important implications in the geomorphological and functional changes of the aeolian sedimentary systems. Changes in land use have varied over time due to the prevailing socioeconomic patterns in each historical moment. We can distinguish two major stages, one historical, prior to tourism development (until 1960s and 1970s) and another one recent and current (after these decades).

30.4.1 Historical Land Uses (<1960–70)

By the eighteenth century these sandy systems lacked value on the part of society, since they were sterile spaces from the agricultural point of view and, in the case of the transgressive dunefields, also from the livestock point of view, due to the mobile dunes. In the vicinity of these dunefields, we must also consider the inconvenience that constituted the sedimentary transport, which caused road cuts or invasion of crops and infrastructures, such as in the Guanarteme dunefield (Santana-Cordero et al. 2014). The dominant uses around of this dunefield, between 1834 and the 1960s, were agriculture, urbanization, recreational uses and sand mining. The first uses detected inner the dunefield, in the nineteenth century, are urbanization and agriculture, which were introduced by the north, south and west, gradually. Already in the twentieth century these uses continue to expand, subtracting surface to the dunefield. Although all uses contributed in different ways to degrade this system, the sand mining and the urbanization were the main causes of its termination. The sale of the lands located within the dunefield and the implementation of a brick factory, due to the need for expansion of the city, eliminated the dunes. The main driver in this process was the development of the port of La Luz and Las Palmas, close to the dunefield, which boosted the island's economy (Santana-Cordero et al. 2016a). In the 1960s came the collapse of the dunefield and its subsequent disappearance.

Trangressive dune sheets, on the other hand, can support agricultural uses, although this implies their deforestation and, therefore, the consequent remobilization of the sediment. The vegetation itself, in any case, also supports exploitation, such as cattle or fuel. Such is the case of La Graciosa, which transgressive dune sheets began to be used massively for livestock since 1730–1736, when the volcanic complex of Timanfaya (Lanzarote) erupted and new lands had to be found to sustain this use. Since then, La Graciosa had a seasonal livestock activity, although human permanent settles were restricted. In 1880 the first permanent settles established in the island, moment since which the population experienced a continuous growth until half-full of century XX, and also the number of houses and other constructions. Also the cattle activity continued increasing, along with the felling of bushes for domestic fuel and for the burning of lime kilns (Santana-Cordero et al. 2016b). There were constant warnings from the authorities that these activities produced the re-mobilization of the sands, invading crops and homes, and legally limiting certain activities. However, most of them continued in time, reaching its maximum intensity in the 1970s. This moment coincides with the development of active processes of aeolian sedimentary dynamics, as it is derived from the existence of some free dunes. The introduction in this decade of domestic gas and electricity decreased deforestation. Also in this decade the last lime kilns were used, so that the pressure on the vegetation decreased significantly. In 1987 the island was declared as protected natural space. Some traditional activities (such as the mining or extensive grazing) were prohibited.

Also the remobilization of the sands, by cutting down the vegetation, supposed, in the case of El Jable (Lanzarote), a threat to the crops located within the system itself, as well as to some surrounding nuclei (Cabrera Vega 2010).

30.4.2 *Recent and Current Land Uses (>1960–70)*

Since the 1960s and 1970s, there has been a change in the perception of coastal sandy systems, as a consequence of their exploitation as tourist resources. Unlike what happened in the previous stage with the transgressive dunefield of Guanarteme, most of the surfaces of the other two transgressive dunefields (Maspalomas and Corralejo), which had been scarcely exploited until then, have been conserved. In fact, a significant part of the aeolian sedimentary systems of the islands have become protected at different scales (regional, national and international). In parallel, the pressure on these spaces has been increasing during the last decades. So much has been the case that most of the changes that have occurred in the aeolian sedimentary systems of the Canary Islands in recent decades have been motivated, directly or indirectly, by urban-touristic development, being the main actions the mining and the construction of buildings and infrastructures (Cabrera Vega 2010; Alonso et al. 2011; Cabrera-Vega et al. 2013; Hernández-Calvento et al. 2014; García-Romero et al. 2016).

The main impacts that have occurred in these systems can be classified into three categories, depending on their nature and the intensity of the disturbances: urban development, management (including beaches services) and user's activities (Table 30.2). With regard to urban development, large volumes of sand have been extracted and the natural systems have been partially occupied by buildings and infrastructures, especially at their margins. Some of these buildings have modified the dynamics of the winds in their surroundings, altering the aeolian transport. This alteration has manifested by two ways: in areas located leeward the buildings, there has been a partial blockage of sedimentary transport, which, in turn, has favored plant colonization and the consequent formation of stabilized dunes; in other areas around the urbanization there has been an acceleration of transport, generating erosive processes and the formation of deflation surfaces, as well as changes in the direction of transport (Alonso et al. 2011; Hernández-Calvento et al. 2014; Smith et al. 2017). In the case of Maspalomas, the alteration of the aeolian sedimentary dynamics by a touristic resort, together with the occupation of part of the system by buildings, has meant significant changes in the vegetation: on the one hand, there has been an increase of most of the plant communities, due to the reduction of the sand contributions and the formation of deflation surfaces in certain zones. On the other hand, there has been a decrease in some plant communities, such as scrub of *Traganum moquinii* (Hernández-Cordero et al. 2012, 2017).

Beach services (kiosks, sunbeds, umbrellas and service vehicles) have generated other types of impact, such as the formation of wind shadows to the leeward of some static elements, altering the pre-existing landforms, especially in the foredune.

Table 30.2 Major causes and consequences of the landforms changes in the aeolian sedimentary systems of the Canary Islands

Causes		Impacts				
Drivers	Pressures	Disappearance	Erosion	Directional changes	Accumulation	Stabilization
I. Urban development	1. Sediment extraction	Quantity of sediment, and dunes				
	2. Dunefield occupation		Deflation surfaces leeward de urbanization	Vertex dunes, alteration of sand transport direction in the entire active dunefield	Windward-echo dunes, bungalow-top dunes, larger dunes	Leeward the urbanization
II. Management	1. Facilities on beach-dune system (kiosks and sunbeds)	Sand sheets	Wind shadows leeward the facilities		Bulldozed dune, camel dune and accumulation round facilities	
	2. Cleaning activities at the beach	Primary dunes				Cobbled base dune, echo dunes
III. User' activities	1. Trampling	Micro-landforms	Remobilization			
	2. Goros	Nebkhas	Wind shadows leeward the goros		Windward and both sides leeward the goros	
	3. Others (felling vegetation, urination)	Nebkhas	Remobilization			
IV. Combination			Deflation areas (II + III)		Accumulation ridge (I + III)	

Modified and updated from Cabrera-Vega et al. (2013)

Table 30.3 Environmental changes in the transgressive dunefields of Maspalomas (Gran Canaria) and Corralejo (Fuerteventura) through evolution of landforms, bare sand and vegetation

	Maspalomas (1961–2003)		Corralejo (1969–2003)	
Landforms	Surface variation (ha)	Variation in the system (%)	Surface variation (ha)	Variation in the system (%)
Barchanoid ridges, barchan dunes and sand sheets	−156.9	−47.8	−602.1	−61.6
Foredune	−3.7	−28	−0.6	−4.4
Nebkhas	−33.0	−73.7	199.4	33.0
Stabilized dunes	68.5	305.8	131.2	27.4
Deflation surfaces	22.1	218.8	3.1	40.3
Other landforms[a]	−32.5	−100	0.0	0.0
Total of the dunefield	−114	−24	−268.7	−12.9
Vegetation	35	38.8	82.9	8.9
Bare sand	−149.1	−61.3	−368.5	−32

Modified from García-Romero et al. (2016)

[a]Only in the dune system of Maspalomas and it includes cliff-top dunes, echo dunes, falling dunes and climbing dunes

The activity of the users has also produced changes in the foredune, mainly by alteration of vegetation, due to the use of plants as urinals or by the construction of stone structures around the plants, named *goros* (Hernández-Cordero et al. 2012). In the case of Maspalomas, synergy has been detected in the impacts, the consequence of which has been the disappearance of some stretches of the foredune and the formation of an accumulation ridge, which advances rapidly towards the inner system (Hernández-Calvento et al. 2014).

Considering jointly the impacts produced on the aeolian geomorphology in the two dunefields (Table 30.3), a significant reduction in the surfaces occupied by most of the landforms is detected, some of which have disappeared. Given the role it plays in the protection of these systems and in the development of their characteristic processes, the case of the foredunes is relevant. Its surface area has decreased by 4.4% in the case of Corralejo, and by 28% in the case of Maspalomas. Also barchanoid ridges, barchan dunes and sand sheets have been replaced by landforms of smaller volume or dynamics. On the contrary, stabilized dunes and deflation surfaces have experienced a considerable increase (305.8 and 218.8%, respectively, in the case of Maspalomas, and 27.4 and 40.3%, respectively, in the case of Corralejo) (Table 30.3 and Fig. 30.5).

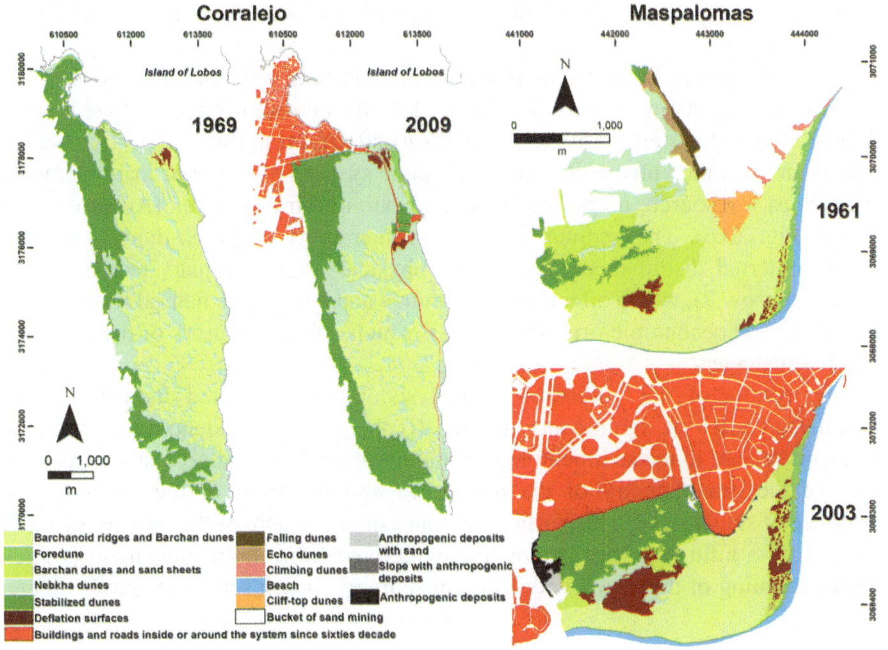

Fig. 30.5 Recent changes in the dune fields of Corralejo (left) and Maspalomas (right). Modified from García-Romero et al. (2016)

30.5 Deficiencies and Challenges in the Management of the Aeolian Sedimentary Systems of the Canary Islands

The arid nature of the aeolian sedimentary systems of the Canary Islands gives them a sensitive nature due to the intensification of natural processes linked to scarce rainfall and high temperature (Hesp 1991; Cabrera-Vega et al. 2013). The anthropogenic actions, continuous throughout the year (there is no seasonality in the arrival of tourists), affect these processes, altering them. The alterations experienced in the last 60 years have meant an increase in their geomorphological fragility, making these systems increasingly vulnerable. Peña-Alonso (2015) and Peña-Alonso et al. (2016) analyzed, from a geomorphological point of view, the vulnerability of these systems, by means of indicators adapted to their characteristics. Vulnerability is understood as the relationship between three groups of main factors: (i) exposure factors, both natural (marine and wind impact) and anthropic (urbanization, management activities and activities carried out by users); (ii) susceptibility factors, or intrinsic factors (geomorphological and vegetation characteristics); and (iii) geomorphological resilience (Cardona 2006; IPCC 2007), defined as the capacity of each system to maintain its functions over time, within its own variability.

The relationships between the biophysical and social processes that occur in the aeolian sedimentary systems of the Canary Islands have allowed to analyze them as socio-ecological systems (Curtin and Prellezo 2010). From this perspective we can understand that the vulnerability of beach-dune systems affects four main elements: systems as a resource, users, managers and public infrastructures (Anderies et al. 2004) (Fig. 30.6). The geomorphological vulnerability of beach-dune systems is due to external disturbances, both of biophysical origin (climate, tectonics, volcanism) (arrow 7), which affects infrastructures and systems as natural resources, as well as of socioeconomic origin (arrow 8), which affects the users of the resources and the managers.

The results of the vulnerability analysis, through adapted indicators, have revealed a series of significant patterns: (i) the marine incidence affects all the elements related to the sedimentological nature of the beach-dune systems, as is the case of the granulometry of the dry beach and the front of the foredune, the presence of shells and the proportion of stones on the beach, or the number of sandy bars; (ii) the influx of users, the activities derived from the management (vehicle traffic, cleaning of the beach with heavy machinery, beach services) and the degree

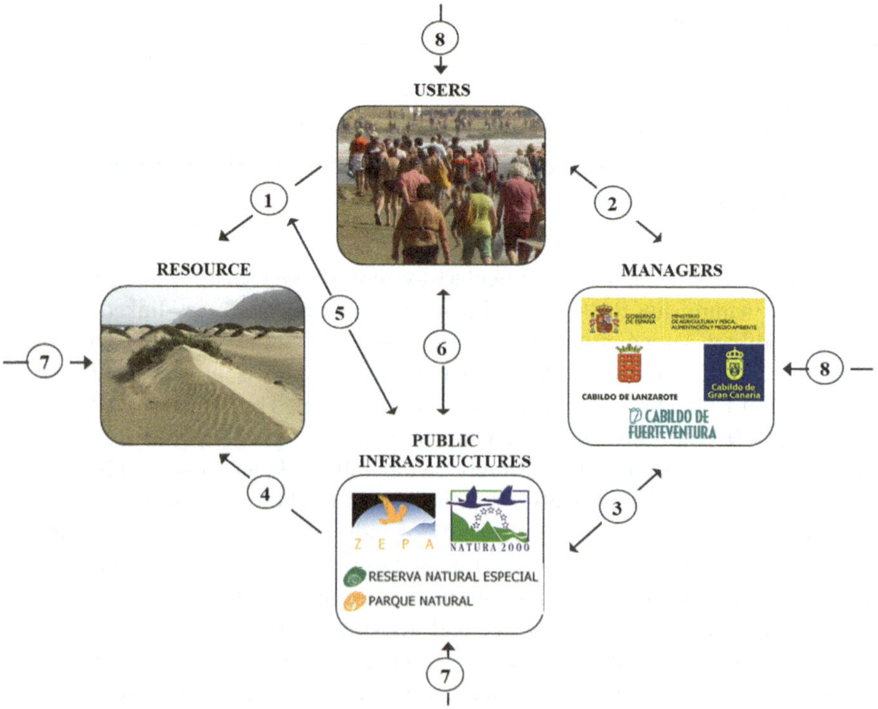

Fig. 30.6 Socio-ecological system of the aeolian sedimentary systems of the Canary Islands

of urbanization affect the state of the vegetation (coverage, height, vigor, and type) and the geomorphology-sedimentology (quantity and distance between nebkhas, presence of embryonic dunes, height of the foredune, granulometry of the front of the foredune and the beach, shells and stones on the beach); and (iii) the geomorphological resilience in the most frequented coastal zones is influenced by the management measures carried out in recent decades, mainly respect to vehicle traffic and beach cleaning with heavy machinery. As a consequence, a loss of systemic equilibrium has been generated, which manifests in the superficial decrease of the foredune and in the increase of the distance between nebkhas. In some cases, as in Maspalomas, the tendency of change results in the loss of the system's capacity to maintain its geomorphological functions as they were before the touristic development.

The analysis of these processes in the beach-dune systems of the transgressive dunefields of Maspalomas and Corralejo, as well as of the transgressive dune sheets of El Jable and Las Conchas (La Graciosa), has allowed to observe factors that determine their weaknesses and strengths geomorphological (Peña-Alonso et al. 2016). On the one hand, high degrees of geomorphological susceptibility (intrinsic weakness) are related to granulometries greater than 2 φ on the front of the foredune and on the dry beach, which in turn is determined by strong waves and/or low input of fine sediments to the systems. High degrees of geomorphological susceptibility are related, in turn, to the absence of embryonic dunes and submerged or emerged sandy or rocky bars, as well as the scarcity of vegetation in the foredune. On the other hand, elements of strength have been identified, derived from the absence of erosion scarps in the foredunes, as well as from the considerable beach widths (between 50 and 100 m at low tide), dissipative modal states and scarcity of deflation surfaces.

More recent studies have allowed to know that the geomorphological processes and the landforms are in the basis of all the biophysical and social relationships that make up these systems (Peña-Alonso et al. 2017). Therefore, the geomorphological resource is the key element of the economic development of the environment, as well as its natural wealth. Consequently, the natural processes that occur in the aeolian sedimentary systems of the Canary Islands are fundamental in the economic development of the archipelago. However, until now there are no management tools adapted to their functional particularities, being the norm the application of tools created to manage similar systems in temperate regions. The Government of Spain has encouraged a series of actions based on studies, recommendations and strategies for managing dune systems. The works carried out by García-Mora et al. (2000) and García-Mora et al. (2001) served as a reference to define the problems of dunes in Spain within the framework of the *Manual de restauración de dunas costeras* (Manual for the restoration of coastal dunes, Ley et al. 2007). However, studies on the vulnerability of the aeolian sedimentary systems of the Canary Islands (Hernández-Calvento 2006; Jiménez et al. 2006b; Government of the Canary Islands 2004, 2006a, b; Medina et al. 2007; Cabrera Vega 2010; Pérez-Chacón et al. 2010; Hernández-Cordero 2012; Viera Pérez 2015; Peña Alonso 2015), have shown that indicators systems, as well as many management measures, contained in the aforementioned *Manual* (Ley et al. 2007) are not directly applicable.

Having observed such deficiencies in the application of evaluation methodologies and adequate management measures, some important management challenges have been identified.

The first one would be related to the analysis of the geomorphological vulnerability of the totality of the aeolian sedimentary systems of the Canaries at a detailed scale. The use of standard evaluation methodologies by means of indicators systems, approached from a multidisciplinary perspective, would be the basis of an adequate and specific management of these systems.

The second challenge is the need to address programs for the public use of these spaces, which allow their sustainable use. This would imply the realization of compatibility proposals between tourist and recreational activities and the conservation of natural processes. For this, the following premises should be taken into account: (i) the need to inform about environmental values and about the importance of conserving them; (ii) the need to report on prohibited and permitted uses; (iii) enforce the regulations established for each protected natural space, maintaining a permanent surveillance; (iv) protect the dunes, especially those stabilized or in the process of stabilization, as well as the foredune, of the alterations derived from the trampling of the users and the transit of vehicles. In this sense, it is proposed: (a) the correct signaling in some languages of the prohibited and permitted uses; (b) the channeling of users to the beaches through controlled accesses; (c) cordon off the most sensitive areas, such as the foredunes; (v) address the sustainable management of beaches, limiting actions that may alter the dynamics and morphology of foredunes or destroy embryonic dunes. This premise would go through some changes, such as: (a) limit the use of heavy machinery in cleaning and conditioning works in the vicinity of the foredunes; and (b) the arrangement of the equipment of the beaches (kiosks, hammocks, etc.), so that their design, location or time of permanence do not induce significant alterations in the aeolian sedimentary transport.

The third challenge is the establishment of suitable environmental restoration methods for the regeneration of degraded systems. This must be done taking into account: (i) the restoration of the foredunes, where they have been destroyed or fragmented, must be undertaken through the generation of nebkhas with specimens of *Traganum moquinii*. In some systems other species could be used, depending on the current and historical structure and dynamics of each case; (ii) the distance between the nebkhas should allow the aeolian sedimentary transport to the inner systems, in order to facilitate the development of mobile dunes, sand sheets or active nebkhas fields; and (iii) in environmental restoration works, the introduction of plant species, whether autochthonous or exotic, different from those indicated for each system, must be prevented.

The fourth of these challenges is the development and implementation of control programs for invasive exotic plant species, lasting over time. Specifically, it would be necessary to control and, if possible, eradicate species such as *Neurada procumbens* and *Sesuvium portulacastrum* in Maspalomas, as well as *Nicotiana glauca* in Maspalomas and *Corralejo*. These species can produce significant ecological alterations, such as the displacement of autochthonous species, changes in the floristic composition of plant communities and alteration of plant succession processes (Hernández-Cordero 2012). Ultimately, these alterations can impact on aeolian sedimentary processes.

Finally, it is essential that follow-up programs of the implemented measures are carried out. The capacity of these systems to show geomorphological changes in short periods of time (weeks-months) after changes in their environment makes periodic evaluation necessary. The new processes that could be identified over time must be taken into account when managing these systems in order to carry out an adaptive management that fits their dynamics.

Acknowledgements This book chapter is a contribution to the REN2003-05947, SEJ2007-64959, CSO2010-18150, CSO2013-43256-R and CSO2016-79673-R projects of the Spanish National Plan for R+D+i (innovation), co-financed with ERDF funds. Also to the program of support for research of the Canary Islands Government, also co-financed with ERDF funds, and to the fellowship program of the University of Las Palmas de Gran Canaria. Some results have been obtained from studies funded by the Organismo Autónomo Parques Nacionales (the 'Isla de La Graciosa' Centre) and by the former Dirección General de Costas, both of the Ministerio de Agricultura, Alimentación y Medio Ambiente, Gobierno de España. The cartography of the current aeolian sedimentary systems of the Canary Islands has been developed for the Dirección General de Protección de la Naturaleza, Gobierno de Canarias, as part of the service "Tipificación, caracterización y correlación con hábitats dunares de interés comunitario de los sistemas dunares de Canarias" (Resolución nº 412/2017 de 09/10/2017), financed by the ERDF Canarias Operational Program (2014–2020).

References

Alonso, I., Hernández, L., Alcántara-Carrio, J., Cabrera, L., & Yánez, A. (2011). Los grandes campos de dunas de Canarias. In E. Sanjaume & J. Gracia (Eds.), *Las Dunas de España* (pp. 467–496). Sociedad Española de Geomorfología.

Anderies, J. M., Janssen, M. A., & Ostrom, E. (2004). A framework to analyze the robustness of social-ecological systems from an institutional perspective. *Ecology and Society, 9*(1), 18. http://hdl.handle.net/10535/2858.

Cabrera Vega, L. L. (2010). *Sedimentología, estratigrafía, dinámica sedimentaria y evolución de El Jable (Lanzarote). Propuesta de gestión* (Ph.D. Thesis). Universidad de Las Palmas de Gran Canaria, 252 pp. https://acceda.ulpgc.es/handle/10553/4852.

Cabrera-Vega, L. L., Cruz-Avero, N., Hernández-Calvento, L., Hernández-Cordero, A. I., & Fernández-Cabrera, E. (2013). Morphological changes in dunes as indicator of anthropogenic interferences in arid dune fields. *Journal of Coastal Research Special Issue, 65,* 1271–1276. https://doi.org/10.2112/si65-215.1.

Cardona, O. D. (2006). A system of indicators for disaster risk management in the Americas. In *Measuring vulnerability to natural hazards. Towards disaster resilient societies.*

Carracedo, J. C., Day, S., Guillou, H., Rodríguez Badiola, E., Canas, J. A., & Pérez Torrado, F. J. (1998). Hotspot volcanism close to a passive continental margin: The Canary Islands. *Geological Magazine, 135*(5), 591–604.

Criado, C. (1987). Evolución geomorfológica y dinámica actual del jable de Corralejo (Fuerteventura, Islas Canarias). *Revista Geografía de Canarias, 2,* 29–52.

Criado, C., Guillou, H., Hansen, A., Hansen, C., Lillo, P., & Torres, J. M., et al. (2007). Mapa geomorfológico del Parque Natural de las Dunas de Corralejo. In *XIII Jornadas de estudio de Lanzarote y Fuerteventura* (pp. 399–427). Tomo II.

Criado, C., Yánez, A., Hernández, L., & Alonso, I. (2011). Origen y formación de los depósitos eólicos de Canarias. In E. Sanjaume & J. Gracia (Eds.), *Las Dunas de España* (pp. 447–466). Sociedad Española de Geomorfología.

Chapman, V. J. (1964). *Coastal vegetation* (p. 242). London: Pergamon Press.

Curtin, R., & Prellezo, R. (2010). Understanding marine ecosystem based management: A literature review. *Marine Policy, 34*(5), 821–830.

Fernández Galván, M., Santos Guerra, A., & Domínguez, F. (1982). El complejo dunar de Corralejo (Fuerteventura). *Gaceta de Canarias, 3,* 73–83.

Ferrer-Valero, N., Hernández-Calvento, L., & Hernández-Cordero, A. I. (2017). Human impacts quantification on the coastal landforms of Gran Canaria Island (Canary Islands). *Geomorphology, 286,* 58–67. https://doi.org/10.1016/j.geomorph.2017.02.028.

García-Mora, M. R., Gallego-Fernández, J. B., & García-Novo, F. (2000). Plant diversity as a suitable tool for coastal dune vulnerability assessment. *Journal of Coastal Research,* 990–995.

García-Mora, M. R., Gallego-Fernández, J. B., Williams, A. T., & Garcia-Novo, F. (2001). A coastal dune vulnerability classification. A case study of the SW Iberian Peninsula. *Journal of Coastal Research,* 802–811. http://www.jstor.org/stable/4300241.

García-Romero, L., Hernández-Cordero, A. I., Fernández-Cabrera, E., Peña-Alonso, C., Hernández-Calvento, L., & Pérez-Chacón, E. (2016). Urban-touristic impacts on the aeolian sedimentary systems of the Canary Islands: Conflict between development and conservation. *Island Studies Journal, 11*(1), 91–112.

García-Romero, L., Hernández-Cordero, A. I., Delgado-Fernández, I., Hesp, P. A., Hernández-Calvento, L., & Viera-Pérez, M. (2017). Evolución reciente de geoformas erosivas inducidas por impacto urbano-turístico en el interior de un sistema de dunas transgresivo árido (Maspalomas, islas Canarias). *Geo-Temas, 17,* 263–266.

Gobierno de Canarias. (2004). *Plan Director de la Reserva Natural Especial de la Dunas de Maspalomas*. Consejería de Medio Ambiente y Ordenación Territorial. Dirección General de Ordenación del Territorio, 85 pp.

Gobierno de Canarias. (2006a). *Plan Rector de Uso y Gestión del Parque Natural del Archipiélago Chinijo: Introducción y memoria informativa*. Consejería de Medio Ambiente y Ordenación Territorial. Dirección General de Ordenación del Territorio, 214 pp.

Gobierno de Canarias. (2006b). *Plan Rector de Uso y Gestión del Parque Natural de Corralejo: Documento informativo*. Consejería de Medio Ambiente y Ordenación Territorial. Dirección General de Ordenación del Territorio, 109 pp.

Hernández Calvento, L. (2006). *Diagnóstico sobre la Evolución del Sistema de Dunas de Maspalomas (1960–2000)*. Las Palmas de Gran Canaria: Cabildo de Gran Canaria. (361 pp).

Hernández-Calvento, L., Jackson, D. W. T., Medina, R., Hernández-Cordero, A. I., Cruz, N., & Requejo, S. (2014). Downwind effects on an arid dunefield from an evolving urbanised area. *Aeolian Research, 15,* 301–309. https://doi.org/10.1016/j.aeolia.2014.06.007.

Hernández Calvento, L., Jackson, D. W. T., Cooper, A., & Pérez-Chacón, E. (2017). Island-encapsulating eolian sedimentary systems of the Canary and Cape Verde archipelagos. *Journal of Sedimentary Research, 87,* 117–125. https://doi.org/10.2110/jsr.2017.6.

Hernández Cordero, A. I., Pérez-Chacón Espino, E., & Hernández Calvento, L. (2006). Vegetation colonisation processes related to a reduction in sediment supply to the coastal dune field of Maspalomas (Gran Canaria, Canary Islands, Spain). *Journal of Coastal Research, 48,* 69–76. (Special Issue).

Hernández-Cordero, A. I. (2012). *Análisis de la vegetación como indicadora de las alteraciones ambientales inducidas por la actividad turística en la Reserva Natural Especial de las dunas de Maspalomas* (Ph.D. Thesis). Universidad de Las Palmas de Gran Canaria, 574 pp. http://hdl.handle.net/10553/9768.

Hernández-Cordero, A. I., Pérez-Chacón Espino, E., & Hernández-Calvento, L. (2012). La investigación como soporte de la gestión: el ejemplo de la duna costera (foredune) de

Maspalomas (Gran Canaria, Islas Canarias). In A. Rodríguez-Perea, G. X. Pons, F. X. Roig-Munar, J. A. Martín-Prieto, M. Mir-Gual, & J. A. Cabrera (Eds.), *Gestión integrada de playas y dunas: Experiencias en Latinoamérica y Europa* (289–306). Palma de Mallorca: Monografies de la Societat d'Història Natural de les Balears 19.

Hernández-Cordero, A. I., Gracia-Prieto, F. J., Hernández-Calvento, L., Pérez-Chacón Espino, E., & Alonso, I. (2015a). Proposal for new EU habitats associated with coastal dune fields of the Macaronesian region. A case study in the Canary Islands (Spain). *Journal of Coastal Conservation, 19,* 213–225. https://doi.org/10.1007/s11852-015-0382-z.

Hernández-Cordero, A. I., Hernández-Calvento, L., & Pérez-Chacón Espino, E. (2015b). Relationship between vegetation dynamics and dune mobility in an arid transgressive coastal system, Maspalomas, Canary Islands. *Geomorphology, 238,* 160–176. https://doi.org/10.1016/j.geomorph.2015.03.012.

Hernández-Cordero, A. I., Pérez-Chacón Espino, E., & Hernández-Calvento, L. (2015c). Vegetation, distance to the coast, and aeolian geomorphic processes and landforms in a transgressive arid coastal dune system. *Physical Geography, 36*(1), 60–83. https://doi.org/10.1080/02723646.2014.979097.

Hernández Cordero, A. I., García Romero, L., Hernández Calvento, L., & Pérez-Chacón Espino, E. (2015). Tecnologías de la información geográfica aplicadas al análisis de la relación entre la vegetación y las tasas de desplazamiento de dunas en Maspalomas (Gran Canaria, Islas Canarias). *SÉMATA, Ciencias Sociais e Humanidades, 27,* 95–114.

Hernández-Cordero, A. I., Hernández-Calvento, L., & Pérez-Chacón Espino, E. (2017). Vegetation changes as an indicator of impact from tourist development in an arid transgressive coastal dune field. *Land Use Policy, 64,* 479–491. https://doi.org/10.1016/j.landusepol.2017.03.026.

Hesp, P. A. (1991). Ecological processes and plant adaptations on coastal dunes. *Journal of Arid Environments, 21,* 165–191.

Hesp, P. A. (2002). Foredunes and blowout: initiation, geomorphology and dynamics. *Geomorphology, 48,* 245–268.

Hesp, P. A., & Walker, I. J. (2013). Coastal Dunes. In J. F. Shroder, N. Landcaster, D. J. Shermann, A. C. W. Baas (Eds.), *Aeolian geomorphology. Treatise on geomorphology* (Vol. 11, pp. 328–355). San Diego: Academic Press.

HIDTMA-IBERINSA. (2005). *Estudio ecocartográfico del litoral de la isla de Fuerteventura.* Dirección General de Costas, Ministerio de Medio Ambiente.

Instituto Geológico y Minero de España, IGME. (2004). *Mapa Geológico de España 1:25.0000 y memoria explicativa, Isla de Fuerteventura.*

IPCC. (2007). *Climate change 2007: the physical science basis. Contribution of Working Group I to the fourth assessment report of the Intergovernmental Panel on Climate Change.* Geneva, Switzerland: IPCC, 104 pp.

Jiménez, J. A., Alonso, I., & Valdemoro, H. I. (2006a). *Estudio del estado actual de los sistemas dunares de Corralejo y Jandía (Fuerteventura). Identificación de problemas y propuestas de actuación. Informe I, Corralejo.* Dirección General de Costas, Ministerio de Medio Ambiente.

Jiménez, J. A., Alonso, I., & Valdemoro, H. I. (2006b). *Estudio del estado actual de los sistemas dunares de Corralejo y Jandía (Fuerteventura). Identificación de problemas y propuestas de actuación. Informe II, Jandía.* Dirección General de Costas, Ministerio de Medio Ambiente.

Ley, C., Gallego-Fernández, J. B., & Pascual, C. (2007). *Manual de restauración de dunas costeras.* Ministerio de Medio Ambiente y Medio Rural y Marino, 251 pp.

Malvárez, G., Jackson, D., Navas, F., Alonso, I., & Hernández-Calvento, L. (2013). A conceptual model for dune morphodynamics of the Corralejo dune system, Fuerteventura, Spain. *Journal of Coastal Research Special Issue, 65,* 1539–1544. https://doi.org/10.2112/si65-260.1.

Máyer Suárez, P., Perez-Chacón Espino, E., Cruz Avero, N., & Hernández-Calvento, L. (2012). Características del viento en el campo de dunas de Maspalomas (Gran Canaria, Islas Canarias, España). *Nimbus, 29–30,* 381–397.

McDougall, I. (1971). Volcanic island chains and seafloor spreading. *Nature Physical Science, 231,* 141–144.

Meco, J., Scaillet, S., Guillou, H., Lomoschitz, A., Carracedo, J. C., Ballester, J., et al. (2007). Evidence for long-term uplift on the Canary Islands from emergent Mio-Pliocene littoral deposits. *Global and Planetary Change, 57*(3), 222–234. https://doi.org/10.1016/j.gloplacha.2006.11.040.

Meco, J., Ballester, J., Betancort, J. F., Cilleros, A., Scaillet, S., & Guillou, H., et al. (2008). *Historia geológica del clima en Canarias*. Las Palmas de Gran Canaria, 296 pp.

Medina, R., Camus, P., Requejo, S., Luque, A., Alonso, I., Hernández, L., et al. (2007). *Estudio integral de la playa y dunas de Maspalomas (Gran Canaria)*. Madrid: Ministerio de Medio Ambiente.

Morales, E., Contreras, H., López, H., & Oromí, P. (2001). Artrópodos de Puntallana (La Gomera): especies de particular interés y su conservación. *Revista de la Academia Canaria de las Ciencias, 13*(4), 153–165.

Olson, J. S. (1958). Rates of succession and soil changes on southern Lake Michigan sand dunes. *Botanical Gazette, 119*(3), 125–170.

Peña-Alonso, C. (2015). *Diseño y aplicación de indicadores de vulnerabilidad y calidad para playas y dunas de Canarias: una propuesta metodológica* (Ph. D. Thesis). Universidad de Las Palmas de Gran Canaria, 445 pp. http://hdl.handle.net/10553/24452.

Peña-Alonso, C., Hernández-Cordero, A. I., & Hernández-Calvento, L. (2016). Estado de conservación de las dunas costeras de Canarias. In F. X. Roig-Munar (Ed.), *Restauración y gestión de sistemas dunares. Estudio de casos* (pp. 177–200). Càtedra D'ecosistemas Litorals Mediterranis.

Peña-Alonso, C., Hernández-Calvento, L., Pérez-Chacón, E., & Ariza-Solé, E. (2017). The relationship between heritage, recreational quality and geomorphological vulnerability in the coastal zone: A case study of beach systems in the Canary Islands. *Ecological Indicators, 82*, 420–432. https://doi.org/10.1016/j.ecolind.2017.07.014.

Pérez-Chacón, E., Hernández-Calvento, L., Hernández-Cordero, A., Máyer-Suárez, P., Romero-Martín, L., & Alonso-Bilbao, I., et al. (2007). Maspalomas: claves científicas para el análisis de su problemática ambiental. Universidad de Las Palmas de Gran Canaria, 38 pp.

Pérez-Chacón, E., Hernández-Calvento, L., Fernández-Negrín, E., Romero, L., Máyer, P., & Hernández-Cordero, A., et al. (2010). *Caracterización del sistema sedimentario eólico de La Graciosa (archipiélago canario)*. Informe final. Centro 'Isla de La Graciosa' (OAPN-Ministerio de MAMRM). University of Las Palmas de Gran Canaria. Las Palmas de Gran Canaria, 151 pp.

Pérez-Chacón, E., Hernández-Calvento, L., Fernández-Negrín, E., Máyer, P., Cabrera-Vega, L., & Cruz, N., et al. (2012). *Evolución reciente del sistema sedimentario eólico de La Graciosa (Archipiélago canario): claves para su diagnóstico ambiental*. Informe final. Centro 'Isla de La Graciosa' (OAPN-Ministerio de MAMRM). University of Las Palmas de Gran Canaria. Las Palmas de Gran Canaria, 151 pp.

Pye, K. (1990). Physical and human influences on coastal dune development between the Ribble and Mersey estuaries, northwest England. In K. F. Nordstrom, N. P. Psuty, & R. W. G. Carter (Eds.), *Coastal dunes. Form and process* (pp. 339–359). Chichester: Wiley.

Santana-Cordero, A., Monteiro-Quintana, M. L., & Hernández-Calvento, L. (2014). Reconstructing the environmental conditions of extinct coastal dune systems using historical sources: The case of the Guanarteme dune field (Canary Islands, Spain). *Journal of Coastal Conservation, 18*, 323–337. https://doi.org/10.1007/s11852-014-0320-5.

Santana-Cordero, A., Monteiro Quintana, M. L., & Calvento, Hernández. (2016a). Reconstruction of the land uses that led to the termination of an arid coastal dune system: The case of the Guanarteme dune system (Canary Islands, Spain), 1834–2012. *Land Use Policy, 55*, 73–85. https://doi.org/10.1016/j.landusepol.2016.02.021.

Santana-Cordero, A., Monteiro-Quintana, M. L., Hernández-Calvento, L., Pérez-Chacón-Espino, E., & García-Romero, L. (2016b). Long-term human impacts on the coast of La Graciosa, Canary Islands. *Land Degradation and Development, 27*, 479–489. https://doi.org/10.1002/ldr.2369.

Santos Guerra, A. (1993). Dry coastal ecosystems of the Canary Islands and the Ilhas Selvagens. In E. Van der Maarel (Ed.), *Ecosystems of the world 2B: Dry coastal ecosystems (Africa, America, Asia and Oceania)* (pp. 51–57). Amsterdam: Elsevier.

Smith, A. B., Jackson, D. W. T., Cooper, J. A. G., & Hernández-Calvento, L. (2017). Quantifying the role of urbanization on airflow perturbations and dunefield evolution. *Earth's Future*. https://doi.org/10.1002/2016ef000524.

Sunding, P. (1972). *The vegetation of Gran Canaria*. Skrifter Utgitt av Det Norske Videnskaps-Akademi i Oslo. I. Matematisk-Naturvidenskapeling Klasse *29*, 1–186.

Viera Pérez, M. (2015). *Estudio detallado de la duna costera de Maspalomas (Gran Canaria, Islas Canarias): interacción "Taganum moquinii"-dinámica sedimentaria eólica en un entorno intervenido. Recomendaciones de cara a su gestión* (Ph.D. Thesis). Universidad de Las Palmas de Gran Canaria, 572 pp. http://hdl.handle.net/10553/17572.

Part V
Case Studies of Coast-Human Interaction Problems and Coastal Management

Chapter 31
Coastal Management in the Basque Coast: A Case Study of Dredging and Dumping Operations Along the Oka Estuary

Manu Monge-Ganuzas, Alejandro Cearreta and Ane García-Artola

31.1 Historical Background

The first data available on the interaction between human beings and the Basque estuaries are situated in the Prehistory, after the last glacial maximum, at around 18,000 years BP (Aguirre et al. 2000). At this time, human activities consisted mainly in coastal mollusc gathering, for both aesthetic and food purposes. Evidences of this activity have been found into the sedimentary record of several caves dispersed along the Basque littoral. These are the cases of Antolinako cave (Oka watershed, 18,800–17,500 years BP; Aguirre et al. 2000), Santimamiñe cave (Oka watershed, 16,300–11,500 years BP; Barandiarán 1962), Santa Katalina (Lekeitio 12,700–9800; Imaz 1994) or Kobeaga II cave (Ispaster, 7500–6500 years BP, López Quintana 2000), among others (Fig. 31.1).

Then, identified Roman settlements (dated on this area between the 1st and 5th centuries) on the banks of the Basque estuaries suggest that these environments were often used as waterways. These are the cases of the Oka estuary, where a Roman settlement was established in Forua (Bizkaia province) (Martínez and Unzueta 1989) or the Roman city of Oiasso located at Irun (Gipuzkoa province) within the Bidasoa estuary (Urteaga 2005).

M. Monge-Ganuzas (✉)
Service of Urdaibai Biosphere Reserve, Directorate of Land Planning, Biodiversity and Climate Change, Basque Government, Madariaga Tower, 48350 San Bartolome, Busturia, Spain
e-mail: manu-monge@euskadi.eus

A. Cearreta · A. García-Artola
Departamento de Estratigrafía Y Paleontología, Facultad de Ciencia Y Tecnología, Universidad Del País Vasco UPV/EHU, Apartado 644, 48080 Bilbao, Spain
e-mail: alejandro.cearreta@ehu.eus

A. García-Artola
e-mail: ane.garcia@ehu.eus

J. A. Morales (ed.), *The Spanish Coastal Systems*,
https://doi.org/10.1007/978-3-319-93169-2_31

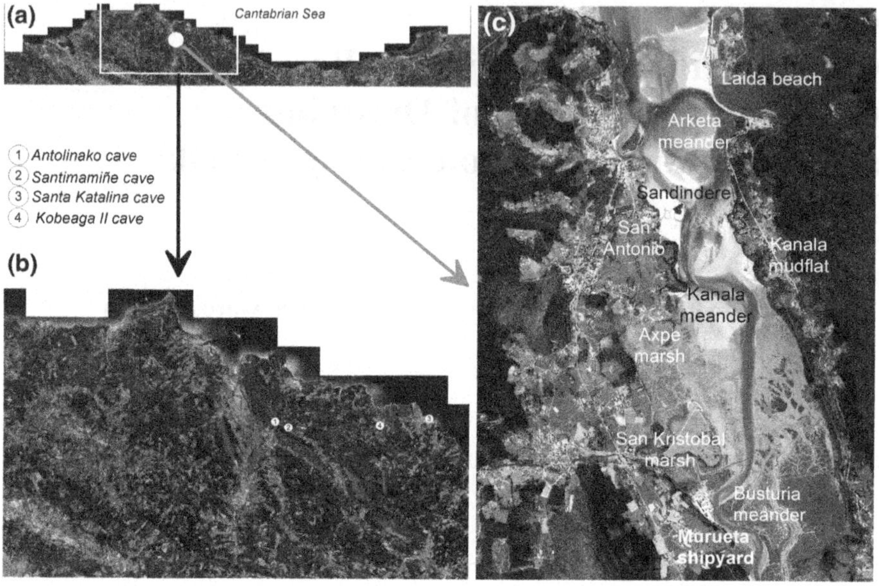

Fig. 31.1 Location of the Oka estuary, the sedimentary macrostructures and those places described in the text

During the early Middle Ages (5th to the 10th centuries) there is not too much information about human activities on the Basque estuaries. Probably, due to the barbarian invasions occurred across Europe at this time, there was an exodus of the human population from rural areas close to the estuaries to the cities situated in the hinterland and human pressure over the estuaries was probably low. However, it is also probable that the estuaries were still used as waterways. It is interesting the progressive salt marsh reclamation occurred during the late Middle Ages in most of the Basque estuaries (Gogeaskoetxea and Juaristi 1997) that caused dramatic sedimentary and environmental consequences on the Basque estuaries (see Chap. 19).

At the same time, the development of new technological knowledge allowed tidal mills to be constructed at the margins of the estuaries modifying them partially. Various examples of these constructions can still be observed for instance at the Oka estuary (Basoinsa 1995).

During the 19th–20th centuries, partial occupation of the intertidal estuarine margins occurred with the widespread development of oyster hatcheries (Basoinsa 1995).

The construction of rigid ports structures inside the estuaries since the Middle Ages also modified their previous contour and altered the hydrodynamic and sedimentary transport patters. In addition, the need for depth at the mouths of these ports during the 20th and 21st centuries resulted in the increase of sand dredging and dumping activities (Uriarte et al. 2004). These activities, in many cases, modified dramatically the environmental characteristics of the estuaries and, in others, generated certain positive effects such as the generation of artificial recreational beaches.

Contemporarily, the regional industrial development of the 19th and 20th centuries provoked the discharges of different types of pollutants to the estuarine waters. These pollutants were progressively accumulated mainly into the muddy upper estuaries. The case of Nerbioi-Ibaizabal estuary has been especially dramatic as the majority of its species and habitats were eliminated or altered (Cearreta et al. 2002). It is also interesting the case of the Oka estuary where pollutant disposal, mainly heavy metals from the industries located at the upper estuary, was performed during at least 30 years (Irabien and Velasco 1999).

However, sediment dredging and dumping can be considered among the main human activities developed at the Basque estuaries with direct influence on their hydrodynamics, sedimentary patterns and environmental quality. The description of these activities and its geological consequences will be described in this chapter considering the Oka estuary as a case study.

31.2 Dredging and Dumping, Morphodynamic Response and Induccd Consequences

31.2.1 Introduction

The Oka estuary is located in the southeastern Bay of Biscay (Fig. 31.1). This drowned fluvial valley (Pritchard 1952, 1960) is a total mixed (Dyer 1973), mesotidal (Hayes 1975) with around 12-hour tides. The tidal range is around 4.5 m on springs and 1.5 m on neaps and it is classified as a tide dominated estuary (Dalrymple et al. 1992). Its highest distance across is around 1000 m, its measurement lengthwise is 12,000 m. It presents an intertidal area of near to 2 km^2. The regional wind strength reaches average day-to-day velocities of 1–2 m/s (period May–October) and its direction is both north and south. Sporadically, within this time, exceptional speed values higher than 6 m/s have been registered. In contrast, during the rest of the year is noticed a prevalence of winds coming from the north that present a mean daily speed nearby to 4 m/s or upper; the highest monthly records sometimes can achieve more than 10 m/s (Monge-Ganuzas et al. 2004).

Although UNESCO declared the Oka estuary as a Biosphere Reserve in 1984, human activities still influence it strongly. During the last 60 years various sediment dredging and dumping activities to deepen and maintain the navigation route from the Murueta shipyard (Fig. 31.2) to the open sea have been performed along the lower Oka estuary. Monge-Ganuzas et al. (2013) analysed these activities for the interlude 1957–2005. They determined a series of human induced

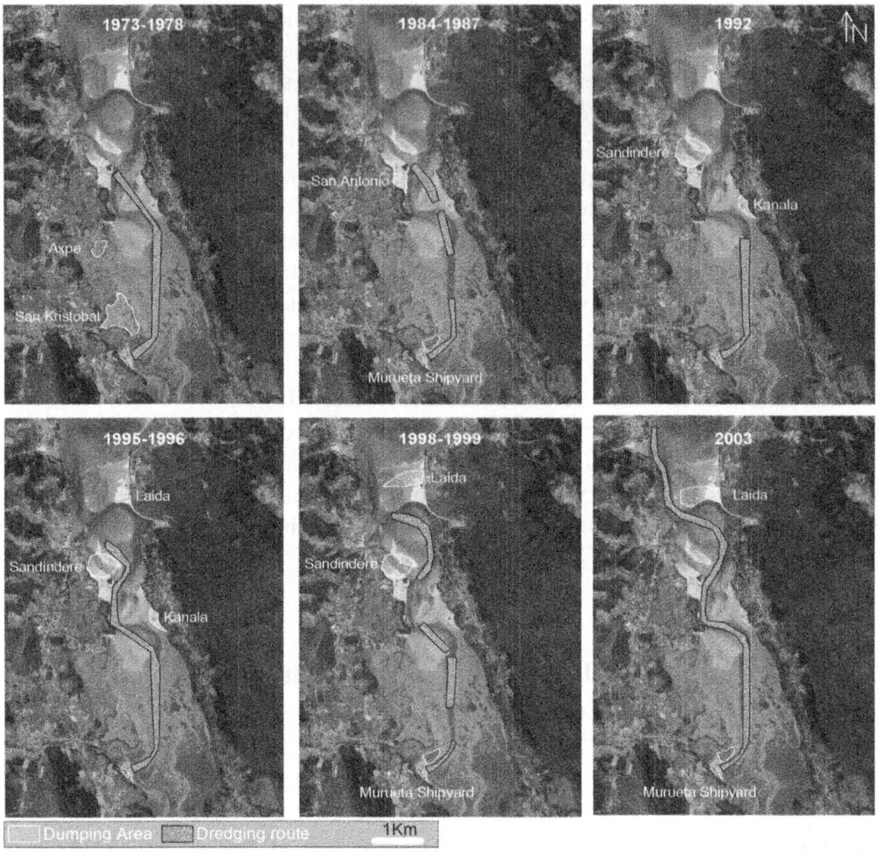

Fig. 31.2 Description of the dredging routes and dumping places developed at the Oka estuary from 1973 to 2003 (modified from Monge-Ganuzas et al. 2013)

morphological variations in Oka estuary and established various morphodynamic reasons in order to achieve its sustainable development by means of vertical aerial photographs and ortophotographs, historical and technical information and bathymetric surveys of the dredged and dumping zones.

31.2.2 Morphodynamic Response of the Oka Estuary

During 1957–1973 interlude very small sand extractions were carried out along the lower estuary. This situation was considered by Monge-Ganuzas et al. (2013) as characteristic of the pristine estuarine situation of Oka estuary. Consequently, it has been considered as a datum for subsequent experimented changes.

During 1973/1978 interval the first dredging and dumping operations were carried out in the Oka estuary. The main ebb channel of the estuary was dredged and 220 × 10^3 m^3 of sand (longitude: 2800 m; wide: 40 m; depth: −2 m -Bilbao port ordnance datum-) were deposited over San Kristobal and Axpe salt marshes (Fig. 31.2). Due to these actions, the meandering channels of Busturia and Kanala turned into a straight channel, thus, they became inactive. Meantime, no dredged areas maintained the features they had during the 1957–1972 period, with the exception of the Laida beach which was eroded. During the following years the shape of this beach varied every year.

In 1984, 125 × 10^3 m^3 of sand (longitude: 1912 m; wide: 40 m; depth: −1 m -Bilbao port ordnance datum-) were dredged again from the main ebb channel and dumped over the San Antonio salt marsh (Fig. 31.2). Later, in 1987 in front of to the Murueta shipyard an unknown volume of mud was dredged and dumped over an adjacent salt marsh.

Five years later, in 1992, 56 × 10^3 m^3 of sand were dumped over the Kanala intertidal mudflat (68%) and the north of Sandindere, over a sandy intertidal flat (32%) (Fig. 31.2). Consequently, the dredged zone, from the Murueta shipyard to Kanala, was converted in a straight channel with tiny creeks draining into it. The Arketa meander was abandoned while the Kanala meander maintained its shape. Consequently, an avulsion occurred towards the main flood delta channel at the main ebb channel. Meanwhile, tidal currents re-worked the sand of the dumping areas and after the dredging of the ebb channel located to the west, in front of San Antonio, the original flood-ebb tidal channel complex became deeper and wider. Consequently, the flood-ebb channel complex was altered.

During the 1995/1996 interval a new dredging was developed at the same areas and around 45 × 10^3 m^3 of sand were dumped on Kanala, Sandindere and the Laida beach (Fig. 31.2). Due to this, the supratidal Laida beach become wider and the Arketa meander was discarded once more as the tidal flow was channelized along this new channel. The area nearby to Sandindere and Kanala became supratidal and the main ebb channel remained divided in two straight segments. The cut induced by the action of the newly excavated channel induced the abandonment of the Kanala. Tidal currents re-worked the edge of the dumping areas of Axpe, San Kristobal and Kanala and the eroded sediment was spread-out towards the inner estuarine area, over the mudflats further inland.

During the 1998/1999 period two new dredging and dumping operations were undertaken (length of 2010 m; width: 40 m; depth of −1 m -Bilbao port ordnance datum-; volume 1998: 50.6 × 10^2 m^3/1999: 42 × 10^3 m^3) (Fig. 31.2). The muddy sediment was moved away and the sand was dumped onto the Laida beach and the San Antonio area. After supratidal Laida beach became wider and the Arketa and Kanala meanders were discarded. The flood-delta increased in extension while the tidal inlet width was reduced. Moreover, the main ebb-tidal channel tried to regain its meandering configuration and moved towards the northwest. Moreover, vegetation colonized the Kanala and Sandindere dumping areas and formed supratidal zones.

The last dredging, up to present, was carried out along the main ebb channel in 2003 (width: 40 m; depth −1 m -Bilbao port ordnance datum-). Extracted sand (240×10^3 m^3) was dumped on the Laida beach and muddy sediments (44×10^3 m^3) were removed (Fig. 31.2). Later, the Kanala meander activated and the main ebb channel migrated towards the north. Furthermore, the main ebb channel and the flood tidal delta moved eastwards and the border between the intertidal sandy and muddy zones progressed around 200 m landwards.

The morphological response of the lower Oka estuary to the 2003 dredging and dumping activities was illustrated by Monge-Ganuzas et al. (2013) by means of bathymetric surveys. Before dredging (March 2003) the main ebb channel showed a variable depth between −1 and −2 m (Bilbao port ordnance datum). Immediately after dredging (June 2003) the whole main ebb channel presented a deepness of around −2 m.

Later (September 2006) the main ebb channel depth was dramatically decreased to between 0 and −1 m. Thus, significant accumulations of sand developed in zones adjacent to the main ebb channel. These morphological changes were associated with the invasion of the previous main ebb channel provoked by the extension of the lateral and front lobes of the flood. Consequently, only three years after the dredging the lower Oka estuary exhibited a diminished depth and had progressed towards a worse situation in terms of navigability.

The sand introduced into the estuary by waves and flood tidal currents was redistributed during a tidal cycle. Van Veen (1936) suggested the use of the terms “flood channel” and “ebb channel” for the cases whereas exist a main tidal direction in certain estuarine zones. These channel types are normally elusive owing to tidal asymmetries (Robinson 1960). In the lower Oka estuary such tidal asymmetries exist (Al-Ragum et al. 2014). During the last period of the ebb and even though the flood currents are still acting, ebb currents continue getting out the estuary for around 1 h (Monge-Ganuzas et al. 2008, 2015). Henceforth, the morphodynamic shape of Oka estuary before 1973 reveals a regular pattern. This pattern consists of reciprocally evasive meandering ebb channels and straight flood channels separated by intertidal shoals. This morphology has been referred to as a multi-channel system (Jeuken and Wang 2010) and has also been defined in other estuaries worldwide (Van Veen et al. 2005; Toffolon and Crosato 2007).

Important morphodynamic changes produce by human can interference on this ebb-flood tidal multi-channel system (Winterwerp et al. 2001). These authors, based on morphological characteristics and numerically computed patterns of tide averaged sand transports, proposed the term cells (a chain of so-called macro- and meso-cells) for these patterns. Each macro-cell comprises an ebb channel and a flood channel, displaying net sediment exchanges between the macro-cells (Jeuken and Wang 2010). Moreover, these authors stated that the complete degeneration of the cells of an estuary can be provoked by dredging (Fig. 31.3). Also, they presented that the effect of dumping over the cells can increase the transport competence of its channels. This means that dumped sediments are normally easily eroded and transferred to another zones. In contrast, when sand-dumping exceeds more

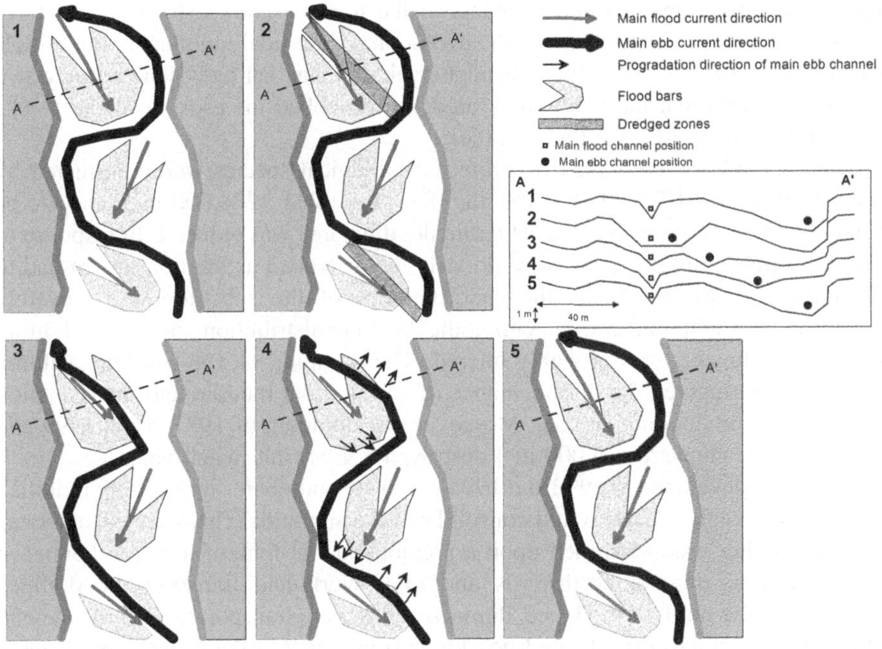

Fig. 31.3 (1–5): Idealized morphodynamic response of ebb/flood channel cells to straight dredging and crosssectional variation of the estuarine bottom morphology, and lateral displacement of the ebb main channel during the ebb/flood channel cells re-equilibrium time-lapse

than 10% of the transport capacity of any channel, current velocity decreases and consequently the cell degenerates.

Once the 1973–1978 dredging was performed, the Busturia and Kanala meanders were deserted. Kanala meander recuperated after the 2003 dredging but the Busturia meander continued inoperative. The presence of the straight waterway after dredging doubtless amplified the transport flood currents competence and caused a precipitous infilling. Van Rijn (2004) affirmed that reductions of deepness take place in areas with transport competence of streams decreased and confirmed that dredging can provoke this. The necessity for supplementary dredging at the infilled channels later 1973–1978 reinforces this assertion.

The ebb/flood channel cells were altered by all dredging as they produced straight channels. In that situation, after each dredging there was a gradually migration of main ebb channel due to the tendency of flood multi-channel cells to recover their previous status. However, the channels could not regain their original position due to consecutive dredging. Moreover, as dredging did not respect the original meanders, these were abandoned and the rate of sedimentation probably increased. In absence of meanders the energy dissipation during the ebb was lower and during low and high tidal stages the main ebb channel filled up rapidly. Due to this facts were necessary further dredging and dumping. This statement is reinforces by the continuous navigation problems in the lower estuary. Moreover, as have been evidenced by examination of the evolution of intertidal mudflats in the middle estuary from 1971 to 2005, the intertidal creek drainage net has been cutting down slowly. It seems that the continuous dredging provoked the erosion of the creeks with the aim of equilibrate their profiles.

Salt marsh areas colonized by protected biological species were eliminated by the dumping of 1973–1978 over the San Kristobal (204,000 m^2) and Axpe (36,000 m^2). Consequently, the estuarine flooding area was reduced. In response to this fact other estuarine areas were flooded and/or eroded as the tidal prism had to accommodate. Moreover, as the characteristics of the substrate were severely modified by dumping, consequently, the benthic distribution along the bottom estuarine sediments was probably altered. The dumping of 1984 at San Antonio (50,000 m^2) triggered the disappearance of an intertidal mudflat and the reduction of the estuarine flooding surface. Moreover, the dumping of 1987, 1998/1999 and 2003 near the shipyard (15,000 m^2) destroyed a great salt marsh area.

Dumping developed in 1992 and 1995–1996 at Sandindere significantly modified the estuary mouth dynamics and consolidated the supratidal character of this area.

As dumping was performed upon a previous lateral lobe of a flood channel, it was forced to migrate northwards and to re-work and transport the dumped materials to the northwest. Hence, dumped sands were transported towards the ebb channel. In response this channel became narrower and at the same flood channel avulsed provoking the abandonment of the Arketa meander. The second dumping at Kanala created a supratidal sandy zone in the place where previously it was a mixed intertidal area. Sand was transported towards the middle estuary through the flood channels and the grain size characteristics were altered. Because of this situation, the ebb channel was forced to modify its position while the curvature of the Kanala meander increased and the intertidal zone area was reduced.

The pattern of the breaking waves was altered by the dumping performed on the Laida beach in 1995, 2001 and 2003. Dumped sand was rapidly eroded by waves and tidal currents and immediately re-introduced into the estuary (Monge-Ganuzas et al. 2008).

Laida beach has continued to exhibit a morphological instability caused by wave energy variability generated by natural processes which modify the coastline and show a general erosive tendency in this location. Because of it, recently local authorities have nourished the eroded supratidal Laida beach (Monge-Ganuzas et al. 2017).

Even though there is not enough bathymetric data on the Oka estuary yet, if dredging, dumping and estuarine morphodynamic responses are analysed, we can easily notice that Oka estuary shows a natural tendency to infill, i.e., to lose its capacity as shown by historical and Quaternary geological data (Cearreta et al. 2005, 2006; Monge-Ganuzas et al. 2006) and that dredging and dumping operations have provoked an imbalance of the ebb/flood channel cells and the modification of the behaviour of the adjacent intertidal zones, in many cases with important consequences (e.g. the disappearance of the famous Mundaka surfing wave in 2003–2005; see Sect 31.2.4).

The data analysed evidence that Laida beach area has varied throughout the study period. During the period 1957–1995 its supratidal shape varied depending on sediment availability, wave and wind action and tidal energy. Conditioned by these processes the ebb-tidal channel situated to the south of Laida adapted to the variable morphology of the Laida beach. The dumped sediments upon the beach during the period 1995–2003 were gradually re-worked by waves and tidal currents and re-introduced into the estuary. In the absence of dredging and dumping (1957–1973 period), the lower Oka estuary had a stable four ebb-flood tidal multichannel system. However, the consecutive dredging and dumping altered it and modified the estuarine sedimentary dynamics. Dredged channels rapidly filled it were necessary consecutive dredging every 5 years.

The majority of humans show a static vision of the littoral. The case described shows that this vision does not fit properly with the reality. Consequently, further alteration of estuarine morphodynamic processes should be avoided can be consider as a lesson to be learned from this case study. Especially, in the case of the Oka estuary that was declared Protected Area (RAMSAR wetland, Natura 2000 site, Global Geosite and core area of the Urdaibai Biosfere Reserve, among others distinctions).

31.2.3 Negative Effects of Dredging and Dumping on the Ecosystems

The dredging and dumping executed at estuaries can cause their damage in many ways. Especially, they cause alterations on the macrobenthos that is located in the base of the trophic network (Boyd et al. 2005). Depending on the grain size of the estuarine bottom that vary from clayey-muddy to sandy, the location and abundance of the communities of macrobenthos that constitute the food of many shorebirds are variable (Colwell 2010). Generally, sandy areas are frequently poorer in shorebird food than mudflats (Burger et al. 1997). Accordingly, habitat loss in very ecologically-sensible habitats, such as intertidal mudflats is often caused by dredging and dumping carried out in estuarine areas (Monge-Ganuzas et al. 2013). Thus, sand dumping in some delicate estuarine areas with active sediment transport could lead to mudflats coverage and, consequently, long-term negative impacts on benthic communities as well as on shorebirds (Piersma et al. 2001).

Arizaga et al. (2017) used retrospective analyses of the consequences of dredging and dumping operations on the abundance and diversity of shorebirds in the muddy intertidal zone of the Oka estuary (Fig. 31.4). These authors used long-term data of shorebird censuses conducted in the Oka estuary affected by the 2003 dredging and dumping episode.

In addition to this study, they as well contrasted caused superficial grain size trends before and after the dredging and dumping events. They also anticipated that the consequences of the dredging and following dumping events should have been more acute on those species that forage regularly or merely on the mudflats. Really,

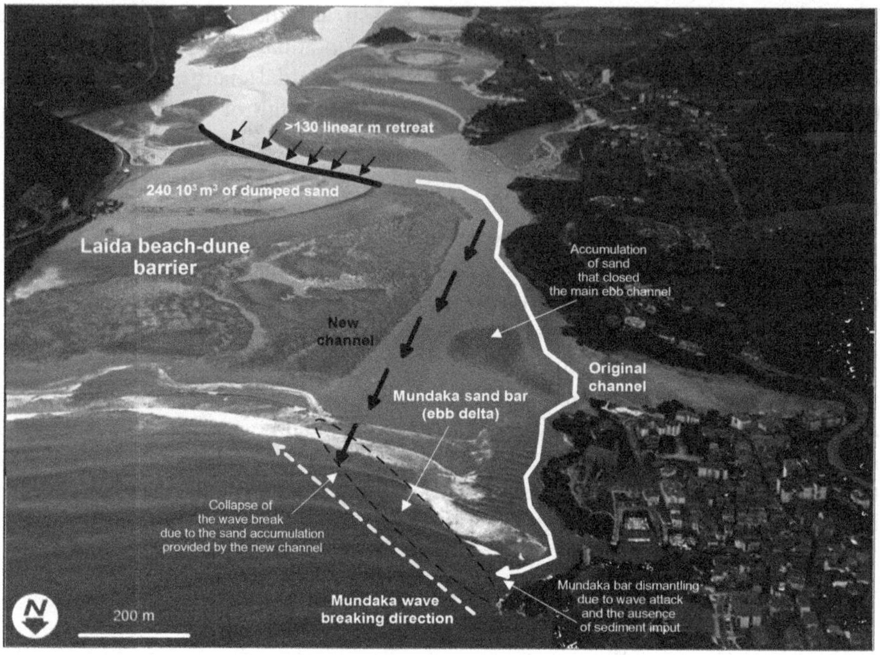

Fig. 31.4 Sand dumping-induced sedimentary responses between June 2003 and May 2005 (modified from Monge-Ganuzas et al. 2008)

they detected a diminished in population amount of numerous shorebird species which hinge on mudflats to feed just one or two years following each event. In a wider perspective, on the other hand, it is acknowledged that dredging can have a critical negative influence on shorebirds as population size of bivalves or other potential target is decreased, moreover because of straight sediment removal at foraging places (Lewis et al. 2001) or for the reason that these eating lands are covered with materials reworked from dumping places that modify invertebrate populations, as certainly happened at the Oka estuary. The circumstance that variety of shorebirds persisted invariable at the Oka estuary in spite of variations in richness following the latter dredging and dumping event of 2003, insinuates that the maximum abundant species were equally altered. The results of this research also indicate that the adverse consequence was extremely quick: the population size of certain species was perceived to diminish just two years later the dredging and dumping events.

Resilience is the capacity of an ecosystem to tolerate perturbation without switching to an alternate state (Standish et al. 2014). The Oka estuary has been subject to recurrent dredging operations during the last four decades. Dredged materials seem to be re-worked by the tide- and wave-induced currents, and this process may allow the recovering of the system morphology after some years (Monge-Ganuzas et al. 2013). However, even if a system could recover after a perturbation, recurrent perturbations may lower its capacity for recovering over the long-term (Díaz-Delgado et al. 2002). Noteworthy, Arizaga et al. (2017) observed that even in 2016, i.e. 13 years after the last dredging and dumping operations carried out in 2003, the percentage of sand within the intertidal mudflat sediment have passed from a mean of 38 to 64%, with this percentage decreasing across a north-south axis (i.e., from the dumping site towards upper estuary areas) (Fig. 31.4). This result suggests that the estuary has been unable to return to the previous state before the dredging and dumping episode and it may be discussed to what extent this effect is reversible, at least in the short- to medium-term. The combined action of the waves and tides, together with the increase of the sea level (assessed to be currently of 2 mm/year) (Leorri et al. 2013), will probably strengthen this sand covering of the existing mudflats during the next years, and hence it is unlikely to expect a recovering of shorebird abundance at these areas in the Urdaibai Biosphere Reserve.

31.2.4 Effects of Dredging and Dumping on the Society and Economy

Following the results of the study by Monge-Ganuzas et al. (2008), it is possible to state that dredging and dumping operations can also cause negative economic effects on the area surrounding the affected estuaries. As it has been explained before, in May 2003, 240,000 m^3 of sand were dredged and dumped on the southern area of the Laida beach adjacent to the estuary mouth. After few months there was strong erosion of the deposited sediments (160,000 m^3 in December 2004) as a consequence of the northwards migration of the final meander of the main estuarine channel before it reached the estuarine mouth (Cearreta et al. 2004). Concurrently, on the external area, the estuarine inlet abandoned its original position on the western flank of the estuary and migrated eastwards, cutting and eroding the sandy ebb delta to a new position on its western flank (Mundaka sand bar) (Fig. 31.4). This provoked a variation in the wave breaking pattern, which had made surfing at this location famous within the international surfing community (Fig. 31.5) and resulted in the suspension of the Billabong Pro Surfing Championship.

Under normal conditions, waves break on a sand bar (ebb tide delta) produced by the tidal currents at the Oka estuary mouth. The wave fronts arrive close to the Mundaka port from a mainly NW direction. In general, the Mundaka sand bar presents an oblique slope to the approaching waves and this induces the eastwards progressive breaking of the wave for approximately 400 m (Monge-Ganuzas et al. 2015). Hence the name of *Mundaka left wave* as when viewed from the sea it breaks progressively from right to left. The most favourable conditions to form breakers suitable for surfing occur around low tide due to the smaller depth of water over the sand bar. Essentially, the formation of the famous left wave requires moderate to intense wave conditions (typical of the autumn-winter), low tide (if possible during spring tides) and southerly or westerly winds.

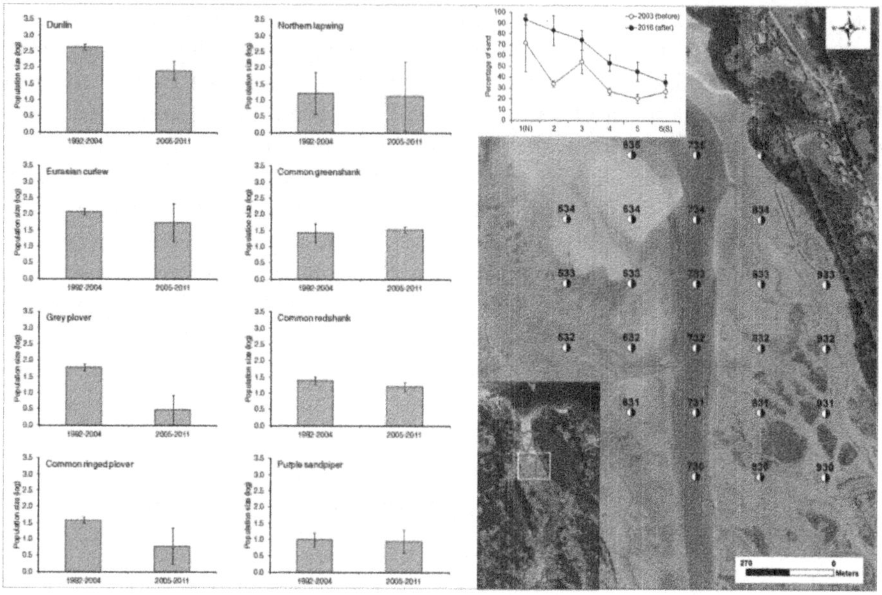

Fig. 31.5 Left side: Mean (±95% confidence interval) population size (log-transformed) of shorebirds before and after the dredging and dumping operations of 2003 at the Oka estuary; Right side: Location of the sampling points for sediment characteristics all along the intertidal mudflats at the Oka estuary and mean (±SE) percentage of sand along a north-south axis (1 stands for the sampling points 635–835; 2 for the points 534–834, etc.) for the same samples (modified from Arizaga et al. 2017)

The quality of the surfing depends upon changes of the bar morphology over the year, with the optimum conditions occurring immediately after the summer, then deteriorating slowly throughout the winter. The attraction of large numbers of surfers each year is an important source of income for the surrounding communities and, consequently, there was much local concern due to the loss of wave-quality for surfing already in 2004.

A morphodynamical study of the Oka estuary mouth was carried out with the purpose of establishing the factors controlling the pattern of wave-breaking in the area and to discuss possible corrective measures to adopt in case modified environmental conditions would not return quickly to the normal situation (Cearreta et al. 2006).

The sedimentary equilibrium of the Oka estuary mouth is controlled by the continuous movement of sand between its different morphological elements (supratidal sand dune, intertidal sandflats and underwater sandy ebb delta). There is a dynamic equilibrium due to the variable intensity with time of the various tidal and wave processes. Since the dumping of sand in May 2003, the volume of eroded sediment on the landward area of the Laida beach increased substantially, reaching a value of 11,000 m^3/month during the second half of 2003. This increase of the erosion rate in such a short period of time introduced an enormous amount of sand into the inlet area. Part of this sand was deposited there, provoked the formation of a new anomalous inlet to the east and modified the ebb delta. Later, the rate of erosion decreased and almost matched the normal values before the artificial sand dumping.

During the summer, on the intertidal area north of the Laida beach, sand ridges migrate to the southwest transporting sediment from the northeastern area of the open bay towards the beach and inlet. In this way, a typical summer beach profile characterized by ridges and runnels is produced in the lower intertidal area (low-tide terrace) and a wide berm is formed. The anomalous inlet, generated as a consequence of the infilling of the original inlet, disappeared in July–August 2005 due to the displacement of the sand ridges mentioned above and the transport of water and sediment induced by the ebb tide. These processes reinforced the development of the Mundaka sand bar and returned the inlet to its normal position close to the western rocky margin of the estuary.

On the other hand, during winter these sand ridges present on the low-tide terrace in summer disappear and part of those sediments is introduced into the estuary. The remainder sand is transported to the proximal area of the Mundaka sand bar and leaves a flat rippled intertidal area. Furthermore, the slopes between the subtidal and intertidal areas and the intertidal and supratidal areas (berm) become more pronounced giving place to a typical winter profile.

Generally, the annual process shows a sediment transport from the northeast to the southwest along the east rocky margin of the open bay and the northern area of the Laida beach to the estuarine inlet. From there, some sediment is moved landwards into the estuary to feed the flood delta and some is transported seawards to feed the Mundaka sand bar (ebb delta). This process operates throughout the year but whereas tidal dynamics are more important during summer, incident wave dynamics are dominant during winter.

In summary, the extra sand introduced artificially into the lower estuarine system caused an upset in the natural estuarine processes that was mitigated by the system after 3 years. No anthropogenic corrective measurements were necessary on the lower estuary during this interval and the environment recovered naturally with the restoration of the usual morphological features. However, during the time the wave breaking was altered, important income losses took place and a dramatic social unrest was noticed and reflected by the regional and national media.

31.3 Conclusions

Dredging and dumping operations performed at the Oka estuary during the last four decades have provoked dramatic consequences on the estuarine environment. Some of these consequences are the destruction of salt marsh environments after the dumping of sand or the change in the distribution and location of flood/ebb channel cells and its subsequent sedimentary and hydrodynamic effects. A more serious problem caused by these activities is the permanent modification of the grain size characteristics of the estuarine bed that have led to the alteration in the ethology of several seabirds. Moreover, some of the estuarine hydrodynamic and sedimentary responses to the dredging and dumping activities performed have resulted in negative effects on the regional economy and the surfing community.

Taking into account that the Oka estuary is part of the declared Biosphere Reserve and the activities to be developed on it must be sustainable, both conservation of the environment and economic and social development must be reconciled. Definition of a multi-year morphodynamic monitoring programme and a well-planned strategy for the whole estuary in order to provide validated and reliable science-based data for management are indispensable. This statement is reinforced by the unquestionable importance of the Oka estuary as provider of significant ecosystem services for humans, and it could be extrapolated to the majority of the estuarine environments in the Bay of Biscay.

Acknowledgements This research was funded by the projects Harea-Coastal Geology Group (Basque Government, IT976-16) and ANTROPICOSTA (MINECO, CGL2013-41083-P). It is Contribution 43 of the Geo-Q Zentroa Research Unit (Joaquín Gómez de Llarena Laboratory).

References

Aguirre, M., López-Quintana, J. C., & Sáenz de Buruaga, A. (2000). Medio ambiente, industrias y poblamiento prehistórico en Urdaibai (Gernika, Bizkaia), del Würm reciente al Holoceno medio. *Illunzar, 4,* 13–38.

Al-Ragum, A., Monge-Ganuzas, M., Amos, C. L., Cearreta, A., Townend, I., & Manca, E. (2014). An evaluation of the Rouse theory for sand transport in the Oka estuary, Spain. *Continental Shelf Research, 78,* 39–50.

Arizaga, J., Amat, J. A., & Monge-Ganuzas, M. (2017). The negative effect of dredging and dumping on shorebirds at a coastal wetland in northern Spain. *Journal for Nature Conservation, 37,* 1–7.

Barandiarán, J. M. (1962). Exploraciones de la caverna de Santimamiñe (Basondo: Cortézubi): 5ª Memoria. Campaña de 1961. Reeditado en Barandiarán, J. M. 1976: Obras Completas. Tomo IX. *Editorial La Gran Enciclopedia Vasca* (pp. 369–403). Bilbao.

Basoinsa. (1995). *Plan de Acción Territorial del Área de Especial Protección de la Ría* (pp. 900) unpublished.

Boyd, S. E., Limpenny, D. S., Rees, H. L., & Cooper, K. M. (2005). The effects of marine sand and gravel extraction on the macrobenthos at a commercial dredging site (results 6 years post-dredging). *Journal of Marine Science, 62,* 145–162.

Burger, J., Niles, L., & Clark, K. E. (1997). Importance of beach, mudflat and marsh habitats to migrant shorebirds on Delaware Bay. *Biological Conservation, 79,* 283–292.

Cearreta, A., Irabien, M. J., Leorri, E., Yusta, I., Quintanilla, A., & Zabaleta, A. (2002). Environmental transformation of the Bilbao estuary, N. Spain: microfaunal and geochemical proxies in the recent sedimentary record. *Marine Pollution Bulletin, 44,* 487–503.

Cearreta, A., Monge-Ganuzas, M., Irabien, M. J., Iriarte, E. (2005). *Herramientas geológicas para la interpretación medioambiental del yacimiento romano en Forua (Reserva de la Biosfera de Urdaibai)* (Unpublished Technical Report for the Culture Department of the Biscay Province Council). pp. 31.

Cearreta, A., Monge-Ganuzas, M., & Iriarte, E. (2004). *Seguimiento y evaluación sedimentológica del proyecto de regeneración dunar en Laida (Reserva de la Biosfera de Urdaibai)* (Informe final (Octubre 2001–Diciembre 2004). Unpublished Report for the Environment and Land Use Department of the Basque Government). pp. 42.

Cearreta, A., Monge-Ganuzas, M., & Iriarte, E. (2006). *Seguimiento y evaluación sedimentológica del proyecto de regeneración dunar en Laida (Reserva de la Biosfera de Urdaibai). Informe final (Octubre 2001–Diciembre 2004)* (Unpublished Report for the Environment and Land Use Department of the Basque Government). pp. 42.

Colwell, M. A. (2010). *Shorebird ecology, conservation and management* (p. 328). Berkeley: University of California Press.

Dalrymple, R. W., Zaitlin, B. A., & Boyd, R. (1992). Estuarine facies models conceptual basis and stratigraphic implications. *Journal of Sedimentary Petrology, 62,* 1130–1146.

Díaz-Delgado, R., Lloret, F., Pons, X., & Terradas, J. (2002). Satellite evidence of decreasing resilience in Mediterranean plant communities after recurrent wildfires. *Ecology, 83,* 2293–2303.

Dyer, K. R. (1973). *Estuaries: A physical introduction* (p. 140). London: Wiley & Sons.

Gogeaskoetxea, A., & Juaristi, J. (1997). Aprovechamientos históricos y privatización de las marismas de Urdaibai. *Lurralde, 20,* 169–189.

Hayes, M. O. (1975). Morphology of sand accumulations in estuaries. In L. E. Croning (Ed.), *Estuarine research 2* (pp. 3–22). New York: Academic Press.

Imaz, M. (1994). Los moluscos marinos de Laminak II. *Kobie, 21,* 221–224.

Irabien, M. J., & Velasco, F. (1999). Heavy metals in Oka river sediments (Urdaibai National Biosphere Reserve, Northern Spain): Lithogenic and anthropogenic effects. *Environmental Geology, 37,* 54–63.

Jeuken, M. C. J. L., & Wang, Z. B. (2010). Impact of dredging and dumping on the stability of ebb-flood channel system. *Coastal Engineering, 57,* 553–566.

Leorri, E., Cearreta, A., García-Artola, A., Irabien, M. J., & Blake, W. H. (2013). Relative sea-level rise in the Basque coast (N Spain): Different environmental consequences on the coastal area. *Ocean and Coastal Management, 77,* 3–13.

Lewis, M. A., Weber, D. E., Stanley, R. S., & Moore, J. C. (2001). Dredging impact on an urbanized Florida bayou: Effects on benthos and algal-periphyton. *Environmental Pollution, 115,* 161–171.

López Quintana, J. C. (2000). El yacimiento prehistórico de la cueva de Kobeaga II (Ispaster, Bizkaia): cazadores-recolectores en el País Vasco atlántico durante el VIII y VII milenio B. P. *Illunzar, 4,* 83–162.

Martínez, A., & Unzueta, M. (1989). Forua: un asentamiento romano alto imperial en la franja cantábrica del País Vasco. En: *El solar Vascón en la antigüedad, Cuestiones de lengua, arqueología, epigrafía e historia. VII Cursos de verano de la* (pp. 37–46). Spain: Universidad del País Vasco.

Monge-Ganuzas, M., Cearreta, A., & Evans, G. (2013). Morphodynamic consequences of dredging and dumping activities along the lower Oka estuary (Urdaibai Biosphere Reserve, southeastern Bay of Biscay, Spain). *Ocean and Coastal Management, 77,* 40–49.

Monge-Ganuzas, M., Cearreta, A., & Iriarte, E. (2008). Consequences of estuarine sand dredging and dumping on the Urdaibai reserve of the biosphere (Bay of Biscay): the case of the "Mundaka left wave". *Journal of Iberian Geology, 34,* 215–234.

Monge-Ganuzas, M., Evans, G., & Cearreta, A. (2015). Sand-spit accumulations at the mouths of the eastern Cantabrian estuaries: The example of the Oka estuary (Urdaibai Biosphere Reserve). *Quaternary International, 364,* 206–216.

Monge-Ganuzas, M., Gainza, J., Liria, P., Epelde, I., Uriarte, A., Garnier González, R. M., et al. (2017). Morphodynamic evolution of Laida beach (Oka estuary, Urdaibai Biosphere Reserve, southeastern Bay of Biscay) in response to supratidal beach nourishment actions. *Journal of Sea Research, 130,* 85–95.

Monge-Ganuzas, M., Iriarte, E., Cearreta, A., & Arana, X. (2004). Sustainable management in the Urdaibai Reserve of the Biosphere (Southern Bay of Biscay). Coastal dune Regeneration. In D.

R. Green (Ed.), *Delivering sustainable coast: Connecting science and policy* (Vol. 2, pp. 725–726). Cambridge Publications. Proceedings VII Internacional Symposium Litoral. Aberdeen (U.K.) 20–22 September.

Monge-Ganuzas, M., Iriarte, E., & Cearreta, A. (2006). Análisis sedimentario y evolución holocena del estuario superior del Oka en el área de Portuzarra (Gernika-Lumo, Bizkaia). *Illunzar, 6,* 39–55.

Piersma, T., Koolhaas, A., Dekinga, A., Beukema, J. J., Dekker, R., & Essink, K. (2001). Long-term indirect effects of mechanical cockle-dredging on intertidal bivalve stocks in the Wadden Sea. *Journal of Applied Ecology, 38,* 976–990.

Pritchard, D. W. (1952). Salinity distribution and circulation in the Chesapeake Bay estuarine system. *Journal of Marine Research, 11,* 106–123.

Pritchard, D. W. (1960). *Lectures on Estuarine Oceanography. Chesapeake Bay Institute and Department of Oceanography* (p. 154). Maryland: John Hopkins University.

Robinson, A. H. W. (1960). Ebb-flood channel systems in sandy bays and estuaries. *Geography, 45,* 183–199.

Standish, R. J., Hobbs, R. J., Mayfield, M. M., Bestelmeyer, B. T., Suding, K. N., & Battaglia, L. L. (2014). Resilience in ecology: Abstraction, distraction, or where the action is? *Biological Conservation, 177,* 43–51.

Toffolon, M., & Crosato, A. (2007). Developing macro-scale indicators for estuarine morphology: The case of the Scheldt estuary. *Journal of Coastal Research, 23,* 195–212.

Uriarte, A., Collins, M. B., Cearreta, A., Bald, J., & Evans, G. (2004). Sediment supply, transport and deposition: contemporary and Late Quaternary evolution. In A. Borja & M. B. Collins (Eds.), *Oceanography and marine environment of the Basque country* (Vol. 70, pp. 97–132). Elsevier Oceanography Series.

Urteaga, M. (2005). El Puerto romano de Irún (Guipúzcoa). In M. Urteaga & M. J. Noain (Eds.), *Mar Exterior: el occidente atlántico en época romana, Actas del Congreso Internacional, Pisa 2003* (pp. 85–106). Rome: Escuela Española de Historia y Arqueología CSIC.

Van Rijn, L. C. (2004). Estuarine and coastal sedimentation problems. In *Proceedings of the 9th international symposium on river sedimentation* (pp. 156–177). Yichang, China.

Van Veen, J. (1936). *Onderzoekingen in the Hoofden* (p. 252). SGravenhage.

Van Veen, J., van der Spek, A., Stive, M., & Zitman, T. (2005). Ebb and flood channel systems in the Netherlands tidal waters. *Journal of Coastal Research, 21,* 1107–1120.

Winterwerp, J. C., Wang, Z. B., Stive, M. J. F., Arends, A., Jeuken, C., Kuijper, C., & Toolen, P. M. C. (2001). A new morphological schematization of the western Scheldt estuary, The Netherlands. In *Proceedings of the 2nd IAHR Symposium on Rivers, Coastal and Estuarine Morphodynamics Symposium* (pp. 525–533). Obihiro, Japan.

Chapter 32
Shoreline Evolution and its Management Implications in Beaches Along the Catalan Coast

José A. Jiménez and Herminia I. Valdemoro

32.1 Introduction

The Catalan coast is located in the NE Spanish Mediterranean (Fig. 32.1). It has about 600 km long coastline, with about 270 km of beaches. It has a wide variety of temperate coastal systems comprising considerable geo- and biodiversity which is represented in cliffs, rocky coasts, sandy beaches, coastal plains, estuaries, and river deltas. In spite of this natural richness, it can be considered as a paradigm of the highly developed areas of the Mediterranean coastal zone. Accordingly, one of its main characteristics is its high susceptibility to change/damage due to the accumulation of human-induced pressures. Thus, using a generic definition of coastal hotspots as regions significantly affected by impacts resulting from the combination of natural hazards and human vulnerability (see e.g. Newton et al. 2012), i.e. situations in which stresses/pressures coincide with a high exposure and low adaptive capacity, the Catalan coast can be used as a good example of hotspotness along the Mediterranean (Jiménez et al. 2017). This condition arises from the combination of local natural and societal factors, which can be summarized in:

- The population is mostly concentrated in the coastal zone. The average density in coastal municipalities is 1518 people/km^2, whereas the mean value for Catalonia is 232 people/km^2.
- The economy of coastal comarcas contributes 65% of the Catalan GDP at market prices whereas they represent 22.8% of the total surface.
- Tourism provides about 11% of the Catalan GDP and, coastal tourism is the major contributor to the sector, with coastal destinies comprising about 63% of tourism overnights (excluding Barcelona city which attract 29%).

J. A. Jiménez (✉) · H. I. Valdemoro
Laboratori D'Enginyeria Marítima, Universitat Politécnica de Catalunya, BarcelonaTech, C/Jordi Girona 1-3, Campus Nord Ed. D1, 08034 Barcelona, Spain
e-mail: jose.jimenez@upc.edu

J. A. Morales (ed.), *The Spanish Coastal Systems*,
https://doi.org/10.1007/978-3-319-93169-2_32

Fig. 32.1 The catalan coast

- There is a large number of infrastructures along the coast modifying the natural dynamics and representing a point-source of stress on the system.
- About 70% of existing beaches are retreating during the last decades (CIIRC 2010).
- Coastal damage has significantly increased during the last 50 years (Jiménez et al 2012).

One of the main characteristic of the above mentioned factors is that most of them take place along the sedimentary part of the coast, i.e. in beaches. Due to this, beaches can be considered as one of the main coastal resources to be preserved and, following a worldwide trend, its present and future states must be treated as a strategic issue from the natural (e.g. Schlacher et al. 2007; Defeo et al. 2009) and economic standpoints (e.g. Houston 2008). These environmental units are characterized by a continuous interaction between the natural, socio-economic and administrative subsystems and, as a consequence, existing conflicts would reflect perturbations in any of them with final consequences on the overall system quality.

The coastal zone in general, and beaches in particular, is a highly dynamic system which is continuously evolving at different time and spatial scales and, due to this, any attempt of long-term management requires to characterize its evolution to forecast its future development. Within this context, the main aim of this work is to characterize the current state of beaches along the Catalan coast taking into account their evolutive behaviour, main factors affecting their evolution and their implications for present and future coastal management. The structure of this chapter is as follows: (i) the second section gives an overview on the evolutive behavior of beaches along the Catalan coast; (ii) the third section analyzes natural and anthropogenic sediment sources for beaches; (iii) the fourth section characterizes one of the most important perturbations affecting beach stability along the Catalan coast, i.e. marinas; and finally, (iv) the fifth section analyzes the implications of coastline stability on functions provided by beaches which will condition their management.

32.2 Shoreline Evolution in Catalan Beaches

As it was mentioned, beaches are one of the most important environments of the Catalan coast from the human perspective. In spite of this, there are few systematic studies characterizing them. The most comprehensive one was done by CIIRC (2010) where beaches were fully characterized in physical terms, existing conflicts and management aspects. In that study, the "representative beach" along the Catalan coast is characterized by an average width of about 40 m, with coarse sediment and a relative steep beachface. However, as it was mentioned before, the main characteristic of these coastal systems is its high dynamics and, in consequence, the description of beaches at a given time is just a fixed state which is clearly insufficient to afford their management. Thus, to complement previous studies and to assess future status of beaches and associated implications, we present here an overview on main aspects on shoreline dynamics along the Catalan coast. To characterize the current decadal-scale behavior, rates of displacement have been calculated by analyzing shorelines extracted from a collection of 10 aerial photographs covering the period 1995–2015. In each beach, shoreline evolution rates have been calculated in control profiles with a spacing of about 100 m by means of the linear regression technique. This method filters out short-term fluctuations and obtained rates can be considered as representative of the evolution at decadal scale. Obtained values for all control profiles within a beach are averaged to obtain a representative shoreline evolution rate.

The statistical distribution of beach-averaged shoreline rates of displacement during the period 1995–2015 is shown in Fig. 32.2. This period corresponds to a situation in which most of major coastal works, including harbors, and beach nourishment operations had been executed and, due to this, it can be considered as representative of the action of the littoral dynamics under current conditions. Obtained results show that about 65% of the sedimentary coastline is retreating, with an average retreat rate of −1.6 m/yr. If we consider all beaches along the Catalan coast (eroding, accreting and stable), the calculated average shoreline rate of displacement is −0.44 m/yr. These values are lower than the ones previously reported by CIIRC (2010) where evolution rates were calculated using the same technique but for a shorter period (1995–2004). This decrease in shoreline rate of displacements can be associated to different factors. First, rates presented here are obtained for a longer period and, in consequence, estimated trends are less sensitive to inter-annual fluctuations. During the covered period, years 2001–2003 can be characterized as "stormy" due to the occurrence of many energetic storms and the Catalan coast was severely affected (Jiménez et al. 2012). Due to this, the contribution of these years to the evolution rates calculated for the period 1995–2004 tended to overweight the erosive behavior. On the other hand, the extension of the evolution study until 2015 permitted the coast to be partially recovered from the effects of that period.

Figure 32.2 shows representative shoreline evolution rates for all comarcas (administrative units comprising a set of municipalities) along the Catalan coast. As it can be seen, although erosion is the dominant behavior, there is a spatial variation in evolution rates along the coast. This spatial pattern reflects the difference in the intensity of littoral dynamics and the importance of local human-induced perturbations along the Catalan coast. Thus, if we do not consider the Ebro delta which has some of the most retreating beaches along the Catalan coast (Jiménez et al. 1997; Sánchez-Arcilla et al. 1998), the largest shoreline retreats are found along the Maresme coast in Barcelona (see also Ballesteros et al. 2018a), where about the 67%

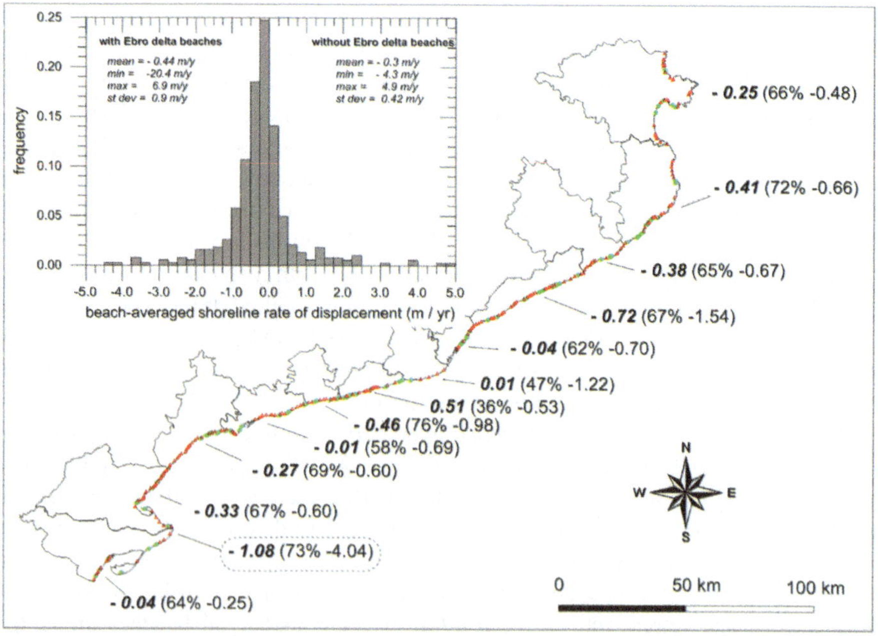

Fig. 32.2 Average shoreline rate of displacement (in bold), percentage of eroding shoreline and average erosion rate (in brackets) for comarcas along the Catalan coast. Inset shows the histogram of beach-averaged shoreline evolution rates during the period 1995–2015 along the Catalan coast (largest erosion rates experienced by some of the delta beaches are not shown in the histogram)

of its shoreline is retreating at an average erosion rate of about −1.5 m/yr. This coastal stretch is characterized by the presence of long rectilinear beaches extending from the Tordera river at the North to Barcelona city at the south. Originally this was an interrupted coast but now, the presence of five marinas have segmented the coast and altered the littoral dynamics which is characterized by a SW directed net longshore sediment transport varying from 25,000 up to 100,000 m^3/yr (CIIRC 2010). The combination of this sediment transport pattern and the presence of these anthropogenic barriers is the main element to control shoreline evolution along this stretch.

On the other hand, there are some comarcas where spatially-averaged evolution rates seem to indicate that beaches are close to equilibrium conditions. However, the percentage of eroding shoreline within the unit as well as the representative erosion rate reflect that, instead of beaches at equilibrium, they are composed by eroding and accreting beaches with a similar contribution to the overall rate. An example of this behaviour is found at the Llobregat delta, where the erosive behaviour of the northern beaches is compensated by the shoreline progradation at the southern end of the littoral cell.

As a final comment, the observed overall erosion of Catalan beaches seems to reflect that, at decadal scale, although alongshore processes can locally control shoreline fluctuations, the integrated behaviour is controlled by offshore sediment losses. Moreover, this erosive behaviour has occurred in spite of different nourishment operations along the Catalan coast amounting more than 6 Mm^3 at an equivalent rate of about 320,000 m^3/yr (see next section), so the real long-term shoreline erosion rate must be larger than the estimated one (−0.44 m/yr).

32.3 Sediment Sources for Catalan Beaches: From Rivers to Humans

One of the main factors determining long-term coastline evolution are sediment sources which is the positive contribution to local sediment budget in beaches. Until recent times, riverine inputs have been the main sediment source for beaches along the Catalan coast. Textural properties of the sediment along the coast reflect their provenance according to river basins characteristics plus their reworking and redistribution by the dominant littoral dynamics. In general, beaches northwards of Barcelona are characterized by the presence of coarse sediments which have been supplied by medium-to-small rivers such as Muga, Fluvià and Ter for Costa Brava beaches and, Tordera river for Barcelona beaches, in addition to ephemeral torrential streams along the entire coast (Liquete et al. 2009; Fernández Salas et al. 2015). Southwards of Barcelona, beaches are composed of fine sediments mainly supplied by the Llobregat and Ebro rivers, and smaller contributions from Besós, Foix, Gaia and Francolí rivers (Fernández Salas et al. 2015). Figure 32.3 shows representative sediment texture parameters in beaches within different littoral cells along the Catalan coast obtained from sediment samples taken in the swash zone (CIIRC 2010). As it can be seen, most of littoral cells are characterized by beaches with a relatively narrow range of textural properties (*d50* and σ). The main exception is the Maresme cell which presents beaches with grain sizes ranging from coarser than 1.5 to 0.4 mm. Along this area, a fining sequence in the direction of the net longshore sediment transport is observed, with the coarsest sediments at the source point (La Tordera river mouth) and progressively finer sediments towards the south. Sediment textural properties at the southern Maresme beaches are also product of artificial nourishment operations with a borrow sediment with a mean size of about 0.4 mm (Peña et al. 1992). The large variation in sediment characteristics in Costa Brava is reflecting the local geomorphology which is dominated by isolated pocket beaches with sediments being controlled by local sources.

These sediment textural characteristics are important from the beach management standpoint since they largely determine morphological aspects such as beachface slope and berm height. As it can be seen in Fig. 32.3, fine sediment beaches are characterized by mild beachface slopes and relative low berm heights (e.g. Ebro and Llobregat deltas, Costa Dorada) whereas coarse sediment beaches have steep slopes and high berms (Maresme and Costa Brava). This sediment—morphology pattern will attract users with particular preference on such characteristics and, in fact, they are typical variables valued to assess beachgoers preferences (e.g. Roca and Villares 2008; Lozoya et al. 2014) and to assess beach quality in terms of recreation (e.g. Ariza et al. 2008b; 2010).

Although rivers are the natural sediment source for beaches along the Catalan coast and, in this sense, they play a crucial role in coastal stability, measurements of riverine sediment fluxes are seldom available and, in most of the cases, existing ones are not representative of all flow conditions, being scarce under the most active situations, i.e. floods. For Catalan rivers, the most relevant existing data have been collected in the Ebro and Tordera rivers which have been used to estimate representative sediment fluxes (e.g. Guillén and Palanques 1997; Rovira and Batalla 2006). In general, the lack of enough field data is usually solved by applying hydrological models that estimate sediment fluxes to the ocean using drainage basin, river characteristics and rainfall rate as input data (e.g. Syvitski and Milliman 2007). In the Catalan coast, Liquete et al. (2009) made an estimation of sediment

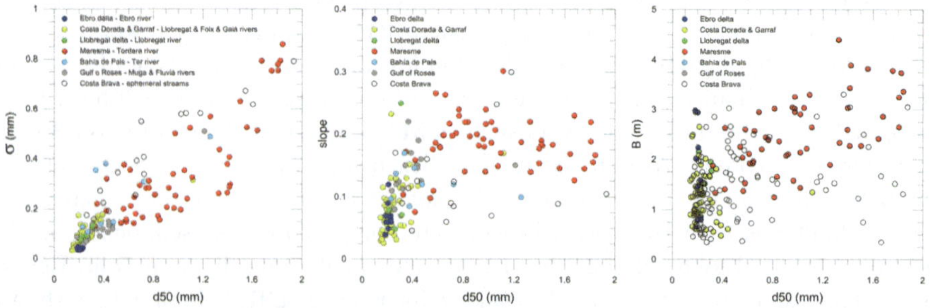

Fig. 32.3 Sediment textural parameters and beachface morphology in beaches within sedimentary cells along the Catalan coast (see Fig. 32.1 for locations). From left to right: standard deviation *vs* mean grain size; beachface slope *vs* mean grain size; berm height *vs* mean grain size

fluxes of main rivers of internal basins (excluding Ebro river) and found that, with independence of the used model, all predicted values were significantly larger (one to two orders of magnitude) than the observed (measured) ones. They concluded that, for most of Catalan rivers, existing measures are sparse and historical base levels are not well known and, this leads to an uncertainty in river sediment transport (and hence, in sediment supplies to the coast) of several orders of magnitude.

In spite of the lack of reliable historical base levels of riverine sediment inputs to the coast, there are strong evidences of their decrease during the last decades, mostly due to anthropogenic factors. The human alteration of drainage basins modifies river water and sediment fluxes and thus their discharges into the sea (e.g. Walling 2006). Main human impacts on rivers within Catalan internal basins have been identified by Liquete et al. (2009) and they cover most of possible human alterations leading to a decrease in sediment fluxes, i.e. damming, water extraction, urbanization, sand and gravel mining. As an example of the current level of anthropization, the Ebro river is heavily segmented by the presence of more than 170 artificial barriers along the main course and its tributaries (MAGRAMA 2012), which has led to a decrease in sediment fluxes larger than 95% (Guillén and Palanques 1997). Also, sediment supplies of the Tordera river has been severely affected by human action due to large-scale sediment minning from the river course from the 60s to early 80s (e.g. Rovira et al. 2005). One of the most evident implications of this reduction in river sediment supplies is the drastic reshaping suffered by beaches along the Ebro (e.g. Jiménez et al. 1997) and Tordera deltaic coasts (e.g. Jiménez et al. 2011).

Under a scenario characterized by the presence of eroding beaches and the decrease of river sediment supplies, these have been substituted by human-managed sources. In this sense, artificial nourishment has become one of the most used ways to try to counteract beach erosion along the Catalan coast during the last decades as it has been the case for nearly worldwide coasts (Hanson et al. 2002; Dean 2003; Cooke et al. 2012; Luo et al. 2016). Figure 32.4 shows annual and cumulative sediment volumes used in beach nourishment in the Catalan coast during the last four decades (1980–2010). Although showed data just correspond to Barcelona and Tarragona beaches, they can be considered as representative of the total volume since this part of the Catalan coast concentrates most of this type of actuations. As it can be seen, about 23 millions of m^3 of sand have been used until 2010 to nourish Catalan beaches, with about the 70% being supplied to beaches along the Barcelona province. This is equivalent to an average artificial sediment supply of about 775,000 m^3/year during that period, although its temporal distribution is quite irregular (Fig. 32.4). Thus, the temporal pattern shows years with large sediment supplies and periods of inactivity. Peaks in the time series correspond to the occurrence of very large nourishment operations (>1–3.5 Mm^3) to recover fully eroded stretches (e.g. Peña et al. 1992; Escartín et al. 1996; Galofré et al. 2003) or to create new artificial beaches, as it is the case along the Barcelona city waterfront (Peña and Covarsi 1994). On the other hand, the period of inactivity during 1995–2001 corresponds to an apparently temporary change in coastal actuation policy together the absence of important shoreline stability problems. Finally, the last period is characterized by different renourishment projects along the coast to counteract local problems mainly driven by storm impacts. This coastline protection strategy, based on the use of artificial nourishment as a main element, is currently maintained although priority criteria are established to define those sites to be protected (e.g. see the proposed Actuation Strategy for the protection of the Maresme coast, CEDEX 2014).

Most of the sediment used to nourish beaches has been obtained by nearshore dredging along the Catalan inner shelf. In addition to associated potential social conflicts (see Sect. 32.5), one of the evident environmental risks associated to beach nourishment is related to potential effects on the existing *Posidonia oceanica* meadows, a highly valuable ecosystem which is very sensitive to burial (Manzanera et al. 1998). Although this nourishment-seagrasses interaction has been detected in some parts of the Spanish Mediterranean (González-Correa et al. 2008; Aragonés et al. 2015), there is no empiric evidence of this impact along the Catalan coast. The only systematic monitoring on *Posidonia* meadows after a nourishment project in the Catalan coast was reported by Manzanera et al. (2014) who did not find any significant effect in the influence area of a 2.5 Mm^3 nourishment in Tarragona.

In addition to these sediment volumes managed by the Spanish Government, other sediment management projects have been implemented along the Catalan coast during the last years by other stakeholders. The most important one was done between 2002 and 2005 associated to the works for the expansion of the Port of Barcelona. Thus, a 3 Mm^3 nourishment to create a new beach just southwards of the Llobregat river mouth was planned to be executed during the period 2002–2006 to compensate physical impacts induced on deltaic beaches. In addition to this, a continuous sediment backpass of 100,000 m^3/y from the southernmost end of the cell (in the surroundings of the Port Ginesta marina) to compensate shoreline retreat

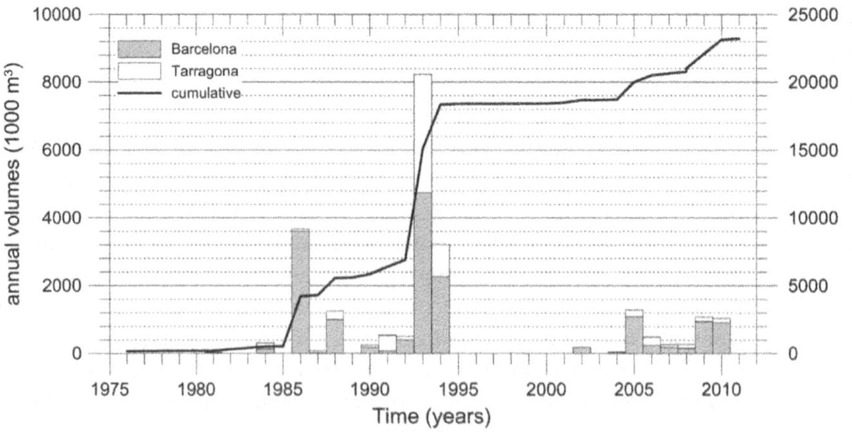

Fig. 32.4 Annual and cumulative sediment volumes used in beach nourishment in Barcelona and Tarragona provinces (data from MAPAMA)

of northern beaches has been implemented until nowadays and it is executed at yearly basis. Finally, smaller amounts of sediment have also been managed in marinas along the Catalan coast. In this case, the main target was to combine maintenance of harbour installations with beach nourishment in adjacent areas to compensate the impact of these installations on sediment transport patterns. They consisted in local dredging with sediment being supplied to adjacent eroded beaches. Most of these actuations took place during the period 2006–2007 and they were managed by the Government of Catalonia and they amounted to a total of 570,000 m^3 in 12 operations in 8 harbours.

To summarize the status of sediment supplies to the Catalan coast, we can conclude that natural sources have been significantly reduced and their role in the coastal sediment stock is limited under present conditions. These sources have been substituted by artificial nourishment at equivalent rates of about 1.3 Mm^3/yr and 0.6 Mm^3/yr during the periods 1981/94 and 2002/10 respectively. However, it has to be considered that even with this sediment supply Catalan beaches are retreating, so current nourishment volumes are insufficient to compensate erosion processes. Moreover, if we want to mimic the role of former riverine sediment supplies, artificial nourishment need to be done in a systematic and programmed manner and, any discontinuity in operations should be equivalent to the observed decrease in natural sediment supplies. This implies to assess the available stock of sediment susceptible to be used in future beach nourishment operations, i.e. to identify and define a strategic sediment reservoir (Marchand et al. 2011).

32.4 The Role of Harbours on Coastline Stability

One of the main characteristics of Mediterranean countries is the presence of small/medium ports providing services to different sectors: tourism, fisheries, commerce, industry and transport. In addition to the contribution to the local economy, the construction of these installations represents a direct (and indirect) pressure on the coastal zone which produces different types of impacts (e.g. Petrosillo et al. 2010). The intensity of this pressure is variable along the Mediterranean countries with Spain being one of the countries with the largest number of marinas per km of coastline, with only 15 km between marinas on average (UNEP 2009). In the case of the Catalan coast, there are 45 small and medium harbours (excluding the ports of Barcelona and Tarragona), with an average distance between them of about 11.5 km (Fig. 32.5), thus converting this area into one of the most densely harbour-populated regions of the Mediterranean coastline. The magnitude of the pressures exerted by these harbours as well as their environmental impact depend on their characteristics and, due to this, the largest pressures are expected to be associated with the largest harbours (Ports of Barcelona and Tarragona). However, the existing large number of small and medium harbours as well as their spatial distribution makes these 'secondary' installations a very important source of pressures on the Catalan coast which is distributed along the entire territory covering all types of geomorphologies, from rocky/cliffy to sandy coastlines. When associating the impact induced by these installations to a specific economic sector, tourism and recreation emerges as the dominant driver. Thus, about the 50% of the existing ports are solely dedicated to yachts. The remaining ones, with the exception of two industrial harbours (dedicated to cement carriers) are mixed, combining yachting with other sectors such as fishing and/or commercial.

As it was mentioned before, these installations generate a significant pressure on the coastal system, inducing a series of direct impacts, such as the modification of the littoral dynamics and the stability of the adjacent coast (e.g. Hsu et al. 1993; Klein and Zviely 2001) and the alteration of coastal habitats (e.g. Di Franco et al. 2011). Moreover, indirect impacts are also produced by ships using the installations, such as pollution (e.g. Piehlet et al. 2002) or physical impacts of boats on submerged habitats—anchoring-(e.g. Lloret et al. 2008). Figure 32.5 shows the spatial distribution of main pressures exerted by these installations inducing direct impacts on coastal stability.

The first considered impact is the direct transformation of a natural coastline (sandy or rocky) to an artificial line unable to react to the coastal dynamics. The new coastline will be formed by harbour installations and, this impact can be simply measured in terms of the *length of the rigidized coastline* by the harbour. The relative importance of the impact of existing installations along the Catalan coast is shown in Fig. 32.5. In total, about 23.7 km of the coastline are rigidized, with an average rigidization length per installation of 550 m, with a maximum value of around 1.5 km (Port of Sant Carles de la Ràpita) and a minimum value of less than 100 m. Minimum values are found in the surrounding of inland marinas such as Ampuriabrava, where only the harbour entrance is strictly placed along the coastline. The second impact is the destruction of the original coastal habitat due to its direct occupation by the installation. This impact can be measured in terms of the

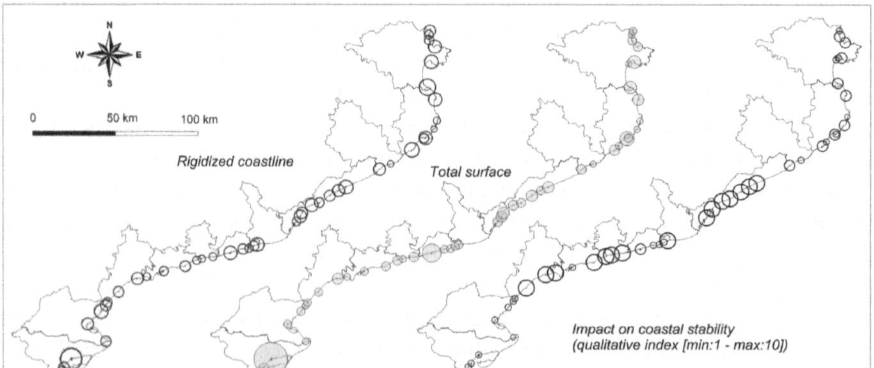

Fig. 32.5 Indicators of physical impacts of harbours along the Catalan coast. The size of the circles is proportional to the relative magnitude of each impact and they are scaled independently and each circle represents a harbour

overall *coastal surface occupied by the harbour* at land and at sea (Fig. 32.5). In total, small and medium harbours occupy a surface of 8.2 million m^2 from which about 5 million m^2 are at the sea and 3.2 million m^2 are on land. The average total surface occupied per installation is about 18.2 ha, with average surface values of 11.5 and 7 ha at sea and on land respectively. Finally, the last indicator of physical induced impacts is the harbour's *potential to modify the littoral dynamics* and, therefore, to alter coastal evolution in the adjacent areas. This impact is the direct driver of most of observed mid-term coastline changes along the Catalan coast (Sect. 32.2). In areas with significant net longshore sediment transport, they induce severe downcoast shoreline erosion and sediment accumulation upcoast of installations. When installations are located in bays or sheltered areas, affectation to coastal stability is of much smaller length scale and they are essentially controlled by local modification of hydrodynamic conditions through wave diffraction. To indicate this impact, installations have been classified into three main types, which, in decreasing order of magnitude of the impact, are: (i) barrier ports in open sandy coastlines where they block or alter the longshore sediment transport, inducing downcoast erosion; (ii) harbours in sedimentary bays where they locally alter the littoral dynamics and beach stability; and (iii) harbours in rocky coasts where their impact on sediment dynamics is minimal. The spatial distribution of this impact along the Catalan coast can be seen in Fig. 32.5. The spatial distribution of induced impact shows that the coasts of the Barcelona and Tarragona provinces are the most affected ones (Fig. 32.5). This is due to the dominance of large and straight sandy coastlines subjected to a significant net longshore sediment transport pattern. The most typical example is the Maresme coast, northwards of Barcelona, where the presence of five marinas has segmented an originally uninterrupted sedimentary coast inducing one of the the largest comarca-averaged shoreline erosion rates (Fig. 32.2).

In summary, the large number and spatial distribution of harbour installations along the Catalan coast convert this element in one of the most important human-induced pressures for coastline stability. Since most of harbours are dedicated to recreational yatching, induced impacts have to be charged to the tourism/leisure sector. Barrier harbours and straight shorelines dominated by significant net longshore sediment transport rates is the worst combination regarding coastline stability being Maresme and Costa Dorada the most affected areas.

32.5 Implications for Beach Management

In order to assess potential implications of coastline stability we focus on the currently most important functions provided by beaches in anthropized environments and which are usually the main target for current management practices, i.e. protection and recreation (e.g. Micallef and Williams 2002). The first one refers to the role of the beach to dissipate wave energy during storm impacts and, thus, to protect the hinterland from direct wave action. The second one involves its use for leisure and it is given by its recreational carrying capacity (RCC) which from the physical standpoint essentially consists in a given available surface per user. To properly provide both functions, beaches must have a given set of physical characteristics which can be parameterized as a function of its width (e.g. Valdemoro and Jiménez 2006; Jiménez et al. 2011; Bosom and Jiménez 2011). Taken this into account and following Jiménez et al. (2017), we have classified beaches in three categories according to their width and the expected level of fulfilling required conditions to support both functions: optimum, medium and low. Table 32.1 shows the selected values for each category and associated implications for each function.

Figure 32.6 shows the distribution of the different beach functional classes along the Catalan coast under current (2015) and future (2035) conditions. At present, about 64% of existing beaches present optimum conditions for providing both functions, with most of the less favoured beaches being located in Barcelona (Maresme) and Tarragona (Costa Dorada) provinces. In any case, it has to be considered that beaches were classified in terms of their average width and, there are a non-negligible number of cases of beaches with some part of their extension being narrower than the optimum value and, thus, presenting localized problems. Identified areas with the highest frequency of beaches below optimum conditions are also critical stretches with respect to potential consequences to failure in both functions. Thus, Maresme is the metropolitan coast close to Barcelona where important infrastructures close to the shoreline exist (e.g. coastal railway) and their beaches are highly frequented by local population and, also by tourists in its northernmost part. The sensitivity of Maresme beaches regarding protection under present conditions is consistent with previous vulnerability and risk assessments to storm-induced hazards (Bosom and Jiménez 2011; Ballesteros et al. 2018a, b; Jiménez et al. 2018). On the other hand, Costa Dorada is one of the most important coastal tourism destinies in Catalonia, with beaches being intensively used during the summer season.

When we consider the future beach status, a significant worsening of functional conditions is observed. Thus, taking into account current shoreline rates of

Table 32.1 Beach classes according to its width and associated implications for recreational and protection beach functions Adapted from Jiménez et al. (2017)

Class	Width (m)	Protection	Recreation
I (low)	<20 m	frequent problems under storms with low *Tr*	Low to very low RCC Overcrowding events
II (medium)	20–40 m	problems under storms with relative long *Tr*	Medium RCC High density
III (optimum)	>40 m	unfrequent problems or under storms with very long *Tr*	High RCC Low density Problems not likely

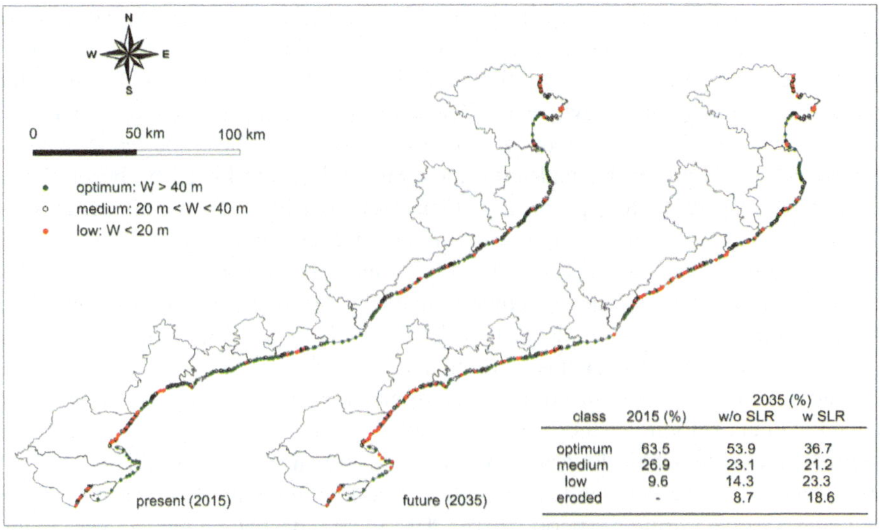

class	2015 (%)	2035 (%) w/o SLR	2035 (%) w SLR
optimum	63.5	53.9	36.7
medium	26.9	23.1	21.2
low	9.6	14.3	23.3
eroded	-	8.7	18.6

Fig. 32.6 Spatial distribution of beaches in terms of their functional width at present (2015) and at the near future (2035) assuming that their evolution is controlled by current shoreline displacement rates of change. Table shows the percentage of length of Catalan sandy beaches of a given class (excluding Ebro delta beaches) at 2015 and at 2035 under different evolution scenarios. SLR is given by the AR5 RCP8.5 scenario

displacement (Fig. 32.2), about 9% of the beaches will be fully eroded, with only the 54% of the sedimentary coastline having beaches with the required width to provide recreation and protection in optimum conditions (Fig. 32.6). Again, the most of affected areas are located in Barcelona (Maresme) and Tarragona (Costa Dorada) provinces, which are characterized by relatively narrow and highly eroding beaches. Due to the before mentioned socio-economic characteristics of both areas, the detected narrowing trend would result in a significant increase in the associated risk (Ballesteros 2017; Ballesteros et al. 2018a). If we also include potential changes due to climate change, the situation clearly worsen due to the sea-level rise (SLR) induced retreat (Jiménez et al. 2017). Thus, under the AR5 RCP8.5 scenario, the enhanced shoreline retreat will lead to a further narrowing of all beaches along the coast in such a way that about 18% of beaches will disappear by 2035. Moreover, this additional background erosion makes that only about 37% of the Catalan beaches would present optimum conditions to provide recreation and protection functions.

Results presented in Fig. 32.6 are in agreement with current problems reported along the Catalan coast. Thus, Fig. 32.7 shows the main beach-related conflicts reported by municipalities along the Catalan coast (CIIRC 2010). As it can be seen, the most frequent problem is related to erosion with the consequent loss of beach surface (about 70% of municipalities). This problem becomes one of the most important management constraints since it will affect the stock of the resource to be managed. One of the consequences of this surface loss is the increase in the exposure of existing infrastructures to wave action during storms and, in consequence, a large part of municipalities reporting beach surface losses also reported damage in infrastructures, which is in fact, the second most frequently identified concern (Fig. 32.7). Thus, about 55% of coastal municipalities reported damage in infrastructures, being promenades the most affected one followed by beach facilities. With respect to this problem, Jiménez et al. (2012) found an increase in damage along the Catalan coast during the last decades in absence of any increase in storminess. They associated this trend to an increase in exposed values along the coast during the last decades together a smaller protection provided by narrowing beaches due to the dominant decadal-scale erosive behaviour. In their analysis, one of the most affected areas by this increase in damage was, again, the Maresme coast. An indicator of the recurrence of this type of problems is the restoration programme called "Plan Litoral" which is launched by the MAPAMA to counteract storm-induced damages along the coast after severe stormy seasons. This plan has as "surname" the year of application and, an example of the increasing sensitivity of our coasts is that during the period 2014–2017 there have been three recovery plans.

With respect to the implications of the beach configuration on the recreational use, Valdemoro and Jiménez (2006) identified long-term shoreline erosion as a critical process to condition RCC in intensively used beaches, being the main factor to determine the beach failure from the recreational standpoint. As a consequence, the predicted significant decrease in optimum beaches by 2035 will induce an increase in beach user density and a decrease in the overall recreational carrying capacity. This could cause a significant problem in highly frequented beaches since the saturation level of 4 m^2/people could be exceeded more frequently and, thus, severely affecting the beach user satisfaction (Roca et al. 2008). To highlight the importance of this problem, it has to be considered that the main objective of beach management implemented by coastal municipalities was to satisfy the recreational use (about 75% of the municipalities identified this as the most important target, Fig. 32.7, see also Ariza et al. 2008a). Also, conflicts related with overcrowding events have been reported under current conditions by some municipalities (Fig. 32.7). A detailed analysis of the influence of erosion on the recreational

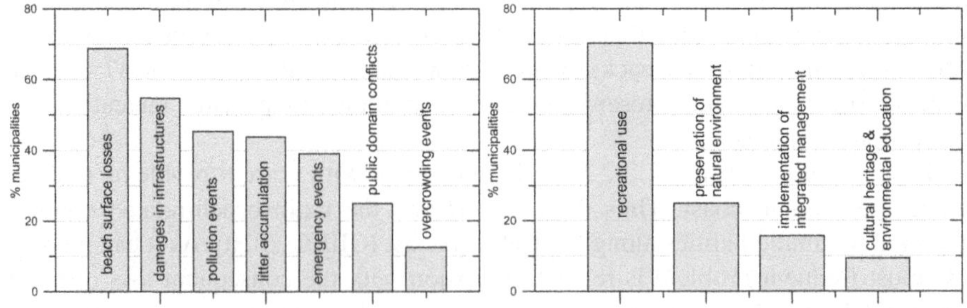

Fig. 32.7 Main beach-related conflicts (left) and main objectives in beach management (right) reported by municipalities along the Catalan coast

function along municipalities of the Catalan coast can be seen in Ballesteros (2017) and Ballesteros et al. (2018a).

The relative high frequency of beach problems and conflicts in some parts of the Catalan coast makes that local stakeholders are well aware about the current risk level and main threats and impacts (Villares et al. 2015; Ballesteros et al. 2016). However, although this is a positive aspect to properly include risk management within long-term coastal management plans, it conditions stakeholder's awareness on lower intensity long-term threats. In this sense, although the expected impact of SLR on the status of Catalan beaches is significant (see table inserted in Fig. 32.6) and, in consequence, the expected impact on functions to be provided is high or very high (Jiménez et al. 2017; Ballesteros 2017), the risk associated with this hazard is almost unappreciated by local community since local perception is more concerned by daily aspects (Villares et al. 2015; Ballesteros et al. 2016).

The response of the Administration to identified problems and conflicts along the Catalan coast can be classified as largely reactive, with most of actuations being driven by the appearance of problems (e.g. Jiménez et al. 2011, 2012). An example of this is the above mentioned beach nourishment works along the Catalan coast which seldom include planned re-nourishment operations. This has resulted in a situation where municipalities with eroding beaches usually claim for works to restore/recover their beaches. In spite of that, some municipalities also show their opposition to beach nourishment using sediment obtained from nearshore dredging as occurred in 2016 by a group of municipalities from the northern part of Maresme. They argued this opposition by claiming that this type of actions would promote unsustainable coastal management although no alternatives on how to restore beaches were given. This dichotomy between asking beach maintenance and, simultaneously, opposing common protection measures, as well as social activism questioning current coastal protection measures due to its potential negative ecological impacts, is an indicative of the increasing social conflict associated to this problem. In fact, it can be considered as one of the potential barriers for coastal adaptation under current conditions and climate change scenarios (Hinkel et al. 2018).

Another coastal-erosion related problem occurs in severely affected areas where local stakeholders have been directly affected by the identified hazards without a rapid response from the Administration or where that response has not been satisfactory to their interests. In some of these cases, they have taken self-protection measures that could be classified as illegal ones since most of them have been executed without any permission from the responsible Administration. An example of this situation is the area surrounding the Tordera delta, where the rapid shoreline retreat during the last 2 decades have affected to different campsites at both sides of the river mouth (Jiménez et al. 2011; 2018). As a consequence of this, and to prevent the continuous damage experienced by these installations, they have taken different actions such as temporary dunes to prevent flooding, sand bags and rock revetments to stop erosion (Jiménez et al. 2018). Although some of them have no major consequences; others have severely impacted the adjacent coast increasing the existing shoreline retreat (Fig. 32.8) in such a way that not only have not solved the existing problem but also has generated or accelerated another one.

Finally, the persistence of erosion problems and the identified consequences have driven the launching of dedicated long-term strategies for protection of the most affected part of the territory considering the different functions provided by beaches. Within this context, the MAPAMA has prepared a Strategy for Actuations in the Maresme coast (CEDEX 2014) where different alternatives for coastal

Fig. 32.8 Downdrift induced erosion due to a revetment built in front of a campsite in the northernmost part of the Maresme coast

protection are considered. It has to be highlighted that the analysis covers present conditions and also includes the potential effects of SLR on future protection needs.

As a final summary, the current status of Catalan beaches reflects the integrated effects of natural dynamics and human influence in the territory. Main human forcings are related to variations in sediment supply to beaches and perturbations in sediment transport patterns due to coastal works. As a result of this, beaches are progressively narrowing and, their capacity to provide protection and recreational functions is decreasing which would induce an increase in the number of conflicts along the coast. At present, the adopted approach to manage existing problems is reactive but the expected increase in the frequency and intensity of the problems in the near future require adopting a different view. Thus, it should be necessary to adopt a long-term, adaptive and sustainable strategy, to define the required actuations and to estimate and make available the required resources for their implementation.

Acknowledgements This work has been done in the framework of the *PaiRisClima* (CGL2014-55387-R) and *M-CostAdapt* (CTM2017-83655-C2-1-R) research projects (MINECO/AEI/FEDER, UE).

References

Aragonés, L., García-Barba, J., García-Bleda, E., López, I., & Serra, J. C. (2015). Beach nourishment impact on Posidonia oceanica: case study of Poniente Beach (Benidorm, Spain). *Ocean Engineering, 107,* 1–12.

Ariza, E., Jiménez, J. A., & Sardá, R. (2008a). A critical assessment of beach management on the Catalan Coast. *Ocean and Coastal Management, 51,* 141–160.

Ariza, E., Sardá, R., Jiménez, J. A., Mora, J., & Ávila, C. (2008b). Beyond performance assessment measurements for beach management: application to Spanish Mediterranean beaches. *Coastal Management, 36,* 47–66.

Ariza, E., Jiménez, J. A., Sardá, R., Villares, M., Pintó, J., Fraguell, R., et al. (2010). Proposal of integral quality index for urban and urbanized beaches. *Environmental Management, 45*, 998–1013.

Ballesteros, C. (2017). *The coastal risk landscape. Application on the Catalan coastal zone* (Ph. D. Thesis). Universitat Politècnica de Catalunya, Barcelona.

Ballesteros, C., Roca, E., Jiménez, J. A., & Villares, M. (2016). Assessing the coastal risk landscape in the Maresme region (NW Mediterranean). In *ECSA 56 Coastal systems in transition: From a 'natural' to an 'anthropogenically-modified' state*. Bremen, Germany.

Ballesteros, C., Jiménez, J. A., Valdemoro, H. I., & Bosom, E. (2018a). Erosion consequences on beach functions along the Maresme coast (NW Mediterranean, Spain). *Natural Hazards, 90,* 173–195.

Ballesteros, C., Jiménez, J. A., & Viavattene, C. (2018b). A multi-component flood risk assessment in the Maresme coast (NW Mediterranean). *Natural Hazards, 90,* 265–292.

Bosom, E., & Jiménez, J. A. (2011). Probabilistic coastal vulnerability assessment to storms at regional scale—application to Catalan beaches (NW Mediterranean). *Natural Hazards Earth System Sciences, 11,* 475–484.

CEDEX. (2014). *Estrategia de Actuación en el Maresme*. Ministerio de Fomento, Madrid: Informe Final.

CIIRC. (2010). *Estat de la zona costanera a Catalunya*. Barcelona: Dept. Política Territorial i Obres Públiques.

Cooke, B. C., Jones, A. R., Goodwin, I. D., & Bishop, M. J. (2012). Nourishment practices on Australian sandy beaches: A review. *Journal of Environmental Management, 113,* 319–327.

Dean, R. G. (2003). *Beach nourishment: theory and practice*. Singapore: World Scientific Publishing Company.

Defeo, O., McLachlan, A., Schoeman, D. S., Schlacher, T. A., Dugan, J., Jones, A., et al. (2009). Threats to sandy beach ecosystems: A review. *Estuarine, Coastal and Shelf Science, 81,* 1–12.

Di Franco, A., Graziano, M., Franzitta, G., Felline, S., Chemello, R., & Milazzo, M. (2011). Do small marinas drive habitat specific impacts? A case study from Mediterranean Sea. *Marine Pollution Bulletin, 62,* 926–933.

Escartín, F. J., Vidaor, A., Castaneda, A. M., & Lupon, N. (1996). Nourished beach control between balis and arenys harbours (Spain). In *Proceedings 25th International Coastal Engineering Conference* (pp. 4014–4023). ASCE.

Fernández-Salas, L. M., Durán, R., Mendes, I., Galparsoro, I., Lobo, F. J., Bárcenas, P., et al. (2015). Shelves of the Iberian Peninsula and the Balearic Islands (I): Morphology and sediment types. *Boletín Geológico y Minero, 126*(2–3), 327–376.

Galofré, J., Montoya, F., Enríquez, J., & Jiménez, J. A. (2003). Calafell-Vendrell beach nourishment evolution: is the "coastal box" closed or open? In *Coastal Sediments 2003* (CD-ROM). Corpus Christi: World Scientific Publishing Corporation and East Meets West Productions.

González-Correa, J. M., Torquemada, Y. F., & Lizaso, J. L. S. (2008). Long-term effect of beach replenishment on natural recovery of shallow Posidonia oceanica meadows. *Estuarine, Coastal and Shelf Science, 76,* 834–844.

Guillén, J., & Palanques, A. (1997). A historical perspective of the morphological evolution in the lower Ebro river. *Environmental Geology, 30,* 174–180.

Hanson, H., Brampton, A., Capobianco, M., Dette, H. H., Hamm, L., Laustrup, C., et al. (2002). Beach nourishment projects, practices, and objectives—A European overview. *Coastal Engineering, 47*(2), 81–111.

Hinkel, J., Aerts, J., Bown, S., Jiménez, J. A., Lincke, D., Nicholls, R. J., et al. (2018). The ability of societies to adapt to 21st century sea-level rise. *Nature Climate Change, 8,* 570–578.

Houston, J. R. (2008). The economic value of beaches: A 2008 update. *Shore and Beach, 76*(3), 22–26.

Hsu, J. R., Uda, T., & Silvester, R. (1993). Beaches downcoast of harbours in bays. *Coastal Engineering, 19,* 163–181.

Jiménez, J. A., Sánchez-Arcilla, A., Valdemoro, H. I., Gracia, V., & Nieto, F. (1997). Processes reshaping the Ebro delta. *Marine Geology, 144,* 59–79.

Jiménez, J. A., Gracia, V., Valdemoro, H. I., Mendoza, E. T., & Sánchez-Arcilla, A. (2011). Managing erosion-induced problems in NW Mediterranean urban beaches. *Ocean and Coastal Management, 54,* 907–918.

Jiménez, J. A., Sancho, A., Bosom, E., Valdemoro, H. I., & Guillén, J. (2012). Storm-induced damages along the Catalan coast (NW Mediterranean) during the period 1958–2008. *Geomorphology, 143–144,* 24–33.

Jiménez, J. A., Valdemoro, H. I., Bosom, E., Sánchez-Arcilla, A., & Nicholls, R. J. (2017). Impacts of sea-level rise-induced erosion on the Catalan coast. *Regional Environmental Change, 17,* 593–603.

Jiménez, J. A., Sanuy, M., Ballesteros, C., & Valdemoro, H. I. (2018). The Tordera Delta, a hotspot to storm impacts in the coast northwards of Barcelona (NW Mediterranean). *Coastal Engineering, 134,* 148–158.

Klein, M., & Zviely, D. (2001). The environmental impact of marina development on adjacent beaches: A case study of the Herzliya marina, Israel. *Applied Geography, 21,* 145–156.

Liquete, C., Canals, M., Ludwig, W., & Arnau, P. (2009). Sediment discharge of the rivers of Catalonia, NE Spain, and the influence of human impacts. *Journal of Hydrology, 366,* 76–88.

Lloret, J., Zaragoza, N., Caballero, D., & Riera, V. (2008). Impacts of recreational boating on the marine environment of Cap de Creus (Mediterranean Sea). *Ocean and Coastal Management, 51,* 749–754.

Lozoya, J. P., Sardá, R., & Jiménez, J. A. (2014). Users expectations and the need for differential beach management frameworks along the Costa Brava: Urban Versus natural protected beaches. *Land Use Policy, 38,* 397–414.

Luo, S., Liu, Y., Jin, R., Zhang, J., & Wei, W. (2016). A guide to coastal management: Benefits and lessons learned of beach nourishment practices in China over the past two decades. *Ocean and Coastal Management, 134,* 207–215.

Manzanera, M., Pérez, M., & Romero, J. (1998). Seagrass mortality due to oversedimentation: An experimental approach. *Journal of Coastal Conservation, 4*(1), 67–70.

Manzanera, M., Alcoverro, T., Jiménez, J. A., & Romero, J. (2014). The large penumbra: Long-distance effects of artificial beach nourishment on Posidonia oceanica meadows. *Marine Pollution Bulletin, 86,* 129–137.

Marchand, M., Sánchez-Arcilla, A., Ferreira, M., Gault, J., Jiménez, J. A., Markovic, M., et al. (2011). Concepts and science for coastal erosion management—An introduction to the CONSCIENCE Framework. *Ocean and Coastal Management, 54,* 859–866.

Micallef, A., & Williams, A. T. (2002). Theoretical strategy considerations for beach management. *Ocean and Coastal Management, 45,* 261–275.

Ministerio de Agricultura, Alimentación y Medio Ambiente. (2012). *Estrategia marina. Demarcación marina Levantino-Balear. Evaluación inicial. Parte II: Análisis de presiones e impactos*. Madrid.

Newton, A., Carruthers, T. J. B., & Icely, J. (2012). The coastal syndromes and hotspots on the coast. *Estuarine, Coastal and Shelf Science, 96,* 39–47.

Peña, C., & Covarsi, M. F. (1994). Project, works and monitoring at Barcelona Olympic beaches. In *Proceeding 24th International Coastal Engineering Conference* (pp. 3564–3578). ASCE.

Peña, C., Carrión, V., & Castañeda, A. (1992). Projects, works and monitoring at Barcelona coast. In *Proceeding 23rd International Coastal Engineering Conference* (pp. 3385–3398). ASCE.

Petrosillo, I., Vassallo, P., Valente, D., Mensa, J. A., Fabiano, M., & Zurlini, G. (2010). Mapping the environmental risk of a tourist harbor in order to Foster environmental security: Objective Versus subjective assessments. *Marine Pollution Bulletin, 60,* 1051–1058.

Piehlet, M. F., Maloney, J. S., & Paerl, H. W. (2002). Bacterioplanktonic abundance, productivity and petroleum hydrocarbon biodegradation in marinas and other coastal waters in North Carolina, USA. *Marine Environmental Research, 54,* 157–168.

Roca, E., & Villares, M. (2008). Public perceptions for evaluating beach quality in urban and semi-natural environments. *Ocean and Coastal Management, 51,* 314–329.

Roca, E., Riera, C., Villares, M., Fraguell, R., & Junyent, R. (2008). A combined assessment of beach occupancy and public perceptions of beach quality: A case study in the Costa Brava. *Spain. Ocean and Coastal Management, 51,* 839–846.

Rovira, A., & Batalla, R. J. (2006). Temporal distribution of suspended sediment transport in a Mediterranean basin: The Lower Tordera (NE SPAIN). *Geomorphology, 79*(1–2), 58–71.

Rovira, A., Batalla, R. J., & Sala, M. (2005). Response of a river sediment budget after historical gravel mining (the lower Tordera, NE Spain). *River Research and Applications, 21,* 829–847.

Sánchez-Arcilla, A., Jiménez, J. A., & Valdemoro, H. I. (1998). The Ebro delta: Morphodynamics and vulnerability. *Journal of Coastal Research, 14,* 754–772.

Schlacher, T. A., Dugan, J., Schoeman, D. S., Lastra, M., Jones, A., Scapini, F., et al. (2007). Sandy beaches at the brink. *Diversity and Distributions, 13,* 556–560.

Syvitski, J. P. M., & Milliman, J. D. (2007). Geology, geography, and human battle for dominance over the delivery of fluvial sediment to the Coastal ocean. *Journal of Geology, 115,* 1–19.

UNEP. (2009). *State of the environment and development in the mediterranean*. Athens: UNEP/MAP-Plan Bleu.

Valdemoro, H. I., & Jiménez, J. A. (2006). The influence of shoreline dynamics on the use and exploitation of Mediterranean tourist beaches. *Coastal Management, 34,* 405–423.

Villares, M., Roca, E., & Oroval, L. (2015). El análisis de la percepción del riesgo costero: una herramienta para la GIZC. In *VIII Jornadas de Geomorfologia Litoral.* Málaga: Universidad Pablo de Olavide.

Walling, D. E. (2006). Human impact on land-ocean sediment transfer by the world's rivers. *Geomorphology, 79,* 192–216.

Chapter 33
Coastal Management in the Balearic Islands

Francesc X. Roig-Munar, José Ángel Martín Prieto, Josep Pintó, Antonio Rodríguez-Perea and Bernadí Gelabert

33.1 Introduction

Coastal dunes represent important natural systems for balancing the beach while providing different functions such as protection against erosion and important ecological niches (Psuty 2004). However dunes are fragile and dynamic environments that are currently threatened both by natural and human factors.

Beaches are the most important natural resource to activity tourism in the Balearic Islands. Several analysis of their attendance and monetary value has shown its importance for the tourism economy (Mas and Blázquez 2005). Anthropogenic influence has supposed the overall degradation and dependence on artificial sand nourishment of many coastal sandy systems (Rodríguez-Perea et al. 2000). As a result, a wrong perception about the coastal areas has been created, no only by users but also by managers.

F. X. Roig-Munar (✉) · J. Á. M. Prieto
Environmental Consulting Carritxaret, 18.6 es Migjorn Gran,
E07749 Minorca, Spain
e-mail: xiscoroig@gmail.com

J. Pintó
Department of Geography, Universitat de Girona, Plaça Ferrater Mora 1,
17071 Girona, Catalonia, Spain
e-mail: josep.pinto@udg.edu

A. Rodríguez-Perea
Department of Geography, Universitat de les Illes Balears, Carretera de Valldemossa,
Km 7,5, E07122 Palma, Majorca, Spain
e-mail: arperea@uib.es

B. Gelabert
Department of Biology, Universitat de les Illes Balears, Carretera de Valldemossa,
Km 7,5, E07122 Palma, Majorcta, Spain
e-mail: bernardi.gelabert@uib.es

J. A. Morales (ed.), *The Spanish Coastal Systems*,
https://doi.org/10.1007/978-3-319-93169-2_33

Tourist activity on the Balearic Islands is focused on a strong demand for sunshine and beaches. Consequently, the activities developed on coastal areas have given to degradation, particularly of the sandy coast. From the beginning of mass tourism development, 50 years ago, the Balearic Islands have experienced a severe process of urbanization with direct consequences above coastal areas, with huge disarrangements between the natural system and the new uses (Mir-Gual and Pons 2011).

Unfortunately, the coastal management applied on the sandy coasts in Balearic Islands along the last decades has been based almost exclusively on the provision of leisure services, including mechanical cleaning, withdrawal of *Posidonia oceanica* and artificial beach nourishment.

Although in recent years there is some concern about the recuperation of dune systems, the dynamics of the system and the relationship with the beach is ignored or unknown. Recreation prevails instead rehabilitation, recuperation and management of the dune system.

The measures applied in the sandy coasts, far from being in consonance with the conservation of beaches and dunes, have been conceived as administrations of services, creation, stabilization and growth of beaches. Their geo-ecologic conservation has been obviated. Economic interests have prevailed over the environmental ones, what has generated serious ecological impacts.

The high coincidence of natural, socio-economic, historical-cultural and legal-administrative agents determines that the coastal zone is an environment with different degrees of fragility and should be managed taking into account all the elements that constitute it. Modifying any factor in a given coastal area could impact on another area and bring about (often unwanted) changes on an indeterminate time scale that can range from the short to the long term.

In recent years there has been an awareness of the constant deterioration of natural ecosystems, such as coastal systems. The stability and balance of dunar systems is determined by several factors such as sand supply (Psuty 1988; Aagaard et al. 2004), sediment transport rate (Namikas et al. 2010; Delgado-Fernández 2011), wave and wind forces (Hesp 1988; Ruz and Allard 1994), the wind flow over dunes (Arens et al. 1995; Hesp et al. 2005), the state of the beach at (Martinez et al. 2001), the occurrence and magnitude of storms (Morton 2002) and vegetation (Miot da Silva et al. 2008; Miot and Hesp 2010). Other factors play a key role in the morphological variability of the dune, such as topography, sediment availability, beach width and geometry, and the presence of species (Hesp 2002; de Vries et al. 2012).

The human impact on these morphologies has been widely studied and described, pointing out the causes of degradation due to massive tourism development, construction of sea walks, high anthropic pressure, the installation of services and incorrect management that generate negative impacts (Carter 1988; Nordstrom 2008; Gómez-Piña et al. 2002; Martín-Prieto et al. 2009; Roig-Munar et al. 2015). For most coastal areas in the Mediterranean, population growth and accelerated development of productive activities have led tensions and conflicts on the coast, where dune systems have been under pressure for 30 years, with a destruction of 75% of Mediterranean dunar systems (Meulen van der and Salman 1996).

Knowledge of beach-dune models can be valid tools for management. In this way, Cleary and Hosier (1979) identified a storm-dune cycle of evolution based in storms. Short and Hesp (1982), studied beach-dune interaction with emphasis on the morpho-dynamic interaction and the response of the beaches to the energy of the wind and waves. The models of Psuty (1988, 2004) provide a starting point to identify the most important variables to predict the morphology and behavior of foredune. Foster and Cheng (2001) provided a substitute for the beach sediment supply used by Psuty (1988, 2004) who uses the erosion/accretion rate to identify a foredune pattern or clusters towards stages of a development cycle from the incipient dunes to foredunes. Hesp (2002) synthesized the models of Hesp (1988), Carter (1988), Arens and Wiersma (1994), in a classification based on morpho-ecological states of the foredune, where stage 1 represents maximum stability and naturalness while stage 5 represents erosion with tendency to disappear. Recently, Houser (2009) and Houser and Hamilton (2009), found that the nature and timing of storm occurrences are critical for the development of dunes. Martín-Prieto et al. (2010), established a relationship between the erosion of the beach and the status of the foredunes line in both natural and anthropogenic states. However, current geomorphological knowledge is far from being applied by managers, and even the most familiar beach-dune interaction models (Short and Hesp 1982; Psuty 1988; Sherman and Bauer 1993; Hesp 2002; Balaguer and Roig-Munar 2016).

In coastal management, human factors are sometimes overlooked, so long-term erosion and sustainability problems can aggravate, creating difficulties in maintaining coastal systems. For this reason, it is necessary to monitor and analyze the management that allows an objective and critical vision of the established plans and the results obtained in the actions, as well as the response of the dune systems. For monitoring, it is necessary to elaborate a specific vulnerability index (Leatherman 1997; Morgan 1999; Laranjeira et al. 1999; Pereira et al. 2003; that must take in account both geomorphological variables of use and management (Roig-Munar 2011).

33.2 General View of Beach-Dune Systems in the Balearic Islands

From western Mediterranean seafloor, where deeps reach 3000 m at Algerian Basin, emerge a great relief: The Balearic Promontory, where the emerged areas of this Promontory are the Balearic Islands. The Promontory is formed by two blocks Menorca-Mallorca to the north shallower less than 200 m and Pitiüses to the south, separated by the Mallorca channel 1000 m deep. The Balearic Promontory is the northwest offshore continuation of the Betic-Rif orogenic arc and represents the northern part of the arc in its first phase of evolution (Bourrouilh 1983; Gelabert

1998; Sàbat et al. 2011). One of the most important features along the archipelago is the lack of rivers or permanent water courses. It is because of the lithology, mostly calcareous, and a scarce precipitation, conditioned by the Mediterranean climate. This aspect, lack of rivers, is an important factor, because there is a null or insignificant input of sediment into de beach-dune system.

The Balearic Archipelago is constituted by four main islands: Mallorca (72.6% of the territory), Menorca (14%), Ibiza (10.79%) and Formentera (1.64%), as well as a set of small islands and uninhabited islands (Fig. 33.1).

The general structure of the beach-dune systems of the Balearic Islands ranges from the submerged zone to the emerged area (Fig. 33.2) and they are based on five zonations described by Servera (1997) and Rodríguez-Perea et al. (2000).

33.2.1 Submerged Zone

In the submerged zone, the offshore sector, which is the deepest and furthest from the coastline (up to approximately −40 m) up to the point where the seagrass meadows grow (*Posidonia oceanica*), can be differentiated from the nearshore sector, which is closest to the shoreline and where sediment is continuously redistributed. It is an area of considerable dynamism, characterized by the formation of underwater sediment bars, where processes of sediment exchange initiate in

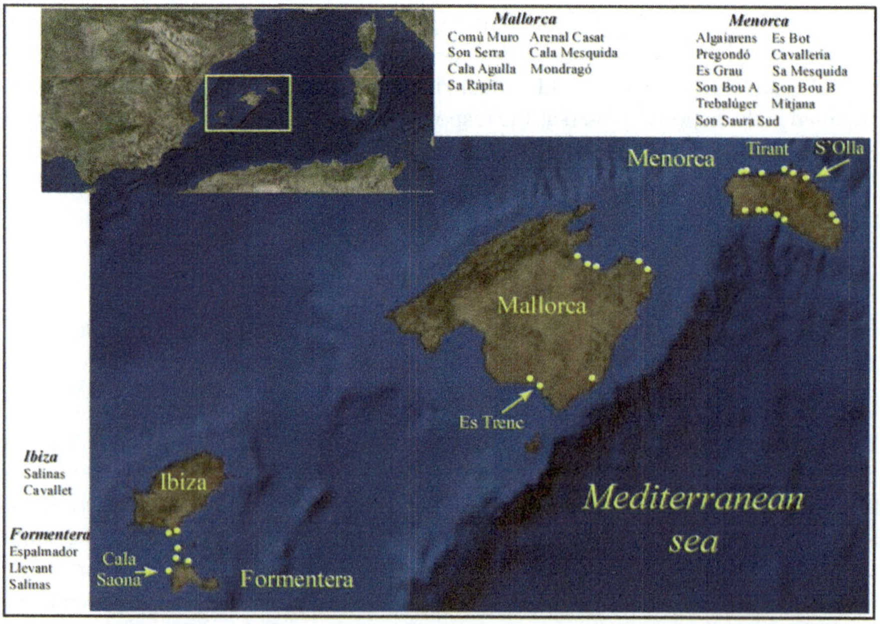

Fig. 33.1 Location of the Balearic Islands and the beach-dune systems studied in this work

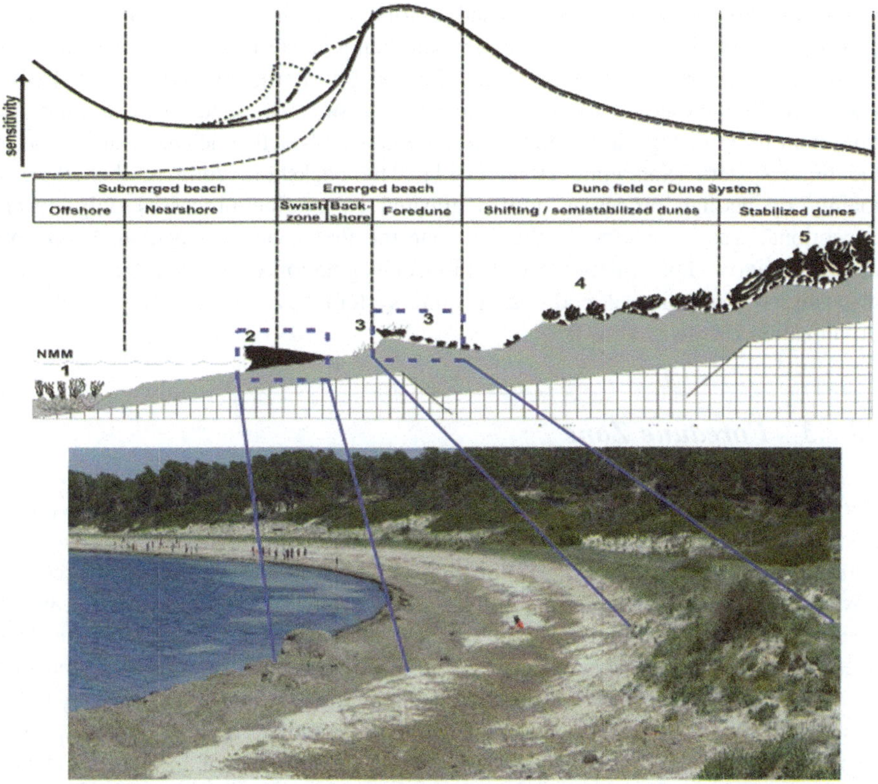

Fig. 33.2 Sectors of the beach-dune and their different degrees of geo-environmental sensitivity. *Source* Own-based and modified from Balaguer and Roig-Munar (2016), Roig-Munar (2004), Rodríguez-Perea et al. (2000) and Brown and McLachan (1990)

order to ensure maintenance of the beach-dune system (Komar 1998) (Fig. 33.2). The origin of the sediment supply is perhaps one of the factors with greatest influence on most coastal dune systems, where *Posidonia oceanica* meadows play an important role in the sediment production, which has a biological origin over 80% of cases (Servera 1997; Rodríguez-Perea et al. 2000; Roig-Munar 2011). This is the area where processes of sediment transfer from the submerged zone to the beach constitute the main variable to be taken into account.

33.2.2 *Beach Zone*

The beach zone is the area most commonly frequented and it is where many of the management and maintenance measures are usually undertaken to ensuring the beach is in good condition for users. It is the zone in which the sedimentary balance of the submerged area with the exposed sectors (subaerial sectors) is most apparent (Woodroffe 2002). The swash zone is where banquettes of *Posidonia oceanica* are deposited, developing a natural barrier formed by alternating sediment and dead leaves. It offers protection to the foreshore in case of storms. The foreshore plays an important role in supplying sediment to the exposed beach zone (backshore) and to the dune system (Sanjaume et al. 2011). The backshore zone is characterized mainly by aeolian processes and is affected by marine processes only during exceptional wave episodes. In this subzone the first ephemeral aeolian landforms appear (shadow dunes, nebkas) with pioneer vegetation (Servera 1997), a feature also observed in many altered coastal dune systems, even on beaches affected by urbanization (Pintó et al. 2013).

33.2.3 Foredune Zone

The foredune zone is the area immediately after the backshore and the zone with the first permanent dune formations. It is formed as sediment is transported by the wind and trapped by herbaceous vegetation (Fig. 33.1). Any alteration to the vegetation cover in this area can lead to its rapid destabilization and the disappearance, which results in transgressive morphologies such as blowouts (Hesp 2002). These dune accumulations, as sediment reservoirs, can ensure the equilibrium of the beach in case of severe storms, when waves reach the dune foot, thus enabling recovery of the beach area. Foredune also absorb the force of the wind and reduce the transport of marine spray into the dunar system, allowing the development of arboreal vegetation in the inner zone (Martín Prieto and Rodríguez-Perea 1996; Servera 1997).

33.2.4 Zone of Mobile and Semi-stabilized Dunes

Between the foredune zone and the mobile and semi-stabilized dune zone there is a depression known as a dune slack. This area is also known as a secondary dune or grey dune zone (Fig. 33.1). The conditions for plant growth are more suitable and the development of vegetation allows for the production of a layer of humus, which helps the formation of soil (Ley et al. 2007). The extension of this zone is not uniform and is usually conditioned to the conservation/degradation state of the foredune and to the topography of the terrain. From this condition depends the development of different types of dune shapes (parabolic, hummock).

33.2.5 Stabilized Dune Zone

The development of edaphic soil increases as we move inland, where the processes of wind deflation are smaller and dunes are stabilized by shrub and arboreal vegetation. There is little contribution of sand and it only takes place during episodes of strong winds (Lynch et al. 2010). Frequently, the vegetation of these areas in the Balearics is affected by human action, associated with processes of cereal sowing and reforestation (Mayol 2006).

Based on these five zonifications (Fig. 33.2), beach-dune interaction models can be evaluable tools for managers in order to implement management strategies (Martín-Prieto et al. 2009). However, this knowledge is far from being used by those responsible for coastal management and planning, who focus on socio-economic tools mainly. Ignoring these interactions can lead to a poor management, where erosion problems are continually aggravated over time, necessary to maintain the natural state of beach-dune systems (Rodríguez-Perea et al. 2000; Roig-Munar 2011; Mir-Gual et al. 2013).

Figure 33.2 shows several curves or points of sensitivity, which have been defined based on the different degree of pressure they are subjected. The first sensitivity curve, associated with the Balearic beach-dune systems, was described by Rodríguez-Perea et al. (2000) and is situated on the *Posidonia oceanica* meadows as sediment-producing habitat, submerged beach stabilizer and wave energy dissipater. The second sensitivity curve, (Roig-Munar and Martín-Prieto 2005), is situated on the accumulated berms of *Posidonia oceanica* on the swash-zone. It is a a sector of sedimentary transfer between the emerged and submerged ones, as contribution of organic matter between the beach and the dune plant communities, basic for stabilization, as well as a buffer element of the force of the storms. The mechanical removal of these berms results in the continuous erosion of the beach. The third sensitivity curve is located on the emerged beach, where the mechanical cleaning actions affect the ephemeral morphologies of the backshore and accelerates the destabilization of slopes in the dune fronts (Roig-Munar 2004). The fourth curve is located on the first dune ridges, and is frequently affected by the urbanization, promenades, users, the presence of bars-restaurants and the degradation of dune vegetation.

In this way, four critical points can be differentiated in the degree of sensitivity along the beach-dune profile, taking in account that the front dune sector is the key in the dune system (Martín-Prieto et al. 2007, 2010, 2016; Mir-Gual et al. 2013). Not knowing these points has supposed both loss of surface and volume of beach and dune, and even disappearance in some examples. Management based on the application of geomorphological criteria allows the recuperation of dunar systems, but working without it can aggravate the erosive processes that are intended to recover (Roig-Munar et al. 2009).

The management of beaches in the Balearics has been characterized by the lack of awareness of environmental sensitivity curves through the application of management measures (Fig. 33.2). These have not taken into account the geomorphological and environmental peculiarities of coastal ecosystems.

33.3 Objectives

The main goal of the present work is to carry out a spatio-temporal analysis of the evolution of 27 beach-dune systems of the Balearic Islands (Fig. 33.2 and Table 33.1). They present different tourist, recreational, service, state, management and conservation uses (Roig-Munar et al. 2006). The analysis allows a sampling of typologies of the beaches of the Balearic Islands (Roig-Munar and Comas-Lamarca 2005) and the morpho-ecological evolution of dune systems through the use of management variables. The purpose is to determine the spatio-temporal evolution of these systems and relate them with its use, planning, and management based on different qualitative variables. Also, four representative dunar-systems (Tirant and S'Olla in Menorca, es Trenc in Mallorca and Cala Saona in Formentera) are studied in order to show their spatio-temporal analysis.

33.4 Methodology

The methodology is based on the revision and modification of different methods adapted to the conditions and realities of the Balearic Islands (Fig. 33.2). A total of 18 variables have been identified and defined (Table 33.1), resulting from a redundancy analysis through an array of correlations on 36 variables used in Menorca (Roig-Munar and Comas-Lamarca 2005) and later simplified along the Balearic Islands

Roig-Munar (2011). These non-redundant variables have been ordered using physical and geo-environmental parameters in order to establish an evolutionary trend from 1956 to 2016 (Hesp 1988, 2002; Mir-Gual and Pons 2011; Corbau et al. 2009), condition, use and management parameters (Williams and Morgan 1995; Leatherman 1997; Laranjeira et al. 1999; Hesp 2002; Roig-Munar and Comas 2005; Martín-Prieto et al. 2016). The variables are qualitative, mostly discontinuous with values between 0 and 5 (being 0 the worst and 5 the best), with the aim to characterize the condition of dune systems and determine the key values in their evolution and/or degradation (Table 33.1). Values have been given for each variable and each case study, through photogrammetric analysis, bibliographic consultation, oral sources of former municipal managers of beaches and field work in recent years.

The variables are:

- Geomorphological: these include the beach-dune system, condition and conservation of foredunes, the presence of ephemeral morphologies at dune foot, vegetation cover and increment of erosive processes.
- Variables of anthropic pressure: presence of urbanizations in the dune system, proximity to the urban zone, installations on the dune system.
- Variables of management or protection, such as Natural Area of Special Interest (ANEI) or Natural Park

Table 33.1 Variables for the spatio-temporal analysis of dune systems in the Balearic Islands

Variables	Code	1	2	3	4	5
Beach	PT	Acretion	Balance	Erosion		
Foredune	FR	Stage 1	Stage 2	Stage 3	Stage 4	Stage 5
Neomorfology dune foot	OD	Very	Half	Little	No	
Deflation channels	BF	No	Little	Half	Very	
Outcrops dune system	UD	No	Little	Half	Very	
Beach vegetation	VP	100–75%	75–50%	50–25%	25–0%	
Plant cover dune system	CV	100–75%	75–50%	50–25%	25–0%	
Burial of arb./shrub vegetation	VE	No	Little	Half	Very	
Root exhumation	ER	No	Little	Half	Very	
Paths in the dunar system	CA	No	Little	Half	Very	
Temp. facilities on foredune	TF	No	Yes			
Perman. facilities on foredunes	PF	No	Little	Very		
Permanent installations dunes	PD	No	Little	Very		
Urbanization	DU	0–25%	25–50%	50–75%	75–100%	
Natural park	PN	Yes	No			
ANEI	AN	Yes	No			
Negative management measures	GN	No	Yes			
Positive management measures	GS	Yes	No			

- Management: removal of accumulations of *Posidonia oceanica*, mechanical cleaning of the beach, permanent or temporary installations in dunes, permanent or temporary installations in foredunes, Natural Area of Special Interest (ANEI), Natural Park, sustainable management technics such as sedimentary traps.

We assumed that the conservation status presented an optimal condition for each beach-dune system in 1956, without use and management. At that time, the only erosive agents were of natural origin.

Ten periods of analysis correspond to the years 1956, 1975, 1989, 1995, 2001, 2006, 2008, 2010, 2012 and 2016. This analysis has been based on aerial photographs to observe the most significant changes of each unit (Roig-Munar et al. 2012; Martín-Prieto et al. 2016), adding a temporal and evolutionary component (Garrote et al. 2001). For each period, use, management and planning variables have been applied to obtain the morpho-ecological response of each system. In order to investigate the relationship between the 18 variables (Table 33.1) and the 27 dune systems (Fig. 33.2) a multivariate factorial statistical analysis was performed to represent a large number of observed variables (270 cases, 180 variables). Finally, statistical analysis was carried out using principal component analysis (PCA).

33.5 Results

33.5.1 Statistic Analysis

An analysis has been carried out for all the 27 systems in the Balearic Islands and in particular the four case studies; S'Olla, Tirant in Menorca, es Trenc in Mallorca and Cala Saona in Formentera (Fig. 33.2), as these are representative of the different types of evolution over time, conditioned to the management in each of them. A main component analysis (PCA) was carried out using the program SPSS v23 to identify the most significant variables to describe and analyze the evolution of dune systems over time. The correlation between variables has been calculated from a matrix defined by 270 cases (27 dune systems for 10 years of observations) and 18 variables (Table 33.1). It has allowed refusing those redundant, as well as those that do not show any type of variability or discriminant information for the space-time study of dune systems.

In all the cases analyzed, the ACP analysis has identified two factors that explain a percentage of the variance that exceeds 59.4% in all dunar systems in the Balearics. In each analysis, the different factors are linked to a greater or lesser extent with any of the two axes generated the by the factorial analysis. Although each of the 27 systems analyzed exhibits a combination of different factors, it can be observed that in most of them the variables FR, OD, PT, AN, GS (Table 33.1) are responsible for the greater variability of the cases studied. Thus, foredune (FR) and dune foot neo-morphologies (OD) determine the stability of the dune system, following the morphological classification of Hesp (2002) and the works applied in Menorca by Roig-Munar et al (2009). These two variables are related to the state of the beach (PT) and its environmental management (GS), following the methodologies applied and evaluated in the dune systems of the Balearic Islands (Roig-Munar et al. 2009).

So, for the analysis of the set of 27 Balearic dune systems in 10 periods, the factors F1 and F2 explain 59.4% of the variance, slightly higher than the 43.2% established by Roig-Munar et al. (2006). The first two correlation factors explain almost half of the variance, where the variables VE (0.832), CV (0.741), OD (0.777) and FR (0.718) are positively related to F1. They are morphological variables that allow the stabilization of the system, especially on the foredune (Fig. 33.2). Meanwhile, variables PF (0.819), PD (0.769) and TF (0.670) do it with F2, that is, as service variables on the front dune system that facilitate the erosion of the dunar system, which is aggravated over time.

In Fig. 33.3, can be observed a concentration of correlations that determines that the vertical axis is related mainly to the management variables, while the horizontal axis does with those that determine the morphological response of the system to these management variables (Table 33.1). Therefore, in the Balearic Islands, the variable relative to the front dune morphologies plays an important role, following the model described in Fig. 33.2 of environmental sensitivity curves. Thus, it is necessary to take into account the management and conservation of the beach-dune systems, according the classifications made by Hesp (1988, 2002), the results in the evolution of dune systems in the Balearic Islands (Roig-Munar et al. 2012; Martín-Prieto et al. 2016), the affectation of services on dunes by Alonso et al. (2002) or the management of the emerged beach and dune fronts (Rodríguez-Perea et al. 2000; Mir-Gual et al. 2013).

In this way the F1 and F2 components define a factorial space with four well-defined quadrants (Fig. 33.3), which allow us to understand the evolution of the dune systems of the Balearics (Roig-Munar et al. 2006, 2009), these quadrants are defined:

- upper right quadrant (F1 and F2 positive): unaltered dune systems without management or protection, in a natural state;
- upper left quadrant (F1 negative and F2 positive): altered dune systems with management and recuperation processes;
- lower left quadrant (F1 and F2 negative): altered and degraded dune systems
- right lower quadrant (F1 positive and F2 negative): non-natural, non-altered dune systems

The representation of the 27 cases studied allows analyzing both global and individual the temporal evolution of all the beach-dune systems. The analysis starts from a theoretical state of naturalness in 1956, where anthropic intervention was minimal or nil. The representation of the annual averages indicates four evolutionary phases:

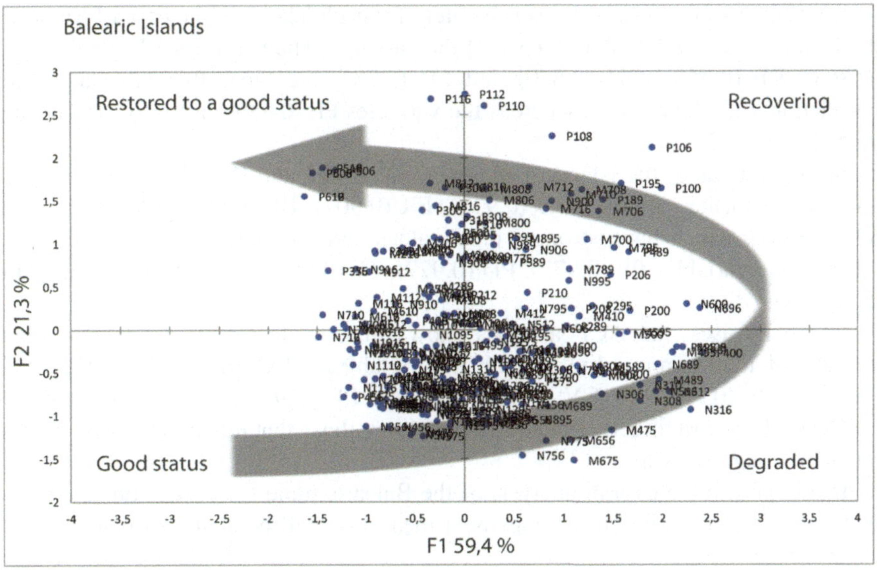

Fig. 33.3 Spatio-temporal evolution of the Balearic dune systems (1956–2016)

1. Between 1956 and 1989: this stage coincides with the beginning of sun and beach tourism, mainly from the 60s, where this tourism is centered on the sandy coast, shifting from a natural state to a degraded state. We start from an optimum situation of naturalness, towards an alteration due to the lack of management, even the absence of management and legal protection figures. The dynamic tends towards the alteration and disappearance of the beach-dune system. there is a lack of awareness about the problem of erosion or loss of nature.
2. From 1995 to 2000: there is a tendency towards a phase of alteration of dune systems due to an increase in degradation processes, associated to a lack of management and regulation of uses and an increase in tourist activities. This erosive phase is slowed due to a management based on geomorphological criteria and the application of protection and/or management figures, attenuating the erosive processes for each island.
3. From 2000 to 2006: there is a general tendency towards the recuperation of dune systems, through management and protection. The application of sustainable management methods and the regulation of public use, allows the stabilization of erosive processes and the restoration of the dunar front and inner morphologies, in Menorca and Formentera mainly.
4. From 2006 to 2016: there is a general tendency towards the recovery of most of dune systems, through management and protection. In some cases, the tendency is the renaturation, migrating towards a more natural, due to a more sustainable management. However, in 2012, the trend moves to negative due to a drop in management attributable to the economic crisis. This situation affects negatively the dune morphologies by the abandonment of measures of control, management and planning.

If we analyze the observations separately for each island, in the case of Menorca the factors F1 and F2 explain 59.3% of the variance. The variables VE (0.903), UD (0.868), BF (0.858), OD (0.840) and FR (0.731) are those that have a greater correlation with the F1 axis, whereas the variables PF (0.889) and ANE (0.817) are with the axis F2.

In the systems of the island of Majorca, factors F1 and F2 collect 65.8% of the variance, with the variables OD (0.913), FR (0.907), BF (0.883), PV (0.859) and VE (0.849)), which have stronger eigenvalues and correlations with the F1 axis, whereas the variables PF (0.977), PD (0.977), TF (0.952) and DU (0.881) are with the F2 axis.

In the Pitiüses, the ACP analysis shows that the factors F1 and F2 explain 73.02% of the variance, linked with the variables FR (0.971), VP (0.953), OD (0.951), VE (0.952 and CV (0.920) (0.936), PF (0.912), TF (0.830) and PN (−0.831). The most directly linked to axis F1, are those that relate to the axis F2, the latter in a negative way.

In addition to the overall analysis of the Balearic dune systems, 4 dune systems that have received different management measures, will be analyzed below.

Fig. 33.4 Panoramic views of the four dunar systems analized (From top left to bottom right: Cala Saona, es Trenc, Tirant and S'Olla)

33.5.2 Cala Saona

The Cala Saona dunar system is located on the west coast of the island of Formentera (Figs. 33.1 and 33.4) and occupies an area of approximately 13.3 Ha, of which 0.7 correspond to foredunes. It shows a good conservation and anthropogenic pressure is concentrated on the beach mainly, the inner dunar system is little anthropized.

In Cala Saona, the two factors explain more than 90% of their variability, where the factors F1 and F2 explain 92.8% of the variance. The variables UD (0.938), OD (0.921) and FR (0.917) are positively related to F1, while PF (0.945), PD (0.945) and TF (0.945) do it with F2. The representation of this system (Fig. 33.5) exemplifies an erosive tendency in the first years of analysis, with a restoration of the system, where we highlight;

1. Between 1956 and 1989: this stage coincides with the beginning of sun and beach tourism associated with a hotel plant located behind the dune system, which estimulates erosion. It is not an accelerated erosive phase, but degradation processes occur.
2. Between 1995 and 2006: there is a trend to degradation with timid recuperation towards a renaturation of the system. This phase is progressively slowed due to the management and application of protection and/or management figures, attenuating the erosive processes.
3. Between 2006 and 2016: there is a general trend toward renaturation and the recuperation of the dune system, through the application of management technics such as of closure and management of uses.

33.5.3 es Trenc

The beach-dune system of es Trenc is located to the south of Mallorca (Figs. 33.1 and 33.4) and it is the most important of the island. It occupies an area of approximately 9 km^2, of which 4.5 km^2 correspond to Holocene morphologies. Foredune is present throughout the beach, although in some sections there are strong discontinuities. The mobile and semi-stabilized dunar sector occupy a surface approximately 8.6 km^2, extending inland.

Currently, the beach-dune complex shows a strong unbalance, amplified in the last decades, mainly due to a considerable increase in the number of users. The vegetation is eliminated due to trampling processes on the foredunes. The partial destruction of foredune results in the formation of blowouts that invade the sector of semi-stabilized dunes.

In this dunar system, F1 and F2 explain 78.7% of the variance. In this case, the variables FR (0.805), BF (0.923), PF (0.917), PD (0.917) and TF (0.917) are positively related to F1. They are variables that display poor management, showing the degradation of the front dune system (FR), with the developing, evolution and consolidation of inland deflation channels (BF), favored by installations on the dune fronts (PF, TF and P.S). Meanwhile, variables PT (0.899) and UD (0.562) do with F2 and show significant degradation with the outcropping of fossil dunes and beach erosion linked with the disappearance of the foredune (Fig. 33.5. Two main periods can be highlighted;

1. Between 1956 and 2000: this stage coincides with an important use and exploitation of the dune system, generating continuous erosive processes due to the lack of management and planning measures. The system shifts to degraded states (Fig. 33.5), since the most affected sector is the foredune and emerged beach.
2. Between 2001 and 2016: there is a tendency towards a recuperation phase. The measures implemented have not had positive results due to the lack of criteria, maintenance and regulation of uses. In general, this beach-dune system is still in an erosive state.

33.5.4 Tirant

It is a longitudinal dunar system, with a north-south orientation, located in a bay exposed to the winds of the northern component (Figs. 33.1 and 33.4). It is one of the largest dimensions of the island of Menorca penetrating almost two kilometers inland. In addition, it is associated to a very important humid zone. It has been exploited through a sand quarrying activity for building purposes between the‘70s and‘90s. The morphological consequences of these actions resulted in an impact on

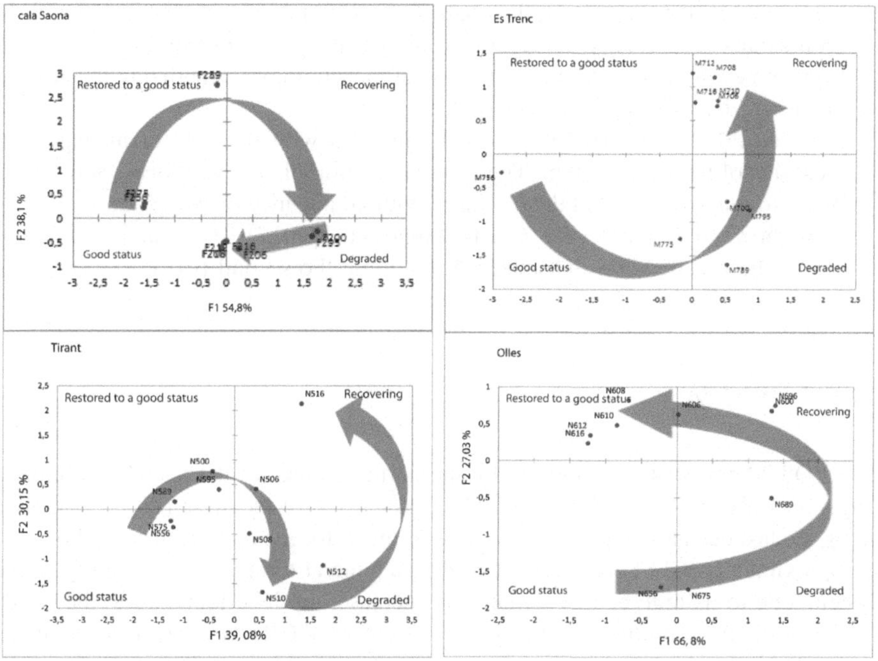

Fig. 33.5 Spatio-temporal evolution of the four dune systems (1956–2016): Cala Saona, es Trenc, Tirant and S'Olla

the semi-stabilized dune system, with a notable sediment loss that was aggravated by the lack of an adequate foredune management.

The results of the Tirant dune system show a correlation between the F1 and F2 axes, which explain 69.2% of the variance, where the variables ER (0.933), CA (0.846), PT (0.781) are related to F1, while that AN (−0.788) is negatively. This variables show the abandonment in the management of the quarry (located in the internal sector of dunar system) that stimulated the reactivation of erosive processes, besides there was an incorrect management of the beach and its dune front. These variables are correlated with FR (0.955), OD (0.947) and CV (0.766), which show in axis F2, due a lack of management. In addition, in the early 2000s a management of the dunar system was initiated, allowing some balance, but it was abandoned later, resulting in degradation of the dune system (Fig. 33.5). Three periods can be highlight:

1. Between 1956 and 2001: this stage is characterized by erosive processes in the inner and front dune system, the later allows the advancement of lobes with the development and extension of transgressive blowout morphologies (Garriga-Sintes et al. 2017).
2. Between 2006 and 2010: during this period there is a general tendency towards the recuperation of the system with the application of management technics on

the beach and the dune front, the control of uses on dunes and the installation of sedimentary traps. This management allows the re-naturalization of the dune front and even the recuperation of blowout morphologies, stabilizing the erosive processes detected in the first period.

3. Between 2012 and 2016: this period coincides with the lack of management measures of the front system. There is reactivation of erosive processes from the beach to the sector of stabilized dunes, with an advance of the deflation lobules between 2 and 7 meters/year (Garriga-Sintes et al. 2017). This reactivation place at risk the recuperation that materialized along the second period.

33.5.5 S'Olla

The s'Olla dunar system is located on the northern coast of the island of Menorca (Figs. 33.1 and 33.4). It has been is altered as a result of the urbanization processes along the last decades. The surface occupied by Holocene and recent morphologies is approximately 0.24 km2 (Servera 1997). In general terms, this is a highly active system, due to its high energy dynamism, increased by its orientation to the winds of the northern component. The dunar system of s'Olla is characterized by its perpendicularity to the coastline. This tendency manifests itself mainly from the development inland of two blowouts; located in both sides. The high energetic dynamism of this dune system and the state of degradation of foredunes over the last decades, are the two agents that explain the scarce surface occupied by mobile or semi-stabilized dunes.

The results obtained in the S'Olla dunar system, show that F1 and F2 axes explain 93.8% of the variance. FR (0.973), OD (0.984), VP (0.990) and BF (0.919), geomorphological variables, are positively related to F1, while DU (0.963) and AN (-0.933) are the variables reflected on the F2 axis, the latter in a negative way. The temporal evolution of this system is closely linked to the management, since it presented an advanced stage of degradation. However, the negative trend was reversed towards the practically renaturation of the beach-dune system, thanks to the application of management measures. This change was achieved by applying measures based on the morphodynamic sectors described in Fig. 33.2. The evolution of the system is marked by two highly differentiated periods (Fig. 33.5):

1. Between 1956 and 1996: this stage coincides with an important process of degradation of the dune front and the advance of important inland deflation channels. These start in the sequence of blowouts of the dune front, that stimulates the degradation of the beach-dune system.
2. From 2000 to 2016: in the year 2000, a integral management of the dunar system was initiated through actions on the beach and the inner dune system, migrating the shifting towards its renaturation. Throughout this period, the system moves from erosive to stable.

33.5.6 Trend of Foredune as a Function of Management (2017–2050)

It is attempted to predict the geomorphological tendency of the front dune system based on the classification of Hesp (2002). For this, we rely on different management scenarios, short, medium and long term, 2017–2050 and on the sensitivity curve of the front dune system (Fig. 33.2), key to the stability of the beach-dune system. We start from the present state of dune systems up to 2016 (Fig. 33.5) where foredune (FR, Table 33.1) represents the key element in the conservation of all the analyzed dune systems (Figs. 33.3 and 33.5). Based on this condition we give values to the foredune from a hypothetical incorrect management between 2016 and 2035, where coastal management would not be applied. This negative trend is reversed in 2035, with the application of correct management until 2050, following the tendency of the period 2006–2016. The criteria for the evaluation of foredune stages have been applied with the knowledge of an expert based on the results obtained in different analyzes of dune systems in the Balearic Islands (Roig-Munar et al. 2006, 2009, 2011; Martín-Prieto et al. 2009, 2010, 2016; Mir-Gual et al. 2013).

33.5.6.1 Cala Saona

In the first phase, the dune system trends towards erosion (2017–2035), due to incorrect coastal management on the foredune. Later these erosive processes stabilize, reaching a stage 3 for several years. The last period, 2036–2050 with the application of management measures, the system retrieves stage 1 in a short period.

33.5.6.2 es Trenc

This dunar system starts from a very degraded stage of the foredune, where despite the application of management measures these have not been positive maintaining in the last years the stage 4 (Figs. 33.5 and 33.6). Figure 33.6 shows the trend of the system towards stage 5 between 2017 and 2035, where there are no signs of recuperation, with processes of reactivation, loss of forms and vegetation. The application of management and control measures on the dune system (2036–2050) gives a tendency of progressive recuperation, shifting from stages from 5 to stages 2.

33.5.6.3 Tirant

It starts from a very degraded stage of the front dune system, where it has undergone important erosive processes in the last decade (Fig. 33.5). The lack of management aggravates these processes between 2017 and 2035 (Fig. 33.6), where the

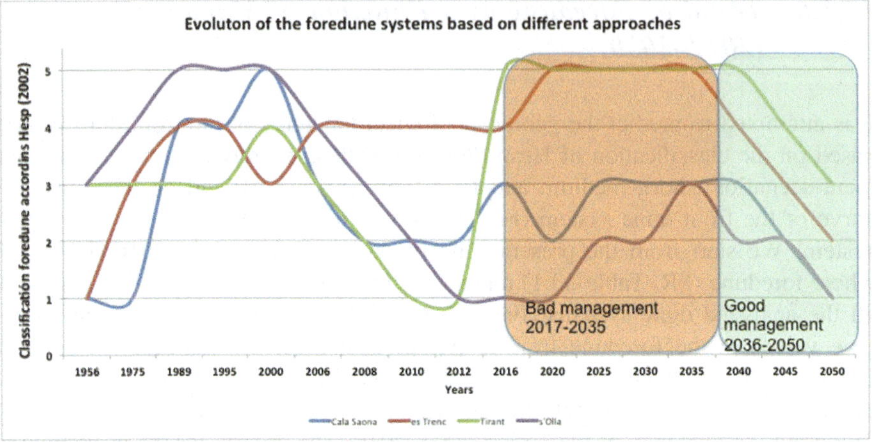

Fig. 33.6 shows the evolution of the front dune systems from their management, showing a different evolution for each system according to its environmental state

dune system not only shows deterioration, but the internal dune system is reactivated (Garriga-Sintes et al. 2017), destabilizing the whole dune system (Fig. 33.2), resulting in negative sedimentary balances. The application of management measures on the dune system (2036–2050) results in a gradual recuperation with initial stages shifting from 5 to stages 3, although erosion processes throughout the system remain active.

33.5.6.4 S'Olla

It stars from a condition of recuperation and balance of the dune system in 2016. The lack of management measures (2017–2035) results in a breakdown of equilibrium with an inertia towards the degradation of the dune front and associated vegetation (Fig. 33.6). In spite of these erosive processes, these do not reach stage 3, because it started from a stage of equilibrium.

Figure 33.6 shows that the behavior of these 4 dune systems in short-term 2016–2035, would be conditioned by the current geo-environmental state (Fig. 33.5). The response of the front system to an incorrect management is aggravated in those that show erosive processes, as Trenc or Tirant, that reach stage 5, the most degraded. In stable systems, such as Saona or s'Olla, there would be incipient processes of erosion towards stage 2. Thus, erosion is aggravated in those systems deteriorated by inadequate management, aggravating and accelerating erosive processes on foredune, destabilizing the balance of the system.

33.6 Conclusions

The evolution of dunar systems in the Balearics must be explained by a combination of natural processes, uses and management that determine the evolution of a beach-dune system, since these are the most important tourist attraction.

In the Balearic Islands, Coastal management has basically consisted in making the beach-dune space functional in order to satisfy the users attendance, what generated conservation difficulties. Problems such as the alteration and elimination of beach-dune neo-morphologies and associated vegetation, elimination of dune formation, alteration and destabilization of beach profiles, increase of wind transport by erosion of dune fronts and reactivation of dunar systems, loss of biological diversity, loss of beach surfaces and volumes, erosive morphologies, advancement of dune lobes, burial of arboreal vegetation, among others. These impacts have displayed with the disappearance of dune morphologies that have put at risk of erosion the dunar systems of the Balearic Islands.

The result of the spatio-temporal analysis (1956–2016) applied to the 27 Balearic dune systems allows to differentiate 4 evolutionary periods. These periods have determined its degree of conservation associated with a tourist model of coastal exploitation, inadequate management measures and lack of management of uses and services and unapplied protection figures. The factorial analysis based on the analysis of main components has been applied through 27 qualitative variables that determine the evolution of dune systems based on the beach-dune profile. From this analysis, seven dominant variables have been detected, that play an important role in the evolution of dune systems, and all of them are geomorphological.

The geomorphological aspects (the morphology of the beach and foredunes mainly), are the key element to manage to recuperate and stabilize the dune system, following the proposals of sensitivity curves established in this work. The results obtained show different patterns of spatio-temporal morphological behavior associated with management and planning measures. In general terms, the trend of dune systems has been towards the renaturation of the beach-dune complex, with measures that have influenced the foredunes (Fig. 33.2) and have allowed the progressive renaturation of the system towards stabilization.

Although the dune evolution shows a negative trend in the first two periods of the analysis, it can be observed a trend of recuperation in most of the systems. There is a positive evolution in general terms within the context of the beach-dune system and with special incidence in the front dune system. However, in the fourth period erosive processes are due to a lack of monitoring in the management.

The trend towards recuperation presents a slow, but progressive and positive evolution in general terms. Procedures that initially facilitate erosion due to extensive procedure have been rectified, while the lack of management criteria gave rise to reactivation of dynamic processes. It is demonstrated that dune restoration requires a significant effort, as well as the combination and integration of different environmental criteria to recover dune morphologies.

The spatial-temporal analysis is presented as a good tool to measure and control the conservation status of dune systems and to establish trends associated with their use, planning and management. The variables used allow establishing qualitatively possible trends in the short and medium term taking into account aspects of management and use that affect the stability of the system. It is evident that the change of management, based on the sensitivity curves, can change the erosive tendency of a system towards its renaturation. But it is also demonstrated that the lack of a continuity of these efforts can reactivate the previously managed erosive processes that facilitates erosive processes.

This analysis is a tool to identify, analyze and manage the sensitive points of the dune system, with special emphasis on the key element, the foredunes. The results obtained give special importance to the dune management and restoration measures. Attempts to dune restoration have been partial in some of the Balearic beach-dune systems, since the geomorphological aspects of each system have not been taken into account.

In general, the implementation of the management measures carried out in the Balearic Islands have been able to get the objectives of regeneration and dune restoration that have allowed the stabilization of the systems. These techniques are successful and widely applied if they are based on geomorphological criteria. Their cost is low, easy to apply, wearable in many cases and they use nature for their purpose.

References

Aagaard, T., Davidson-Arnott, R., Greenwood, B., & Nielsen, J. (2004). Sediment supply from shoreface to dunes: Linking sediment transport measurements and longterm morphological evolution. *Geomorphology, 60,* 205–224.

Alonso, J., Alcantara-Carrió, J., & Cabrera, L. (2002). Tourism resorts and their impacts on the beach erosion at sotavento beaches, Fuerteventura, Spain. *Coastal Research, 36,* 1–7.

Arens, S. M., & Wiersma, J. (1994). The Dutch foredunes: Inventory and classification. *Journal of Coastal Research, 10*(1), 189–202.

Arens, S. M., van Kaam-Pters, H. M., & van Boxel, J. H. (1995). Airflow over foredunes and implications for sand transport. *Earth Surface Processes and Landforms, 20,* 315–332.

Balaguer, P., & Roig-Munar, F. X. (2016). Coastal dune systems and the concept of integrated coastal and marine management (ICMM). En F. X. Roig-Munar & F. J. Quintana (Eds.), *Restauración y Gestión de Sistemas Dunares. Estudio de casos* (pp. 201–220).

Bourrouilh, R. (1983). Estratigrafía, sedimentología y tectónica de la isla de Menorca y del nordeste de Mallorca (Baleares). In *La terminación nororiental de las cordilleras béticas en el Mediterráneo occidental Colección memorias* (Vol. 99, p. 672). Madrid: IGME.

Brown, A., & McLachlan, A. (1990). *Ecology of sandy shores* (p. 572). Hardbound: Elseiver.

Carter, R. W. G. (1988). *Coastal environments: An introduction to the physical, ecological, and cultural systems of coastlines* (p. 617). London: Academic Press.

Cleary, W. J., & Hosier, P. E. (1979). Geomorphology, washover history, and inlet zonation: Cape Lookout, N. C. to Bird Island, N. C. In S. P. Leatherman (Ed.), *Barrier Islands: From the Gulf of St. Lawrence to the Gulf of Mexico* (pp. 237–269). New York: Academic Press.

Corbau, C., Simeoni, U., Archetti, R., Peretti, A., & Farina, M. (2009). Winter sandy protections of the Northern Adriatic Coast against flooding: Preliminary results. *Journal of Coastal Research, 56,* 1194–1198.

de Vries, S., Southgate, H. N., Kanning, W., & Ranasinghe, R. (2012). Dune behavior and aeolian transport on decadal timescales. *Coastal Engineering, 67,* 41–53.

Delgado-Fernandez, I. (2011). Meso-scale modelling of aeolian sediment input to coastal dunes. *Geomorphology, 130,* 230–243.

Foster, E. R., & Cheng, J. (2001). *Shoreline change rate estimates: Gulf County Florida. Florida department of environmental protection, office of beaches and coastal systems* (p. 63). Tallahassee: Florida State University.

Garriga-Sintes, C., Martín-Prieto, J. Á., Roig-Munar y, F. X., & Rodríguez-Perea, A. (2017). Reactivación del sistema dunar de Tirant, N de Menorca, asociada a la falta de gestión y extracciones de áridos. *Geo-Temas, 17*, 167–170.

Garrote, J., Page, J., & Garzón, G. (2001). Morfodinámica en un complejo playa-Ria bajo condicionantes antropicos. Oyambre Cantabria. *Geogaceta, 30,* 55–58.

Gelabert, B. (1998). *La estructura geològica de la mitad occidental de la isla de Mallorca* (p. 129). Madrid: ITGE.

Gómez-Piña, G., Muñoz-Pérez, J. J., Ramírez, J. L., & Ley, C. (2002). Sand dune management problems and techniques, Spain. *Journal of Coastal Research, 36,* 325–332.

Hesp, P. A. (1988). Surfzone, beach and foredune interactions on the Australian south east coast. *Journal of Coastal Research, 3,* 15–25.

Hesp, P. A. (2002). Foredunes and blowouts: Initation, geomorphology and dynamics. *Geomorphology, 48,* 245–268.

Hesp, P. A., Walker, I. J., Davidson-Arnott, R. G., & Ollerhead, J. (2005). Flow dynamics over a vegetated foredune at Prince Edward Island, Canada. *Geomorphology, 65,* 71–84.

Houser, C. (2009). Synchronization of transport and supply in beach-dune interaction. *Progress in Physical Geography, 33,* 733–746.

Houser, C., & Hamilton, S. (2009). Group sensitivity of post-hurricane beach and dune recovery to event frequency. *Earth Surface Processes and Landforms, 34,* 613–628.

Komar, P. D. (1998). *Beach processes and sedimentation* (p. 544, 2nd ed.). New Jersey: Prentice Hall.

Leatherman, S. P. (1997). Beach rating: A methodological approach. *Journal for Coastal Research, 3,* 253–258.

Laranjeira, M. M., Pereira, A., & Williams, A. T. (1999). Comparison of two checklist methods for assessment of coastal dune vulnerability. *Boletin del Instituto Espanol de Oceanografia, 15,* 259–268.

Ley, C., Gallego-Fernández, J., & Vidal, C. (2007). *Manual de restauración de dunas costeras* (p. 244). Rural y Marino: Ministerio de Medio Ambiente.

Lynch, K., Jackson, D. W., & Cooper, J. A. (2010). Coastal foredune topography as a control on secondary airflow regimes under offshore winds. *Earth Surface Processes and Landforms, 35,* 344–353.

Martín-Prieto, J. A., & Rodríguez-Perea, A. (1996). Participación vegetal en la construcción de los sistemas dunares litorales de Mallorca. In A. Grandal d'Angade (Ed.), *IV Reunión de Geomorfología* (pp. 785–798). A Coruña: Sociedad españolas de Geomorfología O Castro.

Martín-Prieto, J. A., Roig-Munar, F. X., & Rodríguez-Perea, A. (2007). Análisis espacio temporal (1956–2005) de la foredune de Mesquida (N Mallorca) mediante el uso de variables geoambientales y antrópicas. En L. l. Gómez-Pujol & J. Fornós (Eds.), *Investigaciones Recientes (2005–2007) en Geomorfología Litoral* (pp. 137–140).

Martín-Prieto, J. A., Roig-Munar, F. X., Pons, G. X., & Rodríguez-Perea, A. (2009). La foredune como elemento estabilizador en sistema playa-duna. En J. A. Morales, M. Cantado, A. Rodríguez y & I. Delgado (Eds.), *Nuevas contribuciones sobre geomorfologia litoral* (pp. 45–48).

Martín-Prieto, J. A., Roig-Munar, F. X., Rodríguez-Perea, A., & Pons, G. X. (2010). Evaluació de la línea de costa de la platja de es Trenc (S. Mallorca). En J. Mayol, Ll. Muntaner, y & O. Rullán (Eds.), *Homenatge a Bartomeu Barceló i Pons, geògraf* (pp. 423–440).

Martín Prieto, J. A., Roig Munar, F. X., Rodríguez, Perea A., Mir, Gual M., Pons Buades, G. X., & Gelabert Ferrer, B. (2016). La erosión histórica de la playa de sa Ràpita (S. Mallorca). *Investigaciones Geográficas, 66,* 135–154.

Martinez, M. L., Vazquez, G., & Salvador, S. C. (2001). Spatial and temporal variability during primary succession on tropical coastal sand dunes. *Journal Vegetation Science, 12,* 361–372.

Mas, L. l., & Blàzquez, M. (2005). Anàlisi de la freqüentació d'ús a les platges i paràmetres de sostenibilitat associats. *Documents d'Anàlisi Geogràfica, 45,* 15–40.

Mayol, J. (2006). Un gran projecte de postguerra: La repoblació forestal de les dunes de Formentera. *Revista Eivisssa, 44–45,* 9–16.

Miot da Silva, G., & Hesp, P. A. (2010). Coastline orientation, aeolian sediment transport and foredune and dunefield dynamics of Moçambique Beach, southern Brazil. *Geomorphology, 120,* 258–278.

Miot da Silva, G., Hesp, P. A., Peixoto, J., & Dillenburg, S. R. (2008). Foredune vegetation patterns and alongshore environmental gradients: Moçambique beach, Santa Catarina Island Brazil. *Earth Surface Proceeding Land, 33,* 1557–1578.

Mir-Gual, M., & Pons, G. X. (2011). Coast sandy strip fragmentation of a protected zone in the N of Mallorca (Western Mediterranean). *Journal of Coastal Research SI, 64–1,* 1367–1371.

Mir-Gual, M., Pons, G. X., Martín-Prieto, J. A., Roig-Munar, F. X., & Rodríguez-Perea, A. (2013). Geomorphological and ecological features of blowouts in a western Mediterranean coastal dune complex: A case study of the Es Comú de Muro beach-dune system on the island of Mallorca, Spain. *Geo-Marine Letters, 33,* 129–141.

Morgan, R. (1999). A novel, user-based rating system for tourist beaches. *Tourism Management, 20,* 393–410.

Morton, R. A. (2002). Factors controlling storm impacts on coastal barriers and beaches a preliminary basis for near real-time forecasting. *Journal of Coastal Research, 1,* 486–501.

Namikas, S., Edwards, B. L., Bitton, M. C. A., Booth, J. L., & Zhu, Y. (2010). Temporal and spatial variabilities in the surface moisture content of a fine-grained beach. *Geomorphology, 114,* 303–310.

Nordstrom, K. F. (2008). *Beach and dune restoration* (p. 22). Cambridge, England: Cambridge University Press.

Pereira, C., Jiménez, J. A., Medeiros, C., & Marinho da Costa, R. (2003). The influence of the environmental status of Casa Caiada and Rio Doce Beaches (NE-Brazil) on beaches users. *Ocean and Coastal Management, 46,* 1011–1030.

Pintó, J., Martí, C., & Fraguell, R. M. (2013). Evaluación de los sistemas dunares de la Costa Brava. In R. Sardá, J. Pintó, & J. F. Valls (Eds.), *Hacia un nuevo modelo integral de gestión de playas* (pp. 73–86). Girona: Documenta Universitaria.

Psuty, N. P. (2004). The coastal foredune: A morphological basis for regional coastal dune development. In M. L. Martinez & N. P. Psuty (Eds.), *Coastal dunes: Ecology and conservation (ecological studies)* (pp. 11–27). Berlin: Springer.

Psuty, N. P. (1988). Sediment budget and dune/beach interaction. *Journal of Coastal Research, 3,* 1–4.

Rodríguez-Perea, A., Servera, J., & Martín-Prieto, J. A. (2000). *Alternatives a la dependència de les platges de les Balears de la regeneració artificial* (Informe METADONA). Universitat de les Illes Balears: Col·lecció Pedagogia Ambiental 10. 108 pp.

Roig-Munar, F. X. (2004). Análisis y consecuencias de la modificación artificial del perfil playa-duna por el efecto mecánico de su limpieza. *Investigaciones geográficas, 33,* 87–103.

Roig-Munar, F. X. (2011). *Aplicació de criteris geomorfològics en la gestió dels sistemes litorals arenosos de les Illes Balears*. Ph.D. thesis, (p. 366). University of Balearic Islands, Palma de Mallorca, Spain. (in Catalan).

Roig-Munar, F. X., & Martín-Prieto, J. A. (2005). Efectos de la retirada de bermas vegetales de Posidonia oceánica sobre playas de las islas Baleares: consecuencias de la presión turística. *Investigaciones geográficas de México, 57,* 39–52.

Roig-Munar, F. X., & Comas Lamarca, E. (2005). Propuesta de un modelo de clasificación para las playas de las islas Baleares mediante el análisis de variables de uso, estado y gestión. *Boletín de la Asociación de Geógrafos Españoles, 40,* 429–448.

Roig-Munar, F. X., Martín-Prieto, J. A., Comas-Lamarca, E., & Rodríguez-Perea, A. (2006). Space-time analysis (1956–2004) of human use and management of the beach dune systems of Menorca (Balearic I.). *Journal of Coastal Research,* (48), 107–111.

Roig-Munar, F. X., Rodríguez-Perea, A., Martín-Prieto, J. A., & Pons, G. X. (2009). Soft management of Beach-Dune systems as a tool for their sustainability. *Journal of Coastal Research SI, 56,* 1284–1288.

Roig-Munar, F. X., Martín-Prieto, J. A., Rodríguez-Perea, A., Pons, G. X., Mir-Gual, M. (2012). Risk assessment of beach-dune system erosion: Beach management impacts on the Balearic Islands. *Journal of Coastal Research, 28*(6), 1488–1499.

Roig-Munar, F. X., Martín-Prieto, J. A., & Rodríguez-Perea, A. (2015). Crítica a las diferentes medidas de restauración en sistemas playa-duna. El caso de las Islas Baleares y la Costa Brava. *Geotemas, 15,* 81–84.

Ruz, M. H., & Allard, M. (1994). Foredune development along a subarctic emerging coastline, Eastern Hudson Bay. *Canada Article in Marine Geology, 117,* 57–74.

Sàbat, F., Gelabert, B., Rodríguez-Perea, A., & Jiménez, J. (2011). Geological structure and evolution of Majorca: Implications for the origin of the western Mediterranean. *Tectonophysics, 510,* 217–238.

Sanjaume, E., Gracia, F. J., & Flor, G. (2011). Introducción a la geomorfología de sistemas dunares. In E. Sanjaume & F. J. Gracia (Eds.), *Las dunas en España* (pp. 13–65). Sociedad Española de Geomorfología: Fundación Empresa-Universidad Gallega.

Servera, J. (1997). *Els sistemes dunars litorals de les Illes Balears* (p. 904). Tesis doctoral: Departament de Ciències de la Terra, Universitat de les Illes Balears.

Sherman, D. J., & Bauer, B. O. (1993). Dynamics of beach-dune systems. *Progress in Physical Geography, 17–4,* 413447.

Short, A. D., & Hesp, P. A. (1982). Wave, beach and dune interactions in South Eastern Australia. *Marine Geology, 48,* 259–284.

Van der Meulen, F., & Salman, A. H. P. M. (1996). Management of mediterranean coastal dunes. *Ocean and Coastal Management, 30*(2–3), 177–195.

Williams, A. T., & Morgan, R. (1995). Beach awards and rating systems. *Shore and Beach, 63*(4), 29–33.

Woodroffe, C. D. (2002). *Coasts: Form, process and evolution* (p. 617). Cambridge: Cambridge University Press.

Chapter 34
The Integrated Coastal Zone Management in the Canary Islands

Emma Pérez-Chacón Espino, Carolina Peña-Alonso, Aarón M. Santana-Cordero and Luis Hernández-Calvento

34.1 Introduction

The coasts of the Canary Islands have undergone a substantial transformation in recent decades. The economy of the Islands, based on agrarian activity and trade, changed radically from the 1960s (Ruiz and Quintana 2001). Since then, the progressive incorporation of an urban-touristic territorial model has dramatically changed the landscapes of the Islands, especially those around the littoral areas (Bianchi 2004). The consequences of these changes have had different signs, but the environmental cost has been notably high (Greenpeace 2009). This situation is particularly serious if we take into account, as Prieto and Navarro (2012: 358) point out, that "the coastal strip is a non-renewable resource in which, from the point of view of sustainability, consumption rates must be minimized and the reuse thereof encouraged". The same authors highlight that, considering the whole of the Canary

E. Pérez-Chacón Espino (✉) · C. Peña-Alonso · L. Hernández-Calvento
Grupo de Geografía Física y Medio Ambiente, Instituto de Oceanografía y Cambio Global, IOCAG, Universidad de Las Palmas de Gran Canaria, ULPGC, Parque Científico-Tecnológico de Taliarte, Calle Miramar 121, 35214 Telde, Las Palmas, Spain
e-mail: emma.perez-chacon@ulpgc.es

C. Peña-Alonso
e-mail: cpena@becarios.ulpgc.es

L. Hernández-Calvento
e-mail: luis.hernandez.calvento@ulpgc.es

A. M. Santana-Cordero
Department for Management of Science and Technology Development, Ton Duc Thang University, Ho Chi Minh City, Vietnam
e-mail: aaron.santana@ulpgc.es

A. M. Santana-Cordero
Faculty of Environment and Labour Safety, Ton Duc Thang University, Ho Chi Minh City, Vietnam

J. A. Morales (ed.), *The Spanish Coastal Systems*,
https://doi.org/10.1007/978-3-319-93169-2_34

Islands, the 20% of the first 500 m of the littoral areas is an altered surface (Prieto and Navarro 2012: 361).

The environmental alteration of the coast of the Canary Islands follows the same trend described for other areas of the Spanish coast. The Observatory of Sustainability of Spain estimates, referring to the Mediterranean coast and a part of the Atlantic and the Cantabrian coasts, that "between 1987 and 2005 an average of more than two hectares was urbanized every day in the first 500 m, which means that in less than one generation, almost half of the Spanish Mediterranean coastline (43%) has become artificial" (Observatorio de la Sostenibilidad 2014: 95). In 2009 more than the 40% of the Spanish population lived in coastal municipalities, and 45 million of foreign tourists visited the coastal areas for holidays (Barragán and Borja 2011: 9). Furthremore, there are many works in the last decades that warn about the coasts in danger (Pilkey and Cooper 2014), the alteration of the dune systems (Bauer and Sherman 1999; Jackson and Nordstrom 2011) or the need to manage the coastline from new approaches (Farinós Dasí 2011; Williams and Micallef 2011; Barragán 2014; Botero et al. 2017).

In the Canary Islands, coastal areas have become the main support for the tourism industry. Specifically, during the year 2017, 16.7 million of tourists visited the islands, while the resident population in the Canary Islands is only 2.16 million (ISTAC 2018). The development of the tourism activity has led the growth of urban-touristic centers, as well as the main infrastructure and territorial facilities. From the administrative point of view, the Canary Islands are a Spanish autonomous community that has a regional government (Canarian Government). In turn, each island has an island government: the islands councils or "Cabildos", each time with more competences. At the local scale, there are 88 municipalities in the archipelago, of which 77 have coastline. The Spanish State retains most of the legal competences over the coastal zone. In this context, Integrated Coastal Zone Management (ICZM) in the Canary Islands is undoubtedly a challenge. However, according to García Sanabria et al. (2011: 240), there is no specific policy in the archipelago of integrated coastal and marine management, only sectoral policies with different incidences on the littoral areas. At the same time, works addressing the problems of the Canarian coasts as a whole are scarce (Bianchi 2004; Pérez-Chacón et al. 2007; García Sanabria et al. 2011).

Taking into account this background, this chapter is a reflection about the Integrated Coastal Zone Management in the Canary Islands, since the perspective that "the Integrated Coastal Zone Management deals with problems and conflicts that take place in coastal marine areas … [and that its] … emphasis is on the impact of human activities on these ecosystems and, above all, on the way to approach solutions" (Barragán 2014: 29). The chapter begins addressing the consequences that the traditional land uses had on the Canary Islands' coastal areas, then the policies and strategies of the coastal management are shown, as well as the problems and the conflicts and, finally, some examples of good practices and search for solutions are addressed.

34.2 The Coast of the Canary Islands: General Settings

The general features of the coasts of the Canary Islands are determined by their volcanic nature, their oceanic and climatic conditions, their biodiversity and the history of their human occupation. Bearing in mind all these factors together, the processes on the coasts of these Islands are different with respect to the rest of the Spanish coast.

34.2.1 *Location of the Canary Islands*

The Canary archipelago is composed of seven islands and a group of islets of volcanic nature. These Islands are located in the eastern sector of the middle Atlantic, on the African plate. The closest distance between the African continent (Cape Jubi-Western Sahara) and the Canary Islands (Fuerteventura) is only 95 km. The archipelago extends 515 km from east to west, the largest archipelago of the Macaronesian region (7447 km^2) (Fernández-Palacios y Dias 2001). The climatic conditions of the Canary Islands zone are directly related to the influence of the Azores anticyclone and its position in the ocean, close to the African continent (Marzol 2001). Thus, in autumn and winter seasons the arrival of oscillation from the Polar Front to which the southwestern ones are sometimes added produces climatic perturbations. On the other hand, on the coasts of the Islands, especially in the eastern ones, a regime with few thermal contrasts and a very irregular rainfall predominate, which does not usually exceed 100 mm per year. In them the climatic conditions can be considered arid. The insular nature determines that the coastal perimeter of the Islands (1553 km in total) is very significant with respect to the surface they occupy, being the second Spanish Autonomous Community in coastline length, after Galicia.

34.2.2 *Coastal Geology and Landforms*

The volcanic formation of the Canary Islands comprises from the Miocene to the present, although each island has materials with different ages (Fig. 34.1). The oldest one is Fuerteventura (20.2 Ma), with sub-aerial eruptive phases, between the Lower and Middle Miocene (up to 12 Ma), and another Plio-Quaternary volcanic reactivation (<5 Ma). On the other hand, the islands of La Palma (1.8 Ma) and El Hierro (1.2 Ma) are the youngest ones of the archipelago, as their phases of shield growth occur during the Quaternary (Hernán 2001). The submarine eruptions that formed the islands are less studied, but in Fuerteventura there have been dated Oligocene submarine basalts (about 33.4 Ma, according to Anguita et al. 2002) and sub-volcanic and plutonic rocks from the Cretaceous (>65 Ma). The age of each

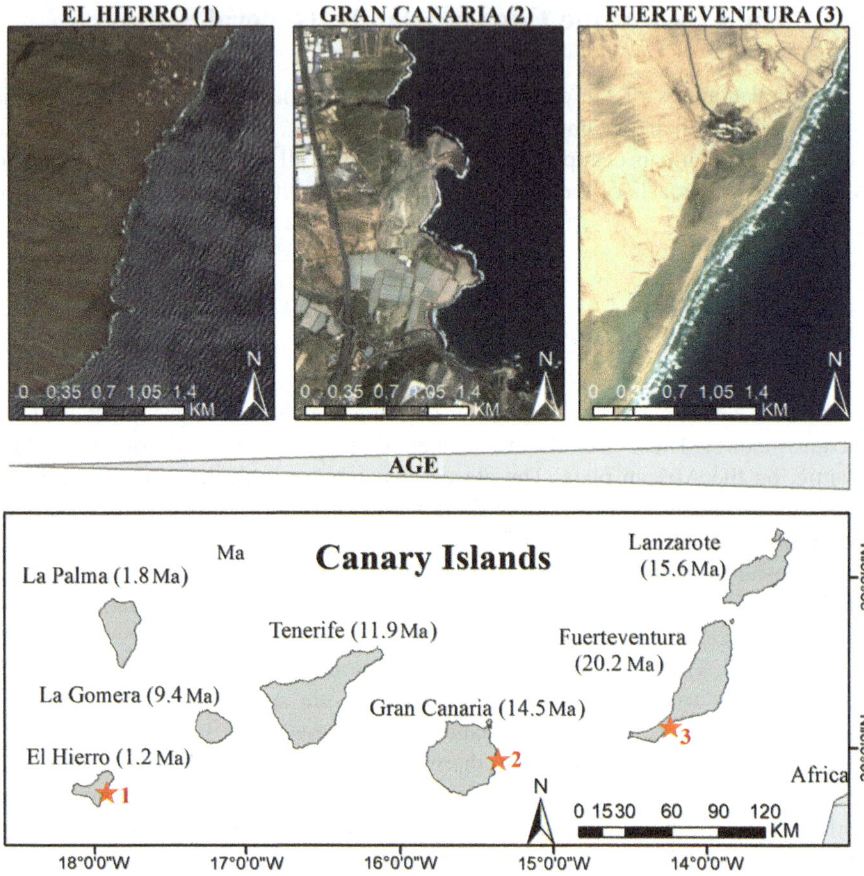

Fig. 34.1 General view of the Canary Islands and age of each island. The age of the islands was consulted in Carracedo et al. (2008)

island determines its morphological characteristics, observing important differences between the eastern and western islands (Fig. 34.1).

The coasts of the western islands, the most recent ones in the geological context of the archipelago, is characterized by being very rugged, as seen in the example shown for the island of El Hierro (Fig. 34.1). The presence of coastal outcrops and the small insular marine platform prevent the existence of significant coastal drifts. Therefore, the beaches are usually small coves (100–300 m long), associated to ravine's mouths or to retreated cliffs. In general, they are formed by thick black sand and boulders or dark blocks of basaltic nature (Table 34.1).

The eastern islands, being geologically older than the western ones, have been exposed for more time to erosional processes. Therefore, they have straighter coastlines and larger coastal and marine platforms, as can be seen, for example, on the island of Fuerteventura (Fig. 34.1). On these platforms there are significant

Table 34.1 Natural characteristics of the canarian coasts

	Natural characteristics				
	Coastal perimeter (Km)[a]	Beaches at coastline (%)[a]	Granulometric predominance in beaches[a]	Cliffs at coastline[a] (%)	Low-lyings[a] (%)
Lanzarote	213.26	15.58	Pebbles/sand	74.27	1.03
Fuerteventura	325.91	23.65	Sand	50.29	25.22
Gran canaria	236.64	23.91	Pebbles	58.43	7.34
Tenerife	398.18	16.86	Pebbles	64.66	12.04
La Gomera	117.65	12.72	Pebbles/sand	84.94	1.27
La Palma	155.75	7.13	Pebbles/sand	81.98	7.64
El Hierro	106.5	4.98	Pebbles/sand	88.03	6.67
Canary Islands	1,553.89	17.08	–	66.89	10.95

[a]Data from the Canary Islands statistics institute (ISTAC 2018)

accumulations of sand-sized grains (Table 34.1) with a high percentage of organ-oclastic components. In these islands there are beaches of larger dimensions, rec-tilinear and associated, in some cases, to dune systems.

In some islands have taken place volcanic manifestations during the Quaternary and in historical epoch, that is, from the fifteenth century. These events of recent volcanic activity, where lava flows reached the coast, have modified some coastal sectors, giving rise to differentiated coastal landforms in the context of each island. In local terminology, lands gained to the sea by recent lava flows are termed "low islands" (Yanes 2006).

34.2.3 Current Human Characteristics

With the arrival of mass tourism to the Canary Islands, the construction of infrastructures, facilities and buildings exponentially increased the demand for labor. For this reason, a generalized immigration flow from inland to the coast was generated, contributing, with the continuous arrival of tourists, to the occupation of the coast. This phenomenon of "littoralisation" is linked to the growth of old urban area as well as to the construction of new ones in coastal areas not previously occupied. This situation has generated important transformation of the coast, especially significant in the islands of Lanzarote, Gran Canaria and Tenerife (Table 34.2).

At present, the arrival of tourists exceeds the number of inhabitants registered in each of the islands (Table 34.2). One of the main characteristics of the Canary

Table 34.2 Human characteristics of the canarian coasts

	Human characteristics				
	Human transformed coastlines[a] (%)	Coastline under environmental protection[b] (%)	Local population[a] (2017)	Tourist arrivals[a,c] (2017)	Tourists per km of coastline (2017)
Lanzarote	9.13	59.25	147,023	3,146,119	14,752.50
Fuerteventura	0.84	55.02	100,299	2,390,980	7,336.32
Gran Canaria	10.32	29.20	843,158	4,587,576	19,386.31
Tenerife	6.43	56.84	894,636	6,181,592	15,524.62
La Gomera	1.06	65.36	20,976	–	–
La Palma	3.24	50.15	81,35	407,708	2,617.71
El Hierro	0.33	49.69	10,679	–	–
Canary Islands	5.08	52.06	2,108,121	16,713,975	10,756.22

[a]Data from the Canary Islands statistics institute (ISTAC 2018)
[b]Own elaboration based on the WMS data of the protected natural spaces of the Canary Islands (www.ide.canarias.es)
[c]Nationals and foreigners

Islands, as a sun and beach touristic destination, is their mild climate during most of the year. In other parts of Europe this type of tourism is seasonal (in summer). In the Canary Islands, the influx of tourists is continuous throughout the year, although two maximum inputs of tourists are identified (Turégano 2005). The first one takes place in summer, and is characterized by Spanish tourists, mainly. The second one occurs in winter, and is played by foreign tourists. During the autumn and spring seasons the influx of visitors remains continuous, although it is reduced in number with respect to the high seasons (ISTAC 2018).

The maturity of some destinations, such as Playa del Inglés (south of Gran Canaria), Puerto de la Cruz (north of Tenerife) or Costa Teguise (east of Lanzarote), has produced, along with other external factors (economic crisis), a phase of economic decline in the touristic model of the Canary Islands. The dependence on large tour operators, the scarce spending on complementary offer, or the subjection to the continuous growth of arrivals (Guerra and Pérez 2008) are elements that facilitate the tendency of these spaces to become obsolete environments. These factors negatively affect the configuration and profitability of the tourist centers, the socio-economy of the islands and the environmental state of the coast. In addition, instead of rehabilitating the degraded tourist areas, the real estate and tourism promoters promote the occupation of new coastal areas not yet altered.

This process finds a limit in the environmental protection of the Canary coasts through legal instruments of supra-European, European, national and regional scopes (Table 34.2). The protection level of each island is around 50% of the coastal perimeter, with the exception of the island of Gran Canaria that does not reach 30%. In the latter case, the lack of protection of the western coastal strip of the municipality of San Bartolomé de Tirajana, at the south of the island, is due to the urban-tourist overcrowding of the coast zone (sealed thanks to the construction of artificial beaches, marinas, dykes and jetties and the construction of the coastline). This situation has generated the only case in the Canary Islands in which a marine Special Area of Conservation (SAC), the marine strip of Mogán (REF: ES7010017), does not include the adjacent coastal zone.

34.3 Traditional Land Uses in the Canarian Coasts

The coasts of the Canary Islands have had several land uses and management mechanisms along their history. Among the traditional land uses mining, firewood extraction, seaweed harvesting, salt pans establishment, agrarian activity and the use of lime kilns can be mentioned. They were developed linked to the previous socioeconomic system of the Islands, in which the agrarian activity predominated (Ruiz and Quintana 2001). It is important to highlight that in the main islands, Gran Canaria and Tenerife, the change of socioeconomic system came with the mass tourist activity that begun in 1960s, but in the other islands this process was different. The example of the La Graciosa island is very illustrative, due to the traditional activities continued existing until the declaration of the island as protected natural area by the Canarian Government, and the prohibition of these activities.

In the rocky coasts, apart from the shell-fishing and the fishing, the *orchilla* collection, set of lichens of the *Roccella* genus that were located on coastal cliffs, was very noTab. 34. in some islands. Practically, from the XV century, and in a marginal form until the XIX, their economic exploitation was of great interest for extracting dyes (Hernández Rodríguez 2004).

Until recent times, in 1940s and 1950s, the seaweed harvesting that arrived to the coasts in spring and autumn was something common. These collected species were of the genus *Glidium* (*G. canariensis*) and *Pterocladia*. From them, agar was obtained, substance used as fertilizer crop in pharmacy and in bacteriology and in certain industries. However, the high costs of the transport and the small amount of raw material collected caused that its exploitation was not renTab. 34. and the industry ceased. Another marine plants, as *Cymodocea nodosa*, was recollected to be used as filling material for mattresses and pillows (Portillo Hahnefeid 2008).

Another activity inherent to the islands was the salt production through salt pans. The Tenefé salt pans (Gran Canaria) constitute a good example. These were created in the XVIII century and continue functioning in the current times. They have an annual production that varies between 180 and 280 tons and have been declared as spot of Heritage of Cultural Interest (Spanish initials: BIC) (FEDAC 2017). Another example is the Fuencaliente salinas, in La Palma, that although are relatively recent (1967), they currently carry out this traditional activity. It is necessary indicate that these Salinas were declared *Sitio de interés científico* (site of scientific interest) by the Canarian Government in 1994 (Grupo Salinas marinas de Fuencaliente 2017).

But the major intensity of the land uses was produced over the sandy coasts. Mining was an important activity documented for several coastal sandy environments, such as the Guanarteme and Maspalomas dune systems (in Gran Canaria), Corralejo and Jandía (in Fuerteventura), El Jable (in Lanzarote) and La Graciosa. This resource was used in construction, mainly for making bricks. Also as filling material, such as the first docks that currently form the navy base, in Las Palmas de Gran Canaria. It was the main material for constructing houses in some neighborhoods, as Guanarteme (Las Palmas de Gran Canaria). The establishment of a brick factory in 1924 into this dune system triggered that the mining acquired an industrial dimension, which was the cause of the disappearance of the entire dune system, according to some oral sources. One management measure in these moments was the forbidden of extracting sand from the beaches (Santana-Cordero et al. 2016a).

In Lanzarote, the sedimentary system of El Jable constituted one of the most productive sectors for the agricultural activity in the island. Likewise, in this system also there was firewood extraction for creating lime kilns. The existence of goat cattle in extensive regime could cause the vegetation reduction (Cabrera Vega 2010).

Firewood extraction, from woody shrubs, was a common practice—for example- in La Graciosa island between 1880 and 1987, moment in which the island was declared as protected natural area by the Canarian Government. This resource was used as fuel for domestic necessities and for lime kilns burning. It is known that the island was totally deforested in the most human pressure period, 1940–1970. With these land uses, the island lived a long period of grazing that extends, at least between 1736 and 1987 (Santana-Cordero et al. 2016b). The impact of the activities around the lime kilns over the dune ecosystems was also very significant in other islands, as Gran Canaria and Fuerteventura.

In other aeolian sedimentary systems, as Maspalomas and Jandía, the agricultural activity and extensive grazing existed as well. In Maspalomas, an export agriculture existed until 1960. Until the 1930s the dominant crops were barley and wheat, being afterwards substituted by tomato, until the beginnings of 1960s. This last crop occupied conditioned lands over aeolian sediments (Hernández-Calvento 2006).

The environmental consequences of these land uses had different intensity, especially significant in the beach-dune systems of some islands. The research performed in historical ecology in the Canary Islands, already mentioned in the previous paragraphs, shows that many of these dune systems experienced drastic ecological changes before the tourism development in the XX century.

34.4 Characteristics of the Current Coastal Management

The traditional land uses of the coastal zone, linked to the agrarian socioeconomic system, significantly transformed some areas, especially those occupied by beach-dune systems. These changes are smaller compared to the process of human transformation of the coastal area that, progressively, experienced many islands since 1960. For example, in Gran Canaria, after the urban-tourist development, the 43% of the coastal landforms have been affected by anthropic impacts, being the most affected ones the sedimentary landforms (coastal dunes, paleo-dunes, beaches and wetlands). In consequence, the geodiversity has undergone a significant reduction (Ferrer-Valero et al. 2017). Even so, the increase in ship traffic and various types of impacts that negatively affect the marine environment should be added.

However, and despite the importance of the coastline in the Canary Islands, public administrations have not been able to design a strategy for an integrated coastal zone management. It is managed from sectoral policies, poorly coordinated, where different administrations converge. The competences (Table 34.3) are shared between the general administration of the State, the regional (Canarian Government), the insular (*Cabildos*) and the municipal. This distribution of competences is similar to that in other Spanish regions, with the exception of the insular administrations. Among the responsibilities of the *Cabildos*, directly or indirectly related to the management of coastal areas, are the regional planning (insular plans of management) and that of protected areas (plans for the management of natural resources and management instruments). They also deal with the management of protected areas, a matter with a high incidence on the coast if we take into account that, in the Canary Islands, 46% of the first 500 meters of coast are protected (Greenpeace 2013: 19). However, when management or planning decisions affect the public domain and areas of conservation, state and regional reports are decisive, as well as requesting their authorizations for coastal actions.

With regard to the marine environment, the waters of the Canary Islands are delimited by the annex of the Law 44/2010, of December 30, of Canarian Waters. These Canarian waters constitute the maritime area of the Autonomous Community of the Canary Islands. It is a delimitation that, for the moment, does not imply changes in the distribution of competences between the State and the regional government.

The Canary Islands have managed to get the International Maritime Organization to recognize the ecological, socio-economic and scientific importance of their waters, because their vulnerability is being increased due to intense maritime traffic, and the probability that accidents involving pollution by hydrocarbons and other hazardous substances can occur. Therefore, they have been declared a Specially Sensitive Marine Area (SSMA) in 2005. Following its entry into force (in 2006), protective measures have been established, including areas to be avoided by the ships, the existence of two obligatory routes for navigation (Fig. 34.2) or the implementation of a mandatory notification system for ships in transit.

Table 34.3 Distribution of competences among the state, regional, insular and municipal administration

	Spanish state	Canarian region	Island	Municipality
Navegation and comunications				
Mineral and energy resources				
Defending				
Biological resources	Territorial sea and Exclusive economic zone (EEZ)	Inland waters and fisheries sector management	Natural resources ordinance Plan (NROP) Insular management plans (IMP)	
Waste disposal		Land-based sources	Land-based sources	Land-based sources
Marina and fishing ports				
Research				
Recreational uses		Coastal zone Internal waters	Coastal zone	Coastal zone
Conservation and protection		Protected marine areas Internal waters	Áreas marinas protegidas Internal waters	
Coastal management	Coastal Law	Territorial planning Tourism management	Territorial planning	Urban planning

The existence in the Canary Islands of three Marine Reserves of Fishing Interest is another indicator of the social, economic and ecological value of its marine environment. One of them is in the surroundings of the island of La Graciosa (declared in 1995), another in El Hierro (since 1996) and the third one in La Palma (2001). The ecological recovery experienced in these waters since they were declared is remarkable, as well as the degree of satisfaction of the fishermen of these environments (López-Ornat et al. 2014).

Another aspect that has had a special impact, directly or indirectly, on the management of the coastal-marine area has been the legislation that, between 1985 and 2003, was developed in the Canary Islands for their territorial organization. A set of laws follow one another to protect more than 40% of the Canarian territory (Law 12/87 of declaration of Natural Areas and Law 12/94 of Protected Natural Areas of the Canary Islands); while incorporating instruments to limit tourism and urban growths (Law 3/85 of Urgent Measures, Law 1/87 of Insular Plans, Law 5/87 on the Urban Planning of Rustic Land), environmental criteria (Law 11/1990 of Prevention of Ecological Impact), integrative approaches (Legislative Decree

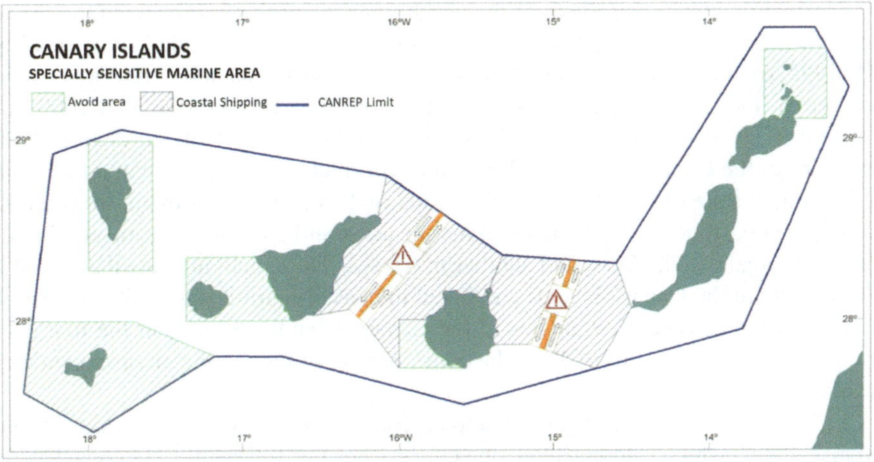

Fig. 34.2 Restricted areas (surfaces with frames) and obligatory routes (orange strips) for navigation in transit through the SSMA of those ships whose origin or destination is not a port in the Canary Islands. *Source* modified from OMI (2003). Document MEPC 51/8, of October 24, 2003, on the designation of the Canary Islands as SSMA, presented by Spain

1/2000) and guidelines (Law 19/03 of Guidelines for General Arrangement and Tourism Management) from which a tourist moratorium was derived. Although urban-touristic growth has been exponential in the first decade of this century, specifically since 1997 (Greenpeace 2013), it is clear that this would have been much greater if the aforementioned legislation not existed. Guadalupe et al. (2016) show the area for touristic use that was declassified on the coast of Famara (Lanzarote) thanks to the enactment of some of these laws (Fig. 34.3).

Fig. 34.3 Simulation of tourism projects planned on the coast of Famara (Lanzarote) between 1960 and 1991. In red the only one project executed and in orange the surface planned to be developed that did not get executed. *Source* modified from Guadalupe et al. (2016)

Meanwhile, the Government of the Canary Islands properly processed the list of "Sites of Community Importance" (SCI) (Council Directive 92/43/ EEC, of May 21, 1992, on the Conservation of Natural Habitats and Fauna and Wild Flora), now converted into "Special Area of Conservation" (SAC) (Decree 174/2009, of December 29, declaring Special Conservation Zones as part of the Natura 2000 Network in the Canary Islands and measures for maintenance these natural areas in a favourable state of conservation). Of a total of 176 SAC, 66 of them, inner areas, border the coastline, 23 of them are in marine areas and 2 of them are mixed, that is, they have a part in the land, and another in the sea. These data indicate that 51.70% of the SAC in the Canary Islands are linked to coastal-marine zones. This implies a change in the protection strategies in the Canary Islands, because in the first declaration of protected areas of 1987 the consideration of the marine environment was scarce.

In 2009, progress was made in implementing Specific Guidelines for Coastal Management, which, among their objectives were the following: (1) consider the need to reduce urban and infrastructures pressure on the coast, (2) coordinate policies and public actions on the coast, (3) define criteria for management of the coastline, among others. It progress was an unsuccessful attempt. It would have been a first document, at regional level, for the Integrated Coastal Zone Management (ICZM) in the Canary Islands, but this document was never processed. In 2003, in the Canary Islands begun a process towards a permissive and discretional territorial arrangement, causing a series of legislative amendments that lead to the current 4/2017 Law on Soil and Protected Natural Areas of the Canary Islands. This law aims to eliminate restrictions, reduce the weight of planning and public control over land, while weakening regional and insular competences. In addition, this Law gives supremacy to the municipalities through a series of competences, but not the material and human means to develop them. From the point of view of the ICZM it is foreseeable that this Law will contribute to the lack of regulation of land in coastal areas, where the real estate pressure is extreme.

As a counterpoint, it contrasts the proliferation of experiences at the municipal or insular level in the absence of a regional policy on ICZM. Many of the experiences are based on processes of citizen participation, which indicate the feasibility of creating spaces to project and manage coastal areas in a participatory manner. Some of them will be exposed in the section dedicated to "good practices".

34.5 Environmental Problematic of the Coast and the Integrated Management

The environmental problematic of the coasts of the Canary Islands is complex, as it happens in other touristic islands (Mimura et al. 2007). In insular ecosystems, due to their own geographical distribution, the synergy of the impacts from each action is intensified. For this reason it is especially important facing these problems from strategies that combine different administration levels and different users.

In the marine environment, the vulnerability increase due to the merchant ship traffic growth, especially with ships transporting hydrocarbons or toxic substances, which oblige to different administrations to coordinate. The Canary Islands have become in an usual route for the companies that explore or exploit fossil fuels in the NW African continental margin, and for the tankers performing routes from Africa (Nigeria, Equatorial Guinea or Angola) towards Europe with the obtained fuel in. An accident of one of these tankers would generate serious consequences to the Canarian ecosystems, and might be a catastrophe for the coasts in which the tourists develop their activities, leaving the Islands without drinking water which is taken from the ocean and desalinized. Thus, contingency plans should be perfectly defined (Canarian Government 2006), the different administrations informed and the population aware of how to act. In the last decades the petroleum platforms have also increased their traffic in the Atlantic and, with they, the introduction of invasive species (algae, crustaceans and fishes, among others) that begin to colonize the ports where these platforms arrive for their provisioning (Pajuelo et al. 2016).

Another example showing the need of the coordination among the administrations is the performing of military maneuvers in the surroundings of the Canarian waters, especially when they are developed close to natural protected areas. Some studies (such as Jepson et al. 2003) show the spatial and temporal relationship between the cretaceous mortality and the use of the military sonar during the maneuvers. In this case, the competent administration to regulate these activities is the State, but the management of the natural protected areas is under the regional and local administrations. The consequences affect to the wild environment and, by extension, to the citizens.

In the transition between the marine and coastal environment, the problematic caused by the construction of ports is especially significant. Along with commercial and fishing ports, the marinas are increasingly in demand by business sectors. The type of port varies according to the competent administration. For this reason it is essential a coordination for the evaluation and minimization of potential impacts. This coordination is of special interest to evaluate the suitability of the creation or expansion of other ports in the future. Likewise, the intensification of recreational activities in the sea, related with the tourism (boat excursions, jet skis, etc.), make also necessary the coordination among the different administration for planning these uses.

Finally, in the coastal environment, the conflicts between development and conservation as well as the need to face the problems from the integrated management intensify. The concentration of economic activities around and on the coast, the expansion without precedents in the edification, the increase of public infrastructures (promenades, canalization of sewage water, etc.), the proliferation of dumping points, the urban-touristic growths in highly dangerous areas (cliffs, ravine mouths, etc.), have increased the vulnerability of the coastal areas, as it happens for example in face of flooding in touristic areas (Máyer and Pérez-Chacón 2006), and especially in face of the processes that could be generated as a consequence of the climate change (Fraile-Jurado et al. 2014); and as it happens in the geomorphological level in the beaches that after disturbances that have occurred in the last 50 years have not been able to maintain their functional landforms (Peña-Alonso et al. 2017). In order to illustrate this problematic two examples have been selected, one bearing in mind the archipelago scale, the dumping points of sewage into the sea, and another one local, the urbanization of a beach-dune ecosystem, Corralejo. Both of them show the impossibility of face the problems from sectoral policies.

34.5.1 Sewage Dumping Points into the Sea

An environmental problem that affects all the islands, although at different degree, is the sewage dumping points into the sea (Table 34.4). A census performed in 2008 shows that there were 504 dumping points, amount that has fallen to 355 in 2017.

Regarding to 2008, Tenerife and Gran Canaria were the islands with more sewage dumping points, 166 and 150, respectively. The following ones were Lanzarote (80), Fuerteventura (59), La Palma (24), El Hierro (13) and La Gomera (12). Regarding the distribution of them, Tenerife concentrated these points in the northeast and the southwest coasts. In Gran Canaria all the coast was affected, except the west coast and a little coastal area in the east, close to the military facilities of Gando and the airport. In Lanzarote they concentrated in the east coast surrounding the capital of the island, and in the south. Likewise, in Fuerteventura they concentrated in the capital and in the touristic zone of the north of the island. Finally, in La Palma there is a concentration of dumping points close to the capital. Of the already mentioned dumping points, 105 are authorized, 372 not authorized, 11 with an expired license and 13 with a license in process.

For 2017, the distribution of dumping points is the following: Tenerife (146), Gran Canaria (109), Fuerteventura (35), Lanzarote (33), La Palma (17), La Gomera (9) and El Hierro (6). Of them, 183 are authorized, 246 not authorized, 12 with an expired license and 71 with a license in process.

From these data some conclusions can be drawn. From the spatial point of view (Fig. 34.4), the areas with more dumping points density are the coastal zones where urban and touristic growth of each island are located. Although there is a reduction in the number of dumping points between 2008 and 2017, the remaining amount is very high. Furthermore in the same period, the difference between authorized and not authorized dumping points is shortened, but the not authorized ones continue being the dominant ones. These dumping points are a threat for the marine biodiversity, the quality of the waters and the hygiene of the coasts. Thus, the not authorized dumping points can cause the arrival of waste to the beaches, including those with a great touristic use, the most important activity of the Canarian economy.

From the point of view of the management, this is an aspect in which the competences are distributed among the regional, insular and local administrations.

Table 34.4 Sewage dumping points into the sea in the Canary Islands in 2008 and 2017

Spills	Gran Canaria		Fuerteventura		Lanzarote		El Hierro		La Palma		La Gomera		Tenerife	
	2008	2017	2008	2017	2008	2017	2008	2017	2008	2017	2008	2017	2008	2017
Total	150	109	59	35	80	33	13	6	24	17	12	9	166	146
Authorized	33	32	9	82	6	6	2	0	3	3	3	3	49	57
Not authorized	115	77	43	27	72	27	11	6	19	14	8	6	104	89
Expired license	0	2	4	1	1	0	0	1	2	0	0	0	4	8
License in process	1	17	1	4	1	19	0	2	0	5	1	5	9	19
Paralyzed	0	0	1	0	0	0	0	0	0	0	0	0	0	0
Out of competence	1	0	0	0	0	0	0	0	0	0	0	0	0	0
License archived	0	0	1	0	0	0	0	0	0	0	0	0	0	0

Source Canarian Government database

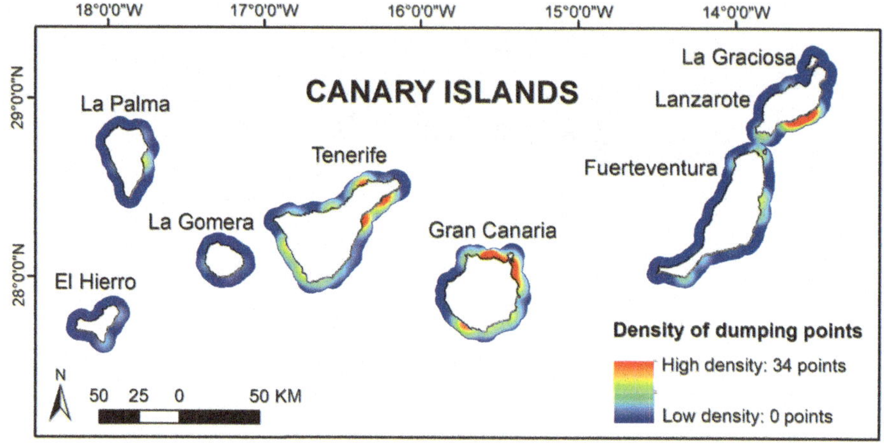

Fig. 34.4 Distribution of the dumping points density in the Canary Islands in 2008. *Source* Canarian Governement database. Prepared by the authors)

If in 2017 246 dumping points remain it is due to the administration does not have known how to solve this problem so far. Meanwhile, as the population becomes aware, a social alarm is producing, especially among the business sectors devoted to tourism. In short, this is a complex problem, whose solution requires the implication of all actors (administrations, businesses, research centers, users, social organizations, etc.) in the design of an integrated management strategy. At the same time, this problem might also be an indicator that in the Canary Islands the loading capacity is being surpassed.

34.5.2 *Problems Between the Touristic Growth and the Conservation of Dune Ecosystems: The Example of Corralejo (Fuerteventura)*

At a local scale we want to illustrate a problem, that with some nuances, has happened in numerous sectors of the coast of the Canary Islands (García-Romero et al. 2016). This shows the complexity when in a given area the competences are distributed among the State, regional government, insular government (*Cabildo*) and municipal council.

The Corralejo dune field is locate at the north of Fuerteventura, in the municipality of La Oliva, occupying in current times an extension of 18.12 km^2. The surroundings of this dune system, towards the west and the south, is formed by volcanic lava flows and volcanic eruption centers. Towards the north and the east the dune field is delimited by the coast. This system is a protected natural area under different legal figures: natural park (Canarian Government), Special Protected Areas

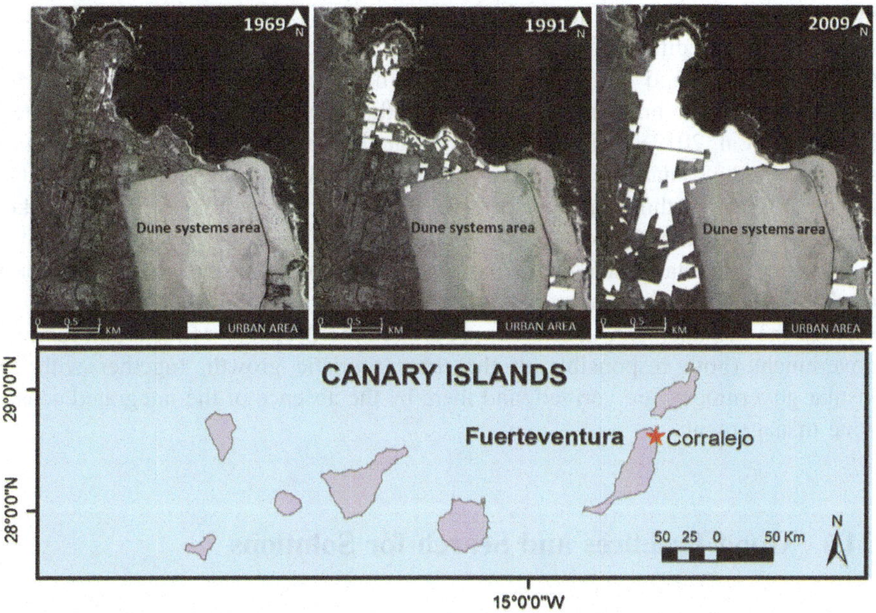

Fig. 34.5 Urban area evolution in the north of Corralejo (Fuerteventura) dune system. Adapted from Fernández-Cabrera et al. (2011)

for Birds (SPAs, European Union) and Special Areas of Conservation (SACs, European Union). However, all these legal figures do not have avoided that in current times this system is being subjected to an intense human pressure.

The beginning of the current urbanization of Corralejo was done from the little traditional village, progressively surrounding the dune system along the north and the west; whereas along the east a road and two hotels were constructed at the end of 1970s. Nevertheless, the massive urban growth of the Corralejo coast was not produced until the 1990s (Fig. 34.5), with the proliferation of hotels and extensive resorts. In 1969 the built surface was 6.72 ha, and in 2009 it was 320.94 ha. During the construction phase, population growth also was very significant: in only 13 years, Corralejo goes from having 3412 inhabitants (1996) to 14,117 ones registered in 2009 (Fernández-Cabrera et al. 2011). This urban development was performed according to municipal subsidiary regulations. From this development, it is significant that in 2011, 19 partial plan obtained unfavorable sentences for the local administration, although some of the urbanizations had already started to construct. This shows the conflicts produced in zones between urban development and conservation and between public and private interests.

Meanwhile, as it is a protected area, this system has a management plan (Master Plan for Use and Management, MPUM). In the study performed by Fernández-Cabrera et al. (2011) it is found that 60%t of the forbidden uses (according to the MPUM) were unfulfilled in 2011; while only 20 actions established by the MPUM were executed. Moreover, this study indicates evidences that in some sectors the edifications have blocked the entrance of sediments from the beach to the dune field, and that some dunes have been stabilized in zones close to the urbanization, thus breaking the natural sedimentary dynamics of the system (Malvárez et al. 2013). This is a relevant issue, because until now the conservation strategies in the Canary Islands have been limited only to protect the visible part of the dune fields, without any perspective about their surroundings and the linked processes between both areas.

In short, it is established a paradox of putting in risk, with the touristic territorial model adopted, some resources essential for the tourist activity. From the point of view of the management, interferences between the decisions of local and regional government (both responsible of the urban-touristic growth, together with the insular government) are verified, and there by the absence of the integrated coastal zone management.

34.6 Good Practices and Search for Solutions

In the Canary Islands, there is a lack of an ICZM regional strategy. Nevertheless, there are numerous experiences of various kinds, which can be considered as "good practices" of integrated management. Some of them have been promoted by the State administration, others ones by the regional administration and even some of them by city councils or by citizen groups. In most of them, social participation, the ICZM's fundamental axis, is a common denominator. Among these experiences, the citizen participation forums stand out, such as the Strategic Plan of Citizen Participation of the island of Lanzarote and La Graciosa; the ecotourism micro-area of Las Canteras beach, in Las Palmas de Gran Canaria (Gran Canaria) and the associated Citizen Participation Council; or the marine protected micro-areas, located in several spots of the Canary Islands, supported by local and regional administrations.

More recently, new proposals at a regional scale have also been presented, such as "Marine eco-areas. A sea for everybody", which is an initiative promoted by the General Directorate of Tourism Management and Promotion of the Canarian Government, with the aim of boosting coastal areas by promoting the touristic uses of the Canary coast, based on the sustainable use of natural heritage. This proposal aims to integrate different groups involved in coastal management (scientists, politicians, citizens, etc.), with the aim of generate collaboration and consensus processes.

Another experience is the development of the "Pilot Tool for the Canary Coast", with the participation of the General Directorate of the Coast and the Sea of the Ministry of Agriculture and Fisheries, Food and Environment (Spanish State scale). Its aim is the creation of a database integrating information on the boundaries of the maritime and terrestrial public domain. The database aims to be a tool for the establishment of political decisions that affect planning. Taking as a starting point the information contained in this database, a work commission is proposed where political decisions related to the planning of the Canarian coast would be addressed.

To illustrate these experiences, three examples have been selected, one related to the creation of a marine reserve, another one about the formation of a citizenship council to manage a small island, and the third one on a local experience of recovering a coastal lagoon, in a touristic area. The common denominator in the three cases is the coordination between institutions, the complicity of different actors involved, and the application of technical and scientific knowledge to design projects that, finally, have been executed and assumed by the affected population.

34.6.1 The Marine Reserve of Punta de La Restinga—Mar de Las Calmas (El Hierro): An Example of Sustainable Management of Marine Resources

Punta de la Restinga—Mar de las Calmas marine reserve is located on the southwest coast of El Hierro, and is an example of good practice in the management of the coasts. The reserve was created in 1996, through the *Ministerial Order of January 24, 1996 (BOE No. 30 of February 3, 1996) and Decree 30/1996 of February 16 (BOC No. 31 of March 11)* (MAPAMA 2017). It has an area of 750 ha and includes areas deeper than 300 m.

The main objective of the reserve is the protection, regeneration and development of resources of fishing interest for the maintenance of sustainable fisheries that allow artisanal fishermen to preserve their traditional way of life (MAPAMA 2017). In addition, the reserve also carries out diving activities through 12 points established for this purpose, since its funds also have the value of being a biodiversity hotspot in the archipelago.

The management of the marine reserve began in 1997 with the establishment of the Joint Commission for Management and Follow-up (General Secretariat of Maritime Fishing and Vice-Minister of Fisheries of the Canary Islands Government). The General Secretariat of Maritime Fishing published posters and triptychs, installed billboards and endowed the reserve with a ship, called "El Guincho". The surveillance service is assumed by the Canarian Government management throughout the year.

Annually, the General Secretariat of Maritime Fishing carries out the control of the dive sites and in 2001 installed buoys at said points (Canarian Government 2017).

This initiative is increasingly valued for the affected population and for those who visit El Hierro. It represents an example of sustainable uses of marine resources, combining fishing exploitation and touristic use.

34.6.2 Council of Citizenship of La Graciosa: A Tool for Integrated Management

La Graciosa is a small island of 29.5 km^2 located northwest of Lanzarote (Fig. 34.1), administratively belonging to the municipality of Teguise (Lanzarote). It has two population centers, Pedro Barba and Caleta de Sebo, where the population is concentrated, which in 2016 amounted to 746 inhabitants (INE 2018). This island is located within the Chinijo Archipelago, accompanied by a series of islets that, as a whole, are listed as a protected natural area through the figures of Natural Park (Legislative Decree, 1/2000, of May 8), Special Area of Conservation (Decree 174/2009 of December 29), Special Bird Protection Area (Order AAA/1260/2014, of July 9), at the same time that it contains the largest marine reserve in Europe, with 700 km^2.

The small size of the island and its few inhabitants, has generated a scenario where historically has been fundamental citizen deliberation on economic, ecological, political, territorial and social aspects. In the 80 s of the last century there was a transition towards the tourist and urban development. The island, as state heritage, was managed by the then Ministry of Finance (Spanish State administration). In 1983, this Ministry proposed a project for the building of hotels with 2500-bed, but the municipal government wanted to limit it to 120 squares on some 10,000 m^2 of land, proposing a preliminary draft in 1984. During these years of confusion over the alternatives for the future of La Graciosa, an assembly was consolidated. It was composed by politicians, environmental associations, members of the fishing sector and citizens, who requested the environmental preservation of the island. This was the reason that led the State authorities to curb urban planning claims, declaring La Graciosa as a Natural Park in 1986 (Socorro 2000).

The citizenship was maintained for decades, and began to take strength through the citizens' initiative called "La Graciosa 8th Island" in 2014, and that even today is still vindicating. At this time, the promoters of this initiative won the support of all the political forces represented at the Parliament of the Canary Islands for the recognition of the singularities of La Graciosa, and the start-up of the Council of Citizenship of La Graciosa that consolidated in the year 2014. This Council functions as an organ for citizen participation in which residents, associations and collectives of La Graciosa are represented, as well as institutions that have competences on the island, such as the Lanzarote's Cabildo, the City Council of Teguise, the Canarian Government or the Government of Spain. This body allows the discussion and agreement of decisions related to education, culture and sports, fishing, tourism, environment and social services.

The representation of the different social groups involved in the island, and their survival throughout the last decades in the form of a "social council", are a real and effective example of the ICZM at a local scale.

34.6.3 Oasis 2000 Project: The Recovery of a Coastal Lagoon in Maspalomas (Gran Canaria)

The Oasis 2000 Project was proposed in 1994 with the aim of recovering, from the environmental point of view, a coastal lagoon located at the Fataga ravine's mouth, in the south of Gran Canaria. This lagoon is part of the Special Natural Reserve, which also includes the beach-dune system of Maspalomas. It is also a Special Area of Conservation (SAC ES7010007), forming part of the Natura 2000 network. It is an area of high value, both from the point of view of geodiversity and biodiversity and a resting place for migratory birds on their migratory flights between Africa and Europe.

From the decade of 1960 the touristic development began in the zone, intensifying in the later decades. The environmental consequences of this process have been widely studied (Hernández-Calvento 2006; García-Romero et al. 2016; Hernández-Cordero et al. 2017). It has been observed, among other issues, a deep alteration of the functioning of the dune system, as well as erosional processes on the beach of Maspalomas, the most important touristic resource of Gran Canaria.

When the Oasis 2000 Project was proposed, the coastal lagoon was deteriorated, after decades of tourist activity in its immediate surroundings. Therefore, the main objective of the project was the recovery of the wet zone, eliminating impacts derived from human activity. The action proposals were the following: restoration of the habitat for threatened flora and fauna; adequacy of the space for a compatible use with respect for the environment and urban rehabilitation of the area of public use of the palm grove and the environment of the coastal lagoon, locally known as *Charca de Maspalomas*.

The project was executed mainly between 1994 and 2000, although some actions had begun earlier. The strategy was based on initiating actions aimed at ensuring that nature itself is responsible for spontaneous colonization processes. The budget was reduced, and the main actions were: the elimination of artificial elements of the mouth of the ravine to favor the infiltration and the spontaneous colonization of hygrophilous species; the close of the road on the left bank of the lagoon and eliminate the parking lots (which will be relocated in an area far from the natural environment), as well as the remains of an old touristic center. Additionally, preparing the promenade along the right margin of the lagoon (contiguous to the urbanized area), and to place in it a lookout and information panels on the fauna and flora of the place was proposed. The management measures applied to the left margin of the lagoon, intended to reduce human pressure and facilitate natural regeneration, while those on the right bank proposed to channel people to the beach and facilitate public use of the area.

From the point of view of management, the team that made possible the execution of this project consisted of the conservative director (biologist) of the protected area, the mayor (geographer) of the municipality in which the lagoon is located, an architect (design of the proposal) and a journalist (dissemination of the actions and results obtained). The joint work of these actors made possible to create

Fig. 34.6 Coastal lagoon of Maspalomas (Gran Canaria) before (1992) and after (2016) of the execution of the Oasis 2000 Project. *Photos* Carlos Suárez Rodríguez (canariashistoriasnaturales.blogspot.com.es) and Emma Pérez-Chacón

a wide support network for the project, conciliating administrations and actors involved, a key issue so that a good part of the proposed actions could be implemented (Fig. 34.6). It is an excellent example in which the ICZM achieves the recovery of a highly degraded environment by human activities.

34.7 Conclusions

The use of coastal resources in the Canary Islands has been changing throughout history. The coastal management has depended on the socio-economic organization prevailing in each stage, being the most drastic change when the islands moved from an agrarian productive model to the urban-tourism one, currently prevailing. The beach-dune systems have been the ones that have registered a degree of anthropogenic alteration more significant before the arrival of the tourist industry: grazing, cultivation, firewood, charcoal and extraction of sand were common uses, with different intensities. Although today the survivals of these ecosystems are protected, there is no doubt that they are still vulnerable. On the other hand, the uses derived from the current productive model have generalized their impacts (disappearance or degradation of ecosystems, human alteration of the coast, etc.) to the whole coast, both in the terrestrial and marine areas of the coastal zone.

Human pressure on the Canary coast is high, as is the value of the coastal ecosystems. Although the islands are a broadly coastal territory, there is not any specific integrated management strategy at the regional scale. The current management is only carried out through sectoral policies, applied with greater or worse fortune according to the time, where different administrations are involved. There are numerous problems arising from interferences between administrations acting at different scales, and from the poor coordination and cooperation between them, with the assignment of competences being confusing in some cases.

On the other hand, there are isolated examples of good practices, some developed by the administration (agreements, consortiums). But greater scope has diverse citizen initiatives that try, from collaborative strategies between administrations and with citizen participation, to develop the ICZM. These are also observed in the fields of research and training. Given the inaction of the authorities, are citizens who are taking the initiative, which is a paradox.

The ICZM must become a regional priority in the Canary Islands, which encourages the coordination of state, regional, insular and municipal policies. For this, it must have adequate financing, and must be able to create complicity among all the actors involved. This is the only way to overcome the current "de-integrated" management and promote the sustainable use of one of its main resources of the islands: the coast.

Acknowledgements This book chapter is a contribution to the REN2003-05947, SEJ2007-64959, CSO2010-18150, CSO2013-43256-R and CSO2016-79673-R projects of the Spanish National Plan for R+D+i (innovation), co-financed with ERDF funds. Also to the program of support for research of the Canary Islands Government, also co-financed with ERDF funds, and to the fellowship program of the University of Las Palmas de Gran Canaria. Some results have been obtained from studies funded by the Organismo Autónomo Parques Nacionales (the 'Isla de La Graciosa' Centre) and by the former Dirección General de Costas, both of the Ministerio de Agricultura, Alimentación y Medio Ambiente, Gobierno de España.

References

Anguita, F., Márquez, A., Castiñeiras, P., & Hernán, F. (2002). Los volcanes de Canarias. *Guía geológica e Itinerarios*. Ed. Rueda. Madrid.

Barragán, J. M. (2014). Política, Gestión y Litoral. Una nueva visión de la Gestión Integrada de Áreas Litorales. Ed. Tébar Flores, Madrid.

Barragán, J. M., & Borja, F. (2011). Capítulo 13. Litorales. *Evaluación de los Ecosistemas del Milenio de España*, pp. 673–739.

Bauer, B. O., & Sherman, D. J. (1999). Coastal dune dynamics: Problems and prospects. In *Aeolian environments, sediments and landforms* (pp. 71–104). Chichester: Wiley.

Bianchi, R. V. (2004). Tourism restructuring and the politics of sustainability: A critical view from the European periphery (The Canary Islands). *Journal of Sustainable Tourism, 12*(6), 495–529.

Botero, C. M., Cervantes, O., & Finkl, C. W. (2017). *Beach management tools-concepts, methodologies and case studies* (Vol. 24). Springer.

Cabrera Vega, L. L. (2010). Sedimentología, estratigrafía, dinámica sedimentaria y evolución de El Jable (Lanzarote). Propuesta de gestión. PhD thesis—Unpublished.

Canarian Government. (2006). *Plan específico de contaminación marina accidental de Canarias*. Consejería de Medio Ambiente y Ordenación del Territorial.

Canarian Government. (2017). http://www.gobiernodecanarias.org/agricultura/pesca/temas/reservas_marinas/rrmm_la_restinga.html. Consulted: 16/10/2017.

Carracedo, J. C., Perez-Torrado, F. J., & Rodriguez-Badiola, E. (2008). Canarias: islas volcánicas intraplaca. In F. Pérez-Torrado & M. C. Cabrera (Eds.), *Itinerarios geológicos por las Islas Canarias, Gran Canaria* (Vol. 4, pp. 11–25). Sociedad Geológica de España, Geo-Guías.

Farinós Dasí, J. (ed. y coord.) (2011): La gestión integrada de zonas costeras, ¿algo más que una ordenación litoral revisada? Publicacions de la Universitat de València, València.

FEDAC. (2017). Grupo salinas marinas de Fuencaliente S.L. (http://salinasdefuencaliente.es/historia_salinas_fuencaliente/); https://lagraciosadigital.wordpress.com/2013/10/09/las-salinas-mas-antiguas-de-canarias/; http://www.cienciacanaria.es/secciones/a-fondo/310-salinas-de-canarias).

Fernández-Cabrera, E., Pérez-Chacón Espino, E., Cruz Avero, N., Hernández-Cordero, A., & Hernández-Calvento, L. (2011). Consecuencias ambientales del crecimiento urbano-turístico en el sistema de dunas de Corralejo (Fuerteventura-Islas Canarias). In V. Gozálvez-Pérez & J. A. Marco-Molina (Eds.), *Urbanismo expansivo de la utopía a la realidad* (pp. 241–252). Asociación de Geógrafos Españoles: Colegio de Geógrafos de España y Universidad de Alicante, Alicante.

Fernández-Palacios, J. M., & Dias, E. (2001). Marco biogeográfico macaronésico. In J. M. Fernández-Palacios & J. L. Martin (Eds.), *Naturaleza de las Islas Canarias* (pp. 45–52). Santa Cruz de Tenerife: Ecología y conservación. Publicaciones Turquesa.

Ferrer-Valero, N., Hernández-Calvento, L., & Hernández-Cordero, A. I. (2017). Human impacts quantification on the coastal landforms of Gran Canaria Island (Canary Islands). *Geomorphology, 286,* 58–67.

Fraile Jurado, P., Sánchez Rodríguez, E., Fernández Díaz, M., Pita López, M. F., & López Torres, J. M. (2014). Estimación del comportamiento futuro del nivel del mar en las Islas Canarias a partir del análisis de registros recientes. *Geographicalia, 66,* 79–98.

García-Romero, L., Hernández-Cordero, A. I., Fernández-Cabrera, E., Peña-Alonso, C., Hernández-Calvento, L., & Pérez-Chacón, E. (2016). Urban-touristic impacts on the aeolian sedimentary systems of the Canary Islands: conflict between development and conservation. *Island Studies Journal, 11*(1).

García-Sanabria, J., García Onetti, J., & Barragán, J. M. (2011). Las Comunidades Autónomas y la gestión integrada de las áreas litorales de España. *Proyecto "Red Española de Gestión Integrada de Áreas Litorales (REGIAL). Fundación Biodiversidad y la Universidad de Cádiz.*

Greenpeace. (2009). *Destrucción a toda costa. Situación del litoral español y sus espacios protegidos*. Greenpeace: Spain.

Greenpeace. (2013). *Destrucción a toda costa. Análisis del litoral a escala municipal*. Greenpeace: Spain.

Guadalupe Aparicio, F., Pérez-Chacón Espino, E., & García Márquez, F. (2016). Planificación territorial y modelo de desarrollo turístico: Famara (Lanzarote), un viaje al pasado…o al futuro. Actas VIII Congreso Internacional de Ordenación del Territorio – VIII Congreso de Derecho Urbanístico. FUNDICOT. pp. 671–684.

Guerra, R., & Pérez, T. (2008). Canarias: entre el desarrollo turístico y la protección al medio. Études Caribéennes, pp. 9–10. http://etudescaribeennes.revues.org/1302.

Grupo salinas marinas de Fuencaliente S.L. (2017) (http://salinasdefuencaliente.es/historia_salinas_fuencaliente/).

Hernán, F. (2001). Estructura geológica. In J. M. Fernández-Palacios & J. L. Martin (Eds.), Naturaleza de las Islas Canarias (pp. 59–64). Santa Cruz de Tenerife: Ecología y conservación. Publicaciones Turquesa.

Hernández-Calvento, L. (2006). *Diagnóstico sobre la evolución del sistema de dunas de Maspalomas (1960–2000)*. Las Palmas de Gran Canaria, Spain: Cabildo Insular de Gran Canaria.

Hernández-Cordero, A. I., Hernández-Calvento, L., & Pérez-Chacón Espino, E. (2017). Vegetation changes as an indicator of impact from tourist development in an arid transgressive coastal dune field. *Land Use Policy, 64,* 479–491. https://doi.org/10.1016/j.landusepol.2017.03.026.

Hernández Rodríguez, G. (2004). *La orchilla en Canarias. Implicaciones socioeconómicas*, Tesis doctoral de la Universidad de Las Palmas de Gran Canaria. ACCEDA. Biblioteca digital de la Universidad de Las Palmas de Gran Canaria. https://doi.org/file:///C:/Users/usuario/AppData/Local/Temp/1860-1.pdf (Consulta 23/02/2018).

INE. Instituto Nacional de Estadística. (2018). http://www.ine.es/nomen2/index.do). Consulted: 1/02/2018.

ISTAC. Instituto Canario de Estadística. (2018). http://www.gobiernodecanarias.org/istac). Consulted: 1/02/2018.

Jackson, N. L., & Nordstrom, K. F. (2011). Aeolian sediment transport and landforms in managed coastal systems: A review. *Aeolian Research, 3*(2), 181–196.

Jepson, P. D., Arbelo, M. D., Deaville, R., Patterson, I. A. P., Castro, P., Baker, J. R., et al. (2003). Gas-bubble lesions in stranded cetaceans. *Nature, 425,* 575–576.

López-Ornat, A., Atauri, J. A., Ruiz, C., & Múgica, M. (2014). *Beneficios sociales y ambientales de las reservas marinas de interés pesquero.* Fundación Fernando González Bernáldez.

Malvárez, G., Jackson, D., Navas, F., Alonso, I., & Hernandez-Calvento, L. (2013). A conceptual model for dune morphodynamics of the corralejo dune system, fuerteventura Spain. *Journal of Coastal Research, 65*(sp2), 1539–1544.

MAPAMA: Ministerio de Agricultura y Pesca, Alimentación y Medio Ambiente. (2017). http://www.mapama.gob.es/es/pesca/temas/proteccion-recursos-pesqueros/reservas-marinas-de-espana/la-restinga/caracteristicas/. Consulted: 15/10/2017.

Marzol, V. (2001). El clima. In J. M. Fernández-Palacios & J. L. Martin (Eds.), *Naturaleza de las Islas Canarias Ecología y conservación* (pp. 87–93). Santa Cruz de Tenerife: Publicaciones Turquesa.

Máyer, P., & Pérez-Chacón, E. (2006). Tourist activity and floods on the southern coast of Gran Canaria. An induced risk? *Journal of Coastal Research, 48,* 77–80.

Mimura, N., Nurse, L., McLean, R., Agard, J., Briguglio, L., Lefale, P., et al. (2007). Small Islands. *Climate Change, 16,* 687–716.

Observatorio de la Sostenibilidad. (2014). [en línea]: *Sostenibilidad en España 2014.* http://www.observatoriosostenibilidad.com/sostenibilidad-en-espana-2014/ Consulted: 17/01/2018.

Pajuelo, J. G., González, J. A., Triay-Portella, R., Martín, J. A., Ruiz-Díaz, R., Lorenzo, J. M., et al. (2016). Introduction of non-native marine fish species to the Canary Islands waters through oil platforms as vectors. *Journal of marines systems, 163,* 23–30.

Peña-Alonso, C., Fraile-Jurado, P., Hernández-Calvento, L., Pérez-Chacón, E., & Ariza, E. (2017). Measuring geomorphological vulnerability on beaches using a set of indicators (GVI): A tool for management. *Journal of Environmental Management, 204,* 230–245.

Pérez-Chacón Espino, E., Hernández Calvento, L., & Yanes Luque, A. (2007). L'impact des aménagements sur les littoraux canariens. In: Etienne y Paris (coord): Les littoraux volcaniques: une approche environnementale. Clermont-Ferrand (France), Ed. Presses Universitaires Blaise-Pascal, Coll. Volcaniques., 173–191 pp.

Pilkey Jr, O. H., & Cooper, J. A. G. (2014). *The last beach.* Duke University Press.

Portillo Hahnefeid, E. (2008). *Arribazones de algas y plantas marinas en Gran Canaria: características, gestión y posibles usos* (p. 88). S.A., Las Palmas: Instituto Tecnológico de Canarias.

Prieto, F., & Navarro, E. (2012). Desarrollo litoral sostenible en España: proceso de litoralización (1960–2012) y escenarios para el 2030. Destrucción de un bien común y pérdida de los servicios de los ecosistemas. Lecciones aprendidas pensando en Latinoamérica. La tragedia de los comunes en la costa española. En: Barragán-Muñoz, JM (coord.) I Congreso Iberoamericano de Gestión Integrada de Áreas Litorales, Cádiz, 357–370 pp.

Ruiz, J. F. M., & Quintana, J. B. (2001). *Geografía de Canarias: sociedad y medio ambiente.* Cabildo de Gran Canaria.

Santana-Cordero, A. M., Monteiro-Quintana, M. L., & Herández-Calvento, L. (2016a). Reconstruction of the land uses that led to the termination of an arid coastal dune system: The case of the Guanarteme dune system (Canary Islands, Spain), 1834-2012. *Land Use Policy, 55,* 73–85.

Santana-Cordero, A. M., Monteiro Quintana, M. L., Hernández-Calvento, L., Pérez-Chacón Espino, E., & García Romero, L. (2016b). Long-term human impacts on the coast of La Graciosa, Canary Islands. *Land Degrad Develop, 27,* 479–489.

Socorro, G. C. (2000). Turismo, decisiones políticas y cambio social en un pueblo de pescadores: la Caleta del Sebo en la isla de La Graciosa. *Coloquios de Historia Canario Americana, 14* (14), 159–174.

Turégano, M. Á. S. (2005). Turismo, empleo y desarrollo. *Papers: revista de sociologia*, *77*, 79–104.

Williams, A., & Micallef, A. (2011). *Beach management: Principles and practice*. Routledge.

Yanes, A. (2006). Caracterización morfológica de las costas de las Islas Canarias. In R. Pie (Ed.), *Vilanova Proyecto de directrices de ordenación del litoral de Canarias*. Y (dirs): Barcelona, Edición de Equip BCpN.

Part VI
Final Remarks

Chapter 35
Future Perspectives of the Spanish Coast

Juan A. Morales

35.1 Introduction

The chapters included in this book demonstrate a high degree of knowledge about the coastal dynamics of many segments of the coast. The highlights of the functioning of wind, waves, tides and currents in each one of the sectors composing the Spanish coast have been studied and described. The main geomorphologic and sedimentary features of, so as the depositional or erosional trends or each coastal track are also defined. According with this knowledge, national, regional and local administrations have in their hands the tools to develop a correct Integrated Coastal Zone Management in each one of these coastal tracks agreeing with the criteria suggested by Evans (1992). This is vital in Spain, especially, take into account that a big part of the Spanish economy is based in the tourism of sun and beaches. Nowadays, the present coastline suffers the errors committed by the coastal managers in the past, before all this knowledge had been acquired.

As was commented in Chap. 1, in the 60s and 70s decades many dams were built along the main Spanish rivers cutting the natural sediment bypass to their mouths and coastal adjacent areas. The presence of hundred of dams caused a sedimentary deficit on the coastal systems and the coast responds with a generalized retreat, observed in every track studied in the chapters of this book. In addition, many hard structures like groins, breakwaters and seawalls, were built in lineal coasts, to avoid this coastal erosion. Other rigid structures like jetties were also built to stabilize estuarine mouths where harbors were established in its interior. A consequence of these structures is the disruption of the natural longshore

J. A. Morales (✉)
Department of Earth Sciences, University of Huelva, Huelva, Spain
e-mail: jmorales@uhu.es

J. A. Morales (ed.), *The Spanish Coastal Systems*,
https://doi.org/10.1007/978-3-319-93169-2_35

transport, compartmenting the coast in small cells and also causing problems of erosion updrift of each rigid structure.

Other problem caused by an incorrect coastal management was the dismantling of entire chains of coastal dunes. Many kilometers of the Spanish coast were urbanized transforming the natural beach dune system in urban beaches attached to promenades or directly to first-line buildings. The presence of these structures caused a rigidization and a leak of flexibility of the coast. The absence of the foredune facilitates the direct attack of storm waves on these buildings and promenades, causing every year tremendous economic damages.

By all of these causes, the present Integrated Coastal Zone Management along the Spanish coasts represent an exciting challenge for the competing administrations, since today the entire Coast is public and protected by laws, but with a big responsibility, because now all the scientific knowledge is available for the coastal managers.

35.2 Future Challenges for the Managers

The framework of the global warming represents a direct menace, not only for the natural coastal systems, but mainly for the human structures installed in them. During next decades a slight but continuous sea level rise and an increment of the frequency and intensity of storms are expectable (Eliot 2016).

In this context, natural environments continue having the flexibility to evolve according these new parameters. In these systems, the tendencies of the future coastal evolution will be conditioned by the quantitative balance between: the velocity of sea-level rise, the rate of sediment supply from the continent and from the shelf and the volumentric capability of coastal agents to rework this sediment supply. As was explained in the Chap. 1 of this book, Global warming acts on the three variables: (1) on a hand, the sea level will rise a mean about 3.0 mm/yr in the Spanish coastal areas, (2) on the other hand, in Spain is expectable a decreasing of the rains, which imply a diminution of sedimentary supply coming to the coast from rivers, (3) finally, an increment of wave energy (and sediment rework capability) will be caused by significant increases of wave dimensions, so as a spike of the number of storms.

Just writing these words, the fourth of a series of extreme storms affected severely the Spanish Coastline. From the 1st of March 2018 these four successive storms affected directly the Gulf of Cadiz in three weeks. Two of these storms, named Emma and Gisele coincided with spring tides. The coincidence of high astronomic tides and a storm surge of 0.60 m elevated the sea altitude to unusual levels unreached since 1981. The duration of various days of these storms provoked also the coincidence on the coast of sea waves and strong winds. In the case of

Emma, the 1st of March, wind reached velocities upper than 30 knots, that incremented the wave dimensions to reach Hs of 6.40 m. The arrival of Gisele, two weeks later, again during spring tide conditions increased the damages in the urban beaches previously dismantled by Emma (Fig. 35.1a, b).

It can be expected that natural systems will be tailored to these new dynamic conditions. So, the coast will evolve to a transgressive situation with beaches and dunes even more eroded and overwashed (Fig. 35.1 c, d) and displaced to upper areas and fluvial mouths invaded by marine waters. On the contrary, the anthropized systems lost the natural flexibility and cannot evolve in an adequate way to this new scenary. The consequences for the Spanish Coast are severe. These, have been described in a study by the EU (Policy Research Corporation 2016), and already exposed in the Chap. 1 of this book.

The measures for the coastal protection in Spain have traditionally used rigid defenses, without taking into account the effects that these installations normally trigger. The evolution of knowledge in the last decades should make managers reflect that relocation and the named soft measures (replenishments) should be best solutions for protection of our coast. In this sense, the managers must use dynamic criteria, generated by geomorphologists and sedimentologists, to define the management policies and maintenance of the coast in the context of Global change.

This is a challenging framework for the coastal managers and the economic development of the entire country will be in their hands in the next decades, since

Fig. 35.1 Damages caused by the storms Emma and Gisele in the Southwestern Spanish Coast. **a** Damages in the promenade of Playa Central (Isla Cristina, Huelva). **b** Damages in the buildings of La Antilla (Lepe, Huelva). **c** Scarp berm in a foredune. **d** Washover fan over a salt marsh by breaching of a foredune

not only the touristic business, but the biggest cities, the richest industries and the main harbors are located just in the coast. For this reason will be necessary a cross-jurisdictional cooperation, taking into account that in Spain there are three levels of administration: national, regional and local.

35.3 Perspectives in the Coastal Science

In recent decades the research efforts that encourage a deeper knowledge of the coastal system are increasing.

The main contributions of the coastal investigators in the last years have been making around three main research lines: (1) contribution to the general knowledge on the coasts from a dynamic, geomorphological, sedimentological and environmental points of view, (2) contributions about pure methodological development and use of new techniques to the coastal knowledge and (3) contributions of environmental diagnosis, showing the effects of some human actions on the coast, mainly focused to planning of coastal management (Andrés and Gracia 2000; Blanco-Chao et al. 2003; Hernández-Calvento et al. 2005; Gómez-Pujol and Fornós 2007; Morales et al. 2009; Montoya et al. 2011; Flor et al. 2013; Malvárez et al. 2015; Pons-Buades et al. 2017). These same three lines mark the tendencies for the next years, especially in the context of the Global change.

35.3.1 Studies on Coastal Dynamics, Geomorphology and Sedimentology

The general guidelines of the knowledge on the dynamics and geomorphologic functioning of the Spanish Coast are already established. However, much remains to be done in different ways. Here are the main trends that must be the future research lines to respond to the immediate challenges.

- *Underground record and architectural studies*: The main part of the studies already developed was centered in the surficial distribution of sediment. Future research must focus the stratigraphic record, especially regarding the 3D architectural disposal of the sedimentary facies. These studies, can contribute to determine geometries of sedimentary bodies and, superimposing a sediment chronology, can help to understand possible slight sea level movements in the last 5000 years.
- *Study of coastal events*: Coastal events are important energetic and destructive punctual processes. It is important as a future guideline to characterize the effects of each one of the storms and tsunamis arrived to the coasts in the future, but also recognize the record of these events in the past in order to establish return periods and make correct previsions of future damages.

- *Extension of the studies to the submarine areas of the coast*: At the moment, the biggest part of studies was focused in the emerged part of the coast, but the shore is an integrated system and to get an adequate knowledge of the dynamic processes it is necessary to extend the studies to the sub-littoral areas. In this sense will be necessary to apply new techniques described as follows.
- *Upscale and downscale the studies*: Once the mesoscale dynamics of our coasts have been characterized it is necessary a two-sense change of scale. In a hand, more detailed and specific studies in a minor scale would be useful to understand small nuances in processes and sedimentary distribution. On the other hand, a connection of studies done in adjacent areas has to be done in a macroscale to understand some problems in a wider context.

35.3.2 Studies on Methodological Development and Use of New Techniques

- *Sensors for process measuring*: Developments in technique incorporated during the last years a new generation of sensors to measuring processes. In this sense, wave gauges, tidal stations and Doppler currentmeters, capable to do time-continuous measures and register them in dataloggers could be the best examples. These datasets will contribute to a better understanding of the environmental knowledge of the coastal systems.
- *Geophisical methods*: Last years, is being a trend the use of acoustic techniques like Side Scan Sonar or Seismic profiles to complete the geomorphological and sedimentological underwater and underground information of coastal systems. These methods were widely and successfully used by the marine geologists in deeper areas. The technical development of cheaper, more reduced and versatile equipments will contribute to extend the studied areas to the subtidal coast.

In the same way, the use of Ground Penetration Radar is being used to characterize the internal structure of sandy barriers and littoral dunes y the land portion of the coast.

- *Topographic and bathymetric methods*: New acoustic and laser techniques like multibeam echo sounds and aerial LIDAR have allowed in the last years to get high resolution bathymetric and topographic records. These records contributed to elaborate digital terrain model (DTM) that integrate terrestrial and underwater very useful in coastal areas to understand some unaboarded geomorphological aspects.
- *Mathematical models*: Modelling of hydrodynamic and sedimentary processes is a new tool which is being used by geoscientists to understand the functioning of the coastal systems. Uncalibrated models were at the moment an excellent method to know the cause of past effects, but also for a good prevision of future actions on the coast. In the future, the use of these models in coordination with

the data obtained by the previously described sensors will allow a correct calibration of the models in order to develop previsions totally reliable which can help the coastal managers to take adequate decisions.

35.3.3 Environmental Studies and Integrated Coastal Zone Management

The beginnings of the studies about the coast were aboarded by the coastal engineers. The first aim of these studies was to adapt the coast to human requirements. When geo-scientists enter to study the coast, a pure scientism was focused to the correct understanding of the natural coastal system and the consequences on coastal dynamics of the structures built by the civil engineers. Now, under the present state of the art, a new focus is necessary: knowledge to be applied to the ICZM. In order to get the best criteria to take decisions will be necessary studies about the followings items.

- *Mapping vulnerability*: Detailed charts of the coastal zone, including maps of degree of vulnerability to coastal events, not only in natural environments, but especially in the human occupied coast.
- *Conceptual modeling*: Conceptual models are conceived as a method of visualizing integrated and interpreted data. There are different conceptual models that would be applied in Coastal Geosciences, but the most interesting for the coastal managers are those that synthesize processes and sedimentary transport as a response of modifications in the parameters that make evolve the coast (natural or human). In this way, conceptual models are a useful way to convey resource information to politics, since must be a system of representation understandable to experts and non-experts.
- *Transversal interdisciplinary studies and debates*: After directives of the European Union like Horizon 2020, scientist would tend to a transversal and interdisciplinary vision of their studies. Take into account these new rules, coastal scientists are obligated to incorporate the vision of geographers, geologists, oceanographers, biologists, archaeologists, coastal engineers, mathematicians, physicians, geochemicals and lawyers to build multi-visionary teams capable to analyze any aspect of the future coastal planification that hardly would be approached by individual specialists.

35.4 Science for the Society

The last challenge that needs to be addressed is to carry the knowledge to the general society. Since the entire population is and will be affected by the coastal processes, in the future, the coastal scientists are strongly forced to communicate our discoveries to people. Therefore, it is not enough to write scientific publications, but write communications comprehensive for a non-technical audience and report it in media, webpages and social networks.

Ultimately, coastal scientists have to focus on the future by addressing challenges that become increasingly difficult, especially in a context of economic crisis that considerably limits investments in research. However, we will address this task with our best weapons: effort and illusion.

References

Andrés, J. R., & Gracia, F. J. (Eds.). (2000). *Geomorfología Litoral Procesos Activos. SEG Monography* (Vol. 7). Madrid.

Blanco-Chao, R., López-Bedoya, J., & Pérez Alberti, A. (Eds.). (2003). *Procesos geomorfológicos y evolución costera: actas de la II Reunión de Geomorfología Litoral*. Univ. Santiago de Compostela. 426 pp.

Eliot, M. (2016). Coastal sediments, beaches and other soft shores. In *Coast adapt information manual* (Vol. 8). National Climate Change Adaptation Research Facility. 69 pp.

Evans, A. W. (1992). The application of geomorphology in coastal management studies. *Ocean and Coastal Management, 17,* 47–55.

Flor, G., Flor-Blanco, G., & Pando, L. A. (Eds.). (2013). Geotemas 13. 226 pp.

Gómez-Pujol, L. l., & Fornós, J. J. (Eds.). (2007). *Investigaciones recientes en Geomorfología litoral*. Univ. Islas Baleares. Mallorca. 224 pp.

Hernández-Calvento, L., Alonso, I., Mangas, J., & Yanes, A. (Eds.). (2005). *Tendencias actuales en Geomorfoplogía litoral*. Univ. Las Plamas de Gran Canaria. 289 pp.

Malvárez, G., Navas, F., & Guisado, E. (Eds.). (2015). Geotemas 14. 236 pp.

Montoya, I., Rodríguez-Santalla, I., & Sánchez-García, M. J. (Eds.). (2011). Avances en Geomorfología Litoral. Tarragona. 174 pp.

Morales, J. A., Cantano, M., Rodríguez, A., & Delgado, I. (Eds.). (2009). Nuevas contribuciones sobre Geomorfología Litoral. Serv. Publ. Univ. Huelva. 152 pp.

Pons-Buades, G., Lorenzo-Lacruz, J., & Gómez-Pujol, L. l. (Eds.). (2017). Geotemas 17. 346 pp.